Essentials of Landscape Ecology

Essentials of
Landscape Ecology

Kimberly A. With

OXFORD
UNIVERSITY PRESS

OXFORD
UNIVERSITY PRESS

Great Clarendon Street, Oxford, OX2 6DP,
United Kingdom

Oxford University Press is a department of the University of Oxford.
It furthers the University's objective of excellence in research, scholarship,
and education by publishing worldwide. Oxford is a registered trade mark of
Oxford University Press in the UK and in certain other countries

© Kimberly A. With 2019

The moral rights of the author have been asserted

First Edition published in 2019

Impression: 1

Published in the United States of America by Oxford University Press
198 Madison Avenue, New York, NY 10016, United States of America

British Library Cataloguing in Publication Data

Data available

Library of Congress Control Number: 2018962521

ISBN 978–0–19–883838–8 (hbk.)
ISBN 978–0–19–883839–5 (pbk.)

Printed in Great Britain by
Bell & Bain Ltd., Glasgow

Links to third party websites are provided by Oxford in good faith and
for information only. Oxford disclaims any responsibility for the materials
contained in any third party website referenced in this work.

Preface

Writing a textbook is rather like being enrolled in a multi-year course on the subject, in which the roles of student and teacher are essentially interchangeable. Since we are all really students at heart, my approach to writing this textbook has been from the standpoint of a student, starting with myself. I know what difficulties I have faced in understanding certain topics and analytical approaches, and after teaching landscape ecology for more than 20 years, I know what difficulties many of my students encounter when first exposed to particular concepts or methods. This textbook is thus for the student of landscape ecology, whatever your academic standing, professional rank, or title. Welcome to our shared course on landscape ecology.

The scientific domain encompassed by landscape ecology is incredibly vast and wide-ranging. Landscape ecology is built upon contributions from various fields in the natural and social sciences, and in turn contributes to both basic and applied problems in each of its constituent fields. No single textbook—including this one—can possibly do it justice. My focus in this textbook is thus on the *ecology* of landscape ecology. I am an ecologist by training, and besides landscape ecology, much of my research has been in the areas of behavioral, population, and community ecology. In addition, I have long been interested in conservation and the applications of landscape ecology for conservation biology. Thus, my perspective herein is inevitably that of an ecologist, with an eye toward applications of landscape ecology for conservation and natural resource management.

Despite this—or maybe even because of this—it is my hope that this book will be of interest to students and professionals in other areas of landscape ecology, such as geographers and landscape planners, who might desire an ecological overview to complement their training or design perspective. Judging from the composition of my own landscape ecology course over the years, I envision an audience that is likely to be quite diverse regardless. In the past, this course has attracted students from all over campus and from fields as diverse as ecology, environmental science, entomology, geography, agronomy, range sciences, forestry, conservation biology, fisheries, and wildlife management. The academic background, training, and expectations of these students are all very different, which demands a varied and diversified approach in the presentation of the material. Some students require an introduction to basic ecological concepts and principles, whereas others are looking for synthesis, critical review, and a deeper treatment of the analytical, methodological, or modeling approaches they are hoping to implement in their own research.

The principal challenge in writing this textbook, as in the classroom, has thus been to develop a curriculum that serves the needs of a diverse audience. To that end, I have tried to provide a mix of basic concepts, examples, and case studies, as well as a more advanced treatment of certain topics. The examples and case studies were selected to emphasize a range of organisms, systems, and geographic locations to the extent possible, although this was often influenced by the idiosyncratic nature of scientific research as well as by my idiosyncratic research into that literature. I therefore offer advance apologies if I appear to have overlooked an important paper or study, perhaps even one of your own contributions to the field. I fear this is inevitable, given the vastness of the literature, the sheer size of the task at hand, and the limited space available to address each topic.

In terms of the book itself, I thank Sharon Collinge, Henri Décamps, Kevin McGarigal, Jean Paul Metzger, John Wiens, and Jingle Wu for their suggestions, support, and enthusiasm for the initial book proposal. I am also indebted to the following colleagues who took the time to review and provide comment on various sections or chapters of the book (and sometimes, on more than one): John Briggs, Thomas Crist, Lenore Fahrig, Olivier François, Janet Franklin, Marie-Josée Fortin, Doug Goodin, Eric Gustafson, Colleen Hatfield, Nancy McIntyre, Jean Paul Metzger, Emily Minor, Rick Ostfeld, Luciana Signorelli, Mark Ungerer, Helene Wagner, Lisette Waits, Jingle Wu, and Patrick Zollner. I would also like to thank the graduate students who have participated in my landscape ecology course at Kansas State University, and who provided input on many of the initial chapter drafts and offered suggestions for improvement, including other examples or topics I had overlooked. Special thanks in this regard are due to Rachel Pigg, Jay Guarani, Mark Herse, Sean Hitchman, Nate Cathcart, Reid Plumb, Emily Williams, and E. J. Raynor.

Of course, this book would never have seen the light of day without a publisher. I very much appreciate the care and oversight provided by Ian Sherman and Bethany Kershaw, senior and assistant commissioning editors, respectively, for biology at Oxford University Press, and Ioan Marc Jones, senior production editor for academic texts at OUP, for managing the book's transition through to its final production phase. Paul Beverley had the unenviable task of copy-editing the entire text, but the book is better for his efforts. This project started out under the auspices of Sinauer Associates (now an imprint of OUP), and I therefore owe a large debt of gratitude to Andy Sinauer for the opportunity and for his support (and patience) during this long process; production editors, Kathaleen Emerson and Stephanie Bonner, as well as art director and production manager, Chris Small, for their guidance, design aesthetic, and attention to detail; Jan Troutt, for her beautiful artwork; Johannah Walkowicz and Michele Bekta for their efforts behind the scenes in obtaining permissions; copy editor Carol Wigg for comments and suggestions on the first set of chapters; and photo researchers, Mark Siddall and David McIntyre, for uncovering many of the images contained within these chapters. The end result is a textbook that is as attractive as I hope it is informative.

Last, but certainly not least, I thank my husband, Gray Woods, for his unremitting support and encouragement throughout the many years it has taken me to complete this project. I dedicate this book to our son, Johnathan, who I fear no longer remembers a time when his mother *wasn't* working on this book, becoming a sort of bibliographic sibling. Wise beyond his years, he had asked me at the start, "How do you know what to write?" I don't recall my answer then, but I have one for him now: figuring out what to write is easy when one has access to such great material; it's figuring out when to stop that is hard. For, as I now realize, the book will never truly be finished, but hopefully, this is a good and worthwhile start.

Kimberly A. With
Division of Biology
Kansas State University

Contents

Chapter 4   Landscape Pattern Analysis   127

1

An Introduction to Landscape Ecology
Foundations and Core Concepts

At first glance, this image (**Figure 1.1**) evokes a pastoral landscape, with cultivated fields of row crops and perhaps some lavender growing in the distance. Fields are neatly arrayed and delineated, some apparently fallow or newly plowed, others bearing regular furrows or bisected by roadways and irrigation canals. The landscape bears the strong imprint of human land use, with all of its lines and orderliness. Upon closer inspection, however, we come to realize that this is not an actual photograph, but an abstraction, a quilted landscape created using colored swatches of silk crepe. Nevertheless, the quilt does depict an actual landscape—the 'fields of salt' in a region of the San Francisco Bay Area where natural wetlands have been converted to salt ponds for industrial use. As with our earlier impression of an agricultural landscape, this image similarly calls to mind the dramatic effects humans can have on landscapes through the alteration of their structure and function. A quilt is thus an apt metaphor for the landscape: as patchworks of

Figure 1.1 *Fields of Salt*, art quilt. © Linda Gass 2007.

Essentials of Landscape Ecology. Kimberly A. With, Oxford University Press (2019).
© Kimberly A. With 2019. DOI: 10.1093/oso/9780198838388.001.0001

different land covers and land uses, landscapes are similarly shaped by humans and, like quilts, they are sometimes in need of restoration and conservation.

Why Study Landscape Ecology?

We are living in a transformative era. Since the start of the Industrial Revolution, humans have issued in a new geological period—the Anthropocene—a 'geology of mankind' (Crutzen 2002). There are few places on Earth that have not been touched, either directly or indirectly, by humans (Vitousek et al. 1997; **Figure 1.2**). A burgeoning global population has increased both our need for land and the mass consumption of resources provided by that land. Modern technologies have made it more efficient and economical for us to exploit the land and its resources, while global transportation networks and the globalization of economies have increased not only the interconnections among diverse regions of the globe, but also the extent of our impact on those regions. With landscapes worldwide being transformed at a rate and scale that rivals even the largest of natural forces, it should come as no surprise that a new and comprehensive science is needed

to tackle the complex ecological and societal consequences of human land use.

The human modification of landscapes is hardly a recent phenomenon, however. Landscapes bear the imprint of past human land uses that in some cases date back centuries or even millennia. We humans have a long history of altering the landscapes around us, whether it be for quarrying stone, felling forests, plowing grasslands, draining wetlands, or damming and diverting rivers. For example, the indigenous people of the Mississippian culture, which flourished more than a thousand years ago throughout the river valleys of the midwestern and southeastern United States, transformed these floodplain landscapes through the creation of huge earthen mounds. One cannot help but marvel at the industry of the people who created these earthen pyramids, the largest of which stands 30 meters tall and covers 6 hectares, solely by packing mud and clay by hand. The largest Mississippian site is Cahokia, located just east of modern-day Saint Louis, Missouri (**Figure 1.3**). During its heyday (around 1100), Cahokia was one of the world's great cities, with a population exceeding that of many European cities of the time, including London. The inhabitants of Cahokia were

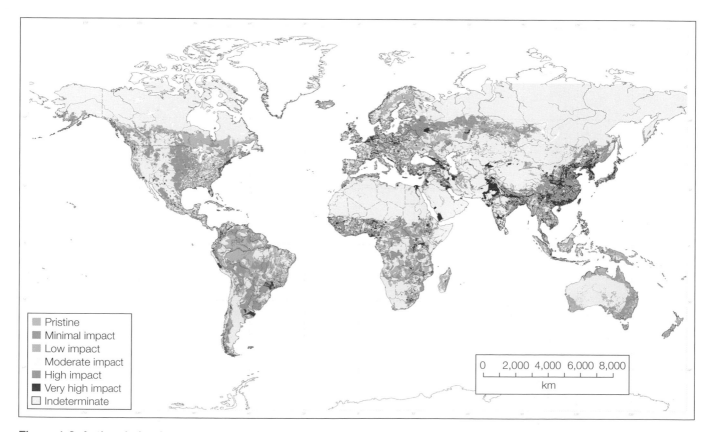

Legend:
- Pristine
- Minimal impact
- Low impact
- Moderate impact
- High impact
- Very high impact
- Indeterminate

0   2,000 4,000 6,000 8,000
km

Figure 1.2 Anthropic landscapes.   Humans have a pervasive influence on landscapes. Worldwide, few areas have not been affected—either directly or indirectly—by our activities. This global anthropic landscape map was generated from the overlay of a global human population density map with a global land-quality map to illustrate the distribution of anthropic tension zones where humans are having the greatest impact, particularly on soil resources.
Source: USDA-NRCS (2000).

(A)

(B)

Figure 1.3 Cahokia.   Cahokia was a cultural mecca in its day (1050–1250), a bustling urban center at the heart of what was then one of the largest cities in the world. (A) Aerial view of Cahokia today, with Monks Mound in the distance. (B) Artist's rendition of what the Central Plaza with Monks Mound might have looked like during Cahokia's heyday (~1100).

Source: (A) © National Geographic Creative/Alamy Stock Photo. (B) Courtesy of Cahokia Mounds State Historic Site, painting by L. K. Townsend.

early landscape architects, having carefully designed and engineered these mounds, which evidently required a great deal of technical expertise (Dalan et al. 2003). The largest mounds were likely built for ceremonial and religious purposes, and may have been topped by large buildings where the elites or rulers lived (**Figure 1.3B**). Although the stonemasonry involved in the construction of the better-known pyramids of ancient Egypt and Mesoamerica is also a marvel of engineering, these earthen mounds, which persist today as small hills dotting the landscape (**Figure 1.3A**), are a testament to the ability of even pre-industrial humans to physically shape the landscape.

The effect of human land use on landscapes—past, present, and future—is a major focus of landscape ecology. As humans alter the landscape, they also invariably alter its ecology, including the ecological flows and myriad ecological interactions that occur within the landscape. All landscapes are heterogeneous in that they are made up of a variety of landforms, ecosystems, vegetation communities (habitat types), and land uses, reflecting the different processes—physical, biological, and anthropogenic—that have shaped them. In turn, the structure of the landscape—its composition and configuration of habitat types or land uses—influences the biological and physical processes that give the landscape its form. The flows of materials, nutrients, and organisms across the landscape are all important for the maintenance of critical ecological functions that contribute to the structure and diversity (i.e. heterogeneity) of the landscape in the first place. **Landscape ecology**, then, is the study of the reciprocal effects of pattern on process: how landscape patterns influence ecological processes, and how those ecological processes in turn modify landscape patterns (Risser et al. 1984; Turner 1989; Pickett & Cadenasso 1995; Turner et al. 2001). From this standpoint, humans are but one of the many forces that shape landscapes, albeit an important one.

In this introductory chapter, we place the emergence of landscape ecology as a relatively new scientific discipline within its historical context by discussing the contributions from its elemental disciplines to its development, as well as the different schools of thought that have subsequently shaped its science and practice. Next, we highlight the core principles and major research themes in landscape ecology, which will be addressed more fully throughout the remainder of this textbook. We conclude with an overview of the book itself to help guide our study of landscape ecology.

Birth of a Discipline

In comparison to many other established fields in the natural and social sciences, or even to other areas of ecology, landscape ecology is a relatively new science. If we mark the birth of a scientific discipline by the establishment of a professional society and/or a scholarly journal dedicated to its study, then landscape ecology emerged only about 35 years ago. Since it is rare to witness the birth of a discipline, a brief overview of its development is warranted, especially since this provides insight into the different perspectives on the study and practice of landscape ecology today.

The International Association for Landscape Ecology (IALE) was officially founded in October 1982 at an international symposium held in the spa town of Piešťany, in what is now western Slovakia. Its conception, however, occurred some 18 months earlier, in April 1981, at the first international congress for landscape ecology at Veldhoven in the Netherlands (Antrop 2007). The formation of IALE thus represented the culmination of a long gestation among European ecologists and geographers, who as far back as the late 1960s and early 1970s had perceived the need for a broader, multidisciplinary science concerned with the management, planning, and design of landscapes (Naveh 2007; Wu 2007a). The IALE now consists of more than two dozen regional chapters from all over the world, representing individual countries as well as collectives of nations (e.g. Africa-IALE and IALE-Europe).

The largest of the IALE regional chapters is the United States chapter (US-IALE).[1] Its inaugural meeting was held in January 1986 in Athens, Georgia (USA), just a few years after the establishment of the international governing body, reflecting an early interest in the field by a number of American ecologists. Indeed, several American ecologists had been among the attendees at the landscape ecology symposia in Europe. In April 1983, 25 ecologists and geographers were invited to a three-day workshop in Allerton Park, Illinois (USA) to discuss opportunities for developing landscape ecology in North America (Risser et al. 1984; Risser 1995). The Allerton Park workshop represents a pivotal moment, not just for landscape ecology in North America, but for the discipline as a whole. The workshop is widely credited with establishing a new paradigm for landscape ecology, one that continues to guide it today—a focus on the reciprocal effects of landscape pattern and ecological process (Wiens 2008; Wu 2013).

At the outset, the idea of landscape ecology sounded appealing.... Urgency and a sense of responsibility for the quality of landscapes motivated us... and we realized that the integration of ecology and human activities was not only necessary but could bring new insights to the study of landscapes.

Forman & Godron (1986)

At around this same time, the publication of two seminal textbooks (one by Zev Naveh and Arthur Lieberman [1984], and the second by Richard Forman and Michel Godron [1986]) further helped to elevate and establish landscape ecology as a new discipline. The discipline's flagship journal, *Landscape Ecology*, published its first issue in 1987, with American ecologist Frank Golley at its helm (Golley 1987). The journal has since developed into one of the top-ranking journals in the fields of geography and ecology (Wu 2007a). Not surprisingly, these early scholars and leading figures in landscape ecology were all participants in the initial European symposia, the Allerton Park workshop, or both.

The mid-1980s thus represented a watershed period for landscape ecology. The rise of landscape ecology at that time can be attributed to a number of factors related to theoretical and conceptual developments in ecology, technological advances, and increasing concern over human impacts on the environment. In the field of ecology, several developments had particular relevance for landscape ecology: (1) island biogeography and metapopulation theory provided a new paradigm

---

[1] The United States regional chapter voted in April 2019 to become IALE-North America, so as to better reflect its multi-national constituency.

for studying the effects of habitat patchiness on eco-logical systems; (2) there was a growing recognition of the importance of spatial scaling in the design and interpretation of ecological research; and (3) there was a shift from viewing ecological systems as closed (iso-lated) and driven toward an equilibrium state to view-ing them instead as open (connected) and dynamic systems. In addition, technological advances in remote sensing, geographic information systems (GIS), and computer processing were making it possible to collect, store, analyze, and display unprecedented amounts of geospatial data over vast spatial extents, allowing for the first time a true study of landscapes at broad spatial scales. Finally, the growing concern over human modification of the environment, especially in the heavily industrialized nations where landscape ecology originated, was a motivating force in the development of the discipline in the mid-1980s. The cumulative effects of human land use were clearly responsible for the loss and fragmentation of habitats that in turn were contributing to a global extinction cri-sis (e.g. Wilcove et al. 1998). Indeed, it is no coincidence that conservation biology also emerged as a scientific discipline during the 1980s (Soulé 1985).

Although landscape ecology may not have emerged as a scientific discipline until the 1980s, its antecedents can be traced back decades and even centuries. Well before there was a journal or a society of landscape ecology, there were societies and journals devoted to human ecology, land-use planning, and design (i.e. land-scape architecture). Journals such as *Landscape Planning* and *Urban Ecology* (which have since been merged into a single journal, *Landscape and Urban Planning*) were first published in the mid-1970s, more than a decade before *Landscape Ecology*. Journals and professional organizations concerned with the management of eco-nomically important landscapes such as forests and rangelands date from the first half of the 20th century. Ecologists and geographers can trace their academic roots back even farther, to the many professional soci-eties and their journals that appeared in the late 19th and early 20th centuries. Indeed, a German geographer, Carl Troll, is widely credited with coining the term 'landscape ecology' (*landschaftsökologie*) in 1939 (Troll 1939). Over the ensuing decades, Troll continued to refine his view of landscape ecology, which he defined as 'the study of the main complex causal relationships between the life communities and their environment' that 'are expressed regionally in a definite distribution pattern (landscape mosaic, landscape pattern)' (Troll 1971). Troll's training in geography, coupled with his early interest in botany, contributed to his unique understanding of interactions between geomorphology and vegetation patterns, particularly in the mountainous regions of the world, where he used aerial photographs to study elevational vegetation gradients. As one of the first landscape ecologists, Troll deftly blended the spa-tial approach of the geographer with the functional approach of the ecologist, a combination that has come to epitomize landscape ecology today (Turner et al. 2001).

Regional Perspectives on Landscape Ecology

Early contributions by European geographers and ecolo-gists to the nascent field of landscape ecology helped define—as well as gave name to—a new discipline. The European perspective pervades the science and practice of landscape ecology in many parts of the world today, although the subsequent establishment of landscape ecology in North America followed a very different tra-jectory during its development. In this section, we con-sider how these different regional perspectives have contributed to the development of landscape ecology.

European Perspective

European landscape ecology has a long tradition that has greatly influenced its subsequent development and character. Europeans recognized early on that society was placing increasing demands on landscapes and that the environmental problems created by those demands were far too complex to be solved individu-ally by existing disciplines, but instead required a new multidisciplinary perspective to address them (Antrop et al. 2009). From its earliest beginnings, then, European landscape ecology has been marked by a strong holis-tic and human-centered perspective (Wu & Hobbs 2007; **Figure 1.4A**).

In densely populated Europe, the main concern is on cultural landscapes and the natural and cultural heritage related to these. Most traditional landscapes lose rapidly their ecological and heritage values, which are considered as "natural and cultural capital." There is a growing need to plan future land-scapes in an increasingly urbanized society and polarised environment in the perspective of sustainable development and participatory planning. Antrop et al. (2009)

The concept of holism, as applied to the landscape, implies that its functioning cannot be understood sim-ply by studying some aspect of it in isolation. The Dutch landscape ecologist Isaak Zonneveld, for example, was a proponent of the **land-unit concept** (Zonneveld 1989). The land-unit concept was first proposed by the Australian land surveyor and mapper Clifford Stuart Christian, who advocated for a systems approach to

(A) European perspective

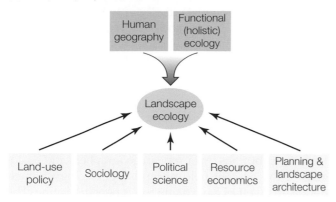

(B) North American perspective

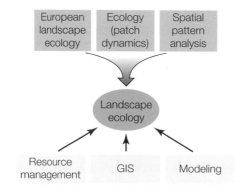

Figure 1.4 Regional perspectives on landscape ecology. The development of landscape ecology in the 1980s followed different trajectories in (A) Europe and (B) North America.
Source: After Wiens 1997.

the study of landscapes as 'hierarchical wholes.' In his view, a landscape is a system of interacting land units, defined as 'parts of the land surface…having a similar genesis and [that] can be described similarly in terms of the major inherent features of consequence to land use—namely, topography, soils, vegetation and climate' (Christian 1958, p. 76). We'll return to this idea of landscapes as hierarchical systems in **Chapter 2**. For now, it is worth noting that the land-unit approach was ultimately an effort to develop a more systematic way of defining, mapping, and integrating multiple attributes of landscapes (e.g. landforms, soils, vegetation, or land uses) that had traditionally been studied individually by different types of scientists. Beyond providing a more holistic approach to the study of landscapes as integrated wholes, however, the land-unit approach was also presented as a more efficient and cost-effective land survey method and mapping tool that could aid in 'the evaluation of the suitability of landscape for any kind of land use' (Zonneveld 1989, p. 68). This approach thus epitomizes the practical and human-centered applications of European landscape ecology, whose primary aim has been to facilitate the evaluation,

mapping, planning, design, and management of landscapes for human land use (e.g. Zonneveld 1972; Naveh & Lieberman 1984; Schreiber 1990).

In many respects, the European view of landscapes, and thus of landscape ecology, is a reflection of the political and economic integration of Europe itself. With the integration of the European Communities in 1967 and the subsequent formation of the European Union in 1993, the rapidly changing face of the European landscape is now being shaped largely by common policies, such as the Common Agricultural Policy. Although the EU does not have a common landscape policy, there has been some progress at the political level in recognizing the broad value of landscapes beyond the economic or resource benefits they provide. In 1995, the EU's European Environmental Agency published its Dobříš Assessment, a report on the state of the European environment, which includes a chapter devoted to landscapes and their importance to the future of the European environment (Stanners & Bourdeau 1995). Most notably, the chapter emphasizes the importance of preserving the unique character and diversity of landscapes as part of the natural and cultural heritage of Europe. The Dobříš Assessment is thus credited with helping to draw the attention of policymakers to the various pressures that are leading to a decline in the diversity, distinctiveness, and value of landscapes throughout Europe (Antrop et al. 2009).

The Dobříš Assessment was influential in the Council of Europe's development of the European Landscape Convention, which went into force in 2004. The convention is an international treaty for the comprehensive protection, management, and planning of landscapes throughout Europe, including all natural, rural, and urban landscapes as well as inland waters and coastal marine areas, regardless of their condition (it includes 'everyday, outstanding and degraded landscapes').

The landscape has an important public interest role in the cultural, ecological, environmental and social fields, and constitutes a resource favourable to economic activity and whose protection, management and planning can contribute to job creation;

…The landscape contributes to the formation of local cultures and…is a basic component of the European natural and cultural heritage, contributing to human well-being and consolidation of the European identity;

…The landscape is an important part of the quality of life for people everywhere;

…The landscape is a key element of individual and social well-being and…its protection, management and planning entail rights and responsibilities for everyone.

Preamble to the European Landscape Convention

Remarkably, the establishment of the European Landscape Convention was principally motivated not by economic concerns over the provisioning services provided by landscapes (although they are seen as one of its benefits), but rather by a concern for societal well-being and the quality of life for European citizens; that is, by the public good. Landscapes are perceived to have strong cultural as well as natural values and, as such, become part of a country's identity and cultural heritage. Because the landscape plays such an important role in the well-being of individuals and society at large, the Convention argues, all Europeans are adversely affected by a deterioration in the quality of their surroundings. The public should thus have some say, as participating stakeholders, in how landscapes are managed, rather than leaving these sorts of decisions solely to those with specialized or economic interests. By agreeing to the terms of the European Landscape Convention, participating nations have acknowledged their collective duty to make provision for the protection, management, and planning of landscapes to fulfill these sociocultural values, especially at the local or regional level. As of February 2019, the convention had been ratified by 39 of the 47 member states in the Council of Europe.

In response to these developments, IALE-Europe was formed in 2009 as a new supranational chapter of the International Association for Landscape Ecology (Antrop et al. 2009). IALE-Europe aims to promote collaboration among members of Europe's diverse academic community of landscape ecologists and to make their collective expertise available to institutional and societal stakeholders as well as the policymakers involved in decisions about landscape management and planning at the European level.

North American Perspective

The emergence of landscape ecology in North America was clearly influenced by its success in Europe, which had excited the interest of a number of ecologists in both the United States and Canada (**Figure 1.4B**). As mentioned previously, the landscape ecology movement in North America can be traced to the workshop held in 1983 at Allerton Park, a former estate now overseen by the University of Illinois at Urbana-Champaign near Monticello, Illinois. Most of the 25 invited participants were from the United States, although individual attendees from Canada and France were also present (Wiens 2008). This small group, most of whom were systems ecologists, formed the nucleus of what became a new movement in landscape ecology (Wu 2013).

The specific developmental pathway that landscape ecology took in North America might thus be attributed to a strong founder effect (Wiens 1997), as it deviated from its European roots as a result of the particular backgrounds and research interests of this small gathering of ecologists and geographers. This is immediately apparent from the definition of landscape ecology given in the report of that workshop, as the study of 'the relationship between spatial pattern and ecological processes [that] is not restricted to a particular scale' (Risser et al. 1984, p. 255). In contrast to the European perspective, this definition is clearly more ecological and explicitly spatial; it also allows for the possibility that landscapes—and thus landscape ecology—need not be concerned solely with broad spatial scales, an issue we'll return to later.

At the Allerton Park workshop, participants were asked to consider how landscape ecology might contribute to four major areas of inquiry (Risser et al. 1984):

1. How heterogeneity influences the flux of organisms, materials, and energy across the landscape.

2. The formative processes, both past and present, that give rise to landscape patterns.

3. How heterogeneity affects the spread of disturbances, such as fire, across the landscape.

4. How natural resource management might be enhanced by adopting a landscape ecological approach.

Thus, the workshop put forward many of the now-recognizable themes of landscape ecology—the importance of heterogeneity, scale, and disturbance dynamics for understanding the reciprocal effects of spatial pattern and ecological process (Wu 2013). In retrospect, this is not surprising, given that many of the participants had also been active in the International Biological Program (1964–1974), whose ambitions included the development of complete systems models to predict the effects of anthropogenic and environmental change on entire ecosystems (e.g. on total productivity), in addition to the more general application of ecosystem science to natural resource management (Boffey 1976). Thus, many of the Allerton Park participants were already, to quote one of them, 'primed…for thinking about landscapes in terms of flows and fluxes, energy and materials, and management implications' (Wiens 2008, p. 127).

Given the ecological and systems modeling backgrounds of many of the Allerton Park participants, it was perhaps inevitable that landscape ecology in North America would take on a more quantitative and spatial modeling character than that of its European congener (**Figure 1.4**). The Allerton Park workshop occurred at a time when microcomputers and GIS were both in their infancy, but already several participants were keen to exploit the opportunities these new technologies afforded landscape ecology, especially for analyzing broad-scale patterns of landscape change (Iverson 2007). The need to quantify and compare the spatial attributes of landscapes spurred the development of new landscape metrics (O'Neill et al. 1988a) as well as spatial modeling approaches to permit the statistical comparison of landscape data with known distributions (i.e. neutral landscape models; Gardner et al. 1987). Patch-based ecological theory, such as the theory of island biogeography, metapopulation theory, and patch dynamics theory, provided a quantitative and predictive framework that helped inform initial landscape ecological research into how patch structure could influence the structure and dynamics of ecological systems (MacArthur and Wilson 1967; Levins 1970; Pickett & White 1985).

The early focus on natural resource management is also quite telling. Whereas Europeans were principally concerned with the management of cultural landscapes, North Americans were more focused on natural or managed landscapes, of which vast tracts still remain. Many of these lands are owned and managed by the federal government. For example, the federal government owns more than a quarter of the land in the United States (28%), amounting to about 2.6 million km² (Gorte et al. 2012). For comparison, this is an area roughly two-thirds the size of the entire European Union. Most federal lands are located in the western half of the United States, where 42% of the land area is federally owned, and in Alaska, where 62% of the land area is federally owned. Federal lands are held in the public trust, and most (96%) are managed for a variety of competing purposes (timber, grazing, mining, recreation, wildlife, conservation) by four different government agencies: Forest Service, National Park Service, Bureau of Land Management, and Fish and Wildlife Service. The US National Forest System, for example, encompasses 780,000 km², which is about the size of the total land area of Germany, UK, Slovakia, Denmark, and the Netherlands combined. In Canada, forest covers about 4 million km² (equivalent to the size of the EU), and 93% of this forested land is publicly owned (crown land) and managed under the purview of either the provincial (77%) or federal (16%) crown (Annual Report, The State of Canada's Forests 2011). Little wonder then that forest and natural resource management should play such a major role in the subsequent development and application of landscape ecology in North America (Boutin & Hebert 2002; Liu & Taylor 2002; Bissonette & Storch 2003; Perera et al. 2007).

Globalization of Landscape Ecology

Although interesting from a historical standpoint, these different regional perspectives should not be taken too literally, at least in terms of how landscape ecology is currently practiced. There are many North American landscape ecologists who are concerned with land-use planning and management issues, just as there are many European landscape ecologists who study the relationship between landscape pattern and ecological process. Thus, to quote Wu & Hobbs (2007, p. 277):

> It is evident that the European and North American approaches to landscape ecology have differed historically. On the one hand, the European approach is characterized by a holistic and society-centered view of landscapes…On the other hand, the North American approach is dominated by an analytical and biological ecology-centered view of landscapes…This dichotomy, of course, is an oversimplification of the reality because neither of the two approaches is internally homogeneous in perspectives and because both have been changing as an inevitable consequence of increasing communications and collaborations among landscape ecologists worldwide.

These two traditional landscape perspectives have been variously adopted—and adapted—in other areas of the world, such as Australia, Latin America, and China. Australian landscape ecologist Richard Hobbs has characterized the Australian approach as taking 'a pragmatic middle road which combines both aspects' (Hobbs & Wu 2007, p. 7). Landscape ecology in Latin America, whose emergence led to the formation of three IALE chapters (Argentina and Brazil in 2005, and Chile in 2016) likewise appears to have embraced both perspectives. For example, the first bulletin for the IALE-Chile chapter (published May 2018) defines landscape ecology as 'an interdisciplinary science that studies the spatial variation of landscapes [across] a wide range of scales,' but also emphasizes the chapter's commitment to applying landscape ecology to the sustainable use of natural resources and 'to work and collaborate on public policies on management issues related to landscape ecology, such as territorial planning, ecosystem services and the effects of global change' (https://www.iale-chile.cl/). The development of landscape ecology in China was similarly influenced by the North American perspective at the outset, but now appears to be embracing the more holistic European perspective that places greater emphasis on landscape planning, design, and environmental management (Fu & Lu 2006, pp. 239–240):

> China is a developing country with [a] large human population and diversified environmental conditions. The drive for socioeconomic development is very strong, and at the same time environmental quality, resource usability and ecological security are also important concerns for the sake of regional sustainable development. Therefore, it is crucial to harmonize the relationships between human population growth, regional economic development and environmental conservation. Consequently, future landscape ecological research in China should take the responsibility of

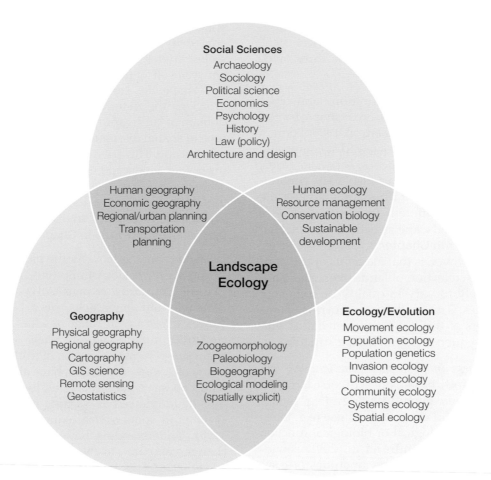

Social Sciences
Archaeology
Sociology
Political science
Economics
Psychology
History
Law (policy)
Architecture and design

Human geography
Economic geography
Regional/urban planning
Transportation
planning

Human ecology
Resource management
Conservation biology
Sustainable
development

**Landscape
Ecology**

Geography
Physical geography
Regional geography
Cartography
GIS science
Remote sensing
Geostatistics

Zoogeomorphology
Paleobiology
Biogeography
Ecological modeling
(spatially explicit)

Ecology/Evolution
Movement ecology
Population ecology
Population genetics
Invasion ecology
Disease ecology
Community ecology
Systems ecology
Spatial ecology

Figure 1.5 Domain of landscape ecology. Landscape ecology is a multidisciplinary science that occurs at the intersection of fields in ecology/evolution, geography, and the social sciences.

exploring the complex interactions between human activities and landscape dynamics under a holistic landscape framework, in which humans are treated as landscape ingredients equivalent to other biotic and abiotic components of the landscapes…Many aspects, including ecological, economic and cultural, should be integrated. The holistic approach is effective in studying the multifunctionality of Chinese landscapes.

Today, the apparent difference in perspectives on landscape ecology is perhaps less a 'European versus North American' distinction than an 'applied science versus basic science' dichotomy found in many fields in the ecological, biological, and physical sciences. Although this dichotomy is a false one in any science, it is especially so in landscape ecology. The relationship between science and practice is reciprocal (Wiens 2005). The two perspectives are in fact complementary, and both are ultimately necessary to the science and practice of landscape ecology (Hobbs & Wu 2007). Landscape ecology is a vibrant, cross-disciplinary science that not only integrates research across the natural and social sciences (**Figure 1.5**), but also transcends the

traditional boundaries between science and practice by promoting the integration of interdisciplinary research with stakeholder concerns about the environmental and societal consequences of human land use (Wu 2006).

Core Concepts of Landscape Ecology

From the start, landscape ecologists have emphasized the structure and function of landscapes and how landscapes change over time (Forman & Godron 1986). Three attributes are characteristic of all landscapes and provide the basis for quantifying and comparing them:

1. **Landscape structure** pertains to the diversity and spatial arrangement of landscape elements (e.g. habitat patches).
2. **Landscape function** refers to the interaction among these spatial elements (e.g. the flow of energy, nutrients, species, or genes among habitat patches).
3. **Landscape change** refers to how landscape structure and function vary over time.

Research in landscape ecology is motivated by several guiding principles or core concepts (Wiens 1997, 2005) that are variously related to these three landscape characteristics.

Landscapes are heterogeneous. By definition, landscapes are mosaics of different landforms, ecosystems, habitat types, or land uses (Forman & Godron 1986; Forman 1995b). While landscapes are often viewed as comprising discrete elements (e.g. habitat patches, corridors, roads, water bodies), heterogeneity can also vary continuously over the landscape, as it does along an ecocline or ecological gradient. Different formative processes give rise to different landscape structures, a topic that we will explore further in **Chapter 3**. One consequence of heterogeneity, however, is that elements of the **landscape mosaic**—the collection of land covers and land-use types—are likely to vary in quality or suitability for different species. This will be a recurring theme in the later chapters of this book, where we deal with the ecological consequences of landscape structure.

Landscapes are diverse in form and function. We can characterize landscapes based on a diverse array of features related to their geomorphology (e.g. mountainscapes), primary land cover or land use (e.g. forest landscape), a specific ecological or biological function (e.g. landscape of fear; Laundré et al. 2001; soundscapes, Pijanowski et al. 2011), the amenities or commodities they provide (e.g. agricultural landscape), or in relation to human occupation and values (urban landscape; cultural landscape). Landscapes need not be landlocked, however. If heterogeneity is a defining characteristic of landscapes, then marine and freshwater systems would similarly qualify, for they also exhibit heterogeneity in the distribution of substrates, habitats, resources, and environmental conditions. Thus, we can define riverine landscapes (Wiens 2002) and marine landscapes or 'seascapes' (Pittman et al. 2011). These are all 'landscapes' in that they comprise spatially heterogeneous areas, whose study might therefore benefit from a landscape ecological perspective.

Landscapes are scale-dependent. Although landscapes have traditionally been viewed on human terms, as areas of broad spatial extent, there has been a persistent movement within landscape ecology to define a landscape simply as a spatially heterogeneous area (Turner 1989) that is scaled relative to the process or organism of interest (Wiens & Milne 1989). Two species that occur within the same habitat, such as grasshoppers and bison (*Bison bison*) on the tallgrass prairies of North America, are likely to have very different perceptions of the landscape in terms of the distribution and availability of their preferred forage, given the different scales at which they each operate. The idea that landscapes are scale-dependent acknowledges that patchiness (i.e. heterogeneity) may exist simultaneously across a range of scales. Multiscale patch structure may reflect the different scales at which various processes or disturbances that shape the landscape operate. Such a view also broadens the definition of 'landscape,' and thus the domain of landscape ecology. From this viewpoint, any type of spatial distribution at any scale could constitute a landscape. This perspective is important, for it recognizes that landscape ecology is not simply 'regional ecology' or 'broad-scale ecology,' but a research paradigm for investigating the effect of spatial pattern on ecological processes at any scale. We will discuss this perspective more fully in **Chapters 2 and 3**.

Landscapes are dynamic. As the biochemist and science fiction writer Isaac Asimov once famously observed, 'the only constant is change.' All ecological systems are dynamic, and landscapes are no exception. Disturbances across a wide range of scales, from those occurring over minutes within a few square centimeters to those operating over tens of millennia across thousands of square kilometers, have all contributed to the landscapes we see around us today. Landscapes are thus better viewed as 'shifting mosaics' than as static systems in some sort of equilibrium (Wu & Loucks 1995). Understanding how anthropogenic landscape change compares to the natural disturbance regime, and whether it fits within the range of historical variation for a particular landscape, is important for evaluating the potential impacts of human activities on the structure and function of landscapes. We will discuss this topic more fully in **Chapter 3**. Furthermore, the rate at which landscapes change may be just as important as—if not more important than—the resulting structural changes (e.g. in the amount or configuration of habitat) for certain ecological responses or landscape functions. For example, species may exhibit a lagged response to rapid habitat loss or fragmentation, such that an assessment of a population's responses to landscape change might underestimate its actual risk of extinction (Schrott et al. 2005). We will consider the effects of landscape dynamics on a variety of ecological responses throughout this book.

Spatial context is important. Given that landscapes are heterogeneous, we can expect that ecological dynamics will vary spatially. For example, if habitats vary in their suitability for a particular species, then population growth rates of that species will likewise vary among habitat types. High-quality habitats should support viable populations, whereas low-quality habitats cannot (Pulliam 1988). Our understanding of the population dynamics within patches thus requires information on habitat quality (in other words, population dynamics are habitat-dependent). However, habitat quality may also be spatially dependent, such that two sites within the same habitat might vary in quality depending on their specific location (i.e. their spatial context). For example, reproductive success for forest-breeding songbirds might be higher for individuals that nest in the center of a habitat patch than at its margins (where nest predation rates tend to be higher), or in a forest patch surrounded by second-growth forest than in one surrounded by agricultural fields (where again, nest predation rates are expected to be higher), or in a landscape that is still predominantly forested than in one that has very little forest cover remaining (where nest predation rates are uniformly greater; Donovan et al. 1997). Thus, spatial context may well be important for understanding what goes on within individual patches (or landscapes), and will definitely be important for understanding what goes on between patches.

Ecological flows are important. Because spatial context is important, we should anticipate that the nature of the patch boundary and the intervening **matrix**—the mosaic of land-cover or land-use types that occurs between habitat patches—can influence the magnitude of ecological flows among patches. Patch boundaries may be porous to the movement of certain organisms but impermeable to others, depending on how different organisms perceive and respond to the structure of the **habitat edge**, which occurs at the juxtaposition of different vegetation communities. The transition between vegetation types may be abrupt, creating what is known as a 'hard edge,' especially if organisms are unwilling or unable to cross that boundary, or may be a more gradual transition from one habitat type to the next, creating a 'soft edge.' Flows may also occur between ecosystems, such as across the land-water interface. Agricultural or stormwater runoff from the surrounding landscape is a familiar source of non-point-source pollution in marine and freshwater systems. However, these flows do not occur in just one direction, from land to water, but may also occur in the opposite direction, from water to land, as in the case of marine subsidies along coastal areas or desert islands (Polis & Hurd 1996). Asymmetrical flows across patch or system boundaries can have profound effects on the dynamics within, as well as between, patches or systems on the landscape. This will be a recurring theme in many of the later chapters of this book, where we consider the effects of asymmetrical flows (individual movement and dispersal, gene flow, and nutrient flows; **Chapters 6, 9, and 11**) on various ecological processes.

Connectivity is important. The notion that organisms, materials, or nutrients flow to varying degrees among patches or systems on the landscape implies that some areas of the landscape are connected, at least functionally, even if they are not obviously connected in a structural sense (e.g. via habitat corridors). Connectivity can be considered an emergent property of landscapes; it emerges as a consequence of the interaction between ecological flows and the landscape pattern (Taylor et al. 1993; With et al. 1997). It could thus be argued that much of landscape ecology is ultimately concerned with the measurement and study of connectivity. Connectivity is important for understanding the propagation of disturbances across the landscape; the movement and redistribution of organisms, materials, and nutrients; the resulting structure and dynamics of populations; gene flow and population genetic structure; the spread of invasive species and diseases; community patterns and dynamics; and ecosystem structure and function (**Chapters 6–11**). Thus, connectivity will be a pervasive theme throughout this book, and an entire chapter (**Chapter 5**) is devoted to this important concept.

Landscapes are multifunctional. Humans are the principal driver of landscape change worldwide, as landscapes are increasingly being transformed and used for a variety of ecological, societal, and economic functions (Ojima et al. 1994b; Vitousek et al. 1997; **Figure 1.2**). Landscapes are thus multifunctional in that they provide humanity with an array of goods and services. Therefore, landscape management requires a means of identifying and resolving the conflicts that inevitably arise in response to competing interests and valuation systems (Mander et al. 2007). Sustainability is a key concept in landscape ecology (Wu 2006) and is the basis for sound ecosystem and natural resource management (Liu & Taylor 2002; Bissonette & Storch 2003). Recall that landscape ecology arose in response to the perceived need to manage resources more holistically and at a broader landscape scale. Although we will consider the management implications of landscape ecology throughout this book, the final chapter (**Chapter 11**) examines in greater detail how principles derived from landscape ecology can be used to meet the environmental and societal challenges that stem from human land use (i.e. landscape sustainability).

Organization of this Book

The demands of time and your personal interests will necessarily dictate how you use this text. Because many diverse fields contribute to landscape ecology, readers of this book are also likely to be quite diverse. Thus, although the book's organization represents a natural ordering of topics (at least from the author's perspective), your individual interests and needs will obviously dictate which topics are emphasized, and in what order.

The first five chapters cover not only the discipline of landscape ecology (introduced in this chapter), but also its major research themes, including issues of spatial and temporal scale (**Chapter 2**), landscape heterogeneity and dynamics (**Chapter 3**), landscape pattern analysis (**Chapter 4**), and landscape connectivity (**Chapter 5**). These chapters provide many of the core concepts that will be emphasized repeatedly throughout the remainder of the book. **Chapter 5**, on landscape connectivity, could be considered a bridge chapter, given that it spans the domains encompassed by landscape pattern analysis in the chapter that precedes it and ecological responses to landscape pattern in the chapters that follow.

Chapters 6–11 consider the ecological consequences of spatial pattern for a wide range of processes and phenomena involving individual movement and dispersal (**Chapter 6**), population distributions and dynamics (**Chapter 7**), population spatial spread (**Chapter 8**), gene flow and population genetic structure (**Chapter 9**), community structure and dynamics (**Chapter 10**), and ecosystem structure and function (**Chapter 11**). Although readers should feel free to focus on chapters that are of particular interest, the chapters have been developed and arranged in a hierarchical fashion, such that material in later chapters builds on concepts and approaches presented in earlier chapters.

Each chapter provides a mix of basic concepts, examples, and case studies along with more advanced topics related to the theoretical foundation, quantitative methods, or modeling applications relevant to a particular area of research. The beginning student, practitioner, or casual reader wishing for an overview of the basic concepts should focus on the introductory sections of the chapter and the chapter summary points at its end. More advanced students and research scientists are likely to benefit most from the more in-depth coverage of methodologies, analyses, and modeling considerations featured in most chapters. However, this book is not an instruction manual. Although issues involved in the collection, analysis, and modeling of spatial data are discussed, the primary objective is to give you the necessary information with which to evaluate these concerns from the standpoint of your own interests, research, or management needs. The intent is to provide an overview of available tools and methods, along with some general guidance as to their use, and then direct interested readers to additional resources where they can obtain more detailed information on the topic.

Finally, as a pedagogical tool, this book has been organized around the way in which these topics are presented in my own course on landscape ecology. It is hoped that the discussion questions at the end of each chapter will challenge the reader to think more deeply or broadly about the topics presented within the chapter. These questions can also help facilitate discussion within a classroom or seminar setting. Some of these questions can be used for class assignments or as essay questions on examinations, and thus might prove especially useful in that regard for instructors—and students—of landscape ecology.

Chapter Summary Points

1. Landscape ecology studies the reciprocal effects of spatial pattern (heterogeneity) and ecological processes. It emphasizes the structure, function, and change in landscapes over time.

2. The ways in which human land-use activities modify landscape structure and function are a major research focus in landscape ecology, which provides a scientific basis for understanding and managing landscapes as well as the goods and services they provide.

3. Landscape ecology is a relatively new scientific discipline, having become established some 35 years ago in Europe before spreading to North America, Australia, and elsewhere. Its historical roots can be traced back much earlier, however. The German geographer Carl Troll first coined the term 'landscape ecology' (*landschaftsökologie*) in 1939 in reference to his study of the interaction between geomorphology and vegetation patterns along elevational gradients.

4. The rise of landscape ecology in the mid-1980s may be attributed to three factors: (i) an increasing concern over human impacts on the environment, especially in the heavily industrialized nations where landscape ecology first originated; (ii) technological advances in remote sensing, geographic information systems, and computer processing, which made the collection, analysis, and modeling of geospatial data over broad regional areas not only feasible, but efficient; and (iii) conceptual and theoretical developments within the field of ecology, especially patch-based theory, a growing recognition of the importance of spatial scale for the design and interpretation of ecological research, and a shift away from equilibrium theories to dynamic views of ecological systems such as landscapes.

5. Different regional perspectives influenced the early development of landscape ecology as a science. In Europe, the focus was more on land-use planning and the management of cultural landscapes; the focus was thus squarely human-centered. In North America, the development of the discipline was more heavily

influenced by ecological theory and spatial modeling applications, and thus the focus was more on the study and management of ecological landscapes. Landscape ecology has benefitted from these diverse perspectives and has matured into a cross-disciplinary science that integrates the natural and social sciences and transcends the traditional boundaries between science and practice.

6. Research in landscape ecology is motivated by eight core concepts: (i) landscapes are heterogeneous, and this heterogeneity is important for understanding spatial processes across the landscape; (ii) landscapes are diverse and are found in aquatic and marine systems as well as terrestrial ones; (iii) landscapes are scale-dependent, in that heterogeneity exists across a wide range of scales, and thus landscape ecology is not restricted simply to the study of broad spatial extents; (iv) landscapes are dynamic, such that the rate of landscape change can be just as important as the magnitude of change for understanding and predicting its consequences; (v) spatial context is important for understanding the distribution and dynamics of ecological systems; (vi) ecological flows are important to many ecological phenomena and are influenced by the differential permeability of landscape elements and the nature of patch boundaries; (vii) connectivity is important and emerges as a consequence of the interaction between landscape pattern and ecological process; (viii) landscapes are multifunctional, which requires that resources and land uses be managed sustainably.

Discussion Questions

1. In what ways might regional differences in how landscape ecology is defined and practiced (e.g. the European versus North American perspectives) contribute positively to the growth and development of the field? In what ways might such regional differences be detrimental?

2. Although landscapes are traditionally viewed as encompassing broad areas of *land*, a different view of landscapes as simply 'spatially heterogeneous areas' defined relative to the organism or process of interest has since emerged within the field of landscape ecology. Discuss how this more general definition of 'landscape' affects the science and practice of landscape ecology.

3. Which of the core concepts in landscape ecology could be applied to the system in which you are currently working? How might a landscape ecological perspective thus prove useful to the study or management of this system?

Scaling Issues in Landscape Ecology

Impressionistic landscape painters of the 19th century were not only masters of light, they were masters of scale. Up close, the painting in **Figure 2.1** is nothing more than daubs of paint that create mosaic-like patches of color. When viewed at a distance, however, the colors come together to create the illusion of shadows, reflections, movement, and the impression of dappled light on shimmering water. Our ability to perceive the overall landscape is influenced by the scale, or distance, at which the painting is viewed. The mastery of impressionist painters lies in their ability to render these visual effects, based on an understanding of how a fine-scale mosaic of color will 'scale up' to create the overall impression of a landscape. Scale is thus important for creating (and appreciating) impressionistic landscape paintings, but how might scale be important for understanding patterns in actual landscapes?

Why is Scale so Important in Ecology?

The evolutionary biologist Theodosius Dobzhansky famously said that 'Nothing in biology makes sense except in the light of evolution.' We can just as easily assert that 'Nothing in ecology makes sense except in the light of scale.' Patterns are ubiquitous in nature, and ecological systems naturally exhibit patchiness across a wide range of spatial and temporal scales. As a consequence, there is no single 'best' or 'natural' scale at which ecological phenomena should be studied—a fact that is reflected in the wide range of disciplines that have emerged in ecology, focusing on different levels of organization across a range of scales, from the behavioral ecology of individuals to the dynamics of ecosystems. By definition, landscapes are inherently patchy or heterogeneous, at least across some range of scale. Our perception of patchiness (and thus of the landscape), as well as our ability to assign the correct mechanisms to observed patterns, is critically dependent on scale. Landscape ecology, perhaps more than any other field of ecology, is expressly concerned with understanding how scale affects the measurement of heterogeneity and the scale(s) at which spatial patterns are important for ecological phenomena. This dependence of pattern on the scale of observation has been called 'the central problem in ecology' (Levin 1992).

It is argued that the problem of pattern and scale is the central problem in ecology, unifying population biology and ecosystems science, and marrying basic and applied ecology.

Levin (1992)

Essentials of Landscape Ecology. Kimberly A. With, Oxford University Press (2019).

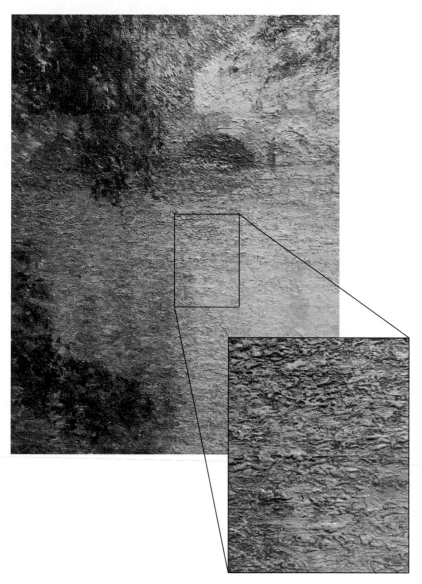

Figure 2.1 Impressionistic landscape painting. *Mill at Limetz* by Claude Monet (1888), on display at the Nelson-Atkins Museum of Art, Kansas City, Missouri (USA).

Source: Art credited in the caption. Photo by Kimberly A. With.

Consider that many pressing environmental problems are related to human land use, which occurs at broad spatial scales and can transform entire ecosystems fairly quickly, with far-reaching consequences. However, most ecological research is short-term and conducted at much finer scales than the regional scale of human land use, producing a scaling mismatch between the scale of the study relative to the scale of the problem. The patterns and processes we measure at fine spatial scales and over short time periods are unlikely to behave similarly at broader scales and extended time periods. Clearly, our ability to predict the consequences of human impacts on the environment requires that we be able to extrapolate information across scales of both space and time. An understanding of pattern-process linkages—the study of landscape ecology—thus requires an understanding of how patterns change with spatial and temporal scale, as well as the development of methods for extrapolating information across scales.

In this chapter, we begin with a discussion of how scale is defined and used in different disciplines that contribute to landscape ecology. We next discuss elements of scale and how different dimensions of scale influence our ability to detect patterns. This then leads to a consideration of how we go about defining the appropriate scale for study, and the various factors that ultimately constrain our choice of scale. Because selecting the appropriate scale for study might be easier if we better understood how the system was structured or organized, we briefly consider the applications of hierarchy theory for landscape ecology, including the question of whether the landscape is in fact a level of organization in the traditional biological hierarchy. Finally, we turn our attention to issues involving the extrapolation of information across scales and levels of organization, and how this ultimately affects our ability to scale up and make predictions about future system behavior (i.e. ecological forecasting).

Uses (and Misuses) of Scale in Ecology

To geographers, cartographers, and most other people, scale is the resolution or degree of reduction of a map. In other words, **map scale** is the ratio of the distance on a map to the corresponding distance on the real landscape. For example, if 1 cm on a map is the equivalent of 100 m (10,000 cm) in the real world, then the map scale is 1/10,000 or 1:10,000. If 1 cm on a map is the equivalent of 10 km (1,000,000 cm) on the ground, then the map scale is 1/1,000,000 or 1:1,000,000. Because the degree of reduction is 100 times greater in the second map, objects will appear much smaller than on the first map. Thus, maps with a high degree of reduction (e.g. a map of North America) are considered **small scale** (1/1,000,000 = 0.000001). Conversely, maps that are **large scale** (e.g. a street map of New York City's Greenwich Village) present much more detail because the degree of reduction is less (1/10,000 = 0.0001). Obviously, the total area of coverage will be less for large-scale than for small-scale maps. There is thus a trade-off between map scale—the resolution or amount of detail presented in a map—and the area of coverage. Nevertheless, the distinction between large versus small scale is somewhat arbitrary; in practice, maps having a scale 1:75,000 or greater (e.g. 1:25,000 or 1:10,000) are considered to be small scale.

The cartographer's definition of scale, however, is the *opposite* of how ecologists define scale. To an ecologist, 'large scale' implies an area of large spatial extent (e.g. an entire mountain range), whereas 'small scale' refers to an area of smaller extent (e.g. a meadow on one peak in the range). Thus, large scale to a cartographer is small scale to an ecologist. Since landscape ecology is a multidisciplinary field, made up of both geographers and ecologists, it is best to avoid confusion in the use of these terms and reserve large and small scale for map scale (i.e. the scale or ratio of spatial reduction), and instead use the terms **broad scale** and **fine scale** when referring to landscapes or other ecological phenomena. This is the convention used throughout this textbook.

Ecological Scale

Ecological scale refers to the measurements that bound the observation (data) set, which is defined in terms of grain and extent (**Table 2.1**).

- The **grain** of the observation set is the finest scale at which data are measured or obtained; it is the resolution of the dataset, and it applies both to the spatial grain and the temporal grain of the study. An example of the **spatial grain** of an ecological study might be the size of the sampling frame used to measure vegetation, the size of a net or sieve used to sample organisms in the water or soil, or the pixel size of remotely sensed imagery. The **temporal grain** of study refers to the frequency or minimum interval over which data are collected (e.g. whether on an hourly, daily, monthly, or annual basis).

- The **extent** of the observation set refers to over how large an area or how long a time period the study is conducted. As with grain, we can specify the spatial and temporal extent of the study. Examples of **spatial extent** include the size of the study plot (ha) or transect (km) over which data are collected, the regional extent of the study (encompassing the sum total of study plots), or the area of coverage in a satellite image. **Temporal extent** pertains to the duration of a study, in terms of over how long a time period the data were collected (e.g. whether over a single season or several years).

Relationship between Grain and Extent

Studies involving small spatial extents usually require sampling at a finer grain in both space and time than those conducted at broader scales. Partly this is due to the fact that processes that operate at finer spatial scales also operate at faster rates, thus necessitating more frequent sampling. This is a consequence of *hierarchical system organization*, which is discussed later in this chapter.

The relationship between the spatial and temporal scaling of ecological phenomena can be depicted as a **space-time diagram**. Space-time diagrams were first developed by the physical oceanographer Henry Stommel to highlight scales bounding important biological and physical phenomena within oceanic systems (Stommel 1963), and space-time diagrams are sometimes referred to as Stommel diagrams in his honor (Vance & Doel 2010). To illustrate the processes involved in the formation of vegetation patterns at different scales, Delcourt and her colleagues (1983) adapted the space-time diagram to include different types of environmental disturbances, each of which operates within a particular **domain of scale** (e.g. the range of scales encompassing a particular type of disturbance; see **Table 2.1**), as well as the scale of biotic responses to those disturbances. Vegetation patterns thus manifest

TABLE 2.1   Some scale-related terms and definitions

Term	Definition	Examples
Grain	Finest scale of the observation or dataset; applies to both spatial and temporal dimensions of the study (spatial grain; temporal grain)	Size of sampling frame; resolution (pixel size) of remotely sensed imagery; frequency of sampling; minimum length of observation period
Extent	Broadest scale that encompasses the observation or dataset; applies to both spatial and temporal dimensions of the study (spatial extent; temporal extent)	Size of the study area (plot size) or total area surveyed; duration of study (number of seasons or years)
Scale	In ecology, the units of measurement that define the grain and extent of the study system or observation (data) set; the scale of observation	A fine-scale study has a limited spatial extent, whereas a broad-scale study encompasses a larger spatial extent
Domain of scale	Range of scales (grain and extent) that bound a particular ecological phenomenon; a scaling relationship that is linear or constant across a particular range of scales	Power-law relationships
Level	A system of interacting components in a hierarchically organized system; a level of biological or ecological organization	The cell represents a level of biological organization, in that cells are made up of organelles, but are also the building blocks that give rise to specific types of tissues or organs
Hierarchical structure	In a nested hierarchy, each level or scale contains, and is made up of, lower levels or scales	Landscapes may exhibit hierarchical patch structure, in which smaller patches are nested within larger patches (**Figure 2.8**)
Hierarchical organization	System of interacting systems in which hierarchical structure emerges, shaped by the relative strength and frequency of interactions among components, both within and between system levels	Populations may exhibit hierarchical organization, in that they are made up of individuals, whose interactions lead to the emergence of social systems and other group behaviors; populations also interact with each other to give rise to source–sink metapopulation dynamics, which influence the likelihood that the species will persist in a given landscape
Scaling up (upscaling)	Extrapolation of data or model predictions obtained at fine scales to infer patterns or system behavior at broader scales in space or time	Studies of insect movements within microlandscapes are assumed to have relevance for understanding the movement responses and dispersal success of larger animals in fragmented landscapes
Downscaling	Extrapolation of data or model predictions from broad scales to infer patterns or system behavior at finer scales	Coarse-scale data on species distributions may be used to predict the local occupancy of a species using occupancy-area relationships

across a wide range of spatial scales, from the fine scale of individual plants or populations all the way up to the physiographic provinces and vegetation formations at continental scales, and across corresponding timescales that range from seasons or years to glacial–interglacial cycles (**Figure 2.2**).

To illustrate how this can be applied to real systems, consider the space-time diagram for various types of disturbances responsible for shaping a boreal forest landscape (Peterson et al. 1998; **Figure 2.3**). Disturbances in this system include physical ones, such as fire and climatic cycles, as well as biotic disturbances resulting from spruce budworm outbreaks and herbivory. In the case of herbivores, the magnitude of the effect is very much dependent on the spatial range encompassing scales of movement, from short-range

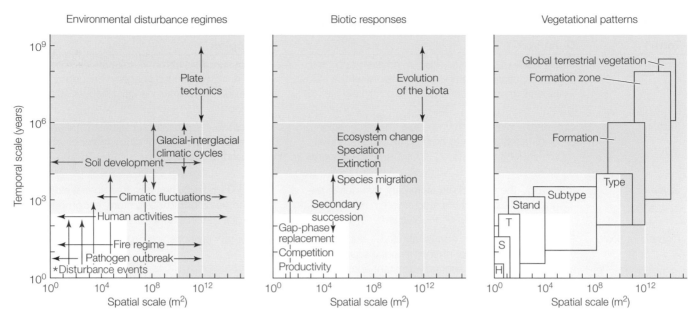

Figure 2.2 Space-time diagram.   The spatial and temporal scales bounding different disturbance regimes, biotic responses, and associated vegetation patterns are represented here by a series of space-time diagrams. Disturbance events include wildfires, wind damage, clearcuts, floods, and earthquakes. In the rightmost graph, abbreviations are H, herbs; S, shrubs; T, trees.
Source: After Delcourt & Delcourt (1988).

foraging decisions to infrequent long-distance dispersal events, which in turn is a function of body size. We'll discuss the effect of different types of biotic and abiotic disturbances—and the scales at which they occur—on the formation of landscape patterns more fully in **Chapter 3**.

From a research standpoint, most studies are conducted at finer scales or are of shorter duration than the ecological phenomena they seek to explain. Inevitably, there is a trade-off between grain and extent in terms of logistics and sampling feasibility. Studies conducted at a broad spatial extent, including many landscape studies, may preclude fine-scale sampling owing to the inordinate demands this would place on time, personnel, and funding. For example, it is a practical impossibility (and likely unnecessary, in any case) to sample a 100 km² landscape using a 1 m² sampling frame on a daily basis. Sampling even a 1 ha plot at this spatial resolution would result in 10,000 data points, which would be overwhelming, especially if attempted on a daily basis. Technology is rapidly closing the gap, however, making the collection of enormous amounts of information over large areas at a fine spatial resolution not only possible, but even cost-effective. For example, Landsat satellite imagery has a resolution of 15 m (panchromatic) or 30 m (multispectral) for the entire planet, with a return interval of 16 days, and is freely available. The latest high-resolution commercial satellite imagery provides a ground resolution of 1 m or less, but is not freely available. Processing such large

amounts of data still remains a daunting and time-consuming task, however.

Effect of Changing Grain and Extent

Two locations on a landscape will tend to become less similar the farther apart they are in space (and probably time as well). This is the basis for **Tobler's first law of geography**, which states that 'everything is related to everything else, but near things are more related than distant things' (Tobler 1970). This is also integral to the concept of spatial autocorrelation in geostatistics, which measures the degree of correlation between two samples taken at different distances apart, and which will be covered in more detail in **Chapter 4**. For now, we will assume on the basis of Tobler's law that spatial variance increases with increasing distance (i.e. scale). In other words, the variability among our samples is expected to increase, the larger the area and the more distant the locations we sample.

How, then, does changing the grain and extent affect spatial variance? If we hold grain size constant (e.g. we use a single sampling frame of a given size), then sampling over a larger area (increasing spatial extent; **Figure 2.4A**) will generally increase the variability contained within the dataset because of an increase in the number of habitat types or other landscape features encountered (i.e. spatial heterogeneity increases). Thus, the between-sample variation (spatial variance) is expected to increase with an increase in spatial extent (Wiens 1989a; **Figure 2.5B**).

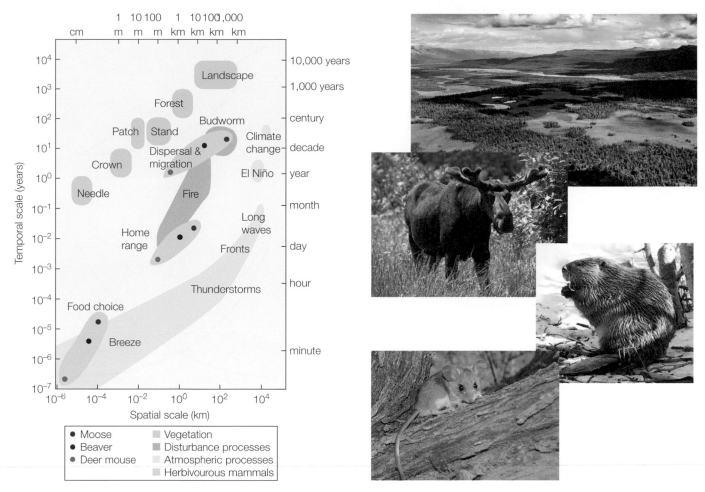

Figure 2.3 Space-time diagram of the boreal forest. A variety of disturbances, both abiotic and biotic, shape the structure of the boreal forest landscape. Herbivores such as the moose (*Alces alces*), beaver (*Castor canadensis*), and deer mouse (*Peromyscus maniculatus*) operate across a range of scales on the landscape, from daily foraging movements to annual dispersal events. The impact of these herbivores will thus depend on their scale of space use, which in turn is a function of body size.

Source: After Peterson, Allen, & Holling (1998). Photos: Moose by Ray Dumas and Deer Mouse by Gregory "Slobirdr" Smith (CC BY-SA 2.0 License). Boreal landscape by Carol M. Highsmith. Beaver photo © Jody Ann/Shuttestock.com.

If we instead hold extent constant, we can increase the grain size of our observation set by increasing the 'sampling window' or 'box size' of our analysis using some type of data aggregation procedure (e.g. averaging data over a larger area to achieve a coarser resolution in the dataset; **Figure 2.4B**). Increasing the grain of observation or measurement (i.e. a decrease in the spatial resolution of the dataset) generally results in a decrease in spatial variance (**Figure 2.5A**). As grain size increases, more of the spatial variability inherent in the system is contained *within* each sample (**Figure 2.5B**). As a consequence, samples become more similar to each other as the grain size approaches the spatial extent of the observation set. Thus, between-sample variation decreases with an increase in grain size (Wiens 1989a; **Figure 2.5B**).

To illustrate, consider a heterogeneous landscape consisting of different land-use/land-cover types (**Figure 2.6**). At the finest scale, corresponding to our data resolution

(a 1 × 1 pixel or window size), the landscape contains numerous patches, both large and small, and appears to have a river running along the lower portion of the map. As we increase the grain or window size of the analysis, a majority rule is used to assign the value of the most-common cover type within each window to that larger pixel; thus, if 60% of the pixels within a 10 × 10 window are agriculture, the entire 10 × 10 pixel would be classified as agriculture. Thus, as we increase the grain size of our analysis, patches take on a more blocky appearance and the smaller scattered patches disappear altogether. At the coarsest resolution (60 × 60), we still have the overall impression of the original landscape, in which the dominant cover types and largest patches persist. Note, however, that we have lost some features—the river no longer spans the landscape but has been reduced to a few 'pixel puddles,' which disappear entirely if we increase grain size further (i.e. to 80 × 80 or 100 × 100) (Wu et al. 2002). Thus,

(A) Changing extent, holding grain constant

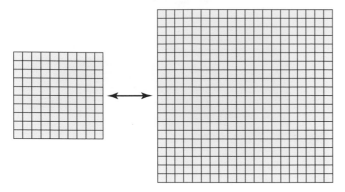

(B) Changing grain size, holding extent constant

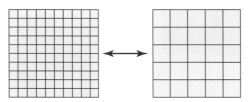

Figure 2.4 Relationship between spatial grain and extent. Changing the spatial grain and extent of an ecological investigation can reveal how patterns change as a function of scale. For instance, we can expand the scope of our study simply by increasing the spatial extent, while maintaining the same sampling grain (A). Alternatively, multiscale analyses of spatial pattern (heterogeneity) typically increase grain size while holding extent constant (B).

Source: Wu et al. (2002).

(A)

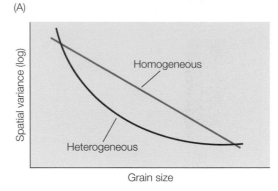

(B)

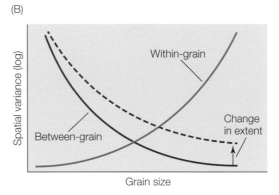

Figure 2.5 Effect of increasing grain size (scale) on spatial variance. (A) Spatial variance tends to decline with increasing grain size. The relationship is expected to be linear when the distribution is uniform or homogeneous, but may be curvilinear or strongly non-linear depending upon the distribution and scale of patchiness in heterogeneous landscapes (see variance staircase, **Figure 2.7**). (B) The effect of changing grain size on heterogeneity within versus between samples (grain). The difference *between* samples decreases with increasing grain size because more of the heterogeneity inherent to the system is contained within the individual samples. The degree of heterogeneity (spatial variance) *within* samples therefore increases with increasing grain size. However, changing the spatial extent of the study will likely lead to an overall increase in heterogeneity, and thus the relationship between spatial variance and scale is shifted upward for the between-grain comparison among samples.

Source: After Wiens (1989a).

increasing the grain size of the analysis leads to a loss of information—a decrease in heterogeneity—owing to the loss of cover types that were already uncommon on the landscape. In particular, these rare cover types are more likely to be lost when they occur as small scattered patches versus as a clumped distribution; if the distribution is clumped, patch sizes are larger, which increases the likelihood that the rare cover type is in the majority at a particular location (Turner et al. 1989b). We will discuss the effects of scale on the analysis of landscape pattern more fully in **Chapter 4**.

Given that increasing grain size may result in a loss of information, why would we ever wish to do this? Systematically increasing grain size is critical to the analysis of how pattern changes with scale. In heterogeneous or patchy landscapes, spatial variance does not scale monotonically with increasing grain size, but instead may exhibit abrupt shifts or transition zones. These **domains of scale** may correspond to different pattern-process relationships within each domain or to changes in process constraints that give rise to different relationships as to how spatial variance scales with increasing distance or grain size (**Figure 2.7**). For example, patch structure may be hierarchically nested, in which patches occur within larger patches embedded

within even larger patches (**Figure 2.8**). Nested patch structure is expected to exhibit abrupt changes in spatial variance with scale, perhaps reflecting different processes that operate within each domain of scale. The transitions between these domains are the 'risers' between the 'steps' (domains of scale), which produces the **variance staircase** that characterizes the scales of nested patch structure in the landscape (**Figure 2.7**). Understanding how spatial variance scales is ultimately important for evaluating the extent to which we can hope to extrapolate information across scales, as from finer scales to broader scales. Patterns that exhibit this sort of variance staircase function have a more complicated scaling relationship than those that scale

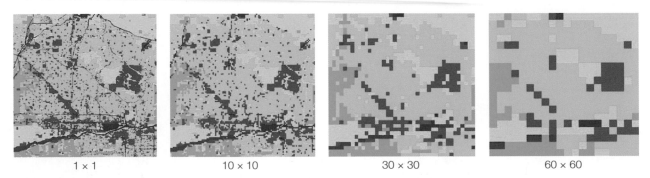

1 × 1	10 × 10	30 × 30	60 × 60

Figure 2.6 Effect of increasing grain size on landscape heterogeneity. As the grain (window size) of the analysis is increased, spatial information is aggregated to produce a map of coarser resolution. Inevitably, this results in the loss of small, scattered patches and a reduction in overall landscape heterogeneity (i.e. the number of land-use/land-cover types). The map shown here is a Landsat image of the central metropolitan area of Phoenix, Arizona (USA).

Source: Wu et al. (2002).

monotonically (e.g. linearly). We'll return to issues involving the extrapolation of information across scales later in the chapter.

Choosing the 'Right' Scale of Study

Choosing a scale appropriate to the question or phenomenon to be studied is crucial to the success of a study, and thus deserves careful consideration prior to embarking on any ecological (or other scientific) investigation. As we shall see in **Chapters 3 and 4**, the choice of scale affects our ability to detect patterns (if that is the goal of the study), or whether environmental heterogeneity is likely to be a 'nuisance' variable in our analysis (if pattern detection isn't the goal), or even if the data are likely to be spatially correlated and thus violate the assumption of independence that underlies parametric statistics. An inappropriate choice of scale can obscure the relationship between pattern and process, leading to interpretations or conclusions that are meaningless or erroneous if, for example, the wrong process is attributed to a particular pattern, when the process doesn't even operate within that particular

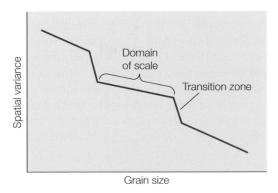

Figure 2.7 Variance staircase. Domains of scale, where the relationship between spatial variance and scale is linear or monotonic, are separated by transition zones. Transition zones are abrupt thresholds, which may signify a change in the underlying process responsible for the pattern. This particular staircase thus has three domains of scale.

domain of scale. A careless or blind choice of scale could also undermine efforts to compare results among studies that end up being conducted at different scales, thus frustrating synthesis and the development of general principles in ecology.

To illustrate the problem, consider that much of the controversy surrounding the importance of competition in structuring ecological communities in the mid-1980s was as much a result of studies being conducted at different scales as it was due to ideological differences among the protagonists in this debate, as well as the fact that other processes, such as predation, also play a role in structuring communities. For example, two species may exhibit a negative association at fine spatial scales (e.g. at the scale of individual territories, ha) because of competitive interactions, but exhibit a positive association at broader spatial scales (km^2) because they have the same general habitat association (Wiens 1989a). Thus, evidence consistent with competition as an organizing force in structuring communities is evident in the first study (a negative association between two species), but not in the second (the association between the two species is positive). It is therefore pointless to argue whether competition is or is not important for structuring communities without first ensuring that we are conducting our investigations at the scale(s) over which competitive interactions actually occur.

So how do we go about deciding on the 'right' scale of study? In planning our study, we obviously need to decide where we are going to conduct our investigation, which leads to the first of many scale-related decisions regarding how large our study area should be, since this will bound the spatial extent of our data or observation set. If we desire replication among study sites, we will need to decide not only how large the individual study sites should be (i.e. plot size), but also how far apart they should be and thus how large an area we will ultimately need in order for the sites to be spatially independent and yet still be similar enough to serve as replicates.

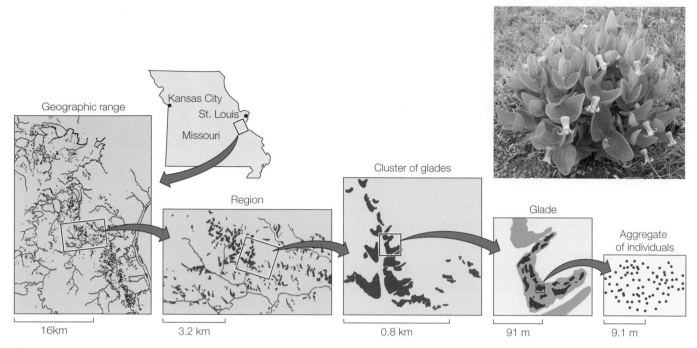

Figure 2.8 Hierarchically nested patch structure. The distribution of Fremont's leather flower (*Clematis fremontii*) exhibits hierarchically nested patch structure across a range of scales in eastern Missouri (USA).
Source: After Erikson (1945).

Once that decision has been made, we next need to decide how we are going to sample within those plots, in terms of the number of sampling points and the distances between them (if sampling at fixed distance intervals, such as along a transect). Then we have to decide on the spatial grain of sampling, in terms of what size sampling frame we should use, or within what radius (neighborhood) of our sampling points we should survey. Finally, we need to decide on the temporal grain, in terms of how often we are going to sample or make observations, and on the temporal extent or duration of our study. Can we perform daily surveys, or will once a week suffice? Is this study being performed over the course of a single season, or will it be conducted over multiple years?

Clearly, the answers to these sorts of questions will depend on a number of factors related to the type of system or organism being studied, the type of data required to address the question of interest, the research conventions and established sampling protocols, and various constraints involving time, funding, personnel, and the type of sampling equipment and research sites available. In practice, the choice of scale often boils down to the following:

- *External constraints.* These are constraints that dictate the scale of study, both in terms of grain and extent, but which lie outside the control of the investigator. For example, study sites may be limited in availability, either in terms of the

distribution of the habitat types, land uses, or patch sizes that occur within a given landscape or region, or because sites are simply inaccessible owing to distance, terrain, or land ownership issues that prevent access. Time, personnel, and funding constraints are ever-present realities that limit the scope of research, both in time and space. We might like to conduct a decades-long study of land-use change in a particular region, but the funding needed to support our research proposal is generally of far shorter duration (<5 years) and cannot be guaranteed. Finally, the availability of equipment or data may also impose constraints beyond investigator control. For example, satellite imagery might be available, but only at a particular resolution. If a finer resolution is needed, it would be nice to obtain high-resolution images of our study site, but perhaps we do not have the means available to purchase them or to acquire these data ourselves. We all work with what we have or can reasonably do within such constraints.

- *Research conventions and sampling protocols.* Science involves the observation and quantification of natural phenomena. Entire monographs have been written about how to sample or measure particular phenomena. For example, there are books and many scholarly articles written on the proper sampling techniques, census, or survey methods to be used for vegetation, birds, fish, or insects. Because our observations are very much influenced by our choice of sampling method, this is not a trivial issue. It does underscore, however, that the sampling protocols or techniques we adopt will also have a large effect on the scale of our investigation, particularly with regard to the grain of our observation set. We are obliged to adopt certain sampling protocols that have been advanced in the literature as 'best' for avoiding known sampling pitfalls—such as sampling error related to insufficient replication, a lack of independence among samples, or detection bias. In other cases, we are bound by tradition in our choice of sampling scale. For example, the Daubenmire frame (a 20 × 50-cm quadrat) is frequently used to survey ground vegetation cover or composition along a transect. By setting a lower limit to our observations, these frame dimensions could influence our ability to detect patterns or compositional changes in vegetation that occur at finer scales. On the other hand, the widespread adoption of such sampling conventions does promote standardization, thus enhancing our ability to make meaningful comparisons among studies that are conducted at the same scale. Unfortunately, however, standardization does not guarantee that research has been conducted at a scale that is appropriate to our question or system.

- *Personal experience or expert opinion.* In the absence of well-established sampling guidelines, we must rely on our own experience or the experience of others to decide how best—and at what scale—to sample and acquire the information needed to address our research questions. For example, if we have experience working with birds in grasslands, we would establish larger study plots than if we were working with birds in forests, because we know that grassland birds have larger territories than forest birds and we want to ensure that we will have a sufficient density of birds (or their nests) for study. If we are inexperienced with the system or there is insufficient information to guide us, we might consult with others as to how best to proceed. For example, graduate student advisory committees and peer reviewers of research proposals routinely provide input on study design and sampling methods, and this could easily extend to a discussion of the appropriateness of the investigator's choice of scale, in terms of both the spatial grain and extent as well as the frequency of sampling and the duration of the study.

- *Arbitrary selection.* Sometimes, it must be acknowledged, the choice of scale is completely arbitrary. We may have given little thought to the particular scale at which we sample, either because we are following established sampling protocols (which may have been established without regard for how the scale of sampling might affect our ability to detect specific patterns), or because we are simply unaware of the importance of scale to our observations (which obviously will not apply to the readers of *this* chapter). More often than not, the choice of scale is informed less by design or an understanding of scaling relationships and pattern-process linkages than by our own perception of scale. For example, a study of population- or community-level responses to landscape pattern will likely require a broader scale of study than that detailing the responses of individuals, although the organisms we choose to study for a behavioral, population-, or community-based study will often depend on whether or not the scales at which those individuals or populations range is convenient to our own (Hoekstra et al. 1991). This has implications for ecology as a discipline, as such preferences can lead to entire areas of study being dominated by research performed at a particular scale, or equivalently, on particular types of organisms or systems (Hoekstra et al. 1991). For example, most experimental research in community ecology has historically been conducted at fine spatial scales (<1 m to 10 m; Kareiva & Anderson 1988), which means that much of it has been on small organisms, such as herbaceous plants, insects, or mollusks (Kareiva & Anderson 1988; Hoekstra et al. 1991).

Sometimes it just isn't all that obvious what scale we should be sampling. We have to start somewhere, however, and thus we have two options: (1) we choose a scale that seems reasonable for addressing our questions (i.e. our choice is informed to some degree by experience) and that we can feasibly sample (because of logistical, funding, or time constraints); (2) we conduct a multi-scale investigation that explicitly considers how the ecological relationships we seek to uncover change as a function of scale. The first option need not be detrimental to the success of our project—we might get lucky and chance on the 'right' scale. Interesting patterns abound in nature, and it is not difficult to find them at whatever scale we might choose to look. This underscores the point that there really is 'no single correct scale' at which to study an ecological system (Levin 1992).

That there is no single correct scale or level at which to describe a system does not mean that all scales serve equally
well or that there are not scaling laws.   Levin (1992)

Some scales are clearly better than others, however. As the debate over the role of competition in structuring communities has taught us, it is possible to find contra-dictory evidence even within the same system, depend-ing on the scale of observation. Competition may be important at some scales but not others, and other pro-cesses can also give rise to patterns of species-avoidance that have nothing to do with competitive interactions. For example, apparent competition occurs when a predator preferentially preys on one of two competing species, giving the appearance that one of the prey spe-cies is competitively dominant over the other, when it is predation rather than competition that is actually responsible for the negative correlation in the density of the two species (Holt 1977).

Perhaps if we better understood the nature of heterogeneity or spatial variability in our system—the scales over which processes operate and patterns emerge—it would be easier to evaluate which scale is 'best' for addressing our particular question of interest. Achieving a better understanding of heterogeneity

within our system will require either a systematic investigation into how pattern varies as a function of scale (our second option above), or a better science of scaling (i.e. the formulation of scaling laws and other methods for extrapolating across scales). Scaling laws are featured in **Box 2.1**, and we will discuss issues pertaining to the extrapolation of information across scales later in the chapter. For now, however, we will turn our attention to how hierarchy theory might be used to provide a better understanding of system structure and organization, based on the scale(s) at which system components interact.

Hierarchy Theory and Landscape Ecology

Hierarchy theory is a theory of system organization that addresses how complex systems arise and how they function. Hierarchy theory asserts that interactions among system components can be ordered based on the frequency and strength of those interactions, and the ordering of these system dynamics gives rise to different levels of hierarchical organization (Allen & Starr 1982; O'Neill et al. 1986). Lower levels are characterized by high-frequency dynamics and strong interactions. Higher levels are made of lower-level

BOX 2.1   Scaling Laws and Fractals

Ecology has occasionally been accused of suffering from 'physics envy' in its quest for universal laws and fundamental principles similar to those that govern physical systems. (Of course, the laws of physics also govern biological and ecological systems!) Even in physics, however, universal laws are not truly universal. Although Newton's laws of motion are useful approximations at the scales and speeds at which most of us operate in everyday life, they do not apply at extremely small scales, at extremely high speeds, or in extremely strong gravitational fields; those conditions require the application of more sophisticated theory, such as general relativity and quantum mechanics. The mathematical physicist Mitchell Feigenbaum has been quoted as saying, 'The only things that can be universal, in a sense, are scaling things.' Ecology may have few laws (just many rules and generalities), but the scaling laws that underpin allometric relationships between body size and other traits and species–area relationships, are among the few that are generally acknowledged to be nearly universal. It is therefore worth exploring the implications of such scaling laws for ecology.

Power-Law Scaling

Most disturbances that occur frequently on a landscape are relatively small, but on rare occasions, a very large

disturbance will occur that impacts much of the landscape. For example, natural disasters such as a 100-year flood, a 200-year fire, a tornado or hurricane, a tsunami, or a volcanic eruption do not happen very often, fortunately. If we plot the number or frequency of events as a function of their size or magnitude, the resulting *power-law distribution* reveals that small event sizes occur with much greater frequency than large ones (**Figure 2.13A**). Power laws have the general form

$$y = \alpha x^k. \qquad \text{Equation 2.1}$$

where x (commonly called the *base*) and y are two related measures, and α and k are constants. The constant k is referred to as the *scaling exponent* and indicates how rapidly the relationship increases (if $k > 0$; **Figure 2.13D**) or decreases (if $k < 0$; **Figure 2.13A**). Power laws are ubiquitous in nature, which is probably just as well, since it might be difficult to survive in a world where large disturbances occurred with the same frequency as small ones.

Fractals: The Geometry of Nature

Fractals are basically a type of power law, in which measured quantities vary as a power of the resolution (scale) at which measurements are made. For example, we

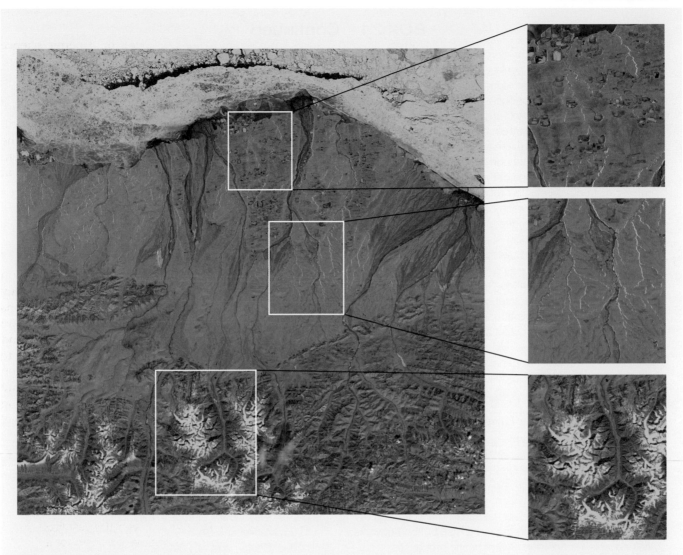

Figure 1 Fractal landscape patterns.   Summer landscapes along the North Slope of Alaska (USA) show various fractal structures, from the pools and ponds that form atop the melting permafrost (top right) to the dendritic network of braided streams (middle right), to the rugged topography of the snow-capped Brooks Range mountains (bottom right).
Source: NASA image created by Jesse Allen, Earth Observatory, using data of the University of Maryland's Global Land Cover Facility.

are all familiar with the fact that the surface area (SA) of an object increases as the square, and volume (V) increases as the cube, of length (l). Thus, SA = l^2 and V = l^3, which we now recognize as the form of a power law. The scaling exponents here are the familiar **Euclidian dimensions** that describe basic geometric shapes: D = 0 is a point, D = 1 is a straight line, D = 2 is a plane, and D = 3 is a volume.

Most properties in nature do not scale as whole numbers, however. For example, trees are not perfect cones, and animals do not move in perfectly straight lines. An animal's wanderings are likely bounded in space (such as by a territory or home-range boundary), but probably do not fully saturate the entire area so as to fill the two-dimensional plane. Thus, the pattern of movements lies somewhere between a straight line (D = 1) and a plane (D = 2); in other words, it occupies

some fractional or **fractal dimension** between 1 and 2 (e.g. D = 1.6).

Fractals are the 'geometry of nature' (Mandelbrot 1983), which is readily apparent not only in the movement pathways of organisms, but also in the complexity of coastlines, the architecture of plants, the rugged terrain of mountain ranges, and in the branching of river networks (**Figure 1**). It is also apparent in the way that populations increase (or decrease), how species diversity changes as a function of area (species–area relationships), and in how the structural and functional traits of organisms scale as a function of body size (allometric relationships). In other words, power laws that describe many ecological and biological relationships are fractal, in that they have a fractal-like design or dynamics and are described with a fractional exponent (i.e. non-integer exponents, such as ¼ = 0.25).

(Continued)

BOX 2.1   Continued

Fractals also have the characteristic of **self-similarity** over an extended range of scales. That is, the properties of the system appear similar regardless of the scale at which they are measured or observed; they exhibit scale invariance. For example, the branching structure of a tree or river is fractal to some degree: branches give rise to branches, which in turn give rise to smaller branches, and so forth. Thus, a branch held upright has approximately the same overall appearance of the larger tree. This hierarchical structure is further reflected in the stream-order system of river networks, which measures the relative size of waterways and also provides a measure of branching complexity (e.g. two first-order streams come together to form a second-order stream, two second-order streams come together to form a third-order stream, and so forth). Rivers and trees do not continue to branch indefinitely, however, and thus exhibit fractal structure (or approximately fractal structure) only across a range of scales.

Fractals thus provide a useful analytical framework for describing complex patterns and system behavior, and may also facilitate the search for general or universal principles that govern the structure and dynamics of complex ecological systems (Brown et al. 2002).

The Universality of Power Laws and Fractals

Some power laws, such as the species–area relationship, appear to be nearly universal, occurring in virtually all taxa and in all types of environments, which suggests that these may share a common underlying mechanism (Brown et al. 2002). At least three mechanisms have been proposed to explain the ubiquity of power-law scaling and fractals: (1) the economy of design in resource-distribution networks; (2) self-organized criticality; (3) stochastic processes.

Design of resource-distribution networks: Relationships that exhibit power-law scaling may result from certain design constraints on resource-distribution networks. Resource-distribution networks tend to exhibit a hierarchical branching pattern ending in fine structures of a given size (e.g. capillaries). The hierarchical branching structure of the network helps to optimize flows and increase the efficiency of resource distribution throughout the network (West et al. 1997; 1999).

Self-organized criticality: Power laws (and thus fractals) are considered the 'fingerprint of self-organized systems' (Solé 2007), where localized interactions among system components spontaneously give rise to higher-order patterns and system behavior (i.e. **self-organized criticality**; Bak et al. 1988). Hierarchy theory is a subset of complex systems theory. Intriguingly, complex systems theory posits that although interactions occur locally, there exists a critical point at which these local interactions can propagate to affect the behavior of the entire system (i.e. they have a cascading or 'domino' effect). In this critical domain, power-law relationships are produced between the frequency and size distribution of events.

For example, an ecosystem is made up of interacting species, which may go extinct locally within a patch for any number of reasons (competition, predation, disease, or some environmental catastrophe). Local extinctions happen episodically throughout the landscape, but are balanced by recolonization. The addition of a new species to the system, such as through speciation or the introduction of a non-native (exotic) species, is a type of local perturbation that may cause a reshuffling of species at the point of introduction, but generally has little effect on the ecosystem as a whole. As the number of species added to the system increases, however, there may exist a critical point (the **critical biodiversity threshold**) at which the introduction of a new species has cascading effects across the entire system, such as when a superior competitor evolves or an invasive species spreads throughout the landscape, triggering a wave of species extinctions (Kaufman et al. 1998). The structure of such systems can be defined by a power law near the critical threshold. For example, the frequency distribution of extinction events of various sizes is a power law: most extinction events are small (e.g. only 1–2 species go extinct across the landscape) but a few are very large (mass extinction events). The system lacks a characteristic scale in the vicinity of the critical threshold; extinction events of *any* size may occur in this domain. At the critical biodiversity threshold, however, the system may be particularly vulnerable to mass extinction and suffer a system collapse (a transition from one system state to another).

To what extent are landscapes self-organized critical systems? Many disturbances that shape landscapes, from the gap-size dynamics of tropical rainforests (Solé & Manrubia 1995; Manrubia & Solé 1997) to the size distribution of landslides (Restrepo et al. 2009) and forest fires (Malamud et al. 1998; Ricotta et al. 1999), exhibit power-law scaling that may reflect a process of self-organization. However, forest fires in western North America may conform to a power-law distribution only across a finite range of scales (i.e. they do not exhibit scale invariance), and even then, only within some regions (**Figure 2.13A**; Reed & McKelvey 2002).

Regardless of whether ecological systems self-organize to a critical state, it is the opposing nature of key processes (births and deaths, speciation and extinction, extinction and recolonization) that is postulated to give rise to power-law or fractal scaling in the structure and dynamics of these systems (Schneider 2001).

Stochastic processes: Power laws may also result from a kind of central limit theorem for multiplicative growth processes. Many different kinds of stochastic processes, with no connection to critical phenomena, have power-law correlations, for example. If stochastic processes with exponential growth are stopped (extinguished) at random, then the distribution of that state will likely exhibit power-law behavior (Reed & Hughes 2002).

systems, but the interaction among these systems is weaker and occurs less frequently.

Hierarchical Organization of Biological Systems

The biological hierarchy is a familiar example of a hierarchically organized system (**Figure 2.9**). Cells are nested within tissues, different types of tissues make up organs, and organs are organized into systems that interact and perform various functions that ultimately contribute to the structure and vital functions of the organism. Individual cells within a given tissue type interact with each other much more frequently and intimately than they do with cells in other types of tissues, which is not to say that cells in different tissues do not interact at all. Hormones produced by one type of cell in a given tissue or organ (e.g. the pituitary gland) can affect other organs (such as the thyroid gland) that in turn may secrete additional hormones that regulate cell functions elsewhere in the body. Many of these types of systems operate through **negative feedback** mechanisms that help maintain homeostasis and thus the functioning of the higher-level system, which is the organism in this case.

The organism, in turn, may interact with other individuals of its kind to fulfill certain needs or functions (resource acquisition or reproduction), giving rise to higher-order group behavior (societies, cultures, politics, economies, religions) that in turn may have broader implications for the type and magnitude of effects that a species has on other species or levels of organization within the biological hierarchy. Such effects typically involve negative feedbacks that help to stabilize these systems, such as when a predator species keeps the population growth of its prey in check, which may have important implications for community or ecosystem function as a whole. For example, the presence of a top predator in a system can help to maintain herbivore populations, contributing to higher diversity and productivity of the entire ecosystem.

The importance of these sorts of negative feedbacks on system stability and function is illustrated by the reintroduction of gray wolves (*Canis lupus*) during the mid-1990s to Yellowstone National Park (USA), from which they had been extirpated some 70 years earlier. During that time, elk (*Cervus elaphus*) populations in the park soared, contributing to overbrowsing of woody vegetation, particularly aspens (*Populus tremuloides*), cottonwoods (*Populus* spp.), and willows (*Salix* spp.).

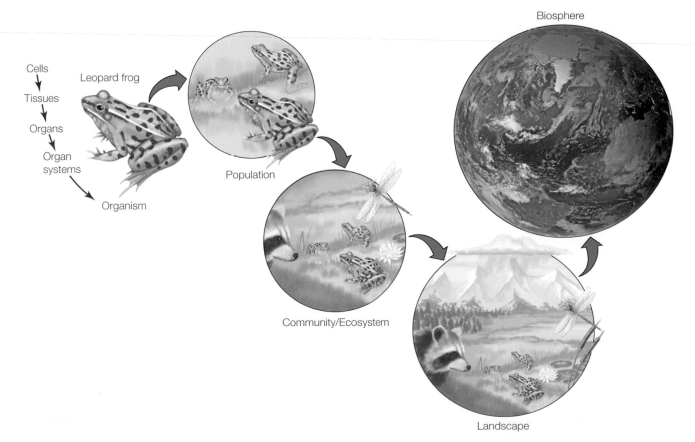

Figure 2.9 Traditional biological hierarchy. In the traditional biological hierarchy, landscapes are a level of organization beyond ecosystems but below the level of the biome or biosphere.

Source: After Sadava et al. (2012).

The reintroduction of wolves led to the discovery of a **trophic cascade** between wolves, elk, and woody vegetation; in the years since wolves have been reintroduced, elk populations have declined and woody vegetation, especially cottonwoods and willows, has increased (Ripple & Beschta 2012). The restoration of a top predator to the Yellowstone ecosystem has benefitted other species as well, including the beaver (*Castor canadensis*), an important ecosystem engineer whose numbers have likewise increased since the reintroduction of wolves, probably as a result of the increase in riparian habitat (i.e. cottonwoods and willows). Wolves thus have a number of direct and indirect effects on the diversity and productivity of the Yellowstone ecosystem.

In other cases, however, **positive feedbacks** may occur among species, which can lead to the amplification of system dynamics, causing destabilization that may lead to altered system structure and function, or even to system state changes, as when overgrazing causes the desertification of grasslands. Similarly, the spread of an invasive species or disease pathogen may occur quite suddenly once a suitable vector has been introduced, a type of positive feedback in which one introduced species facilitates the establishment or spread of another introduced species. If this process of species facilitation continues, whereby each new species introduced makes it that much easier for the next species to invade, then the end result may be an **invasional meltdown** (Simberloff & Van Holle 1999). Invasional meltdowns are the result of multiple positive feedbacks among introduced species that lead to further increases in the rate at which other introduced species become established, as well as the growing impacts these introduced species have on community structure and ecosystem function (Simberloff 2006, 2011).

Because humans have been responsible, whether intentionally or accidentally, for moving and introducing species to new areas across the globe, these sorts of positive feedbacks are often symptomatic of anthropogenic disturbance or ecosystem degradation. For example, the combined introduction of the invasive yellow crazy ant (*Anoplolepis gracilipes*) and a honeydew-secreting scale insect (*Coccus* spp.) to Christmas Island in the Indian Ocean resulted in positive feedback between the two that allowed the formation of high-density ant 'supercolonies' (O'Dowd et al. 2003). These supercolonies now threaten the native red land crabs (*Gecarcoidea natalis*), which the ants are able to overwhelm and kill. A reduction in the land crab population has facilitated the invasion of native rainforests by the giant African land snail (*Achatina fulica*), which benefits from increased resources (native seedlings and leaf litter) as well as the now predator-free space (Green et al. 2011). Thus, a mutualistic interaction between two invasive species has contributed to an invasional meltdown by indirectly facilitating the invasion of African land snails that in turn may be having an impact on the diversity and structure of the tropical rainforests on Christmas Island. We will explore the role of positive feedbacks and catastrophic state changes in ecosystems more fully in **Chapter 11**.

Structure of a Hierarchical System

As a consequence of the interactions and feedbacks that occur among system components, the system may spontaneously self-organize and generate more complex structures and behaviors that are considered emergent, in that they are not simply an additive or constitutive property of the system (more on that in a bit). For example, a hierarchically organized system is made up of at least three levels—a system of systems embedded in a larger system (i.e. a *triadic system*; **Figure 2.10**). We might thus consider the implications of triadic system organization in terms of how we go about studying the system.

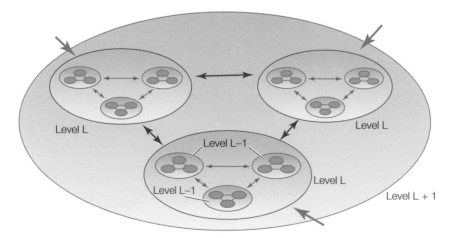

Figure 2.10 Organization of a hierarchical system. A hierarchically organized system is composed of at least three nested levels (triadic structure).
Source: After King (1997).

The *focal level* (L) is the level of primary interest to the investigator. It embodies the range of scales (both spatial and temporal) that bound the particular phenomenon or question we are studying. However, the patterns or dynamics of the focal level that we observe come about because of interactions that occur among subsystems at a *lower level* (L − 1), and are ultimately bounded by constraints imposed by a *higher level* (L + 1). The lower level (L − 1) contributes the mechanistic explanations for the patterns and dynamics we observe within our focal level, while the upper level (L + 1) provides broader context and gives significance to the phenomenon under study. For example, the significance of reproduction cannot be understood simply by studying the individual organism, but becomes apparent only within the context of population dynamics at the next higher level.

To illustrate hierarchical organization more fully, consider that a **metapopulation** is a population of interacting populations, and could itself be embedded within a larger system of interacting metapopulations to give rise to regional source–sink dynamics for a given species. The extinction risk for a given metapopulation (L) depends on the balance of recolonization and extinction dynamics among individual populations (L − 1). Some populations occur in high-quality habitat patches that are capable of producing a surplus of individuals; these patches are **population sources**. Surplus individuals might then recolonize **population sinks**, patches where populations have died out, such as within lower-quality habitats that are incapable of supporting populations by themselves. The relative proportion and configuration of source and sink habitat on the landscape will dictate the interplay between extinction and recolonization, and thus whether the metapopulation is likely to persist. Different landscapes within a region or within the range of the species (L + 1) will have different proportions of good- and poor-quality habitats, and thus metapopulations with different probabilities of extinction risk. The relative proportion of source versus sink landscapes, and the degree to which metapopulations interact through long-range dispersal, determine whether the species is viable at a region-wide scale (With et al. 2006).

Landscape Scale or Landscape Level?

The terms 'landscape level' and 'landscape scale' are often used interchangeably. To what extent is the landscape a level in the traditional biological hierarchy, however, and why might this distinction be important for the science and practice of landscape ecology? The answer is that this distinction ultimately relates to how landscapes are defined and thus what qualifies as the appropriate domain for the study of landscape ecology.

HIERARCHICAL PATCH STRUCTURE   Landscapes frequently exhibit patchiness across a range of scales, and this patchiness is sometimes hierarchically nested, in which small patches are nested within larger patches that in turn are nested within even larger patches (Kotliar & Wiens 1990; **Figure 2.8**). Nested patch structure comes about either because of changes in the types of processes that give rise to patchiness within each domain of scale (different processes, different domains), or because of changes in constraints on a particular process at different scales (same process, different constraints). Adjacent levels in a patch hierarchy are more likely to interact and to have a greater influence on each other than those that are separated by intervening levels of patchiness.

Examples of nested patch structure in ecological systems abound. Streams and rivers are the canonical example of a hierarchically structured system (Frissell et al. 1986; **Figure 2.11**). Microhabitats are nested within larger pools and riffles that in turn make up a stream reach. Reaches are nested within stream segments, and all of the segments together make up the stream system. Different processes operate at each scale and have different consequences for the structure and dynamics of stream habitats and their biota over time (**Figure 2.11**).

Another example is given by the hierarchical structure of East African savannas, in which vegetation dynamics can be understood by organizing processes that influence tree density into five nested scales (Gillson 2004; **Figure 2.12**). At the microscale, tree density is influenced by competitive interactions (e.g. between trees and grasses), fine-scale disturbances such as selective browsing and trampling, and microclimatic factors that affect the availability of germination sites. The interaction among these processes in turn gives rise to local-scale patterns in tree density that are determined by more extensive disturbances, such as fire and browsing damage by elephants, which influence tree recruitment. Differential recruitment in turn contributes to patch dynamics at the landscape scale, in which asynchronous transitions between grasslands and woodlands occur over time to give rise to the mix of trees and grass that is the savanna landscape. Geomorphic and climatic factors, including topography, geology, hydrology, and rainfall, also influence the distribution of habitat types across the savanna. This hierarchy can be extended to the regional scale, where broad-scale changes in herbivore abundance due to massive mortality from disease (e.g. rhinderpest) or hunting (e.g. for ivory) can have a profound effect on the regional dynamics of savanna landscapes. Ultimately, the savanna biome is limited in extent at a global scale by climatic factors (i.e. the macroscale; **Figure 2.12**). Thus, by linking patterns and processes across a wide range of scales, a hierarchical framework provided a useful way of thinking about vegetation patterns—and the processes that shape them—in this system.

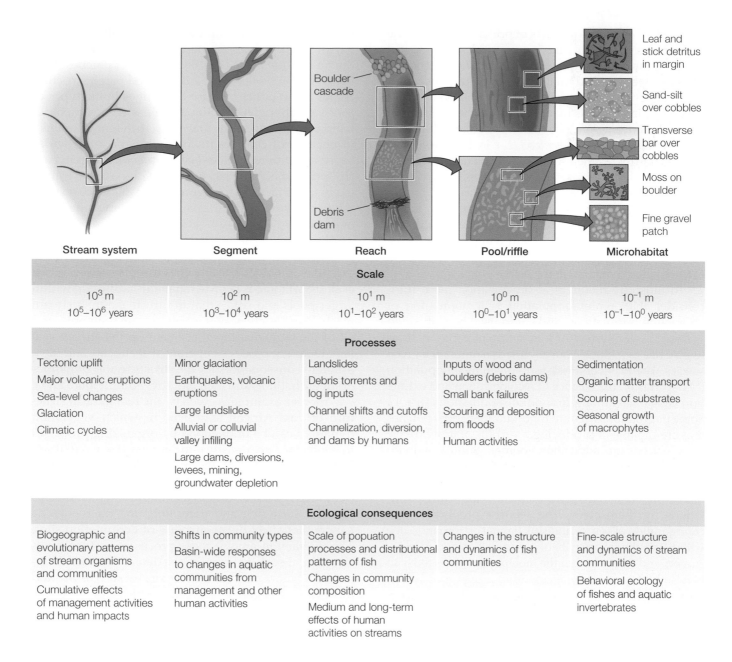

		Scale		
10^3 m 10^5–10^6 years	10^2 m 10^3–10^4 years	10^1 m 10^1–10^2 years	10^0 m 10^0–10^1 years	10^{-1} m 10^{-1}–10^0 years

		Processes		
Tectonic uplift Major volcanic eruptions Sea-level changes Glaciation Climatic cycles	Minor glaciation Earthquakes, volcanic eruptions Large landslides Alluvial or colluvial valley infilling Large dams, diversions, levees, mining, groundwater depletion	Landslides Debris torrents and log inputs Channel shifts and cutoffs Channelization, diversion, and dams by humans	Inputs of wood and boulders (debris dams) Small bank failures Scouring and deposition from floods Human activities	Sedimentation Organic matter transport Scouring of substrates Seasonal growth of macrophytes

		Ecological consequences		
Biogeographic and evolutionary patterns of stream organisms and communities Cumulative effects of management activities and human impacts	Shifts in community types Basin-wide responses to changes in aquatic communities from management and other human activities	Scale of popuation processes and distributional patterns of fish Changes in community composition Medium and long-term effects of human activities on streams	Changes in the structure and dynamics of fish communities	Fine-scale structure and dynamics of stream communities Behavioral ecology of fishes and aquatic invertebrates

Figure 2.11 Hierarchical stream structure. Streams exhibit hierarchical structure in habitats, associated processes, and ecological consequences across a range of spatial and temporal scales.

Source: After Frissell et al. (1986).

HIERARCHICAL STRUCTURE VS HIERARCHICAL ORGANIZATION

If a landscape exhibits hierarchical patch structure, might we then conclude that it is also a hierarchically organized system? Unfortunately, evidence that the landscape exhibits nested patch structure is not sufficient to conclude that it is also hierarchically organized, although it might be. Recall that levels in a hierarchically organized system come about because of interactions that occur at the level below; the level is itself an *emergent property* of these lower-level interactions. Higher levels are thus not simply a collection of their constituent parts (i.e. the whole is greater than the sum of its parts), especially

since not all hierarchical systems are nested. For example, a food chain is hierarchical, but is not nested. Predators are not actually made up of prey.

The critical distinction here, in terms of assessing whether the system is hierarchically organized or just hierarchically structured, is to consider whether the behavior of the system could be predicted simply as some function of its constituent parts (e.g. as an additive function). For example, could we predict higher-order system behavior, such as the kind of social or mating system of a species, simply by studying isolated individuals (as in a lab or zoo), or going further, from its skeleton or fossilized remains? The behavior of

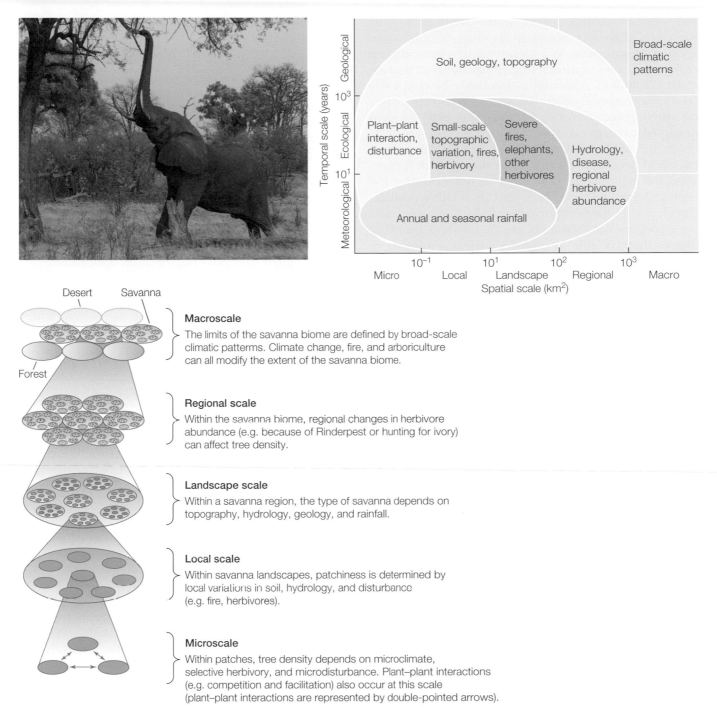

Figure 2.12 Scaling and hierarchical structure of an East African savanna. The distribution of trees is influenced by a combination of geology, topography, and soils; biotic and abiotic disturbances; and weather-related and climatic factors, as depicted in the space-time diagram. These factors all contribute to the hierarchically nested patch structure of trees across a range of scales, from micro to macro scales.

Source: After Gillson (2004). Photo by Charles J. Sharp and published under the CC BY-SA 3.0 License.

individuals is more than just the sum of their individual systems and cannot be predicted simply by understanding the mechanics of how those systems function. Behavior is thus an emergent property of the integrated and interacting system of systems, which can best be understood in the context of the population and its environment.

DISTINCTION BETWEEN SCALE AND LEVEL   Although higher levels of system organization necessarily occur at broader scales in space and time, scale is not synonymous with level (King 1997). *Scale* pertains to the physical or temporal dimensions of the object being measured (e.g. m², yr⁻¹), and is synonymous with the *scale of observation* (i.e. grain and extent). Scale thus has units of measurement. In contrast, *level* refers to level of organization in a hierarchical system; it is an entity or system of interacting components in its own right (**Table 2.1**). Although a level may be characterized by a temporal or spatial scale, the level itself has no units of measurement. For example, a *cell* is a level of biological organization (**Figure 2.9**). A cell may be characterized by its small size and rapid rate of biochemical processes (i.e. by the scale at which it operates), but is ultimately defined by its functions (e.g. the type of hormone it secretes) and its relative position (level) within the biological hierarchy. Cells are made up of organelles, but cells are themselves building blocks that are able to interact with one another in ways that give rise to different types of tissues (e.g. nervous tissue).

IS THE LANDSCAPE A LEVEL IN THE ECOLOGICAL HIERARCHY, OR A SCALE OF OBSERVATION?

The term 'landscape level' implies that landscapes represent a level of ecological organization. Landscapes are generally placed above ecosystems in the traditional biological hierarchy, and were defined early on in the seminal landscape ecology text by Forman and Godron (1986, p. 11) as 'a heterogeneous land area composed of a cluster of *interacting ecosystems*' (emphasis added). If landscapes are made up of interacting ecosystems, then we might reasonably argue that the landscape represents a true level of organization in the traditional biological or ecological hierarchy (**Figure 2.9**). This reinforces the notion that landscapes are 'large scale,' which is certainly the conventional perception.

If, however, landscapes are defined as a 'spatially heterogeneous area' (Turner 1989), then a landscape can be defined at *any* scale relevant to the question, process, or organism under investigation. This does not imply anything about how landscapes are formed or organized, or where they exist (if at all) in the traditional biological hierarchy. From this viewpoint, the landscape is scale-dependent and is defined in terms of its spatial (and temporal) grain and extent. In other words, the landscape it is not necessarily bigger than a community or ecosystem, but is instead the stage on which ecological interactions and processes play out. 'Landscape scale' is thus shorthand for the *scale of landscape observation*. Landscape in this context provides a 'criterion for observation' (Allen & Hoekstra 1992), and is a matter of perspective or a methodological approach that underlies our scientific inquiry, involving the explicit recognition and incorporation of spatial pattern (i.e. the landscape perspective).

Thus, we should not assume a priori that landscapes represent a level of system organization. The distinction between *landscape as level* versus *landscape*

as scale is perhaps best explained by King (1997, p. 204–205):

> Reference to the landscape level implies that the landscape exists or behaves as a consequence of interactions among components at the next lower level and that it interacts with other landscapes as part of a system at the next higher level…A landscape is often conceptualized as an assemblage of patches. Do these patches interact to form the landscape? Do the holistic or aggregate properties or behaviors of the landscape (the larger extent) change if the pattern of patches is altered but the composition remains the same?…If they do, then it may be appropriate to refer to the level of the landscape….Otherwise, if the landscape is simply the collection or set of patches, it is probably more appropriate to simply refer to the (appropriately qualified) scale of the landscape.

Given the definition in this textbook of landscape as a spatially heterogeneous area, we will follow King's advice and refer to 'landscape scale' rather than 'landscape level' throughout this book.

> [G]ood landscape ecology can be practiced without reference to hierarchy theory as a theory of system organization. It probably cannot be done without a theory of scale, especially spatial scale.   King (1997)

Implications of Hierarchy Theory for Landscape Ecology

Whether or not landscapes represent a true level of organization in the traditional biological hierarchy is admittedly less important to the study and practice of landscape ecology than the discipline's focus on spatial patterns and processes. As we have seen, pattern is scale-dependent and may exist at many different, possibly nested, scales on a landscape. Regardless of whether landscapes or other ecological systems are or are not hierarchically organized, hierarchy theory provides a useful framework for thinking about the interaction between patterns and processes, and about the spatial and temporal scales at which these operate. Hierarchy theory also has heuristic value in scaling up information from fine to broad scales, as described in the next section. Extrapolating information across spatial scales often involves translating across levels of organization, such as translating measurements on leaf stomatal conductance to estimate canopy conductance of an entire forest stand (Urban et al. 1987). Such translations may benefit from an explicit consideration of the hierarchical organization of the system (King 1997), as it may not be possible to simply extrapolate across levels of organization, let alone scales. Issues involving the extrapolation of information across scales will thus be discussed next.

Extrapolating Across Scales

Extrapolation involves making predictions about patterns and system behavior beyond the extent of our observations or dataset. Most ecological research is conducted at fairly fine spatial scales and over short time periods (e.g. Kareiva & Anderson 1988), and yet ecologists are increasingly being called on to make forecasts about the long-term ecological consequences of societal impacts on the environment, which have broad-scale regional and even global ramifications (Underwood et al. 2005). Thus, ecologists are frequently confronted with the problem of **scaling up** information obtained at fine spatiotemporal scales to make inferences about the larger system or its predicted behavior (**Table 2.1**). Ideally, this is done explicitly, as when attempting to develop scaling relationships or methods for extrapolating across scales, as we'll discuss later in this section. More usually, however, scaling up is done in a very general or conceptual way, as when we extrapolate the findings from our site-based study to the entire region, or when we argue for the broader implications of our study to other systems.

The essence of the ecological scaling dilemma in forest ecology is this: we know a lot about trees, a fair bit about forest stands, but not very much about forested landscapes or regions; yet it is at these larger scales that management and policy apply.   Urban (2005)

Scaling up is also implicit in the use of experimental model systems. Experimental model systems are frequently used in ecology to simplify the complexities of nature and afford a controlled environment in which to test our ideas about how nature works (Carpenter et al. 1995; Lawton 1995). Even landscape ecology has embraced this approach in the form of experimental model landscapes (microlandscapes, Wiens et al. 1993a; Jenerette & Shen 2012). Experimental model landscapes are small (e.g. 25–256 m^2) plots in which a patchy resource or habitat distribution is created to explore how different patch configurations (e.g. the number, size, and arrangement of patches) influence individual movement, population, and community responses to spatial heterogeneity (e.g. Wiens et al. 1995; With et al. 1999; With & Pavuk 2011). Microlandscape studies have typically focused on small organisms, such as arthropods and small mammals, which might reasonably be expected to operate within the scales encompassed by these small experimental plots. Nevertheless, an assumption inherent to all experimental model landscape studies is that they have relevance for understanding how other species might respond to landscape patterns, such as habitat fragmentation, at broader spatial scales (e.g. Ims et al. 1993). In other words, it is assumed that fundamentally similar processes are at work, whether we are observing how a grasshopper moves through a patchy grass mosaic within an experimental microlandscape (25 m^2) or how a bison roams across a mosaic of burned and unburned areas within a tallgrass prairie landscape (25 km^2).

Even if the responses are fundamentally different across scales, experimental model systems still have value by demonstrating that scale is indeed important to understanding how the process interacts with spatial pattern. As the British ecologist John Lawton has said in defense of experimental model systems (1995, p. 330–331):

> *Model laboratory systems are an ecological tool. Like all tools they do some things well, some things badly, and other things not at all. At worst, they are analogs of how we imagine nature to work, not how it actually works. At best, the task of assembling, maintaining, and predicting the behavior of even moderately complex ecosystems in the laboratory tests our understanding to the limit. More than anything else, model systems act as a bridge between theory and nature. They are not a substitute for studying the real thing, but by simplifying the complexities of nature, model systems can sharpen our understanding of natural processes.*

Thus, experimental model landscapes are useful if they enable us to achieve a better understanding of how spatial patterns influence ecological processes in a general way. From that standpoint, it is immaterial whether we can scale up directly from microlandscape studies to make specific predictions about patterns and processes within larger landscapes. Experimental model landscapes should thus be considered yet another tool in the landscape ecologist's toolbox.

Alternatively, **downscaling** takes information obtained at broader scales (coarse-scale information) to make inferences about local-scale patterns or processes (**Table 2.1**). This is often done, for example, in studies that seek to quantify or predict the effects of land management or climate change on fine-scale ecological processes (**Table 2.1**). A classic example of downscaling is the application of general circulation models (GCMs), which are used to predict patterns or change in the global climate, to forecast effects on climatic conditions at a regional scale (Hewitson & Crane 1996; Wu 2007b). Advanced GCMs involving coupled atmospheric-ocean models are used to simulate the El

Niño Southern Oscillation, from which predictions can be made as to the likelihood of extreme weather conditions occurring within a particular region, such as in Peru or southern California.

One ecological example of downscaling involves taking coarse-scale data on species distributions to predict the occurrence of a species at a more local scale using fine-scale environmental data (i.e. downscaling species distribution models; Keil et al. 2013). For example, many countries around the world conduct bird atlas projects in which information on the distribution and abundance of breeding or resident bird species are obtained and mapped over large geographic areas. The region to be mapped is usually represented as a grid, with grid-cell dimensions varying between 1 and 50 km, depending on the size of the region to be mapped. Trained observers attempt to record all of the species present in their assigned grid cell, by visiting a certain number of locations within that cell. Depending on the resolution of the grid, however, this can amount to some very coarse-resolution data on species occurrences, which might nevertheless be sufficient for getting a general idea of a species' distribution at a continent-wide scale, for example.

Although grain size usually sets a lower limit to the dataset, such that any information below that scale is unknown and unobservable, the existence of certain scaling relationships may permit extrapolation even below the resolution of our data. In the case of species distributions, for example, occupancy patterns (the area or proportion of grid cells occupied) scales in a predictable fashion with grain size (i.e. as a power law or fractal distribution; Box 2.1), which has been referred to as the **scale-area relationship** (Kunin 1998) or **occupancy-area relationship** (He & Condit 2007). Thus, it is possible to use these sorts of relationships in developing methods of extrapolation that permit the downscaling of coarse-scale distribution data to predict the occurrence of a species at a finer scale of resolution (Keil et al. 2013).

Central to the discussion of upscaling and downscaling is whether or not heterogeneity or spatial variance scales in a predictable way. In other words, does the relationship remain relatively constant or change in a monotonic (e.g. linear or exponential) fashion across a range of scales, or are different relationships evidenced at different scales (e.g. domains of scale)? Problems associated with extrapolation within versus between domains of scale are therefore considered next.

Extrapolating Within Domains of Scale

Extrapolating across scales is not a problem if the pattern exhibits scale-invariance; that is, the pattern exhibits a constant and predictable relationship across the range of scales considered (Figure 2.5A). In this case, the phenomenon or pattern has no single or characteristic scale. A power law is one such example of a scale-invariant relationship (Figure 2.13; Box 2.1). Because they describe scaling relationships that remain constant, often over many orders of magnitude, power laws permit extrapolation and prediction over a wide range of scales (Brown et al. 2002).

The use of power laws and other scaling functions represents a form of direct scaling or extrapolation (Denny & Benedetti-Cecchi 2012). For example, simple linear regression can be used to fit empirical data and make extrapolations across space or over time (Figure 2.13), and other approaches for modeling non-linear relationships are also available (e.g. generalized additive models, Guisan et al. 2002). As long as the relationship is monotonic, extrapolation is relatively simple. For example, power functions are monotonically increasing or decreasing in their empirical form (Figure 2.13C); taking the log of the relationship makes it possible to use linear regression methods to obtain scaling exponents (i.e. from the slope of the regression line; Figure 2.13D).

The problem, however, is that no ecological or biological system remains constant over an infinite range of scales. It may be that scale-invariance applies across a particular range of scales, but then there is a transition or critical threshold, owing to different processes or process constraints, that result in a different scaling relationship across another range of scales (i.e. domains of scale; Figure 2.7). Information at a particular scale may be extrapolated to other scales within a domain, but extrapolation across domains of scale should be done advisedly, if at all. We therefore discuss the problem of extrapolating across domains of scale next.

Extrapolating Across Domains of Scale

Because patterns in nature rarely exhibit scale-invariance across an infinite range of scales, the challenge of identifying domains of scale, as well as the limits to extrapolating information across scales, is a major challenge in ecology. This problem was highlighted by American landscape ecologist John Wiens in his classic essay on spatial scaling in ecology (1989a, p. 393):

> Measurements made in different scale domains may therefore not be comparable, and correlations among variables that are evident within a domain may disappear or change sign when the scale is extended beyond the domain…Explanations of a pattern in terms of lower-level mechanisms will differ depending on whether we have reduced to a scale within the same domain, between adjacent domains, or across several domains.

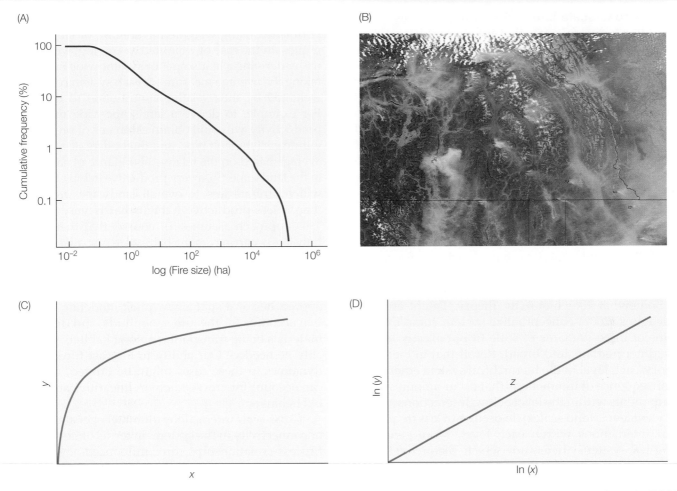

Figure 2.13 Power-law distributions.   (A) The cumulative-size distribution of forest fires in northeastern Alberta, Canada (1961–1998) is approximately a power law. Most fires are small, but a few are very large. Note logarithmic scale on both x and y axes. (B) Satellite image of fires burning in British Columbia and Alberta, Canada in 2003. (C) Empirical form of a power-law function, depicting a positive relationship between the measured quantities. (D) Linear regression of the relationship on a logarithmically scaled plot is typically used to derive the slope (z) as an estimate of the scaling exponent.

Source: (A) After Reed & McKelvey (2002) (B) Image courtesy of Jacques Descloitres, MODIS Rapid Response Team at NASA GSFC and taken from NASA's Earth Observatory.

Direct extrapolation is therefore not recommended across domains of scale.

To delineate domains of scale, and thus the limits to direct extrapolation, we require information on how spatial patterns change as a function of scale, which will be covered in more detail in our discussion of multiscale analysis in **Chapter 4**. For now, however, we might at least define the **scope** of the scale domain over which the power law holds, by examining the relationship between the grain and the extent of measurements (Schneider 2001). Scope can be computed as

$$\left(\frac{Q(m)}{Q(m_0)}\right) = \left(\frac{m}{m_0}\right)^k.$$ Equation 2.2

where the Qs are the quantities associated with measurements obtained at the fine scale (m_0) and at some broader scale (m), respectively, and k is the scaling exponent that relates these two measures. The scope is thus the ratio of the extent to the resolution or grain of the dataset. For example, the scope of a meter stick with a resolution of 0.01 m (1 m = 100 cm) is 1 m/0.01 m = 10^2 which is a narrower scope then if the meter stick has a resolution of 0.001 m (1 m = 1000 mm), where the scope is 1 m/0.001 m = 10^3. The first set of measurements ranges over two orders of magnitude, whereas the second ranges over three orders; the second set thus has a greater scope than the first. Power laws basically relate one scope to another according to the scaling exponent, k. Abrupt changes in k denote a change in scope and can thus aid in the detection of scaling domains.

CROSS-SCALE INTERACTIONS   Extrapolating across scales is further complicated if interactions occur across scales. **Cross-scale interactions** are processes at one spatial or temporal scale that interact with processes at another scale (Peters et al. 2004, 2007). For example, broad-scale drivers can interact with fine-scale processes to determine system dynamics, such as when land-management practices alter the local cycling of soil nutrients (**Chapter 11**). Cross-scale interactions can have important influences on ecological systems, and they pose a significant challenge for understanding and forecasting system behavior.

The challenge of cross-scale interactions is that they can generate emergent behavior that cannot be predicted based on observations at a single scale or even at multiple independent scales. Such emergent behavior may occur suddenly and unexpectedly, such as when a typically local disturbance such as a fire is able to spread across an entire landscape, perhaps transforming it to another state. How might this occur?

Somewhat like hierarchy theory, Peters and her colleagues (2007) conceptualize system complexity in terms of three domains of scale (if not exactly levels): fine, intermediate, and broad. Recall that in hierarchy theory, each level in the hierarchical system emerges as a consequence of interactions that occur among system components within the level below. Interactions among adjacent levels (and scales) do occur, but less frequently than interactions within each level. More generally, complex systems theory (of which hierarchy theory is a part) predicts that the interaction between local-scale processes can sometimes produce long-range correlations and have system-wide effects that can result in non-linear dynamics and system state shifts, much as proposed by the cross-scale interaction framework. In the cross-scale interaction framework, however, it is some type of **transfer process** (e.g. water, wind, animal dispersal) operating at intermediate scales that is ultimately responsible for facilitating interactions between fine and broad scales (Peters et al. 2007). In other words, interactions do not just occur between adjacent domains of scale (e.g. between fine and intermediate scales), but may also cut across scaling domains if such interactions are enhanced by transfer processes that span the intervening scales. Spatial heterogeneity may interact with, and subsequently affect, the behavior of these transfer processes, which only adds to the challenge of understanding how cross-scale interactions might affect the dynamics of the system.

How does this ultimately affect our ability to extrapolate across scales? Consider that if cross-scale interactions do not occur (i.e. spatial heterogeneity and intermediate-scale transfer processes are not important), then direct or linear extrapolation might be used to scale up information. Conversely, if spatial heterogeneity is important but transfer processes are not (the system is essentially a 'closed' one), then extrapolation can be performed using weighted averaging or similar techniques. In the case of weighted averaging, we can scale up by deriving a landscape- or system-wide measure by taking the average measure for each system component, weighted by their relative contribution to the whole. For example, to derive a landscape-wide measure of productivity, we could obtain estimates of productivity within each habitat type, and then calculate a weighted average based on the relative abundance of each habitat in the landscape. So, a very productive habitat that is rare will contribute less to overall landscape productivity than a less-productive habitat that is very common. This approach assumes, of course, that there are not transfers occurring among habitats (e.g. spatial subsidies; **Chapter 11**). When cross-scale interactions are expected to be important (transfer processes occur among different habitats of the landscape), then non-linear statistical approaches or a spatially explicit modeling approach that accounts for the rate, magnitude, and direction of materials being transported among habitats will probably be needed. Our ability to forecast future system dynamics in these cases might be limited, unless we can account for cross-scale non-linearities and threshold behaviors.

Cross-scale interactions ultimately get at the nature of connectivity in the system, in terms of how pattern-process relationships are influenced by transfer processes operating across domains of scales. Highly connected (or overconnected) systems have the highest probability of exhibiting non-linear spatial dynamics, where local effects or processes can propagate throughout the entire system precipitating **thresholds** and possibly catastrophic state shifts (**Chapter 11**). For example, the spread of disease due to increased transport (air travel or shipping) leading to a global pandemic is an example of how enhanced connectivity across a range of scales can facilitate unexpected—and undesirable—events (**Chapter 8**). We will explore issues pertaining to landscape connectivity further in **Chapter 5**.

Uncertainty, Predictability, and Ecological Forecasting

The renowned Danish physicist and Nobel laureate Niels Bohr is credited with saying 'prediction is very difficult, especially about the future.' Extrapolation is fraught with uncertainty, and this uncertainty compounds over space and time, especially if we are attempting to extrapolate across domains of scale or levels of biological organization. Ecological systems are complex systems, in which levels of organization are defined on the basis of local interactions that occur among system components. Thus, it can be difficult to understand or predict the structure or dynamics of the system simply from a very detailed, mechanistic understanding of how the system components are distributed in space or the nature of the interactions among them. Complex systems typically exhibit emergent behavior and sensitivity to initial conditions that make extrapolation from local scales to broad (global) scales difficult.

A critical test of our understanding of complex systems is our ability to make reliable predictions   Gardner (1992)

To use a familiar example, consider the science of meteorology. The weather is very much a complex system, and yet weather forecasters attempt to keep us informed of the expected weather conditions for the coming day or week. Weather forecasts are couched in uncertainty. For example, if there is an 80% chance of rain this afternoon, we would be wise to carry an umbrella, although there is a 20% chance that we won't need it. Weather forecasts become increasingly uncertain the farther out in time meteorologists try to extend their predictions. Weather forecasts rarely extend more than 10 days out, and are often unreliable in any case. They might foretell impending changes based on major weather systems that are tracking across the continent or ocean, but they are unlikely to be able to predict with precision how severe an effect it will have on your particular location until a day or two in advance. A forecast of rain for next Tuesday may well have evaporated by the time Sunday rolls around. Any number of factors can influence the trajectory of the system and thus alter the outcome, requiring updates and refinements to the initial predictions.

Now consider the improbability of trying to forecast the likelihood of rain a year from next Tuesday, let alone 10 or 100 years from next Tuesday. Such long-term weather forecasts would be so unreliable as to be ludicrous. We might still make a general prediction, based on how often it has rained on that particular date in the past. For example, if it rained 40 of the last 100 years on that date, we give it a 40% chance of raining on the same date next year. This is essentially an average, however, which doesn't account for recent climatic changes, such as a drying trend that makes it far less likely to rain on this date in the future. Thus, although long-term weather data may give us some average system behavior, such as the average daily temperature or monthly precipitation, complex systems rarely exhibit 'average' behavior at any given point in time (or space). This fact highlights the sort of

challenge we face in **ecological forecasting**, which underlies much of our environmental, land-management, and conservation policies.

Our ability to make predictions depends on the nature of the relationship between spatial and temporal scales of variation (i.e. on the nature of heterogeneity). As discussed earlier in the chapter, an increase in spatial scaling generally brings about an increase in temporal scaling (space-time diagrams; **Figures 2.2 and 2.3**). Studies that are conducted at the relevant scale in space and time should therefore have the highest predictive power (**Figure 2.14**). As mentioned previously, however, most ecological investigations are conducted at relatively fine spatial scales and over fairly short time periods, which challenges our ability to extrapolate to broader scales in order to make reliable predictions about system behavior. With the advent of landscape ecology and the widespread availability of satellite imagery, we can now obtain information at very broad spatial scales with relative ease. Still, satellite imagery has only been available for about 45 years or so, and other sources of historical landscape data based on aerial photography or land surveys are generally limited in their availability (**Chapter 4**). Thus, most broad-scale landscape studies can document change over a relatively narrow period of time. This is a concern only if these sorts of studies are being used to generate predictions that give the appearance of high predictability in forecasting landscape change, when they simply have not covered a sufficient time period to observe the longer-term dynamics of the system.

Any predictions of the dynamics of spatially broad-scale systems that do not expand the temporal scale are pseudopredictions. The predictions may seem to be quite robust because they are made on a fine time scale relative to the actual dynamics of the system, but the mechanistic linkages will not be seen because the temporal extent of the study is too short. It is as if we were to take two snapshots of a forest a few moments apart and use the first to predict the second.

Wiens (1989a)

The apparent predictability of broad-scale studies conducted over short time periods are really **pseudopredictions** (Wiens 1989a; **Figure 2.14**). Pseudopredictions appear robust, but are only trivially so, given that the temporal scale of the study is not sufficient to capture the dynamics of the broader-scale system. While we would not expect to put much stock in such predictions, this is easier said than done, given that much of the research in landscape ecology (as in other areas of ecology and environmental science) is conducted over much shorter time periods than the dynamics of the systems we study. This is of particular concern in resource management, where policies that impact large areas are often based on short-term studies (Wiens 1989a).

This does not mean that predictions generated by ecological forecasting are so unreliable as to be useless, however. We can infer much about the process of forest succession from studies of the gap dynamics of individual forest stands and the disturbance dynamics of landscapes, without the benefit of a centuries-long study that encompasses climatic changes and the occasional catastrophic fire or volcanic eruption. Nevertheless, our understanding of the role of these sorts of climatic or disturbance processes in shaping the landscape might be incomplete if we are unable to observe the system across the full range of conditions or magnitude of disturbances likely to occur in this system, which could obviously affect the reliability of our predictions. One way to circumvent that problem is to substitute space for time in our study.

SPACE-FOR-TIME SUBSTITUTIONS   Given the difficulty of conducting broad-scale studies over correspondingly long time periods, we might instead identify study sites or landscapes that collectively represent the range of conditions or sorts of dynamics that we expect to occur in our system. For example, if we are interested in the effects of habitat loss and fragmentation, we might identify a series of landscapes with different amounts of habitat or varying degrees of fragmentation. So, rather than observe individual landscapes undergoing habitat loss and fragmentation over time, we instead stitch together a hypothetical time-series based on the range of landscape configurations that are present in the region. Such studies are thus making a **space-for-time substitution**, in that they use replication in space as a proxy for replication in time (Hargrove & Pickering 1992).

This of course assumes that space and time are interchangeable, but is that a reasonable assumption to make? In the case of our habitat fragmentation study, we are assuming that all of the landscapes lie on the same trajectory of change; that is, that all landscapes undergo habitat loss and fragmentation in the same way (**Figure 2.15A**). While that might be a reasonable assumption (especially if all landscapes are located within the same region and subject to the same sorts of climatic factors, disturbances, and land-use pressures), it is also possible that historical contingencies and differences in the rate at which habitat is lost and fragmented could have a profound effect on our ability to forecast the ecological consequences of landscape change.

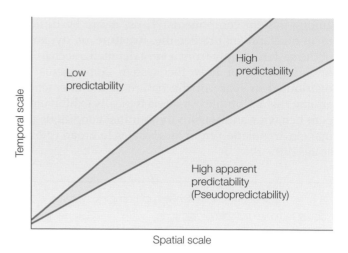

Figure 2.14 Scale and predictability.   An increase in the spatial extent of an investigation generally requires an increase in the temporal extent (i.e. its duration). Studies that are conducted at the appropriate spatiotemporal scales will have high predictive power. However, most ecological investigations are conducted at finer scales than the phenomena they seek to study, even if the research is long-term. Such studies are likely to have low predictive power, in terms of extrapolating the findings to other areas or over broader scales. Conversely, short-term studies conducted at broad spatial scales may appear to have high predictability, but in fact do not. These studies run the risk of psuedopredictability.

Source: After Wiens (1989a).

For example, two landscapes might have the same amount of forest cover (say, 50%) and level of fragmentation, but could have arrived at their present states by different means over variable lengths of time: either slowly as a result of a chronic, low-level disturbance such as selective logging (e.g. 0.5% forest lost per year) or rapidly owing to deforestation (5% loss per year; **Figure 2.15B**). Although the end result is the same (50% habitat is lost in both cases), the ecological consequences of the two scenarios may be very different if, for example, a species is able to track landscape changes in the first scenario (when the rate of habitat loss is slow) but not in the second (when habitat loss occurs more rapidly). It takes time—perhaps years or decades—for population numbers to reach a new equilibrium with the landscape (if they do), such that the species exhibits a lagged response to rapid landscape change. This could lead to the mistaken conclusion that the species can tolerate far more habitat loss and fragmentation than it actually can (i.e. its extinction risk is being underestimated in this scenario; **Figure 2.15B**). Ignoring the rate or history of landscape change can thus have a profound effect on our conclusions or predictions about the effects of habitat loss and fragmentation.

(A)

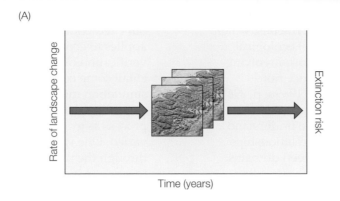

(B)

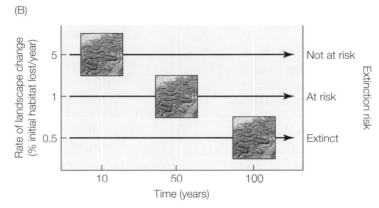

Figure 2.15 Space-for-time substitution.   Most fragmentation studies assume landscapes with the same degree of habitat loss and fragmentation all lie on the same trajectory of change, and are thus ecologically equivalent, such as for assessing a species' risk of extinction (A). However, landscape history, such as how quickly habitat loss occurred, may influence this assessment of extinction risk (B). Populations can track landscape changes that occur gradually (0.5%/year), in which case, the species may already have gone extinct after a 50% loss of habitat (i.e. within 100 years). If habitat is lost too rapidly (e.g. 5%/year), populations may lag behind the rate of landscape change, such that the species may not even be assessed as 'at risk' of extinction even though the same amount of habitat loss has occurred. Extinction risk would thus be underestimated in the very scenario where it matters most. Landscapes with the same level of habitat loss and fragmentation may lie on different trajectories of change, such that space cannot always serve as a substitute for time.

Source: After With (2007). Photo © B Brown/Shutterstock.com

Future Directions

Although the concept that "scale matters" is a central concern of landscape ecology,
we have only fragments of a theory of scaling.   Wiens (1999)

Two decades later, Wiens's sentiment still rings true. Scale is a central concept in ecology, but could it also be a unifying theory? Power laws and fractals might provide some theoretical unity to ecology (Schneider 2001), and hence to landscape ecology. Landscape patterns and spatial phenomena are only fractal-like or scale-invariant across a finite range of scales, however, and non-linearities and threshold behaviors may exist among domains of scale. Thus, we need a more comprehensive approach to the study of scaling relationships than provided by just power laws or fractals. Future research should therefore strive to do the following:

- *Adopt a multiscale research focus*. We need to make the study of scale-dependent changes in ecological

patterns an explicit focus of research efforts. Studies that are conducted at several scales, in which grain and extent are systematically varied independently of one another, are still needed to help identify domains of scale and the pattern-process linkages responsible for various ecological phenomena, as well as the interrelationships that occur among scales (Wiens 1989a). We thus need to move toward a multiscale or hierarchical research approach as a matter of course (Wu 2007b).

- *Advance the theory of scaling*. More progress is needed on the development of scaling theory that is mechanistically based and generates testable hypotheses. Scaling theory needs to progress beyond power laws and fractals, which are not adequate for explaining all ecological scaling phenomena, especially those involving cross-scale interactions that produce non-linearities (Wu 2007b). As Wiens (1989a, p. 394) pointed out, 'Our ability to arrange scales in hierarchies does not mean that we understand how to translate pattern-process relationships across the non-linear spaces between domains of scale.' We need to understand why scaling relationships occur and why relationships scale the way they do, and not just content ourselves with documenting them empirically.

- *Develop new methods for data extrapolation.* A major research challenge will continue to involve the development of methods for extrapolating across scales and the conditions under which these methods are advisable. This will involve the development and application of scaling functions, non-linear statistical methods, spatially explicit modeling approaches, and other types of analytical and modeling approaches (Urban 2005; Wu 2007b).

- *Test and verify scaling relationships and methods for extrapolation*: Beyond research and development,

we need to assess the accuracy and reliability of our predictions, particularly as this assessment applies to ecological forecasting. The testing and verification of scaling relationships will likely entail some combination of experimentation and simulation modeling to explore the range of variability that occurs between pattern and process as grain and extent are systematically varied. One way this could be accomplished is through the use of **neutral landscape models** (discussed in **Chapter 5**), which have been used to explore the effect of changing scale on the analysis of landscape pattern (Turner et al. 1989b); how different resource distributions affect the scale of resource utilization by foragers (O'Neill et al. 1988b); how the scale of heterogeneity affects the redistribution of individuals across different landscape configurations (With et al. 1997); and, to model hierarchical patch structure (O'Neill et al. 1992) and how the availability of habitat at different nested scales affects species coexistence (Lavorel et al. 1994, 1995).

In conclusion, extrapolation need not be viewed just as an approach to forecasting, or predicting patterns or system behavior beyond the extent of our data, but also as a means of assessing the robustness of underlying pattern-process relationships (Miller et al. 2004).

Chapter Summary Points

1. Scale is a central problem in landscape ecology, and in ecology more generally.

2. In ecology, scale is defined by the *grain* and *extent* that bound the study or observation (data) set. Grain is the finest scale of resolution in the data-set, and extent is the broadest scale encompassed by the study. The scale of any study includes both spatial and temporal dimensions that can be defined in terms of their grain and extent.

3. Patterns are scale-dependent. Our choice of scale influences what patterns we detect and whether we are able to assign the correct processes or mechanisms to the observed phenomenon. This in turn will likely influence our interpretation of the results and the conclusions of the study.

4. Although there is no single correct scale at which to study ecological systems, some scales are clearly better than others. Our choice of scale will depend upon the question, process, organism, or system being studied. Ultimately, the scale of study is set by external constraints; established research conventions and sampling protocols; personal experience and expert opinion; or sometimes, arbitrary decisions.

5. Landscapes exhibit patchiness or heterogeneity across a range of scales, which may be hierarchically nested (patches are nested within larger patches). Landscapes may thus exhibit hierarchical patch structure.

6. Landscapes may also exhibit hierarchical system organization. Hierarchy theory addresses how complex systems arise and how they function. According to hierarchy theory, hierarchical system organization emerges as a consequence of interactions among lower-level system components that give rise to higher-order system structure and dynamics. The levels are defined by their constituent parts and position within the hierarchy (higher levels contain lower levels), but also by the strength and rate of interactions (lower-level components interact more strongly and frequently than components of higher levels).

7. Power laws and fractals are common in nature. The existence of power-law behavior and fractal-like structure across a range of scales enables the direct extrapolation of information across scales because the relationship is linear (scale-invariant) within that domain.

8. Extrapolation across scales may involve either the scaling up (upscaling) of fine-scale information to make inferences about broader-scale patterns or system behaviors, or the downscaling of broad-scale information to make predictions about patterns and processes at finer scales.

9. Most patterns in nature exhibit some degree of scale dependency, such that different scaling relationships are associated with different domains of scale. Because the scaling relationship may be non-linear, extrapolation across domains of scale should be approached with a great deal of caution (and a small dose of courage!).

10. Cross-scale interactions can occur if some type of intermediate-scale transfer process (fire, water) is able to facilitate interactions between fine and broad scales. However, feedbacks among these different domains of scale can produce non-linearities and threshold behaviors in which the system unexpectedly shifts from one state to another (i.e. a catastrophic state change).

Discussion Questions

1. Explain why scale is considered the central problem in ecology. Why is scale particularly important to the study of landscape ecology?

2. Identify the grain and extent that bounds the observation set either of your own research or of a study featured in one of your assigned readings. Don't forget to identify the temporal grain and extent, as well as the spatial scale of the study! What factors might have influenced or constrained the choice of scale? How might the choice of scale influence the results or conclusions of the study? What changes would you advise making to the study design to remedy these scaling issues?

3. What is the scale of a landscape? In other words, at what spatial and temporal scales should we study landscapes? Be ready to defend your position by explaining how this would ultimately influence the science and practice of landscape ecology.

4. Try to demonstrate how your study system (or an example from the literature) could be conceptualized or modeled as a hierarchically organized system. Recall that a minimum of three levels is required for study. What other conditions are necessary for demonstrating that this system is hierarchically organized, as opposed to just hierarchically structured?

5. How might a landscape be considered a hierarchically organized system? Discuss whether it matters to the study of landscape ecology if landscapes are not in fact a true level in the ecological hierarchy (i.e. at a level above ecosystems).

6. Using your own research or an example from the literature, identify in what ways information is being extrapolated across scales, both spatially and temporally. Note that extrapolation is implied when extending the results of the study to other locations, systems, or time periods. To what extent might your ability (or that of the author, if this is an example from the literature) to extrapolate information or study results be constrained by the design of the study (i.e. its spatial or temporal grain and extent)? What other factors limit the extrapolation of information in this study?

Landscape Heterogeneity and Dynamics

Few places on Earth bring opposing geological forces together in such stark relief as the landscapes of northern Arizona in the American Southwest. Here we see evidence of a violent past, where eruptive volcanoes gave rise to towering mountain peaks that vie with the power of water that carved a mile-deep gash in the planet's surface. Such dramatic variation in topography generates a similarly spectacular range in environmental conditions, which are reflected in the diverse ecological communities that form along this gradient. One can thus travel from alpine tundra atop the San Francisco Peaks, through spruce-fir and ponderosa pine forests, to pinyon-juniper woodlands and grasslands, down to desert cacti near the bottom of the Grand Canyon, all in a distance spanning less than 130 km (**Figure 3.1**).

Beyond the geological forces that have contributed to this unique landscape, environmental disturbances in the form of decades-long droughts and periodic fires have altered ecological communities and shaped patterns of human settlement and migration throughout the region. The human footprint is clearly evident, from the crumbling remains of ancient dwellings constructed by the Sinagua who inhabited this land more than a millennium ago, to the modern communities that sit in the shadow of the San Francisco Peaks today. Over the past 150 years, land-use practices involving livestock grazing and fire suppression have triggered a shift toward greater tree densities, which in turn have contributed to more-destructive forest fires. In June 2010, for example, an abandoned campfire ignited a fire that scorched more than 61 km² of forest on the slopes of the San Francisco Peaks (**Figure 3.1A**). Pumice mines scar the flanks of the Peaks, and smaller volcanic peaks nearby are being leveled entirely for their pumice. The mighty Colorado River has been tamed and drained through the construction of dams and canals along its length, which power and irrigate the major urban and agricultural centers that paradoxically have arisen from the desert. A river that once carved a canyon grand now no longer reaches the sea, reduced to a mere trickle that ends some 80 km from its delta in the Gulf of California, thereby disrupting the ecology of one of the largest desert estuaries in the world.

Disturbances—both natural and anthropogenic—are a ubiquitous feature of any landscape, contributing to its structure and dynamics, as well as those of the ecological systems embedded within the landscape. This chapter considers the formative processes that shape landscapes and contribute to their heterogeneity and dynamics.

Heterogeneity and Disturbance Dynamics as Core Concepts in Landscape Ecology

Perhaps no other ecological construct is as closely tied to the concept of heterogeneity as is the landscape. Although heterogeneity is a characteristic of all ecological

Essentials of Landscape Ecology. Kimberly A. With, Oxford University Press (2019).
© Kimberly A. With 2019. DOI: 10.1093/oso/9780198838388.001.0001

(A)

(B)

Figure 3.1 Landscapes of northern Arizona are a study in geological contrasts.   (A) In this region of the southwestern USA, a violent past gave rise to mountain ranges such as the San Francisco Peaks (3851 m elevation), the remnants of an extinct stratovolcano. More-recent violence is evidenced by fire scars along its flanks (blackened and bare areas) from a wildlife in June 2010, as well as the scars of a pumice mine (light gray areas, far right). (B) A bit farther to the north, uplifting and erosional forces were responsible for carving the 1800-m deep Grand Canyon, laying bare nearly 2 billion years of geologic history. The environmental gradient from the top of the San Francisco Peaks to the bottom of the Grand Canyon encompasses no fewer than seven life zones, from alpine tundra to Sonoran desert (**Figure 3.2**).

Source: Photos by Kimberly A. With.

systems (and of complex systems more generally), the landscape is explicitly defined as 'an area that is heterogeneous in at least one factor of interest' (Turner et al. 2001). As such, landscapes can be defined at any scale and within any system—aquatic, marine, or terrestrial—because all may exhibit heterogeneity. Heterogeneity is thus a central concept in landscape ecology; in fact, it could be argued that it is *the* central concept. Landscape ecology seeks to understand what factors give rise to heterogeneity, how that heterogeneity is maintained or altered in the face of natural and anthropogenic disturbances, and how heterogeneity ultimately influences ecological processes and flows across the landscape. Because heterogeneity is expressed across a wide range of spatial scales, the landscape perspective can be applied to address these sorts of questions at any level of ecological organization.

It is the revelation of the importance of heterogeneity that makes the landscape perspective so pervasively relevant to ecology at different organizational levels as well as earth sciences across a broad range of spatial scales. Heterogeneity may be regarded as an essential cause and consequence of diversity and complexity in both natural and social systems, and thus plays a key role in dealing with complexity in theory and practice.   Wu (2006)

Although the focus in landscape ecology is typically on **spatial heterogeneity**, disturbance dynamics obviously produce changes in landscape structure over time as well as in space. Landscapes are thus characterized in terms of spatial and **temporal heterogeneity**: how their composition and configuration are shaped by different types of disturbances operating over a range of scales in both space and time (**Figure 2.2**). Heterogeneity and disturbance dynamics are thus inextricably linked, which is why we will tackle both concepts together in this chapter.

To better appreciate the importance of heterogeneity and disturbance dynamics to landscape ecology, we begin this chapter with an overview of how these concepts have traditionally been viewed by ecologists, and the research developments that have led to a more explicit consideration of the role of heterogeneity and non-equilibrium dynamics within the field of ecology. Next, we tackle the various ways in which heterogeneity has been defined, in terms of both its spatial and temporal dimensions. That leads us to consider how disturbance dynamics are defined, by assessing what factors characterize the disturbance regime. We then highlight two disturbance concepts that have received considerable study within ecology: (1) Intermediate Disturbance Hypothesis (IDH), and (2) Large, Infrequent Disturbances (LIDs). In the second half of the chapter, we consider the various geomorphological processes that play a role in landscape formation, as well as the wide range of abiotic and biotic disturbances that contribute to landscape heterogeneity and dynamics. We conclude with a consideration of how anthropogenic disturbances compare to natural disturbances, and the implications of historical variability concepts for the management of landscapes.

Emergence of Heterogeneity and Dynamical Concepts in Ecology

Science seeks to uncover patterns and relationships in the hopes of bringing some orderliness and comprehension to the complexity of the systems we study. To that end, we sometimes invoke metaphors or use simple models to abet our understanding of this complexity. For example, ecological systems have long

TABLE 3.1   Stability concepts

Concept	Definition
Resistance	The capacity of a system to resist change in the face of an external perturbation (i.e. disturbance); it is measured by the degree to which a variable is changed from its equilibrium value following a perturbation
Resilience	Rapidity with which a system returns to a previous equilibrium after a perturbation
Persistence	Ability of a system to remain within defined limits despite perturbations; length of time system remains within a defined state (i.e. within some accepted 'normal' range of variability)
Variability	Degree of change in system properties over a given period of time
Constancy	Inverse of variability; invariance in system properties over a given time period

Source: After Wu & Loucks (1995).

been held to exist within a balanced state, one that is sustained by various regulatory mechanisms that function to maintain the relative stability of the system (**Table 3.1**). There is a comfortable logic to the idea that ecological systems are driven to achieve some sort of natural balance or equilibrium. It fits well with our ordered sense of the universe and is consistent with the various regulatory functions that govern our own homeostasis. The guiding paradigms of ecology have thus traditionally been steeped in 'balance-of-nature' ideology, in which the themes of homogeneity, order, stability, and equilibrium prevail (DeAngelis & Waterhouse 1987; Wu & Loucks 1995). In this section, we examine the various ways in which ideas pertaining to homogeneity and equilibrium have given way to an acceptance of—and an appreciation for—heterogeneity and dynamical concepts in ecology.

Heterogeneity at Broad Geographic Scales: Biogeography and Life Zones

Naturalists and geographers recognized early on that the distribution of plants and animals exhibited some well-defined and predictable patterns. The pioneering work of the 19th-century German naturalist Alexander von Humboldt laid the foundation for the field of plant biogeography (phytogeography)—the geographic study of plant distributions. Not content merely to describe biogeographic patterns, Humboldt advocated for a more quantitative approach to the mapping and study of vegetation, having recognized a connection between elevation, temperature, and the formation of different types of vegetation communities.

This relationship between climatic factors and the distribution of particular communities is ultimately the basis for the **life-zone concept**, developed in 1889 by American naturalist Clinton Hart Merriam to describe the general similarities in plant and animal communities that may be found along either an elevational or latitudinal gradient (**Figure 3.2**). Merriam's work was inspired by the elevational gradient that runs from the San Francisco Peaks to the bottom of the Grand Canyon (**Figure 3.1**). Recall too that it was the study of such elevational zonation in vegetation communities that led German geographer Carl Troll (1939) to first propose landscape ecology as a new ecogeographic study of the relationship between communities and their environment (**Chapter 1**).

The life-zone concept was later revised by the American botanist Leslie Holdridge (1947) to encompass more bioclimatic information (average mean temperature above freezing, total annual precipitation, potential evapotranspiration ratio), thereby producing a more comprehensive classification scheme of vegetation zones that could be applied worldwide. Whereas Merriam identified only seven life zones, most of which were specific to western North America, Holdridge's system defined more than 30 life zones circumscribed by different bioclimatic domains that included the influences of elevation, latitude, and humidity provinces (**Figure 3.3**). Holdridge's life-zone system can thus be applied to tropical and subtropical regions as well as to the more arid regions that formed the basis of Merriam's life-zone classification. Today, the life-zone approach continues to figure prominently in ecosystem mapping, particularly in the context of forecasting how terrestrial ecosystems might shift in response to climate-change scenarios (Lugo et al. 1999; Yue et al. 2011).

> We consider the life zone approach to have many strengths for ecosystem mapping because it is based on climatic driving factors of ecosystem processes and recognizes ecophysiological responses of plants;…it is a relatively simple system based on few empirical data; and it uses objective mapping criteria.
>
> Lugo et al. (1999)

The concept of a life zone implies that vegetation communities are predictable, that they form reasonably stable assemblages that can be classified as distinct vegetation zones. To what extent are vegetation assemblages stable and predictable, however? We consider this next in the context of how plant communities are organized.

Dynamics of Plant Community Assembly: Climax State or Independent Assembly?

The notion that ecological systems exist in some sort of equilibrium or stable state is also evident in early ideas about how these systems might be organized. For example, the American plant ecologist Frederic Clements (1916) proposed that plant communities naturally went through **ecological succession**, a series of stages akin to the developmental stages of an organism, which culminated in a **climax state**, the particular suite of plant species best suited to local conditions (e.g. a beech-maple forest association).

The idea that vegetation communities might be viewed as some sort of 'superorganism,' driven to attain a particular equilibrium state (the mature or climax state) clearly resonates with balance-of-nature ideology. It is also very much at odds with how vegetation communities appear to be organized, however, and Clements's view of community organization was

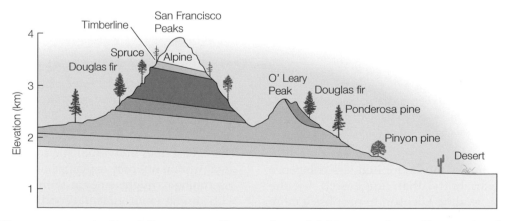

Figure 3.2 The life-zone concept. Vegetation communities vary in predictable ways along either an elevational or latitudinal gradient, largely as a result of differences in bioclimatic factors. The elevational gradient depicted here is located in northern Arizona (**Figure 3.1**), where C. Hart Merriam first developed the concept of life zones. Notice that the relative elevation at which different life zones occur also depends on aspect, the direction the slope faces.

Source: After Bailey (1996).

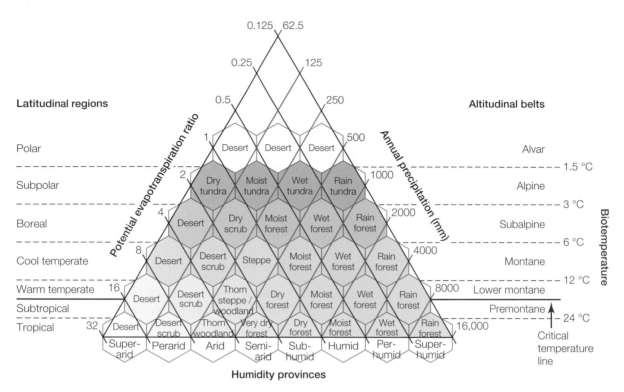

Figure 3.3 Holdridge life-zone diagram. Different vegetation zones can be mapped based on bioclimatic factors related to annual precipitation, the potential evapotranspiration ratio (the amount of water that could be evaporated and transpired by plants if sufficient water is available), and humidity provinces within different altitudinal belts and latitudinal regions.

Source: Image created by Peter Halasz and published under the CC BY SA 2.0 License.

challenged by his contemporary, Henry Allan Gleason, another American plant ecologist, in part because it assumed a great deal of uniformity in the attainment of a particular climax state (Gleason 1926, p. 13):

> [I]t seems that we are treading upon rather dangerous ground when we define an association as an area of uniform vegetation, or, in fact, when we attempt any definition of it. A community is frequently so heterogeneous as to lead observers to conflicting ideas as to its associational identity, its boundaries may be so poorly marked that they can not be located with any degree of accuracy, its origin and disappearance may be so gradual that its time-boundaries can not be located; small fragments of associations with only a small proportion of their normal components of species are often observed; the duration of a community may be so short that it fails to show a period of equilibrium in its structure.

Vegetation communities thus exhibited less coherence and far more variability than suggested by the Clementsian view, leading Gleason to propose a more individualistic concept of community organization (Gleason 1917, 1926). Gleason viewed vegetation communities as highly variable assemblages in which plants species sorted themselves independently of one another, owing to the vagaries of dispersal, time, and

the intrinsic heterogeneity of the physical environment (Gleason 1926: p. 19):

> Heterogeneity in the structure of an association may be explained by the accidents of seed dispersal and by the lack of time for complete establishment. Minor differences between neighboring associations of the same general type may be due to irregularities in immigration and minor variations in environment. Geographical variation in the floristics of an association depends not alone on the geographical variation of the environment, but also on differences in the surrounding floras, which furnish the immigrants into the association. Two widely distant but essentially similar environments have different plant associations because of the completely different plant population from which immigrants may be drawn.

The Gleasonian individualistic concept of community organization not only explains how differences in plant assemblages might emerge within and among areas with similar environmental conditions, but also found empirical support in the apparently independent distribution of plant species along environmental gradients, such as those studied in the Great Smoky Mountains of Tennessee (USA) by the eminent plant ecologist Robert Harding Whittaker, who went on to popularize gradient analysis and ordination techniques in community ecology (Whittaker 1956, 1967).

As with most of the polarizing debates in ecology, there is support for both sides, depending on the scale at which vegetation patterns are viewed. Clearly, broad-scale vegetation zones can be identified and are the basis for ecological land classification (e.g. biomes, life zones, ecoregions, floristic provinces). Considerable heterogeneity nevertheless exists within these regions, owing to a wide range of biotic and abiotic disturbances that operate at various scales in space and time to influence the composition of local plant communities.

Patch Dynamics: A Paradigm Shift in Ecology

That disturbances might be important for the structure and dynamics of plant communities was recognized by at least some early ecologists. Indeed, Clements himself recognized this, pointing out that '[e]ven the most stable association is never in complete equilibrium, nor is it free from disturbed areas in which secondary succession is evident' (Clements 1916, p. 4). In his classic paper on pattern and process in plant communities, Scottish botanist Alexander Stuart Watt described the nature of the 'regeneration complex' (Watt 1947). Watt essentially viewed plant communities as shifting patch mosaics, in which the patches were different successional stages that were dynamically linked. The regeneration sequence exhibited a certain degree of orderliness in its progression, but Watt clearly appreciated the role that disturbance played in disrupting this orderly progression, as demonstrated by his description of the gap-phase dynamics in a beech forest (Watt 1947, pp. 13–14):

Aside from minor fluctuations over periods of time in, say, the number of deaths among old trees, there are exceptional factors of rare or sporadic occurrence, such as storms, fire, drought, epidemics, which create a lag phase of exceptional dimensions . . . the relative areas under the age classes (as a super refinement of phasic subdivision) need bear no relation to current meteorological factors but be explicable in terms of some past event which happened, it may be, 200 or 300 years ago. A series of sporadic exceptional events will obviously increase the difficulties of 'explaining' current relative areas . . . Fortuitous fluctuations of climate and other causes may thus bring about major departures from the normal or ideal.

So, although disturbances were acknowledged to be important in contributing to heterogeneity, they were nevertheless viewed as unusual or uncommon events. That is, disturbance was viewed more as an impediment to attaining the natural equilibrium state, by causing 'major departures from the normal or ideal.' Still, what Watt found interesting was that this distur-

bance-mediated patchiness persisted rather than becoming randomized or homogenized with time, thereby highlighting the potential for long-past disturbances to influence current landscape structure and species distributions (i.e. 'the ghosts of landscapes past'; With 2007).

A more dynamic view of nature began to emerge during the 1970s and 1980s, fostered by a growing recognition that if disturbance events are common, then most ecological systems are unlikely ever to reach any sort of equilibrium or balanced state (Levin & Paine 1974; Wiens 1976; Pickett & White 1985). This thinking is evident in a now-classic paper by Simon Levin and Robert T. Paine on the role of disturbance dynamics in shaping molluskan communities in the rocky intertidal zone along the coasts of western North America (Levin & Paine 1974, p. 2744):

Disturbance, often in the form of extinctions due to natural catastrophe, competition, or predation-related agents, interrupts the local march to and survival of equilibrium (local climax) . . . Disturbance operates in two ways to increase environmental heterogeneity: by providing the opportunity for local differentiations through random colonization and a kind of founder effect ensuring persistence, and by constantly interrupting the natural successional sequences. . . . Such short circuits may prevent local patches from ever achieving equilibrium.

By the 1980s, disturbance was viewed as integral to the development and maintenance of heterogeneity within ecological systems. Disturbances of different sizes, intensities, and frequencies create gaps or patches of disturbed areas, which are then subject to different successional trajectories, resulting in **patch dynamics** (Pickett & Thompson 1978). In the case of landscapes, disturbance and succession are considered to be so ubiquitous that landscapes may be viewed as 'a mosaic of successional patches of various sizes . . . whose diversity depends partially on patch dynamics' (Pickett & Thompson 1978: 29). Patch dynamics have thus contributed to our current view of landscapes as heterogeneous patch mosaics.

Because patch dynamics implies constant change, we might question when—if ever—landscapes could be said to be in equilibrium. Although equilibrium implies a balance between opposing forces, notice that this does not mean that the system is static. The system could, for example, achieve a bounded steady state; that is, a **dynamic equilibrium**. For example, we might observe little change in the overall composition of a landscape over time, despite the fact that certain patch areas on the landscape do change (**Figure 3.4**). If the rate at which disturbances occur is offset by successional changes within previously disturbed patches,

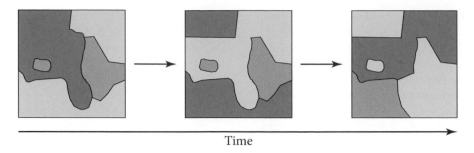

Time

Figure 3.4 Landscapes as shifting patch mosaics. The balance between disturbance and successional dynamics may give rise to a shifting-mosaic steady state, a type of dynamic equilibrium in which the different seral stages or land-cover types attain a more-or-less constant proportion over time (a landscape steady state) despite ongoing disturbances (the shifting mosaic). Thus, although the configuration of the landscape changes over time, the composition of the landscape remains fairly constant.

then the proportion of the landscape within each successional stage should remain fairly constant over time. Thus, landscapes might exhibit a **shifting-mosaic steady state** (Bormann & Likens 1979), possessing a relative, if not absolute, constancy in certain attributes.

As with everything else in ecology, the concept of a steady state is ultimately scale-dependent. The landscape steady state emerges only when a large enough area is observed over a long enough time period. From our perspective, a small area experiences all manner of disturbances and stochastic fluctuations that are averaged out when considered over a larger area and longer time period. If, however, we consider a very large geographic area and adopt the perspective of geologic time periods, the region may not be so stable after all, for it now experiences broad-scale, millennia-long events such as glacial-interglacial cycles. Thus, much of what we perceive as landscape dynamics occurs at intermediate spatial scales and over intermediate time periods (**Chapter 2**). Although larger areas are more likely than smaller areas to exhibit a stable mosaic, it is unclear at exactly what scale that occurs, for it depends on the type of system and its disturbance dynamics. Still, the interaction between landscape heterogeneity and disturbance dynamics means that even relative constancy may be illusory and difficult to find in real landscapes. Equilibrium landscapes should thus be viewed as the exception rather than the rule (Turner et al. 1993).

Toward a Non-Equilibrium View of Ecology

If equilibrium landscapes are the exception, then are non-equilibrium landscapes the rule? We have only to think of how a large disturbance event, such as a massive wildfire, alters landscape structure to envision how an ecological system could be destabilized and pushed far from its supposed equilibrium state. In some cases, the system may be pushed so far that it is unable to recover and undergoes a complete reorganization. Indeed, such disturbances are often classified as

'catastrophic' because of the impact they have on the system. Yet classifying disturbances as catastrophic implies that there is something unusual or even unnatural about them; that is, that they lie outside the natural disturbance regime and thus are best viewed as an anomaly. Although that might be true in some cases (especially where human actions are concerned), it is not true for all disturbances that could be considered catastrophic. Lest we forget, disturbances naturally occur across a wide range of scales in space and time (**Figure 2.2**), and may interact to varying degrees, contributing to feedbacks that can cause instability and uncertainty. Large, infrequent disturbances are very much a part of the natural disturbance regime, as we will discuss later in this chapter.

> The classical equilibrium paradigm has usually implied that historical effects, spatial heterogeneity, stochastic factors, and occasional environmental perturbations play a small or negligible role in governing the dynamics of ecological systems, and these systems therefore are reasonably predictable. Numerous studies have demonstrated that history, heterogeneity, stochasticity, and disturbance all can be very important to the structure and dynamics of ecological systems.
>
> Wu & Loucks (1995)

The ability of a system to return to its equilibrium state (e.g. its recovery time) has sometimes been equated with the system's resilience to disturbance. Rather than a single steady state, however, ecological systems may possess multiple stable states (alternative steady states), in which some perturbation or forcing agent causes the system to 'flip' from one state to another (Holling 1973). Such system-state changes, or **regime shifts**, are found in a wide array of terrestrial, aquatic, and marine systems (Scheffer et al. 2001; Folke et al. 2004; Groffman et al. 2006). For example, overgrazing of arid rangelands may cause the system to shift from a grassland to a shrubland or desert state (Laycock 1991; **Chapter 11**).

For ecological systems that possess alternative steady states, there is no single stable state to which the system returns when perturbed. The concept of resilience thus needs to be redefined. In his seminal 1973 paper on the resilience and stability of ecological systems, Canadian ecologist C. S. 'Buzz' Holling differentiated system stability (i.e. system recovery time, what he later referred to as 'engineering resilience') from the concept of **ecological resilience**, the amount of disturbance necessary to push the system from one state to another (Holling 1973, 1996; Gunderson 2000).

Ecological resilience emphasizes that landscapes or ecosystems might have multiple stable states. The threshold that marks the transition from one system state to another is the sort of non-linear, non-equilibrium dynamic characteristic of complex systems, which is why the effects of disturbance on ecological systems can be so difficult to predict (Folke et al. 2004). We will return to a discussion of system state changes in relation to landscape function in **Chapter 11**.

Homogeneity: The Frictionless Plane of Ecological Theory

Along with equilibrium assumptions, much of the foundational theory in ecology assumes a homogeneous environment, the ecological equivalent of the frictionless plane in physics (**Figure 3.5**). Homogeneity greatly simplifies mathematical models in ecology, allowing for closed-form solutions that are elegant in their simplicity. Such mathematical models also have the advantage of generality in that they could apply—in a very general or abstract way—to any ecological system, although not necessarily to any particular system. Thus, a simple mathematical model, such as for logistic population growth (the Verhulst equation, first published in 1838) or predator–prey interactions (the Lotka–Volterra equations, published in the early 1920s), are still useful in that they provide a baseline as to how populations are expected to grow, or the conditions under which two species could coexist, before adding the complexities of spatial (or temporal) heterogeneity. By working in homogeneous environments (or at least in environments that appear so), ecologists might then better apply or test these theoretical expectations within their own systems.

Cracks in the foundation soon appeared, however. Consider the Lotka–Volterra predator–prey model as an example. In its most basic form, the model predicts a coupled dynamic: predator populations rise and fall in response to the population size of their prey, whose numbers in turn are kept in check by the predator. This exemplifies the balance of nature, a stable oscillation or regular cycling of predator and prey populations. However, the model assumes that predators can react

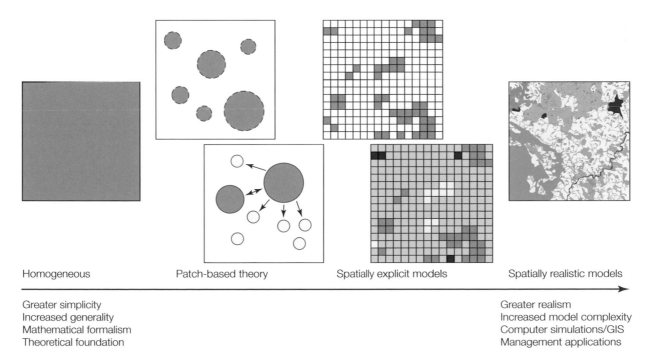

Homogeneous   Patch-based theory   Spatially explicit models   Spatially realistic models

Greater simplicity   Greater realism
Increased generality   Increased model complexity
Mathematical formalism   Computer simulations/GIS
Theoretical foundation   Management applications

Figure 3.5 Representation of spatial heterogeneity in ecological theory and modeling. The development of ecological theory and modeling represents a gradient in terms of how spatial complexity is incorporated, from not at all (the environment is assumed to be homogeneous), to the abstract (patch-based theory, in which patch structure is implied), to the more spatially explicit and realistic.

instantaneously to increases in numbers of their prey. Because population growth rates of predators tend to be slower than those of their prey, however, there is an inevitable time lag to this dynamic. If we add that time lag to the mathematics of the predator–prey dynamic, we find that these fluctuations begin to amplify over time, producing wild oscillations in the numbers of predators and prey as the population dynamics of the two species become decoupled, eventually leading to the extinction of both. Chaotic population fluctuations that lead to extinction do not conform to anyone's idea of stability or balance.

The Lotka–Volterra predator–prey model also assumes that the environment is homogeneous, such that predators encounter prey purely at random, and prey have nowhere to hide from predators. The idea that environmental heterogeneity might be important for understanding predator–prey dynamics, however, was clearly illustrated by Carl Huffaker's classic experiments with mites (Huffaker 1958; Huffaker et al. 1963). Huffaker created an experimental system in which oranges (food patches) and rubber balls (non-food patches) were arrayed to create a patchy resource landscape for six-spotted mites (*Eotetranychus sexmaculatus*), which served as the prey for a predatory mite (*Typhlodromus occidentalis*). By altering the availability and spacing of oranges, Huffaker was able to demonstrate the conditions under which both predator and prey could coexist. Notably, Huffaker experimented with both spatial heterogeneity, by clumping oranges together or interspersing them among rubber balls, and patch connectivity, which he achieved through various means to reduce the dispersal abilities of predators relative to their prey (e.g. 'handicapping' the predatory mite by applying petroleum jelly to the matrix between patches so as to impede their dispersal, and by inserting toothpicks into oranges from atop which six-spotted mites could disperse via their silken threads on air currents).

In homogeneous landscapes where food patches were clumped, or in any landscape where predators were able to disperse freely, neither species persisted for long, as predators quickly located and eradicated all prey populations—thereby eradicating themselves. It was only by creating a spatially complex environment, in which prey were given a dispersal advantage over their predators, that Huffaker was able to achieve a stable coexistence of both species, at least for a while. By adding spatial refugia from predation, the prey species was able to persist through extinction-colonization dynamics among patches; that is, as a **metapopulation**.

On their own, time lags and strong interspecific interactions can destabilize ecological systems, which led theoretical ecologists to investigate how mech-anisms such as disturbance and spatial heterogeneity (patchiness) could overcome these destabilizing feedbacks and increase system stability (DeAngelis & Waterhouse 1987). Mathematical models like metapopulation theory can give rise to system-wide stability (metapopulation persistence) in spite of the transient dynamics of individual patches or populations. We thus consider the emergence of patch-based ecological theory next.

Patch-Based Theory in Ecology

Patch-based ecological theory emerged in the late 1960s and early 1970s, as exemplified by the development of the theory of island biogeography (MacArthur & Wilson 1967; **Chapter 10**) and metapopulation theory (Levins 1969, 1970; **Chapter 7**; **Figure 3.5**). Even then, spatial heterogeneity was incorporated in a fairly abstract way, by making certain assumptions about how patch size or distance ought to influence the colonization and extinction dynamics of species in patches, whether these be oceanic islands or habitat fragments. In other words, such models did not explicitly consider the arrangement of patches in space; they simply formalized a mathematical relationship between aspects of patch structure and certain ecological rates (i.e. the probability of extinction or colonization as a function of the relative size or distance of a patch from a potential source of immigrants).

The emergence of patch-based theory in ecology represented a significant departure in the way spatial heterogeneity was viewed, not only in ecological theory, but also in ecological research. The study of habitat patchiness and its effects soon became a research paradigm in ecology, heightened by the growing concern over human impacts on the environment, especially those contributing to the loss and fragmentation of natural habitats. This concern was highlighted in a landmark publication by John T. Curtis (1956), an American plant ecologist who described the process of land clearing following European settlement of a small township in southern Wisconsin (**Figure 3.6**).

Long before orbiting earth satellites and computer-based mapping software, Curtis created by hand—using land-survey records and his own observations—a time-series of maps that depicted a landscape in transition. Over the course of a century, the Cadiz Township was transformed from nearly continuous deciduous forest to a predominantly agricultural landscape in which the forest had been reduced to a patchwork of small fragments dotting the landscape (**Figure 3.6A**). So compelling was this illustration that Robert MacArthur and Edward Wilson—the architects of island biogeography theory—used it as the very first figure in their own landmark monograph, to illustrate how the principles of island biogeography could also apply

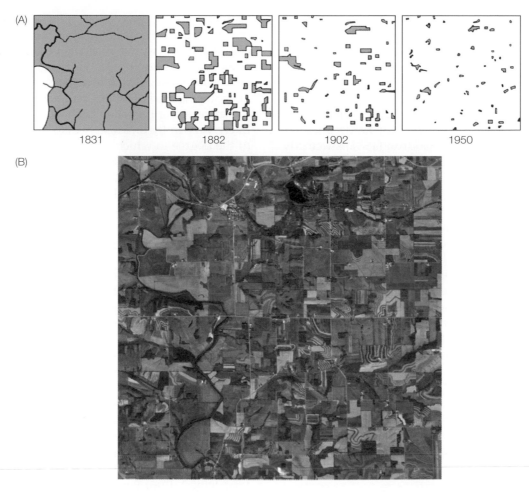

Figure 3.6 **Habitat loss and fragmentation.** (A) The process by which land clearing leads to the loss and fragmentation of natural habitats was famously illustrated by John T. Curtis (1956) in this time-series depicting the changing landscape of Cadiz Township in Green County, Wisconsin (USA). The 36 mi² (93 km²) township was settled by Europeans in 1834, and within 100 years, almost all (96%) of the native deciduous forest had been converted to farmland and pasture. By 1950, the formerly continuous forest had been reduced to 55 small fragments (averaging 14.3 acre = 5.8 ha in size). (B) Cadiz Township today. Some of the same woodlots present in 1950 have persisted to the present day. Some of the smaller woodlots are gone, but others along waterways have actually increased in size.

Source: (A) After Curtis (1956). (B) Google Earth, data provided by USDA Farm Service Agency.

to habitat islands created by habitat loss and fragmentation (MacArthur & Wilson 1967). So compelling is this figure even today, that it has become one of the most reproduced figures in ecology (Wiens et al. 2007). We would therefore be remiss if we did not bow to tradition and include it in this textbook as well (**Figure 3.6**).

> Many of the principles graphically displayed in the Galápagos Islands and other remote archipelagos apply in lesser or greater degree to all natural habitats.... The same principles apply, and will apply to an accelerating extent in the future, to formerly continuous natural habitats now being broken up by the encroachment of civilization.
>
> MacArthur & Wilson (1967)

Over time, spatial heterogeneity has been incorporated with ever-greater complexity into the mathematical formalism of ecological theory, largely due to advances in metapopulation theory and spatial ecology (e.g. Pulliam 1988; Tilman & Kareiva 1997; Hanski 1999). In addition, computer simulation models have contributed to our understanding of how spatial complexity influences ecological processes and dynamics in heterogeneous landscapes (**Figure 3.5**). **Spatially explicit models**—those that consider the location and other spatial attributes of patches explicitly—have therefore been indispensable in the development of spatial and landscape ecological theory, modeling, and ecological forecasting, particularly for assessing ecological responses to different scenarios of landscape change (e.g. Dunning et al. 1995; Turner et al. 1995). We will explore applications of spatially explicit models in many of the later chapters of this book, especially those dealing with population, community, and ecosystem responses to landscape pattern (**Chapters 7–11**).

Toward a Landscape-Mosaic View of Environmental Heterogeneity

The island-biogeographic perspective clearly influenced the early patch-matrix view of landscapes. As Richard Forman and Michel Godron (1981) succinctly put it: 'The structure of a landscape is primarily a series of patches surrounded by a matrix.' Patches can be characterized by a number of attributes, such as their size, shape, and configuration, which can then be quantified to provide an analysis of landscape structure (**Chapter 4**). Patches can also be classified according to the means by which they are created, whether by various types of natural disturbances, by humans, or in response to resource availability or other environmental factors (Forman & Godron 1981). We will return to the discussion of how natural and anthropogenic disturbances contribute to landscape heterogeneity in the second half of this chapter.

The arrangement or structural pattern of patches, corridors, and a matrix that constitute a landscape is a major determinant of functional flows and movements through the landscape, and of changes in its pattern and process over time.

Forman (1995a)

Because landscapes comprise different patch types, the island-biogeographic view of landscapes has given way to the **landscape-mosaic concept** (Wiens 1995b; Forman 1995b). Landscapes are not just patches embedded in a homogeneous matrix, but rather, heterogeneous mosaics of different patch or habitat types (i.e. ecosystems or biotic communities) that vary in their suitability for a given species. Land covers not used for feeding or breeding purposes might thus be considered part of the matrix from that species' perspective, but the matrix could still be traversed to varying degrees during dispersal rather than avoided altogether (more on that in a moment).

Thus, habitat patches or fragments may be like ocean islands, but they are not truly islands in some important respects. Habitat patch boundaries are not always clearly defined, and may represent more of a gradient than a sharp transition between the habitat patch and the surrounding matrix (a soft versus hard edge; **Figure 6.9**). Recall that this was one of the points made by Henry Gleason (1926) in arguing against the prevailing view that vegetation communities represented well-defined assemblages: '… its boundaries may be so poorly marked that they can not be located

with any degree of accuracy.' Further, the intervening matrix between habitat patches is unlikely to be as completely inhospitable as the ocean is for land animals or plants attempting to disperse among islands. This is not to suggest that dispersal through a heterogeneous matrix is without risk or difficulty, only that the intervening habitats that make up the matrix will likely vary in their 'resistance' to movement, in terms of the degree to which matrix habitats either facilitate or impede dispersal. The effect of landscape resistance on movement, dispersal, and other ecological flows is a recurring theme in landscape ecology, and will be addressed in several different contexts throughout this text (e.g. **Chapters 5, 6, and 9**).

Landscape mosaics may contain other elements besides habitat or resource patches, such as habitat corridors, roads, areas of human habitation, lakes, rivers, and streams. Although we tend to view rivers and streams as linear features within a broader terrestrial landscape, they too contain a diversity of spatial elements and thus might be viewed as landscape mosaics in their own right (Fausch et al. 2002; Ward et al. 2002). Spatial elements such as surface water bodies (lakes, tributaries), alluvial aquifers (the hyporheic zone), riparian systems (gallery forests, swamps, marshes), and various geomorphic features (e.g. bars and islands, terraces, deltas, channel networks) are all characteristics of riverine landscapes. River corridors can be viewed as an alternating sequence of floodplains along their length, from narrow canyon-constrained reaches to broad floodplains, like 'beads on a string' (**Figure 3.7**). Floodplain reaches are expansive, containing multiple channels and deep alluvial deposits. The alluvium is thus akin to the matrix, within which the other elements of the riverine landscape are embedded, and among which hydrological exchange occurs along longitudinal, lateral, and vertical dimensions (**Figure 5.17**; Ward et al. 2002). We will discuss the formation of river networks a bit later in this chapter, as this is a major driver of landscape evolution. We will also consider further the connectivity of riverine systems (hydrological connectivity) in **Chapter 5**.

River corridors consist of a dynamic mosaic of spatial elements and ecological processes arrayed hierarchically. As such, they fit comfortably within the framework of landscape ecology.   Ward et al. (2002)

The recognition that rivers and others freshwater systems can be viewed as landscape mosaics, to which the principles of landscape ecology can be applied, has led to the development of **landscape limnology** (Soranno et al. 2010). Landscape limnology is 'the

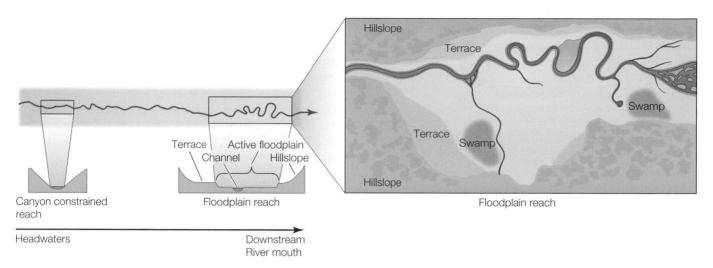

Figure 3.7 River corridors as landscape mosaics. A river corridor may be conceptualized as an alternating sequence of canyon-constrained and floodplain reaches (left). Within the floodplain reach, the alluvium acts as a matrix in which other spatial elements that characterize riverine landscapes are embedded (right).

Source: After J. V. Ward, K. Tockner, D. B. Arscott, et al. (2002).

spatially explicit study of lakes, streams, and wetlands as they interact with freshwater, terrestrial, and human landscapes to determine the effects of pattern on ecosystem processes across temporal and spatial scales' (Soranno et al. 2010, p. 442). In this framework, the freshwater landscape is viewed as a collection of interacting patches or systems that vary in their physical, chemical or biological characteristics, and which are embedded within a broader terrestrial and human landscape context that ultimately affect properties of the freshwater system (e.g. water chemistry, species diversity, primary productivity). Processes operate within particular domains of scale within each landscape, the components of which may be hierarchically arrayed (e.g. factors operating at broader scales have an impact on lower levels in the hierarchy). Landscape limnology thus views freshwater systems as one component of the larger multisystem landscape (i.e. the landscape mosaic, comprising freshwater, terrestrial, and human landscapes), which again emphasizes the integrative nature of freshwater systems, as well as the imprudence of trying to manage them in isolation from their terrestrial and human landscape context. We will return to this idea of interacting ecosystems again in **Chapter 11**. For now, however, the main point to emphasize is that freshwater systems are heterogeneous, spatially structured systems, and thus the principles and approaches of landscape ecology can be applied to their study and management as well (Fausch et al. 2002; Wiens 2002).

The freshwater environment is not the only new frontier to have been colonized by landscape ecology,

however. The concepts and techniques of landscape ecology are increasingly being applied to the study of coastal marine environments, giving rise to the field of **seascape ecology** (Boström et al. 2011; Pittman 2013). A seascape is defined as 'a spatially heterogeneous area of coastal environment (i.e. intertidal, brackish) that can be perceived as a mosaic of patches, a spatial gradient, or some other geometric patterning quantified from either benthic or pelagic environments' (Boström et al. 2011, p. 192). Seascapes encompass a wide variety of semi-terrestrial, subtidal, and intertidal environments, such as salt marshes, mangrove forests, seagrass meadows, coral reefs, and oyster beds. Patch structure is evident across a range of scales in these different seascapes, which can be mapped and quantified using landscape metrics and other spatial analytical tools, just as for terrestrial landscape patterns (Wedding et al. 2011; Pittman 2013). Like its terrestrial counterpart, seascape structure is expected to influence a wide variety of ecological processes that contribute to the overall diversity and ecosystem functioning of these coastal environments, which in turn support a wealth of ecosystem goods and services (Boström et al. 2011).

Because coastal ecosystems are so economically valuable, they tend to be densely populated and heavily impacted by humans through development, pollution, the introduction of non-native species, and overexploitation (Boström et al. 2011). By their very definition, coastal environments occur at the interface between terrestrial and marine systems, and thus often function as critical transition zones that control the flows of organisms, nutrients, detritus, sediments, and other materials across the land-sea interface (Levin et al. 2001), which may subsidize the productivity of terrestrial and marine environments (Polis et al. 1997). Such **spatial subsidies** between systems will be discussed in more detail in the context of interacting ecosystems in **Chapter 11**. In the meantime, we can add seascapes to the purview of landscape ecology.

Emerging evidence indicates that animals in these seascapes respond to the structure of patches and patch mosaics in different ways and at different spatial scales, yet we still know very little about the ecological significance of these relationships and the consequences of change in seascape patterning for ecosystem functioning and overall biodiversity.

Boström et al. (2011)

Given that heterogeneity and disturbance dynamics are such defining characteristics of landscapes, we will now turn our attention to defining the spatial and temporal dimensions of heterogeneity. Because environmental patterns manifest in different ways within landscapes, heterogeneity can have different connotations depending on what is being measured or considered.

How is Heterogeneity Defined?

As mentioned at the beginning of this chapter, landscape heterogeneity includes both spatial and temporal dimensions. We therefore begin by discussing spatial heterogeneity, before turning our attention to temporal heterogeneity.

Spatial Heterogeneity

Spatial heterogeneity can refer either to a landscape's composition (its diversity of land-cover types) or to the degree of spatial variability exhibited by some landscape attribute (its degree of patchiness). **Landscape diversity** refers to the variety of land-cover or land-use types present and their relative abundance. For example, landscapes with high habitat diversity are more heterogeneous than those dominated by a single habitat type (i.e. a homogeneous landscape). We will discuss compositional measures of landscape structure, including landscape diversity, more fully in **Chapter 4**.

Alternatively, landscapes that exhibit a high degree of patchiness, such as in the distribution of habitat or soil properties, might also be considered spatially heterogeneous. Spatial variability (spatial variance) is a statistical concept that examines the degree to which locations are related (either positively or negatively) across the landscape; it is thus a measure of spatial

dependence, or **spatial autocorrelation**. Although spatial variance between sites is expected to increase as a function of distance (Tobler's first law of geography), how spatial variance actually changes with distance ultimately depends on the nature of patchiness in the landscape. Locations sampled within a patch should be more similar to one another (low spatial variance = high spatial autocorrelation) than to locations outside of the patch (high spatial variance = low spatial autocorrelation). By measuring how spatial variance changes as a function of distance, we can therefore determine at what scale(s) patchiness exists in our landscape. Indeed, this is exploited in many types of spatial analysis, to be discussed in **Chapter 4**. In the present context, however, it will suffice to understand that spatial variance is related to patchiness, and as such, spatial heterogeneity could imply a patchy landscape as well.

To address these different facets of spatial heterogeneity, Kotliar and Wiens (1990) proposed defining heterogeneity in terms of *contrast* (different patch types, or diversity) and *aggregation* (the degree to which patches are clumped, which relates to spatial variance and autocorrelation). Landscapes can exhibit heterogeneity along only one of these axes and have either a high degree of aggregation (patchiness; Case 2) or high contrast (diversity; Case 3 in **Figure 3.8**). Landscapes can also exhibit heterogeneity along both axes (Case 4 in **Figure 3.8**). In addition, heterogeneity can be expressed at different scales, such as by having high contrast within patches (Case 4b) as well as between patches (Case 4a).

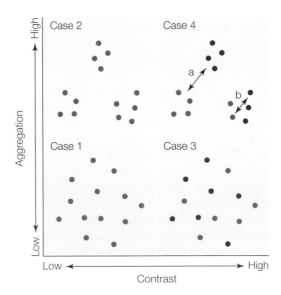

Figure 3.8 Dimensions of spatial heterogeneity. Spatial heterogeneity is defined in terms of both contrast (diversity) and aggregation (degree of spatial dependence). In landscapes with hierarchically nested patch structure (Case 4), landscapes can exhibit heterogeneity between patches (a) and well as within patches (b).

Source: After Kotliar & Wiens (1990).

The concept of heterogeneity is thus a multifaceted one. To complicate matters further, we should also distinguish between **measured heterogeneity** and **functional heterogeneity** (Kolasa & Rollo 1991). Measured (or structural) heterogeneity involves the application of some landscape metric or spatial analysis to quantify heterogeneity. Measured heterogeneity is obviously important for characterizing landscape structure. Less obvious, however, is whether measured heterogeneity is important for functional heterogeneity—the sorts of ecological relationships we seek to uncover between pattern and process. Because landscapes may exhibit patchiness (heterogeneity) across a wide range of scales (**Figure 2.8**), we should anticipate that only patch structure at some scales will be relevant to our investigation. For example, an animal cannot respond to heterogeneity that it does not perceive or never encounters. The lower limits of its perceptual range may thus be set by its sensory abilities or rate of movement, whereas the upper limits are set by its dispersal range (**Figure 6.4**). The existence of patch structure either above or below these limits is irrelevant for understanding how landscape pattern influences foraging or dispersal behavior. One of the main goals of landscape ecology is thus to assess functional heterogeneity, to match landscape structure with ecological function.

Arbitrary measures of heterogeneity are tempting and popular, but their ability to reflect the relevant properties of the system of interest is unclear and questionable.

Kolasa & Rollo (1991)

Temporal Heterogeneity

Temporal heterogeneity describes the pattern that occurs within a data time series. Just as the degree of spatial dependence provides a measure of spatial heterogeneity, we can similarly characterize temporal heterogeneity in terms of temporal dependence, or **temporal autocorrelation**, as the extent to which observations or events are correlated across different lengths of time. Although observations made close in time are likely to be more similar than those separated by longer time intervals, the degree and pattern of temporal autocorrelation will ultimately depend on the nature of the phenomenon being observed. For example, a large earthquake may set off a swarm of smaller earthquakes or aftershocks (an example of a positive correlation), but fires are less likely to occur after a large wildfire has swept through a forest and exhausted the fuel load (an example of a negative correlation). Still, it depends on the time scale over which we view this,

as the frequency with which aftershocks occur decays rather quickly following the main quake, and the conditions that gave rise to a catastrophic wildfire may soon reoccur under a prolonged period of drought if fuels are allowed to build up again. Thus, disturbances may shift from a positive to a negative correlation, and back again, depending on the scale of temporal autocorrelation.

Such cycles are most apparent when examining how environmental conditions fluctuate over time, which may reveal seasonal or multiyear dependencies in the data time series. For example, the minimum air

(A)

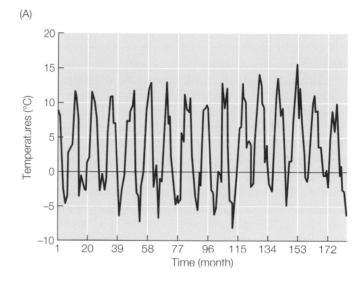

(B)

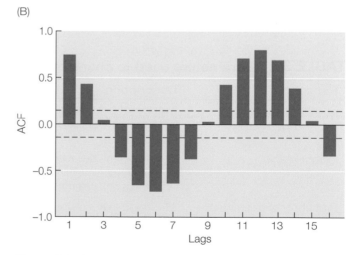

Figure 3.9 Analysis of temporal heterogeneity. Patterns in environmental fluctuations over time (temporal heterogeneity) may emerge when assessed at different temporal scales. (A) The minimum air temperature at this particular location in a temperate region fluctuates in a regular fashion over a 15 year period. (B) The temporal autocorrelation function (ACF), which measures the degree of temporal dependence among observations at different time intervals (lags, in months here), reveals the seasonality in temperature fluctuations.

Source: After Castellarini et al. (2002).

temperature at a given location in the temperate zone is likely to be positively correlated to minimum temperatures the previous month, but will be negatively correlated to minimum temperatures some 6 months later, reflecting the seasonality of temperature fluctuations at middle latitudes (**Figure 3.9**). These sorts of temporal fluctuations in environmental conditions, especially in temperature (either air or sea surface) and precipitation, may in turn affect the frequency or intensity of some types of disturbances that are important drivers of landscape patterns and dynamics, such as flooding, landslides, hurricanes, and fires (e.g. Swetnam & Betancourt 1998; Elsner et al. 1999; Witt et al. 2010). We will discuss the role of these disturbances in contributing to landscape heterogeneity a bit later in this chapter.

Although the focus in landscape ecology is typically on spatial heterogeneity, the nature of temporal heterogeneity is clearly important for characterizing the disturbance regime of a landscape, as we will discuss next. The application of methods involving time-series analysis to the disturbance dynamics of landscapes could thus prove useful not only in characterizing the temporal pattern of disturbance and patch dynamics within a landscape (e.g. Meyer et al. 2007), but also in identifying what qualifies as an unusual disturbance event (i.e. a large, infrequent disturbance). The analysis of time-series data can also help to inform both retrospective and prospective analyses of climate-

change effects on landscapes and other ecological systems (e.g. ecological forecasting; Sabo & Post 2008; Brown et al. 2011).

How are Disturbances Defined?

Disturbances of various types give rise to landscape heterogeneity. A disturbance is defined as 'any relatively discrete event in time that disrupts ecosystem, community, or population structure and changes resources, substrate availability or the physical environment' (White & Pickett 1985). Disturbances are defined in terms of both spatial and temporal dimensions, as you will no doubt recall from our discussion of space-time diagrams (**Figure 2.2**). This once again reinforces the notion that spatial and temporal heterogeneity are inextricably linked.

The Disturbance Regime

The **disturbance regime** of a landscape is described by a number of parameters that encompass both spatial and temporal dimensions (**Table 3.2**); it is sometimes referred to as the **disturbance architecture** of the landscape (Moloney & Levin 1996). Although some of these parameters do not have an explicit spatial or temporal dimension (e.g. disturbance frequency), we must still specify the spatial extent or time period being considered. For example, annual fire frequency—the number of fires that occur in an average year—will

TABLE 3.2 Parameters used to characterize the disturbance regime of a landscape.

Parameter[†]	Description	Example of application*
Disturbance frequency	Number of disturbances that occur within an area over a specified time period	The annual fire frequency, the mean number of fires per year, is 5.6 fires/year in this landscape
Return interval	Length of time between successive disturbance events at a given location; the inverse of its disturbance frequency	The mean fire return interval is about 40 years
Probability of disturbance	Likelihood that a disturbance (or disturbance of a given size) will occur within a particular time period. Although sometimes based on the disturbance frequency (frequency of occurrence/total time period), this is also derived as a multivariate assessment of various risk factors	The probability of a catastrophic forest fire occurring in any given year is 0.005 (1 catastrophic fire/200 years). The actual probability of such a fire occurring is ultimately dependent on the fuel load, fuel moisture level, humidity, temperature, drought index, wind speed, location, time of year when ignition occurs, etc.
Disturbance rate	Number of disturbances that occur within a given area per unit time; a product of both frequency and area	Treefall gaps are created when individual trees die, giving a local disturbance rate of 0.004 (1 tree/250 years). At a landscape scale, however, the disturbance rate from treefall-gap formation may only be 0.0001 (10 trees/100,000 trees/km²/year)

Parameter[†]	Description	Example of application*
Disturbance size (spatial grain and extent of disturbance)	Area affected by an individual disturbance event (the spatial grain of the disturbance), or the entire area affected by a particular disturbance within a given time period (spatial extent of the disturbance)	Fires burned 20% of the landscape this past year, as a result of many small fires that were ignited by lightning but which were quickly extinguished
Disturbance-size frequency distribution	Number of disturbances of a particular size that occur within a given time period	Over the last 100 years, small fires occurred with far greater frequency than large fires, but this past decade has been atypical in the greater occurrence of large fires
Rotation period	Time required to disturb an area of a certain size; not all of the area need experience the disturbance, however	The landscape has a fire rotation period of about 500 years (e.g. the fire cycle)
Spatial pattern of disturbance	Distribution of disturbances in space; the degree of spatial autocorrelation or clustering, which may vary as a function of scale	The pattern of lightning strikes is random, but fires tend to be associated with a particular forest type, resulting in a non-random distribution across the landscape
Temporal pattern of disturbance	Distribution of disturbances over time; the degree of temporal autocorrelation or clustering, which may vary as a function of scale	Although lightning is a random occurrence, fires tend to occur at certain times of the year when lightning storms are common, resulting in a seasonal correlation. Among years, fire frequency is greater during drought years, which may be cyclic, resulting in decadal correlations
Intensity	Magnitude of a disturbance, assayed in terms of physical force (e.g. wind speed, energy released during a fire, height of storm surge)	Fire intensity can be assayed by measuring the radiative energy emanating from burned areas in images obtained via remote sensing
Severity	Impact of the disturbance on the system (e.g. the degree to which organic matter has been removed or destroyed, trees or shrubs killed, or substrates altered)	Fire severity ranges from light surface fires to deep ground and stand-replacing fires in which all trees are killed
Duration	Length of time that disturbance lasts; the disturbance may be a brief or short-term event (a pulse disturbance) or long-lasting in its duration (a press disturbance)	Under conditions of low fire danger, lightning-ignited fires will self-extinguish and are therefore short-lived (i.e. they burn out on their own within hours). However, under extreme fire-danger conditions, fires may spread quickly and encompass a large area, such that they takes months to burn themselves out

Sources: White & Pickett 1985; Moloney & Levin 1996; Lake (2000).

[†]Parameter values in these examples have been fabricated for illustrative purposes and, while realistic, are not representative of forest fires in general or of any forest type in particular.

*Examples describe how these measures could be applied to describe the fire regime within a forested landscape.

likely be higher if we base our calculation on an entire region rather than a particular landscape within that region (unless it is just that one landscape that burns time and again). Similarly, a large forest fire assessed at the scale of the overall landscape may nevertheless contain unburned areas or stands that experienced only minimal fire damage when assessed at a more local scale. Thus, as with most measures in ecology, parameters that characterize the disturbance regime are scale-dependent. Some of these measures do explicitly examine how disturbances scale in space or time, however, such as those concerned with the spatial or temporal distribution of disturbances (i.e. analysis of spatial or temporal autocorrelation; **Figure 3.9**).

Although disturbances occur across a range of spatial and temporal scales, two concepts have received a great deal of attention in ecology and therefore deserve special consideration here: intermediate disturbances and large infrequent disturbances.

The Intermediate Disturbance Hypothesis

The Intermediate Disturbance Hypothesis (IDH) posits that diversity (and, by extension, heterogeneity) should be maximized at an intermediate scale or degree of disturbance (**Figure 3.10**). In a challenge to the prevailing equilibrium view of communities of the time, American community ecologist Joseph H. Connell (1978, p. 1303) observed that:[1]

> *Organisms are killed or badly damaged in all communities by disturbances that happen at various scales of frequency and intensity. Trees are killed or broken in tropical rain forests by windstorms, landslips, lightning strikes, plagues of insects, and so on; corals are destroyed by agents such as storm waves, freshwater floods, sediments, or herds of predators. This hypothesis suggests that the highest diversity is maintained at intermediate scales of disturbance.*

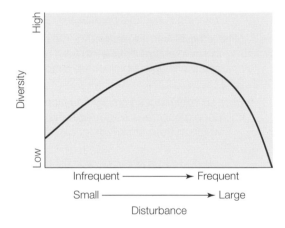

Figure 3.10 The Intermediate Disturbance Hypothesis. Under the Intermediate Disturbance Hypothesis, diversity (e.g. heterogeneity) is expected to be maximized at intermediate levels of disturbance.

Source: After Connell (1978).

---

[1] It should be noted that although Connell first introduced the term 'Intermediate Disturbance Hypothesis,' and thus is usually credited as the originator of the idea, the relationship between intermediate disturbance and diversity was not unknown to ecologists of the time (Wilkinson 1999). Indeed, the British plant community ecologist John Philip Grime had published a paper 5 years earlier that included several figures depicting the same hump-backed relationship between plant species density (number of species/ area) and increasing 'environmental stress' (e.g. drought or soil nutrient content) or management intensity (e.g. mowing, grazing, or trampling), which he assessed in terms of species' relative competitive abilities (Grime 1973a). Grime in turn referenced the great systems ecologist Eugene P. Odum for his observation made a decade earlier that 'the greatest diversity occurs in the moderate or middle range of a physical gradient' (cited in Grime 1973a, pp. 345–346).

The basis for the IDH is intuitively appealing: If disturbances are too frequent, recurring before many species can become established, then the community will be dominated by just those few species capable of rapid growth and good dispersal abilities; that is, species that are good colonizers, what are often referred to as pioneer or ruderal species. At the other extreme, if disturbance is a rare event, then those few species with superior competitive abilities will come to dominate—the so-called climax or late-successional species. Thus, disturbances that occur with intermediate frequency or size should allow for coexistence of the greatest mix of species.

As with many intuitively appealing ideas in ecology, the IDH, while still popular, has been challenged on both theoretical and empirical grounds (Mackey & Currie 2001; Fox 2012). Although some studies have found support for the IDH, many have not (Mackey & Currie 2001). There is also the issue of what exactly qualifies as an intermediate scale, frequency, or intensity of disturbance. Clearly, this determination is system-specific, and must be assessed across the full range of scales over which the disturbance occurs—that is, with respect to the disturbance regime. Because disturbance can influence, as well as be influenced by, landscape structure, there is also the possibility that the landscape itself may influence the relationship between diversity and disturbance.

Consider how the IDH might be used to predict the effect of a disturbance, such as the spread of fire or an invasive species, on landscape heterogeneity. Imagine a completely forested landscape, in which human settlers (the invasive species) arrive and begin clearing land at one corner, transforming the forested landscape over a period of years into a single type of agricultural land use (e.g. pasture; **Figure 3.11A**). Initially, landscape heterogeneity will increase because a new land-cover type is being added, but only up to a point. Once half the forest has been cleared, heterogeneity begins to decline as the landscape is once again dominated by a single cover type (pasture, in this case). Maximum heterogeneity is thus achieved at an intermediate level of disturbance (i.e. when half of the landscape has been disturbed).

What if the landscape being disturbed is already heterogeneous, however? Imagine that each quarter-section of our landscape contains a different land-cover type (**Figure 3.11B**). As before, land clearing starts at one corner and initially contributes to a small increase in heterogeneity as disturbance spreads throughout this first quarter-section. However, continued land clearing eventually leads to a decline in heterogeneity back to its original level as pasture comes to dominate this first quarter. Beyond that, additional land clearing into the other quarter-sections results in a continual decline in heterogeneity, as each of the original land-cover types is converted to pasture in turn.

In this particular heterogeneous landscape, heterogeneity is maximized at a low level of disturbance rather than at the intermediate level predicted by the IDH. Interestingly, the first part of the curve (**Figure 3.11B**) is simply a smaller version of the hump-backed curve from the homogeneous landscape (**Figure 3.11A**), which makes sense given that in both cases the disturbance initially begins and spreads through a single, homogeneous land-cover type. However, this suggests that IDH might only apply at certain scales, based on the scale at which the disturbance occurs relative to the scale at which heterogeneity (or diversity) is expressed in the system (Aronson & Precht 1995).

Regardless, this example underscores the sensitivity of disturbances to initial conditions, in terms of both the composition and distribution of land-cover types on the landscape, as well as to the site of initiation and subsequent pattern of spread by the disturbance throughout the landscape (**Chapter 8**). As stated in **Chapter 1**, predicting the effect of disturbance on landscape heterogeneity, or any aspect of landscape structure, is a major research challenge in landscape ecology. Complex, non-linear responses are to be expected, a matter we shall consider in the next section and again when discussing analysis of landscape patterns in **Chapter 4**. The world is thus more complicated (and interesting!) than predicted by the IDH.

Large Infrequent Disturbances

Throughout history, natural disasters such as floods, wildfires, earthquakes, volcanic eruptions, hurricanes, drought, famine, and disease have all played an important role in the disturbance dynamics of landscapes as well as the ecological and human systems embedded within landscapes. By virtue of their size, large infrequent disturbances (LIDs) exert a powerful influence over a wide area and thus have the potential to effect a complete restructuring or reorganization of ecosystems. Subsequently, LIDs create an enduring legacy on landscapes that may persist for centuries or even millennia (Turner et al. 1997; Foster et al. 1998a). Although disastrous from our perspective, LIDs are part of the normal, long-term dynamics of many ecological systems, and thus we need to understand the effect of LIDs on these systems, especially if we hope to manage them (Dale et al. 1998).

As with intermediate disturbances, what qualifies as a large infrequent disturbance is specific to the system, but it can be defined in terms of how the event compares to the average size, duration, or intensity of that disturbance type in a given system (Turner & Dale 1998). For example, an extreme flood event can be defined as one in which the water depth (stage) or flow volume (discharge, m^3/s) is beyond two standard deviations of the average depth or flow that has been assessed over a period of several decades (Resh et al. 1988). Or, we might evaluate the event in the context of a frequency distribution of disturbance sizes for that system, classifying those at the extreme end of the distribution (the top 1–10%) as LIDs (Turner & Dale 1998; Romme et al. 1998). Recall that the scaling of disturbance-size frequency is often given by a power law (i.e. it exhibits a fractal distribution; **Box 2.1**). Thus, there are a great many more small forest fires that occur within a given landscape over a period of time than large fires (**Figure 2.13A**). In the end, however, we tend to evaluate LIDs from the perceived impact that the disturbance has had on the area and relative to our own lifespans (Turner & Dale 1998), as when we refer to a '100-year flood' event to convey the rarity of the event; a flood of that magnitude is not likely to be seen again in our lifetime.[2] Thus, LIDs generally have a return-time measured in centuries (100–1000 years), and are infrequent from a human perspective. However, it is precisely because of their apparent rarity and unpredictability that we know so little about them (Romme et al. 1998).

Historically, the study of LIDs has not received as much attention from ecologists as the small frequent disturbances that give rise to patch (or gap) dynamics (Turner et al. 1997). Thus, one of the first questions that needs to be addressed is how the effects of LIDs compare to that of smaller, more frequent disturbances (Romme et al. 1998). In other words, are the effects of a single large disturbance equivalent to those of several small ones? Is the difference simply one of degree (i.e. is the difference quantitative, in terms of the total area affected), or do LIDs differ qualitatively in the kinds of effects they have on landscapes and ecosystems? If they differ by degree, then we should be able to extrapolate from what we know about the effects of small disturbances to predict or manage the effects of larger disturbances. From a management or mitigation standpoint, we might choose to implement a series of small disturbances (which are more easily controlled than a single large disturbance) to mimic or stave off the effects of a larger disturbance. For example, we might temporarily restore high water flows to a river regulated by a dam to mimic a seasonal flood-pulse in order to inundate a floodplain (thereby creating a controlled flood) for the purposes of restoring wetland habitat. Or we could set a series of small controlled burns in order to reduce the fuel load and thus minimize the risk of a severe forest fire.

---

[2] While we may not witness a 100-year flood again in our lifetime, this does not mean that such a flood only occurs once every 100 years. A 100-year flood is one that has a 1/100 (0.01) chance of occurring in a given year, and thus there is always some probability (albeit small) that such a flood will reoccur within that 100-year period.

While we recognize these as legitimate management practices, we would do well to remember the lessons learned in the previous chapter regarding the perils of blindly extrapolating across scales, especially in the event of cross-scale interactions, in which transfer processes such as water flows or fire spread interact with spatial heterogeneity in complex, unpredictable, and surprising ways (**Chapter 2**). If LIDs differ from small disturbances not just in degree, but also in the *kinds* of effects they have on landscapes and ecosystems, then we need to be aware of what those differences might be and try to plan for contingencies. To quote Romme et al. (1998, p. 525):

> ...*if LIDs introduce fundamentally new kinds of phenomena into ecological systems, then we must be prepared for some ecological surprises in the aftermath of large events, and we may need to devise strategies for incorporating the large disturbances into our management framework.*

Recall that these sorts of ecological surprises may come in the form of **threshold responses**, abrupt changes in how the system responds to a disturbance past a given size, frequency, duration, or intensity that can lead to system-state changes, as when overgrazing leads to the desertification of grasslands (Scheffer et al. 2001; **Figure 11.9**). We'll talk more about system state changes in **Chapter 11**, but for now, suffice it to say that such

threshold responses make extrapolation from small to large disturbances unwise, or at least unreliable.

Another surprise, given their large size, is that LIDs do not produce a uniform pattern of disturbance across landscapes. Rather, LIDs typically create a heterogeneous mosaic of disturbance severity across a range of scales (Turner et al. 1997). Thus, large islands of green, unburned forest were scattered throughout the blackened, charred landscape of Yellowstone National Park (USA) following the severe fire season of 1988, in which nearly a third of the park burned (Turner et al. 1997; **Figure 3.31**). Improbably, these forest stands were spared, perhaps due to a slight change in wind direction that caused the fast-moving fire to skip over some stands while alighting and igniting others. Because LIDs are capable of generating a range of disturbance intensities over a large area, they ultimately have the potential to create a greater degree of landscape heterogeneity more quickly than a number of smaller disturbances spread over time (Turner & Dale 1998).

Although the study of LIDs and their effects is important from the standpoint of understanding the long-term disturbance dynamics of landscapes and ecosystems, this area of research has taken on an added urgency in the face of global-change issues involving human land use and climate change. Humans are capable of transforming landscapes on a scale comparable to that of LIDs, which we'll explore more fully later in this

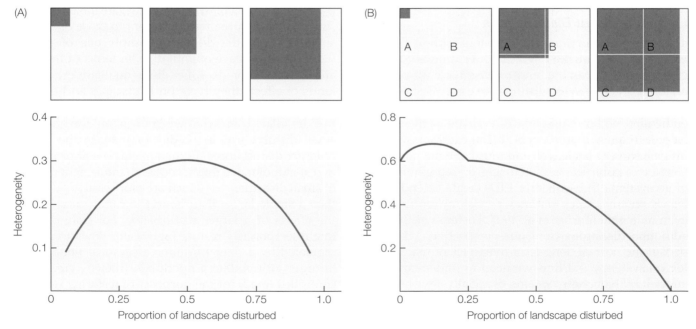

Figure 3.11 Disturbance interacts with landscape heterogeniety.   Maximum heterogeneity is not always attained at intermediate levels of disturbance. (A) Heterogeneity (number of land-cover types) is maximized in a homogeneous landscape when half of the landscape has been disturbed, consistent with the Intermediate Disturbance Hypothesis (**Figure 3.10**). (B) If the landscape is initially heterogeneous, however, intermediate levels of disturbance (50% landscape disturbed) could result in lower heterogeneity, as shown here. In this example, heterogeneity is thus maximized at low levels of disturbance.

Source: After Kolasa & Rollo (1991).

chapter. Global warming and other climatic changes are projected to lead to an increase in the intensity, duration, and/or frequency of disturbances like hurricanes, fire, drought, and flooding over the next century. Thus, some of these LIDs may soon become LFDs—large frequent disturbances—that may not only exceed the ability of ecological systems to respond and recover (i.e. thresholds are exceeded, leading to system state changes), but also our own socioeconomic systems for dealing and coping with natural disasters (IPCC 2012). The management implications of LIDs therefore extend beyond resource and ecosystem management to include disaster risk management, socioeconomic development, public safety, and food security (Dale et al. 1998; IPCC 2012). Land-use planning and ecosystem protection are extremely important for managing disaster risk, especially in terms of hazard mitigation, by affording long-term protection from climate extremes (IPCC 2012). For example, careful land-use management and the protection of environmentally sensitive areas can help mitigate a variety of extreme disturbance events and help avert natural disasters caused by landslides, flooding, or storm surges, as we'll see later in this chapter.

Implications of the Disturbance Regime for Landscape Dynamics

As pointed out at the start of this section, disturbance regimes include both spatial and temporal dimensions, which can interact with landscape heterogeneity (both spatially and temporally) to produce a wide range of dynamics. Although spatial and temporal heterogeneity may be linked, they are not equivalent (Kolasa & Rollo 1991). Two locations in space that are otherwise identical will diverge over time if subjected to different disturbance regimes (Figure 2.15). Temporal heterogeneity thus helps to create spatial heterogeneity. Spatial heterogeneity in turn may influence how those dynamics play out on the landscape, affecting the trajectory the landscape takes through time. For this reason, two locations that differ in their initial spatial heterogeneity, but otherwise experience the same disturbance dynamics, may continue to evolve along different trajectories over time. Not only does this underscore the inherent challenge of predicting landscape change, but it also helps to explain why it is so difficult to infer the disturbance dynamics that contributed to landscape formation from an analysis of current landscape structure (i.e. its measured heterogeneity). Different disturbance dynamics can potentially give rise to the same spatial patterns, and the same type of disturbance can play out differently in different landscape contexts.

To characterize the different sorts of landscape dynamics that might emerge under different disturbance regimes, Turner and her colleagues (1993) adopted a simulation modeling approach to explore the interplay between the spatial and temporal dimensions of disturbance relative to those of the landscape. In their model, disturbances of a fixed size were initiated at regular time intervals, but at random locations, within the landscape (which initially was composed entirely of the mature seral stage; i.e. the landscape was homogeneous). Disturbances 'reset' the area(s) being disturbed to the earliest seral stage. Disturbed areas then underwent a deterministic process of recovery (succession through a series of eight seral stages), and thus may or may not have fully recovered before the next disturbance occurred. The size and timing of disturbances were adjusted so as to encompass a full range of spatial and temporal scales, which were summarized by two parameters: S, the ratio of the size of the disturbance relative to the size of the landscape (i.e. its extent); and, T, the ratio of the disturbance interval relative to the time it takes a site to recover from that disturbance (e.g. the time it takes to go from an early-successional to mature seral stage).

Based on their simulation results, Turner and colleagues (1993) were able to identify six domains that encompassed a range of landscape dynamics (Figure 3.12). Equilibrium or steady-state landscapes are found only when disturbances are relatively small and infrequent (low S and high T); that is, when landscapes experience little disturbance. At the other extreme, landscapes that experience frequent large disturbances, in which the landscape never has a chance to recover (high S and low T), are completely unstable. Such landscapes may undergo a complete transformation (i.e. a system-state change), as when anthropogenic disturbance rapidly clears or radically alters native habitats across entire landscapes. In between these extremes, however, there exists a wide range of non-equilibrium conditions that can produce complex dynamics, ranging from a quasi-equilibrium state (stable, but with some small degree of change) to landscapes that are highly, but predictably, variable (e.g. where disturbance affects more than 50% of the landscape, but occurs infrequently). Thus, a given landscape might be driven to *any* of these states, depending on the nature of its disturbance dynamics.

Although this state-space diagram may not characterize the dynamics of all landscapes, and not every landscape dynamic will necessarily fall within a particular domain exactly as depicted here, this diagram has nevertheless been useful for illustrating how the scaling of disturbances—in both space and time—are likely to give rise to qualitatively different dynamics, even for the same landscape. For example, Turner and colleagues (1993) compared the fire regime of Yellowstone National Park before and after the severe fires of 1988, in which about a third of the park burned

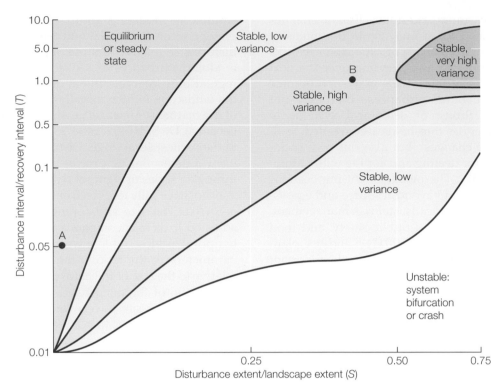

Figure 3.12 Scaling domains of landscape dynamics. The different spatial (S) and temporal (T) dimensions that characterize disturbance regimes are predicted to give rise to different domains of landscape dynamics. For the Yellowstone landscape, the fire regime prior to the catastrophic wildfires of 1988 (1972–1987) suggested a landscape in equilibrium (A), whereas the nature of the 1988-type fire (which occurs only once every 300 years or so in this system) is more consistent with a non-equilibrium landscape that is highly variable, but stable (B). Thus, the same landscape may appear to exhibit qualitatively different dynamics, depending on the temporal (or spatial) extent of the analysis.

Source: Turner et al. (1993).

(~321,000 ha). In the 15 years prior to 1988, more than 200 fires were ignited by lightning, but rarely burned more than 1 ha before dying out on their own; this amounted to about 1% of the entire park area (8300 ha). Thus, far more area burned in a single year (1988) than in the preceding 15 years combined.

To assess the magnitude of this shift, they calculated the spatial and temporal dimensions of the fire regime prior to 1988: $S = 0.01$, since 1% of the park burned during this 15-year period; and $T = 0.05$, since the return interval for small, lightning-ignited fires is about 15 years in this landscape and the recovery time for this forest type is about 300 years (15/300 = 0.05). Thus, the Yellowstone landscape appeared to exhibit equilibrium conditions over this time period, at least with respect to the fire regime (A in **Figure 3.12**). Large fires on par with those that occurred in 1988 are fortunately rare; the last fire of this magnitude occurred around 1700 (about 288 years earlier) and the return interval for such fires is estimated to be around 300 years, so perhaps Yellowstone was due for another very large fire. In any case, if $T \sim 1$ and $S = 0.36$ (because 36% of the park burned during this fire), then the fire regime—when assessed over a longer time scale—now

falls within the region of a stable landscape with high variance (i.e. a highly dynamic landscape; B in **Figure 3.12**). This may explain why previous research failed to find evidence of a shifting-mosaic steady state within Yellowstone National Park (Romme 1982).

Importantly, this characterization of landscape dynamics is very much influenced by the temporal extent of our analysis. The landscape appeared to shift from equilibrium to a highly dynamic state as a consequence of altering the time period over which we assessed the magnitude of the fire regime. We can anticipate that similar sorts of changes will occur if we alter the spatial extent over which disturbances are evaluated (e.g. if we assess fire within an area of a few thousand hectares as opposed to the entire park). Again, the effect of disturbance on landscape dynamics is scale-dependent, which means that a landscape is not inherently stable or unstable, but rather, may exhibit any of these dynamical states depending on the spatial and temporal scale at which it is observed. We might then consider to what extent various land-use or land-management practices would alter landscape dynamics, such as by driving landscape to an unstable state simply through a rescaling of the natural disturbance regime (e.g. through overgrazing or fire suppression). We'll return to that thought at the end of the chapter when we consider natural variability concepts in the context of anthropogenic disturbance and land management. Next, however, we consider the range of disturbances of that contribute to natural variability and heterogeneity within landscapes.

Formation and Evolution of Landscapes: Geomorphological Processes

A variety of processes operating over a wide range of scales in both space and time contribute to the formation of landscapes, and thus to landscape heterogeneity (**Figure 3.13**). Landscape formation is a dynamic process, which geomorphologists attempt to understand through the study of **landscape evolution** (Pazzaglia 2003; **Figure 3.14**). A landscape is therefore a complex system with emergent structure, one that arises as a consequence of interactions among various geomorphological processes and types of disturbances (abiotic, biotic, and anthropogenic) within a particular area over time. We will consider each of these formative processes in turn, starting with an overview of geomorphological processes in this section.

The character of a landscape is most obviously given by its geomorphology, the type and distribution of **landforms** that determine its physical structure. Landforms contribute to the terrain or 'lay of the land,' as reflected by its topography. Landforms are characterized by the processes that create them and provide the distinctive features that we tend to associate with different types of landscapes, such as mountain ranges, canyons, plains, and river deltas (**Table 3.3**). Freshwater and marine landscapes are likewise characterized by their terrain (or rather, their bathymetry), which comprise different types of formations. Opposing geologic forces give rise to the topographic diversity of Earth's surface (e.g. **Figure 3.1**), as well as to many of its unique landforms (**Figure 3.15**). To illustrate, we consider the formation of montane landscapes and river networks, both of which epitomize the antagonistic nature of competing geological forces at work. We will then turn our attention to how such geomorphological processes affect the formation and distribution of soil types, which are important not only from the standpoint of mapping terrestrial ecosystems, but also for understanding human land-use and settlement patterns.

Formation of Montane Landscapes

More than a quarter of Earth's land surface is mountainous terrain (Restrepo et al. 2009). Mountains tower over many landscapes and are one of the most prominent features of these landscapes. By virtue of their height, mountains create a wide range of climatic conditions within a fairly localized area, producing elevational gradients in temperature and precipitation that in turn give rise to ecological gradients, such as the different life zones discussed previously (**Figure 3.2**). Because of their geomorphic and ecological importance, the various means by which mountains are formed deserve at least a brief overview here. In general, mountain building, or **orogenesis**, involves a variety of geophysical processes that deform Earth's crust and upper mantle (the lithosphere) through folding, faulting, or volcanic activity (**Table 3.3**; **Figure 3.16**). We consider each of these in turn, beginning with the role of volcanism in mountain building.

Many of the iconic and culturally significant mountains of the world are **stratovolcanoes**, composite volcanoes that formed as a consequence of repeated eruptions over very long periods of time (tens to hundreds of thousands of years). Some well-known stratovolcanoes include Mount Fuji in Japan, Mount Kilimanjaro in Tanzania, Mount Pinatubo in the Philippines, Mount Vesuvius in Italy, and Mount St Helens in the USA. Volcanoes typically form at the fluid boundary between tectonic plates, where a rupture in Earth's crust allows magma to escape, sometimes with explosive force and catastrophic consequences for the surrounding biota—including humans. Nearly 300,000 people are believed to have been killed as a consequence of volcanic eruptions over the past 500 years (Simkin et al. 2001), and today more than 500 million people are living within the shadow of an active volcano (i.e. within 100 km; Small & Naumann 2001). Volcanoes have played a dramatic role not only in the formation of landscapes, but also in triggering global climatic changes that have contributed to biotic upheavals, ranging from mass extinctions in terrestrial and marine ecosystems to the shaping of human history (Oppenheimer 2011).

Volcanoes may form as part of a larger mountain range that rises along continental margins in relation to a **subduction zone**, where one or more oceanic plates are pushed beneath a continental plate (i.e. a convergent plate boundary; **Figure 3.16**). The Andes, the longest continental mountain range in the world (~7000 km long), were formed by the subduction of the Nazca and Antarctic Plates beneath the South American Plate. Not only do the Andes contain some of the tallest mountains in the world, but the range also includes numerous volcanoes, several dozen of which are currently active. Similarly, the Cascade Range in western North America, which contains more than a dozen large volcanoes including the aforementioned Mount St Helens (which erupted in spectacular fashion in 1980), is the result of the subduction of the Juan de

Figure 3.13 Disturbances large and small contribute to landscape heterogeneity.   Geomorphological processes, such as the volcanic eruption that gave rise to this cinder cone a millennium ago (background), create distinctive landforms that characterize the overall physical structure of the landscape; that is, its terrain. At finer scales, biological processes, such as this conical mound (foreground) created by harvester ants (*Pogonomyrmex* sp.), may also contribute to landscape heterogeneity. The nest mound, which has the appearance of a miniature cinder cone, is constructed of fine gravel within a large disk (2 m diameter) from which the ants have cleared all vegetation.

Source: Photo by Kimberly A. With.

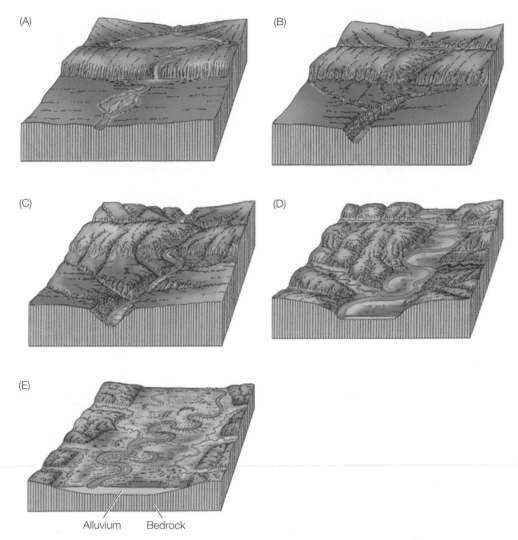

(A)

(B)

(C)

(D)

(E)

Alluvium   Bedrock

Figure 3.14 Landscape evolution. Initially, the 'young' landscape contains lakes and falls (A), which eventually give way and increase stream power, resulting in the carving of deeper channels (B). Over time, the channel widens and increases the drainage area, capturing tributaries to form a branching network (C). Eventually, however, the increased sediment load leads to a greater rate of sedimentation, eventually producing a graded channel or floodplain (D). Finally, the lateral migration of the channel results in a vast, complex floodplain (E).

Source: Ward et al. (2002).

TABLE 3.3 Landforms created by geomorphological processes.*

Process	Description	Landform types
Tectonic	Movement of the earth's crust, resulting in either its uplift, folding, faulting, or subsidence. Volcanism is often but not always associated with tectonic plate movements, and is thus treated as a separate category below	*Uplift and folding:* mountain, mountain range, escarpment, mid-ocean ridge *Faulting and subsidence:* rift valley, structural basin, lake (basin or rift), ocean trench
Volcanic	Deposition of lava, ash, and other pyroclastic materials via explosive or effusive eruption; subsidence of volcanic landforms also creates unique structures	*Depositional:* stratovolcano, shield volcano, cinder cone, volcanic dome, lava plain (field), seamount, oceanic plateau, volcanic island, deep-sea vent *Subsidence:* volcanic crater (caldera), crater lake
Glacial	During glacial expansion, ice sheets scour surface rock layers, and underlying bedrock, producing erosional landforms Glacial retreat deposits eroded materials, along with melting ice, creating depositional landforms	*Erosional:* cirque, glacial horn (type of mountain peak), glacial valley (u-shaped valley), fjord, glacial lake *Depositional:* moraine, esker, kame, drumlin, glacial till, kettle lake, outwash fan

(Continued)

TABLE 3.3   Continued

Process	Description	Landform types
Fluvial	Movement of sediments by water, leading to the creation of both erosional and depositional landforms	*Erosional*: canyon, gorge, ravine, cliff, butte, mesa (table or tableland), hoodoo (rock columns), bench, cave, river channel, fluvial terrace (cut or strath terraces), fluvial valley, oxbow lake, marine terrace, gulf, bay, cove *Depositional:* river delta, alluvial (flood) plain, fluvial fill terrace, beach, sandbar, shoal, spit, barrier island, lagoon
Aeolian	Erosion (abrasion), movement and deposition of sediments or sand by the wind	*Erosional:* cliff (especially coastal cliffs), blowout, yardang (rock formations), desert pavement *Depositional:* dune, sandhill, loess, erg (dune field), dry lake (playa)

*Most landforms are created through the interaction of multiple processes, involving both erosion and deposition, and thus this table should be viewed as illustrative rather than fully comprehensive.

Figure 3.15 Landscape formation. Wind, water, and the freeze-thaw cycle contribute to the differential erosion and weathering of rock layers. Softer rock layers, such as sandstone, are eroded or weathered more quickly than harder layers, such as those containing limestone. This can lead to the creation of unusual—and strikingly beautiful—formations. This formation near Oak Creek Canyon in northern Arizona (USA) is carved from a distinctive red sandstone (colored by iron oxide), which formed from wind-swept sand dunes lining the coast of an ancient sea that covered this region during the Permian (~270 mya).

Source: Photo by Kimberly A. With.

Fuca Plate beneath the North American Plate. Both the Andean and Cascade volcanic arcs are part of the 'Ring of Fire,' a seismically and volcanically active area that encircles the Pacific Basin.

However, some volcanoes occur far from plate boundaries and mid-ocean ridges. For example, the San Francisco Volcanic Field in northern Arizona, which gave rise to the San Francisco Peaks (**Figure 3.1A**), is located well within the interior of the North American Plate. Nevertheless, this field has produced more than 600 volcanoes during its 6 million year history (Duffield 2005). These volcanoes were produced along a conveyor belt of sorts, as the North American Plate slid slowly westward (~80 km in 6 million years, or about 1.3 cm/year). The youngest volcanoes are thus located on the eastern edge of the field, where the most recent eruption occurred only about a thousand years ago and gave rise to the Sunset Crater cinder cone (**Figure 3.13**). Such mid-plate volcanic fields are thought to be the product of 'hotspots' (Wilson 1963) that are fed by unusually hot regions located deep within the mantle, from which **mantle plumes** of molten rock rise to the surface (Morgan 1971).[3] Because these hotspots are relatively stationary, the inexorable movement of the plate over the hotspot creates an age-progression of volcanoes along the track, much as observed in the San Francisco Volcanic Field. Some volcanic island chains, such as the Hawaiian archipelago in the middle of the Pacific Plate, are also thought to have arisen via hotspot volcanism (Wolfe et al. 2009; Rychert et al. 2013; **Figure 3.16**).

Other tectonic forces besides volcanism give rise to mountains. As mentioned previously, the convergence of tectonic plates results in the thrusting and folding of Earth's crust (a **fold-thrust belt**), creating mountain ranges

[3]Geoscientists continue to debate whether volcanic hotspots are fueled by mantle plumes, and even whether mantle plumes exist at all (Foulger 2010; Anderson & Natland 2014). As with most polarizing debates, both sides have their supporters, and both sides claim that the weight of evidence backs their position.

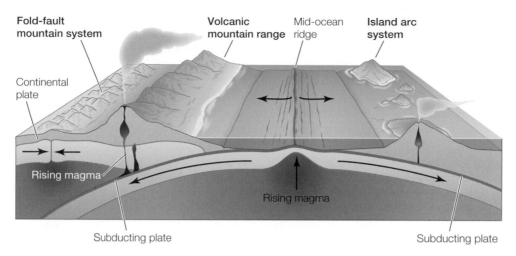

Fold-fault
mountain system

Volcanic
mountain range

Mid-ocean
ridge

Island arc
system

Continental
plate

Rising magma

Rising magma

Subducting plate

Subducting plate

Figure 3.16 Mountain formation (orogenesis).   Volcanic mountain ranges tend to be formed along continental margins as a result of a subduction zone where ocean and continental plates collide. Volcanic island arcs may form at the subducting zone away from mid-ocean ridges. The compression forces of colliding continental plates also produce mid-continental fold-fault mountain ranges.

in addition to volcanoes (**Figure 3.16**). The subduction of oceanic plates beneath continental plates produces mountain ranges along continental margins, as in the case of the Andes and Cascade ranges. In some cases, however, the resulting mountain range may be far removed from the continental margin. For example, the Rocky Mountains in the USA are located about 1000 km from the west coast of North America, but may have formed as result of a very shallow-angled subduction zone that caused 'ripples' (the Rockies) to form far within the interior of the continental plate. Admittedly, this 'flat-slab hypothesis' is not a wholly satisfying explanation, and the origin of the Rockies continues to be something of a puzzle to geomorphologists (e.g. Jones et al. 2011). More usually, mid-continental mountain ranges are produced by the collision of two continental plates, as in the case of the Pyrenees between Spain and France, which were created when Iberia (the landmass making up today's Spain and Portugal) collided with the Eurasian Plate. In fact, many of the mountain ranges in Europe and Asia, including the Alps, the Carpathians, and the Himalayas, were formed along a single fold-thrust belt (the Alpide Belt) as a result of the convergence of the northward-moving African, Arabian, and Indian Plates with the Eurasian Plate.

Alternatively, tension created by tectonic forces (extensional tectonics) can lead to a thinning and fracturing of Earth's crust, producing faults. Although we tend to associate fault zones with seismic activity, certain types of fault movements, especially those resulting in the vertical movement of adjacent blocks, can contribute to mountain formation. **Fault-block mountains**, for example, are produced when part of the block is thrust upward or downward along the fault

line; this often occurs at an angle to create a tilted fault block. As a result, one side of the uplifted block, the **horst**, will have a sheer face or escarpment, whereas the backside of the block mountain will have a more gradual slope. The Sierra Nevada range in California (USA) is an example of a tilted fault-block mountain range, in which the eastern edge of the range is demarcated by the steep Sierra escarpment, the uplifted part of the block, which tapers off gradually to the west toward California's Central Valley. The Sierra escarpment stands in sharp relief to the surrounding Owens Valley, which is the **graben**, or 'dropped' portion of the fault block. Such tilted fault-block mountains are common throughout the Basin and Range region of the western USA and northwestern Mexico, giving rise to the unique and variable topography of this large physiographic province.

Extensional tectonics are also behind continental **rifting**, which may result in elongated dropped valleys (grabens) with steep uplifted sides (horsts), as in the case of the East African Rift Valley. Continental rifting leads to divergent plate tectonics, which were responsible for the break-up of the supercontinent Gondwana some 180 million years ago. More recently, the East African Rift is working to divide the African continent along its Nubian and Somalian Plates, which it should accomplish in another 10 million years or so. Interestingly, mantle plumes have been hypothesized to trigger the initial rupture of the lithosphere, which then sets the stage for the extensional tectonics that cause rifting; such is believed to have initiated the East African Rift, for example (Chorowicz 2005).

Ironically, as mountains rise, they also produce the means of their own destruction. Glaciers form atop mountains as they attain their great heights, which are then capable of dissecting their slopes through the sculpting of deep valleys. Yosemite Valley in the Sierra Nevada, which has been immortalized in the works of landscape painters and photographers such as Albert Bierstadt and Ansel Adams, respectively, was formed in large part by glacial action. In addition, downcutting occurs through the action of fluvial erosion. Water runoff increases as slopes become steeper, thereby increasing the power of streams or rivers to carve into the mountain, eventually triggering landslides that further contribute to the mountain's erosion. The steep terrain of mountain landscapes thus reflects a balance between the opposing forces of uplift and erosion, which in turn involves the dynamic coupling of river incision and landslide erosion (Larsen & Montgomery 2012). In mountain ranges undergoing rapid uplift, such as the Namache Barwa range in the eastern Himalayas, there appears to be a threshold angle in slope steepness (30°) beyond which the rate of landslide erosion increases by an order of magnitude, thereby maintaining slopes at a certain angle of repose, and ultimately, limiting the height attained by the mountain (Larsen & Montgomery 2012; Roering 2012).

While opposing forces may be at work in shaping montane landscapes, these processes operate at very different spatiotemporal scales: uplifting is a slow gradual process, perhaps occurring at a rate of 1 cm/year, whereas the rates of fluvial erosion and landsliding are influenced by more frequently occurring and highly variable factors, such as rainfall events that contribute to flooding, the denudation of hillslope vegetation by humans that increases erosion, or earthquakes that ultimately destabilize these saturated (and perhaps denuded) hillsides, causing them to slide (Restrepo et al. 2009). Thus, mountains may take millions of years to rise but can come tumbling down in a matter of seconds when the hillslope threshold is exceeded. The role of landslides in contributing to the disturbance dynamics of landscapes will be discussed a bit later, but for now, we will shift our attention to a different erosional process, and explore how fluvial erosion interacts with topography in the formation of river networks.

Formation of River Networks

Dendritic river networks are a ubiquitous feature of many landscapes, and thus might be viewed as 'both the skeleton and the circulatory system of Earth's landscapes' (Perron et al. 2012), in that they contribute to landscape structure and serve as transportation networks that carry and deliver materials throughout the landscape. The hierarchical branching structure of river networks exhibits the fractal properties of resource-distribution networks more generally (i.e. power-law scaling), which may reflect a general efficiency that naturally emerges in any type of branching network serving a particular area (or volume; West et al. 1997, 1999; Banavar et al. 1999). Recall that power laws are the hallmarks of fractal structures and critical processes (**Box 2.1**), which in turn are the fingerprints of self-organized systems (Solé 2007). However, to say that the drainage network of river basins exhibits fractal properties or self-organization (Rodríguez-Iturbe & Rinaldo 1997) does not explain how branching occurs (in terms of the specific mechanism or process), or even why branching should occur when and where it does. In other words, how do erosional mechanics give rise to the observed fractal structure of dendritic networks? How does process give rise to pattern?

[S]tudies based on principles other than erosional mechanics… have provided new insights into the structure of natural river networks, but cannot directly relate river network form to the erosional mechanisms that shape the topography: we know what the skeleton of a landscape looks like, but not how it grows.

Perron et al. (2012)

Not only are antagonistic forces at work, but there may exist a critical threshold in the interplay between these forces that ultimately governs when branching occurs. Rivers are formed through incision, an erosional process by which water cuts into the underlying bedrock. The rate of river incision is a function of stream power (a function of slope and the volumetric flow rate), the amount and size distribution of suspended sediments, and the strength or resistance of the rock being incised (Sklar & Dietrich 2001). River incision creates a channel that grows deeper and wider with time through positive feedback: as the river cuts its way through the landscape, it increases the drainage area and discharge rate, which in turn increase the rate of channel incision, which then leads to a further deepening and enlarging of the drainage area. As the channel grows and the drainage basin enlarges, it may eventually impinge on smaller neighboring channels, capturing these to form branching tributaries (**Figure 3.14**).

As with most fractal structures in nature, however, this branching process does not go on indefinitely. At some point, negative feedbacks in the form of depositional processes begin to fill the channels and slow incision rates. For example, soil creep—the gradual slumping that occurs on hillsides due to the force of

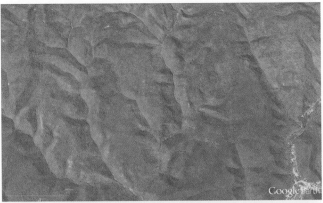

Figure 3.17 Formation of river networks. The formation of river networks reflects the antagonistic processes of erosion and deposition. In branching river networks where soil creep exceeds the channel incision rate, the result is a less rugged landscape with larger valleys (e.g. Allegheny Plateau, Pennsylvania, USA; top) compared to river networks where the channel incision rate exceeds that of soil creep, which results in a finer scale of branching that contributes to a more heavily dissected landscape (e.g. near Salinas Valley, California, USA; bottom).

Source: (A) Google Earth, data provied by USDA Farm Service Agency. (B) Google Earth, data provided by AMBAG.

gravity—works against incision in some drainage systems. There thus appears to be a critical ratio in the antagonistic rates of incision and soil creep that acts as a threshold for the creation of river branching (Perron et al. 2012). To test this, Perron and colleagues (2012) compared dendritic networks in the hills surrounding California's Salinas Valley versus those in the Allegheny Plateau in southwestern Pennsylvania (**Figure 3.17**). They found that rivers in California had four times as many branches as those in Pennsylvania, presumably because its softer rock layers and higher runoff favor incision over soil creep, whereas the older, harder rock layers and wetter soils of the Allegheny Plateau are instead more susceptible to soil creep.

Formation and Diversity of Soils

Erosion is not just important in the formation of mountains and river networks. The weathering of the underlying rock layer (parent material) is also responsible for the formation of different soil types. Soil formation (pedogenesis) is a complex process that is further modified by topography (e.g. slope), climate (temperature, precipitation), and biological activity (soil decomposers, burrowing animals, rooting plants, and humans) (White 2006). The topic of soil formation thus encompasses all of the sections in this chapter pertaining to the various geomorphological, abiotic, and biotic processes that contribute to landscape heterogeneity. Nevertheless, we will discuss soil formation in this section on geomorphological processes because soil formation typically involves the action of slow erosional processes, and thus is more akin to the formation of the other landforms we have been discussing here.

> The landscape displays a remarkable range of soil types, resulting from an almost infinite variation in geology, climate, vegetation and other organisms, topography, and the time for which these factors have combined to influence soil formation.
>
> White (2006)

Soils are dynamic and evolve over time, creating characteristic layers (horizons) that can be used to classify different soil types based on their physical, chemical, and biological properties (e.g. texture, permeability, alkalinity, fertility). For example, the United States Department of Agriculture (USDA) has developed a taxonomy of soil types, based on a hierarchical system of classification that recognizes 12 different orders of soils (e.g. aridisols, mollisols). Other countries, such as Canada, France, and Australia, have adopted similar systems of soil classification. An international standard for soil classification based on these various systems has been developed: the World Reference Base for Soil Resources, which recognizes 32 soil groups worldwide (IUSS Working Group WRB 2006).

Analysis of the **pedodiversity**—the diversity of soil types—of a landscape or region provides important information on the different types, relative abundance, and distribution of soils (Ibáñez et al. 1995, 1998; Guo et al. 2003). A soil map thus gives another view of landscape heterogeneity that is useful for evaluating potential vegetation, land cover, or land use within a region. For example, a variety of soil types are associated with the different physiographic regions in the state of Pennsylvania in the eastern United States (**Figure 3.18**). These distinct regions (provinces) are

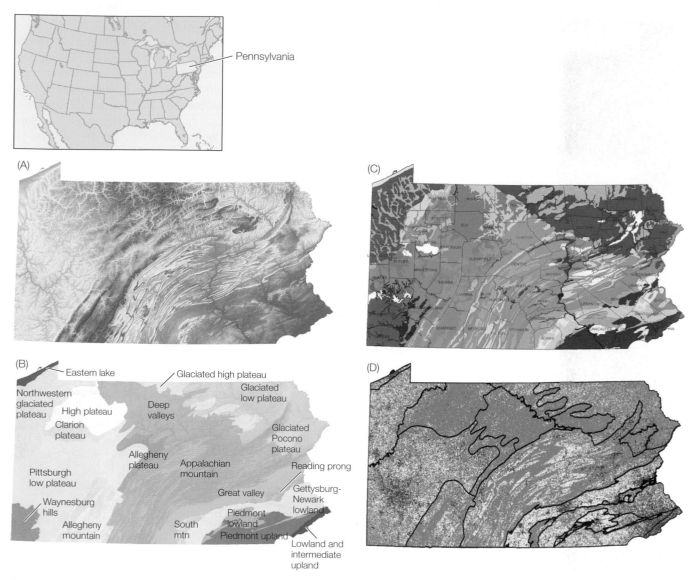

Figure 3.18 Geomorphology and landscape heterogeneity.   Various geomorphological processes are responsible for the complex terrain (A) of a region like Pennsylvania in the northeastern USA, which in turn gives rise to distinct physiographic regions throughout the state (B). The interaction of geomorphology and climate results in the formation of different soil types, which in many cases are unique to a particular physiographic region (C). Topography and soils influence the formation of different vegetation types (D), as well as human settlement and land-use patterns (red = developed areas, yellow = cropland, green= forest).

Source: Courtesy of the Pennsylvania Spatial Data Access Map Gallery.

the product of a complex geomorphology a billion or more years in the making, ranging from repeated periods of glaciation in both the northeastern and northwestern parts of the state; mountain-forming events (orogenies), including the Appalachian event that gave rise to the distinctive Ridge and Valley region of today; the Piedmont region, the eroded, hilly remains of a once-great mountain range; fluvial deposits that built up the western half of the state into the Pittsburgh and Allegheny Plateau regions; and more-recent floodplains along the Atlantic coast in the southeastern region of the state (Barnes & Sevon 2002).

Topography and the distribution of soil types are both important determinants of human land-use and settlement patterns (Amundson et al. 2003). For example, agricultural fields and settlements tend to be concentrated in areas of low relief, such as within coastal plains, river valleys, and atop plateaus or tablelands, where soils tend to be deeper and richer or not so steep and rocky as to preclude their cultivation (**Figure 3.18**). Land-use patterns are thus typically a reflection of the terrain and the geomorphological processes that shaped it, and may mirror the dendritic structure of a drainage basin, the directional flow of

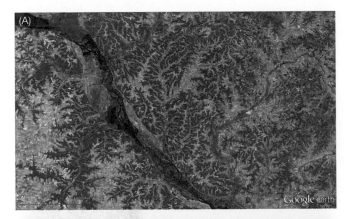

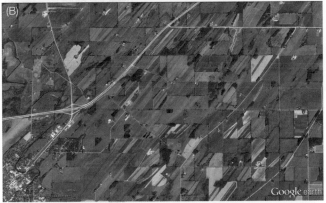

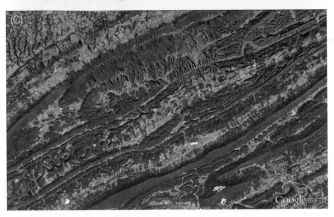

Figure 3.19 Topography strongly influences human land-use patterns. Agricultural fields and settlements (tan or light-colored areas) are concentrated in areas of low relief, such as atop plateaus and within fluvial plains (A), oriented along glacial till (B), or in lowlands and river valleys (C). **(A):** Driftless plain region along the upper Mississippi River that forms the Wisconsin–Minnesota border (USA); **(B)** Drumlin field near Columbus, Michigan (USA); **(C)** Ridge and valley region, southern Appalachians, Tennessee (USA).

Source: Google Earth, data provided by A) Landsat/Copernicus B) USDA Farm Service Agency C) GeoEye, USDA Farm Service Agency, Digital Globe.

glaciers, or the alternating ridges and valleys formed by the fold-and-thrust belt of an ancient mountain range (**Figure 3.19**). The specific land uses, agricultural

crops, or wine varietals that can be grown within a region are as much a function of its soils as its climate.

> [C]ultivated soils might be viewed as domesticated versions of their natural counterparts, with widely differing properties and functions.   Amundson et al. (2003)

Most soil classification maps do not assess the extent to which soils have been manipulated or disturbed, however (Amundson et al. 2003). For example, nearly 20% of the USA is under intensive agriculture, concentrated in the Midwest, Great Plains, Mississippi River Valley, and California's Central Valley. As a consequence, these regions have undergone a significant loss (>50%) of their pedodiversity (particularly of rare soil types) or are in danger of doing so, which has negatively affected the diversity of rare plants as well as the soil microbial diversity that is important for nutrient cycling, and thus, soil fertility (Amundson et al. 2003). Even when land use does not result in the loss or extinction of unique soil types, it still changes soil properties and biogeochemical functioning, such that it is no longer representative of that soil type in its undisturbed state. Undisturbed soils are also important for the sequestration of carbon, and can function as carbon sinks. Many land-use practices associated with agriculture (e.g. tillage, slash-and-burn) have released massive amounts of carbon into the atmosphere as CO_2, contributing to the problem of global climate change. For example, the historical loss of soil carbon as a result of converting native grasslands to agricultural land use may be as high as 5000 Tg in the USA alone (DeLuca & Zabinski 2011).[4] Fortunately, changing agricultural practices (e.g. reduced tillage) and the application of soil conservation measures can help to mitigate increasing atmospheric CO_2 levels through increased carbon sequestration in soils (Post et al. 2012; **Chapter 11**).

Not surprisingly, then, the severity of land degradation, such as from deforestation, overgrazing, or agricultural mismanagement, is ultimately assessed with respect to soil resources and the extent to which these have suffered erosion, acidification, contamination, desertification, or salinization. Soil erosion in particular is one of the most significant environmental and land-management issues we currently face. To illustrate, we consider the North American Dust Bowl of the 1930s (the 'Dirty Thirties'), which underscores the ecological and socioeconomic consequences of mismanaging soil resources (**Figure 3.20**).

[4] A teragram is 10^{12} grams.

Figure 3.20 American Dust Bowl.   Poor agricultural practices coupled with a decade-long drought resulted in the wind erosion of millions of tons of soil during massive dust storms that choked or buried major portions of the Great Plains region during the 1930s. The Dust Bowl is the greatest human-induced environmental disaster to occur within the USA.

Source: Image is a work of the USDA and is in the public domain.

CASE STUDY   The North American Dust Bowl

The Great Plains are an immense, wind-swept, drought-prone region of grassland and steppes in the center of the North American continent. Early European settlers christened the region the 'Great American Desert,' for in their view, this vast, arid, treeless region was little more than a wasteland that was ill-suited to farming.[5] The region was nevertheless settled, due in large part to the granting of land by the federal government to encourage homesteading.[6]

A period of unusually high rainfall, however, revealed that the land could be made highly productive cropland once sufficient water was applied. Indeed, the apparent relationship between this increased rainfall and crop production led some to the mistaken belief that cultivation was actually responsible for this shift in climate, making the region wetter and therefore more favorable for agriculture ('rain follows the plow').[7] The advent of mechanized agriculture, coupled with a booming wheat market in Europe following World War I, high wheat prices, and generous federal farm policies, helped to turn the Great American Desert into America's Breadbasket.

This initial prosperity was short-lived, however. The economic collapse of the Great Depression and an overproduction of wheat caused wheat prices to plummet, compelling farmers to enlarge their acreage to stay solvent and pay for the expensive new farm machinery that enabled them to farm so expansively. As a consequence, over 300 million acres (120 million ha) of native prairie were plowed under and converted into wheat and other crop fields during the 'Great Plow-Up' of the Great Plains (Sampson & Knopf 1994). When drought inevitably returned to the region, it turned out to be a 'drought of record,' lasting for the better part of a decade through the 1930s.

[5] Which seems ironic in retrospect, given that much of the region sits atop the massive Ogallala or High Plains Aquifer (one of the world's largest at 450,000 km²) that now irrigates one of the most productive agricultural regions in the world. The technology to access the aquifer was not available until after World War II, however, and has since given rise to center-pivot agriculture, which creates the 'crop circles' seen by airline passengers flying over the region (**Figure 3.37B**). Once thought inexhaustible, the aquifer has since been seriously depleted and pumped dry in places; it is estimated that a third of the southern High Plains will be unable to support irrigation within 30 years (Scanlon et al. 2012). Nor is the problem of groundwater depletion limited to just the High Plains Aquifer. Groundwater supplies drinking water and irrigation to billions of people around the world. The size of the global groundwater footprint (the area required to sustain current groundwater use) is currently about 3.5 times the actual area of aquifers; thus, an estimated 1.7 billion people live in areas where groundwater resources are under threat (Gleeson et al. 2012).

[6] The Enlarged Homestead Act of 1909 granted 320 acres (1.3 km²), double the amount of land offered in the original Homestead Act (1862), to settlers willing to farm marginal lands in areas like the Great Plains; much of the prime farmland had already been homesteaded under previous acts.

[7] Although the theory was discredited in this particular case, there is evidence that agricultural land use does have an effect on local and regional weather patterns, including in the Great Plains (Raddatz 2007).

Record drought, coupled with high temperatures and winds, killed off the wheat and left behind a region that was largely barren and denuded of the deep-rooted native grasses that had formerly held the highly erodible soils in place. Winds picked up the desiccated soil, producing massive dust storms ('black blizzards') that enveloped the region, with some storms carrying dust as far away as the Atlantic seaboard and beyond. All told, the dust storms of the 1930s carried away several hundred million tons of topsoil, soil that had taken thousands of years to accumulate but was swept away in fewer than ten. The 'Black Sunday' storm of 14 April 1935 was by far the worst, as vividly described by Tim Egan in his book on the Dust Bowl entitled, *The Worst Hard Time* (2006):

> The storm carried twice as much dirt as was dug out of the earth to create the Panama Canal. The canal took seven years to dig; the storm lasted a single afternoon. More than 300,000 tons of Great Plains topsoil was airborne that day.

Ironically, land degradation not only contributed to the dust storms of the drought-stricken 1930s, but also likely amplified the drought itself by creating land-surface feedbacks. The widespread loss of vegetation (both in term of the native grasslands as well as desiccated cropland), coupled with elevated atmospheric loadings of dust, led to even warmer and drier conditions, turning an otherwise modest drought into one of the worst environmental disasters the USA has ever experienced (Cook et al. 2009).

The North American Dust Bowl is not just an accident of history, a 'perfect storm' of global events that could never happen again. Rather, the Dust Bowl should serve as a cautionary tale. There are other areas in the world today where human-induced land degradation coupled with drought, which is likely to be made worse by global warming, have the potential to interact and cause future dust bowls (Cook et al. 2009). An estimated 30% of the world's arable land has become unproductive in just the past 40 years due to erosion, and about 80% of the world's agricultural land currently suffers moderate-to-severe erosion (Pimentel 2006). Many countries, such as the USA, China, and India, are losing soil some 10–40 times faster than the rate at which soils naturally form (Pimentel 2006). As a consequence, world food production may decline by 30% in the coming decades owing to a loss of soil fertility, even as the human population continues its exponential growth, thereby imperiling our future food security (Pimentel 2006).

Landscape Dynamics: Abiotic Disturbances

The geomorphological processes discussed in the previous section are predominantly abiotic, involving as they do the forces of wind, water, and ice. However, the extended time periods and vast spatial scales over which geomorphological processes operate typically lie outside the ecological scales studied by landscape ecologists. In the framework of hierarchy theory, geomorphological processes operate more slowly and with less interaction strength than those that give rise to the structure and dynamics of ecological systems. As such, geomorphology exists one level or more above the focus of the ecologist. Recall, however, that the upper level of a hierarchically organized system provides the context in which the focal-level dynamics occur (**Chapter 2**). Although this imposes constraints on the behavior of the system of interest, it can also give significance to the phenomenon we are studying. Thus, for most landscape ecologists, the geomorphology of a region provides a context—both historical and geographical—for understanding the ecology of landscapes, but the study of landscape formation itself is generally left to the province of geologists and geomorphologists.

This does not mean that landscape ecologists are uninterested in the abiotic forces that shape landscapes. Far from it. Abiotic processes that operate within the spatial and temporal scales of ecological systems, such as fires, windstorms, floods, and even volcanic eruptions, are certainly of interest, as these tend to occur as relatively discrete events—disturbances—that alter the structure of the system and contribute to its dynamics (**Table 3.2**). We thus examine a range of abiotic disturbances in terms of how they contribute to the structure and dynamics of landscapes.

Volcanic Eruptions

There are probably 20 or so volcanoes erupting around the world at this very moment, out of some 50–70 volcanoes that will likely erupt this year, which is still only a small fraction of the ~600 volcanoes whose eruptions have been documented historically (i.e. over the last 500 years) and which may erupt again someday. Although volcanism is a geomorphological process that has contributed to landscape formation, it can also be viewed as a disturbance event, especially in the case of highly explosive eruptions.

As with any disturbance, volcanic eruptions vary in type, magnitude, and intensity, from the continuously effusive flows or fountains of lava characteristic of Hawaiian volcanic eruptions, to the very large explosive-type eruptions (Plinean and ultra-Plinean eruptions),[8] such as the 1883 eruption of Krakatoa in the Sunda Strait of Indonesia, or the 1991 eruption of

Mount Pinatubo in the Philippines. Volcanic eruptions have the potential to act as disturbance events that profoundly alter the structure and ecology of landscapes over broad spatial scales and within a very short period of time. Among the best-studied volcanic eruptions, in terms of documenting the ecological impacts and subsequent recovery, was the May 1980 eruption of Mount St Helens in the Pacific Northwest, one of the largest in recent history.

[8] Plinean and ultra-Plinean eruptions are named for Pliny the Younger, a Roman administrator who had witnessed and later documented the eruption of Mount Vesuvius that famously buried Pompeii and its neighboring towns in 79 CE. Vesuvius has erupted nearly three dozen times since then and is considered one of the most dangerous volcanoes in the world, especially now that the surrounding area is inhabited by more than three million people.

CASE STUDY   The 1980 Eruption of Mount St Helens

After months of seismic unrest, Mount St Helens finally erupted on 18 May 1980, following a moderate earthquake. The earthquake triggered a massive landslide—one of the largest in recorded history—on the north flank of Mount St Helens, which filled a nearby river valley with avalanche debris up to 200 m deep. Seconds later, a cataclysmic explosion blasted a kilometers-wide crater in the side of the mountain and reduced its height by 400 m in an instant (Lipman & Mullineaux 1981). Nearly everything in the blast

arc (a 550 km² area) was obliterated, and the resulting eruption column reached a height of 27 km (~17 miles), raining ash down over a vast region as far away as the Great Plains some 1500 km distant (Nash 2010; **Figure 3.21**). For hours to days after the initial eruption, falling tephra (pumice), pyroclastic flows, and lahars (volcanic mudflows) continued to bury the surrounding landscape and its river channels over an enormous area. Volcanic eruptions are thus not just a single disturbance event, but rather, a series of related disturbances that transform the landscape in different ways and to varying degrees.

Since the 1980 eruption, Mount St Helens has undergone an equally dramatic transformation, from a sterile grey 'moonscape' in the immediate aftermath of the eruption, back to a living green landscape in a matter of years (**Figure 3.22**). The process of recovery through ecological succession has occurred far more quickly than anyone might have imagined, given the extent of the devastation.[9] Although the explosive blast of the eruption leveled or seared much of the surrounding forest in the blast zone, this seemingly stark and lifeless landscape nevertheless held scattered pockets of life, survivors that somehow managed to escape either through sheer luck or because they were protected during the initial eruption and subsequent fallout. For example, many small mammals that live underground or have subterranean burrows managed to survive within areas of the blast zone, as did thousands of small tree saplings that were still covered by the winter snowpack (Crisafulli et al. 2005). Herbaceous plants with belowground buds also survived (Dale et al. 2005a), as did entire aquatic communities that were protected under a layer of ice (Crisafulli et al. 2005). It thus helped that the eruption occurred in early spring, when many seasonal residents, such as

Figure 3.21 Eruption of Mount St Helens.   On 18 May 1980, Mount St Helens in the Pacific Northwest (Washington state, USA) underwent an explosive eruption that reduced the height of the mountain by 400 m (1312 ft) in an instant.

Source: Image is a work of the USGS and is in the public domain.

[9] Nature also had some help. Part of the forest blasted by the eruption was privately owned by a timber company, which undertook a major reforestation effort in the years immediately following the eruption, eventually planting more than 18 million seedlings over a six-year period. Today, these replanted areas are thickly forested, with trees already >20 m. In contrast, areas within the Mount St Helens National Volcanic Monument, which was established in 1982, are undergoing a process of natural regeneration and are thus still sparsely treed.

Figure 3.22 The recovery of Mount St Helens. Landsat satellite imagery reveals the 'greening' of the Mount St Helens landscape in the 30 years following its May 1980 eruption. (A) June 1984, (B) July 2011.
Source: Photos from NASA Earth Observatory.

migratory songbirds and anadromous fishes, had not yet returned to the area; bud break had not yet occurred at higher elevations; and significant snow and ice still covered many areas, including lakes. The ecological consequences would undoubtedly have been far worse had this eruption occurred a bit later in the season.

The presence of these surviving organisms (**biological legacies**) greatly accelerated the process of recovery through ecological succession (Franklin 1990). Recall from our discussion regarding community assembly earlier in this chapter that the Clementsian view of succession envisions an orderly process, one that advances through a series of stages of increasing complexity toward a mature climax state. In the case of primary succession, such as that following a volcanic eruption where the landscape is essentially wiped clean, we might therefore expect a slow progression of species: from early colonizers, such as lichens that help to break down the volcanic rock to form soil, which is then able to support early-successional forbs and grasses, which in time give way to shrubs, and finally, deciduous or coniferous trees many decades later. Because of the existence of biological legacies, however, signs of life were evident within months—not years—after the eruption. These survivors formed the nucleus of recovery, as other species or their propagules (such as fungal spores and plant seeds) began to arrive from outside the blast zone to recolonize disturbed areas. As a consequence, community structure and ecological complexity (i.e. the network of interacting species) could be reestablished fairly quickly in some areas.

The future at Mount St Helens will be one of continuing change. The pace of ecological change will be determined by complex processes of ecological succession influenced by landscape position, topography, climate, and further biotic, human, and geophysical forces. The current volcanic activity at Mount St Helens attests to its dynamic character…Even so, many biotic, landform, and soil legacies of the 1980 eruption will influence ecological processes for centuries to come.   Dale et al (2005b)

Far from being an orderly process toward a single endpoint, however, ecological succession on Mount St Helens has proven to be a complex process, one that is sensitive to landscape position (e.g. proximity to forest edges), initial conditions, and chance events, and which has proceeded at different rates along many different successional pathways (Dale et al. 2005b). Further, successional pathways are continually being modified by secondary disturbances, such as small landslides, flooding, and mudflows, which reset succession to an earlier stage or alter its trajectory altogether. The colonization of disturbed areas by various species is subject to the vagaries of dispersal, making it difficult to predict what community patterns will form on a particular site. Partly this may be due to **priority effects**, the order in which species arrive at a site, which may then influence successional trajectories if already established species then facilitate—or inhibit—the establishment of later-arriving species. Even after arrival, species interactions, such as competition, allelopathy, and herbivory all interact to shape successional trajectories in unpredictable ways. In their nearly 30-year study of primary succession on the massive lahar created by the eruption of Mount St Helens, del Moral and his colleagues (2009, p. 187) were left to conclude that

> [S]tochastic, not deterministic, factors are more closely associated with species patterns. Thus, significant assembly rules to form vegetation units are not yet evident.... Vegetation develops not from adherence to rules, but in response to dispersal limitations and the resistance of established species to invasion.

Recovery on the slopes of Mount St Helens is far from complete. Vegetation is still sparse on the most devastated areas, such as within the debris-avalanche zone, and is likely to remain so for decades to come (**Figure 3.22**; Dale et al. 2005a). In the meantime, Mount St Helens continues to serve as a natural laboratory for studying ecological succession, providing a benchmark not only for understanding how ecosystems recover in the aftermath of volcanic eruptions, but also for how ecological systems respond more generally to broad-scale environmental disturbances.

Landsliding

Landslides play an important role in both landscape formation and the formation of landscape heterogeneity. As we saw earlier in this section, landsliding is an important erosional process responsible for carving mountain ranges and river networks. Landslides involve the mass movement of rocks, mud, soil, or other debris and vegetation down a slope. As such, they are variously characterized as rockfalls, mudflows, slumps, or debris avalanches, depending on the material and mode of movement (**Figure 3.23**). Landslides may be triggered by a variety of natural forces, such as earthquakes, volcanic eruptions, or heavy precipitation events. Recall that one of the more devastating consequences of the 1980 eruption of Mount St Helens was the massive landslide that was triggered by the initial earthquake that set off the eruption. Landslides are thus among the most severe of abiotic disturbances, given that all of the biomass

and soil may be removed in a single event (Myster et al. 1997).

Landsliding is also important from an ecological standpoint, as a type of natural disturbance that has long-lasting effects on ecosystems, and which contributes to overall landscape heterogeneity (Restrepo et al. 2009). As with volcanic eruptions, primary succession commonly occurs on landslides, and the process of ecological recovery is similarly influenced by landscape context and position (Myster et al. 1997). Human land-use activities may exacerbate the risk of landsliding in mountainous areas, through the construction of roads or the deforestation of mountain slopes, which increase erosion and destabilize hillsides (Restrepo et al. 2009). From a human hazard-risk standpoint, landslides pose a deadly threat to communities that lie in their path, having been responsible for more than 32,000 deaths worldwide in one 7-year period between 2004 and 2010 (Petley 2012). In addition, the frequency and magnitude of landsliding appears to be growing in many parts of the world as a result of climate change that is contributing to an increase in heavy precipitation events as well as warming temperatures that are hastening the melt of montane glaciers, both of which can lead to slope failures that produce landslides (Stoffel & Hugel 2012). Landsliding thus is likely to become an increasingly important form of disturbance in the coming decades.

Predicting when and where landslides will occur is no simple matter, however. Although the primary triggers of landsliding generally involve seismic activity or precipitation (or both), the point at which the slope becomes destabilized and gives way (i.e. the hillslope threshold; Larsen & Montgomery 2012; Roering 2012) depends on many factors, such as the slope's angle (its incline), its aspect (the direction the slope faces), the type of geological substrate or soil, the extent and type of vegetation cover, and the type and intensity of human land use on the slope, all of which interact with climate and the geographic location of the mountain

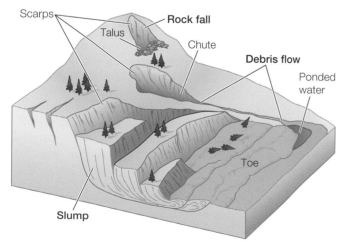

Figure 3.23 Anatomy of a landslide. Landslides are classified by the type of material moved (soil, rock, or debris), as well as their mode of movement downslope (slump, fall, or flow).

Source: After an image from http://geologycafe.com/

range (e.g. proximity to major seismic or precipitation zones) to influence the occurrence of landslides (Myster et al. 1997; Restrepo et al. 2009). Multiple factors thus operate within and across domains of scales (cross-scale interactions) to influence the process of landsliding, resulting in a diversity of landslide types and severity (Restrepo et al. 2009; **Figure 3.23**). As a consequence, landslides may be expressed across a range of scales, from the individual landslide at the slope scale, to the population of landslides triggered by the same seismic or precipitation event within a given mountain range, or as an assemblage of landslide populations that overlap in space and time within a region (Restrepo et al. 2009). Thus, landsliding leaves behind a spatial signature of disturbance, a landscape legacy of disturbance events from hundreds or even thousands of years past (Hewitt et al. 2008), and must therefore be considered a major contributor to the heterogeneity and dynamics of mountainscapes.

Individual landslides are also internally heterogeneous, exhibiting different disturbance zones related to the initial site of slope failure (the headwall or scarp), the transport zone or chute, and the deposition zone (at the foot or 'toe' of the slide; **Figure 3.23**). Although we tend to think of landslides as creating a vertical disturbance gradient down the face of a mountain, there is also a distinct horizontal gradient in the degree of disturbance across the width of the slide. Indeed, the gradient in both abiotic and biotic conditions (e.g.

light, soil nutrient levels, seed pool) may be far greater across the width of the landslide than along its length (Myster & Fernández 1995). For example, the level of disturbance is usually milder along the edges than at the center of the landslide, which tends to be more deeply and frequently scoured of soil and vegetation (Restrepo et al. 2009). In terms of ecological recovery, then, plant propagules usually arrive first and become more abundant along the edges than at the center of the slide, both because of dispersal limitation and because of the greater degree of disturbance within the chute.

As for the vertical disturbance gradient, the deposition zone at the foot of the slide receives vast quantities of soil and organic matter, including plant remnants and seeds, which may hasten re-vegetation and recovery in these areas relative to other areas of the slide. Nevertheless, the reorganization of ecological communities following a landslide is often unpredictable (Myster & Walker 1997), given that successional trajectories vary widely as a consequence of initial conditions, the vagaries of dispersal, priority effects, and the dynamic nature of species interactions, just as we saw in the case of ecological succession following the 1980 eruption of Mount St Helens. To illustrate further, we consider the effect of landslides on ecological succession in the Ninole Hills, a series of prominent ridges along the southeastern flank of Mauna Loa—the largest active volcano in the world—on the island of Hawai'i.

CASE STUDY   Landslides in the Hawaiian Islands

With a mean annual rainfall of around 4 m, the Ninole Hills on the island of Hawai'i are especially prone to landsliding. Ironically, the hills may themselves be the product of a catastrophic landslide (part of the headwall) that occurred when the southern flank of the ancient Mauna Loa 'failed' and slid into the sea some 100,000 years ago (Lipman et al. 1990).[10] Regardless of how they formed, landslides continue to shape the Ninole Hills and are a recurring disturbance that contributes to a diversity of landforms, substrates, and thus vegetation types in this region. For this reason, Carla Restrepo and her colleagues investigated how landslides affected colonization and the successional development of plant communities across a landslide chronosequence (i.e. across landslides that varied in age; Restrepo & Vitousek 2001; Restrepo et al. 2003).

Landslides are fairly common in the Ninole Hills, occurring at a rate of 15% per century. This translates into

an enormous loss of biomass from these systems (53 t/ha/100 y), along with the productive ash-derived soils upon which these vegetation communities developed over hundreds of years (Restrepo et al. 2003). Subsequently, the exposed basalt of recent landslides (4–17 years post-slide) supports a very different type of plant community than the mesic forests found on neighboring undisturbed sites, thereby contributing to the heterogeneity of these montane landscapes (Restrepo & Vitousek 2001). About half (46%) of the 28 species colonizing recent landslides in the Ninole Hills were non-native or invasive species, however, whereas few non-native species occur in undisturbed sites. Adding insult to injury, non-native species prevent the establishment of native species within these disturbed areas, and also stunt the growth of the endemic 'ōhi'a lehua tree (*Metrosideros polymorpha*) that would otherwise dominate these hillsides (Restrepo & Vitousek 2001). Thus, landslides may now be facilitating the invasion of landscapes by non-native species, thereby transforming these communities in different—and possibly irreversible—ways than in the past.

[10] The origin of the Ninole Hills continues to be debated. More recent evidence suggests that basaltic lava flows from a now-buried rift zone may have given rise to the Ninole Hills, either independently or in conjunction with landsliding (Morgan et al. 2010).

On several of the Hawaiian Islands, including the main island of Hawai'i, an introduced invasive species is actually contributing to a greater incidence and severity of landsliding. The velvet tree (or miconia; *Miconia calvescens*) is native to the wet montane forests of Central and South America. As the velvet tree is rather attractive, with enormous green-and-purple leaves, it has been introduced to numerous tropical islands throughout the Pacific. The velvet tree was first introduced to the Hawaiian Islands in the early 1960s and 1970s as a specimen plant in several botanical gardens and parks across the island chain, but was not widely considered a threat until the early 1990s, after it had become naturalized in several areas and begun to spread (Medeiros & Loope 1997).

Like many invasive species, the velvet tree succeeds through excess: it grows and matures rapidly (in ~4 years), and when mature, each tree is capable of producing millions of tiny seeds several times a year, which are spread far and wide by fruit-eating birds (which themselves have been introduced, such as the Japanese white-eye, *Zosterops japonicus*, and common myna, *Acridotheres tristis*) or even by humans (e.g. on mud-caked shoes or tires). The velvet tree is thus able to completely inundate native forests, forming dense monotypic stands with an umbrella-like canopy that shades out native plants. As a result, the groundcover beneath miconia stands is virtually eliminated, leaving nothing but bare ground that is then highly susceptible to erosion, especially as the velvet tree's shallow root system provides little in the way of soil stability (Nanko et al. 2013). Entire hillsides covered with miconia eventually fail and slide away, making the velvet tree a doubly serious threat to the montane ecosystems of the Hawaiian archipelago, where it is officially designated a noxious weed (and unofficially, as the 'purple plague'). Elsewhere in the South Pacific, such as in Tahiti, French Polynesia, where it has replaced more than 75% of the native forests (Meyer 2010), the velvet tree has been dubbed the 'green cancer.' This 'cancer' has since metastasized to several other Society Islands and beyond, to New Caledonia and the wet tropics of Australia, to as far away as Sri Lanka in the Indian Ocean and the Caribbean islands of Jamaica and Grenada. For this reason, the velvet tree ranks among the top 100 of the world's worst invasive alien species (Lowe et al. 2000).

The velvet tree is admittedly a special case. Forested hillsides—and vegetated slopes more generally—usually reduce the likelihood of landsliding, as witnessed by the dramatic effect that clearcutting has on increasing surface erosion, and thus, the frequency of landslides (Guthrie 2002). Indeed, deforestation appears to lower the rainfall threshold at which landslides occur. For example, moderate rainfall events (characterized as those having a short return interval of 3–10 years) were capable of triggering landslides in logged montane forests of central Japan, whereas virtually no landslides occurred in regenerated forests (>25 years post-harvest) during these same rainfall events (Imaizumi & Sidle 2012). Thus, forest harvesting not only increases the frequency of landslides during heavy rainfall events, but can also increase the likelihood that landslides will occur during more moderate rainfall events.

Flooding and the Natural Flow Regime

Landslides are not the only disturbance triggered by extreme precipitation events. Flooding is another possible outcome of heavy or prolonged precipitation, and like landsliding, has the power to erode and mobilize large amounts of material across the landscape. Flooding causes significant environmental and economic damage, and thus poses a substantial threat to human health and safety as well. Flooding accounts for 40% of all natural disasters worldwide as well as half of all human fatalities from natural disasters, either directly due to drowning or indirectly as a result of water-borne diseases that occur in the aftermath of the flood (Ohl & Tapsell 2000). For this reason, a variety of flood-control measures are typically implemented in flood-prone areas (e.g. floodgates, levees, dams) in an effort to protect communities from catastrophic flooding. Given that global climate change is predicted to alter precipitation patterns, resulting in fewer but more intense storms in some areas, and increased snow and glacial melting as a consequence of warming temperatures, the frequency and intensity of flooding is only likely to increase in coming decades, which will seriously challenge the flood-control capacity of these systems (both natural and human-constructed) in flood-prone regions around the world. Indeed, we are perhaps already seeing evidence of this, as devastating and record-breaking floods have occurred in recent years all over the world. Major flooding also occurs along coastal areas during hurricanes or typhoons (tropical cyclones) as a result of storm surge (the wind-driven swell of water onto land), which will be covered in the next section.

From an ecological standpoint, however, periodic or seasonal flooding is integral to the dynamics, complexity, productivity, and functioning of river systems (Parsons et al. 2005). Riverine landscapes are created and maintained by fluvial disturbance dynamics, such as the **flood pulse** or **pulse flows** that lead to the periodic expansion and contraction of these systems (Junk et al. 1989; Tockner et al. 2000). Flooding or peak flows may occur predictably, such as following monsoon rains or during the spring melt of winter snowpacks atop mountains, causing rivers to swell and rage. Only peak flows that occur outside the norm (±2 standard deviations of the mean discharge for that time of the year) might then be considered unusual in rivers that experience seasonal flooding (Resh et al. 1988; **Figure 3.24A**). This is not to suggest that seasonal flood or flow pulses are not disturbances, however, because even predictable disturbances are still a perturbation to the system (Poff 1992). Nevertheless, defining disturbance with respect to the norm might then enable us to identify extreme

flow events—whether river discharge is unusually high or low—as discussed previously in regards to what qualifies as a large infrequent disturbance (LID).

Flooding is just one extreme of the **natural flow regime**, which describes the pattern of variability in water flow (streamflow or discharge) over time and is characterized in terms of the magnitude, frequency, duration, timing, and rate of change in water discharge levels (Poff et al. 1997; **Table 3.4**). Different types of

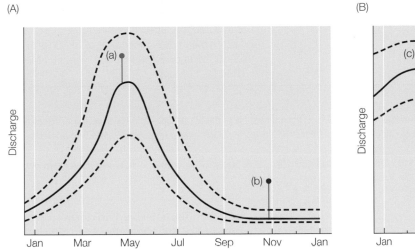

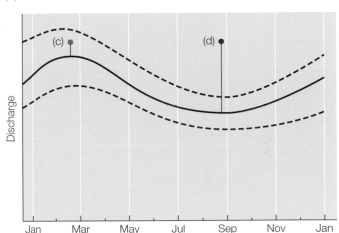

Figure 3.24 Disturbance in riverine landscapes. For montane rivers (A) that experience dramatic flood or flow pulses seasonally, elevated discharge levels may actually constitute less of a disturbance than in rivers that maintain more constant flow rates throughout the year (B). If a disturbance is defined in terms of whether the discharge level occurs outside the norm for that time of year (i.e. beyond the 95% confidence interval; dashed lines), then unusually high (or low) flow rates that occur out-of-season, such as (b) and (d), would qualify as significant disturbance events in these two systems. Note, however, that the same high discharge level, (a) and (d), would either qualify as a disturbance (d) or not (a), depending on the natural flow dynamics of the river.

Source: After Resh et al. (1988).

TABLE 3.4 Components of the natural flow regime used to describe the fluvial disturbance dynamics of riverine systems.

Parameter	Description	Example*
Magnitude of discharge	The volume of water that moves past a fixed location (e.g. a stream gauging station) over a period of time; typically quantified as either a near-instantaneous flow rate (ft^3/s or m^3/s) or volume over a longer time period (e.g. acre-feet/yr). Used to assess high and low flows (i.e. maximum and minimum magnitudes), particularly in regards to the potential for flood or drought conditions	The largest peak discharge (streamflow) of 69,800 ft^3/s was observed at this gauge station on 27 May 1984; the lowest peak streamflow of 5080 ft^3/s was observed on 10 June 1977
Frequency of occurrence	How often the flow is above (or below) a given magnitude within a particular time period. The frequency of occurrence is inversely related to the magnitude of the flow (e.g. the likelihood of a 100-year flood occurring in any given year is 1/100 = 0.01)	During the past six decades (1951–2012), streamflow at this site exceeded the median annual flow rate (5683 ft^3/s) in about half of those years (30/61 years = 0.49)
Duration	Period of time associated with a specific flow condition (e.g. number of days streamflow exceeds a particular flow rate). A pulse disturbance is a short-term event (flood, drought), whereas a press disturbance maintains a certain flow condition over of a longer period of time. A ramp refers to steadily increasing or decreasing flows (**Figure 3.25**)	In 1984, the mean daily streamflow at this site exceeded 60,000 ft^3/s for four consecutive days (25–28 May), which coincided with the seasonal flood-pulse following melting of the winter snowpack at higher elevations

(Continued)

TABLE 3.4 Continued

Parameter	Description	Example*
Timing or predictability	The regularity with which flows of a defined magnitude occur	Peak streamflows at this site occur at the height of the spring melt of the winter snowpack in the Colorado Rocky Mountains and from other high mountains within the drainage area, sometime in late May or early June
Rate of change or flashiness	How quickly the streamflow changes from one magnitude to another (e.g. from low to high flow); streams that change rapidly are termed 'flashy'	At the start of the spring melt following a record winter snowpack (~160% above average) in the Colorado and Gunnison River Basins (the drainage areas that most directly affect flows at this gauge)[1], streamflow increased six-fold (from a mean daily discharge of 8800 to 53,400 ft³/s) over the first two weeks of May 1984 (1–17 May)

Sources: Poff et al. (1997); Lake (2000)

*The example given to illustrate the application of these concepts to regime flow is from a section of the upper Colorado River near the Colorado–Utah state line (USA). Data obtained from USGS streamgauge station number 09163500: http://nwis.waterdata.usgs.gov/nwis/nwisman/?site_no=09163500&agency_cd=USGS).

[1]Data on the annual basin-wide percentages of snowpack in Colorado obtained from the USDA Natural Resources Conservation Service (www.nrcs.usda.gov)

(A) Pulse disturbances

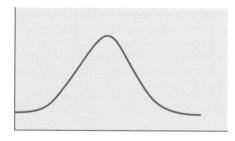

(B) Press disturbances

Strength of disturbing force

(C) Ramp disturbances

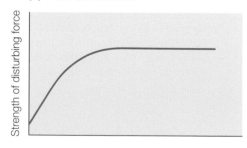

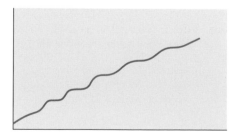

disturbances may be characterized in terms of how quickly they cause water discharge levels to change, as well as how long those changes persist. **Pulses** represent an abrupt short-term change in streamflow (e.g. a flood pulse); **presses** may occur suddenly but persist for an extended period of time (e.g. sediment loads following a landslide or after a catastrophic fire in a forested watershed); and **ramps** involve a steady increase in the disturbance over time (**Figure 3.25**; Lake 2000). For example, drought represents this sort of 'creeping disaster' that gets worse over time. Other examples of ramps include the steady accumulation of sediments as a forested watershed is logged, or the incremental spread of a non-native species that has adverse effects on the ecosystem.

Because flows vary over space as well as time, riverine landscapes can be viewed as shifting patch mosaics in much the same way as terrestrial landscapes are (Pringle et al. 1988; Townsend 1989; Lake 2000; Fausch et al. 2002; Winemiller et al. 2010). Flow dynamics contribute to patch-disturbance dynamics, from the fine-scale disturbances associated with normal flows ('the turning of a stone, the loss of a leaf

Figure 3.25 Three types of river disturbance. (A) **Pulse disturbances** are those in which the magnitude or intensity of the disturbance peaks (or drops) for brief periods, which may occur predictably (e.g. seasonal or annual flood pulses). (B) **Press disturbances** increase (or decrease) abruptly, but are sustained over a long period of time. (C) **Ramp disturbances** gradually increase (or decrease) over time (e.g. a drought that gets worse over time).

Source: After Lake (2000).

pack, and the removal of individuals'; Townsend 1989, p. 44) to major disturbances associated with extreme flow events involving either drought or flash floods (spates). In the case of spates, the river channel is scoured and loses much of its biota, not unlike the effect of a landslide on a hillside. Nevertheless, recovery after even an extreme flow event can be fairly rapid if there exist spatial refugia, such as deep pools in drought-stricken rivers or along banks and other backwater areas during a spate. Organisms may survive within such refugia until normal flows are restored and they are able to recolonize newly available habitats (Townsend 1989; Lake 2000).

By contributing to the redistribution of sediments, woody debris, nutrients, and organisms, the natural flow regime is ultimately responsible for much of the heterogeneity found within rivers (**Figure 3.7**), which in turn exerts an influence on practically every ecological process, from nutrient cycling to food-web dynamics, as well as the nature of interactions and flows that occur among elements of the riverine landscape (e.g. river–floodplain exchanges; Pringle et al. 1988). Flow therefore acts as a 'master variable' in terms of its influence on the ecological integrity of river systems (Poff et al. 1997).

Rivers in their natural, free-flow state thus possess a great degree of heterogeneity and connectivity (**Figure 3.7**). Most (~60%) of the world's large river systems are no longer free-flowing, however, but are fragmented and regulated by dams (Nilsson et al. 2005). There are more than 48,000 large (>15 m high) dams worldwide, regulating river flows in order to meet our ever-increasing water, irrigation, and energy needs. In the USA, Graf (2006) found that very large dams (those capable of storing 1.2 km^3 or more of water) reduced the annual peak discharge by an average of 67%, and up to 90% in the case of some rivers. Dams not only regulate flows, but can also hold back one or more years of discharge in those designed for storage. From this perspective, the most highly regulated rivers in the world are the Volta River in Africa (428% of the annual discharge held back); the Colorado River in the USA (>250%); the Rio Negro in Argentina (140%); and, the Tigris-Euphrates river system in the Middle East (124%) (Nilsson et al. 2005).

Dams are engineered to regulate water flows, and thus they prevent or significantly reduce seasonal or periodic flood pulses. Although flooding is considered by most people to be a destructive force to be prevented at all costs, it is also a constructive force that plays an important role in shaping the geomorphology of landscapes and contributes to the habitat diversity and productivity of riverine systems and their floodplains. As we mentioned in our earlier discussion of landscape evolution and the formation of river networks, flood pulses contribute to the erosional mechanics that help shape riverine landscapes and give rise to the different sorts of geomorphic structures found within these systems, including depositional features such as floodplains and river deltas (**Figures 3.7, 3.14, and Table 3.3**).

Apart from the river itself, the habitats most impacted by the alteration of the natural flow regime are riparian areas and floodplains. 'Riparian' refers to the area bordering a river, stream, lake, or other wetland. Riparian areas are thus an example of an ecological transition zone, occurring at the interface between adjacent terrestrial and aquatic systems (the land-water interface). Riparian areas share attributes of both systems, while possessing certain unique characteristics and dynamics that emerge as a consequence of the interactions and flows between them (Naiman & Décamps 1997). The riparian zone encompasses a diverse array of habitats, such as riparian forests (including gallery forests, or bosques), marginal wetlands, and riparian floodplains, and thus can be difficult to define or delineate precisely. The riparian zone is generally defined as the region of the stream channel between the low- and high-water marks of a river and that area of the surrounding landscape affected by flooding or elevated water tables (i.e. the floodplain; Naiman & Décamps 1997). Thus, the width of the riparian zone varies according to the size of the stream, its position within the drainage network, the hydrologic regime, and the local geomorphology, in addition to various ecological factors (e.g. extent of riparian vegetation). Because of high seasonal and annual variation in streamflow, the riparian zone can be an unusually harsh environment for the establishment of plant and animal species. For example, in the case of plants, Naiman and Décamps (1997, p. 624) note that:

> Seasonal variations in discharge and wetted areas create environmental conditions that challenge even the most tolerant species. Nearly every year, most riparian plants are subjected to floods, erosion, abrasion, drought, freezing, and occasionally toxic concentrations of ammonia in addition to the normal biotic challenges; the life-history strategies of most riparian plants are such that extreme conditions are either endured, resisted, or avoided.

Besides providing primary habitat for many species, riparian areas contribute to a number of other important ecosystem functions and services. Riparian areas may function as: ecological corridors that facilitate dispersal or migration of species—including invasive ones—across the landscape; a major source of allochthonous inputs (e.g. leaf litter, woody debris, terrestrial insects) into aquatic systems, thereby contributing to their productivity (**Figure 5.18**); filtration or buffering systems that trap sediments and take up nutrients, including pollutants such as nitrates and phosphorus commonly found in agricultural runoff; and as refugia for species during prolonged dry periods, thereby helping to maintain landscape and regional diversity (Naiman & Décamps 1997). Riparian areas are particularly sensitive to variation in the hydrological cycle, and thus serve as good indicators of environmental changes caused by alteration of the natural flow regime, such as by dams (Nilsson & Berggren 2000).

Riparian and floodplain areas are also important from an economic standpoint. For example, river floodplains are among the most productive and diverse ecosystems on Earth (Tockner & Stanford 2002). The global value of the ecosystem services provided by floodplains has been estimated at around $4 trillion/year (about $6.3 trillion/year in 2019 dollars), which is about 20 times the area-based value ($/ha/year) of forests and more than 200 times that of cropland (Costanza et al. 1997). Although floodplains cover only 1.4% of Earth's land surface, they contribute 25% of all terrestrial ecosystem services, especially in terms of provisioning water, natural flood regulation, and waste-treatment services to human communities (Tockner & Stanford 2002). Thus, floodplains are also among the most highly developed and densely settled regions on the planet, which is why flood damages have increased in recent years and will continue to do so if flooding risk increases as projected in response to climate change. Not surprisingly, then, floodplains also rank among the most altered landscapes in the world, resulting in either the complete loss or degradation of these systems through the disruption and control of the flow regime (Tockner & Stanford 2002). Because hydrology is the single most important driver of floodplain dynamics, changes to the flow regime can significantly alter the extent, duration, and frequency of floodplain inundation, which has some far-reaching ecological consequences. To illustrate, we consider the plight of the Colorado River in the southwestern USA and its delta where it meets the Gulf of California in Mexico.

CASE STUDY   The Colorado River and Its Delta

As pointed out in the introduction to this chapter, the natural flow of the Colorado River has been seriously disrupted, such that in most years the river no longer reaches its delta in the Gulf of California (**Figure 3.26**). From its humble beginnings as a small mountain stream arising in a wet meadow along the western divide of the Colorado Rockies, the Colorado River travels 2300 km on its journey to the sea, carving numerous gorges and canyons along its length, including the famed Grand Canyon (**Figure 3.1**). The Colorado River and its tributaries drain an enormous watershed that encompasses much of the extremely arid southwestern USA, in addition to portions of northwestern Mexico (**Figure 3.26**). The Colorado has a very long history of water diversion, as evidenced by the ancient irrigation canals constructed by the indigenous Hohokam some 1400 years ago (**Box 3.1**). Today, almost all of the water (99%) in the Colorado River is diverted to supply the water needs of more than 30 million people, to irrigate some 4 million acres (1.6 million ha) of farmland, and to provide hydroelectric power on both sides of the USA–Mexico border (Gerlak et al. 2013). The Colorado is thus one of the most regulated and controlled rivers in the world (Nilsson et al. 2005), thanks to several hundred km of canals[11] and ten major storage dams, including the Hoover[12] and Glen Canyon Dams, in addition to dozens of other smaller dams that have been constructed along its length.

> The number-one problem facing the Colorado River Delta today is insufficient water supply.   Gerlak et al. (2013)

> Dams are ubiquitous on rivers in the USA, and large dams and storage reservoirs are the hallmark of western U.S. riverscapes.   Sabo et al. (2012)

The Colorado River Delta once encompassed about two million acres (8100 km²) of floodplain forest and wetlands, which in turn supported a 'legendary richness' of birds, fish, and other wildlife (Luecke et al. 1999). In his book *A Sand County Almanac*, Aldo Leopold (1949) recounted with great nostalgia a canoe trip he took with his brother through the Colorado River Delta in 1922.[13] He described it as a vast

---

[11] Including the All-American Canal, one of the longest irrigation canals in the world and which travels more than 130 km to water the Imperial Valley of California, a major agricultural production center (**Figure 3.26**).

[12] The Hoover Dam is considered one of the Seven Wonders of the Industrial World (Cadbury 2004). The Grand Canyon, which is just upstream of the Hoover Dam, is generally considered one of the Seven Wonders of the Natural World. The juxtaposition of these two Wonders of the World is certainly not without irony.

[13] It is worth noting that the Leopold brothers embarked on their journey in the same year—1922—that the Colorado River Compact was signed. The Compact governs the allocation of water from the Colorado River among the seven US states within the Colorado River Basin, which facilitated the rapid development of state and federal water works projects under the US Bureau of Reclamation, including the construction of Hoover Dam (completed in 1936). For 6 years, virtually no fresh water

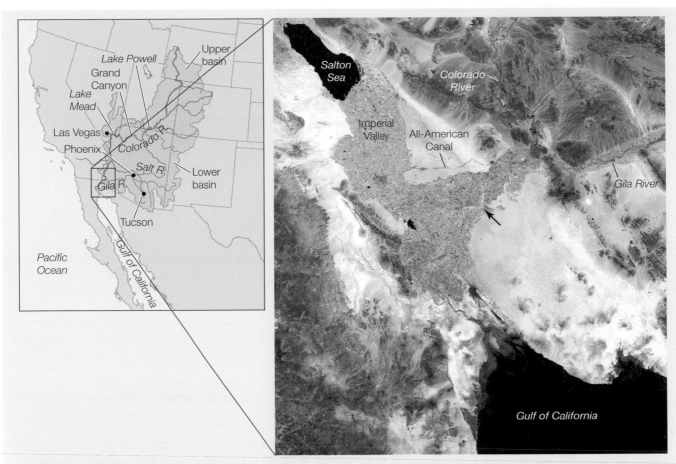

Figure 3.26 The Colorado River Basin and Delta.   The Colorado River and its major tributaries (left) historically drained into the Gulf of California (inset). Almost all of the water flow is now diverted or held behind storage dams (e.g. Hoover and Glen Canyon Dams) to supply municipal water and hydroelectric power to more than 30 million people, and irrigation for 4 million acres of agricultural land in the desert (lime green areas on Landsat image, right). The end result is that the Colorado River no longer reaches its floodplain delta in most years (the black arrow points to a section that ran dry during a drought in May 1990), disrupting the connectivity and ecology of one of the largest desert estuaries in the world (right).

Source: Image used with permission from Alejandro Hinojosa, CICESE (Center for Scientific Research and Higher Education) in Ensenada, Baja California, Mexico.

wilderness, full of green lagoons and twisting channels lined with gallery forests of cottonwood (*Populus* spp.) and willows (*Salix* spp.), and thick mesquite bosques (*Prosopis* spp.). Once considered among the most productive deltas in the world, the Colorado River Delta today covers less than 10% of its former extent, and much of what remains is now fed by agricultural wastewater and other effluent (**Figure 3.26**; Gerlak et al. 2013). In recognition of its global ecological importance, as well as its need for protection, the Colorado River Delta was made an UNESCO Biosphere Reserve in 1993 (the Alto Golfo de California Reserve in Mexico), and a Ramsar site by the Convention on Wetlands of International Importance (the Ramsar Convention) in 1996 (Humedales del Delta del Río Colorado, Mexico). Although the disruption of water flow is largely to blame

reached the Delta as Lake Mead filled behind Hoover Dam (Luecke et al. 1999). Thus, the Delta was already in decline by the time *A Sand County Almanac* was published in 1949.

for the severe contraction of the Colorado River Delta over the past 80 years, altering the flow regime of the Colorado River has had a number of other effects that are also contributing to the loss and degradation of the Delta. The Delta was formed through the deposition of river-borne sediments; prior to being dammed, the Colorado River carried some 135 million tons of sediment and silt to the Gulf annually (Milliman & Meade 1983). The nature of this seasonal flood-pulse, and its hydromorphic and geomorphic effects on the Colorado River, is vividly described by Cohn (2001, p. 999):

Before people started tinkering with it, the Colorado often flooded in spring and early summer, when rain and runoff from melting snow turned the river into a raging giant. The floods swept away salts, plants, and ground litter. In the process, they scoured old sandbars and beaches and laid down new ones; created backwater marshes, oxbows, and sloughs; and raised groundwater levels. As summer turned into fall and winter, the river gradually shrank, in some

*years to a mere trickle. How much water flowed varied,
often widely, from season to season and year to year. Those
days are gone….By preventing the Colorado from flooding,
the dams and reservoirs have altered the river. With its
more uniform, year-round flow, no floods now sweep down
the river. No longer are salts or plant debris washed away,
nor are any new backwaters created or old ones recharged.
The dams have also changed the Colorado from a warm-
to a cold-water river, held back sediment that would
otherwise create new sandbars, and kept nearly all
freshwater from reaching the delta in Mexico.*

Depending on one's perspective, the fact that the Colorado River no longer experiences such mood swings might be considered a good thing. However, from an ecological standpoint, the long-term consequences of altered water flows along the Colorado River have been catastrophic. Very little sediment is now being deposited into the Delta, such that erosion by tidal flows from the Gulf exceeds the rate of accretion by sediments from the Colorado, a highly unusual condition for a river delta (Luecke et al. 1999). The lack of freshwater flowing into the Delta has substantially increased salinity, making the brackish waters uninhabitable for a number of estuarine species, including the critically endangered totoaba (*Totoaba macdonaldi*), a large fish (up to 2 m long and weighing 100 kg) native to the upper Gulf, and the endemic vaquita porpoise (*Phocoena sinus*), which is the most endangered cetacean in the world and may already be extinct by the time you read this.

The riparian areas of the Delta that provide critical habitat for migrating and nesting birds have also been reduced to as little as 1.5% of their original extent (Cohn 2001). Not only have riparian habitats been cut off hydrologically from the main river channel, but their soils have become highly salinized in the absence of periodic flooding, which helped to flush salts from the soil. These highly saline soils are toxic to the native cottonwoods and willows, but are well-tolerated by the non-native tamarisk (saltcedar, *Tamarix* spp.), which has come to dominate waterways throughout the southwestern USA (Merritt & Shafroth 2012).[14] In addition to being salt-tolerant, tamarisk also owes its success to a

highly prolific nature and deep root system, which enables it to access a lower water table beyond the reach of the native cottonwoods and willows (Glenn & Nagler 2005).

In the absence or declining availability of native riparian habitat, many breeding birds are now using tamarisk habitat, although it is unclear whether tamarisk affords high-quality habitat for any of these species (Sogge et al. 2008). Still, this presents a restoration dilemma. Although the removal or control of tamarisk is viewed as integral to the restoration of riparian areas along the Colorado River and its Delta, this needs to be done advisedly, in view of the fact that, in some areas, tamarisk may provide the only riparian habitat left for breeding birds. Regardless, restoration of native riparian habitat in the lower Colorado River and its delta region will require far more than simply removing tamarisk and replanting with native cottonwoods and willows; it will also require a concomitant restoration of natural disturbance processes, such as periodic flooding (Cohn 2001; Merritt & Shafroth 2012).

Although flooding is usually considered a type of natural or abiotic disturbance, it is actually the *absence* of flooding that is a disturbance in these types of river systems. While a return to a natural flow regime is not an option for the Colorado River or its delta, clearly more water is needed. It has been estimated that at least 50,000 acre-feet (61 million m³) of water would be needed annually, along with the occasional flood (260,000 acre-feet, or 320 million m³, every 4 years) to sustain and restore riparian forests and wetlands within the delta (Luecke et al. 1999). In November 2012, the USA and Mexico agreed to a historic 5-year water-sharing agreement (the 'Minute 319' pact[15]), which guarantees that the Colorado River will once again flow to the Gulf (Gerlak et al. 2013). Under the agreement, the two countries aim to provide 52,696 acre-feet (65 million m³) of base flow annually to the Colorado River Delta, in addition to a one-time pulse flow of 105,392 acre-feet (130 million m³). On 23 March 2014, the pulse flow was initiated by opening the gates of the Morelos Dam on the USA–Mexico border, and nearly 2 months later, the Colorado River made its way to the Delta for the first time in many years. Although the pulse flow amounted to less than 1% of the river's pre-dam annual flow, it is nevertheless a hopeful sign that such short-term water releases could permit the restoration of riparian and marshland habitat in the Delta.

[14] In fact, tamarisk may be able to facilitate its own invasion and dominance within these riparian areas, given that it concentrates and exudes salts from its leaves, which end up littering the ground, thereby increasing soil salinity. This apparently makes more of a difference in facilitating tamarisk invasion along the upper Colorado River, however, as soil salinity within the Delta region is already so high due to low water flows and a lack of flooding (Merritt & Shafroth 2012).

[15] https://www.ibwc.gov/Files/Minutes/Minute_319.pdf

Large river deltas the world over—including those of the Nile, Ganges, Indus, Niger, and Mekong—face many of the same threats as the Colorado River Delta. Although anthropogenic soil erosion has increased sediment transport in rivers by an estimated 2.3 billion metric tons annually, less than half of that sediment actually reaches the ocean (only about 0.9 billion metric tons/year), largely because of dams and reservoirs. Over 100 billion metric tons of sediment (and 1–3 billion metric tons of carbon) are now sequestered in reservoirs, most of which were constructed in the past 50 years (Syvitski et al. 2005). The reduction in sediment transport is contributing to coastal retreat in delta regions, where a large fraction of the human population lives, and this at a time when climate-driven rises in sea level also threaten these low-elevation landscapes. Hurricanes (tropical cyclones), which are also projected to increase in frequency and intensity in response to climate change, may further increase erosion rates in some large delta regions (Bianchi & Allison 2009). We consider the effects of hurricanes and storm surges on coastal landscapes next.

Windstorms, Hurricanes, and Storm Surges

High winds are associated with a variety of meteorological phenomena that encompass windstorms (derechos), tornadoes, or hurricanes (tropical cyclones or typhoons), depending on wind speed and formation. As a natural agent of disturbance, wind is particularly important in contributing to the dynamics and heterogeneity of forested landscapes (Mitchell 2013). Wind may serve as a chronic low-level disturbance that causes adaptive or acclimative growth, resulting in thickening and strengthening of stems and roots that increase the tree's resistance to wind damage. For example, trees exposed to chronic high winds atop mountains may become twisted and stunted, a growth pattern known as Krummholz. More intense windstorms can strip the leaves off trees, break off limbs, and snap or topple trees (**windthrow**).

> Windthrow is all too often looked at as an exceptional, catastrophic phenomenon rather than a recurrent driver of ecosystem patterns and processes that falls within the spectrum of chronic and acute effects of wind on forests.
>
> Mitchell (2013)

The extent of wind damage generally increases with wind intensity, and may be expressed across a range of scales. At an individual tree level, windthrow can affect stand and soil developmental trajectories as a result of uprooting, which helps to invert and expose deeper soils, thereby facilitating pedogenesis and nutrient cycling (Mitchell 2013). Uprooting also creates a pit where the tree once stood, which may persist for centuries after the windthrow event, and an adjacent mound from the soil that slumps off the upended root mass, which introduces heterogeneity at a mircosite level (pit and mound microtopography; Mitchell 2013). At a stand level, windthrow creates gaps in the canopy in which **secondary succession** occurs, resulting in a shifting mosaic of disturbance and recovery that increases heterogeneity across the landscape (i.e. **gap dynamics**). Wind damage is also greatest along habitat edges, particularly those resulting from human land

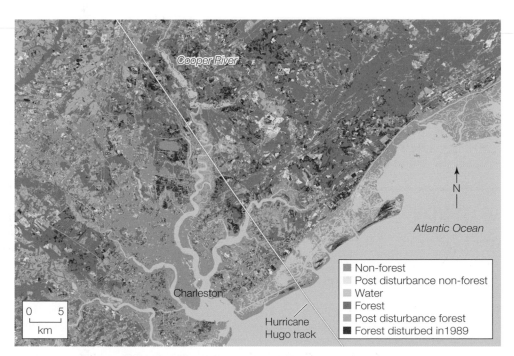

Figure 3.27 Hurricane disturbance mosaic.   Despite their intensity, hurricanes create a heterogeneous disturbance mosaic, as evidenced by the pattern of forest disturbance created by Hurricane Hugo, which made landfall near Charleston, South Carolina (USA) as a Category 4 hurricane on 21 September 1989. Most of the disturbance (in red) is concentrated on the right side of the hurricane's track. In the northern hemisphere, wind speeds are greatest near the eyewall in the right-front quadrant along the hurricane's track, owing to the hurricane's counterclockwise motion and northerly direction of travel. Forests also show signs of prior disturbance from fire and logging (light green areas), and thus these areas were less affected by the hurricane.

Source: NASA Eart Observatory image by Robert Simmon, based on data from Chad Rittenhouse, Department of Forest and Wildlife Ecology, University of Wisconsin-Madison.

use that contributes to the clearing and fragmentation of forested landscapes (**Box 7.1**). The severity and pattern of wind damage across the landscape is a function not only of the intensity, duration, and frequency of windstorms, but also depends on topography, site exposure, the type and depth of soil in which trees are anchored, the structure and composition of the forest or stand, and its past disturbance history (Foster & Boose 1992; Boose et al. 1994; Mitchell 2013). In the case of extreme wind events, such as hurricanes, there are gradients of wind speed across the hurricane's path, which in turn produces a heterogeneous pattern of damage across the landscape (Busby et al. 2008; **Figure 3.27**).

Hurricanes (also known as tropical cyclones or typhoons, depending where you are in the world) bring extreme winds (>118 kph, or >74 mph), torrential rain, and storm surges, and thus pose a multifaceted threat to coastal areas around the world. That threat is compounded by the fact that human population densities tend to be highest along the coast (Michener et al. 1997). In addition, global warming is predicted to produce hurricanes of greater intensity in the coming decades, as well as rising sea levels, which will increase the hazard risk from storm surges—one of the deadlier consequences of hurricane landfall (Bender et al. 2010; Knutson et al. 2010; Lin et al. 2012). The deadly reach of hurricanes is not limited to coastal areas, however. Although hurricanes begin to weaken once they make landfall, the path of the storm can travel a hundred miles (160 km) or more inland. For example, half of the hurricane fatalities in the USA since 1970 occurred inland along the main storm path, mostly (75%) due to drowning deaths in the floodwaters of rivers and streams (a consequence of high rainfall) and wind-induced injuries (Czajkowski et al. 2011).[16] Prior to 1970, however, 90% of hurricane fatalities in the USA were from storm-surge drowning (American Meteorological Society 1973). Clearly, more advanced early-warning systems and the evacuation of coastal areas ahead of hurricane landfall have been successful in mitigating fatalities due to storm-surge drowning. Besides being deadly, hurricanes are also costly and damaging to property and infrastructure, with damage costs and insurance losses likely to increase over the coming decades.

Hurricanes also have devastating impacts on ecological systems. Storm surges and waves erode coastlines and dunes, inundate estuaries and coastal wetlands, and batter coral reefs (Michener et al. 1997). Hurricane-force winds lay waste to forests, and the accompanying torrential rains can cause flooding and landsliding far inland, as discussed previously. Because of the multifaceted nature of hurricane impacts on ecological systems, we consider the effects of hurricanes on two types of coastal systems—tropical forests and coral reefs—both of which are found in the Caribbean and are thus subjected to hurricanes on a relatively frequent basis.

[16] Hurricane-related fatalities along coastal areas in the United States have generally decreased since 1970, with the major exception of Hurricane Katrina, which made landfall (for the second time) as a Category 3 hurricane near New Orleans, Louisiana on 29 August 2005. The resulting storm surge caused a catastrophic failure of the city's levees and floodwalls, producing widespread flooding up to 5.6–7.6 m (15–25-feet) deep throughout much of the city of New Orleans (Andersen et al. 2007). Approximately 1800 deaths were attributed to Katrina, exceeding all other hurricane-related fatalities incurred in the USA between 1970 and 2007 by an order of magnitude. The atypicality of the event, which had societal as well as natural causes, is likely why Hurricane Katrina was omitted from the Czajkowski et al. (2011) comparative analysis of hurricane-related fatalities in coastal versus inland areas.

CASE STUDY   Hurricane Impacts on Coral Reefs and Tropical Forests in the Caribbean

Coral reefs are often called the 'rainforests of the sea,' which suggests certain parallels between the two systems. Both systems are structurally complex, productive environments that support high levels of diversity. Both types of systems can also be found in tropical areas, such as the Caribbean, where hurricanes and tropical storms are a common and recurring disturbance. Over the past 150 years, there have been at least 1000 tropical storms and over 200 Category 2–5 hurricanes to hit the Caribbean (Scheffers et al. 2009). The average return-time for hurricanes in this region is about 9–12 years (Gardner et al. 2005). Subsequently, the impacts of severe storms on tropical forests and coral reefs have been studied more thoroughly in the Caribbean than anywhere else in the world (Lugo et al. 2000). It thus makes sense to consider the dual impacts of severe storms on both coral reefs and rainforests (or tropical forests, more generally), since they frequently co-occur in tropical regions throughout the world.

On coral reefs, the relation between disturbance and diversity is similar to that in tropical forests.   Connell (1978)

Despite these parallels, tropical forests and coral reefs experience hurricanes in very different ways. Coral reefs are impacted by the wind-driven waves generated by the hurricane, whereas tropical forests are primarily impacted by the wind itself. This distinction is important when viewed in terms of the relative amount of energy that is dissipated during a hurricane (i.e. its intensity; **Table 3.2**). Given that waves represent the work of the wind over a large area, they are more powerful than the wind itself, by up to several orders of magnitude (Lugo et al. 2000). Thus, coral reefs are generally subjected to greater stress forces during a hurricane than tropical forests, and likely for a longer period of time, as wave action builds well ahead of the hurricane's arrival and subsides days after its departure. Although coral reefs are essentially autoengineered to withstand wave action, a hurricane is an extreme disturbance event and is therefore capable of causing significant damage to the reef. The wave energy generated by hurricanes is capable of smashing and breaking apart coral structures and reducing reefs to rubble, which may then be scattered about over a large area or eventually washed ashore as beach deposits.

As with any disturbance, however, the impact of hurricanes on coral reefs and tropical forests ultimately depends on the intensity and frequency of occurrence (recall that these were the very systems—albeit in northeast Australia rather than the Caribbean—that inspired Connell's formulation of the Intermediate Disturbance Hypothesis; Connell 1978). And, like other broad-scale disturbances described in this chapter, the damage caused by a hurricane may be variable, devastating some areas completely while leaving other areas relatively undamaged and seemingly undisturbed (Hughes & Connell 1999; Lugo et al. 2000). Such variability in the pattern of disturbance may be due to the position or location of the site in relation to the hurricane, its initial structure and composition, local topography, size or depth (in the case of corals), time since the last major hurricane event, as well as interactions with other types of disturbances.

In the Caribbean, the average hurricane reduces coral cover by about 17%, although the magnitude of this loss increases with hurricane intensity and with the time elapsed since the last impact (Gardner et al. 2005). Interestingly, the overall degree of hurricane damage in coral reefs is largely determined by the time interval between hurricanes (i.e. the return interval; **Table 3.2**). Despite their unpredictable nature, hurricanes are not truly random. In the Caribbean Sea, hurricanes exhibit a significant degree of temporal clustering, in which periods of high hurricane activity alternate with relatively quiet periods (Mumby et al. 2011). Temporal clustering of hurricanes may come about through the Atlantic Multidecadal Oscillation in sea-surface temperatures, which has been linked to long-term variability in hurricane activity (Chylek & Lesins 2008). Perhaps counterintuitively, a strongly clustered hurricane regime may actually allow reef systems to remain in a later successional state (i.e. with higher diversity) for a greater proportion of the time (Mumby et al. 2011). Although hurricanes have a devastating impact on coral reefs, if the next hurricane occurs before much recovery has taken place (as when hurricanes are clustered in time),

then the second hurricane will have far less of an impact because most of the damage was already done the first time around and few species will have had a chance to recolonize the reef in the interim. Thus, the degree of clustering of hurricane events may provide exactly the sort of intermediate disturbance that has been found to drive the disturbance-diversity relationship in corals. If hurricanes were not as clustered, severe storms would be more frequent and would maintain the system in a highly degraded (and depauperate) state. Conversely, if hurricanes are too infrequent, then the reef would rarely experience this type of severe disturbance that contributes to reorganization and helps to maximize diversity in this system.

In the aftermath of a hurricane, coral reefs may undergo a variety of different successional trajectories, but their ability to recover fully may now be compromised by the sheer number of stressors that are contributing to coral declines worldwide, such as warmer ocean temperatures, anthropogenic impacts (pollution, sedimentation), disease (coral bleaching), and invasive species. In the Caribbean, coral cover on reefs has decreased by about 80% in the past 30 years (Gardner et al. 2003). Thus, the impact of hurricanes on coral reefs must be assessed against a backdrop of ongoing declines, which is averaging −2.1%/year in the Caribbean. Unfortunately, coral reefs in the Caribbean no longer appear able to mount a full recovery after a hurricane, and instead are declining more quickly (−7%/year; Gardner et al. 2005). Although this may signify that hurricanes are interacting synergistically with other stressors (Hughes 1994), such accelerated declines in the Caribbean could also mean that the cumulative impacts of these disturbances have exceeded the threshold for recovery by coral reefs. The ability of corals to recover is a function not only of disturbance type and intensity, but also of duration. In a survey of coral reefs worldwide, corals were found to have recovered in 69% of cases involving acute short-term disturbances (e.g. a hurricane), but in only 27% of cases involving chronic long-term disturbances (e.g. disease, pollution; Connell 1997). Thus, the cumulative impacts of ongoing or long-lasting disturbances, most of which are directly or indirectly related to human impacts on the marine environment, may now be largely responsible for the regional decline of Caribbean corals (Connell 1997; Gardner et al. 2005), such that hurricanes no longer function as an extreme disturbance event, in terms of their effect on coral cover and diversity.

As with coral reefs, tropical forests experience a wide range of disturbance severity in response to hurricanes (Bellingham et al. 1995; Lugo et al. 2000; Ostertag et al. 2005). The severity of disturbance is influenced by numerous factors, including tree size, landscape position, and past disturbance history, in terms of whether the tree has suffered previous hurricane damage (Ostertag et al. 2005). For example, larger trees are more likely to be uprooted and suffer heavier damage than smaller trees; hurricane damage is likely to be greater on ridge tops, where wind exposure is greatest, than on hillslopes; and, trees previously damaged by hurricanes are more likely to be damaged even more severely during the next hurricane event (Ostertag et al. 2005).

Still, there are no absolutes, and the pattern of hurricane damage on the landscape tends to be complex and vary within, as well as among, forest types. For example, the severity of wind damage in some tropical forests may be wholly unrelated to tree size (Bellingham 1991; Bellingham et al. 1995). In others, tree damage may be less atop ridges if species are capable of developing tree 'unions' via root grafts among individual trees, giving them greater purchase on ridge tops, which helps to increase their resistance to high winds (Basnet et al. 1993). Hurricane damage on hillslopes may be mitigated by aspect if the slope is oriented away from the direction of the oncoming hurricane (Bellingham 1991). In addition, species-specific traits, such as tree architecture, wood density, and growth rates will also affect a tree's susceptibility to hurricane damage. For example, faster-growing tree species were found to experience greater hurricane damage than slower-growing species, which perhaps reflects certain trade-offs between growth rates and wood density or tree architecture (Ostertag et al. 2005). Thus, some tree species are more likely to incur hurricane damage than others, regardless of landscape position, in much the same way that some types of corals are more fragile and therefore more likely to be damaged by hurricane-driven waves (e.g. branching corals).

Such species-specific responses to hurricanes led Bellingham and his colleagues (1995) to identify four 'response syndromes,' defined in terms of the degree of hurricane damage a tree species generally incurred (their resistance to disturbance) and by their relative rates of growth, sprouting, and recruitment following the hurricane (their 'responsiveness' or capacity for regeneration and recovery; see also Batista & Platt 2003). Some tree species suffer severe damage and are slow to recover from the effects of a hurricane; these species are considered highly **susceptible** to hurricane damage, and include the tree fern *Cyathea pubescens* (Bellingham et al. 1995). Species that experience high damage but rapid recovery following a hurricane, such as the cigarbush (*Hedyosmum aborescens*), are considered to be **resilient**. However, many tropical trees appear to be **resistant** to the effects of hurricanes, in that they suffer relatively little damage but are slow to recover from whatever damage they do incur (Bellingham et al. 1995). Finally, there are species like the achiotillo (*Alchornea latifolia*) that are both resistant to hurricane damage and recover quickly; these are the **usurpers**, which also characterize some introduced invasive species, such as the mock orange (*Pittosporum undulatum*) that is not native to the Caribbean (Bellingham et al. 1995). If hurricane frequency or intensity does increase as predicted due to climate change, then the composition of these tropical forests will likely change in favor of species that are more resistant to damage (i.e. resistant or usurper species; Ostertag et al. 2005). Indeed, an exposed forest atop the Blue Mountains of Jamaica was found to be dominated by hurricane-resistant species (Tanner & Bellingham 2006). Furthermore, the spread of the invasive mock orange (a usurper) across Jamaica appears to have been favored by past hurricane disturbance (Bellingham et al. 2005).

Coral reefs and tropical forests of the Caribbean have thus been shaped in large part by the hurricane disturbance regime. As we have seen, hurricanes create a shifting mosaic of disturbance and recovery within both systems (i.e. gap dynamics). The mechanism by which species persistence is achieved involves resistance, rapid repair, or both (Lugo et al. 2000). As trees or corals are damaged and die, gaps are created within the canopy or reef, which then allows for the recruitment of new species that may subsequently alter the structure and composition of the forest or reef (i.e. secondary succession). Although some aspects of recovery may be fairly rapid, others may require a much longer time interval. For example, defoliated trees may recover within a few months of the storm, but tree mortality may remain elevated for years after a hurricane, as trees slowly succumb to their injuries (Ostertag et al. 2005). Lagged responses to hurricane disturbance are thus to be expected, such that landscapes and reefs inevitably bear the legacy of past hurricane disturbances. To quote Lugo and his colleagues (2000: p. 112) regarding the recovery of tropical forests following a hurricane:

> [T]he path to recovery can lead to innovation and surprise as species importance values change, new species enter at the patch scale, and others exit patches. Full recovery may take as long as 60 years, and the timing of the recovery varies with the forest component or function under consideration.

The same may be said of coral reef recovery. Unfortunately, the long-term recovery of these systems may now be compromised by human actions that create chronic disturbances, such as pollution and overfishing on reefs, or the clearing of tropical forests, which cause dramatic changes in the structure and species composition of these systems. Ultimately, these sorts of changes may be less reversible than those resulting from hurricanes, especially if they end up compromising recovery and affect the ability of these systems to resist future storm damage (Lugo et al. 2000).

The mitigation of damage and human fatalities from storm surge and hurricanes also provides strong incentive for protecting or restoring coastal habitats, such as marshes, dunes, mangrove forests, seagrass beds, and coral reefs. For example, coral reefs can reduce wave energy onshore by up to 90% (Lugo-Fernandez et al. 1998), and coastal wetlands can attenuate surge levels on the order of 1 m for every 4–60 km of intact habitat (Wamsley et al. 2010). In the Caribbean, a multinational insurance facility found that the restoration of reefs and mangrove forests was among the most cost-effective means of protecting these island communities from flooding and wind damage, far beyond human-engineered solutions such as sea walls and other flood-control measures (CCRIF 2011). In the USA, 67% of the coastline is currently protected by natural habitats, but 16% of the total coastline is nevertheless considered at high risk of hazards from storms and sea-level rise (Arkema et al. 2013). More than 1.3 million people live in these high-risk areas, but the loss of coastal habitats would double the amount of coastline exposed to flooding from storm surge and expose an additional 1.4 million people to these sorts of threats, particularly along the Gulf of Mexico and eastern seaboard (Arkema et al. 2013). Globally, the areas most vulnerable to rising sea levels and storm damage are found within the world's densely populated megadelta regions, low-lying coastal areas (especially in Asia and Africa), and among small islands in the Caribbean, Pacific, and Indian oceans (Nicholls et al. 2007, 2011). Populations within coastal areas (≤100 km) are projected to increase from some 2 billion to 5 billion people by 2080 (Nicholls et al. 2007), which would put a significant portion of the global population at risk and cause the displacement of up to 187 million people ('environmental refugees') unless protective measures are taken (Nicholls et al. 2011).

Drought

Extreme precipitation events include not just excessive rainfall and flooding, but also the opposite extremes of insufficient rainfall and protracted drought. Flood events tend to receive greater attention because they occur rapidly and have dramatic consequences that are clearly visible, whereas drought events develop more slowly and may go unnoticed for quite some time, which is why drought is often referred to as a 'creeping

disaster' (Van Loon 2015). Drought can have a variety of direct and indirect effects on both human and ecological systems, which makes drought an especially complex natural disturbance to understand and manage.

Drought is a sustained period of below-normal water availability. It is a recurring and worldwide phenomenon, with spatial and temporal characteristics that vary significantly from one region to another.   Tallaksen & Van Lanen (2004)

Drought is defined as a period of below-normal precipitation, but what is deemed 'normal' depends on what aspect of water use (agriculture, drinking water, hydropower, navigation) or the water cycle (precipitation, evapotranspiration, snow accumulation, soil moisture, streamflow, lake levels, groundwater) we consider. Thus, our assessment of drought severity depends on the type of water deficiency and its impacts. There are four major types of drought (Dai 2011; Van Loon 2015):

1. *Meteorological drought* involves a precipitation deficiency lasting months or years (**Figure 3.28**). Meteorological drought is caused by anomalies in precipitation produced by changes in atmospheric circulation patterns, which in turn are usually triggered by anomalous sea-surface temperatures (e.g. El Niño Southern Oscillation or ENSO). Meteorological drought can propagate through the hydrological cycle and give rise to other types of drought.

2. *Agricultural drought* focuses on soil water deficiencies that occur as a consequence of below-

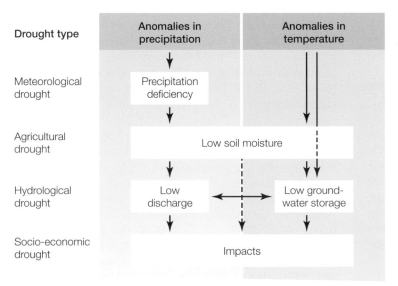

Figure 3.28 Drought types and factors contributing to their development.
Source: Van Loon (2015).

normal precipitation and higher temperatures (**Figure 3.28**), leading to reduced vegetation growth and crop production. In addition, dry soils and high temperatures contribute to reduced humidity and higher soil evaporation rates, which may create positive feedbacks at a regional scale (i.e. the region gets progressively drier) that further exacerbate the drought, much as happened during the North American Dust Bowl (Cook et al. 2009; **Figure 3.20**).

3. *Hydrological drought* is defined by reduced groundwater recharge and streamflow (**Figure 3.28**). Hydrological drought is particularly insidious, for although it is also caused by below-normal precipitation, it lags behind other drought types. Within the hydrological cycle, groundwater is usually the last to be affected, such that only major meteorological droughts ('megadroughts') will result in significant groundwater deficits, following a lag time of months to years (Tallaksen & Van Lanen 2004). Groundwater may take years, centuries, or even millennia to recharge, which means that the effects of hydrological drought may persist long after precipitation patterns return to normal.

4. *Socioeconomic drought* is concerned with the economic and societal impacts of drought and drought-mitigation strategies (Tallaksen & Van Lanen 2004; **Figure 3.28**). Drought has the greatest economic impact on large populations, especially in industrialized countries, whereas the greatest societal impacts of drought are experienced in developing countries that lack the infrastructure to cope with drought (e.g. irrigation; Tallaksen & Van Lanen 2004).

Droughts are defined by more than just a reduction in precipitation (Van Loon 2015). Like any disturbance, drought can be characterized by its intensity, spatial extent, and duration (Dai 2011). **Drought intensity** is assessed by the magnitude of the precipitation, soil moisture, or water-storage deficit. For example, the Palmer Drought Severity Index (PDSI) is widely used for the monitoring and analysis of drought conditions in the USA and elsewhere. The PDSI is based on precipitation, temperature, and soil moisture, and categorizes drought conditions across a gradient from extremely wet (PDSI > 4) to extreme drought (PDSI < −4). In addition to meteorological data, the assessment of drought intensity may also include the economic, environmental, or societal impacts, especially those involving agricultural losses, the depletion of water resources, and effects on human health. For example, the recent multiyear drought in California has been determined to be the most severe drought the region has experienced in some 1200 years (Griffin & Anchukaitis 2014).[17] In 2014 alone, the drought

is estimated to have cost the state $2.2 billion, primarily due to agricultural revenue losses (totaling $1.5 billion) and additional groundwater-pumping costs for crop irrigation (Howitt et al. 2014).

Drought severity is assessed not just in terms of intensity, but also by how widespread it is. Given a measure of drought intensity such as the PDSI, the **drought spatial extent** can be mapped and analyzed. The Dust Bowl of the 1930s was caused by a particularly severe drought in terms of both intensity and extent. At its worst, about 80% of the USA was under moderate-to-extreme drought (PDSI ≤ −2.0). For comparison, the most extensive drought to affect the USA since the 1930s occurred in 2012 and affected about 65% of the country. Based on its intensity and spatial extent, the 1934 drought thus stands out as the single-year benchmark for assessing drought severity in the USA, at least in the years since 1900 (Cook et al. 2010).

Severe drought is the greatest recurring natural disaster to strike North America. A remarkable network of centuries-long annual tree-ring chronologies has now allowed for the reconstruction of past drought over North America covering the past 1000 or more years in most regions. These reconstructions reveal the occurrence of past "megadroughts" of unprecedented severity and duration, ones that have never been experienced by modern societies in North America. There is strong archaeological evidence for the destabilizing influence of these past droughts on advanced agricultural societies, examples that should resonate today given the increasing vulnerability of modern water-based systems to relatively short-term droughts.   Cook et al. (2007)

In addition to its intensity and spatial extent, the 1934 drought was notable in another respect: it was the worst in a series of bad drought years that spanned the decade between 1929 and 1940. The duration of the drought that gave rise to the Dust Bowl is estimated to have lasted from 7 to 12 years, marking it as one of the longest droughts in the USA over the past 150 years (Cook et al. 2010). **Drought duration** is ultimately difficult to quantify, owing to the inherent spatial and temporal variability of drought conditions. Still, droughts of much longer duration have certainly occurred within the USA—and elsewhere—over the past millennium. For example, the southwestern USA experienced several megadroughts during the period spanning 900–1300, which may have contributed to the decline of the Ancestral Puebloans (Anasazi) in the Colorado Plateau (Cook et al. 2007), as well as to the collapse of the

---

[17] Climate scientists use a variety of means to reconstruct past climate (paleoclimate), including the use of climate proxies such as tree-ring data, from which fluctuations in precipitation

levels can be inferred from the thickness of annual growth rings (dendrochronology). These data complement more recent meteorological data obtained via instrumentation, and which are used to calculate drought indices such as the PDSI.

Hohokam, the ancient water engineers of the Salt River Valley in Arizona (**Box 3.1**). Furthermore, several decadal droughts of unusual severity in the central USA during the 14th and 15th centuries may have contributed to the decline of the Mississippian culture, leading to the abandonment of cities such as at Cahokia (Cook et al. 2007; **Figure 1.3**). Here we consider the role of drought, agricultural expansion, and deforestation in the decline and fall of civilizations with the case study of the Classic Maya in Mesoamerica.

BOX 3.1   Ancient Water Engineers of the Colorado River Basin

The Hohokam, who inhabited what is now south-central Arizona for more than 1500 years, began construction of an elaborate canal system some 1400 years ago (Bayman 2001). These canals drew water from the Salt and Gila Rivers, major tributaries of the Colorado River, and irrigated more than 110,000 acres (44,500 ha), turning the desert into productive farmland that may have supported 40,000–80,000 inhabitants at its zenith (Hill et al. 2004; Joseph 2010). The Hohokam's engineering feat is made all the more remarkable by the fact that these canals were dug entirely by hand, using digging sticks and stone hoes that lacked handles. In spite of this, the Hohokam irrigation system comprised more than 500 miles (800 km) of canals, some of which were 20 miles (32 km) long, 85 feet (26 m) wide and 20 feet (6 m) deep (Howard 1992).

The traces of this ancient irrigation system persist today, and in fact, may have contributed to the founding of the city of Phoenix in the 1860s, as these canals were exploited by early European settlers to provide water to the growing desert community.[22] Even today, part of the Phoenix Metropolitan Area's modern canal system is built upon those constructed by the Hohokam more than a millennium ago (**Figure 1**). Nor was the magnitude of what these early engineers accomplished lost on Phoenix's first civil engineer, Omar Turney. On his 1929 map of the Hohokam canal system in the Salt River Valley, Turney noted that the area encompassed 'the largest single body of land irrigated in prehistoric times in North or South America and perhaps in the world.'[23] Then, in a footnote, he paid further homage to the technological achievements of the Hohokam: 'These were the original engineers, the true pioneers who built, used and abandoned a canal system when London and Paris were a cluster of wild huts.'

So what happened to the Hohokam? Where did they go? The Hohokam seemingly disappeared from the region by 1450, abandoning their intricate canal system that had sustained them for centuries. While the various factors that may have contributed to their decline and disappearance are debated within archaeological circles, the Hohokam may simply have been a victim of their own success. The extensive canal network that enabled them to cultivate the desert may also have made them highly vulnerable to environmental and societal changes (Nelson et al. 2010). Putting this in the context of resilience theory and cross-scale dependencies, Nelson et al. (2010) postulated that 'the large-scale irrigation technology and its great capacity to supply agricultural surpluses was highly robust to local fluctuations in rainfall, which contributed to the creation of a regional-scale economy. However, it did so with increased vulnerability to social and ecological perturbations at specific localities, which, because of the crosscutting interdependencies, could be felt across the region.'

In support of this argument, several extreme climatic events were found to have occurred in the late 1300s, at a time when the Hohokam population (and its associated resource needs) were at a peak. Based on a reconstruction of historic water flows in the lower Salt River, this translated into two unusually high-flow years that were each preceded and followed by years of low water flow (Graybill et al. 2006). While a protracted drought might have caused sections of the canal system to run dry or contributed to sedimentation of the canals, interfering with canal function, major floods could cause extensive damage to the canal infrastructure itself (Graybill et al. 2006). Further, from a societal standpoint, the population had become large and concentrated within a single area, putting their local environment at risk for depletion and degradation. It's not called the Salt River for nothing. Irrigating fields on such a massive scale using water from the Salt River may have contributed to the salinization of the fertile valley soils over time, reducing productivity. Given the Hohokam's dependence on agriculture and water, they were necessarily dependent on a social system capable of maintaining the extensive network of canals, and further relied on a regional exchange network to provide additional resources. In speculating on how environmental and societal events might cascade through Hohokam society and eventually led to its collapse, Nelson et al. (2010) envisioned that 'Once that extensive social network became fragmented, resources brought from afar were no longer available and the social relations that supported the canal systems changed. With these changes, the potential vulnerabilities were realized: people suffered, institutions collapsed, and the region was depopulated to such a level that the remaining population was not archaeologically visible.'

No civilization lasts forever, but the fact that the Hohokam were able to manage their water needs via this intricate canal system for almost ten centuries is a remarkable achievement, something that today's inhabitants of the region are unlikely to match given the current rate of water withdrawal from the Colorado River (Wildman & Forde 2012). Water use in

[22] Indeed, the city was named 'Phoenix' because, much like the mythical bird, it arose from the ashes of a previous civilization. Phoenix is now the sixth largest city in the United States, with a population size of about 1.6 million people and over 4 million people in the greater Phoenix Metropolitan Area. Phoenix ranks as one of the fastest-growing cities in the United States.

[23] Turney was apparently unaware of the Chimú along the coast of northern Peru, who engineered their own extensive canal system around the same time as the Hohokam (ca. 900–1480); some of these canals ran 50 miles (80 km) between valleys (Ortloff 1995).

(Continued)

BOX 3.1   Continued

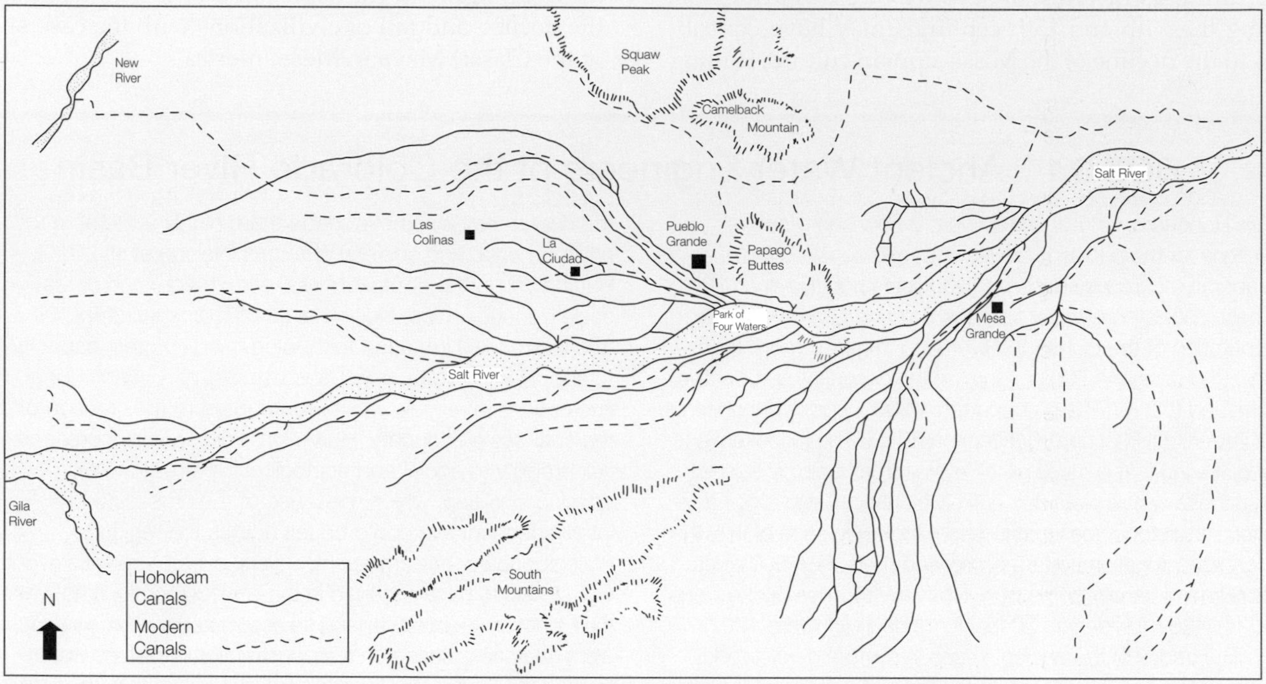

Figure 1 Hohokam canal system in the Salt River Valley of central Arizona.   This reproduction of Omar Turney's 1929 map shows the distribution of irrigation canals constructed by the Hohokam some 1400 years ago, in what is now the Phoenix Metropolitan Area (for reference, Phoenix's international airport is located near the point marked Pueblo Grande). Even today, some of Phoenix's 'modern' canals were built upon those constructed by the Hohokam more than a millennium ago.

Source: After Hunt et al. (2017).

the Colorado River Basin is operating at a deficit, for not only does the river not reach the sea in most years, but even the largest storage reservoirs behind Hoover and Glen Canyon Dams (Lake Mead and Lake Powell, respectively) have reached record-low levels in recent years (USBR 2013a,b). Nor is the water supplied by the Colorado River likely to increase any time soon, given that climate change has contributed to unprecedented declines (relative to the past millennium) in the winter snowpack across the Rocky Mountains over the last few decades, which in turn has reduced the amount of spring runoff that feeds the Colorado River, a trend that is projected to continue throughout the 21st century and beyond (Pederson et al. 2011). Continued low runoff,

coupled with a steadily rising demand (recall that Phoenix is one of the fastest growing cities in the USA) in one of the hottest and most arid parts of the country,[24] will likely exhaust the water storage capacity of these reservoirs within the next few decades (Barnett & Pierce 2008). If that happens, the ensuing water crisis would be devastating to the people and economy of the region, in much the same way it was for the Hohokam: 'people suffered, institutions collapsed, and the region was depopulated.' As the philosopher and poet George Santayana observed: 'Those who cannot remember the past are condemned to repeat it.'

[24] The average high temperature for the Phoenix area is >100°F (38°C) for three months of the year (June-August), and receives an average of 8.5 inches (22 cm) precipitation per year (www.weather.com).

CASE STUDY   Deforestation and the Drought-Mediated Collapse of the Classic Maya Civilization

Drought has contributed to the decline and fall of civilizations throughout human history. Prolonged drought can cause widespread famine, which tends to incite civil unrest, conflict, and mass exodus, thereby contributing to

sociopolitical upheavals that can topple empires. Indeed, a major shift in climate some 4200 years ago (the 4.2 kiloyear event) produced unusually severe drought conditions that lasted a century and caused the societal collapse of several

Old World civilizations, including the Akkadian empire of Mesopotamia, the pyramid-building Old Kingdom of ancient Egypt, and the Indus Valley civilization of northwest India (Weiss & Bradley 2001). A millennium later, a centuries-long drought was again implicated in the 'Late Bronze Age Collapse,' which saw the cultural collapse of several civilizations in the eastern Mediterranean, including the Egyptian empire of the New Kingdom period, the Mycenaean civilization of Greece, and the Hittite empire in what is now Turkey (Kaniewski et al. 2015).

In the New World, a series of droughts—including a megadrought spanning two centuries—is believed to have contributed to the collapse of the Classic Maya civilization in Mesoamerica some 1200 years ago (Gill 2000; Haug et al. 2003; Kennett et al. 2012; Douglas et al. 2015). This ancient civilization spans some 3,000 years of history, during which time the Maya developed an extraordinarily complex and sophisticated society. Today, the ancient Maya are remembered for their monumental stone architecture, including step pyramids, palaces, and observatories that still tower above the tropical jungles of their homeland more than a millennium after these cities were abandoned.

At its height during the Classic Period (250–900 C.E.), the Maya civilization consisted of numerous city-states, the largest of which was located at Tikal in the Petén Basin of present-day Guatemala. The Petén Basin makes up part of the southern Maya Lowlands, which were once home to an estimated 5 million people, making it one of the most densely populated regions in the world at that time (Scarborough et al. 2012). In support of this large population, extensive slash-and-burn agriculture was practiced throughout much of the region, resulting in widespread deforestation (Douglas et al. 2015). Ironically, deforestation may partially account for the shifting of climate toward drier conditions, by amplifying developing drought conditions within the region (Oglesby et al. 2010; Cook et al. 2012). Given the tropical wet-dry climate cycle, the Maya were critically dependent on seasonal rainfall, not only for their crops but also for their own water needs, especially given the scarcity of perennial surface waters in the Petén Basin. In places like Tikal that lacked access to a permanent water source,[18] the ancient Maya initially adapted to drying conditions by constructing extensive water catchment and storage systems in which they could collect and store seasonal runoff (Scarborough et al. 2012). Even the most cleverly engineered system cannot withstand a centuries-long drought, however. The combination of drought and deforestation may thus have contributed to the collapse of the Classic Maya civilization, as described by Oglesby and colleagues (2010, p. 2):

>...the continuing deforestation over hundreds of years slowly put more and more stress on water availability; however for much of this time, the Maya were able to cope through continuous adaptive strategies, even during occasional periods of drought. Once deforestation became near total and a natural drought of sufficient severity came along, the Maya could no longer adapt, and the resulting water shortages lead quickly to extreme social unrest and political instability that in turn induced almost complete collapse of their civilization.

[18] Tikal appears to have been founded on the summit of a ridgetop for strategic reasons, and also because the site held a spring that initially provided a reliable water source. The water demands of the growing population soon exceeded that supplied by the spring, however, especially as city construction progressed and the area became increasingly paved in stone for plazas, ball courts, and causeways, thereby preventing the spring from recharging. The inhabitants at Tikal thus became increasingly reliant upon their engineered water-management system in order to meet their potable water needs (Scarborough et al. 2012).

Droughts are not just a scourge of past civilizations. In recent years, severe and even historic[19] drought conditions have been reported from all over the world, including East Africa, southeast Australia (the 'Millennium Drought'), China, southeast Asia, eastern Brazil, and the western half of the USA. Physical water scarcity[20] currently affects about 20% of the global population, a number that is projected to increase with continued population growth and increased food demand (Molden 2007). Although water is obviously essential for human health and well-being, water is also important for sustaining our agricultural, energy, and industrial sectors. In particular, drought is a major threat to agricultural production, and thus to our global food security (Li et al. 2009). More than 80% of agriculture worldwide is rain-fed (Gornall et al. 2010), and agriculture currently accounts for 70% of global freshwater withdrawals (FAO 2015).

[19] Such claims are usually made in reference to the instrumental record, the period during which reliable meteorological records have been kept, which date back only 150 years or so. The ten warmest years on record have all occurred since 1998 (NOAA National Climatic Data Center, State of the Climate, Global Analysis: www.ncdc.noaa.gov/sotc/global/). Although the instrumental record is limited to the recent past, temperature reconstructions using climate proxies (e.g. tree-ring data, fossil pollen, or sediment cores) indicate that these are also the warmest years of the past several centuries, and in some cases, of the past millennium.

[20] Physical water scarcity, as opposed to economic water scarcity, occurs when there is insufficient water available to meet demands imposed by human water use as well as those of the ecosystem, leading to severe groundwater depletion and environmental degradation (Molden 2007). Although arid regions are most often associated with physical water scarcity, overallocation and mismanagement of water resources can also create physical water scarcity in regions with otherwise abundant water resources. Economic water scarcity, by contrast, pertains to a lack of investment in infrastructure that then limits access to water resources.

In addition, climate change is projected to cause shifts in agricultural production that will reshape the agricultural landscapes of the world (Gornall et al. 2010). By the end of the 21st century, the cropland drought-disaster risk (based on the probability of severe drought conditions[21] and its potential impacts on crop yields) is projected to double, which translates into drought-related yield reductions of 90% for the major crops of the world (barley, maize, rice, and wheat; Li et al. 2009). Climate-change scenarios portend not only a reduction in crop yields around the world (e.g. maize and soybeans in the USA, Schlenker & Roberts 2009), but also substantial regional shifts in lands that are potentially arable (Zhang & Cai 2011). Regions such as Africa and South America, which currently comprise 40% of the world's potentially arable land, could lose as much as 20% of their arable land area by 2100, owing to higher temperatures, reduced rainfall, or both. In contrast, Russia, China, and the USA all have the potential to see increases in arable land area, as rising temperatures contribute to a northward shift in areas suitable for agriculture (Zhang & Cai 2011). Whether such increases in arable land area are actually realized, however, will depend on whether future population growth and land use will allow for agricultural expansion into these areas. For example, increased urbanization (urban sprawl) typically results in the conversion and loss of agricultural lands, which then leads to a process of agricultural intensification, such as the use of irrigation to increase crop yields in remaining agricultural areas (e.g. Alauddin & Quiggin 2008).

Economic, social and political crises have been emerging at an accelerated rate. Although often described individually – the 'food' crisis, the 'energy' crisis, the 'financial' crisis, the 'human health' crisis, or the 'climate change' crisis, to name but a few – these crises are all inter-related though their causes and consequences. Their underlying causes often boil down to the ever-increasing competition for a few key – often-limited – resources, **of which water is common to all**. (emphasis added)   Balaji et al. (2012)

Besides impacting our agricultural interests, drought affects other ecosystem goods and services on which we depend, such as those provided by forests. Deforestation may have amplified the drought that led to the collapse of the Classic Maya civilization, but drought by itself can lead to deforestation through massive forest die-offs. In recent decades, forests all over the world have experienced drought-related declines, which parallel the rise in global temperatures

(Allen et al. 2010). Even tropical rainforests are not immune (Phillips et al. 2009). Trees die en masse during 'hot droughts' (droughts compounded by high temperatures), either as a direct result of water stress (Choat et al. 2012), or indirectly, as a consequence of climate-related increases in pest outbreaks, disease, and severe wildfires (Allen et al. 2015).

For example, tree mortality rates in coniferous forests throughout the western USA have increased exponentially over the past few decades, doubling every 17–29 years, depending on the region (e.g. mortality rates are increasing faster in California and the Pacific Northwest than in the Rocky Mountains; van Mantgem et al. 2009). After factoring out other potential causes of tree mortality (e.g. fire-exclusion policies that increased tree density, and thus mortality from competition and disease), van Mantgem and his colleagues (2009) concluded that warming temperatures and concomitant water deficits are most likely responsible for the observed increase in tree mortality rates. In California alone, nearly a billion trees—an area encompassing 10.6 million ha of forest land—have experienced dangerous levels of water stress, measurable using remote-sensing technologies, following 4 years of severe drought (Asner et al. 2016). In the southwestern United States, the rapid die-off of piñon pine (*Pinus edulis*) has been linked to the protracted drought and associated bark-beetle infestations that have plagued this region since the late 1990s (Breshears et al. 2005). Even in higher-elevation forests, which are an entirely different life zone (**Figure 3.2**), tree species such as the ponderosa pine (*P. ponderosa*) and Douglas fir (*Pseudotsuga menziesii*) have also experienced unusually high levels of mortality due to an increase in fire severity and bark-beetle outbreaks tied to the drought (Williams et al. 2010). The intensity of drought conditions in the southwestern United States over the past two decades is already the most severe of the past 500 years (since the 1500s megadrought) and is on track to become the most severe of the past millennium (Williams et al. 2013). As such, we are witnessing not just a dramatic shift in forest structure and composition at a landscape scale, but a change in the dominant vegetation type at a regional or even continent-wide scale. In that case, the effects of the current drought will likely persist for decades, if not centuries, to come.

Fire

As the foregoing illustrates, drought may be accompanied by additional hazards, such as heat waves and wildfires. The risk of wildfires increases during dry periods, especially during hot droughts (Overpeck 2013). When combined with high temperatures, low relative humidity, and strong winds, drought is an extremely effective priming agent, turning terrestrial ecosystems into tinderboxes that are quick to ignite and combust. Climatic patterns that contribute to extreme

---

[21] Defined in terms of both the frequency and duration of severe droughts, in which the expected PDSI < –3 (severe drought) during the growing season for a region.

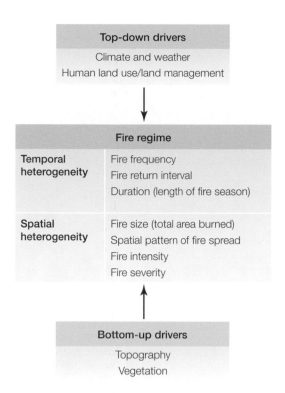

Figure 3.29 Top-down versus bottom-up drivers of fire regimes.

Source: After Falk et al. (2007).

fire-weather conditions, such as the El Niño Southern Oscillation,[22] are thus considered a major 'top-down' driver of **fire regimes** (Falk et al. 2007; **Figure 3.29**; **Table 3.2**). For example, most (90%) of the forest area burned by wildfires historically has been caused by a few, very large fires that ignited under extreme weather conditions (1% of all fires; Gedalof 2011). Given the global trend toward warming temperatures and persistent drought conditions in many parts of the world, extreme fire-weather conditions are an increasingly common phenomenon, as are the large fires they ignite (Westerling et al. 2006).

Fire is a spatial and temporal process, driven by controls acting across a range of scales.   Falk et al. (2011)

Besides climate, fire regimes are also regulated from the 'bottom up' by landscape factors, namely topography and vegetation (**Figure 3.29**). The interaction between topography and vegetation affects the amount and distribution of fuels across the landscape, resulting in a heterogeneous **fuel mosaic** that influences ignition patterns and the frequency, size, distribution, and

severity of fires (Turner & Romme 1994; Falk et al. 2007; McKenzie et al. 2011). Vegetation provides the fuel that carries and sustains fire, but different types of vegetation afford different fuel types (e.g. grass, leaf litter, downed woody debris) that vary in their propensity to burn, and thus, in their resultant fire properties. Vegetation types also differ in their structure,[23] productivity, and post-fire recovery rates, which affect the amount, distribution, and connectivity of fuels across the landscape. Topographic complexity created by variation in aspect, slope, and elevation influence bioclimatic factors, which in turn affect the distribution and productivity of different vegetation types, as well as the availability and moisture content of fuels across the landscape. For example, fire frequency typically exhibits an inverse relationship with elevation (i.e. fires occur more frequently at lower elevations), largely because warmer temperatures at low elevations contribute to drier fuels and a longer fire season, especially if higher elevations also experience a deeper snowpack and later snowmelt.

In western North America, for example, fires occur less frequently in high-elevation mixed-conifer and spruce-fir (*Picea-Abies*) forests, which have dense, high-moisture fuel beds that impede the ignition and spread of fires,[24] than in low-elevation ponderosa pine forests where drier surface fuels, consisting of long-needled litter and cured grasses, tend to promote fire spread (Falk et al. 2011; **Figure 3.2**). Slope aspect can similarly influence fire frequency, given that south-facing slopes have greater sun exposure and thus tend to be warmer and drier than north-facing slopes. For example, fires in the Blue Mountains of eastern Oregon and Washington occurred more frequently on southern aspects, which support ponderosa pine forests, than on northern aspects, which support more mesic forest types (Heyerdahl et al. 2001). Nevertheless, there are no absolutes here, for the effect of slope aspect on fire frequency was only observed in watersheds with steep terrain that had a strong topographic barrier to fire spread (e.g. a river). Similarly, elevation has little effect on fire frequency if watersheds lack a strong altitudinal gradient in fuel moisture, fuel loads, or fuel bed density (Heyerdahl et al. 2001; Taylor & Skinner 2003; Scholl &

---

[22] The El Niño Southern Oscillation is the most important driver of global climatic variability. Its effect on wildfire thus spans the globe, with significant effects on every continent except Antarctica (Gedalof 2011).

[23] Vegetation structure entails both vertical and horizontal dimensions. Vertical structure can be assayed in terms of canopy cover or foliage height among different layers or vegetation strata (e.g. canopy, subcanopy, understory, shrub, or herbaceous layer), which is sometimes expressed as 'foliage height diversity.' Horizontal structure refers to the spacing or spatial pattern of vegetation, which may include density or size-based measures (e.g. different tree-diameter classes) as well as various spatial-statistical approaches used to quantify distributional patterns (**Chapter 4**). The availability and connectivity of fuels should ideally be assessed in both dimensions, in order to gain a better understanding of fire behavior in different vegetation contexts.

[24] Dense litter beds require a longer period of drying to become flammable, especially if the fuels have a high moisture content. The effect is to shorten the length of the fire season, as well as to inhibit fire spread.

Taylor 2010). In other words, topography has little effect on fire frequency if the distribution of fuels is essentially uniform across the landscape.

Even where topography exerts strong controls on fire frequency, topographic effects can be overwhelmed by drought and fire-suppression policies, both of which can create uniform fuel conditions that enhance fuel connectivity across the landscape (Gill & Taylor 2009). The interaction between fuel connectivity and fire frequency is complex, and may well represent a threshold response to factors that are influenced by topographic complexity, such as fuel loads and fuel moisture content (Miller & Urban 2000; Taylor & Skinner 2003). Within coniferous forests of the Sierra Nevada range in California, for example, fuel connectivity may be related to fire frequency only under conditions of intermediate fuel moisture (e.g. 5%), where fuel moisture levels are not so high as to impede fire spread or so low as to promote widespread fire regardless of time since previous fire (Miller & Urban 2000). Fire suppression during the 20th century promoted high fuel connectivity by allowing fuels to accumulate, thereby creating a more uniform distribution of flammable fuels across the landscape, regardless of topographic position (Taylor & Skinner 2003; Gill & Taylor 2009; Scholl & Taylor 2010). In this system, more frequent fires akin to the historical fire regime would thus help to maintain fuels below the threshold levels that promote landscape-wide fuel connectivity, except in years of extreme fire-weather conditions (i.e. drought years; Miller & Urban 2000).

The relationship between landscape factors and fire can also change depending on what aspect of the fire regime we consider. For example, fires might occur less frequently in mesic forests, but when they do occur, they tend to be of greater severity. Returning to the Blue Mountains study, mesic forests historically sustained moderate- to high-severity fire regimes, in which some or all of the trees were killed owing to their long crowns[25] and thin bark, whereas more arid ponderosa pine forests historically sustained low-severity fires, in which most trees were not killed, owing to their high crowns and thick bark (Heyerdahl et al. 2001). **Fire severity**[26] thus characterizes the degree to which vegetation or the soil organic layer has been consumed or killed by fire (Keeley 2009). Fires of low severity are typically light **surface fires** that consume litter, grasses, and herbaceous plant cover, but do not burn into the soil organic layer or affect canopy trees (Turner et al. 1994). As a consequence, low-severity fires tend to occur frequently in some vegetation types (e.g. ponderosa pine

Figure 3.30 Crown fires.   Crown fires exhibit the greatest severity, causing widespread tree mortality that can lead to stand replacement, as in the case of the 1988 Yellowstone fires. Crown fires typically start as surface (ground) fires but then spread vertically up lower branches to engulf the tree canopy (laddering), from which the fire can then spread rapidly among adjacent trees.

Source: National Park Service (Jeff Henry; Yellowstone National Park 1988).

forests), and are therefore characterized by short return intervals (i.e. years to decades; Turner et al. 2003). The most severe forest fires are **crown fires**, which burn through the upper canopy and therefore have the potential to kill trees, possibly leading to stand replacement (**Figure 3.30**).[27] Fires of such high severity have a long return interval (several decades to centuries) and thus occur infrequently, but are nevertheless characteristic of some forest types (e.g. high-elevation forests; Turner et al. 2003). Crown fires also tend to occur in regions that experience severe drought, as such extreme weather conditions are generally necessary to ignite fires of this severity (Turner & Romme 1994).

---

[25] Crown length is a measure of vertical forest structure. Trees with low branches (long crowns) are more likely to carry fire from the ground up to the crown (**Figure 3.30**).

[26] Although fire severity is sometimes used synonymously with fire intensity, the latter is really a measure of the energy output by the fire (Keeley 2009).

[27] Stand replacement fires occur in a variety of ecosystems, from forests to grasslands, and thus may be caused by severe surface or ground fires, in addition to crown fires, depending on the type of system.

Different types of forests are thus characterized by different types of fires (surface vs crown fires) and fire regimes, which has implications for the management of these forests. For example, fire-suppression policies in forests that historically experienced frequent surface fires of low severity (e.g. ponderosa pine forests) have shifted the fire regime of these forests to one of greater fire severity (e.g. Gill & Taylor 2009; Scholl & Taylor 2010).

Understanding how fire regimes might shift in the face of climatic changes that have increased the frequency and severity of droughts is important not only for the future management of these forests, but also for evaluating the potential for dramatic changes in the dominant vegetation type (i.e. system state changes). We consider the effect of climate change on fire regimes and forested landscapes in western North America in the following case study.

CASE STUDY   Are Large Frequent Forest Fires Becoming the 'New Normal' in Western North America?

Forest fires in the western USA have been increasing in size (Westerling et al. 2006) and severity (Miller et al. 2009) in recent decades. For example, the 1988 Yellowstone fires were the first in a series of large, severe fires that have occurred throughout western North America during the past quarter-century (Romme et al. 2011). Recall that the 1988 Yellowstone fires were unusually large and burned about a third of the National Park; a fire of this magnitude had not occurred in Yellowstone for nearly 300 years (**Figure 3.12**). The Yellowstone fires occurred during a year of unusually severe drought (the driest summer then on record), which was also accompanied by high winds and dry lightning, and as such, appear to have been a largely climate-driven event (Turner et al. 2003). Vegetation and topography ultimately had little effect on impeding fire spread in the face of such severe fire conditions.

Ecologists were genuinely surprised by the amount of heterogeneity observed within the perimeter of the 1988 fires, but numerous subsequent postfire studies have confirmed that **heterogeneity is the rule, not the exception**. (emphasis added)   Romme et al. (2011)

Despite their severity, the 1988 Yellowstone fires did not cause uniform devastation across the landscape. One of the many 'surprises' that emerged in the aftermath of the Yellowstone fires was the striking heterogeneity of the resulting burn landscape, and how variable fire severity was even within burned areas (Christensen et al. 1989; Turner et al. 1994; Turner et al. 2003; **Figure 3.31**). Much of the pre-fire landscape (~80%) had been dominated by lodgepole pine (*Pinus contorta*) forests, which formed nearly continuous stands in many places throughout the National Park (**Figure 3.32**). Despite the seeming continuity of fuels across the landscape, the vagaries of the fires' behavior produced a complex mosaic of burned and unburned areas, resulting in a heterogeneous landscape of varying burn severity (**Figure 3.31**). Subsequently, most severely burned areas could be found in fairly close proximity (<200 m) to lightly burned or unburned areas, increasing the likelihood that such areas could serve as a source of propagules

to facilitate the recolonization and recovery of severely burned areas (Turner et al. 2003). Another surprise, however, was that most post-fire recovery occurred via the regrowth or re-sprouting of plants or their underground structures (rhizomes, roots) that had survived the fire, rather than via colonization from unburned sites (Turner et al. 2003). Biological legacies were thus important to fire recovery, much as they were in the case of the 1980 eruption of Mount St Helens. Thus, even severely burned areas were awash with green and flowering plants only a year after the fire (**Figure 3.33**).

The Yellowstone fires were stand-replacing—the most severe type of fire—in which entire stands of lodgepole pines were killed. Forest recovery was therefore expected to be a slow process, in which trees would need to recolonize burned areas, such that it might be decades before forests once again blanketed the Yellowstone landscape. Lodgepole pines are a **fire-adapted species**, however. Populations of lodgepole pine were found to have varying degrees of serotiny throughout the National Park (<1% to 65% of lodgepole pines; Turner et al. 2003). For serotinous trees,

Figure 3.31 Complex fire mosaic created by the 1988 Yellowstone fires.   Stands of charred and fire-killed trees (black and brown areas) are adjacent to forest that was seemingly untouched by the fire. Fire severity also varied within burned areas, as evidenced by islands of green within some burned patches.

Source: National Park Service (Jim Peaco; 24 July 1989).

Figure 3.32 Extent of the 1988 Yellowstone fires.   The fires that burned within Yellowstone National Park during the summer of 1988 burned about a third (36%) of the total park area and was the largest fire to occur in Yellowstone in almost 300 years (burned areas are in red in the 1989 imagery). More than 20 years later, the fire scars from the 1988 fires were still visible (light pink areas in the 2011 imagery), although new fires have occurred in the interim (red areas).

Source: Image by Eric Sokolowsky (2011) and taken from the NASA, Goddard Space Flight Center.

high temperatures, as from a severe fire, are needed to open the mature, closed cones and release seeds. Pine seedlings were thus evident in burned areas the year following the fire, and within 10 years had formed dense stands of young trees (**Figure 3.33**). Nevertheless, pine-seedling density was affected by fire severity, burn-patch size, and topographic position, which contributed to spatial variability in fire recovery. Pine-seedling density was highest in large burn areas, where seedlings thrived in the high-light environment, and in areas with severe surface rather than crown-fire burns, presumably because the latter fire type reduced seed viability (Turner et al. 1994, 2003). Pine-seedling density was also greatest at lower elevations (<2400 m) in which a higher percentage of trees were serotinous, although even then, serotiny varied as a function of stand age (i.e. time since fire), being highest in mature stands (>100 years; Schoennagel et al. 2003).

Factors influencing post-fire stand density, such as pre-fire serotiny and the size and severity of the fire, thus generated considerable spatial variation in the post-fire landscape. Although the spatial variability of post-fire vegetation was greater than expected, the primary effect of the 1998 Yellowstone fires was to alter vegetation structure, especially of post-fire lodgepole pine density,

rather than plant species richness or composition (Romme et al. 2011). Nevertheless, the Yellowstone fires produced a spatially complex pattern of succession and recovery that persists to the present day, and is likely to do so for decades to come (Turner et al. 2003; **Figure 3.32**).

Climate is a major top-down driver of fire activity, but humans can also exert a top-down effect on fire regimes. Although fire-suppression policies do not appear to have contributed to the 1988 Yellowstone fires (Turner et al. 2003; Romme et al. 2011), this has not been the case elsewhere in the western USA, especially in drier forest types that historically experienced frequent surface fires. Using fire-proxy data (i.e. sedimentary charcoal and fire scars)[28] and historical records, Marlon

[28] Charcoal is formed from the incomplete burning of fuels. Charcoal particles may be carried aloft during the fire, transported to lakes, and incorporated into lake sediments where they are preserved. Sediment cores are thus good repositories of information on past fires (paleofires), and the analysis of charcoal and pollen preserved in sediments is widely used to reconstruct landscape history, as well as the changing character of climate-vegetation-fire linkages (Iglesias et al. 2015). Dendrochronological methods are used to date fire scars, which are lesions caused by fires that appear in the growth rings of long-lived trees (Falk et al. 2011).

Figure 3.33 Fire recovery.   Ecological succession occurred rapidly in the aftermath of the 1988 Yellowstone fires. (A) Grasses, fire-weed (*Chamerion angustifolium*), and other forbs quickly colonized or sprouted within burned areas a year after the fires swept through Yellowstone National Park . (B) Some conifers, such as lodgepole pines (*Pinus contorta*), have serotinous cones that only release their seeds after a fire, leading to a rapid sprouting of pine seedlings in the year following the fires. (C) Ten years later, young lodgepole pines already formed dense stands within areas that had suffered the greatest burn severity (i.e. a stand-replacing fire).

Source: Photos A and C by Jim Peaco and Photo B by Ann Deutch.

and her colleagues (2012) reconstructed the fire record of the past three millennia for the western half of the USA. During the first half of the 20th century, fire activity declined to one of its lowest levels of the past 3,000 years, coinciding with the introduction of livestock grazing (which reduces fuel loads, leading to fewer fires) and fire-suppression policies on federal lands.[29] Fire activity thus declined even as successional and climatic changes increased the fire potential of these forested landscapes. Subsequently, a growing 'fire deficit' created a system that is now dangerously out of equilibrium with current climatic conditions (Marlon et al. 2012). If true, then large fires of unusual severity and frequency

will continue to ignite western forests until the deficit is resolved. For example, fires consumed about 32,000 km² across the western half of the USA in 2017, making it one of the most devastating (and costly) years on record for wildfires (NIFC 2017).[30]

To what extent might such changes in the forest-fire dynamics of western North America portend a landscape in transition? Recall that the frequency distribution of historical forest fires typically exhibits power-law scaling: most fires are small or of limited extent, with very large fires being a rare occurrence (i.e. LIDs; **Figure 2.13**). It appears, however, that historical fire records exhibit power-law scaling only when

[29] Recall from **Chapter 1** that 42% of the land area in the western half of the conterminous United States is federally owned.

[30] For comparison, the total area burned is roughly the size of Belgium or the state of Maryland (USA).

forest fires are strongly controlled by landscape factors, such as the distribution of fuels (vegetation) and topography—that is, by bottom-up drivers (McKenzie & Kennedy 2012). Such power-law scaling is thus evident in watersheds with complex topography, where topographic constraints have the greatest effect on fire frequency and spread, as discussed previously. Importantly, power-law behavior appears to occur within a fairly narrow range of conditions related to the ability of fire to spread across the landscape (i.e. a function of fuel connectivity, given by the **percolation threshold**; see **Chapter 5**). In this domain, there appears to be a balance between top-down and bottom-up controls on fire spread, such that fire spread is usually limited (resulting in frequent small fires) except under extreme fire-weather conditions (infrequently, when conditions allow for large, landscape-wide fires; McKenzie & Kennedy 2012).

In less complex or uniform landscapes, however, power-law behavior no longer occurs, and the balance shifts toward top-down controls on fire spread (e.g. climate). Fire regimes may thus exhibit threshold behavior, in which the relative importance of bottom-up versus top-down controls shift, producing very different fire dynamics. Consider the implications of this in the context of the recent fire history and forested landscapes of the western USA, however. If climate change produces more extreme fire weather, such as the recent drought experienced by much of the region, top-down controls on fire spread are expected to strengthen, such that fire spread will no longer be controlled by landscape factors (i.e. by bottom-up controls), even in topographically complex areas. In that case, the system would move toward more frequent large fires, not unlike what has been occurring in the western USA in recent decades.

Thus, although climate change is gradual, it can trigger rapid changes in fire regimes (Romme et al. 2011). For example, climate change is projected to reduce the fire-return interval in the Yellowstone region by an order of magnitude (from the current 100–300 years to less than 30 years by the middle of the 21st century (Westerling et al. 2011). Such frequent fires would be incompatible with most of the conifer species that have dominated the Yellowstone landscape for the past 10,000 years, including the lodgepole pine. There is thus a strong likelihood that Yellowstone's forests will be converted to some other vegetation type (e.g. shrubland) by mid-century. Such a shift in the dominant vegetation type would decrease the amount and condition of fuels, especially as the time for vegetation recovery between fires declines. Perhaps ironically, then, given the role of climate in driving this state change, the system would switch from being a climate-limited to a fuel-limited fire regime, thus eventually limiting fire activity (Westerling et al. 2011).

Fire is not just a disturbance of forested landscapes, however. Virtually all vegetation has the potential to burn, and thus fire is a ubiquitous disturbance in many—if not most—terrestrial ecosystems (McKenzie et al. 2011). Nevertheless, some ecosystems are especially fire-prone, and indeed, may owe their existence to fire (i.e. they are **fire-dependent ecosystems**). Although climatic factors are primarily responsible for global vegetation patterns and life zones (**Figure 3.3**), several of the world's major biomes appear to be controlled more by fire than climate (Bond et al. 2005). In particular, most of the world's mesic grasslands and savannas have the climate potential to be forests, as evidenced by the propensity of these systems to convert to forests when fire is suppressed (Bond et al. 2005).[31]

To illustrate, we'll consider the plight of the tallgrass prairie ecoregion of the eastern Great Plains, which occupies a 'tension zone' between more mesic forests to the east and more arid grasslands to the west. Tallgrass prairie once covered some 68 million ha across central USA, but most was plowed under and converted to row-crop agriculture during the 'Great Plow-Up' of the Great Plains at the turn of the 20th century (recall our discussion earlier in the chapter of the socioeconomic drivers that led to the plow-up and the Dust Bowl of the 1930s). Less than 4% of the historical tallgrass prairie remains today (Sampson & Knopf 1994). Although agricultural conversion still poses a threat in some areas, the expansion of trees and woody shrubs into this now fragmented grassland represents a greater threat at present (Briggs et al. 2005). Historically, frequent and widespread fire limited woody plant expansion, such that the region was once largely devoid of trees (remember, this was part of the 'Great American Desert' that was largely eschewed by early European settlers). As the region became more settled and developed,

---

[31] Although fire may well have contributed to the evolution and spread of grasslands some 6–8 million years ago, it has also been suggested that lower atmospheric CO_2 levels (which benefit the C_4 grasses that make up these fire-prone grasslands and savannas) favored grasslands over shrublands and forests, especially in warmer regions like the tropics (e.g. in Africa and South America; Ehrlinger et al. 1997). Indeed, there is some evidence that the encroachment of woody plants into grasslands, which has been documented in recent decades all over the world, may be due—at least in part—to rising atmospheric CO_2 levels, especially in areas

with higher rainfall amounts (i.e. mesic grasslands and savannas; Knapp et al. 2008; Eldridge et al. 2011; Buitenwerf et al. 2012).

however, fire suppression near population centers has allowed tree species such as the native red cedar (*Juniperus virginiana*) to spread rapidly and form dense, closed-canopy stands in as little as 40 years (Briggs et al. 2002).

Even when fire does occur, it may no longer be effective in preventing woody establishment (Briggs et al. 2002; Briggs et al. 2005). For example, much of the remaining tallgrass prairie is found within the Flint Hills of eastern Kansas, which supports a major cattle industry, resulting in widespread grazing and annual burning of rangelands. Although fire and grazing are critical to maintaining the structure and diversity of the tallgrass prairie (Collins 1992; Knapp et al. 1999), these disturbances historically occurred as a **shifting mosaic** (**Figure 3.4**). Large herds of American bison (*Bison bison*) once roamed the landscape in search of high-quality forage, such as the tender young shoots of grasses that re-sprouted after a fire. Grazing was thus concentrated within recently burned areas (**pyric herbivory**), allowing fuels to accumulate within ungrazed areas, which were then more prone to burn in subsequent years (Fuhlendorf & Engle 2001; Fuhlendorf et al. 2009). The historical tallgrass landscape was thus a dynamic, heterogeneous mosaic of burned and unburned, grazed and ungrazed areas.

Today, annual prescribed fire and widespread cattle grazing create a far more homogeneous landscape within the Flint Hills (Fuhlendorf & Engle 2001). The chronic grazing pressure reduces fuel loads, such that they are unable to carry a hot-enough fire to kill tree or shrub seedlings (Briggs et al. 2005). Once established, shrubs such as the rough-leaf dogwood (*Cornus drummondii*) become resistant to fire, growing ever larger and providing 'safe sites' for the germination of more fire-susceptible tree species, resulting in multiple foci for woody invasion across the landscape (**Chapter 8**). As a consequence, the tallgrass prairie—like many mesic grasslands around the world—is undergoing a conversion to shrubland and forest that may be difficult to avoid at this point (i.e. a system-state change; **Chapter 11**).

In conclusion, the influence of landscape and climatic factors on fire operate across a wide range of spatial and temporal scales, producing varying degrees of top-down versus bottom-up controls on fire regimes, which vary regionally and across systems (Falk et al. 2007; Parisien & Moritz 2009). The result is a complex disturbance process that both creates and responds to landscape heterogeneity (Turner & Romme 1994). Fire thus behaves much like a living organism, in terms of how it consumes vegetation, grows, and interacts with environmental heterogeneity while spreading across the landscape. Fire therefore provides an ideal segue from our discussion of abiotic disturbances to the role of biotic disturbances in contributing to landscape heterogeneity and dynamics.

Landscape Dynamics: Biotic Agents of Landscape Formation and Disturbance

Organisms can also be agents of landscape disturbance, and may even contribute to the development (or conversely, the erosion) of landforms. Indeed, we have discussed several examples already:

- Trees stabilize hillsides by reducing erosion and the landslide potential of montane landscapes. Conversely, the spread of the invasive velvet tree (miconia) destabilizes hillsides and contributes to increased landsliding on the Hawaiian Islands.
- Trees uprooted by windstorms invert soils and create microtopographic features (pits and mounds) that affect nutrient cycling and plant establishment.
- Coral reefs—the rainforests of the sea—are bioengineered structures created by colonies of tiny, polyp-like animals.
- The invasive saltcedar (tamarisk) has lowered the water table and increased soil salinity throughout the Colorado River Basin.
- The grazing activities of the American bison, in conjunction with fire, once helped to create a shifting disturbance mosaic that enhanced heterogeneity and controlled woody plant expansion onto tallgrass prairie landscapes.

As the last example illustrates, biotic disturbances can interact with abiotic disturbances to enhance or alter the extent, severity, or occurrence of these disturbances, often in complex and unpredictable ways (Simard et al. 2011). Even on their own, however, biotic disturbances can match or exceed the scale of abiotic disturbances. For example, more than 470,000 km² of coniferous forest (an area larger than Sweden) have been affected by bark beetles over the past decade in western North America (Raffa et al. 2008). The annual area affected by bark beetles is thus on par with the average area burned by forest fires (Kurz et al. 2008).[32]

[32] Because both fire activity and bark beetle outbreaks have increased in recent decades, it was thought that these disturbances were perhaps interacting synergistically, in which beetle-killed stands increased the susceptibility of forests to large, severe fires (e.g. Logan & Powell 2001). Instead, beetle-killed stands may actually reduce the incidence of severe fires in some forest types, such as the lodgepole pine forests of the Greater Yellowstone ecosystem, owing to a reduction in canopy fuel loads (beetle-killed trees drop their needles, which rapidly decompose and thus afford little fuel for a ground fire, let alone a canopy fire; Simard et al. 2011). Bark beetle outbreaks and fire *are* linked disturbances,

In this section, we consider the role of organisms as agents of landscape disturbance and dynamics, particularly in terms of their ability to contribute to landscape formation across a range of scales (organisms as geomorphic agents), and to engineer landscape features or even entire landscapes (organisms as ecosystem engineers) in ways that are often unique (the keystone role of species). Note that our focus in this section is principally on how organisms generate or help shape landscape structure. The reciprocal effects of landscape structure on organisms is the basis for the second half of this textbook (**Chapters 6–11**).

Organisms as Geomorphic Agents

The study of the interaction between organisms and landscape formation was originally the province of **biogeomorphology** (Viles 1988). Initially, biogeomorphology largely focused on plants (phytogeomorphology), but geomorphologists have since expanded their focus to investigate the geomorphic effects of animals (zoogeomorphology; Butler 1995; Butler & Sawyer 2012). As geomorphic agents, animals and plants contribute to landscape formation by influencing erosional and depositional processes, such as by altering runoff and sedimentation rates or via the redistribution of materials about the landscape. For example, burrowing animals, such as earthworms, ants, and rodents, rework soils through their tunneling activities and also transport materials between the soil and surface, a process known as **bioturbation** (Gabet et al. 2003; Meysman et al. 2006). Although the effects of such animals may appear to be small and highly localized, their cumulative effects can add up over time and space, especially when a large number of bioturbators are present (e.g. soil invertebrates).

For example, harvester ants (*Pogonomyrmex* spp.) actively maintain a cleared disk area around their nest mounds, from which workers clip and remove all vegetation (**Figure 3.13**). Although the surface area of individual nests may cover only a few square meters, there may be as many as many as 20–150 colonies/ha, each persisting up to 15–50 years (MacMahon et al. 2000). Subsequently, the total area cleared by harvester ants may comprise 3% or more of the landscape, resulting in a significant degree of disturbance that is detectable

even in satellite imagery.[33] During nest construction, in which ants excavate chambers and shafts down to a depth of 2 m or more, upwards of 80–280 kg of soil may be moved per hectare annually (MacMahon et al. 2000). Besides mobilizing soil, harvester ants bring back seeds and other organic matter to their nests (hence the reason they are called 'harvester' ants), which results in nutrient enrichment over time that greatly contributes to heterogeneity in the soil nutrient distribution (Wagner et al. 2004). Subsequently, these enriched sites are colonized by plants after the colony dies, often supporting a distinctly different plant community from the surrounding area, thereby enhancing the diversity and heterogeneity of vegetation as well (MacMahon et al. 2000).

Other examples of geomorphic features created by animals, and across a wide range of scales in space and time, include termite mounds in tropical grasslands, some of which have persisted on the landscape for thousands of years; the slow accretion of coral reefs over the millennia; the creation of coral sands via the bioerosion of coral reefs by foraging parrotfish (Scaridae); Mima mounds—small grassland hillocks—created by the burrowing activities of pocket gophers (Geomyidae); and the wallows created by large mammals such as elephants (Butler 1995; Montaggioni 2005; Haynes 2012; Gabet et al. 2014; Erens et al. 2015; Martin et al. 2018).

Organisms as Ecosystem Engineers

Many of the aforementioned organisms are also considered to be ecosystem engineers. **Ecosystem engineers** are 'organisms that directly or indirectly modulate the availability of resources (other than themselves) to other species, by causing physical state changes in biotic or abiotic materials. In so doing they modify, maintain and/or create habitat' (Jones et al. 1994, p. 374). Ecosystem engineering thus overlaps to some degree with biogeomorphology, a consequence of these two fields having arisen independently within disparate disciplines (ecology versus geomorphology; Butler & Sawyer 2012). Thus, although ecologists may also explore how ecosystem engineers give rise to geomorphological features on the landscape, this is usually done within the context of understanding the

therefore, just not in the way usually assumed (i.e. the relationship is a negative one, rather than positive). Like fire, bark beetle outbreaks are also driven by climate and weather-related factors (e.g. Preisler et al. 2012). Warmer temperatures in particular are allowing bark beetles to increase and expand their ranges into forests that occur at higher elevations and latitudes, such that bark beetles are now able to affect a greater number of conifer species and over a much greater spatial extent than in the past (Logan & Powell 2001).

[33] Don't believe me? Go to Google Earth and type the coordinates 35°22′17.36″ N 111°33′18.00″ W, which is the approximate location where the photo in **Figure 3.13** was taken. Zoom into an altitude of about 2.25 km. Those reddish-brown circles with a pale center that dot the landscape are harvester ant mounds. If you then zoom out and move over this open area, you can gain a better appreciation for the density of ant mounds and the degree to which they constitute a broad-scale disturbance on the landscape.

ecological consequences of such environmental changes (Jones 2012).

There are two broad classes of ecosystem engineers (Jones et al. 1994, 1997). **Autogenic engineers** change the environment via their own physical structure. Examples of autogenic engineering include forests, coral reefs, giant kelp (*Macrocystis pyrifera*) forests, and mollusk beds, reefs, or shell middens (Gutiérrez et al. 2003). Autogenic engineers do not need to be alive to perform their engineering functions, as even their structural remnants (logs, dead coral, shells) can change the environment and regulate the distribution and abundance of other resources or species. For example, fallen and uprooted trees contribute to the formation of large woody debris jams ('log jams') in forested river ecosystems, which influence not only streamflow and channel morphology, but also provide pool habitat and shelter for aquatic species (e.g. salmon) and for the establishment of riparian vegetation on bars and islands that form below the debris jam (Abbe & Montgomery 1996).

> Two particular, distinguishing features of ecosystem engineers are that they affect the physical space in which other species live and their direct effects can last longer than the lifetime of the organism—engineering can in essence outlive the engineer.
> Hastings et al. (2007)

In contrast, **allogenic engineers** change the environment by transforming living or non-living materials from one physical state to another, whether mechanically or by some other means. The changes wrought by allogenic engineers can have long-lasting effects, which may persist for decades or even centuries, and are thus likely to outlive the engineers themselves (Hastings et al. 2007). Examples of allogenic engineers include everything from earthworms to elephants (Jones et al. 1994; Jouquet et al. 2006). Nevertheless, beavers are the archetypal allogenic engineer, as we discuss in the following case study.

CASE STUDY Beavers as Ecosystem Engineers

Beavers are large, semi-aquatic rodents found along rivers, streams, and wetlands throughout the temperate and boreal regions of North America and Europe (North American beaver, *Castor canadensis*; Eurasian beaver, *C. fiber*). They are famous for their industry and ability to gnaw down mature trees, build dams, and impound rivers and streams, completely transforming the landscape in the process. Their role as ecosystem engineers is neatly summarized by Robert Naiman and his colleagues (1988, p. 753):

> *Beaver . . . provide a striking example of how animals influence ecosystem structure and dynamics in a hierarchical fashion. Initially beaver modify stream morphology and hydrology by cutting wood and building dams. These activities retain sediment and organic matter in the channel, create and maintain wetlands, modify nutrient cycling and decomposition dynamics, modify the structure and dynamics of the riparian zone, influence the character of water and materials transported downstream, and ultimately influence plant and animal community composition and diversity . . . In addition to their importance at the ecosystem level, these effects have a significant impact on the landscape and must be interpreted over broad spatial and temporal scales.* (emphasis added)

The broad-scale impact of beavers is evidenced by the degree of landscape transformation that occurs following their return to areas from which they had previously been extirpated. By the turn of the 20th century, beavers had been trapped nearly to extinction throughout North America and Europe (Naiman et al. 1988; Rosell et al.

2005).[34] In northern Minnesota, on the Kabetogama Peninsula in Voyageurs National Park, beavers began a slow comeback starting around 1925. At that time, only about 1% of the peninsula (a 298 km^2 area) had been converted by beavers to ponds and meadows. Over the next 60 years, however, the number of beaver impoundments increased, with the most rapid rate of change taking place during a single 20-year period (1940–1960; Naiman et al. 1994). By 1988, about 14% of the landscape had been impounded by beavers, and an additional 12–15% of the landscape was modified as a result of their foraging activities (Naiman et al. 1988). Beavers are able to influence the composition and dynamics of riparian forests by selectively foraging and felling mature trees of certain preferred species (e.g. the trembling or quaking aspen, *Quercus tremuloides*). This then creates disturbance gaps that are colonized by other woody species (Johnston & Naiman 1990a; Rosell et al. 2005).

In modifying riparian areas and converting terrestrial ecosystems to aquatic ones, beavers increase landscape heterogeneity by creating a complex mosaic of aquatic (ponds) and semi-aquatic (marshes, bogs, forested wetlands) habitats within an otherwise forested landscape (Naiman et al. 1988, Johnston & Naiman 1990b; Wright et al. 2002; **Figure 3.34**). The foraging and dam-building activities of beavers give

[34] Beavers were trapped for their pelts, which were used to produce felt hats, including the ubiquitous top hat of the era (late 1700s to mid-1800s). In addition, beaver were harvested for castoreum, a substance secreted by specialized sacs under their tail, which they use for scent marking. Castoreum is used in perfumes, where it supposedly lends leather 'notes' to the fragrance. It has been suggested that the demand for beaver pelts (and castoreum) fueled the exploration and early settlement of the western frontier in North America (e.g. Innis 1999).

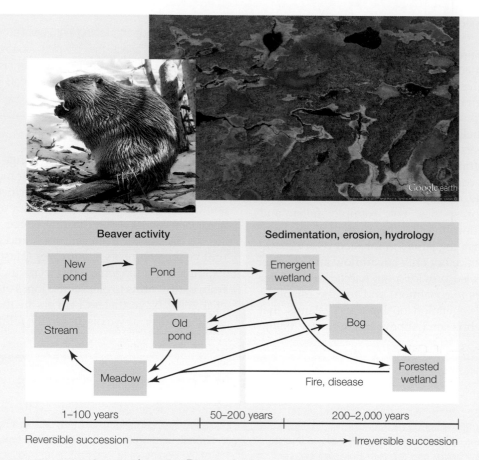

Figure 3.34 Beavers as ecosystem engineers.   Beavers increase the heterogeneity of landscapes, such as here on the Kabetogama Peninsula in northern Minnesota (USA). The result is a complex mosaic of ponds, meadows, and bogs in an otherwise forested landscape. Beaver activity produces multi-successional pathways, creating landscape features that may persist for decades, or even, centuries.

Source: After Naiman et al. (1988). Beaver landscape taken from Google Earth, data provided by Landsat/Copernicus. Beaver photo © Jody Ann/Shuttestock.com

rise to multi-successional pathways, in which the specific trajectory is dependent on a variety of factors including topography, hydrology, vegetation type, and past disturbance history (Naiman et al. 1988). The impacts of beavers on landscapes are also long-lasting. Landscape features created by beavers can persist for decades, or even, centuries (Naiman et al. 1988; Johnston & Naiman 1990a; **Figure 3.34**).

Beavers can also enhance connectivity among wetlands, as well as the connectivity between wetland and upland habitats. Beavers typically construct a system of canals in association with their ponds, by digging channels that can extend up to 500 m into the surrounding landscape, although most are only about 20 m long (Hood & Larson 2015). These channels, which are used by beavers to access foraging areas and to ferry trees and branches back to their ponds, provide a physical link to adjacent wetlands and forests, which may then be used by other organisms (e.g. amphibians) as movement corridors. In some landscapes, the canal system dug by beavers is not only extensive in terms

of total area affected, but may also exceed the impact of dam-building. For example, streams are few but wetlands are plentiful within the Cooking Lake Moraine district[35] of central Alberta, Canada. Beavers there tend to occupy existing kettle lakes ('prairie potholes'), which they modify via the construction of channels all around the edges of these shallow lakes. In one 13 km² landscape, the combined length of all beaver channels totaled about 40 km, a density of about 3 km/km² (Hood & Larson 2015). These channels also increase the length and complexity of the shoreline, which provides beneficial habitat for waterfowl and other species. In addition, channels enhance wetland resilience by increasing drainage into beaver ponds, resulting in larger and deeper ponds that are more resistant to drought and less likely to freeze to the bottom during winter, thereby enhancing survival for beavers as well as other aquatic organisms (Hood & Larson 2015).

[35] Because of its 'knob and kettle' terrain, this region is also referred to as 'Beaver Hills,' which seems especially apt in the present context. The 'hills' or 'knobs' are really just small mounds of glacial deposits (moraines), creating a hummocky landscape.

Beavers, being ecosystem engineers, are among the few species besides humans that can significantly change the geomorphology, and consequently the hydrological characteristics and biotic properties of the landscape. In so doing, beavers increase heterogeneity, and habitat and species diversity at the landscape scale. Beaver foraging also has a considerable impact on the course of ecological succession, species composition and structure of plant communities, making them a good example of ecologically dominant species (e.g. keystone species).

Rosell et al. (2005)

Landscapes are thus fundamentally different when beavers are present than when they are absent. Few species besides humans have this sort of broad-scale, long-lasting, multifaceted impact on landscapes (Rosell et al. 2005). The role of beavers in converting terrestrial ecosystems to aquatic ones is also unique, in that no other species within these landscapes can duplicate the beaver's impact (Kotliar 2000). For this reason, beavers are also considered a **keystone species**, one whose unique effects on community structure and ecosystem function far exceed its own abundance or biomass in the community (Paine 1969; Power et al. 1996).

The Keystone Role of Species

The concept of a keystone species was first introduced by American ecologist Robert T. Paine in reference to the role played by a predatory sea star (*Pisaster ochraceus*) in reducing populations—and thus the competitive dominance—of California mussels (*Mytilus californianus*) in the rocky intertidal zone of the Pacific Northwest (Paine 1969). The bivalve community was far more heterogeneous when sea stars were present than when they were absent. Subsequently, the keystone concept has been applied to a wide range of species, including herbivores, insect pests, and even certain diseases (Mills et al. 1993). For example, the American bison is considered a keystone herbivore of the tallgrass prairie, as we discuss in greater detail next.

CASE STUDY   American Bison as Keystone Herbivores of the Tallgrass Prairie

The American bison (*Bison bison*) is considered a keystone herbivore because its grazing activities 'increase habitat heterogeneity and alter a broad array of plant, community, and ecosystem processes' within the tallgrass prairie (Knapp et al. 1999: 39; **Figure 3.35**). By preferentially grazing on the dominant grass species, bison increase the diversity of herbaceous flowering plants (forbs), which are a major component of plant species diversity in tallgrass prairie ecosystems. Furthermore, bison concentrate their grazing activities within distinct grazing patches (20–50 m²), which may form extensive grazing lawns (400 m²) within areas that have been recently burned (pyric herbivory; Fuhlendorf et al. 2009), thus contributing to greater spatial heterogeneity than would be created by fire alone (Knapp et al. 1999).

In addition, bison engage in a variety of non-grazing activities that likewise increase spatial heterogeneity. For example, bison create wallows by repeatedly rolling on the ground, forming shallow depressions over time (up to 3–5 m wide and 10–30 cm deep). Bison wallows leave an imprint on the landscape that may persist for a century or more and may come to support a very different type of plant community than the surrounding prairie (Knapp et al. 1999; McMillan et al. 2011). Wallows further enhance habitat heterogeneity by functioning as ephemeral wetlands when they fill with water during the rainy season. These small pools are used as breeding habitat by a variety of semi-aquatic organisms, including insects and frogs (e.g. Gerlanc & Kaufman 2003; Pfannenstiel & Ruder 2015).

Finally, bison play a critical role in the cycling and redistribution of nitrogen, a nutrient that directly limits plant productivity in this system **(Figure 11.3)**. Bison obtain nitrogen from the grasses they consume, and then return it in a more labile form through the excretion of urine and feces during their lifetimes, and via the decomposition of their bodies after death (Knapp et al. 1999). By increasing spatial heterogeneity in the availability of nitrogen, bison alter patterns of plant productivity and species composition across the tallgrass prairie landscape. Bison are thus considered a keystone species of the tallgrass prairie because their various and diverse effects on the landscape collectively enhance spatial and temporal heterogeneity across a range of scales (Knapp et al. 1999; **Figure 3.35**).

Figure 3.35 Bison as keystone herbivores. The American bison (*Bison bison*) is considered a keystone herbivore of the tallgrass prairie. Grazing, in conjunction with fire, contributes to greater spatial heterogeneity and biodiversity within the tallgrass prairie landscape.
Source: Photo by Eva Horne.

The American bison, like the beaver, was hunted nearly to extinction by the late 1800s. Unlike the beaver, however, there is little possibility of the bison either naturally recolonizing or being reintroduced into much of its former range. Most of the Great Plains has since been converted to agriculture and other land uses or is now used as a grazing range for cattle. We might thus wonder whether the bison's role as a keystone species of tallgrass prairie is still ecologically relevant, or whether that role has now been assumed by cattle. Cattle are the **ecological equivalents** of bison in many respects: they are large herbivores that preferentially consume grasses, and likewise concentrate their foraging activities within recently burned areas of the landscape (Knapp et al. 1999; Allred et al. 2011a). Grazing by either bison or cattle increases spatial heterogeneity and promotes plant species richness relative to ungrazed grasslands (Towne et al. 2005). But there are some important differences between cattle and bison in terms of their foraging ecology and space use. Cattle are generally stocked at higher densities and graze more uniformly than bison (Towne et al. 2005). In addition, cattle spend more time than bison near water and within riparian and other wooded areas, which can lead to bank erosion, affect water quality, and alter vegetation within areas of heavy use

(Allred et al. 2011a). Nevertheless, the bison's role as a keystone herbivore has largely been replaced by cattle in most grasslands (Allred et al. 2011a).

The importance of grazing and fire for enhancing the diversity and productivity of tallgrass prairie has led to calls for restoring heterogeneity on cattle-grazed rangelands by mimicking the historical bison grazing-fire interaction (Fuhlendorf & Engle 2001; Fuhlendorf et al. 2009). Traditional rangeland management practices seek to minimize environmental heterogeneity so as to provide uniform forage conditions for livestock (which, you'll note, is consistent with an equilibrium-based view of ecosystems; Fuhlendorf & Engle 2004). In the case of tallgrass prairie, this entails prescribed burning of the entire management unit or pasture, often on an annual basis. Rather than burning all of the rangeland in a given management unit, land managers could burn only a third of their pasture each year, in rotation. Such 'patch burning' creates a shifting disturbance mosaic, especially given that cattle preferentially graze within these burned areas (Fuhlendorf & Engle 2004). Subsequently, patch-burn grazing is effective at increasing spatial heterogeneity and native biodiversity within tallgrass prairie rangelands (Fuhlendorf et al. 2006; McGranahan et al. 2012).

The use of **historical** or **natural variability concepts** for managing ecosystems and landscapes embraces a non-equilibrium view of ecological systems. The natural or historical range of variability is defined as the spatial and temporal variation in ecological conditions that occur (or did) within a particular area and time period, in the absence of human activities (Landres et al. 1999). As such, natural variability is seen as a benchmark for evaluating the effects of human land-use and management practices on landscapes, as well as a more effective way of maintaining biological diversity or restoring degraded ecosystems (Landres et al. 1999). The application of natural variability concepts as a management tool is not without challenges, however, as this presupposes that we have information on the historical disturbance regime and understand the landscape-scale effects of that disturbance regime. Further, natural disturbance regimes may include extreme disturbances that may be viewed as impractical or unacceptable in the current landscape context. For example, the restoration of bison throughout much of the Great Plains is no longer possible, and natural grassfires (wildfires) are not allowed to burn unchecked. Thus, the restoration of fire and grazing to these grassland systems is done within a certain degree of 'management variability,' rather than the full range of natural or historical variability (Landres et al. 1999).

Approximating historical conditions provides a coarse-filter management strategy that is likely to sustain the viability of diverse species, even those for which we know little about…Similarly, because of limited understanding about ecosystems, approximating past conditions offers one of the best means for predicting and reducing impacts to present-day ecosystems…   Landres et al. (1999)

Relative Impact of Species on Landscapes

It could be argued that virtually all species have an effect on the distribution and abundance of resources at some scale, and thus all might be considered biotic agents of disturbance. However, the designation of a species as a geomorphic agent, an ecosystem engineer, or a keystone species implies that some species have a greater impact than others. Indeed, some species, such as the beaver, have such a profound effect on the landscape that they have been variously classified as a geomorphic agent (Butler 1995), as ecosystem engineers (Jones et al. 1994), *and* as a keystone species (i.e. a keystone habitat-modifier; Mills et al. 1993). To what extent, then, are these concepts essentially interchangeable when discussing the role of organisms as agents of landscape disturbance?

We have discussed how the designation of species as either geomorphic agents or ecosystem engineers overlaps to some degree, given that these concepts arose independently in disparate disciplines. In contrast, keystone species and ecosystem engineers may appear to be similar in concept but are not in fact synonymous (Jones et al. 1997). Recall that ecosystem engineers provide or create new habitats and resources by modifying the physical structure of the environment (e.g. bioturbation), whereas the effects of a keystone species typically involve trophic or competitive interactions. Although keystone species can contribute to system engineering, they do not always do so, nor is this necessary for engineering to occur (Jones et al. 1997, p. 1949):

> When a beaver cuts down the tree to make the dam it does not have to eat any part of the tree in order for the dam and resulting pond to have an effect. And while disturbance…and engineering can often have similar effects, not all engineering is disturbance (e.g. tree growth) and not all disturbance is engineering (e.g. a hurricane). Many "keystone species"…are engineers (e.g. beavers), but others (e.g. sea otters) are not; some engineering has big effects on other species (e.g. beaver dams), while other impacts may be relatively trivial (e.g. a cow hoofprint).

This quotation highlights three important points: (1) not all keystone species are ecosystem engineers (i.e. these concepts are not synonymous); (2) not all forms of engineering have large impacts on the landscape; and, (3) not all forms of engineering qualify as a disturbance.

As with any type of disturbance, engineers differ in their impacts. Although some engineers have large, varied, and long-lasting impacts on the landscape (e.g. beavers), others have only small, localized effects of short duration (e.g. many songbirds construct nests that last only a single season; Jones et al. 1994). Thus, ecosystem engineers are not always agents of landscape disturbance and dynamics, although they obviously can be. The biggest effects are attributable to those species having 'large per capita impacts, living at high densities, over large areas for a long time, giving rise to structures that persist for millennia and that modulate many resource flows' (Jones et al. 1994, p. 373). It should be readily apparent that no species fulfills all of these criteria—except our own. We humans are thus ecosystem engineers par excellence (Jones et al. 1994).

Humans as the Primary Driver of Landscape Change

Of all the disturbances discussed here, the transformation of landscapes by humans is the most pervasive and globally significant (Vitousek et al. 1997; Foley et al. 2005). The **human footprint**, which maps the degree of human influence on landscapes, is estimated to affect 83% of Earth's total land-surface area (Sanderson et al. 2002).[36] About half of this land area is used for agricultural production (Foley et al. 2005). As a consequence, humans now appropriate about a quarter of the global net primary productivity of terrestrial ecosystems, much of which has been achieved through landscape transformation and other land-use activities, such as prescribed fire and intensive agricultural practices (Haberl et al. 2007). It should come as no surprise, therefore, that human land-use activities are causing worldwide declines in biodiversity. Globally, local species richness has been reduced by an average of 13.6% over the last 500 years, with the worst-affected areas of the world averaging a 77% loss of species as a result of intensive land use and related human pressures (Newbold et al. 2015). By 2100, we could see an average decline in local species richness of 20%, with the greatest losses concentrated in the most biodiverse—but economically poor—countries of the world (Newbold et al. 2015). This may not sound like much, but a 20% loss of species locally is likely to disrupt ecosystem function, and thus the ecosystem goods and services upon which all human societies depend.

In view of the human domination of Earth's ecosystems (Vitousek et al. 1997), there has been a growing movement to designate the epoch we live in as the 'Anthropocene,' a new geological period in which humans have surpassed nature to become the dominant environmental force on Earth (Crutzen 2002). Although the start of the Anthropocene was originally proposed as the beginning of the Industrial Revolution, both earlier and later starting points have been suggested (**Table 3.5**). Many geologists would prefer a formal stratigraphic designation, a 'golden spike' that is widely detectable in the geologic record (Ruddiman et al. 2015).[37] It has therefore been suggested that the

mid-20th century offers a better baseline for demarcating the Anthropocene, as this marks the beginning of the 'Great Acceleration,' a period of widespread and rapidly increasing human global impacts on the environment resulting from the unprecedented burning of fossil fuels, the rapid rise in atmospheric CO_2, deployment of nuclear weapons, and pollution from agricultural and industrial sources (Lewis & Maslin 2015; Swindles et al. 2015; Zalasiewicz et al. 2015). These activities have left a globally coherent anthropogenic signal in the stratigraphic layer that meets the criteria for a 'golden spike.'

> Does it really make sense to define the start of a human-dominated era millennia after most forests in arable regions had been cut for agriculture, most rice paddies had been irrigated, and CO_2 and CH_4 concentrations had been rising because of agricultural and industrial emissions?
>
> Ruddington et al. (2015)

Others argue that a focus on such recent changes overlooks the profound effect that humans have had on the environment for thousands of years, and which have likewise affected the landscapes, atmosphere, climate, and biodiversity of the planet. In particular, it has been argued that humans first started exerting a global impact some 11,000 years ago during the Neolithic agricultural revolution (Ruddiman et al. 2015; **Table 3.5**). The widespread clearing of forests, the spread of agriculture and domestication of livestock, and the concomitant rise in greenhouse gases (atmospheric CO_2 and CH_4) all demonstrate that pre-industrial humans were already altering landscapes and climate on a global scale (Ruddiman 2013). Regardless of the exact start date, however, there can be little doubt that we have had a profound influence on the global environment, the impacts of which will be observable for centuries, and perhaps millennia, to come.

Stages of Anthropogenic Landscape Transformation

It could be argued that all of the global impacts by humans are ultimately rooted in the agricultural transformation of landscapes, which started between 8500 and 11,000 years ago. Wherever and whenever it begins, landscape transformation by humans tends to

[36] The degree of human influence was estimated from geographic proxies such as human population density; the relative accessibility of an area to humans, based on the availability of roadways, railways, and waterways; the degree of landscape transformation, based on the degree of land-cover change, urbanization, and density of transportation networks; and the level of technological development, as assayed by the use of electrical power, which implies fossil-fuel use and thus a greater ability to modify the environment (Sanderson et al. 2002).

[37] A 'golden spike' refers to the Global Boundary Stratotype Section and Point (GSSP) that marks the 'type' location of a particular stratigraphic layer, and thus serves as a standard point of reference for geologists. The lower boundary of a stratigraphic layer (i.e. the beginning of a particular geologic stage) is defined by its unique biological (fossil), chemical, magnetic, and/or climatic profile, which should be consistent in rock layers of that type worldwide. The stratigraphic designation and location of the GSSP must be approved by the International Commission on Stratigraphy (ICS), and then a metal stake or plaque is placed on the rock face to mark the boundary of the geological section at the approved location. The idea is that geologists or other researchers wishing to study that particular

stratigraphic layer would then be able to travel to the location of its GSSP to do so (ease-of-access being one of the considerations the ICS debates when approving a particular location as a GSSP). The first GSSP, marking the Silurian-Devonian boundary, was established in 1972 in Klonk near the village of Suchomasty, about 35 km southwest of Prague in the Czech Republic. As of July 2018, the ICS has approved 71 GSSPs, most of which are in Europe (http://www.stratigraphy.org/ICSchart/ChronostratChart2018-07.pdf).

TABLE 3.5 Some potential start dates for the Anthropocene.

Event	Date[†]	Origin and geographical extent of impacts	Stratigraphic evidence
Neolithic agricultural revolution: Origins of farming	~11,000 YBP	Southwest Asia origin, but multiple independent origins worldwide	Fossil crop pollen; phytoliths;[1] charcoal
Extensive farming: Majority of humans become agriculturalists; conversion of forests to crops and grazing lands; associated fire impacts; increased atmospheric carbon dioxide (CO_2)	~8000 YBP–present	Eurasian origin, global impact	CO_2 inflection in glacier ice; fossil crop pollen; phytoliths; ceramic materials
Rice cultivation and domestication of livestock: Increased methane (CH_4) production	6500 YBP–present	Southeast Asian origin, global impact	CH_4 inflection in glacier ice; stone axes; fossilized remains of domestic livestock
Columbian Exchange: First global trade networks; era of European colonialism and colonization of the Americas; disease spread and reduced human population; reduced farming and fire use leading to forest and grassland regeneration (increased carbon uptake); globalization of crops; cross-continental exchange of biota	1492–1800	Eurasian-Americas event, global impact	Decreased CO_2 in glacier ice (the 1610 CO_2 minima); fossil crop pollen (especially of maize, *Zea mays*); phytoliths; charcoal reduction; increased CH_4
Industrial Revolution: Accelerating fossil fuel use; rapid societal changes; globalized trade	1760–present	Northwest Europe origin, local impacts but eventually spread worldwide	Fly ash[2] from coal burning; increased CH_4 and nitrate (NO_3^-) from fossil fuel burning
Great Acceleration: Rapid increase in human population; development of industrially produced chemicals; detonation of nuclear weapons	1945–present	Local events, global impact	Radionuclides (^{14}C) in tree rings and glacier ice; enriched plutonium (^{239}Pu); compounds from cement, plastic, lead, and other metals in glacier ice or sediments

Source: After Lewis & Maslin (2015)

[†]YBP = years before present

[1]Phytoliths are microscopic, silica-based structures found in some plant tissues, which are deposited in the soil after plants die and decompose. An analysis of phytoliths in the soil can thus help to reconstruct what vegetation types were present in the historical landscape.
[2]Fly ash is an atmospheric pollutant generated by coal combustion, as opposed to 'bottom ash' that falls to the bottom of the boiler. Fly ash is made up of a number of fine particles, including silicon dioxide, aluminum oxide, and calcium oxide, as well as trace concentrations of arsenic, lead, mercury, and other heavy metals.

progress through a series of predictable stages, which can be characterized as (1) pre-settlement, (2) frontier, (3) subsistence, (4) intensifying, and (5) intensive land use (Foley et al. 2005; **Figure 3.36**).

Pre-settlement landscapes are characterized by their natural vegetation or land-cover types, which are converted to other land uses as the region becomes progressively settled, moving from clearings and homesteads within the frontier landscape, to subsistence agriculture and small-scale farming, to intensive agriculture and urbanization. Notice that protected areas are usually not part of the landscape until the later stages of the land-use transition, when there is relatively little

natural vegetation left to protect (**Figure 3.36**). Different parts of the world are at different stages along this landscape-transformation gradient, and not all areas will move through these stages in a consistent or linear fashion (Foley et al. 2005). Some regions may persist in a given stage for long periods of time, whereas others will move rapidly through these stages, transitioning from pre-settlement to intensive land use in a matter of decades.

A case in point is the conversion of the Brazilian Cerrado from a vast tropical savanna to a predominantly agricultural landscape with several large urban centers, a transformation that happened in fewer than 60 years

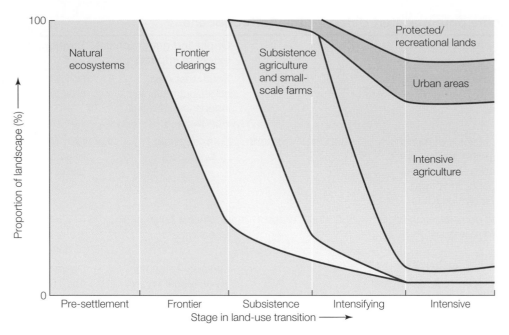

Figure 3.36 Stages of landscape transformation.
Source: After Foley et al. (2005).

(Klink & Moreira 2002).[38] The opening of the agricultural frontier in the Cerrado began in 1960, with the move of the country's capital from Rio de Janeiro to Brasília in 1960,[39] which necessitated the development of an extensive highway system to connect the region with Brazil's other major cities. Agricultural expansion within the Cerrado, which previously had supported only subsistence farming and ranching, was actively promoted through government incentives to settle and farm the region in the 1970s, with the goal of increasing Brazil's agricultural production to self-sustainable levels (Barros 2014). The government also created the Brazilian Agricultural Research Corporation (Empresa Brasileira de Pesquisa Agropecuária, EMBRAPA) in 1973 to speed the growth of the agricultural sector through investments in research, technology, and knowledge transfer. Indeed, it was the technological advances developed and implemented by EMBRAPA that permitted the broad-scale agricultural expansion and intensification of the Cerrado. In particular, the nutrient-poor, acidic soils of the Cerrado were amended on a massive scale, using phosphate fertilizers and lime, to enable the production of cash crops like soybean, corn, cotton,

and sugarcane (the latter is mostly grown as a biofuel). Besides agriculture, the Cerrado is also an important cattle-production region, producing a third of Brazil's beef cattle (Castro 2015). Cattle ranching has thus been a major contributor to land-use change within the region, not only through deforestation but also from the planting of pastures in non-native African grasses to increase forage production for cattle. As a consequence, some 60% of the Cerrado is now directly affected by human land use, and about a third has been converted to crops and planted pastures (Oliveira & Marquis 2002; Klink & Machado 2005). The result, however, has been that the Cerrado has nearly tripled its contribution to the total agricultural production of Brazil within the past four decades (contributing to 19.5% of the national agricultural GDP[40] in 2009; Castro 2015). During this same period, Brazil has successfully transformed itself from a major agricultural importer to one of the world's largest exporters of agricultural commodities (OECD 2015), due in no small part to the agricultural conversion of the Cerrado.

Types of Human Land Use

As the example of the Cerrado illustrates, the transformation of landscapes is driven by a combination of social, economic, and political factors, which can vary over space and time, even within the same region. The types of human land use that contribute to landscape

[38] The Cerrado, which makes up 21% of the land area of Brazil, is also considered a global biodiversity hotspot (Myers et al. 2000), a designation that acknowledges the degree of threat, in terms of the amount of primary vegetation lost and the high degree of endemism within the region.
[39] In terms of distance, this would be akin to moving the United States capital from Washington, DC along the east coast, to the Midwestern city of Chicago, Illinois (a distance of about 927 km, or 576 miles, as the crow flies).

[40] Agricultural GDP is the portion of the gross domestic product (GDP) that comes from the agricultural sector, which includes the value of all agricultural goods and services.

Figure 3.37 Geometry of agricultural land use.   (A) Agricultural kaleidoscope (near Lanchester, Pennsylvania, USA). (B) Crop circles produced by center-pivot irrigation (near Colby, Kansas, USA). (C) Agricultural patchwork quilt (near Bowling Green, Ohio, USA). (D) Ribbon farms (near Leamington, Ontario; Canada). (E) Agricultural jigsaw puzzle (near Summerville, Pennsylvania; USA). (F) Agroforestry checkerboard (near Priest Lake, Idaho; USA).

Source: Google Earth, data provided by A) Landsat/Copernicus B) USDA Farm Service Agency C) USDA Farm Service Agency D) GeoEye E) Landsat/Copernicus F) Landsat/Copernicus.

transformation are varied, but some of the more common and globally significant are: (1) agricultural expansion and deforestation; (2) urbanization; (3) transportation corridors (e.g. roadways); and (4) reforestation.

AGRICULTURAL EXPANSION AND DEFORESTATION In terms of landscape transformation, the conversion of forests to agricultural land use (croplands and pastures) stands out as one of the more striking examples of anthropogenic landscape change (**Figure 3.37**). Croplands and pastures have become one of the largest terrestrial biomes on the planet, occupying 40% of the land surface (Foley et al. 2005). Furthermore, the global demand for agricultural commodities is projected to double as another 2.4 billion people are added to the human population by mid-century (Tilman et al. 2011). If current land-use trends continue, this would amount to an additional 1 billion hectares of land cleared by 2050, which is an area about the size of the USA. Land clearing on this scale for agricultural production—**agricultural extensification**—would result in marked increases in greenhouse gas emissions and global nitrogen-fertilizer use (Tilman et al. 2011), thereby compounding the environmental impacts of land clearing far beyond the local or landscape scale. Further, most of this new agricultural land would come at the expense of tropical forests, where much of the world's biodiversity is concentrated (Gibbs et al. 2010).

Although the potential for rapid agricultural extensification is cause for concern, the clearing of forests to create more land for agriculture is an ancient practice that harkens back to the origins of agriculture itself. Agriculture spread from the Fertile Crescent of southwest Asia into the forested regions of south-central Europe some 7000–8000 years ago, resulting in widespread deforestation from slash-and-burn agriculture, especially after the advent of the plow 5000–6000 years ago (Ruddiman 2003). Elsewhere in Asia and Europe, the paleoecological record reveals a similarly long history of deforestation linked to agricultural expansion (Bhagwat 2014). Indeed, deforestation appears to have been one of the most pervasive drivers of landscape transformation throughout Europe during the Holocene. In other parts of the world, such as the Americas and Australia, broad-scale deforestation generally occurred only after European settlement in the late 15th and 18th centuries, respectively. However, the case study we covered earlier in this chapter, on how deforestation and drought may have contributed to the collapse of the Classic Maya civilization, demonstrates how even pre-Columbian societies were capable of broad-scale deforestation. Thus, although the timing varied widely within and among regions, the process of deforestation has generally followed a similar pattern, with most deforestation taking place initially for the development of agricultural land use (**Figure 3.36**).

It has been estimated that at least half of the world's forests have disappeared as a result of human activity, with three-quarters of that occurring since 1700 (Alberti et al. 2003). Currently, forests cover 31% of the total land area, more than half (53%) of which is located in five countries: Russia, Brazil, Canada, USA, and China (FAO 2010; **Figure 3.38A**). Although the global deforestation rate has declined since 1990, it is still alarmingly high: 13 million ha/year (an area roughly the size of Greece) has been lost each year in the first decade of this century.[41] Again, much of this forest loss is occurring in the tropical forests of the world, such as in South America, Africa, and southeast Asia (**Figure 3.38B**). Currently, only about one-third (36%) of the world's forests can be considered primary or old-growth forest—that is, native forests that have no visible sign of human activity (FAO 2010). Unfortunately, primary forest is being lost at a rate of about 4 million ha/year, such that 40 million ha (an area about the size of Paraguay) was lost between 2000 and 2010 (FAO 2010). Furthermore, remaining forests are likely to suffer degradation from habitat fragmentation. More than 70% of the world's remaining forests are now within 1 km of a forest edge, which puts the vast majority of forests within the range where negative edge effects (human disturbance, altered microclimate, ecological impacts from non-forest species; **Box 7.1**) can have a significant impact on forest ecosystems (Haddad et al. 2015).

Instead of converting more forests to agricultural land, we might instead increase the productivity of current agricultural lands, especially in areas of the world that are currently underyielding. **Agricultural intensification**, involving the use of high-yielding crop varieties, irrigation, fertilizers, and pesticides, has not only increased agricultural production worldwide (the 'Green Revolution'), but is also credited with saving some 15 million km² of forest, an area about 1.5 times larger than the USA (Borlaug 2007). Agricultural intensification is not without environmental consequences, however, in terms of the collateral damage caused by agricultural runoff, increased greenhouse gas emissions, and increased water use (Matson et al. 1997). Still, it has been suggested that the 2050 global crop demand could best be met by adopting a moderate degree of intensification, through the development and transfer of high-yield technologies to existing cropland areas in underyielding nations (Tilman et al. 2011). Although such a targeted strategy would involve some increase in global nitrogen use (albeit less than if we were to stay on our current trajectory), it would greatly reduce the projected amount of greenhouse gas emissions (by 67%) as well as the projected amount of land cleared (by 80%) relative to current agricultural land-use trends (Tilman et al. 2011).[42] Indeed, agricultural intensification has been proposed as a way of mitigating future greenhouse gas emissions, given that clearing forested land for agriculture would release far more carbon into the atmosphere, especially as agricultural expansion is occurring primarily in tropical forests, which are major carbon sinks (Burney et al. 2010).

Clearly, the relationship between agriculture and deforestation is a complex one (Villoria et al. 2014). Although it is often assumed that there is a trade-off between agricultural intensification (developing and adopting yield-enhancing technologies) and agricultural extensification (land clearing) that can be optimized, the reality is far more complex. For example, increases in productivity as a result of the development and transfer of high-yield technologies can increase the profitability of agriculture relative to other land uses (e.g. forests), thereby encouraging further expansion into the frontier (Villoria et al. 2014). Thus, agricultural intensification and agricultural extensification need not be mutually exclusive. Regardless, it seems clear that agricultural expansion is inevitable if we are to meet the future global food demand, and that much of this will come at the expense of tropical forests and savannas.

URBANIZATION For the first time in human history, more people live in urban areas than in rural areas globally.[43] More than half (54%) of the human population currently lives in urban areas, of which there are now 28 'mega-cities' worldwide (cities with more than 10 million inhabitants; United Nations 2014). In the coming decades, we will see further changes to the size and spatial distribution of the human population, with some two-thirds of the global population (66%) projected to be living in urban areas by 2050 (United Nations 2014).

[41] Although most of this deforestation involved the conversion of tropical forests to agricultural land, some forest loss was due to natural causes, such as the severe drought and associated fires that have consumed forests in places like Australia (FAO 2010).

[42] This assumes that the yield gap can be closed using available technologies and genetic materials (e.g. genetically modified organisms or GMOs), which may not be the case in some underyielding countries where farmers do not have access to the technical knowledge and skills required to increase production, the finances to invest in higher production (e.g. irrigation, fertilizer, machinery), and/or the crop varieties that could maximize yields (Godfray et al. 2010).

[43] Because different countries define 'urban' versus 'rural' differently, there is no single or standardized definition of these terms. Although 'urban' implies a high population density, different criteria are in fact used to define urban areas, typically based on some combination of population size or density, administrative, and economic criteria (e.g. percent of population employed in non-agricultural sectors). For example, the US Census Bureau defines urban areas as ≥2500 people (www.census.gov/geo/reference/ua/urban-rural-2010.html), whereas in Canada, the minimum population threshold to qualify as an urban area is 1000 people (www.statcan.gc.ca/eng/subjects/standard/sgc/notice/sgc-06).

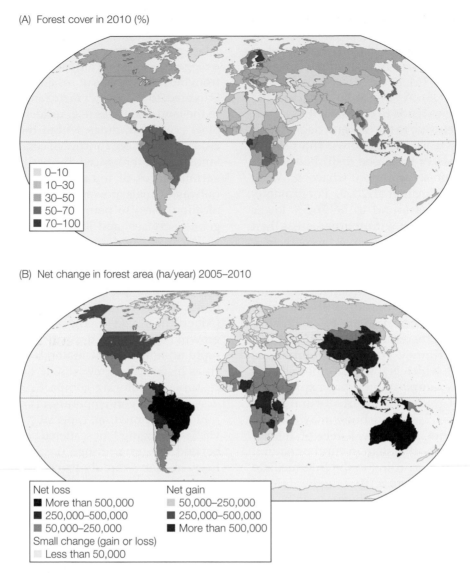

(A) Forest cover in 2010 (%)

▨	0–10
▨	10–30
▨	30–50
▨	50–70
■	70–100

(B) Net change in forest area (ha/year) 2005–2010

Net loss
■ More than 500,000
■ 250,000–500,000
▨ 50,000–250,000
Small change (gain or loss)
▨ Less than 50,000

Net gain
▨ 50,000–250,000
■ 250,000–500,000
■ More than 500,000

Figure 3.38 Global deforestation and reforestation.   (A) Percent forest cover in 2010; (B) Net change in forest area (ha/year) from 2005 to 2010.

Source: Data from FAO (2010).

Although urban areas cover less than 3% of the world's land-surface area, urbanization is a major driver of global environmental change (Alberti et al. 2003; Grimm et al. 2008). Urban expansion and its associated land-cover changes drive habitat loss, threaten biodiversity, alter hydrological systems and biogeochemical cycles, and contribute to climate change through increased greenhouse gas emissions and the urban heat-island effect[44] (Grimm et al. 2008). Furthermore, if current trends in population density and urban expansion continue, urban land cover will increase by 1.2 million km^2 (an area roughly the size of South Africa) by 2030, which would be nearly triple the amount of urban land area present in 2000 (Seto et al. 2012). Since most of this growth is projected to take place within countries that are located in tropical/subtropical regions of the world (e.g. India, Nigeria), urban expansion will result in a considerable loss of habitat within several key biodiversity hotspots as well (e.g. the Western Ghats in India, the Guinean Forests of West Africa; Seto et al. 2012).

Although urbanization represents landscape transformation in the extreme, urban landscapes are far more complex than evoked by the notion of a 'concrete jungle.' Urbanization has typically been viewed along a gradient in human population and building density, both of which tend to decrease away from the urban

[44] In the absence of trees and other vegetation, the concrete and asphalt surfaces of cities absorb more shortwave radiation during the day, such that cities are warmer (by 1–3°C), and take longer to cool down in the evenings, than surrounding suburban and rural areas (Arnfield 2003).

core. Urbanization is thus a continuum rather than a dichotomy. As such, it has been proposed that this **urban-rural gradient** should be studied like any ecological or disturbance gradient to assess human impacts on the biota and environment (McDonnell & Pickett 1990; McDonnell & Hahs 2008). As might be expected, the number of native species tends to decline with increasing urbanization, largely due to the loss and degradation of natural habitats, increased predation or competition with non-native species, or increased disturbance from human activity (e.g. noise pollution; Chace & Walsh 2006; McKinney 2008; Francis et al. 2009). For example, local species richness was found to be 76.5% lower (and abundance 39.5% lower) within urban areas than in areas minimally affected by human activity (e.g. mature forests; Newbold et al. 2015). Urban environments favor a small subset of disturbance-tolerant or highly adaptable species, such that urban areas worldwide are dominated by a similar group of species (and, in some cases, the same species), many of which have been introduced and are therefore not native (so-called **urban exploiters**; McKinney 2002). The ubiquitous pigeon (*Columba livia*), which is found in cities all over the world, is a prime example of an urban exploiter.

As with ecological gradients, the urban-rural gradient is not always linear. Species richness may actually be higher in areas with moderate levels of human development (i.e. the suburbs) than in rural or natural areas (McKinney 2002). According to the Intermediate Disturbance Hypothesis (**Figure 3.10**), a moderate level of disturbance can increase habitat heterogeneity, and thus species diversity, especially if the landscape was largely homogeneous to begin with (e.g. mostly agriculture or forest; **Figure 3.11**). Not all species will benefit from intermediate levels of urban development, however. Disturbance-sensitive species (e.g. forest-interior species) tend to be **urban avoiders**, and thus will be rare or absent from such landscapes. The species most likely to benefit from suburban development are therefore **urban adapters**, which are typically early-successional or habitat-edge species that do well in moderately disturbed or fragmented landscapes. Many of these species also excel at exploiting resources provided by humans, whether this resource provisioning is intentional (e.g. bird feeders) or unintentional (e.g. the bounty provided by vegetable and flower gardens, ornamental shrubs, pet food, or household refuse).

Even if diversity peaks at intermediate levels of urbanization, the urban-rural gradient is less a single, uniform gradient in space than a characterization meant to capture the range in variability of human population density and land-use intensity. As with any landscape, urban landscapes can be viewed as heterogeneous mosaics (Forman 1995a; Cadenasso et al. 2007), consisting of different land uses (industrial, commercial, residential) and open spaces of natural or seminatural areas (e.g. lakes, parks). Urban areas exhibit heterogeneity across a range of scales that are hierarchically nested, ranging from the megalopolis or urban region in which the city is located, to the greater metro area surrounding the city, to the land-use types (e.g. residential), neighborhoods, blocks, and buildings within the city itself (Forman 2014). In addition, functional connectivity of the urban landscape is enhanced through the construction of various types of infrastructure corridors (roadways, railways, canals, power lines, water pipelines, communication lines) to permit the flow of people, material goods, services, and information throughout the city. Urban areas are dynamic; they grow and change over time in response to the interaction between individual human agents (e.g. homeowners, developers, city planners) and the biophysical, social, economic, and political environments in which they live and work (Alberti et al. 2003). The pattern and pace of urban growth has changed dramatically in the current era of rapid urbanization (Ramalho & Hobbs 2012). Whereas cities historically grew slowly over many decades or centuries, exhibiting compact or concentric rings of development, contemporary urban growth is far more rapid and sprawling, especially when growth occurs via 'leapfrogging'[45] or scattered pockets of development beyond the urban fringe (i.e. exurban development). Even in many rapidly growing cities, suburban and exurban areas are growing much faster than other zones, which is a concern given that per capita land consumption tends to be higher in these areas (e.g. the rise of the 'ranchette' in the western USA). In the USA, for example, exurban development is projected to grow at a rate some 6.5 times faster than that of urban or suburban growth over the next decade (Theobald 2005).

Human-dominated landscapes have unique biophysical characteristics. Humans redistribute organisms and the fluxes of energy and materials…Relative to non–human-dominated systems, urban ecosystems have low stability, different dynamics (complex and highly variable on all temporal and spatial scales), more non-native species, different species composition (often simplified, always changed), and unique energetics (anti-entropic in the extreme). They have rich spatial and temporal heterogeneity—a complex mosaic of biological and physical patches in a matrix of infrastructure, human organizations, and social institutions.   Alberti et al. (2003)

[45] Leapfrog development is a pattern of urban sprawl in which developers skip over currently developed areas to obtain land at a lower price farther out, despite the absence of utilities and infrastructure, which then must be extended to service these areas (Heim 2001).

Urban areas thus meet all of the criteria used to define landscapes, in terms of their structure, function, and change, and therefore fit within the purview of landscape ecology. Landscape ecology, and ecology more generally, is increasingly being called upon to assist in developing sustainable solutions to the sorts of challenges faced by urban planners and designers, landscape architects, community leaders, and decision-makers confronted with managing and mitigating the impacts of rapid urban growth. **Landscape design** can help to extend the science of landscape ecology, by putting into practice its pattern-process paradigm within urban areas to enhance ecosystem services and sustainability while continuing to meet societal needs and values (Nassauer & Opdam 2008). For example, investments in green infrastructure (e.g. street trees, green roofs, rain gardens, greenbelts, parks, and wetland restoration) can help to enhance or restore ecological functioning within the urban landscape, which might then confer greater resilience to the urban socioecological system in the face of a major perturbation, such as drought or flooding (Pickett et al. 2011). For their part, ecologists are increasingly drawn to the study of **urban ecology**, recognizing that urban areas are a unique 'hybrid ecosystem' that emerges from the complex interactions between humans and ecological processes (Alberti 2008; see also Marzluff et al. 2008; Gaston 2010; Niemelä et al. 2011; Forman 2014). Given the ascendancy and global impact of urbanization, ecologists can no longer afford to study only 'natural' landscapes if we wish to understand and manage human impacts on the environment, especially given the extraordinarily large ecological footprint of most cities and in view of current urbanization trends (Grimm et al. 2008). Recall that most landscapes now bear the stamp of the human footprint (Sanderson et al. 2002), such that virtually all landscapes are anthropic landscapes or **anthromes** to some degree (Ellis & Ramankutty 2008; **Figure 1.2**).

TRANSPORTATION CORRIDORS Urbanization and urban growth are ultimately made possible by the development of infrastructures capable of moving people, goods, services, and information across the landscape. In particular, the construction of transportation corridors—whether in the form of roadways, railways, or waterways—is both a precursor to and consequence of urbanization. The construction of transportation corridors leads to development, and development leads to the further construction of transportation corridors. We can therefore count the millions of kilometers of transportation corridors that now crisscross the globe as one of the major global impacts of urbanization. At present, there are some 1.1 million km of railways, 2.3 million km of waterways, and 64 million km of roadways worldwide (World Factbook 2013).

Regardless of the type, transportation corridors cut linear swaths through landscapes, and thus contribute to habitat loss and fragmentation. As an example, consider the impact of roads on the pattern of deforestation within the state of Rondônia in Brazil (**Figure 3.39**). Here, as in the Cerrado, the construction of a major interstate highway, along with rural development projects launched by the Brazilian government in the 1970s and 1980s, opened this part of the Amazon Basin to settlement and eventual urbanization (Pedlowski et al. 1997; Ferraz et al. 2005). Forest clearing for timber, agriculture, and pasture by small landholders (<100 ha) initially occurred along major roads, and then expanded along secondary roads to produce this unique 'fishbone' pattern of deforestation over time. These small clearings eventually coalesced to produce larger clearings, especially as landholdings became consolidated into larger units and converted to pastures for cattle ranching (Alves 2002). The population of Rondônia has grown from about 111,000 in 1970 to an estimated 1.76 million in 2018, nearly three-quarters (73.5%) of which lives in urban areas (Instituto Brasileiro de Geografia e Estatística, IBGE).[46] Subsequently, Rondônia has experienced some of the highest rates of deforestation in the Brazilian Amazon, resulting in the loss of 43% of its forested area over the past 40 years (total area of forest lost, as of 2017: 92,086 km²; INPE).[47]

Because of the impact that roadways in particular have on habitat loss and fragmentation, road density (km road/km² land area) is sometimes used as an index of landscape fragmentation, especially in evaluating the extent of forest fragmentation (e.g. Reed et al. 1996; Heilman et al. 2002). Most deforestation occurs in fairly close proximity to roads, as evidenced by the pattern of deforestation in Rondônia. In the Brazilian Amazon as a whole, nearly 60% of deforestation occurs within 25 km of a major road, with almost all (87.5%) deforestation occurring within 100 km of roads (Alves 2002). Even in the USA, where the overall road density is on the lower end among developed countries (68 km/km²),[48] roads still affect a significant portion of the total land area: an estimated 83% of the land area is

[46] IBGE is the Brazilian Institute of Geography and Statistics (www.ibge.gov.br).

[47] INPE is Brazil's National Institute for Space Research (Instituto National de Pesquisas Espaciais; www.inpe.br). So, for comparison, the amount of forest lost in Rondônia is roughly equivalent to the size of the US state of Indiana. The Brazilian state of Rondônia, in turn, is approximately the size of the US state of Oregon (~240,000 km²).

[48] For comparison, many European countries have much higher road densities: UK = 172.4, France = 172.5, Germany = 180.5, The Netherlands = 372, and Belgium = 498.7 km/km² (World Development Indicators, Transportation Infrastructure Data, most recent data available 2005–2007). In terms of overall length, however, the USA has the most roadway of any country (~6.6 million km, 65% of which are paved; World Factbook 2013).

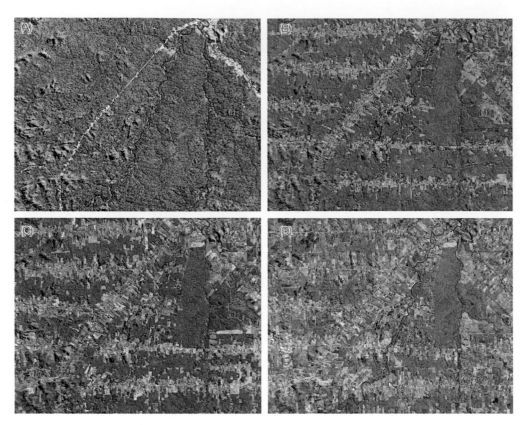

Figure 3.39 Fishbone deforestation along roadways in Rondônia, Brazil. The municipality of Ariquemes is located in the upper-right corner of each image, which is at the intersection of two highways. Deforestation proceeds out from the highways and along secondary roads in a fishbone pattern over time.

Source: Landsat Imagery obtained from USGS, Earthshots: Satellite Images of Environmental Change.

within a kilometer of a road, whereas 20% is within about 100 m of a road (Forman 2000; Riitters & Wickham 2003).

Unfortunately, the environmental impacts of transportation corridors are not limited to just the width of the corridor. Similar to habitat edge effects (**Box 7.1**), the environmental impacts of transportation corridors may extend many hundreds of meters beyond the edge of the rails, road, or bank (Forman & Alexander 1998). For example, the increased runoff from roads may increase the rate and extent of hillside erosion, contributing to landslides. Within a watershed, stormwater runoff from roads can increase stream discharge rates, which in turn may alter channel morphology and restructure riparian areas (**Figure 3.7**), exacerbating flood risk downstream. In addition, the transport of sediments, toxic chemicals (e.g. de-icing salt), and other pollutants are increased along roadways, which affect not only the adjacent terrestrial habitats, but can also find their way into aquatic systems through surface and subsurface flows. Disturbances such as fire, deforestation, disease, and invasive species may all be initiated, introduced, or spread along transportation corridors (e.g. Hulme 2009; **Chapter 8**). These disturbances may

act in concert to enhance the overall environmental impact of transportation corridors.

To illustrate, we once again return to the Brazilian Amazon, where the risk of contracting malaria was found to be a function of road density, which in turn relates to selective logging practices[49] (Hahn et al. 2014). Almost all cases of malaria in Brazil are from the Amazon Basin, where deforestation increases the forest-edge habitat favored by the primary mosquito vector (*Anopheles darlingi*) of the malaria parasite, as well as the contact between humans and the mosquito vectors of malaria (Castro et al. 2006). Selective logging may not appear to be as great a disturbance as deforestation, but it matches or exceeds the total area affected by deforestation in some areas of the Brazilian Amazon (Asner et al. 2005). Road densities are higher in areas undergoing selective logging, as roads are necessary for gaining access to the forest, as well as for transporting logs from those forests (Hahn et al. 2014). Subsequently, selective logging extends deeper into intact forest and

[49] Selective logging involves the targeted removal of certain trees or tree species, as opposed to clear-cutting (deforestation), in which all trees are removed.

creates more forest-edge habitat relative to deforestation (Broadbent et al. 2008). Road density and selective logging are thus interrelated disturbances associated with opening up the Amazon frontier, which in turn has contributed to a higher incidence of malaria among the non-indigenous human population (Hahn et al. 2014).

From an ecological standpoint, transportation corridors fragment the landscape and disrupt the connectivity of populations, which can have negative effects on population viability and population genetic structure (**Figure 9.12**). Because transportation corridors carry fast-moving vehicles, they typically pose a barrier (or at least, a filter) to animal movements across landscapes, either through direct mortality as a result of wildlife-vehicle collisions, or indirectly because of behavioral avoidance of traffic-related disturbances (e.g. noise, light, pollution; Forman & Alexander 1998; Forman et al. 2003; Coffin 2007). For example, a meta-analysis of published research found that transportation corridors generally reduced bird and mammal abundances up to a distance of 1 km and 5 km from the corridor, respectively (Benítez-López et al. 2010). Clearly, a finding of reduced species abundances near roadways or other transportation corridors demonstrates an ecological effect, but whether this is due to behavioral avoidance, road-induced mortality, or both, is impossible to say without further study into the specific mechanism behind the response.

Behavioral avoidance of roadways, for example, has been convincingly demonstrated by animal-movement studies using radio- or satellite-tracking technology (**Chapter 6**). Such studies reveal not only how individuals respond when encountering the road edge (**Figure 6.10**), but also whether such encounters occur less often than expected (i.e. that individuals are actively avoiding roads). For instance, female woodland caribou (*Ranger tarandus caribou*) in Alberta, Canada, were significantly less likely to cross roads than might be expected if roads were encountered at random (Dyer et al. 2002). In the midwestern USA, several reptile species (the massasauga rattlesnake, *Sistrurus catenatus*; eastern box turtle, *Terrapene carolina*; and ornate box turtle, *T. ornata*) were found to cross roads less often than expected, thereby demonstrating that even reptiles can exhibit a behavioral avoidance of roadways (Shepard et al. 2008). Figuring out *why* the turtle (or snake or caribou) failed to cross the road is another matter, however.

Demonstrating that animals cross roadways less often than expected is only the first step in trying to uncover what specific aspect(s) of the roadway they are avoiding (e.g. noise, traffic speed or volume, openness of roadway). To address this, one study went so far as to create a 'phantom road' by blaring traffic noise from speakers arrayed along a roadless stretch of forest to document the response by migrating songbirds (McClure et al. 2013). Nearly 60% of bird species (13 of 22 species) exhibited a negative response to the phantom road, and bird abundance there declined by about a third compared to a nearby control area that lacked roads and experimental traffic noise. Furthermore, birds that stayed in spite of the noise suffered a decline in their physical condition, apparently because they were too disturbed to forage (Ware et al. 2015). Given that fat reserves are needed to fuel migration, traffic noise poses an 'invisible source of habitat degradation' within stopover sites for migratory songbirds (Ware et al. 2015). Such findings of the impact that traffic noise can have on wildlife populations is significant in view of how much land area is potentially within earshot of a road in many parts of the world. For this reason, **soundscape ecology** has emerged as a field to understand the contribution of anthropogenic noise to the natural soundscape, as well as the ecological consequences of that noise (Pijanowski et al. 2011).

More obviously, road-killed animals (roadkill) provide graphic evidence of the effect that roadways have on some species, not just at an individual level, but also at the level of the population (Fahrig & Rytwinski 2009). For example, road mortality was found to be the 'tipping point' contributing to population declines in the bare-nosed wombat (*Vombatus ursinus*), a large burrowing herbivore of southeastern Australia, whose populations are also under siege from introduced predators (wild dogs and the red fox, *Vulpes vulpes*) and sarcoptic mange (Roger et al. 2011). Even in protected areas, road mortality may account for nearly 14% of the total population annually (Roger et al. 2012). Wombats may be particularly susceptible to becoming roadkill because they are attracted to roadsides, which provide suitable habitat for grazing, especially if in close proximity to forest cover or thorny shrubs (such as the introduced and invasive blackberry, *Rubus fruticosus*, which often grows along roadsides) that offer protection from predators (Roger et al. 2007; Roger & Ramp 2009).

The pattern of roadkill is rarely random, tending to be concentrated along certain sections or types of roadways (roadkill 'hotspots'), and perhaps then only during certain times of the year (e.g. during seasonal migrations between breeding and wintering ranges). Various landscape factors, such as the presence of ponds or vegetation alongside the road, in addition to aspects of the roadway itself, such as type of road bed (whether raised or level), road width, traffic volume and vehicle speed, can all influence the incidence of wildlife-vehicle collisions (e.g. Fahrig et al. 1995; Trombulak & Frissell 2000; Clevenger et al. 2003). Certain species also just seem to be more vulnerable to road mortality. Some species, like the wombat, may actually be attracted to roads if roadsides provide foraging opportunities or breeding habitat. In some agriculturally dominated landscapes, for example, roadsides (road verges) may well contain the only native vegetation left on the landscape, and thus can provide breeding habitat for grassland or farmland birds (Morelli et al. 2014). Among birds and mammals, species that are herbivorous or omnivorous are particularly vulnerable to vehicle collisions, probably because these species frequently forage along roadsides (e.g. kangaroos, *Macropus* spp.; white-tailed deer, *Odocoileus virginianus*; moose, *Alces alces*; Cook & Blumstein 2013). Other species may lack behavioral avoidance of roads or are simply too slow-moving, increasing the likelihood they will be hit by vehicles (Fahrig & Rytwinski 2009). As with most responses to landscape structure, species differ in their individual responses to roadways, such that roadways have differential effects on species' populations. Although these effects are often negative, some may be neutral or even positive at the population level, as when roadside verges provide habitat that is otherwise lacking in the broader landscape (which may prove beneficial only so long as these species also have the ability to avoid colliding with vehicles; Rytwinski & Fahrig 2013; Morelli et al. 2014).

The growing concern over the environmental and ecological impacts of roadways has given rise to the field of **road ecology** (Forman & Alexander 1998; Forman et al. 2003). Road ecology seeks 'to quantify the ecological effects of roads, with the ultimate aim of avoiding, minimizing, and compensating for their negative impacts on individuals, populations, communities, and ecosystems' (van der Ree et al. 2011). With some 6.6 million km of roadway and more than 250 million motor vehicles registered in the USA alone, the potential for wildlife-vehicle collisions represents a significant and growing threat to wildlife and the public safety. It is estimated that 1–2 million collisions between vehicles and large animals (mostly deer) occur annually in the USA (Huijser et al. 2008). Further, road mortality is among the major threats to 21 federally listed threatened and endangered species, including the Florida panther (*Puma concolor coryi*), desert tortoise (*Gopherus agassizii*), and California tiger salamander (*Ambystoma californiense*). Reducing vehicle collisions is thus particularly important to the conservation and recovery of these species (Huijser et al. 2008).

Mitigative measures to reduce wildlife-vehicle collisions may include fencing roadways to prevent animals from crossing, installing signage or animal-detection systems to warn drivers of animals on the roadway (**Figure 3.40**), and constructing wildlife-crossing structures to permit safe passage either under or over the roadway. Wildlife-crossing structures are an obvious (albeit costly) means of mitigating wildlife-vehicle collisions and restoring connectivity to wildlife populations (**Figure 5.1**), but their efficacy ultimately depends on how the target species respond to the design and place-

Figure 3.40 Mitigating the impacts of roadways on wildlife. The installation of wildlife warning signs is meant to alert drivers to the possibility that wildlife may be present along certain stretches of the roadway. This sign along a road in Australia warns passing motorists to be on the lookout for wombats, which have a penchant for foraging along roadsides and may thus wander out onto the road, with sometimes unfortunate consequences.

Source: Photo by Phil Whitehouse and published under the CC BY 2.0 License.

ment of the crossing structure (Clevenger & Waltho 2005). Just as with species' responses to roadways, species vary in their use of different types of crossing structures. For example, the ideal crossing structure for a grizzly bear (*Ursus arctos*), elk (*Cervus elaphus*), or wolf (*Canis lupus*) is very different from that for a cougar (*Puma concolor*) or black bear (*Ursus americanus*): the former prefer wide, open crossing structures (e.g. landscaped overpasses), whereas the latter prefer more constricted passageways (e.g. underpasses). Road-mitigation measures should therefore strive to incorporate a diversity of crossing structures in order to reduce the barrier effect of roadways and maximize population connectivity for the greatest number of species (Clevenger & Waltho 2005; Mata et al. 2008; **Chapter 5**).

REFORESTATION The widespread planting of trees, along with natural reforestation in some regions, has helped to slow the rate of deforestation at a global scale. In the last decade, the net loss of forest area was reduced by about 3 million ha/year, from 8.3 million ha/year during the last decade of the 20th century to 5.2 million ha/year during the first decade of the 21st century (FAO 2010; **Figure 3.38B**). Planted forests now make up about 7% of the total forest area globally (FAO 2010). Further, the area of planted forest has been increasing by 5 million ha/year, largely due to **afforestation**, the planting of trees in areas that have not been forested in recent times (FAO 2010).

China in particular has undertaken large-scale afforestation projects in recent years, in an effort to reverse decades of deforestation and degradation that have contributed to soil erosion and catastrophic flooding (Zhang et al. 2000). The Natural Forest Conservation Program (NFCP) and the Grain for Green Program (GFGP) are two of the largest and most ambitious programs undertaken by China, which were initiated in 1998 and 1999, respectively (Zhang et al. 2000; Liu et al. 2008). The overall goal of the NFCP is to protect and restore natural forests through a combination of logging bans, reduced timber harvests, and afforestation programs, which includes payments to forest industries and forestry workers to offset their economic losses (Liu et al. 2008). In a similar vein, the GFGP uses economic incentives to encourage rural farmers to convert croplands on steep slopes to forests by giving them grain and cash subsidies. The Chinese government has thus made an enormous investment in these two reforestation programs that is projected to exceed $100 billion by 2020 (Liu et al. 2008; Ren et al. 2015; Ahrends et al. 2017). The NFCP has already produced significant gains in forest cover (defined as a ≥20% increase) over a large portion of China, totaling about 157,315 km^2 (Viña et al. 2016), whereas the GFGP is credited with contributing another 28 million hectares of forest nationwide (Hua et al. 2016).[50]

Reforestation may also occur naturally as a consequence of agricultural abandonment. For example, the reforestation of the northeastern USA is a well-documented case of forest recovery following changing land-use and settlement patterns that led to widespread agricultural abandonment (Foster et al. 1998b; Foster 2002). Prior to European settlement, the region was largely forested and governed by natural disturbances and successional dynamics until about 1650 (Thompson et al. 2013). Over the next two centuries, some 60% of the forest was logged and cleared for agricultural land use in response to European colonization and expansion throughout the region (Cogbill et al. 2002). Then, around the mid-1800s, industrialization and increased urbanization, coupled with the shift in agricultural production to the midwestern USA, resulted in widespread abandonment of agricultural fields in the east. These fields were left to undergo a process of natural reforestation and forest growth

(Foster et al. 1998b), and today some 80% of the region is once again forested, although less than 1% is considered old-growth forest (Thompson et al. 2013).

Despite the apparent recovery, to what extent are modern forests reminiscent of the historical forests prior to European settlement? In colonial North America, early land surveyors marked the corners of land parcels using the nearest tree ('witness trees'). In so doing, they unwittingly provided future ecologists with information on the composition of the historical forest landscape, which is useful for making comparisons to the modern-day landscape (Cogbill et al. 2002; Thompson et al. 2013; **Chapter 4**). Although the region was and is largely forested, there are numerous forest types (ecotones) distributed across a broad geographic gradient that exhibits a gradual transition from coniferous forests in the north to deciduous forests in the south (Cogbill et al. 2002; **Figure 3.41**). The changes that have occurred must therefore be placed in the context of these different forest ecoregions. For example, beech (*Fagus*) and hemlock (*Tsuga*) originally comprised about 40% of the forest taxa in the northern and central Laurentian and Adirondack-New England forest provinces, whereas oaks (*Quercus*) and pines (*Pinus*) dominated (~50%) the more southern or coastal forests of the Eastern Broadleaf and Central Appalachian provinces (**Figure 3.42A**).

Although almost all of the original taxa are still present in the modern-day landscape, about a third have undergone significant declines, whereas another third have increased in relative abundance across the region (Thompson et al. 2013). Beech is no longer a major component of any forest type, and oaks have declined significantly, especially in the Eastern Broadleaf forest province where they were once the dominant tree (**Figure 3.42B**). Instead, about a third of all trees are now maples (*Acer*), regardless of forest type, reflecting a general homogenization of forests across the region. These changes reflect a shift from late-successional species such as beech and hemlock to early-successional ('pioneer') species such as maples. Thus, after more than 150 years of reforestation, the forests of the northeastern USA are still in the early stages of recovery, as eloquently summarized by Thompson and his colleagues (2013, p. 9):

> ...after 400 years of land use [the eastern forest] is at once largely unchanged and completely transformed. It is unchanged insomuch as all the major arboreal taxa remain. With few exceptions, the same taxa that made up the forest in the pre-colonial period comprise the forest today, despite ample opportunities for species invasion and loss. In this sense, the regional ecosystem has been quite resilient and the recovery of the eastern forests has been quite real in extent and composition. Yet, at the same time, the forest has been radically transformed. The relative abundance and distribution

[50] Not all gains in forest cover equate to forest recovery, however. In southwestern China, for example, forest cover increased by 32% between 2000 and 2015, but this increase was due almost entirely to the conversion of croplands to tree plantations, which are usually monocultures of non-native tree species (Hua et al. 2016, 2018). In fact, native forest cover declined by 6.6% during this same period, as a result of landowners clearing native forests to make way for tree plantations, driven by the perceived profitability of planting trees on those lands (Hua et al. 2018). The gains in China's forest cover may thus be more modest than reported here, for it ultimately depends on how 'forest' is defined (Ahrends et al. 2017).

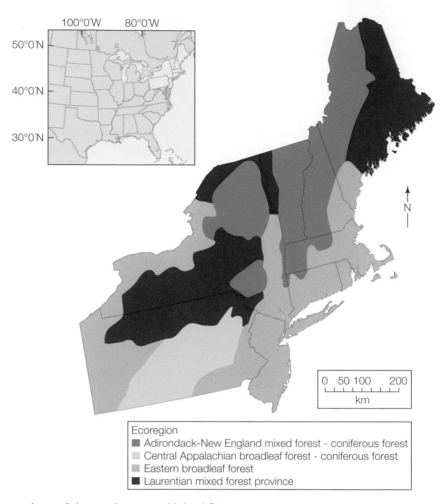

Figure 3.41 Forest ecoregions of the northeastern United States.
Source: After Thompson et al. (2013).

of most taxa have shifted dramatically; the relationship between forest composition and the environment has been weakened; and variable patterns of land use have imposed a mosaic of impacts whose legacies are evident centuries later.

Land-Use Legacy Effects

As the foregoing example illustrates, the legacy of past land use may continue to influence the composition and structure of landscapes today. In other words, the 'ghosts of landscapes past' haunt the landscapes of the present (With 2007). The human footprint on the landscape is typically large, deep, and long-lasting. Human land use exerts a strong legacy effect that may persist for decades or centuries, even after the landscape has been abandoned and appears to have recovered (Foster et al. 2003). Such appearances can be deceiving, because the recovered landscape may only superficially resemble its pre-settled state (**Figure 3.42**). The distribution of species may therefore reflect the

land-use history of past landscapes more than the present ones.

The forests of the northeastern USA retain other land-use legacies from America's Colonial period. The herbaceous understory of woodlands that recovered after agricultural abandonment still bears the legacy of past cultivation: they contain more weedy species and fewer ericaceous shrubs than woodlands that were never cleared and plowed (Foster et al. 2003). Many of these forest herbs have low seed production, lack persistent seed banks, and are dispersal-limited (e.g. their seeds are ant-dispersed or lack morphological structures—such as the 'wings' of maple seeds—that facilitate long-distance dispersal), presumably because they evolved within a relatively stable forest ecosystem characterized by fine-scale patch disturbances (Bellemare et al. 2002). Given the broad-scale disturbance of human land use, species that lack long-range dispersal abilities are often unable to recolonize following agricultural abandonment (Flinn & Velland 2005). In addition, past agricultural land use

(A)   Pre-settlement landscape

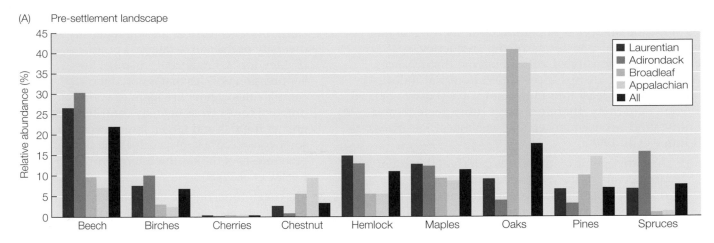

(B)   Current landscape

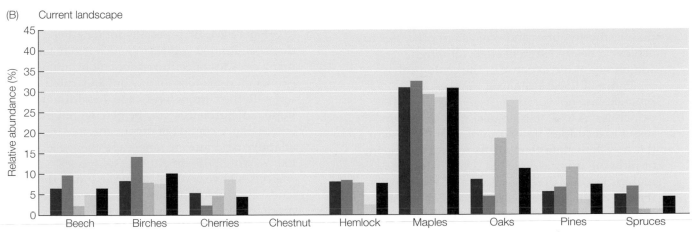

Figure 3.42 Reforestation of the northeastern United States. Following European settlement, much of the forest in the northeastern USA (**Figure 3.41**) was logged and cleared for agricultural land use. Industrialization and increasing urbanization, coupled with a shift in agricultural production to the Midwest, led to widespread farm abandonment in the mid-19th century. After more than 150 years of natural reforestation and forest growth, the region is once again largely forested and even contains most of the same tree taxa that made up the pre-settlement forest. Nevertheless, there have been some significant changes in the relative abundance of these taxa, with a shift from late-successional species, such as beech and hemlock, in favor of early-successional species (maples). The American chestnut (*Castanea dentata*) died out by 1950 due to a fungal disease (chestnut blight) that was accidentally introduced from SE Asia.

Source: After data from Thompson et al. (2013).

may affect the availability and distribution of soil nutrients, which can persist for many decades, centuries, or millennia, and thus permanently alter the composition and diversity of forest ecosystems (Dupouey et al. 2002; Fraterrigo et al. 2005). In that case, it may not be possible to recover the historical vegetation, even if dispersal limitation is eventually overcome in time.

Past land use also influences the biological diversity of aquatic systems. Land-use practices throughout the watershed may affect a wide range of conditions, such as hydrology, organic inputs, temperature, and water chemistry, and are thus capable of contributing to strong legacy effects (Allan et al. 1997). For example, patterns of fish and invertebrate diversity within streams of the southern Appalachians were better explained by the intensity of agricultural land use some 40 years

earlier than by current land use (Harding et al. 1998). Thus, some forest streams contained fish and invertebrate species that were more typical of agricultural streams; a legacy of the past agricultural land use that had occurred within those watersheds. Reforestation over the past half-century thus produced little recovery of these forest-stream communities. As in terrestrial systems, the recovery of aquatic biota from deforestation or agriculture may take many decades to achieve, if ever.

Human land use is certainly not the only type of disturbance that leaves a legacy on the landscape. Indeed, we have previously discussed in this chapter how other types of natural disturbances, such as fire, hurricanes, and landslides, can create legacies that influence successional pathways in complex and often unpredictable ways. How, then, is human land use different from any of these other types of disturbances? By now, the answer is hopefully clear, as it again has to do with scale, in terms of the frequency, spatial extent, and intensity of anthropogenic disturbances. Anthropogenic disturbances differ from natural disturbances in at least four important ways (Foster et al. 1998b), and may therefore have unexpected—and unintended—consequences (e.g. ecological surprises, Romme et al. 1998):

- Anthropogenic disturbances can rival the extent or intensity of the most severe natural disturbances, but may occur with far greater frequency (large frequent disturbances, LFDs) and exceed the recovery time of the system (**Figure 3.12**).

- Land-management practices typically alter disturbance regimes and ecosystem processes to achieve a desired system state, although sometimes at the expense of the ecological resilience of these systems if management practices are not based on sound ecological principles. A loss of ecological resilience may result in a sudden transformation of the system (i.e. a system state change), as when overgrazing or plowing up grasslands in drought-prone regions leads to desertification and Dust Bowls (**Figure 3.20**). Further, as we have seen in this chapter, agricultural expansion and intensification have profoundly altered the availability and quality of water resources throughout the world, as a result of deforestation, water withdrawals for irrigation, and nutrient runoff from fertilizer use. Agricultural modifications of hydrological flows have thus contributed to a number of documented system state changes, including the eutrophication of freshwater systems, the development of hypoxic 'dead zones' along coastal areas, and the salinization of soils (Gordon et al. 2008).

- Land use and land-management practices may be applied in a fairly uniform manner across a landscape or region, which ignores (or is even intended to eliminate) the inherent variability in topography, climate, natural disturbances, and environmental heterogeneity. For example, annual burning is prescribed in order to create a more homogeneous forage distribution for cattle, but also reduces the heterogeneity and diversity of the native tallgrass prairie (Fuhlendorf & Engle 2001). Blanket fire-suppression policies that ignored the complexities of how fire regimes varied among different forest types actually contributed to greater fire risk and larger, more severe fires in some forest ecosystems, such as the ponderosa pine forests of the southwestern USA (e.g. Keeley et al. 2009).

- The magnitude, rate, or extent of human land use may exceed the capacity of species to respond to such landscape changes. Indeed, landscape legacy effects are often the result of lagged species' responses to landscape change (e.g. **Figure 2.15B**), especially for long-lived or dispersal-limited species in which it may take many decades or centuries for colonization-extinction dynamics to play out and 'catch up' with the current landscape. In addition, human disturbances may be new within the evolutionary context of organisms, making adaptive responses to landscape change less likely (With 2015). In either case, species' responses have become decoupled from the landscape, which means that uncovering pattern-process linkages (i.e. the study of landscape ecology!) will prove all the more difficult.

An understanding of land-use legacies is important for putting the current ecosystem or landscape in the context of its past trajectory of change (Foster et al. 2003; **Figure 2.15B**). This includes not only human land use, but also past climate, natural disturbances, and successional processes. All of these 'ghosts of landscapes past' help to shape the current landscape, but also constrain its future response. From a management standpoint, an understanding of historical variability is often deemed essential for providing the reference or 'baseline' conditions that managers might then seek to emulate (Landres et al. 1999; Swetnam et al. 1999; Keeley et al. 2009). Using the past to manage for the future is only possible if we understand the nature of such legacy effects on the ecological systems we seek to manage. In the meantime, our ongoing land-use activities are leaving a landscape legacy for future generations.

Future Directions

Because heterogeneity is such a pervasive theme in landscape ecology, understanding how different processes give rise to spatial heterogeneity—and at what scales—will continue to be an ongoing research challenge, especially as broad-scale disturbances caused by human land use and climate change become more frequent, and the need to predict, manage, and mitigate these types of disturbances becomes more urgent. In particular, the potential for **cross-scale interactions**, in which disturbances at one scale interact with those at another scale, must be explicitly addressed in future research on disturbance dynamics (Peters et al. 2004; 2007). Recall from **Chapter 2** that cross-scale interactions may be important for understanding the relative importance of pattern-process relationships, such as those that give rise to landscape heterogeneity and dynamics. Because cross-scale interactions often produce non-linear dynamics and ecological thresholds, they can generate ecological surprises and system state changes that pose a particular challenge to ecosystem management and ecological forecasting. Human land use and the other types of disturbances we have discussed in this chapter (hurricanes, fire, grazing) are the very sorts of environmental drivers that can simultaneously influence both broad-scale patterns and finer-scale processes in ways that reinforce, modulate, or amplify the effects of the disturbance, producing new and surprising ecological responses.

There is also the issue of **compounded disturbances**. Most ecological systems are subjected to more than one type of disturbance. These disturbances may have a purely additive effect, or they may have a synergistic effect. Compounded disturbances also have the potential to yield ecological surprises (Paine et al. 1998).

Although an ecosystem may recover from a single large disturbance (a volcanic eruption, fire, or hurricane), the combined effects of multiple disturbances within that same recovery period may lead to a long-term—and perhaps permanent—alteration of the system (i.e. an irreversible state change). This is particularly true if the recovery potential of the system has already been eroded, perhaps due to the loss of species or important ecosystem processes as a result of past anthropogenic disturbance. Thus, a forest might recover from a drought, a fire, or a bark-beetle outbreak, but combine two or three of these disturbances, in conjunction with past land-use or management practices that have homogenized the forest (in terms of its tree species, stand ages, or fuel loads) or increased tree densities, and it seems unlikely that the system will retain many characteristics of its previous state, or even, remain a forest. To quote Paine and his colleagues (1998, p. 535):

[I]n a world of ever-more-pervasive anthropogenic impacts on natural communities coupled with the increasing certainty of global change, compounded perturbations and ecological surprises will become more common. Understanding these ecological synergisms will be basic to environmental management decisions of the 21st century.

Chapter Summary Points

1. Heterogeneity is one of the core concepts of landscape ecology. Given that a landscape is defined as a spatially heterogeneous area, it could be argued that heterogeneity is *the* central concept of landscape ecology, which can be defined at any scale and within any system, whether aquatic, marine, or terrestrial. It is the prevalence of heterogeneity that makes landscape ecology so relevant to such a wide range of scales, systems, and disciplines.

2. Over the past century, ecology has undergone a paradigm shift from an equilibrium to a non-equilibrium view of ecological systems, one that embraces the dynamical nature of these systems, as well as their spatial complexity. Concepts of spatial heterogeneity in ecology have evolved over time, from the simple patch-matrix view informed by island biogeography and metapopulation theory, to the current landscape-mosaic concept in which landscapes are viewed as heterogeneous assemblages of different patch types or land covers that vary in suitability, use, and resistance to ecological flows.

3. Heterogeneity includes both spatial and temporal dimensions. Spatial heterogeneity incorporates

the dual concepts of composition (the number or diversity of land-cover types) and configuration (e.g. the spatial arrangement or degree of patchiness of different land covers). Temporal heterogeneity refers to the pattern within a data time series, such as whether events are aggregated or randomly distributed over time. Regardless of how it is defined, we should take care to distinguish between measured or structural heterogeneity (the quantitative assessment of heterogeneity) and functional heterogeneity, the heterogeneity that is relevant to the ecological process or phenomenon under investigation.

4. Disturbances of various types—abiotic, biotic, and anthropogenic—contribute to landscape heterogeneity. The disturbance regime is characterized by spatial and temporal parameters that define the size, frequency, and intensity or severity of a disturbance.

5. The Intermediate Disturbance Hypothesis posits that the greatest diversity or heterogeneity is found at intermediate scales of disturbance, in which disturbances are neither too large or too small, nor too frequent or too rare.

6. Large infrequent disturbances (LIDs), such as a wildfire or a volcanic eruption, have a profound and lasting impact on landscapes. The ecological effects of LIDs are fundamentally different from those of other types of disturbances, and may exceed the cumulative impacts of many small disturbances. LIDs thus pose a particular management challenge, as the consequences of these LIDs cannot simply be scaled up or predicted from the known effects of small disturbances, leading to 'ecological surprises.' This is especially salient, given that human land use and climate change are causing shifts in the disturbance regime toward more frequent large disturbances in many landscapes.

7. Landscape formation is a dynamic process, involving the interaction between various geomorphological and disturbance processes, which causes the landscape to 'evolve' over time. Opposing geologic forces, such as tectonic uplifting versus faulting and subsidence, give rise to the topographic diversity of Earth's surface. Erosional and depositional processes, involving the action of wind, water, and ice, continue to sculpt the landscape and contribute to the formation of unique landforms, river networks, and the diversity and distribution of soils, which are an important determinant of vegetation and land use (e.g. agricultural production and productivity).

8. Abiotic disturbances that shape landscapes and thus contribute to landscape heterogeneity include volcanic eruptions, landsliding, flooding, windstorms, hurricanes, storm surges, drought, and fire.

9. Like any disturbance, volcanic eruptions vary in severity, although the most powerful and spectacular—such as the 1980 eruption of Mount St Helens in the Pacific Northwest—are the sorts of large infrequent disturbances that have contributed to our understanding of the complexities and unpredictability of ecological recovery (e.g. ecological succession) following a severe disturbance event.

10. Ecological recovery is a complex process that depends on landscape position, initial conditions, and chance events. Biological legacies—the surviving species or their propagules—are important for ecological recovery and influence the rate of recovery following a disturbance. Successional trajectories are also dependent on priority effects, the order in which species arrive at a site, which may then influence the ability of other species to become established.

11. In montane landscapes, landsliding is important to the formation of landscapes and landscape heterogeneity. Landsliding may be triggered by a variety of forces, such as earthquakes, volcanic eruptions, heavy precipitation, deforestation, and other human land-use activities that erode or destabilize hillsides. Landslides exhibit both vertical and horizontal disturbance gradients that contribute to variation in disturbance severity, and therefore, in ecological recovery rates.

12. Flooding is part of the natural flow regime of riverine landscapes. Seasonal variation in precipitation and snowmelt contribute to periodic flood and flow pulses, causing the expansion and contraction of these systems during the year, as well as variation in streamflow between years. Flooding helps shape riverine landscapes through erosional and depositional processes, which give rise to the geomorphology and habitat heterogeneity within these systems. By regulating water flows, dams disrupt flood pulses and alter the hydrology, morphology, and productivity of riverine systems, including their riparian areas and floodplains.

13. Damage from windstorms affects the structure and composition of forests, especially along coastal areas where hurricanes may make landfall. The disturbance effects of hurricanes may be felt hundreds of kilometers inland, however. In addition to high winds, hurricanes are also accompanied by storm surges, which are devastating to coastal and estuarine areas, but also damage coral reef systems, which are a type of seascape.

14. Drought is considered a 'creeping disaster' because its effects tend to develop over time and are more likely to go unnoticed until one or more types of water use are affected. Drought is divided into four major types, reflecting the impact that water deficiencies have on human systems: meteorological drought, involving a period of below-normal precipitation, is the basis for all other drought types; agricultural drought, refers to soil water deficiencies that affect crop production; hydrological drought, which is defined by reduced streamflow and groundwater recharge, and thus affects water availability; and, socioeconomic drought, which considers the economic and societal impacts of drought.

15. Hot droughts—drought accompanied by high temperatures—increase the risk of wildfires. Fire regimes are controlled by top-down drivers involving climate, weather, and land-use/land-management practices, and bottom-up

drivers that include topography, fuels, and their interaction. Climatic changes that are contributing to hotter and drier conditions are contributing to larger and more frequent fires in some forest types.

16. Organisms are also agents of landscape disturbance, and may even contribute to the formation of geomorphic features across a range of scales (e.g. ant or termite mounds, coral reefs). In particular, ecosystem engineers play a critical role in the creation and maintenance of landscape heterogeneity. Autogenic engineers alter the landscape via their own physical structure (e.g. forests, coral reefs), whereas allogenic engineers transform landscapes through the physical modification of living or non-living materials (e.g. beaver dams). Ecosystem engineers are therefore often (though not always) considered to be keystone species, which enhance the diversity and heterogeneity of the landscapes they inhabit.

17. Humans are ecosystem engineers par excellence. Of all disturbance types, the transformation of landscapes by humans is the most pervasive and globally significant, which has issued in a new geological era, the Anthropocene.

18. The global impact of human land use has historically been rooted in the agricultural transformation of landscapes, which started some 8500–11,000 years ago and proceeds through a series of predictable stages involving a pre-settlement phase, the development and settlement of the agricultural frontier, subsistence agriculture, intensifying land use, and finally, intensive land use. Different parts of the world progress through these stages at different rates, and are thus at different stages along this landscape-transformation gradient. Agricultural expansion has mostly come at the expense of the world's forests.

19. In addition to deforestation, urbanization represents another major form of landscape transformation. For the first time in human history, more people live in urban areas than in rural areas globally. Although the effects of urbanization are typically viewed along an urban-rural gradient, urban areas are heterogeneous landscapes in their own right, representing a hybrid ecosystem that emerges from the interactions between humans and ecological processes.

20. Transportation corridors, such as roads, are both a precursor to and consequence of urbanization. Roadways contribute to habitat loss and fragmentation, and the effects of roads (the road-effect zone) may extend hundreds of meters beyond the road corridor itself. Besides contributing directly to wildlife mortality through collision with vehicles (roadkill), roads may constitute a barrier to movement through behavioral avoidance, leading to the fragmentation of populations in some species. The emerging field of road ecology seeks to understand the environmental and ecological consequences of roadways, and to develop measures to mitigate the impacts of roads on wildlife populations (e.g. wildlife-crossing structures).

21. Intensive land use need not be the endpoint in the landscape-transformation gradient. The abandonment or restoration of agricultural lands has led to the reforestation of landscapes in some regions of the world. Nevertheless, the legacy of past land use may persist for decades, and even centuries, following land abandonment. The ghosts of landscapes past may thus continue to shape and influence the heterogeneity of the landscapes around us, even those that appear to be relatively undisturbed by human land use today.

Discussion Questions

1. If heterogeneity is the defining characteristic of a landscape, then in what way(s) is your study area heterogeneous? Besides the heterogeneity of cover types or land uses, what are some other ways in which your study area might be viewed as a landscape? How many 'landscapes' are there within your study system, therefore?

2. What types of disturbances shape landscape patterns in your study system? Construct a space-time diagram (**Figures 2.2 and 2.3**) that depicts the scaling domains over which these disturbances are expected to operate in your system. Which of these overlap with the scaling domain of your study?

3. Given that disturbances are ubiquitous within landscapes and other types of ecological systems, when does a disturbance no longer qualify as a disturbance? For example, if a river floods seasonally each year, can flooding really be considered a disturbance within that system?

Consider the flip side to this question: how can the *absence* of a disturbance (e.g. fire suppression) qualify as a disturbance? How should we thus define 'disturbance'?

4. How would you characterize the dynamics of the landscape in which you work (or alternatively, in a landscape example provided by your instructor)? Is the landscape stable or unstable? Calculate the spatial and temporal dimensions of the disturbance (S and T), and evaluate this with respect to the state diagram depicted in **Figure 3.12**. How might your assessment of landscape dynamics change if you were to evaluate a different type of disturbance within this same landscape (e.g. grazing rather than fire)?

5. In characterizing landscape disturbance dynamics, we are evaluating the time it takes the system to recover relative to the disturbance frequency or interval between disturbances. How is disturbance intensity—the severity of the ecological impacts of the disturbance—expected to alter this relationship? Similarly, how might the duration of the disturbance (e.g. a low-level chronic disturbance vs an intense, but short-lived disturbance) additionally modify our assessment of landscape dynamics?

6. River disturbances may be classified as a pulse, press, or ramp (**Figure 3.25**). To what extent can these different types of river disturbance be applied to terrestrial systems? That is, are there similarities in how terrestrial and riverine landscapes respond to the intensity and duration of disturbances?

7. Although many disturbances, such as drought, are viewed principally in terms of their negative impacts on ecological (and human) systems, how might such disturbances actually have a *positive* effect on landscape structure and function? Can you envision an intermediate disturbance effect for drought (or fire or wind or landsliding), for example?

8. Keystone species are important agents of disturbance that have a major effect on the structure and dynamics of the landscapes they inhabit. How might an abiotic disturbance, such as fire, be considered a 'keystone process'?

9. How do abiotic disturbances differ from biotic disturbances, in terms of their impacts on the landscape? Are these differences primarily qualitative (they differ in kinds of effects) or quantitative (they differ in frequency, intensity, or extent)? Which type of disturbance is likely to have a greater or longer-lasting impact on the landscape? Which type of disturbance might therefore be easier to manage?

10. Consider how anthropogenic disturbances differ from so-called natural disturbances. As in the previous question, discuss whether these differences are primarily qualitative (they differ in the kinds of effects they have on landscapes) or quantitative (they differ in their frequency, intensity, or extent). Which type of disturbance is likely to have a greater or longer-lasting impact on the landscape?

Landscape Pattern Analysis

Deforestation—and the loss of natural habitats more generally—is one of the most pressing global-change issues of our time. Forests provide a wealth of ecosystem services, from the timber and forest products used by humans, to providing habitat for a diverse array of species, to the regulation of climate and the protection of soil and water resources. Over the past 25 years, global forest cover has declined by 3.1% while the human population has increased some 30%, with a concomitant increase in the demand for forest and agricultural products (Keenan et al. 2015). Because most of this forest loss is a result of clearing for agriculture and other land uses, it is obviously important to be able to assess and monitor changes in forest cover over time.

This is a surprisingly difficult task (MacDicken et al. 2015). National agencies around the world have historically used a variety of inventory methods, including ground-based and aerial surveys, to assess and monitor their forest resources. Since 1948, the Food and Agricultural Organization of the United Nations (FAO) has published its Global Forest Resources Assessment (FRA) based on an analysis of available national forest inventories. Over time, the FRA has become more comprehensive, especially as satellite imagery is increasingly being incorporated in national forest inventories (70% of countries in the FRA 2015 report; MacDicken 2015). It is because of these sorts of global assessments that we can quantify how forest trends change over time (e.g. the global rate of deforestation has actually decreased by 50% over the past decade), as well as identify where deforestation rates are still high or have increased in recent years (e.g. the tropical forests of South America and Southeast Asia; Keenan et al. 2015).

In addition to quantifying global and regional declines in forest cover, the fragmentation of forest landscapes is also of concern from an ecological standpoint. The loss and fragmentation of habitats are widely regarded as the greatest threat to biodiversity (Hanski 2011), which demands that we have some way of quantifying such landscape changes. While it may seem obvious how we should quantify habitat loss, how should we measure habitat fragmentation? Will a single measure suffice, and if not, then what measures should we use? At what point *do* landscapes become fragmented? The analysis of landscapes and other spatial data is a major focus of landscape ecology, being essential to the study of how landscape patterns affect ecological processes, but it also has great relevance for conservation, resource management, and sustainable land use.

On the Importance of Landscape Pattern Analysis in Landscape Ecology

By now, the following points regarding landscapes and landscape ecology should be abundantly clear: heterogeneity is a defining characteristic of landscapes; heterogeneity

Essentials of Landscape Ecology. Kimberly A. With, Oxford University Press (2019).
© Kimberly A. With 2019. DOI: 10.1093/oso/9780198838388.001.0001

is thus one of the core concepts of landscape ecology; heterogeneity is a multifaceted concept, involving both spatial and temporal dimensions; heterogeneity can be expressed across a range of (potentially nested) scales; and, landscape ecology is expressly concerned with understanding the relationship between spatial patterns and ecological processes, at whatever scale. It thus follows that the description of landscape pattern is a requisite for understanding pattern-process linkages in ecology.

Not surprisingly, then, the analysis of landscape heterogeneity has been a major research focus in landscape ecology, becoming one of its defining characteristics to the point that it has sometimes been conflated with its geospatial technologies and tools (e.g. remote sensing and geographical information systems or GIS). To be sure, a certain preoccupation with landscape pattern analysis did emerge early on in the discipline, due in large part to the newly afforded view of landscapes provided by satellite imagery and the resultant need to quantify and compare landscape patterns at such broad, geographic scales (O'Neill et al. 1988a; Turner & Gardner 1991). There are now hundreds of quantitative measures that have been proposed for quantifying various aspects of landscape heterogeneity (Gustafson 1998). Landscape pattern analysis has also benefitted in recent years from technological advances in computing (especially mobile computing), remote sensing, GIS, and the development of landscape modeling and analysis software. In particular, the internet revolution has increased both the availability of, and access to, geographical datasets and the software tools needed to analyze them, many of which are now free and open-source (Steiniger & Hay 2009).

Landscape pattern analysis has become a *sine qua non* for environmental monitoring, natural resource management, landscape planning, sustainable development, and species conservation (Hunsaker et al. 1994; O'Neill et al. 1997; Klopatek & Gardner 1999; Botequilha Leitão & Ahern 2002; Bennett et al. 2006; Heinz Center 2008). The intensification of human land use is a major global-change issue, requiring an indepth analysis of how landscapes are being transformed over time, in terms of the rate, magnitude, and direction of those changes, as well as how anthropogenic disturbances compare to the natural disturbance regime (e.g. Kupfer 2006; **Chapter 3**). Nevertheless, the quantification of landscape heterogeneity remains a significant research challenge, owing to the complexity of the interacting processes that give rise to landscape patterns, as well as to the multidimensional nature of the heterogeneity concepts themselves (Kolasa & Rollo 1991).

> Studying spatial structures is both a requirement for ecologists...and a challenge. It is the new paradigm for field ecologists.
> Legendre (1993)

In this chapter, we begin with an overview of the various types and sources of landscape data, from historical land surveys to modern remote-sensing technologies. Then, we discuss the different types of data models (raster versus vector), including the different ways of representing spatial elements in a dataset. We need some way to manage, manipulate, analyze, and display these spatial data, which is why we will review the subsystems, functions, and capabilities of geographical information systems next. Once we have a better idea of how spatial data can be obtained, processed, and manipulated, we will turn our attention to the analysis of landscape patterns, starting with landscape metrics and then proceeding to a discussion of some common methods in spatial statistics for the detection and description of spatial structure. We will also cover tests for evaluating significant departures from complete spatial randomness (statistical inference) and the estimation of spatial data values via kriging (spatial interpolation). This chapter should thus set the stage for our subsequent discussions on how landscape pattern interacts with ecological processes in the second half of the textbook.

Sources of Landscape Data

Given our definition of a landscape as a spatially heterogeneous area that is scaled relative to the organism or process of interest (Turner 1989; Wiens & Milne 1989; **Chapter 1**), the potential sources of landscape data are essentially infinite, encompassing a wide variety of environmental and ecological attributes, systems, and scales that can be visualized, mapped, and analyzed using the quantitative methods of landscape ecology and spatial statistics. In other words, we are not limited to using only aerial photographs or satellite imagery as our main source of landscape information. Any ecological pattern that varies in its occurrence or intensity across space can be assigned a physical location (**spatially referenced** or **georeferenced**) and thus mapped to create a 'landscape.'

Because of the many potential sources and types of spatial information, we cannot possibly cover all of these here in this one section. However, many of these other data sources will be discussed in the chapters that follow, as we explore various ways of characterizing ecological responses to spatial pattern. We will therefore restrict our discussion here to the types and availability of landscape data defined by the scale of human land-use activities, and which are obtained via historical land surveys or remote sensing.

Historical Land Surveys

The aerial photographs and satellite images that are so indispensable to landscape analysis today are fairly recent innovations, providing a record of the landscape that is at best maybe 100 years old in the case of aerial photographs, and only about 45 years old in the case of satellite imagery. Before then, the 'lay of the land' was surveyed on foot (or by some other conveyance) and mapped by hand by various explorers, surveyors, travelers, or settlers, who were either just passing through or seeking to lay claim to the land. The maps

and journals they left behind are often the first pictorial or written records we have of the landscapes they traversed, providing a rare glimpse into how these historical landscapes might have looked in comparison to the present-day landscape (Whitney 1994; Vellend et al. 2013; Silcock et al. 2013).

Still, early maps and written records are bound to be subjective, especially if these are individual accounts that were not part of a systematic land survey. Systematic land surveys have proven to be especially useful for reconstructing historical landscapes, especially those that pre-date the settlement and land-clearing that inevitably followed European colonization and expansion in countries like the United States, Canada, and Australia (Whitney 1994; Vellend et al. 2013). In places where

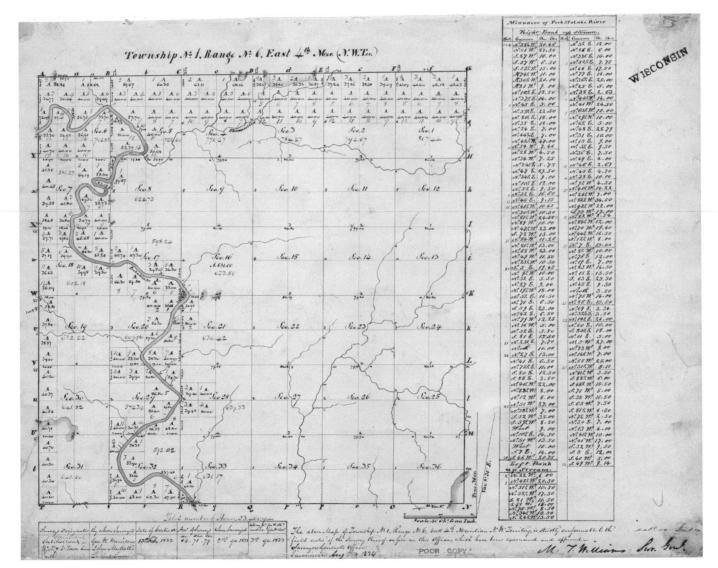

Figure 4.1 Land-survey map. Original federal land survey of the Cadiz Township (cf. **Figure 3.6**) in Green County, Wisconsin (USA) that was undertaken in 1833 for the US Public Land Survey.

Source: Image obtained from the US Department for the Interior, Bureau of Land Management, General Land Office.

European settlement has been fairly recent (<200 years), these early survey maps provide some of the best information available as to the state of the pre-European settlement landscape, particularly in terms of the distribution of its forests. Recall that one of the classic studies of landscape change following European settlement is that depicting the loss and fragmentation of forest in the Cadiz Township of Green County, Wisconsin (**Figure 3.6**), which was based on the original federal land survey of the area from 1833 (**Figure 4.1**).

In the United States, the **US Public Land Survey** was a result of the Land Ordinance of 1785, which set forth how the nascent republic should go about surveying, subdividing, and selling off the largely unmapped territory that lay northwest of the Ohio River (i.e. the Northwest Territory).[1] The US Public Land Survey commenced on 30 September 1785 at a point (the aptly named 'Beginning Point') on the north bank of the Ohio River along the border between Ohio and Pennsylvania, in an area that became known as the 'Seven Ranges' townships. In time, much of the continental United States (with the exception of the original 13 states, Kentucky, Tennessee, and Texas) were surveyed and mapped under the methods established for the US Public Land Survey.

Under the US Public Land Survey, surveyors established a coordinate grid system consisting of meridians (north-south lines) and range lines (east-west lines). Each Public Land Survey commenced from an initial point at the intersection of a baseline and a principal meridian (**Figure 4.2**). From there, the land was divided into townships measuring 36 square miles (6 × 6 miles = 9.7 × 9.7 km). Each township was further subdivided into 36 1-square-mile sections (1.6 × 1.6 km). A section can then be partitioned more finely (e.g. quarter-sections), depending on land ownership.

To delineate these boundaries, surveyors placed a wooden post or stone[2] to mark each section, quarter-section, and meander corner,[3] and then 'blazed' two to four nearby trees as 'witness' or 'bearing' trees on which they inscribed the township, range, and section information (Schulte & Mladenoff 2001; **Figure 4.2**).[4] Because surveyors also entered the species of tree, its diameter, compass bearing, and distance from the

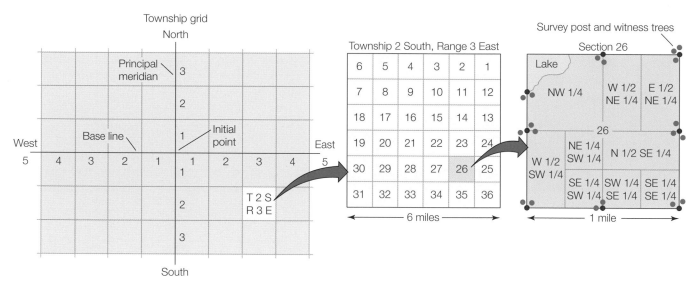

Figure 4.2 Public Land Survey System. In the United States, most public (federal) lands were initially surveyed using a rectangular survey system. Working from a principal meridian (north-south line) and baseline (east-west line), the surveyor divided the land into a grid of townships. Each township is 36 square-miles (6 × 6 miles = 9.7 × 9.7 km), and is further subdivided into 36, 1 square-mile sections (1.6 × 1.6 km). Sections (640 acres = 259 ha) may then be partitioned into quarter-sections (160 acres = 65 ha), half-quarter (80 acres =32.4 ha) or quarter-quarter (40 acres = 16.2 ha) sections. Posts were placed at all section, quarter-section, and meander corners, and two to four witness trees were blazed and inscribed with the coordinates.

Source: After an image taken from the US Department for the Interior, Bureau of Land Management.

[1] All the land west of the original 13 states on the eastern seaboard to the Mississippi River, including the Northwest Territory, was acquired from Great Britain in the Treaty of Paris (1783). The Treaty of Paris marked the end of the American Revolutionary War (1775–1783), and thus the birth of the United States as a sovereign nation. The US Public Land Survey System parcelled out this newly acquired 'public' territory for sale to private landowners, which encouraged settlement of these frontier lands while helping to retire the nation's war debt.

[2] A more substantial marker consisting of a metal pipe with either a metal or plastic cap, on which the survey coordinates are printed, is now used to mark survey corners.

[3] Surveyors also marked where section lines crossed rivers or lakes, which are referred to as 'meander' corners (**Figure 4.2**).

[4] In grasslands, where trees are scarce, section corners were marked with an earthen mound and trenches around the post.

section corner into their field notes, these records provide an unusually comprehensive survey of the occurrence and distribution of various tree species in the pre-European landscape (Schulte & Mladenoff 2001). Little wonder, then, that this dataset has been so widely used to reconstruct these historical landscapes, often serving as a baseline for assessing how forest communities have changed over time, and for evaluating current land-management practices and the restoration potential of landscapes in view of their past disturbance regimes and historical range of variability (e.g. Delcourt & Delcourt 1996; Radeloff et al. 1999; Manies & Mladenoff 2000; Schulte et al. 2002; Bolliger et al. 2004; Rhemtulla et al. 2009).

As valuable as these land-survey data have been for the reconstruction of historical landscapes, they have some limitations (Schulte & Mladenoff 2001; Williams & Baker 2010). Though systematic, the US Public Land Survey was not a scientific survey, but a cadastral survey.[5] Therefore, in addition to the sorts of measurement and recording errors that can plague any ecological survey, there is also the added potential for selection bias and ambiguities in tree-species identification (Manies et al. 2001; Mladenoff et al. 2002; Kronenfeld & Wang 2007). For example, surveyors preferred trees that were easy to blaze and inscribe when selecting witness trees (Manies et al. 2001). Such preferences could clearly influence the types and sizes of trees chosen, as species with thin bark and trees of moderate size (large enough to be inscribed but not so large as to risk being logged for timber) would likely be over-represented in the witness-tree dataset. Even then, individual surveyors might have exhibited different preferences for tree species that met these criteria, as was found in an analysis of variability among the public land surveyors of northern Wisconsin (Manies et al. 2001). In other cases, the surveyor may have been less than precise in recording the specific species of witness tree, simply using a generic term such as 'pine' or 'oak' (Mladenoff et al. 2002; Williams & Baker 2010).

Land-survey records thus do not represent a random or unbiased sample of available trees, which may compromise our ability to reconstruct the composition and structure of forested landscapes prior to European settlement. Nevertheless, many of these sources of bias can be tested for, and if significant, resolved statistically prior to analysis (e.g. Manies & Mladenoff 2000; Mladenoff et al. 2002; Kronenfeld & Wang 2007; Bouldin 2008; Williams & Baker 2010). Not all uncertainty or bias that exists in the data is important from an ecological standpoint. Significant biases in tree species and size become less important when evaluated at coarser scales (at the township or above) or over a broader spatial extent (Manies et al. 2001; Schulte et al. 2002; Weng & Larsen 2006; Liu et al. 2011). For example, although significant selection bias was found among surveyors of hardwood forests in northern Wisconsin, ultimately this bias did not have a great effect on the resulting landscape reconstruction (Manies et al. 2001). Mature hardwood forests are dominated by a few core species, which would constrain the species present at each site, such that large deviations due to selection bias were uncommon. Therefore, at the broader landscape or regional scale, any differences that emerge in forest composition are more likely to be a consequence of environmental heterogeneity (different forest types) than surveyor bias (Manies et al. 2001; Liu et al. 2011).

The US Public Land Survey has served as a model for land-survey systems used by other countries, such as the public land surveys in Australia (Fensham & Fairfax 1997) and the crown land surveys in Canada (Jackson et al. 2000). In addition, the legacy of the US Public Land Survey is still evident today in the gridlike layout of public roadways (especially in flatter rural areas), and in the rectangular agricultural fields that are familiar to airline passengers flying over the middle of the country (**Figure 3.37B and C**). Short of a time machine, land-survey records offer our only view into the recent past, extending the period over which we can evaluate landscape change by an additional 100–150 years beyond existing remotely sensed imagery. To help put this in perspective, we next turn our attention to the various advances in imaging technology that have given rise to the remote sensing of landscapes.

Remote Sensing

Remote sensing is simply the gathering of information from a distance, which can include anything from astronomy to bird watching. For our purposes here, we will adopt the following definition of remote sensing (Campbell & Wynne 2011, p. 6):

> Remote sensing is the practice of deriving information about the Earth's land and water surfaces using images acquired from an overhead perspective, using electromagnetic radiation in one or more regions of the electromagnetic spectrum, reflected or emitted from the Earth's surface.

This definition may appear rather technical, but is simply meant to emphasize that we can obtain far more information about Earth's surface than just that observable within the visible light spectrum; that is, beyond what we can discern by looking at a color aerial photograph. For example, the Advanced Very High Resolution Radiometer (AVHRR), which is carried onboard the

---

[5]Cadastral surveys are concerned with the delineation of land-parcel boundaries (e.g. property lines). A plat map is an example of such a survey, showing how the land has been subdivided for legal and/or taxation purposes.

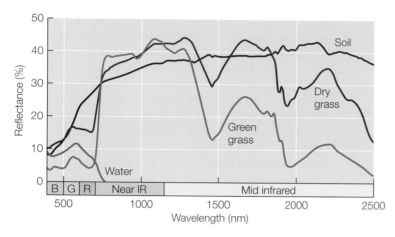

Figure 4.3 Spectral signatures for different land-cover types. Different land covers reflect radiant energy at different wavelengths within the electromagnetic spectrum (**Box 4.1**), which gives them a characteristic spectral signature that is useful for distinguishing different cover types or vegetation conditions (healthy or actively photosynthesizing vegetation versus senescent vegetation).

NOAA[6] polar-orbiting satellites, measures reflectance from Earth's surface in the visible, near-infrared, and thermal infrared portions of the electromagnetic (EM) spectrum (**Box 4.1**). We'll discuss the spectral properties of different types of remotely sensed images a bit later, but for now, it will suffice to know that different land covers or vegetation conditions reflect EM radiation at different wavelengths and therefore can be distinguished by their **spectral signatures** (**Figure 4.3**). Infrared images, for example, can detect drought or disease stress in plants long before it is visible to the human eye. The spectral signature of vegetation is affected by factors such as the leaf water content: healthy (green) plant tissues contain more water and thus will absorb more radiation and have lower reflectance values in the mid-infrared range of the EM spectrum than dry vegetation (e.g. compare the spectral signatures for green grass versus dry grass in the mid-infrared range of **Figure 4.3**).

Many research areas in landscape ecology—but especially those dealing with land-use and land-cover changes, disturbance dynamics, and sustainable land or resource management—have benefitted from the acquisition and analysis of remotely sensed data. Indeed, it could be argued that the discipline of landscape ecology owes its existence to remote sensing (Groom et al. 2006). Recall from **Chapter 1** that the term 'landscape ecology' was coined by the German geographer Carl Troll in reference to his studies of elevational gradients in vegetation patterns, which were largely based on the interpretation of aerial photographs from montane regions around the world. Aerial photography thus represents both an early form of remote sensing and the early use of remote sensing in landscape ecology. Aerial photography continues to play an important role in

landscape analysis, and in ecosystem management more generally (Morgan et al. 2010). A review of papers published in the journal *Landscape Ecology* during a recent 5-year period found that nearly half (46%) of all remote-sensing studies made use of aerial photographs, whereas slightly fewer (42%) used satellite imagery, mostly Landsat satellite imagery (Newton et al. 2009). We will therefore focus on the types of remotely sensed imagery that can be obtained via aerial photography and satellite imagery, although we include a discussion of some other types of remote sensing that are commonly used in topographic mapping (**Table 4.1**).

AERIAL PHOTOGRAPHY   Photography (at least the kind involving chemical processing) has been around only since the 1830s, and the earliest aerial photographs, taken from hot-air balloons, date back to the mid-1800s (Campbell & Wynne 2011). The French photographer Gaspard-Félix Tournachon (also known as 'Nadar') is credited with taking the very first aerial photograph from a balloon over a French village in 1858, although this photo has not survived to the present day. The earliest surviving photograph is believed to be one taken over Boston, Massachusetts (USA) in 1860 by the American photographer James Wallace Black. These aerial photographs enabled humans to obtain, for the first time, a true 'bird's-eye view' of the landscapes around them. Initially, the complexity of the early photographic process[7] limited aerial photography to the use of tethered balloons, which greatly constrained the vantage point

---

[6]The National Oceanic and Atmospheric Administration (NOAA) is a bureau within the U.S. Department of Commerce. NOAA includes a number of agencies, such as the National Environmental Satellite, Data and Information Service (NESDIS), which operates several environmental satellites and provides data and information services related to Earth-system monitoring (www.nesdis.noaa.gov/content/about).

[7]Photography in the 1850s involved using the wet plate or collodion process, a rather inconvenient process that required photographers to work quickly, as the entire process—from preparing the photographic material to exposing and developing it—had to be completed within a span of about 10–15 minutes. The photographer thus needed to carry around a portable darkroom, which is not very convenient for field work, let alone aerial photography from a balloon! The wet plate collodion process was replaced a couple decades later by the dry-plate (gelatin) process. Dry plates were simpler and easier to work with, as the photographer could take the exposed plates back to their darkroom and develop them later, thus simplifying taking of photographs in the field, especially aerial photographs of landscapes.

that could be attained. These earliest aerial photographs are thus better regarded as curiosities rather than an accurate source of landscape information (Campbell & Wynne 2011).

Over the next few decades, the development of dry-plate processing made free-flight aerial photography possible. Balloons no longer needed to be tethered, so aerial photographs could be taken from greater heights. Free-flight aerial photography was not limited to balloons, however. By the turn of the 20th century, kites, rockets, and even pigeons were being outfitted with cameras for the purposes of aerial photography. The turn of the 20th century also saw the development of the first powered, fixed-wing aircraft following the successful flights by Americans Orville and Wilbur Wright in 1903, and by the Brazilian aviation pioneer Alberto Santos-Dumont in 1906. The use of powered aircraft for aerial photography took off during World War I (1914–1918), when reconnaissance and surveillance missions were flown to record the position and movements of enemy forces on both sides (Campbell & Wynne 2011).

After the war, the value of aerial photography for conducting broad-scale land surveys soon became readily apparent. Although aerial photography was already being used for mapping land forms and vegetation patterns in the 1920s, it quickly found major applications in agriculture and forestry during the 1930s and 1940s, especially in the United States. Recall that in the 1930s, the United States suffered an extraordinarily devastating environmental crisis, the Dust Bowl, which was exacerbated by the global economic crisis of the Great Depression (**Figure 3.20; Chapter 3**).

In the wake of these twin disasters, a number of federal programs and agricultural policies were established to ensure better management of soil resources and a more balanced production of agricultural commodities in the future. Compared to field surveys, aerial photography offered a far more effective and efficient means of mapping soils, agricultural land use, and crop conditions; in other words, for monitoring and administering these federal programs. Federal agencies such as the Soil Conservation Service (now the USDA Natural Resources Conservation Service), the Agricultural Adjustment Administration (which was combined with a number of other agencies to form the USDA Farm Service Agency), and the US Forest Service were therefore responsible for commissioning most of the aerial photography of that era. The USDA Aerial Photography Field Office now maintains one of the largest collections of historical aerial photography in the United States (Mathews 2005; **Table 4.2**).

Within the United States, various other federal agencies, such as the National Aerial Photography Program of the US Geological Survey (USGS), also have large holdings of aerial photographs (**Table 4.2**). The USGS is responsible for producing a nationwide series of topographic maps, including the 7.5-minute quadrangle format (1:24,000 scale) or 'topo quad map' that is well-known to outdoor enthusiasts in the United States (**Figure 4.4**). This format is unique to the United States, as most other countries that have national mapping programs (**Table 4.3**) have adopted a standard metric 1:25,000 or 1:50,000 scale, and produce maps of even larger scale in some cases (e.g. 1:10,000 in the

TABLE 4.1 Advantages and disadvantages of different types of remotely sensed imagery (after Morgan et al. 2010).

Type of imagery	Advantages	Disadvantages
Traditional aerial photographs (film-based)	• Versatile—images can be taken according to the specified needs of the user or program (e.g. area of coverage, time of year, temporal frequency, photographic scale, etc.) • Relatively easy to capture, in terms of equipment needed • Long time series (dating back to the 1930s–1950s) may be available for some areas (**Table 4.2**) • High spatial resolution	• Processing requires considerable time and effort (film development, orthorectification) • Availability of aerial photographs typically dependent on specific program needs and spatial coverage (**Table 4.2**) • Manual interpretation of aerial photographs can be subjective • Photographs may be of variable or insufficient quality, owing to weather, environmental variability, positional variability in flight path, etc. • Standardizing image contrast and rectification may therefore be difficult • Metadata limited or inconsistent (mostly in the case of historical aerial photographs)
Digital aerial photographs	• Same advantages as traditional (film-based) photography, plus: • Immediate image access (viewable during flight) can optimize exposure conditions in flight	• Shorter time series available (since 1990s) • Availability of digital aerial photographs still dependent on specific program needs and spatial coverage (**Table 4.2**) • Photographs may still be of variable or insufficient quality

(Continued)

TABLE 4.1 Continued

Type of imagery	Advantages	Disadvantages
	• Enhanced positional accuracy (many digital cameras have GPS) • Images can be copied repeatedly without loss of data or image quality, and storage media can be re-used (memory devices) • Extensive radiometric calibration procedures are available	• Often coarser in resolution than film-based photographs • Large amounts of digital storage space may be required to store high-resolution digital images
Satellite imagery	• Broad spatial coverage (global) • Systematic collection of imagery with high temporal frequency • Sensors cover a broader spectral range than cameras • Many methods for image analysis have been developed • Newer high-resolution imagery broadens the time period over which landscape change can be studied (i.e. facilitates comparison with historical aerial photographs) • Metadata precise and easily obtained • Imagery is easily accessible (online) and many images are available for free	• Shorter time series available than for traditional (film-based) aerial photographs (1970s onward) • Generally coarser resolution than film-based photographs • Higher resolution imagery is expensive • Given the spatial coverage (global) and frequency of image collection, satellite-image datasets are huge and require large amounts of digital storage space, especially for high-resolution images • Image processing is still time-intensive • Equipment costs are prohibitive and generally limited to federal agencies or commercial companies, which limits public access to imagery • Sensors may not be serviceable, due to their location in space • Image quality may be affected by atmospheric conditions and weather (e.g. cloud cover)

TABLE 4.2 Some sources of aerial photographs for the analysis of landscapes and land-use change over time in the United States.

Source	Description	Time period	Resolution or Scale	Online
United States National Archives and Records Administration (NARA)	Historical black-and-white aerial photographs Obtained by the Soil Conservation Service (now the National Resources Conservation Service), Tennessee Valley Authority, US Forest Service, the US Geological Society (USGS), among others Black-and-white and color aerial images of natural disasters (tornadoes, hurricanes, flooding, wildfires)	Historical aerial photographs of landscapes prior to 1955 (mostly 1930s–1950s) Aerial images of areas affected by natural disasters up through the present	Variable	www.archives.gov/research
USDA Farm Service Agency Aerial Photography Field Office (APFO)	Mostly black-and-white (BW), with some natural color, and color infrared (CIR) images Archival images obtained through the Agricultural Stabilization and Conservation Service (now the Farm Service Agency) and the US Forest Service, among other agencies Replaced by NHAP, but provides duplicate storage for aerial imagery obtained through NHAP and NAPP	1947–present Most of the film holdings are from 1955–1979	Variable, most are: 1:20,000–1:40,000 (BW) 1:40,000–1:60,000 (CIR)	www.fsa.usda.gov/programs-and-services/aerial-photography/

Source	Description	Time period	Resolution or Scale	Online
USGS Earth Resources Observation and Science (EROS) Center Long-Term Archive (LTA)				www.usgs.gov/ centers/eros/ science/ usgs-eros- archive- products- overview
National High Altitude Photography (NHAP)	Black-and white and color infrared images Obtained from an altitude of 40,000 feet (12.2 km) Centered over USGS 7.5-minute quadrangles Available as medium or high-resolution digital images	1980–1989 (length of program)	1:80,000 (BW) (11 × 11 miles) 1:58,000 (CIR) (8 × 8 miles)	www.usgs.gov/ centers/eros/ science/ usgs-eros- archive-aerial- photography- national-high- altitude-pho- tography-nhap
National Aerial Photography Program (NAPP)	Black-and-white and color infrared images Obtained from an altitude of 20,000 feet (6.1 km) over the conterminous United States Centered over quarters of the USGS 7.5-minute quadrangles Available as medium or high resolution digital images Replaced NHAP	1987–2007 (length of program)	1:40,000 (5 × 5 miles)	www.usgs.gov/ centers/eros/ science/ usgs-eros- archive-aerial- photography- national-aerial- photography- program napp
Digital Orthophoto Quadrangle (DOQs) and Quarter-Quadrangle (DOQQ)	Black-and-white, natural color, or color infrared Computer-generated images of aerial photographs that combine the image characteristics of the photo with the georeferenced qualities of a map Two types: 3.75-minute (quarter-quad) DOQs and 7.5-minute (full-quad) DOQs; the latter are currently only available for a few western states	1987–present	1 m	www.usgs.gov/ centers/eros/ science/ usgs-eros- archive-aerial- photography- digital-ortho- photo-quad- rangle-doqs
High Resolution Orthoimagery (HRO)	Orthoimagery includes black-and-white, natural color, color infrared, and color near-infrared Orthorectified aerial photographs; combines the image characteristics of an aerial photograph with the geometric qualities of a map May be created from several images from a variety of sources, so resulting resolution, area of coverage, and projection will vary among datasets Foundation for most public and private geographical information systems	2000–present	1 m or finer	www.usgs.gov/ centers/eros/ science/ usgs-eros- archive-aerial- photography- high-resolu- tion-orthoim- agery-hro

(Continued)

TABLE 4.2 Continued

Source	Description	Time period	Resolution or Scale	Online
USDA Farm Service Agency National Agricultural Imagery Program (NAIP)	Digital orthophotographs in natural color or color infrared imagery Aerial imagery for the conterminous US during the peak of the growing season ('leaf on') Acquired on a 3-year cycle (since 2009); 5-year cycle before 2009 Focus is on agricultural areas, but NAIP partners with other federal agencies to acquire full state coverage Centered on a 3.75-minute quarter-quadrangle Available as Compressed County Mosaic (CCM) or Digital Ortho Quarter Quad (DOQQ); also available through USGS EROS in JPEG2000 (JP2) format	2003–present	1 m	www.fsa.usda.gov/programs-and-services/aerial-photography/imagery-products/index www.usgs.gov/centers/eros/science/usgs-eros-archive-aerial-photography-national-agriculture-imagery-program-naip

Netherlands or 1:5,000 in Germany). Recall from **Chapter 2** that these are considered large-scale maps because the representative fraction is large relative to that of a small-scale map (e.g. 1/10,000 = 0.0001 vs 1/100,000 = 0.00001). A map unit corresponds to a smaller distance on the ground in large-scale maps, and thus large-scale maps show more detail, but over a smaller areal extent, than small-scale maps. This explains why relatively small countries, such as the Netherlands or Germany, tend to produce larger-scale maps than very large countries (e.g. Australia, Canada, or the United States). For example, it takes some 57,000 individual topo quad maps to cover the United States at the 1:24,000 scale (i.e. all of the contiguous US, plus Hawaii and areas of Alaska).

It may at first seem surprising that topographic maps, which reflect the three-dimensional terrain or relief of the landscape, can be produced with such accuracy from two-dimensional aerial photographs. That this is at all possible is due to the science of **photogrammetry**, which uses analytical methods based on triangulation to obtain measurements from photographs and produce a geometric reconstruction of the image. In aerial photogrammetry, multiple overlapping aerial photographs (each having at least a 60% overlap with adjacent images in the forward direction) are taken along a flight path. Cartographers process these images using a stereoplotter that combines overlapping photographs (producing a stereo or three-dimensional effect) and enables them to determine elevation by analyzing the degree of 'spatial shift' (the parallax) between the stereo pairs. In effect, this creates a three-dimensional model of the landscape's terrain,

on which the cartographer can draw contour lines for the desired elevations to produce a topographic map of the area. This all used to be done by hand using projection stereoplotters, but is now done on computers with stereoplotting software.

The first aerial photographs were **panchromatic**: black-and-white images in which the visible spectrum is represented as a single 'channel' or band that does not separate white light into its different spectral colors (**Figure 4.5**), and thus captures only the relative brightness (luminance) of different landscape features (**Figure 4.6A**). However, because panchromatic film is sensitive to all wavelengths of light across the visible spectrum (panchromatic means 'across colors'), it has the ability to produce images of higher resolution than color film. Thus, panchromatic images may actually be preferred in aerial landscape photography (and by landscape photographers) where spatial detail is more important than color for distinguishing ground features or patterns. For example, most USGS topographic maps have been produced from panchromatic images.

Color photography was attempted almost from the moment photography first developed in the mid-1850s, but it didn't become commercially viable until the French brothers Auguste and Louis Lumière began marketing their Autochrome process in 1907. Color film was invented following a series of innovations in color processing beginning in the 1930s, but it didn't enjoy mass market appeal until the mid-1960s. All color processing, whether chemical or digital, is based on the three-color method, in which the visible light spectrum is separated into three bands corresponding to red, green, and blue (RGB) light (**Figure 4.5**). Color

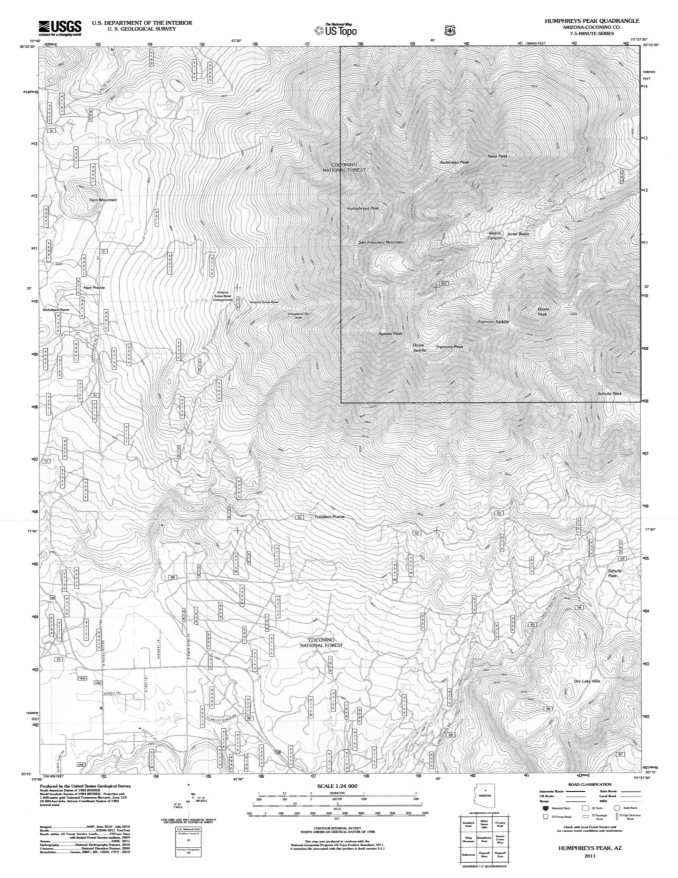

Figure 4.4. Topographic map. A USGS topographic map of the Humphreys Peak quadrangle (7.5-minute format = 1:24,000 scale), which encompasses the San Francisco Peaks in northern Arizona, USA (**Figure 3.1**). Humphreys Peak is the highest (3852 m elevation) of several peaks formed by this now-extinct volcano. The northeast quarter-quadrangle has also been highlighted for reference (e.g. **Figure 4.7**).

Source: Image courtesy of the U.S. Geological Survey.

TABLE 4.3 National mapping programs for various countries around the world.

Argentina	
Instituto Geográfico Nacional (IGN)	www.ign.gob.ar
Australia	
Geoscience Australia	www.ga.gov.au
Belgium	
Institut Géographique National (IGN)	www.ign.be
Brazil	
Instituto Brasileiro de Geografia e Estatística (IBGE)	www.ibge.gov.br
Canada	
Natural Resources Canada Canada Center for Mapping and Earth Observation (CCMEO)	www.nrcan.gc.ca/earth-sciences/geomatics/10776
China	
Surveying and Mapping, Ministry of Natural Resources	www.mnr.gov.cn
Czech Republic	
Český úřad zeměměřický a katastrální (ČÚZK)/Czech Office for Surveying, Mapping and Land Registry	www.cuzk.cz
Denmark	
Geodatastyrelsen/Danish Geodata Agency	www.gst.dk
Finland	
Maanmittauslaitos (MML)/National Land Survey of Finland (NLS)	www.maanmittauslaitos.fi/asioi-verkossa/karttapaikka
France	
Institut Géographique National (IGN)	www.ign.fr
Germany	
Bundesamt für Kartographie und Geodäsie (BKG)/Federal Agency for Cartography and Geodesy	www.bkg.bund.de
India	
Survey of India	www.surveyofindia.gov.in
Iran	
National Cartographic Center (NCC)	www.ncc.org.ir
Ireland	
Ordnance Survey Ireland (OSi)	www.osi.ie
Italy	
Istituto Geografico Militare (IGM)	www.igmi.org
Japan	
Geospatial Information Authority of Japan (GSI)	www.gsi.go.jp
Mexico	
Instituto Nacional de Estadística y Geografía (INEGI)	www.inegi.org.mx
Netherlands	
Kadaster/Land Registry	www.kadaster.nl
Portugal	
Instituto Geográfico do Exército (IGEOE)	www.igeoe.pt
South Africa	
National Geospatial Information (NGI)	www.ngi.gov.za
Spain	
Instituto Geográfico Nacional (IGN)	www.ign.es
Sweden	
Lantmäteriet	www.lantmateriet.se

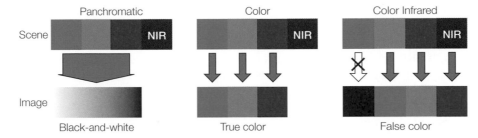

Figure 4.5 Color assignment models for different types of imagery. In **panchromatic** imagery, the visible light spectrum is represented as a single channel across all colors, giving a black-and-white image that records only the relative brightness of different landscape features. For **color** imagery, the three primary colors (blue, green, and red) are represented by three channels or bands. The combination of these three color bands produces natural-color images (true color). In **color infrared** imagery, the blue band is filtered out and the remaining three channels are reassigned to different colors so that objects reflecting in the near-infrared (NIR) range appear red, producing a false-color image.

Source: After Campbell & Wynne (2011).

Figure 4.6 Aerial photographs. Comparison of black-and-white (A) and color infrared (B) aerial photographs of an agricultural landscape just west of Garden City, Kansas (USA). Images have been rotated and aligned to show the same areal extent (bounded by the yellow box in the black-and-white photo). These photographs were taken at the same time (9 May 1983) by two different camera systems aboard the same aircraft, so as to simultaneously capture landscape images in both black-and-white (1:80,000) and color infrared (1:58,000), as part of the USGS National High Altitude Photography (NHAP) program (**Table 4.2**). The color-infrared photo (B) is a false-color image, in which healthy or actively growing crop fields are dark red, fallow fields appear as various shades of blue, and bare fields are dark blue or purple. The Arkansas River running just below the middle of each image seems to divide the landscape into two types of agricultural—or more precisely, irrigation—methods: rain-fed or traditional irrigation north of the river and center-pivot irrigation south of the river. Today, almost all of the agricultural fields in this region are irrigated by the center-pivot method.

Source: Image courtesy of the U.S. Geological Survey, Earth Resources Observation and Science Center.

film has three layers (emulsions), each of which is sensitive to a different color range: the top layer is sensitive to blue, the middle layer is sensitive to green, and the bottom layer is sensitive to red. The combination of these three primary colors (RGB) produces images that look naturally colored to our eyes,[8] which is why these are referred to as **natural-color images**. Natural-color images are useful in mapping applications that benefit from color separation, such as the mapping of vegetation communities or land-cover types (**Figure 4.7A**).

Color-infrared (CIR) film is sensitive to both visible light and longer wavelengths in the near-infrared (NIR; **Box 4.1**). Although NIR adds a fourth spectral channel, we can't actually see this part of the spectrum (it lies just outside the visible range) and we can only see (and therefore use) three primary colors (RGB) in any case. We therefore need to filter out one of the color bands (blue, as it turns out) and re-assign the remaining channels (G, R, and NIR) to our three primary colors if we want to make use of NIR in our imagery. Thus, green becomes blue, red becomes green, and NIR becomes red; anything blue appears black (**Figure 4.5**). Because we have reassigned colors, the colors in the image are no longer true, and thus the

resulting image is called a **false-color image** (**Figure 4.6B**).

Despite the unusual color scheme, CIR imagery does have some advantages over both natural-color film and black-and-white film. First, by dropping the blue band, CIR film is less sensitive to atmospheric scattering (haze),[9] which degrades natural-color images taken at high altitudes, and thus produces images that are sharper and clearer than color film. Second, from a landscape-mapping perspective, CIR images are especially useful in identifying healthy versus diseased vegetation. Healthy, photosynthetically active plants reflect strongly in the NIR and thus will appear red in CIR images (**Figure 4.6B**). It was for precisely this reason, in fact, that the technology was initially developed during World War II (1939–1945), as a means of detecting camouflaged military equipment (Campbell & Wynne 2011). The human eye may be fooled by green-painted surfaces or equipment concealed by cut vegetation, but these forms of camouflage are readily visible in CIR imagery given that living (photosynthetically active) vegetation appears red in contrast to the camouflaged objects, which would show up as blue or green. CIR imagery is now

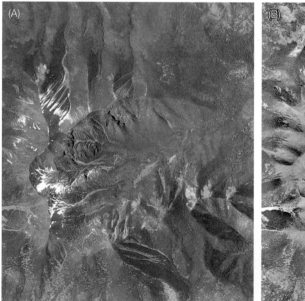 

Figure 4.7 Digital orthophotographs.   These orthorectified images of the San Francisco Peaks in northern Arizona (USA) are centered over the northeast quadrangle of the USGS topographic map (**Figure 4.4**). The digital color orthophoto (A) was taken with a color near-infrared sensor as part of the USDA National Agricultural Imagery Program (NAIP; **Table 4.1**). The black-and-white USGS digital orthophoto quarter-quadrangle (DOQQ) is really a composite of nine photographs that were taken by the USGS National Aerial Photography Program (NAPP; **Table 4.1**).

Source: (A) courtesy of the USDA, and (B) courtesy of the USGS.

[8]The images look natural to our eyes because we only have three types of photoreceptors (cone cells) that respond to wavelengths of light in either the red, green, or blue region of the visible spectrum.

[9]The short wavelengths of blue light are particularly susceptible to atmospheric scattering, which helps to explain why we perceive the sky to be blue.

BOX 4.1   Some Basic Remote Sensing Concepts

To understand remote sensing, we first need to understand the properties of the electromagnetic (EM) spectrum and how physical objects emit and reflect that EM radiation.

As the name suggests, EM radiation consists of both electric and magnetic components that are propagated as an energy wave through space and time at a constant speed (i.e. the speed of light ≈ 300,000 km/s). Energy waves are characterized by their amplitude, frequency, and wavelength. The **amplitude** is the height of the wave and is a measure of the amount of energy carried by the wave. The **frequency** is the number of wave crests that pass a fixed point in a given period of time (e.g. hertz = 1 cycle/s).

The **wavelength** is the distance from one wave crest to the next (e.g. μm). Wavelength is inversely proportional to frequency; shorter wavelengths have higher frequencies than longer wavelengths. Thus, EM energy can be specified using either wavelength or frequency, but the convention in remote sensing is to use wavelength.

Much of the EM spectrum is invisible to us (**Figure 1**). The visible part of the spectrum occupies only a very narrow range (400–700 nm), producing the familiar colors of the rainbow, with blue and violet light (400–475 nm) on one end and red light (650–700 nm) on the other end. Our visual system does not permit us to see EM radiation that lies

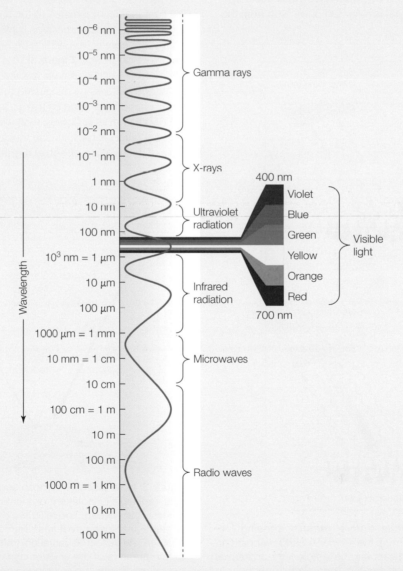

Figure 1   Electromagnetic spectrum.   Different types of sensors are capable of detecting the radiant energy that is reflected or emitted at particular wavelengths of the electromagnetic spectrum. The sensors used in remote sensing operate primarily in the visible, infrared, and microwave portions of the spectrum.

(Continued)

BOX 4.1   Continued

outside this range because our eyes are tuned only to these particular wavelengths. Sensor technologies thus permit us to 'see' EM radiation across a much greater range of the spectrum than we would otherwise be able to detect. In remote sensing, most observation is done within the visible and infrared regions of the spectrum (wavelengths = 0.4–15 µm), although microwave radar is also used in remote-sensing applications. The longer wavelengths of microwave radiation can penetrate cloud cover, which clearly is beneficial for the remote sensing of the Earth's surface.

Where does EM radiation come from? In remote sensing, the sun is the main source of EM radiation (solar radiation) that is measured by **passive remote sensors,** which detect radiation that is reflected or emitted from the

Earth's surface (**Figure 2**). Cameras used in aerial photography are one type of passive remote sensor. Processed film or digital image sensors capture different wavelengths of light (primarily red, green, and blue) within the visible spectrum (and sometimes in the near-infrared) to create images. Satellite imagery, such as those acquired by the Landsat satellites, is another example of passive remote sensing. This contrasts with **active remote sensors**, which emit a form of artificial radiation and measure its reflectance back from the surface (e.g. radar or lidar). The sun produces the full spectrum of EM radiation, which first must pass through the Earth's atmosphere where some of this radiation is scattered or absorbed (e.g. by cloud cover). The light that is transmitted to the surface is then scattered, absorbed, or reflected back as light and/or thermal energy to varying degrees, depending on the spectral properties of the object or surface (**Figure 3**). The fundamental premise of remote sensing is that we are able to identify different landscape features by measuring the radiation that is reflected and/or emitted by them (Campbell & Wynne 2011). Different land covers, land uses, vegetation types, and vegetation conditions can thus be identified on the basis of their unique **spectral signatures** (**Figure 4.3**).

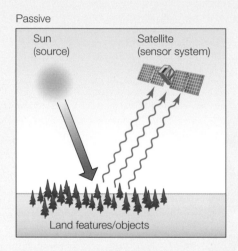
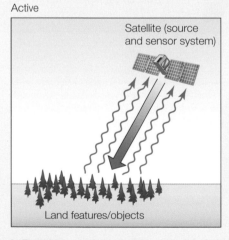

Figure 2 Passive versus active remote sensing. In passive remote sensing (top), the sensor is carried aboard an aircraft or satellite and measures the visible light or thermal energy being reflected or emitted from Earth's surface. In **active remote sensing** (bottom), the energy source and sensor are both carried aboard the aircraft or satellite.

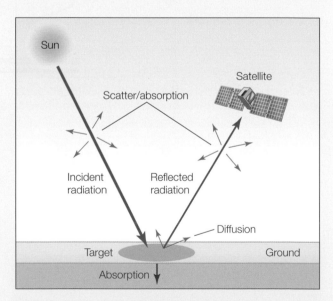

Figure 3 How passive remote sensing works. Incoming solar radiation ('insolation' or incident radiation) is variously scattered, absorbed, and reflected by the atmosphere and by different landscape features. Remote sensors detect and measure the reflected radiation from Earth's surface across certain portions of the electromagnetic spectrum (e.g. visible and infrared), which give unique spectral signatures that can be used to identify different land-cover types or vegetation conditions (**Figure 4.3**).

used in a wide range of remote-sensing applications, particularly to assess vegetation conditions related to productivity (plant biomass, vegetation density, crop yields) and plant health (disease outbreaks, water stress) in forestry and agriculture (**Figure 4.6B**).

Historically, then, most aerial photography was captured on film. Increasingly, however, aerial photographs are being acquired using digital cameras, which first became commercially available in 1999. The image sensors of digital cameras possess certain advantages over film (**Table 4.1**). Digital image sensors can be configured to have very narrow spectral sensitivities, enabling them to record only specific regions of the spectrum, which may contribute to greater precision in measuring and reproducing the spectral characteristics of the scene (Campbell & Wynne 2011). Beyond that, the ability to capture images electronically via digital cameras means that images can be viewed almost instantly, and greatly facilitates the ease with which these images can be transferred, copied, displayed, processed, analyzed, and stored. Aerial photographs processed from film can still be scanned into a digital format, but not without some loss of image quality. Once digitized, however, these aerial photographs can then be processed further using remote sensing software or brought into a GIS for analysis.

A common type of digital aerial photograph is the **orthophoto** or **orthoimage**. A digital orthophoto is a computer-generated image of an aerial photograph, in which the image distortion created by topographic relief and camera tilt has been removed. This process of geometric image correction is called **orthorectification**. Orthorectification is necessary so that distances can be accurately measured within the image via photogrammetry, and so that mapped object locations are correct relative to each other (i.e. spatial fidelity is retained). The resulting orthoimage possesses a uniform scale, like that of a planimetric map,[10] and thus combines the characteristics of the original photograph with the spatial accuracy of a map. For example, the USGS produces two sizes of digital orthophotos, one based on its 7.5-minute quadrangle format (digital orthophoto quads or DOQs) and the other a quarter of this size (a 3.75-minute quarter-quadrangle, sometimes abbreviated DOQQs; **Table 4.2**; **Figure 4.7**). Orthophotos form the basis of orthophoto maps, which are used in many digital map products, such as those available within a GIS.

One common approach to orthorectification is to use a **digital elevation model (DEM)** to correct terrain

distortion in aerial (and satellite) imagery. A DEM[11] represents continuous variation in elevation as a raster dataset (**Figure 4.8**). We'll talk more about raster data a bit later in the chapter, but for now, it will suffice to know that a raster is a grid-based representation of the landscape (i.e. a lattice). A raster dataset is thus a computer data file that consists of rows and columns of data representing the landscape (the xy coordinates of the grid). In the case of the DEM, the raster dataset consists of the elevation values (the z values) within each grid cell or pixel, which is the smallest resolution of the imagery or the finest scale at which the data were collected (i.e. the spatial grain). DEMs are commonly used in GIS and now provide the basis for most digitally produced topographic and relief maps. For example, the USGS produces the National Elevation Dataset, a seamless DEM based on its 7.5-minute quadrangles (**Table 4.2**).

AIRBORNE LIDAR   Increasingly, airborne **lidar**[12] is replacing aerial photography for creating topographic maps and DEMs. Lidar involves **active remote sensing**, in contrast to aerial photography, which is a form of **passive remote sensing** (Box 4.1). Although lidar technology was developed in the late 1960s, it has been used in topographic mapping only since about 2000. Lidar uses a laser scanner to generate a narrow beam of monochromatic light ('pure' light, consisting of a very narrow range of wavelengths, usually in the NIR), which is transmitted as a series of rapid pulses (up to 150,000 pulses/s) directed at the landscape, and then analyzes how long it takes to receive the return signal that is reflected back from the surface. Because the laser pulse travels at a constant speed (i.e. the speed of light), the amount of time between the transmission and reception of the signal is related to the distance the signal has travelled (the range distance). The range distance is combined with information on the aircraft's

---

[10]Planimetric maps represent only the horizontal locations of features, whereas topographic maps represent both the vertical and horizontal positions, showing the relief as contour lines or relief shading (e.g. **Figure 4.4**). A street map is an example of a planimetric map.

[11]There is a good deal of confusion and overlap in the usage of the terms Digital Elevation Model (DEM), Digital Terrain Model (DTM), and Digital Surface Model (DSM). A DTM is a three-dimensional representation of the terrain surface, which includes not only elevations, but also other geographical elements and natural features of the landscape (e.g. rivers, ridgelines, etc.). A DTM is essentially a DEM that has been augmented by additional features in the landscape, but which are not present in the original data; a DEM is thus considered to be a 'bare ground' DTM. A DSM depicts the elevations of all reflective surfaces, such as buildings and vegetation; these signals are sometimes filtered out to leave just the bare-ground signal (i.e. a DEM). We won't worry about distinguishing differences among these different sorts of models in this text, and will just stick to calling these all 'DEMs' throughout the book.

[12]The word 'lidar' (pronounced lie-dar) is considered to be either an acronym (LIDAR or LiDAR) for 'Light Detection and Ranging' (Campbell & Wynne 2011) or, per the Oxford English Dictionary, a portmanteau of 'light' and 'radar.' Given that the etymology apparently supports the latter claim, we will use the typography 'lidar' throughout this text.

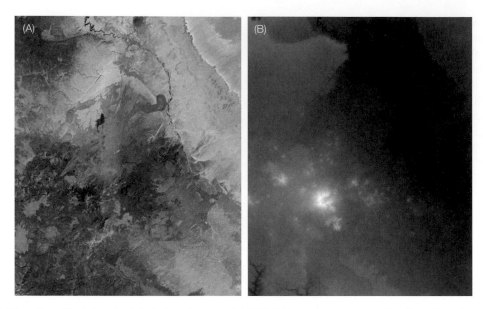

Figure 4.8 Digital Elevation Model. A digital elevation model (DEM) that encompasses the San Francisco Peaks in northern Arizona (Figure 3.1). The San Francisco Peaks show up as the large, brightly illuminated area in the southwest quadrant of the DEM image (B). Smaller volcanic cones are also evident as small white spots surrounding the Peaks. The dark area in the north-eastern quadrant is a wide valley and gorge carved by the Little Colorado River, which joins the main stem of the Colorado River—and the larger and deeper Grand Canyon—some 20 km to the north (compare with the Landsat image; (A)). This particular DEM is based on data obtained from synthetic aperture radar (SAR) data from the NASA Shuttle Radar Topography Mission (1 × 1 degree tile, 1 arc-second = 30 m resolution).

Source: (A) Image from Google Earth, data provided Landsat/Copernicus; (B) DEM image is the work of the USGS.

position and attitude (obtained from the onboard GPS and inertial measurement unit, respectively) to calculate the precise geographic location and elevation of each ground point (i.e. in xyz space). Because multiple returns are received for each pulse, a given lidar dataset may comprise many millions of measurements (data points) of the terrain. The result is an extremely high-density 'cloud' of data points that is essentially a three-dimensional digital scan of the terrain. Images created from lidar data are often of such high resolution that they can be mistaken for black-and-white photographs.

Compared with traditional mapping systems based on photogrammetry, lidar provides a fast and accurate alternative to mapping large areas at high resolution and is gradually being adopted by several national mapping programs as the primary technique for generating DEMs (e.g. the USGS is now using lidar data to create its National Elevation Dataset; Meng et al. 2010). The intensity of the reflected energy can also be analyzed to provide additional information on surface characteristics, such as differences in vegetation density or tree canopy height, and thus can play a role in forest inventory and monitoring forest changes (e.g. as a result of crown expansion or dieback; Reutebuch et al. 2005; Antonarakis et al. 2011). Lidar does have its limits, however, as it is not able to penetrate dense forest canopies (e.g. rainforests) or cloud cover. Although lidar cannot penetrate dense forest cover, the data

point coverage is usually sufficient to enable ground measurements to be taken through small holes in the canopies of most forest types, making it possible, via various ground-filtering algorithms, to filter the signal and isolate the topographic signature (Meng et al. 2010). Lidar is therefore preferred over photogrammetric techniques for topographic mapping because it can 'see through' the canopy to map the ground below vegetation better than most other technologies.

Lidar is also useful in the bathymetric mapping of rivers, lakes, and coastal waters. For bathymetric mapping, a laser in the blue-green wavelengths is used to penetrate the water, but because lidar is based on the transmission and reception of laser signals, its use is limited to relatively shallow waters where light can penetrate (e.g. up to 50 m). Airborne lidar is now the primary method for measuring river topography over large areas, with resolutions of 1 m or better (Carbonneau et al. 2012). Other bathymetric mapping technologies, such as sonar (SOund Navigation and Ranging), are generally better suited for deeper waters, such as for mapping the topography of the ocean floor. Some sources of lidar data in the United States include the NOAA Coastal Services Center ('Digital Coast') and the United States Interagency Elevation Inventory, a joint project led by NOAA and the USGS, with contributions from the Federal Emergency Management Agency.

SYNTHETIC APERTURE RADAR Another form of active remote sensing involves the use of **synthetic aperture radar** (SAR) to obtain digital elevation information. Synthetic aperture radar is a type of imaging radar[13] in which pulses of EM radiation within the microwave range (wavelengths of 1–300 mm) are directed at Earth's surface (**Box 4.1**). The resolution attained by early airborne radar systems was ultimately limited by the length of the antenna—a long antenna permits the acquisition of higher-resolution radar images, but there are limits to the size of an antenna that can be carried by an aircraft. This is also a practical concern with space shuttles or orbiting Earth satellites today, as their greater distance from Earth's surface would necessitate a very long antenna indeed. With SAR, a higher resolution is achieved, in effect, by synthesizing a large antenna via the forward motion of the actual antenna on the radar platform (Chan & Koo 2008). A pulse is transmitted at each location along the flight path, and the return echoes are recorded and their Doppler shift variation analyzed. The Doppler shift provides a unique signature for each location, which upon processing can be used to create high-resolution images. This is the type of SAR used on most operational SAR satellites.

Alternatively, with SAR interferometry (InSAR), two SAR images of the same region are acquired simultaneously from different positions to synthesize a very long antenna, either from separate systems aboard two platforms on different flight (or orbital) paths, or by carrying two widely spaced radar systems aboard the same platform, so as to create signals that are slightly out of phase. This phase difference (phase shift) between signals produces a stereoscopic effect between the two SAR images (much like that in aerial photographs) that can be analyzed via interferometry to produce an **interferogram**, which is then used to create the DEM. SAR interferometry was first used for topographic mapping in 1974, but really gained widespread recognition when it was used aboard the Space Shuttle Endeavor during its Shuttle Radar Topography Mission (SRTM) of February 2000.[14] One set of antennas was located in the shuttle's payload bay and the other set was located at the end of a 60-m mast that was extended from the bay, thereby enabling the SRTM

to obtain SAR data simultaneously from two, slightly different positions in a single pass. The data from this mission have been used to produce a high-resolution, digital topographic database for about 80% of Earth's land area (between 60°N and 56°S). The DEMs produced from the SRTM C-band data are distributed through the USGS EROS Data Center (**Table 4.2**), whereas the finer-resolution SRTM X-band radar data are being processed and distributed by the German Aerospace Center's (DLR) Earth Observation Center.

Admittedly, SAR systems are more expensive to operate than aerial photography, lidar, or even other types of radar-imaging systems (e.g. real aperture systems, such as SLAR or 'side-looking airborne radar'). The interferometric analysis of SAR imagery is rather complicated and requires special expertise to interpret interferograms and generate images. Still, SAR has certain advantages over other remote-sensing technologies used for topographic mapping. Like lidar, it has the ability to measure terrain elevation day or night, because it is not dependent on receiving reflected light energy from an illuminated surface, as in aerial photography. Unlike lidar, however, microwave signals are able to penetrate cloud cover, giving a clear view of Earth's surface regardless of local weather conditions. SAR interferometery provides such detailed digital information on topography that it can distinguish small differences or changes in surface geometry, such as hillside deformations that might presage a landslide or volcanic eruption, which is why it is often used in risk assessment of earthquake, volcanic, and landslide hazards. Oceanographers are using SAR data to study surface and internal waves, the rate at which polar ice sheets are shrinking, and for monitoring shipping lanes for icebergs (e.g. Rignot et al. 2008). Ecological applications of SAR include the estimation of net primary production in ecosystems; mapping forest biomass; forest, crop or land-cover classifications; delineation of floodplain inundation and water-level changes; three-dimensional mapping of species' habitats; forest or crop inventories; and monitoring soil moisture and vegetation condition (e.g. Waring et al. 1995; Kasischke et al. 1997; Imhoff et al. 1997; Bergen & Dobson 1999; Alsdorf et al. 2000).

SATELLITE IMAGERY Remarkably, it took only about 100 years to go from the first aerial photograph, taken aboard Nadar's tethered hot-air balloon in 1858, to the first satellite images of Earth's surface via the Explorer 6, which was launched in 1959 by the United States.[15]

[13]The word 'radar' is actually an acronym for 'RAdio Detection and Ranging,' which was coined by the US Navy in 1940. As with most remote-sensing technologies, radar was quickly coopted for military and strategic reconnaissance, which contributed to the rapid development of radar system technology in the years leading up to World War II (Campbell & Wynne 2011).

[14]The SRTM was an international project spearheaded by the United States National Geospatial-Intelligence Agency (NGA) and the National Aeronautics and Space Administration (NASA), with participation by the German Aerospace Center (DLR) and the Italian space agency (ASI) [Campbell & Wynne 2011; www2.jpl.nasa.gov/srtm/mission.htm].

[15]The first images of Earth from space were actually captured some 13 years earlier, in 1946, during a sub-orbital V-2 rocket flight launched by the United States. The first artificial Earth satellite, Sputnik 1, was launched by the Soviet Union (USSR) in 1957, a feat that spurred the 'Space Race' between the USSR and the USA during the Cold War era following World War II, and which

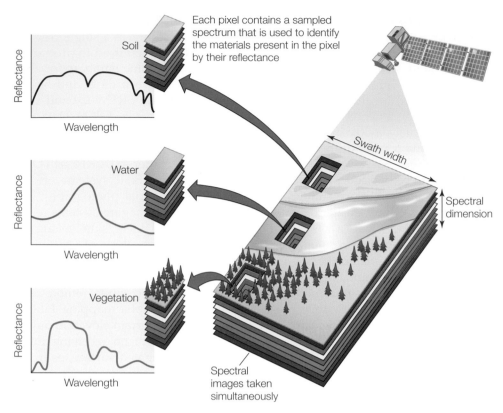

Each pixel contains a sampled spectrum that is used to identify the materials present in the pixel by their reflectance

Soil

Reflectance

Wavelength

Water

Reflectance

Wavelength

Vegetation

Reflectance

Wavelength

Swath width

Spectral dimension

Spectral images taken simultaneously

Figure 4.9 Multispectral images.   Most environmental satellites, including Landsat, use some sort of multispectral scanner, in which several spectral ranges (bands) are measured simultaneously as the sensor sweeps across and along the satellite's ground track. The spectral signatures (**Figure 4.3**) can then be analyzed to identify different landscape features, and multiple images can be combined to create multispectral images (**Figure 4.10**).

Source: After Shaw & Burke (2003).

Although the first Earth observation satellites were actually for monitoring the weather,[16] the first satellite dedicated solely to the observation of Earth's land surface was launched by NASA in 1972 as the Earth Resources Technology Satellite (ERTS), which we know today as Landsat 1 (Campbell & Wynne 2011). There have now been a total of eight Landsat missions, the most recent of which was launched in February 2013, making this the longest-running program for the acquisition of satellite imagery (45+ years). Millions of images have been taken and are now freely available online through either the USGS EarthExplorer (earthexplorer.usgs.gov), Global Visualization (GloVis) Viewer (glovis.usgs.gov), or the Landsat LookViewer (landsatlook.usgs.gov).

[T]he value of any sensor must be assessed not only in the context of its specific capabilities but also in the context of these characteristics relative to other sensors.

Campbell & Wynne (2011)

Over its long history, Landsat has undergone numerous upgrades to its sensor instrumentation. The first Landsat vehicles (Landsats 1, 2, and 3) carried two sensor systems. The first, the Return Beam Vidicon (RBV), provided a camera-like perspective to replicate the sort of imagery obtained using aerial photography with color infrared film (i.e. it had three spectral channels in the red, green, and NIR; **Figure 4.5**). The second, the **Multispectral Scanner System (MSS)**, recorded data in four spectral bands (green, red, and two NIR bands). Landsat 3, launched in 1978, added a fifth channel in the far-infrared (thermal) range. The multispectral scanner measures reflected radiation coming off Earth's surface within these 4–5 spectral ranges while the sensor systematically sweeps across and along the satellite's ground track (defined by the satellite's orbit, the sensor's swath width, and swathing pattern; **Figure 4.9**).

contributed to the rapid development of satellite technology, unmanned planetary and deep-space probes, and manned missions to the moon (Earth's only natural satellite).

[16]The TIROS (Television and Infrared Observation Satellite) series of weather satellites were first launched in 1960 by NASA and its partners.

Thus, this represents a form of passive remote sensing (**Box 4.1**). A separate image (scene) is produced by each spectral channel, each emphasizing the landscape features that reflect most strongly within that particular part of the spectrum (**Figure 4.9**). These scenes can then be combined to create a single composite multispectral image (**Figure 4.10**).

The MSS proved superior to the RBV sensor and was thus retained in the upgraded Landsats 4 and 5, which were launched in 1982 and 1984, respectively. New to these missions was the addition of the **Thematic Mapper (TM)** sensor, which was basically an enhanced MSS (Campbell & Wynne 2011). The TM acquired data within seven spectral bands (the previous five bands, plus two new bands in the mid-infrared region of the spectrum), and thus provided far greater spatial resolution (30 m for all but Band 6, which measured in the far infrared and had a resolution of 120 m), spectral separation, geometric fidelity, and radiometric accuracy than the MSS. In 1993, Landsat 6 was launched but failed to reach orbit; it carried the Enhanced Thematic Mapper (ETM) sensor, which included a panchromatic or 'sharpening' band (15 m spatial resolution) for a total of eight channels. The loss of Landsat 6 might have dealt a serious blow to the continuity of the Landsat program. Fortunately, Landsat 5 continued

to operate well beyond its 3-year design life,[17] preventing a data gap until the successful launch of Landsat 7 in April 1999.

The sensor onboard Landsat 7 is the **Enhanced Thematic Mapper Plus** (**ETM+**), which measures eight spectral bands that span the visible, NIR, mid-infrared, and far (thermal) infrared wavelengths, providing spatial resolutions of 15 to 60 meters (**Table 4.4**). Like its predecessor, Landsat 7 has continued to operate long past its design life (it was designed to last a minimum of 5 years) and remains operational as of 2019 (20 years), if not entirely problem-free. The sensor's scanline corrector (SLC), which compensates for the forward motion of the satellite and ensures that scan lines are properly oriented in the processed imagery, failed on 31 May 2003 (Campbell & Wynne 2011). Without an operating SLC, images have wedge-shaped gaps near the edges (reflecting the zig-zag pattern traced by the sensor along the satellite's ground track), which amounts to a loss of about 22% of the data for any given scene. Put another way, some 78% of the scene is still clearly visible and unaffected by the SLC failure. Thus, Landsat 7 continues to acquire images in the 'SLC-off' mode, and various methods are used to fill in the gaps of SLC-off images, from creating mosaics from several images to data interpolation of image data using geostatistics (e.g. Maxwell et al. 2007; Roy et al. 2010; Chen et al. 2011; Wulder et al. 2011; Zhang et al. 2014).

Landsat 7 has a repeat interval of 16 days; that is, it completes a full scan of Earth's surface about every two weeks, following 233 orbits by the satellite (**Figure 4.11**). During each orbit, the sensor scans an 185-km-wide swath along the satellite's ground track. Image data are framed in 170 km increments (scenes) along the ground tracks; each scene thus covers an area of approximately 170 km (north-south) by 183 km (east-west). It takes the satellite 98.8 minutes to complete one orbit, and so Landsat 7 orbits Earth 14.6 times each day. The ground tracks for that day's orbits are each shifted 2752 km (24.7°) to the west of the previous orbit (when measured at the equator). The intervening ground tracks are filled in during subsequent orbits by the satellite over the next 16 days, but not in numeric order. Because each successive day's track is shifted west by 1204 km (10.8°), scenes located on adjacent tracks are actually separated in time by a week. For example, if Path 1 is scanned on Day 1, the track immediately west (Path 2) would not be scanned until a

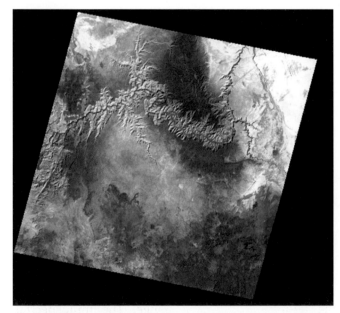

Figure 4.10 Multispectral image. A false-color composite of the Grand Canyon in northern Arizona (USA), which was created by combining reflectance within four spectral bands (green, red, and two NIR channels) by the Multispectral Scanning System (MSS) onboard Landsat 2.

Source: Image courtesy of the NASA Landsat Program.

---

[17]Landsat 5 was finally decommissioned on 5 June 2013, and holds the record for the longest-operating Earth observation satellite (nearly 30 years!). It orbited the planet more than 150,000 times and transmitted over 2.5 million images (www.usgs.gov/land-resources/nli/landsat/final-journey-landsat-5-a-decommissioning-story).

TABLE 4.4 Spectral bands measured by sensors aboard Landsats 7 and 8. Landsat 7 carries the Enhanced Thematic Mapper Plus (ETM+) sensor, whereas Landsat 8 carries the Operational Land Imager (OLI) and Thermal Infrared Sensor (TIRS).

Landsat 7				Landsat 8				Properties/Mapping Applications[2]
Band	Name[1]	Wavelength (μm)	Spatial resolution (m)	Band	Name[1]	Wavelength (μm)	Spatial resolution (m)	
				1	Coastal/ Aerosol	0.43–0.45	30	Deep blue band; Useful for studies of aerosols and coastal waters
1	Blue	0.45–0.52	30	2	Blue	0.45–0.51	30	Penetrates clear water; Useful for bathymetric mapping, mapping of coastal waters, and distinguishing soil from vegetation and deciduous from coniferous vegetation
2	Green	0.52–0.60	30	3	Green	0.53–0.59	30	Records green radiation from healthy vegetation; Assesses peak vegetation, which is useful for assessing plant vigor; also useful for assessing water turbidity
3	Red	0.63–0.69	30	4	Red	0.64–0.67	30	Chlorophyll absorption in this part of the spectrum useful for plant-type discrimination and the discrimination of slopes with vegetation
4	NIR	0.77–0.90	30	5	NIR	0.85–0.88	30	Absorbed by water; Useful for delineating shorelines and for assessing plant biomass
5	SWIR-1	1.55–1.75	30	6	SWIR-1	1.57–1.65	30	Indicative of moisture content; Useful for mapping soil and vegetation moisture, and for penetrating thin clouds; Used to detect plant stress and delineating burned areas

Landsat 7				Landsat 8				Properties/Mapping Applications[2]
Band	Name[1]	Wavelength (μm)	Spatial resolution (m)	Band	Name[1]	Wavelength (μm)	Spatial resolution (m)	
6	TIR	10.40–12.50	60* (30)	**10**	TIRS-1	10.60–11.19	100* (30)	Useful for mapping thermal differences in water currents, monitoring fires, and estimation of soil moisture
				11	TIRS-2	11.50–12.51	100 (30)*	Improved mapping of thermal differences in water currents, monitoring fires, and estimation of soil moisture
7	SWIR-2	2.09–2.35	30	**7**	SWIR-2	2.11–2.29	30	Useful for discrimination of rock types (hydrothermally altered rocks associated with minerals); Provides improved discrimination of moisture content in soil and vegetation, and penetration of thin clouds; Used to detect drought stress, burned and fire-affected areas, including active fires (especially at night)
8	Panchromatic	0.52–0.90	15	**8**	Panchromatic	0.50–0.68	15	Useful in 'sharpening' multispectral images
				9	Cirrus	1.36–1.38	30	Improved detection of cirrus clouds

[1]NIR = near-infrared, SWIR = short wavelength infrared, TIR = thermal infrared (far infrared)
[2]Sources: Campbell & Wynne (2011); USGS Landsat (http://landsat.usgs.gov/best_spectral_bands_to_use.php)
*ETM+ Band 6 is acquired at a 60 m resolution, but images acquired after 25 February 2010 are resampled to 30 m. Similarly, TIRS bands are acquired at 100 m but are resampled to 30 m.

week later (on Day 8). On Day 2, the satellite's ground track has shifted west by 1204 km, such that Path 8 is scanned on the day after Path 1. So, the ordering of the ground tracks over the 16-day interval in this example would be: 1, 8, 15, 6, 13, 4, 11, 2, 9, 16, 7, 14, 5, 12, 3, 10. Not surprisingly, there is some overlap between adjacent tracks; this 'sidelap' ranges from 7.3% at the equator to nearly 84% near the poles.

Since the early days of Landsat, these images have been indexed using the **worldwide reference system (WRS)**. The WRS is a coordinate system based on the specific orbital track of the satellite (i.e. the longitudinal paths; **Figure 4.11**), which are then subdivided into latitudinal rows that correspond roughly with the center of each scene. The combination of path and row numbers uniquely identifies each scene. Because of an altitude change in the satellite orbits between the first three Landsat missions and subsequent missions, two different systems are used: WRS-1 to reference scenes acquired by Landsats 1 to 3, and WRS-2 for scenes acquired during later missions. Under the WRS-2 system, there are 233 paths (corresponding to the 233 orbital tracks needed to complete a whole-Earth scan) numbered from east to west (1–233; **Figure 4.11**). Path

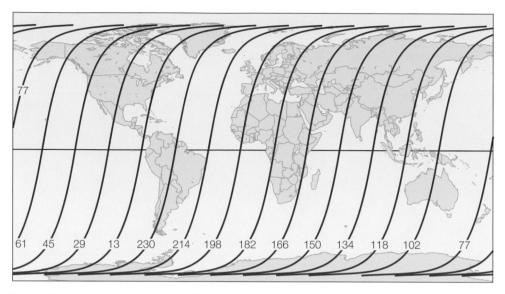

Figure 4.11   Orbital paths for Landsats 4–8.   Paths are numbered from 1 to 233, corresponding to the number of orbits it takes Landsat to scan the entire Earth's surface. This forms the longitudinal coordinate (path number) of the worldwide reference system (WRS-2) used to index Landsat images.

Source: After an image from the USGS.

1 crosses the equator at 64.6°W longitude, near Calanaque in Amazonas state, Brazil.[18] Rows were established so that Row 60 would correspond to the equator (0° latitude) and broken up into a total of 119 rows from there. Thus, Row 1 is at 80.8°N latitude and Row 119 is at 80.8°S latitude. Coverage of a particular landscape may require multiple scenes, representing some combination of paths and rows. For example, the state of Virginia in the eastern United States requires a total of 15 scenes (tiles) for complete coverage, involving portions of six orbital paths (Paths 14–19) and three rows (Rows 33–35; **Figure 4.12**).

A new era of Landsat began with the launch of Landsat 8 in February 2013. Originally called the Landsat Data Continuity Mission (LDCM), Landsat 8 is a joint initiative between NASA and the USGS. Like Landsat 7, Landsat 8 completes a full Earth scan every 16 days, but with an 8-day offset from Landsat 7. Landsat 8 carries two sensors: (1) the **Operational Land Imager** (**OLI**), which uses next-generation sensor technology[19] to collect spectral data consistent with previous Landsat missions, but with improved measurement capabilities and the addition of two new spectral bands in the deep blue and shortwave-infrared ranges, for a total of nine spectral bands; and (2) a **Thermal Infrared Sensor** (**TIRS**), which has two channels (TIRS-1 and TIRS-2) and a 100 m resolution (**Table 4.4**). The two sensors will simultaneously collect multispectral digital images to afford seasonal coverage of the global land surface, including its surface waters, for a minimum of five years. The launch of Landsat 9 is planned for December 2020, and will carry upgraded sensor technologies (OLI-2 and TIRS-2) that will further increase the imaging capacity of the program over previous Landsat missions (landsat.usgs.gov/landsat-9-mission).

The knowledge gained from 40 years of continuous data contributes to research on climate, carbon cycle, ecosystems, water cycle, biogeochemistry and changes to Earth's surface, as well as our understanding of visible human effects on land surfaces. Building off that research, the Landsat imaging data set has, over time, led to the improvement of human and biodiversity health, energy and water management, urban planning, disaster recovery and agriculture monitoring.

NASA Landsat Overview (www.nasa.gov)

---

[18]The WRS was based on a grid system designed by the Canadian Centre for Remote Sensing to manage their data archive after Landsat 1 was launched. According to Landsat Program lore, Path 1 was designated as such because this was the first orbital path to cross any part of Canada, comprising a pair of scenes (Rows 27–28) that encompass the southeastern edge of the island of Newfoundland (NASA/USGS, personal communication).

[19]The OLI instrument uses what is called a 'push-broom' sensor, which aligns a long imaging detector array (containing over 7000 detectors per spectral band and over 13,000 detectors for the panchromatic band) along the satellite's focal plane, such that it can view the entire swath (185 km) at once, as opposed to the 'whisk-broom' sensors employed by earlier Landsat satellites, whose scan mirrors sweep across the field of view. The push-broom design offers greater sensitivity and improved performance over previous whisk-broom sensors (landsat.gsfc.nasa.gov/operational-land-imager-oli/).

Figure 4.12 Landsat scene coverage for the state of Virginia (USA). Scenes are catalogued using the worldwide reference system (WRS-2) according to the orbital path number (**Figure 4.11**) and a row number that is aligned along the approximate center of scenes at a given latitude (e.g. the equator = Row 60).

The enhanced imagery acquired by Landsat 8 is already contributing to the study of rapid environmental changes and seasonal effects on ecosystem productivity at an unprecedented scale and level of detail. For example, you may recall the case study on the Colorado River Delta from **Chapter 3**, in which we discussed how the disruption of the natural flow regime, through the damming and diversion of water along the length of the Colorado River, has dramatically altered this formerly productive estuary, resulting in the loss and contraction of much of the riparian habitat throughout the lower Colorado and its delta region. Recall, too, that an international compact between the United States and Mexico, called Minute 319, was to restore a minimum base flow, as well as allow a one-time pulse flow to occur downriver from the Morelos Dam, which is the last of the many dams on the Colorado River. On 23 March 2014, the Morelos Dam floodgates were opened, and 130 million m³ of water was allowed to course below-dam over a period of 8 weeks. Although much of the water soaked into the ground for the first 60 km, helping to recharge groundwater along the riparian zone, some water did make it to the Colorado River Delta for the first time in more than a decade. The resulting rapid increase in riparian vegetation along the lower Colorado could actually be observed and quantified through an analysis of data obtained by Landsat 8 (**Figure 4.13**).

This 'green-up' in riparian vegetation along the lower Colorado River was evidenced by comparing the **Normalized Difference Vegetation Index** (**NDVI**) between years (August 2014 versus August 2013). The NDVI is the most widely used index for identifying vegetated areas and their condition from multispectral remote-sensing data, having been developed during the early days of Landsat 1 (Tucker 1979). The NDVI is a ratio that expresses the difference between the reflectance in the NIR versus the red wavelengths as:

$$NDVI = (NIR - Red)/(NIR + Red). \quad \text{Equation 4.1}$$

The NDVI is normalized in this fashion, as a proportion of the total reflectance in the red and NIR bands, to give a bounded index between −1.0 and +1.0. The NDVI exploits the fact that healthy, photosynthetically active vegetation has a unique spectral signature compared to other land-cover types (**Figure 4.3**). The photosynthetic machinery of green vegetation absorbs more heavily in the red part of the visible-light spectrum (0.62–0.70 μm) than in the NIR (0.76–1.1 μm), whereas the opposite is true for non-vegetated areas (bare soil, rock, pavement), water, and snow or ice cover. Thus, NDVI ≤ 0 for non-vegetated areas and NDVI > 0 for vegetated areas, with higher NDVI values associated with areas of greater plant biomass and productivity (e.g. tropical forest NDVI → 1.0).

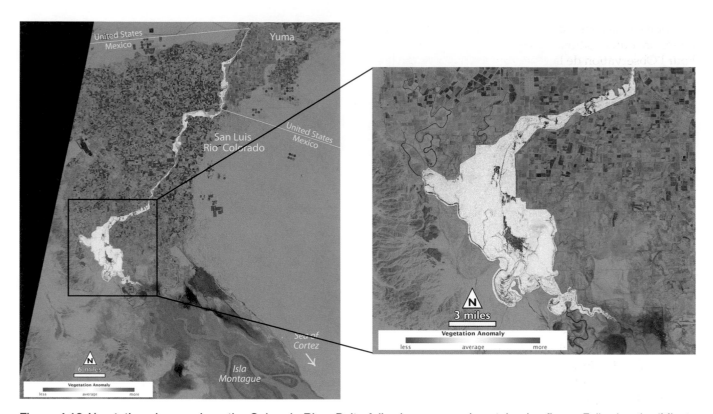

Figure 4.13 Vegetation change along the Colorado River Delta following an experimental pulse flow. Following the 'Minute 319' agreement between the United States and Mexico, some 130 million m³ of water were released below Morelos Dam on the Arizona-Mexico border in March 2014. For the first time in more than a decade, water flowed to the Colorado River Delta in the Gulf of California, which is also known as the Sea of Cortez (**Figure 3.26**). The increase in riparian vegetation (dark green-blue) as a consequence of this pulse flow is clearly evident, as shown here by this comparison of the degree of change ('vegetation anomaly') in the normalized difference vegetation index (NDVI) between 15 August 2014 and 12 August 2013, using Landsat 8 data.

Source: Images courtesy of NASA's Goddard Space Flight Center (Visualization of the NDVI data is by Jesse Allen for NASA's Earth Observatory and Footage of the Colorado River courtesy of Andrew Quinn and Owen Bissell).

Because the NDVI is calculated from reflectance values measured within the red and NIR spectral bands, we obviously need to ensure that we are using the correct band numbers when calculating this index. The band numbers assigned to the red and NIR channels are not standardized and thus vary among satellite programs, and even between missions within a given program (e.g. Landsat). For example, the red and NIR reflectance data are given by Bands 3 and 4 for Landsat 7 images, but these spectral ranges correspond to Bands 4 and 5 in the Landsat 8 imagery (**Table 4.4**). In addition, the spectral ranges for a given channel are narrower for Landsat 8 data, raising concerns about how well NDVI values might compare between Landsat 7 and 8 datasets. Such effects due to 'spectral band differences' between these two datasets are obviously a concern if we are interested in analyzing land-cover change over time, for example. Fortunately, initial spectral band comparisons between these two sensor types (ETM+ vs OLI) for various land-use/land-cover types have revealed only subtle differences in reflectance data, as well as in the performance of several vegetation indices, including the NDVI (Li et al. 2014).

Although Landsat is the longest-running satellite program dedicated to land observation, it is by no means the only such program. The Moderate-Resolution Imaging Spectroradiometer (MODIS) instrument was launched by NASA on the Terra satellite in 1999 and on the Aqua satellite in 2002. Both instruments have a swath of 2330 km and image the entire Earth every 1–2 days. The instruments capture data in 36 spectral bands (between 0.405 and 14.385 μm), and acquire data at three spatial resolutions (250 m, 500 m, and 1 km). Although designed to provide broad-scale readings of global dynamics (e.g. cloud cover, sea- and land-surface temperatures), its imagery is also used by the US Forest Service for detecting and mapping wildfires. MODIS also produces global images of the dominant land-cover types and vegetation dynamics, albeit at a low resolution (at either a 500 m or 5600 m spatial resolution; lpdaac.usgs.gov).

Landsat has served as a model for satellite programs in other nations, such as the SPOT (Satellite Pour l'Observation de la Terre) program developed by the French Space Agency (Centre National d'Etudes Spatiales, CNES) in conjunction with the Belgian Scientific, Technical and Cultural Services (SSTC) and the Swedish National Space Board (SNSB). SPOT 1 was launched in 1986, and the most recent satellite in the series (SPOT 7) was launched in June 2014, despite the agency's decision to terminate the program at the conclusion of the SPOT 5 mission.[20] The continuation of the SPOT program is thus attributable to a commercial venture by Astrium (a subsidiary of Airbus Defence and Space, which was formerly known as the European Aeronautic Defence and Space Company, EADS). In December 2014, SPOT 7 was sold to Azerbaijan's space agency Azercosmo, which renamed the satellite *Azersky*. As the second-longest running satellite program, SPOT has acquired more than 30 million images, at resolutions of 1.5–20 m, over its 30+ years of operation. Another well-known commercial vendor of high-resolution (sub 1 m) satellite imagery is DigitalGlobe,[21] which now offers a range of imagery obtained from its fleet of satellites (IKONOS, GeoEye, QuickBird, and WorldView) that may also be purchased online.

UNMANNED AERIAL VEHICLES Although satellite imagery is a boon to many landscape studies, such data may not be useful for all ecological investigations, especially those conducted at finer spatial scales where the cost of high-resolution satellite imagery is prohibitive. In recent years, the development of lightweight, unmanned aerial vehicles (UAVs or 'drones') have the potential to 'revolutionize spatial ecology' (Anderson & Gaston 2013). In the past, UAVs have been rather large (on par with fixed-winged aircraft) and expensive to operate, but the increasing miniaturization and cost reductions associated with GPS units, sensors, cameras, and embedded computers have all contributed to the rapid development, availability, and affordability of smaller UAVs (i.e. UAVs that are small enough for a person to carry).

Small, cost-effective, fixed-winged drones are already being used for a variety of landscape and conservation applications, from land-use/land-cover mapping and the characterization of forest structure to wildlife population monitoring and biodiversity surveys (e.g. Koh & Wich 2012; Vermeulen et al. 2013; Getzin et al. 2014; Fornace et al. 2014). For smaller field studies, rotary-wing UAVs, such as the various 'quadcopters' available on the commercial market, might prove especially

useful, given that these can be made to fly in any direction (including up and down) as well as hover, making them ideal for detailed inspection or survey work. Indeed, flexibility of use is one of the main advantages of small UAVs for remote sensing. The user can control where, when, and how often the vehicles are flown (although there are limits; see below). Further, although small UAVs are normally flown manually by radio control, many are also capable of autonomous flight and can be programmed by entering a predetermined flight path (e.g. the GPS waypoints) using software on a desktop computer or even a mobile phone app, thereby affording even greater precision for spatial sampling and survey work.

> UAVs fill a niche but do not replace existing remote-sensing methods.
> Fornace et al. (2014)

Although small UAVs hold great promise for filling the 'data gap' between ground surveys and traditional remote-sensing methods, they do have limitations. Their small size limits their payload capacity, which in turn constrains not only the imaging system that can be used (more on that in a moment), but also the endurance and thus flying time of the craft (e.g. <25 min for quadcopters). The total flight distance the UAV can cover is limited not only by its endurance, but also by obstacles in the environment that interfere with the control or operation of the craft (e.g. dense tree canopies, power lines). Inclement weather, such as wind, driving rain, or high temperatures, may limit use of the UAV or ground it completely. It goes without saying that small UAVs are necessarily limited to very low altitudes (typically <100 m), if not by design then by federal aviation regulations. Obtaining permits to fly the UAV over a particular area can also be an issue (Fornace et al. 2014). Finally, the images (or video) obtained by small UAVs are mostly limited to color (RGB) at present, although some have been outfitted with lightweight thermal systems and NIR cameras (Anderson & Gaston 2013; Fornace et al. 2014). Continued advances in the miniaturization of sensor technology may allow the collection of multispectral data by small UAVs at some point in the future. Until then, however, the use of UAVs is perhaps better viewed as ancillary to the sorts of remote-sensing methods we've discussed previously.

From Landscape Data to Landscape Data Analysis

There are many different types and sources of spatial data, but these data will typically be stored digitally in

---

[20]SPOT 5 was launched in May 2002 and stopped functioning in March 2015, after more than a decade in service.

[21]DigitalGlobe merged with its main competitor, GeoEye, in 2013.

FID	Name	Value	Public land?
1	Water	4	Yes
2	Grassland	3	No
3	Deciduous forest	2	No
4	Coniferous forest	1	Yes

Values	Name	Count
1	Coniferous forest	10
2	Deciduous forest	9
3	Grassland	2
4	Water	4

Figure 4.14 Vector versus raster data models. A vector data model (left) is better for representing discrete features with complex geometries, whereas a raster data model (right) is usually preferred for representing continuous surfaces or features that lack well-defined boundaries. The attribute tables for each data model are also presented below each map for comparison.
Source: After a figure created by Jeffrey Dunn, contributing author to https://blogs.lib.uconn.edu.

either a raster or vector format (**Figure 4.14**). Although we can convert from one data form to another, some landscape features naturally lend themselves to one or the other of these formats. For example, some spatial features, such as the locations of individual trees, the trace of a river, or the simple geometry of agricultural fields are easily delineated on the landscape. Such spatial data can be mapped as points, lines, or polygons, respectively, using a **vector data model** (Figure 4.15):

- **Points** represent objects that occur at discrete locations on the landscape; that is, these data essentially have no dimension relative to the scale at which they are being mapped (D = 0). Points are thus represented by a single node. Examples of spatial point data may include the locations of individual trees, small ponds, or bird nests.

- **Lines** are one-dimensional (D = 1) features that occur on the landscape (e.g. streams or roads), or may represent a series of linked data points (e.g. elevation contour lines or animal movement pathways). Lines consist of at least two data points (nodes), defining the endpoints of the line.

- **Polygons** are two-dimensional (D = 2) features on the landscape (e.g. forest stands, agricultural fields, lakes), or may represent a sequence of points defining a bounded area (e.g. an animal's home range or an administrative area). Polygons are basically an enclosed line and thus require at least three points (nodes) to define the shape of the polygon.

Note that the representation of landscapes as a collection of polygons fits neatly within the traditional patch-mosaic view of landscapes (**Chapter 3**). Vector data are good for representing the size and shape of discrete landscape features whose boundaries are reasonably well defined (e.g. political or administrative boundaries, roads, rivers). From a data-management standpoint, vector data files are also smaller and more easily manipulated than raster data files (to be described next) because only the point coordinates that delineate landscape features are stored rather than a value for every grid cell in the raster. Vector data can also depict the relationships among features, such as where flows are likely to occur and in what direction (e.g. water). This typology is epitomized by a network or graph data structure (**Figure 4.15**), which we will cover in more detail in **Chapter 5** when we discuss applications of graph theory for quantifying landscape connectivity.

Raster is faster, but vector is correcter.   Old GIS adage

Yes raster is faster, but raster is vaster, and vector just seems more correcter.   Tomlin (1990)

In contrast to the vector data model, the **raster data model** represents the landscape as an array of regularly spaced grid cells (i.e. as a lattice; **Figure 4.14**). The raster format is often used to represent data that vary

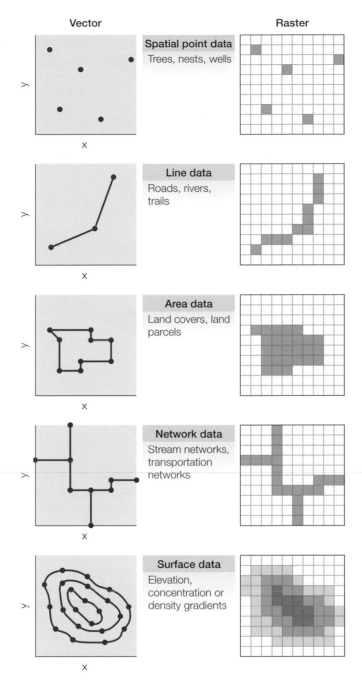

Figure 4.15 Comparison of spatial data models. Vector and raster data models represent various spatial features of the landscape (point patterns, lines, areas, networks, or surfaces) using different data structures. Vector data represent landscape features as points, lines, or polygons. Raster data represent the landscape as a grid, with individual cells containing information about that location on the landscape (e.g., land cover type, species richness).

continuously over the surface of the landscape, such as elevation, NDVI, or soil pH (**Figure 4.15**). Each grid cell (pixel)[22] represents a location of fixed size on the landscape (e.g. 30 m), and thus the raster format is emi-

[22] A 'pixel' is a contraction of 'picture element,' the finest resolution of a digital or graphics image (a type of raster), and thus, represents the grain size of a raster image or data set. For raster imagery, the resolution is expressed as 'pixels per inch' (PPI) and as 'dots per inch' (DPI) in the case of printers.

nently compatible with remotely sensed imagery. Each grid cell is assigned a specific numeric value (e.g. elevation) or class (e.g. land-cover type) corresponding to a particular attribute of interest. If more than one attribute is of interest, then different **spatial data layers** are created, one for each attribute (**Figure 4.16**).

Raster data may thus be numerical or categorical in type. **Categorical data** are characterized by the classes or categories used to define different cell types (e.g.

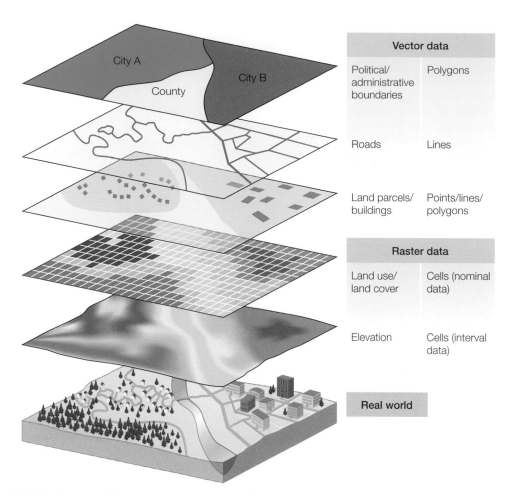

Figure 4.16 Spatial data layers.   Each data layer contains information on a different landscape feature, which may be represented using different data structures (vector vs raster) obtained from different data sources. Spatial data layers can be combined (overlay) for analysis within a GIS.

Source: After an image obtained from NOAA/NWS.

land use or land cover). Numerical data can be converted to categorical data by assigning classes to different numerical ranges (e.g. 1–15 years since disturbance = 'early-successional forest'). Although the raster data format is typically used to represent continuous distributions, discrete data can also be represented as raster data. For example, a point feature would be represented as a single cell, but lines and polygons can be constructed by grouping contiguous cells (**Figure 4.15**). The simple structure of raster datasets also enables certain analyses and operations, such as overlays and buffers, to be performed more quickly and efficiently than with vector datasets, as we'll discuss in the next section.

For any given project, it is quite possible that the spatial data layers will involve a mix of both vector and raster formats (**Figure 4.16**). Depending on the needs of the project, we may have to convert our vector data to a raster format, or vice versa. Thus, there is usually a good deal of data processing, manipulation, and conversion that is required before we can proceed to

the pattern-analysis stage. Because the manipulation and analysis of spatial data are frequently performed within a **geographical information system** (**GIS**), we'll cover the various components, subsystems, and capabilities of a GIS next.

Geographical Information Systems

The first operational GIS was developed in the 1960s for the Canadian government by the English geographer, Roger Tomlinson, who is thus regarded as the 'father of GIS.' Tomlinson developed the Canada Geographic Information System (CGIS) to collect, store, analyze, and manipulate geographical data collected as part of the Canada Land Inventory, a 30-year federal-provincial project charged with evaluating and mapping the capability of rural lands (over 2.5 million km²) to sustain agriculture, forestry, recreation, and wildlife (sis.agr.gc.ca/cansis/nsdb/cli/index.html). The CGIS developed by Tomlinson pared this initial undertaking down from an estimated three years to a matter of weeks, and for about a quarter of the cost (www.ucgis.org/roger-tomlinson).

[A] GIS is designed to accept large volumes of spatial data, derived from a variety of sources, including remote sensors, and to efficiently store, retrieve, manipulate, analyze and display these data according to user-defined specifications.

Marble & Peuquet (1983)

To many, GIS is equated with the software that is used to store, manage, analyze, and display spatial information. There are a number of commercial GIS software providers, as well as several free and open-source GIS platforms (e.g. GRASS, Neteler et al. 2012; see Steiniger & Hunter 2013 for others). Software is only one component of a GIS, however. The software is obviously running on some platform, and so the computer hardware, its operating system, web server, network, and any associated mobile GIS devices are also considered to be a component of the GIS. Some degree of expertise is needed to manage and utilize the GIS, and thus the various administrators, data managers, GIS technicians, application managers, and end users represent yet another component of the GIS. The methods for the input, management, and analysis of spatial data, as well as guidelines for their use and application, represent an additional component of the GIS. Finally, none of this matters without some sort of spatial data that can be obtained, processed, and made available for our use. Thus, a GIS really consists of five, integrated components: (1) spatial data, (2) software, (3) hardware, (4) methods, and (5) personnel (Heywood et al. 2011).

The architecture of a GIS can also be viewed in terms of its different functional subsystems or capabilities, which involve (1) data input, (2) data processing and database management, (3) data manipulation and analysis, and (4) data output and display (Marble & Peuquet 1983). Because of the integrated nature of a GIS, there is some degree of overlap among these various subsystems, and thus these subsystems are not as separate and distinct as suggested here. Nevertheless, this list provides a useful way to characterize and discuss the different functions and capabilities of a GIS, and we will discuss each of these subsystems in turn.

Data Input Subsystem

The data input subsystem involves the capture, collection, or conversion of spatial information into a digital format that can be used by the GIS. As we discussed earlier in this chapter, spatial data can be obtained from a variety of sources, such as digitized or scanned maps and photographs, satellite imagery, field or land surveys,

and from administrative or census data (**Tables 4.2, 4.3**). Although spatial data are obviously a key component of a GIS, these data are usually combined with non-spatial information known as **attribute data**. Attribute data describe the characteristics of the spatial data, such as land-cover type, stand age, or tree density at each location. Attribute data may be classified into four types of measurement:

- **Nominal data** represent discrete categories or types, and thus are identified by name or class only (e.g. land-cover types).
- **Ordinal data** are organized hierarchically and thus have a rank ordering. For example, roads may be organized into a hierarchy based on size and traffic volume into primary highways, secondary highways, local roads, and unpaved roads.
- **Interval data** have numeric values obtained from a measurement scale with an arbitrary (non-zero) reference point (e.g. elevation is measured as distance above mean sea level).
- **Ratio data** also have numeric values, but these are obtained from a measurement scale with an absolute reference point at zero (e.g. stand age or tree density).

To understand how or why the distinction between interval and ratio data matters, consider that a temperature reading of 0°C does not imply an absence of temperature, whereas a density estimate of 0 trees/km² does indicate an absence of trees. If the temperature should warm from 10°C to 20°C, we would not say that it is now twice as warm as it was, but we can state that tree density has doubled if we go from 10 trees/km² to 20 trees/km².

Attribute data are stored in a table that is linked to the specific spatial feature by a unique identifier (e.g. a FID or Feature Identification Number; **Figure 4.14**). An attribute table is thus an example of a **relational database**, in which one or more datasets (tables) can be linked based on their shared attributes.[23] The attribute table consists of columns of data (attributes) corresponding to the individual spatial features (rows), which may be created and stored as spreadsheet files, tab- or comma-delimited text (ASCII) files, dBASE files (*.dbf), or INFO database tables (used by ESRI ArcInfo™ to create coverages). Various operations are performed

[23]In a relational database, every table shares at least one attribute with another table in either a one-to-one, one-to-many, many-to-one or many-to-many relationship. For example, in a one-to-many relationship, a particular spatial feature (i.e. a table record or row) in one dataset is linked to many other datasets that also contain information about that particular feature. In this way, data can be organized and accessed in virtually an unlimited number of ways, making it possible to build and analyze some very complex datasets to uncover some equally complex relationships among spatial attributes.

using the attribute data during the data processing or data manipulation and analysis stages. For example, we can reclassify nominal or ordinal data into fewer categories if we wish (e.g. creating a single category of 'forest' by combining deciduous and coniferous forest types), or we can perform various arithmetic operations on ratio or interval data (e.g. adding the number of tree species surveyed to get a single measure of tree species richness at each site). Although we cannot perform arithmetic operations on nominal or ordinal data (recall that these are not based on a measurement scale), we can sort interval or ratio data into ordinal categories for thematic-mapping purposes (e.g. by assigning stand ages to different categories, such as 'early-successional' or 'old-growth' forest).

Data Processing and Database Management Subsystem

Once we have acquired or captured the necessary landscape data for our project, the accuracy and precision of these data must be evaluated and any data-quality issues addressed. **Data accuracy** defines how close the data come to the actual ('true') value in the real landscape (**Figure 4.17**). Errors in data accuracy include **attribute errors**, in which the characteristics of the landscape features have been applied incorrectly; or **positional errors**, in which the locations of spatial features are incorrect or have been omitted altogether (Bolstad & Smith 1992). Attribute errors may occur because of observer bias (e.g. different observers perceive and record attributes differently), sampling errors, problems with the instrumentation (e.g. sensors have not been properly calibrated and so give erroneous readings), or arise from data-entry errors. Positional errors that affect data accuracy may arise as a consequence of careless measurement and operator error in the field, data-entry errors (if spatial data are entered manually), inadequate sensitivity of the equipment or sensor, registration errors, and other inaccuracies that creep in during the data conversion or processing stage (e.g. digitizing or scanning hard-copy maps and aerial photographs to a digital format).

By contrast, **data precision** refers to the repeatability or reproducibility of the data when obtained under similar conditions (**Figure 4.17**). Precision is indicated by the range of variability or uncertainty associated with the readings obtained for a given location; inconsistent or uncertain readings have low precision. For example, the position of some features in our dataset may be difficult to determine with great precision, either because they lack discrete boundaries or because the resolution of the dataset is too coarse relative to the feature being mapped. In the latter case, we can determine the specific pixel in which a given feature occurs (the spatial data are accurate), but we are

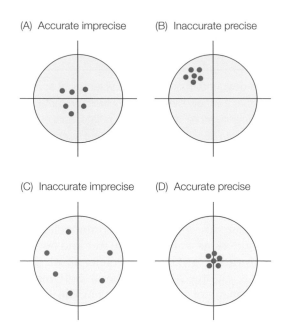

Figure 4.17 Accuracy versus precision in spatial data. If the crosshairs are centered on the true location of a point feature on the landscape (e.g. a tree), then the distribution of points indicate the relative degree of accuracy and precision in the data obtained, for example, by different observers. Data that are accurate (A and D) lie very near to the true coordinates of the point, whereas data that are precise (B and D) show very little variation among different observers, but may (D) or may not (B) be accurate. Ideally, we would like to obtain spatial data that are both accurate and precise (D).

Source: After Heywood et al. (2011).

unable to determine exactly where within that pixel the feature occurs, given the resolution of our dataset (i.e. the spatial data lack precision). Note, therefore, that data can be accurate but lack precision, or vice versa (**Figure 4.17**). Ideally, we want to obtain spatial data that are both accurate *and* precise (i.e. high-quality data).

If we are obtaining landscape data from other sources, we can evaluate data quality by reading the metadata file that accompanies the dataset, if available. The metadata file contains important information about the data; that is, it contains data about the data. Depending on the format or standard used, the metadata file typically includes information on the data source (e.g. the agency responsible for its collection, such as the USGS); the type of spatial data, the method of collection (e.g. lidar), and level of processing; the date the data were acquired; the bounding geographic coordinates (the spatial extent of the dataset); the spatial resolution (i.e. the spatial grain of the dataset), which is usually a function of the sensor technology; the data format (raster or vector); the projection or coordinate system used;[24] a full description of the dataset and its attributes; any constraints on the use of the data; and, how to cite and give credit for the use of the data (attribution information).

Assuming we are satisfied with the quality of our landscape data, there are still a number of operations we may need to undertake before we can begin data analysis. Indeed, data processing is usually the most time-intensive and expensive part of a GIS project, and the importance and complexity of this stage cannot be overstated (Heywood et al. 2011). Some important data-processing tasks include:

- **Reprojection**, to ensure that the projection and coordinate systems are the same for all of our spatial data layers;

- **Registration or georeferencing**, in which any scanned images are digitally aligned or 'anchored' to the appropriate geographic coordinates;

- **Resampling**,[25] to alter (coarsen) the resolution of a raster image and ensure that the spatial resolution is consistent among images to be analyzed;

- **Reclassification**, in which the pixel values in a raster dataset are changed to different values, usually to provide a broader classification scheme (e.g. several forest types may be lumped into a single category of 'forest');

- **Edge matching**, to ensure that adjacent images are properly aligned (i.e. that spatial features along the edges match up);

- **Vectorization** or **rasterization**, to convert spatial data from one format to another (from raster to vector, or vector to raster, respectively);

- **Coordinate thinning**, in which data points are removed from lines and polygons in a vector dataset, giving shapes a simpler or 'smoother' appearance.

In addition, we should seek to identify and edit any remaining spatial errors in our database. This includes verifying our spatial data and landscape maps through some combination of visual inspection (if we are familiar with the study area), comparison with a reference image (e.g. an aerial photograph or DOQ), or **ground-truthing**, in which field data are obtained on-site to verify our classification of remotely sensed imagery (e.g. to verify land-cover types).

Spatial databases and their attribute tables are typically large and complex, and thus data need to be organized in such a way as to permit the quick and efficient storage, editing, and retrieval of information. A database management system (DBMS) helps streamline the chore of editing, updating, and manipulating both spatial and attribute data. The GIS may have either an embedded DBMS or a link to a commercial DBMS (e.g. dBASE™, Oracle®, Access®). The organization of spatial data usually follows a thematic approach, in which different themes are represented as individual spatial data layers that contain information on the feature of interest. For example, one spatial data layer might contain information on soil classifications, another may contain roads, another has information on land-cover types, and still another could have elevation data (**Figure 4.16**).

Data retrieval typically takes the form of a **query**, in which the user can have the GIS program sort, search, and locate spatial features of interest, based on specific attributes or by location.[26] For example, we might want to select all habitat patches of a certain size or cover type for subsequent analysis. Spatial queries are thus closely linked to the data manipulation and analysis subsystem of the GIS, which is often considered the core of a GIS, and yet, would not be possible without the availability of high-quality spatial data that can be efficiently stored, managed, and retrieved as a result of the data processing and database management subsystem.

Data Manipulation and Analysis Subsystem

The manipulation and analysis of spatial data is the *raison d'être* of a GIS, and is what separates it from other types of information systems. Entire books are devoted to spatial analysis within a GIS environment, so just a brief overview of some of these functions will be given here. Specifically, we cover **area and distance functions, neighborhood functions, overlay functions, interpolation functions, topographic functions**, and **network analysis** in the following sections.

AREA AND DISTANCE FUNCTIONS   Area and distance functions provide the most basic of spatial measures

[24]Many different geographic projections have been devised to convert the Earth's curved surface to a flat plane for mapping purposes (e.g. the transverse Mercator projection). Coordinate systems are used to provide a spatial or geographic reference (georeference) so that locations can be mapped, and distances measured, accurately. The most commonly used geographic coordinate system involves the use of latitude and longitude (defined in either degrees-minutes-seconds or decimal-degrees). Some coordinate systems, such as the Universal Transverse Mercator (UTM) coordinate system, are based on geographical projections. The UTM is a global coordinate system that is both accurate and precise (within a meter), which is why it is commonly used in GPS and GIS mapping.

[25]Note that resampling in this context is very different from its usage in statistics, where resampling methods such as bootstrapping are used to estimate the precision of a sample statistic, or Monte Carlo simulations are used to perform randomization tests.

[26]Boolean operators, such as AND, OR, NOT, and XOR (either/or, but not both), are frequently used for spatial and non-spatial queries in a GIS. Boolean operators are based on Boolean algebra, which was developed by the English mathematician, George Boole (1815–1864), and consists of binary logic (1 = true, 0 = false). Thus, if we want to combine two spatial datasets, as when creating map overlays, we can use the AND operator (both criteria must be true) or the OR operator (at least one criterion must be true).

Points	Lines	Polygons

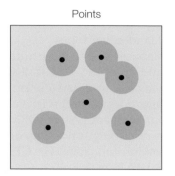 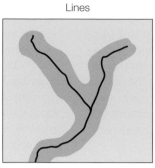 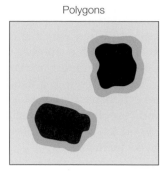

Figure 4.18 Neighborhood functions: buffering. A buffer is created as either a fixed or variable distance around the spatial entity (e.g. points, lines, or polygons), and can be performed on either vector or raster data.

and are a key feature of many landscape metrics and spatial analyses. We will cover these measures in greater detail when we address those topics later in the chapter.

NEIGHBORHOOD FUNCTIONS These functions incorporate the influence of neighboring points or cells on a given location. This can be done for mapping purposes or as part of a proximity analysis. For example, **buffering** is used to create a zone within a fixed (or sometimes, varying) distance around the spatial entity (**Figure 4.18**). Buffering can be used for a wide range of applications, such as to determine the potential accessibility of certain resources or habitat to an individual or population (e.g. by highlighting all habitat within a 1 km radius), or in evaluating the potential impact of certain types of disturbances at varying distances from a source (e.g. a volcanic hazard or flood-zone risk map), or even to delineate sensitive areas around wetlands or waterways for protection (e.g. riparian buffer zones).

Another type of neighborhood function involves **spatial filtering**,[27] which is often used when processing remotely sensed images (e.g. resampling to smooth out otherwise 'noisy' data), but can also be used for spatial analysis. The 'filter' is a data-processing window or search neighborhood of a certain size and shape (usually a square 'box' for raster data or a circle for spatial point data) that is moved systematically over the entire landscape to ensure that each cell or point is processed. The focal location is then assigned a new value based on the mean, mode, or some other function of data within the filter (**Figure 4.19**). Filter

windows of different sizes are often used to investigate how landscape patterns change as a function of changing (i.e. coarsening) the resolution of the landscape (**Figure 2.6**), which may serve as a prelude to understanding scale-dependent relationships between landscape patterns and ecological processes (e.g. Wu et al. 2002). Spatial filtering has also been used to investigate whether 'holes' in the dataset represent true absences (data values = 0), or whether these are a result of under-sampling or under-reporting in that area (e.g. Curtis 1999).

OVERLAY FUNCTIONS Perhaps the most important type of function performed by a GIS (Haywood et al. 2011), overlay functions govern how spatial data layers are integrated. It was this capability that motivated the development of GIS in the first place (i.e. the CGIS developed by Roger Tomlinson). Overlays are fairly easy to accomplish with raster data, as mathematical operators or 'map algebra' can be used to combine cell values among input data layers in various ways to produce the output map layer or dataset. For example, if we are trying to integrate data on the distribution of a species (species present = 1, species absent = 0) with the distribution of its preferred habitat (preferred = 1, not preferred = 0), we can simply add or multiply the cell values to obtain an output map that displays the overlap between the two (**Figure 4.20**).

Whether this output map will prove useful in uncovering the spatial relationship of interest, however, depends on how we code the initial data and the specific overlay functions we perform. Using our previous example, if the objective is to highlight suitable habitat sites that are currently occupied by the focal species, then it would be better to multiply rather than add the cell values so that the resulting output map is clear and unambiguous (**Figure 4.20**). However, if we wish to differentiate preferred habitat sites that

---

[27]Spatial filtering in this context differs from its application in regression analysis, where spatial filtering is done to separate out the spatial effects (i.e. spatial autocorrelation) from a variable's total effects so as to permit the use of conventional linear regression models (e.g. Getis & Griffith 2002).

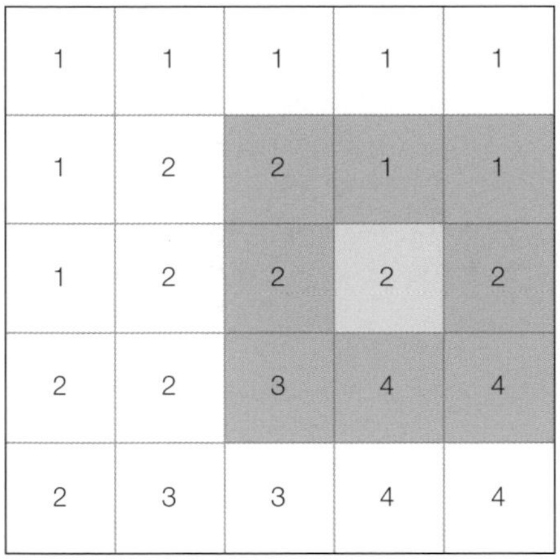

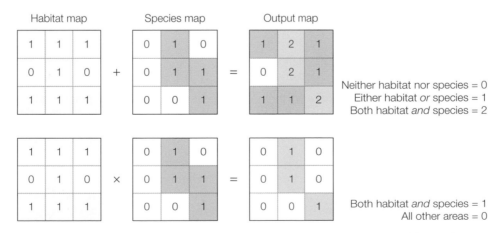

Filter type	Description	Value
Mean	Average value of cells in neighborhood	2.4
Modal	Most common value among cells in neighborhood	2
Minimum	Lowest value among cells in neighborhood	1
Maximum	Highest value among cells in neighborhood	4
Diversity	Number of different class types in neighborhood	4

Figure 4.19 Neighborhood functions: spatial filtering. Filtering assigns a different value to the focal cell within a prescribed neighborhood, as a function of the particular operation employed. For the highlighted neighborhood in the sample landscape above (which is the same raster landscape depicted in **Figure 4.14**), the focal cell could be assigned a value ranging from 1 to 4, depending on which filter is used.

Source: After Heywood et al. (2011).

are currently unoccupied from occupied sites that contain sub-optimal (non-preferred) habitat, then we need to consider whether some other overlay function would better resolve that difference in our output, or whether we perhaps need to apply overlay functions sequentially.

Overlay functions involving vector data are admittedly more difficult and time-consuming to perform (Heywood et al. 2011). There are three main types of vector overlays: point-in-polygon, line-in-polygon, and polygon-on-polygon (**Figure 4.21**). **Point-in-polygon overlays** are used to uncover associations between spatial point data and land-cover types, which are depicted as polygons. For example, we might wish to analyze the distribution of songbird nests with respect to the distribution of forest on a landscape (**Figure 4.21A**). A **line-in-polygon overlay** is a bit more complicated to perform, as the line(s) are broken into smaller segments to reflect where they intersect the polygon(s); topological information is thus retained in the output map (Heywood et al. 2011). For example, if we overlay a road line map over a forest polygon map, perhaps as a prelude to analyzing the effects of roads on nesting songbirds, the lines denoting roads are divided into additional line segments to delineate which sections of the road actually lie inside the forest (**Figure 4.21B**).

Finally, **polygon-on-polygon overlays** can combine polygon map data in several different ways to produce output maps. For example, if we overlay the proposed boundaries of a conservation reserve with a land-cover map by combining both features, we can more easily visualize and quantify what fraction of the total forest

Figure 4.20 Overlay functions: raster data. Various mathematical operations can be performed on raster data layers to produce the output map. In this example, the presence (value = 1) or absence (value = 0) of suitable habitat and the focal species in each cell provide the data input layers. Different output maps can be produced using different operations on the same data, which may be useful in uncovering different types of spatial relationships.

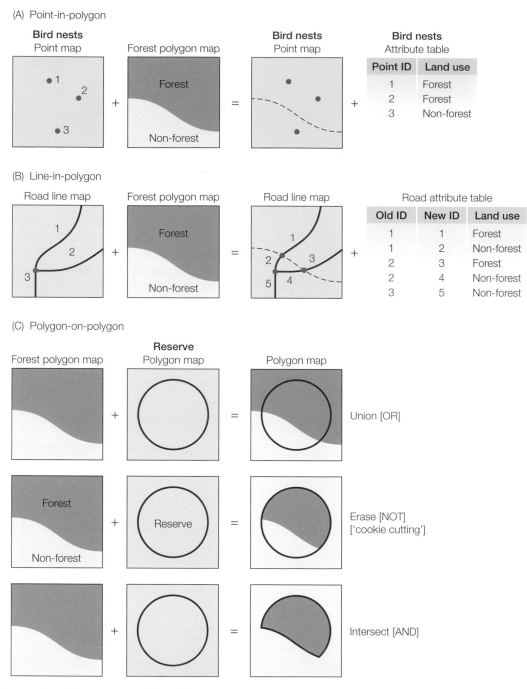

Figure 4.21 Overlay functions: vector data. There are three main types of vector overlays: point-in-polygon, line-in-polygon, and polygon-on-polygon. Several types of operations (union, erase, or intersect) can be performed on polygon-on-polygon operations to produce different types of output maps.

Source: After Heywood et al. (2011).

cover will be protected by using the Union function[28] in our GIS overlay toolset. Union behaves like the Boolean operator OR, by highlighting areas that are either forest *or* reserve, and thus gives us both features of interest (**Figure 4.21C**). Perhaps we are only inter-

ested in the amount of forest cover within the proposed reserve itself, however. In that case, we can overlay the two input maps using a 'cookie-cutting' function that excludes the forest area outside the reserve boundaries using the Erase function. Erase acts like the Boolean operator NOT, and highlights all areas that satisfy our first criterion *but not* the second (i.e. all areas that lie

[28]Operations such as union and intersection are derived from mathematical set theory.

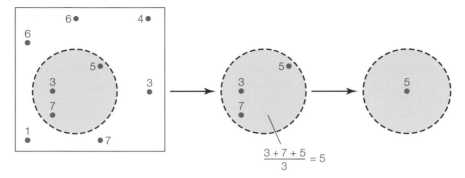

Figure 4.22 Interpolation based on spatial moving average. As a spatial filter (dashed circle) is moved systematically over the landscape surface, the average of the data values that fall within the filter is assigned to that neighborhood. Spatial moving average is often used for data smoothing.
Source: After Heywood et al. (2011).

within the reserve boundary, but not outside the boundary; **Figure 4.21C**).

In the example given, the reserve functions as an **identity feature**, as the boundary of the reserve both delimits and defines the area of protection (i.e. gives it an associated 'identity'), which is why it is retained in the output map (Heywood et al. 2011). But what if we want to limit our analysis to *just* the protected forest area within the reserve? Perhaps we need to determine whether we have sufficient forest habitat to sustain a forest-nesting songbird. In that case, we can use the Intersect overlay function to identify those areas of our landscape that are both forest *and* lie inside the reserve's boundaries. Intersect thus functions as the Boolean operator AND, in which *both* criteria must be satisfied.

The equivalent of point-in-polygon, line-in-polygon, and polygon-on-polygon overlays can also be performed with raster data. Recall that in raster datasets, point data are represented as single cells, line data are represented as a linear array of cells, and polygons are defined as a group of adjacent cells (**Figure 4.15**). So, the cell values for both spatial point features (e.g. bird nests) and lines (roads) could be denoted in terms of their presence or absence on the input map (e.g. road present = 1, no road = 0). The cells that make up polygons can take different values to denote different areal features on the landscape, such as different land-cover types (e.g. **Figure 4.14**). Again, we just need to ensure that the coding of our input maps will provide meaningful results when these are combined (overlaid) to produce the output map. For example, if we are interested in mapping out a scenic drive through the forested areas of the raster landscape depicted in **Figure 4.14**, we might want to reconsider our initial code values for roadways. If we overlay maps by simply adding the cell values for roads (value = 1) and

deciduous forest (value = 2), for example, the resulting output map would instead depict these areas as grassland (value = 3). The solution is thus to recode these features to ensure that their sum will produce a unique value that can be unambiguously attributed to 'forested roads.'

INTERPOLATION FUNCTIONS These are used to generate continuous landscape surfaces, by applying various techniques to estimate values of unsampled locations based on the available spatial data (i.e. the sampled locations). Interpolation methods, especially those based on **triangulated irregular networks (TINs)**,[29] are commonly used for producing DEMs and contour maps. Interpolation may also be used for resampling spatial datasets or images, and for converting a continuous surface from one representation to another (e.g. from a contour to a grid).

There are many different methods for interpolation, but the **spatial moving average** is perhaps the most commonly used in GIS (Heywood et al. 2011). As the name implies, a neighborhood filter (box or circle) is moved in a systematic fashion over the landscape and the mean value is assigned to each location, based on the values of the sampled sites within that filter (**Figure 4.22**). This method works best when a fairly large number of data points are available, as otherwise

[29] A triangulated irregular network, or TIN, is a vector-based data model that is constructed by triangulating a set of data points (nodes) to produce a continuous landscape surface. Methods such as Delauney triangulation are used to connect spatial data points, resulting in a network of irregularly shaped triangles (i.e. as opposed to the regular arrays of raster datasets). Because nodes can be placed irregularly over a surface, TINs can have a higher resolution in areas where a surface is highly variable or where more detail is desired, and a lower resolution in areas that are less variable.

the mean value may not provide a good estimate of that particular location (de Smith et al. 2018).

Ideally, the size of the spatial filter is based on the distance over which local variability occurs or is deemed to be important (Heywood et al. 2011). An **inverse distance-weighting function** can then be employed to give greater influence to points that are nearer than those farther away (another application of Tobler's first law of geography; Tobler 1970). A simple way to do this is to obtain the value of the focal location (z_j) by dividing the value of each sampled location (z_i) by its distance from the focal location (d_{ij}), raised to a power of α based on the expected rate of distance decay:

$$z_j = k_j \sum_{i=1}^{n} \frac{1}{d_{ij}^{\alpha}} z_i. \qquad \text{Equation 4.2}$$

where k_j is an adjustment to ensure that the weights ($1/d_{ij}$) add up to 1 (de Smith et al. 2018). The rate of distance decay may simply be the straight-line (Euclidean) distance ($\alpha = 1$), but faster rates may also be employed, such as the squared distance between points ($\alpha = 2$).

Spatial moving average and inverse distance-weighting are both examples of **deterministic interpolation methods**, in which the weighting function is determined by the method, algorithm, and/or user-defined inputs. This is in contrast to **geostatistical interpolation methods**, such as kriging, which first analyze the spatial data and then derive a model and preferred set of parameters to interpolate the landscape's surface; in other words, these are not predetermined nor selected by the user a priori (de Smith et al. 2018). We will return to this topic when we discuss kriging later in this chapter.

TOPOGRAPHIC FUNCTIONS Topographic functions can be applied to DEMs to permit a three-dimensional visualization of the terrain (**Figure 4.23**). Slope (the steepness of the terrain) and aspect (the direction the slope faces) can be calculated from a DEM using a variety of methods (Wilson & Gallant 2000). A knowledge of slope curvature, in terms of whether the terrain is concave (negative curvature, such as a valley) or convex

Figure 4.23 Three-dimensional rendering of a digital elevation model for Mount St Helens in Washington State (USA). This shaded relief model of Mount St Helens (**Figure 3.22**) was created entirely from elevation data acquired during the Shuttle Radar Topography Mission (SRTM). The topographic slope is shaded so that northeast slopes appear bright and southwest slopes appear dark. Topographic height was color-coded, with lower elevations in green, rising through yellow and tan, to white at the highest elevations. The view here is looking to the southeast, where two other potentially active volcanoes can also be seen in the background (Mount Adams on the left and Mount Hood on the right).

Source: Image courtesy of NASA/JPL/NGA.

(a positive curvature, such as a ridge), is important for understanding hydrology and various geomorphic processes, such as weathering, runoff, erosion, transport, and deposition (**Chapter 3**; Heywood et al. 2011). This in turn might facilitate the development of hazard risk maps, such as for landslides, avalanches, and flooding. **Analytical hillshading** calculates the location of shadows based on the sun's position in the sky (represented by altitude and azimuth), and illuminates the terrain accordingly to create a virtual relief of the landscape (**Figure 4.23**; Heywood et al. 2011). Apart from its visual appeal, analytical hillshading is useful for estimating the amount of sunlight received by different parts of the landscape over time, which can then be used to model snow melt, rates of photosynthesis, and suitability for different vegetation or crop types (Heywood et al. 2011). Finally, visibility analyses can highlight what aspects of the terrain are visible from a particular location, which can then be presented as a **viewshed** map in which areas that are visible are highlighted (Heywood et al. 2011). More sophisticated computing approaches are used for creating animations or visualizations that permit one to explore the terrain ('fly-throughs'). For example, we can explore the entire length of the Grand Canyon by taking a virtual flight over the Colorado River in Google Earth.

NETWORK ANALYSES   Network analyses are used to assesses connectivity within landscapes. A network is defined as a set of nodes (points) connected by lines, which may represent actual connections (roads or waterways) or virtual connections (e.g. flows of information or individuals) (**Figure 4.15**). Network analyses may be used to calculate the shortest or 'least-cost' path distance between two points, the most efficient route within a network (i.e. the 'travelling salesman problem'), or to trace the flow of information, individuals, materials, or services throughout a network (route tracing; Heywood et al. 2011). Graph-theoretic approaches (**Chapter 5**) are commonly used to assess the connectivity of vector data networks, particularly if done in a GIS (de Smith et al. 2018), but connectivity within raster datasets can be explored via applications of percolation theory and other modeling approaches (With 2002a; **Chapter 5**). Connectivity is a core concept in landscape ecology and will be covered in greater detail in **Chapter 5** and elsewhere in the text.

Data Output Subsystem

Finally, the GIS data output subsystem is responsible for visualizing or mapping spatial information and for displaying the results of the spatial analyses or simulations. Although GIS is best known for displaying spatial information in the form of maps, data output can also include plots, tables, lists, directions, animations, three-dimensional terrain models (**Figure 4.23**), and

'fly-throughs' that can take the audience on a virtual flight over (or through) the landscape's terrain.

Now that we have a better idea of how spatial data are processed, stored, and manipulated within a GIS, we can turn our attention more fully to the analysis of those spatial data. To a large degree, spatial analysis is informed by the type of spatial data we have. For example, point pattern data, such as the distribution of bird nests or trees, are typically analyzed using spatial statistics. Thematic data, such as land-cover polygons or grid-cell patches, are analyzed using landscape metrics. We thus consider each of these approaches in turn, starting with landscape metrics.

Landscape Metrics

Although remote sensing and GIS have played a major role in contributing to the availability and analysis of landscape data, it was the release of the FRAGSTATS landscape analysis program (McGarigal & Marks 1995)[30] that really helped to 'revolutionize the analysis of landscape structure and firmly entrench landscape pattern indices or "landscape metrics" in the minds and statistical tool boxes of many landscape ecologists' (Kupfer 2012, p. 401). Although FRAGSTATS may have facilitated the use of landscape metrics by a much wider audience, the development and use of landscape metrics was an early hallmark of landscape ecology (e.g. Forman & Godron 1986; O'Neill et al. 1988a). Thus, many of these metrics have since been incorporated into other widely used stand-alone and GIS-integrated landscape analysis packages (Kupfer 2012). For example, the R package *landscapemetrics* calculates most of the commonly used metrics that are currently available in FRAGSTATS, as well as some other newer metrics (Hesselbarth et al. 2019).[31]

> One of the trademarks of landscape ecology, especially in North America, has been its extensive use of landscape metrics, among numerous methods for spatial pattern analysis.
> Li & Wu (2004)

[30]The current release of FRAGSTATS (v4.2) and its associated documentation are available for download at http://www.umass.edu/landeco/research/fragstats/fragstats.html (McGarigal et al. 2012).

[31]R is an open-source programming language and environment for statistical analysis and graphics, and can be downloaded for free from one of the many Comprehensive R Archive Network (CRAN) mirrors: https://cran.r-project.org/mirrors.html. Select the institution closest to your location and follow the instructions for downloading. Although R comes with a basic suite of pre-installed packages for data management, graphing, and analysis, there are literally thousands of other user-generated packages that have been developed to perform various types of analyses and tasks, and which are likewise freely available for download. For those seeking guidance, there are now numerous books devoted to data analysis and visualization within R.

TABLE 4.5   Common landscape metrics. Landscape metrics measure either the composition or configuration of landscapes, and may be variously calculated at patch, class, or landscape scales as indicated.

Landscape metric	Patch	Class	Landscape
Composition measures			
Landscape richness (R)			X
Proportion of each land-cover class (p_i)		X	X
Landscape diversity measures (Shannon, evenness, dominance)			X
Configuration measures			
Number of patches		X	X
Patch size (area, mean, area-weighted mean)	X	X	X
Perimeter (edge)	X	X	X
Patch shape (perimeter-to-area ratio, fractal dimension)	X	X	X
Edge contrast/like-adjacencies (cell-based)	X	X	
Interspersion/juxtaposition (patch-based)		X	X
Aggregation/clumpiness index		X	X
Contagion			X

Landscape metrics are applied to the analysis of **categorical maps** (also known as **thematic** or **chloropleth maps**). Categorical maps represent the landscape as a patch-based mosaic. Depending on the data format, the patches of the landscape are either represented as polygons (vector format) or as a collection of grid cells (raster format), which have been classified into discrete categories (e.g. land uses or land-cover types; **Figure 4.14**). Recall, however, that quantitative data can also be categorized into classes. For example, forest stand age—a continuous quantitative variable— might be used to distinguish different successional stages, such as 'old-growth forest' (forest stands >200 years) from 'early-successional forest' (forest stands <15 years). The classification scheme and the way in which patches are defined can have major effects on the resulting landscape pattern, and thus the landscape metrics describing that pattern, as we will see later in this section. For now, realize that landscape representation—how we delineate and classify patches—can potentially have as much effect on our analysis of landscape pattern as the geomorphological forces and natural disturbances that shaped those landscape patterns in the first place!

Literally hundreds of metrics have been developed for the purposes of landscape pattern analysis (Gustafson 1998), and this diversity can seem overwhelming to the uninitiated. Further complicating matters is the fact that many measures are redundant, whereas others measure several components of landscape structure simultaneously, which can make them difficult to interpret (O'Neill et al. 1988a; Li & Reynolds 1993; Riitters et al. 1995; Hargis et al. 1998; Cushman

et al. 2008). To keep it simple, we will focus here on a set of standard metrics that will generally suffice for most landscape studies. Our discussion of landscape metrics will be organized around the aspect of landscape structure that they are designed to measure (composition vs configuration), as well as by the organizational scale[32] to which they are usually applied (patch, class, or landscape).

To begin, landscape metrics can be partitioned on the basis of whether they quantify the composition or configuration of landscapes (**Table 4.5**). **Landscape composition** refers to the number and relative abundance of different land covers or patch types on the landscape. Measures of landscape composition thus quantify the overall diversity or heterogeneity of the landscape. In contrast, **landscape configuration** refers to the spatial arrangement of land covers or patch types on the landscape. Measures of landscape configuration thus assess the number, size, shape, aggregation, or interspersion of different patch types or land covers on the landscape. Both types of measures— composition and configuration—are needed to describe and compare landscape patterns. In addition, landscape metrics can be calculated at different scales of organization: the patch, class, or landscape scale (**Table 4.5**). With these distinctions in mind, we will now explore some examples of each type of metric.

---

[32]Landscapes are viewed here as hierarchically structured arrays, consisting of different land-cover classes that are made up of one or more patches (patches → classes → landscape). This is referred to here as 'organizational scale' to distinguish this from the scale of landscape analysis.

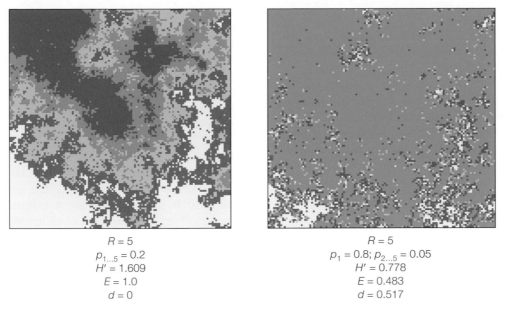

$R = 5$
$p_{1...5} = 0.2$
$H' = 1.609$
$E = 1.0$
$d = 0$

$R = 5$
$p_1 = 0.8; p_{2...5} = 0.05$
$H' = 0.778$
$E = 0.483$
$d = 0.517$

Figure 4.24 Assessment of landscape diversity.   Both landscapes have the same number of land-cover types ($R = 5$), but differ in the relative proportions of those cover types (p_i). All cover types have the same abundance ($p_i = 0.2$) on the left, but one cover type dominates the landscape on the right ($p_1 = 0.80; p_2 = p_3 = p_4 = p_5 = 0.05$). The landscape on the left therefore possesses greater evenness (E), and thus, higher landscape diversity, than the landscape on the right.

Measures of Landscape Composition

Measures of landscape composition are always calculated at the landscape scale; that is, across the entire spatial extent of the landscape. As discussed in **Chapter 2**, the spatial extent of a landscape is best defined by the ecological phenomenon being studied. More often than not, however, landscape extent is defined by political, administrative, or land-ownership boundaries, or is based on the study's logistical or sampling constraints (i.e. the maximum area that can be feasibly surveyed given the limitations of time, personnel, and funding). Regardless of how the landscape is ultimately defined, we can use these measures to quantify and compare the composition of different landscapes. These measures are affected by spatial scale, however, and thus we should ensure that we are comparing landscapes of the same or similar size when conducting a comparative analysis. Many of these measures of landscape composition are also used to assess species diversity within communities (**Chapter 10**), and thus have the same basis in information theory.

- *Landscape richness* **(R)**—the number of different land-cover types (the total number of classes or categories) present within the landscape. This provides a basic measure of landscape diversity or heterogeneity.
- *Proportion of land cover* **(p_i)**—For each land-cover type, the relative proportion of the total landscape occupied by that cover type is obtained as

$p_i = \sum_1^N n_i / N$, where n_i is the area or number of cells of land-cover type i and N is the total area (spatial extent) or number of cells within the landscape. A related measure is the **total class area (CA)**, which is assessed over all cells (or patches) of a given land-cover type and is given in some unit of area (e.g. ha or km²).

- *Landscape diversity (H′, E, and d)*—Although richness (R) gives the number of different land-cover types present, it does not provide any information on the relative abundances of those different cover types. All else being equal, we would consider a landscape to be more heterogeneous (to have greater diversity) if all cover types are equally abundant, than if only one of those cover types dominates the landscape (**Figure 4.24**).

We can calculate a diversity measure in which each land-cover type is weighted by its proportional abundance on the landscape, using the Shannon Diversity Index (H'):[33]

---

[33]Developed by Claude Shannon (1916–2001), an American mathematician considered to be the 'father of information theory,' the Shannon Diversity Index is probably the most widely used index of species diversity in ecology, where it is sometimes referred to as (1) the Shannon–Wiener Index (apparently because Shannon credited mathematician Norbert Wiener for his contributions to communications theory, although Wiener was not actually involved in developing the index itself); (2) the Shannon–Weaver Index (which is a misapplied reference to an earlier publication by Shannon and Weaver); or simply, (3) the Shannon Index (Spellerberg & Fedor 2003).

$$H' = -\sum_{i=1}^{R} p_i \ln(p_i). \qquad \text{Equation 4.3}$$

where p_i is the proportion of land-cover type i, and R is the total number of land-cover types in the landscape (i.e. the landscape richness).

Because H' is unbounded, it is usually desirable to scale this index to the maximum H' (H'_{max}) expected for a given level of richness, R. The H'_{max} for a given level of richness is calculated simply as $\ln(R)$. This gives us the maximum value of H' expected if all of the R cover types are equally abundant on the landscape. Thus, we can calculate landscape evenness (E) as

$$E = H'_L / H'_{max} = H'_L / \ln(R). \qquad \text{Equation 4.4}$$

where H'_L is the Shannon Diversity Index calculated for a given landscape (Equation 4.3). Evenness thus measures the proportion of maximum diversity (heterogeneity) attained by the landscape with R cover types. Because it is a proportion, it is a bounded index: $0 < E \leq 1$. If all cover types are equally abundant, then $H'_L = H'_{max}$ and $E = 1$.

Given that evenness assesses the degree to which land covers are equally abundant on the landscape, the converse of this would be the degree to which the landscape was dominated by a single land-cover type. Dominance (d) can thus be defined as

$$d = 1 - E. \qquad \text{Equation 4.5}$$

The less equitable the relative abundance of the R land-cover types ($E \to 0$), the greater the degree of dominance by one of those cover types ($d \to 1$).

Example 4.1 **Comparison of heterogeneous landscapes using landscape composition metrics**. To illustrate the application of these landscape composition metrics, let's compare the two landscapes depicted in **Figure 4.24**. Both have the same richness ($R = 5$), but differ in the relative proportions of the five land-cover types on the landscape. Although most landscape analysis programs calculate diversity measures (e.g. FRAGSTATS, McGarigal et al. 2012; the R package *landscapemetrics*, Hesselbarth et al. 2019), it is simple enough to calculate these types of measures in a spreadsheet, given that we know what R and p_i are for each of these landscapes (try it!).

For the landscape in which all land-cover types have the same relative abundance ($p_i = 0.2$), it should come as no surprise that $E = 1$, given that this is the very definition of evenness. For a landscape with five equally abundant cover types ($p_i = 1/5 = 0.2$), this is the maximum level of heterogeneity (diversity) that can be attained ($H'_L = H'_{max}$). Conversely, the second landscape is clearly dominated by one cover type ($p_1 = 0.80$), although the other four cover types have the same

relative abundance ($p_2 = p_3 = p_4 = p_5 = 0.05$). Subsequently, the landscape appears more homogeneous, which is borne out by its relatively low diversity index ($H'_L = 0.778$) relative to the maximum level of diversity expected [$H'_{max} = \ln(R) = \ln(5) = 1.609$]. Thus, this landscape has lower evenness ($E = 0.778/1.609 = 0.483$) and higher dominance ($d = 1 - 0.483 = 0.517$), consistent with our initial impression of this landscape.

Because H', E, and d are all related mathematically, they provide redundant information on landscape composition. As mentioned previously, E is generally preferred because it is scaled to the maximum heterogeneity expected for a given level of richness. This means that E is easily interpretable, and can be compared among landscapes with different numbers of cover types. In that case, however, it is important to report richness (R) for each landscape, in conjunction with E, so as to provide a basis for comparison. For example, if two landscapes are equally diverse in terms of E, we still might view a landscape with a greater R as being more heterogeneous than a landscape with a lower R.

Measures of Landscape Configuration

Unlike landscape composition metrics, measures of landscape configuration span the entire range of organizational scales (**Table 4.5**). Some of these metrics can even be calculated at all three organizational scales (i.e. patch, class, and landscape scales). We'll thus tackle measures of landscape configuration by organizational scale, starting with patch-scale metrics, and then discuss how these measures might be applied to other scales, as appropriate. Before we begin discussing configuration metrics at any scale, however, we should first decide how we will define patches for analysis. This is more of an issue with raster data than vector data, because vector-based landscapes represent patches as polygons and thus patches have already been defined. Our discussion here will therefore center on how patches are defined in raster landscapes.

DEFINING PATCHES USING NEIGHBORHOOD RULES The most obvious way to define patches within a raster landscape is on the basis of **like-adjacencies**; that is, by combining all adjacent cells of the same type into a single patch. That seems straightforward enough, but we then have to decide which cells are to be considered 'adjacent.' Cells that share an edge (nearest neighbors) are clearly adjacent, but what about cells that lie along the diagonal (the next-nearest neighbors)? Might these be considered adjacent also?

It depends on our choice of **neighborhood rule** (**Figure 4.25**). Neighborhood size is defined by the number of cells that are considered to be adjacent to the focal cell. For example, the **4-cell neighborhood rule** (or more

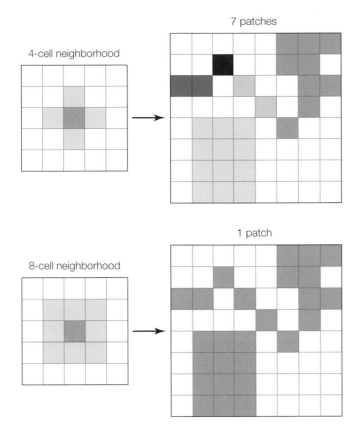

4-cell neighborhood

7 patches

8-cell neighborhood

1 patch

Figure 4.25 Defining patches with neighborhood rules. The choice of neighborhood rule (e.g. either the 4-cell or 8-cell rule) in a landscape analysis will influence the number of patches, and thus, the patch structure (patch size and shape) of the landscape.

simply, the 4-cell rule) includes only those four cells that share an edge with the focal cell (i.e. the nearest neighbors). The **8-cell neighborhood rule** (8-cell rule) includes the four nearest neighbors, plus the four cells that are diagonal to the focal cell (the next-nearest neighbors). Any cell within the neighborhood that is of the same type as the focal cell is considered adjacent, and therefore part of the same patch. If only a single like-cell is adjacent to the focal cell, then that particular patch would consist of only two cells. The smallest possible patch size is thus a single cell, when there are no like-cells within the defined neighborhood of the focal cell.

Of the two neighborhood rules, the 4-cell rule is obviously the most conservative, in that it imposes a stricter definition on how patches are defined: cells are only considered adjacent if they share an edge. Because our choice of neighborhood rule influences how patches are defined, this decision will affect the number, size, and shape of patches on the landscape. In other words, how we define patches ends up affecting our analysis of patch structure, and thus landscape configuration. To illustrate, consider the simple landscapes depicted in **Figure 4.25**. If we use the 4-cell rule

to define patches, the landscape has seven patches: one patch of 12 cells, one patch of 8 cells, one patch of 2 cells, and four patches with only a single cell each. If we use the 8-cell rule, however, the landscape now consists of a single, large, irregularly shaped patch. All 26 cells are considered adjacent, either along an edge or on the diagonal, and are thus part of the same patch.

Regardless of whether we use the 4-cell or 8-cell neighborhood rule, there is no guarantee that our analysis of landscape structure will relate to how some species or process moves through the landscape. Recall from **Chapter 3** that we need to distinguish measured (structural) heterogeneity from functional heterogeneity. Ideally then, our choice of neighborhood rule is based on an understanding of how some process or organism moves through the landscape; that is, we should ideally adopt a functional definition of landscape structure. For example, if we are interested in studying animal movement responses to landscape structure (**Chapter 6**), an animal moving from cell to cell is likely to do so by crossing an edge, rather than exactly at the point that defines the diagonal between two cells. Animals may appear to move along the diagonal, but are often 'cutting the corner,' such that movement between cells still occurs across cell edges. In that case, perhaps the 4-cell rule would best represent the movement process and thus how patch structure is likely to affect movement across the landscape. But what if we are interested in studying plant dispersal? If most seeds are dispersed within a certain radius of the plant, then perhaps an 8-cell rule would better reflect the resultant distribution of safe sites[34] within the dispersal neighborhood of the plant. Some seeds are bound to fall within those diagonal cells if seeds are dispersed in a circular area around the plant.

Then again, perhaps seed dispersal occurs over greater distances, beyond the immediate neighborhood around the plant. In that case, a larger neighborhood would be needed to assess the distribution of safe sites that fall within the plant's dispersal range, which may include cells that are ultimately unsuitable for seed germination. For example, we could expand our neighborhood rule to encompass all cells within a two-cell radius of the focal plant (i.e. a 5 × 5-cell block), which would give us a 24-cell neighborhood. A larger neighborhood rule could also apply to animal dispersal; individuals may be willing to cross small gaps in the distribution of their preferred habitat when dispersing across the landscape. Small gaps—say one or two cells wide—might therefore be so inconsequential that they are essentially perceived to be part of the larger habitat patch. Once again, a larger neighborhood rule, one that encompasses the size of habitat

---

[34]A 'safe site' (*sensu* Harper et al. 1961) is a site that is suitable for seed germination and plant establishment (Fowler 1988).

gaps individuals are willing to cross, would be preferred for our analysis of patch structure and landscape configuration.

These examples highlight an important point: We are not limited to defining patches based solely on physical adjacency (a measure of structural connectivity). Because habitat cells may be functionally connected by dispersal, the **dispersal distance** or **gap-crossing ability** (i.e. the gap-size distance individuals are willing to cross) may be used to identify the appropriate neighborhood size for defining patches and analyzing landscape structure. This is central to determining **patch isolation**, an important landscape metric that is often used to assess fragmentation effects on populations and communities. This is such an important point, that it deserves to be highlighted: *Distance alone does not indicate whether a habitat cell or patch is isolated. Rather, patch isolation depends on the degree to which habitat is functionally connected, whether by dispersal, gene flow, or some other process.*

The functional connectivity of patches and landscapes is a key concept in landscape ecology, and thus will be covered more fully in **Chapter 5**. Functional connectivity will be a recurring theme in many of the chapters that follow. In the meantime, we will restrict our focus here to structural measures of connectivity and heterogeneity (i.e. the 4- and 8-cell rules), since these are most often used for the analysis and comparison of landscape structure (e.g. the 8-cell neighborhood rule is the default in Fragstats, although the 4-cell rule is available as an option; McGarigal et al. 2012).

PATCH-SCALE METRICS   Patch-scale metrics describe the properties of individual patches, such as their size and shape. They are usually calculated for a particular land-cover class, but can also be calculated across all cover types for the entire landscape (**Table 4.5**).

- *Patch area* **(A)**—Size is the most fundamental patch-scale metric, and is expressed in terms of unit-area (e.g. in ha or km²). Patch area has important implications for many—if not most—ecological processes. For example, patch area is often related to a species' risk of extinction, as well as the total number of species within that patch (**Chapters 7 and 10**).

- *Patch perimeter* **(P)**—The perimeter of a patch is the distance along its outer edge (e.g. in m or km). The amount of edge may be important for determining the likelihood of movement or flows across patch boundaries (either into or out of patches; **Chapter 6**), as well as the potential influence of the surrounding matrix on patch properties and dynamics (e.g. negative edge effects; **Chapter 7**).

- *Patch shape* **(P/A, P/A*,** and **D)**—Patch shape (patch geometry) is important for evaluating the potential for patch-boundary effects on animal movement, gene flow, and other ecological flows (e.g. **Chapter 6**). Patch shape can be quantified in a number of different ways, of which we'll consider three approaches here.

The simplest patch-shape measure is the **perimeter-to-area ratio** (P/A). For two patches of the same area ($A_1 = A_2$), the patch with the most edge ($P_2 > P_1$) will have the greater P/A ratio, and thus, a more complex shape. To illustrate, let's compare the two patches in **Figure 4.26**. Both have the same area (each habitat cell = 1 km², so $A_1 = A_2 = 9$ km²), but differ in shape, with the first patch configured as a simple square. Because of the more complicated shape of the second patch, it has more edge (each cell edge = 1 km), and thus, a greater P/A ratio than the first patch (2.2 vs 1.3). All else being equal, then, patches that are elongated or have a more complex shape will have a greater P/A ratio than simple, geometric patch shapes (i.e. square or round patches).

Because the P/A ratio varies with the size of the patch (A), however, it is usually best to standardize this by the P/A for a simple square of 1 unit-area (e.g. P/A of a single cell = 4/1 = 4). Thus, the **standardized P/A** (P/A*)[35] for a patch can be obtained as

$$P/A^* = \frac{0.25(P)}{\sqrt{A}},\qquad \text{Equation 4.6}$$

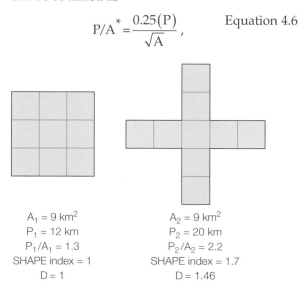

A₁ = 9 km² · A₂ = 9 km²
P₁ = 12 km · P₂ = 20 km
P₁/A₁ = 1.3 · P₂/A₂ = 2.2
SHAPE index = 1 · SHAPE index = 1.7
D = 1 · D = 1.46

Figure 4.26 Effect of patch geometry on patch metrics. Patches with simple square shapes (left) have both their standardized perimeter-to-area (P/A) ratio (P/A* = SHAPE index) and fractal dimension (D) equal to 1. Given the same total area, more complex patch shapes (right) have proportionately more edge (perimeter), and so, have a SHAPE index and D greater than 1.

[35]This is the SHAPE index in Fragstats (McGarigal et al. 2012).

which returns a value of 1 for a patch made up of a single habitat cell $[0.25(4)/\sqrt{1} = 1/1 = 1]$. This gives us a way of comparing patch shapes of different sizes in a meaningful way. For example, if we had a square patch that was four-times larger than the one depicted in **Figure 4.26** (i.e. A = 36 km²), this would appear to have a smaller P/A ratio despite being the exact same shape (P/A = 24/36 = 0.7). However, if we standardized P/A using Equation 4.6, we now obtain the same index for both patches, describing a simple square (i.e. P/A* = 1).

Given the standardized P/A, we can also calculate the **fractal dimension** (D) of the patch. In the context of patch geometry, the fractal dimension describes the complexity of the patch's shape, which is fractional because it occupies a dimension in between the traditional (Euclidean) dimensions (i.e. $1 \le D \le 2$; **Box 2.1**). For example, patches that have been created by natural processes are rarely perfectly square or circular; in fact, the presence of such simple geometric shapes on the landscape is usually a tell-tale sign of human land-use activities (**Figure 3.37**). We can calculate the fractal dimension (D) of a patch as

$$D = \frac{2\ln[(P)0.25]}{\ln(A)}. \qquad \text{Equation 4.7}$$

which, as you've no doubt noticed, is based on the standardized P/A ratio (P/A*; Equation 4.6). Thus, if the P/A* index for a simple square is 1, we should expect the same value for the fractal dimension (D) of a square patch, when this is calculated using Equation 4.7 (try it!).[36] Although this may seem like a redundant measure, the fractal dimension has the desirable property of being a bounded index (unlike the P/A* index), given that it encompasses the range $1 \le D \le 2$. Thus, patches with more complex shapes take up more space in two dimensions than simple shapes, all else being equal (**Figure 4.26**).

- *Habitat contiguity within patch* (**C$_p$**)—The extent to which habitat cells within patches are contiguous (i.e. adjacent) can be assessed by calculating habitat contiguity.[37] For example, two patches might have the same areal extent and overall shape (equivalent perimeters), but differ in how habitat is distributed within the patch. A patch with uniform habitat (from edge-to-edge) has greater contiguity than one that is riddled with many small gaps throughout. Assessing habitat contiguity is somewhat complicated, involving a moving-window algorithm (**Figure 4.27**),

and thus will not be covered in detail here (see the explanation given in McGarigal et al. 2012). Fortunately, contiguity is far more easily interpreted than it is calculated, given that it is a bounded index: $0 \le C_p \le 1$. Contiguity equals 0 for a one-cell patch, and $C_p \to 1$ with increasing contiguity of habitat within the patch.

- *Patch isolation* (*Euclidean nearest-neighbor distance*; **ENN**)—The straight-line (Euclidean) distance between patches is perhaps the simplest measure of patch isolation, although it must be emphasized that this is a structural rather than functional measure of isolation. We will discuss patch-connectivity measures in more detail in **Chapter 5**, but it is still worth mentioning the ENN measure here, for this is often used as a measure of habitat fragmentation and can affect a great many ecological processes and dynamics (e.g. dispersal or colonization success; **Chapter 6**). The ENN distance is simply the shortest distance

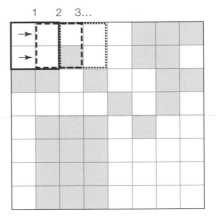

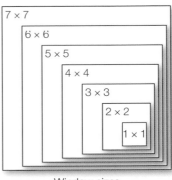

Window sizes

Figure 4.27 Moving-window analysis. A moving window of different sizes is used to assess landscape pattern at different scales. The density or presence of a given cover type is assessed within each window or 'box' of a given size (lower). The window is moved either one window-unit over (moving window) or one column over (gliding box), depending on the specific algorithm (upper). The gliding-box algorithm is used in the lacunarity analysis of landscape structure, for example (**Figure 4.38**).

[36] This is the FRAC index in FRAGSTATS (McGarigal et al. 2012).

[37] This is the CONTIG index in FRAGSTATS (McGarigal et al. 2012).

(m or km) to the nearest neighboring patch of the same type, measured edge-to-edge. If the 4-cell neighborhood rule is used, the minimum possible distance between two neighboring patches is equal to the distance along the diagonal, whereas the minimum ENN distance is twice the cell size if the 8-cell neighborhood rule is used (**Figure 4.25**). For example, if the cell size is 30 m, then the minimum ENN distance would be 42.4 m if patches are being defined by the 4-cell rule (calculated using the Pythagorean theorem), but the minimum ENN distance is 60 m if patches are defined using the 8-cell rule (2 × 30 m = 60 m).

Example 4.2 **Using patch-scale metrics to compare the largest patch in a clumped landscape versus the largest patch in a fragmented landscape**. To illustrate the application of these patch-scale metrics, let's compare the two landscapes depicted in **Figure 4.28**. Although both landscapes have the same amount of habitat (50%), they differ in the degree to which that habitat is fragmented (clumped vs fragmented). Because of the sheer number of patches, especially in the fragmented landscape, we will just apply these patch metrics to the largest patch in each landscape.

First off, we can see that the largest patch in the clumped landscape is much larger than the largest patch in the fragmented landscape; in fact, almost all of the habitat in the clumped landscape is contained within that largest patch (judging by the largest patch index, LPI, a class-scale metric that will be defined in the next section; **Table 4.6**). Note, too, that the size of the largest patch is little changed by the choice of neighborhood rule in the clumped landscape, owing to the high degree of habitat contagion (C_p or CONTIG; **Table 4.6**). Thus, patch structure is affected by our choice of neighborhood rule to a greater

degree in the fragmented landscape than in the clumped landscape.

Another major difference between the clumped and fragmented landscapes lies in the amount of edge of the largest patch (largest patch perimeter, P; **Table 4.6**). The largest patch on the fragmented landscape has considerably more edge, and thus a more complex shape (based on both the P/A* index and D), than the largest patch on the clumped landscape (**Table 4.6**). Finally, it is worth noting that the largest patch is somewhat more isolated on the clumped landscape than in the fragmented landscape, when using the 8-cell rule (based on the ENN distance; **Table 4.6**). This might at first seem counterintuitive, especially since patch isolation is often used as a measure of fragmentation. However, the ENN is a measure of the gap size between patches. If much of the habitat is concentrated within a few, widely spaced patches, as it is here in the clumped landscape, then the gaps between patches tend to be larger, especially when compared to the more dispersed pattern of habitat on the fragmented landscape (**Figure 4.29**). Thus, patch isolation (as indexed by ENN) may not be a particularly good measure of habitat fragmentation after all.

CLASS-SCALE METRICS   As with patch-scale metrics, measures of landscape configuration are usually obtained for one or more land-cover classes individually. Given that we are still quantifying attributes of patches, many of the patch-scale metrics discussed previously will also apply at the class scale, but are just now being assessed over the entire distribution of patches for that class (e.g. mean patch size or shape). The following are some common metrics obtained for individual classes:

- *Number of patches* (**NP**)—The number of patches is a reflection of how the habitat or land cover is dis-

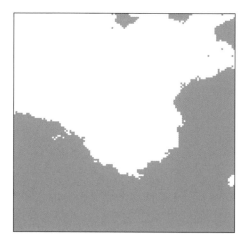
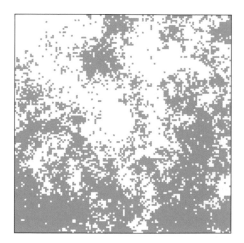

Figure 4.28 Landscape fragmentation.   Both landscapes (128 × 128 cells) have the same proportion of habitat (50%), but differ in their degree of habitat fragmentation, with the landscape on the right exhibiting greater fragmentation.

TABLE 4.6 **Effect of fragmentation on patch-scale metrics.** Patch-scale metrics for the largest patch within a clumped landscape versus the largest patch within a fragmented landscape (**Figure 4.28**). Each landscape (128 × 128 grid cells = ~14.7 km² areal extent given a 30 m resolution) comprises 50% habitat and was analyzed using different neighborhood rules in FRAGSTATS. The FRAGSTATS acronyms are also provided for metrics where these differ from the text.

Landscape Metric	Clumped		Fragmented	
	4-cell rule	8-cell rule	4-cell rule	8-cell rule
Largest patch size (A; AREA; km²)	7.03	7.03	2.56	5.77
Largest patch perimeter (P; PERIM; km)	19.4	20.2	53.2	144.2
Largest patch index (LPI)[1]	47.8	47.8	17.4	39.3
Perimeter-area ratio (P/A; PARA)	2.8	2.9	20.8	25.0
Standardized P/A ratio (P/A*; SHAPE)	1.83	1.90	8.28	14.9
Fractal dimension (D; FRAC)	1.08	1.08	1.29	1.35
Contiguity Index (CONTIG)	0.975	0.975	0.835	0.804
Euclidean nearest neighbor distance (ENN; m)[2]	42.4	180	42.4	60

[1]The largest patch index (LPI) is actually a class-scale metric in FRAGSTATS, but is presented here so as to give a better indication of the size of the largest patch relative to that of the entire landscape. Bear in mind that only 50% of the landscapes being analyzed here is classified as 'habitat,' and thus LPI → 50 indicates that much of the habitat present on the landscape is contained within that largest patch.
[2]Note that the minimum ENN distance for a 30-m cell is 42.4 m (i.e. the distance along the diagonal) when defined using the 4-cell rule; it is 60 m when defined using the 8-cell rule (2 × 30 m = 60 m).

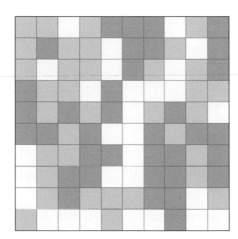

Figure 4.29 Heterogeneous landscape. This landscape (10 × 10 cells) was generated as a random distribution of three equally abundant cover types.

tributed on the landscape. For a given amount of land cover (p_i), fragmented landscapes will have more patches than unfragmented landscapes.

- *Largest patch index* **(LPI)**—The percentage of the landscape that is occupied by the largest patch. The largest patch index is considered a class-scale metric because the largest patch can only be identified via comparison with other patches of the same class. The LPI is equivalent to the proportion of land cover (p_i) if the largest patch contains all cells of a particular class type (LPI = p_i). Thus, LPI = 100 when the landscape is made up entirely of a single large patch (i.e. 100% of a given cover type).

- *Total edge or patch perimeter* **(TE)**—The sum total of all edge lengths or patch perimeters (e.g. m or km) for a given class. The total edge (TE) gives an indication of how fragmented the cover type is on the landscape. Because fragmented landscapes have more patches, they have more total edge than unfragmented landscapes for a given p_i.

- *Mean patch size* **($\overline{A}$)**—The mean patch size is simply the arithmetic mean, calculated as the total class area (CA) divided by the number of patches ($\overline{A}$ = CA/NP). Although mean patch size provides some information on the expected size of an average patch for a given class type, this measure is not actually influenced by the individual patch sizes used to calculate the mean. To illustrate, imagine we have a landscape consisting of six patches of equal size (A = 25 km²), such that the mean patch size is the same as individual patch size [$\overline{A}$ = 6(25)/6 = 150/6 = 25 km²). Now imagine a second landscape with the same total patch area (150 km²), but in which one patch is very large but the rest are small ($A_{1\cdots5}$ = 10 km², A_6 = 100 km²). Despite the wide range in patch sizes, the mean patch size is the same as in the first landscape ($\overline{A}$ = [5(10) + 100]/6 = 150/6 = 25 km²). Although mean patch size provides a good description of patches found on the first landscape, it doesn't do a very good job of capturing the characteristics of an 'average' patch in the second landscape.

One way to deal with this is to weight patches by their relative size when calculating the mean patch size; that is, to calculate an **area-weighted mean patch size** ($\bar{A}_w$). In calculating the arithmetic mean, all patches were given equal weight: it didn't matter whether patches were of equal size, as only the total class area and number of patches were used to calculate mean patch size. To calculate a weighted mean, we weight each patch (A_i) by its proportional contribution to the class (A_i/CA) and then sum over all patches ($i = 1..NP$) as:

$$\bar{A}_w = \sum_{i=1}^{NP}\left[A_i\left(\frac{A_i}{CA}\right)\right]. \qquad \text{Equation 4.8}$$

So, using the patch-size distribution from the second landscape in the above example, we have five small patches (10 km²), each of which makes up 6.7% of the total class area (10/150 = 0.067), and a single large patch (100 km²) that makes up 66.7% of the total class area (100/150 = 0.667). The area-weighted mean patch size for this landscape is therefore:

$$\bar{A}_w = 5(10)(0.067)+(100)(0.667) = 3.35 + 66.7 = 70.1 \text{km}^2.$$

The area-weighted mean patch size for this second landscape is clearly larger than the arithmetic mean we calculated previously (70.1 km² vs 25 km²). In general, the area-weighted mean is biased toward larger patches; larger patches do contribute more to the total class area, after all. From an ecological standpoint, larger patches are also likely to be more important and exert greater influence on various sorts of patch dynamics, such as those driving metapopulation or community dynamics (**Chapters 7 and 10**). Thus, the area-weighted mean patch size is probably a more useful or relevant descriptor of an 'average' patch than the arithmetic mean in most ecological applications (Li & Archer 1997).

- *Patch size variability* (**SD**, **CV**, or **range**)— Regardless of which mean is used, some measure of variability around the mean should also be reported, such as the **standard deviation (SD)** or **coefficient of variation (CV)**. These so-called 'second-order statistics' are often more informative than the first-order statistics upon which they are based (i.e. the mean).[38] For example, we would have instantly detected the difference in the patch-size distribution for the two landscapes above if the mean patch size (which was identical for both landscapes) had been accompanied by a standard deviation, or better yet, the coefficient of variation, since the latter provides a standardized measure of variability (the ratio of the standard

deviation to the mean, CV = SD/$\bar{A}$). The higher the CV, the more variable the patch sizes within the landscape. Conversely, the lower the CV, the more similar the patch sizes. Our first landscape, in which all patches were the same size exhibited zero variability and thus CV = 0, which is the minimum value that the CV can take. For the second landscape, we can calculate the SD as the square root of the average squared differences from the mean patch size (using the arithmetic mean here):

$$SD = \sqrt{\frac{\sum_i^{NP}\left(A_i - \bar{A}\right)^2}{NP-1}}. \qquad \text{Equation 4.9}$$

So, for our second landscape, the SD would be

$$SD = \sqrt{\frac{5(10-25)^2+(100-25)^2}{6-1}} = \sqrt{\frac{5(-15)^2+(75)^2}{5}}$$
$$= \sqrt{\frac{6750}{5}} = \sqrt{1350} = 36.7\text{km}^2.$$

The CV of the mean patch size would thus be 36.7 km²/25 km² = 1.47. The theoretical maximum value for CV is given by $\sqrt{(NP-1)}$ (i.e. all values but one are equal to 0). Given that $CV_{max} = \sqrt{5} = 2.24$ for a landscape with six patches, we can more easily evaluate the degree of variability observed in the second landscape (i.e. it's about 66% of maximum variability: 1.47/2.24 = 0.656).

If we instead use the area-weighted mean for patch size ($\bar{A}_w$), we need to calculate the **weighted standard deviation (SD$_w$)** rather than the arithmetic standard deviation given in Equation 4.9. The SD$_w$ is calculated as

$$SD_w = \sqrt{\frac{\sum_{i=1}^{NP}\left[\frac{A_i}{CA}\left(A_i - \bar{A}_w\right)^2\right]}{NP-1/NP}}. \qquad \text{Equation 4.10}$$

For the second landscape, in which $\bar{A}_w$ = 70.1 km², we thus have

$$SD_w = \sqrt{\frac{\sum_{i=1}^{NP}\left[5\left(0.067(10-70.1)^2\right)\right]+\left[0.667(100-70.1)^2\right]}{5/6}}$$
$$= \sqrt{\frac{1210+596}{0.8333}} = \sqrt{2167} = 46.6\text{km}^2.$$

Now, the CV of the area-weighted mean patch size is 46.6 km²/70.1 km² = 0.66. Patch sizes thus appear to be less variable when the area-weighted mean and standard deviation are used (0.66 vs 1.47), because the weighted mean is better at capturing the size distribution of patches (i.e. the deviation of patch sizes from the area-weighted mean is less than the deviation from the arithmetic mean).

Finally, it might also be worthwhile to report the **range** in patch sizes, which is simply obtained as the difference between the largest and smallest patches on

---

[38]In statistics, 'order' refers to the power of the term used to describe the data distribution. For example, the mean is a first-order statistic ($\bar{X}$); the variance of the mean (s^2) and its derivatives (SD or CV) are considered second-order statistics (Equation 4.9); and, the skewness of the data distribution is a third-order statistic defined as $\sum_{i=1}^{N}\left(X_I - \bar{X}\right)^3/(N-1)s^3$, where s is the standard deviation and N is the sample size.

the landscape (range = maximum A_i – minimum A_i), as well as what those minimum and maximum patch sizes are. Thus, for the second landscape, the minimum patch size is 10 km² and the maximum patch size is 100 km², which gives a patch-size range of 90 km². In the end, such simple descriptors may communicate as much—if not more—to an audience than the statistical measures of patch-size variability (SD, SD$_w$, or CV), which are far less intuitive to most people.

- *Mean patch shape ($\overline{P/A}$, $\overline{P/A^*}$, or $\overline{D}$)*—Now that we have an idea of how to calculate the mean or area-weighted mean patch size, as well as a measure of variability around that mean, we can also use these formulae to calculate other sorts of class means and measures of variability that are based on patch area, such as patch shape (e.g. P/A). If patch area is being assessed in terms of area-weighted means, then it makes sense to be consistent and use the area-weighted patch mean when calculating the mean perimeter-area ratio ($\overline{P/A}_w$) or its standardized form ($\overline{P/A^*}_w$), as well as the fractal dimension based on the P/A$_w$ ratio ($\overline{D}_w$). We should obviously take care to report which mean is being used, especially if we are using an area-weighted mean, as otherwise the arithmetic mean will be assumed.

- *Mean patch isolation (Euclidean Nearest Neighbor, $\overline{ENN}$)*—Mean patch isolation, as used here, is simply the arithmetic mean of the nearest-neighbor distances (in m or km) obtained for each patch of a given land-cover type. If we are comparing landscapes with the same amount of cover, mean patch isolation can give us an idea of the distribution of patches on the landscape, in terms of whether patches are generally far apart ('isolated') or close together (aggregated or clumped). By comparing the mean patch isolation to the mean dispersal distance or gap-crossing ability of a species, we might also have an idea of whether the landscape is likely to be perceived as connected by that organism (**Chapters 5 and 6**). Regardless, any inferences about the distribution or connectivity of a particular land-cover type will necessarily be limited if based solely on mean patch isolation. There are a variety of other metrics that are better suited for evaluating the degree to which a particular land-cover type is aggregated, which we'll consider next.

- *Aggregation measures [**P(a_{ii})**, **AI**, and **CI**]*—We can derive a simple measure of aggregation based on the proportion of like-adjacencies; that is, the likelihood that two adjacent cells belong to the same cover type. It stands to reason that if the probability of like-adjacencies for a given cover type is high, then that cover type is more aggregated than if the probability of like-adjacencies is low. So, we first need to calculate the **probability of like-adjacencies**

[**P(a_{ii})**], which is simply the proportion of like-adjacencies (a_{ii}) relative to the total number of adjacencies for that cover type (a_{ij}), given R cover types:

$$P(a_{ii}) = \frac{a_{ii}}{\sum_{j=1}^{R} a_{ij}}. \qquad \text{Equation 4.11}$$

We can then multiply this probability by 100 if we wish to convert this to a percentage.[39] Adjacencies are obtained from an adjacency matrix, which contains the frequencies of each adjacency type. For example, if we have three cover types, the adjacency matrix would look like this:

Cover type	Cover type		
	a_1	a_2	a_3
a_1	$a_{1,1}$	$a_{1,2}$	$a_{1,3}$
a_2	$a_{2,1}$	$a_{2,2}$	$a_{2,3}$
a_3	$a_{3,1}$	$a_{3,2}$	$a_{3,3}$

The like-adjacencies lie along the diagonal and are highlighted here in boldface type; these are the number of cells in the landscape that are adjacent to another cell of the same cover type. Although adjacencies may be calculated using either the 4-cell or 8-cell rule, we'll assume a 4-cell rule here. Thus, interior cells of a landscape (those not along the boundary edge of the landscape) have four adjacent cells; those along the boundary edge have three adjacent cells, except for the corners, which have two adjacent cells.

To illustrate, consider the landscape depicted in **Figure 4.29**, which has three cover types of roughly equal abundance ($p_{green} = 0.36$; $p_{tan} = 0.35$; $p_{yellow} = 0.29$). Starting with the cell in the upper-left corner of the landscape (a tan cell), we can see that this is adjacent to two like-cells (i.e. the cell immediately to the right and below this cell are both tan). If we move one cell to the right (which is also a tan cell), this cell is adjacent to three other cells (one to the left, one to the right, and one below), but only one of these is a like-cell (i.e. the previous cell to its left). If we count up all of the different types of adjacencies, the adjacency matrix for this landscape would look like this:

Cover type	Cover type		
	green	tan	yellow
green	**58**	44	28
tan	44	**44**	40
yellow	28	40	34

Thus, there are 360 adjacencies in this landscape (an 8 × 8 inner landscape = 64 cells × 4 edges = 256 adjacencies; 8 cells along the outer boundary of the landscape × 4 sides = 32 cells × 3 edges = 96 cells; and, 4 corner cells × 2 edges = 8 cells; thus, 256 + 96 + 8 = 360).

[39] FRAGSTATS reports like-adjacencies as a percentage (PLADJ; McGarigal et al. 2012).

If the cover types are randomly distributed, then $a_{ii} = p_i$; that is, the like-adjacency of a given cover type should be equivalent to its proportional abundance on the landscape (Gardner & O'Neill 1991). Thus, if the percentage of like-adjacencies is less than p_i, the cover type is more dispersed than expected for a random landscape (i.e. it is overdispersed), but if the percentage of like-adjacencies is greater than p_i, then the cover type is more aggregated than expected (underdispersed). When calculated as a percentage, the value of $P(a_{ii}) = 0$ when the cover type is completely disaggregated, such that there are no like-adjacencies (i.e. every cell of the landscape is a separate patch). Conversely, $P(a_{ii}) = 100$ when all cells are of the same cover type (i.e. the entire landscape is made up of single cover type). For the landscape depicted in **Figure 4.29**, P(green, green) = 0.4, P(tan, tan) = 0.31, and P(yellow, yellow) = 0.29. Thus, only the yellow cover type is randomly distributed [P(yellow, yellow) = p_{yellow} = 0.29]. The green cover type is a bit more clumped than expected [P(green, green) > p_{green} = (0.4 > 0.36)] whereas the tan cover type is a bit more dispersed than expected for a random distribution [P(tan, tan) < p_{tan} = (0.31 < 0.35)].[40]

Adjacencies can be calculated in different ways: sometimes the double-count method is used, in which a cell side may be counted twice (as described above), and sometimes the single-count method is used, such that each cell side is only counted once. This use of the double-count versus the single count method can vary among metrics, so it is important to note which method is being used to ensure that the appropriate calculations or equivalent comparisons are being made among studies (Riitters et al. 1996). In Fragstats, for example, the percentage of like-adjacencies is calculated using the double-count method, whereas the **aggregation index** (**AI**) is calculated using the single-count method (McGarigal et al. 2012).

The AI is standardized by the maximum like-adjacencies possible, which occurs when all cells of a given cover type are found within a single patch. Thus, the AI is a ratio of the actual shared edges relative to the maximum possible shared edges for a given cover type (He et al. 2000). For any p_i, AI = 0 when the cover type is completely disaggregated (i.e. there are no like-adjacencies), and AI = 100 when the cover type is aggregated into a single large patch. As before, AI should approach p_i for a randomly distributed cover type (He et al. 2000). For the random landscape depicted in **Figure 4.29**, AI_{green} = 48.3, AI_{tan} = 37.9, and

AI_{yellow} = 36.2, which once again underscores the fact that the green cover type is more aggregated than expected, even after accounting for its slightly greater abundance relative to the other cover types.

We have just explored one way in which aggregation can be adjusted for p_i (i.e. AI). Another approach is to standardize like-adjacencies by that expected under a random distribution ($a_{ii} = p_i$), which gives a **clumpiness index** (CI).[41] Thus,

CI = 0   when the cover type is randomly distributed,

CI = 1   when the cover type is maximally clumped (i.e. in a single large patch), and,

CI = –1   when the cover type is maximally disaggregated (i.e. every cell is a separate patch).

Because this index is standardized to a random distribution (where $a_{ii} = p_i$), it provides a useful measure of fragmentation independently of cover (p_i; McGarigal et al. 2012). For comparison, the random landscape we've been discussing (**Figure 4.29**) has a CI_{green} = 0.2, CI_{tan} = 0.1 and CI_{yellow} = 0.05. Although these are all fairly close to random (CI = 0), as might be expected given that these were all generated as a random distribution, the green cover type is still somewhat more clumped than the other two, even after adjusting for differences in p_i.

Example 4.3 **Using class-scale metrics to compare a fragmented versus a clumped landscape**. To illustrate the application of these class-scale metrics, let's once again compare the two landscapes depicted in **Figure 4.28**. Both landscapes have 50% habitat cover ($p = 0.5$), but differ in the degree to which that habitat is fragmented. This is perhaps most evident in the number of patches present on the landscape (**Table 4.7**). Depending on the neighborhood rule used to define patches (4-cell vs 8-cell rule), the clumped landscape has 10–17 patches, whereas the fragmented landscape has 217–465 patches. The fragmented landscape thus possesses some 22–27 times more patches than the clumped landscape. Not surprisingly, then, the largest patch in the clumped landscape contains 95% of the habitat present (LPI/p_i = 47.7%/50% = 0.95), whereas the largest patch on the fragmented landscape contains either 36% or 78% of the total habitat, depending on whether the 4-cell or 8-cell rule is used to define patches, respectively (**Table 4.7**). Twice as much habitat is found in the largest patch when the larger neighborhood rule is used on the fragmented landscape. The choice of neighborhood rule thus makes a bigger difference to the patch structure of the fragmented landscape than it does to the patch structure of the clumped landscape.

The clumped landscape may have fewer patches, but these average 1.5–3.6 times larger than patches in

---

[40]Interestingly, this landscape pattern was actually generated as a random distribution of all three cover types, in which all had equal probability of being assigned to a given cell (p_i = 0.333). The deviation of the individual cover types from that expected—both in terms of p_i and $P(a_{ii})$—is caused by the relatively small spatial extent of the landscape (10 × 10 = 100 cells); a larger landscape would thus produce p_i closer to the initial probabilities.

[41]This is the CLUMPY index in Fragstats (McGarigal et al. 2012).

TABLE 4.7 Effect of fragmentation on class-scale metrics. Comparison of class metrics between a clumped and fragmented landscape (**Figure 4.28**), each comprising 50% habitat and analyzed using different neighborhood rules in FRAGSTATS. The FRAGSTATS acronyms are also provided for metrics where these differ from the text; the standard deviation or weighted standard deviation, as appropriate, is also given in parentheses for class means.

Landscape Metric	Clumped		Fragmented	
	4-cell rule	8-cell rule	4-cell rule	8-cell rule
Percent cover (p_i; PLAND; %)	50	50	50	50
Number of patches (NP)	17	10	465	217
Largest patch index (LPI)	47.7	47.7	18.1	39.1
Total edge (TE; km)	15.7	15.7	227.6	227.6
Area-weighted mean patch size ($\bar{A}_w$; AREA_AM; km²)	6.8 (1.65)	6.8 (2.10)	1.9 (0.17)	4.6 (0.39)
Area-weighted mean patch shape ($\overline{P/A}_w$; SHAPE_AM)	1.81 (0.227)	1.89 (0.301)	6.72 (0.578)	12.58 (1.112)
Area-weighted mean patch fractal dimension ($\bar{D}_w$; FRAC_AM)	1.08 (0.027)	1.08 (0.035)	1.24 (0.043)	1.32 (0.065)
Mean patch isolation (Euclidean nearest neighbor, $\overline{ENN}$; m)	84.06 (60.67)	171.09 (118.40)	54.22 (20.45)	71.83 (20.57)
Percentage of like-adjacencies (PLADJ)	97.4	97.4	75.8	75.8
Aggregation index (AI; %)	98.5	98.5	76.7	76.7
Clumpiness index (CI; CLUMPY)	0.97	0.97	0.54	0.54

the fragmented landscape, depending on the neighborhood rule ($\bar{A}_w$; **Table 4.7**). Patches also tend to have simpler shapes in the clumped landscape, compared to the more complex patch geometries of the fragmented landscape (based on $\overline{P/A}_w$ and $\bar{D}_w$; **Table 4.7**). Patches are 1.5–2.4 times farther apart, on average, in the clumped versus the fragmented landscape ($\overline{ENN}$; **Table 4.7**). However, fragmentation does increase the total amount of edge (TE); the fragmented landscape has 14.5 times more edge than the clumped landscape (**Table 4.6**). Finally, the habitat is almost maximally clumped (CI = 0.97) in the clumped landscape, whereas it is more randomly distributed in the fragmented landscape (CI = 0.54). Nevertheless, habitat in the fragmented landscape is still more clumped than expected for a completely random distribution (AI > p_i; 76.7% > 50%; **Table 4.7**).

The analysis of these two landscapes (**Figure 4.28**) thus serves to illustrate some of the general effects of habitat fragmentation on landscape structure. For a given amount of habitat, fragmentation tends to produce many more small patches with greater edge and more complex geometries, than if habitat were less fragmented. However, patch isolation may not be greater in fragmented landscapes. As demonstrated here, inter-patch distances are actually greater in the clumped landscape. Once again, mean patch isolation is primarily a measure of distribution, the degree to

which patches are dispersed or clumped, rather than habitat fragmentation per se.

LANDSCAPE-SCALE METRICS Even though class-scale metrics are assessed across the entire spatial extent of the landscape, these metrics are still calculated for each class individually. In the case of landscape-scale metrics, metrics are calculated across all patches, regardless of cover type. Thus, most of the same metrics described previously at the class scale can be calculated at the landscape scale, which just involves averaging over a larger population of patches (i.e. all patches over all cover types combined; **Table 4.5**). So, to avoid repetition, we will just focus here on one more metric that is unique to the landscape scale, and then apply landscape-scale metrics to the analysis of the landscapes depicted in **Figure 4.28**.

• *Landscape contagion* (**LC**)—Landscape contagion is a measure of both aggregation and interspersion (O'Neill et al. 1988a; Li and Reynolds 1993). Whereas aggregation refers to the extent to which like-cells are adjacent (like-adjacencies), **interspersion** pertains to the likelihood that cells of one cover type are next to a different cover type (i.e. the *un*like-adjacencies). All else being equal, a landscape in which all cells of different cover types are interspersed will have lower contagion than one in which all cells of a given type are aggregated into a single patch. The metric is based on the proportional abundance of each cover type (p_i)

TABLE 4.8 **Effect of fragmentation on landscape-scale metrics.** Comparison of landscape-scale metrics between a clumped and fragmented landscape (**Figure 4.28**), each comprising 50% habitat and analyzed using different neighborhood rules in FRAGSTATS.

Landscape Metric	Clumped		Fragmented	
	4-cell rule	8-cell rule	4-cell rule	8-cell rule
Percent cover (p_i; PLAND; %)	50	50	50	50
Number of patches (NP)	31	16	911	449
Largest patch index (LPI)	50.4	50.5	39.8	41.5
Area-weighted mean patch size ($\overline{A}_w$; AREA_AM; km²)	7.1 (1.78)	7.1 (2.39)	3.3 (2.30)	4.8 (3.97)
Area-weighted mean patch shape ($\overline{P/A}_w$; SHAPE_AM)	1.77 (0.213)	1.83 (0.33)	8.77 (0.605)	12.72 (1.083)
Area-weighted mean patch fractal dimension ($\overline{D}_w$; FRAC_AM)	1.07 (0.025)	1.08 (0.047)	1.23 (0.04)	1.32 (0.063)
Mean patch isolation (Euclidean nearest neighbor, $\overline{ENN}$; m)	100.2 (182.82)	193.15 (248.83)	54.2 (21.49)	70.7 (22.80)
Percentage of like-adjacencies (PLADJ)	97.6	97.6	76.1	76.1
Aggregation index (AI; %)	98.7	98.7	76.9	76.9
Landscape contagion (CONTAG; %)	44.1	44.1	10.8	10.8

and the proportion of adjacencies between all combinations of cell types (i.e. the adjacency matrix, using the double-count method) as:

$$LC = 1 + \left[\frac{\sum_{i=1}^{R} \sum_{j=1}^{R} \left(p_i \times \frac{a_{ij}}{\sum_{j=1}^{R} a_{ij}} \right) \ln \left(p_i \times \frac{a_{ij}}{\sum_{j=1}^{R} a_{ij}} \right)}{2 \ln(R)} \right] \times 100.$$

Equation 4.12

Because LC is expressed as a percentage, the index gives a value of 0 when all cell types are maximally disaggregated and interspersed, and approaches 100% when all cells of a given type are aggregated into a single patch.[42]

Example 4.4 **Using landscape-scale metrics to compare a fragmented versus a clumped landscape.** Although we've been concerned with the distribution of 'habitat' on these landscapes, the landscapes could be viewed as comprising two cover classes that differ in their suitability for a given species (e.g. suitable vs unsuitable habitat, or high-quality vs low-quality habitat). Thus, if we combine patches across both cover types, we have a total of 16–31 patches in the clumped landscape and 449–911 patches in the fragmented landscape, depending on which neighborhood rule we use

to define patches (**Table 4.8**). Notice that patches tend to be larger than when the single class ('habitat') was analyzed previously, but patch geometry is little affected, presumably because like-adjacencies—and the aggregation index, which is based on like-adjacencies—remain unchanged (compare **Tables 4.7 and 4.8**). In the clumped landscape, patches average 16–22 m farther apart than when 'habitat' alone was analyzed (**Table 4.7**), depending on which neighborhood rule is used. Although we've been referring to these landscapes as 'clumped' versus 'fragmented' based on appearances, we should be reassured by the landscape contagion measure (LC), which indicates that the cover types on the fragmented landscape are minimally aggregated and approaching maximum interspersion of the two cover types (LC = 10.8 %), whereas cover types on the clumped landscape have a higher (and intermediate) degree of aggregation with less interspersion (44.1%; **Table 4.8**).

Effects of Pattern and Scale on Landscape Metrics

Now that we have some idea of what landscape metrics are available, we should bear in mind a number of known issues that affect the behavior and values attained by these metrics. An understanding of these issues is essential if we are to obtain a meaningful interpretation of landscape metrics (Hargis et al. 1998). Thus, we will consider here how the following affect the behavior of landscape metrics: (1) landscape pattern, in terms of both the amount and aggregation of land covers, (2) thematic resolution, (3) spatial resolution,

[42]Landscape contagion (LC) is inversely related to edge density (Hargis et al. 1998). Edge density (or ED in FRAGSTATS; McGarigal et al. 2012) is the total edge (TE) per unit area within the landscape (e.g. km/ha).

(4) the Modifiable Areal Unit Problem (MAUP), and, (5) spatial extent.

Effect of landscape pattern—Given that landscape metrics are designed to quantify the composition and configuration of landscapes, we naturally expect that these metrics will be sufficiently sensitive to detect such differences or track changes in landscape pattern over time. Nevertheless, many of these landscape metrics may change in unexpected ways as a function of the proportion of land cover (p_i) or the distribution of patches on the landscape (Gustafson & Parker 1992; Hargis et al. 1998). We therefore need to be aware of how these metrics behave in response to changes in landscape pattern, so as to ensure that we are interpreting these metrics correctly.

A meaningful interpretation of landscape metrics is possible only when the limitations of each measure are fully understood, the range of attainable values is known, and the user is aware of potential shifts in the range of values due to characteristics of landscape patches.   Hargis et al. (1998)

The study of how metrics behave in different landscape contexts has benefitted greatly from the development and application of **neutral landscape models** (Gardner et al. 1987; Gardner & O'Neill 1991; With & King 1997). Neutral landscape models (NLMs) are simulated landscape patterns, in which the amount and distribution of land covers are created using a simple statistical distribution or algorithm. For example, the landscapes depicted in **Figures 4.28 and 4.29** are examples of landscape patterns that were created as either a fractal or random distribution of habitat, respectively, using a NLM generator (e.g. the R package *NLMR*, Sciaini et al. 2019). We will discuss these different classes of neutral landscape models in more detail in **Chapter 5**. For now, however, we'll highlight the utility of NLMs for exploring the behavior of landscape metrics. Because NLMs are generated using a known spatial distribution, they provide a baseline or null model for evaluating how specific changes, such as in the amount or aggregation of habitat, will affect the value of the metric. By understanding the expected behavior of the metric, we can then gauge whether observed changes in real landscapes are due to the specific biotic or abiotic factors in question (**Chapter 3**), or whether such changes are expected purely on the basis of changes in the proportion of land cover (Gustafson & Parker 1992; Remmel & Fortin 2013).

By way of illustration, let's consider how altering the amount of habitat (p_i) affects some basic metrics, such as the number of patches (NP), the largest patch

index (LPI), and the total amount of edge (TE). By creating a series of random NLMs that represent a gradient in habitat abundance (p_i), we can see that the number of patches exhibits a right-skewed distribution that peaks at $p_i = 0.3$ (**Figure 4.30**). As the proportion of habitat increases beyond this point ($p_i > 0.3$), the number of patches decreases as habitat eventually coalesces into a single large patch, whereas below this point ($p_i < 0.3$), patches are so small (many comprise just a single cell) that any further habitat loss results in a loss of patches as well. Given the nature of this relationship, however, we cannot gauge where we are on the spectrum of landscape disturbance based on just the number of patches alone. For example, there are approximately 1000 patches in landscapes that have either 10% or 50% habitat. Put another way, we cannot infer how fragmented the landscape is simply from the number of patches present.

It would thus help to have some additional information on how patch size—such as the largest patch index (LPI)—changes as a function of the amount of habitat (p_i) on the landscape. Interestingly, the LPI exhibits a threshold at about $p_i = 0.5$, below which the largest patch represents only a tiny fraction of the landscape (**Figure 4.30**). This sort of threshold response is related to the **percolation threshold**, the threshold amount of habitat at which a single large patch no longer spans the landscape. Because percolation thresholds have been used to define landscape connectivity, we'll reserve further discussion on this topic until **Chapter 5**.

Finally, while the number of patches peaked at $p_i = 0.3$, the total amount of edge (TE) exhibits a more symmetrical distribution with a vertex at $p_i = 0.5$ (**Figure 4.30**). Thus, more patches do not necessarily equate to more edge habitat. Instead, the greatest amount of edge is found in landscapes with an intermediate amount of habitat. As before, however, we cannot use TE alone to define the degree of landscape fragmentation because landscapes near either end of the spectrum (e.g., $p_i = 0.2$ or $p_i = 0.8$) have an equivalent amount of edge (TE = 300 km; **Figure 4.30**).

Habitat aggregation—in conjunction with habitat amount—can also influence the behavior of landscape metrics. For example, we can create more aggregated habitat distributions using algorithms derived from fractal geometry, in which the degree of aggregation (denoted by the Hurst Dimension, H) can be adjusted independently of the amount of habitat (p_i). We'll discuss this further in **Chapter 5**, but for now, let's compare the effect of p_i on these same landscape metrics in fractal landscapes that possess a moderate degree of aggregation (H = 0.5; **Figure 4.31**). As we saw earlier from our comparison of fragmentation effects on class metrics, habitat aggregation results in fewer and larger

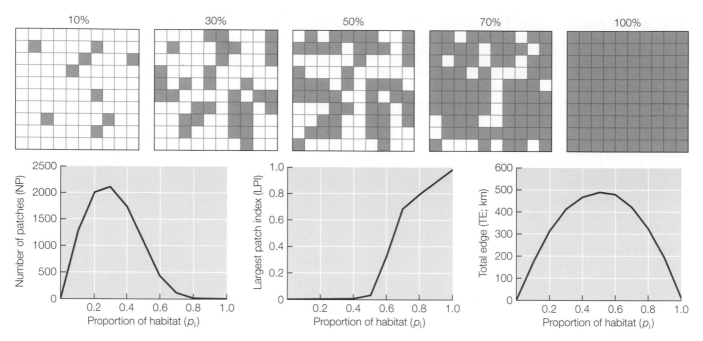

Figure 4.30 Effect of habitat amount on landscape metrics in a random landscape. The behavior of several metrics (number of patches, largest patch index, and total edge) can be evaluated by varying the amount of habitat (p_i) in a simulated series of random landscapes (128 × 128 cells). Smaller versions of these random landscapes (10 × 10 cells) are depicted here for reference (top). Patches were defined in this analysis using the 4-cell neighborhood rule.

patches with less total edge than more fragmented distributions (**Table 4.7**). Thus, it should come as no surprise that both the number of patches (NP) and total edge (TE) are lower in these fractal landscapes—with their more aggregated habitat distribution—than in the random landscapes (**Figure 4.30**). Although TE is still maximized at $p_i = 0.5$, there is now less of a peak to the NP distribution: the greatest number of patches occurs across a range of habitat abundance ($0.2 \leq p_i \leq 0.6$) and drops off fairly quickly on either end of this range. The greatest difference, however, lies in the response of the largest patch index (LPI). The largest patch no longer exhibits a threshold response to p_i, but instead decreases in direct proportion to the amount of habitat on the landscape (i.e. linearly; **Figure 4.31**).

A far more comprehensive analysis of the relative effects of habitat amount (p_i) and aggregation (H) on landscape metrics has been carried out using fractal neutral landscapes by Neel and her colleagues (2004). This analysis uncovered three broad 'behavioral groups' of metrics, based on whether the metrics are primarily influenced by (1) habitat amount (p_i); (2) habitat aggregation (H); or, (3) the interaction between the two ($p_i \times$ H). Nearly a third of the metrics (30%) are influenced primarily by habitat amount (p_i), whereas only a small fraction (7/50 = 14%) are directly related to habitat aggregation (H; **Table 4.9**). Thus, a majority of landscape metrics (56%) are influenced by both habitat amount and aggregation, and exhibit either a para-

bolic or non-linear relationship with p_i (**Table 4.9**). In other words, many of the metrics that are typically used to quantify habitat fragmentation (e.g. NP, TE, AI, ENN) are confounded with the amount of habitat on the landscape, and thus do not provide a wholly unambiguous analysis of fragmentation per se. We should thus ensure that we select metrics that are primarily influenced by habitat aggregation if our goal is to assay landscape fragmentation. Ideally, however, our landscape analysis will incorporate multiple metrics from each of these three behavioral groups, so as to provide a more complete assessment of landscape pattern.

Effect of thematic resolution—As mentioned previously, the **thematic resolution**—how finely we choose to categorize land covers—can have a major effect on the analysis of landscape pattern (Buyantuyev & Wu 2007). Consider the landscapes depicted in **Figure 4.32**. The landscape on the left is the same as that in **Figure 4.24** and has five land-cover types ($R = 5$). We'll designate these covers as (1) water, (2) coniferous forest, (3) deciduous forest, (4) savanna (sparsely treed grassland), and (5) open grassland (no trees). Perhaps all that really matters for our study, however, is whether we have forest or grassland, and thus we can simplify the representation of this landscape by combining coniferous and deciduous forest into a single category ('forest') and by combining savanna and grassland into another category ('grassland'). Our landscape now has

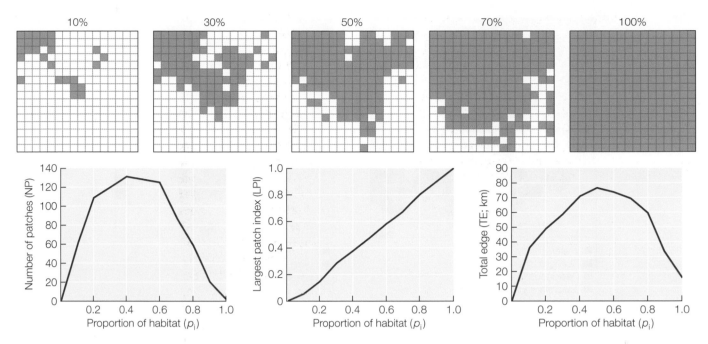

Figure 4.31 Effect of habitat amount on landscape metrics in a fractal landscape.   The behavior of several metrics (number of patches, largest patch index, and total edge) as a function of habitat amount (p_i) in a series of fractal landscapes (H = 0.5; 128 × 128 cells). Smaller versions of these fractal landscapes (16 × 16 cells) are depicted here for reference (top). Patches were defined in this analysis using the 4-cell neighborhood rule.

TABLE 4.9   Relative effects of habitat amount (p_i) versus fragmentation (H) on various class-scale metrics (Table 4.7). After Neel et al. (2004).

I. Metrics affected primarily by habitat amount (p_i)

- Area-weighted mean patch area ($\bar{A}_w$; AREA_AM)
- Largest patch index (LPI)

II. Metrics affected primarily by habitat aggregation (H)

- Standard deviation of fractal dimension (FRAC_SD)
- Coefficient of variation of fractal dimension (FRAC_CV)
- Clumpiness Index (CI; CLUMPY)

III. Metrics affected by both habitat amount (p_i) and aggregation (H)

—Parabolic response as a function of habitat amount (p_i); magnitude of the vertex decreases with increasing aggregation (H)
  - Area-weighted mean perimeter-area ratio (P/A_w^{*} ; SHAPE_AM)
  - Standard deviation of $\overline{P/A_w^{*}}$ (SHAPE_SD)
  - Coefficient of variation of $\overline{P/A_w^{*}}$ (SHAPE_CV)
  - Area-weighted mean fractal dimension ($\bar{D}_w$; FRAC_AM)
  - Number of patches (NP)[1]
  - Total edge (TE)[1]
—Slightly non-linear trend (from high p_i and H to low p_i and H)
  - Aggregation index (AI)
  - Percentage of like-adjacencies [$P(a_{ii})$; PLADJ]
—Strongly non-linear relationship (at high H and low p_i)
  - Area-weighted mean of Euclidean nearest-neighbor distances (ENN_AM)
  - Standard deviation of ENN (ENN_SD)
  - Coefficient of variation of ENN (ENN_CV)

[1]Although not examined by Neel et al. (2004), this metric also exhibits a parabolic response (**Figures 4.30 and 4.31**).

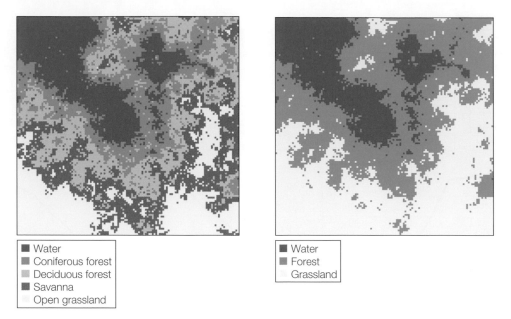

- ■ Water
- ■ Coniferous forest
- ■ Deciduous forest
- ■ Savanna
-    Open grassland

- ■ Water
- ■ Forest
-    Grassland

Figure 4.32 Effect of thematic resolution on landscape metrics.   Changing how land-cover classes are defined, by aggregating cover types into fewer categories, can have a major effect on measures of landscape composition and configuration.

three cover types ($R = 3$), and thus lower richness. Since the cover types are no longer completely equal in abundance ($p_{water} = 0.2$, $p_{forest} = 0.4$, $p_{grassland} = 0.4$), landscape diversity has decreased a bit as well ($H' = 1.05$, $E = 0.96$). **The same landscape can thus appear more—or less—diverse depending on how we classify land covers**. Thematic resolution clearly affects landscape composition metrics, but it can also affect configuration metrics because cells of different cover types are now being combined (Buyantuyev & Wu 2007). The effect of the classification scheme on landscape metrics is most pronounced—and least predictable—for landscapes with <10 classes (Huang et al. 2006).

Effect of spatial resolution—Another issue we might encounter concerns the spatial grain or resolution of the landscape. As the spatial resolution of remotely sensed imagery continues to increase (i.e. high-resolution data), we are increasingly confronted with the problem of dealing with far more landscape data than we strictly need or can easily handle. Remember that not all heterogeneity is necessarily relevant to the pattern-process relationships we seek to uncover (**Chapter 3**). In that case, it might be better to aggregate these fine-scale data to provide a resolution that is better aligned with the scale of the ecological data we have collected.

To illustrate the implications of doing so, however, consider the heterogeneous landscape that was first depicted in **Figure 4.29**. To produce a landscape with a coarser resolution, we could apply a **majority rule** within a moving 2 × 2 window over the landscape, as illustrated in **Figure 4.33**. Starting at the top left corner of the land-

scape image, we find that three of the four cells within this window are tan, and thus assign the category of 'tan' to the larger pixel of our new landscape. The window is then moved to the next 2 × 2 group of cells, a new majority cover type is assigned, and so on. In the event that two cover types are equally abundant within the window, we can choose one or the other at random to represent the class of the larger pixel. As you'll recall from **Chapter 2**, coarsening the resolution increases the spatial grain of the landscape, which can lead to a loss of heterogeneity, particularly of rare cover types (Turner et al. 1989b; **Figure 2.6**). Furthermore, the resulting landscape pattern is more blocky, which will have an effect on all of our configuration metrics as well. **Thus, landscape pattern analysis is strongly influenced by the resolution or spatial grain used to represent the landscape**.

Modifiable Areal Unit Problem—Both of the preceding issues—the reclassification of cover types and the coarsening of the spatial resolution—illustrate how landscape pattern is sensitive to the way in which we define the units of our analysis. In other words, the same basic data yield different results depending on how pixels are aggregated to define patches. This is referred to as the **Modifiable Areal Unit Problem (MAUP)**, which was first described in 1934 (Gehlke & Biehl 1934) and later codified by British geographer Stan Openshaw (Openshaw & Taylor 1979; Openshaw 1983). The areal units of the landscape (i.e. cells or patches) are considered to be modifiable in that they can be defined or depicted in different ways through reclassification or rescaling, as discussed above. Similarly, the way in which we define

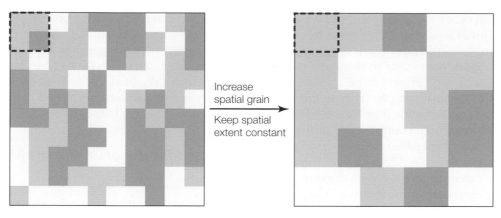

Figure 4.33 Effect of spatial resolution on landscape metrics. Coarsening the resolution of the landscape, such as by assigning a majority rule within a 2 × 2 window as demonstrated here, can alter the apparent composition and configuration of the landscape. Both landscapes have the same spatial extent, but differ in grain size, with the landscape on the right having a coarser resolution (larger grain size) than the landscape on the left.

patches in a raster landscape (i.e. different neighborhood rules) or whether we choose to represent patches in a raster or vector format can also modify the areal units of the landscape. The MAUP comes about for a variety of reasons related to the presence of spatial autocorrelations (either positive or negative) in the dataset; the scale at which various processes operate to give rise to the underlying spatial patterns (e.g. local versus regional processes); and neighborhood effects (e.g. the influence of the surrounding area on the data).

Wherever areal data are used and aggregated to define patch structure, the modifiable areal unit problem may occur. It is critical to understand how the results of landscape analysis may be affected by both scale and zoning problems…the MAUP in general has widespread implications for landscape ecological studies.   Jelinski & Wu (1996)

The problem with modifiable areal units in the analysis of landscape pattern is thus two-fold, involving the interrelated problems of (1) the scale effect, in which different results are obtained when spatial data are combined into larger areal units (e.g. by increasing grain size; **Figure 4.33**), and (2) the zoning effect, whereby different results are obtained because of how patches are defined or classified within the landscape (**Figure 4.32**; Jelinski & Wu 1996). Note that the zoning effect occurs when the scale of analysis is otherwise fixed, but the way in which data are being aggregated is changed (i.e. how the landscape is 'zoned' or divided up). If the spatial units are aggregated or specified differently, we might observe very different patterns and relationships. This poses a problem for landscape analysis because the results may be due, at least in

part, to the classification or aggregation scheme used, rather than the landscape pattern per se. From a landscape ecological perspective, the concern is that the representation of the landscape might not reflect the underlying processes or relationships of interest.

The scale effect, on the other hand, may offer insights into the hierarchically nested patch structure of landscapes and how spatial patterns change as a function of scale (**Figures 2.7 and 2.8**). Thus, as pointed out by Jelinski & Wu (1996: 138):

> the MAUP is not really a 'problem', per se; rather, it may reflect the 'nature' of the real systems that are hierarchically structured. It does not, therefore, present any real impediment to understanding spatial phenomena if recognized and dealt with explicitly. On the contrary, it carries critical information we need to understand the structure, function and dynamics of the complex systems in [the] real world. (emphasis added)

This is a salient point, especially for uncovering pattern-process linkages (i.e. landscape ecology!). The scale effect is purposely exploited in many multiscale spatial analyses, as we will see a bit later in the chapter.

Effect of spatial extent—Another scaling issue that can have an effect on landscape metrics involves changing the spatial extent of the landscape. As we saw in **Chapter 2**, heterogeneity tends to increase with increasing spatial extent (**Figure 2.5B**). As we broaden the areal extent of our landscape, we are bound to encounter new land-cover types and other landscape features, thereby increasing landscape richness and diversity (i.e. heterogeneity). Depending on how these land-cover types are distributed, however, sampling a smaller area within the larger landscape is unlikely to produce a 'sub-landscape' with a similar composition or configuration. In other words, if we reduce the spatial extent of our initial landscape, we cannot be certain that we will retain the same number of cover types in the same relative abundances and in the same configuration as the original (larger) landscape.

Take **Figure 4.34** as an example. This is the same heterogeneous landscape depicted in **Figure 4.29** (just 'flipped' horizontally), which comprises a random distribution of three cover types that have roughly equal abundances (p_{green} = 0.36, p_{tan} = 0.35, p_{yellow} = 0.29). If we reduce the spatial extent to focus just on the one quadrant, however, we can see that the relative abundances of these cover types are now highly skewed (p_{tan} = 0.52, p_{green} = 0.28, and p_{yellow} = 0.20; **Figure 4.34**). Given the shift in landscape composition (p_i), we can expect that landscape configuration metrics (e.g. mean patch size) will likewise be affected.

The potential for this sort of scaling effect should obviously be kept in mind when attempting to compare landscapes of different spatial extents. In general, the more clumped or aggregated the landscape pattern or cover type, the greater the effect of changing spatial extent on the resulting metrics of landscape pattern (Saura & Martínez-Millán 2001). Although the effect of changing spatial extent tends to be less predictable than that of changing spatial grain, some simple scaling relationships (involving either a linear, logarithmic, or power-law function) have been developed that permit the extrapolation (or interpolation) of at least a few of these metrics across landscapes of different extents (e.g. NP, TE, H'; see Wu et al. 2002; Wu 2004).

Unfortunately, most other metrics do not exhibit such consistent patterns to permit the development of simple scaling relationships. This requires an evaluation of both the underlying spatial structure and metric behavior when attempting to compare landscapes of different spatial extents. In that case, it may be better to depict how metrics change as a function of scale when attempting to characterize or compare landscapes (via scalograms; Wu et al. 2002). Such multiscale landscape analyses are routinely performed in the context of spatial analysis, as we'll discuss in a bit.

Use and Misuse of Landscape Metrics

The validity of landscape analysis will increase if indices are selected according to their ecological relevance rather than the convenience of computer programs.   Li & Wu (2004)

The previous section outlined the variable and sometimes unpredictable effects of pattern and scale on landscape metrics, thereby identifying some of the limitations and challenges surrounding their application and interpretation (Li & Wu 2004). There are a number of other issues that contribute to the improper use of landscape metrics, however, which we should guard against:

- *Indiscriminate use of landscape metrics*—The widespread availability and usability of landscape analysis software makes it all too easy (and tempting) for users to calculate dozens of metrics all at once, sometimes without understanding what the metrics are actually measuring and thus whether they are even appropriate for the objectives of the analysis. Needless to say, the blind and mindless calculation of landscape metrics is not considered a 'best practice' in landscape ecology! Although there is little

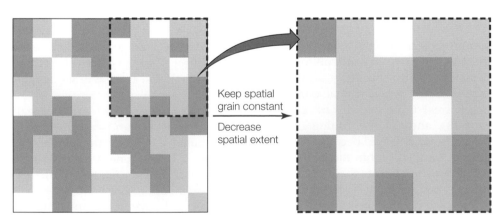

Figure 4.34 Effect of spatial extent on landscape metrics.   Depending on how spatial data are clustered, changing the spatial extent or comparing landscapes of different spatial extents can have a major effect on measures of landscape composition and configuration. The landscape on the right is from the upper right quadrant of the landscape on the left, and so actually covers a smaller spatial extent (5 × 5 cells), but has been rescaled here for the purposes of comparison. The grain size is the same in both landscapes.

consensus as to which landscape metrics—or how many—are ultimately needed to describe landscape patterns (Riitters et al. 1995; Cain et al. 1997; Cushman et al. 2008), there are a number of basic landscape metrics that will generally suffice for most studies, such as those highlighted in this chapter.

- *Landscape pattern analysis as an endpoint*—All too often, landscape pattern analysis is treated as an end, rather than as a first step in understanding how structural changes in landscapes might influence ecological processes (Li & Wu 2004). The very foundation of landscape ecology is built on the premise that spatial patterns affect ecological processes, and ecological processes in turn give rise to spatial patterns (Turner 1989; Pickett & Cadenasso 1995). Even if the relationship between pattern and process is not reciprocal, there is still an expectation that spatial pattern matters for the ecological process or phenomenon being investigated. It thus behooves us to establish that linkage between spatial pattern and ecological process.

- *Fishing with landscape metrics*—On the other hand, fishing expeditions that seek to uncover any and all significant correlations between landscape metrics and some ecological process or other empirical measure likewise cannot be considered a best practice. The statistical use of metrics should always be guided by an a priori hypothesis, preferably informed by theory or observation as to how that aspect of landscape pattern is expected to influence the ecological response being studied.

- *No link exists between pattern and process*—If spatial pattern does not matter to the ecological response, or if the process of interest is not responsible for the observed landscape pattern, then landscape pattern analysis is of questionable value. Admittedly, it is not usually clear whether or at what scale(s) landscape pattern might be important for the ecological response of interest until one has conducted a landscape or spatial pattern analysis. In that case, we should still be wary of ascribing significance to any untested correlations between landscape metrics and ecological responses.

- *Ecological irrelevance*—Landscape metrics, as descriptors of landscape pattern, are assumed to have relevance for the ecological process or phenomenon being studied. Rarely is this assumption tested, however. For example, we might expect dispersal success to decline as a consequence of habitat loss and fragmentation,

and yet Schumaker (1996) found that some common measures of landscape pattern (e.g. NP, ENN, CI) were only weakly correlated with the success of simulated dispersers on real landscapes that had 15–22% habitat. This is significant, especially given that the nearest-neighbor distance (ENN) is the most commonly used measure of patch isolation, and yet is often poorly correlated with ecological processes such as dispersal and gene flow (**Chapters 5, 6 and 9**).

[T]he assumption that metrics which quantify landscape structure capture functional landscape properties is perhaps too uncritically accepted, and the ecological relevance of many landscape indices is often unproven and questionable.
Kupfer (2012)

Adding greater functionality to landscape metrics can enhance their ecological relevance. Landscape measures that incorporate the **functional connectivity** of patches, based on the relative ease with which individuals can move through the intervening matrix (landscape resistance), are generally better predictors of dispersal success and gene flow than structural measures (ENN), for example (**Chapters 5 and 9**). Or, by accounting for edge effects in our analysis of patch sizes, based on how far such effects are expected to extend into the patch (**Box 7.1**), we can obtain an estimate of available habitat that is perhaps more relevant for assessing the patch occupancy or population viability of a forest-interior species (i.e. based on the amount of core habitat).[43] All of this underscores the importance of deriving 'ecologically scaled landscape indices' that explicitly incorporate information on how organisms or other processes respond to spatial heterogeneity (Vos et al. 2001).

- *Ignoring scaling relationships*—Because many landscape metrics change in complex ways with increasing spatial grain or extent of the landscape, direct comparison between landscape metrics obtained at different scales may not be prudent (Wu et al. 2002). In that case, we have three options: (1) restrict comparisons to landscapes having the same grain (resolution) and spatial extent; (2) derive scaling relations that permit the extrapolation/interpolation of metrics between landscapes that differ in scale, which is possible for some but not all metrics (Wu et al. 2002; Wu

---

[43]Core-area metrics for a specified edge depth (m) can be calculated within FRAGSTATS (McGarigal et al. 2012) and the R package *landscapemetrics* (Hesselbarth et al. 2019), for example.

2004); or (3) adopt a multiscale analysis of the landscape metric, as opposed to just a single-scale measure. As discussed previously, one way to do this is to construct 'scalograms,' in which the metric is computed over progressively broader scales (either in terms of increased grain or extent), and then compare these scaling relationships among landscapes (Wu 2004).

Alternatively, we can use multiscale statistical methods to evaluate the scaling of landscape patterns. We therefore consider such methods next.

Spatial Analysis

Spatial analysis involves the application of spatial statistics or geostatistical techniques that have been developed over the past half-century in a variety of disciplines, including plant ecology, epidemiology, human geography, and mining engineering (Fortin & Dale 2005). Despite the diversity of methods, spatial statistics may be grouped according to their primary purpose and objectives (Fortin & Dale 2005):

1. detection and description of spatial structure (**pattern exploration**)

2. estimation of parameters from a spatial distribution or the testing of hypotheses regarding spatial dependence (**statistical inference**)

3. estimation and the production of continuous surface maps from sampled locations (**spatial interpolation**)

4. analysis of connectivity among locations (**network typology**)

The key purpose in performing spatial analysis is to determine whether the data lack independence: and if so, what is the nature of the spatial dependence.

Fortin & Dale (2005)

Although these categories are not mutually exclusive, our focus in this section will primarily be on the use of spatial statistics for the detection and description of spatial structure (pattern exploration), and testing whether patterns depart from the null hypothesis of complete spatial randomness (statistical inference). We will also touch briefly upon data estimation and spatial interpolation using kriging. The description of network typology and the analysis of landscape connectivity will be left to **Chapter 5**, and is also covered in **Chapter 9** on landscape genetics. Spatial analysis is ubiquitous in landscape ecology, and thus applications of these and other types of spatial statistics and spatial

modeling techniques will be presented throughout the textbook.

A Primer to Spatial Statistics

Before we begin our study of spatial statistics in earnest, we should first discuss some of the underlying assumptions and sampling considerations related to the use of these statistics: 1) the relationship between spatial point patterns and spatial point processes, 2) sampling design, and 3) the assumption of stationarity and its implications for spatial statistics.

SPATIAL POINT PATTERNS AND SPATIAL POINT PROCESSES   One common class of spatial statistics involves the analysis of **spatial point patterns** (point pattern statistics). A point pattern is simply a collection of data points that are spatially referenced (i.e. a landscape). At a minimum, each data point includes information as to its location on the landscape (i.e. its xy coordinates). Such data are referred to as location-based or **point-referenced data** (Perry et al. 2002).

For example, if we are analyzing the distribution of a particular tree species, we can record the coordinates for each tree on the landscape (**Figure 4.35A**). While we are at it, we might as well measure the canopy area of each tree, which will give us an idea of how tree size classes are distributed across the landscape. Our dataset now includes tree size as well as location (**Figure 4.35B**). Location-based data may thus possess attributes (**point-referenced data with attributes** or 'marks') that give the identity, magnitude, or intensity of the variable at that location (x, y, and z; Perry et al. 2002). Perhaps we are working in a forest that is experiencing an outbreak of bark beetles, and we want to analyze the pattern of damage (i.e. the locations of dead and dying trees). We could thus record whether individual trees are healthy or not (i.e. trees are in one of two possible states; **Figure 4.35C**). Then again, there is more than one type of tree in the forest, and so we might wish to quantify how the pattern of tree diversity changes across the landscape. In that case, we could record the number of tree species encountered at sampling points arrayed at regular intervals along transects (i.e. a survey grid; **Figure 4.35D**). Our dataset thus consists of the coordinates and the number of tree species present at each survey point.

A spatial point pattern is assumed to be the result of a **spatial point process**. A spatial point process is any type of random process that generates a point pattern (Perry et al. 2006). For example, lightning strikes could be considered a spatial point process, resulting in a spatial point pattern of ignitions across the landscape. Although the distribution of ignitions across the landscape may be random, they need not be, as the distribution and condition of flammable

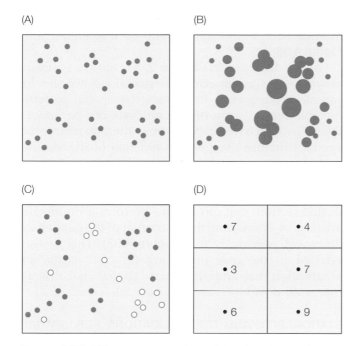

Figure 4.35 Different types of spatial point data. Spatial point data may take the form of (A) point-referenced data with just the locations (*xy* coordinates) of individuals or objects; (B) point-referenced data with numeric attributes, such as information on the magnitude or intensity (size, abundance, concentration) at those locations; (C) point-referenced data with discrete attributes, such as the identity or state of the individual at each location (e.g. species type or presence/absence); or (D) area-referenced data, in which a systematic survey (contiguous samples or a raster data set) is converted to point-referenced data with attributes (e.g., number of species at each sampling point).

Source: After Perry et al. (2002).

fuels ultimately determines whether a fire is ignited by the strike.

The fact that spatial point processes generate spatial point patterns is important from the standpoint of statistical inference. As we shall see, spatial statistics assume stationarity to test for departures from complete spatial randomness. **Complete spatial randomness** is the realization of a spatial Poisson process, the simplest type of point process, in which the locations of points are independent and follow a Poisson distribution (i.e. the spatial point pattern is random). Subsequently, the statistical parameters used to characterize the distribution (the mean and variance) do not vary across the landscape (i.e. they are stationary). It is because of these underlying assumptions of stationarity and complete spatial randomness inherent to a spatial Poisson process that we can then test for spatial structure (departures from randomness) in our data (Wagner & Fortin 2005).

The upshot of all this is that patterns are not processes. Spatial processes give rise to spatial patterns,

but not all spatial patterns are the result of spatial processes. For example, the interaction between fire, fuels, and topography gives rise to a spatially heterogeneous disturbance mosaic (**Figure 3.31**). In this case, fire is considered a *spatially structured* process, rather than a spatial process (*sensu* Legendre 1993; Hawkins 2012), even if the ignition pattern of those fires was created by a spatial process, such as lightning. This helps to explain, yet again, why we cannot infer spatial process from an analysis of pattern, especially if other types of processes are operating (e.g. non-spatial or spatially structured processes).

SAMPLING DESIGN CONSIDERATIONS   As the examples in the preceding section illustrate, point patterns encompass many different types of data that are obtained via different sampling methods (Perry et al. 2002; Fortin & Dale 2005). For example, if we obtain the location of every tree on the landscape, we will have conducted an **exhaustive survey** of the landscape (**Figure 4.35A–C**). In some cases, however, it might be more practical or cost-effective to survey at regular intervals along individual transects or across a sampling grid, such that our sampling units are contiguous (**Figure 4.35D**). Note that if sampling is done across a grid, we are representing the landscape as a lattice (i.e. a raster dataset). Spatial statistics can thus be applied to raster datasets in which grid cells contain categorical or numeric information about each sampling location (e.g. a species' presence or absence, abundance, or the total number of species present).

Sampling along a transect or across a survey grid is a type of systematic survey (the distance between sampling points is at a fixed distance interval). However, many studies adopt a random or stratified random sampling design, such that samples are variably spaced across the landscape (i.e. the sampling units are not contiguous). The type of sampling scheme is important for two interrelated reasons: first, it affects our ability to detect and describe spatial structure in the data; and second, it determines what types of spatial statistics we can perform on our point pattern data. For example, some approaches assume exhaustive sampling of all point locations (e.g. Ripley's *K*-function), whereas others require that sampling units be contiguous (e.g. lacunarity analysis).

In fact, all aspects of the sampling design, such as the number of samples (sample size), the grain and extent of sampling, the spatial lag (distance between samples), and the sampling strategy (systematic, random, stratified random, etc.) can affect our ability to detect and describe spatial patterns (Fortin & Dale 2005; **Chapter 2**). Sample size is an important consideration in any study, but takes on added significance when the goal is to detect and describe spatial patterns. For example, it has been suggested that at least 30 survey locations are necessary to detect significant spatial autocorrelation (Legendre & Fortin 1989), but reliable estimation of spatial pattern may require information from 100 or more sampling locations, depending on the nature and scaling of spatial structure (Fortin & Dale 2005).

Although we have previously discussed how spatial grain and extent affect the appearance and analysis of landscape patterns (**Figures 2.5, 2.6, 4.33, and 4.34**), **spatial lag** is also important to our ability to detect spatial structure. To illustrate, imagine we are surveying the relative abundance of some species across a transect (a one-dimensional landscape), with the goal of identifying areas of high versus low abundance (i.e. abundances that are significantly greater than or less than the mean abundance, respectively). A spatial lag that is too coarse relative to the scale of patchiness will fail to detect spatial structure. In the example given, if samples are taken at 5 m intervals, abundance would appear uniformly high because we are sampling entirely inside the patches (**Figure 4.36, a**). A somewhat shorter lag distance (4 m) would at least enable us to detect the existence of spatial structure, for we are now encountering variability in species abundance among our samples because our sampling sites fall both inside and outside of patches (**Figure 4.36, b**). An even finer lag distance (3 m) would be ideal, however, as this would enable us to better characterize the scale of patchiness, by examining how the variability in species abundance changes

across the transect (i.e. going from high, to low, to intermediate, to low values, and back to high again; **Figure 4.36, c**).

As we discussed in **Chapter 2**, it is not always clear at what scale (grain, extent, or spatial lag) we should sample if our goal is to characterize spatial pattern. For this reason, a multiscale analysis can be undertaken, in which we explore how the appearance of spatial structure changes as a function of altering the spatial grain or lag distance at which the pattern is measured. Although multiscale analysis is intrinsic to some of the spatial statistics that we will be discussing in this section, one can always perform a multiscale analysis by coarsening the resolution of the dataset, as discussed previously (e.g. **Figure 4.33**). This is also advised in the case of testing for stationarity, an assumption that underlies many spatial statistics, as discussed next.

STATIONARITY AND ITS IMPLICATIONS FOR SPATIAL STATISTICS   Sampling design is not the only issue that affects our ability to detect and describe spatial patterns. Many spatial statistics assume **stationarity,** such that the statistical parameters used to characterize the spatial pattern (e.g. the mean, variance, or covariance) are invariant across the landscape (note the single horizontal line for mean abundance across the one-dimensional landscape in **Figure 4.36**). The assumption of stationarity is to spatial statistics what the assumptions of normal distribution and equal variance (homoscedasticity) are to parametric statistics (Fortin & Dale 2005). Stationarity can be tested by examining how the

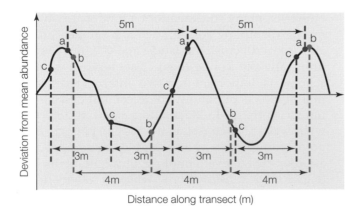

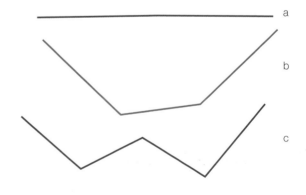

Figure 4.36 Effect of spatial lag on the detection of spatial pattern.   In this example, the degree to which a species' abundance deviates from the mean (horizontal line) is surveyed along a transect (a one-dimensional landscape). If the spatial lag (the distance interval between sampling points) is too coarse (5 m), then abundance will appear uniformly high across the transect because all samples are taken at the 'peaks' of the distribution (at the **a** points on the graph). If we sample at a shorter lag distance (4 m), we can at least detect that there is variation in abundance, as samples are taken at both high and low points of the distribution (**b** points on the graph). An even shorter lag distance (3 m) would be preferable, however, as this would enable us to better resolve the scale of patchiness in this distribution (e.g. the fact that there are three abundance 'hotspots' along this transect; **c** points on the graph).

Source: After Fortin & Dale (2006).

mean and variance of the pattern change as function of scale (e.g. within sliding windows of different sizes; Fortin & Dale 2005). Recall from our previous discussion, however, that stationarity is actually a property of the spatial point process, and not of the data. Stationary processes can generate patterns that look non-stationary (e.g. the data possess a trend or gradient; Fortin & Dale 2005). In that case, however, data can be detrended prior to analysis to conform with the stationarity assumption (Wagner & Fortin 2005).

In other cases, stationarity may be scale-dependent, as a result of different processes operating at different scales or within different areas of the landscape. Non-stationarity is common in broad-scale datasets such as those used in biogeographic studies (Hawkins 2012). All is not lost if the assumption of stationarity cannot be met, however. We might at least be able to identify homogeneous regions within our landscape in which stationarity can be reasonably assumed, and then calculate our spatial statistics for those regions separately.

The basic problem of spatial analysis of landscapes is that several processes creating heterogeneity often operate at the same time. These processes may interact, so that the parameters of one process change with the heterogeneity resulting from other processes. This means that the observed pattern can rarely be attributed to a single, stationary process, as many methods in spatial analysis assume.

Wagner & Fortin (2005)

Alternatively, it may be that the data are so heterogeneous that stationarity cannot be assumed even within different regions of the landscape. In that case, **local spatial statistics,** which measure the degree of spatial dependence only within a local neighborhood (as opposed to averaging across the entire landscape or a region within a landscape), are more appropriate for detecting clusters or patchiness in the data (i.e. hotspots; Getis & Ord 1996). Local spatial statistics include a variety of approaches, which may be grouped under the general umbrella of Local Indicators of Spatial Association (LISA; Anselin 1995), such as second-order neighborhood analysis (Getis & Franklin 1987); local spatial autocorrelation statistics (the Getis-Ord Index, G*: Getis & Ord 1992; Ord & Getis 1995); and even, local versions of global statistics (e.g. local Moran's I), among others.

To keep our discussion in this section (relatively) simple, we will be focusing primarily on **global spatial statistics**, in which stationarity is assumed for the entire landscape, or at least, the individual regions within a landscape if separate analyses are performed by region.

Spatial Statistics

We will now discuss some of the more common types of spatial statistics that you are likely to encounter or perhaps use in your own research. Most of these methods have been developed for the detection and description of spatial structure, and/or testing whether those spatial patterns are significantly different from the null expectation of complete spatial randomness. As with any branch of statistics, spatial statistics is a vast field and there are many more types of analyses than can be covered in this one section, or even, in a single chapter. For those desiring a more comprehensive and rigorous treatment on spatial data analysis, there are a number of excellent textbooks devoted specifically to these methods and their applications in ecology (e.g. Dale 1999; Haining 2003; Fortin & Dale 2005), along with books and online resources for implementing these analyses within a variety of programs, such as the ArcGIS® Spatial Analyst and Spatial Statistics extensions, and various R packages[44] such as *splancs, spatstat, gstat, and geoR* (Baddeley et al. 2015; Brunsdon & Comber 2015; Bivand et al. 2017; Pebesma & Graeler 2018; Ribeiro & Diggle 2018).

POINT COUNT DATA: AGGREGATION INDICES A simple measure of distribution is given by the **variance-to-mean ratio (VMR)**, in which we compare the sample variance (s^2) to the sample mean ($\bar{x}$) across the entire study area to obtain the ratio ($s^2/\bar{x}$) :

$$s^2/\bar{x} = \left(\frac{\sum_{i=1}^n \left(x_i - \bar{x} \right)^2}{n-1} \right) / \bar{x} .$$

Thus, VMR = 1 when the variance equals the mean ($s^2 = \bar{x}$), which is the hallmark of a Poisson distribution (i.e. a random distribution). The VMR has thus been interpreted as an index of spatial dispersion (**Figure 4.37**): a clumped or underdispersed distribution is suggested by a VMR >1 ($s^2 > \bar{x}$), and an overdispersed or regular distribution is given by VMR < 1 ($s^2 < \bar{x}$). Note, however, that we are only inferring the spatial pattern of the data distribution. The VMR does not actually use (or even require) spatially referenced data. In other words, the VMR only makes use of the data values and not the specific locations at which those data were obtained (i.e. this is attribute-only data).

---

[44]Within R (see footnote 31), you can often find related packages organized by 'task views' (https://cran.r-project.org/web/views/). For example, there is a task view for the Analysis of Spatial Data, which summarizes all of the packages currently available for reading, writing, manipulating, analyzing and visualizing spatial data (https://cran.r-project.org/web/views/Spatial.html).

Let's apply this approach to the landscape depicted in **Figure 4.35D**, in which we'll assume that we have obtained species richness (number of species) at these six locations. We wish to determine whether species richness is randomly distributed throughout the landscape, or whether richness is more (or less) dispersed than expected under the assumption of a random distribution. The mean richness is 6 species (7+4+3+7+6+9 = 36/6 = 6), with a sample variance (s^2) of 4.8 species. Thus, the VMR for species richness in this landscape is 0.8 (4.8/6 = 0.8), which suggests that there is a bit of overdispersion in the distribution of species richness (VMR < 1). That is, the probability of finding high (or low) species richness at any two adjacent sites is less than expected if the distribution were completely random.

The simplicity of the VMR is certainly appealing, but unfortunately, is also problematic. First, if the data are spatially autocorrelated, then the sample variance (s^2) does not provide a very good estimate of the *spatial variance* in the dataset (Dale et al. 2002). Sample variance would not capture the variance staircase expected of nested patch structure, for example (**Figure 2.7**). Second, and most critically, VMR has been shown to be essentially 'useless as a measure of departure from randomness' (Hurlbert 1990).[45] Although a random distribution is Poisson, and a Poisson distribution has equal mean and variance, we should not assume that the converse is also true (Dale 2000; Dale et al. 2002). That is, distributions with equal mean and variance are not necessarily Poisson, and thus, equal mean and variance (VMR = 1) should not be taken as evidence of randomness (Hurlbert 1990). This seriously undermines the use of VMR to infer patterns of spatial dispersion.

A better measure of dispersion is given by **Morisita's Index (I_M)**, which superficially resembles VMR, but is more sensitive to departures from randomness and is not as strongly affected by sample size as some other aggregation indices (Hurlbert 1990). Morisita's index gives the probability that two, randomly selected samples (e.g. individuals or species) come from the same site as:

$$I_M = \frac{s^2 - \bar{x}}{\bar{x}^2 - (s^2/n)} + 1, \quad \text{Equation 4.13}$$

where n is the number of sampling sites (Morisita 1962, 1971). So, for a Poisson distribution in which the variance equals the mean ($s^2 = \bar{x}$), we would have $I_M = 1$ (because the numerator, ($s^2 - \bar{x}$), would then be 0 in Equation 4.13). Using our previous example of species richness (**Figure 4.35D**), we would have

[45]Hurlbert's (1990) treatise on the spatial distribution of the montane unicorn (*Monoceros montanus*) is that rarest of beasts, a highly technical scientific paper that is also fun to read.

$$I_M = \frac{(4.8-6)}{(6^2-(4.8/6)} + 1 = (-1.2/35.2) + 1 = -0.034 + 1 = 0.966.$$

Since $I_M < 1$, this suggests that the distribution of species richness across the landscape is somewhat less than expected for a random distribution (i.e. it is overdispersed; **Figure 4.37**). That is, it is slightly less likely that two species drawn at random come from the same site than if the distribution were completely random. Higher values of I_M ($I_M > 1$) suggest a more clumped or aggregated distribution, meaning that the probability of randomly drawing two samples that come from the same site is greater than expected if these were randomly distributed. Thus, $I_M = 2.5$ indicates that the distribution is 2.5 times more aggregated than a random distribution; in other words, the probability of two randomly selected samples coming from the same site is 2.5 times greater than if the distribution were random.

Aggregation indices offer an easy way to characterize distributions, but they are not generally the best way to do so. Spatial patterns are sensitive to scale, and thus a single measure of aggregation will rarely suffice (Dale 2000). In that case, a multiscale analysis of spatial pattern is needed.

CONTIGUOUS SAMPLES: LACUNARITY ANALYSIS AND BLOCKED-QUADRAT VARIANCE METHODS

Lacunarity analysis is inherently a multiscale analysis of spatial pattern or, more accurately, of the gap structure of landscapes. A *lacuna* is a space or gap. Although it may seem odd to focus on the distribution of spaces or gaps (absences) between patches or events rather than the patches or events themselves, it turns out not to matter from the standpoint of implementation, as we can just as easily apply lacunarity analysis to habitat or other types of patches. Nevertheless, some ecological processes, such as dispersal, are influenced more by the distances between patches than by patch structure per se (With & King 1999b). In that case, our search for meaningful pattern-process relationships may prove more fruitful if we relate such responses to the 'gappiness' of landscapes, rather than just their patchiness.

Lacunarity can be calculated in different ways, but we will focus here on the **gliding-box method** (Plotnick et al. 1993). Recall that a gliding box is a type of moving-window analysis, in which boxes (windows) of different sizes are moved across the landscape, one sampling unit (e.g. grid cell) at a time (**Figure 4.27**). This, then, requires that our sampling units are contiguous (e.g. **Figure 4.35D**). Within each box of size l, the number of occurrences are counted (e.g. the number of empty or filled cells, depending on whether we are focusing on absences or presences, gaps or patches, respectively). For example, if we wish to quantify the gap structure of the landscape depicted in **Figure 4.27**, we begin with a 2×2 box ($l = 2$) at the upper left corner and count the number of empty cells ($c = 4$), slide the box over one column and count the number of empty cells ($c = 3$), and so on, until we have covered the entire landscape. The procedure is then repeated for the next box size (e.g. $l = 3$) up to the maximum box size, which is usually taken to be one-half the linear dimension of the landscape ($0.5L$) so as to ensure there are a sufficient number of samples (at least four) to calculate a mean and variance of the box counts obtained at that scale.

The **lacunarity index** (Λ) is basically a VMR of the box counts at each l scale, which is calculated as:

$$\Lambda(l) = s^2(l) / (\bar{x}(l))^2 + 1. \qquad \text{Equation 4.14}$$

Note the similarity to Morisita's index (Equation 4.13). The two indices are closely related, but lacunarity has the advantage of permitting an examination of pattern across a range of scales (Dale 2000). The lacunarity index is usually interpreted graphically, by plotting Λ values as a function of scale (i.e. the box or window size, l). As box size (l) increases, the counts within boxes become more similar (the variance among boxes decreases), and thus, lacunarity decreases (recall that between-sample variance is expected to decrease as a function of increasing grain size; **Figure 2.5**). In theory, $\Lambda \to 0$ as $l \to L$; that is, the lacunarity index goes to zero as the box size approaches the size of the landscape. At the finest scale ($l = 1$), which is equivalent to the size of the sampling unit (e.g. the grain size or resolution of a grid cell), the lacunarity index (Λ) is equal to $1/p$, the proportion of filled sites on the landscape (e.g. the proportion of habitat on the landscape; Plotnick et al. 1993). Thus, sparsely filled landscapes (low p) will have greater lacunarity (more open space) than those with a greater number of filled sites (high p). Therefore,

it is usually desirable to normalize the lacunarity index (Λ^*), by dividing lacunarity values by the maximum lacunarity attained at the finest scale (i.e. $l = 1$, where $\Lambda = 1/p$), so that $0 \leq \Lambda^* \leq 1$. A bounded lacunarity index facilitates comparisons between landscapes that differ in the amount or fragmentation of habitat, for example.

To illustrate, let's perform a lacunarity analysis of the two landscapes depicted in **Figure 4.28**. Recall that both landscapes have the same amount of habitat (50%), but differ in their degree of fragmentation (clumped vs fragmented). Although we have previously compared aspects of their patch structure using landscape metrics (**Tables 4.6–4.8**), we will now explore how the gap structure of these two landscapes differs, and over what range of scales. Both curves show the expected decline in the normalized lacunarity index (Λ^*) as a function of scale; however, lacunarity is generally higher on the clumped landscape than on the more fragmented one over most of this range ($l = 5$–100; **Figure 4.38A**). That is, the clumped landscape is more 'gappy,' in that it possesses gaps that are larger and more variable in size than those in the fragmented landscape. The fragmented landscape is more patchy, but the clumped landscape is more gappy. Notice that the greatest difference in lacunarity between these two landscapes is observed at $l = 9$ (**Figure 4.38B**). Such differences in lacunarity could thus help identify the critical scale(s) at which certain ecological processes—such as those sensitive to large gaps or matrix effects—are most likely to be influenced by landscape pattern.

Lacunarity analysis is somewhat similar to the **blocked-quadrat variance** methods developed by plant ecologists. The original blocked-quadrat variance (BQV) method (Greig-Smith 1952) combines quadrats into non-overlapping blocks (i.e. boxes or windows), and calculates the variance at each block size, where block sizes increase as a power of 2 (e.g.

Random	Clumped (underdispersed)	Regular (overdispersed)
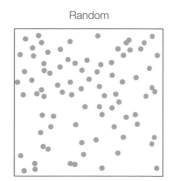	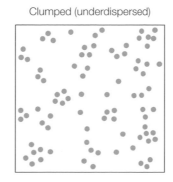	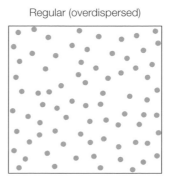

Figure 4.37 Patterns of spatial dispersion. Points are independent of one another in a random (Poisson) distribution, but there is a greater probability of finding another point nearby in a clumped (underdispersed) distribution, whereas there is a decreased probability of finding another point nearby if the pattern is regular (overdispersed).

Source: After Dale (1999).

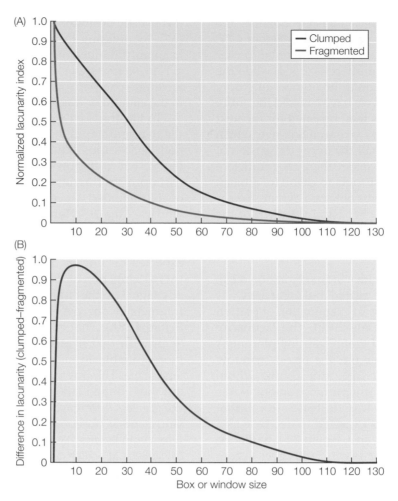

Figure 4.38 Lacunarity analysis. (A) The normalized lacunarity index (Λ^*) is plotted against the size of the gliding window (**Figure 4.27**) for the clumped and fragmented landscapes depicted in **Figure 4.28**. (B) The difference in the standardized lacunarity index between these two landscapes is greatest at a box or window size of 9 units (e.g. 270 m × 270 m = 7.29 ha).

$2^0 = 1 \times 1$, $2^1 = 2 \times 2$, $2^2 = 4 \times 4$, $2^3 = 8 \times 8$, etc.).[46] A plot of how variance changes as a function of block size can reveal the scale(s) at which spatial pattern (patchiness) occurs, which appears as a peak (or multiple peaks) in the distribution (i.e. a random distribution would have constant variance as a function of block size). Because blocks do not overlap, this method is sensitive to the block starting position, and thus could miss or mischaracterize the scale at which the variance actually peaks (Dale 1999). Further, because blocks double in size, it is not possible to determine at exactly what scale the variance might actually peak, especially at the largest block sizes (e.g. high variance at block size 256 and the next block size at 512 would only tell us that spatial structure exists somewhere in between those two scales; Dale 1999).

Because of these issues, **local quadrat variance** methods are favored that use a complete range of block sizes and average over all possible starting positions (i.e. these are moving-window techniques; Hill 1973b). In the **two-term local quadrat variance (TTLQV)**, the squared difference between all adjacent pairs of blocks of a given size are averaged over all possible positions (Dale 1999, 2000). By contrast, the **three-term local quadrat variance (3TLQV)** uses three adjacent boxes of a given size, in which a difference is obtained by subtracting twice the value of the middle box from the sum of the two outer boxes (Dale 1999, 2000). These differences are then squared and averaged over all positions to obtain the variance term. For both methods, the variance is calculated over a range of block sizes, and when plotted as a function of block size, peaks in the variance are again indicative of spatial structure (Dale 2000). The three-term form is generally recommended over the two-term form because it is less sensitive to trends in the data (Lepš 1990).

[46]The BQV method has also been adapted to transects (Kershaw 1957).

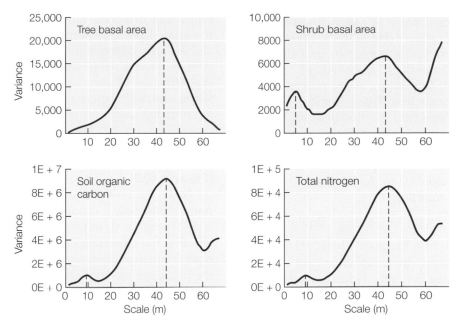

Figure 4.39 Detection of spatial scaling using three-term local quadrat variance (3TLQV). The distribution of trees and shrubs influence carbon and nitrogen pools in a savanna landscape in southern Texas (USA). All show a peak around 43–45 m, which corresponds to the average distance between woody patches and neighboring grasslands. In addition, understory shrubs appear to be contributing to finer scale patchiness (5–9 m) in both soil organic carbon and nitrogen.

Source: After Liu et al. (2010).

To illustrate an analysis of local quadrat variance, we'll consider a study by Liu and colleagues (2010), who examined how the woody invasion of a grassland in southern Texas (USA) has affected the distribution of soil organic carbon (SOC) and total nitrogen pools. Most of the carbon and nitrogen in arid and semi-arid systems reside below ground, and although the shift from grassland to a woody plant-dominated system could increase carbon storage (i.e. creating a greater carbon sink), there is little consensus on what effect woody invasion might actually be having on carbon storage in grasslands (e.g. whether positive, neutral, or negative), especially given the interactive effects of climate, topography, type of vegetation, and herbivory. Woody encroachment has increased over the past century, and thus Liu and colleagues (2010) sought to compare the scaling relationships between SOC, total nitrogen, woody plant cover, grass cover, forb cover, and litter across a hillslope gradient that encompassed both grass and woody plant-dominated communities. Using 3TLQV, they found a strong association between the spatial distribution of tree and shrub cover and spatial variation in soil carbon and nitrogen (**Figure 4.39**), whereas no such relationship existed between soil carbon and nitrogen and the cover of grass or forbs. The variation at the broader scale (43–45 m) corresponds to the average distance between the centers of woody plant communities and adjacent grasslands, whereas the finer scale variation (10 m)

appears to reflect the local influence of shrub cover on SOC and total nitrogen. There was also a strong correlation between litter amount, root biomass, shrub and tree cover with SOC and total nitrogen, which suggests that invasive woody plants are indeed altering both the storage and spatial distribution of soil nutrient pools, likely via root turnover and litter production, in this grassland (Liu et al. 2010).

POINT PATTERN ANALYSIS: RIPLEY'S *K*-FUNCTION
What if, rather than contiguous samples, our landscape data take the form of a mapped point pattern, in which we have recorded the spatial coordinates of every individual or object of interest (i.e. an exhaustive survey; **Figure 4.35A**)? One of the most commonly used multiscale analyses of point-referenced (coordinates only) data is the **Ripley's *K*-function** (Ripley 1976; Fortin & Dale 2005). Conceptually, implementation of the Ripley's *K*-function, $K(t)$, involves centering a circle of radius t on each point,[47] and then counting the number of other points (neighbors) that fall within that circle (Haase 1995). The procedure is then repeated for circles of different radii. If the spatial point pattern is consistent with a homogeneous Poisson process, such that it exhibits complete spatial randomness at all

---

[47]Circles rather than windows, boxes, or blocks of different sizes are thus used to assess how pattern changes as a function of scale in this approach.

scales, then the expected value of $K(t) = \pi t^2$; that is, the area of a circle having radius t (Haase 1995; Dixon 2002).

The $\hat{K}(t)$ statistic is an estimate of $K(t)$, and although various derivations have been proposed for calculating $\hat{K}(t)$, these all entail counting every pair of points separated by a distance less than t for each radius (we'll skip the equations for how this is done, but the interested reader is invited to consult one of the cited references for details, e.g. Haase 1995; Dixon 2002; Fortin & Dale 2005). We can compare the observed $\hat{K}(t)$ statistic to the expected $K(t)$ to derive a standardized measure called the **L-function**, $\hat{L}(t)$:

$$\hat{L}(t) = \sqrt{\hat{K}(t)/\pi} - t. \qquad \text{Equation 4.15}$$

The spatial point pattern is random when $\hat{L}(t) = 0$ for a given t, clumped (underdispersed) when $\hat{L}(t) > 1$, and evenly distributed (regular or overdispersed) when $\hat{L}(t) < 1$.[48]

By plotting the L-function against t, we can determine how pattern changes as a function of scale. A point pattern that exhibits complete spatial randomness at all scales would thus have $\hat{L}(t) = 0$ for all values of t; that is, the graph of the L-function versus t would be a horizontal line ($y = 0$). To determine whether $\hat{L}(t)$ deviates significantly from a random distribution, the critical values for rejecting the null hypothesis with 95% certainty (a significance level of $\alpha = 0.05$) can be approximated as $\pm 1.42\sqrt{A}/n$, or with 99% certainty ($\alpha = 0.01$) as $\pm 1.68\sqrt{A}/n$ (Ripley 1976). Nowadays, most spatial statistics programs typically use Monte Carlo methods to achieve the acceptance envelope for rejecting the null hypothesis (Fortin & Dale 2005; Manly 2006).[49]

To illustrate, let's compare the distribution of trees in two landscapes (**Figure 4.40A**). The landscapes cer-

tainly *look* different, but how would we characterize the spatial point pattern in each? Are these patterns significantly different from a random distribution, and if so, at what scale(s)? By plotting $\hat{L}(t)$ as a function of scale (the lag distance, t), we can see that both landscapes are overdispersed (trees are evenly distributed) at short lag distances (e.g. $t = 3$). At greater lag distances (e.g. $t > 10$), however, trees are randomly distributed in the first landscape whereas they are clumped (underdispersed) in the second landscape (**Figure 4.40B**).

Not only have we been able to describe the spatial pattern and how it changes as a function of scale (pattern exploration), but we have also been able to test whether the pattern is significantly different from random (statistical inference). Recall that these are two of the primary objectives of spatial statistics. By identifying the range of scales over which tree distributions are (or are not) random, we can begin searching for the processes that might be responsible for the patterns at those scales. For example, trees might exhibit a uniform distribution over shorter distances because of competition (self-thinning) or negative allelopathy that results in a certain degree of spacing between individual trees (e.g. Kenkel 1988). At broader scales, wind dispersal of seeds could explain the random distribution of trees in some landscapes, whereas secondary seed dispersal by small mammals (seed caches) might be responsible for a more clumped tree distribution in other landscapes. Although spatial statistics like Ripley's K-function (or its equivalent, the L-function) cannot tell us which of these hypothesized mechanisms is actually responsible for the observed pattern, we at least now have a much better idea of where—and at what sorts of processes—we should be looking.

Ripley's K-function can be extended to the analysis of bivariate or multivariate (multi-type) spatial point patterns, in which the data also contain information about the type, status, or size of each location (i.e. point-referenced data with attributes or 'marks'). For example, we might have data on the distribution of two tree species, in which each point also includes information on the type of species at that location (**Figure 4.35C**). Although we could use the methods just described to perform separate analyses on each species' point pattern, or even on both species combined (i.e. by ignoring species type), it might be more interesting to see whether these species appear to be influencing each other's distributions. That is, are we more (or less) likely to find one species of tree within a certain distance (t) of a randomly chosen tree belonging to the other species?

The bivariate form of the Ripley's estimator is $\hat{K}_{ij}(t)$, where i and j are the different types of points (tree species, in this example). If the spatial process is stationary,

---

[48]Some authors obtain $\hat{L}(t)$ as $t - \sqrt{\hat{K}(t)/\pi}$ (e.g. Fortin & Dale 2005). This affects the interpretation of the resulting $\hat{L}(t)$ values: overdispersed (regular) patterns give positive values, whereas underdispersed (clumped) patterns give negative values. It is obviously important to understand which way $\hat{L}(t)$ is being calculated, so as to ensure that results are being interpreted correctly.

[49]Although this is often referred to as a 'confidence interval' or 'confidence envelope,' this is not strictly accurate for we are not trying to evaluate how well our estimate compares to the true value of $K(t)$; that is, we are not attempting to say that we have a 95% level of confidence that our estimate is 'correct.' Rather, we are attempting to compare our estimate to the value of $K(t)$ expected under the null hypothesis of complete spatial randomness. If our estimated value falls within the bounds of that envelope, there is a 95% probability that the pattern we've analyzed is random; that is, we accept the null hypothesis. In the former interpretation (as a confidence interval or envelope) we are getting at the precision of the estimate, but in the latter, we are interested in testing the 'goodness of fit' to the hypothesized model, which is why this is referred to here as an 'acceptance envelope' (or 'non-rejection envelope'): values outside this envelope would lead to the rejection of the null hypothesis that the pattern exhibits complete spatial randomness (Baddeley et al. 2015).

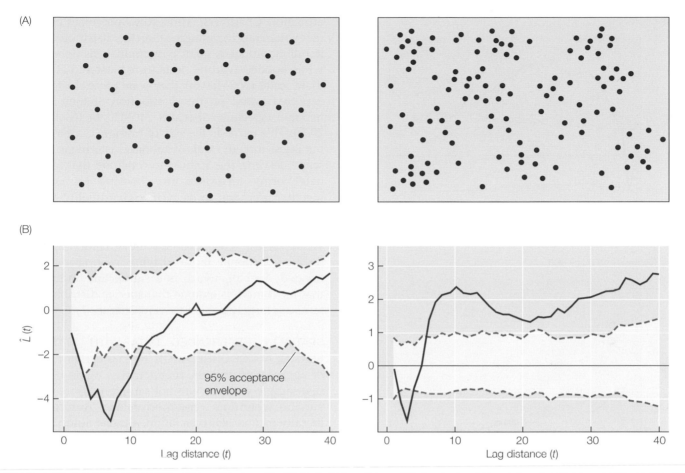

Figure 4.40 Analysis of spatial point patterns with Ripley's *K*-function.   (A) Analysis of these two landscape patterns identifies the range of scales over which the point pattern is either regular (overdispersed), random, or clumped (underdispersed). (B) The *L*-function, $\hat{L}(t)$, is the standardized form of Ripley's *K*, and is interpreted with respect to the hypothesis of complete spatial randomness (the 95% acceptance envelope). The pattern is overdispersed (a regular or uniform distribution) when the *L*-function lies below the envelope, but underdispersed (clumped or aggregated) above the envelope.

then the corresponding pairs of **cross-*K* functions** are equal: $K_{ij}(t) = K_{ji}(t)$. If we assume that the spatial process(es) that gave rise to the two types of point patterns are acting independently, then the expected value of the cross-type *K*-function is $K_{ij}(t) = \pi t^2$, which means we can use the corresponding $\hat{L}_{ij}(t)$ function (Dixon 2002).

To illustrate, we'll consider an analysis that was performed on the distribution of bald cypress (*Taxodium distichum*) and black gum (*Nyssa sylvatica*) trees in a swamp hardwood forest of South Carolina, USA. Based on a univariate analysis of their individual distributions, cypress trees were found to be strongly clustered (underdispersed) at lag distances of 4–27 m (**Figure 4.41A**), whereas gum trees were randomly distributed (Dixon 2002). In the field, the two tree species appeared to be spatially segregated, which was borne out by the bivariate analysis: the *L*-function comparing the distribution of cypress to gum trees, $\hat{L}_{ij}(t)$, is significantly overdispersed across a distance range of 3–35 m (**Figure 4.41B**). That is, the number of cypress

trees in the neighborhood of black gum trees (or vice versa) is less than expected if the distributions of these two tree species were completely independent. As to why they exhibit a negative association is another question. Whether this is due to allelopathy or resource competition between species, or whether the two species simply occupy distinct habitats within this swamp forest, are all questions for further study. Again, linking spatial pattern to process—and vice versa—is the very definition of a landscape ecological investigation!

[B]irds can be territorial at fine scales, clustered at other scales, and randomly distributed at still further scales, so *methods that quantify spatial patterns and resolve their probable causes are needed at multiple spatial scales.* (emphasis added)   Melles et al. (2009)

As an example of how alternative hypotheses regarding causal processes can be tested using point pattern analy-

(A) Cypress

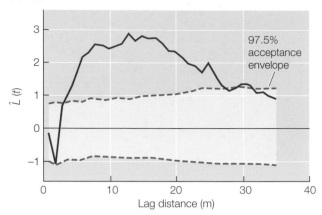

(B) Cypress and black gum

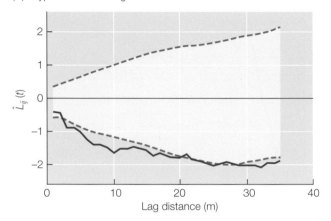

Figure 4.41 Bivariate analysis with Ripley's *K*-function. (A) Univariate analysis of the distribution of bald cypress (*Taxodium distichum*). Note that trees are significantly clustered at distances 4–27 m. (B) The bivariate analysis of both bald cypress and black gum (*Nyssa sylvatica*) reveal that the two species are not randomly distributed with respect to each other. Instead, their joint distribution is over-dispersed between 3 and 35 m, meaning that there are fewer cypress trees in the neighborhood of black gum trees than expected, and vice versa. The 97.5% acceptance envelope is for the null hypothesis of complete spatial randomness.

Source: After Dixon (2002).

ses, Melles and her colleagues (2009) used a combination of univariate and bivariate methods to tease apart the influence of habitat availability versus social attraction on the distribution of hooded warbler (*Wilsonia citrina*) nests within a forested landscape in southern Ontario, Canada. Hooded warblers nest within small forest gaps, and thus nest locations tend to be clustered within the landscape. Are nests clustered simply because the hooded warbler's breeding habitat is clustered, however? Like many songbirds, the hooded warbler is territorial, but the presence of actively nesting individuals within an area may signify favorable breeding habitat, causing other individuals to settle there (conspecific

attraction; **Chapter 6**). Thus, social processes might better explain nesting patterns than the distribution of habitat alone. Indeed, point pattern analysis revealed that while hooded warbler nests were clustered at intermediate scales (360–480 m), they were more clustered than expected based on the distribution of their breeding habitat, especially at scales 240–420 m (Melles et al. 2009). This exceeds the size of a single male's territory, but is less than the scale of habitat clustering in the landscape, leading the authors to conclude that females in neighboring territories are choosing to nest closer together, perhaps to be near higher-quality resources and/or mates (extra-pair paternity by neighboring males is not uncommon in hooded warblers, evidently). Whatever the reason, conspecific attraction appears to be an additional factor driving nesting patterns in hooded warbler, which is a significant finding given that most models assume that species distribution patterns are largely habitat-driven (**Chapter 7**).

SPATIALLY REFERENCED DATA WITH ATTRIBUTES: SPATIAL AUTOCORRELATION In our earlier discussion of aggregation indices, mention was made of the fact that sample variance was not a good measure of spatial variance when data were autocorrelated. Autocorrelation literally means 'self-correlation.' **Spatial autocorrelation** is thus the degree to which data for a given variable (e.g. tree height, abundance, species richness) are correlated in space. Recall from **Chapter 2** that near points on a landscape tend to be more similar than points that are farther apart; this is the basis for Tobler's first law of geography (Tobler 1970). We thus expect that the degree of similarity or autocorrelation between locations will decrease with increasing lag distance. That is only an expectation, however, because the relationship is typically not linear: the variable of interest may respond in complex ways to the underlying heterogeneity of the landscape, in addition to whatever intrinsic processes are responsible for its distribution in the first place (Fortin & Dale 2005).

The most common coefficient for spatial autocorrelation is **Moran's *I*** (Moran 1950). As you may recall from your basic statistics course, the Pearson's correlation coefficient (*r*) measures the degree to which two variables (*x* and *y*) covary; that is, whether their relationship is positive (*r* > 0), independent (*r* = 0), or negative (*r* < 0). The correlation coefficient can be thought of as a type of standardized covariance,[50] in which the sample covariance of two variables (s_{xy}) is divided by the product of their standard deviations ($s_x s_y$), so that the resulting coefficient is bounded: $-1 \leq r \leq 1$. Thus, two variables

[50]Recall from your basic statistics course that the sample covariance between two variables *x* and *y* is given as $s_{xy} = \dfrac{\sum_{i=1}^{n}(x_1 - \bar{x})(y_i - \bar{y})}{n-1}$, where *n* is the sample size.

exhibit a positive linear relationship when $r = 1$, but a negative linear relationship when $r = -1$.

In the case of spatial autocorrelation indexed by Moran's I, we are interested in the degree to which the values of two locations within a certain distance (d) covary. Since we are dealing with only a single variable (x), however, we calculate **spatial autocovariance** ($s_{x_i x_j}$) as the summed products of the deviations of each point in the pair (x_i and x_j) from the global mean ($\bar{x}$) obtained over the entire landscape for that variable $\left[\sum_{i=1}^{n} \sum_{j=1}^{n} (x_i - \bar{x})(x_j - \bar{x}) \right]$ for all pairs located within a given distance, divided by the total number of pairs at that distance (Fortin & Dale 2005). Moran's I is then obtained by dividing the autocovariance ($s_{x_i x_j}$) by the standard deviation for that variable (s_x) so that the resulting coefficient is bounded: $-1 \le I \le 1$. Moran's I is interpreted in much the same way as the Pearson's correlation coefficient, but by adding a spatial dimension to the analysis, we can also make inferences about the pattern of distribution and how it varies as a function of scale (d):

$I_d \rightarrow 1$ Values at locations separated by distance d are similar, and thus exhibit positive spatial autocorrelation. The distribution is aggregated or clustered, which may occur if the pattern is patchy or cyclic;

$I_d = 0$ Values at locations separated by distance d are independent, such that the variable exhibits no spatial autocorrelation. The distribution is thus random;

$I_d \rightarrow -1$ Values at locations separated by distance d are diametrically different (i.e. one high, the other low), and thus exhibit negative spatial autocorrelation. The distribution may be overdispersed or consist of an alternating pattern.

To illustrate the application of Moran's I in detecting and interpreting spatial structure, we'll consider a study by Diniz-Filho and colleagues (2010), who were interested in determining to what extent variation in the metacommunity structure of terrestrial-breeding frogs in the Brazilian Amazon could be attributed to environmental variation (heterogeneity), spatial variation (distance effects), or some combination of the two. Species abundance data were obtained across a large study area (72 km²) for some two-dozen frog species, and Moran's I coefficients were calculated at different lag distances (1–9 km) for each species. The individual coefficients were averaged over all 28 species for each distance class ($\bar{I}_d$), and then plotted as a function of lag distance (d) to produce a **spatial correlogram (Figure 4.42)**. Species abundance patterns are strongly clustered ($\bar{I}_d > 0$) up to a distance of about 2.5 km, at which point, abundances generally become independent ($\bar{I}_d = 0$) and increasingly

more variable among species (note that the standard error bars increase with distance; **Figure 4.42**). Using simulation models, the authors demonstrated that this is the pattern of spatial autocorrelation expected if neutral (random) processes rather than environmental variation drive species abundances. Species abundances tend to be similar at short distances because frog dispersal mainly occurs within this range, whereas species abundances are largely independent at greater distances because species are randomly distributed within the study area (Diniz-Filho et al. 2010). This study thus demonstrates one way in which spatial autocorrelation analysis can be used to disentangle the relative contributions of spatial versus environmental variation to species distribution patterns, which can help determine whether intrinsic processes (e.g. dispersal limitation) or habitat heterogeneity are largely responsible for observed spatial structure.

Another common coefficient of spatial autocorrelation is **Geary's** c (Geary 1954). Although Geary's c tends to be inversely related to Moran's I, it is not identical. The calculation of spatial variance in Geary's c is based on the sum of the squared differences $\left[\sum_{i=1}^{n} \sum_{j=i}^{n} (x_i - x_j)^2 \right]$ between points (x_i and x_j) at various distances, d, divided by the standard deviation of that variable across the landscape (s_x). It thus behaves like a distance measure, and takes on very different values than Moran's I (Fortin & Dale 2005):

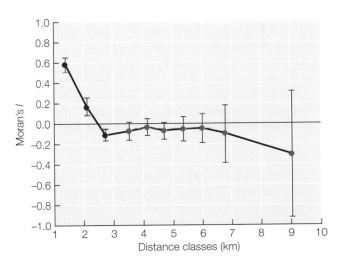

Figure 4.42 Spatial autocorrelation with Moran's I. The abundance patterns for 28 frog species in the Brazilian Amazon are spatially structured. Abundances for all species are positively autocorrelated up to about 2.5 km, after which they are largely independent, becoming more variable as the distance between sites increases (>6 km). Symbols are the mean Moran's I based on the abundances of 28 species; red symbols are significantly different from random ($I = 0$), given that their error bars do not overlap 0.

Source: After Diniz-Filho et al. (2011).

$c \to 0$ strong positive correlation

$c = 1$ spatial independence (complete spatial randomness)

$c \to 2$ strong negative correlation

Because it is not based on differences from the global mean $(\bar{x})$ like Moran's I, the Geary's c coefficient might be expected to be more sensitive to local variation in spatial autocorrelation. For example, the presence of data outliers (either very high or low values) can bias the global mean, which in turn could lead to the under- or overestimation of the degree of spatial autocorrelation under Moran's I (Fortin & Dale 2005). Geary's c is not immune to the effects of outliers, however, given that differences between points are squared. Moran's I is more commonly used in ecology because its values and interpretation are similar to those of the Pearson correlation coefficient, r (Fortin & Dale 2005).

Besides the detection and description of spatial structure, autocorrelation coefficients can be compared to their expected value to test for spatial independence (i.e. complete spatial randomness: Moran's $I = 0$; Geary's $c = 1$). However, if we are interested in testing for significant departures from randomness at each distance class (d), we need to adjust the probability level (α) of our test, given that we are performing multiple significance tests on the same data. The most widely used adjustment is the Bonferroni correction, in which α is divided by the number of distance classes (d) to give a more conservative test of significance (Fortin & Dale 2005). For example, if we initially set the significance level of our test to $\alpha = 0.05$, but we are testing for the significance of departures from a random distribution at nine distance classes, then our adjusted $\alpha = 0.05/9 = 0.006$. Thus, we would reject the null hypothesis for a given distance class only if there is a <0.6% probability that the observed autocorrelation coefficient could have come from a random distribution, as opposed to a <5% probability prior to applying the Bonferroni correction. We are thus making it more difficult to reject the null hypothesis, so as to guard against the possibility of making an incorrect inference.[51] Alternatively, a progressive Bonferroni correction can be applied, in which the test for significance becomes increasingly more stringent the more tests performed (Legendre & Legendre 2012; Dale & Fortin 2014). That is, the first test (for distance class 1) would be at the $\alpha = 0.05$ level, the second (distance class 2) would be at the $\alpha = 0.05/2 = 0.025$ level, the third

(distance class 3) at the $\alpha = 0.05/3 = 0.017$, and so on, up to the total number of tests (distance classes).

INTERPOLATION OF SPATIAL DATA: VARIOGRAMS AND KRIGING   The field of geostatistics has given rise to a variety of statistical methods for the estimation of spatial data based on **interpolation** techniques; that is, data estimation given an incomplete sampling of the landscape. In this case, however, the spatial variance of a variable is estimated by the **semivariance** function, $\hat{\gamma}(h)$, which is a distance-based function that is similar in form to Geary's c (i.e. the calculation of spatial variance is based on the sum of squared pairwise differences). Semivariance is obtained by dividing the spatial variance by 2 (hence, semi- or 'half' the variance)[52] and not by the standard deviation as in Geary's c. As with autocorrelation, we can plot the semivariance as a function of distance (h)[53] to create a **semivariogram** (or simply, **variogram**) to explore how spatial variance changes as a function of scale (**Figure 4.43**).

Among the parameters that can be obtained from the variogram are the nugget, range, and sill (**Figure 4.43**). The **nugget** is the value of the semivariance at $h = 0$ (i.e. at the y-intercept). Although this should be zero in theory, the intercept is typically greater than zero in practice, owing to the presence of fine-scale heterogeneity that we have not measured (our sampling grain is too large) or because of sam-

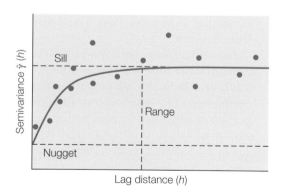

Figure 4.43 Semivariogram.   An idealized semivariogram, with the locations of the nugget, sill, and range identified.

Source: After Karl & Maurer (2010).

---

[51]This is referred to as a Type I error in statistics, and α is thus the Type I error rate. Setting the significance level of a statistical test at $\alpha = 0.05$ means that we are willing to accept a 5% probability of incorrectly rejecting the null hypothesis (i.e. rejecting the null hypothesis when it is in fact true).

[52]The autocovariance is divided by 2 because each pair of points is considered twice in the calculation of the semivariance function (Fortin & Dale 2005). In any case, semivariance is a bit of a misnomer. The term was originally applied to the variance of pairwise differences between points, but in a sense we only want 'half' of this; that is, 'the variance per point when the points are considered in pairs' (Webster & Oliver 2007: 54). What we are really calculating is the variance of a measured variable at a given lag distance, h (Bachmaier & Backes 2008). Thus, variance and variogram are the more apt terms.

[53]In geostatistics, the lag distance is denoted as h, rather than d as in spatial statistics (Fortin & Dale 2005). The notation h is thus used here for consistency with the literature.

pling or measurement errors resulting from differences among observers (observer bias), instrumentation errors, and so forth (Fortin & Dale 2005). The **range** is the distance (*h*) over which the measured variable is spatially autocorrelated. The range is identified by where the semivariance values start to level off (i.e. the sill). The **sill** refers to the part of the variogram where the semivariance levels off and forms a plateau. In this domain, distance no longer affects the spatial structure of the data (i.e. the data are spatially independent). Not all spatial patterns have a sill, however.

Although we could fit a line to the empirical data by eye to obtain these parameters, it is advisable to use a model to obtain the best fit. This **theoretical variogram** could be modeled as a simple linear or exponential relationship, neither of which have sills. In landscape ecology, however, we are interested in capturing the sill because it helps to determine the range of spatial autocorrelation in our data. In that case, a bounded theoretical variogram, in which the semivariance levels off (thereby capturing the sill), is preferred. There are many different types of bounded variograms, but the most common are the spherical model (which looks like the idealized semivariogram in **Figure 4.43**), the Gaussian model (a sigmoidal or 'S-shaped' curve), and the linear + sill model, which is a type of nested model in which the semivariance increases linearly up to the sill but then levels off (Fortin & Dale 2005). Which model ultimately provides the best fit to the empirical data might again be determined by eye, but the model goodness-of-fit is more usually evaluated using generalized least-squares and maximum likelihood (e.g. Akaike information criterion, AIC) methods nowadays (Fortin & Dale 2005). Given that spatial dependence is greatest at short distances (i.e. up to the range), we should ensure that the model fit is a good one within at least this region of the semivariogram. The reliability of $\hat{\gamma}(h)$ also decreases as a function of distance because there are fewer pairs of points at those longer lag distances. Ideally, there should be at least 30–50 pairs of points in each lag distance class, so as to increase the statistical reliability of the estimate (Rossi et al. 1992).

One of the reasons for deriving the model fit of the semivariogram analytically is so that we can predict values at unsampled locations on our landscape using kriging. **Kriging**[54] is a type of spatial interpolation method, which uses parameters obtained from the theoretical semivariogram (i.e. the nugget, range, and sill). We previously discussed interpolation functions under the data manipulation and analysis subsystem of a GIS. Recall that one of the most commonly used methods of interpolation is a spatial moving average (**Figure 4.22**),

as when the search neighborhood or filter is based on an inverse distance-weighting function (Equation 4.2). Kriging is basically a weighted moving average, in which the sampling points that fall within the range (estimated from the semivariogram) have greater influence than those farther away (Krige 1966). Ideally, we should have 15–20 sampling points or more within the search neighborhood, so as to obtain reasonably robust estimates of the data values being interpolated (Fortin & Dale 2005). There are several different types of kriging, depending on whether interpolation is performed for specific locations on the landscape (punctual kriging) or for an area (blocked kriging). If the mean of the variable being interpolated is constant across the landscape (i.e. globally constant), then this calls for 'simple' kriging, whereas if the mean is locally constant, then 'ordinary' kriging is used. It gets more complicated from there, and those desiring a more in-depth coverage of methods should refer to a geostatistics textbook (e.g. Isaaks & Srivastava 1989; Chilès & Delfiner 2012).

For our purposes here, an illustration of how semivariance and kriging can be used in an ecological context will suffice. As we discussed previously, the soil landscape is a heterogeneous mosaic, in which the distribution of soil nutrient pools is likely to be patchy across some range of scales (e.g. **Figure 4.39**). Nitrogen availability limits primary production in most ecosystems. Although nitrogen is abundant in the atmosphere (~78% of our atmosphere is made up of nitrogen gas, N_2), it cannot be used by most organisms in its gaseous form, but can be converted to more usable forms (ammonium, NH_4^+ and nitrate, NO_3^-) through the biological action of bacteria and fungi.[55] The relative abundance of NH_4^+ to NO_3^- in soils is determined by the amount of soil organic matter, as well as factors that affect soil microbial activity, such as temperature, moisture, pH, and aeration (Britto & Kronzucker 2002).

Although plants respond in a variety of ways to resource heterogeneity, in terms of their establishment, growth, and reproductive output, they can also exert an influence on the chemical, physical, and biological properties of the soil. For example, soils that develop under tree canopies are often richer in organic matter, nutrients, and water availability, which should thus influence the distribution and availability of nitrogen across the landscape. In an oak savanna in southwestern Spain, Gallardo and colleagues (2000) compared the distribution of soil nitrogen pools beneath the tree canopy (largely made up of holm and

[54]Kriging is named for Danie G. Krige (1919–2013), a South African mining engineer who pioneered the field of geostatistics.

[55]Nitrogen fixation is the process whereby atmospheric nitrogen (N_2) is converted to ammonium (NH_4^+). Besides bacteria, legumes and a few other plants are also capable of fixing nitrogen, thanks to the presence of root nodules that contain nitrogen-fixing rhizobia bacteria. The conversion of ammonium to nitrate (NO_3^-) is known as nitrification, and is also performed by bacteria.

cork oaks, *Quercus ilex* and *Q. suber*, respectively) relative to the open grassland between trees. By fitting a spherical model to their empirical semivariogram, they demonstrated that soil ammonium levels (NH_4^+) were spatially autocorrelated up to a range of 9 m, which is approximately the width of a tree's canopy (**Figure 4.44A**). In contrast, the range of soil nitrate (NO_3^-) was only about 2.5 m, indicating a much finer scale of patchiness. Indeed, the kriged soil surface map around a single holm oak illustrates the much greater heterogeneity in the distribution of soil nitrate levels compared to ammonium levels (**Figure 4.44B**). Whereas the highest levels of soil ammonium are concentrated under the tree canopy, where soil organic matter is also highest, soil nitrate levels are largely depleted in this same zone, except near the trunk where stemflow may be responsible for locally high nitrate concentrations. Nitrate is water soluble and can travel farther in the soil, where it can then be taken

up by the tree's roots. Oak trees are thus a major source of spatial heterogeneity in the distribution of soil nitrogen within this semi-arid system, which has implications for the distribution of other plant species. For example, perennial grasses are largely found beneath the tree canopy, whereas annual grasses are typically found in the open areas between trees in this system (Gallardo et al. 2000).

ANISOTROPIC PATTERNS: DIRECTIONAL CORRELOGRAMS AND VARIOGRAMS Up to now, we have been assuming that spatial patterns are the same regardless of direction. In other words, the degree and range of spatial autocorrelation obtained from a transect along the north-south direction would be the same as that obtained from a transect along the east-west direction. Spatial patterns can vary depending on direction, however. For example, the distribution of seeds from a wind-dispersed tree may possess a strong directional

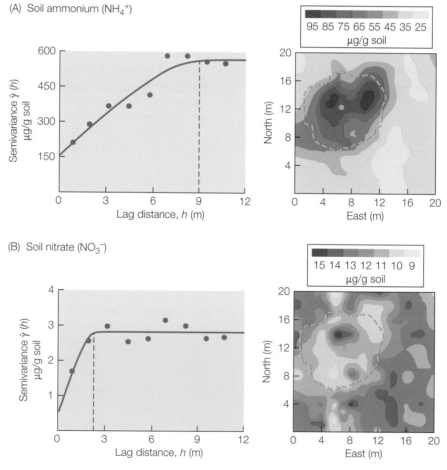

Figure 4.44 Spatial interpolation based on kriging. A spherical model was fit to the empirical semivariograms for soil ammonium (A) and soil nitrate (B) levels taken in an open oak forest in southwestern Spain. The range of spatial autocorrelation is indicated by the vertical dashed line in each semivariogram. To the right of each semivariogram is a kriged map showing the distribution of soil ammonium (top right) and soil nitrate (bottom right) relative to the canopy (dashed line) of a single holm oak (*Quercus ilex*).

Source: After Gallardo et al. (2000).

component, consistent with the direction of prevailing winds. Our ability to accurately determine and define the spatial pattern of seeds will thus depend on whether we are taking this directional component into consideration.

The statistical approaches discussed thus far have all been 'omnidirectional;' that is, the similarity between locations is assessed within a given distance, regardless of direction. This is fine for spatial patterns that are **isotropic**, which vary in a similar way in all directions, but is not well-suited to **anisotropic** patterns, which vary as a function of direction in addition to distance. To address this, we can implement spatial analyses using a 'window' or filter that compares points within a certain direction as well as distance. That is, the lag distance measure is now a vector.

To illustrate, we'll consider an application of directional variograms involving the distribution of first-year (age 0) Atlantic cod (*Gadus morhua*) in the Barents Sea, which lies north of the Arctic Circle between Norway and Russia (Ciannelli et al. 2008; **Figure 4.45**). The Barents Sea supports the largest remaining Atlantic cod fishery in the world. Although located in the Arctic, the Barents Sea is highly productive because the North Atlantic Current brings warmer waters to this region, creating phytoplankton blooms in the spring

that feed zooplankton, and in turn, young fish like the Atlantic cod and capelin (*Mallotus villosus*), an important forage fish of the adult cod. Most of the young cod appear to be concentrated in the western half of the Barents Sea, where the warm Atlantic waters enter the Sea and major spawning grounds are found along the northern Norwegian coast (**Figure 4.45A**). An omnidirectional variogram indicates that the abundance of young cod are no longer spatially autocorrelated beyond a distance of about 800 km (**Figure 4.45B**). A directional variogram, however, reveals that there is strong anisotropy to the distribution of young cod in the Barents Sea (**Figure 4.45C**). Spatial variability is greatest along the east-west (90°) and northeast-southwest (45°) axes, resulting in a spatial range of 950–1200 km in these directions as opposed to only 600–650 km in the other directions. The omnidirectional variogram thus essentially averaged out this variability in spatial structure. Whether this matters or not will depend on our questions and objectives, but accurately describing spatial pattern should always be the implicit goal of any spatial analysis.

As we conclude this section on spatial analysis, it is worth reiterating that this section of the chapter is only intended to provide an introduction to some of the

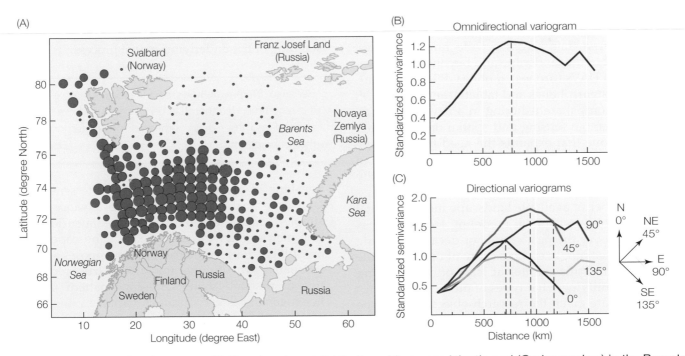

Figure 4.45 Directional variograms. (A) The abundance distribution of first-year Atlantic cod (*Gadus morhua*) in the Barents Sea above the Arctic Circle (circle size is proportional to abundance). Although the omnidirectional variogram (B) suggests a range of spatial autocorrelation up to about ~800 km, directional variograms (C) reveal strong anisotropy in the distribution. Spatial variance (standardized semivariance) is greater along the northeastern (45°) and eastern (90°) axes of the distribution, than along the northern (0°) and southeastern (135°) axes. Subsequently, the range of spatial autocorrelation varies, from ~700 km up to ~1200 km, depending on direction.

Source: After Ciannelli et al. (2008).

important concepts and common methods of spatial statistics. We will be continuing this discussion in the next chapter on landscape connectivity, and indeed, throughout the remainder of the textbook, as new spatial analytical and modeling techniques are introduced. But, just as this chapter cannot be considered a complete treatment of spatial pattern analysis by itself, the introduction of these other spatial analytical methods and models elsewhere in the book does not mean that these are only appropriate for addressing those particular applications. Part of the challenge (and fun!) of spatial analysis is to identify how different types of spatial data can best be analyzed to address a particular question or objective, irrespective of the specific research area to which it is being applied.

Future Directions (and Some Caveats)

Spatial variance of observed measures...is no longer viewed as a statistical annoyance. It is now treated as a biologically important quantity that changes value depending on the scale of measurement.
Horne & Schneider (1995)

If spatial autocorrelation is part of nature, and we are trying to understand nature, it makes little sense to claim that spatial autocorrelation in data represents some sort of bias, artefact or distortion.   Hawkins (2012)

Ecologists have come a long way since viewing spatial heterogeneity (spatial variance) as simply a nuisance parameter that must be eliminated or accounted for in the study design and statistical analysis of ecological data. Spatial variance is now recognized as an important attribute of ecological systems, making it worthy of study in its own right (Legendre 1993; Horne & Schneider 1995; Hawkins 2012). Technological advances in remote sensing, information technology, and mobile computing have made it easier (and cheaper) than ever before to collect and analyze spatially referenced data across a wide range of scales, from the plot to the landscape and beyond. In particular, the rise of UAVs (drones), mobile GPS and computing devices (smartphones and tablets), and mobile GIS applications (apps) are ushering in a new era of individualized remote sensing and spatial data collection in the field, one that expands the need for landscape and spatial analytical tools to a much wider range of ecologists outside those working at traditionally defined 'landscape scales.'

The number of available landscape metrics, spatial analytical tools, and spatial modeling techniques has proliferated over the past couple of decades, offering ecologists a seemingly endless array of options. While it is hoped that this chapter at least offers a starting point for identifying what types of metrics or analyses are appropriate given the research objective and type of spatial data available, the ready availability of software capable of performing multiple analyses with a click of the mouse or the addition of a few lines of code makes it tempting to just run everything and see what sorts of interesting results emerge at the end. Although a comparative analysis of methods is to be recommended (Dale 2000), we should save the fishing expeditions for our favorite lake or stream, and try to approach landscape and spatial analysis more thought-fully, by giving due consideration to the underlying assumptions and limitations of the various methods or models. Nor is there any reason to make the analysis more complicated than need be. Just because we can calculate dozens of landscape metrics doesn't mean that we should do so, especially when only a few metrics will generally suffice to characterize the composition and configuration of most landscapes. Start simple, and incorporate additional metrics or greater model complexity only when necessary to achieve a better description of the spatial pattern or an understanding of the underlying spatial process.

Because different methods react to different features in data, it is recommended that data be analysed by more than one method and the results compared for greater insight into their characteristics.   Dale (2000)

Speaking of greater complexity, remote sensing and landscape analysis have started to advance beyond simple thematic mapping and two-dimensional pattern metrics (Newton et al. 2009). Landscape metrics are still patch-based metrics, which imposes a particular spatial structure that may not be truly representative of the underlying landscape pattern if spatial heterogeneity is a continuous variable rather than a categorical one. Categorizing a continuous variable into patches 'discretizes' the spatial data, which may subsume internal heterogeneity that may be important for understanding the relationship between pattern and process. For that reason, three-dimensional surface analysis, or **surface metrology**, which has been developed and applied in the fields of microscopy and photogrammetry, represents the next generation of surface metrics in landscape ecology (McGarigal et al.

2009). Given that remote sensing has the potential to provide a three-dimensional characterization of landscapes (e.g. lidar), the use of surface metrics seems far more apt, and would eliminate some of the issues we've discussed throughout this chapter, such as the effect of thematic resolution on landscape analysis.

If space was once ecology's 'final frontier' (Kareiva 1994), then space-time is its newest frontier (Cressie & Wikle 2011). The idea that 'space matters' is so fundamental and deeply ingrained in landscape ecology that the analysis of spatial pattern has largely overshadowed the analysis of temporal patterns, even though the study of disturbance dynamics and landscape change are both major areas of research (**Chapter 3**). The analysis of landscape change—how landscape patterns change over time—is generally carried out as a comparative analysis of the landscape at different points in time (i.e. a chronosequence), using any of the various landscape metrics or spatial statistics discussed in this chapter. Thus, the analysis of landscape change is a static one, performed individually on each landscape in the time-series. Increasingly, however, the emphasis is on developing spatiotemporal statistical approaches that describe temporal variability in space *dynamically* (e.g. Cressie & Wikle 2011). We have entered the big-data era, in which large spatiotemporal datasets are being amassed, archived, and made available through online repositories, including a number of citizen-science sites. Thus, in addition to the availability of high-resolution imagery capable of providing information on vegetation condition on a weekly (or even daily) basis, we now have large amounts of data accessible online for a variety of spatiotemporal phenomena such as the movements of individual animals (e.g. Movebank), the abundance and distribution of species (e.g. ebird), and the incidence and spread of invasive species or disease (Bonney et al. 2009; Kays et al. 2015; Meentemeyer et al. 2015). Although methods for the analysis of these sorts of spatiotemporal phenomena will be discussed in **Chapters 6–9**, the opportunities and challenges of analyzing large spatiotemporal datasets will only increase in the future.

At this point in time, pattern description is probably still less of a challenge than understanding the process(es) that give rise to the pattern. The relationship between pattern and process is rarely clearcut. Different processes can give rise to similar patterns, and the same process may generate different patterns in different landscape contexts. In addition, processes may interact and create a 'mosaic of intermingled and confounded spatial patterns' (Fortin & Dale 2005, p. 4), which in turn affect the intensity and types of ecological processes that operate upon the landscape. Although the feedbacks between pattern and process can be difficult to distinguish, untangling these sorts of pattern-process relationships represents one of the grand challenges of landscape ecology, and indeed, of ecology more generally. The rest of the book is devoted to the investigation of these pattern-process relationships.

Chapter Summary Points

1. The description of landscape pattern, and of spatial structure more generally, is a requisite for understanding the relationship between pattern and process in ecology. Landscape pattern analysis is thus a major focus of landscape ecology.

2. At the traditionally defined 'landscape scale,' historical land surveys and remote sensing provide the major sources of landscape data, but ecological data can be collected, mapped, and analyzed as 'landscapes' across a range of spatial scales.

3. Historical land surveys are often the earliest records of the pre-settlement landscape. In the United States, the Public Land Survey System was a systematic survey that began in 1785, and provides an unusually comprehensive survey of tree species ('witness trees' that marked survey boundary points) present in the pre-European landscape. The US Public Land Survey record is thus often used to reconstruct the historical landscape, and to provide a basis for comparison with the current forest composition.

4. Remote sensing simply means to gather information from a distance. From a landscape data perspective, remote sensing includes the use of aerial photography, airborne lidar, synthetic aperture radar (SAR), satellite imagery, and unmanned aerial vehicles (drones). Remarkably, it took only about a century to go from the first aerial photograph, taken from a tethered hot-air balloon, to the first satellite images of Earth's surface.

5. The longest-running program for the acquisition of satellite imagery (45+ years) is the Landsat program. There have been a total of eight Landsat missions, the most recent of which launched in February 2013; a ninth mission is planned for December 2020. Millions of images have been taken and are now freely available online. Other satellite programs complement the Landsat program by offering either more frequent (MODIS) or higher-resolution imaging (e.g. SPOT and other commercial vendors).

6. Spatial data can be mapped using either a vector data model (as points, lines, or polygons) or as a raster data model (as a grid or lattice). The vector data format is used for depicting discrete structures or features (e.g. individual trees, administrative boundaries, roads, agricultural fields, or forest management units). The raster data format is generally preferred for representing continuous surfaces, in which grid cells (pixels) contain specific numeric or categorical values (e.g. land-cover types).

7. A geographical information system (GIS) is designed to accept large volumes of spatial information from a variety of sources, and to efficiently store, retrieve, manipulate, analyze, and display those data. A GIS is made up of a number of subsystems, including the data input subsystem (data files and relational databases); the data processing and database management system (e.g. reprojection, registration, or reclassification of images); the data manipulation and analysis subsystem (e.g. area and distance functions, neighborhood functions, overlay functions, and interpolation functions); and the data output subsystem, for visualizing, mapping, and displaying the results of the spatial analyses or simulations.

8. The analysis of categorical landscape data involves the use of landscape metrics. Although there are hundreds of landscape metrics, they can be divided into two classes, in terms of whether they measure landscape composition or landscape configuration.

9. Measures of landscape configuration are based on patch-definition rules (neighborhood rules), and may be applied at the scale of the individual patch (e.g. patch area, patch perimeter, or patch shape); to the entire land-cover class (e.g. number of patches, mean patch size, aggregation index); or, to the entire landscape (landscape contagion). By contrast, landscape composition metrics (proportion of land covers, landscape richness, or diversity) are always assessed at the scale of the entire landscape.

10. Landscape metrics are sensitive to a number of factors related to the modifiable areal unit problem: the 'units' (patches) that make up the landscape can be modified by both the spatial resolution (grain) and thematic resolution of the image or dataset. In that case, a multiscale analysis of the landscape metrics can uncover the nature of these scaling relationships and help to facilitate comparisons among landscapes.

11. Landscape pattern analysis is meant to be the first step in uncovering the relationship between pattern and process. Not all metrics are equally good at the task, however, and there is no guarantee that a given metric will have relevance for the ecological process of interest. The goal is obviously to adopt ecologically relevant landscape metrics; that is, landscape metrics that assess functional properties of the landscape.

12. Spatial analysis refers to a broad class of methods in spatial statistics and geostatistics that have the primary aims of detecting and describing spatial structure. Many of these methods also test whether the observed pattern departs from complete spatial randomness, as well as the scale(s) at which such departures occur.

13. Spatial point patterns are location-based data (xy coordinates only), although locations may also possess attributes ('marks') as to their identity, intensity, or magnitude (xy and z data). Different spatial statistics are indicated, depending on the type of spatial point data and sampling design. Some analyses require exhaustive sampling of the point pattern or that samples be contiguous, for example.

14. Spatial autocorrelation is a ubiquitous feature of landscapes and spatial ecological patterns. Although spatial autocorrelation is expected to decline with distance, in keeping with Tobler's first law of geography (i.e. near locations tend to be more closely related than locations farther apart), the reality is often more complicated, depending on how spatial pattern changes as a function of scale. Thus, spatial correlograms and semivariograms, in which the degree of spatial autocorrelation or semivariance, respectively, is plotted as a function of lag distance, can reveal the scale(s) at which significant spatial structure occurs.

15. Kriging, derived from the field of geostatistics and based on the semivariogram, is used in spatial interpolation and surface mapping. It thus illustrates how spatial analysis can also be used for data estimation in cases where sampling of the landscape is incomplete.

Discussion Questions

1. Technological advances in remote sensing, geographic information systems, and computer processing have all contributed to the study and analysis of landscapes at broad spatial scales. Little wonder, then, that some perceive landscape ecology as essentially synonymous with remote sensing and GIS. Should we define a discipline by its tools? Put another way, can landscape ecology be done in the absence of such tools?

2. Human land-use, such as that involving row-crop agriculture, creates strong geometric patterns on the landscape. With reference to the aerial photographs depicted in **Figure 4.6**, what can we infer about the agricultural practices in this region? For example, what information can we glean from the size, shape, or layout of fields? Are both of these images really necessary? What information can be obtained from the color infrared image that cannot be obtained from the black-and-white image, or vice versa?

3. How many Landsat scenes would be required to cover the state, province, or country in which you live (e.g. **Figure 4.12**)? Give the orbital path and row numbers of these scenes (i.e. the WRS index).

4. What spatial data layers might you use to represent the landscape in which you live or work (**Figure 4.16**)? Explain whether you would use vector or raster data structures (**Figure 4.15**) for a given layer, and then see whether you can find what remotely sensed images or products are actually available for a given type of spatial data layer (i.e. identify the specific agency or source).

5. If we can't obtain landscape or spatial data that possess a high degree of accuracy and precision, then should we prioritize data that are at least accurate (even if not precise) or data that have high precision (even if not very accurate)? Which sort of error will have the greatest effect on the analysis of landscape pattern? Which sort of error might be easiest to correct prior to analysis?

6. What landscape metrics would you use to assess the degree of habitat fragmentation within a landscape? Be sure to identify whether the metrics you select are measures of landscape composition or configuration, and explain what values of these metrics would enable you to determine that the landscape is fragmented. Can landscape fragmentation be assessed independently, or is it a relative measure that can only be evaluated in comparison to another landscape or relative to a particular ecological response?

7. Changing the spatial grain, spatial extent, or thematic resolution of a landscape can all affect the analysis of landscape pattern. Discuss how changing each of these would likely influence the following landscape metrics, and explain your reasoning. Be sure to consider how the degree of landscape richness (R) or fragmentation might alter your expectations.

   a) Proportion of habitat (p_i, PLAND)

   b) Number of patches (NP)

   c) Mean patch size ($\bar{A}$, Area_MN)

   d) Total edge (TE)

   e) Clumpiness index (CI, CLUMPY)

   f) Mean patch isolation ($\overline{ENN}$)

   g) Shannon's evenness index (E, SHEI)

   h) Landscape contagion (LC, CONTAG)

8. Identify a spatial point pattern relevant to your study system, or in an example provided by your instructor. How would you characterize this point pattern: as coordinates only or coordinates with attributes? How would you go about sampling this pattern (e.g. exhaustive survey, systematic survey, or random or stratified random sampling)? Based on the nature of the point pattern data and your sampling design, what sorts of spatial analyses would you suggest to assess whether the data are spatially structured, and if so, whether the pattern exhibits patchiness at one or more scales?

Landscape Connectivity

5

Habitat corridors seem such an obvious way to restore connectivity among populations that have become fragmented by habitat loss. Species unable to cross the intervening land-use matrix might at least be willing to disperse along habitat corridors through that matrix to reach other habitat fragments. Similarly, populations fragmented by other types of anthropogenic barriers (e.g. dams, roads, fences) might also be reconnected through the construction of wildlife-crossing structures, such as a landscaped overpass built above a busy highway (**Figure 5.1**). Whether or not such corridors 'work' in restoring population or landscape connectivity, however, depends on the species and landscape in question, as well as the type of corridor involved (Beier & Noss 1998).

By definition, habitat corridors are long, relatively narrow strips of habitat, which means they possess a disproportionate amount of edge relative to the habitat area they contain (i.e. corridors have a high perimeter-to-area ratio; Rosenberg et al. 1997). Corridors might therefore increase predation risk for dispersing organisms, assuming dispersers are able to locate and utilize these as movement corridors in the first place. Clearly, corridors will not be effective at restoring connectivity if they are not used or do not result in safe passage. In other cases, corridors may be too effective, especially if they facilitate the spread of non-native species (Resasco et al. 2014) or diseases (Hess 1994). In such cases, corridors could have detrimental effects on the very species they are supposed to benefit.

How can we gauge how well corridors actually function in preserving or establishing connectivity among populations? Should we try to reconnect all populations, or perhaps just the largest or most remote? Is there some optimum level of connectivity? Perhaps we can dispense with corridors altogether if individuals are at least willing to traverse certain types of habitats in the landscape matrix. In that case, maybe we could promote or restore connectivity by managing the matrix? Then again, perhaps it would be better if we took a more proactive approach and prevented populations from becoming isolated in the first place. But, at what level of habitat loss does connectivity become disrupted on the landscape? In order to address these issues, we turn our attention in this chapter to the important concept of landscape connectivity.

What is Landscape Connectivity and Why is it Important?

The concept of landscape connectivity was first introduced by Canadian landscape ecologist, Gray Merriam, who emphasized the importance of the interaction between behavior and landscape structure for understanding a species' movements among habitat patches (Merriam 1984). Subsequently, landscape connectivity has been defined as the 'degree to which the landscape facilitates or impedes movement among resource patches' (Taylor et al. 1993, p. 571) and as 'the functional relationship among habitat patches, owing to the spatial contagion of habitat and the

Essentials of Landscape Ecology. Kimberly A. With, Oxford University Press (2019).
© Kimberly A. With 2019. DOI: 10.1093/oso/9780198838388.001.0001

Figure 5.1 Wildlife corridor.   A landscaped overpass facilitates connectivity by providing wildlife with a safer means of crossing a four-lane highway.

Source: Skyward Kick Productions/ Shutterstock.com

movement responses of organisms to landscape structure' (With et al. 1997, p. 151). Both of these definitions highlight the functional nature of connectivity, by emphasizing the interaction between movement and landscape structure.

Although the original definition of landscape connectivity dealt with the movement of organisms, the concept has since been extended to other types of movement or flows across the landscape. Landscape connectivity is important for many ecological and biophysical processes, including those responsible for shaping landscape patterns, as well as those that are themselves shaped by landscape patterns. Processes as diverse as dispersal; gene flow; the flow of water, materials, or nutrients; the spread of an invasive species, disease, or pest; and even the spread of a disturbance, such as fire, are all potentially influenced by the connectivity of habitat or resources across the landscape. For this reason, landscape connectivity has been called 'a vital element of landscape structure' (Taylor et al. 1993). Because landscape connectivity provides a measure of how landscape pattern affects ecological processes, it might also be said to be vital to the study of landscape ecology itself.

Given the importance of landscape connectivity for ecological flows, the disruption of connectivity may have some important ecological consequences. For example, the wholesale loss and fragmentation of natural habitats are disrupting landscape connectivity and thus the function of many ecological systems, including aquatic and marine ones. Dispersal among habitat patches is an important determinant of recolonization and patch occupancy rates, population size, and genetic diversity, all of which contribute to population viability and the extinction probability of a species. The maintenance or restoration of connectivity is an important issue in reserve design, land management, and land-use planning, especially in the debate over corridor efficacy and matrix management. The ecological consequences of connectivity are therefore explored throughout this book.

How best to define or measure landscape connectivity? Unfortunately, there is no single definition or measure that is best in all cases. In this chapter, we will review the various ways in which connectivity is defined and examine how it can be measured or assessed. We begin with a discussion of what landscape elements contribute to connectivity, as well as the distinction between structural and functional connectivity, and between patch and landscape connectivity. We then turn our attention to different approaches for assessing landscape connectivity in terrestrial, as well as aquatic systems. We conclude with a discussion of how these ideas on connectivity can be extended to multiple scales or levels of ecological organization, followed by some final thoughts on whether landscape connectivity should be treated as an independent or dependent variable in ecological investigations.

Elements of Landscape Connectivity

What elements of the landscape confer connectivity? It depends on the type of landscape we have in mind. If landscapes are viewed simply as patches embedded within an ecologically neutral matrix, then we might first evaluate connectivity in terms of whether habitat patches are generally large or small. If patches are large, then there is at least some degree of habitat connectivity within those patches (habitat contiguity, C_p; Chapter 4), but it is also possible that a patch might be so large as to span the entire landscape, thereby providing habitat connectivity at the landscape scale.

If patches are small, then we are more concerned about how many patches there are and their relative proximity to one another. If patches are few and far between, then we might assess connectivity in terms of what linkages exist between them. Patches linked by habitat corridors are obviously connected, at least in a structural sense (**Figure 5.2A**). If habitat corridors occur among some patches but not all, we might then wonder what level of connectedness is necessary or optimal for preserving overall connectivity of the landscape. Perhaps some types of corridors are better or more effective than others? For example, certain features of the landscape may naturally function as corridors, as in the case of streams, rivers, and their associated riparian areas; mountain passes and longitudinal valleys; and, along the habitat boundaries that define ecotones. In other cases, however, corridors are specifically created to restore or maintain connectivity, such as through the preservation of habitat strips and greenways; the installation of fish ladders around locks and dams; and the construction of wildlife-crossing structures over or under major highways (**Figure 5.1**). Habitat corridors may also be created unintentionally, incidental to human land-use activities, such as the planting of tree windbreaks and hedgerows along property lines; clearings for utility right-of-ways; and the construction of roads, trails, and canals for transportation. Clearly, we need some way to quantify the quality or strength of connections among patches, especially if flows tend to occur primarily in one direc-

tion, as well as some way of evaluating the overall connectivity of the landscape.

Corridors are an obvious way to create or restore connectivity, but they are not the only way. It may be that we are concerned about maintaining landscape connectivity for a species of conservation concern in the face of the ongoing loss and fragmentation of its breeding habitat. If the species has sufficient dispersal and habitat-detection abilities to cross at least small habitat gaps, then corridors may not be needed if habitat patches are sufficiently close together and fall within the dispersal range of the species. Even if the inter-patch distances of breeding habitat exceed the gap-crossing abilities of the species, it might still be possible to restore connectivity by creating **stepping stones**: small patches that are strategically placed within the dispersal range of the species, for the sole purpose of encouraging movement across larger habitat gaps (**Figure 5.2B**). However, this requires that we have some basic understanding of a species' movement or dispersal behavior, in terms of how far it moves, how likely it is to cross habitat gaps of a certain size, its ability to detect another habitat patch from a distance (the **perceptual range**), and how effective it is in searching and finding stepping-stone habitat within the matrix.

Rarely are landscapes managed for a single species, however. Another species may have no difficulty crossing even large gaps, but its willingness to do so may depend on the type of matrix it must cross. For example, a species may move readily through secondary forest, but avoid crossing open fields, whereas large rivers and major highways are complete barriers to dispersal. Thus, the species may have to undertake a more circuitous route through preferred habitats to traverse the landscape, which is not without cost, but may afford less risk in the end. To assess landscape connectivity for this species, we thus need information on the species' relative rates of movement in different habitat types, as well as its willingness to cross particular types of patch boundaries or landscape features. We also need more information on the landscape itself, such as the types and distribution of various habitats or land uses, as well as the location of potential barriers or movement corridors.

If we define movement corridors as those areas of the landscape that facilitate movement or dispersal, we should be aware that such areas may not be a distinct or obvious feature of the landscape. Instead, movement corridors may emerge from the juxtaposition of certain habitat types or land uses that end up 'funneling' individuals across the landscape (e.g. agricultural fields adjacent to secondary forest as opposed to suburban development; Gustafson & Gardner 1996). We thus need to know how different cover types vary

(A)   (B)   (C)

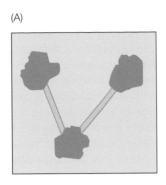
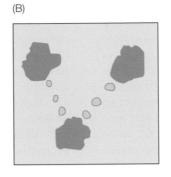
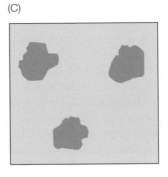

Figure 5.2 Elements of landscape connectivity.   Habitat corridors are an obvious way to maintain or restore connectivity on a landscape (A), but habitat stepping stones (B) can be just as effective for species that are able to cross short distances across the matrix. Because corridors and stepping stones are meant to promote dispersal, these habitats need not be of high quality, so as to discourage organisms from settling within them. Alternatively, managing the matrix (C) to ensure that the intervening habitat or habitats are permeable to dispersing organisms may also help to promote landscape connectivity, especially in heterogeneous landscapes where dispersal may be more likely to occur within some matrix habitats than others.

in their resistance to movement to evaluate whether the landscape is connected for the species.

Clearly, the assessment of landscape connectivity is more difficult in a heterogeneous landscape, especially given that the same landscape may be connected for some species and processes, but not others. In other words, whether a landscape is connected or not depends upon the species (or process) in question. Landscapes are not inherently connected or disconnected; connectivity is ultimately determined by the scale with which the species or process interacts with landscape structure. In addition, landscape connectivity is not defined solely by structural measures or features (e.g. the presence of habitat corridors), but ideally by the functional relationship between landscape pattern and process (e.g. dispersal or gene flow). We therefore consider this distinction between structural and functional connectivity next.

Structural versus Functional Connectivity

Structural connectivity refers to the adjacency of habitat patches (spatial contagion) or the existence of actual linkages among patches (i.e. habitat corridors). Structural connectivity is therefore fairly easy to assess, especially given the widespread availability of spatial analysis tools in GIS and landscape analytical software (**Chapter 4**). Although these metrics may provide an initial assessment of connectivity—a first approximation of the potential connectivity of the landscape—structural measures of connectivity may not be related to function. For example, Schumaker (1996) found that nine commonly used indices of landscape pattern (e.g. number of patches, patch area, nearest-neighbor distance, contagion, mean perimeter-area ratios) were

only weakly correlated with dispersal success in real and simulated landscapes (the latter with random habitat distributions).

In contrast, **functional connectivity** incorporates the response of the process or organism to landscape structure. This can be subdivided further into *potential* versus *actual connectivity* (Calabrese & Fagan 2004). For **potential connectivity**, we can use some very basic information on dispersal or movement, such as the average or maximum dispersal distance, to assess connectivity. To measure **actual connectivity**, more detailed movement data are required, such as the response of the organism to habitat edges or its relative rate of movement in different habitat types. Bélisle (2005) suggests equating functional connectivity with *travel costs*, which is what resistance-surface models, circuit theory, and some applications of graph theory that incorporate weighted edges as part of a least-cost path analysis attempt to do, and which will be covered later in this chapter.

> We should abandon the common belief that each landscape is associated with a certain connectivity value. It is not. Connectivity has two dimensions: landscape and the organism considered. Only a combination of these two will yield a meaningful value of connectivity.
>
> Kindlmann & Burel (2008)

The primary difference between potential and actual connectivity lies in the amount of information available on the response of the organism or process to landscape pattern. Although assessing the actual connectivity of the landscape might be the goal, we may not have sufficient empirical information on how landscape structure influences movement behavior or ecological flows across the landscape to permit this level of assessment, or to test our hypotheses as to how landscape connectivity (or rather, the disruption of landscape connectivity) might influence dispersal or other processes across the landscape. Although significant progress in the study of behavioral responses to landscape structure is being made within the area of **behavioral landscape ecology** (Chapter 6), most analyses of landscape connectivity assess the structural or potential connectivity of the landscape, rather than of the actual or realized connectivity. Whether or not this is a problem depends on the goals and objectives of the analysis, and how closely structural (or potential) connectivity relate to the actual or realized connectivity of the landscape.

Structural and functional connectivity can be synonymous if, for example, a species is confined to move through preferred habitat (it lacks gap-crossing abilities and does not move through the matrix) but there is sufficient habitat available to permit it to move freely about the landscape. We should not assume that structurally connected landscapes are necessarily functionally connected, however. Habitat corridors provide structural connectivity, but may not promote functional connectivity if they are not used by the target organisms or their dispersal success is low. As Tischendorf and Fahrig (2000a) point out, corridors might be a component of landscape connectivity, but they do not determine its connectivity.

Corridors in a landscape may therefore be a component of its connectivity if they promote movement among habitat patches, but they do not determine its connectivity. The degree to which corridors contribute to landscape connectivity depends on the nature of the corridors, the nature of the matrix and the response of the organism to both.

Tischendorf & Fahrig (2000a)

Given the importance of landscape connectivity for ecological flows, a high level of functional connectivity is generally considered a desirable property of landscapes, something to be managed and enhanced when possible. This is understandable given that habitat loss and fragmentation caused by human land-use activities typically result in a disruption of landscape connectivity, and thus restoring connectivity is generally the goal. High connectivity is not always desirable, however. For example, a highly connected landscape may permit the rampant spread of a destructive disturbance, such as fire, a noxious pest, or a virulent disease (Hess 1994; Resasco et al. 2014).

Furthermore, management recommendations to promote high connectivity could actually jeopardize conservation efforts if based solely on identifying those habitats or landscape elements in which individuals tend to move most quickly (i.e. with little resistance). For example, individuals might move quickly through an area that it perceives to be risky (e.g. high predation risk) or because it offers few resources (i.e. low-quality habitat). An increase in these sorts of habitats, ostensibly to promote connectivity, could have a detrimental effect on species management. More rapid movements through less-desirable habitats could result in excessive dispersal and low philopatry, causing individuals to suffer lower breeding success and/or higher mortality (Bélisle 2005).

The pertinent question in managing for landscape connectivity should thus be: 'How much movement is necessary to maintain a given process?' The answer to that question is not a simple one, and obviously depends upon the process in question. This represents the single most important challenge in quantifying and managing for landscape connectivity.

Patch Connectivity versus Landscape Connectivity

Although landscape ecologists view connectivity as a property of the entire landscape, connectivity is viewed as an attribute of the habitat patch in other areas of ecology, most notably in metapopulation ecology (Tischendorf & Fahrig 2000b, 2001; Moilanen & Hanski 2001). Although our primary focus in this chapter is on landscape connectivity, we begin with an overview of patch-connectivity measures and their applications, so as to provide a basis for evaluating the different scales at which connectivity can be assessed.

The notion that habitat patches are connected via dispersal is a key feature of both island biogeography and metapopulation theory. The number of species on an island or habitat patch represents a trade-off between colonization and extinction rates. Although extinction is assumed to be inversely related to island or patch area, colonization rates are affected by the degree of isolation (distance) from the mainland or source patch. Patch isolation is thus the inverse of connectivity. Isolated islands or patches receive few to no immigrants (colonists) and are therefore more vulnerable to local extinctions. Patch-connectivity measures appear prominently in the metapopulation literature and in studies of habitat fragmentation effects, and are therefore discussed next.

Patch-Based Connectivity Measures

There are three main classes of patch-connectivity metrics: (1) nearest-neighbor distance, (2) neighborhood

(A)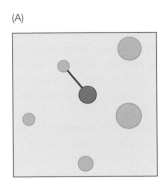 (B)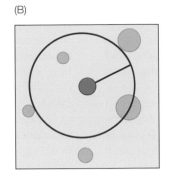 (C)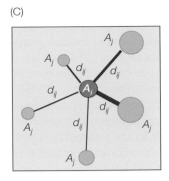

Figure 5.3 Patch-based connectivity measures. (A) The nearest-neighbor measure is simply the distance to the nearest patch, or ideally, the nearest occupied patch. (B) The neighborhood (buffer) measure includes all patches within a certain radius of the focal patch (two patches here). Ideally, the buffer distance is set to the maximum dispersal range of the species. (C) The incidence-function model measure weights the contribution of each patch to the focal patch (A_i), based on their individual areas (A_j) and distances (d_{ij}). Large, near patches are thus more likely to be a source of immigrants to the focal patch than small patches located farther away.

(buffer) measures, and (3) the incidence-function model measure (a distance-weighted, area-based measure). Other patch-based measures derived from graph theory are discussed a bit later in the chapter.

NEAREST-NEIGHBOR DISTANCE Of the patch connectivity measures, the simplest and most commonly used is the **Euclidean nearest-neighbor (ENN) distance**, which was presented in **Chapter 4**, and gives the straight-line distance from the focal patch to the nearest patch (ideally, an *occupied* patch; see Prugh 2009; **Figure 5.3A**). The nearest-neighbor distance contains very little information about the patch network and no information about the dispersal abilities of the species. It does not consider patches beyond the nearest one, but since outlying patches might also influence patch occupancy or population size, it is perhaps not surprising that ENN is generally a poor predictor of colonization patterns (Moilanen & Nieminen 2002). Also, because inter-patch distances are usually calculated from patch centers, the sizes and shapes of patches can strongly influence apparent connectivity. For example, two large patches may appear far apart (i.e. isolated) when measured from center to center, and yet their edges may in fact be quite close together. One solution is thus to calculate patch distances from edge to edge. Another suggestion is to use a neighborhood buffer measure, which sums the amount of habitat within a given radius of the focal patch, and is discussed next.

NEIGHBORHOOD (BUFFER) MEASURES Neighborhood measures are easy to implement within a GIS by setting a 'buffer' around patches, and calculating the amount of habitat within the buffer (a fixed radius or neighborhood; **Figure 4.18**, **Figure 5.3B**). From a biological perspective, this approach makes sense because it provides a direct measure of how much habitat or

potential sources of immigrants are in close proximity to a patch. However, one must parameterize this metric by selecting a threshold distance that determines how large the buffer will be, and thus buffer measures are sensitive to the size of the radius selected (Moilanen & Nieminen 2002). Ideally, the buffer width should reflect the maximum dispersal distance of the organism, but is sometimes set arbitrarily (Bender et al. 2003b).

INCIDENCE-FUNCTION MODEL MEASURE This measure is basically a distance-weighted, area-based measure that assesses the area and distance to all potential source patches (**Figure 5.3C**). It can take a number of forms, but we'll consider here the measure obtained from the incidence-function model (IFM) originally proposed by Finnish ecologist, Ilkka Hanski (1994), and which is generally presented as:

$$S_i - A_i^c \sum_{j \neq i} D(d_{ij}, \alpha) A_j^b \qquad \text{Equation 5.1}$$

where S_i is the connectivity of patch i; A_i is the area of patch i (the target patch); A_j is the area of patch j (the source patch); the exponents b and c scale the emigration and immigration rates to the size of the source and target patches, respectively; $D(d_{ij}, \alpha)$ is the dispersal kernel, which describes how the species' probability of dispersal decays with distance, where d_{ij} is the distance between patches i and j, and α is a species-specific parameter (or vector of several parameters) describing the dispersal ability of the species (Moilanen & Hanski 2001). The IFM measure thus sums the potential contribution of all patches, weighted by their area and distance, to the connectivity of the focal patch.

Because it includes information on dispersal ability, the IFM measure is more biologically realistic and thus was found to perform better than the ENN and buffer measures (Moilanen & Nieminen 2002). This is further supported by a simulation model analysis, in which area-based metrics in general (such as the buffer and IFM measures) were better predictors of immigration over ENN measures (Bender et al. 2003b). However, in a survey of two-dozen metapopulations representing a wide array of taxa (arthropods, amphibians, and reptiles), the distance to the nearest *occupied* patch was as good a predictor of colonization and patch occupancy as the best buffer and IFM measure (Prugh 2009).

Greater realism comes at a cost. The IFM measure requires more information to parameterize than either the ENN or buffer measures. Increased biological realism is a potential drawback if parameter estimates are not readily available (i.e. for a, b, and c). Although the performance of the IFM measure is relatively insensitive to the scaling constants b and c (Moilanen & Nieminen 2002), performance was strongly affected by a (Prugh 2009). It is not always clear what a is, and thus, how it should be parameterized. Various definitions have been given in the literature, including: mean daily movement rates, distance-based survival rate, migration rate, colonization rate, and more generically, a 'species-specific parameter describing the dispersal ability of the species' (Moilanen & Hanksi 2001: 149), resulting in a lack of consistency in the way a, and thus $S_{i'}$ are calculated among studies (Prugh 2009).

Which Patch-Connectivity Measure to Use and When?

In practice, patch isolation—as with any measure of connectivity—can be difficult to quantify because it should ideally be based on biological information related to the dispersal range or movement abilities of the species. Different methods can produce conflicting estimates of patch connectivity, and thus choosing an appropriate measure can be difficult and must be done with care. The best patch-connectivity measure, however, is not necessarily the most complex or biologically realistic one. The definitive question is: 'How well does the patch-connectivity measure relate to function, such as dispersal or colonization success?'

Which patch-connectivity measure is 'best' may also depend on how much information is required to address the question or process of interest. It is a general modeling axiom that one should start simple and add complexity only as needed. Although ENN measures lack the biological detail of IFM measures, they may be sufficient for predicting patterns of patch occupancy and colonization for some species, especially in simple landscapes where the matrix is homogeneous. Nearest-neighbor distance measures have the advantage of being easy to compute; they require a minimum amount of spatial or biological information, and thus may permit more straightforward comparisons among

species and studies. However, if the goal is to assess the metapopulation dynamics of a species of conservation concern in a given landscape, or to evaluate the contribution of specific habitat patches to the population viability of a target species, more biologically realistic measures may ultimately be required (Prugh 2009).

None of these patch-connectivity measures may satisfactorily explain patch occupancy or colonization in all cases. For example, Winfree and her colleagues (2005) found that none of these patch-connectivity indices performed particularly well in predicting population responses (immigration, colonization, or abundance) of voles, bees, or butterflies from different landscapes. In this case, it was suggested that patch-connectivity measures may function better as measures of habitat structure than as measures of habitat connectivity from the organism's point of view (Winfree et al. 2005), which again emphasizes the important distinction between structural and functional connectivity.

The relative influence of matrix habitat on inter-patch movements, which will differ among species, likely accounts for much of the difference among studies as to whether (and which) patch-based measures are good predictors of immigration rates. If the matrix structure affects dispersal behavior or mortality, then patch-connectivity measures that ignore this may overestimate immigration rates, leading to overly optimistic estimates of patch occupancy and persistence in metapopulation models. In a simulation study of dispersal within heterogeneous landscapes, in which the contagion of matrix habitat (two types) was varied, isolation metrics were found to be poor predictors of patch immigration rates, especially for habitat specialists, which exhibited the greatest sensitivity to matrix structure (Tischendorf et al. 2003). In a related simulation study that additionally varied the grain and contrast of matrix habitats, both patch size and isolation were found to be poor predictors of inter-patch movement when the matrix consisted of many cover types in a coarse-grained pattern (Bender & Fahrig 2005). This was supported by field studies with eastern chipmunks (*Tamius striatus*, a habitat specialist) and white-footed mice (*Peromyscus leucopus*, a habitat generalist).

Patch-connectivity measures may thus perform best in landscapes in which the matrix is homogeneous (or the disperser behaves as if it were), or when adjustments are made to patch-connectivity measures to incorporate the **effective isolation** of patches (Ricketts 2001), which weights distances among patches by the perceived resistance of the matrix to dispersal. For example, two patches may be the same distance from a source patch, but have different **patch contexts**, in terms of the surrounding habitat matrix. Patches embedded in matrix habitats that offer greater resistance to

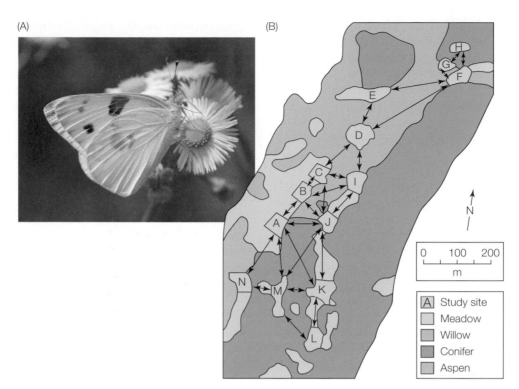

Figure 5.4 Effective isolation of habitat patches. In an alpine valley in the Colorado Rocky Mountains, butterflies like this checkered white (*Pontia protodice*) (A), were more likely to move through willow habitat than coniferous forest as they flitted among meadows (B). Conifer habitat thus had a greater resistance to movement than willow habitat for most butterfly taxa.

Source: After Ricketts (2001). Photo by Megan McCarty and published under the CC BY 3.0 License.

movement are effectively more isolated than those embedded in matrix habitats that offers less resistance. For example, butterflies are far more likely to traverse willow (*Salix* spp.) thickets than coniferous forests as they flit from one montane meadow to another (Ricketts 2001; **Figure 5.4**). Coniferous forests thus provide greater resistance to butterfly movement, such that butterflies are less likely to be found in meadows surrounded by conifers. Meadows within a forest matrix are effectively more isolated than meadows surrounded by willows. **Matrix resistance** underlies the basis for calculating least-cost paths and resistant surfaces, which are explored later in the chapter when we discuss approaches for assessing connectivity in heterogeneous landscapes.

From Patches to Landscapes

Although metapopulation ecologists and landscape ecologists tend to define and measure connectivity at different scales (i.e. at the patch scale vs the landscape scale, respectively), connectivity is determined by the same fundamental process—movement in a spatially structured landscape (Moilanen & Hanski 2001). Why, then, do we need to concern ourselves with assessing connectivity at the landscape scale? Can't we simply aggregate or extrapolate patch-connectivity measures to obtain an overall measure of landscape connectivity? First, we know from **Chapter 2** that landscape patterns

may change as a function of scale. Thus, patches can be highly connected (or clustered) across a range of scales, and yet the overall landscape may not be connected. If landscapes possess hierarchical patch structure (**Figure 2.8**), this will be important for understanding how organisms or processes are influenced by habitat connectivity at different scales, and how landscape connectivity emerges as the aggregate of movement responses to patch structure across scales.

Second, processes or organisms operate within particular domains of scale, both in space and time (**Figures 2.2 and 2.3**). For processes or species that operate at the scale of the entire landscape, connectivity might not be well predicted from measures of patch connectivity. In other words, it may not be possible to scale up from patch-based connectivity measures to obtain an overall measure of landscape connectivity (Tischendorf & Fahrig 2001). This alone requires that we adopt different measures or approaches in assessing landscape connectivity.

Finally, landscapes generally exhibit greater heterogeneity or complexity than the traditional view of habitat patches embedded within a homogeneous matrix. This may complicate the assessment of distance-based patch connectivity measures if organisms exhibit different behavioral responses to patch boundaries or move at different rates within different habitats. Although adjustments to patch-isolation measures can be made to include matrix resistance (e.g. effective isolation), and some landscape connectivity measures also fail to consider differential movement responses in heterogeneous landscapes, this is still an area where landscape approaches have traditionally had the advantage over patch-based approaches.

Methods for Assessing Landscape Connectivity

Various methodological approaches have been developed for assessing landscape connectivity beyond simple landscape metrics. These include (1) percolation-based neutral landscape models; (2) graph-theoretic approaches; (3) simulation models of movement on heterogeneous landscapes; (4) least-cost paths and resistance-surface modeling; and (5) applications of circuit theory. We will study each of these approaches in turn. Because connectivity is a multidimensional concept in riverine landscapes, we will consider those methods separately. Finally, we will conclude this section with a discussion of some issues to consider when deciding which landscape-connectivity approach to use in a given study.

Neutral Landscape Models

Neutral landscape models (NLMs) were first presented by American landscape ecologist Robert Gardner and his colleagues (Gardner et al. 1987). Derived from percolation theory, NLMs are grid-based (raster) maps in which habitat patterns are generated as a statistical distribution or by a mathematical algorithm (**Figure 5.5**). NLMs are also variously referred to as artificial landscapes, simulated landscapes, or virtual landscapes in the literature. NLMs have two general purposes: (1) to determine how the structure of real landscapes deviates from that produced by some theoretical or hypothesized spatial process (e.g. departure from a random distribution; **Chapter 4**); and (2) to generate hypotheses

about how ecological processes are affected by known spatial structure (With & King 1997). The first purpose investigates how a given process might give rise to landscape pattern, whereas the second investigates how landscape pattern might affect some ecological process. Although NLMs were first introduced to make inferences about factors that might have given rise to specific landscape patterns (Gardner et al. 1987; Gardner & Urban 2007), they have since been used to explore ecological responses to landscape structure in a wide variety of applications (e.g. With & King 1999b; With et al. 2002; With & Pavuk 2011).

Neutral landscape models have also been used to define landscape connectivity, and perhaps most importantly, when landscapes become disconnected. Because NLMs are derived from percolation theory, landscape connectivity is defined by whether or not habitat (and presumably the species or process within that habitat) 'percolates' across the landscape. The **percolation cluster** (or spanning cluster) is defined as a habitat patch that spans the entire extent of the landscape (**Figure 5.6**). Within the percolation framework, patch structure emerges depending upon the neighborhood or movement rule used to define patches (e.g. a Rule 1, 4-cell neighborhood vs a Rule 3, 12-cell neighborhood; **Figure 5.6**). In other words, patches may include habitat cells that are immediately adjacent (structural connectivity) as well as those that can be accessed by virtue of the dispersal or gap-crossing abilities of the species or process in question (functional connectivity). Patches are thus not defined a priori as required of the patch-connectivity measures discussed previously, or as with the graph-theoretic approaches to be discussed a bit later.

TYPES OF NEUTRAL LANDSCAPE MODELS   Although any theoretical spatial distribution could be used to generate a NLM (even a checkerboard!), the most commonly used are either simple random maps or are

(A) Simple random

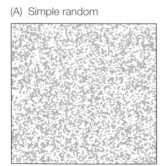

(B) Hierarchical random

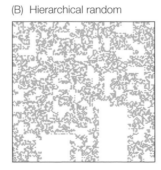

(C) Fractal

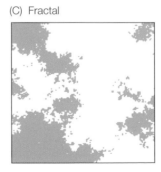

Figure 5.5 Neutral landscape models.   Examples of neutral landscape models generated as either a simple random distribution of habitat (A) or by a fractal algorithm involving either curdling (hierarchical random, B) or midpoint-displacement (fractal, C). All three landscapes contain the same amount of habitat ($p = 0.33$).
Source: After With (1997).

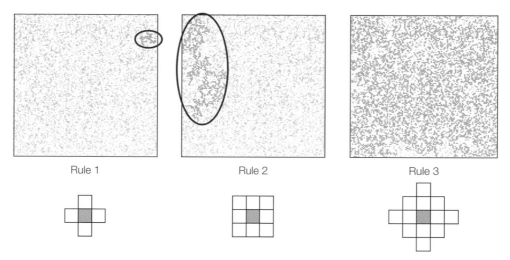

Figure 5.6 Landscape connectivity as a percolation process.   The presence of a percolation or spanning cluster is used to define landscape connectivity within the neutral landscape framework. Whether a landscape with a given amount of habitat 'percolates', however, depends on the neighborhood rule used to define patch structure. Although these three landscapes are identical ($p = 0.33$), the size of the largest patch varies depending on which neighborhood rule is used (the largest patch is highlighted in dark green in each landscape, and is also circled in red where the largest patch is small). Note that the largest patch spans the landscape (i.e. percolates) only for Rule 3 (a 12-cell neighborhood), which might equate to a species or process capable of traversing small gaps.

Source: After With (1997).

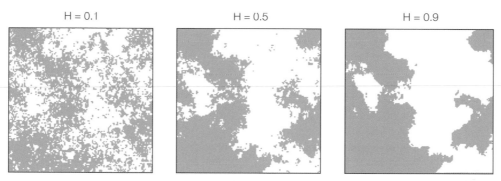

Figure 5.7 Fractal landscapes.   Each map has the same amount of habitat ($p = 0.5$), but varies in the spatial contagion (Hurst Dimension, H) to produce a gradient in habitat fragmentation.

Source: After With (1997).

based on some fractal algorithm (**Figure 5.5**).[1] At first glance, a **random landscape** (**Figure 5.5A**) may appear to be a poor model of landscape structure, in that real landscapes are unlikely to be formed by random processes and exhibit a random distribution of habitats or land uses (**Chapter 3**). That is precisely the point, however. The use of random NLMs is less about whether or not real landscapes are random, then the extent to which they behave as if they are (i.e. they exhibit properties similar to those generated by a random distribution of land covers and land uses). Random NLMs can therefore provide a baseline (a null model) for evaluating how and when landscape pattern matters (i.e. when do we need a more complex model of landscape

pattern?), as well as for assessing how simply changing habitat abundance (p) affects landscape structure and overall connectivity (e.g. how much habitat can be lost before landscape connectivity is disrupted?).

Fractal landscapes (**Figure 5.5C**) exhibit spatial contagion and thus have a greater resemblance to natural landscape patterns. Landforms and vegetation associations may be shaped by various processes that give rise to complex geometries and distributions that can be described using fractal measures (**Box 2.1; Chapter 3**). Spatial contagion can be specified by the Hurst dimension (H), which provides a measure of spatial autocorrelation. For purely random patterns, $H = -0.5$ (Keitt 2000). Maximum contagion is attained at $H = 1.0$, when much of the habitat is aggregated into a single patch on the landscape (**Figure 5.7**). We can use fractal NLMs to simulate both the loss of habitat (p) as

[1] Neutral landscape models can be created using freely available software, such as the R package *NLMR* (Sciaini et al. 2019).

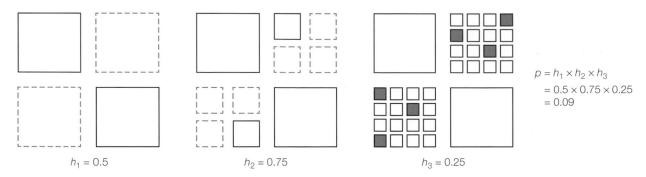

$$p = h_1 \times h_2 \times h_3$$
$$= 0.5 \times 0.75 \times 0.25$$
$$= 0.09$$

$h_1 = 0.5$ $h_2 = 0.75$ $h_3 = 0.25$

Figure 5.8 Illustration of the curdling algorithm used to generate hierarchical random landscapes. The presence of habitat at the coarsest scale (h_1) constrains the availability of habitat at the next level (h_2), which in turn constrains the availability of habitat at the next level (h_3), and so on. The total amount of habitat on the resulting landscape (p) is given by the product of the habitat proportions at each level.

Source: After Pearson & Gardner (1997).

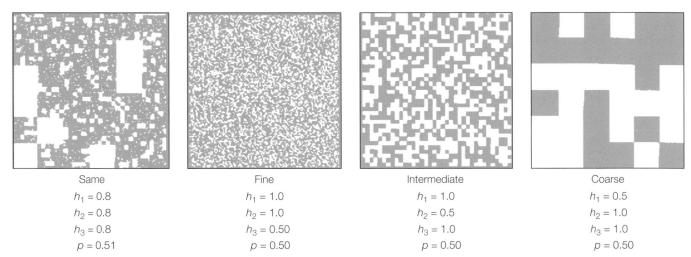

Same	Fine	Intermediate	Coarse
$h_1 = 0.8$	$h_1 = 1.0$	$h_1 = 1.0$	$h_1 = 0.5$
$h_2 = 0.8$	$h_2 = 1.0$	$h_2 = 0.5$	$h_2 = 1.0$
$h_3 = 0.8$	$h_3 = 0.50$	$h_3 = 1.0$	$h_3 = 1.0$
$p = 0.51$	$p = 0.50$	$p = 0.50$	$p = 0.50$

Figure 5.9 Hierarchical random landscape patterns. Although all of these landscapes have approximately the same total amount of habitat ($p = 0.5$), the graininess of the pattern can be adjusted by varying habitat availability at a single level, whether at the finest scale ($h_3 = 0.5$) or at the coarsest ($h_1 = 0.5$).

Source: After Plotnick et al. (1993).

well as different patterns of fragmentation (H) on the landscape. The two parameters (p and H) are adjusted independently of each other, and thus fractal NLMs are useful for disentangling the relative effects of habitat loss from fragmentation on landscape connectivity and on other ecological responses (With 1997; With & Pavuk 2011).

Finally, another NLM that has been used and which is also generated using a fractal algorithm called *curdling* is the **hierarchical random landscape** (O'Neill et al. 1992; **Figure 5.5B**). Although habitat is randomly distributed, the amount of habitat is adjusted independently at each of L levels (i.e. $h_1, h_2, \ldots h_L$ where L is the number of levels). This permits the creation of patchy habitat or resource distributions at different scales, mimicking the hierarchically nested patch structure found in many landscapes. To illustrate how hierarchical random maps are created, imagine a landscape in which patchiness occurs at three different scales

(**Figure 5.8**). If half of the landscape contains habitat at the broadest scale ($h_1 = 0.5$), and 75% of *those* habitat cells contain habitat at the next scale ($h_2 = 0.75$), and finally, 25% of the second-level cells also have habitat at the finest scale ($h_3 = 0.25$), then the total amount of habitat on the landscape, p, is $h_1 \times h_2 \times h_3 = (0.5 \times 0.75 \times 0.25) = 0.09$ or 9% (**Figure 5.8**). This is a recursive procedure, so if habitat is not present at the coarsest scale, it will not be present at the finest scale either.[2] Habitat may thus percolate at some levels but not others, such that overall landscape connectivity is determined by the combined habitat probabilities and the way in which those habitat probabilities are assigned to the various levels. By adjusting the number of levels and the proportion of habitat within each, it is possible to generate some fairly complex—if blocky—landscape

[2] This can be remedied by seeding an additional fraction of cells at subsequent levels with habitat (adding 'wheys' to the curds).

patterns that vary in the grain of the resulting habitat distribution (**Figure 5.9**). Such blocky patterns might be used to simulate the geometry of human land-use patterns, for example, given their resemblance to the square or rectangular motifs reflected in agricultural or managed forest landscapes (**Figure 3.37**).

CRITICAL THRESHOLDS IN LANDSCAPE CONNECTIVITY   For a simple landscape consisting of a single habitat type, landscape connectivity exhibits a **critical threshold** in response to the amount of habitat on the landscape (**Figure 5.10**). The critical threshold in landscape connectivity (i.e. **the percolation threshold, p_c**) is the level of habitat (p) at which the landscape becomes disconnected. By convention, the critical threshold is defined as the specific habitat level (p) at which there is at least a 50% probability that landscapes with that amount of habitat possess a percolating cluster (i.e. where the percolation probability curve crosses the horizontal line in **Figure 5.10A**). Thus, if only 2 out of 10 landscapes with 55% habitat percolate (a 20% probability of percolation), then in general, landscapes with 55% habitat would be considered disconnected because there is ≤50% probability that landscapes with that amount of habitat possess a percolating cluster.

If we assume a random distribution of habitat and the most conservative rule of habitat connectivity (i.e. Rule 1 or a 4-cell neighborhood), then $p_c = 0.59$, meaning that the landscape is predicted to become disconnected once the amount of habitat falls below 59% habitat (in other words, when 41% of the habitat has been removed). As **Figure 5.10A** illustrates, a small loss of habitat initially has very little effect on the overall connectivity of the landscape. As the amount of habitat loss approaches the threshold, however, landscape connectivity suddenly becomes disrupted. Small losses of habitat in this domain thus lead to a dramatic change in landscape connectivity, from a connected to a disconnected state.

The prediction of critical thresholds in landscape connectivity—that small losses in habitat may lead to a rapid and potentially unexpected decline in landscape function—is arguably the single most important contribution of NLMs. However, it is important to understand that no single threshold value defines when landscapes become disconnected. For example, threshold values are affected by the spatial contagion of habitat on the landscape or the pattern of habitat loss (whether disturbances are small and dispersed or large and concentrated in space). In terms of structural connectivity, landscapes with a high degree of spatial contagion (H = 1.0) are able to maintain overall connectivity across a greater range of habitat loss (compare the critical threshold values for the fractal, H = 1.0 landscapes vs. the random landscapes in **Figure 5.10A**). In other words, if habitat loss is concentrated within a

few areas of the landscape, so as to preserve large blocks of habitat elsewhere, then overall landscape connectivity can potentially be preserved at higher levels of disturbance.[3]

Critical thresholds in landscape connectivity are also sensitive to the neighborhood rule that is used to define habitat patches (**Figure 4.25, Figure 5.10**). If the neighborhood rule represents the maximum dispersal distance of a species, then different species will have different perceptions of whether a given landscape is connected or not. This permits us to assess species' perceptions of landscape connectivity. For example, a bird willing to cross gaps, such as clearcuts in a forest, would perceive the landscape as connected across a much greater range of habitat than one unwilling to do so. In the case of the first bird, habitat does not need to be contiguous (structurally connected), but instead can be functionally connected via dispersal.

From the standpoint of functional connectivity, however, the dispersal or gap-crossing abilities of the target species also determine which pattern of fragmentation is most conducive to maintaining overall landscape connectivity. For example, a species incapable of crossing gaps would benefit from high habitat contagion (**Figure 5.10A**) whereas a species capable of crossing small gaps (Rule 3) might, paradoxically, benefit from a more dispersed or fragmented pattern of disturbance, at least in terms of its ability to move about on the landscape (compare the shift in the percolation threshold between Fractal, H = 1.0 and random landscapes in **Figure 5.10B** relative to those in **Figure 5.10A**). This is because the creation of small gaps enables the species to use habitat patches as stepping stones that can then facilitate its movement across the landscape. Note that this does not imply that there is sufficient habitat to sustain viable populations of the species, only that habitat is sufficiently connected to sustain movement or dispersal across the landscape.

APPLICATIONS OF NEUTRAL LANDSCAPE MODELS Neutral landscape models have heuristic value in theoretical landscape ecology, especially for generating hypotheses and testing the effect of landscape structure and critical thresholds on movement and other ecological phenomena. In practice this has involved a combination of computer simulation modeling, field experimentation, or both. Because NLMs offer a systematic way of modeling habitat loss and fragmentation,

---

[3] Unless the percolation cluster is specifically targeted, in which case this might prove to be an extraordinarily efficient means of disrupting overall landscape connectivity. This might be done, for example, to disrupt the spread of an invasive species or disease pathogen, or possibly to minimize the risk of spread of a disturbance like fire across the landscape (i.e. strategically placed firebreaks).

(A) Rule 1

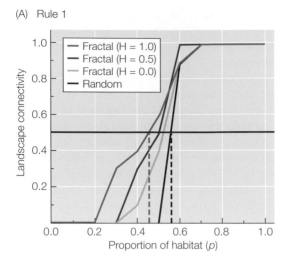

(B) Rule 3

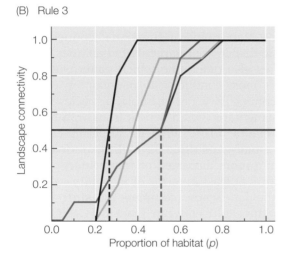

Figure 5.10 Thresholds in landscape connectivity. The probability that a percolating cluster occurs on a landscape with a given amount of habitat (p) is a function of the neighborhood rule used to assess habitat connectivity (**Figure 5.6**) and the degree of habitat fragmentation (H, **Figure 5.7**). (A) Under Rule 1, the percolation threshold (p_c) occurs at about $p = 0.59$ or 59% habitat in random landscapes (the black vertical dashed line), but at a lower level ($p = 0.43$ or 43% habitat) in clumped fractal (H = 1.0) landscapes (the blue vertical dashed line). (B) Under a larger neighborhood rule (Rule 3), however, the thresholds are reversed, occurring at higher habitat levels in clumped fractal landscapes ($p_c = 0.51$) than when habitat is randomly distributed ($p_c = 0.28$). Gap sizes are smaller in random than in clumped fractal landscapes, and thus a species or process capable of crossing only small gaps (one cell wide under Rule 3) would be able to percolate across a greater range of p values if habitat is randomly distributed.

Source: After With & King (1999a).

and for quantifying when landscapes became disconnected, NLMs have potential applications for conservation biology as well (With 1997). We consider next a few of the ways in which NLMs have been used in landscape ecology.

Effect of landscape connectivity on the spread of disturbances: The spread of a disturbance, such as fire, will depend on both landscape pattern and properties of the disturbance regime itself, such as the frequency and intensity of the disturbance. In an early application of random NLMs, Turner and her colleagues (1989a) modeled the effect of disturbance on landscape structure by exploring how variation in the frequency and intensity of a hypothetical disturbance interacted with the amount of disturbance-prone habitat on the landscape. Disturbances can respond to—as well as create—landscape pattern. Thus, Turner et al. (1989a) found that disturbance should have little effect on landscape structure when the disturbance occurs infrequently and susceptible habitat on the landscape is uncommon (i.e. $p < p_c = 0.59$ for nearest-neighbor spread on a random landscape). On such landscapes, susceptible habitat is fragmented, occurring as relatively small, isolated patches, which does not permit the disturbance to spread far beyond its initial point of initiation. If much of the landscape is susceptible ($p > p_c$), however, then disturbance can spread throughout the landscape even at low levels of disturbance, owing to the greater overall connectivity of the landscape. Put another way, landscapes that are dominated by a single susceptible habitat may become fragmented (disconnected) by disturbances of only low-to-moderate intensity, depending upon how frequently disturbances occur. Even low-level disturbances can have a large impact on the landscape if initiated often enough. Still, predicting the effect of disturbance on landscape structure remains a challenge. As Turner et al. (1989a) point out, it will also depend on the type of disturbance, in terms of whether the disturbance is mostly confined to spread *within* a given habitat type (as in their analysis) or whether it can spread *between* habitats (such as from fields to adjacent forests). For disturbances that spread between habitats, the adjacency of these habitat types (a_{ij}) will likely be an important determinant of whether and when a disturbance will spread. An increase in landscape heterogeneity could thus increase the likelihood of spread, and thus landscape connectivity, for these types of disturbances.

Percolation in experimental model landscapes: One of the more obvious consequences of disrupting landscape connectivity should be a disruption in the movement or dispersal of organisms across the landscape. Inspired by NLMs, Wiens and his colleagues (1997) tested the effect of landscape structure on the movement patterns of darkling beetles (*Eleodes obsoleta*). They ran individual beetles on a small, experimental landscape grid (5 × 5 m) that had a random distribution of habitat (grass turf in a sand matrix), varying the specific amount of grass cover among trials (0, 20, 40, 60, and 80%) to encompass the theoretical percolation threshold for random maps ($p_c = 0.59$ by the 4-neighbor rule). Thresholds in a number of movement parameters (e.g. net displacement, mean step length) occurred, not at the expected percolation threshold, but between 0% and 20% habitat. Beetles are not restricted to move just through grass, but also moved (and did so more quickly) across the sand matrix. Thus, beetle movement behavior violates the basic assumption of nearest-neighbor percolation, and so it is perhaps not surprising that their movement thresholds did not match this particular percolation threshold. The results of this experiment remind us: (1) the oft-cited threshold of 59% is specific to certain assumptions regarding how movement occurs across the landscape, and so should not be used as a generic threshold of landscape connectivity; (2) critical thresholds in landscape connectivity may not coincide with thresholds in movement metrics even so; and (3) the definition of landscape connectivity ultimately depends upon both landscape pattern and how individuals move among patches. That movement may exhibit a threshold response to landscape pattern is an important finding, however, which may have consequences for evaluating the effect of habitat loss and fragmentation on foraging success, territoriality, mate finding, and other behaviors that influence individual fitness.

Species' perceptions of landscape connectivity in heterogeneous landscapes: Whether or not a given species perceives a landscape as connected will depend not only on the amount and arrangement of habitat on the landscape, but also on the species' own habitat affinities and movement responses to landscape structure. For example, With and her colleagues (1997) conducted a simulation modeling experiment using both random and fractal NLMs (**Figures 5.5 and 5.7**) to explore how species' interactions with landscape heterogeneity influenced the distribution of populations across landscapes. In this study, landscapes were modeled as heterogeneous mosaics that varied in the relative abundance and quality of three habitat types (high, medium, or low carrying capacity), and for which the species exhibited different preferences (based on the probability of remaining in a given type of habitat cell). Recall that fractal landscapes have greater spatial contagion than random ones, and thus one habitat always percolated at the levels of habitat considered in this study ($p \geq 0.33$), assuming nearest-neighbor movement in landscapes with an intermediate degree of aggregation (H = 0.5), whereas habitat in random landscapes percolated only when habitat exceeded the percolation threshold ($p > 0.59$; again, assuming nearest-neighbor movement).

Whether or not these landscapes were actually connected, however, depended on species' habitat affinities and the quality of those habitats. Populations were more patchily distributed (i.e. fragmented) when the species had a high affinity for high-quality habitats that were rare or uncommon on the landscape ($p < p_c$), even in the seemingly well-connected fractal landscapes. Thus, structural connectivity does not guarantee functional connectivity, but even functionally connected landscapes may possess structural differences that could have important ecological consequences. For example, functionally connected random landscapes have more edge habitat than functionally connected fractal landscapes that have a high degree of habitat contagion. If the amount of edge is related to increased rates of predation or competition, then survival rates should be lower in such landscapes. Thus, functionally connected landscapes should not be assumed to be functionally equivalent from an ecological standpoint.

Plant dispersal and species coexistence on hierarchical landscapes: In a pair of papers, Lavorel and her colleagues (1994, 1995) used hierarchical random NLMs to simulate the distribution of suitable colonization sites at three nested scales of patchiness, to explore how differences in the dispersal and dormancy strategies of annual plants (how plants escape disturbance either in space or time, respectively) interacted with landscape pattern to influence species coexistence. Although the entire landscape was potentially suitable for colonization ($h_1 = 1.0$), the availability of colonization sites was constrained at the two finer scales (h_2 and h_3) by varying the fraction of suitable habitat at each level (all six possible combinations of h_2 and h_3 where $h_i = 0.32$, 0.60, or 0.88; cf. **Figure 5.9**). Such differences in habitat suitability at finer scales might arise in real landscapes because of variation in the distribution of soil nutrients, microtopography, or local disturbances. Much like real landscapes, habitat suitable for colonization tended to be more aggregated (patchy) in these hierarchical landscapes than in simple random NLMs (especially when $h_2 < h_3$).

As a consequence, Lavorel et al. (1995) found that short-range seed dispersal was the most effective strategy for promoting population spread across these landscapes. Long-range dispersal is risky when resources are patchily distributed, as suitable sites for colonization are more likely to be clustered in the vicinity of the parent plant. All else being equal, the better colonizer (with short-range dispersal) will eventually win the competition for space by saturating all available sites, making species coexistence unlikely (Lavorel et al. 1994). However, which type of short-range disperser might win will depend upon its response to local disturbance, and the frequency or intensity of that disturbance. For example, if disturbance is rare, producing many seeds that then go dormant is not a winning strategy; a species that produces fewer

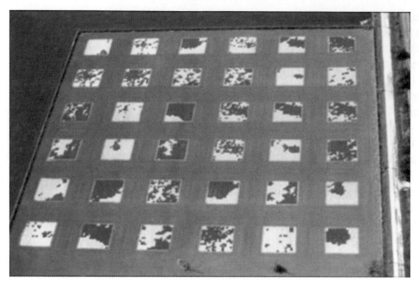

Figure 5.11 Aerial view of an experimental landscape system founded on fractal neutral landscapes.   Red clover (*Trifolium pratense*) was planted to cover 10, 20, 40, 50, 60, or 80% of the landscape plot (plot = 16 m × 16 m) as either a clumped (H = 1.0) or fragmented (H = 0.0) distribution in this experimental model landscape system.
Source: Photo by Kimberly A. With.

dormant seeds will quickly dominate the landscape. At the other extreme, frequent disturbances would favor dormancy, as the species could then escape in time and wait to germinate when conditions are favorable. If dormancy is actually broken by disturbance, however, then all bets are off. Then, it generally pays to produce fewer dormant seeds of this type. Instead, a strategy of 'risk spreading,' in which some intermediate fraction of seeds go dormant each generation but germination is independent of disturbance, appears to be the most successful at hedging one's bets across a wide range of conditions, and promotes coexistence. The results of these simulations thus demonstrate that differences in the ways species exploit resources in space and time can potentially mediate coexistence through storage effects (Chesson 1994). In other words, no species does best under all environmental conditions, but given a temporally or spatially variable (heterogeneous) environment, some species may be able to 'store' gains made in good years or in good environments, which enable them to weather the bad years or buffer against population losses in poor environments.

The relative effect of habitat area versus fragmentation on community patterns: Because habitat loss may lead to fragmentation, the effects of habitat loss and fragmentation are inextricably linked and therefore confounded in many ecological investigations. To disentangle the effects of habitat area from fragmentation, With and Pavuk (With et al. 2002; With & Pavuk 2011) constructed an experimental landscape system in the field, in which they planted clover (*Trifolium pratense*) within individual plots (16 × 16 m) according to a fractal dis-

tribution defined by both the amount (10, 20, 40, 50, 60, or 80%) and fragmentation (clumped: H = 1.0; fragmented: H = 0.0) of habitat (**Figure 5.11**). Their approach is unique among experimental fragmentation studies because it adopted a 'top-down' approach to create landscape patterns, by setting the habitat amount and overall pattern of fragmentation (H) at the scale of the entire landscape. This contrasts with the traditional 'bottom-up' approach in which patches of varying sizes are arrayed at different distances to create fragmented landscape patterns.

The fractal clover plots were colonized by insects and other arthropods from the surrounding agricultural landscape. Insects are common study subjects in experimental fragmentation studies because they are small and thus tend to operate at spatial and temporal scales that ecologists find especially convenient. In general, habitat area had a far greater—and more consistent—effect on insect diversity than fragmentation (With & Pavuk 2011). In particular, predators and parasitoids were more sensitive to habitat area than their herbivorous prey. Interestingly, patch-occupancy patterns of one of the major predators in this system, the ladybird beetle *Harmonia axyridis*, exhibited a threshold at low levels of habitat (<20%), matching similar thresholds in the distribution of its aphid prey (*Acyrthosiphon pisum*) that in turn reflected **lacunarity thresholds** in the distribution of habitat (With et al. 2002). Recall that lacunarity measures the variability and size of inter-patch distances (the gap structure) of landscapes (**Chapter 4**). Thus, lacunarity increases in landscapes with little habitat cover (<20%), especially when habitat is clumped, because both the size and variability of gap sizes increases. Aphids are fairly sedentary (except for winged morphs) and thus populations became more patchy and isolated as their clover habitat declined. Habitat fragmentation can interfere with predator search efficiency, preventing them from locating and concentrating within areas of high prey density, which could undermine biological control efforts and lead to an increased risk of pest outbreaks (Kareiva 1987). The non-native *H. axyridis* had been introduced specifically as a biocontrol agent of pea aphids in this region, and given its greater mobility (relative to a native ladybird beetle), it was apparently able to track aphid distributions even as they became scarce and increasingly fragmented, as evidenced by the coincident thresholds in these predator–prey distributions (With et al. 2002). So, although individual species may be sensitive to fragmentation effects, we are reminded that different species will respond in different ways and at different scales to landscape structure, with the net result being an overwhelming effect of habitat area rather than fragmentation on the overall response of the community to landscape pattern.

In summary, landscape connectivity is generally far more complex than suggested by percolation thresholds. For starters, the relationship between percolation, structural connectivity, and functional connectivity is not always obvious. As Metzger and Décamps (1997) point out, it is possible that a landscape might possess a high degree of habitat connectivity and yet not percolate (as when a landscape is dominated by two large habitat patches that are separated by a barrier that is impassable), just as it is possible to have a percolating landscape with a low overall degree of habitat connectivity (as when a narrow corridor bisects the landscape but is otherwise not connected to habitat patches elsewhere on the landscape). For this reason, they suggest percolation be defined only for interior habitats and by quantifying the degree of buffering (habitat expansion) necessary to connect these patches of interior habitat. This should help guard against cases where only a narrow corridor is present on the landscape, as this is unlikely to contain interior habitat that would promote high ecological functioning on the landscape. They also reiterate the distinction between structural and functional connectivity by pointing out that matrix permeability, corridor quality, and other landscape elements such as habitat stepping stones can influence ecological flows across the landscape, and must therefore be considered when evaluating landscape connectivity. Percolation thresholds might thus be viewed as a first approximation of the structural connectivity of a landscape, but may or may not relate to the functional connectivity of that landscape, in terms of its capacity

to sustain ecological flows or other ecological functions. As with most general models, the concept of percolation and its neutral landscape derivatives are abstractions of reality. Their primary value lies in detecting thresholds in landscape structure, which in turn might have important ecological consequences, and for exploring when landscape structure might be important for understanding ecological responses to landscape pattern.

Graph-Theoretic Approaches

Graph theory (or network analysis) is becoming an increasingly popular way of assessing landscape connectivity, especially for landscapes that are made up of discrete patches such as forest fragments, alpine meadows, or ponds. Graph theory effectively blends patch-based and landscape approaches in assessing connectivity, with a host of measures that can be applied at either the patch or landscape scale (Minor & Urban 2008). At a minimum, this requires information on the spatial coordinates of habitat patches and the maximum dispersal distance of the organism, but it is also possible to incorporate additional information on the size or quality of habitat patches, and on the resistance of the matrix to movement among patches, to provide greater biological realism (Urban et al. 2009).

WHAT IS A GRAPH?   We discussed in **Chapter 4** the different types of spatial data, including raster data formats, which represent landscapes as an array of pixels (lattices, such as the NLMs discussed above), or vector data formats, which represent landscapes as a collection of points, lines, or polygons, such as the information contained within a GIS shapefile. Basically, vector formats use mathematical relationships between points and the paths connecting them to describe the landscape. This is called a **graph**, in which habitat patches or populations (and sometimes individuals) are represented as points or polygons (**nodes**) that are connected by lines (**edges**) representing the actual or presumed linkages (i.e. functional connectivity) among specific nodes (**Figure 5.12**). Linkages reflect the likelihood of movement or flows among nodes, which may be distance-based (assuming some maximum dispersal distance) or weighted to reflect resistance of the intervening matrix to dispersal, providing a cost to movement. In this context, the 'cost' may represent the unwillingness of individuals to move through a particular environment, the physiological or energetic cost of movement, higher mortality incurred by individuals moving through the habitat, or some combination of these factors (Zeller et al. 2012). A **path** is a sequence of edges joining nodes. For example, the path may represent the possible route(s) an individual could take while moving through the landscape. Graphs are connected if

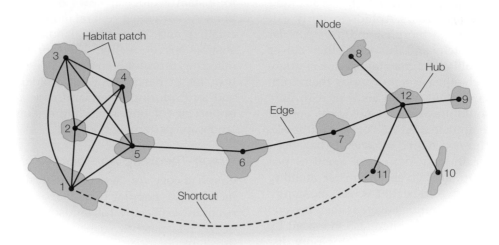

Figure 5.12 A landscape graph, with key elements highlighted.
Source: After Minor & Urban (2008).

there exists a path between each pair of nodes (i.e. every node is reachable from some other node). However, a disconnected graph may include several connected **components** or **subgraphs**, reflecting different scales of patchiness or clustering within the landscape.

GRAPH METRICS Many graph metrics have been developed to define connectivity at both the patch (node) and landscape (network) scales (**Table 5.1**). Node-based metrics assess how connected an individual node is to others. High-degree nodes (**hubs**) are likely to be important for overall network connectivity (i.e. **critical nodes**). On the other hand, network connectivity measures are based on some aspect of the size or extensiveness of the largest component on the landscape. For example, graph diameter, which is defined as the 'longest shortest path' on the network, reflects the total inter-patch distance an organism would have to cross in order to traverse the largest component. This is similar to the percolation cluster of NLMs and can be interpreted as the degree to which the landscape is potentially traversable (Bunn et al. 2000).

Connectivity may come about either because habitat is very abundant on the landscape ($p \rightarrow 1.0$), or because patches are close together, or some combination of the two. Thus, **graph diameter** may or may not indicate a high degree of overall connectivity for the landscape network. For example, a short graph diameter can mean either that (1) the graph is highly connected, and most nodes (patches) are connected to each other; or (2) the graph is very poorly connected and broken into multiple, small components. Thus, Ferrari et al. (2007) introduced an area-based metric (F) that measures the relative proportion of habitat in the largest

patch (A_{LP}) relative to the proportion contained within the largest component (A_{LC}): $F = A_{LP}/A_{LC}$. If all habitat is connected in one large component, then $A_{LC} = 1.0$ and $F = A_{LP}$, which represents the proportion of the largest component that is found in a single large patch. Recall that components are identified by a graph analysis of inter-patch connectivity, which is different from simply assessing the largest patch size (cluster) using GIS methods; thus, $A_{LC} \geq A_{LP}$. This area-based metric is thus a complementary measure to graph diameter, and can be used as an additional criterion to identify landscapes that have undergone habitat loss without substantial fragmentation, which are not easily identified simply through a comparison of graph diameters.

CRITICAL THRESHOLDS IN NETWORK CONNECTIVITY A network consists of nodes and links, and thus the loss of either type of element, such as through habitat loss or fragmentation, may jeopardize the structural and functional integrity of the landscape. Some nodes may be more important than others (high-degree nodes or hubs) for maintaining network structure. Similarly, certain links will likewise prove more consequential for facilitating flows through the network (e.g. shortcuts or habitat corridors). Thus, we can examine the targeted removal of either patches or links to evaluate their relative importance on network connectivity, and identify critical thresholds in the proportion of links or nodes that can be removed before network connectivity is disrupted. This also permits an evaluation of **network robustness**, the resilience of the network to the removal of nodes or links before overall connectivity is altered.

TABLE 5.1 Graph-theoretic measures used to assess connectivity at the patch (node) and landscape (network) scales.

Measure	Definition	Ecological interpretation[1]
Patch-scale (node) attributes		
Node degree	Measure of node connectivity; the number of connections to other nodes (equivalent to number of patches within a neighborhood or dispersal range). If edges are weighted, then this is a measure of 'source strength'. A hub is a high-degree node (i.e. a node connected to many other nodes)	High-degree nodes (hubs) may be particularly important from a conservation standpoint (critical habitat nodes). Small patches with a low degree may be vulnerable to extinction if their neighbors are removed
Centrality (eigenvector centrality)	Similar to node degree, but accounts for the fact that not all connections are equally well-connected; a measure of the relative position of a node or edge in terms of connectivity	Connections to well-connected nodes will be more influential than less well-connected nodes
Clustering coefficient	Probability that two nodes connected to another particular node are themselves connected; measures the average fraction of the node's neighbors that are also neighbors with each other	Highly clustered nodes facilitate dispersal and spread of disturbances, and can contribute to network resilience owing to redundant pathways
Betweenness	Number of shortest paths that a particular node or edge lies on	High betweenness implies linkages or 'stepping stones' between subgroups; these patches may control flows across a network and be critical for maintaining connectivity
Network (landscape) attributes		
Minimum spanning tree	A 'tree' is a path that includes no cycles; a 'spanning tree' contains every node in the graph. The spanning tree with the shortest length is the 'minimum spanning tree'	Minimum spanning tree is analogous to a percolation cluster ('connectivity backbone of landscape')
Graph diameter	Longest shortest path between the two most distant nodes in the network	Total inter-patch distance an organism would have to traverse to span the largest cluster; analogous to percolation backbone
Characteristic path length (CPL)	Mean of all pairwise graph distances connecting nodes; measure of the average shortest path length over the network	A short CPL indicates that all patches are easily reached, implying a patchy population rather than a metapopulation or subpopulations
Compartmentalization (connectivity correlation)	Measures the relationship between node degree and the average node degree of its neighbors	High compartmentalization slows movement through the network and may isolate the potentially cascading effects of disturbance
Node-degree distribution	Distribution of node degree values in a network	Provides a diagnostic of network types and other properties of the network
Network robustness (resilience)	Number of nodes or edges that can be removed without altering network connectivity. Depends strongly on the node-degree distribution; networks with significant variance in node connectivity are most robust to the random removal of nodes	Useful for identifying critical habitat nodes or edges that are essential for maintaining the overall connectivity of the network. Compartmentalization may increase overall network robustness

[1]Sources: Bunn et al. (2000); Urban & Keitt (2001); Minor & Urban (2007, 2008); Garroway et al. (2008).

As an illustration of the importance of habitat nodes for network connectivity, Keitt and his colleagues (1997) performed a network analysis of suitable forest habitat for the Mexican spotted owl (*Strix occidentalis lucida*), which is a federally listed threatened species that occurs within forested canyons of the southwestern USA. They identified a single habitat node—a small forest district in New Mexico—that was critically important for preserving overall network connectivity, in that removal of this node would fragment the network (and potentially the Mexican spotted owl metapopulation) into two separate regions. Notably, this habitat node did not support a viable owl population, and thus had been previously deemed unimportant for the conservation of this species. Pursuant to this analysis, the forest district's importance as a critical node for Mexican spotted owls resulted in a re-evaluation of timber harvesting plans in this area.

Similarly, the loss of critical linkages among populations, as a result of habitat loss or other landscape changes that increase the effective isolation of patches, can also disrupt network connectivity. Using different edge thresholds, such as dispersal distances or resistance values, we can explore the effect of 'dropped edges' on overall network structure. For example, van Langevelde (2000) found that a threshold distance of 2.4–3 km best explained colonization patterns of the Eurasian nuthatch (*Sitta europaea*) in forested landscapes of The Netherlands. The threshold distance thus gives an indication of the distances covered by dispersing nuthatches that lead to successful colonization. Based on this analysis, the ability of nuthatches to colonize forest patches would be seriously compromised in fragmented landscapes where the average inter-patch distances exceeded the critical threshold of 3 km.

Graph metrics can also exhibit critical threshold behavior equivalent to percolation thresholds because they are measuring the same phenomenon: the likelihood of traversing the network or landscape. For example, thresholds in the graph diameter of fractal NLMs occur at the same critical thresholds in habitat abundance (p_c) predicted by percolation analysis (Ferrari et al. 2007). Thus, the analysis of network connectivity is really just percolation analysis in a different landscape context or format.

TYPES OF GRAPHS   Just as there are different types of NLMs, there are different types of graph structures. Understanding graph structure (**network typology**) is of interest because it can help in the characterization of different landscape configurations, which may give insight into the formative processes responsible for those patterns (Cantwell & Forman 1993). Because network structure is an emergent property, identifying the type of network can enhance our understanding of ecological flows across the landscape or of properties such as the vulnerability of the network to disturbance (Minor & Urban 2008). Four common types of networks are presented in **Box 5.1**.

BOX 5.1   Common Network Structures

Planar networks: In planar networks (**Figure 1A**), nodes are connected just to their immediate neighbors and connect to more distant nodes only by passing through other nodes (as stepping stones; Minor & Urban 2008). Planar networks tend to have long path lengths because there are no shortcuts; they may or may not have a high clustering coefficient. A lattice (NLM) is basically a planar network in which percolation occurs through sites (site percolation) rather than through links (bond percolation, as in networks).

Random networks: In the case of random networks (**Figure 1B**), linkages among nodes are entirely random. This is reflected in the node-degree distribution, which exhibits a normal (Gaussian) distribution. Each node thus has approximately the same degree of connectivity (i.e. the network lacks hubs or highly connected nodes), and the characteristic path length increases monotonically regardless of whether node removal is random or targeted. If a real landscape exhibits random graph properties, this suggests that dispersal among nodes is basically random

and that nodes are separated by short paths. The loss of nodes (patches) would steadily decrease the ease with which flows occur across the landscape. Just as with random NLMs, random networks provide a useful null model for determining whether observed properties of a network are a consequence of some non-random process or are simply a byproduct of random linkages among nodes.

Scale-free networks: In scale-free networks (**Figure 1C**), most nodes have relatively few connections while a few nodes are highly connected hubs. The node-degree distribution is thus a continuously decreasing function (i.e. a power-law or fractal distribution). Removal of a high proportion of nodes is expected to have little impact on the network because most nodes are not well connected. However, the targeted removal of the most connected nodes leads to a rapid increase in the characteristic path length and network fragmentation. Identifying these critical nodes is thus of great concern from a conservation or management standpoint.

Small-world networks: Small-world networks (**Figure 1D**) are common to many real-world networks. Small-world networks are characterized by shortcuts that allow rapid and direct movement between distant nodes, resulting in a small graph diameter relative to the number of nodes (Minor & Urban 2008). They also tend to have many more hubs and a high clustering coefficient compared to random graphs. Each node has approximately the same influence on the characteristic path length if removed, but the added feature of clustering might create alternate paths between nodes (i.e. redundancy), such that the impact of node removal could be less than on random networks (Garroway et al. 2008). Small-world networks are more vulnerable to random disturbance than scale-free networks because shortcuts make spreading through the network relatively quick and easy.

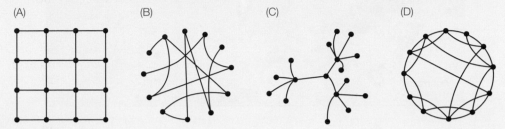

(A)   (B)   (C)   (D)

Figure 1 Different theoretical network structures.   (A) Planar (B) Random (C) Scale-free (D) Small world
Source: Minor & Urban (2008).

Beyond a taxonomy of graph types, what possible difference does it make whether a forested landscape is structured like a random or small-world network? It might matter from the standpoint of network resilience. For example, if a forest songbird can disperse up to 1.5 km, then a landscape might be viewed as a highly connected network (e.g. 83% of forest patches occur within a single component; Minor & Urban 2008). If we know that the forest network has a larger diameter than expected for either random or scale-free networks, this implies that there are few shortcuts and that movement through the network is relatively slow. That might not be a problem, however, as this level of connectivity might still be sufficient for maintaining dispersal or gene flow among populations, while slowing the spread of diseases such as avian influenza or West Nile Virus across the landscape. Further, if we find that the node-degree distribution is intermediate to that of random and scale-free networks (showing a heterogeneous node degree), this would indicate a certain resilience against random node removal, such as through forest harvesting. This is especially true if the clustering coefficient is also high, indicating the presence of many redundant pathways that confer resilience to the network. Thus, an understanding of network typology may be important for evaluating network resilience, as well as assisting in land management and conservation planning for target species.

APPLICATIONS OF GRAPH THEORY: EXTENSIONS AND CASE STUDIES   We conclude this section with a selection of case studies that illustrate applications of graph theory, especially for conservation.

Connectivity in marine systems: Connectivity is not just a property of terrestrial systems. For many marine species, such as those found on coral reefs, populations are connected primarily through the passive dispersal of larvae by ocean currents, often over great distances. Using a biophysical model with passive dispersal, Treml and colleagues (2008) estimated connectivity among reef islands of the Tropical Pacific for coral larvae (**Figure 5.13**). The scale of population connectivity for corals across the Pacific was on the order of 50–150 km. Patterns of connectivity among some reef islands were predominantly driven by climatic events (e.g. El Niño), such that connectivity was only realized in certain years, which underscores that connectivity may possess a temporal as well as spatial dimension. This type of analysis can aid in identifying and prioritizing areas for marine conservation, by considering the importance of specific island stepping stones to the overall connectivity of coral reef systems. Further, this approach can be used to identify key dispersal routes as well as barriers to dispersal, which may assist in highlighting those species most susceptible to dispersal limitation and population isolation, thereby facilitating conservation efforts for critical species.

Species' perceptions of network connectivity: Different species may have different perceptions as to whether a particular landscape is connected or not. For example, the connectivity of a landscape consisting of forested swamps and wetlands in the southeastern USA became

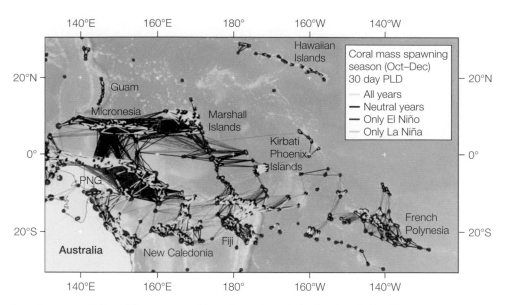

Figure 5.13 Coral reef connectivity. The connectivity of coral reefs in the South Pacific was evaluated in terms of graph theory and a biophysical model that included information on ocean currents and the amount of time marine larvae typically remain in the water column. Although connectivity among neighboring islands is consistently high (yellow lines), El Niño years provide an opportunity for long-range connectivity among distant island groups (red lines).
Source: Treml et al. (2007).

disrupted at an edge threshold distance of 19 km (Bunn et al. 2000). This particular landscape would therefore be perceived as connected for any species with a dispersal range of at least 20 km (e.g. American mink, *Mustela vison*), but not for species with more limited dispersal (e.g. 5 km; prothonotary warbler, *Protonotaria citrea*). Thus, an analysis of the habitat network in this landscape suggests that species such as the prothonotary warbler may function as a set of largely independent populations, whereas species such as the mink will likely behave as one large patchily distributed population (i.e. a metapopulation). Network analysis can thus provide a first step in evaluating the extent to which populations might be isolated, with all the attendant demographic and genetic consequences that isolation brings, in an initial conservation-needs assessment.

Source–sink dynamics in landscape networks: Beyond just location, nodes in a graph can be assigned other attributes related to the size or quality of patches to reflect habitat heterogeneity inherent in the landscape. For example, in an analysis of forest connectivity for the wood thrush (*Hylocichla mustelina*), Minor and Urban (2007) examined the physical structure of the patch network, and also assessed habitat quality of individual forest patches: birds nesting in high-quality forest patches should produce more offspring (potential population sources) than birds nesting in patches of lower quality (population sinks). Nest parasitism by brown-headed cowbirds (*Molothrus ater*) and predation

by general predators are the major factors limiting breeding success in wood thrushes, and both of these rates tend to be higher along forest edges (**Box 7.2**). Thus, small forest fragments, which have a high edge-to-area ratio and are mostly edge habitat, are expected to have low reproductive potential and provide poor breeding habitat for thrushes. Minor and Urban (2007) thus indexed patches in terms of quality-weighted area, which was simply the patch size multiplied by patch quality (the average distance to a non-forest edge for every pixel within the habitat patch). From a management standpoint, it is obviously important to identify and protect high-quality patches that are likely to have high reproductive value for the target species. Nevertheless, we should not overlook the potential value of even low-quality habitats in contributing to the overall connectivity of the network (critical nodes; Keitt et al. 1997).

Directional flows and the implications for population connectivity: In population source–sink dynamics, the net flow between patches occurs in one direction, from sources to sinks. Other types of directional or asymmetrical flows may occur in response to prevailing wind or water currents. Linkages (edges) among patches can thus reflect the net directionality of movement among nodes in the form of a **directed graph** (or digraph, for short). For example, the population connectivity of chinook salmon (*Oncorhynchus tshawytscha*) in dammed rivers of California's Central Valley in the USA was evaluated using weighted digraphs, based on the level of recruitment between source (donor) and sink (recipient) populations (Schick & Lindley 2007). Compared to the pre-dam riverscape, the current population network has far fewer source populations (net exporters), many of which are now either extinct or have become population sinks (net importers). By analyzing the effect of sequential dam addition to the river network, Schick and Lindley (2007) demonstrated how a single dam can impact almost the entire salmon population network in this region. Further, by incorporating recruitment as a measure of connectivity, the authors were able to extend the usual spatial aspects of graph analysis, which are based on dispersal assumptions, to consider the demographic consequences of altering network structure for the long-term viability of these populations.

Assessing Connectivity in Heterogeneous Landscapes

Landscapes are heterogeneous and thus it may seem overly simplistic to represent them as grids or networks consisting of habitat patches embedded in a matrix. That may be a fair description of some landscapes or how some organisms perceive their landscape (e.g. extreme habitat specialists), but it clearly does not suffice for others. As a rule, we should start simple and incorporate complexity into our models or landscape representations only as needed and not simply because they promise greater realism. Such promises may prove false and may carry costs in terms of greater model uncertainty. Still, there are times when greater complexity is required, such as when simple mathematical models fail to adequately capture the interaction between pattern and process, or when we are evaluating different management plans for specific species in a particular landscape. Although we have previously explored some ways in which heterogeneity can be incorporated within the framework of graph-theoretic and neutral landscape modeling approaches, the assessment of landscape connectivity necessarily becomes more complicated when organisms exhibit different movement responses within different elements of the landscape. The challenge is greater still if the landscape exhibits continuous variation rather than discrete patches. In this section, we consider other approaches that have been used to assess connectivity in heterogeneous landscapes: grid-based simulation models, least-cost path and resistance-surface modeling, and applications of circuit theory.

GRID-BASED SIMULATION MODELS An increase in behavioral and landscape complexity may necessitate a numerical simulation of individual movement on a heterogeneous landscape to determine if (or within what region) the landscape is connected. Individual-based simulation models vary in complexity, from simulating dispersal simply as a diffusion process (e.g. random walk models) to the incorporation of detailed movement algorithms in which individuals move at different rates or suffer different rates of mortality within different habitat types, or differ in the likelihood of moving from one habitat type to another, for example. The landscape is usually represented as a raster map or grid, and often within a GIS framework if the goal is to model dispersal and assess connectivity for some target species in an actual landscape. Grid-based simulation models have the power to yield insights into functional connectivity, but tend to have higher data requirements and may be more difficult to implement (requiring expertise in computer programming) and to parameterize than other approaches.

For example, connectivity among populations of the European badger (*Meles meles*) in the central Netherlands was assessed using a grid-based random walk model (Schippers et al. 1996). Movement within a given habitat type was simulated as a stochastic process, in which individuals were assumed to follow a correlated random walk (i.e. movement direction was determined probabilistically but favored continued movement in a forward direction), resulting in more of a straight-line or directed pattern of movement than given by a purely random walk (**Chapter 6**). Because the landscape is heterogeneous, different habitat types offer different incentives for badgers. Thus, badgers are not only more likely to move into habitats they prefer, but are likely to spend more time there once they arrive. Life is not without risk, especially if you are frequently on the move, and thus badgers accrue mortality risk as they move about the landscape, especially in regions of high road density (traffic accidents are the main source of badger mortality). Badgers would prefer not to cross major motorways, and canals are apparently deathtraps for badgers (who can swim, but often are unable to climb the steep banks to get out), and so various features of the landscape serve as barriers to movement. The movement algorithm of this simulation model therefore incorporated information on the likelihood of movement between cover types and across barriers, as well as movement and mortality rates within different habitats types. Connectivity was then assessed as the fraction of simulated individuals (out of 5000) that arrived at various target populations from a given source population. This was repeated for every population, producing a connectivity matrix that depicted the degree of connectivity among 34 badger populations.

This study illustrates the computationally intensive nature of this type of modeling approach, although computational efficiency and run time is rarely an issue any more given the much faster processors of today's computers. Still, if we are concerned about efficiency, another possibility is to consider a network analysis of least-cost paths and resistance surfaces, which we consider next.

LEAST-COST PATHS AND RESISTANCE SURFACES
The analysis of least-cost paths involves identifying the optimal route between two locations, based on the assumption that organisms will attempt to minimize their travel costs through a heterogeneous matrix that offers differential resistance or risk to movement (Bunn et al. 2000). For example, individuals might use habitat patches of a certain type as stepping stones (low cost), rather than taking a shortcut across an inhospitable matrix (high cost). Least-cost paths thus combine habitat quality and Euclidean (straight-line) distances in determining the effective distance or cost between two patches. In graph-theoretic parlance, the links or edges between nodes are weighted, depending upon the probability of movement between them. Tools that enable the calculation of least-cost paths are now a common feature of most GIS packages (e.g. Spatial Analyst in ArcGIS™).

Resistance or cost surfaces depict the relative costs of movement across a heterogeneous landscape (Zeller et al. 2012). A resistance value is assigned to each habitat type, which indexes the expected costs of movement through that cover type. For example, a resistance value of 1.0 indicates minimal resistance (i.e. movement through preferred habitat), a resistance value of 2.0 means that individuals are expected to move successfully only half as far as within the preferred habitat, all the way up to the maximum resistance value for habitats that pose an absolute barrier to movement. A least-cost path analysis is then used to find the shortest functional distance between each cell to every other cell in the landscape within a maximum dispersal distance. The least-cost 'kernel' is a surface that represents the probability that an individual dispersing from the focal cell arrives at any other point in the landscape.

Using this resistant-kernel modeling approach, Compton and his colleagues (2007) assessed the connectivity of vernal pools for mole salamanders (*Ambystoma* spp.) throughout the state of Massachusetts in the eastern USA (**Figure 5.14A**). Although ephemeral pools represent critical breeding habitat, these salamanders actually spend much of their lives in the surrounding upland forests. Thus, connectivity was assessed at multiple scales encompassing local connectivity between breeding pool and upland habitats; neighborhood connectivity of pools within the dispersal range of these salamanders; and the regional connectivity of pools within a specified distance (pool clusters). Although there is some empirical information on salamander dispersal and mortality in different habitats, the authors ultimately relied upon expert opinion to parameterize resistance values of land-

cover types in this landscape. Thus, preferred habitats, such as vernal pools and upland forests, received a resistance value of 1.0; old-fields, pastures, and agricultural lands received intermediate resistance values (3–10); and large streams, urban areas, and major highways were assigned the largest resistance values

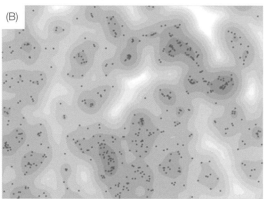

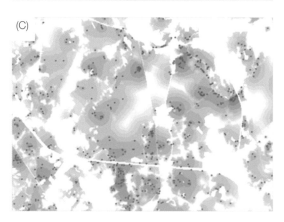

Figure 5.14 Resistant-kernel estimator of landscape connectivity for woodland salamanders. (A) Marbled salamander (*Ambystoma opacum*). (B) Connectivity of vernal pools (dark green dots) using a standard-kernel estimator; darker contours indicate a higher probability of salamanders arriving at the target pool, and thus a higher degree of connectivity. (C) Same landscape, but connectivity assessed with the resistant-kernel estimator, which includes the relative resistance of roads and other habitats and land uses to salamander dispersal.

Source: (A) © Jason Patrick Ross/Shutterstock.com. (B) and (C) provided by Kevin McGarigal.

(25–40), with salt marshes representing an absolute barrier to dispersal. Some 30,000 potential vernal pools throughout the state were then ranked based on their overall degree of connectivity (i.e. averaged across local, neighborhood, and regional scales), since pools that are not well connected at any one scale would be less likely to contribute to metapopulation viability. This approach thus enabled the investigators to identify pools with the highest connectivity scores, which can then assist with prioritizing pools for more focused monitoring and conservation (**Figure 5.14B and C**). Otherwise, individually surveying some 30,000 pools for salamanders would be a daunting task, the associated costs and time being prohibitive, and it still would not give us a comprehensive assessment of connectivity at the broader landscape or regional scales that are ultimately impacted by land development that threatens salamander (and other amphibian) populations.

Although resistance-surface models provide a foundation for assessing population and landscape connectivity, they do not by themselves identify specific movement corridors that might be essential for preserving connectivity and the long-term viability of populations. Thus, Cushman and his colleagues (2009) used an enhanced resistance-mapping approach by adding a 'source-destination' least-cost path analysis to identify potential corridors and barriers to movement for a target species, the black bear (*Ursus americanus*), in the northern Rocky Mountain region of the USA. In this case, genetic information was used to map resistance of the landscape to gene flow rather than the movement of individuals per se, an approach that we will discuss in more detail in **Chapter 9**. For now, a high degree of genetic relatedness among locations implies high connectivity, because individuals must have successfully dispersed from the source location to reproduce at the destination location for there to be such high genetic similarity between the two locations. By modeling least-cost paths between a large number of source locations near the Canadian border and destination locations along the northern border of Yellowstone National Park (25,600 potential paths in all), Cushman et al. (2009) identified three major corridors running through this region, as well as nearly two dozen potential barriers that might impede movement (and thus gene flow) along these corridors, either because they represent areas that are vulnerable to future human development (owing to gaps in federal land ownership) or contain major highways. At present, major highways intersect these movement corridors in at least six areas, which may require mitigative actions, such as the construction of wildlife overpasses or underpasses to permit safe passage. Indeed, a wildlife overpass was constructed at one of the major dispersal barriers identified in this analysis (the Evaro Hill overpass on Highway 93 in Montana), and there are more than 40 underpasses that were installed along this stretch of highway alone.

APPLICATIONS OF CIRCUIT THEORY Another approach for identifying movement corridors and evaluating connectivity in heterogeneous landscapes has been adapted from electrical circuit theory (McRae 2006; McRae et al. 2008). In this context, the landscape is represented as a circuit, consisting of a discrete network of nodes (e.g. habitat patches or populations) through which dispersal or gene flow occurs (**Figure 5.15**). In other words, the landscape is depicted as a graph (**Figure 5.12**), with resistors between nodes that reflect the relative resistance of the intervening habitat to movement or gene flow. Functionally, this is equivalent to weighted edges that have been used in graph-theoretic applications to calculate least-cost paths, which, as you'll recall, is measured along a single optimal pathway. Unlike least-cost path analysis, however, the **resistance distance** (otherwise known as the **effective resistance**) derived from circuit theory also includes information on the availability of alternative pathways (i.e. redundancy), in addition to information on the shortest or least-cost pathway.

To illustrate how these compare, consider that if two nodes (node **a** and node **b**) are connected by a single pathway consisting of two identical segments having the same resistance and length (unit length = 1), then the resistance distance (d_R) is equivalent to the least-cost path distance (d = 2 segments/pathway = 2; **Figure 5.15A and D**). If, however, there are two identical pathways between these two nodes, the least-cost path remains unchanged (d = 2) whereas the resistance distance will now be reduced by half owing to the added redundancy (d_R = 2 segments per pathway/2 pathways = 1; **Figure 5.15B and E**). Taking this a step further, we can compare the path distance between two nodes connected by three identical pathways (**Figure 5.15C and F**). Although this once again remains unchanged for the least-cost pathway (d = 2, because the shortest path is still only two-segments long), the resistance distance for the circuit has been reduced to d_R = 2/3 (2 segments per pathway/3 pathways). Thus, resistance distance decreases as the redundancy of the network increases. In general, the resistance distance provides a measure of isolation (or really, effective isolation) between nodes: isolation is low (and thus connectivity is high) when the resistance distance is small, such as when there are many connections of low resistance, whereas isolation is high (and connectivity is low) when the resistance distance is large, such as when there is a single connection of high resistance (McRae et al. 2008).

Circuit theory has largely been used to evaluate the effects of landscape heterogeneity on gene flow and the genetic structure of populations (i.e. landscape genetics, **Chapter 9**), but could also be applied to other

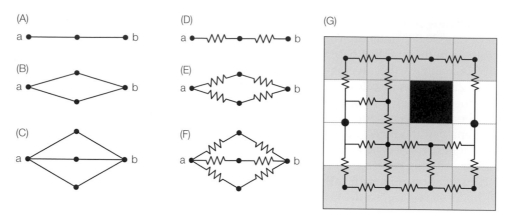

Figure 5.15 Representation of landscape structure in circuit theory. (A–C) Three graphs, each having the same edge weights. Despite the differences in graph structure, the shortest or least-cost path distance (*d*) is the same in each (*d* = 2). (D–F) Three circuits, analogous in structure to the graphs in A–C, but the edges have now been replaced with unit resistors. Because circuit theory provides an integrated measure of resistance that adjusts the least-cost distance by the availability of alternative pathways, each of these structures has a different resistance distance (2, 1, and 2/3, respectively). Note that while the resistance distance is equivalent to the least-cost distance in both A and D, the resistance distance is half that of the least-cost distance when comparing B and E. (G) Cell-based landscapes can also be represented as a circuit. Here, two patches (white) are separated by an intervening habitat matrix of some finite resistance (green); a barrier (black square) to dispersal or gene flow is also present within the matrix. Flows between the two patch nodes (large dots) can be quantified by integrating across all of the available pathways to compute the resistance distance.

Source: After McRae et al. (2008).

types of flows, such as dispersal. Circuit theory takes an 'isolation by resistance' approach to the study of dispersal and gene flow, which has been found to outperform traditional models of genetic differentiation based on isolation by distance (i.e. the Euclidean or straight-line distance between populations) or least-cost path analysis (McRae 2006; McRae & Beier 2007). Genetic distance—the degree to which two populations differ genetically—is expected to be better correlated with resistance distance (isolation by resistance) than either the Euclidean distance (isolation by distance) or least-cost path because the resistance distance better accounts for the multiple pathways (redundancy) by which gene flow may occur across the landscape. The existence of multiple dispersal pathways would equate to higher connectivity, which would then lead to less genetic differentiation among populations than might otherwise be expected based on isolation by distance or least-cost path analyses.

For example, gene flow between northern and southern populations of the big-leaf mahogany (*Swietenia macrophylla*) in Central America was better explained by a circuit-based modeling approach that incorporated isolation by resistance, then either isolation by distance or least-cost path distance (i.e. the R^2 value is greatest in **Figure 5.16A**, indicating that more of the variation in the data is explained by this model; McRae & Beier 2007). In particular, the better model fit given by circuit theory is evidenced by pairwise comparisons between seven populations located in the northern and central portions of Central America to the most distant

population located to the south in Panama, within the narrow isthmus that connects North and South America (blue circles in the graphs; **Figure 5.16**). More of these points tend to lie along the regression line under the isolation-by-resistance model than for the other models, which suggests that the circuit-based approach gives a better picture of gene flow among populations in this landscape.

We will discuss landscape effects on gene flow more fully when we study landscape genetics in **Chapter 9**. In the meantime, we consider some issues pertaining to the parameterization of resistance surfaces.

ESTIMATING LANDSCAPE RESISTANCE TO MOVEMENT A major challenge in employing cost weights or resistance surfaces is determining the costs or resistance values of different landscape elements (Koen et al. 2012; Zeller et al. 2012). Cost weights and resistance values are either obtained from empirical data or are based on expert opinion, or some combination of the two (Spear et al. 2010; Zeller et al. 2012). We consider the pros and cons of each here.

Ideally, empirical data are used to estimate resistance. Empirical estimates may be based on actual movement data obtained through direct observation or telemetric studies of individuals or their propagules (**Chapter 6**); genetic data obtained from genotyping individuals or their propagules (landscape genetics, **Chapter 9**); detection data (obtained either as presence-absence or presence-only data, and often used in conjunction with resource-selection functions or other

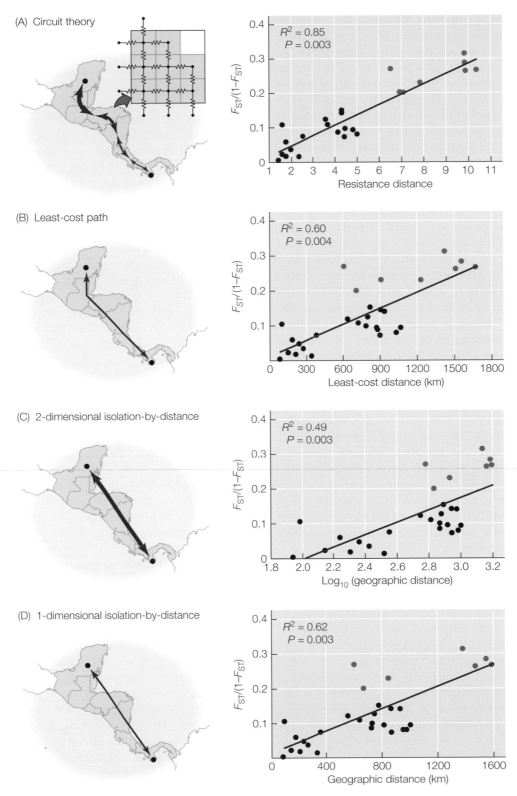

Figure 5.16 Using circuit theory to assess population connectivity. Gene flow among big-leaf mahogany (*Swietenia macrophylla*) populations in Central America is better explained by an isolation-by-resistance model derived from circuit theory (A) than least-cost path (B) or isolation-by-distance (Euclidean) models (C and D). The adjacent graphs show the linear relationship between a measure of genetic distance (the linearized F_{ST}; Equation 9.17) and resistance distance. The blue data points in each graph are the pairwise comparisons between seven populations and the most southern (distant) population; the black data points are the pairwise comparisons among the remaining seven populations. Linear regression lines are based on all data points.

Source: After McRae & Beier (2007).

species distribution models; **Chapter 7**); or some combination of these approaches (Zeller et al. 2012). Quantifying movement behavior is a challenge in practice (**Chapter 6**), and thus the actual costs of movement (in terms of energetics, survival, or individual fitness) are not usually known. Furthermore, animals may engage in different scales of movement (foraging, home-range movements, dispersal), which requires an understanding of the appropriate environmental variables that confer resistance at whatever movement scale is being studied (Zeller et al. 2012; **Figure 6.2**). Detection data are more easily obtained than movement data, but as detections usually represent individual point locations, it is unclear how these might translate (if at all) into movement and thus landscape resistance (Beier et al. 2008). Resistance is thus being inferred rather than directly measured when detection data are used (Zeller et al. 2012). Genetic data would at least seem to offer irrefutable proof that individuals have moved successfully in the past, but this again provides an indirect rather than direct measure of resistance (Zeller et al. 2012). Further, there is always the possibility that resistance estimates based on past gene flow may not match the current landscape of interest (Landguth et al. 2010).

In the absence of empirical data, many studies rely instead on expert opinion to obtain resistance values (43% of 96 studies in one review; Zeller et al. 2012). Usually, information from the literature forms the basis of the opinion, although occasionally input is solicited from other experts in the field or is based on the researcher's own experience. Expert opinions are subjective and are difficult to evaluate in terms of performance, especially since opinions may be based on the sorts of habitats in which the species occurs and not on resistance to movement per se (Zeller et al. 2012). Nevertheless, expert opinion may be the only data available for estimating resistance surfaces, and given the urgency of conservation action in many cases, it at least provides an interim solution until empirical data can be obtained (e.g. Compton et al. 2007; Zeller et al. 2012).

Regardless of the approach taken, there will always be varying degrees of uncertainty that inevitably enter into the parameterization of resistance surfaces, especially when expert opinion is involved. Therefore, a sensitivity analysis of resistance estimates to relative cost weights should be conducted (Rayfield et al. 2010; Koen et al. 2012; Zeller et al. 2012). This may be done by systematically varying cost weights (resistance values) and assessing the effect on the resulting resistance surface, or on some other attribute of connectivity, such as the effective resistance or least-cost path. Not only would this permit an evaluation of model uncertainty in the evaluation of landscape connectivity, given that the true movement costs are unknown, but it might then reveal for which landscape elements we require better information on resistance.

Assessing Connectivity in River Networks

As with terrestrial systems, humans have significantly altered the connectivity of river networks by disrupting flows and the movement of organisms (**Chapter 3**). Barriers may be physical, such as through the installation of dams; chemical, through alterations to water chemistry or an increase in acidity as a consequence of runoff, industrial effluent, or acid mine drainage, which render entire sections of the river uninhabitable for certain species; or hydrological, such as through diversions and drawdowns that contribute to low flow rates or permanently drain sections of the river. At the other extreme, we have also artificially enhanced connectivity by providing routes around natural barriers, either through species introductions—both intentional and accidental—via recreational fishing and boating, the construction of navigation canals, or the discharge of ballast water at major shipping ports, all of which have facilitated the spread of non-native and invasive species through river systems and among water bodies (**Chapter 8**). Again, not all connectivity is necessarily good or desirable!

Rivers and streams are often seen as the epitome of connectivity, as so much of what goes on is tied to water flow and hydrology.
                                                              Wiens (2002)

Although connectivity is common to both terrestrial and river systems, the main difference is that flows tend to occur longitudinally, as well as laterally and vertically, in river systems (Ward 1997). Connectivity is thus a multidimensional concept in riverine landscapes. **Longitudinal connectivity** refers to the degree to which upstream and downstream sections of a river are connected (**Figure 5.17A**). Although this is the more obvious dimension of riverine connectivity, not all aspects of longitudinal connectivity are obvious, since this involves more than just the flow of water. There also exists a gradient in physical and chemical (and thus biological) processes along the length of river, which is loosely associated with its width or stream order. This is the **river continuum concept** (**Figure 5.18**), first proposed by Vannote and colleagues (1980). Rivers tend to be heavily subsidized by inputs from the surrounding landscape at their source or in sections of low stream order, where stream primary production (P) is less than that required by stream organisms (R, respiration). Thus, $P/R < 1$ in these sections (**Figure 5.18**). Farther downstream, productivity may meet or exceed that required by stream organisms ($P/R > 1$). Excess energy and materials may be transported downstream, which might then subsidize higher-order sections in large rivers where productivity once again is insufficient to meet the energy

(A) Longitudinal connectivity

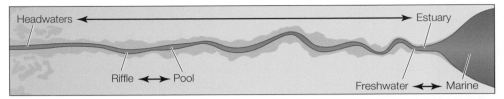

(B) Lateral connectivity

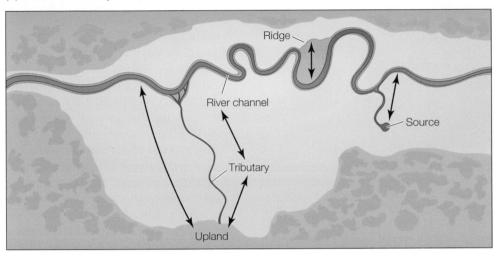

(C) Vertical connectivity

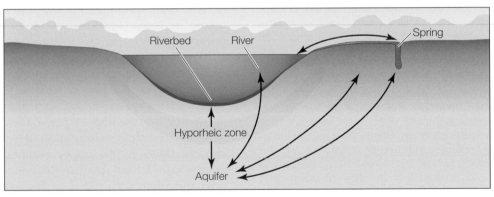

Figure 5.17 Dimensions of connectivity in riverine systems. (A) Longitudinal connectivity (B) Lateral connectivity (C) Vertical connectivity.

Source: After Wiens (2002).

needs of stream organisms (P/R < 1). Thus, the creation of dams along rivers not only obstructs water flow and the movement of organisms, but also the free flow of sediments, organic matter, and nutrients downstream.

Lateral connectivity refers to flows that occur beyond the banks of the river, as a result of periodic flooding of the floodplain (**Figure 5.17B**). Once again, this periodic or seasonal exchange of water, sediment, organic matter, nutrients, and organisms may provide critically important subsidies that benefit both terrestrial and aquatic systems. For example, deforestation

and the construction of levees along major rivers in the Pacific Northwest of North America has interfered with the recruitment of large woody debris. Historically, large woody debris created massive 'log jams' that helped to shape these river systems, giving rise to a complex and highly dynamic braided structure consisting of multiple channels with forested islands, deep pools, and floodplain sloughs (Collins & Montgomery 2002). Lateral connectivity has thus been disrupted as rivers have become increasingly isolated from their forested floodplains. River restoration in these areas must focus not just on the river itself, but also on

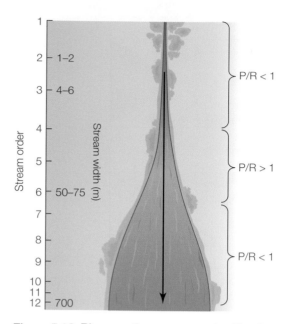

Figure 5.18 River continuum concept.   The river continuum concept depicts the relative importance of allochthonous (external) to autochthonous (internal) inputs to the system. When the ratio of primary production (P) to ecosystem respiration (R) is less than 1 (P/R < 1), primary producers (e.g., benthic algae or phytoplankton) are not able to meet all of the energetic demands of species within the aquatic community or ecosystem, and thus external inputs from the surrounding landscape assume greater importance for ecosystem functioning. The relative importance of allochthonous inputs (subsidies) from the surrounding landscape can change along the length of the river, based on light levels and the amount of shading relative to the width or order of the stream. Thus, small forested streams (1–2 order) may rely predominantly on allochthonous inputs (P/R < 1), whereas somewhat larger segments of the stream may be predominantly autochthonous (P/R > 1). Very large streams or rivers may once again rely heavily on allochthonous inputs (P/R < 1), including any excess production from upstream. Rivers thus exhibit longitudinal connectivity, not only in terms of water flow and the movement of organisms (which can move both upstream and downstream), but also in terms of the flow of energy and organic matter downstream.

Source: After Vannote et al. (1980).

removal of levees and on reforestation of the surrounding floodplain, in order to restore processes critical to the function and the dynamics of these systems.

Vertical connectivity refers to the exchanges that occur between the river, the atmosphere, and the groundwater or sediments below (i.e. the hyporheic zone; **Figure 5.17C**). Vertical flows involve not only precipitation that falls directly into the river or as runoff from the surrounding landscape, but also as groundwater that percolates up through the sediments into the river. For some rivers, groundwater may even contribute the main source of water flow at certain times of the year. Even in rivers without a substantial

contribution of groundwater, however, the exchange of water in the hyporheic zone is still important for nutrient cycling, temperature modulation, and filtration in these systems, and it also provides habitat for many aquatic insects and the eggs and larval stages of some fishes (e.g. salmon). The hyporheic zone runs parallel to the river channel, and varies in size from a few centimeters deep to several kilometers wide, depending upon geology, channel morphology, and the size of the river. Thus, much of the river may actually be running underground and far beyond the visible banks of the river, which explains why human alterations of the landscape elsewhere in the watershed can have such an impact on water quality. In particular, paving over landscapes not only increases surface runoff, but also disrupts vertical connectivity by blocking the flow of water into the ground that would otherwise be able to slowly recharge riparian aquifers. Loss of vertical connectivity also occurs when water wells draw down the water table, such as for irrigation and municipal water supplies. A loss of vertical connectivity can thus lead not only to a reduction in water availability, but also to declines in water quality when the ecosystem services provided by the hyporheic zone—such as water filtration and temperature modulation—are compromised.

Understanding how riverine habitats are connected, both spatially and temporally, is thus a challenge, but is also key to addressing many questions of interest to river and stream ecologists (Fullerton et al. 2010). Compared to terrestrial and marine systems, riverine connectivity is much more vulnerable to disruption because the flow of water is unidirectional and so few pathways exist for dispersal and other flows to occur. Further, the importance of the *type* of connectivity tends to shift from longitudinal (upstream to downstream) flows in the upper reaches of the river, to increasingly lateral (mainstem versus floodplain habitats) and vertical (surficial versus hyporheic) flows in lower sections of the river (Fullerton et al. 2010). These lateral and vertical dimensions of connectivity blur the distinction between the aquatic and terrestrial realms, making it difficult to demarcate the boundaries of the river. Connectivity is also more obviously a dynamic concept in riverine systems, which may well change depending on flow levels and precipitation, such that the system is connected only at certain times of the year or maybe only in certain years (**Figure 3.24 and 3.28; Table 3.4**).

Thus, connectivity in river systems is more complex and difficult to analyze than in other ecological systems. Efforts have been made either to adapt traditional landscape metrics or graph-theoretic approaches, or to devise new approaches that explicitly incorporate the network structure of rivers (Fullerton et al. 2010).

For example, Cote and his colleagues (2009) developed a dendritic connectivity index, which assesses the probability that fish can move between two randomly chosen points in a river network. Dendritic connectivity is thus affected by how many barriers exist between the two points and the degree to which these barriers are passable. Cote et al. (2009) found that longitudinal connectivity suffers most in the early stages of fragmentation, because there are fewer alternative paths available in a river network, especially if the main branch of the river is impacted. This is in contrast to two-dimensional terrestrial systems, where overall landscape connectivity is generally maintained until later periods of fragmentation, since there are usually alternative paths for movement available (i.e. greater redundancy and thus greater network resilience). Diadromous species—such as the spring-run Chinook salmon mentioned earlier in the section of graph-theoretic approaches to landscape connectivity—are the most sensitive, since they must access freshwater environments from the ocean to breed. Thus, barriers such as dams can potentially render all upstream habitats unusable, especially if barriers are placed near the mouths of river.

Which Landscape Connectivity Approach to Use and When?

As with patch-connectivity measures, the decision of which landscape connectivity measure or approach to use will depend on a number of factors related to the research objective, the type of landscape (whether discretely patchy or continuously heterogeneous) or type of landscape data available, the availability of information on movement behavior or dispersal range of the target species, perceived ease of implementation or computational efficiency, and personal preference. If the research objective is primarily heuristic, to generate general hypotheses or insights into how landscape structure affects some ecological process or to tease apart the relative effects of habitat loss from fragmentation on some ecological response, then NLMs are tailor-made for these sorts of applications, especially when coupled with other types of ecological models to develop spatially explicit ecological theory (e.g. fractal NLMs coupled with metapopulation models to assess extinction thresholds for different types of species; With & King 1999a). If the objective is more applied, in terms of developing management recommendations for maintaining landscape connectivity for a species of conservation concern, then we might employ a simulation modeling approach using remotely sensed land-cover data within a GIS (i.e. a grid-based simulation model) instead. Then again, we might opt for a graph-theoretic or circuit-based approach, or perhaps even a resistant-kernel modeling approach, depending on the type of landscape and movement data available.

Some landscapes lend themselves more naturally to one approach than another. For example, the connectivity of plant populations distributed among rocky outcrops within a montane forest landscape might best be modeled using a graph-theoretic or circuit-based approach because patches can be clearly delineated and perhaps the genetic relatedness of individuals can be used as a measure of gene flow among different populations to give us a sense of the overall connectivity of this network. If the intervening matrix is heterogeneous, however, and differentially affects the movements of pollinators (such as butterflies or hummingbirds) or seeds (if dispersed by ants or small mammals), then we will need more information on the likelihood of movement through different types of habitats (and for pollen versus seeds) by these different animal vectors in order to analyze least-cost paths or resistance distances that affect gene flow or to identify important stepping stones for maintaining overall connectivity of the plant population network. Or, we might instead adopt a resistance-surface modeling approach to evaluate connectivity and make predictions about how different forest management plans might be expected to affect gene flow in this system. Alternatively, if the landscape exhibits less well-defined patchiness, or it is not immediately clear how the species might be perceiving patchiness, then a grid-based simulation modeling approach may be appropriate for evaluating landscape connectivity. This may incorporate just some basic information on movement (e.g. maximum dispersal distance) or entail the development of very detailed movement rules governing the response of individuals to different elements of the landscape (e.g. movement rates within different habitats, response to edges, barriers to movement).

Although there may not be a single best way to assess landscape connectivity, since what is deemed 'best' may ultimately be a matter of opinion, the examples given in this chapter illustrate a variety of ways by which we can model or evaluate landscape connectivity. However, these examples should not be taken to mean that the particular approach used to assess landscape connectivity is the only one that could (or should) be used in that context. For example, gene flow can be (and has been) modeled as a percolation process (e.g. Green 1994; Ezard & Travis 2006), and thus can be studied using NLMs and not just graph-theoretic or circuit-based approaches. In a similar vein, theoretical graph structures can be used as neutral models for assessing how the alteration of network structure (through the removal of a certain proportion of nodes or links) affects overall landscape connectivity and the occurrence of critical thresholds. Other ways in which these landscape connectivity methods have been used are scattered throughout this textbook. So although this chapter provides some general guidelines as to which methods to use and when, you are encouraged to evaluate the relative merits of these various approaches within the context of your particular system and research needs.

Beyond Landscape Connectivity

Although landscape connectivity is considered a system-wide property, landscapes may exhibit patchiness—and thus connectivity—across a range of scales. Our discussion of patch-based versus landscape measures of connectivity illustrated two scales at which connectivity may be assessed, but other scales of patchiness may be nested between as well as beyond these scales. For example, we discussed how hierarchical neutral landscapes could be used to simulate nested patch structure at different scales within a landscape. In the case of graph-theoretic applications, edge thresholding can be used to explore how the network structure changes at different scales (edge distances), revealing graph components (subgraphs) that persist across a wide range of scales, and which therefore may contribute to the robustness of the network or be crucial for preserving network resilience. We may choose to define connectivity at various scales that have biological or management relevance, as in the case of the resistant-kernel modeling approach that assessed the connectivity of vernal-pool amphibians at local, neighborhood, and regional scales. Finally, ecological systems themselves are interconnected. Terms such as **metapopulation** (a system of interacting populations linked by dispersal); **metacommunity** (a collection of local communities linked by the dispersal of multiple, potentially interacting species; Leibold et al. 2004), **metaecosystem** (a set of ecosystems connected by the spatial flows of energy, materials, and organisms, Loreau et al. 2003); and, **metalandscape** (a system of interacting landscape populations connected by dispersal; With et al. 2006) have all been added to the ecological lexicon to emphasize the connections among more traditional units of ecological study. Nor do the connections stop at traditional disciplinary boundaries. The importance of the land-water interface for managing water quality is increasingly dissolving the boundary between terrestrial and aquatic fields of study by emphasizing the linkage between these systems. For example, water chemistry and the species composition of aquatic systems are profoundly affected by the surrounding land use, such as through the addition of nitrogen and phosphorus into lakes and streams from agricultural runoff. We will return to the topic of connections between the land-water interface and other metaecological systems in the second half of the book.

Should Landscape Connectivity be a Dependent or Independent Variable?

Because landscape connectivity has the potential to influence many ecological processes, such as dispersal success, gene flow, and the redistribution of individuals or materials across the landscape, it might reasonably be argued that it could be treated as an independent variable in a study of landscape effects on process (Goodwin 2003). For example, we might wish to relate some connectivity measure such as the percolation threshold to a particular ecological response (e.g. dispersal success). In a review of the literature, Goodwin (2003) found that most (78%) studies in fact treat landscape connectivity as an independent variable, which addresses the question: 'How does landscape connectivity affect ecological processes?' This question obviously assumes that the measure of connectivity we use accurately represents landscape connectivity for the organism (or process) in question. As mentioned previously, however, structural measures of connectivity are not always meaningful from an ecological standpoint. Further, this seems to contradict the view of landscape connectivity as an emergent property that comes about through the interaction of movement responses to landscape structure.

If landscape connectivity depends on the scale of the process or species considered, then perhaps landscape connectivity should be viewed as a dependent variable in our analysis? This perspective considers questions such as: 'How does movement interact with landscape structure to influence connectivity?' or 'Is this landscape connected from the perspective of this species?' Interestingly, Goodwin (2003) found that studies that viewed connectivity as a dependent variable tended to adopt a simulation modeling approach in order to assess landscape connectivity, perhaps reflecting the perceived complexity of the interaction between species movement responses and landscape pattern. Recent studies using circuit theory, least-cost modeling, and resistance surfaces are also clearly dealing with landscape connectivity as a dependent variable.

Ultimately, both approaches are necessary for a comprehensive understanding of landscape connectivity. Landscape connectivity is both a 'vital element of landscape structure' (Taylor et al. 1993) that can influence ecological flows, and a property that emerges as a consequence of the interaction between landscape pattern and ecological process (With et al. 1997). Landscape connectivity thus epitomizes the reciprocal nature of pattern-process linkages, which defines the study of landscape ecology.

Future Directions

Future research must continue to advance our understanding of how movement interacts with landscape structure to influence landscape connectivity. This is essentially the core of connectivity research, and thus represents an ongoing need, especially in the context of conservation and land-use planning. Nevertheless, connectivity research would benefit from more empirical research on species' movement responses (or the movement of their propagules) to landscape structure, especially in terms of movement costs, not only so that we may incorporate more realistic movement behavior in our models, but also to determine which aspects of movement behavior (and at what scales) are important to the assessment of functional connectivity. This might then lead to a better understanding of when such detailed information on movement responses to landscape structure is really necessary to assess connectivity. For example, it would be good to know when structural connectivity is a reasonable proxy for functional connectivity, especially since it is far easier to assess the structural connectivity of a landscape.

Along those lines, we should strive for more testing or validation of connectivity assessments, especially if these are being conducted for the purposes of land-use planning or as part of a conservation-needs assessment. Given some general information on a species' dispersal distance or its presumed cost of moving through different landscape elements, it is now all too easy (and commonplace) to generate maps depicting the supposed connectivity of habitat or populations on the landscape. While this might be an important first step in an analysis of landscape connectivity, especially if data (and time) are limited, it should not be viewed as the final product or end-stage of the analysis. Ideally, these connectivity maps represent hypotheses to be tested, which then lead to further refinements as additional movement data are acquired and putative linkages or critical habitat nodes are evaluated further.

We also need a better understanding of the relationship among various connectivity measures. Although this once again reiterates the need for understanding the relationship between structural and functional connectivity, it also highlights the deepening concern that different connectivity measures may give different results as to whether or how landscapes are connected. Few studies have directly compared multiple measures of connectivity (Goodwin 2003), and thus it would be interesting to know whether an assessment of connectivity for a given landscape is robust to different representations. This applies equally well to the assessment of landscape connectivity based on different types of processes (movement versus gene flow). We do not know at this point how sensitive connectivity assessments might be to the use of different types of movement data, let alone movement versus genetic data.

Finally, the development and application of connectivity measures in aquatic systems, especially for dendritic systems such as rivers, lag far behind those in terrestrial systems (Fullerton et al. 2010). Some progress is being made in this area, such as the application of graph-theoretic approaches to evaluate the disruption of connectivity among fish populations (Schick & Lindley 2007) or in the development of unique metrics for quantifying the connectivity of riverine systems (Cote et al. 2009). Still, there are a number of challenges unique to riverine systems that defy the simple translation of connectivity concepts or measures from terrestrial to riverine systems, and which will therefore require the development of new and innovative approaches.

Chapter Summary Points

1. Landscape connectivity is important for many ecological flows that occur on landscapes: dispersal; gene flow; the spread of disturbances, such as fire; the spread of disease or an invasive species; water; and the redistribution of matter, nutrients, and energy.

2. Landscape connectivity can be defined *structurally*, in terms of the spatial contagion or adjacency of habitat, and *functionally*, in terms of the interaction of some process (dispersal, gene flow) with landscape pattern.

3. Functionally connected landscapes need not be structurally connected, but structurally connected landscapes are not always functionally connected, although that is the usual assumption (or hope).

4. Landscape connectivity is a landscape-wide property, although landscapes may exhibit connectivity across a range of scales.

5. Landscape connectivity can be inferred or assessed using landscape metrics, percolation-based neutral landscape models, graph theory, circuit theory, grid-based simulation models, and the use of resistance-surface and least-cost path modeling. Each approach has its associated costs and benefits, as well as degree of data requirements.

6. Landscape connectivity may exhibit threshold behavior in response to habitat loss and

fragmentation. Although thresholds in landscape connectivity may have important ecological consequences, ecological thresholds (e.g. dispersal thresholds, extinction thresholds) may not occur exactly at the landscape connectivity threshold. Different processes operating at different scales are responsible for these different threshold phenomena.

7. Different species will have different perceptions of landscape connectivity, depending upon the scale of dispersal relative to landscape pattern.

8. Assessment of landscape connectivity may differ depending upon the type of information used (dispersal versus genetic distances).

9. Landscapes are not inherently connected or disconnected, for it depends upon the scale, process, or organism being considered.

10. Connectivity is a multidimensional concept in riverine landscapes, which exhibit longitudinal, lateral, and vertical flows. Connectivity is more easily disrupted in riverine than terrestrial landscapes (e.g. a dam across the main stem of a river). Connectivity is also a more dynamic concept in these systems, as connectivity may only manifest in certain seasons or years, depending on streamflow.

11. Landscape connectivity can be either a dependent or independent variable in landscape ecological research. Landscape connectivity is treated as an independent variable in ecological investigations that seek to uncover how landscape connectivity (or a disruption of that connectivity) influence ecological flows across the landscape. Landscape connectivity is treated as a dependent variable in studies that assess whether the landscape is connected from the perspective of different species or processes, based on how those species or ecological flows interact with landscape structure. Both approaches are ultimately needed to permit a comprehensive understanding of landscape connectivity.

Discussion Questions

1. Under what circumstances might structural connectivity be a reasonable proxy for the functional connectivity of a landscape?

2. Consider the various approaches that have been used to assess landscape connectivity (e.g. landscape metrics, neutral landscape models, graph theory, circuit theory, grid-based simulation models, resistant-kernel modeling) and discuss whether these approaches primarily measure the structural, potential, or actual connectivity of the landscape. What are the trade-offs or limitations to each approach?

3. If landscape connectivity exhibits a threshold response to habitat loss, what information is needed to identify when or at what level of habitat such thresholds might occur? What are the implications for land management or conservation if thresholds in landscape connectivity do occur?

4. If different species have different perceptions of landscape connectivity, how might landscapes be managed so as to preserve connectivity for the greatest number of species?

5. Is landscape connectivity both necessary and sufficient for successful management or conservation? If not, what other information or factors must be considered?

6. Assessments of connectivity may differ for a given landscape, depending upon the information used (e.g. dispersal versus genetic distances). Ideally, what type of information should be used to assess landscape connectivity? What do different types of information convey about connectivity, and is one type of information inherently better than others when assessing landscape connectivity? For example, would it be better to manage for dispersal connectivity or genetic connectivity?

7. Is landscape connectivity always a desirable management goal? If a little connectivity is beneficial, is more connectivity necessarily better? Consider under what circumstances a high degree of landscape connectivity might be undesirable. Discuss ways in which landscapes might be managed so as to balance both desirable and undesirable flows across the landscape, such as the dispersal of a species of conservation concern versus the spread of an introduced predator or disease of that species.

8. Using the examples and case studies provided in this chapter or from your assigned readings, determine whether landscape connectivity is viewed primarily as a dependent variable or as an independent variable in each study. Identify what connectivity approaches or metrics are used in each case, and discuss their relative merits and shortcomings in quantifying connectivity. In your opinion, what is the best way to measure landscape connectivity?

6

Landscape Effects on Individual Movement and Dispersal

Behavioral Landscape Ecology

The migration of animals has always been something of a mystery. Where do they go? What routes do they follow? Do individuals from the same region all migrate to the same place? Through mass-marking programs, we have learned that some species are capable of truly extraordinary journeys, travelling between continents, hemispheres, and across oceans over the course of a single year. With recent advances in animal-tracking technologies, however, we can now follow the migration of individuals via satellite in real time, which is rapidly transforming our understanding of animal movements and the degree to which different landscapes or regions are connected.

For example, the leatherback turtle (*Dermochelys coriacea*) spends much of its life at sea, with females coming ashore only briefly to lay their eggs in the warm soft sands of tropical beaches. In the Pacific Ocean, there are two regional nesting populations: the West Pacific population, which nests in Indonesia, and the East Pacific population, whose nesting grounds are scattered along the coasts of western Mexico and Central America (**Figure 6.1**). Where do these turtles go after nesting? Although leatherbacks are found throughout the Pacific basin, we might naturally assume that a turtle foraging off the coast of California is one that nested a bit farther south on a beach in Mexico. Through satellite tracking, however, we now know that leatherbacks foraging off the California coast are actually from Indonesia, having traversed the entire ocean basin to forage in these highly productive coastal areas (**Figure 6.1**; Bailey et al. 2012). As with their nesting grounds, the foraging grounds of these two regional populations are also segregated. The West Pacific population that nests in Indonesia feeds in many different areas, from the South China Sea to southeastern Australia, all the way across the ocean to the west coast of the USA. By contrast, the East Pacific population has a more localized foraging range in the South Pacific, feeding in offshore upwellings where their jellyfish prey are concentrated.

An analysis of the migratory pathways of leatherback turtles has thus identified not only the linkages among nesting and foraging ranges, reinforcing that these are indeed two separate regional populations, but also the areas of concentrated use within these foraging areas where leatherbacks may be particularly vulnerable to threats posed by human fishing activities. Given that the leatherback turtle is critically endangered and bycatch is a major source of

Essentials of Landscape Ecology. Kimberly A. With, Oxford University Press (2019).
© Kimberly A. With 2019. DOI: 10.1093/oso/9780198838388.001.0001

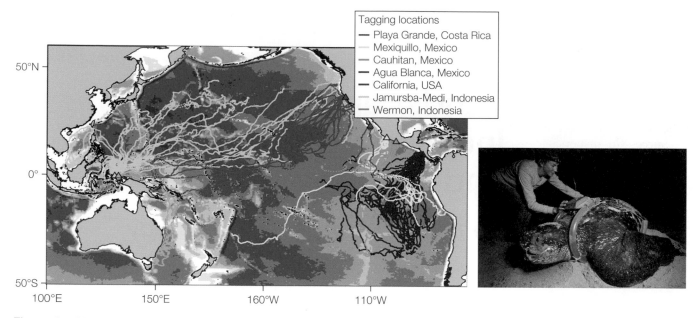

Figure 6.1 Movement patterns of leatherback sea turtles (*Dermochelys coriacea*). Individual turtles were tracked via satellite telemetry from their nesting grounds (Indonesia and western North America) to their respective foraging areas in the Pacific Ocean.

Source: Figure from Bailey et al. (2012). Leatherback turtle © Mark Conlin / Alamy Stock Photo

mortality, an understanding of its movement behavior and migratory patterns in relation to different oceanographic conditions throughout its range is absolutely essential to the conservation of this marine species. Nor is the leatherback turtle unique in this regard, as most species disperse at some point in their lifecycle. We therefore turn our attention to the landscape ecology of movement behavior and dispersal in this chapter.

Why are Movement and Dispersal Important from a Landscape Ecological Perspective?

Movement is the essence of life. All living organisms need to move about to find food, shelter, or mates, either for themselves or for their offspring. Thus, even sedentary life forms, such as plants and corals, have a motile life stage or dispersive propagules that enable these otherwise sessile organisms to colonize new areas and exploit new resources. Because environmental conditions, habitat, food, and mates are not evenly distributed across the landscape, we can anticipate that the patchy distribution of these resources will affect the movement and redistribution of individuals (or their propagules) across the landscape. The effect of resource or habitat patchiness on individual movement thus represents the finest scale of organismal response to landscape pattern.

Individuals engage in a diverse array of movements, covering a wide range of behaviors and scales that reflect the motivational state, sensory abilities, and movement mode of the individual, as well as the nature of the landscape through which they move. Thus, we might wish to understand how an individual's search behavior is modified by landscape structure when searching for food versus when searching for suitable habitat in which to breed. Different rules of movement may govern individual responses to a patchy resource distribution depending on the object of the search (food versus territory) or the scale traversed (within-habitat movements while searching for food vs between-habitat movements when searching for a territory). Then again, perhaps all movement is fundamentally similar, regardless of purpose, once we account for differences in scale and the nature of patchiness. There are basic constraints on how organisms use space and acquire energy, at whatever scale these processes occur.

Movement is a fundamental ecological process that functions not only in individual resource acquisition but also in **dispersal**, the movement of individuals from one habitat patch or population to another. As discussed in **Chapter 5**, dispersal is central to functionally based definitions of landscape connectivity, as well as for evaluating the relative isolation of individual habitat or resource patches on the landscape (patch-based connectivity measures). Dispersal also affects gene flow, which influences the genetic structure of populations across landscapes (**Chapter 9**). Thus, gene flow can also be used to assess landscape connectivity, particularly in terms of determining historical patterns of connectivity among populations. Dispersal has therefore been called 'the glue that keeps populations together' (Hansson 1991), and its measure provides an index of population connectivity.

Successful dispersal can lead to the colonization of new areas or the recolonization of areas where the species occurred previously. The degree to which patch colonization rates are able to offset local extinction rates affects metapopulation dynamics and the persistence of spatially structured populations in fragmented landscapes (**Chapter 7**). Dispersal also plays a role in the spread of species across landscapes and regions, enabling some species to track changing environmental conditions and shift their distributions in response to climate change, while allowing for the invasive spread of others (**Chapter 8**). Shifting distributions of species, coupled with the dispersal of disease vectors or their hosts, figure prominently in the emergence and spread of infectious diseases (**Chapter 8**). Communities and ecosystems are linked by the movements of species between systems, which can help to subsidize the productivity of these systems, and demonstrates how dispersal can affect the structure and function of higher-order ecological systems, beyond that of just the individual or population (**Chapters 10 and 11**).

In this chapter, we thus turn our attention to the importance of movement and dispersal in a landscape ecological context. We begin by discussing the different scales in space and time over which movement is expressed, from the fine-scale foraging behaviors of individuals to the dispersal of individuals among populations, communities, and ecosystems. Because different species—and even different life stages within a species—operate at different scales, we can use information on movement responses to environmental heterogeneity, at whatever scale it occurs, to identify species' perceptions of landscape structure. Next, we explore movement responses to patch structure, in terms of what factors influence whether or when an individual should leave a patch, how elements of the landscape influence movement between patches, and what factors influence the selection of a new patch. Ultimately, our understanding of movement responses to patch structure will require a study of individual movement, and thus we consider the various methods available for tracking animals, as well as the mathematical and simulation modeling approaches that have been developed to analyze movement pathways. Because the movements of most animals are bounded in space and ultimately in time, this sets an upper limit to the range of heterogeneity that individuals may encounter, and thus the sorts of habitats or resources they might be able to use. We therefore consider the estimation of space utilization and home-range size in this chapter. Finally, although much of this chapter revolves around animal movement responses to landscape structure, we conclude with a discussion of the various approaches to measuring plant dispersal, which, as with animals, is important for evaluating how the movements of individuals (or their propagules) translate into the redistribution or spread of populations across the landscape.

Scales of Movement

Because individuals engage in an array of movement behaviors, they are likely to be affected by patch structure and heterogeneity across a range of scales. We can therefore conceptualize movement responses to patch structure as a spatially and temporally nested hierarchy, in which different types of movement manifest within particular domains of scale (Ims 1995; **Table 6.1**; **Figure 6.2**). Indeed, a variety of seemingly disparate topics in ecology dealing with foraging behavior, predation, dispersal, habitat selection, or gene flow (including pollinator behavior) are really all dealing with same phenomenon—individual movement responses to environmental heterogeneity—but are investigating the nature of this interaction at different scales relevant to the behavior of interest (Mayor et al. 2009). Of course, individuals may respond to different environmental cues or to different types of resources or habitat patches depending on whether they are foraging for food versus searching for a territory in which to breed. Different constraints may thus govern different types of movement, such that a detailed understanding of foraging behavior probably won't tell us very much about how that same individual selects a territory or breeding site within the landscape.

Nevertheless, individual fitness is the common currency that links all of these different types of movement. An individual must be successful at foraging and finding suitable breeding habitat and a mate if it is to survive and reproduce, which means that it must somehow offset the costs of movement and dispersal, such as increased predation risk, with the benefits gained through the acquisition of food, shelter, or mates (Zollner & Lima 2005). Cost-benefit trade-offs, such as between foraging and predation risk, may also change as a function of scale (Mayor et al. 2009). We should thus anticipate that the response to patch structure across a range of scales is likely to be non-linear, which could complicate the extrapolation of fine-scale movement responses to predict broader-scale patterns of habitat occupancy and range distributions (**Figure 6.2**), a topic we shall address a bit later in the chapter.

TABLE 6.1 **Parameters commonly used to describe animal movement pathways.** Refer to **Figure 6.19** for additional information.

Movement parameter	Description
total path length (L)	The total distance travelled by the individual, obtained as the summation of all line segments (steps, ℓ) making up the movement pathway as $\sum_{1}^{n}\ell$, where n is the total number of steps
mean step length $(\overline{\ell})$	The average distance moved between sequential locations, obtained as $\dfrac{\sum_{1}^{n}\ell,}{n}$ where n is the total number of line segments (steps, ℓ)
net displacement (R_n)	The straight-line distance travelled by the individual between the start and end points of its movement path. This is sometimes converted to a rate of net displacement, in which R_n is divided by the total tracking time, in order to provide a more standardized measure of net displacement for individuals that have been observed for different lengths of time
mean turning angle ($\overline{\theta}$)	Turning angles are circular quantities, and thus cannot simply be averaged arithmetically. For example, 0° and 360° are identical angles, and thus their arithmetic average (180°) would not be correct. Instead, the mean turning angle is derived as the trigonometric mean as $\overline{\theta} = \arctan(\overline{\sin\theta} \,/\, \overline{\cos\theta}\,)$
mean vector length (r)	A vector expresses both the distance and direction travelled from some reference point. The mean vector length is a unit vector measure (i.e. the normalized vector) of the dispersion of turning angles, ranging from a perfectly uniform distribution (non-directional or completely random movement; r = 0) to perfectly directional (r = 1.0). It thus gives a measure of path tortuosity (r → 0 for very convoluted pathways), and is calculated as $r = \sqrt{\overline{\sin\theta}^{2} + \overline{\cos\theta}^{2}}$

Source: Crist et al. (1992).

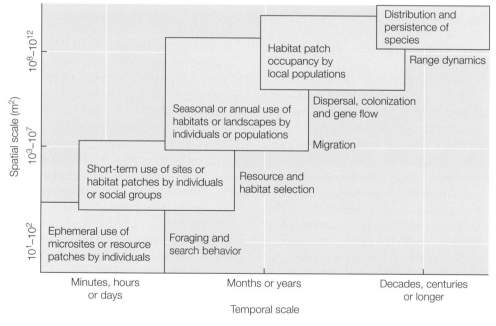

Figure 6.2 **Space-time diagram of animal movement.** Various types of movement phenomena are associated with different kinds of space use across a wide range of spatial and temporal scales. The spatiotemporal domains depicted here are loosely based on the elk (*Cervus canadensis*, **Figure 6.3**), and thus are not intended to encompass the relevant scales of movement or space use for all organisms.

Source: After Mayor et al. (2009).

Movement Responses to Hierarchical Patch Structure

Although different movement behaviors may be expressed at different scales, the same movement behavior (e.g. foraging) could also span different domains of scale. For example, a **foraging hierarchy** for a large herbivore, such as an elk (*Cervus canadensis*), may encompass varying degrees of patchiness in the distribution of forage that correspond to different scales of movement (Senft et al. 1987; **Figure 6.3**). At the

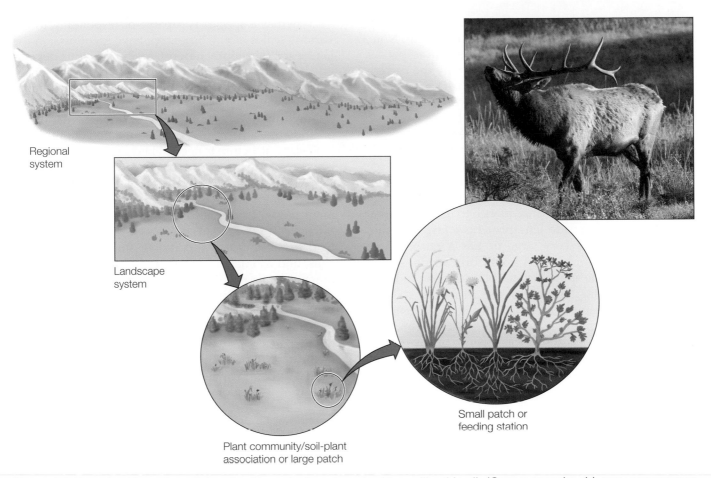

Regional
system

Landscape
system

Plant community/soil-plant
association or large patch

Small patch or
feeding station

Figure 6.3 Foraging hierarchy. A foraging hierarchy for a large herbivore, like this elk (*Cervus canadensis*), may encompass patch structure across a range of scales, from that of the individual feeding station to the regional distribution of landscapes through which the elk migrates during its seasonal movements from its high-elevation summer range to its wintering range in low-lying valleys.

Source: After Senft et al. (1987); photo by Kimberly A. With.

finest scale, foraging decisions are influenced by the distribution and abundance of preferred grasses and forbs at a microsite or feeding-station scale. These finer-scale patches may in turn be nested within larger patches comprising the particular plant community or habitat in which the elk is foraging. At a broader scale, foraging habitat is distributed unevenly throughout the landscape, which is made up of different habitat types, not all of which provide suitable forage but which may nevertheless afford other amenities, such as forest shelter from predators or the elements. Finally, elk may track the seasonal availability of forage among landscapes, by moving down from their high-elevation summer range to low-lying valleys during the winter. Consistent with the notion of hierarchy theory (**Chapter 2**, **Figure 2.8**), foraging hierarchies are defined not just by the scaling of nested patch structure, but also by the forager's movement responses and the rates at which it interacts with these different scales of patchiness. Foraging decisions are made far more frequently at finer scales than at broader scales in

the hierarchy: an elk may make millions of foraging decisions at a microsite scale over the course of a year, but may travel between different landscapes only a couple times a year. Decisions made at a broad scale, however, will constrain all other decisions available to the elk at finer scales in the foraging hierarchy. A preference for a particular forb will go unmet if the elk is not in a habitat patch or landscape where that forb species occurs.

In defining foraging hierarchies, or the range of scales governing movement responses to hierarchical patch structure more generally, we must identify the **perceptual grain and extent** over which the species interacts with its environment (Kotliar & Wiens 1990). The finest scale at which an organism responds to patch structure is referred to as the resolution or grain of its movement response (**Figure 6.4**). Below this, the organism perceives the environment to be homogeneous, or at least, behaves as if it does. Because it is not possible to know how another species actually perceives its environment, we are assuming that the finest scale at which an organism responds to environmental heterogeneity is equivalent to its perceptual grain. Note, however, that the spatial grain of movement responses may reflect a coarser grain than that set by the perceptual abilities of the organism (**Figure 6.4B**, Species C). That is, the organism may well perceive finer-scale structure but does not respond to it, either because that level of detail is irrelevant to its current objectives or because it represents background 'noise' and is ignored. For example, we have the visual acuity to see that there are cracks in the sidewalk, but these are generally ignored as we walk down the street. Sidewalk cracks are usually of no consequence to us, in that they do not affect our stride length or how fast we walk, unless the crack is very large or has a raised edge, causing us to stumble and break our stride. At the other extreme, the spatial extent bounding an individual's movements determines the broadest scale at which it may respond to patch structure. Whether it then perceives broader scale heterogeneity beyond this cannot usually be known in the absence of a response. Thus, the grain and extent of an organism's movement response are usually determined behaviorally rather than perceptually.

Whether a species is capable of responding to the various scales of patchiness we have identified on the landscape (**structural heterogeneity**) will thus depend on the grain and extent of its movement responses to that patch structure (**functional heterogeneity**). Foraging hierarchies are defined functionally, which requires that we adopt a species-centered definition of landscape structure. A species-centered definition of landscape structure is needed because different organisms are likely to perceive and respond to patch structure across different ranges of scale (**Figure 6.4**), owing to differences in body size, perceptual abilities, movement mode, or developmental stage (Wiens & Milne 1989; With 1994a,b). For example, larger organisms tend to move faster and farther than smaller organisms, and as such, are likely to have either a broader range of response (**Figure 6.4**, compare Species A and B) or a range that is shifted toward broader spatial scales (**Figure 6.4**, compare Species A and C). Body size is also expected to correlate with the **perceptual range**, the distance over which species can detect landscape features (Lima & Zollner 1996). Once again, what a species appears capable of perceiving is gauged by the nature of its response, in terms of whether it orients

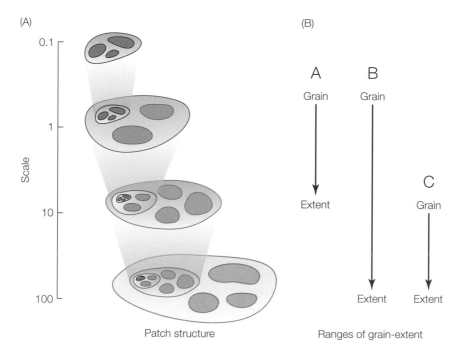

Figure 6.4 Range of movement responses to hierarchical patch structure. (A) A nested hierarchy of habitat or resource patches. (B) Species may not respond to all scales of hierarchical patch structure. The spatial range (grain and extent) encompassing the movement responses of each species (A–C) to this patch structure is indicated by the corresponding arrow. Thus, Species A and B both respond to fine-scale patchiness (they share a similar perceptual grain), but Species B operates over a broader range of scales than Species A, and is thus influenced by patch structure at broader spatial scales (the two species have different perceptual extents). Conversely, Species B and C operate at the same broad spatial extent, but Species B responds to patchiness at a finer scale than Species C (they have a different perceptual grain). A species' perception and response to patch structure may well change throughout its lifetime (i.e. different life stages may exhibit different spatial ranges), and may also be modified by landscape context (i.e. the species perceives and responds to a narrower or broader range of patch structure in certain landscape contexts).

Source: After Kotliar & Wiens (1990).

toward some feature, like a forest edge. Among small mammals in fragmented forest, for example, the comparatively large fox squirrel (*Sciurus niger*) was able to perceive and orient toward forest edges from a distance of almost 500 m when placed in the adjacent field matrix, which was nearly five times farther than the much smaller white-footed mouse (*Peromyscus leucopus*; Mech & Zollner 2002).

Allometric Scaling of Movement

Allometric scaling relationships, such as between body size and movement rate, dispersal distance, or home-range area, can help to identify the spatial (and temporal) scales over which different species operate, and therefore, how they are likely to perceive and respond to the distribution and abundance of habitat or other resources within a given landscape. Recall from **Chapter 2** that allometric relationships exhibit power-law scaling (Equation 2.1); in other words, the scaling relationship is fractal. For example, home-range size (H) is thought to scale with body mass (M) as $H = \alpha M^k$, where α is a constant and k is the scaling exponent (the slope of the relationship on a log-log plot), which varies between 0.75 and ~1.0 for cold-blooded and warm-blooded vertebrates, respectively (Hendriks et al. 2009). Allometric relationships presumably reflect the energetic constraints imposed by body size, which places an upper limit on home-range size and other space-use requirements. In turn, such space-use requirements will likely determine the species' dispersal range, perhaps more so than body size alone (Bowman et al. 2002). For example, mammals with large home ranges disperse proportionately farther than those with smaller home ranges. A similar relationship exists between bird territory size and dispersal distance (Bowman 2003), although it appears that birds disperse proportionately farther than mammals, perhaps owing to their greater vagility.

Whatever the explanation, these sorts of allometric relationships can provide a useful tool for identifying the scope of the scaling domain (Equation 2.2) that bounds individual movement and space use for different species. Since the distribution of resources or habitat within the landscape also influences energy acquisition and space use, allometric scaling relationships should interact with the scaling of landscape patterns to influence movement and the redistribution of individuals. For example, Milne and his colleagues (1992) used a variety of allometric relationships to regulate the foraging energetics and movement behavior of simulated foragers, which included mass-dependent relationships for animal density (number of foragers of a given size that could be supported by a given-sized landscape), home-range size (amount of foraging area needed for each animal of a given size),

ingestion rate, metabolic rate, movement rate, and transport costs (energy expended by a given-sized animal when moving between home ranges). Their simulation model demonstrated how 'allometric herbivory' (scale-dependent foraging) could interact with scale-dependent (fractal) resource distributions to influence foraging success, and how this in turn influenced dispersal and species' distribution patterns across the landscape. For example, foragers initially concentrated their foraging activities within areas of high resource abundance within their home ranges, and thus exhibited relatively sedentary behavior. As these resource areas were depleted, however, foragers began moving about in search of suitable range elsewhere, until by the end of the simulated time period, most individuals in the population were moving about in search of food and were highly dispersed.

Interestingly, this shift in movement behavior—from concentrated foraging to increased dispersion—was modified by landscape pattern and was only evident when forage was patchily distributed across the landscape, not when the forage distribution was uniform. Patchy resource distributions thus tended to 'mobilize' foragers, causing them to shift the scale at which they interact with the landscape pattern to broader scales. This is consistent with previous work, based on percolation theory, which attempted to define the resource utilization scales at which species should operate, given constraints on movement across random landscapes (i.e. the species should be able to move freely across the landscape when the critical resource or habitat occupies ≥59% of the landscape; O'Neill et al. 1988b). Basically, species should operate at broader scales when resources are either scarce or aggregated on the landscape. Put another way, only species capable of traversing large distances would be able to make use of resources that are sparse or patchily distributed, as this would enable them to move among patches and locate new resources.

Movement Responses to Patch Structure

Movement through a patchy landscape typically entails three stages: (1) emigration, leaving a patch; (2) movement among patches, typically through a heterogeneous matrix of variable resistance; and (3) immigration, finding and selecting a new patch to enter (Bowler & Benton 2005). A variety of intrinsic (behavioral) and extrinsic (environmental or landscape) factors influence movement decisions at each stage (**Figure 6.5**). Our discussion here encompasses a wide range of movement behaviors, from foraging behavior to dispersal and habitat selection, so as to highlight the generalities that

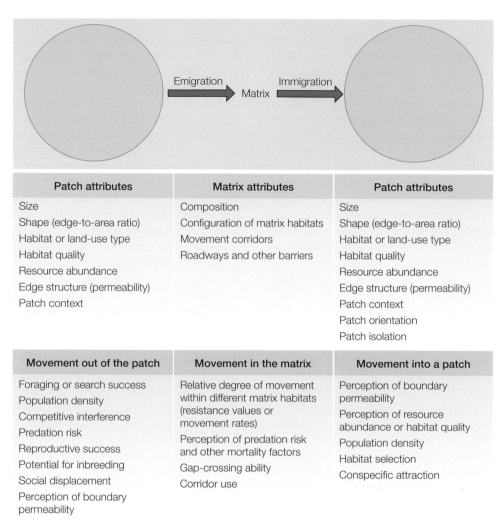

Patch attributes	Matrix attributes	Patch attributes
Size	Composition	Size
Shape (edge-to-area ratio)	Configuration of matrix habitats	Shape (edge-to-area ratio)
Habitat or land-use type	Movement corridors	Habitat or land-use type
Habitat quality	Roadways and other barriers	Habitat quality
Resource abundance		Resource abundance
Edge structure (permeability)		Edge structure (permeability)
Patch context		Patch context
		Patch orientation
		Patch isolation

Movement out of the patch	Movement in the matrix	Movement into a patch
Foraging or search success	Relative degree of movement within different matrix habitats (resistance values or movement rates)	Perception of boundary permeability
Population density		Perception of resource abundance or habitat quality
Competitive interference		
Predation risk	Perception of predation risk and other mortality factors	Population density
Reproductive success		Habitat selection
Potential for inbreeding	Gap-crossing ability	Conspecific attraction
Social displacement	Corridor use	
Perception of boundary permeability		

Figure 6.5 Landscape and behavioral factors affecting movement within and among patches.
Source: After Baguette & Van Dyke (2007).

influence movement responses to patch structure at whatever scales these interactions occur. Further, the term 'dispersal' is applied to many different types of movement across a range of scales (Bowler & Benton 2005). For example, 'dispersal' in some quarters may be defined as the movement from the natal patch to the breeding patch (natal dispersal) or among patches of breeding habitat from one season to the next (breeding dispersal), which works fine for territorial species like birds. In other cases, however, dispersal might be defined as the movement from one population to another, or even more generically, as the movement from one habitat patch to another. Without getting bogged down in semantics, the presentation in this section reflects the terminology of the literature, although an effort is made to point out where movement theory developed in one context (e.g. optimal foraging) could also be used to assess the relative costs and benefits of movement in other contexts. Although the specific costs or

benefits may change depending on the scale of movement being considered (foraging among resource patches within a home range versus moving between seasonal home ranges), the consequences for the individual (survival and reproductive success) all share a common currency measured in terms of fitness.

Movement out of Patches (Emigration)

The decision to leave a patch is predicated on the decision to remain in a patch; that is, whether the benefits of staying within the patch outweigh the costs. We therefore begin by discussing what factors influence whether individuals should remain in a patch, before turning our attention to what factors might then motivate them to leave.

HOW MUCH TIME TO SPEND IN A PATCH? SEARCH EFFORT AND RESIDENCE TIME It may seem obvious that individuals should spend more time in areas where resources are abundant than in areas where resources are scarce. To do so, however, individuals need to adjust their movement behavior to maximize their time within a patch once it has been located. For example, foraging individuals tend to slow down and turn more frequently within a resource patch than when moving between patches. This produces a pattern of **area-restricted search** that increases encounter rates with the resource, thereby enabling foragers to exploit patchily distributed resources more efficiently (Kareiva & Odell 1987).

The amount of time spent in an area should correlate with search effort, which again is expected to be greater in areas of high resource abundance. One way to assess search effort is by the mean **first-passage time**, which is the time required for an animal to traverse a given distance (actually, a circle of a given radius; Johnson et al. 1992). If an animal exhibits concentrated search in certain areas of the landscape (area-restricted search), then it will take longer to traverse a given distance compared to areas where the animal moves quickly and spends little time. Mean passage time will also increase as the distance (radius) used increases. Thus, first-passage time provides a scale-dependent measure of movement behavior and search effort, and can be used to assess the scale(s) of patchiness to which individuals respond. For example, an analysis of first-passage times for foraging Antarctic petrels (*Thalassoica antarctica*) revealed a hierarchical search strategy, in which petrels concentrated their search efforts at a scale of 20–50 km that was nested within a broader area of restricted search on the order of several hundred km (Fauchald & Tveraa 2003).

Consistent with the idea that individuals should spend more time within areas of high resource abundance, we could also calculate an index of **residence time** (Turchin 1991). A low residence time would indicate that individuals are moving quickly through a particular habitat, whereas a high residence time implies that they are spending more time in that area. If residence times vary greatly among different patch or habitat types, this may well influence the distribution of individuals across the landscape. Intuitively, we expect that individuals will accumulate in areas with a higher residence index, which may lead to a patchy population distribution. Thus, an index of residence time can provide a link between individual movements and the spatial distribution of populations (Turchin 1991).

To explore this potential, With and Crist (1995) calculated a residence index for two grasshopper species, based on their relative rates of movement in different habitats within the shortgrass steppe of eastern Colorado (USA). One species (*Psolessa delicatula*) had a higher residence time in homogeneous habitats dominated by grass, which corresponded to a low probability of moving out of this habitat. The other species (*Xanthippus corallipes*) exhibited high residence times within homogeneous as well as very heterogeneous habitats containing a mix of bare ground, grass, cactus, and shrubs. Very little of the landscape (8%) comprised homogeneous grass cover, and thus we might expect that *P. delicatula* in particular would exhibit a patchy distribution, given its high residence time within this habitat. However, this habitat type appears to have been too scarce to cause a similarly patchy distribution in the grasshopper, especially given that this species was able to move and forage within the other two habitat types. Thus, this species ended up having a more widespread distribution across the landscape, whereas the other species (*X. corallipes*), with its higher residence times within habitat types that comprised a third of the landscape, attained a patchy distribution. Our ability to translate individual movements into patterns of species distribution thus requires information about the composition and configuration of the landscape, in addition to the relative rates of movement or residence times within different habitats that make up that landscape.

Further, if movement behavior is simultaneously influenced by patch structure over more than one scale, as in a foraging hierarchy (**Figure 6.3**), then the broader spatial context in which patches are embedded could also influence patch residence times. For example, Searle and colleagues (2006) examined the influence of patch context on residence times in foragers, by creating a hierarchically nested resource in which they manipulated both the size and distribution of feeding stations to create a hierarchical design of stations nested within patches, which in turn were grouped together to form even larger patches. By altering the distances among feeding stations and patches, they could then test whether residence times for large herbivores, such as the mule deer (*Odocoileus hemionus*) and grizzly bear (*Ursus arctos*), could be explained by attributes of the feeding station alone (the local resource level), or whether additional information on patch context (the distance among neighboring stations or patches) was required. For both species, residence time—whether within a local feeding station or patch—was always better explained when information on the broader patch context was included in the model. Inter-patch distances had a greater effect than inter-feeding station distances on residence time in grizzly bears, whereas both were about equally important in explaining residence time in mule deer. Thus, patch context may well influence foraging decisions in these and other animals, such that movement behavior evidenced at broader scales in a foraging hierarchy cannot be derived simply as the summation of finer-scale movement behaviors.

Patch residence time is related to the 'giving-up time,' the optimal amount of time a forager should spend within a patch before leaving to find a more-profitable patch elsewhere. We consider the concept of optimal foraging next.

WHEN TO LEAVE A PATCH? OPTIMAL FORAGING AND THE MARGINAL VALUE THEOREM A patch residence time implies that individuals will eventually leave and move to a different patch. The decision to leave will likely be determined by how quickly resources or prey within the patch are depleted relative to the likelihood of encountering another resource patch elsewhere. If we assume that individuals forage optimally, meaning that they seek to maximize their rate of energy intake[1] while minimizing their time spent foraging, then the decision to leave a patch would ideally represent a trade-off between the cost of staying in a depleted patch assessed against the risk of searching for a new patch but then possibly finding a more profitable patch.

Even organisms like [the nematode worm] *C. elegans*, which lack the spine to make decisions, can potentially adjust the rate at which they leave a patch in response to deteriorating conditions.   Adler & Kotar (1999)

The **marginal value theorem** predicts that individuals should leave a resource patch when their net intake rate drops to the overall average for that habitat (Charnov 1976; **Figure 6.6**). Graphically, this corresponds to the point of diminishing returns on the resource intake curve, which is a function of habitat quality. Thus, individuals should spend more time in patches when foraging in high-quality habitats than when foraging in low-quality habitats. Within a given habitat, the amount of time spent in patches (the **giving-up time**) should be greater for higher quality patches than for lower quality patches.[2] Finally, the giving-up time is expected to increase as the distance between patches increases. Thus, the farther apart patches are, the longer the expected travel time, and thus the longer foragers should remain within the patch once they get there (compare the red line to the blue line in **Figure 6.6**). Note that theory predicts that the forager should leave the patch well before all resources have been consumed.

[1]Energy intake is expected to be related to survival and reproduction, and thus, individual fitness.

[2]Note that the patch residence time increases with the giving-up time; residence time and giving-up time are thus related—but not necessarily equivalent—concepts. Residence time within a patch may increase for reasons that have nothing to do with patch quality and food availability (resource intake), as when movement through a patch is reduced owing to the structural complexity of the habitat (habitat is physically difficult to move through) or because of social inhibition (high density resulting in aggressive interactions that prevent individuals from leaving patches), as we discuss in the next section.

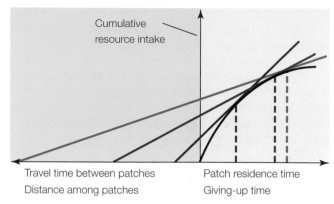

Figure 6.6 Marginal value theorem.   The marginal value theorem predicts that a forager should leave a patch before the point of diminishing returns in its resource intake (black curve). The optimum time spent within the patch is a function of how profitable the patch is and the travel time between patches (i.e. the distance among patches). Thus, an individual should spend more time (patch residence time) or take longer to give up (giving-up time) when the travel time (or distance) among patches is greater (e.g. compare the blue dashed line to the red dashed line).

For example, if a predator attempted to stay until all prey were captured, it would spend ever more time searching as prey become increasingly scarce, and the search costs would exceed any benefit that could possibly be gained once those last few prey were finally captured. Such a predator could hardly be said to be foraging optimally. The marginal value theorem thus predicts that all patches should be depleted to an equivalent prey density (the **giving-up density**), regardless of the initial patch quality (Brown 1988).

Giving-up densities may thus reflect how foragers perceive patch quality, in terms of both resource availability and predation risk. A high-quality patch may go under-exploited if the perceived predation risk is simply too great to permit the forager to linger there. Higher giving-up densities may thus be associated with a higher perceived risk of predation. For example, Druce and colleagues (2009) experimentally manipulated resource availability for klipspringers (*Oreotragus oreotragus*), a small, nimble antelope of rocky outcrops (koppies) in southern and eastern Africa (**Figure 6.7**). Artificial food patches (trays of alfalfa pellets) were positioned across a range of scales (microhabitat, intermediate, and landscape scales) to evaluate how klipspringers perceived foraging risk, based on the giving-up densities within patches. Although foraging klipspringers were unaffected by the availability of cover at the microhabitat scale (0–4 m), they apparently perceived an increasing risk of predation as they foraged farther away from rocky outcrops when this was assessed at an intermediate scale (10–60 m). At the broadest scale (4.41 ha), their pattern of giving-up densities demonstrated that klipspringers very clearly perceived areas of high versus low predation risk on

Figure 6.7 Klipspringer (*Oreotragus oreotragus*).
Source: Photo by Kimberly A. With.

the landscape, corresponding to a **landscape of fear** (*sensu* Laundré et al. 2001; **Box 6.1**).

The marginal value theorem—and optimal foraging theory more generally—has provided an important foundation for predicting patch use, albeit at finer spatial scales encompassing resource patches (e.g. Pyke 1984; Stephens & Krebs 1986; Stephens et al. 2007). As we've discussed previously, however, foraging behavior can be expressed across a hierarchy of spatial and temporal scales, from fine-scale movements within food patches to broader-scale movements between seasonal home ranges (**Figures 6.2 and 6.3**). Principles of optimal foraging could therefore be applied to movement expressed over broader spatial and temporal scales, by relating the benefits derived from habitat use at a particular scale to the associated costs (Owen-Smith et al. 2010). For example, shifts in movement behavior or intensity of space use may reflect not just how animals perceive resource distributions across the landscape, but also potential threats to their survival (i.e. the landscape of fear; **Box 6.1**). It could even be argued that the sorts of movement behaviors expressed at these broader spatial and temporal scales are ultimately more important to survival and reproduction than fine-scale foraging behaviors (Owens-Smith et al. 2010). Regardless, the increasing use and availability of satellite telemetry for studying animal movements means that we can now obtain high-resolution data on animal movements over much longer time periods, thereby expanding the degree to which movement responses to landscape structure can be linked to individual and population performance.

Nor are animals the only creatures to forage optimally. Plants have likewise been shown to obey the predictions of the marginal value theorem (McNickle

---

BOX 6.1 The Landscape of Fear

A species' perception of landscape structure is a function not just of resource distribution, but also of predation risk. Foraging theory predicts that animals should sacrifice feeding effort in order to reduce their risk of predation, and thus animals should spend more time in areas where they have a lower risk of predation than in high-risk areas (Lima & Dill 1990). This means that an animal's foraging behavior or habitat use may not be a simple reflection of the abundance and distribution of habitat or food resources across the landscape, but instead may involve another layer of constraint imposed by their perception of predation risk. Predation risk can likewise be thought of as a spatially and temporally variable landscape, one that is created not only by the abundance and distribution of predators but also by their perceived lethality, a species' encounter rate with these predators, and their ability to detect them (Brown 1999), all of which may vary among habitats and over time to create a

dynamic 'landscape of fear' (Laundré et al. 2001). The landscape of fear may thus shape animal movements and space use as much as—if not more than—the distribution of habitats or other resources that are more typically mapped to define a species' resource landscape.

In a given landscape, species could suffer predation at the talons, claws, or fangs of a whole suite of different predators, each of which may require predator-specific behaviors for mitigating that risk. The landscape of fear may thus be multidimensional as well as dynamic. For example, vervet monkeys (*Chlorocebus pygerythrus*, formerly *Cercopithecus aethiops*) are a small, diurnal primate that live in social groups and are native to southern and eastern Africa (**Figure 1A**). They are abundant and widespread, occurring in a range of different habitat types, and spend time both in the trees and on the ground. They thus encounter a diverse array of predators that include baboons

(Continued)

BOX 6.1   Continued

(Chacma baboon, *Papio ursinus*), eagles (crowned eagle, *Stephanoaetus coronatus*, and Verreaux's eagle, *Aquila verreauxii*), leopards (*Pantharus pardus*), and, potentially, snakes (e.g. African rock python, *Python sebae*). In response to a perceived threat, vervets emit predator-specific alarm calls that elicit different types of behaviors in other members of the troupe; for example, a 'leopard' alarm call causes monkeys on the ground to scurry up the nearest tree, whereas an 'eagle' alarm call causes monkeys to look up (Seyfarth et al. 1980). These alarm calls are distinctive

even to human researchers, and by recording when and where these different alarms are given, it might then be possible to map out the landscape of fear for vervets.

At one site in the Soutpansberg mountain range of northern South Africa, Willems and Hill (2009) set out to do just that within the context of understanding how various landscape factors affected home-range use by a group of vervet monkeys. Within the heterogeneous landscape encompassed by their home range (114 ha), vervets clearly exhibited a greater intensity of use within some portions of

(A)

(B)

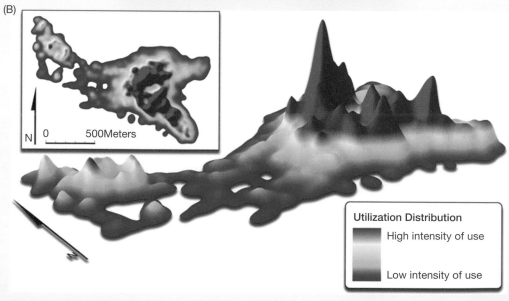

Utilization Distribution

High intensity of use

Low intensity of use

Figure 1 (A) Vervet monkey (*Chlorocebus pygerythrus*). (B) Home range space use.   Intensity of space use within the home range (114 ha) of a group of vervet monkeys in the Soutpansberg Mountain Range, South Africa.

Source: (A) Photo by Alexander Landfair; (B) Willems & Hill (2009).

their range than others (**Figure 1B**). But, do these regions of greater use coincide with the distribution of preferred habitat, areas of abundant food, or perhaps the availability of some other critical resource, such as surface water or sleeping trees? Or, are these areas of concentrated use a reflection of higher perceived predation risk elsewhere? Perhaps home-range use represents some combination of resource distribution and the landscape of fear?

Using a spatial regression modeling approach, Willems and Hill (2009) demonstrated that the intensity of vervet home-range use was negatively related to the landscapes of fear for baboons and leopards (but not eagles and snakes), as well as the distance to sleeping trees and surface water

(**Figure 2**). In other words, vervets tended to avoid areas where they perceived a high risk of predation from baboons and leopards, and did not like to stray too far from their preferred sleeping trees or open water sources (e.g. <200 m). In addition, intensity of space use was also positively correlated with habitat types that had the most abundant food resources (riverine forests and thickets; **Figure 2**). Much of the spatial variation in home-range use by vervets (>60%) could thus be explained by a combination of habitat selection, the distribution of critical resources (sleeping trees and surface water), and predator-specific fear landscapes.

Leopards and baboons apparently represent the greatest threat to vervet longevity. Snakes are not a major

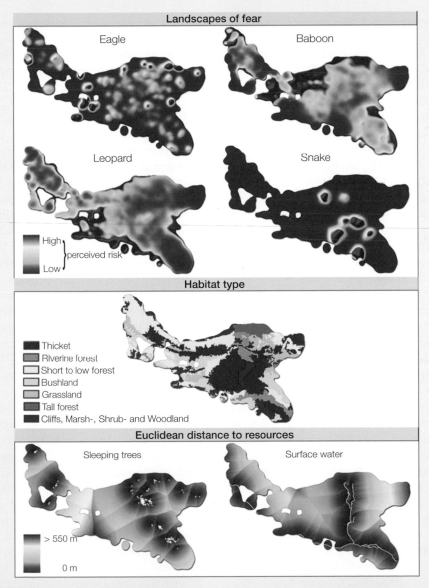

Figure 2 Landscapes of fear relative to the distribution of habitat types and resources. Top: Landscapes of fear for different predators. **Middle:** Distribution of habitat types within the vervet home range. **Bottom:** Distribution of critical resources.
Source: Willems & Hill (2009).

(Continued)

BOX 6.1   Continued

predator of monkeys, but eagles are, so why wasn't the eagle fear landscape an important determinant of vervet space use? One possible explanation is that eagles operate over a broader range than that encompassed by the vervet home range, and thus the risk of eagle predation was essentially uniform across the landscape. Further, since eagles are aerial predators, vervets may have an easier time detecting them from a greater distance, and then tend to escape vertically (within trees) rather than horizontally in space (by switching habitat use). Thus, these factors may collectively produce a 'flat' landscape of fear where eagle predation risk is concerned across the vervet home range.

Most home-range analyses attempt to link the intensity of animal space use with the distribution of habitats or other resources (see **Space Use and Home-Range Analysis**). By including predator-specific landscapes of fear in their analysis, however, Willems and Hill (2009) were able to explain more fully space use by vervets in the context of both resource distribution and predation risk. As demonstrated by their analysis, the effects of fear may even exceed those of the resource distribution in shaping space use. Thus, home-range analyses that ignore the fear landscape may well be neglecting an important determinant of space use for many animals.

& Cahill 2009; Cahill & McNickle 2011). For example, plant roots will slow their growth rate in order to remain in patches of higher soil resources for longer periods of time, consistent with the expectations of the marginal value theorem. We will return to the study of plant dispersal at the end of the chapter.

WHEN TO LEAVE A PATCH? DENSITY-DEPENDENT DISPERSAL AND THE SOCIAL ENVIRONMENT Although the marginal value theorem predicts that individuals should leave a patch when foraging conditions are no longer optimal, this is not the only factor that could cause individuals to leave a patch. If patches vary in quality or resource levels, then population density will likely vary among patches as well, with high-quality patches capable of supporting higher population densities. In many species, dispersal has been found to be density dependent, such that individuals are more likely to leave patches the more crowded they become (e.g. insects, Harrison 1980; birds and mammals, Matthysen 2005). Although density-dependent emigration might be attributed to a per capita reduction in resource availability (**exploitative competition**) or to an overall deterioration of conditions within the patch, it could also result from an increase in agonistic interactions with other individuals (**interference competition**; Bowler & Benton 2005). Interference competition ranges from overtly aggressive interactions between individuals (of either the same or different species) over a common resource to the usurpation of high-quality habitat by socially dominant individuals. Such competitive asymmetries can produce biased patterns of dispersal, in terms of which individuals are most likely to disperse in response to increasing density (e.g. sex-biased or natal dispersal; Greenwood 1980). Thus, aspects of the social environment, such as the sex ratio or kin structure of a population, can also affect the decision to leave a patch. In particular, dispersal should be favored in environments where there is a high probability of relatedness, because not only would this serve

to reduce competition among close kin, but it would also help to prevent inbreeding (Bowler & Benton 2005). The following case study illustrates the role the social environment plays on the probability of emigration in the great tit (*Parus major*).

The great tit (*Parus major*) is perhaps the most widely studied bird in the world, having been the subject of many now-classic studies in behavioral ecology, especially in regards to foraging ecology and territoriality, as well as on the evolution of avian life-history traits, such as brood size (e.g. Perrins 1965; Royama 1970; Krebs 1971; Cowie 1977). A handsome and familiar bird of gardens throughout Europe and Asia, the great tit generally inhabits forests where it establishes and defends territories throughout the breeding season. Great tits are secondary cavity nesters, meaning that they do not excavate their own nest cavities within trees, but instead move into the nest cavities excavated by other species (e.g. woodpeckers) or use natural cavities. Nest sites are therefore a limiting resource for great tits. As a consequence, they will readily use nest boxes, which is convenient for avian biologists, who can then easily manipulate breeding densities, clutch or brood sizes, and even offspring sex ratios, providing an unparalleled opportunity to disentangle the relative effects of density and sex ratio on dispersal. Nicolaus and colleagues (2012) experimentally adjusted both nestling density and sex ratio to evaluate how the social environment affects natal dispersal (the dispersal of juveniles from their natal area to a breeding territory) in populations of great tits in the Netherlands. Juvenile great tits were more likely to emigrate (leave their natal patch) from areas with high breeding densities than from areas with low densities. Overall, juvenile females were more likely than males to emigrate (as with most birds, natal dispersal is female-biased in great tits), but the effect of density on emigration was greatest in juvenile males. Thus, density-dependent effects on natal dispersal were greater for males than females. Since males compete for territories and juveniles are subordinate to adults, it would appear that increased male–male competition for territories in high-density populations is likely responsible for the differences between sexes in the role that density dependence plays in the natal dispersal of great tits.

To add another twist to the evaluation of density dependence on dispersal, consider that the structural complexity of the habitat itself can influence population density independent of resource availability, thereby modifying the outcome of competitive interactions and thus the strength of density dependence (Chesson & Rosenzweig 1991). Imagine two patches that vary in the structural complexity of their habitats, but which are otherwise identical in terms of their resource levels and population size (Levin et al. 2000). Because individuals encounter each other less frequently in the more structurally complex habitat patch, this leads to fewer aggressive interactions among individuals, which in turn reduces competitive interference, which should therefore decrease the likelihood of dispersal from this patch type. In other words, structural complexity should weaken density dependence, all else being equal. In a similar vein, habitat loss and fragmentation may likewise alter population densities and competitive interactions in ways that promote density-dependent dispersal. For example, we know that habitat loss produces smaller patches with more edge habitat. Competitive interactions among species may be edge-mediated and therefore greater within small patches owing to a greater frequency or intensity of interactions among individuals or to the movement of new species from the surrounding matrix into the patch (Fagan et al. 1999). Further, individuals within small patches are more likely to encounter the edge, which may increase their likelihood of leaving the patch (Stamps et al. 1987). We discuss this more fully in the next section on how edge permeability affects movements into or out of patches.

Up to now, we have been focused on how high population density can drive emigration from a patch (positive density-dependent dispersal), but we should also consider that it is possible for density to have a negative effect on dispersal. Although this is less common, negative density-dependent dispersal[3] does occur in some species (Matthysen 2005). For example, patches with abundant or high-quality resources should retain and even attract individuals, resulting in limited emigration even at high population densities (i.e. there is no need or motivation to leave the patch), at least up to the point where resources become limiting. Conversely, it may be that population density is so high that emigration becomes physically impossible owing to increased agonistic interactions with surrounding neighbors, which effectively block individuals from leaving (the social fence hypothesis; Hestbeck 1982). Finally, individuals might be forced to leave a patch at low rather than high population densities if suitable mates are scarce and become difficult to find (**Allee effect**; Stamps 1988).

EDGE PERMEABILITY Leaving (or entering) a patch entails crossing some sort of boundary, a discontinuity in the habitat or environment. A large part of an organism's movement response to landscape structure is therefore dictated by its perception of patch boundaries (Wiens et al. 1993a). The greater the contrast between habitats or environmental conditions that define the boundary, the more likely the boundary will be perceived as such by an organism (**Figure 6.8**). Perception of the patch boundary also depends on the dispersal ability or movement rate of the organism, as very mobile organisms are less likely to perceive or be affected by patch boundaries, unless they are quite distinct (**Figure 6.8**). Whether or not an individual leaves or enters a patch is therefore conditioned on the perceived permeability of the patch edge. **Edge permeability** is not just a function of the structure or environment at the patch edge, but also of how the organism perceives and interacts with that structure (**Figures 6.9 and 6.10**).

Edge permeability is ultimately determined by the degree to which individuals cross a patch boundary (Stamps et al. 1987). **Soft edges** are easily traversed and

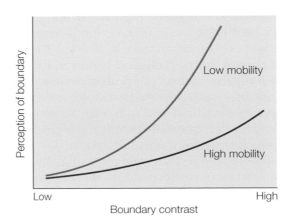

Figure 6.8 Perception of patch boundaries. The perception of patch boundaries depends on the degree of contrast between the patch and matrix, and the relative mobility of the organism. Species with low mobility are more likely to perceive patch boundaries as 'high contrast' compared to species with higher mobility.

Source: After Wiens (1992).

---

[3]Density has a positive effect on dispersal (positive density-dependent dispersal) when the likelihood of dispersal or emigration increases with increasing population density (i.e. the slope of the relationship between the likelihood of dispersal and population density is positive). Conversely, density has a negative effect on dispersal (negative density-dependent dispersal) when the likelihood of dispersal or emigration decreases with increasing population density (i.e. the slope of the relationship between the likelihood of dispersal and population density is negative).

Figure 6.9 Edge contrast affects species' perceptions of edge permeability. (A) High-contrast edges, such as at the boundary between an agricultural field and forest, are likely to be perceived as 'hard edges' that are relatively impermeable to movement. (B) Edges with lower contrast, such as between this forest edge and abandoned crop field, are more likely to function as 'soft edges' that permit movement between adjacent habitat or land-use types.

Source: Photos by Kimberly A. With.

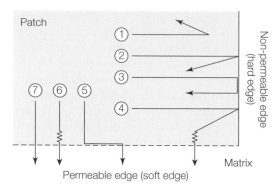

Figure 6.10 Edge permeability is influenced by the type of behavioral response to patch edges. Hard edges are not permeable and are either avoided completely (1) or are deflecting and not crossed (2–4). Soft edges are differentially permeable to movement (4–6), with some species being completely unaffected (7). The matrix (patch context) also influences the degree to which a patch edge might be permeable to movement, and a given patch might have more than one edge type. Behavioral responses are as follows: (1) avoidance of edge; (2) deflecting edge with return to patch interior; (3) edge following before return to patch interior; (4) deflected by hard edge, but subsequent cautious approach and crossing of a softer edge elsewhere in the patch; (5) edge following before emigrating from patch; (6) cautious approach followed by emigration from patch; and, (7) emigration without inhibition.

Source: After Lidicker & Koenig (1996).

are thus permeable to movement (**Figures 6.9B and 6.10**). In contrast, a **hard edge** is a boundary that is never or rarely traversed because it imposes a physical (structural), physiological, or psychological barrier to movement (**Figure 6.10**). Examples of physical barriers include ecotonal boundaries, such as at the land-water interface; the margin between adjacent land uses, such as between a crop field and forest (**Figure 6.9A**); roads and highways; or structures such as fences, dams, and weirs.

Some barriers might instead pose more of a physiological constraint to movement, owing to abrupt changes in moisture, temperature, salinity, or pH levels that do not always coincide with an obvious transition between habitats. For example, the salt line in river estuaries—the boundary between freshwater and seawater—may constitute a hard edge for many fish species, whereas others like the mummichog (or banded killifish, *Fundulus heteroclitus*), which occurs along the Atlantic coast of North America and can tolerate a wide range of salinity (euryhaline), treats this as a soft edge. Similarly, while the

land-water interface might be construed as a hard edge to most fish species, some such as the freshwater eel (either the European eel, *Anguilla anguilla*, or the American eel, *Anguilla rostrata*) are able to travel overland in search of new habitat on rainy nights, and may even come onto sodden agricultural fields at night to forage for worms and snails (Tesch 2003). Thus, barriers to movement (or conversely, corridors or opportunities for movement) may only occur at certain times or under certain conditions, and are therefore more phenomenological than structural in some cases.

In other cases, the edge boundary appears to be more of a psychological barrier, created by an unwillingness to cross certain types of edges into habitats that the organism is otherwise physically able or willing to traverse. For example, woodland salamanders might be inclined to cross forest edges adjacent to open fields or residential areas, but not forest edges adjacent to other types of clearings, such as roadways (Gibbs 1998). In other cases, species may exhibit a complete avoidance of forest edges, despite an apparent willingness to traverse open habitats in other contexts. This was found to be the case in red-spotted newts (*Notophthalmus viridescens*), a woodland salamander that is otherwise capable of dispersing long distances through open habitats (>1 km) during its terrestrial (eft) stage, but is nevertheless reluctant to cross from forests into fields (Gibbs 1998). That individuals are actively avoiding forest edges was convincingly demonstrated in spotted salamanders (*Ambystoma maculatum*), in which radio-telemetry studies showed individuals reversing course upon encountering forest edges adjacent to open grassland (Rittenhouse & Semlitsch 2006; **Figure 6.10**).

Patch context thus affects the degree of edge contrast, and therefore has the potential to alter edge permeability. Depending on the nature of the surrounding matrix, a given patch might have edges that differ in contrast and thus in edge permeability (i.e. a soft edge on one side and a hard edge on another side; **Figure 6.10**). Movement into (immigration) or out of (emigration) a given patch may not be equivalent across all edges. If so, then immigration or emigration rates cannot be predicted simply on the basis of patch size or shape (e.g. edge-to-area ratio) without additional information on patch context. This was initially demonstrated in a simulation model, which revealed that emigration was influenced more by edge permeability than by patch shape in patches with relatively hard edges (Stamps et al. 1987). Conversely, patch shape was the more important determinant of emigration in patches that had soft edges. Thus, the effect of patch shape on immigration and emigration is likely to depend on the surrounding matrix (edge contrast), as well as a species' perception of edge permeability.

To test this, Collinge and Palmer (2002) designed a field experiment in which they trapped ground-dwelling beetles (Coleoptera: Tenebrionidae and Carabidae) as they moved through grass patches that differed in shape (square vs rectangle) and edge contrast (hard vs soft, created by mowing grass at the patch periphery to different heights). As predicted, the effect of patch shape on beetle movement was more pronounced in low-contrast patches that had tall grass margins, than in the high-contrast patches delineated by short grass margins. Beetles were most likely to move into rectangular patches (which have more edge relative to square patches of the same area) with a low edge contrast. In fact, the patches with high edge contrast experienced net emigration (more beetles left than entered), whereas low-contrast patches experienced net immigration (more beetles entered than left). These differences in patch immigration and emigration may have been a result of **edge effects**: the shorter grass of high-contrast edges had less cover and higher surface temperatures, and these microclimatic effects may have permeated into the patch interior, creating an unfavorable environment that beetles sought to escape. Whatever the reason, beetles clearly responded to edge contrast, which influenced their movements between patches, and ultimately, affected their abundance and distribution in this experimental grassland system.

High edge contrast may also contribute to asymmetrical flows across patch boundaries. For example, planthoppers (*Prokelisia crocea*) exhibit different edge behavior depending on whether patches of their host plant, prairie cordgrass (*Spartina pectinata*), are on mudflats or embedded in a matrix of smooth brome (*Bromus inermis*), a non-native grass in this system (Haynes & Cronin 2006). The cordgrass-mudflat edge is a high-contrast edge that is also highly impermeable to movement; it could thus be considered a hard edge. The cordgrass-brome edge is low contrast (cordgrass and brome are structurally similar) and effectively invisible to planthoppers, which readily cross this type of boundary as they move within their preferred cordgrass habitat. However, assessment of edge permeability depends on which direction you are looking at it. Although planthoppers were reluctant to cross the cordgrass-mudflat edge to *leave* a patch, they readily crossed this boundary to *enter* cordgrass patches from the mudflat matrix (Reeve et al. 2008). Immigration rates (movement into the patch) thus exceeds emigration rates (movement out of the patch), resulting in asymmetrical flows across the patch boundary for cordgrass patches found on mudflats. In contrast, flows are symmetrical across patch boundaries for cordgrass patches embedded in swards of smooth brome. This translates into higher movement rates—and thus connectivity—between cordgrass patches when in brome.

Beyond edge contrast, the shape of the edge itself (**edge tortuosity**) can also influence edge permeability. Although anthropogenic habitat edges tend to be fairly straight and simple in form, natural habitat edges may exhibit more complex and irregular shapes, in which fingers of habitat extend out into the matrix (convex edges) in some places or the edge cuts more deeply into the patch (concave edges) elsewhere. The effect of edge shape on edge permeability was demonstrated experimentally in the field with meadow voles (*Microtus pennsylvanicus*), by mowing grass patches so as to create complex edge shapes consisting of straight, convex, and concave edges (Nams 2012; **Figure 6.11A**). Meadow voles prefer to tunnel through the dead vegetation of grassy areas, and so mowed areas create a semi-permeable barrier to movement. How permeable that barrier actually is to vole movements depends on edge shape, however. Voles were more likely to enter the matrix at concavities than at convexities or straight edges (**Figure 6.11B**). Because concavities represent areas where the edge cuts into the habitat patch, voles may have a greater tendency to cross the matrix (mowed areas) at these locations because this edge type maximizes protection, by allowing voles to remain in closer proximity to the patch habitat (e.g. there is grass cover on three sides), at least initially. Thus, a habitat patch with many edge concavities is more permeable to vole movements than a patch with many straight or convex edges.

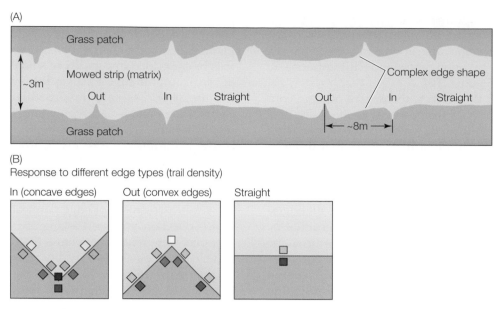

Figure 6.11 Edge shape affects the permeability of edges for meadow voles (*Microtus pennsylvanicus*). (A) Grass fields were mowed so as to create complex edge shapes, in which straight edge sections were alternated with convex ('Out') and concave ('In') edges. (B) More voles crossed into the matrix at concave edges. Track plates (small squares) were used to quantify trail densities along the edge (a measure of vole edge-crossings), with darker colors corresponding to higher densities. Thus, concave edges are more permeable to vole movements than other edge types.

Source: Modified from Nams (2012).

Movement Between Patches

Whether or not an individual will move between patches may depend not only on its perception of edge permeability, but also on its perception of the distance to the next patch (the inter-patch distance or gap size), its ability to move through the intervening matrix, and the perceived risk of doing so. For example, we expect movement to occur freely between patches that fall within the dispersal distance of a species, which is reflected in the various forms of the patch-connectivity measures discussed in **Chapter 5**. Recall, however, that patch connectivity is also a function of the resistance of the intervening matrix to movement. Thus, patches may be effectively isolated—even though they fall within the dispersal range of the species—if individuals are unwilling or unable to traverse the matrix (Ricketts 2001; **Figure 5.4**). We thus begin with a discussion of how the gap-crossing abilities of a species influence movement between patches.

GAP-CROSSING ABILITIES The **gap-crossing abilities** of species can be determined through behavioral observation or by tracking individuals as they move about the landscape. An ingenious method developed for forest passerines involves the use of recorded 'mobbing' calls to see if birds can be lured across forest gaps (Desrochers & Hannon 1997). Mobbing calls are given by some bird species when a predator is sighted, which attracts other birds of its own—as well as other—species. Playbacks of mobbing calls should thus signify actual predation risk (i.e. a predator has been sighted

nearby), which goes beyond whatever psychological barriers birds might have to crossing large open areas (i.e. a perception of higher predation risk). Using this methodology, Creegan and Osborne (2005) showed that several forest passerines in Scotland differed in their willingness to cross gaps, and furthermore, differed in the size of gaps they were willing to cross (**Figure 6.12**). Larger species, such as the chaffinch (*Fringilla coelebs*), were willing to cross bigger gaps (150 m wide) than were smaller species (e.g. goldcrest, *Regulus regulus*), presumably because their correspondingly longer wings made it easier—and safer—to do so. Whatever the reason, these differences in gap-crossing abilities may well translate into different perceptions of habitat availability. Within a fragmented forest landscape, for example, more habitat would likely be accessible to chaffinches than to goldcrests or European robins (*Erithacus rubecula*), given their different gap-crossing abilities (**Figure 6.12**).

The ability of a species to gauge the distance between patches—and thus whether it will cross a particular gap or not—likely depends on its **perceptual range**, the ability to detect habitat from a distance (Zollner & Lima 1997). As illustrated by the previous example, the perceptual range may well correlate with body size (Mech & Zollner 2002). To a large degree, however, the perceptual range will depend on the senses by which an organism perceives its environment. For example, the prickly pear cactus bug (*Chelinidea vittiger*) locates its *Opuntia* host through olfaction. However, their ability to detect *Opuntia* cactus

(A) No gap-crossing ability

(B) Goldcrest (46 m)

(C) Robin (60 m)

(D) Coal tit (92 m)

(E) Chaffinch (150 m)

Figure 6.12 Perception of habitat availability for forest birds with different gap-crossing abilities. Within a fragmented forest in Scotland, songbirds differed in the gap sizes they were willing to cross. Different species would thus have different perceptions of habitat availability. If we focus on a single patch (dark green), a species that lacked gap-crossing abilities would be restricted to this one patch (A). However, a species like the goldcrest (*Regulus regulus*) that is willing to cross gaps up to 46 m wide would be able to access a couple of neighboring patches (shown in light green, B). The European robin (*Erithacus rubecula*), with its slightly greater gap-crossing ability (60 m), could potentially access a third additional patch (C), whereas many more patches would lie within the gap-crossing abilities of the coal tit (*Periparus ater*; D) and chaffinch (*Fringilla coelebs*; E). Thus, species with better gap-crossing abilities would be able to access more habitat within this woodland.

Source: After Creegan & Osborne (2005).

from a distance (their perceptual range) is very much influenced by the size of the target patch (bigger *Opuntia* patches are 'smellier'), the structure of the intervening matrix (odor plumes travel farther in structurally simple habitats), and the location of the target patch relative to prevailing winds (cactus bugs generally orient toward *Opuntia* patches that are located upwind; Schooley & Wiens 2003).

Species may use more than one sensory modality to perceive habitat from a distance, depending on the environmental conditions and structural complexity of the habitat. For example, the ability of northern flying squirrels (*Glaucomys sabrinus*) to orient toward their preferred old-growth forest habitat was positively influenced by wind, which suggests that they are able to locate habitat via olfaction (Flaherty et al. 2008). That they also rely upon vision, however, is evidenced by the fact that flying squirrels had a much larger perceptual range in areas with fewer visual obstructions (100–150 m in clearcuts versus 20–50 m in second-growth forests). In addition, flying squirrels released into clearcut or second-growth forests immediately climbed the nearest stump or tree, apparently to scan the horizon to orient themselves, before climbing back down and scampering off. Thus, a species may also be able to alter their perceptual range behaviorally to gain a better vantage point, as in the case of northern flying squirrels.

A species' perceptual range is not fixed, but may be modified by the environment, including the perceived risk of predation (Zollner & Lima 2005). For example, the perceptual range of white-footed mice (*Peromyscus leucopus*), in terms of their ability to orient toward forest edge, increased five-fold (from 15 m to 75 m) under full moonlight relative to moonless nights (Zollner & Lima 1999). If predation risk is greater under full moonlight, however, then mice would do well to limit their dispersal to shorter distances or to moonless nights. Once again, dispersal involves trade-offs, in this case between perceptual range and predation risk. Hungry mice are another story, however, and thus we consider next how energy reserves or the motivational state of the individual might alter these sorts of trade-offs when moving through the matrix between patches.

MOVEMENT IN THE MATRIX: WHEN TO EAT AND RUN?
The need to feed provides a strong motivation to move, and thus a hungry animal might be willing to accept a higher risk of predation than it otherwise might. As we discussed previously, foraging behavior involves trade-offs between energy intake and predation risk, reflected in such measures as the giving-up density within patches. As a consequence, hungry animals might interact differently with landscape structure than those that do not need to forage en route. What's the best strategy for moving between patches, then? Should

animals move as quickly as possible to reduce their risk of predation? Or, would it be more prudent to slow down, so as not attract the attention of potential predators? Hungry animals might also need to slow down to forage, but doing so could increase predation risk if it interferes with their ability to watch for predators (i.e. the time spent searching for food versus scanning for predators represents yet another trade-off). Thus, slowing down to forage may not be a good strategy unless the animal is low on energy reserves and faces a low risk of predation (Zollner & Lima 2005).

An experimental study on darkling beetles (*Eleodes* sp.; Coleoptera: Tenebrionidae) demonstrates how motivational state (whether the individual is hungry or not) and perceived predation risk influence movement behavior. Darkling beetles are the stereotypical black beetle. They are found worldwide, but are especially adapted to arid environments, such as the western plains of North America. The adults are highly conspicuous as they meander about the landscape, which suggests that they have little to fear from predators. Indeed, darkling beetles have a fairly effective deterrent against would-be predators: when threatened, they elevate their abdomens, from which they exude a foul-smelling liquid that is usually sufficient to deter most predators (and quite a few meddlesome humans). Thus, predation risk is expected to be low for adults, in which case, they should be able to slow down and forage at their leisure whenever their energy reserves run low. This was tested experimentally by depriving beetles of food for a month, and pitting them against well-fed beetles in arenas containing different arrays of food patches (McIntyre & Wiens 1999a). Consistent with expectations, starved darkling beetles slowed down and covered less ground than well-fed beetles, primarily because the hungry beetles stopped to feed when they encountered food patches. The distribution of food patches also affected movement behavior, in that hungry beetles showed evidence of an area-restricted search when food was clumped or uniformly distributed, but not when the distribution of food was unpredictable (i.e. randomly distributed). Thus, hungry beetles exhibited a more intensive—and efficient—pattern of search than well-fed beetles, at least when foraging in a patchy landscape.

On the other hand, it might pay to move through the matrix as quickly as possible, either to avoid predation or simply because it is more efficient and easier to do so. For example, recall that planthoppers (*Prokelisia crocea*) were found to move more rapidly through the matrix when patches of their host plant (prairie cordgrass) were on mudflats than when immersed in swards of smooth brome (Haynes & Cronin 2006). As mudflats are sparsely vegetated, it is possible that planthoppers may perceive a higher risk of predation and therefore attempt to move through them quickly to

minimize that risk. However, their main predators are spiders, which are actually more abundant in brome. Thus, mudflats could simply offer a lower resistance to movement, as smooth brome is structurally very similar to prairie cordgrass. Indeed, their overall rate and pattern of movement in smooth brome is similar to that in their preferred cordgrass habitat (Haynes & Cronin 2006). Not only do planthoppers move faster, but they also move in a more directed fashion across mudflats, which is consistent with the notion that mudflats offer little resistance to movement. Unfortunately, their movements are apparently not directed *toward* their cordgrass hostplant, which are patchily distributed throughout the mudflat. Perhaps surprisingly, then, planthoppers actually enjoy greater dispersal success (in terms of finding cordgrass patches) when moving through brome, as their more convoluted movements in this matrix type increase the likelihood of encountering their patchily distributed hostplant. Thus, although high **matrix resistance** (the degree to which the matrix reduces movement) is generally thought to reduce connectivity, we have here an example where connectivity of cordgrass patches is actually *greater* in the high-resistance matrix.

CORRIDOR USE   The fact that some species move faster in the matrix might even be exploited in conservation planning and reserve design to facilitate successful dispersal among habitat patches. For example, the rare Fender's blue butterfly (*Icaricia icarioides fenderi*) survives in a dozen or so prairie remnants scattered throughout the Willamette Valley of western Oregon (USA). Prairie remnants contain the butterfly's larval host plant (lupines, *Lupinus* sp.) as well as the nectar sources required by adults. Fender's blue butterflies do make occasional forays outside these prairie remnants, but as the matrix lacks any resources they require, they end up moving rather quickly through these areas (Schultz 1998). Given that populations of Fender's blue butterflies are now relatively isolated from one another, would a habitat corridor help to restore population connectivity? Restoring prairie habitat to create corridors between populations would require a good deal of effort and expense, but is certainly possible. Would habitat corridors be a wise investment in this case (Simberloff et al. 1992)?

Apparently not. Given the slow rate of movement within their preferred prairie habitat, it is highly unlikely that individual butterflies would be able to traverse the distances required to link populations (3–30 km) over the course of their short lifetimes (i.e. butterflies were estimated to travel <0.75 km in their 10-day lifespan; Schultz 1998). However, a Fender's blue might disperse more than twice that far (>2 km) when moving through the matrix. Thus, population connectivity might instead be achieved through the creation of habitat stepping stones that are placed strategically within the perceptual range of the butterfly (**Figure 5.2B**). Butterflies might then be motivated to move more quickly among habitat patches, with the stepping stones supplying nectar resources to fuel their travels, but of insufficient size to encourage them to stay long.

Dispersal observations...are useful in comparing alternative reserve designs. Although they do not provide a blueprint specifying how land managers should proceed, they do help reject strategies that are likely to fail.   Schultz (1998)

This is not to suggest that corridors are never effective in restoring or maintaining connectivity, however. The efficacy of corridors should be assessed in terms of whether or not they actually facilitate movement or dispersal between habitat patches or populations (i.e. whether corridors restore functional connectivity for the target species; Rosenberg et al. 1997; **Chapter 5**). For corridors to be effective, individuals must be able to locate them and be willing to use them in preference to simply moving through the matrix. More importantly, however, corridors should enhance movement rates and dispersal success; individuals that use corridors should realize a lower rate of predation or travel costs compared to moving through the matrix. Ideally, we would like to know whether the target species is likely to benefit from a habitat corridor *before* we go to the trouble and expense of creating one.

To what extent can we predict corridor use on the basis of a species' movement behavior, such as its response to habitat edges? A corridor, by definition, is a narrow strip of habitat; it is thus basically two long edges. As we've discussed, many species are 'deflected' by habitat edges (**Figure 6.10**), which helps to keep the individual within the patch. In a corridor, this sort of edge-deflecting behavior could help to direct movement along the corridor (Haddad 1999). Given our previous discussion on edge permeability, we might therefore predict that species with a strong reluctance to cross habitat boundaries, particularly high-contrast edges, would benefit most from corridors. Indeed, corridor use was accurately predicted for several butterflies based solely on their movement behaviors at habitat boundaries: two open-habitat butterfly species (the sleepy orange, *Eurema nicippe*, and cloudless sulphur, *Phoebis sennae*), which were unwilling to cross forest edges, used corridors of open habitat more than expected if their movements were purely random, whereas a habitat generalist (the pipevine swallowtail, *Papilio troilus*), whose movements were little affected by habitat edges, showed no preference for moving along corridors (Haddad 1999).

From this we might conclude that, for species that use corridors, the bigger the corridor the better. Not only would bigger (wider) corridors be easier to find from within the patch, but they might also provide more cover and food resources during transit, thereby increasing the odds that individuals would survive the journey through them. Further, wider corridors might help to mitigate some of the negative edges effects associated with narrow habitat strips, such as increased predation risk. On the flip side, however, corridors that are too wide might not be perceived as corridors so much as simply habitat in which to settle. If edge effects ultimately degrade habitat quality, individuals that settle within corridors may suffer serious fitness consequences, in terms of lower breeding success or higher mortality (i.e. a **population sink**). Is there thus an optimum width that would promote movement through corridors, so that they function as links rather than sinks?

To test this, Andreassen and colleagues (1996) experimented with corridor width on the movement behavior of the root or tundra vole (*Microtus oeconomus*), a medium-sized rodent that inhabits wet meadows throughout the Holarctic region. Root voles were forced to run a gauntlet of sorts, in which they were provided with a 310 m long corridor of different widths (0.4, 1 and 3 m wide) that also contained simulated competitors (caged voles) or predators (fresh fox scat) at fixed intervals to reinforce the perceived costs of moving along the corridor. Corridor width ultimately had a greater effect on corridor use than the presence or absence of competitors and predators. Voles moved farther and faster in corridors of intermediate (1 m) width than either extreme. In fact, voles rarely entered narrow corridors at all, which were little wider than the size of their runways (which are 0.1 m in width). Although voles readily used the widest corridors (3 m), these were ultimately less effective in facilitating movement between patches because voles engaged in more exploratory behavior and moved in a 'zig-zag' fashion, perhaps because they had difficulty perceiving the edges of the corridors, which would otherwise help direct movement along its length. Thus, corridors that are too narrow will not be used (too risky), whereas corridors that are too wide may be perceived simply as elongated habitat rather than as movement corridors.

[T]he value of corridors as landscape connectors may be dynamic, reflecting variation in the degree of contrast between patch and matrix environments. Therefore, given a choice of a pathway through a corridor or through the matrix, optimal behavior is a conditional, not an obligate, response.

Rosenberg et al. (1997)

Nor are animals the only organisms to benefit from habitat corridors. Plants can also benefit, especially if their propagules are vectored by animals that use corridors. For example, Tewksbury and his colleagues (2001) made use of a large-scale landscape experiment to evaluate the efficacy of corridors in promoting both pollen and seed movement (**Figure 6.13**). They used a deciduous holly (*Ilex verticillata*) that is dioecious and thus incapable of self-pollination, and planted several male (pollen-producing) plants in the center patch and a number of female (fruit-producing) plants in each of the four peripheral patches, one of which was connected to the central patch via a corridor. As flowers of this holly are pollinated by a variety of insects, and the only pollen source available was in the central patch, any differences in fruit set (i.e. successful pollination) among the experimental holly populations would have to be due to differences in how pollinators moved between the central and peripheral patches. Consistent with expectations, fruit set was 69% greater for hollies in the corridor-connected patch than in any of the uncon-nected patches. In another experiment, the movement of seeds from the center patch for the bird-dispersed holly (*Ilex vomitoria*) was measured in peripheral patches by inviting birds to alight—and subsequently defecate—on artificial perches. More than twice as many seeds were defecated in connected than unconnected patches. By facilitating the movements of animal dispersal vectors, habitat corridors can thus help to maintain important plant-animal associations involving pollination and seed dispersal. We will return to the measurement of plant dispersal again at the end of the chapter.

Movement into Patches (Immigration)

The fact that animals are able to discriminate among habitat or resource patches that differ in quality or profitability means that they might then choose which patches to enter. We have already discussed some fac-tors that influence immigration into a particular patch, such as the ability of species to detect habitat from a distance (perceptual range and gap-crossing abilities), whether or not they orient toward habitat or make use of corridors, and the relative permeability of the patch boundary itself. Here, we turn our attention to the pro-cess of habitat selection and how this influences the distribution of individuals across the landscape.

HABITAT SELECTION   Habitat selection refers to the disproportionate use of habitat relative to its availabil-ity in the landscape. In other words, individuals are found within certain habitats to a much greater (or lesser) degree than expected if they were simply using habitat in proportion to its availability (i.e. at random). Habitat selection involves a series of behavioral responses that occur across a range of scales, in which

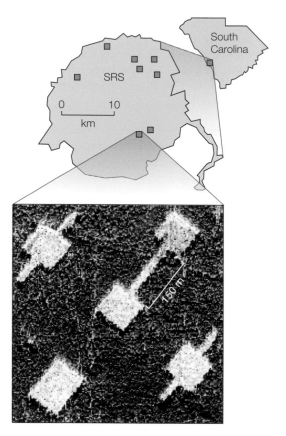

Figure 6.13 Experimental landscape used to test corridor use by different species. At the Savannah River Site (SRS) in South Carolina (USA), a replicated series of fragmented landscapes was created via clearcutting. As shown in the false-color image, each experimental landscape consists of five patches of open habitat (green) embedded in a forest matrix (red) of loblolly (*Pinus taeda*) and longleaf pine (*P. palustris*). The central patch (1 ha) is connected to a second patch by a 150 m long corridor. It is surrounded by three other patches ('unconnected patches'), each of which is 150 m from the center patch and equal to the size of the center patch plus its corridor (1.375 ha) to control for any increased area effects of the corridor. Two of these unconnected patches have that extra area added as extensions ('wings') on either side of the patch. The third unconnected patch had this additional area added to its back end, resulting in a rectangular shape.

Source: Tewksbury et al. (2001). Copyright (2002) National Academy of Sciences, U.S.A.

the habitat available at a given level is constrained by the decisions made at previous levels. Thus, habitat selection can be viewed as a nested hierarchy, much as we discussed previously in the context of foraging hierarchies (**Figure 6.3**). For example, within a specific landscape or region of the geographical range of a species (first-order selection), second-order selection determines the location of the home range of the indi-vidual or social group (Johnson 1980; **Figure 6.2**). Within the home range, third-order selection pertains to the disproportionate use of various habitats, such as for foraging. Finally, the preferential movement among resource patches within a given habitat (e.g. foraging behavior) can be viewed as a fourth-order selection. Although we have thus far discussed selection at the level of habitat and resource patches, we will return to habitat selection at the home-range level a bit later in this chapter and also in **Chapter 7**.

(A) Ideal free distribution

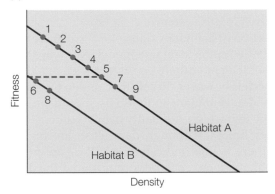

(B) Ideal despotic distribution

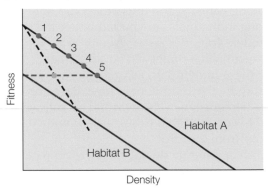

Figure 6.14 Density-dependent habitat selection. Fitness is assumed to decrease linearly with increasing population density in two habitat patches that vary in quality (Habitat A = good-quality habitat; Habitat B = poor-quality habitat). (A) Under the ideal free distribution, individuals (green dots) distribute themselves equally among the two habitat types so as to maximize their individual fitness. Individuals settle initially in Habitat A (individuals 1–4) up to a critical density when fitness is equivalent in both patch types (individuals 5 and 6), at which point they begin settling in Habitat B. (B) Under the ideal despotic distribution, fitness is expected to decline more quickly in the high-quality habitat for subordinate individuals (black dashed line), owing to displacement by dominant individuals (despots). Thus, after the initial settlement of Habitat A by a couple of dominant individuals (1 and 2, green), a subordinate (yellow) would achieve higher fitness by settling in Habitat B, even though there is higher-quality habitat still available. In the case of the ideal despotic distribution, individuals are selecting lower-quality habitats at lower densities (i.e. sooner) than predicted by the ideal free distribution (which, in the example here, ordinarily would not occur until at least five individuals had settled within Habitat A, represented by the blue dashed line).

Source: Modified from Johnson (2007).

Regardless of the scale at which it occurs, habitat selection has important fitness consequences (Morris 1992). For example, we have discussed previously the trade-off between foraging and predation risk; individuals must balance their need to eat against their risk of being eaten. Dispersal among habitat patches might also reflect a series of trade-offs between population density, resource availability, and social pressures that must likewise be assessed against the risk of predation or failing to find suitable habitat or a mate elsewhere if an individual decides (or is forced) to leave a patch. Because habitat varies in quality, we can also expect that this will contribute to spatial variation in population density, reproductive success, predation risk, competition, and thus in the strength of density-dependent dispersal across the landscape. Given all this, how should individuals select habitat in a patchy landscape? How do these sorts of behavioral decisions regarding patch or habitat choice influence settlement patterns and thus the distribution of individuals across the landscape?

In a now classic paper, Fretwell and Lucas (1970) proposed that habitat selection should be density-dependent, reflecting an **ideal free distribution** of individuals among patches that differ in habitat quality. Much like the marginal value theorem we discussed previously, this model predicts that individuals should distribute themselves among patches in such a way as to maximize their average fitness. As we have discussed, individual fitness is expected to decline with increasing population density owing to resource depletion and competition (all individuals here are assumed to have the same competitive ability). If two habitat patches vary in quality, then individuals should choose the better-quality habitat, but only up to the point where they could achieve the same level of expected fitness in the lower-quality patch (**Figure 6.14A**). In other words, increasing population density within good-quality habitats will eventually lower the expected gain in individual fitness to the point where an individual would do just as well in a lower-quality patch that is less crowded. At equilibrium, all patches are filled in proportion to habitat quality and every individual does equally well. The resulting distribution of individuals among patches is referred to as 'ideal free' because we are assuming that individuals behave ideally, meaning that they are able to correctly assess habitat quality, that they have complete information on the quality of all patches, and that they are free to move to whichever patch will give them the highest fitness return.

Individuals might not always be free to move among patches, however, if socially dominant individuals ('despots') prevent subordinate ones from settling within high-quality habitats. In that case, subordinate individuals are forced to select poorer-quality habitats sooner than under the ideal free distribution, even though higher-quality habitat is still available (**Figure 6.14B**). This type of preemptive distribution is referred to as the ideal despotic distribution and is to be expected among highly territorial species (Fretwell 1972). In this case, a few individuals control the best resources, such that dominants settle disproportionately in the highest-quality habitats, resulting in a large discrepancy in fitness between dominant and subordinate individuals.

The distinction between these models is important because they make contrasting predictions about how population density is related to habitat quality (Van Horne 1983; Johnson 2007). Under the expectations of the ideal free distribution, density is a reflection of habitat quality: even though all individuals have the same average fitness at equilibrium, density is ultimately greater in high-quality habitats. In that case, we could identify high-quality habitats as those that supported the highest population densities. Under an ideal despotic distribution, however, the equilibrium density among good and poor-quality habitats depends on the relative competitive abilities of individuals. Because subordinates are more greatly influenced by competition than dominant individuals, density could be highest in poor-quality habitats if subordinates are forced to settle there. In that case, density would be a misleading indicator of habitat quality (Van Horne 1983), but we may not realize that, unless we base our determination of habitat quality on some measure of individual performance (e.g. survivorship, reproductive success).

The difficulty in assessing how population density relates to habitat quality, and thus habitat selection within patchy environments, is illustrated by the following case study. Intuitively, small habitat patches should have fewer resources and thus a greater intensity of competitive interactions than larger patches, which should lead to individuals settling preferentially within large patches (i.e. larger patches should support a higher population density than small patches). This was tested by Levin and his colleagues (2000) in a study of the settlement patterns of juvenile three-spot damselfish (*Stegastes planifrons*), a common species found on coral reefs throughout the Caribbean. Damselfish vigorously defend their small territories (<1 m²) year-round, and thus Levin and his colleagues sought to compare the density and behavior of damselfish on small patch reefs to those occupying territories on a large continuous reef habitat. Counter to expectations, damselfish densities were actually highest on small reef patches, with neighbors twice as close as those on the continuous reef habitat. Based on density alone, we might conclude that small patches were of better habitat quality (but see Van Horne 1983).

However, the closer proximity of neighbors on small reef patches resulted in a greater frequency of agonistic interactions (2–4× greater than among damselfish on continuous reef habitat), which detracted from the time available for foraging, such that damselfish on the continuous reef habitat foraged twice as much as those on small patch reefs. This in turn translated into faster growth rates for damselfish on the large reef. Thus, higher densities are not associated with better habitat quality, at least in this system.

The selection of habitat by damselfish was therefore not consistent with the ideal free model. Nor was their pattern of settlement entirely consistent with the ideal despotic model, however. Recall that under the ideal despotic model, dominant individuals should preferentially settle in the highest-quality habitats where they enjoy higher fitness. Although damselfish are highly territorial, there is little indication that they are minidespots. Settlement on territories occurs under the dark of night, and thus individuals are neither prevented from settling nor are they displaced later by more-dominant individuals. Further, short-term survivorship did not vary among habitats, and thus fitness did not vary greatly among individuals as expected under the ideal despotic model. True, the faster growth rates on the continuous reef habitat might accrue additional fitness benefits for these individuals, as in greater fecundity[4] and thus reproductive output, over time. Based on the information at hand, however, Levin and his colleagues were forced to conclude that habitat selection by damselfish was far from the ideal.

Damselfish apparently do not satisfy several of the assumptions required by these models: for example, they may lack an ability to rate habitat quality accurately, especially given that they are locating and settling on reefs at night when their sensory abilities are more limited; they likely do not have complete knowledge of all habitat patches in any case, since settlement appears to be more passive than active (more on that in a moment); and/or they do not move freely among patches after settling, presumably because dispersal is risky (e.g. predation), and so are unable to shift to a better-quality patch and achieve an ideal distribution. Instead, the distribution of damselfish may simply reflect the passive sampling of individuals: individuals encounter small patch reefs more frequently because these are abundant and have a more-dispersed or fragmented distribution in this system. The continuous reef habitat may be large, but it is restricted to a single

[4]In fish, fecundity is related to body size, which in turn is given by the length-weight relationship $L = aW^b$, where b is the growth rate of the fish and is usually close to 3; that is, fish length is expected to increase as the cube of its body mass or weight (i.e. the 'cube law;' Froese 2006). You will no doubt recognize this as yet another example of an allometric relationship of the form $y = ax^k$ (Equation 2.1).

location, and thus may end up 'sampling' less of the population in the end.

This highlights how both the amount and relative distribution of good- versus poor-quality habitats could influence habitat selection, and thus the distribution of individuals across a landscape (Pulliam & Danielson 1991). For example, if the two habitat types are randomly interspersed, then the targeted removal of one will lead to an increase in the distance among patches of the other. Species with a limited dispersal range would thus end up sampling fewer patches. If the habitat being destroyed happens to be of particularly high quality, in which individuals enjoy high reproductive success and survivorship, then more individuals will end up settling in the marginal habitat, simply because this appears to be the best choice among the available options. Once again, density or settlement patterns may not correlate with habitat quality (Van Horne 1983). Even though some of these individuals breed successfully in the marginal habitat, they produce fewer offspring than if they had been able to select a site of higher quality. Far from being ideal, the apparent selection of these marginal habitats not only leads to a reduction in individual fitness, but also results in a lower population size and higher extinction risk across the landscape as a whole. Clearly, then, not all habitats that are utilized are necessarily valuable to the species. The preservation of marginal habitats that appear to support high population densities, at the expense of lower reproductive success or survival within those habitats, may lead to lower population sizes that compromise the long-term viability of the species. We will address this further in **Chapter 7**, when considering landscape effects on source–sink population dynamics.

Whether or not animals behave according to the ideal is arguably less important than understanding what factors induce them to settle in the first place. We consider the role of conspecific attraction on patch choice next.

CONSPECIFIC ATTRACTION   Earlier in the section, we considered how social factors might prevent individuals from leaving a patch (Allee effects and the social fence hypothesis), but it is also possible that the presence of other individuals could be quite attractive to potential immigrants, increasing the likelihood that they will settle within a patch. Why might individuals preferentially settle in habitats that are already occupied by others? For starters, this may increase the likelihood of finding a mate, which is obviously a requisite for breeding success. Or, the fact that others have already settled there may be an indicator of habitat quality or abundant resources (i.e. a source of public information; Valone 1989, 2007). There may be greater safety in numbers, such that an individual can reduce

its own risk of predation owing to enhanced vigilance (the many eyes hypothesis; Powell 1974) or simply through a dilution effect (the selfish herd hypothesis; Hamilton 1971). Ultimately, then, conspecific attraction acts to reduce Allee effects, as well as the costs associated with searching and settling within habitat (Stamps 2001). It is thus worth considering what role conspecific attraction plays in habitat selection.

Before we can document that conspecific attraction is a major factor in habitat selection, however, we first need to control for any other effects due to patch structure or habitat quality. Although this can be done statistically, it is also possible to test this experimentally, as in a study of juvenile settlement patterns in the bronze anole (*Anolis aeneus*), a small lizard found throughout the Lesser Antilles in the Caribbean (Stamps 1988). Habitat arenas were carefully constructed in the field to ensure that the distribution of habitat and resources were identical between sites. Juveniles were then given a choice of potential 'homesites' (basically, a pile of sticks) that were placed adjacent to an area already occupied by lizards, or next to a similar area that was completely vacant. Juveniles showed a clear preference for settling near established territory-holders, providing unequivocal evidence that conspecific attraction can play a role in habitat selection.

For conspecific attraction to be advantageous during settlement, the presence of conspecifics must be an honest indicator[5] of habitat quality (Doligez et al. 2003). Animals often take their cues from other individuals— of either the same or different species—to locate food and other resources (Sumpter 2010). For example, the frenzied activity of aerially diving seabirds signals the location of schooling fish on the open ocean to other seabirds, in much the same way that the presence of other birds at a garden feeder or ducks on a pond might signify that a food source or predator-free space has been found.[6] The presence of conspecifics in an area communicates high-quality habitat during settlement only when population density is relatively high, however. Using a spatially explicit simulation model to explore various strategies of habitat selection, Fletcher (2006) demonstrated that conspecific attraction should be most effective at moderate-to-high population densities. At low densities, there are simply too

[5]'Honesty' in this context simply means that the cue is a reliable one, and does not imply anything about individual motivation (i.e. whether the signaler or receiver both stand to gain from the exchange of information or not). The examples given here may thus arise through 'copying behavior,' in which individuals gain by exploiting the finds of others (Sumpter 2010).

[6]Indeed, duck hunters purposely exploit this sort of conspecific attraction through the use of inanimate duck decoys to lure ducks into shooting range. Decoys have also been used in conservation to attract colonial nesting seabirds, such as alcids and terns, in an effort to recover nesting colonies that have been extirpated from islands.

few conspecifics to provide a reliable guide to habitat selection, but at extremely high densities, aggressive interactions among territorial individuals would likely prevent settlement.

Conspecific attraction might also help mitigate some of the adverse effects of habitat degradation, such as through habitat loss and fragmentation. For example, Fletcher (2006) found that a habitat-selection strategy based on conspecific attraction led to higher fitness gains for individuals in moderately fragmented landscapes (i.e. fractal landscapes with an intermediate degree of spatial contagion, H = 0.5; **Figure 5.7**) or in landscapes that contained large amounts of low-quality habitat, because individuals could better locate and aggregate within high-quality areas (i.e. relative to a random search). Furthermore, conspecific attraction generated the sort of patch-size and (to a lesser degree) edge-avoidance effects on settlement patterns that are typically observed in empirical studies, where individuals settle away from habitat edges and in large patches rather than small ones. Conspecific attraction may thus help to explain some of the observed effects of habitat loss and fragmentation on population distributions and dynamics, a topic we will explore more fully in **Chapter 7**.

PATCH SHAPE AND ORIENTATION   Previously, we discussed how patch shape could influence movements into and out of patches within the context of edge permeability (Stamps et al. 1987; Collinge & Palmer 2002). Recall that patches with more complex shapes have proportionately more edge relative to area, thereby increasing the likelihood that individuals will encounter these patches. Whether they then decide to enter these more complex patches, however, depends upon that species' perception of edge permeability, habitat quality, and density of conspecifics within the patch, as we've discussed. For example, patches with a high amount of edge habitat may be dominated by negative edge effects, such as higher predation rates, and thus may not afford high-quality habitat in which individuals choose to settle.

The orientation of the patch can also influence immigration rates and settlement patterns in landscapes where the movement of individuals is largely directional, such as during the seasonal migration of birds or the passive dispersal of marine organisms via ocean currents (Gutzwiller & Anderson 1992; Tanner 2003). All else being equal, elongated patches oriented perpendicular to the direction of movement should intercept more individuals than either more compact (e.g. square) patches or elongated patches oriented in line with the movement of individuals. To test this experimentally, Tanner (2003) constructed a series of artificial patches (all 1 m²) that were either square or elongated, and which were positioned either parallel or perpendicular

to the direction of tidal currents in a seagrass (*Zostera muelleri*) community off the coast of southern Australia. These artificial seagrass patches were colonized by a variety of marine organisms, such as polychaete worms, mollusks, amphipods (shrimp-like crustaceans), decapods (true shrimp and crabs), and fish. Not surprisingly, the importance of patch shape and orientation depended on the relative dispersal abilities of the species. Patch shape and orientation had little effect on the settlement patterns of active swimmers like fish, and no effect at all on groups with only limited dispersal ability (adult polychaetes) or those that do not rely on currents for dispersal (adult mollusks and decapods). Thus, patch shape and orientation had the greatest effect on amphipods, which have poor swimming ability and are passively dispersed via ocean currents. Elongated patches oriented perpendicular to the current intercepted more amphipods than either square patches or elongated patches oriented parallel with the current.

Throughout this section, we have explored how patch structure can influence the different stages of movement, from the factors associated with leaving a resource or habitat patch, to those affecting movement through the matrix, and finally, the patch attributes associated with habitat selection and settlement within patches. Central to our discussion has been the interplay between species' dispersal abilities and the patch structure of the landscape. Given that the particular movement pathway taken by an individual provides a spatial record of how it has interacted with environmental heterogeneity, an analysis of movement pathways should prove useful in defining or comparing how different organisms perceive and respond to landscape structure. We therefore consider the measurement and analysis of animal movement pathways next.

Analysis of Movement Pathways

Before we can analyze movement pathways, we must first obtain data on animal movements. As we shall see, our choice of tracking method and study design will have a large effect on the analyses we might employ and our ability to make inferences about movement responses to landscape structure. We thus consider some issues in tracking and measuring animal movements before turning our attention to the analysis of movement pathways.

Analysis of movement data is inextricably intertwined with modeling movement, and collecting data without a clear idea of how it will be analyzed will almost certainly be a waste of time.

Turchin (1998)

Tracking Animal Movements

Various methods and technologies have been developed for the study of animal movement (tracking): direct observations of focal individuals; the capture, release, and recapture or resighting of marked individuals; the use of weather radar to track patterns of migrating birds, bats, and insects; and remote sensing of individuals outfitted with transmitters, which can be tracked on foot or by vehicle, from the air, or even from satellites in space. Each approach to tracking has trade-offs that ultimately influence the grain and extent of movement data that can be obtained. For example, focal observations generally allow for the study of fine-scale movements, such as foraging or ovipositing behavior, but it may be difficult to follow or keep the animal in view without disturbing it, and thus observations are generally limited to short time periods (i.e. the duration of an observation bout, which is often measured in minutes). Mapping the movement pathway then becomes a challenge, requiring some sort of spatial marking technique (e.g. numbered flags to mark the animal's location at set time intervals or whenever it stops) that can be surveyed later to provide spatial coordinates (e.g. using a hand-held GPS unit). Alternatively, a numbered grid system can be superimposed on the study area and the individual's coordinates recorded as it moves through the area.

Regardless of the method employed, it may not be possible to make repeated observations of the same individual unless individuals are uniquely marked, limiting the scope of movement behaviors that can ultimately be studied through direct observation. The study of marked individuals can thus provide movement data over broader scales, such as dispersal or migration distances, in addition to estimates of survivorship, home-range size, and population size.

For mark-recapture or mark-resighting studies, different types of markers are commonly used, depending on the type of organism studied (dyes; powders; beads; streamers; fin, ear, or wing tags; leg bands or rings; **Figure 6.15**). Marks vary in their degree of visibility and permanence, which is an important consideration when deciding which type of marker to use. One should also consider, however, whether the type of marker used could adversely affect the movements, social interactions, or survivorship of the marked individuals. The process of capturing and handling individuals to mark or identify them can itself cause stress or injury that might alter their subsequent behavior or survivorship, especially if this is being done repeatedly to determine their pattern of movements from recaptures.

Marking is obviously done with the expectation that at least some individuals will be recaptured or resighted at a later date. Rates of return or resighting

Figure 6.15 Examples of marking and tracking devices used in animal movement studies. (A) patagial tag on a California condor (*Gymnogyps californianus*); (B) colored beads inserted above the tail of a collared lizard (*Crotaphytus collaris*); (C) tracking collar on a black-tailed prairie dog (*Cynomys ludovicianus*); (D) a paper wing tag on a monarch butterfly (*Danaus plexippus*); (E) colored leg bands on a burrowing owl (*Athene cunicularia*); and (F) tracking collar on an African wild dog (*Lycaon pictus*).

(A)

(B)

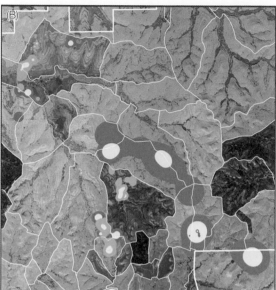

Figure 6.16 Telemetry studies provide spatial information on the movements, space use, and home-range size of individuals.   (A) A hand-held antenna is used to triangulate and locate the unique radio frequency being transmitted by a tagged snake (B) A kernel-based analysis of home-range sizes based on radio-tracking individual snakes across a tallgrass-prairie landscape (green = yellow-bellied racer, *Coluber constrictor flaviventris*; brown = Great Plains rat snake, *Pantherophis emoryi*; yellow = high-use areas within the ranges of each species). Thin white lines denote watershed boundaries, which are subjected to different fire and grazing treatments (dark areas are recent spring burns) at the Konza Prairie Biological Station (Kansas, USA), a National Science Foundation Long-Term Ecological Research site.

Source: Map and photo courtesy of Page E. Klug.

are often low, however, which means that many individuals must be captured and marked in the hopes that a few will be recovered at a later date (e.g. mass-marking programs, such as bird-banding or ringing programs). For those individuals that do not return, it is impossible to know whether they died or whether they simply settled elsewhere, beyond the search area, making it difficult to tease apart mortality from permanent emigration (see Sandercock 2006 for a review of methods). Further, individuals may lose their marks over time, making it difficult to recover data on those individuals. Thus, little fine-scale information on movement is likely to be obtained using mark-recapture/resighting methods unless the animal is territorial or has a high degree of site fidelity, such that there is a reasonable expectation of obtaining repeated sightings or captures over some period of time (e.g. over the course of a season).

The use of radio- and satellite-telemetry has therefore been a real boon for the study of animal movement, as once the transmitter is implanted or affixed to the animal, it is possible to locate the individual again and record its location repeatedly (from every few minutes or hours to daily or weekly intervals) over a period of time (measured in days, weeks, months, or even years), as determined by the type of transmitter, its size, and its power source (whether battery or solar). With telemetry, we can thus track

animal movements across a broad range of spatial scales, from fine-scale movements within the territory or home range, to seasonal movements along dispersal corridors between landscapes, to the specific routes taken by species that migrate between continents or across oceans. Furthermore, some transmitters have sensors that record temperature, depth (for aquatic or marine species), proximity to other individuals, and whether the animal is hibernating or has died. The use of VHF radio-transmitters involves manual tracking by a researcher, and thus requires considerable search effort to find and determine the location of the transmitting individual (**Figure 6.16**). In contrast, satellite transmitters can provide continuous, automated tracking over an unlimited geographical range, and even pinpoint the individual's location to within a few meters. The longevity of solar-powered satellite transmitters allows long-term study over multiple years, potentially encompassing the lifetime of the individual.

With so many advantages to satellite tracking, one might wonder why VHF transmitters or other tracking methods are still used in animal movement studies. Two reasons: size and cost. Transmitters in general are costlier than tags and other marking materials, and until recently, their size has limited their use to fairly large animals (e.g. sea turtles, birds of prey, large mammals). However, advances in technology are ever reducing the size of transmitters, such that small GPS transmitters are now available for use in smaller birds, bats, and reptiles. Nevertheless, this comes at considerable cost and at the expense of battery life, which is limited to a few days or months, depending on how frequently the unit is set to 'fix' the location of the individual (more frequent fixes will drain the battery more quickly). Inevitably, then, the use of transmitters will impose a limit on the number of individuals that can be feasibly (and affordably) studied, which will be far fewer than in observational or mark-recapture/resighting studies. Regardless of the type of transmitter used, we still face the same sorts of issues discussed previously in regards to the study of marked animals, namely whether the stress of capture and handling or the device itself is likely to alter the behavior or survival of tagged individuals. There are thus ethical considerations in the capture, marking, and tracking of animals, which necessitates the review and permitting of such studies (especially those involving vertebrates) by academic institutions and wildlife management agencies in many countries.

Indirect measures of movement, involving the use of stable isotopes and molecular techniques, have also helped expand the scope of movement studies. The analysis of population genetic structure reveals historical as well as contemporary patterns of dispersal and gene flow, which may be shaped by landscape structure, and which will be discussed more fully in **Chapter 9**. Although the analysis of **stable isotopes** can uncover

linkages among ecosystems and the importance of spatial subsidies to those systems (a topic we will return to in **Chapter 11**), they can also be used to assess the connectivity among migratory populations that breed and overwinter in different locations, perhaps even on different continents (**migratory connectivity**; Hobson 1999; Webster et al. 2002). Various biogeochemical, biological, and anthropogenic processes give rise to distinct gradients and landscape signatures in the environmental ratios of stable isotopes, which are subsequently reflected in the diets and body tissues of individuals (e.g. in hair, feathers, claws, or muscle), and which can therefore be used to determine the likely origins of these individuals when sampled elsewhere, such as on their wintering grounds.

For example, the ratio of deuterium (a stable hydrogen isotope) in precipitation exhibits a continent-wide pattern across North America, which tends to increase along a gradient running from the northwest to the southeast (**Figure 6.17**). The flight feathers of migratory songbirds are usually grown on or near their breeding grounds prior to migration, and thus an analysis of deuterium ratios within these feathers can help identify the breeding origins and migratory connectivity between breeding and wintering populations of Neotropical migrants (those that breed in North America but overwinter in Central or South America; Hobson & Wassenaar 1997). Note, however, that the use of deuterium is less useful for assessing the

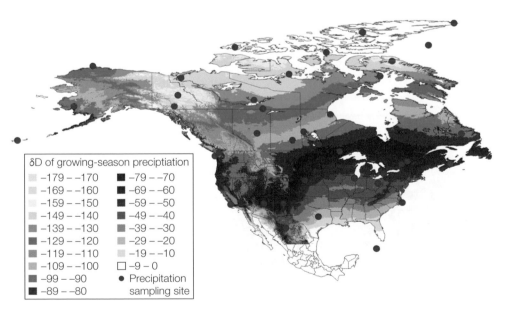

Figure 6.17 Latitudinal gradient in the stable hydrogen isotope ratio (δD) of precipitation during the growing season across North America. Precipitation becomes more 'enriched' in deuterium (²H) along a gradient running from north to south as a result of increasing temperatures, although this pattern is modified by elevation and proximity to the coast. Nevertheless, the values for δD (‰) tend to increase (become less negative) with decreasing latitude. A stable isotope analysis of animal tissues (insect chitin, bird feathers, mammal fur), which reflect the conditions of the local growing season where the tissues were formed, can then be compared to this sort of isotopic gradient to infer the connectivity of breeding and wintering populations for migratory species.

Source: Meehan et al. (2004).

migratory connectivity of Afro-Palearctic migrant species (long-distance migrants that breed in Europe and winter in sub-Saharan Africa), because the precipitation gradient there runs largely east to west and thus lacks a strong latitudinal signature (Rubenstein & Hobson 2004). In that case, other stable isotopes ($\delta^{13}C$, $\delta^{15}N$) can be used to assess migratory connectivity.

Until recently, stable isotope analysis afforded a rare glimpse into the seasonal movements of small migratory bird species over very large geographic distances. Nor are birds the only migratory species to have benefitted from stable isotope analysis: the seasonal movements of butterflies, fish, bats, sea turtles, and even whales have also been inferred from an analysis of stable isotopes (Rubenstein & Hobson 2004). Nowadays, however, light-level geolocators are more typically used to map migratory pathways in small passerines (Stutchbury et al. 2009; Fudickar et al. 2012). A geolocater (or geologger) is a small tracking device that periodically records ambient light levels to determine the bird's position (i.e. based on day length, which varies latitudinally, and the time of solar noon, which varies longitudinally). An individual's entire migration journey can thus be recorded over the course of a year. However, individuals must be recaptured in order to recover the tags and download the data, which means recovery rates may be low (e.g. 10-36%, Stutchbury et al. 2009). Nevertheless, geolocators are able to map bird migration routes and wintering locations with far greater accuracy than is currently possible with stable isotope analysis.

Assessing migratory connectivity has been helpful not only in determining the linkage among seasonally disjunct populations, but also in terms of evaluating the extent to which individuals from a single breeding population all migrate to the same non-breeding location or whether they spread out more widely across the entire wintering range (Webster et al. 2002). This clearly has implications for the conservation of migratory species, in terms of the scale of habitat protection required to secure populations of a given species.

Scaling Issues in Tracking Animal Movements

The pathways of some animals may naturally consist of a series of individual moves punctuated by stops, such as when a bee or butterfly flies among flowers and pauses briefly to forage from the inflorescence of each. For other animals, however, movement is more continuous and it is less clear how often readings should be taken to give a reasonable facsimile of their movement pathway. Animal movement pathways are invariably represented as a series of discrete movement segments (**steps**) and turning angles for analysis or to permit comparison with theoretical models of movement, as we'll discuss in subsequent sections. Depending on the temporal or spatial grain we use, however, information on the finer-scale details of movement will inevitably be lost, which means that the probability distribution of step lengths and turning angles used to describe the movement pathway will be influenced by our measurement scale. For example, if we take readings on the location of a tagged individual once a day, we will not know how it moved in the interim and thus risk missing out on some potentially important behavioral interactions with landscape structure (**Figure 6.18**).

Taking readings too often (oversampling), however, might be a waste of our time and resources if this adds little additional information on how the animal moves, or breaks the pathway down into a trivial series

Figure 6.18 The effect of scale of measurement on animal movement pathways. Taking less-frequent readings on the location of an individual results in a loss of information, as demonstrated here by the increasingly coarse representation of the pathway (from left to right). Some statistical procedures, such as renormalization, intentionally aggregate data to give a coarser resolution of the pathway in order to assess the scale at which structure of the movement pathway begins to break down and significant information is lost. These pathways were generated as a simple random walk made up of 5000 steps (left) that was then renormalized with either 555 (middle) or only 61 increments (right).

Source: After Wiens et al. (1993b).

of steps and turning angles that ultimately have little to do with how individuals are actually interacting with their environment. This is also a concern from the standpoint of testing statistical models of movement. Too-frequent sampling will result in a higher degree of correlation in the direction of movement, giving the appearance of a correlated random walk (to be discussed later in this section), even if the animal is actually making random turns (Hawkes 2009). Thus, some careful thought must be given to how often we record an animal's location, which will obviously be related to the tracking methods used, as well as the questions being asked. If our interest is in mapping home ranges or identifying dispersal corridors, then we probably do not need to take a reading every 5 minutes, but would it suffice to take readings every hour or even just once a day?

The increasing use of satellite telemetry to track animal movements, which can obtain positional data every few seconds or minutes that can be retrieved through a computer while sitting at one's desk, means that oversampling rather than under-sampling is likely to become the greater concern in animal movement studies. Practical concerns over battery life aside (frequent sampling drains transmitter batteries more quickly), the real issue is whether having such fine-scale spatial data on movement is really all that informative or even useful, especially if this exceeds the resolution of the resource or habitat data available to us (Hebblewhite & Haydon 2010). We might have information on movements that occur at the scale of a few meters, but if the spatial resolution of our map data is much larger (e.g. 250 m for MODIS imagery), we will be unable to link these fine-scale movement behaviors to habitat distributions at those scales, given the degree of scaling mismatch between our datasets.

As with most scaling issues, there is no simple solution to the problem of what scale or how often we should take readings on an animal's movements that will give us only the essential details of its interactions with the landscape. Having familiarity with the species to be tracked at least provides some initial guidance as to what scale or frequency of readings is likely to be most useful in capturing the movement behavior of interest. Beyond that, statistical approaches can be used to assess the degree of autocorrelation among movement steps to determine the scale(s) at which movements might be trivially correlated, and which therefore add little information to our understanding of animal interactions with the patch structure or resource distribution. For example, we might resample the pathway at increasingly coarser resolutions to determine the scale(s) at which the structure of the pathway abruptly changes using renormalization procedures or by assessing serial correlations in turning angles, under the assumption that this will help us determine when oversampling is no longer an issue (Wiens et al. 1993b; Turchin 1998; **Figure 6.18**). Of course, this assumes that we have collected movement data at a sufficient resolution to capture the 'real' structure of the pathway in the first place.

This also ignores the possibility that animals may exhibit different movement behaviors or responses to heterogeneity at different scales. As discussed earlier, we expect to see scale-dependent shifts in the structure of movement pathways, corresponding to different behavioral domains of scale (Nams 2005; **Figure 6.2**). This also assumes that serial correlations in turning angles are not of interest and just represent 'noise' that is an artifact of our sampling scale. This is not always the case, however. The degree of correlation among moves can be used to evaluate which of various models of animal movement might apply to our study organism, which would provide insights not only into how they move, but also into how we might then be able to translate individual movements into predicting the distribution and viability of populations in different landscape contexts (e.g. Heinz et al. 2006).

Measuring Movement Pathways

For most tracking methods, we will end up with a series of spatial coordinates (points) that mark the location of the individual as it moves about the landscape. The points are then connected by line segments (steps) to form the movement **path** (**Figure 6.19**). There are a number of properties of the movement pathway that are particularly useful for describing movement patterns, which are summarized in **Table 6.1**.

A study of movement patterns in darkling beetles serves to illustrate how these measures can be used to compare animal movements in different habitats or among different species (Crist et al. 1992). In the short-grass steppe of eastern Colorado (USA), these ground-dwelling beetles must navigate a grassy landscape interspersed with patches of bare ground, cactus, or shrubs as they wander about in search of food, which consists of plant litter and other detritus. A simple inspection of their movement pathways suggests that beetles may experience greater difficulty in moving through structurally complex habitats, such as those containing shrubs, given that they travel considerably farther in grassy areas within a fixed period of time (**Figure 6.20A**). Or, perhaps beetles encounter so many resources (food, shade, shelter) in shrubby habitats that they end up slowing down, which reduces their distance travelled in these areas? Whatever the reason, net displacement (R_n, **Figure 6.19B**) was significantly lower in shrub-dominated plots for all three species studied (**Figure 6.20B**) as a result of shorter mean step lengths ($\bar{\ell}$) and mean vector lengths (r), compared to movement

(A)

(B)

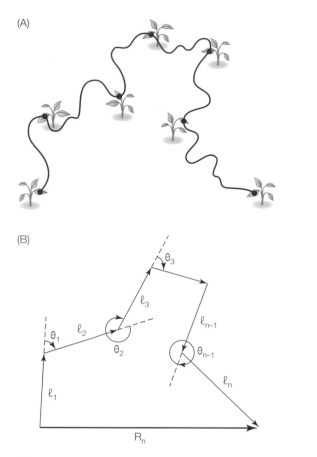

Figure 6.19 Representation of an animal's movement pathway. (A) Moves may represent the individual's location at a given point in time, or actual stops it makes while foraging or ovipositing. (B) The spatial coordinates (points) of the animal's locations are then connected by line segments (ℓ_n), and the change in direction from its subsequent location is given by the turning angle (θ_n). The animal's net displacement (R_n) is the straight-line distance travelled between its initial location and the endpoint of the path.

Source: After Kareiva & Shigesada (1983).

pathways in predominantly grassy or bare-ground areas (mean turning angle $\bar{\theta}$ was essentially zero for all three species, meaning that these beetles exhibited directed movement regardless of habitat type). In addition, the largest beetle (*E. hispilabris*), which is 2–4× larger than the other two species studied, appeared little affected by cactus cover, although this cover type reduced the net displacements of the other two species. Indeed, this species was observed crawling over and through small cactus clumps, whereas the smaller beetle species rarely did so. Thus, although habitat structure clearly affected the movement pathways of all three species, species responded in different ways to the structural complexity of these habitats, which may reflect differences in body size or movement ability and the scale(s) at which they interact with environmental heterogeneity.

The potential for scale-dependent movement responses to patch structure is clearly of interest, especially if different species are perceiving or responding to environmental heterogeneity across different domains of scale (**Figure 6.4**). The **fractal dimension** (D) of the movement pathway gives a measure of the complexity of the movement pathway (its **tortuosity**). Fractal analyses of movement pathways have been used to assess species' perceptions of landscape structure (With 1994a), and to identify different domains of scale bounding animal movements (Nams 2005). We thus consider this measure in greater detail next.

FRACTAL ANALYSIS OF ANIMAL PATHWAYS Larger animals tend to move faster and farther per unit time than smaller animals. Thus, distance-based movement parameters, such as net displacement or mean step length, will obviously differ between individuals or species of different sizes. Although individuals may differ in the absolute distance or rate at which they move, they may nevertheless exhibit similar responses to the patch structure or resource distribution of the landscape. If so, then movement pathways of larger individuals would just be scaled-up versions of those of smaller individuals.

Fractal analysis provides a means of quantifying the spatial complexity of movement pathways, by assessing how the structure of the pathway varies as a function of the scale at which it is measured. Recall from **Chapter 2** that a fractal structure is one that exhibits the same degree of complexity at whatever scale it is viewed. Thus, if the movement pathways of small individuals are basically scaled-down versions of those produced by larger individuals, then they would both have the same fractal dimension, suggesting a similar pattern of movement or response to environmental heterogeneity, despite the absolute differences in movement length. The fractal dimension of the movement pathway can be obtained using the 'dividers method' (Dicke & Burrough 1988), whereby the overall path length (L) is measured at different scales (ruler lengths) and regressed on a log-log plot (**Figure 6.21**). The total path length is greater at finer scales because more of the details of the pathway are included in the measurement, whereas these are lost if the path is measured at a coarser resolution (i.e. a larger ruler). For example, the total path length is usually greater than the net displacement because the latter ignores the meanderings made by the animal between the endpoints of the path. Thus, the comparison between these two measures is essentially one of scale: the use of a small ruler (the individual step lengths) versus a larger ruler placed at the endpoints of the pathway.

A fractal analysis of movement pathways measures the total path length (L) at different ruler sizes (δ) as

$$L(\delta) = k\delta^{1-D} \qquad \text{Equation 6.1}$$

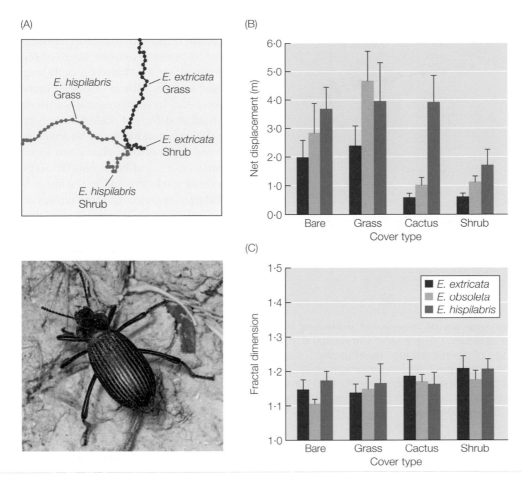

Figure 6.20 Effect of habitat structure on movement pathways of darkling beetles.   (A) Representative pathways for two species of *Eleodes* beetles as they move through either grass- or shrub-dominated plots (5 × 5 m). The individuals' locations were marked at 5-second intervals (dots) using numbered toothpick flags, for a total of 100 time steps (500 s). The pathways of four individuals from two different plot types (grass vs shrub) have been superimposed here for comparison. (B) Mean net displacement for three species of *Eleodes* beetles in different habitat types along a gradient of increasing structural complexity. (C) Fractal dimension of movement pathways for these same three species in different habitat types.

Source: Modified from Crist et al. (1992). Photo by Judy Gallagher and published under the CC BY 2.0 License.

in which k is the intercept of the regression line. Note that the slope of the regression line is equal to $1 - D$, which we can rearrange to solve for D as $1 -$ slope. Since the slope is negative (the total path length decreases with the increasing size of the ruler used to measure it), this gives us $D = 1 - (-\text{slope})$. So, if the slope is -0.09, then $D = 1 - (-0.09) = 1.09$. For movement in two dimensions, $D \to 1$ for a linear path and $D \to 2$ for a very convoluted pathway (i.e. random movement). Thus, our hypothetical pathway with $D = 1.09$ would be fairly linear, but with a few more twists and turns than a perfectly straight line. It should also be noted that because D is an exponent in this scaling relationship, small differences in D reflect large differences in overall movement (Milne 1997).

To illustrate the application of fractal analysis to movement pathways, we once again consider the darkling beetles mentioned above. Although one species (*E. hispilabris*) was larger and moved some 2–8× farther than

the smallest species (*E. extricata*), depending on habitat, fractal analysis revealed that the overall structure of their movement pathways was in fact quite similar ($D \approx 1.15$; **Figure 6.20B**). The similarity in the fractal structure of their movement pathways suggests that all three beetle species are responding similarly to the patch structure of the environment (Crist et al. 1992). This was not the case for other insects in this same grassland system, however. Grasshopper species exhibited size-related differences in their movement responses to patch structure (**Figure 6.21**), which extended even to different life stages (nymphs vs adults) in one grasshopper species (*Opeia obscura*; With 1994b). Among adult grasshoppers, the larger species (*Xanthippus corallipes*) moved 6× faster, and was apparently less affected by fine-scale heterogeneity than two smaller species, as indicated by its more linear movement pathways (With 1994a; **Figure 6.21**). However, adults of one of the small grasshopper species (*O. obscura*)

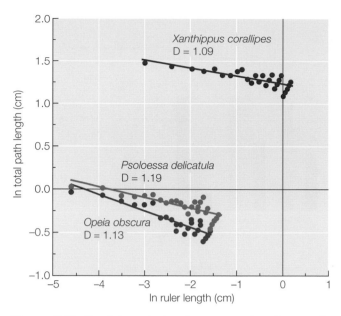

Figure 6.21 Fractal analysis of movement pathways for three grasshopper species. The fractal dimension (D) is obtained as 1 – slope of the regression line of the total path length (L) versus the size of the ruler length (length scale, δ). In this case, one species of grasshopper (*Xanthippus corallipes*) not only moves farther, but also in a more-linear manner (D ≈ 1), than the other two species.

Source: After With (1994a).

were more strongly affected by habitat structure than their even-smaller nymphs, despite the fact that adults moved 2–6× faster and farther than nymphs, depending on the structural complexity of the habitat (With 1994b). Nymphs of this species tend to cling to individual grass blades, and thus did not really interact with the two-dimensional patch structure of the environment in the same way (or, at least, not at the same scale) as adults. Compared to beetles and grasshoppers, harvester ants (*Pogonomyrmex occidentalis*) in this grassland had more convoluted movement pathways (D = 1.3), which may reflect differences in forage distribution and search strategies among these different groups of insects (Wiens et al. 1995). For example, grasshoppers and beetles are immersed in a widely available resource (grass or plant litter, respectively), which may account for their more linear movements. In contrast, harvester ants forage for patchily distributed seeds, and thus their more convoluted movement pathways may reflect the greater uncertainty associated with searching for a more variable and less-predictable resource.

Although fractal analysis can compare the structure of movement pathways within a particular domain of scale (thereby providing a scale-independent measure of movement), it can also be used to assess the potential for scale-dependent shifts in movement responses between domains of scales. We've been discussing the former application, in which a single fractal dimension (D) is

calculated for the entire movement pathway. This is reasonable if individuals are exhibiting the same general type of response across the range of scales being studied (i.e. the relationship between path length and the measurement scale is a linear one, see **Figure 6.21**). If individuals respond in fundamentally different ways to patch structure at different scales, however, then we expect this to be reflected in the fractal analysis of the movement pathway as an abrupt change in D at those scales. For example, Nams and Bourgeois (2004) performed a fractal analysis of tracks created in the snow by American martens (*Martes americana*), a large weasel of northern coniferous forests, and found a transition in the fractal dimension of those pathways at a scale of about 3.5 m (**Figure 6.22**). They interpreted this to mean that American martens were responding in different ways to habitat features within each domain. Although the fractal dimension of marten pathways increased with scale, pathways were essentially linear—corresponding to directed travel—at scales below 3.5 m, which perhaps differentiated finer-scale movements within a particular habitat type from those of longer movements within the home range.

Now that we have obtained some information on movement pathways, we can use these measures to parameterize or test models of animal movement. Not only might such measures give us additional insight into the behavioral processes or environmental cues that influence movement, but this is also an essential first step in translating individual movement behavior into the spatial distribution and spread of populations. We turn to models of animal movement next.

Models of Animal Movement

In this section, we will explore two general frameworks for modeling animal movement: 1) mathematical models, and 2) spatially explicit models.

Mathematical Models of Animal Movement

The early mathematical framework for modeling animal movement was founded on applications of diffusion theory (Skellam 1951). Although a brief overview is given here, a more comprehensive treatment of random walk and diffusion models can be found in one of the texts (Okubo 1980; Turchin 1998; Okubo & Levin 2002) or review articles (Codling et al. 2008) devoted to the topic.

SIMPLE RANDOM WALKS We begin by assuming that individuals move about aimlessly on the landscape. Not that we believe that animals actually move this way, but the assumption of a random walk is one of the simplest movement behaviors imaginable, the ultimate in 'behavioral minimalism' (Lima & Zollner 1996). Even though real animals might not move exactly in

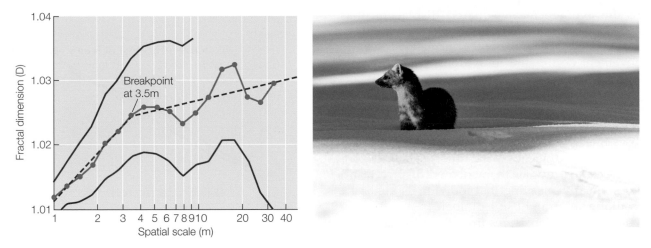

Figure 6.22 Fractal analysis can reveal domains of scale in movement behavior.   In a fractal analysis of pathways created by American martens (*Martes americana*) in the snow, a shift in the fractal dimension (D) occurs at a scale of 3.5 m, which coincides with the domain of foraging activities. Dashed lines (red) represent two different regression lines, fit separately to each domain. Each point is the mean fractal dimension obtained over all movement pathways, with the solid red lines representing the 95% confidence intervals.
Source: After Nams and Bourgeois (2004). Photo by Yellowstone National Park.

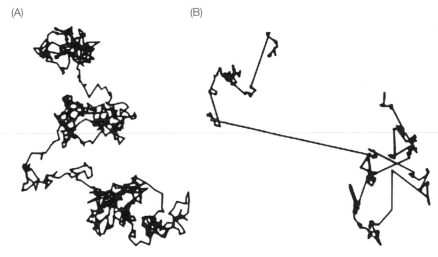

Figure 6.23 Comparison of movement models.   (A) Simple random walk (B) Lévy walk. The Lévy walk is a fractal pattern of movement, in which long-range movements occur more frequently than expected for a random distribution of move lengths.

this fashion, our interest here is in examining the extent to which they can be modeled as if they do. As we shall see in **Chapter 8**, we can model the spread of an invasive species or disease as a diffusion process, in which we are basically assuming that individuals move about at random. Thus, although the behaviors of individuals may be unpredictable, there may nevertheless be properties of the larger population that *are* predictable, such as the rate at which it spreads, in much the same way that properties of a gas are predictable (such as its pressure or volume at a given temperature) but the movements of individual gas molecules are not.

In a **simple random walk**, movements are uncorrelated and unbiased (**Figure 6.23A**): uncorrelated, in that the direction of movement is completely independent of the previous direction moved; and unbiased in that

the direction taken is completely random (i.e. there is no preferred direction). If an individual moves with equal probability in any direction, then this process is essentially Brownian motion and the behavior of a large number of random walkers produces the standard diffusion equation (Equation 6.2), which predicts that the **mean squared displacement** ($\overline{R}_t^2$; the average squared straight-line distance travelled by a population of walkers) increases linearly over time, t. This relationship is important, for it enables us to estimate the **diffusion coefficient**, D, that is used in models of population spread (**Chapter 8**).[7] Thus,

$$\overline{R}_t^2 \sim t^\mu \qquad \text{Equation 6.2}$$

[7] To help avoid confusion between the fractal dimension, D, and the diffusion coefficient, D, the latter is italicized in this text, although context should usually make it clear which 'D' is being discussed.

where the average mean squared displacement is expected to increase linearly as some power μ of time, which gives the slope or rate of this increase. So, if $\mu = 0$, no movement occurs and individuals simply remain in place during the observation period, which is certainly disappointing if we are interested in studying movement. For animals that are taking simple random walks, $\mu = 1$, because mean squared displacement increases linearly with time. If $\mu = 2$, then movement is increasing quadratically with time, such that absolute displacement (rather than squared displacement) increases linearly, meaning that individuals are basically walking in straight lines (albeit in a randomly chosen direction). Departures from the expectations of a random walk ($\mu = 1$) are thus informative, and indicate that individuals are moving either slower ($0 < \mu < 1$) or faster ($1 < \mu < 2$) than if movement were purely a diffusion process. For example, environmental heterogeneity and patch structure could slow movement, especially if individuals spend time within patches or encounter other obstacles during movement. Individuals might instead move faster than expected if they occasionally exhibit directed or straight-line movements, such as orienting toward a resource or habitat patch. We consider next a modification of the simple random walk model to include short-term directed movements.

CORRELATED RANDOM WALKS Far from wandering aimlessly, most animals have a tendency to maintain the same heading, at least over short distances, a behavior that gives rise to short-term directed movements (directional persistence). The direction of movement between successive steps may thus be correlated, even though there is no overall directional bias to the pattern of movement. In a **correlated random walk**, movement direction is still uniformly distributed, meaning that turning angles are basically drawn from a random (Gaussian) distribution centered on zero degrees (i.e. no directional bias), even though the heading of each step depends on the direction of the previous step. Departures from the expectations of a correlated random walk may thus indicate more complex movement behaviors in response to environmental heterogeneity or patch structure. For example, the flight patterns of the ubiquitous cabbage white butterfly (*Pieris rapae*) were well described by a correlated random walk when laying its eggs (ovipositing) on host plants in a homogeneous habitat of collards, but deviated considerably from these expectations when foraging for nectar in a more heterogeneous goldenrod field (Kareiva & Shigesada 1983; **Figure 6.24**).

Greater directional persistence, such as orienting toward a resource or habitat patch, can also be accommodated within this mathematical framework by increasing the probability of moving in a certain direction, thereby adding a directional trend to the walk

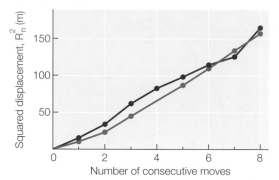

(A) Ovipositing

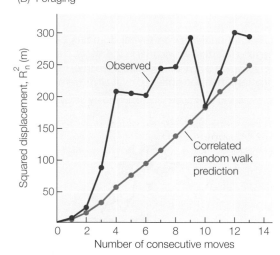

(B) Foraging

Figure 6.24 Correlated random walk. Movement pathways of cabbage white butterflies (*Pieris rapae*) fit a correlated random walk when ovipositing in a homogeneous habitat (A), but not when foraging in a heterogeneous one (B).
Source: After Kareiva & Shigesada (1983). Photo by Kimberly A. With.

(i.e. drift). A **biased correlated random walk** is basically a correlated random walk with drift. For example, random walkers have no directional bias and move with equal probability in any direction; the expected position of the individual does not shift over time (no drift). In a biased correlated random walk, however, the individual's expected position shifts in the direction toward which movement is biased (drift). An animal's movement might thus be characterized as a biased correlated random walk if its movements are correlated (there is a tendency to move in the same direction as the previous step) but they exhibit a bias toward (i.e. an attraction to) certain habitat or landscape features, such as a particular host plant or habitat type.

Butterfly flight patterns, such as the cabbage white butterflies mentioned above, might actually be better described by a biased correlated random walk in patchy or heterogeneous landscapes. Many butterflies engage in 'foray loops' at patch edges, in which they fly out from a habitat patch for a short distance before looping back and returning to the original habitat patch. This clearly non-random behavior would seem to demand a more complex model of movement behavior. However, Crone and Schultz (2008) have demonstrated that this sort of looping behavior can be generated from a biased correlated random walk, in which movement is biased toward habitat patches. Butterflies can apparently detect host plants or changes in habitat structure (e.g. patch boundaries) at a distance of about 15–30 m, and thus this sort of looping behavior, which results from a bias toward patch edges, can only occur over short distances that lie within their perceptual range. Nevertheless, the fact that a fairly simple movement model can mimic this sort of behavioral response to patches reinforces our assertion that complex behaviors need not require complex models to understand or predict.

LÉVY WALKS Over longer time periods and spatial scales, many animal pathways display a combination of movement behaviors, involving concentrated search within a resource or habitat patch separated by longer movements between patches (**Figure 6.23B**). Thus, while random walks consist of step lengths drawn from a Gaussian distribution, **Lévy walks** are characterized by step lengths that are drawn from a 'heavy-tailed' distribution, in which long step lengths occur more frequently than expected from a normal distribution.

As such, the frequency distribution of step lengths (l) can be described by a power-law relationship, which as you recall, is associated with fractal patterns:

$$P(l) \sim l^{-\mu} \qquad \text{Equation 6.3}$$

The exponent or slope of this relationship varies as $1 < \mu \leq 3$. In particular, Lévy walks in which $\mu = 2$ may represent an optimal search strategy in landscapes where prey or resources are scarce and unpredictable, because this pattern of movement tends to increase encounters with new patches over a simple random walk (Viswanathan et al. 1996). Nevertheless, the direction of movement in a Lévy walk is still uniformly distributed (i.e. turning angles are randomly distributed around a mean of 0°), as with the previous random walk models we have discussed.

Lévy walks have been reported in animals as diverse as honey bees (*Apis mellifera*; Reynolds et al. 2007), wandering albatrosses (*Diomedea exulans*; Viswanathan et al. 1996), jackals (*Canis adustus*; Atkinson et al. 2002), spider monkeys (*Ateles geoffroyi*; Ramos-Fernández et al. 2004), fallow deer (*Dama dama*; Focardi et al. 2010), and various marine predators (Sims et al. 2008; **Figure 6.25**). The fact that Lévy walks appear to be so ubiquitous among so many different types of animals could indicate that this behavior is adaptive and has evolved across different taxa because it is an efficient search strategy for exploiting patchily distributed resources in heterogeneous and unpredictable environments. Conversely, Lévy walks may simply arise as a consequence of how individuals are forced to interact with patchy resource distributions (i.e. the search behavior is emergent rather than adaptive; Viswanathan et al. 1996).

To test this, Sims and his colleagues (2008) examined the scaling properties of prey distributions (krill) in relation to the movement patterns of krill-eating basking sharks (*Cetorhinus maximus*) and Magellanic penguins (*Spheniscus magellanicus*). Intriguingly, predator movements and krill distributions exhibited similar

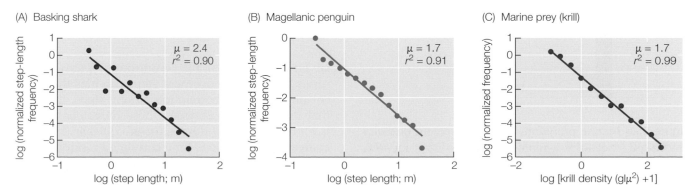

Figure 6.25 Analysis of movement as Lévy walks. The movement patterns of the basking shark (*Cetorhinus maximus*, A) and Magellanic penguin (*Spheniscus magellanicus*, B) exhibit similar scaling properties that match the distribution of their prey (C), which suggests that predators are adopting an optimal search strategy ($\mu = 2$).

Source: After Sims et al. (2008).

fractal scaling properties, which were very close to that expected ($\mu \approx 2$) for optimal search success in patchy, unpredictable environments (**Figure 6.25**). In simulations, the foraging success of Lévy walkers was indeed greater when searching for prey that exhibited a fractal (Lévy) distribution rather than a random distribution (Sims et al. 2008). Further, many of these marine predators have sufficient behavioral flexibility to switch foraging patterns depending on the type of habitat or prey distribution they encounter. Predators exhibited Lévy behavior in the deeper, less-productive waters of the open ocean, where prey are expected to be scarce and patchily distributed, but then switched to random (Brownian) movements in highly productive environments, such as along shallow frontal-shelf waters, where prey are more abundant (Humphries et al. 2010; **Figure 6.26**). Again, we should anticipate that individuals will exhibit different behavioral responses or patterns of movement in different habitat or landscape contexts. The fact that species may well exhibit different types of movement depending on scale, however, means that extra care must be taken when attempting to fit movement data to a particular model, a point we shall discuss next.

SCALE DEPENDENCE OF RANDOM WALK MODELS By attempting to relate patterns of movement to a particular model, we are hoping to gain insight into the nature of the processes that underlie behavioral responses to resource or habitat distributions. However, a given movement pathway could potentially fit more than one of the models discussed above, depending on the time period and spatial scale involved, or the statistical approach used to test model fit. For example, short-range correlations in movement may disappear when assessed over longer time periods, such that a correlated random walk becomes a random walk at broader scales (Bartumeus et al. 2005).

Likewise, some movement patterns that have been described as Lévy walks, such as those of the wandering albatross (Viswanathan et al. 1996), might simply be random walks when this fit is tested statistically (using model-selection procedures based on maximum likelihood estimates) rather than the usual graphical approach that consists of regressing the frequency distribution of move lengths as a function of scale on a log-log plot (Edwards et al. 2007). The key to distinguishing a random walk from a Lévy walk is the extent to which the frequency distribution of step lengths better fits an exponential (Poisson) versus a power-law (fractal or Lévy) distribution, but this can be difficult to determine on a log-log scale, with the result that many movement patterns that look like Lévy walks may not in fact be generated by a fractal process. Indeed, a composite random walk model, in which step lengths were randomly drawn from two different exponential distributions (capturing the expected behavior of individuals foraging within patches versus when moving between patches) generated movement patterns that were indistinguishable from a Lévy walk (Benhamou 2007). Further, in a comparison of random walk models to Lévy walks, it was shown that autocorrelation inevitably leads to a Lévy walk (Reynolds 2010).

> The many different types of models applied to such data underscore the fact that movement data often look similar…and models that might fit equally well could involve different assumptions or, at least, be interpreted in different ways.
> Schick et al. (2008)

> Emergent movement patterns should not be confused with the processes that gave rise to them.   Benhamou (2007)

Given that representing a continuous pathway as a series of discrete movement steps can bias the apparent degree of autocorrelation among steps, it may be that the difference between these movement models is more a matter of degree (or scale!) than kind. The analysis of continuous-time correlated random walks, which avoids the aforementioned problem of discretization of continuous movement pathways, demonstrates that Lévy walks are an inevitable byproduct of autocorrelation, which may explain the apparent ubiquity of Lévy walks in nature (Reynolds 2010). Hence, we would do well not to associate movement patterns

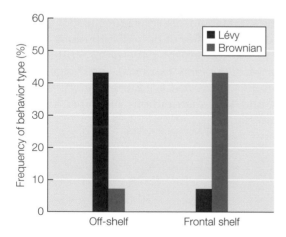

Figure 6.26 Marine predators switch movement behavior depending on habitat context. In the Northeast Atlantic, predators foraging in less-productive habitats (off-shelf), where prey is likely to be scarce and patchily distributed, were more likely to adopt a Lévy search strategy, whereas predators in more-productive habitats (frontal shelf) switched to a random (Brownian) search that may be more efficient when prey are widespread and abundant.

Source: After Humphries et al. (2010).

with a particular process (Poisson vs Lévy), given that different processes can potentially give rise to the same pattern, without first considering the scale and underlying assumptions inherent in each model. Further, we should test the potential fit of our movement data to different models, and not just for consistency with a particular model that happens to be of interest.

Spatially Explicit Models of Animal Movement

Although we can make some inferences about how species are interacting with landscape structure from just an analysis of their movement pathways (e.g. whether they perceive the landscape to be patchy), we might eventually wish to explore how individuals interact with spatial complexity directly, through the use of a spatially explicit simulation model. We discussed how grid-based simulation models could be used to assess landscape connectivity in heterogeneous landscapes in **Chapter 5**. Recall that we can simulate movement of individual dispersers through a heterogeneous landscape, using either a neutral landscape model (**Figure 5.5**) or actual landscape map. We can imbue our virtual dispersers with various movement rules, ranging from how far and in what direction they are likely to move with each time step; to how fast and through which habitats they are most likely to move; to the potential costs, in terms of energetics and mortality risk, of dispersing or in moving across particular landscape elements (e.g. roadways). In short, we can program movement any way we wish, focusing on the fates of individual dispersers (individual- or agent-based models) or on the likelihood of dispersal success for a population of dispersers (e.g. dispersal kernels), which gives us the complete freedom and unparalleled flexibility to explore the interaction between individual movement behavior and landscape structure. For example, this might enable us to gain a basic understanding of how habitat loss versus fragmentation influences foraging or dispersal success, the redistribution of individuals across the landscape, or the probability of patch colonization, and thus can be used to generate hypotheses that we can then test experimentally or empirically. Alternatively, we might be interested in parameterizing the movement algorithm of a metapopulation or other type of spatially explicit population model (SEPM) that is being used to assess management options for a species of conservation concern (**Chapter 7**).

To illustrate applications of the spatially explicit simulation modeling approach, we consider next a couple of examples related to quantifying dispersal success in both fragmented and heterogeneous landscapes.

DISPERSAL SUCCESS IN FRAGMENTED LANDSCAPES

Habitat loss and fragmentation are expected to have a negative effect on dispersal success. Intuitively, dispersal success should decline as habitat becomes scarcer. If dispersal is completely random, such that the disperser is able to move to any randomly selected point on the landscape, then the probability of successful dispersal is given by

$$P(\text{success}) = 1 - u^m \qquad \text{Equation 6.4}$$

where $u = 1 - h$, in which h is the amount of habitat on the landscape, and m is the number of dispersal steps (of random length and in a randomly selected direction) that an individual can take while searching for habitat (Lande 1987). So, if habitat covers 80% of the landscape, then a disperser has a 80% chance of success, even if it only has a single chance at dispersal $[1 - (1 - 0.8)^1 = 1 - (0.2)^1 = 0.8]$. If it has $m = 2$ chances, then its success rate climbs to 96% $[1 - (0.2)^2 = 1 - 0.04 = 0.96]$.

If dispersal is truly random, then the arrangement of habitat on the landscape does not matter for predicting dispersal success; only the amount of habitat is important. Remember that the disperser is moving to a randomly selected location with each step. Thus, the more suitable habitat there is on the landscape, the greater the likelihood that a randomly selected location will contain habitat. However, if animals do not disperse as a series of completely independent steps, but instead exhibit some degree of spatial autocorrelation in their movements (i.e. dispersal occurs as a series of small, sequential steps), then the distribution of habitat will matter for predicting dispersal success. But does the distribution of habitat always matter, or does it matter only when habitat is scarce (e.g. McIntyre & Wiens 1999b)? At what critical level of habitat loss might fragmentation (the distribution of habitat) be important for predicting dispersal success?

To investigate this, With and King (1999b) developed a spatially explicit dispersal model, in which individuals were permitted to move through adjacent cells (one of the four nearest-neighbors) on fractal landscapes that varied in both the amount (h) and fragmentation (H) of habitat (**Figure 5.7**). Dispersal was initiated from within a randomly selected habitat cell, and while the direction of travel was random, the movement step length was fixed to a single cell (i.e. individuals could only move one cell left or right, forward or backward). Individuals were constrained to move a certain number of steps (m), in which they either succeeded in finding a suitable habitat cell or died trying. Dispersal success was thus assessed as the proportion of dispersers that succeeded in finding habitat on a given landscape type.

As expected, dispersal success increased with search effort (m) as well as the amount of habitat (h) on the landscape, but was also affected by the degree of habitat fragmentation (H; **Figure 6.27**). For example, dispersal

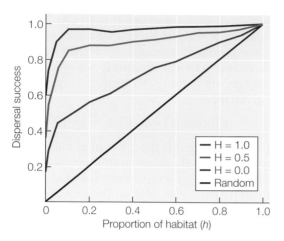

Figure 6.27 Effect of landscape structure on dispersal success. Dispersal success is higher in clumped fractal landscapes (H = 1.0) than in more fragmented fractal landscapes (H = 0.0) when dispersers are constrained to move only through adjacent cells (**Figure 5.7**). Dispersal success declines precipitously when ≤10% habitat remains. Dispersal success for random dispersal is also shown for comparison. All comparisons here are for a single dispersal attempt (m = 1).

Source: After With & King (1999b).

success is only 10% in landscapes with 10% habitat when dispersal is random and limited to a single dispersal attempt (m = 1; Equation 6.4). Conversely, success is virtually guaranteed (97%) when that habitat is clumped (H = 1.0), even with only a single dispersal attempt that is restricted to the local neighborhood (nearest-neighbor dispersal). Interestingly, dispersal success declines precipitously below about 10–20% habitat (a **dispersal threshold**), which may coincide with the threshold in gap sizes that occur in this domain (**lacunarity threshold**; Figure 4.38), leading to an increased isolation of patches and a decline in dispersal success. The gap structure (inter-patch distances) of the landscape may thus be a more important determinant of dispersal success than the patch structure (the size distribution of patches or total amount of habitat) when habitat is limiting (<10–20%) on the landscape.

DISPERSAL SUCCESS IN HETEROGENEOUS LAND-SCAPES As we saw in the first part of this chapter, the composition of the landscape matrix can influence animal movements. Different matrix types offer varying degrees of resistance to movement, in terms of the rate or likelihood with which animals move through them. Recall from our discussion of landscape resistance in **Chapter 5** that grid-based simulation models, least-cost paths, resistance-surface models, and circuit theory have all been used to evaluate the functional connectivity of heterogeneous landscapes, based on the perceived resistance or relative costs to individuals moving through different land covers (Bunn et al. 2000; McRae et al. 2008; Zeller et al. 2012). Because movement

is expected to occur more readily through some types of habitats than others, heterogeneous landscapes present an added complication to our assessment of landscape effects on dispersal success.

To evaluate the potential for matrix heterogeneity to influence dispersal success, Gustafson and Gardner (1996) developed an individual-based simulation model, in which they simulated the movement of a large number of dispersers on heterogeneous landscapes. Movement behavior was modeled as a self-avoiding walk,[8] in which the individual was unable to return to a previously visited cell until it had moved >2 steps, so as to prevent individuals from simply moving back and forth between two cells. They used a combination of neutral landscape models (heterogeneous random and hierarchical random maps; **Figure 5.5**) and actual landscape images, in which forest patches were embedded in a heterogeneous but predominantly agricultural matrix (**Figure 6.28**). Much of the variation in dispersal success (89%) on these landscapes could be explained simply by patch size and isolation (e.g. A_i and d_{ij}; Equation 5.1). Large, close patches exchanged more individuals than small, isolated patches. Nevertheless, changes in matrix heterogeneity sometimes had significant—and unpredictable—effects on the rates of dispersal among patches. In particular, dispersal among patches tended to be asymmetrical, in that the exchange of individuals was often greater in one direction than another (i.e. $d_{ij} \neq d_{ji}$). To illustrate, consider patches 1 and 3 in **Figure 6.28**. There were nearly 7× more individuals that emigrated from patch 1 to patch 3, as there were that immigrated from patch 3 into patch 1 ($d_{13} = 0.13$, $d_{31} = 0.02$). Patch 1 is fairly isolated, and thus only receives dispersers from patch 3. Patch 3, however, has four other patches in close proximity with which it also exchanges individuals. Thus, differences in patch configuration can account for dispersal asymmetries among patches.

Landscape structure is in fact likely to produce barriers to movement in certain directions and thereby force dispersing individuals to concentrate movement within restricted corridors… that may not be obvious to human observers.

Gustafson & Gardner (1996)

Further, the specific dispersal routes taken by individuals moving between patches were rather amorphous and did not correspond with any obvious features or

---

[8]A self-avoiding walk is simply a random walk through adjacent cells in a lattice, with the constraint that individuals are not permitted to visit a cell more than once. In the Gustafson and Gardner (1996) study, individuals were prevented from backtracking (i.e. revisiting their most recent location) and thus moved randomly into one of the other seven cells in the immediate neighborhood (i.e. Rule 2 movement; **Figure 5.6**).

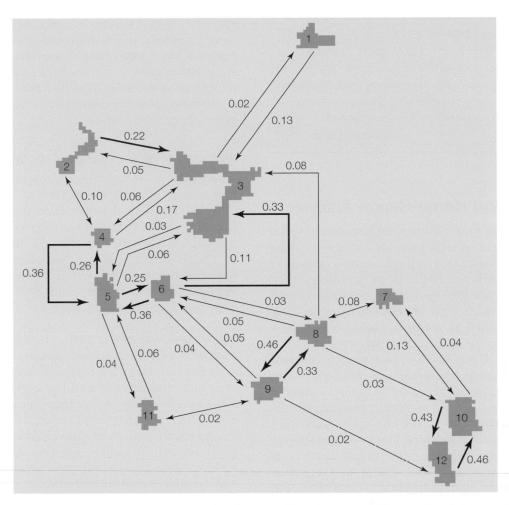

Figure 6.28 Dispersal success in a heterogeneous landscape. Transfer rates between forest fragments for simulated dispersers in an agriculturally dominated landscape. The complexity of the agricultural matrix is not depicted here, so as to better highlight the differential transfers between forest fragments.

Source: After Gustafson & Gardner (1996).

land-cover types on the landscape. In all of our previous discussions on habitat corridors or dispersal barriers, these have always been assumed to be discrete, structural features of the landscape. The results of this simulation model, however, demonstrate how regions of the landscape that function as dispersal corridors may not be all that obvious within a heterogeneous matrix. As a consequence, these conduits may go undetected and unprotected. For example, areas of less-preferred habitat may be used more frequently during dispersal when adjacent to favorable habitat, which again underscores the potential value that less-suitable habitat may have as dispersal corridors.

Before we leave the topic, we should bear in mind that the seemingly greater realism of spatially explicit simulation models must not be confused with greater accuracy or precision. Whether a spatially explicit dispersal model is a 'good' model depends on the purpose of the model, the objectives of the analysis, and the type and quality of movement data available. For example,

if the intent is to use the simulation model as a planning or assessment tool to evaluate different management scenarios for a species of conservation concern, then we will need some fairly detailed information on how that species moves about the landscape. Errors in parameter estimation will propagate throughout the model run and lead to errors in our model predictions. For example, Ruckelshaus and her colleagues (1997) demonstrated how errors in estimating dispersal mortality and how far dispersers can travel resulted in substantial errors in predicting dispersal success, even within a simple spatially explicit population model. Further, these prediction errors were higher in landscapes that had a low percentage of suitable habitat, which are precisely the sorts of landscape scenarios in which species would be at greatest risk, and where accurate population assessments are most crucial. They thus conclude that less-detailed models should be considered when the quality of available data does not match the complexity of the model.

> While digitized maps of habitat might be accurate and suggestive of great realism, the models of animal movement that are used to connect species to their landscape often rest on a very flimsy data base. Either more energy must be put into learning how organisms disperse and die along the way, or alternative modeling approaches must be explored.
>
> Ruckelshaus et al. (1997)

Space Use and Home-Range Analysis

The movements of most animals are bounded in space and ultimately in time, which sets an upper limit to the range of heterogeneity that individuals may encounter and thus the sorts of habitats or resources they might be able to use (the spatial extent of movement; **Figure 6.4**). The sum total of movements over long periods of time, such as within a season, a year, or a lifetime, circumscribes an area that gives the individual's **home range**. The concept of home range was first defined by American mammalogist William Henry Burt (1943) as 'that area traversed by the individual in its normal activities of food gathering, mating and caring for young.' The home range concept is somewhat different from that of a **territory**, as it need not involve an area that is defended, but rather is just a pattern of space use that is area-restricted. The home range thus links the movement of an individual to the distribution of resources necessary for their survival and reproduction, providing a fundamental measure of individual space use (Börger et al. 2008). An analysis of space use within the home range can reveal habitat preferences (the disproportionate use of habitats relative to their abundance), whereas quantifying the size and distribution of home ranges provides information on the degree of spatial overlap among individuals or species, and can be used to estimate population size or carrying capacity in different habitats (Downs & Horner 2008).

Of course, not all areas within an animal's home range are used equally—or at all—and thus the location and frequency with which various areas are used gives rise to the **utilization distribution** (*sensu* Van Winkle 1975). The utilization distribution has contributed to the current view of the home range as a probability density surface that depicts the relative intensity of use of various areas. The utilization distribution can be used to estimate the home-range boundary (the smallest area within which an individual spends 95% of its time) or of the core area of use (the area encompassing 50% of space use; White & Garrott 1990). These attributes of the home range can then be used to evaluate variation in space use among individuals within a population (between sexes or life stages), or for a given

species between seasons (winter vs summer ranges) or even among different species (Nilsen et al. 2008; **Figure 6.16**). More recently, various statistical modeling techniques have been developed that make direct use of the utilization distribution in estimating home-range size (Kernohan et al. 2001). We discuss home-range estimators next.

Methods of Home-Range Estimation

Two of the most common techniques for estimating home-range size are the minimum convex polygon (MCP) and the kernel density estimation (KDE) method (Laver & Kelly 2008), and we therefore focus on these two approaches here. Regardless of the approach used, we should be cognizant of the potential biases of different methods for estimating home ranges and the implications of these biases in any subsequent analyses based on these estimators.

MINIMUM CONVEX POLYGON   The MCP is the smallest area of a convex polygon[9] that contains all of the recorded locations for a particular individual. Because of its historical precedent and ease of implementation, it has long been used to estimate home-range size and is still often used to permit comparisons among studies, especially with older ones (Nilsen et al. 2008). The MCP suffers from a number of known issues, however, such as a sensitivity to sample size, sampling method and duration, serial autocorrelation among movement locations, outliers, and so forth (Swihart & Slade 1985a,b; Seaman et al. 1999). Further, the MCP provides no measure of space use within the home range and often contains large areas of unoccupied space if home ranges are not in fact convex (Worton 1989). Thus, the MCP is generally considered an unreliable estimator that is subject to unpredictable bias, leading Börger and colleagues (2006) to proclaim that it should never be used in the estimation of home-range size. Nevertheless, the MCP continues to be popular in home-range studies, where it is often combined with other modeling approaches (such as KDE) to estimate home-range size (Laver & Kelly 2008). Where the goal is to make comparisons among species that exhibit very large differences in home-range size, the MCP is perhaps as good as any other estimator, especially since such large differences in range size among species will likely swamp any differences resulting from the choice of estimator (Nilsen et al. 2008).

KERNEL DENSITY ESTIMATION   In using kernel methods to estimate the utilization distribution, the home range is viewed as a three-dimensional surface, in which

[9]A convex polygon has all interior angles ≤180°, such that its vertices all point outward.

the 'height' corresponds to the amount of time the animal spent in a given area within the home range (Worton 1989). In practice, implementation of the KDE method involves placing a **kernel** (a bivariate normal probability density function) over each recorded location of the individual (**Figure 6.29A**). Our dataset is thus a collection of individual locations recorded over some period of time (i.e. a spatial point pattern), from which density estimates (the intensity of use) are obtained across the entire home range by averaging the densities of overlapping kernels at each spatial coordinate (in xy space). The width of the kernel (or bandwidth, h) determines the relative contribution of more distant points to the density estimate at a given coordinate. For example, a wider kernel (a large bandwidth) allows more distant points to influence density estimates at a particular location, resulting in a greater 'smoothing' of the distribution. For this reason, h is also referred to as a smoothing parameter. Thus, narrow kernels reveal greater fine-scale detail in the data structure, whereas wide kernels reveal the overall shape of the distribution. As will be discussed in the following, the selection of an appropriate

kernel width is one of the main challenges to applying KDE techniques. Choosing a kernel width that is too small will yield 'noisy' estimates with potentially spurious structure, whereas choosing a kernel width that is too large will yield overly smoothed estimates that obscure important structure (Keating & Cherry 2009). Assuming an appropriate kernel width has been used, however, then once the utilization distribution has been estimated, density can be converted into a home-range estimate by connecting areas of equal density to describe a particular usage area of the home range (e.g. 95% of the utilization distribution).

Kernel density estimation generally outperforms other estimation techniques (Worton 1989; Seaman & Powell 1996) and is thus the most widely accepted method of home-range estimation (Kernohan et al. 2001). There is the potential problem of a lack of consistency among studies, however, because there are so many different ways in which KDE can be implemented (Laver & Kelly 2008). We thus consider some of these issues in implementation that may affect KDE-based analyses of home ranges.

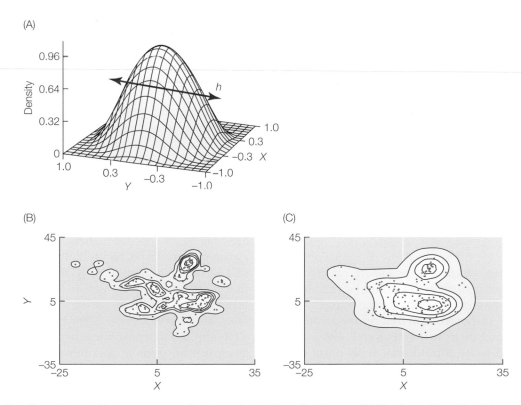

Figure 6.29 Estimation of animal home ranges using kernel density estimators. (A) The kernel is a bivariate probability distribution, in which the volume under the curve integrates to 1. The width of the kernel (h) adjusts the scale over which recorded animal locations influence the density estimate at a given point. Different types of kernel density estimators can give estimates of the size and shape of the home range. (B) The estimator based on the least-squares cross-validated (LSCV) kernel generally provides a more accurate representation of the home range compared to many other procedures, such as the reference bandwidth kernel (C). Contours depict the 95%, 72.5%, 50%, 27.5%, and 5% intervals of the estimated home range.
Source: After Seaman & Powell (1996).

Selection of a kernel width (h): The kernel width (*h*) has great influence on kernel density estimators (Silverman 1986). As mentioned above, kernel widths that are too large will oversmooth the distribution and overestimate home ranges, whereas those that are too small have the potential to underestimate home ranges (Kernohan et al. 2001). The most widely recommended—and thus widely used—procedure is the least-squares cross-validation technique (LSCV), which examines various bandwidths (*h*) and selects the one that minimizes the estimated error (the difference between the unknown true density function and the kernel density function; Seaman & Powell 1996).

Despite its widespread endorsement (Seaman et al. 1999; Millspaugh et al. 2006), LSCV does not perform well in all cases. Although a large sample size is usually considered desirable for increasing the precision of home-range estimates (see *Sample size and serially correlated data*, below), LSCV may fail for very large data sets (e.g. those containing hundreds of individual locations), such as those obtained through satellite tracking (Hemson et al. 2005). As we discussed earlier in this chapter, oversampling becomes a problem when fixes on an animal's location are taken too frequently because this increases the likelihood that locations will be spatially correlated (if not identical). In terms of the utilization distribution, these high-density areas give the appearance of high-intensity use (see *Intensity of habitat use*, below), which then affects the success of LSCV (the selected *h* is likely too small) and biases the estimation of home-range size (Hemson et al. 2005). The obvious solution might be to obtain or use fewer data, but how many fewer is difficult to gauge, especially in advance. The issue of 'how much data is enough?' is a thorny one and must be considered carefully in any home-range analysis, regardless of whether LSCV-KDE or some other approach is used. There are a number of alternative selection methods that can be used if needed (e.g. Gitzen et al. 2006).

Effect of home-range shape and boundary effects: The KDE approach assumes that the animal's activity is concentrated in the center of its home range, with activity decreasing outward from that center according to a bivariate normal distribution (**Figure 6.29A**). Most home ranges tend to be multimodal with more than one center of activity, however, and thus KDE may overestimate home-range size when the spatial pattern of locations does not conform to a bivariate normal distribution (Downs & Horner 2008). For example, KDE performs poorly when ranges are linear (as for riverine species), or that have large amounts of unused area in their interior (as in disjunct ranges), or that are demarcated by a sharp boundary produced by some topographic barrier to movement, such as a river or lake (Blundell et al. 2001; Getz & Wilmers 2004; Row &

Blouin-Demers 2006). In these cases, KDE may extend the 'tails' of the utilization distribution beyond the observed data, past perceived barriers, or into non-habitat areas that are not used by the individual (Getz & Wilmers 2004).

Various approaches have been developed to deal with these issues. For example, Getz and Wilmers (2004) introduced the local convex hull (LoCoH or LCH) method to address the perceived inability of KDE to adequately resolve narrow movement corridors, interior 'holes' (unused areas) in the home range, and sharp boundaries. Their approach involves constructing a local convex polygon (the local hull) using a specified number of nearest neighbors at each recorded location, which produces a set of non-parametric kernels that overlap to produce the utilization distribution (Getz et al. 2007). Thus, the LCH method uses kernels that arise from the data, unlike parametric kernels whose form is specified a priori (e.g. a bivariate normal kernel with width *h*). However, traditional kernel methods can certainly produce utilization distributions with irregular surfaces and multiple peaks (multimodal distributions; **Figure 6.29B, C**). In fact, Lichti and Swihart (2011) found that while the LCH method was better at excluding unused areas outside the home-range boundary, the KDE method performed as well or better in all other respects. They thus advise that researchers continue using KDE for home-range estimation, unless they are especially concerned about boundary effects.

Sample size and serially correlated data: Estimation of home-range size is sensitive to sample size, the number of recorded locations obtained for the individual (Seaman et al. 1999; Börger et al. 2006). Substantially fewer locations are generally needed to provide accurate estimates when KDE methods are used, however. For example, 100–300 locations may be necessary if using MCP (Harris et al. 1990), but only 30–50 locations are minimally needed when using kernel methods (Seaman et al. 1999). Nevertheless, larger sample sizes still increase the ability of KDE (as well as LCH, for that matter) to fit the structure of the utilization distribution more precisely, as well as to exclude unused areas at the periphery (i.e. the boundary effect). Thus, KDE applied to a small number of locations (<50) will generally be poor at identifying fine-scale structure and will have a tendency to overestimate home-range size (Seaman & Powell 1996).

Despite this, most home-range studies have been based on <50 locations (Laver & Kelly 2008). Rather than basing sample size on some published minimum, the preferred approach is to perform an asymptote analysis of the data, in which the sensitivity of the estimator to a range of sample sizes is explored. This is typically done via visual inspection, by plotting the resulting home-range size as a function of sample size (based on the sequential addition of locations until the total number of locations is reached) and then identifying the sample size at which home-range size appears to asymptote. A more quantitative approach involves identifying the sample size at which the 95% confidence interval of the bootstrapped home-range estimate is within some percentage (e.g. 5–10%) of the total home-range size (using all locations) for some number of consecutive locations (e.g. 5; Laver & Kelly 2008).

The feasibility of tracking animals using satellite-based transmitters, which permit the automatic logging of individual locations at fixed time intervals, means that sample-size issues are likely to be less of an issue in future home-range studies. Instead, serial autocorrelation leading to a lack of independence among data points is likely to be the greater concern. As discussed previously, serial correlation is usually considered a 'statistical flaw' in movement studies rather than a source of information, which as Benhamou and Cornélis (2010) point out, is rather paradoxical given that home ranges are spatial distributions that emerge as a consequence of movement processes rather than by some sort of spatial point process. By itself, serial correlation does not matter where KDE is concerned, as the main goal is really to obtain a representative sample of locations, as discussed above (Swihart & Slade 1985a,b). Still, applying a minimum smoothing constraint (the kernel width, h) to locations that are serially correlated can lead to the overestimation of home-range size if this ends up including areas that were never visited. The key is thus to increase spatial resolution of the utilization distribution. To that end, Benhamou and Cornélis (2010) exploited the serially correlated nature of movement data to create a kernel that is locally biased in the direction the individual moves between successive locations. The result is a narrower anisotropic kernel than that of the classical kernel, which as you recall, assumes a probability density distribution that is isotropic (i.e. it tapers off uniformly from the center in all directions; **Figure 6.29A**). This movement-based kernel also succeeded in modeling sharp home-range boundaries, as it accurately represented the barrier posed by the Niger River to the movements of an African buffalo (*Syncerus caffer*) herd, thereby giving a more reliable estimate of their range size (Benhamou & Cornélis 2010).

Intensity of habitat use: Although the utilization distribution is supposed to give the intensity of space use within a home range, the inclusion of locations associated with resting bouts or times when the animal is not actively moving may give a false impression of relative habitat use or habitat preferences. For example, if we include multiple readings of locations obtained at a nest or den site, these repeated fixes will inflate the apparent use of whatever habitat is associated with

that site, thereby minimizing the importance of other habitats that are used by the individual for foraging or other activities (Millspaugh et al. 2006). The obvious solution is thus to use only data from when animals are actively moving, such as during foraging, to obtain a biologically representative estimate of the utilization distribution.

Dynamic home ranges: For many species, the utilization distribution changes over time, such as in the diel movements of an aquatic species (vertical migration) or the seasonal use of different habitats or landscapes by a migratory species. Home ranges are thus dynamic, and yet, most home-range studies focus solely on the spatial dimensions of the utilization distribution (intensity of space use in two dimensions) rather than on its temporal dimension (variability in the intensity of that space use over time; Laver & Kelly 2008). Utilization distributions might thus be defined instead as the 'relative frequency distribution of an animal's occurrence in space and time' (Keating & Cherry 2009). Time can be added directly as a covariate in the model to produce temporally varying utilization distributions, in which the probability of occurrence varies continuously over the course of a day or throughout the year, thus eliminating the need to divide location data into discrete (and subjective) time periods for analysis (e.g. winter vs summer home-range use).

Although time can be treated as a continuous linear variable, as when modeling occurrence over a period of years, time may also be represented as a continuous circular variable, as when modeling seasonal occurrence throughout the year (Keating & Cherry 2009). The latter requires a different sort of kernel that accounts for the circularity of the data distribution to ensure that locations are weighted appropriately. For example, an individual's location at 23:55 hours is likely to be more similar to its location at 00:00 hours (a difference of 5 minutes) than to its location at 22:00 hours (a difference of 1 hour and 55 minutes). To illustrate this approach, Keating and Cherry (2009) adopted a kernel for circularly distributed data (a wrapped Cauchy distribution) in their analysis of the utilization distributions for two groups of Rocky Mountain bighorn sheep (*Ovis canadensis*) during the course of a year within Glacier National Park, Montana, USA (**Figure 6.30**). Although bighorn sheep have been described as having up to four seasonal ranges, their seasonal use and movements among these areas may in fact be far more fluid than generally appreciated. For the groups studied by Keating and Cherry (2009), ewes moved freely among a number of winter and late-winter/spring ranges during these periods. Although ewes tended to be concentrated within a few areas during the summer, individual ewes sometimes traversed the entire extent of their year-long range,

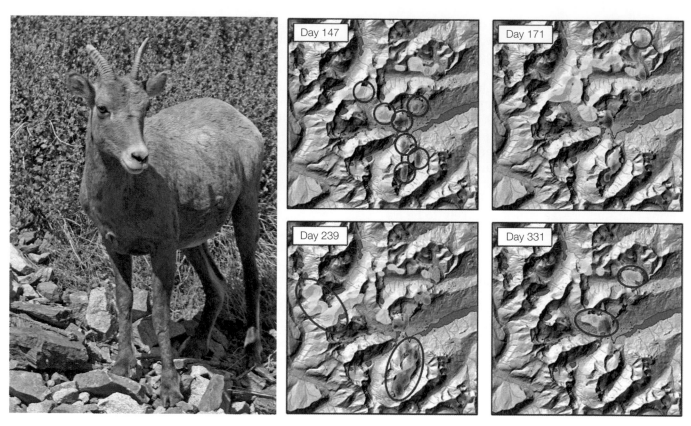

Figure 6.30 Home-range dynamics of Rocky Mountain bighorn sheep (*Ovis canadensis*). Over the course of the year, the concentration of space use for two groups of ewes (identified in green and red) in Glacier National Park (Montana, USA) shifted in continuous, but systematic, ways. Red circles depict concentrated use associated with lambing areas (day 147), a major salt lick (day 171), late-summer habitat (day 239), and rutting grounds (day 331).

Source: Keating & Cherry (2009). Photo by Kimberly A. With.

and so might be found anywhere during this period of time. Thus, although there were predictable patterns of space use during certain times of the year (**Figure 6.30**), the utilization distributions of bighorn sheep were ultimately stochastic in both space and time, such that the probability of occurrence peaked in different locations during different seasons. The use of more traditional kernel methods to derive seasonal utilization distributions tended to average across the entire range during a given period, thus obscuring finer details of space use that were confined to only a brief period in time. This might be important when evaluating the habitat needs of a given species, as traditional KDE methods might otherwise overlook the importance of certain areas (such as salt licks) that are used only briefly but are nevertheless an essential component of the species' home range.

Measuring Plant Dispersal

So what about plants? Up to now, we have focused almost exclusively on the analysis of animal movement in this chapter. Not all animals are mobile (e.g.

corals), but those that are sessile tend to be aquatic or marine and have larvae or propagules that are dispersed passively by water currents (e.g. **Figure 5.13**). Clearly, plants are also sessile, but dispersal can be accomplished through a variety of means, involving the movement of seeds, spores, plant fragments (e.g. pieces of stolons or roots), bulbils, or through clonal growth. Some plants can disperse in more than one way, such as by seeds and plant fragments. Although plant dispersal may entail passive movement via wind or water, dispersal can also be achieved through various adaptations of the plant itself (e.g. explosive pods that shoot seeds a short distance from the 'parent' plant) or facilitated by animal vectors (including humans) that are capable of transporting seeds or plants over much longer distances. For example, some plant seeds are 'sticky' or have spines or hooks that enable them to adhere to animal fur (or the socks and pant legs of unwary hikers), others produce fleshy fruits that invite animals to consume them in order that their seeds might be regurgitated or defecated at a later date somewhere else, and still others are carried off to be cached and discarded or forgotten.

Plant dispersal can thus be a complex process, involving different stages of movement by different means operating at different scales. For example, many trees in the Brazilian Cerrado, such as the monkey pepper tree (*Xylopia aromatica*), produce fleshy fruits that are eaten and dispersed primarily by birds (Christianini & Oliveira 2010). Birds are messy eaters, however, and a substantial portion of the fruit crop (25%) ends up on the ground beneath the trees. Nevertheless, much of this fallen fruit (up to 83%) is ultimately gathered up by ants, and although some of these do consume seeds (an act of seed predation), others take the seeds back to their nests where the seeds may eventually germinate (**directed dispersal**). In addition, some ants even remove the processed *Xylopia* seeds found in bird feces, resulting in a second bout of dispersal when these seeds are then carted off to the ant's nest (**secondary dispersal**). Thus, ants contribute almost as much as birds to *Xylopia* seed dispersal (21% vs 32%, respectively), albeit at very different scales. Birds are clearly important for the long-distance dispersal of seeds, but ants provide both short-range dispersal benefits (rescuing fallen seeds and moving them beyond the canopy of the parent tree, thereby giving them another shot at dispersal and establishment) as well as secondary directed dispersal that may increase germination success by 'fine-tuning' seed placement following their initial displacement by birds (nests of ground-dwelling ants tend to provide a safer and more-suitable environment for seed germination, since they are underground, than wherever birds happen to deposit them). Birds and ants thus provide complementary seed dispersal services in this system.

As with animals, dispersal is important for understanding the spatial distribution, dynamics, and spread of plant species across the landscape. Seed dispersal in particular is an important driver of plant population dynamics because most plants are dependent on seeds for regeneration, and because germination success, seedling establishment, and recruitment are all critically dependent on the distribution of suitable microsites, which tend to be patchily distributed across the landscape (Schupp & Fuentes 1995). Although these sorts of local population processes may require detailed information on dispersal on the order of a few meters, we may ultimately require information on dispersal at much larger scales if our goal is to understand metapopulation dynamics or the rate of invasive spread (Bullock et al. 2006). We should therefore consider the various means by which plant seed dispersal can be measured, which are divided here into direct and indirect methods.

Direct measurement of seed dispersal typically entails either tracking individual propagules to see where they end up, or by inferring dispersal from the number of propagules that end up in seed traps at a given location (Bullock et al. 2006). Each approach has its advantages and disadvantages. Which approach is better—tracking versus trapping—will therefore depend on the size and dispersal mode of the seed (or plant part), as well as the type of movement information required by the analysis or model to be developed.

Tracking seed dispersal, or the transport of plants or plant parts more generally, is considerably more laborious and time-intensive than trapping. As with animal tracking, we will either need to follow the movement of individual seeds or their animal vectors. Thus, many of the issues discussed previously in the context of animal tracking apply here as well, especially if it is the animal vector that is being tracked. Ideally, animal vectors are tracked while actually carrying seeds, although even then, it is not always possible to observe when seeds have been deposited and thus dispersed (Bullock et al. 2006). In the case of human-vectored dispersal of seeds or plants, identification of the major transportation routes, travel patterns, and behavior of people must be studied. Many plant propagules can 'hitchhike' on the boots, clothing, or equipment of humans and be spread all over the world. For example, Eurasian watermilfoil (*Myriophyllum spicatum*) is a popular aquarium plant that has become widely established throughout North America, where it is considered a highly invasive nuisance species that is easily spread among water bodies by recreational boaters (Johnson et al. 2001). Even the remote, inhospitable landscape of Antarctica is under siege from alien plant invasions. Seeds are inadvertently carried in by the tens of thousands of tourists, scientists, and support staff that disembark on the continent every year (Chown et al. 2012). The average visitor to Antarctica was found to carry about 10 seeds on their person, with about half of these seeds coming from plant species that hail from similarly cold regions of the world (e.g. alpine or arctic regions), and which can therefore survive and establish in the ice-free areas near the coastal regions and research stations of Antarctica.

Mark-recapture techniques are also used to study seed dispersal, using a variety of materials (paint, fluorescent powder, radioisotopes) to mark and subsequently relocate dispersed seeds. However, it can be difficult to find marked seeds in dense litter or within structurally complex habitats, especially at increasing distances from the source. For large seeds, such as acorns, metal brads or nails can be pounded into the seed and then recovered later using a hand-held metal detector (Sork 1984). For example, tagged acorns (*Quercus* spp.) were found to be eaten or dispersed by rodents up to 20 m from their source (Sork 1984; Steele et al. 2001). More recently, tiny coded metal tags (1 mm long), which were originally developed for use in fish, have been used to track ant-dispersed seeds over short distances (<1 m; Canner & Spence 2011). The wire tags can be found with a metal detector even when buried underground (up to 4 cm; Canner & Spence 2011). Most plant seeds are too small to mark, however, and those dispersed by wind or water may be impossible to follow in the field, especially if they travel long distances. Thus, seed trapping or modeling air flows or water currents provide better avenues for measuring or predicting dispersal for these sorts of propagules.

Trapping is generally much easier and less time-intensive than tracking individual propagules. Various types of seed traps are available, depending on the size or type of plant propagule being sampled, and include sticky traps (like those used for insects), ground trays, pitfall traps, or nets (for larger or aquatic seeds). Alternatively, one may dispense with the trap and sample the substrate directly, using a sieve to separate seeds from the substrate or allowing the seeds to germinate within the substrate itself (so seedlings rather than seeds are counted, giving a measure of successful dispersal). Trapping is also likely to give a broader view of dispersal than tracking individual propagules, since it is not restricted to a particular type of movement vector.

To be successful, trapping does require a carefully designed study and adequate sampling to ensure that the area within the potential dispersal range is sufficiently covered (by samples or traps), and that the seed source itself is (or can be) isolated to reduce confounding effects from neighboring sources. In turn, this may require some a priori information on the pattern of dispersal in order to design the layout of traps (or samples) in the first place. For example, long-distance dispersal events may be missed if traps or samples are not placed far enough out from the source (or at a sufficient density or size to capture what may well be a rare event in any case). Similarly, the degree to which the pattern of seedfall is clumped or anisotropic could be overlooked if traps are arrayed improperly along transects, such that the degree of spatial dependence among dispersal distances is missed (e.g. the spatial grain of sampling is too coarse) or the transect itself is

not aligned with the direction of seedfall. Clearly, this could bias our assessment of the dispersal kernel. Given the usual constraints and diminishing returns of increased sampling, the ideal trap design will need to balance these sorts of trade-offs in terms of the area surveyed and the total number of traps against the frequency of sampling, as well as in the number of areas that can be surveyed (site replication). Regular or random grids of traps might work if there is uncertainty about the shape of the dispersal kernel or pattern of dispersal. Trap grids are also commonly employed in areas where the distribution of sources is uniform (Bullock et al. 2006). Where sources are isolated (a point seed source), however, concentric arrays are more commonly used, in which traps are arranged either in rings emanating out from the source or within particular sectors, which attempt to sample the same proportion of area as distance increases from the source (Bullock & Clarke 2000).

Finally, plant dispersal can be studied indirectly, using molecular approaches (Bullock et al. 2006). Genetic markers, for example, can provide information on long-distance dispersal, which might be difficult or impossible to obtain through more direct methods. Genetic analysis can provide evidence of dispersal across a range of temporal scales, from the movement of individuals within a generation to movement among populations over several generations. Parentage analysis, using either allozyme or microsatellite data, can aid in identifying the putative source of seeds or seedlings, and thus can be used to generate dispersal kernels (Nathan et al. 2003). Further, since genetic markers vary in how they are inherited, we can use different type of markers (nuclear vs organellar) to parse the relative contributions of both pollen (using nuclear microsatellite markers) and seeds (using maternally inherited chloroplast microsatellite markers) to patterns of plant dispersal (Nathan et al. 2003; Austerlitz et al. 2004; **Table 9.1**). Although we will discuss these sorts of genetic approaches in more depth in the chapter on landscape genetics (**Chapter 9**), it is still worth noting here that such approaches are available to assist with quantifying dispersal in plants.

Future Directions

Because the successful movement and dispersal of organisms is essential for the maintenance of so many important ecological processes, an understanding of individual movement responses to landscape structure is in many ways foundational to the study of landscape ecology, the study of the effect of spatial pattern on ecological flows. An understanding of landscape effects on movement and dispersal will

certainly be key to our ability to manage species in the face of global changes wrought by an increasingly warmer climate; the wholesale destruction, fragmentation, and degradation of native habitats; and the rampant spread of introduced species and diseases. Although dispersal influences the extent to which species will be able to cope with these different dimensions of environmental change, we still have a long way to go toward understanding how most species perceive and respond to environmental heterogeneity (Lima & Zollner 1996). Recent advances in tracking technology and molecular genetics have greatly facilitated this task, however.

We know remarkably little about the sorts of information available to animals at the scale of ecological landscapes, and we know even less about how such information is used in decisions regarding movement and patch/habitat selection.

Lima & Zollner (1996)

As we continue to overcome many of the logistical problems that have previously hindered our ability to study animal movements or plant dispersal, we must now shift our attention to address the gaps in our understanding of how landscape structure affects individual movement and dispersal. As many factors—both behavioral and environmental—influence movement and are likely to act in concert, future research should seek to understand how different factors are integrated within as well as among different stages of movement (i.e. emigration, inter-patch movement, and immigration). For example, most research tends to focus on a single stage of the movement process (e.g. emigration) rather than consider the simultaneous effects of landscape structure on the entire process (Bowler & Benton 2005). Nor has there been sufficient study of the domains of scale—in both space and time—that bound different types of movement responses to landscape structure (e.g. **Figure 6.2**). As we've discussed throughout this chapter, the costs and benefits of movement are predicted to vary as a function of scale, and thus future work should seek to understand how these costs affect an individual's decision to leave a patch, how fast and how far they (or their propagules) travel, the routes taken while travelling through the matrix, and their decision to select or settle within a particular patch, all of which may be modified by landscape structure (Bowler & Benton 2005). An understanding of how landscape structure influences movement behavior is thus important not only for understanding the proximate causes of dispersal, but also for being able to translate movement responses to understand or predict patterns of space use and population distributions across the landscape, especially in the context of rapidly changing environments. Uncovering the linkages between movement behavior, dispersal, and population processes thus represents the future of behavioral landscape ecology (Hawkes 2009).

Chapter Summary Points

1. The movement behavior of individuals, or their propagules, represents the finest scale of organismal response to landscape pattern.

2. Movement encompasses a wide range of behaviors including resource acquisition (foraging behavior); space use; habitat selection; dispersal; gene flow; the spread of invasive species, pests, and pathogens; and species' range shifts in response to environmental change.

3. Movement responses to patch structure may be hierarchically nested, reflecting a range of behaviors (foraging, space use, habitat selection, dispersal) in response to the distribution of habitats and resources across a range of spatial and temporal scales (e.g. foraging hierarchies). Decisions made at a broader scale, such as in the selection of habitat, constrain what resources are then available at a finer scale.

4. Different species are likely to have different perceptions of landscape structure. The finest scale at which an organism responds to environmental heterogeneity is its perceptual grain; the broadest scale bounding its movement is its perceptual extent. On the basis of allometric scaling relationships, larger animals are expected to operate across a broader range of scales, or have their perceptual resolution shifted to broader scales, because they tend to move faster and farther and have larger home ranges than smaller animals.

5. Movement across the landscape involves at least three stages: movement out of patches (emigration), movement through the matrix, and movement into patches (immigration). Different landscape and behavioral factors affect movement at each stage.

6. The decision to leave a patch (emigration) may be influenced by the rate of resource intake (a function of resource abundance, patch quality, and/or distance travelled); perception of predation risk (giving-up densities); density-dependent dispersal; and the nature of the social environment (sex ratio, kin structure).

7. Edge permeability—the extent to which individuals cross a patch boundary—is a function of the edge contrast (structural differences between the habitat edge and surrounding matrix), edge shape, and the movement behavior of the species in question. Edges that are easily

traversed are permeable to movement and are called soft edges. Hard edges are patch boundaries that are rarely if ever crossed, because they impose a physical, physiological, or psychological barrier to movement.

8. Gap crossing refers to the ability or willingness of individuals to cross the intervening matrix (gaps) between habitat patches. Gap-crossing ability depends on the perceptual range of the species (the farthest distance at which the species is able to perceive habitat); its movement abilities; the size of the gap; and the perceived risk of crossing the matrix, which will vary depending upon environmental conditions (e.g. whether the moon is full or not) or the motivational state of the individual (e.g. whether its energy reserves are running low).

9. Matrix resistance—the degree to which the matrix impedes movement between patches—generally reduces patch or population connectivity, leading to the effective isolation of patches, even if patches are otherwise within the gap-crossing range of the species.

10. For corridors to be effective in restoring patch or population connectivity, individuals must be willing to use them, and realize lower mortality or travel costs while doing so, relative to moving through the matrix. For species that are able to move rapidly through the matrix, connectivity might be better achieved through the use of habitat stepping stones, rather than corridors of continuous habitat.

11. Movement into a patch (immigration) may reflect density-dependent habitat selection (an ideal free or ideal despotic distribution), conspecific attraction, or attributes of the patch itself (e.g. size, shape, orientation, habitat quality).

12. A variety of tracking methods are available to study animal movement. Although recent technologies such as satellite tracking permit the near-continuous recording of an individual's location, more data are not necessarily better if the landscape data do not match the resolution of the movement data.

13. Animal movement has been variously modeled as a random walk, correlated random walk, or Lévy flight (i.e. fractal scaling). The movements of a wide range of animals, from insects to marine mammals, appear to exhibit fractal scaling. The pervasiveness of fractal movement may indicate that this sort of search behavior is particularly efficient for exploiting patchily distributed resources within an unpredictable environment.

14. More complex movement responses to landscape structure can be explored through the use of spatially explicit simulation models, in which movement behavior is simulated on either an image of an actual landscape or a simulated landscape (i.e. a neutral landscape model).

15. Dispersal success—the probability of finding a habitat patch—is only influenced by the amount of habitat on the landscape if dispersal is truly a random process. For animals that have more localized dispersal, both the amount and distribution of habitat are often important in predicting dispersal success. Dispersal success in heterogeneous landscapes is influenced by the composition of the matrix and its differential resistance to movement. Differential movement through matrix habitat can result in asymmetrical dispersal, in which flows between patches are greater in one direction than the other.

16. The movements of most animals are bounded in space. The sum total of movements over time circumscribes the individual's home range. As not all areas within the home range are used equally, we can characterize the home range as a probability density surface that reflects the relative frequency or intensity of space use within different areas (the utilization distribution). Home ranges may vary in size among individuals as a function of sex or life stage, season (winter vs summer range), the habitat or landscape context, as well as among different species. Home-range estimation procedures thus use statistical or other modeling techniques to quantify home-range size so as to permit comparisons among individuals or species.

17. Plant dispersal may occur via passive transport (wind or water currents) or be actively vectored through the movements of animals. As many plants are dispersed by more than one means, plant dispersal can be a complex process involving different stages of dispersal that occur at different scales. As with animals, dispersal is important for understanding the spatial distribution and dynamics of plant species across the landscape.

18. Plant dispersal may be studied through the direct measurement of seed dispersal (e.g. tracking individual seeds or their animal vectors) or through the use of seed traps and mark-recapture techniques. Increasingly, indirect measures based on molecular approaches are being used to infer patterns of dispersal among populations or regions. Different types of molecular markers can be used to distinguish the movement of pollen

(using nuclear microsatellite markers) from that due to seeds (using maternally inherited

chloroplast microsatellites), both of which contribute to patterns of plant dispersal.

Discussion Questions

1. Construct a foraging hierarchy, either for a species you work with or an example provided by your instructor. Be sure to identify the spatial and temporal scales of patchiness that correspond to each level of the hierarchy, and discuss what movement behaviors are expressed at each level.

2. For a given landscape, either one you work in or an example provided by your instructor, compare the perceptual ranges of two different species (e.g. a granivorous ant versus a granivorous rodent or bird) and consider how differences in the scales at which these species operate would likely affect the scales of patchiness with which they are able to interact. What might these differences in landscape perception mean in terms of competition for a common resource (e.g. seeds)?

3. Although habitat preferences or patch quality might be inferred based on how much time an individual spends within a given habitat or patch type (e.g. relative movement rates, residence times, or giving-up times), is this really the best measure of preference? For example, we assume that individuals will spend more time in areas where resources are abundant than where they are not. Is that a fair assumption? How else might we interpret slower rates of movement through a particular habitat or patch type? Why else might individuals spend more time in less-preferred areas of the landscape?

4. Foraging behavior—and movement behaviors in general—are typically viewed in terms of cost-benefit trade-offs. Consider the foraging hierarchy you developed for Discussion Question 1 (or use the one illustrated in **Figure 6.3**). What are the different costs and benefits associated with the movement behaviors at each level of the hierarchy? Are these trade-offs the same at each level, or do they change as a function of scale? What are the fitness consequences at each level? At what scale, then, do these trade-offs ultimately have the greatest effect on survival and reproduction for this species?

5. Landscapes of fear depict spatial variability in predation risk that influence the movement behavior or space use of the species (**Box 6.1**). What other types of 'landscapes' might influence animal movement or space use? In other words, what other types of constraints on foraging or

movement behavior could be mapped to give a fuller understanding of animal responses to landscape structure?

6. Does the idea of plants foraging seem too great a stretch, or does the fact that they can alter their growth form in response to environmental heterogeneity provide an appropriate analog to animal foraging behavior? Looking beyond the obvious differences in the type of response (physical versus behavioral), how might plant foraging responses be similar to that of animals in landscapes with a heterogeneous resource distribution? Does the fact that some plants are carnivorous, like *Philcoxia* of the Brazilian Cerrado that traps nematodes in its sticky underground leaves (Pereira et al. 2012), alter your perspective on plant foraging behavior?

7. What types of organisms are most likely to benefit from habitat corridors? Under what conditions might stepping stones be a better solution for facilitating movement or dispersal across the landscape? Would managing the intervening matrix to minimize landscape resistance be a more—or less—effective strategy for promoting connectivity between habitat patches or populations?

8. How would you design a movement study to characterize a foraging hierarchy for a species, such as the one you highlighted in Discussion Question 1? What sort of methods or tracking technologies would be appropriate for the study of movement responses to landscape structure in this species? Consider what sampling frequency and study duration would be required, as well as the number of individuals that might have to be tracked in order to provide sufficient information on movement behavior across a range of scales. What other scaling issues do we need to consider?

9. Based on an analysis of the movement data obtained from your study outlined in response to Discussion Question 8, what attributes of the movement pathway would suggest whether individuals were responding to patch structure at a particular scale? In other words, how would you expect different movement parameters (**Table 6.1**) to vary depending upon whether individuals are perceiving the landscape as homogeneous or heterogeneous?

10. True or False: The best movement model is the simplest one that makes the fewest assumptions. Be prepared to explain and defend your position.

11. Using a species you work with or an example provided by your instructor, outline the various landscape factors that are likely to influence the size and utilization intensity of an animal's home range. How might this change over time, such as throughout the year or lifetime of the individual?

12. Design a study of seed dispersal for a plant species of your choosing, or one suggested by your instructor. Consider the various modes of dispersal and therefore the scales over which dispersal might occur (don't forget about secondary dispersal!) in your study design. Is it sufficient to just study seed dispersal, or might it also be necessary to include pollen dispersal in order to understand the resulting pattern of distribution for this species?

Landscape Effects on Population Distributions and Dynamics

7

If a piece of the sky should fall, it would resemble nothing so much as the cerulean warbler (*Setophaga cerulea*; **Figure 7.1**). A neotropical migrant, it is known in South America as Reinita Cielo Azul, which means 'little queen of the blue sky.' Lovely to behold, it is unfortunately the fastest declining warbler in North America, having lost some 83% of its total population over the past 40 years (Butcher & Niven 2007). The reasons for its decline are most likely a combination of habitat loss and fragmentation on its breeding and wintering grounds. On its wintering grounds, which are located in the forested mountains of Venezuela and Colombia and extend down through the Andes of Peru, its survival is threatened by deforestation and the economic pressures that are driving the change from traditional shade coffee to higher-yield sun coffee. On its breeding grounds in the eastern USA, it is likewise threatened by deforestation of its preferred hardwood forests, such as by mountain-top removal mining. In addition, fragmented forests may host a suite of nest predators as well as the brown-headed cowbird (*Moluthrus ater*), an obligate brood parasite, which collectively reduce the reproductive success of cerulean warblers that do manage to find suitable habitat in which to nest.

Given that cerulean warblers are declining at a rate of 4.5%/year, we might wonder whether there are still population strongholds somewhere within their range that could be targeted for protection. But how do we identify these population strongholds? Can we stabilize population declines elsewhere? If so, would it be better to increase the amount of protected forest, or to improve the quality of the bird's remaining forest habitat? Would populations recover more quickly if we were to focus on measures aimed at improving their nesting success on their breeding grounds, or on increasing their overwinter survival on the wintering grounds?

To answer these sorts of questions, as well as to monitor and assess the rate of population decline in this—or any other—species, we are going to need a population model. If habitat loss and fragmentation are the primary factors responsible for the cerulean warbler's decline, then we will need to understand the interaction between landscape structure and demography that give rise to population declines. In short, we are going to need a spatially explicit population model. We therefore consider landscape effects on demography in this chapter, as part of our broader discussion concerning the landscape ecology of population distributions and dynamics.

Essentials of Landscape Ecology. Kimberly A. With, Oxford University Press (2019).
© Kimberly A. With 2019. DOI: 10.1093/oso/9780198838388.001.0001

Figure 7.1 Cerulean warbler. The cerulean warbler (*Setophaga cerulea*) is a neotropical migratory songbird that breeds in the temperate hardwood forests of North America, and overwinters in the montane forests of South America. A species of conservation concern, it is listed as Vulnerable on the International Union for Conservation of Nature (IUCN) Red List.

Source: Photo by Mdf and published under the CCA-SA 3.0 License.

Why Should Landscape Ecologists Study Population Distributions and Dynamics?

The spatial distribution and dynamics of a population result from the interplay of dispersal and demography with landscape structure. Although the fact that a population is declining could be assessed without specific reference to the landscape, it is impossible to determine what aspects of landscape structure—in terms of the amount, quality, or configuration of habitat—are contributing most to population decline, and thus what land-management practices would be most likely to effect a recovery, without putting this into a spatial or landscape context. For example, should we focus our efforts on restoring connectivity among existing populations, or should we instead try to improve the quality of remaining habitat on the landscape? Our answer will depend on whether landscape structure is having a greater effect on dispersal or demographic processes, and how the population is distributed among habitats on the landscape.

An understanding of population distributions and dynamics is thus fundamental to species management and conservation. Wildlife and fisheries management require information as to how the amount, distribution, and quality of habitats affect population numbers, in order to set harvest levels or to manage landscapes (or seascapes or riverscapes) to improve habitat quality for target species. Human land use leading to habitat

loss and fragmentation is generally regarded as the greatest threat to biological diversity (e.g. Wilcove et al. 1998). Assessment of a species' extinction risk thus requires the use of spatially explicit population models to determine the population's current status as well as its future viability under various land-use change scenarios. Global climate change, in conjunction with changes in human land use, will force some species to shift their ranges in order to track environmental changes that affect the distribution of their habitats. Whether species will be able to shift their distributions quickly enough, however, will likely depend on their dispersal ability and reproductive capacity relative to the types of geographic and landscape barriers they encounter on the modern landscape, such as dams, roadways, and major metropolitan areas, and the loss and fragmentation of habitats.

In this chapter, we focus on the effects of landscape structure on the distribution and dynamics of populations. We begin with an overview of the general effects of habitat loss and fragmentation, taking care to distinguish between the two processes, as they are not equivalent in their expected effects on populations. Next, we discuss the spatial analysis of species distributions and the different approaches for modeling species distributions, especially in forecasting species' range shifts in response to projected landscape- or climate-change scenarios. Then, because population assessment figures so prominently in evaluating a species' extinction risk to landscape change, we consider the different classes of population models that are used to estimate population growth rates and population viability along a gradient of increasing spatial complexity. For the uninitiated, we include a brief primer on traditional population models to introduce the basic concepts, before launching into a discussion of how spatial complexity can be incorporated within these general modeling frameworks, which leads to new insights and powerful tools for evaluating the effect of landscape structure on patch occupancy and population dynamics.

Overview: Effects of Habitat Loss and Fragmentation on Populations

Given the rate at which natural habitats have been—and continue to be—destroyed or degraded worldwide, it is inevitable that this will have a negative impact on the populations of many—if not most—species. Habitat loss not only reduces the absolute amount of habitat on the landscape, but may also reduce the effective area of remaining habitat through fragmentation effects. We thus need to distinguish between effects on populations that are primarily due to **habitat loss** from those that are due to **habitat fragmentation**. Because habitat fragmentation generally occurs through a process of habitat loss, the terms are often used interchangeably. They are not synonymous, however, and may have very different implications for the management or conservation of populations.

Habitat Loss versus Fragmentation

Habitat loss and fragmentation are typically viewed as having four main effects on landscape structure: (1) a reduction in the total amount of habitat on the landscape; (2) an increase in the number of remaining habitat patches; (3) a decrease in the size of remaining habitat patches; and, (4) an increase in patch isolation (Fahrig 2003). Each of these effects on landscape structure can be quantified and each has been used—either singly or in combination—to characterize habitat fragmentation (**Chapter 4**). However, as pointed out by Fahrig (2003), many of these effects are principally a result of habitat loss and not fragmentation per se.

Habitat loss results in the reduction of total habitat area on the landscape, a decrease in patch size, and an increase in patch isolation (**Figure 7.2**). Although habitat fragmentation typically occurs through a process of habitat loss, fragmentation more properly refers to the *pattern* of habitat loss and results in an increase in the number of patches on the landscape. Fragmentation may or may not result in an increased isolation of patches, and may actually *decrease* patch isolation, depending on the pattern of habitat loss. For example, a fine-grained pattern of habitat loss, in which many small gaps are created, may leave behind small patches of habitat that can serve as 'stepping stones' to dispersing organisms across the matrix. Further, patch isolation may actually reflect the amount of habitat area in the surrounding landscape (inter-patch distances tend to be greater in landscapes with low levels of habitat), and thus patch-isolation measures may be confounded with the total amount of habitat on the landscape (Bender et al. 2003a). Thus, one of the primary measures usually associated with habitat fragmentation—patch isolation—is actually more a product of habitat loss than fragmentation. Of all the effects normally associated with habitat fragmentation, only one—an increase in the number of remaining habitat patches—can be attributed unambiguously to habitat fragmentation.

Why does this matter? Beyond a desire to make correct inferences about what aspect of landscape change—habitat loss or fragmentation—is negatively affecting the target species, consider the consequences from a management or conservation standpoint. If we believe that habitat fragmentation is generally to blame for population declines and increased extinction risk, then we might reasonably seek to mitigate such effects by attempting to restore connectivity among isolated patches through the creation of habitat corridors. If,

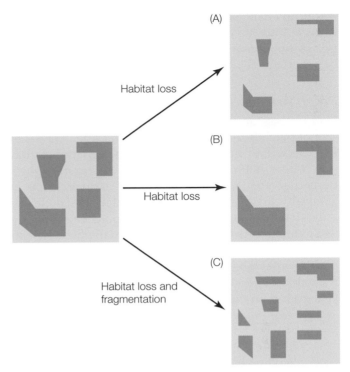

Figure 7.2 Differential effects of habitat loss versus habitat fragmentation on landscape structure. Habitat loss by itself can reduce the size of patches (A) or increase the distance between remaining habitat patches (B). In contrast, fragmentation as a consequence of habitat loss results in an increase in the number of patches as well as a decrease in patch size. Despite the use of patch isolation as a measure of fragmentation in many studies, habitat fragmentation may reduce the degree of patch isolation (C). As shown here, the average inter-patch distance is less in the landscape undergoing habitat loss and fragmentation (C) than in those undergoing pure habitat loss (A, B).

Source: Fahrig (1997).

however, it is the sheer loss of habitat that is to blame, then our efforts are misguided as we cannot hope to stabilize the population simply by restoring the connectivity of existing habitat (Fahrig 1997). Instead, we need to restore a sufficient amount of habitat—and of sufficient quality—to ensure long-term **population viability**. We will address 'how much habitat is enough?' later in this chapter in the context of extinction thresholds.

> The fragmentation literature provides strong evidence that habitat loss has large, consistently negative effects on biodiversity. This implies that the most important question for biodiversity conservation is probably 'How much habitat is enough?' Fahrig (2003)

Regardless of whether it is habitat loss, habitat fragmentation, or some combination of the two that is ultimately responsible for population declines, the resulting changes in patch size and configuration are expected to have important consequences for local population dynamics. A brief overview of those patch-based effects is given next.

Patch Area and Isolation Effects

Habitat loss and fragmentation typically reduce the size of patches remaining on the landscape. Patch size is expected to have a positive effect on population size: larger habitat areas can support larger populations than smaller areas. Given that extinction risk is generally inversely related to population size, it follows that populations within large patches should be less extinction-prone than those in small patches. Such positive patch-area effects are likely to be strongest for **interior species** (those sensitive to habitat edge) and for habitat specialists, especially for species that live year-round on the landscape (resident species; Bender et al. 1998). Species with large area requirements, in terms of their territory or home-range sizes, are also expected to exhibit a positive association with patch size. Conversely, **edge species** are expected to be negatively affected by patch size (Bender et al. 1998). Recall that the proportion of edge (the patch perimeter) decreases as patch area increases (P/A ratio; **Chapter 4**), and thus proportionately less edge habitat is available to edge species in large patches. **Patch occupancy**—the probability that the species is present within a patch—may also be influenced by patch size. Larger patches are more likely to be occupied than smaller patches for any of the following reasons: they support larger populations that are less prone to extinction; they present a larger 'target' to dispersers; or they are perceived to be more attractive to colonizers (e.g. large patches might be perceived as having more resources or higher habitat quality).

Habitat loss and fragmentation[1] can also increase the isolation of remaining patches on the landscape. Patch isolation is a function not only of inter-patch distances but also of the dispersal or gap-crossing ability of the species and the nature of the intervening matrix (**Chapters 5 and 6**). Intuitively, isolated patches are less likely to be colonized than those that are in close proximity to a large population source. Recall, however, from our discussion of landscape resistance surfaces in **Chapter 5**, that the intervening matrix can substantially increase the **effective isolation** of patches, if it provides a hostile environment for dispersers (Ricketts 2001). Thus, the straight-line (Euclidean) distance between patches is not always a good measure of patch isolation (**Chapters 4 and 5**). Unfortunately, isolated populations

are less likely to be 'rescued' from extinction. We will discuss the effects of landscape structure on source–sink population dynamics a bit later in the chapter.

In spite of this consideration of patch-based effects on populations, it must be emphasized that habitat loss and fragmentation are processes that occur at a landscape-wide scale. Although many studies attempt to infer fragmentation effects from the study of the size or isolation of individual patches (McGarigal & Cushman 2002), this is generally not advisable, especially given that patch size and isolation are often more a consequence of habitat loss than habitat fragmentation per se (Fahrig 2003). This means, however, that replication for fragmentation studies must take place at the scale of entire landscapes, not patches. Landscape replicates can be difficult to define, but even when a landscape series is identified across a gradient of habitat amount or fragmentation, we cannot assume that these all lie on the same trajectory of landscape change (**Figure 2.15**). Landscape history and the rate at which habitat is lost or fragmented may have different consequences for extinction risk, which will be discussed later in the chapter.

Fragmentation and Edge Effects

Of all the supposed effects of habitat fragmentation, perhaps the most clear-cut is an increase in the amount of edge habitat that accompanies the subdivision of existing habitat blocks into many small patches. Although habitat loss by itself can increase the amount of edge on a landscape, fragmentation amplifies this effect, resulting in greater edge for a given amount of habitat (**Figures 4.30 and 4.31**). While that is true at the landscape scale, recall that at the patch scale, small patches have proportionately more edge than large patches (based on the patch P/A ratio). Thus, patch area and edge may be confounded at a local scale unless care is taken to separate the two effects (Fletcher et al. 2007). For studies that have controlled both, however, edge effects are often stronger and more frequently observed than patch-area effects.

So, what are edge effects? **Edge effects** refer to the change in some ecological response as a function of distance to a habitat edge. The magnitude and direction of that response (whether positive or negative) typically depends on the nature of the edge or edge type (Yahner 1988; Saunders et al. 1991). Edges represent the interface between different habitats and their associated biota. Naturally occurring edges between different habitats or ecosystems are called **ecotones**, and typically reflect an underlying physical gradient or **ecocline**. Ecotones may occur as an abrupt transition between ecosystems, or as a gradual one in which there is considerable overlap or blending between the two systems (e.g. the transition between grasslands and

---

[1] Roadways, weirs, fences, and other physical barriers to dispersal can also fragment habitats and isolate populations.

forest, resulting in savanna). Recall that this creates a 'soft edge' between adjacent habitat types (**Figure 6.9**).

Disturbance-induced edges, such as those produced by land-clearing, often stand in sharp contrast to their surroundings, however (i.e. a 'hard edge'; **Figure 6.9**). Disturbance typically produces structural changes in the vegetation along the edge, resulting in a more open canopy that allows for greater light, and subsequently, a different microclimate that is warmer, drier, and windier than the interior (Saunders et al. 1991). This has implications for habitat-interior species, especially species that are susceptible to desiccation, such as amphibians or insects. Because habitat edges are at the juxtaposition of different habitats, they may provide a mix of resources that are attractive to many

species (**landscape complementation**; Dunning et al. 1992; **Figure 7.3A**), and subsequently, may enhance interactions among species. Unfortunately, many of these interactions are negative ones, in terms of their impact on interior species. Competition, predation, and parasitism may all be greater along habitat edges, thus limiting survival, reproductive success, or population recruitment of interior species (Yahner 1988; Saunders et al. 1991). Nor are these negative effects restricted to just the edge margins. The microclimatic and biotic responses to the habitat edge may extend 200 m or more into the patch (Laurance et al. 2011). Different types of edge effects that occur within forest fragments are discussed in **Box 7.1**.

To illustrate the possible negative effects of habitat edges on the abundance and distribution of populations, consider the Pacific trillium (*Trillium ovatum*), a small herbaceous plant with a white or pink tricorn flower, which grows in the moist, shady understory of coniferous forests in western North America (**Figure 7.4**). Forest clearcutting for timber produces a strong, persistent edge effect on trillium. Trillium populations within 65 m of a clearcut-forest edge have had almost no recruitment in the years since the adjacent land was clearcut, which in some cases was 30 years earlier (Jules 1998). In contrast, recruitment rates and population sizes of trillium are higher within the forest interior. Population declines along forest edges thus appear to be driven by negative edge effects, perhaps as a consequence of adverse changes in the microclimate (especially due to increased

(A) Landscape complementation

(B) Landscape supplementation

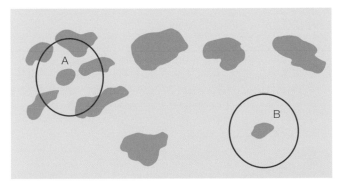

Figure 7.3 Landscape complementation versus landscape supplementation. (A) Landscape complementation occurs when two (or more) resources are required by the species (i.e. resources are not substitutable), and the different patch types are in close proximity. These landscapes will support larger populations (point A) than do landscapes in which the two patch types are far apart (point B). (B) In the case of landscape supplementation, the resource required by the organism is substitutable, and so individuals can supplement their resource intake by using nearby patches or making use of another resource in another patch type. Populations will be larger in landscapes where individuals can move to patches providing the same or similar resource (point A), as long as these fall within the dispersal range of the species (red circles).

Source: After Dunning et al. (1992).

Figure 7.4 Pacific trillium (*Trillium ovatum*).

Source: Photo by Kimberly A. With.

BOX 7.1 How Far Do Edge Effects Penetrate into the Forest Interior?

The **Biological Dynamics of Forest Fragments Project** (BDFFP) is a long-term study that was initiated by the renowned conservationist Thomas Lovejoy in 1979 to study the effects of fragmentation in the Amazonian rainforest. Located just north of Manaus in Brazil, it is the largest (1000 km²) and longest-running fragmentation study in the world. Over the past several decades, these experimental forest fragments (ranging in size from 1 to 100 ha) have experienced a wide array of ecological changes, and edge effects have been a dominant driver of these changes (Laurance et al. 2002, 2011). Many of the microclimatic effects penetrate 100 m into the forest fragment, and some effects penetrate up to 400 m, causing the edge to be warmer, drier, and windier than the interior (**Figure 1**). The result is a higher incidence of tree mortality up to 300 m beyond the edge boundary, and a host of other changes to forest structure in between. Some trees simply drop their leaves and die standing, unable to withstand the light intensity and dry conditions; others are ripped apart or felled by winds; those that remain may be slowly strangled by lianas. The increased leaf fall and drier soil conditions near edges slow decomposition and create a dense litter layer that smother germinating seeds and seedlings, further preventing recruitment of rainforest trees. In their place, opportunistic pioneer species such as

Cecropia and other early-successional species take advantage of the newly created treefall gaps, altering vegetation structure and composition along edges.

Although some animals respond favorably to these changes, many others do not. Changes in the ant, beetle, and butterfly communities can be detected up to 200 m from the forest edge, and insectivorous understory birds appear to be particularly sensitive to edge effects. Edge effects also appear to be cumulative, such that the presence of two or more nearby edges increases the severity of the effect, which may explain why small fragments (<10 ha) are so severely altered by forest fragmentation.

The intensity of these edge effects varies in space and time, however, and is influenced by factors such as the time since the edge was created, the number of nearby edges, and the type of matrix surrounding the fragment. For example, the matrix has changed markedly since the study was initiated, evolving from large cattle pastures to mosaics of abandoned pasture and regrowth forest. As secondary forest reclaims pastures, some of the earlier edge effects have been mitigated: several insectivorous birds, various monkeys, forest spiders, and euglossine bees have recolonized fragments. Forest edges are 'softened' by the regrowth of secondary forest, which may enhance the

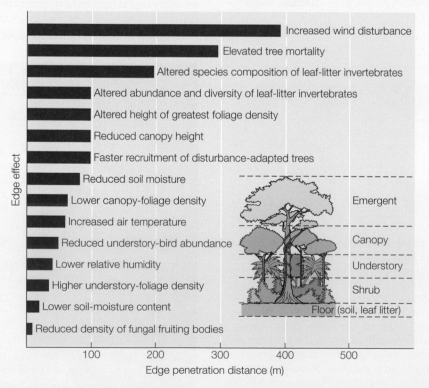

Figure 1 Edge effects within the Amazonian rainforest. Penetration distances for various types of edge effects into forest fragments of the Biological Dynamics of Forest Fragments Project.

Source: After Laurance et al. (2002).

movements of pollinators and seed dispersers, affecting the seed rain into forest fragments.

No two fragments are alike, however, even if they share similar land-use histories. Identically sized fragments exhibited very different trajectories of change, leading to ecological divergence among landscapes. The effects of deforestation are not usually immediate, and may take decades to play out, especially in the case of rare or long-lived species. It may take a century or more for dynamics to stabilize even within small fragments, owing to the long lifespan of many tropical trees. Even then, there are likely to be long-lasting effects on the composition and structure of these forest fragments. Such insights into the multifaceted nature and enduring legacy of edge effects on the Amazon rainforest would have been impossible without benefit of long-term study, and the BDFFP has unquestionably been unique in both its scope and duration.

wind) and increased seed predation by deer mice (*Peromyscus* spp.), which have been found to occur at higher densities within clearcuts and along forest edges. In turn, these negative edge effects translate into demographic changes within trillium populations, involving reduced seed set and survivorship of seeds and seedlings near edges. As a consequence, forest fragmentation has significantly altered both the distribution and abundance of trillium, which are now largely restricted to uncut forest in the study region.

Edge effects are thought to be the primary driver behind many ecological responses to habitat fragmentation. Fragmented landscapes, by definition, have many small patches that are likely to be dominated by edge effects, and so it stands to reason that edge effects should be greater overall in fragmented landscapes. For example, predation is the main cause of nest failure in most bird species, accounting for some 80% of nest losses, on average (Martin 1993). Nest-predation rates may thus have significant demographic consequences for breeding bird populations, even driving population declines in some species. Nest-predation rates were found to be greater for forest birds that nested near forest-field edges, as opposed to the interior of forest fragments (Gates & Gysel 1978). In addition, nest-predation rates tend to be greater in fragmented forest landscapes than in landscapes with more continuous forest cover (Robinson et al. 1995; Donovan et al. 1997; **Box 7.2**). Nest-predation rates are thus typically viewed as a type of negative edge effect that is expected to be most prevalent in fragmented landscapes.

Are the negative effects of forest fragmentation on bird populations the result of negative edge effects that are merely being amplified at the broader landscape scale, as suggested above, or are negative edge effects in fact being driven by factors operating at the broader landscape scale? Consider this: although some studies have indeed found evidence for lower nesting success near habitat edges, a majority of studies have not (Lahti 2001). Habitat edges are supposed to enhance nest-predation rates by supporting or attracting higher numbers of nest predators, and yet a meta-analysis found little evidence of this for most edge types, excepting those created by agriculture (Chalfoun et al. 2002a). Indeed, the relationship between nest predators and fragmentation was far greater when assessed at the landscape scale, especially in landscapes being fragmented by agriculture (Chalfoun et al. 2002a). The amount of agriculture in the landscape may be associated with an increased abundance or diversity of generalist predators, which can increase predation rates along habitat edges (Andrén 1995). Although this may give the appearance of a negative edge effect in agriculturally dominated landscapes, predators may be responding (numerically or functionally) to the increase in agricultural land cover rather than to an increase in habitat edges per se (Ryall & Fahrig 2006). So, not only is landscape context important for evaluating the strength of edge or fragmentation effects, but edge effects may actually be confounded with those of landscape context. This is not to suggest that negative edge effects do not occur or are not important, only that care must be taken in parsing out at which factors are driving the response, and at what scales.

Furthermore, it can be difficult to scale up from local responses to the entire landscape to predict species' responses to fragmentation. Even if edges have a negative effect on local nest success, the species may be able to compensate in other ways. For example, Flaspohler and his colleagues (2001) found that although nest predation increased near edges (<300 m) for a forest songbird, the ovenbird (*Seiurus aurocapillus*), birds nesting near edges were able to offset higher predation rates by increasing the number of eggs they laid. Productivity of edge-nesting ovenbirds was thus similar to ovenbirds nesting in the forest interior. For other species, however, the cumulative effects of edge-related declines in productivity may well translate into population declines at the landscape scale in fragmented landscapes (Donovan et al. 1995; With & King 2001).

Of course, not all species have negative responses to habitat edges; many more species exhibit either no response or actually have a **positive edge response**, in which abundances are highest along edges (Ries et al. 2004). Positive edge effects may occur either because the adjacent habitats offer complementary (different, but non-substitutable) resources the species requires and the edge offers convenient access to both, or

BOX 7.2   Edge Effects on the Nesting Success of Songbirds

Predation is responsible for most nest failures in songbirds, and nest predation rates tend to be higher near habitat edges (Martin 1993). The reasons for this are likely multifaceted (Chalfoun et al. 2002a). Disturbance may reduce the structural complexity and vegetative cover along the edge, making nests more visible and thus easier for predators to find. Nesting songbirds might be attracted to habitat edges if they provide greater food availability (such as near human habitation), in which case nest predation could simply be a density-dependent response to the number of nests, making edges an **ecological trap** for nesting songbirds (Gates & Gysel 1978). Or, the edge may be attractive to the nest predators themselves (i.e. predators exhibit a positive edge response; **Figure 7.5**), causing either an increase in foraging activity along edges (a functional response) or an increase in the number or types of predators present (a numerical response), leading to greater predation pressure on nesting songbirds near habitat edges (Chalfoun et al. 2002b). The type of edge is also a factor in the magnitude of the edge effect on nest success. For example, nesting success for grassland birds in tallgrass prairie fragments declined within 50 m of shrubby edge habitats (**Figure 1**), but was unaffected by proximity to agricultural edges or roads (Winter et al. 2000). Conversely, nest predation rates in forest birds are often higher near agricultural edges, which may be related to a higher abundance, diversity, or activity of predators along these edges (Chalfoun et al. 2002b).

In fragmented forests and grasslands of North America, brood parasitism by brown-headed cowbirds (*Moluthrus ater*) is an additional concern for nesting songbirds (Askins 1995; **Figure 2**). Cowbirds do not build their own nests but instead lay their eggs in the nests of other species, leaving the host bird to rear the cowbird's offspring, often to the detriment of the host's own reproductive fitness (**Figure 2**). For example, cowbirds sometimes remove host eggs before laying their own eggs in the nest (a type of predation), and cowbird young usually hatch sooner and grow more rapidly than the host bird's young, outcompeting them for food or even ejecting them outright from the nest. Cowbirds originally occurred in the open grasslands of central North America, but have since spread into forested landscapes to the east and west as these were cleared and became increasingly fragmented. Like predation, rates of brood parasitism may be higher near edges, especially if cowbirds occur at higher density there (Chalfoun et al. 2002b).

Edge effects on nest predation and parasitism rates are not universal (Lahti 2001; Chalfoun et al. 2002a). As with most ecological responses to landscape pattern, it depends on the scale, landscape context, and predator species in question. Different predators respond to fragmentation in different ways and at different scales, and thus it is difficult

to make generalizations (Chalfoun et al. 2002b). For example, if a major nest predator is itself negatively impacted by fragmentation and avoids edges, then nest predation rates may well be lower along edges. The predator community may be influenced by the nature of the landscape matrix as well, especially if the type of habitat adjacent to an edge has an additive effect on predation, in which both habitats contribute predators. In sum, the local-scale effects of habitat edges on nest success are extremely variable and depend on the landscape context and region in which the edges are embedded (Donovan et al. 1997; Chalfoun et al. 2002a). In other words, a landscape perspective is ultimately needed for evaluating edge effects.

Figure 1 Nest predation rates tend to be higher near habitat edges. Nests are vulnerable to a suite of nest predators, including snakes. Top: A black rat snake (*Elaphe obsoleta*) prepares to depredate the nest of a Bell's Vireo (*Vireo bellii*). Bottom: A racer (*Coluber constrictor*) caught in the act of depredating a Dickcissel (*Spiza americana*) chick, through the use of a 'camera trap.'

Source: Top photo courtesy of Karl Kosciuch and bottom photo courtesy of Page E. Klug.

Figure 2 Nests near habitat edges may suffer higher rates of brood parasitism.   The dickcissel (top right), a sparrowlike bird that breeds in North American tallgrass prairie, is a frequent target of the brood-parasitic brown-headed cowbird (*Moluthrus ater*, top left). Here, a parasitized nest contains two blue dickcissel eggs and five speckled brown-headed cowbird eggs (bottom).

Source: Photo of brown-headed cowbird by Rodney Campbell, and published under the CC By 2.0 License. Dickcissel photo by Rita Wiskowski. Photo of dickcissel nest courtesy of William E. Jensen.

because resources are concentrated along habitat edges (Dunning et al. 1992; Ries & Sisk 2004; **Figure 7.5b, d, e**). If resources are similar and substitutable (supplementary) between adjacent habitats (**Figure 7.3b**), then a neutral edge response is expected (**Figure 7.5c**). Negative edge effects are expected only when habitat quality differs greatly between adjacent patches (e.g. a hard edge) and the resources are otherwise supplemental, but are concentrated in one of those patches (i.e. the high-quality habitat; **Figure 7.5a**). Note that a negative edge effect for one species (e.g. lower nesting success in forest songbirds) may come about because of a positive edge response by another species (a generalist predator from the matrix that exhibits a numerical or functional response[2] to habitat edges).

We have barely scratched the surface as to the possible effects of habitat loss and fragmentation on popu-

lations, which is easily a book in itself (e.g. Lidenmayer & Fischer 2006; Collinge 2009). We will therefore continue this discussion throughout the remainder of this chapter, and indeed, throughout the book. Next, however, we turn our attention to landscape effects on the abundance and distribution of species.

Species Distribution Patterns

Information on the abundance and distribution of a species is a requisite for assessing what environmental factors are most important—and at what scale—to the spatial dynamics of populations. Such information might also enable us to forecast how a species' distribution could change under scenarios of future climate and landscape change. Species distributions can be assessed across a range of scales, from that of a single study area or landscape to a broader regional or continent-wide scale encompassing the entire range of the species.

[2] A numerical response involves an increase in predator density with increasing prey density, whereas a functional response involves an increase in prey consumed by predators as a function of increasing prey density.

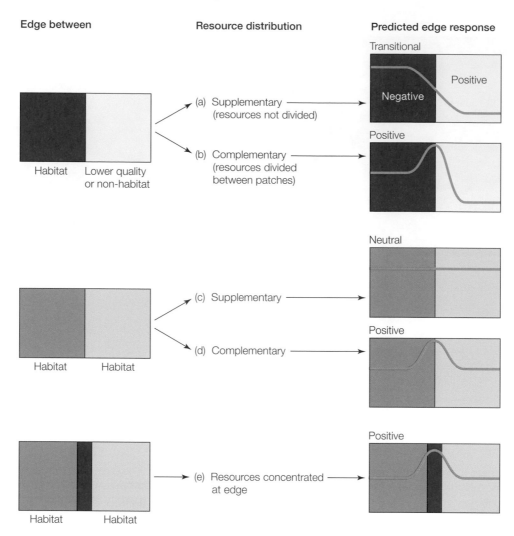

Figure 7.5 Population responses to habitat edges. A negative response to habitat edge is expected when habitat quality varies greatly between patches and resources (which are otherwise supplementary) are concentrated in one of those patches (the higher quality patch; a). Positive edge effects occur when resources between patches are complementary (different and non-substitutable; b and d), or when resources are concentrated at the edge (e). No response to the habitat edge (neutral) is predicted when there is little difference in habitat quality and resource concentrations are similar (supplementary) between patches (c).
Source: Ries & Sisk (2004).

Spatial Pattern Analysis of Species Distributions

As a starting point, we might be interested in assessing the spatial pattern of the species' distribution, in terms of whether individuals exhibit a random, uniform, or clumped (patchy) distribution (**Chapter 4**). From this, we might be able to infer what mechanisms (e.g. competition, allelopathy, social behavior) or other environmental factors are influencing the distribution of individuals. Patterns of distribution may well change as a function of scale, however, with individuals exhibiting a clumped or patchy distribution at some scales but a more uniform or random distribution at others. It is also possible that different life stages within a species could exhibit different distributional patterns. To further complicate matters, species' distributions may

represent the superposition of processes acting at different scales (e.g. cross-scale interactions), giving rise to hierarchically nested patterns of distribution (**Chapter 2**).

For example, the critically endangered tropical tiniya tree, *Shorea congestiflora*, which is endemic to the rainforests of Sri Lanka, is one such species that exhibits clustering at multiple scales and for different size classes. Using a point pattern analysis that incorporated a double-clustering process, Wiegand and his colleagues (2007) demonstrated how this species' distribution exhibited a nested pattern of clustering at two scales, encompassing all live trees at the broadest scale (26 m, **Figure 7.6A**), and within those clusters, a finer scale of clustering (8 m) for seedlings and saplings (**Figure 7.6C**). Moreover, the distribution of *S. congesti-*

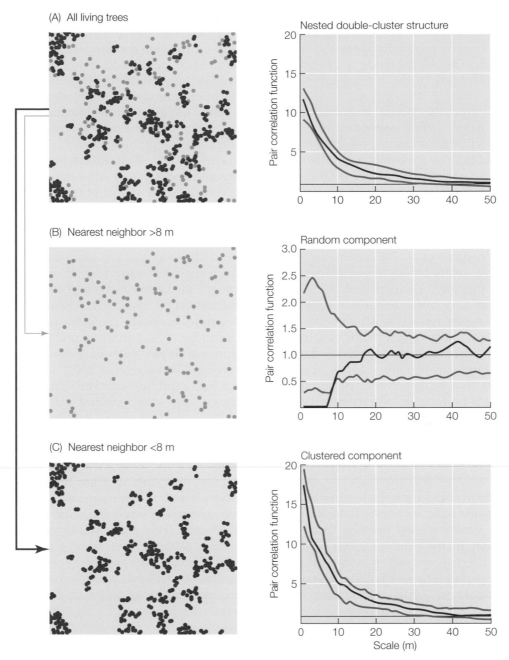

Figure 7.6 Distribution of the tropical tiniya tree   (*Shorea congestiflora*). (A) The distribution of living trees across the landscape (500 m × 500 m) is consistent with a hypothesized double-cluster process (i.e. the red line in the graph was fit to the data and falls within the 95% confidence envelope depicted in blue) in which neighboring trees (<8 m) are assumed to be nested within larger clusters (26 m). Note that the pair correlation function is a variation of the Ripley's *L* function (**Figure 4.40**) and is interpreted similarly. A substantial number of trees (12%) had no nearest neighbor within 8 m, and thus the distribution of *S. congestiflora* may represent the superposition of two independent patterns: a random component that is made up of isolated trees (>8 m; B), and a double-clustered component created by neighboring trees (<8 m) that form larger clusters (up to 26 m; C). Clustering at the broadest scale is thought to reflect dispersal limitation in the species, whereas finer-scale clustering (8 m) likely represents self-thinning among seedlings and saplings, given that this scale corresponds to the gap size of dead canopy trees.

Source: After Wiegand et al. (2007).

flora appeared to result from the superposition of two independent processes that simultaneously gave rise to both randomness and a clustered distribution of trees (**Figure 7.6B, C**). This may indicate that the distribution of *S. congestiflora* is produced by two different modes of dispersal operating at different scales: primary dispersal of seeds in the vicinity of adult trees, which results in clustering (a case of dispersal limitation); and, secondary dispersal through surface runoff that occasionally carries seeds farther downslope and adds a

random component to the distribution. The fact that
adult trees did not show fine-scale clustering suggests
that self-thinning occurs, especially since the size of
fine-scale clusters (8 m), which are made up of seedlings
and saplings, corresponds to the typical gap size of
dead canopy trees.

Although an analysis of species' distribution patterns can help us gain a better understanding of the
ecological processes that might be responsible for these
patterns, we should take care whenever we attempt to
infer process from pattern. For example, the point
pattern analysis used above is not mechanistic. Such
analyses are descriptive and inductive, in that they can
be used to test whether an observed pattern is well
described by a given model (e.g. the double-cluster
process used by Wiegand et al. 2007). Although they
might suggest causal relationships, these relationships
must be tested empirically or experimentally to demonstrate a link. It is also worth bearing in mind that
current distributions may reflect the action of past processes, in which case, using spatial pattern analysis to
gain an understanding of ecological processes is essentially a space-for-time substitution (Law et al. 2009).
Our inferences may thus be incorrect if space and time
are not perfectly substitutable, as when past processes
no longer operate in the present.

Patch versus Landscape Effects on Species Distributions

Because species distribution patterns tend to be clustered or patchy at some scale, especially if the habitat in
which the species occurs is also patchily distributed
across the landscape, many techniques and modeling
approaches have been developed to estimate patch
occupancy. This is particularly evident in metapopulation models, which we will discuss a bit later in the
chapter. In the meantime, we consider here scale-
dependent effects on patterns of patch occupancy. To
what extent is patch occupancy influenced by the properties of the patch and its immediate surroundings
(patch context) versus characteristics of the landscape
in which the patch network is located (landscape context)? In other words, is the occurrence of a species
more a function of patch-scale characteristics or of
broader landscape attributes?

Local patch properties, such as patch size or habitat
quality, are often important determinants of species
occurrence. In a review of the literature, Mazerolle
and Villard (1999) found that patch-scale variables
(e.g. patch area, P/A ratio, habitat type) were significant predictors of a species' presence or abundance in
virtually all of the studies surveyed. However, landscape variables that related to the amount and configuration of habitat (e.g. distance to nearest patch, degree

of habitat fragmentation) were a significant predictor
in more than half (60%) of the studies. Importantly,
both patch-scale and landscape factors were required
to predict species occurrence in at least half of these
studies, suggesting that species distribution or patch-
occupancy models based solely on patch-scale characteristics will be insufficient for many—if not most—taxa.
In particular, vertebrates such as birds, mammals, reptiles, and amphibians, were more sensitive to landscape context than invertebrates (80% vs 20% of
studies, respectively). Although this might be expected
given the finer spatial scales at which most invertebrates operate relative to vertebrates, this may also
reflect a bias in how the landscape is defined in these
sorts of studies.

If the extent of the landscape is scaled appropriately (e.g. relative to an organism's dispersal or foraging
range), then even insects and other arthropods may be
influenced by broader landscape characteristics. For
example, landscape context was found to have a scale-
dependent effect on the abundance and occurrence of
different types of bees in northwestern Germany
(Steffan-Dewenter et al. 2002). Solitary bees and bumble bees exhibited a greater correlation between abundance and the amount of seminatural habitat
(grasslands, hedgerows, fallow lands, gardens) in the
surrounding landscape at spatial scales less than 1 km
(250 and 750 m, respectively), whereas honey bees
(*Apis mellifera*) were most correlated with landscape
context at a scale of 3 km (**Figure 7.7**). Interestingly,
honey bee abundance appeared to *decline* with increasing amounts of seminatural habitat in the surrounding
landscape. This counterintuitive relationship may be
related to the wide-ranging foraging habits (up to 3
km) of honey bees and their keen ability to home in on
the most-profitable resource patches. Bee abundance
in this study was based on the number of visits to
experimental flower patches placed at the center of
each landscape. Honey bees would thus appear more
abundant in landscapes with less seminatural habitat
because they were more attracted to the experimental
flower patches, given the relative scarcity of other nectar sources in those landscapes. Different types of bees
may therefore have different perceptions of landscape
structure. The broader effects of landscape context on
bee abundance may be explained by their habitat
specificity and foraging range. For example, solitary
bees are more specialized in their habitat requirements,
being restricted to seminatural habitats such as calcareous grasslands and orchards, and have a shorter foraging range than honey bees. Thus, solitary bees may
be particularly sensitive to the loss and fragmentation
of seminatural habitats, which could have important
consequences for pollination services in agricultural
landscapes.

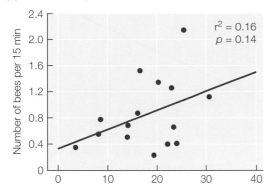

(A) Bumble bees, radius 750 m

$r^2 = 0.16$
$p = 0.14$

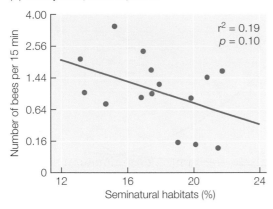

(B) Honey bees, radius 3,000 m

$r^2 = 0.19$
$p = 0.10$

Figure 7.7 Scale-dependent responses to landscape context in different groups of bees. Experimental flower patches (12 pots consisting of four different flower species) were placed at the center of landscapes having different amounts of seminatural habitat in Lower Saxony, Germany. The rate at which bees visited these experimental flower patches was then used as a proxy for bee abundance within each landscape. Bumble bee abundance was most influenced by the amount of seminatural habitat within a 750 m radius of the experimental patches (A), whereas honey bee abundance declined as the amount of seminatural habitat increased within a 3 km radius of patches (B).

Source: After Steffan-Dewenter et al. (2002).

Species Distribution Models

Species-habitat relationships are among the most fundamental of ecological relationships, and these are now encapsulated in an array of modeling approaches, including **resource-selection functions**, **ecological niche models**, and **climate envelope models**. Although these approaches may differ in their conceptual frameworks or analytical methods, they all have the same general goal of describing and predicting the occurrence or abundance of species in heterogeneous landscapes. An entire book (Franklin 2010a) and several review articles have been published on mapping and

modeling species distributions, and so only a general overview of these approaches is given here.

Resource-Selection Functions

Resource-selection functions (RSFs)[3] are popular for modeling the relative use of habitat and the expected distribution of species across a heterogeneous landscape. RSFs are defined operationally by relating patterns of species occurrence to various measured attributes of the habitat, and the function itself provides values that are proportional to the probability of that habitat being used by the target species (Boyce et al. 2002). RSFs are usually estimated from presence-absence data (i.e. habitat that is used versus unused) using a binomial generalized linear model, such as logistic regression (Guisan & Zimmermann 2000), although other modeling approaches and selection procedures have been advanced in the literature. RSFs can be implemented within a GIS, thereby enhancing their value for land-management planning and population viability analysis of target species in a given (or future) landscape (Boyce et al. 2002). To illustrate this approach, we consider the development and application of RSFs in the following case study involving the selection of landscapes during winter by woodland caribou (*Rangifer tarandus caribou*) in different parts of its range within Québec, Canada.

Woodland caribou are declining throughout North America, and are thus a high-priority species for conservation and management; it is classified as threatened or endangered in various parts of its range by the Committee on the Status of Endangered Wildlife in Canada. As a consequence, Canadian legislation requires that the critical habitat of the woodland caribou be defined and protected. Although woodland caribou are known to prefer mature coniferous forests, peatlands, and open areas with lichen, there are some broad regional gradients in these habitat attributes across its range. Fortin and his colleagues (2008) developed an RSF, based on aerial surveys of pathways made by caribou through the snow, to characterize the habitat characteristics of landscapes selected by caribou in Québec during winter. The top-ranked model highlighted the importance of regional

[3] Resource-selection functions (RSFs) share some similarities with utilization distributions, which were discussed in the context of home-range estimation (another pattern of space use) in **Chapter 6**, especially when utilization distributions are coupled with the environment to explain intensity of resource use (resource-utilization functions, RUFs). A major difference between the two, however, is that the utilization distribution is estimated prior to fitting the RUF (e.g. the method of kernel density estimation), whereas the RSF is estimated explicitly as a function of spatial covariates (Hooton et al. 2013). There are other differences as well. In examining the properties of both, Hooten et al. (2013, p. 1153) concluded that 'generally the RSF is preferred because it is slightly easier to implement and yields unbiased inference about selection coefficients when no measurement error exists in the telemetry data. However, we note that when there is location uncertainty in the data, a modified version of the RUF can outperform the traditional RSF in terms of less bias in the estimation of selection coefficients.'

differences in how caribou were selecting winter land-scapes, based on the amount of coniferous forests and lichen, and in relation to road density (**Figure 7.8**). Perhaps surprisingly, the association between caribou and roads was positive in the southern part of the range (**Figure 7.8A**). This should not be taken to mean that cari-bou have an affinity for roads, however; their positive association with roads was perhaps unavoidable given the high road density in this region (Fortin et al. 2008).

Such regional differences in landscape selection by caribou reflect the broader-scale gradients in climate (precipitation is twice as high in the eastern than west-ern part of the region), as well as in road density and forest-harvesting trends, which are higher in the south-ern than northern part of the range. When the RSF was used to map the probability of woodland caribou occur-rence across the boreal forest of Québec, woodland caribou are more likely to occur in areas having lakes, coniferous forest, and greater lichen cover, but are less likely to occur in areas with mixed and early-seral stage forests or high road density. Given that woodland cari-bou are associated with different landscape characteris-tics in different parts of their range, however, the conservation needs of this species are unlikely to be well-served by a single definition of critical habitat that is applied uniformly throughout its range.

As with any modeling approach, RSFs are not with-out their limitations and there are a number of issues that should be considered and addressed in the con-struction or application of RSF models. These concerns are not unique to RSF models, however, but may apply more generally to species distribution models or other modeling approaches that rely upon presence-absence data (e.g. patch occupancy models, such as metapopu-lation models). We consider a few issues of particular interest here, but a more in-depth treatment can be found in one of the texts devoted to the topic (e.g. Manly et al. 2002; MacKenzie et al. 2006).

FALSE ABSENCES, FALSE PRESENCES, AND DETEC-TION BIAS Because species distribution models, including RSFs, rely on presence-absence data, the reli-ability of 'presences' and 'absences' becomes a key issue in fitting these models. Most models assume that spe-cies are present in suitable habitat ('true presences') and are absent from unsuitable ones ('true absences'), but the reality is far more complex (**Table 7.1**; Hirzel & Le Lay 2008). In the case of population source–sink dynamics, for example, the species may occur in mar-ginal habitats that are not really suitable (a 'fallacious presence'; **Table 7.1**). Evaluating absences is admit-tedly more difficult. Is the species really absent from a particular habitat patch, or has it just been missed dur-ing surveys (i.e. missed presences; **Table 7.1**)? Mobile or cryptic species could easily be overlooked, and it may be difficult to detect a species in some habitats

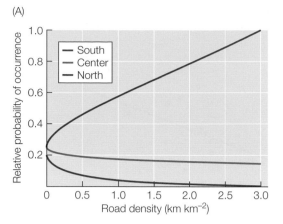

(A)

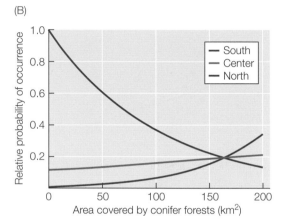

(B)

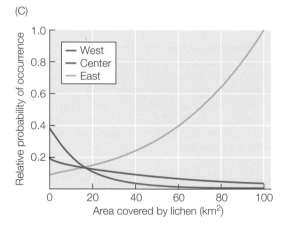

(C)

Figure 7.8 Landscape selection in woodland caribou (*Rangifer tarandus caribou*) varies by region. Within the boreal forests of Québec, Canada, the occurrence of cari-bou with respect to road density (A), amount of coniferous forest (B), and lichen cover (C) is predicted to vary region-ally. Separate resource-selection functions were derived for individual landscapes (201 km²) sampled along a latitu-dinal (south, center, north) and longitudinal (west, center, east) gradient across the study region (a 161,920 km² region that includes the southern half of the caribou's range in Québec).

Source: After Fortin et al. (2008).

TABLE 7.1 Relationship between species occurrence data (presence vs absence) and habitat suitability, illustrating potential sources of error.

			Habitat	
			Suitable	Unsuitable
Species	Presence	Recorded	True presence (TP)	Fallacious presence (FP)
		Unrecorded	Missed presence 1 (MP1)	Missed presence 2 (MP2)
	Absence		Fallacious absence (FA)	True absence (TA)

Source: Hirzel & Le Lay (2008).

more than others (if the habitat is very dense, for example). Or perhaps we have not surveyed the area sufficiently to encounter the species; the more we search, the more likely it is that we will detect the species if it is present. Such biases in search effort and detection can—and should— be adjusted for in the development of RSF and other types of patch-occupancy models (e.g. MacKenzie et al. 2002). Another potential source of error occurs because species may not occupy all of the available, suitable habitat because of dispersal limitation or local extinctions. In that case, absences from otherwise suitable habitat are considered 'fallacious' (**Table 7.1**; Hirzel & Le Lay 2008). Determining whether an absence is true, fallacious, or simply a presence that was missed is clearly problematic, and yet so very crucial to the development and interpretation of species distribution models.

USE VERSUS AVAILABILITY It is difficult to define what habitats are 'unused' if only the location or presence of the organism is recorded, as is the case with telemetry data that track the location of individuals over time ('presence-only data;' Boyce et al. 2002). We might then characterize what habitat is potentially available to the organism, based on a random sample of locations from the surrounding area (presence-available data). The difficulty there lies in defining what is 'available' to the organism, however. Habitat sites are available only if the individual can actually get to them, which means we should evaluate availability in terms of the movement scale, dispersal range, or home-range size of the species, depending on what type of habitat use is being modeled by the RSF (e.g. foraging vs habitat selection).

SCALE-DEPENDENT OR HIERARCHICAL HABITAT USE Spatial scale is especially important if the organism selects habitat in a hierarchical fashion (decisions at the broadest scale constrain availability of habitat at finer scales), or if factors affecting habitat use operate at different scales. We can incorporate scale-dependence either by constructing a set of scale-specific predictors that are likely to affect the distribution of a species at various scales, or by adopting a hierarchical model-selection approach, which entails selecting the parameters from the best model at each scale and then combining

these into an overall model to predict species occurrence (Elith & Leathwick 2009). Habitat use may also vary depending upon landscape context, as demonstrated above in the example of the woodland caribou (**Figure 7.8**), which again argues for constructing different RSF models for different regions or landscapes.

TEMPORAL DYNAMICS Habitat use may also be dynamic or time-dependent, perhaps reflecting seasonal or annual differences in resource availability (McLoughlin et al. 2010). In the case of seasonal differences, we could develop separate RSF models for the species at different times of year, much in the same way we can develop different scale-dependent models of RSF, as noted above. We should anticipate that population size may also fluctuate among years, perhaps because of changes in resource availability. Population density may interact with habitat selection (e.g. source–sink population dynamics), and thus could be driving or contributing to annual variability in habitat use. Methods for incorporating ecological dynamics in RSF models, such as those linked to density-dependent habitat selection, are currently at the forefront of development (McLoughlin et al. 2010).

EXPLANATION VERSUS PREDICTION Species distribution models are used to both explain and predict patterns of abundance and distribution. For many, however, the most important function of RSFs is their predictive power (Boyce et al. 2002). Predictive models may be used either to predict the occurrence or abundance of a species within a particular landscape context (i.e. within the range of habitats or time periods sampled) or to forecast species occurrences in a new landscape context, such as in response to projected climate or land-use changes (Elith & Leathwick 2009; Franklin 2010b). The RSF developed for the woodland caribou is an example of the former (**Figure 7.8**). Beyond enhancing the robustness and predictive ability of the model, the incorporation of the sorts of spatial and temporal dynamics we've just discussed should also help provide a better understanding of the ecological processes underlying patterns of habitat use and species distributions. If so, this would go a long way toward strengthening the linkage between ecological theory and the operational use of RSFs in conservation and land management (McLoughlin et al. 2010).

As many applications of RSF models have conservation and management implications...the most important consideration for evaluating RSF models is prediction. If a model reliably predicts the locations of organisms, it is a good model.

Boyce et al. (2002)

Ecological Niche Models

Before we talk about ecological niche modeling, we should first review the concept of the **ecological niche**. American naturalist Joseph Grinnell first coined the term 'niche' to refer to the habitat required by a species for foraging and reproduction (in this case, for a bird, the California thrasher, *Toxostoma redivivum*; Grinnell 1917). This initial concept of niche was subsequently expanded by British ecologist Charles Elton to include the role of the species within its community or ecosystem, and by limnologist George Evelyn Hutchinson, who viewed the niche as a multidimensional space (the 'n-dimensional hypervolume') that encompassed all of the environmental conditions under which the species could persist. This gave rise to the distinction between the **fundamental niche**, the potential range of the species, and the **realized niche**, which is the subset of the fundamental niche that is actually occupied by the species, owing to constraints on its distribution resulting from the presence of other species (competitors, predators) or other factors (e.g. geographical barriers to dispersal). For example, a non-native or invasive species may come to occupy more of its fundamental niche once introduced and established within a new location, where the pressures of competition or predation that had formerly limited its distribution in its native land have now been relaxed.

Ecological niche modeling entails the use of various techniques that relate species occurrence data to ecological, climatic, or environmental features of the landscape in order to define the species' ecological niche (Peterson 2003). The ecological niche model (ENM) is then projected onto the same or different landscapes to identify geographic regions that meet the species' niche requirements—and those that do not—thereby producing a potential or hypothesized distribution for the species. Although the goal is to model the fundamental niche of the species, the use of occurrence data (which often are in the form of presence-only data) should at least correlate with the realized niche of the species (Elith & Leathwick 2009). To illustrate the application of an ENM, we highlight the following case study involving an analysis of the predicted distribution of an endangered parrot in Mexico.

Figure 7.9 Yellow-headed amazon parrot. The yellow-headed amazon parrot (*Amazona oratrix*) is endangered by habitat destruction and illegal capture for the pet trade in its native Mexico.

Source: Photo by Kimberly A. With.

The yellow-headed parrot (*Amazona oratrix*) is highly valued in the pet-bird trade because it is intelligent and attractive, with a reputation for being a good 'talker' (**Figure 7.9**). Although the species ranges into the northern parts of Central America, its broadest distribution occurs within Mexico, where it is found in dry and semideciduous tropical forests along the Pacific and Gulf coasts. The yellow-headed parrot has been extirpated from parts of its range as a result of habitat destruction and capture for the pet-bird trade, and is classified as Endangered on the IUCN Red List, meaning that it faces a very high risk of extinction in the wild. Thus, the trade of wild-caught birds is strictly prohibited by international treaty under the Convention on International Trade of Endangered Species (CITES). Despite this, the status and distribution of yellow-headed parrots in Mexico are largely unknown, particularly along the Pacific coast where the rates of deforestation have been greatest.

Using ecological niche modeling, Monterrubio-Rico and colleagues (2010) compared the current distribution of the yellow-headed parrot to its historical distribution to assess the magnitude of range reduction. Location data obtained from museum specimens and published records, in conjunction with vegetation maps, digital elevation data, and maps of Mexico's biogeographic regions, were used to reconstruct the historical distribution of the parrot. For the current distribution, field surveys provided presence-absence data, which were then combined with information on the ecological niche of the species (it is generally found at elevations <900 m and in a variety of forest types) and land-cover data from a national forest inventory map. Subsequently, Monterrubio-Rico and colleagues estimated that yellow-headed parrots currently occupy only 21% of their historical range along the Pacific coast in Mexico. Several populations within this range appear to be isolated, and much of its preferred semideciduous tropical forest is unprotected. Thus, the yellow-headed parrot is a conservation priority and has become a flagship species for parrot conservation within the region.

Among the approaches currently used in ecological niche modeling, GARP (Genetic Algorithm for Rule-set Prediction; Stockwell & Peters 1999), which was used in the parrot example above, and MaxEnt (Maximum entropy modeling of species' geographical distributions; Phillips et al. 2006, Elith et al. 2011) are among the most popular. Popularity comes at a cost, however. The ready availability of user-friendly software, in addition to the widespread availability of geospatial data layers and biodiversity databases, has greatly contributed to the rise in the use of ecological niche modeling to predict species distributions. Unfortunately, availability and ease of implementation also increase the risk that such modeling will be done unadvisedly, without due consideration of the limitations of these approaches.

For example, Lozier and colleagues (2009) demonstrated how ENMs can be used to generate very convincing—albeit dubious—distribution maps, based on reported sightings of the mythological creature (or cryptozoid, depending on your beliefs) known as Sasquatch, or more colloquially as 'Bigfoot,' in western North America. Bigfoot is a large, hairy, human-like biped that supposedly inhabits the forests of the Pacific Northwest. The authors illustrate a rather convincing niche model that significantly predicts the current distribution of the Sasquatch, as well as its projected distribution under a scenario of climate change(!) in which the Sasquatch is expected to shift its distribution to higher latitudes and elevations in the future. They also—just as convincingly—demonstrate the overlap in the current distribution map of the Sasquatch with that predicted by a niche model for the American black bear (*Ursus americana*). Thus, a skeptic might conclude that many Bigfoot sightings are really that of black bears—a mere case of mistaken identity (Lozier et al. 2009).

Far from being a criticism of ENMs—or species distribution models more generally—this example is meant to illustrate the importance of data quality and the ease with which such models can be developed and subsequently misinterpreted. Again, careful consideration

must be given to the underlying data requirements and methodological limitations of each approach (Elith et al. 2006; Peterson et al. 2007; Phillips & Dudík 2008). Although we consider a few issues here in the context of ENMs, these apply to other types of species distribution models as well.

PRESENCE-ONLY DATA AND OTHER DATA-QUALITY ISSUES Biodiversity databases are increasingly available online in the public domain, ranging from the collections of herbaria and natural history museums to continent-wide standardized surveys and citizen-science datasets for some groups (e.g. birds). These databases may contain errors in the accuracy of species identification or in their georeferencing of species' locations, however (Lozier et al. 2009). In addition, occurrence data may have been obtained haphazardly, without benefit of a planned or systematic sampling scheme, such that absences cannot be inferred with certainty (Elith et al. 2006). Although presence-only data can be used to reliably model species' distributions (e.g. Elith et al. 2006), care must obviously be exercised in evaluating the quality of the data and thus in interpreting model results.

SAMPLE SIZE AND PREDICTION BIAS Ecological niche models typically 'train' and 'test' the model on the same landscape by dividing the dataset (i.e. the same data are not being used to both train and test the model); a sufficient amount of data must thus be available to develop a robust model as well as to provide a rigorous test of that model. How much data might be needed, however, depends upon the modeling approach and intended use of that model (Wisz et al. 2008). Although model performance generally improves with the addition of new information, there is usually a point of diminishing returns, where additional data or variables add little to the accuracy of the model's predictions (Stockwell & Peterson 2002). In fact, model performance may even *decrease* with increasing information, usually because the additional variables produce models that are too specific—models that have been optimized for a particular case and which therefore perform poorly on new data—a problem referred to as **prediction bias** or overfitting of the model (Stockwell & Peterson 2002). Some methods, such as GARP and MaxEnt, appear to be less sensitive to small sample sizes (e.g. 10–30 locality points) than other approaches, but no method performs consistently well with <30 samples (Wisz et al. 2008). More data are needed if the nature of the ecological relationship is a complex one (i.e. interactions occur among variables) or if the species is widespread, where more data are required to adequately characterize the species' distribution (Wisz et al. 2008). Widespread species may exhibit regional differences in their ecological characteristics, however, and thus modeling all of these populations together may overestimate the breadth of the species' ecological niche, thus reducing model accuracy (Stockwell & Peterson 2002).

EXTRAPOLATION VS INTERPOLATION   One of the main challenges of ecological niche modeling, and species distribution models more generally, is the problem of **transferability**, projecting species distributions onto different landscapes from those used to generate the model. Transferability (also known as the generality of the model) has important implications for forecasting how species distributions are likely to change in response to climate change and for predicting the potential for invasive spread of non-native species (Peterson et al. 2007; Elith et al. 2010). This represents yet another problem in **extrapolation** or translating across scales (**Chapter 2**), here involving the challenge of predicting species distributions in different landscapes across geographic space or at different points in time (Elith et al. 2010). This is different from the problem of **interpolation**, the accuracy of the model in predicting species' distributions within the sampled landscape, although this is no less important an issue. Not all of the commonly employed ecological niche modeling approaches perform equally well at both interpolation and transferability (Peterson et al. 2007; but see Phillips 2008), which requires some care by the user in considering the goal and intended use of the ENM. There is often a trade-off between the generality of a model and its ability to be applied to specific cases. Which type of model is better, therefore, depends on the intended application.

Climate Envelope Models

Climatic constraints, such as precipitation or temperature, limit the distribution of many species. Climate envelope models (CEMs) depict the geographic distribution of a species' climatic niche, and may provide insights into the factors limiting a species' range (Graham et al. 2010). CEMs have been used to evaluate the suitability of habitat beyond the current range of a species; if the CEM predicts little or no suitable habitat beyond the current range, then climatic factors may be limiting the species' distribution. If the CEM predicts that suitable habitat does occur beyond the current range, then factors other than climate may be limiting the species' distribution (e.g. dispersal limitation due to geographic barriers).

To evaluate the potential for climatic constraints on species distributions, Svenning and Skov (2004) examined the extent to which the current ranges of European trees filled their potential range, as predicted by a CEM. The ranges of temperate tree species are strongly constrained by physiological thresholds in response to winter cold, heat during the growing season, and drought. The authors thus estimated potential species'

ranges based on three bioclimatic variables related to these thresholds: the number of growing degree days; absolute minimum temperature, estimated from the mean temperature of the coldest month; and water balance (calculated as the yearly sum of the monthly differences between precipitation and potential evapotranspiration). As a measure of how well each species filled its potential range, Svenning and Skov (2004) calculated a realized-to-potential range size ratio. Based on their model analysis, they estimated that about two-thirds of European trees (n = 55 species) presently occupy <50% of their potential range, with most occupying only about 28% (**Figure 7.10**). Thus, a majority of European tree species are missing from a large part of their potential range. Given the widespread naturalization of many of these tree species outside of their present range, the low degree of range-filling may be a result of dispersal constraints on postglacial expansion. One implication of this, however, is that many European tree species may be unable to track the more rapid climatic changes occurring today.

Nor are trees and other plants the only types of species whose ranges are likely to be limited by climate. The ranges of vertebrates may also be constrained by climatic factors related to temperature or precipitation. For example, many montane bird species within the Andes Mountains of Colombia have restricted ranges, contributing to the high degree of bird diversity and endemism found in this region (Graham et al. 2010). In Colombia, the Andes consist of three cordilleras, each of which has its own unique assemblage of birds, which implies that dispersal limitation may be important in setting range limits for these species. Within a cordillera, however, species tend to occur at particular elevations, which may reflect environmental constraints on species distributions at a finer scale. Graham and her colleagues (2010) employed a CEM to determine the relationship between potential and realized distributions, both across and within cordilleras, for 70 bird species. As in the European tree study, birds in the Colombian Andes occupied <50% of their potential range, which is perhaps not unexpected, given that these particular species were known to be range-restricted in the first place. Within a cordillera, however, species generally occurred across the entire area predicted to be climatically suitable for that species, which mostly corresponded to environmental (rather than geographic) gradients. Once again, we see that species' distributions may be influenced by different factors at different scales.

Beyond evaluating the extent to which climatic factors appear to be limiting the current distribution of species, CEMs are also important for predicting the impact that climate change might have on the future distribution of species (Franklin 2010b). Although this is a common desire in species distribution modeling, we should be aware that this involves extrapolating

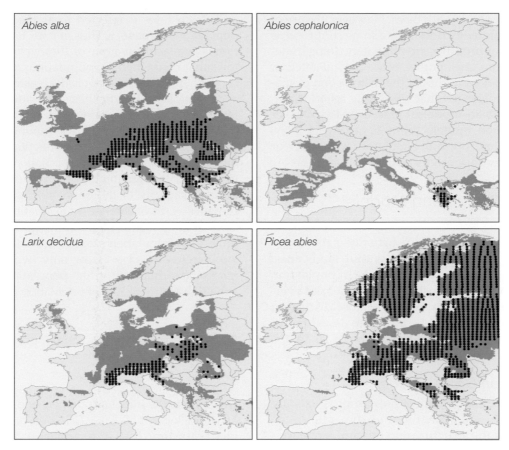

Figure 7.10 Figure 7.10 Realized versus potential distribution of four European trees.   Climatic envelope models were used to predict the potential distribution (dark green shaded areas) for different species, based on their physiological thresholds for cold, heat, and drought tolerance. The current distribution (black points) of these trees thus does not fill all of the potential range, possibly because of dispersal constraints that limited postglacial expansion.

Source: After Svenning & Skov (2004).

model results to different landscapes and time periods from those used to generate the relationship in the first place. Thus, all the issues we discussed previously in regards to the predictability and extrapolation of models applies here as well. Forecasting range shifts in response to climate change is a growing concern, however, and thus merits a bit of discussion here.

TESTING PREDICTIONS ABOUT FUTURE DISTRIBUTIONS   One of the key assumptions of species distribution models is that species are at an equilibrium with their environment, which clearly is not the case under a scenario of climate change, whether we are looking back into the past or forward into the future (Elith & Leathwick 2009). We thus have to assume that our current species records will cover the range of conditions that will (or did) occur, so that we can extrapolate our model to the novel landscape and time. The use of species distribution models, such as CEMs, for prediction rather than just explanation ultimately has implications for the way that models are fitted and evaluated (Elith & Leathwick 2009). Although CEMs are widely used to forecast shifts in species ranges under future

climate change scenarios, these models are rarely validated against independent data, and their fundamental assumption that climate limits species' distributions is rarely tested (but see Duncan et al. 2009). Despite all this, species distribution models are really one of the few practical approaches currently available for forecasting (or hindcasting) distributions in response to changing environmental conditions.

Summary

All of these methods achieve the goal of modeling species distributions, but in different ways using different methods. Clearly, there is considerable overlap among these different types of models. Modeling resource selection is akin to modeling a species' niche, and both RSFs and ENMs may include climate data like CEMs. Modeling shifts in species' distributions to climate change is not limited to just CEMs, but is also being investigated within the context of ENMs and RSFs. Indeed, MaxEnt has been used within the context of both ecological niche modeling (Peterson et al. 2007) and climate envelope modeling (Graham et al. 2010). Thus, it probably matters less what we end up calling these models than what we do with them. However, the use of such diverse terminology does create some confusion as to whether these approaches are ultimately doing something different from one another. Since they all attempt to use occurrence data to model the distribution of species, it has been suggested that we simply refer to them all as **habitat suitability models** (Hirzel & Le Lay 2008) or **species distribution models** (Elith & Leathwick 2009; Franklin 2010a), and not quibble about the manner or scale at which they supposedly achieve this aim.

A Primer to Population Models

Despite their utility, species distribution models do not indicate whether populations are actually viable or are capable of long-term persistence within landscapes. For that determination, we need a model of population dynamics. We thus begin this section with a basic overview of population growth models, before discussing how spatial complexity can be added to assess population viability and extinction risk within a landscape context.

Basic Population Growth Model

Fundamental to understanding how populations are affected by landscape structure and dynamics is an understanding of the structure and dynamics of populations themselves. At its most basic, population dynamics involve the change in population size (ΔN) from one time period (N_t) to the next (N_{t+1}): $\Delta N = N_{t+1} - N_t$. The change in population size is a result of processes that contribute to the population-wide birth rate (B) and death rate (D), as well as the rates at which individuals move into (immigration, I) and out of (emigration, E) the population. Thus, population change (ΔN) can be described as

$$\Delta N = (B + I) - (D + E). \qquad \text{Equation 7.1}$$

These are commonly referred to as the **BIDE factors**. Intuitively, populations will increase if the number of individuals being added to the population (through births and immigration) exceeds the number of individuals lost (through death or emigration). Thus, $\Delta N > 0$ when $(B + I) > (D + E)$.

To simplify things a bit more, let's assume that the population is 'closed', meaning that it is isolated and not connected to other populations by dispersal. Individuals thus cannot enter (I) or leave (E) the population as before. Population dynamics can now be attributed to the intrinsic processes of birth and death: $\Delta N = B - D$. The difference between birth and death rates gives us an immediate understanding of the population trajectory: if births exceed deaths, the population is increasing; if deaths exceed births, the population is declining; and, if births balance deaths, then the population is stable. The difference between birth rates

and death rates is referred to as the **intrinsic rate of population increase**, r. Thus, if $r = 0.1$, then the population is increasing at a rate of 10%/year. It therefore follows that $r > 0$ when the populations is increasing, and $r < 0$ when the population is decreasing.

To simplify things further, let's assume that the population is capable of growing without bounds, exponentially. In other words, we are assuming that there is no resource limitation, disease, competition, predation, or other factors that might limit population growth. Though unrealistic, this assumption allows us to examine the intrinsic capacity of the population to grow, and thus provides a critical baseline for evaluating to what degree these or other factors might limit population growth. To project the population size at the next time step (N_{t+1}), we therefore need to know only two things: (1) how many individuals are in the population at time t (N_t), and (2) the intrinsic rate of population growth, r. Thus,

$$N_{t+1} = rN_t. \qquad \text{Equation 7.2}$$

In general, we can project the population size at any future point in time, t, as

$$N_t = N_0 e^{rt} \qquad \text{Equation 7.3}$$

where N_0 is the initial population size and e is the base of natural logarithms (e = ~2.718). We can think of natural logs as the time it takes to achieve a certain amount of exponential growth, and thus all exponentially growing systems are simply scaled versions of a common rate, e. That is, it doesn't matter whether we are talking about population growth or the interest compounded on our investments, these are all just different forms of exponential growth (or decay, as the case may be), which just increase (or decrease) at some rate depending upon the specific value of r (the scaling exponent). Thus, the basic equation for population growth is a **power law**, as discussed previously in **Chapter 2 (Box 2.1)**. For example, if we want to project how large an initial population of 100 individuals will be in 10 years given $r = 0.1$, then we can solve for $N_t = 100e^{0.1 \times 10}$, which gives us an ending population of about 272 individuals (Try it! What would the population size be if $r = 0.05$? What about if $r = -0.1$?). Like compounded interest, the population in this scenario would continue to grow at a constant rate indefinitely (in an ideal world), enabling us to calculate the size of the population at any point in time.

Matrix Population Models

The basic model of population growth discussed above assumes that the population is homogeneous; in other words, all individuals are assumed to have the same probability of dying or giving birth in a given time period. In reality, individuals are not demographically equivalent. The early life stages we variously refer to as seeds, fingerlings, larvae, hatchlings, or newborns do not reproduce, tend to be extremely vulnerable to predators or the elements, and thus incur high mortality (death) rates. In contrast, adult life stages do reproduce (perhaps at different rates or degrees of success depending on age or experience) and may be larger, faster, more mobile, or simply more adept at avoiding predators and weathering environmental extremes, and thus have lower mortality rates. For some species, then, birth and death rates vary according to the life stage or age of the individual.

Populations are thus heterogeneous, and we can incorporate such age- or stage-structured demography through the construction of a **matrix population model**. The British mathematician, Patrick Leslie, is usually credited with the development of matrix models, and thus these are often referred to as **Leslie models** in his honor. Although Leslie models are based on age-structured demography, the related **Lefkovitch models**, developed by Canadian statistician Leonard Lefkovitch, are based on stage-structured demography. A focus on life stage might be more useful than age in species where it is difficult to age individuals accurately or where fecundity or survival are more closely related to life stage (or some other attribute, such as size) than to age. Since then, matrix population models have undergone further development and have been applied to a wide range of problems in ecology, conservation, and management (Caswell 2000).

To illustrate a simple matrix model, imagine a migratory songbird that nests in the temperate forests of North America and spends the rest of the year in South America, like the cerulean warbler introduced at the start of the chapter (**Figure 7.1**). The lifespan of a migratory songbird is fairly short, but we'll assume here they live 4 years to keep things simple. A **life-cycle graph** depicts the different stage or age classes, as well as the transitions between them (**Figure 7.11**). For the songbird, we might identify three stage classes: (1) juveniles or first-year birds, which do not breed; (2) second-year birds, which either do not breed (because they are not successful at acquiring territories or mates) or are just inexperienced at nesting and rearing offspring, resulting in low fecundity for this stage; and (3) mature adults, which are experienced breeders. There are two types of transitions: (1) survival (S_{ij}), the probability of surviving from stage i to the next stage j;[4] and, (2) fecundity (F_{i1}, sometimes referred to as *fertility*), assessed as the average number of female offspring produced per female at stage i; it is thus the contribution to the first life stage.[5]

In the case of our migratory songbird, the different stages mostly correspond to age, except for the adult stage where individuals may remain for two years (assuming a maximum four-year life span). In general,

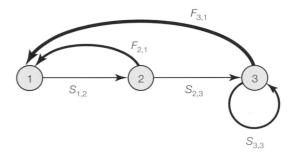

Figure 7.11 Life-cycle graph for a hypothetical migratory songbird. The three life stages are: (1) first-year birds or juveniles, (2) second-year birds, which are inexperienced breeders, and (3) adults. The transitions between stages are shown by arrows indicating survival (S_{ij}) and fecundity, the contribution to the first life stage ($F_{i,1}$).

survival is low for juvenile or first-year birds; fledglings are particularly vulnerable to predators, starvation, and inclement weather in the weeks after they leave the nest and learn how to fend for themselves. Assuming they survive this gauntlet, they must then contend with their first migration, winging their way many thousands of miles to South America on their own. The dangers are myriad, and so perhaps a third of first-year birds will survive to return to breed the next year ($S_{1,2} = 0.3$). Among breeding adults, survival might be twice as high ($S_{2,3} = 0.6$), although they too must contend with the perils of migration, as well as disease, predation, and old age. Older adults might therefore have slightly lower survival than younger adults ($S_{3,3} = 0.5$). In terms of fecundity, we'll assume that older females (3 years) are more experienced, and produce about one more female offspring per female than do second-year females ($F_{2,1} = 1.55$ female young/second-year females; $F_{3,1} = 2.40$ female young/adult females). Note that nest success (and thus fecundity) is usually low for most songbirds, particularly in disturbed habitats and fragmented landscapes that support an abundance of generalist predators that prey upon eggs and nestlings (**Box 7.2**).

We can represent these stage- or age-specific demographic rates within a matrix, in which the first row is reserved for fecundity rates ($F_{i,1}$), survival rates (S_{ij}) are listed on the diagonal, and the columns correspond to the specific age class or stage. For the generic songbird above, the matrix, **A**, would look like this:

$$\mathbf{A} = \begin{bmatrix} 0 & F_{2,1} & F_{3,1} \\ S_{1,2} & 0 & 0 \\ 0 & S_{2,3} & S_{3,3} \end{bmatrix} \qquad \text{Equation 7.4}$$

in which the various demographic rates are obtained empirically from the study population, such as through observation of nest fate and fledging success to estimate age-specific fecundity, and mark-recapture techniques to estimate age-specific survival. To project the population size forward in time [n(t+1)], we also need

---

[4] For long-lived species, individuals may stay within the same stage for more than one year (S_{ii}).

[5] It is a convention of population models to focus only on females because it is easier to assess female fertility than male fertility, and female fertility often sets a limit to how fast the population can grow.

information on the numbers of individuals within each stage, which is represented by a vector, $\mathbf{n}(t)$:

$$\mathbf{n}(t) = \begin{bmatrix} n_1 \\ n_2 \\ n_3 \end{bmatrix}.$$ Equation 7.5

Subsequently, population projection with a stage- or age-structured model is given by the equation:

$$\mathbf{n}(t+1) = \mathbf{A}\mathbf{n}(t).$$ Equation 7.6

Thus, for our hypothetical songbird population, the elements for this matrix and vector would be:

$$\mathbf{n}(t+1) = \begin{bmatrix} 0 & 1.55 & 2.40 \\ 0.3 & 0 & 0 \\ 0 & 0.6 & 0.5 \end{bmatrix} \begin{bmatrix} 200 \\ 120 \\ 60 \end{bmatrix}.$$

Note that $F_{1,1} = 0$ since first-year birds are juveniles and do not breed. In this example, we are assuming that the population consists of 200 juveniles, 120 second-year birds and 60 third or fourth-year birds. Matrix algebra is then used to calculate the discrete-time or **finite population growth rate** (λ), which is derived as the dominant eigenvalue of the matrix, $\mathbf{A}$.[6] The finite population growth rate, λ, describes the stepwise (i.e. geometric) increase in the population from one year to the next, as typically occurs in species with discrete breeding seasons like our migratory songbird (i.e. breeding does not occur continuously throughout the year, as with exponential growth). The calculation of λ assumes that the age- or stage-structured demographic rates do not change over time, and that the population exhibits a stable age or stage distribution (i.e. the numbers or proportion of individuals among the different classes do not change over time). For our hypothetical songbird population matrix, $\lambda = 0.957$.

So, what does this value of λ mean? Because λ is the change in population size from one year to the next, it is basically a ratio of the relative population size:

$$\lambda = \frac{N_{t+1}}{N_t}.$$ Equation 7.7

Thus, if the population size remains unchanged from one year to the next, then $N_{t+1} = N_t$ and $\lambda = 1.0$, which provides a criterion for stability or **population viability**. Above this threshold, $\lambda > 1$ and the population increases. Below this threshold, $\lambda < 1$ and the population decreases. Thus, if $\lambda = 0.957$, we know immediately that the population is decreasing (because $\lambda < 1$), and further, that it is decreasing at a rate of 4.3%/year ($0.957 - 1.0 = -0.043 \times 100 = -4.3\%$). As you may recall from the introduction to this chapter, this is the

[6] The matrix, A, has several solutions, which are called eigenvalues. The largest or leading eigenvalue is the most important or 'dominant' one.

annual rate of decline exhibited by the cerulean warbler, one of the fastest declining songbirds in North America.

Regardless of whether population growth is exponential or geometric, r and λ are related mathematically as

$$r = \ln(\lambda)$$ Equation 7.8

and

$$\lambda = e^r.$$ Equation 7.9

Thus, if $\lambda = 0.957$, then $r = \ln(0.957) = -0.043$. For a population in which $r = -0.043$, we can obtain λ as $e^{-0.043} = 0.957$. We can also project population size at some future point in time t as

$$N_t = N_0 \lambda^t$$ Equation 7.10

which, as you'll note, is equivalent to Equation 7.3 because $\lambda = e^r$ (Equation 7.9). Therefore, if our hypothetical songbird population begins with a population size $N_0 = 380$ and a population growth rate $\lambda = 0.957$, will it be able to persist for another 100 years? Maybe, but the population is likely to be perilously close to extinction by then, with only a handful of females remaining [$N_t = 380(0.957)^{100} = 4.7$ females].

Source–Sink Population Dynamics

Although we have discussed heterogeneity within populations, in terms of age- or stage-structured demography, population growth rates may vary *among* populations as a consequence of differences in habitat quality or patch context. Thus, population growth rates may be spatially dependent if fecundity, mortality, or both vary among habitats. Intuitively, we expect that populations should do well and increase over time in high-quality habitats, whereas the opposite should occur in poor-quality habitats.

If we think of this in terms of the BIDE factors, then populations in high-quality habitats should have a higher birth than death rate, resulting in a positive population growth rate (B − D = r > 0). As a consequence, populations in high-quality habitats may produce a surplus of individuals, such that emigration rates exceed immigration rates (E > I). Thus, we expect high-quality habitats to be **population sources** and function as net exporters of individuals. In contrast, poor-quality habitats are likely to function as **population sinks**. Sink populations are not intrinsically viable (D > B), but persist because of immigration from nearby source populations (I > E); they are thus net importers of individuals and are subsidized by source populations.

In a seminal paper, American ecologist Ron Pulliam formalized the theory of **source–sink population dynamics**, based on this relative assessment of BIDE rates in different habitat types (Pulliam 1988). In this same paper, Pulliam went on to classify source–sink populations in terms of habitat-specific demography as

$$\lambda = S_A + \beta S_J \qquad \text{Equation 7.11}$$

in which S_A and S_J are survivorship estimates for the adult and juvenile life stages, respectively, and β is the habitat-specific per-capita birth rate (assuming a simple, two-stage life cycle in which only adults reproduce). Thus, $\lambda > 1$ in habitat sources (populations are increasing) and $\lambda < 1$ in habitat sinks (populations are declining).

Unfortunately, this has led to the general practice of defining population sources and sinks solely on the basis of within-habitat reproductive and survivorship rates, which ignores the relative contributions of emigration and immigration, which are the other defining criterion of source–sink habitats. For example, a population with $\lambda > 1$ (B > D) doesn't necessarily mean that it is functioning as a source; those surplus individuals could simply be dying off (i.e. E = 0). For a population to be a source, it must provide emigrants to at least one sink population. Similarly, populations with $\lambda < 1$ are not automatically sinks, unless they are also being sustained through immigration (i.e. I > E). Central to the idea of source–sink populations is that two or more populations are dynamically linked by asymmetrical dispersal.

How Best to Identify Population Sources and Sinks?

Identifying habitats that are population sources is a challenge in practice. Many studies simply use population size (N) or density (individuals/unit-area) to identify sources and sinks (Runge et al. 2006); habitats that support large populations are assumed to be sources and those with small populations are branded sinks. Unfortunately, population size or density is not always a reliable indicator of habitat quality (Van Horne 1983). Population density is sometimes higher in poor-quality habitats, as when a temporary increase in resource levels leads to an overproduction of young (or seeds or eggs, as the case may be), resulting in a 'spill-over' of individuals into poor-quality habitats. In species that exhibit social dominance, high-ranking individuals may saturate available high-quality habitat, forcing less-dominant individuals to occupy marginal habitats (i.e. a preemptive distribution; **Chapter 6**). Or, higher densities in poor-quality habitat could result from a lagged response to recent habitat changes, especially for species that exhibit a high degree of site fidelity and which therefore may not respond immediately to environmental or anthropogenic changes that degrade habitat quality (Wiens et al. 1986).

In other cases, marginal habitats might actually be *more attractive* to individuals, because these habitats appear to offer suitable habitat or have more abundant resources, and thus may have higher population densities in spite of their lower rates of reproductive success and survivorship. Attractive low-quality habitats represent **ecological traps** and are also thought to occur as a consequence of rapid habitat changes, resulting in the decoupling of the perceived attractiveness of habitat and its quality (Robertson & Hutto 2006). Conversely, species may avoid settling in high-quality habitat if they perceive it to be marginal habitat (**perceptual traps**; Patten & Kelly 2010). Thus, population density is a misleading indicator of habitat quality for a number of reasons, such that habitat quality should be assessed on the basis of intrinsic demographic rates (fecundity and survivorship) rather than population size.

Having said that, it can very difficult to obtain population vital rates for most species, let alone for multiple populations of the same species in different habitat types. For example, reproductive rates can be obtained more easily in some organisms than others, depending on the ease of finding and counting the seeds, eggs, or young produced in each habitat. Survivorship can be difficult to estimate for mobile organisms. For example, mark-recapture techniques are frequently used in vertebrate populations to estimate apparent survivorship, the probability of surviving and returning to the same locale the following year. However, any individuals failing to return are assumed to have died. Mortality and emigration rates may thus be confounded in such measures, which can lead to biased assessments of habitat quality: areas that export many individuals (high emigration rate) might be misclassified as sinks (because of low apparent survival) when in fact they are sources (Runge et al. 2006).

Further, habitats capable of supporting viable populations may nevertheless experience a net influx of dispersers, and if higher population density leads to increased mortality or decreased reproduction, then this can have a negative effect on population vital rates, making sources appear to be population sinks. These **pseudo-sinks** can be difficult to distinguish from genuine sinks without information on whether populations are capable of persisting in the absence of immigration (Watkinson & Sutherland 1995). Although relative rates for all four BIDE factors might ultimately be important for evaluating whether a habitat patch is indeed functioning as a population source, emigration and immigration rates were the least likely to be measured in empirical source–sink studies (Runge et al. 2006), undoubtedly because of the difficulty of quantifying dispersal in most organisms (**Chapter 6**).

Ideally, the identification of source and sink habitats should be based on long-term demographic study that is capable of distinguishing births from immigration and deaths from emigration (Dias 1996). For example, reproduction and/or survival may be high within marginal habitats during years of exceptional resource productivity (boom years), which could lead to the erroneous classification of these areas as sources if not evaluated over a longer time period. Similarly, a study conducted during a year of abnormally low resource levels (bust years) because of drought or some other environmental extreme might result in low survival or reproduction even within normally productive sources. For example, a twenty-year study of Siberian jays (*Perisoreus infaustus*) in northern Sweden tracked annual changes in the source–sink dynamics of individual territories (Nystrand et al. 2010). While most territories in an undisturbed forest functioned as sources on average, the investigators still found that most of these exhibited random fluctuations among years as to whether they represented a demographic source or sink. Thus, this study underscores the importance of long-term data for identifying high-quality habitat: a single snapshot in time is unlikely to provide a reliable picture of the source-sink status of a population.

Over time, sink habitats may convert to habitat sources if populations eventually adapt to the new conditions (a **source–sink inversion**; Dias 1996). This is believed to have happened in the case of the blue tit (*Parus caeruleus*) on the island of Corsica off the southern coast of France. Mainland populations of blue tits nest in both deciduous (*Quercus pubescens*) and evergreen (*Q. ilex*) oak forests in southern France, but breeding success (proportion of eggs resulting in surviving offspring) is far lower in evergreen forests owing to a mismatch between the timing of reproduction—a time of peak food demand—and peak food availability within this habitat (Blondel et al. 2006). Thus, deciduous forests function as population sources and evergreen forests are population sinks for blue tits on the mainland, a supposition that was supported by genetic assays that revealed patterns of relatedness consistent with asymmetrical dispersal from deciduous sources to evergreen sinks. On Corsica, however, forests are dominated by evergreen oaks, and breeding success of blue tits is similar in both forest types. Thus, evergreen forests no longer function as population sinks on Corsica, where blue tits have shifted the timing of their reproduction to coincide with peak food availability in this habitat. Although source–sink inversions are possible in theory, we probably should not count on adaptive responses arising and rescuing populations in landscapes where changing climatic conditions or human land-use activities are rapidly degrading habitats (With 2015).

Are Habitat Sinks a Drain on Habitat Sources?

As we have seen, different habitat types can make different contributions to the overall size and growth rate of the population. From the standpoint of preserving viable populations, would it thus be better to protect a large amount of poor-quality habitat or a smaller amount of high-quality habitat? What is the optimal amount and arrangement of source versus sink habitat on the landscape that will enhance population persistence? As usual, it depends on the species in question. If the species is very mobile and preferentially settles within high-quality habitat (according to a preemptive distribution), then too much sink habitat can dilute the contribution of high-quality habitats, reducing the size and growth rate of the population on the landscape (Pulliam & Danielson 1991; **Figure 7.12**).

In other cases, habitat sinks can have a direct negative effect on source populations, as was demonstrated in an experimental study of root voles (*Microtus oeconomus*) in southern Norway (Gundersen et al. 2001). In a simple two-patch system, voles were released into one of the habitat patches (the source) and the other patch was maintained as a mortality sink, by removing any individuals that colonized that patch. Over the course of the breeding season, source populations in these experimental plots decreased relative to those in control plots (where voles were not being removed from the second patch), despite the fact that reproduction and per capita recruitment was nearly twice as high within source patches of the treatment plots. Population declines in this source–sink system were due to density-dependent dispersal: voles continued to disperse from source (high density) to sink (low density) populations in experimental plots, but not within control plots where densities were similar in the two patches.

Clearly, it is important from a management or conservation standpoint that population sources be identified and high-quality habitats be protected or restored. That is not to suggest that lower-quality habitats, even those that support population sinks, are not important, however. Sink habitat can sometimes help stabilize the dynamics of source populations. In the case of the reed warbler (*Acrocephalus scirpaceus*) in the Netherlands, small marshes (ditches along roads) support lower population densities, and were thus assumed to be of poorer quality, than larger marshland areas (Foppen et al. 2000). With the aid of a simulation model, Foppen and colleagues (2000) explored how small marshes (1–5 ha) in close proximity (2–5 km) to a large marshland (10 ha) could actually enhance overall population size on the landscape. Further, this configuration of sink habitat enabled sources to recover more quickly

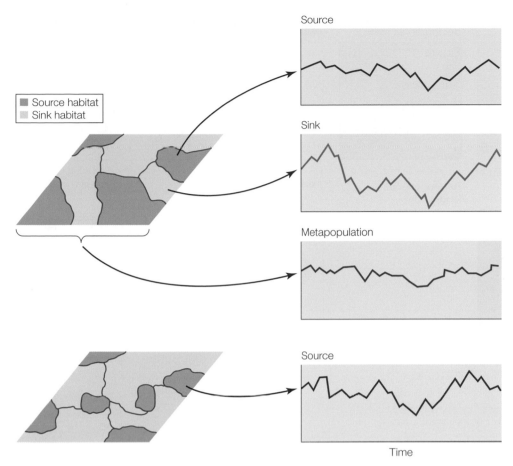

Figure 7.12 Source–sink population dynamics.   Populations in source habitats tend to have more stable dynamics than those in sink habitats. In sink habitats, death rates exceed birth rates, and the rate of emigration exceeds immigration, resulting in large fluctuations in population size over time. Note that the population size in sink habitats can sometimes exceed that of source habitats, and thus population size by itself is not a reliable indicator of whether a population is a source or a sink. In heterogeneous landscapes with sufficient source habitat (top left), the net effect of source–sink population dynamics is to stabilize metapopulation dynamics at the landscape scale. Increasing the proportion of sink habitat (bottom left), however, can destabilize the dynamics of populations within source habitats (compare source graph at bottom right to source graph at top right), especially if there are density-dependent effects on reproduction and survival.

Source: After Wiens (1989b).

following a population crash, suggesting that sink habitats may sometimes be able to function as refugia from which individuals can later recolonize source populations after an environmental catastrophe (assuming that these sink habitats haven't likewise been hit by the catastrophe). Sufficient connectivity among source–sink populations may thus enhance colonization and reduce extinction rates at the landscape scale, a phenomenon that we will explore in the next section on metapopulation dynamics.

Metapopulation Dynamics

As we have seen, basic population growth models typically assume a closed population and treat the landscape as if it were homogeneous, in that environmental heterogeneity and patch structure are ignored. These models are thus non-spatial (**Figure 7.13**). The inclusion

of spatially structured demography relaxes the assumption of homogeneity, and acknowledges that demographic processes may vary spatially because of differences in habitat quality. In the case of source–sink population dynamics, there is an explicit recognition of the importance of dispersal among populations (from sources to sinks). Thus, populations are not usually closed, as even seemingly isolated populations may be connected via dispersal to other populations.

A collection of populations that are linked by dispersal is referred to as a **metapopulation**, a 'population of populations' (Levins 1970). Metapopulation models are **patch-occupancy models**, in which the proportion of habitat patches occupied is assessed rather than the rate of population growth. Although individual populations may go extinct, these habitat patches may be recolonized by immigrants from other populations. Metapopulation persistence is thus dependent upon

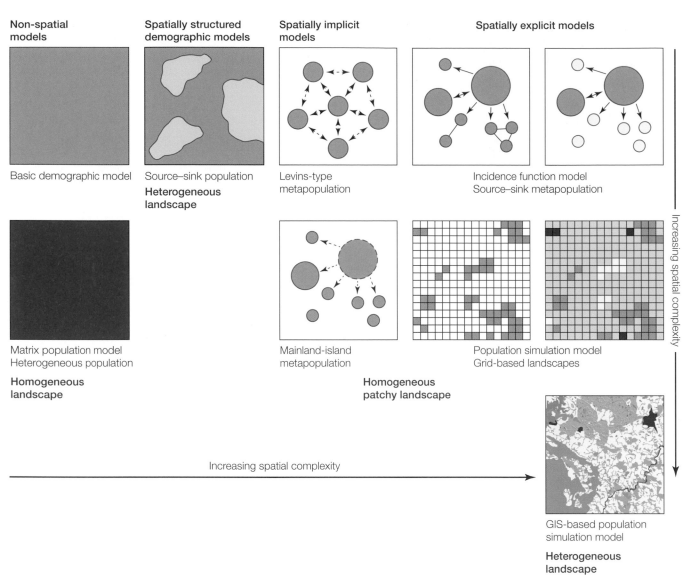

Figure 7.13 Different classes of population models, along a gradient of increasing spatial complexity. Non-spatial models include basic demographic and matrix population models; in neither case is demography spatially dependent, and so the landscape is assumed to be homogeneous. The original formulation of the source–sink population dynamic involves spatially structured demography, in which population vital rates vary among habitats. Habitat sources have birth and immigration rates that exceed death and emigration rates. Metapopulation models were initially spatially implicit, in that patch structure was implied by the presumed effects of colonization–extinction dynamics on patch occupancy. Incidence-function models are a class of metapopulation model that are spatially explicit, in that the size and location of patches influences colonization and extinction dynamics, and thus metapopulation persistence. These models can also accommodate differences in habitat patch quality, allowing for the modeling of source–sink dynamics in heterogeneous landscapes. Finally, spatially explicit population simulation models couple theoretical or real (GIS-based) landscape maps with a population simulator in which dispersal, reproduction, and mortality may all vary among habitats in heterogeneous landscapes.

the colonization and extinction dynamics of the entire population of populations, much as population dynamics are dependent upon the collective births and deaths of individuals within the population. Because many species have patchy distributions, and habitat fragmentation creates the appearance of a metapopulation, the metapopulation paradigm has come to epitomize the spatial dynamics of populations and underpins the management and conservation of both plants and animals in a diverse array of systems that include aquatic and marine environments, in addition to terrestrial ones.

Classical Metapopulation Models

What we now view as the classic form of the metapopulation model was first presented by American mathematical ecologist Richard Levins to evaluate whether different crop planting strategies (planting crop fields all at the same time or at staggered intervals) might be able to control pest outbreaks (Levins 1969, 1970). Agricultural fields were viewed as discrete patches in which crop pests were either present or absent. The dynamics of the pest population, in turn, was characterized in terms of local extinctions (e) and recolonizations (m) by pests dispersing from other fields in the region. The change in the proportion of fields (patches, p) occupied over time (t) can thus be described simply as the difference between colonization and extinction dynamics. Intuitively, the only patches that can go extinct (at a rate, e) are those that are already occupied (ep). Likewise, colonists can only come (at rate, m) from occupied patches (mp), and the only patches eligible for recolonization are those that are unoccupied ($1 - p$). Thus, the difference between colonization and extinction dynamics can be expressed mathematically as

$$\frac{dp}{dt} = mp(1-p) - ep. \quad \text{Equation 7.12}$$

If we assume that recolonization (m) and extinction (e) rates do not change over time, then patch occupancy (p) also will not change over time (t), and thus $dp/dt = 0$ (i.e. patch occupancy is at an equilibrium). If we solve Equation 7.12 for equilibrium (i.e. we set the left-hand half of the equation to 0), we get

$$\hat{p} = 1 - \frac{e}{m} \quad \text{Equation 7.13}$$

which gives us the expected proportion of habitat patches occupied at equilibrium ($\hat{p}$). Note that equilibrium does not mean that the recolonization and extinction rates are equal! Indeed, what would happen if the rate of recolonization exactly balanced the rate at which populations went extinct? If $e = m$, then $e/m = 1$ and $\hat{p} = 1 - 1 = 0$. In other words, no patches are occupied ($\hat{p} = 0$) and the metapopulation is extinct. Thus, metapopulation persistence in this context occurs only when the rate of recolonization exceeds the rate at which populations go extinct ($m > e$).

As with all models, the classical form of the metapopulation model makes a number of simplifying assumptions. These assumptions are: (1) the population of patches is large, and thus stochastic effects on colonization and extinction rates are minimal; (2) all patches are equally accessible (m is a constant and is therefore not influenced by inter-patch distances or the type of matrix between patches); (3) all patches have the same probability of extinction (e is a constant and is not influenced by patch size or quality); (4) patches are either at carrying capacity or unoccupied (the dynamics of populations within patches are ignored); and (5) the dynamics of individual populations are independent and asynchronous. Synchronous dynamics could come

about because of strong environmental forcing (all populations within the region are subjected to a severe drought) or because of excessive dispersal that effectively produces a single panmictic population, either of which could lead to correlated extinctions and hasten metapopulation extinction. Indeed, Levins concluded in his seminal (1969) paper that applying control practices synchronously across all agricultural fields (planting all fields at once, and then harvesting them simultaneously) would be the best strategy for eradicating crop pests, because there would be no refugia where pest populations could escape and later spread to recolonize fields from which pests had been eradicated.

Despite its simplicity, small tweaks to the model can produce surprising—and surprisingly complex—results. For example, we might envision a landscape in which there is a secure core population that never goes extinct and which serves as a perennial source of colonists to outlying populations. This sort of **mainland-island metapopulation** requires only a small adjustment to the colonization term of the classic model (Equation 7.12), in which we now assume that the fraction of island patches occupied (p) is no longer relevant for assessing colonization because colonists are continually coming from the large mainland population. This is also equivalent to a persistent propagule rain or a seed bank, where the current patch occupancy is not important for predicting future patch colonization. Thus, for a mainland-island metapopulation,

$$\frac{dp}{dt} = m(1-p) - ep \quad \text{Equation 7.14}$$

and the equilibrium solution is given by

$$\hat{p} = \frac{m}{m+e}. \quad \text{Equation 7.15}$$

Now if recolonizations balance extinctions ($m = e$), then $\hat{p} = 0.5$, meaning that 50% of island patches will be occupied at equilibrium.

An example of a mainland-island metapopulation is given by the bay checkerspot butterfly (*Euphydryas editha bayensis*), a federally listed threatened species under the US Endangered Species Act (**Figure 7.14**). A habitat specialist, the bay checkerspot prefers native host plants (especially dwarf plantain, *Plantago erecta*) that thrive in serpentine soils, which tend to be patchily distributed along ridgetops and outcroppings of the coastal mountain range within the San Francisco Bay Area in California (USA). Unfortunately for the butterfly, much of its former range came to be viewed as prime real estate, and thus the bay checkerspot now occurs in only a handful of core areas within the heavily populated—and heavily developed—San Francisco Bay Area. At the time of its listing in the late-1980s, the largest extant population of bay checkerspots was located near Morgan Hill and was estimated to contain

Figure 7.14 Bay checkerspot butterfly (*Euphydras editha bayensis*). The bay checkerspot butterfly exhibits mainland-island metapopulation dynamics in the serpentine grasslands of the San Francisco Bay Area in California (USA).

Source: Photo by Josh Hull, USFWS and published under the CC by 2.0 License.

hundreds of thousands of butterflies. Although smaller, outlying areas sometimes had butterfly populations, these would often go extinct as a result of drought and other environmental fluctuations. The persistent Morgan Hill population may have served as a source of colonists that rescued these outlying populations from extinction (Harrison et al. 1988): the pattern of recolonization was found to be distance-dependent, such that only habitat areas closest to the Morgan Hill population (1.4–4.4 km) were likely to be recolonized. Thus, this system has the appearance of a mainland-island metapopulation, which the astute reader will note is basically a source–sink population. Without the large, persistent Morgan Hill population, it is unlikely that the smaller populations would be able to persist as a classic metapopulation on their own. Clearly, identification of mainland populations (i.e. source populations) is important for the conservation or management of such metapopulations.

Despite the emphasis on patch occupancy and where colonists are coming from, the classic metapopulation model and its derivations are **spatially implicit population models** (Figure 7.13). The specific location, size, and configuration of individual patches are ignored and the effect of patch structure on colonization–extinction dynamics is not explicitly considered in the construction of these models. Instead, the notion that the landscape is patchy is merely implied through the phenomenological considerations of patch colonization and extinction rates and the focus on patch occupancy. As we shall see next, it is possible to include these spatial considerations explicitly using incidence function models.

Incidence Function Model

Intuitively, large patches should be able to support populations that are larger, and less prone to extinction,

than populations in small patches. It also seems likely that patches that are close to other (especially large) patches will have a higher probability of being recolonized than those that are more isolated. We can include the effect of spatial structure on colonization–extinction dynamics explicitly using an **incidence function model**, which was popularized by Finnish ecologist Ilkka Hanski (Hanski 1994). The probability that patch i is occupied at a given point in time is given by its incidence (J_i) as

$$J_i = \frac{C_i}{C_i + E_i} \qquad \text{Equation 7.16}$$

in which C_i is the probability that the patch becomes colonized and E_i is the probability that the population in patch i goes extinct in that time period. The astute reader will notice that Equation 7.16 is very similar in form to Equation 7.15, the equilibrium patch occupancy for a mainland-island metapopulation. The difference here is that we are focused on the occupancy of a single patch, which is likely to be affected by the occupancy of surrounding patches, which in turn is based on the relative size and proximity of those patches. Thus, we cannot simply scale up from patch-based occupancy to metapopulation occupancy without considering these patch-specific rates of extinction and colonization.

To obtain a patch-specific rate of extinction (E_i), we will assume that extinction is inversely proportional to patch area (A_i), which is a reasonable proxy for population size (and, it is usually more convenient to measure patch size than population size). Thus, the probability of extinction increases as patch size decreases as

$$E_i = \frac{e}{A_i^x} \text{ if } A_i > e^{1/x} \qquad \text{Equation 7.17}$$

where e is a species-specific probability of extinction in an area of fixed size (e.g. 1 ha) or the minimum patch size (A_0) below which extinction is certain ($E_i = 1$). Populations are assumed to go extinct because of random environmental fluctuations (e.g. drought), and thus x reflects the relative strength or intensity of environmental stochasticity. When $x > 1$, there exists a range of patch sizes where extinction is highly unlikely (i.e. patches are very large and environmental stochasticity is weak). When $x < 1$, however, even populations in large patches are likely to go extinct because of strong environmental stochasticity. Note that $E_i = 1$ for $A_i \leq e^{1/x}$, which means that extinction is certain to occur in patches below a critical size (A_0) relative to the intensity of environmental stochasticity (Hanski 1994).

The probability that patch i will be colonized (C_i) is a function of the number of immigrants arriving at the patch (M_i). We can derive a patch-based measure of colonization as

$$C_i = \frac{M_i^2}{M_i^2 + y^2} \qquad \text{Equation 7.18}$$

such that the rate of colonization increases in a sigmoidal (s-shaped) fashion, quickly at first and then gradually decreasing as the maximum is reached; y thus gives the rate at which the colonization probability reaches unity with an increasing number of immigrants, M_i. Intuitively, M_i should be influenced by the number, size, and proximity of neighboring populations. In other words, M_i can be equated to *patch connectivity* (S_i), which was presented in **Chapter 5** as one of the patch-based measures of connectivity (the incidence-function measure; Equation 5.1). For review purposes, recall that S_i is weighted by the area and distance of neighboring patches j, under the rationale that larger, closer patches provide more immigrants to the target patch i than smaller patches that are more distant. If $M_i = S_i$, we can rewrite Equation 7.18 as

$$C_i = \frac{1}{1 + \left[\dfrac{y}{S_i}\right]^2}. \qquad \text{Equation 7.19}$$

If we substitute Equations 7.17 and 7.19 for E_i and C_i, respectively, in Equation 7.16, we get the following for the expected occupancy of patch i

$$J_i = \frac{1}{1 + \left(1 + \left[\dfrac{y}{S_i}\right]^2\right)\dfrac{e}{A_i^x}}. \qquad \text{Equation 7.20}$$

The incidence-function model (Equation 7.20) can be fitted to empirical patch-occupancy data using a non-linear regression with maximum likelihood estimation for a given dispersal range (a species-specific parameter, α, which is a component of S_i; Equation 5.1) and the observed patch state p_i (occupied, $p_i = 1$; unoccupied, $p_i = 0$) as the dependent variable rather than the expected patch occupancy J_i (assuming patch occupancy is at an equilibrium; Hanski 1994).

The incidence-function model is an example of a **spatially explicit population model** (Figure 7.13). Although more realistic in its representation of the patch structure of the landscape, the addition of such spatial information on the size and location of patches complicates the estimation of patch occupancy quite a bit from a mathematical standpoint. Empirically, data on the location and size of patches are fairly easy to obtain, assuming habitat occurs as discrete patches in the environment. With respect to the species, the most critical information concerns its pattern of patch occupancy and dispersal range (α), from which the connectivity of patches (S_i) will be assessed. Although patch occupancy can be based on a single survey, it would obviously be

better to have data from several surveys so as to obtain more robust estimates (e.g. at least three years; Etienne et al. 2004). Further, Hanski (1999) has suggested that the system of patches should be sufficiently large (at least 30 patches), with a sufficient number of occupied and unoccupied patches (at least 10 each), to provide good information for parameter estimation. As in classic metapopulation models, colonization–extinction dynamics are assumed to be in a quasi-equilibrium, such that patch occupancy remains relatively constant over time. Admittedly, it can be difficult to obtain all of the necessary species-specific parameters empirically, although some can be estimated indirectly by solving an equation for the unknown parameter.

Despite these caveats, this approach has been successfully applied to estimate patch occupancy in a number of species, including the well-studied Glanville fritillary (*Melitaea cinxia*). We discuss how this approach can be extended to evaluate the overall metapopulation capacity of fragmented landscapes for the Glanville fritillary in **Box 7.3**. Although a full treatment of incidence-function models lies beyond the scope of this text, the interested student is referred to Hanski's text on metapopulation ecology for further inspiration (Hanski 1999).

Source–Sink Metapopulations

One of the simplifying assumptions of the classic metapopulation model is that all patches are of uniform size and quality. Yet, as we have just discussed, patches clearly vary in size and may also vary in habitat quality, such that populations within these patches will have different population growth rates. Previously, we discussed how spatially structured demography gave rise to source–sink population dynamics. Although the focus is typically on demographic rates, the definition of population sources and sinks also involves consideration of the relative rates of immigration and emigration. Population sources produce a surplus of individuals that can then colonize unoccupied habitat patches or rescue declining populations (sinks) from extinction. The very notion of a source–sink population dynamic is thus consistent with that of a metapopulation, a collection of populations linked by dispersal, although dispersal in this case is asymmetrical (i.e. from source populations to sink populations).

Some of the metapopulation models we have discussed in this section might thus be viewed as source-sink metapopulations. For example, the concept of a mainland-island metapopulation involves the existence of a large habitat patch that provides a perennial source of immigrants to outlying habitat patches that otherwise could not persist by themselves (i.e. they are sinks). The incidence-function model explicitly

BOX 7.3 The Metapopulation Capacity of Fragmented Landscapes for the Glanville Fritillary

Butterflies are frequent subjects of metapopulation studies because they often occur in patchy habitats that are relatively easy to delineate and map. For example, the Glanville fritillary (*Melitaea cinxia*) occurs in open grassy meadows throughout Europe (**Figure 1**). Its patch occupancy patterns in the Åland Archipelago of southern Finland have been studied for more than 20 years by the Finnish ecologist Ilkka Hanski and his colleagues.

Beyond the occupancy of individual patches, the bigger question concerns the probability of persistence for the entire metapopulation. This requires summarizing across the colonization and extinction rates of individual patches. To some extent, this is like summarizing across the birth and death rates of individual stage or age classes in a matrix population model (Equation 7.4). The patch structure of a landscape can thus be represented as a matrix **M** of individual patch elements, m_{ij}, characterized by their extinction (E_i from Equation 7.17) and pairwise colonization rates [S_i from Equation 4.1] as

$$m_{ij} = A_i A_j \exp\left(-\alpha d_{ij}\right) \qquad \text{Equation 1}$$

where A_i and A_j are the areas of patches i and j, respectively, d_{ij} is the distance between patch i and patch j, and α is the species' dispersal ability, where $1/\alpha$ is the average dispersal distance. Each element m_{ij} of the landscape matrix **M** gives the contribution that patch j makes to the colonization rate of patch i, multiplied by the expected lifetime of that patch (Hanski & Ovaskainen 2000). In other words, m_{ij} measures the fraction of time patch i would be occupied if patch j were the only source of immigrants (Ovaskainen & Hanski 2004). Thus, for a simple three-patch system, the landscape matrix **M** would look like this:

$$\mathbf{M} = \begin{bmatrix} 0 & A_2 A_1 e^{-\alpha d_{2,1}} & A_3 A_1 e^{-\alpha d_{3,1}} \\ A_1 A_2 e^{-\alpha d_{1,2}} & 0 & A_3 A_2 e^{-\alpha d_{3,2}} \\ A_1 A_3 e^{-\alpha d_{1,3}} & A_2 A_3 e^{-\alpha d_{2,3}} & 0 \end{bmatrix}. \qquad \text{Equation 2}$$

The **metapopulation capacity** of the landscape, λ_M, is given as the dominant eigenvalue of the matrix **M**. Metapopulation capacity thus captures the effect of landscape structure—the size and configuration of patches on the landscape—on the likelihood of metapopulation persistence. It can be thought of as the sum of all the contributions from individual patches, which might be useful in ranking patches as to their importance for metapopulation persistence. For long-term persistence, λ_M has to exceed a threshold set by the species' ability to occupy patches, δ, as

$$\lambda_M > \delta \qquad \text{Equation 3}$$

where $\delta = e/m$, the ratio of its extinction rate (e) to colonization rate (m). Thus, overall patch occupancy of a fragmented landscape at equilibrium is given by

$$\hat{p}_\lambda = 1 - \frac{\delta}{\lambda_M}. \qquad \text{Equation 4}$$

Again, the astute reader will notice that similarity between Equation 4 and the equilibrium solution for patch occupancy in Levins' classic metapopulation model (Equation 7.13). Indeed, if we present that result in terms of

Figure 1 Glanville fritillary (*Melitaea cinxia*).

Source: Photo by Marie-Lan Nguyen and published under the CC By 2.0 License.

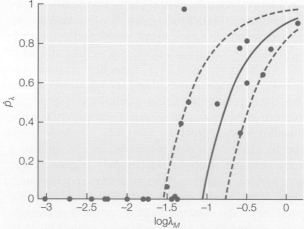

Figure 2 Metapopulation capacity. The metapopulation capacity (λ_M) of 25 patch networks for the Glanville fritillary (*Melitaea cinxia*) in the Åland archipelago of southwestern Finland. The solid line is based on the average of estimated δ values for patch networks in which $\hat{p}_\lambda > 0.3$. The envelope (dashed lines) around this are the minimum and maximum estimates after outliers are removed.

Source: After Hanski & Ovaskainen (2000).

the fraction of suitable habitat on the landscape (*h*) that is occupied, we get

$$\hat{p} = 1 - \frac{\delta}{h}.$$ Equation 5

Thus, the metapopulation capacity of the landscape, λ_M, plays exactly the same role that *h* does in the simple Levins' model in which only the amount of habitat is important because all patches are assumed to be identical in size and quality and are equally accessible. Here, we acknowledge that the metapopulation capacity of a landscape is not based solely on the amount of suitable habitat because not all patches are equal in size or are equally accessible. Patch occupancy at the landscape scale thus depends on the interaction between species' characteristics and the spatial configuration of patches on the landscape.

The threshold value of λ_M required for metapopulation persistence can be obtained by rewriting Equation 4 as

$$\delta = \lambda_M \left(1 - \hat{p}_\lambda\right).$$ Equation 6

This was done for about two-dozen patch networks within the range of the Glanville fritillary in the Åland Islands of Finland, in which δ was estimated for networks in which $\hat{p}_\lambda > 0.3$ (Hanski & Ovaskainen 2000; **Figure 2**). Below this threshold, the butterfly was absent (or nearly so) from the network. This approach can also be used to rank patches within the network as to their relative importance to the overall persistence of the metapopulation in the landscape, which has obvious implications for targeting populations for conservation or management.

---

incorporates patch-area effects on occupancy patterns, and to the extent that habitat area is related to patch quality (and it may not be), then this too evokes the idea of a source–sink metapopulation. Larger patches are an important source of immigrants that help to preserve the integrity and persistence of the entire network of populations. Besides habitat area, it is possible to incorporate additional information on habitat quality into the incidence-function model, as illustrated once again with the Glanville fritillary.

Cattle grazing reduces habitat quality for the Glanville fritillary by denuding vegetation, which in turn leads to warmer and drier conditions that increase mortality of the butterfly's host plants as well as the butterfly larvae that feed upon them (Moilanen & Hanski 1998). Thus, grazing reduces the *effective area* of the patch: large grazed patches may be demographically equivalent to smaller patches that are ungrazed. Habitat quality can have similar effects on rates of immigration and emigration. For example, information on the relative abundance of flowers used for nectaring was also found to be an important measure of habitat quality in the Glanville fritillary: butterflies were attracted to patches with abundant flowers (increased immigration) and stayed longer within these patches (decreased emigration; Moilanen & Hanski 1998). Habitat quality may therefore influence all of the BIDE factors that contribute to source–sink dynamics. Although habitat quality in the Glanville fritillary system ultimately had less of an effect than patch area and isolation on patterns of patch occupancy, we should not assume that this is generally true for all systems.

Bigger isn't always better. Just as we should not identify population sources from a survey of population size, we should not assume that large habitat areas are necessarily functioning as population sources without some independent measure of habitat quality.

For example, the Flint Hills region of eastern Kansas in the USA contains the largest remnant of the tallgrass prairie ecosystem, a biome that once covered some 160 million hectares throughout central North America. Because almost all of the native grasslands in North America have now been converted to agriculture and other land uses, the Flint Hills may be critically important for the continued persistence of grassland species at a continent-wide scale. Grassland birds in particular have exhibited steep declines in numbers, presumably as a result of the widespread destruction of their grassland habitat. With nearly 2 million hectares of largely native grassland, the Flint Hills has been considered a population stronghold for grassland birds. However, a demographic analysis of several grassland bird species revealed that none was regionally viable within the Flint Hills ($\lambda < 1$), probably because of intensive land-management practices throughout the region, which entail widespread cattle grazing and annual spring burning to improve forage conditions for cattle (With et al. 2008). Although fire and grazing are necessary for maintaining tallgrass prairie (which otherwise would convert to forest), the increased frequency and scale at which these disturbances are now applied to the entire region have effectively homogenized and degraded grassland habitat for nesting grassland birds. The largest remaining tallgrass prairie landscape thus does not appear to be functioning as a demographic source for grassland birds, and it remains to be seen whether there is sufficient habitat at even the continental scale to maintain viable populations, assuming these species are able to function as a regional metapopulation.

Metapopulation Viability and Extinction Thresholds

The plight of grassland birds raises an important question: How much habitat is sufficient to maintain viable populations of grassland birds, or butterflies, or of any other species? Is there some minimum amount of habitat, some threshold, below which populations will likely go extinct? American evolutionary biologist and population ecologist Russell Lande extended Levins' classic metapopulation model to assess the critical amount of habitat necessary for metapopulation persistence (Lande 1987). In nature, species do not saturate all available habitat and thus can go extinct even when some suitable habitat remains. Is there some minimum level of habitat necessary to ensure the continued persistence of the metapopulation? The implications for conservation and management are obvious, especially considering that habitat loss poses the greatest threat to species (Wilcove et al. 1998).

The minimum amount of habitat required for metapopulation persistence is the **extinction threshold**, which is defined as the proportion of habitat (h) at which all populations have become extinct on the landscape (i.e. $\hat{p} = 0$). The prediction of an extinction threshold originates from a demographically based metapopulation model, in which the patches now are assumed to represent territories or home ranges (all of the same size and quality) and dispersal and life-history processes affect colonization–extinction dynamics. Territories are assumed to be randomly or evenly distributed throughout the landscape, and are either inherited (with probability ε) or acquired during dispersal (in which m sites are searched at random) by juveniles that are produced at a rate R_0' (the mean lifetime production of female offspring by females, which sums across the age-structured birth and survival schedule for the species). If we are pessimistic about the juvenile's chances of finding a suitable territory, then the likelihood of failure is a combination of not inheriting its natal territory $(1 - \varepsilon)$ and not finding a suitable, unoccupied site in m attempts, where $\hat{p}h$ is the proportion of occupied suitable habitat and $1 - h$ is the proportion of unsuitable habitat:

$$(1-\varepsilon)(\hat{p}h+1-h)^{m}.\qquad \text{Equation 7.21}$$

If we are feeling optimistic, then the probability of successful dispersal is the probability of unsuccessful dispersal (Equation 7.21) subtracted from unity $[1-(1 - \varepsilon (\hat{p}h + 1 - h)^m]$. If we assume demographic equilibrium ($\lambda = 1$), then the characteristic equation for geometric population growth (assuming reproduction occurs at discrete yearly intervals) is given by the combined probabilities of successful dispersal and the reproductive rate, R_0') as

$$\left[1-(1-\varepsilon)(\hat{p}h+1-h)^{m}\right]R_0' =1.\qquad \text{Equation 7.22}$$

Solving for the equilibrium territory occupancy, $\hat{p}$, we find that

$$\hat{p} = \begin{cases} 1-\dfrac{1-k}{h} & \text{for } h > 1-k \\ 0 & \text{for } h \le 1-k \end{cases}\qquad \text{Equation 7.23}$$

where k is the **demographic potential** of the population,

$$k =\left[\left(1-1/R_0'\right)/\left(1-\varepsilon\right)\right]^{1/m}.\qquad \text{Equation 7.24}$$

The demographic potential, k, is thus a composite parameter consisting of information on the species' reproductive and dispersal abilities. More practically, it gives the proportion of suitable habitat occupied ($\hat{p}$) when the landscape is entirely suitable ($h = 1.0$). For example, if $k = 0.6$ (**Figure 7.15**), then the species would be able to occupy 60% of a landscape that is entirely suitable ($\hat{p} = 0.6$ when $h = 1.0$). The species is therefore predicted to go extinct ($\hat{p} = 0$) in landscapes with ≤40% habitat ($1 - k = 1 - 0.6 = 0.4$). In other words, the species could withstand a loss of 60% habitat before crossing the extinction threshold (**Figure 7.15**). Species with low demographic potentials, owing to a combination of poor dispersal ability and low reproductive rates, will thus be more sensitive to the effects of habitat loss than species with higher demographic potentials (**Figure 7.15**). In practice, k is determined indirectly by assessing the observed patch occupancy (assuming demographic equilibrium, $\hat{p}$) in a landscape with a given amount of habitat (h) as

$$k = 1 - (1 - \hat{p})h.\qquad \text{Equation 7.25}$$

The application of this approach for evaluating the critical amount of old-growth forest required for

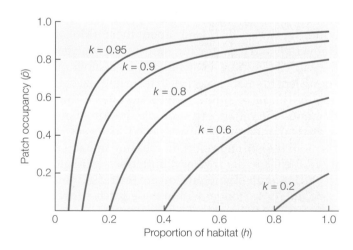

Figure 7.15 Extinction thresholds. Patch occupancy ($\hat{p}$) declines non-linearly with decreasing habitat (h) for species with different demographic potentials (k). The extinction threshold is the critical level of habitat (h_{crit}) at which $\hat{p} = 0$. Species with low demographic potentials ($k = 0.20$) have higher extinction thresholds ($h_{crit} = 0.80$), and thus are more sensitive to habitat loss, than species with higher demographic potentials (e.g., $k = 0.95$, $h_{crit} = 0.05$).

Source: After Lande (1987).

persistence of the northern spotted owl (*Strix occidentalis caurina*), a federally listed threatened species under the US Endangered Species Act, is highlighted in **Box 7.4**.

This extinction threshold model makes a number of important predictions. First, as mentioned previously, it suggests that we should not expect that all available habitat will be occupied, even when the landscape is

---

BOX 7.4 How Much Habitat is Needed to Conserve Northern Spotted Owls?

The northern spotted owl (*Strix occidentalis caurina*, **Figure 1**) is a habitat specialist of old-growth coniferous forests in the Pacific Northwest of the USA. Prior to 1800, old-growth forest (>200 years old) comprised 70–80% of the forests in this region (Franklin & Spies 1984). These ancient trees are a valuable commodity, however, and heavy logging subsequently destroyed much of the old-growth forest throughout the region, especially on private lands. By the mid-1980s, only 38% of national forest land in the Pacific Northwest contained old-growth forest. Under the guidelines of the forest management plan then proposed by the US Forest Service, the amount of old-growth forest deemed necessary for supporting a minimally viable population of northern spotted owls (550 pairs) on national forest lands in the Pacific Northwest was estimated at 7–16%. Would this be enough?

Figure 1 Northern spotted owl   (*Strix occidentalis caurina*).
Source: Photograph courtesy of John and Karen Hollingsworth.

Using a demographically based metapopulation model, Lande (1988) estimated the extinction threshold, the critical amount of habitat necessary to maintain population persistence, for northern spotted owls. Recall that the amount of old-growth habitat on federal lands was estimated to be 38% ($h = 0.38$). A three-year survey of spotted owls in the national forests of this region revealed that about 44% of suitable sites were occupied each year, on average ($\hat{p} = 0.44$). From this, the demographic potential of northern spotted owls was estimated to be $k = 0.79$ (Equation 7.25). Thus, the extinction threshold for northern spotted owls could be obtained as $h_{crit} = 1 - k = 1 - 0.79 = 0.21$. In other words, the northern spotted owl is predicted to go extinct when old-growth forest covers <21% of the landscape. As the proposed Forest Service plan called for 7–16% old-growth forest, these guidelines would likely have resulted in the extinction of the northern spotted owl from the region.

The northern spotted owl was eventually listed as a threatened species under the US Endangered Species Act in 1990, citing logging of old-growth forest as the primary threat to the owl's recovery. Protection of the owl, under both the Endangered Species Act and the National Forest Management Act, has led to significant changes in forest practices in the Pacific Northwest, which was met with much controversy and quite a few lawsuits. Most notably, a comprehensive Northwest Forest Management Plan was adopted in 1994 by the federal agencies that administer forest lands within the range of the northern spotted owl, with the aim of protecting owls and other species dependent on old-growth forest while ensuring a sustainable harvest of timber and non-timber resources on federal lands. Although logging has been substantially reduced throughout the region, an analysis of long-term demographic data (over 20 years) indicates that the northern spotted owl has continued to decline throughout much of its range, by an average –2.9%/year ($\lambda = 0.971$; Forsman et al. 2011). Although this may be a legacy of the widespread clearing of old-growth forests that occurred prior to 1990, there is now a new threat to the owl's recovery, posed by the invasion of a larger, more aggressive congener, the barred owl (*Strix varia*), into the range occupied by spotted owls. The reason behind the barred owl's westward expansion is unknown, but since the species is opportunistic and occurs in a variety of forest types, it is at least possible that landscape disturbance from logging and human development could be partly to blame.

entirely suitable ($h = 1.0$). This is an important expectation from a land-management or restoration standpoint, as we may require substantially more habitat to ensure (meta)population persistence or even to achieve a high probability of occurrence of the target species on the landscape. Second, species are expected to vary in their sensitivity to habitat loss. Some species are intrinsically rare, by virtue of their dispersal abilities and life-history traits, and these are often the species of greatest concern to conservationists. Third, the approach to extinction may be non-linear, occurring as a threshold response to habitat loss. Although initial losses of habitat may have small consequences for patch occupancy, even a small loss of habitat near the extinction threshold (h_{crit}) will have disproportionately large effects on patch occupancy. If populations do exhibit non-linear threshold responses to habitat loss, then it is obviously important to anticipate when such thresholds might occur, as otherwise our management recommendations might seriously overestimate how much habitat loss a species can tolerate before going extinct.

EFFECTS OF HABITAT FRAGMENTATION ON EXTINCTION THRESHOLDS Landscapes undergoing habitat loss may also experience fragmentation. Given that fragmentation could lead to decreased dispersal success (**Chapter 6**) or a decline in reproductive success (**Box 7.2**), it stands to reason that habitat fragmentation should have an additional, negative effect on the extinction threshold. Thus, we might predict that the extinction threshold for a given species will be higher (meaning extinction occurs sooner, at lower levels of habitat loss) in landscapes that are being fragmented than in those that are not (i.e. fragmented h_{crit} > clumped h_{crit}).

To test our hypothesis, we can first examine how relaxing the assumption of global dispersal affects the extinction threshold. Recall that the extinction threshold model (Lande 1987) assumes that habitat is randomly distributed (or is being destroyed at random) and that dispersal is global, meaning that individuals can access any potential site on the landscape (up to m attempts). Although some highly mobile species might be capable of global search, most exhibit a more localized search within a dispersal neighborhood. If dispersal is limited to a local neighborhood, then individuals will be more sensitive to the spatial arrangement of habitat than if dispersal is global, where only the total amount of habitat on the landscape (h) affects dispersal success (**Figure 6.27**). Colonization rates will thus be heavily influenced by the proportion of sites occupied within the local neighborhood, which in turn is affected by how much habitat has been destroyed ($h_d = 1 - h$). Using a **spatially explicit simulation model** of local dispersal on random landscapes, Bascompte and Solé (1996) demonstrated how patch occupancy ($\hat{p}$) was gen-

erally lower than that predicted by Lande's extinction threshold model, at least beyond a certain level of habitat destruction (h_d). The more habitat destroyed, the smaller and more isolated habitat patches become, and the lower the likelihood that patches will be occupied even when suitable habitat is available. Thus, for a given amount of habitat destroyed (h_d), fewer patches should be occupied than under an assumption of global dispersal. Extinction thresholds therefore occur sooner—at lower levels of habitat destruction (h_d)—and more abruptly if dispersal is limited to a local neighborhood.

To explore the effect of fragmentation—the pattern of habitat destruction—on extinction thresholds, With and King (1999a) developed a spatially explicit simulation model of local dispersal in fractal landscapes (**Figure 7.16A**). Dispersal success was much higher in landscapes with clumped habitat distributions (H = 1.0) than in more fragmented distributions (H = 0.0), especially when dispersers were constrained to search within a small neighborhood (i.e. small values of m; **Figure 6.27**). As a consequence, extinction thresholds were shifted to lower habitat levels (meaning thresholds occurred later, at a greater loss of habitat) in landscapes with clumped habitat distributions. In the example given in **Figure 7.16B**, $h_{crit} = 0.5$ in highly fragmented landscapes (H = 0.0), $h_{crit} = 0.33$ for moderately fragmented landscapes (H = 0.5), and $h_{crit} = 0.13$ in clumped landscapes (H = 1.0). Thus, the species in this scenario would require nearly four times more habitat to ensure metapopulation persistence in landscapes undergoing both habitat loss and fragmentation (H = 0.0) than in those undergoing primarily habitat loss (H = 1.0), a habitat difference of 37% ($0.5 - 0.13 = 0.37$). Also note that extinction thresholds occur *sooner* in fragmented landscapes than predicted by Lande's extinction threshold model (assuming global dispersal). In the example given in **Figure 7.16B**, the extinction threshold predicted by Lande's model is $h_{crit} = 0.21$ ($h_{crit} = 1 - k = 1 - 0.79 = 0.21$). Coincidentally, this was also the estimated extinction threshold for the northern spotted owl (**Box 7.4**). Thus, we might conclude that extinction thresholds are likely to be higher (occurring at lower levels of habitat destruction) in landscapes where habitat is also being fragmented. If true, then considerably more old-growth forest would be needed to maintain owl populations in the Pacific Northwest.[7]

Not all demographic potentials are created equal, however. Because of the added complexity of incorporat-

[7] Northern spotted owls appear to be more affected by the amount of remaining old-growth forest, however, than fragmentation per se. Spotted owls have large home-range requirements and thus site selection is influenced primarily by the amount of old-growth forest in the surrounding landscape (200 ha), as well as by the size of old-growth patches (Meyer et al. 1998). As discussed previously, patch size is really more of a habitat-area effect than a fragmentation effect.

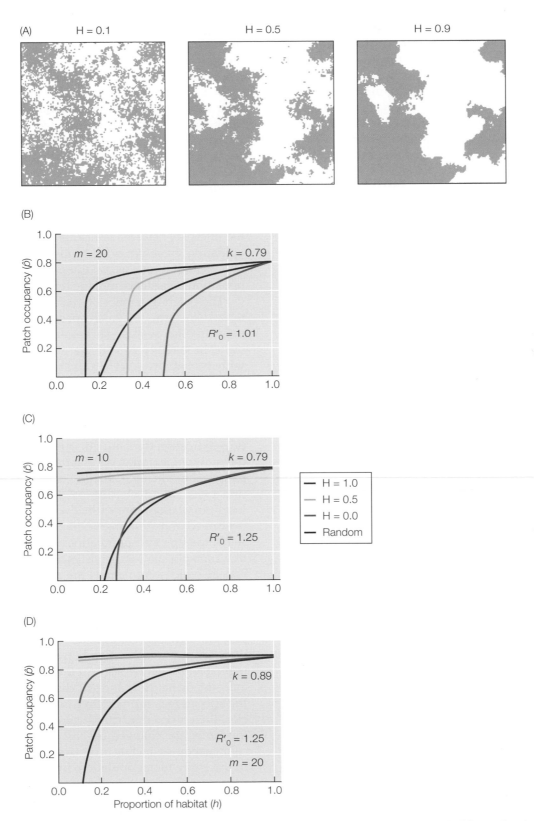

Figure 7.16 Effects of habitat fragmentation on extinction thresholds.   (A) Fractal landscapes with different levels of fragmentation (B) Extinction thresholds occur sooner in highly and moderately fragmented landscapes than in random landscapes, but occur later in highly clumped landscapes ($R'_0 = 1.01$, $m = 20$). (C) Extinction thresholds for a species with the same demographic potential (k) as in B, but with a higher reproductive rate ($R'_0 = 1.25$) and lower dispersal ($m = 10$). (D) Extinction thresholds for a species with the same dispersal ability (m) as in B, but with a higher reproductive rate ($R'_0 = 1.25$) and thus demographic potential (k).

Source: After With & King (1999a).

ing spatial information explicitly into the extinction threshold model, it was necessary to look at the individual contributions of dispersal (m) and reproductive rate (R_0') to the demographic potential (k) in order to evaluate fragmentation effects on extinction thresholds. Different combinations of m and R_0' can give rise to the same demographic potential (k), but have different consequences for a species' vulnerability to habitat loss and fragmentation (which is why the extinction threshold is no longer simply $h_{crit} = 1 - k$). For example, we can compare two species (**Figure 7.16B and C**) with the same demographic potential ($k = 0.79$), but with slightly different reproductive rates ($R_0' = 1.01$ or 1.25) and correspondingly different dispersal abilities ($m = 20$ or 10). Despite having the same demographic potential, the species with the higher reproductive rate ($R_0' = 1.25$) is less affected by fragmentation, even though it has half the dispersal range ($m = 10$) of the other species ($m = 20$). If dispersal range is kept constant ($m = 20$), but reproductive rate is increased (**Figure 7.16B and D**), thresholds no longer occur and the species is able to maintain near-maximum patch occupancy ($\hat{p} = k$) across a wide range of landscape scenarios (in terms of h and H).

Thus, small changes in reproductive rate can be as important—if not more so—than increasing dispersal on extinction thresholds. Put another way, species with high reproductive rates should be able to persist in landscapes with less habitat. This was demonstrated empirically for forest-breeding songbirds in North America, in which a negative relationship was found between the annual reproductive output (clutch size × number of broods) and the minimum habitat requirements (the amount of forest cover at which the species had a 50% probability of presence over a 10-year period) of 41 species (Vance et al. 2003). On average, forest birds with low reproductive output required twice as much forest cover as bird species with double the reproductive output. From a management standpoint, increases in reproductive output could also be achieved by increasing habitat quality, such as through improved habitat management and restoration, the direct supplementation of food or breeding resources (e.g. addition of nest boxes), or even predator-control programs (in cases where habitat loss and fragmentation has artificially increased predation pressure by generalist or non-native predators, such as feral cats). Although much of the focus has traditionally been on mitigating the adverse effects of habitat loss and fragmentation on dispersal, these results suggest we shouldn't lose sight of the importance of habitat quality for mitigating the effects of habitat loss and fragmentation on reproductive success.

EXTINCTION THRESHOLDS IN DYNAMIC LANDSCAPES Although the previous extinction threshold models have been couched in terms of evaluating the effects of habitat destruction and fragmentation on metapopulation persistence, the reality is that these models are only making comparisons across a set of landscapes with different amounts or configurations of habitat. Such models are implicitly making a **space-for-time substitution**, in which the landscapes (and populations within them) are all assumed to lie on the same trajectory of change (**Figure 2.15**). Habitat loss and fragmentation are dynamic processes, however, and thus the rate at which habitat is destroyed may be critically important for evaluating extinction thresholds. Further, in some systems, habitat or resource patches may be extremely dynamic or ephemeral. An understanding of the interplay between landscape dynamics and metapopulation dynamics would thus contribute to a better understanding of the conditions necessary for metapopulation persistence in these types of systems.

The patch dynamics of a landscape can be characterized in terms of patch-specific 'birth' and 'death' rates. In other words, patches arise or are created with some probability h_b and are exhausted or destroyed with probability h_d. Thus, patches have an expected 'lifespan' $L = 1/h_d$, which gives the rate of patch turnover (e.g. long-lived patches result in slow rates of turnover and thus, landscape change). The metapopulation in turn is characterized by the sorts of colonization–extinction dynamics discussed previously, except that now colonization is influenced by the rate at which patches become available (h_b), and population extinction rates are affected by patch longevity, L (i.e. the local population goes extinct when the patch is either destroyed or its resources exhausted). The dynamics of the landscape and population are thus coupled, such that metapopulation persistence is determined by the rate at which populations are able to respond to the rate of landscape change. Using a spatially explicit metapopulation model, Keymer and his colleagues (2000) demonstrated that metapopulation persistence—and thus extinction thresholds—in dynamic landscapes depend on the interaction between (1) the amount of habitat in the landscape, (2) the rate of change in the amount of habitat, and (3) the life-history traits of the species (e.g. reproductive output). For a given species, there existed a critical value for L, beyond which the landscape changed too fast relative to the scale of colonization–extinction processes, resulting in extinction of the entire metapopulation. In highly dynamic landscapes, extinction thresholds were thus found to be affected more by the rate of landscape change than by the amount of habitat destroyed. Metapopulation persistence is more difficult when habitat loss is dynamic, especially if it is occurring rapidly, and thus more suitable habitat may ultimately be required to avert extinction in such landscapes (i.e. species' extinction thresholds are expected to occur sooner, at lower levels of habitat loss).

That the history of landscape change (such as the rate of habitat loss) might be important for assessing extinction risk is not generally appreciated. Admittedly, we do not always have good historical records that enable us to reconstruct past landscapes (**Chapter 4**), although more recent landscape transformations can—and must—be documented, especially where this is happening rapidly. Still, most population viability analyses use the current landscape either to assess the conservation status of the population (is the population stable or declining?), or as a point of departure for evaluating how different land-management scenarios might affect extinction risk for the target species. Again, this assumes that all landscapes arrived at their current state in the same way, or at least, at the same rate. As Schrott and his colleagues (2005) demonstrated, however, rapid landscape change can lead to a decoupling of population and landscape dynamics, such that populations may exhibit a **lagged response** to landscape change. Initially, the population does not decline in response to rapid landscape change because it takes time for the population and demographic consequences of habitat loss and fragmentation to be reflected in viability measures (e.g. $\hat{p}$ or λ). Thus, we may be underestimating extinction risk in the very scenarios—rapid landscape change—where it matters most.

EVIDENCE FOR EXTINCTION THRESHOLDS   Although theory predicts the occurrence of extinction thresholds, what evidence do we have that species actually exhibit non-linear declines in patch occupancy or occurrence as a result of habitat loss and fragmentation? Few studies have investigated extinction thresholds empirically, and most of those have focused on birds (Swift & Hannon 2010). For example, Villard and his colleagues (1999) documented a non-linear relationship between the probability of occurrence and the amount of forest cover for several migratory songbirds in southern Ontario, Canada. Generally, birds did not exhibit a sharp threshold at some critical level of habitat, below which the species was never present. Rather, the probability of occurrence declined gradually with decreasing forest cover. There was also considerable variability in species occurrence within landscapes with similar amounts of habitat; for example, some landscapes with only 10% forest cover still contained ovenbirds (*Seiurus aurocapilla*) or chestnut-sided warblers (*Setophaga pensylvanica*), even though this was past the putative threshold for these species. How then do we define thresholds? Should the threshold be the point at which the curve first begins to decline, or at the midpoint of the curve where the rate of decline is most rapid?

In practice, extinction thresholds have been defined in different ways by different investigators (**Figure 7.17**). What constitutes 'extinction' also varies among studies. Extinction may be assessed either in terms of the species' prevalence (presence, patch occupancy, or population size) or its probability of persistence in landscapes with different amounts of habitat cover (Swift & Hannon 2010). In the extreme, an extinction threshold could manifest as a sharp transition, in which the species is either present or not depending on the amount of habitat on the landscape (**Figure 7.17A**). Such an extreme threshold response seems unlikely beyond the theoretical thresholds discussed previously. Instead, thresholds are more likely to manifest as a gradual decline in patch occupancy or in the probability of persistence (**Figure 7.17B**), as illustrated in the case of the forest songbirds discussed above. This makes identification of an exact habitat threshold more problematic, and the threshold in this case is less an absolute point than a range, and should therefore be used only as a guideline for assessing what minimum habitat levels might

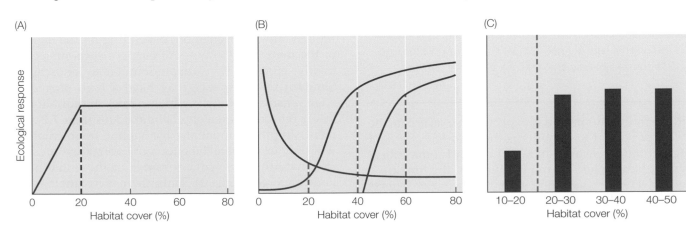

(A)   (B)   (C)

Figure 7.17 Definition of ecological thresholds.   Thresholds may manifest either as (A) an abrupt decline at some critical level of habitat (red dashed line), (B) a gradual decline in response to habitat cover, in which the threshold may be defined as the midpoint of the curve, where the rate of change shifts direction (blue dashed lines), or (C) a categorical response, in which the response is significantly lower below some habitat level (green dashed line).

Source: After Swift & Hannon (2010).

ensure the occurrence or persistence of a species on the landscape. If fewer landscape data are available, then a much coarser resolution of thresholds might still be possible, again by detecting declines in patch occupancy or occurrence among different habitat-level categories (**Figure 7.17C**). Typically, thresholds are defined visually rather than statistically, and thus do not adequately evaluate whether the response is indeed non-linear (Swift & Hannon 2010).

Theory also predicts that extinction thresholds should occur sooner—at lower levels of habitat loss—if habitat is also being fragmented (e.g. With & King 1999a). However, one study that sought to examine fragmentation effects on the extinction threshold found little evidence for thresholds, let alone in the direction expected for populations in fragmented landscapes. Among eight tree species in peninsular Spain, only the European beech (*Fagus sylvatica*) exhibited a threshold at lower levels of forest cover within more fragmented landscapes (**Figure 7.18**; Montoya et al. 2010). Although the occurrence of all species declined with decreasing forest cover, some did not have an 'extinction' threshold (where the probability of occurrence = 0) or had a higher probability of occurrence within more fragmented landscapes, counter to the extinction threshold hypothesis. For example, the sessile oak (*Quercus petraea*) exhibited an extinction threshold at about 30% forest cover in landscapes with a low degree of fragmentation (**Figure 7.18**). However, this species was present within fragmented landscapes that had as little as 10% forest cover. Thus, fragmentation actually seems to benefit some of these trees, especially those that are 'edge species' and can tolerate the higher light or temperatures along edges.

In most landscapes the total area of suitable habitat will be of greater importance than its spatial arrangements for species living in this particular habitat. This is especially true for landscapes with a high proportion (more than 30%) of the suitable habitat left... The negative effects of [fragmentation] on the original sets of species may not occur until the landscape consists of only 10–30% of the original habitat.

Andrén (1994)

Although fragmentation is expected to exacerbate the effects of habitat loss, perhaps these added effects of fragmentation are only apparent beyond some critical level of habitat loss? From a review of the literature, Andrén (1994) concluded that the negative effects of fragmentation[8] generally affected the occurrence of birds and mammals only below a threshold of

about 10–30% habitat. Dramatic changes in landscape structure are expected to occur within this domain, such as an increase in the degree of patch isolation, which increases exponentially below 20% habitat. As discussed in **Chapter 6**, thresholds in the gap structure of landscapes may occur ≤20% habitat (lacunarity thresholds), which can interfere with dispersal success (dispersal thresholds; With & King 1999b). Similarly, Fahrig (1997) demonstrated (based on the results of a spatially explicit simulation model, see below) that the effect of habitat amount far outweighed that of fragmentation on the probability of extinction, at least when habitat cover exceeded 20% of the landscape.

These results should not be construed to mean that fragmentation is never important for assessing a species' extinction risk, or that maintaining at least 20% or 30% habitat will be sufficient for averting extinction risk in all species. First, recall that species will likely require far more habitat than might be predicted on the basis of extinction thresholds, especially if habitat is being lost at a rapid rate and the population exhibits a lagged response to landscape change. Second, species characteristics, such as dispersal behavior and minimum area requirements, and how these are influenced by landscape structure, will ultimately determine a species' sensitivity to habitat loss and fragmentation. The effects of fragmentation on extinction thresholds are expected to be greatest for species with a limited dispersal range and strong habitat preferences (i.e. habitat specialists) in landscapes where there is a clear difference between habitat types (Luck 2005).

Intuitively, a species with good dispersal or gap-crossing abilities is expected to be relatively unaffected by habitat loss and fragmentation, unless the intervening matrix is exceedingly hostile and individuals suffer high mortality during dispersal. In that case, even a good disperser might be very sensitive to habitat loss and fragmentation, depending on the quality of the matrix or the types of barriers encountered in highly modified or human-dominated landscapes. Similarly, some species may have a high degree of edge sensitivity, suffering higher rates of mortality or lower reproduction along habitat edges (**Box 7.2**). By increasing the amount of edge habitat, fragmentation will only exacerbate the effects of habitat loss on species that are highly edge sensitive or those that have large area requirements. Thus, there is no single habitat threshold that guarantees persistence in all cases. By the same token, we should not expect that species will always exhibit a non-linear or threshold response to habitat loss and fragmentation (**Figures 7.15–7.18**); this is a hypothesis that must be assessed on a species-by-species basis.

---

[8] Fragmentation effects here were defined in terms of decreased patch size and increased patch isolation, but recall that these measures may reflect habitat loss rather than fragmentation per se.

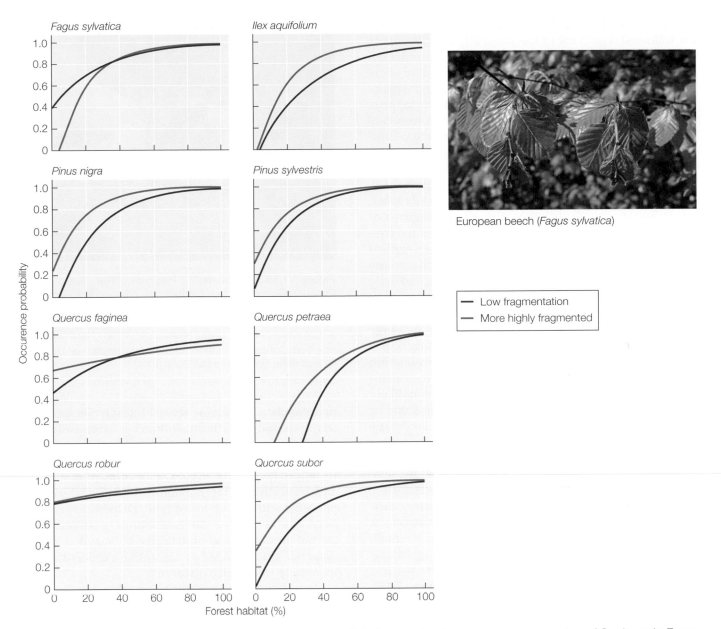

European beech (*Fagus sylvatica*)

— Low fragmentation
— More highly fragmented

Figure 7.18 Thresholds in the occurrence of tree species within fragmented landscapes.   In a region of Spain, only *Fagus sylvatica* (top left) shows a threshold in the expected direction, in which the threshold at which the species becomes absent occurs at higher habitat levels in more fragmented landscapes. Note that most other species have a higher probability of occurrence in fragmented landscapes.

Source: After Montoya et al. (2010).

Spatially Explicit Population Simulation Models

More complex representations of the landscape generally require numerical simulation of how population processes interact with different elements of landscape structure. In such **spatially explicit population simulation models**, the population model is coupled with a landscape map, which is typically grid-based (**Figure 7.13**). At a minimum, the grid cells are characterized by habitat type, but may also contain information pertaining to habitat quality, stand age, or time since disturbance. The

population simulator is a computer program in which different modules (subroutines) execute the various rules, mathematical functions, or algorithms that determine how, when, and where dispersal, reproduction, and mortality occur. Grid-based population simulation models were first used in landscape ecology to explore the effects of landscape structure, such as the effects of habitat area and fragmentation, on population persistence (Fahrig 1991, 1997).

Spatially explicit population simulation models are as diverse as the applications for which they are developed. The population model may be individual (agent) based, in which the behavior and fates of individuals are followed within the model (basically a BIDE model), or population based, in which demographic or dispersal rates and other species attributes (e.g. area- or edge-sensitivity) are applied to individual populations within the landscape (basically a simulated metapopulation model). The landscape itself may be a theoretical construct (a neutral landscape model), an actual landscape map obtained from remotely sensed imagery, or a simulated landscape depicting different management or land-cover change scenarios (**Figure 7.13**). Population dynamics can also be simulated on a dynamic landscape, in which different disturbance regimes or harvesting strategies are simulated and alter the configuration of the landscape over time. Spatially explicit population simulation models are thus central to many population viability analyses, in which a spatially explicit population model is developed or parameterized for a particular target species and implemented within a GIS for a given landscape or landscape-change scenario. These models thus give land managers and conservation biologists a powerful tool for evaluating which strategies are most likely to achieve their management aims, given that this sort of assay is unlikely to be accomplished through more traditional study or experimentation, especially given the broad landscape or regional scales over which these sorts of assessments are generally conducted.

One of the first management applications of a spatially explicit population simulation model was for the threatened Leadbeater's possum (*Gymnobelideus leadbeateri*) in southeastern Australia (Lindenmayer & Possingham 1996). The Leadbeater's possum is a small, nocturnal, communally nesting, tree-dwelling marsupial that occurs within mountain ash (*Eucalyptus regnans*) forests of the Central Highlands in Victoria (**Figure 7.19**). The Leadbeater's possum is an old-growth species that requires large trees that are several hundred years old for nesting, but uses younger trees (*Acacia* spp.) for foraging, and thus benefits from uneven-aged forests. Large trees are a valuable commodity, however, and more than 75% of the possum's remaining range (80 × 60 km) was slated for timber harvest. Lindenmayer and Possingham (1996) simulated several timber-management and reserve-design scenarios in a dynamic landscape context (i.e. the habitat suitability of forest stands varied over time owing to regrowth following timber harvest or periodic wildfires) to assess the relative effects on population viability of the Leadbeater's possum. The population model contained basic life-history information related to the minimum home-range size and maximum population density, age-structured reproduction and mortality rates, and density-dependent dispersal and mean distance moved by dispersing individuals. The patch structure of the existing landscape was characterized by the distribu-

Figure 7.19 Leadbeater's Possum. The Leadbeater's Possum (*Gymnobelideus leadbeateri*) is endemic to Victoria in southeastern Australia and has become a faunal emblem of the state. Once thought to be extinct, a small population was rediscovered in the Central Highlands in 1961. The species is listed as Critically Endangered by the International Union for Conservation of Nature (IUCN).

Source: Photo courtesy of Dan Harley, Zoos Victoria.

tion, number, and size of several forest types, including stands of old-growth forest. Patches were assigned a suitability index, which was related to the availability of nest trees (large trees with hollows) and food resources (primarily gum exudates from *Acacia* trees), and basically set the maximum breeding density of possums within each forest type. Dispersal was simulated as straight-line movement among suitable habitat patches, and the probability of survival during dispersal was modeled as a negative exponential with increasing inter-patch distance.

Based on these simulation scenarios, Lindenmayer and Possingham (1996) found that the amount of protected habitat was insufficient to ensure the persistence of Leadbeater's possum, and that the effects of intensive logging operations—coupled with wildfires—had created a patch structure that was incompatible with the long-term conservation of this species. After extensive sensitivity analyses of various model scenarios, Lindenmayer and Possingham (1996) concluded that the best conservation strategy for Leadbeater's possum would be to set aside several small (50–100 ha) reserves within every wood-production area and to ban post-fire salvage operations when these areas burned in a wildfire. The creation of several small reserves provides a risk-spreading strategy that may prevent a wildfire or other catastrophe from wiping out all of the protected populations (as opposed to protecting the species in a single large reserve), whereas salvage-logging tends to remove large, fire-damaged trees that might otherwise support suitable nesting cavities for possums for many decades before the tree eventually dies and collapses.

This example illustrates two very important contributions of spatial population simulation models and of population viability analyses more generally: the ability to rank various management strategies, and the ability to test the sensitivity of those rankings to changes in parameter values or assumptions. As exhaustive as this modeling approach for the Leadbeater's possum was, no model can possibly foresee every contingency. Despite their recommendation to create a number of small reserves to prevent a single catastrophic event from wiping out the entire metapopulation, Lindenmayer and Possingham (1996) could not possibly have foreseen the catastrophic wildfires of February 2009 (the 'Black Saturday' bushfires) that scorched some 450,000 ha throughout Victoria, including about half of the possum's remaining habitat, and reducing the surviving population to less than 1500. The Leadbeater's possum thus faces a precarious future and is at high risk of becoming extinct within the next 20–40 years, especially as logging continues to threaten what little old-growth forest now remains (only 1.5–3% of its historical range; Lindenmayer et al. 2013).

Although more realistic in its portrayal of population processes and landscape structure, spatially explicit population simulation models do have their limits. Consider that every parameter included in the model requires information. Spatially explicit models, in general, have higher data demands than population models that are not spatially structured. For many species, especially those that are rare and thus of conservation concern, the necessary life-history data to parameterize the model may not be readily available, or if they are, it may only be a species-specific rather than habitat-specific demographic rate. Information on dispersal is also necessary to program how individuals move through different elements of the landscape, but unfortunately, our understanding of dispersal is poor for most species (**Chapter 6**). Thus, 'dispersal rules are often coarse caricatures of biological reality due to difficulty of empirically determining dispersal distances, age of dispersers, and mortality during dispersal' (Beissinger & Westphal 1998, p. 830).

The quality and amount of available data are thus an issue in parameterizing spatially explicit population models, as this will have a direct bearing on the reliability of the results; if we put 'garbage in' we should anticipate that we will get 'garbage out.' With every parameter we add comes added uncertainty or error in those parameter estimates. Sensitivity analysis, in which we quantify how changing parameter values over some range might influence our model results, can give us an idea of whether or when this uncertainty might matter. Uncertainty propagates in both space

and time, however, if we are thinking to extend our model to larger areas or other landscapes from which the data were obtained, or if we are interested in forecasting population responses over time in response to future land-use or climate-change scenarios (**Chapter 2**). At the extreme, propagation error can be so severe as to compromise the use of such models as management tools (Ruckelshaus et al. 1997). Nor is the population model the only source of error in spatially explicit population models. The classification and spatial representation of the landscape is another potential source of error. For example, the various sorts of GIS operations that we routinely perform on spatial data, such as aggregation and interpolation, are all subject to varying degrees of error (**Chapter 4**). Such errors can also be expected to propagate in a highly complex and variable fashion in our model runs.

The best current applications of [spatially explicit population models] are in making comparative and qualitative statements about the likely population responses to a set of potential or real landscape alterations.....Even with the best knowledge, however, complex landscape structures are not deterministic entities, and we cannot expect to forecast new futures with a high degree of precision.   Dunning et al. (1995)

The uncertainty in spatial population estimates can—and should—be reported as part of the model output and can even be mapped. For example, Pidgeon and her colleagues (2003) mapped the reliability of habitat suitability estimates for the black-throated sparrow (*Amphispiza bilineata*) in a desert scrub landscape of the southwestern USA (**Figure 7.20A**). Nesting success of this species was obtained from different habitat types, and some habitats were clearly more productive for nesting sparrows than others. Variability in habitat quality could then be mapped onto the landscape, with high-quality habitats having higher nest success than poor-quality habitats (**Figure 7.20B**). Not all habitat types had sufficient nest data or sampling coverage to provide reliable estimates of habitat quality, however, and thus Pidgeon and her colleagues (2003) derived a spatial reliability score to indicate the degree of uncertainty across the landscape. Although a third of the population nested in mesquite (*Prosopis glandulosa*), this habitat ultimately contributed only 10% of successful nests, and thus appears to be a population sink, or even an ecological trap, for black-throated sparrows. Conversely, breeding densities were low in black grama (*Bouteloua eriopoda*) grasslands where nest-success rates were reliably high. Once again, density is not a reliable indicator of habitat quality, but even demographic measures, such as nest-success rates, may be less reliable in

Figure 7.20 Mapping spatial reliability in habitat quality. (A) Black-throated sparrows (*Amphispiza bilineata*) breed within several different habitats of a desert scrub landscape (New Mexico, USA). (B) Reliability of nest success estimates. Darker colors indicate where estimates of nest success can be considered reliable. Hotter colors (red) show poor-quality habitats for breeding black-throated sparrows.

Source: Map after Pidgeon et al. (2003). Photo by Caleb Putnam and printed under the CC BY-SA 2.0 License.

some locations on a landscape than in others. Thus, it is important that our designation of habitat quality be qualified in terms of our confidence in the data.

Which Population Model to Use and When?

Given the plethora of population modeling approaches, it can seem overwhelming at first to decide which approach might be appropriate in a given context. As always, the best model to use depends on one's research objective. Because basic demographic and metapopulation models contain so many unrealistic assumptions and ignore the complexity of the real world, one might wonder when—if ever—such models could possibly be useful. The utility of the tool lies in the hand of the craftsman, however. To some, a useful model is one that can be applied to a particular species or system. To others, a useful model is one that leads to new or unexpected insights into system behavior. For example, the idea that metapopulations can persist even in the face of local population extinctions, because of recolonization from neighboring populations, is not something that can be predicted from consideration of a simple demographic model that assumes the population is closed. The prediction of extinction thresholds in species' responses to habitat

loss came from a fairly simple, demographically based metapopulation model. Thus, the simplicity of these sorts of models is also their strength; they have the advantage of *generality* and potentially can apply to any species or system. That does not mean that every species *will* exhibit an extinction threshold, but it does create an awareness—a prediction—that species *might* exhibit a non-linear response to habitat loss. These types of models thus have heuristic value. Ecological and evolutionary theory is based on many such general models, and therefore have utility.

On the other hand, we may need to assess the extinction risk or population viability of some target species in a particular landscape. For this, we will likely require a more specific, spatially explicit modeling approach that is parameterized for the species (and landscape) in question. Again, our choice of model will depend upon the nature of the habitat or species distribution, the type and availability of population data at our disposal, and perhaps our own modeling proclivities. If the species occurs within habitat that is distributed as discrete patches on the landscape or within habitat that has been extensively fragmented, then a metapopulation modeling approach, such as the incidence function model, might prove useful. The incidence function model is best applied to very patchy or fragmented landscapes in which habitat occupies a small fraction of the landscape (e.g. <20%), such that only inter-patch distances (rather than patch geometry) are influencing colonization rates (Hanski 1999). If the landscape has been fragmented recently, however, the colonization–extinction dynamics of the species may not yet have attained equilibrium, perhaps owing to lagged population responses to rapid landscape change. In other cases, species simply may not be able to function as a metapopulation when their habitat becomes fragmented, especially if their former distribution was essentially continuous and they lack the dispersal capabilities to traverse long distances to locate habitat, or the intervening matrix is too hostile for those who try, resulting in high dispersal mortality. Spatial subdivision may thus be a necessary, but not sufficient, condition for metapopulation persistence, and we should not assume that all fragmented populations are capable of being managed—or modeled—as a metapopulation. If the landscape is more complex, then a spatially explicit simulation model is generally the preferred approach. The increased availability of commercially available population modeling software, often coupled with a GIS, increases the appeal and ease of use of spatially explicit simulation models. This also increases the ease of misuse if the model functions and limitations are not fully understood by the end-user, however.

Because spatially explicit population models are more 'realistic' in their portrayal of population processes and landscape structure, we might naturally assume that they are better, especially if we are planning to use the model as a management or decision-support tool. We should guard against equating greater model complexity, or apparent realism, with superiority, however. Recall that a model is only an abstraction of reality; a model is meant to simplify the complexity inherent in the real world and thus no model will include everything, nor should it (else why would you need a model?). It is a general modeling axiom to start simple and add detail only as necessary to explain or predict the phenomenon of interest. Further, every parameter or function in the model requires information, information we may not have in terms of good-quality data over a long-enough time period to derive robust estimates to parameterize the model. Thus, the most complicated models are not necessarily the best ones, and some model-selection procedures (such as those founded on Akaike's Information Criterion, AIC) actually rank models based on the principle of parsimony: all else being equal, in terms of model goodness-of-fit, the simplest model is the best one. Thus, although spatially explicit population models have the appearance of greater realism, this comes at the cost of generality, and potentially, reliability.

When using a population model as a decision-support tool, it is obviously important that our model predictions are reliable and accurate, and can thus help inform what management action will be most effective (McCarthy et al. 2003). Given that many such model analyses are inevitably hampered by a lack of data, however, we might wonder whether population models are really all that useful in practice for estimating species' extinction risk. Indeed, when the precision and reliability of estimates from a generic matrix population model were investigated, it was found that the amount of data needed for reliable predictions was 5–10 times the desired timeframe of the predictions (Fieberg & Ellner 2000). So, if we want to make reliable predictions of, say, whether a species is likely to go extinct 100 years from now, we would need 500–1000 years of data on the relevant population parameters (e.g. population size, birth rates, death rates) from which to derive reliable estimates! Put another way, if we have 40 years of data, then our population projections are expected to be reliable only to about 4–8 years into the future. Although absolute predictions, such as the resulting population size or long-term extinction probability, may not be very reliable, we might still be able to use population models to make relative comparisons among different proposed management actions or projected scenarios of landscape change (Beissinger & Westphal 1998). For example, McCarthy and colleagues (2003) found that population models were useful for making relative comparisons among different management actions over a 100-year horizon, based on only 10 years of data. Although 10 years still qualifies as a long-term study in ecology, it is certainly a more attainable goal than five or ten centuries of data.

As with any model, [population viability analysis] should and always will be an imperfect description of reality, and it is misguided to use the model testing to indicate whether PVA is 'right' or 'wrong.' Clearly, some models will make poor predictions and others will make better predictions. The role of model testing should be to identify the weakest aspects of the model so that its predictions can be improved.... **The process of parameter estimation, model construction, prediction, and assessment should be viewed as a cycle rather than a one-way street.** (emphasis added)   McCarthy et al. (2003)

In conclusion, population models are indispensable tools in the context of population viability analysis for species of conservation concern, but we would do well to heed the advice of Beissinger and Westphal (1998) regarding best practices, namely that we should (1) evaluate relative rather than absolute rates of extinction, (2) use short time periods for making population projections, (3) start with simple models and choose an approach that the data can support, and (4) use models cautiously to diagnose causes of decline and potential for recovery.

Future Directions

Although investigation into the effects of habitat loss and fragmentation has been an active area of research for several decades now, there is still much to learn regarding the relative effects and scale(s) at which fragmentation—as opposed to habitat area—differentially affect dispersal and the demographic processes that give rise to the distribution and dynamics of populations. Future research should thus strive to tease apart these relative effects to assess which has the greater effect on dispersal and demography, and thus which should be targeted in any mitigative measures we might propose or pursue. Further, we need a better understanding of the mechanisms by which habitat loss and fragmentation influence population processes, such as through the creation of edge effects, and how these effects might scale up from the patch scale to influence species distributions and population dynamics at a broader landscape scale. Fragmentation research needs to move beyond an analysis of patch area and isolation effects, however, in recognition of the fact that both habitat loss and fragmentation are primarily landscape-wide phenomena. Thus, researchers should adopt more of a landscape-wide than purely patch-based perspective in the future, one that emphasizes the relationship between whole-landscape measures of heterogeneity and structural complexity with population processes, and that appreciates the importance of landscape context for understanding and predicting population responses to landscape change.

Of all the negative population consequences of habitat loss and fragmentation, perhaps the most disquieting is the prediction of extinction thresholds, which may lag in response to landscape change. Identifying when and where such thresholds might occur is a challenge in practice, however, and thus we need to develop better ways of defining thresholds, both statistically and empirically. In general, empirical support for extinction thresholds currently lags far behind theoretical predictions.

Models of species distribution and population dynamics are essential tools for predicting species' responses to landscape change. Metapopulation theory has long been the paradigm for modeling and managing spatially distributed populations, but increasingly, spatially explicit population models are becoming the approach of choice for assessing population viability and extinction risk within heterogeneous landscapes. If we are to be able to use population models as decision-support tools, however, we should seek to demonstrate a link between model predictions and any management options we might pursue, perhaps by conducting field tests of the underlying model assumptions and through validation of any secondary model predictions where possible (Beissinger & Westphal 1998). Uncertainty is a fact of life and so we must learn to confront uncertainty in our models, by developing ways of characterizing it, reducing it, and assessing its impact on our decision-making process. The model should not be viewed as an end in itself, but rather, we should consider the full cycle of development, implementation, and evaluation (including experimental testing) as part of the model exercise. This is especially true if predictions regarding species' distributions or population viability and extinction risk are being extended to other landscapes or to other scales in space or time (e.g. ecological forecasting; **Chapter 2**). In that case, we need methods for evaluating how model performance varies across different spatial and temporal scales, especially in the context of whatever management decisions might stem from the model's predictions.

Chapter Summary Points

1. Habitat loss and habitat fragmentation are not synonymous. Habitat loss results in a reduction of the total amount of habitat on the landscape. Although habitats become fragmented through a process of habitat loss, fragmentation more properly refers to the spatial pattern of habitat loss.

2. From a population standpoint, it matters whether it is habitat loss or fragmentation that is having the greater effect on population declines. We cannot hope to remedy population declines by mitigating fragmentation effects (e.g. by attempting to restore habitat connectivity) if it is the sheer loss of habitat that is to blame.

3. The sheer loss of habitat is often to blame. Habitat loss has been found to have large, consistently negative effects on populations and overall species diversity. The effect of fragmentation is generally weaker, and may only be significant at certain levels of habitat.

4. Habitat fragmentation increases the amount of edge habitat on the landscape. Edge effects may penetrate many tens to hundreds of meters into the habitat interior, resulting in more extreme environmental conditions (wind, light, temperature) and higher rates of competition, predation, parasitism, or disease transmission in proximity

to the habitat edge. Consequently, these sorts of edge effects may have a negative effect on some populations, especially interior species. Conversely, many other species have a positive response to habitat edges, either because patches contain different resources that are complementary or because resources are concentrated within the edge habitat. Negative edge effects are therefore expected only when the difference in habitat quality is very great and resources (that are otherwise supplementary) are concentrated in one patch.

5. Species distribution models describe and predict the occurrence of species at a broad landscape or regional scale. The most common approaches include resource-selection functions, ecological niche models, and climate envelope models. These models variously relate environmental and climatic factors to species-occurrence data to derive relationships that can then be used to predict the likely distribution of the species in other landscape contexts or how species' distributions may shift in response to climate change.

6. Population responses to habitat loss and fragmentation are commonly assessed using metapopulation theory and spatially explicit population models. These types of population models incorporate the effects of patch structure on dispersal and demography to varying degrees. Such spatially explicit models are often used in population viability analyses to evaluate the likelihood that populations will persist under different scenarios of landscape change.

7. Species may exhibit a threshold response to habitat loss and fragmentation, where the probability of a species' occurrence (e.g. patch occupancy) or persistence declines rapidly below some critical amount of habitat. Population models can help identify where and under what conditions extinction thresholds might occur.

8. Variation in habitat quality can produce source-sink population dynamics, in which populations in low-quality sink habitats (where death rates exceed birth rates) are subsidized by immigrants from high-quality source habitats (where birth rates exceed death rates). The ratio of sink to source habitat can have a destabilizing effect on overall species persistence at a landscape scale. From a management or conservation standpoint, we want to ensure that we can identify, manage, and protect habitats that have the potential to function as population sources.

9. Where used as a decision-support tool, the best population model is ultimately one that makes accurate and reliable predictions, rather than the most complex or seemingly 'realistic' one.

Discussion Questions

1. From your assigned readings or a quick survey of the recent literature, identify how investigators have characterized fragmentation effects on populations. Have the investigators differentiated between the effects of habitat loss from those of fragmentation? If not, to what extent might this affect the interpretation of their results or the conclusions/recommendations of the study? If the authors were careful to distinguish between the two, did habitat loss or habitat fragmentation have the larger effect? In each case, discuss why distinguishing between the relative effects of habitat loss and fragmentation might be important for understanding population responses to landscape change, and how this could affect the sorts of land-management decisions or conservation actions we might take.

2. Fragmentation is considered a landscape-scale phenomenon. From your readings or survey of the recent literature, determine whether fragmentation effects on populations were studied at the scale of the entire landscape versus at the patch scale. In other words, were landscape properties (total habitat area and degree of fragmentation) or patch-based properties (patch area and isolation) used to define fragmentation effects? If fragmentation effects were assessed at the patch scale, to what extent can we extrapolate these effects to understand population responses at a landscape scale? In general, what factors might complicate our ability to scale up from population responses at the patch scale to make inferences about population dynamics at the landscape scale?

3. Habitat fragmentation is often assumed to have a negative effect on populations. Can you think of some instances in which fragmentation might actually be beneficial or have a positive effect on populations or on some population process?

4. Using an example provided by your instructor or from your own research, identify and discuss how various environmental and landscape factors are likely to influence the distribution of the target species. Based on the sorts of relationships you derive, explain what pattern of distribution

is expected in the study landscape and how this might change under hypothesized scenarios of land use or climate change. Discuss what factors might limit the reliability or accuracy of your predictions.

5. The metapopulation paradigm underlies the management and conservation of many species. Should all species be managed as metapopulations (i.e. as spatially subdivided populations that are linked by dispersal), however? What types of species are most likely to benefit from a metapopulation approach? What are some potential consequences (positive and negative) of managing species as metapopulations?

6. Species may exhibit a threshold response to habitat loss. Different species have different sensitivities to habitat loss and fragmentation, however, which may result in a wide range of extinction threshold values for species within a given community. Discuss what land-management strategies you would advise so as to benefit the greatest number of species.

7. As a test of your understanding as to how landscape structure influences population processes, outline the structure of a spatially explicit population model for the species you study, or one suggested by your instructor. That is, consider how the distribution and quality of habitat is expected to influence dispersal, reproductive success, and survivorship of your species. It might help to begin by considering how the model will be used and the availability of data that could be used to parameterize the model. Be sure to identify any potential sources of uncertainty that might influence the accuracy or reliability of your model.

8

Landscape Effects on Population Spatial Spread

Range Shifts, Biological Invasions, and Landscape Epidemiology

In February 2006, hibernating bats covered with a mysterious white powder were discovered in a cave in the northeastern United States. The following winter, bat populations from four other caves in the area exhibited similar symptoms, and hundreds of dead and dying bats were reported. By the end of 2018, white-nose syndrome—named for the appearance of affected bats—had spread to bat colonies in 33 states and 7 Canadian provinces, including the state of Washington in the Pacific Northwest (**Figure 8.1**). Caused by a cold-adapted fungus (*Pseudogymnoascus destructans*; Lorch et al. 2011), the disease invades and destroys the skin, including the bat's membranous wings, and is able to spread readily within hibernating colonies through close contact with infected individuals. Within a decade, the disease had killed nearly six million bats and decimated bat populations throughout the region, including several species already on the brink of extinction (Foley et al. 2010). The demise of bat populations is likely to hit agricultural interests hard, given the valuable service bats provide in controlling many crop and forest pests, which could lead to agricultural losses of $3.7 billion/year (Boyles et al. 2011). Thus, bats are of great ecological and economic importance, and the massive mortality and rapid rate with which this disease has spread has contributed to one of the greatest wildlife disease crises in recorded history. But where did the disease come from and how does it spread so quickly among populations? Was its emergence and spread facilitated by humans, perhaps transported from Europe where the fungus occurs but apparently does not cause mass-mortality events? If we understand the rate and pattern of spread, perhaps we can slow or even prevent further spread through targeted control programs, such as closure of caves to recreational caving (**Figure 8.1**)? The adoption of a landscape epidemiological approach might thus be helpful in mapping out and predicting the pattern, rate, and direction of spread in this—and other—emerging infectious diseases.

Spatial Spread: The Good, the Bad, and the Ugly

From a population standpoint, spatial spread could be considered the epitome of success. Not content to merely persist, a population that is spreading is one that is growing and actively establishing new colonies, sometimes in far-away lands where it may exert great influence on the communities it invades. Range expansion and

Essentials of Landscape Ecology. Kimberly A. With, Oxford University Press (2019).
© Kimberly A. With 2019. DOI: 10.1093/oso/9780198838388.001.0001

(A)

(C)

(B)

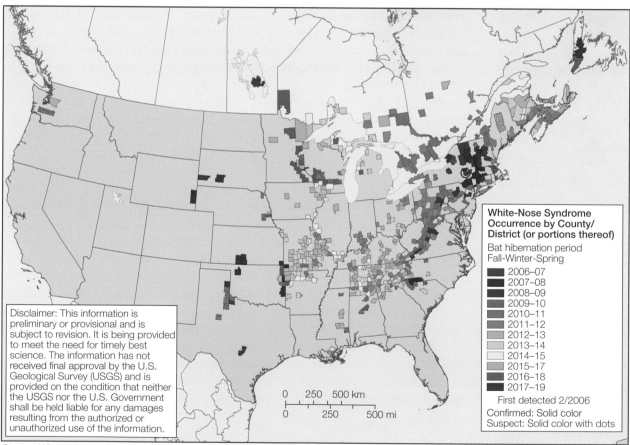

White-Nose Syndrome Occurrence by County/ District (or portions thereof)

Bat hibernation period Fall-Winter-Spring

2006–07
2007–08
2008–09
2009–10
2010–11
2011–12
2012–13
2013–14
2014–15
2015–17
2016–18
2017–19

First detected 2/2006

Confirmed: Solid color
Suspect: Solid color with dots

Disclaimer: This information is preliminary or provisional and is subject to revision. It is being provided to meet the need for timely best science. The information has not received final approval by the U.S. Geological Survey (USGS) and is provided on the condition that neither the USGS nor the U.S. Government shall be held liable for any damages resulting from the authorized or unauthorized use of the information.

0 250 500 km
0 250 500 mi

Citation: White-nose syndrome occurrence map - by year (2018). Data Last Updated: 07/02/2018.
Available at https://www.whitenosesyndrome.org/resources/map.

Figure 8.1 Spread of white-nose syndrome among bat colonies across North America. (A) Little brown bat (*Myotis lucifugus*) infected with white-nose syndrome, which is caused by a fungus (*Pseudogymnoascus destructans*); (B) Map showing the occurrence and spread of white-nose syndrome, following its initial discovery in central New York (epicenter circled in red) in February 2006; (C) A sign from a state park in Indiana (USA) advising of cave closures to help slow the spread of white-nose syndrome.

Source: (A) Photo by Marvin Moriarty/USFWS. (C) Photo © Tom Uhlman / Alamy Stock Photo.

biological invasions, including disease spread, are inherently spatial processes, involving the successful introduction or colonization, establishment, and dispersal of organisms (or their propagules) into new areas. As such, spatial spread involves the interaction of both dispersal and demography with landscape structure, two ecological processes that we have studied in previous chapters, and which we will now bring together here in order to consider landscape effects on the patterns and rates of population spatial spread.

A landscape ecological perspective benefits the study of spatial spread in a number of ways. The spread of species and disease has been greatly enhanced through human transport and land-use intensification. The introduction and spread of non-native species pose a major threat to native biota worldwide, in some cases transforming entire landscapes and disrupting ecosystem function in terrestrial as well as aquatic and marine systems. In addition, land-use change, especially that contributing to the loss and fragmentation of native habitats, is increasing both the incidence and spread of non-native species and diseases that threaten our food supplies, our industries, our economies, and our very health and well-being. Human activities thus create an array of disturbances that may directly or indirectly facilitate the spread of disease and non-native species, both within and among landscapes. A better understanding of how landscape structure influences spatial patterns of spread might also offer new opportunities for managing landscapes to prevent or reduce the spread of an invasive species, pest, or pathogen.

In this chapter, we first explore natural patterns of spread involving species' range shifts and evaluate to what extent species will be able to shift their distributions in response to future land-use and climate-change scenarios. Although the expansion of native species into new areas could be considered invasive in some contexts, we typically think of invasive spread as the introduction of a non-native species that has the potential to cause a great deal of ecological or economic harm. We therefore examine what effect landscape structure might have on the various stages of the invasion process. We then briefly review spatial models of invasive spread, which are used to predict whether, when, and how fast a non-native species is likely to spread. Finally, we end this chapter with a discussion of disease spread in a landscape context, in which we explore how the field of landscape epidemiology is contributing to a better understanding of the interaction between pathogens, their vectors, and their hosts with environmental heterogeneity, thereby influencing the incidence and persistence of disease on the landscape.

Landscape Effects on Species' Range Shifts

Range shifts are a population-level phenomenon, representing the differential survival, reproduction, and dispersal of individuals over many generations, which collectively results in a change in the distribution of the species. If positive population growth is coupled with successful colonization, then the species' range may expand. **Range expansion** is perhaps best illustrated by the spread of an invasive species or disease into new regions, which we will discuss a bit later in this chapter. Conversely, the extinction of local populations along the periphery of the range can lead to a **range contraction**, which, unless accompanied by range expansion elsewhere, will lead to an absolute reduction in the total range of the species. An example of this sort of range expansion-contraction dynamic is given by the Edith's checkerspot butterfly (*Euphydryas editha*) in western North America in response to climate change (**Box 8.1**).

The distribution of most species is ultimately limited by climatic factors, such as temperature or precipitation. Increasingly, we are seeing dramatic changes in the occurrence and distribution of species across landscapes and regions, which in many cases represent

---

BOX 8.1   Are Species Shifting their Ranges in Response to Climate Change?

The world is heating up. The average global temperature has increased by about 0.8°C (1.4°F) since 1880, with much of that increase occurring in just the last 40 years. As temperatures rise, many species in the temperate zone will need to shift their distributions to higher latitudes or to higher elevations in order to track the resultant environmental changes. Such range shifts do not happen overnight, however, but are instead a multi-generational phenomenon that occurs because of differential colonization–extinction dynamics at the range margins. That is, the range 'shifts' because of the successful colonization and increased persistence of populations at the expanding range front, which is then offset by higher extinction rates along the contracting range margin.

One of the best examples of how climate-driven range shifts occur via differential extinction rates along latitudinal

(Continued)

BOX 8.1 Continued

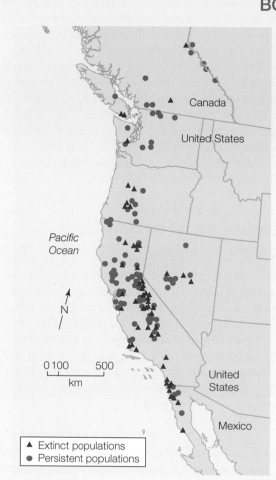

Figure 1. Distribution of Edith's checkerspot populations.
Source: From Parmesan (1996).

Legend:
▲ Extinct populations
● Persistent populations

Map labels: Canada, United States, Pacific Ocean, N, 0 100 500 km, United States, Mexico

or altitudinal clines is given by the work of Camille Parmesan on the Edith's checkerspot butterfly (*Euphydryas editha*). The Edith's checkerspot is broadly distributed across western North America, from Mexico to Canada, where it occurs within a wide range of habitats that include coastal chaparral, grassy ridgetops, and alpine tundra (**Figure 1**). The bay checkerspot (*Euphydryas editha bayensis*), which was mentioned in the previous chapter (**Figure 7.14**), is one of its endangered subspecies that occurs in the serpentine grasslands of the San Francisco Bay Area.

Over the past century, Edith's checkerspot has shifted its range by about 100 km northward, as well as by about 100 m upwards in elevation (Parmesan 1996; Parmesan et al. 2005). A disproportionate number of populations have gone extinct along the southern edge of the range (>70% extinct), as well as at lower elevations (>40% of populations extinct at elevations <2400 m). In comparison, <20% of populations went extinct at the northern periphery of the range, and <15% of populations went extinct at higher elevations (>2400 m). A warming trend has occurred during this same period, resulting in isothermal shifts of some 105 km northward and upward, as well as a lighter snowpack at lower elevations (the snowpack is now 14% lighter and melts a week earlier than a century ago). The Edith's checkerspot is known to be sensitive to extreme climatic events, such as drought, which disrupt the synchronicity between the butterfly and its host plant. The observed shift in the distribution of the Edith's checkerspot is thus consistent with the expectations of a climate-driven range shift, produced by differential colonization–extinction dynamics at the range margins.

range shifts in response to climate change, as in the case of Edith's checkerspot (Parmesan 2006). Many species formerly restricted to lower latitudes have been steadily moving toward higher latitudes, such that some 'tropical' species (and diseases) are now being seen with increasing regularity in more temperate regions of the world. Conversely, some high-latitude as well as high-altitude species are at increased risk of extinction in a warmer world and are currently undergoing range contractions (Thomas et al. 2006).

In the previous chapter, we examined how species distribution models can make use of climate data and species' habitat associations to predict where species ought to occur within a landscape or region (**Chapter 7**). We'll now consider the use of species distribution models to predict how species might shift their ranges in response to climate change, especially within the context of an increasingly human-modified and fragmented landscape.

Modeling Species Distributions in Response to Climate Change

Because climate is an important factor that limits the distribution of many species, bioclimatic models that predict how species' distributions might change in response to climate-change scenarios are becoming increasingly popular for understanding the consequences of global climate change on species diversity. For example, Peterson and his colleagues (2002) demonstrated the utility of species distribution models in predicting the effects of climate change on nearly 1900 species of birds, mammals, and butterflies in Mexico (**Figure 8.2**). Environmental niche models (developed using GARP) were projected onto future landscapes modeled under different climate-change scenarios obtained from a general circulation model, and with different dispersal assumptions (the species could access all suitable habitat anywhere within its range, the species could move only through contiguous habitat,

Colonizations

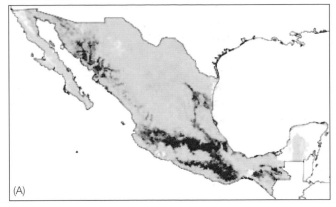

(A)

Extinctions

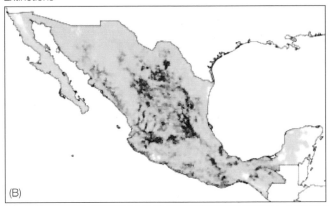

(B)

Species turnover

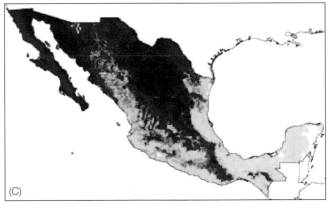

(C)

Figure 8.2 Projected changes in the distribution of birds, mammals, and butterflies in response to climate change throughout Mexico. Areas where most species will likely be able to colonize new habitats (A), or conversely, where most species are likely to lose habitat (B) are highlighted in red (darker shades correspond to larger numbers of species). (C) The cumulative change in the predicted number of species as a function of these sorts of climate-driven range shifts can then be indexed by the degree of species turnover. Species turnover is projected to be greatest within Baja California, where almost half of the current species are expected to be replaced by others.

Source: After Peterson et al. (2002).

or the species could not disperse at all), to assess the predicted change in habitable area relative to the species' current distribution. Although few species are predicted to lose all habitat, some are expected to suffer severe range contractions and more fragmented distributions, while still others may gain habitat and expand into new areas. These predicted changes in species' distributions were unevenly distributed across Mexico, however, with colonization of new areas most likely to occur along the major mountain ranges and local extinctions primarily concentrated within the Chihuahuan desert region and northwestern coastal plain area (**Figure 8.2**). In sum, the regions with the greatest expected change in species (i.e. areas with high **species turnover**) are in the Chihuahuan desert in northern Mexico, interior valleys extending south to Oaxaca, and along the Baja California peninsula, where predicted turnover rates are as high as 45% (**Figure 8.2**).

This assumes, of course, that species are sorting themselves independently of one another. In using species distribution models to predict the effects of climate change, we are ignoring the potential for biotic interactions to either facilitate or impede the ability of species to colonize new areas or persist in old ones. For example, the ability of species to colonize new areas may depend on the presence of a particular host or prey species, or on the availability of 'enemy-free space' devoid of predators or pathogens. Thus, the resulting distribution of a species may be constrained more by the sorts of novel species interactions it must contend with as communities are restructured by climate change, rather than by climate change per se (Davis et al. 1998). The counter-argument, however, is that biotic interactions are unlikely to be as important as climate in limiting species' ranges at the broad geographical scales at which these species distribution models are being applied (Pearson & Dawson 2003). While it may not be possible to include all of the biotic and abiotic factors that ultimately define the realized niche of a species, species distribution models can at least provide a first approximation of how species' ranges might shift in response to climate change.

Will species actually be able to track such rapid changes in climate, however? Species in temperate and mountainous regions of the world have certainly had to overcome such challenges in the past, as their ranges contracted or expanded during the last glacial period. For example, the study of tree migration rates (the rate at which tree ranges have shifted), based on an analysis of the fossil pollen record preserved in sediments, suggests that many species were apparently able to track climatic changes in the past. Historical tree migration rates have been estimated at 100–1000 m/yr across North America and Europe during the last postglacial warming period, which began some 10,000–20,000 years ago (Pearson 2006). Of course, past tree migrations occurred over many thousands of years. It is estimated that trees would have to migrate at far greater rates (>1000 m/yr) over the next century to keep pace with the current rate of climatic warming (Malcolm et al. 2002). For example, Iverson and Prasad (1998) used species distribution models to map and predict the potential range shifts of 80 tree species in the eastern USA in response to climate-change scenarios involving a doubling of atmospheric CO_2 levels and warmer temperatures. Based on their analysis, changing environmental conditions would require nearly half of these species to shift either the northern or southern limits of their range by at least 100 km northward. Since different environmental factors often constrain the northern versus southern limits of a species' range, shifts in the northern limits of a species' range may not be offset by shifts in its southern limits (i.e. ranges will ultimately contract).

Although the fossil pollen record suggests that such high migration rates must have occurred in the past, it is difficult to explain how tree populations could possibly have achieved these rapid rates of migration given our current understanding of tree demography and seed dispersal, a problem referred to as **Reid's paradox** (Clark et al. 1998).[1] Is it possible that fossil pollen records might be overestimating historical rates of tree migration, therefore? Based on molecular evidence obtained from the DNA in chloroplasts, McLachlan and colleagues (2005) found that the postglacial expansion of some tree species in the eastern USA, such as the American beech (*Fagus grandifolia*) and red maple (*Acer rubrum*), may have occurred much slower (<100 m/yr) than suggested by pollen reconstructions. Instead, it appears that at least some of these tree species may have been able to persist as small isolated populations in glacial refugia very near the northern limits of (or at higher elevations within) their current distribution, from which they could then move out to recolonize northern portions of their range as temperatures warmed and the glaciers retreated (**Figure 8.3**). Such 'cryptic' northern populations would not be well preserved by the fossil pollen record, which is less reliable for identifying the occurrence of small populations (small populations produce less pollen that can be preserved in the record), as well as the routes of population spread. However, if postglacial colonization from isolated refugia occurred, then tracking changing environmental conditions

(A)

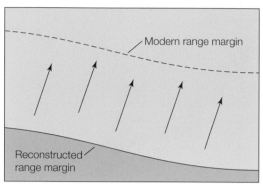

(B)

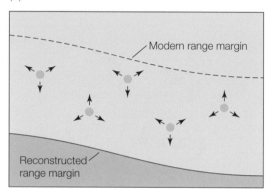

Figure 8.3 Alternative mechanisms for range expansion. Species may track changing environmental conditions, such as climate change, through broad-scale shifts in their range limits (A) or from a series of spatially distributed populations (e.g. glacial refugia) beyond the range limit (B).

Source: After Pearson (2006).

through broad-scale range shifts along a front might not be the only way in which species' range shifts occur (Pearson 2006). Thus, we should not assume that modern species will either be able to migrate as quickly as in the past (if pollen records are accurate) or, alternatively, that they may not need to migrate at all (if refugia populations are able to persist) in order to keep pace with rapid environmental change, especially as the landscapes through which they must now move have been extensively transformed by humans.

Although not all species will be able to spread far enough or fast enough to track climatic changes, it appears that most temperate species have thus far been able to keep pace with rising temperatures. In a meta-analysis of a wide range of species groups (plants, butterflies, birds, and mammals), Chen and colleagues (2011) found that species distributions have shifted to higher latitudes at a rate of 17 km/decade, or to higher elevations at a rate of 11 m/decade, with the fastest range shifts occurring in areas where warming has been greatest. Although two-thirds of the nearly 1400

[1] Clement Reid (1853–1916) was a British geologist and paleobotanist, who estimated that it would have taken oaks (*Quercus* spp.) nearly a million years to expand their range northward 600 miles (966 km) following the last glacial retreat to achieve their current northern range limit in Great Britain (cited in Clark et al. 1998). Empirical evidence suggests that tree migration rates were some three orders of magnitude faster than estimated by Reid, giving rise to the paradox that now bears his name.

species they examined had shifted their distribution toward the poles or to higher elevations, one-third had not. Why? Are these species simply poor colonists, unable to move far and fast enough to track environmental change? Or are they habitat specialists whose habitats disappeared or become too fragmented for them to locate or survive? We consider the effect of landscape fragmentation on the potential for species to shift their ranges next.

Range Shifts in Fragmented Landscapes

Regardless of the role that climate plays in setting range limits, the distribution of many species is also being affected by the widespread loss and fragmentation of their habitat as a consequence of human activities. As we have seen in previous chapters, both habitat loss and fragmentation have a negative effect on colonization success and population persistence (**Chapters 6 and 7**). Even a species with very good dispersal abilities and population growth potential may have a difficult time tracking rapid climate change if confronted by sprawling metropolitan areas, impenetrable dams, or vast agricultural 'deserts' that have come to dominate so many Anthropocene landscapes.

To evaluate what effect habitat fragmentation might have on tree migration rates, and thus on the potential for species to track changing climatic conditions within fragmented landscapes, Collingham and Huntley (2000) coupled a spatially explicit population model (SEPM) with neutral landscape models to simulate tree migrations across landscapes with varying degrees of suitable habitat and levels of fragmentation. Their SEPM was parameterized for a wind-dispersed tree, the small-leaved lime (*Tilia cordata*), which is native to most of Europe and western Asia, and is widely cultivated throughout North America. Importantly, the parameters of this SEPM were initially tuned to give a migration rate comparable to that of the early postglacial migration of *T. cordata* (0.4–0.5 km/yr). Hierarchical random landscapes (**Figure 5.5**) were generated to produce landscapes with varying habitat amounts (p = 1, 5, 40, or 90%) and degree of fragmentation; basically, they modified habitat availability at three nested levels to produce landscapes in which fragmentation was modeled as either a blocky fine-grained (small gaps) or coarse-grained (large gaps) pattern of disturbance (e.g. **Figure 5.9**). Based on their simulations, Collingham and Huntley found little effect of habitat availability on migration rates until the amount of suitable habitat fell below 25%. Below this threshold, migration rates dropped sharply (i.e. non-linearly) and the degree of fragmentation became important for predicting migration rate, with the fastest rate of spread seen in landscapes with numerous small gaps that can serve as 'stepping stones' to migration. Nevertheless, it

may be difficult for this species—and others like it—to maximize their full migration potential, regardless of how that habitat is arrayed, given that many landscapes now contain significantly less than 25% habitat suitable for colonization (indeed, there is <12% woodland habitat in all of the UK). Still, it is obvious that landscapes will need to contain a sufficient amount of functionally connected habitat to support both persistence as well as migration if species are to track environmental changes and shift their ranges in response to climate change.

Although habitat fragmentation may interfere with the ability of some species to keep pace with rapidly changing environmental conditions, other species may actually experience an increase in the availability of suitable habitat as temperatures warm, which could possibly enhance the connectivity of their habitat and lead to faster rates of migration and thus range expansion. To explore this possibility, Wilson and his colleagues (2009, 2010) adopted a metapopulation approach (using the incidence function model discussed in **Chapter 7**) to test how climate change has affected habitat use and the regional distribution of the Holarctic grass skipper (also known as the silver-spotted skipper in Europe, *Hesperia comma*). In Britain, this butterfly is typically found in calcareous grasslands, but an increase in summer temperatures (+2.8°C) over the past two decades has apparently enabled it to broaden its use of habitats (which now include all grasslands regardless of slope aspect) as well as its distribution. In that time, it has increased its area of occupancy 10-fold and tripled the number of grassland patches it occupies. Although we might expect that wide-ranging species and habitat generalists would be able to shift their distributions far more readily than more dispersal-limited species with localized habitat distributions, this study demonstrates how even a habitat specialist may be able to expand its distribution if climate change increases the availability of suitable habitat. Nevertheless, it appears that further range expansion may be constrained by habitat fragmentation (modeled as an increase in patch-isolation effects on the colonization rate, Equation 5.1), which may explain why this butterfly has failed to colonize new habitat patches as quickly as might have been expected given the increased habitat availability, even after allowing for its limited dispersal abilities. Even under the most favorable climate-change scenario, the skipper is predicted to occupy only about half of the available habitat within the next 100 years. Notice that, here again, we have an example of a species whose range shift is expected to occur through metapopulation processes involving colonization–extinction dynamics, rather than expansion along a continuous front (**Figure 8.3**).

Of course, not all range expansions are driven by climate change. In many cases, the introduction of a species to a new environment, especially if free of the competitors or predators that formerly limited its distribution and abundance elsewhere (**enemy release**), may now permit it to spread far and wide, perhaps even aggressively, to the detriment of native species. Such species are frequently viewed as aliens or exotics, as weeds or pests, or as problem species that require control rather than conservation. We thus turn our attention to the spread of invasive species next.

Landscape Effects on Invasive Spread

Invasive species are changing the landscapes of the world. As with many other global-change phenomena, such as land clearing and climate change, the introduction and spread of non-native species is likewise facilitated by humans. Along with the loss and fragmentation of native habitats, invasive species are among the leading threats to biodiversity (Wilcove et al. 1998). The potential link among these threats has not generally been appreciated, however (With 2002b). Human land-use patterns may enhance the invasibility of landscapes (Hobbs 2000), by creating disturbances such as those resulting in habitat loss and fragmentation. Although disturbance is generally assumed to promote invasive spread (Fox & Fox 1986), in some cases it is the suppression or alteration of natural disturbances (fire, floods, grazing) that opens the door to invasion (Hobbs & Huenneke 1992). For example, the suppression of fire leads to the woody invasion of some grasslands,[2] such as the tallgrass prairie of North America (Briggs et al. 2002). Invasive species themselves may modify natural disturbance regimes, such as fire cycles, providing a positive feedback that further enhances their spread (D'Antonio & Vitousek 1992). An example of this is given by the invasion of cheatgrass (*Bromus tectorum*) in the Great Basin region of the western USA, where it has transformed the native shrubsteppe into an annual grassland (**Box 8.2**).

Landscape transformation, such as that illustrated by the cheatgrass example, is one of the most severe outcomes of a biological invasion. Although not all invasive species transform landscapes, some landscapes can transform an introduced species into an invasive one, depending on the type, intensity, and distribution of disturbances on the landscape. Thus, understanding how landscape structure influences the various stages of the invasion process may help identify the species

and landscape scenarios in which invasive spread is most likely to occur, and perhaps assist with the development of land-management strategies for controlling current as well as future invasions. We discuss the landscape ecology of invasive spread next.

Landscape Ecology of Invasive Spread

The invasion process can be broken down into several different stages that include: (1) introduction, (2) colonization, (3) establishment, and (4) dispersal to new sites, which may then give rise to (5) spatially distributed populations (multiple foci) and (6) invasive spread. Landscape structure may affect any or all of these stages (**Figure 8.4**).

STAGE 1: INTRODUCTION The introduction of a non-native species (or pathogen) may be done intentionally, such as when a parasitoid wasp is released as a biocontrol agent of an agricultural pest, or accidentally, as when marine invertebrates are transported across oceans on the hulls or in the ballast water of cargo ships before being discharged into the waters of unsuspecting ports all over the world. The globalization of economies, coupled with our own global travel and transport, has greatly increased the rate of introduction of non-native species at major ports-of-entry. Although the spread of non-native species has clearly been facilitated by our globe-trotting ways, thereby enabling such species to overcome geographical barriers and dispersal constraints that formerly limited their distribution, it is still possible for landscape structure to exert some control over where such far-flung introductions might occur.

For example, geography and topography play a major role in shaping human land-use patterns, and thus influence where introductions of non-native species are most likely to occur. Land use generally has the greatest influence on the distribution of non-native species (Hobbs 2000). Intensity of land use usually decreases with increasing elevation, however, and thus lower elevations tend to be more heavily invaded than higher elevations, even after the elevational gradient in climate has been factored out (Stohlgren et al. 2002). International shipping ports tend to occur within natural bays or deep rivers, where they are surrounded by settlements or are in close proximity to natural waterways that can facilitate spread. If not, then the necessary transportation infrastructure is developed to efficiently move cargo (and any non-native stowaways) farther upstream or inland. Human settlements are themselves hot zones for invasion, from ornamental plants that escape cultivation to feral livestock and pets that escape domestication and wreak havoc on neighboring biological communities. The invasion ecology literature is thus replete with examples that demonstrate a relationship between the numbers of non-native species (especially plants) with proximity to human

---

[2] Although the transition from grassland to forest is part of the natural ecological succession in some systems, it is considered a form of invasion in tallgrass prairie because it results from the disruption of the natural disturbance regime (fire) that would otherwise prevent trees and shrubs from colonizing these grasslands. Recall that in disturbance-mediated systems, it is the absence of disturbance (e.g. fire) that is actually the disturbance, which may then permit the establishment of species that are not native to the system.

BOX 8.2   How the West Was Lost...to Cheatgrass

Other than humans, perhaps no single species has had a greater effect on the western landscapes of North America than cheatgrass (*Bromus tectorum*) (**Figure 1**). By the middle of the 19th century, grazing by cattle and sheep was already widespread throughout the open rangelands of the western USA, which may have contributed to the spread of cheatgrass, a Mediterranean grass that was unintentionally sowed as either a contaminant in grain or perhaps in straw used for packing material (Knapp 1996).

In the arid Great Basin, which is particularly vulnerable to overgrazing, cheatgrass has succeeded in converting more than half of the native shrubsteppe habitat to annual grassland. Cheatgrass is able to 'cheat' native grasses owing to a suite of life-history traits that seem tailor-made for an invader: prolific seed production (up to 18,000 seeds/m^2); rapid, early-season growth, which gives cheatgrass a competitive advantage over native species; formation of dense stands (12,000–15,000/m^2) that crowd out other grasses and forbs; and a high tolerance for disturbances,

such as fire and grazing. In fact, cheatgrass has altered the historical fire regime of the native shrubsteppe, from a fire return interval of 45–75 years to now only 3–5 years. The dense, near-continuous stands of cheatgrass provide a highly combustible fuel that is easily carried throughout a landscape. Once burned, these areas are then recolonized by cheatgrass, by virtue of its superior colonization and competitive abilities, enabling it to continue its domination of these grasslands and shrublands.

Cheatgrass thus promotes its own successful invasion by increasing both the frequency and intensity of fire. As a consequence, cheatgrass has fundamentally altered the structure and function of these systems, from shrubsteppe formerly dominated by long-lived perennial shrubs (e.g. *Artemisia tridentata*) interspersed with native bunch grasses, to annual grasslands consisting of near-continuous stands of cheatgrass, punctuated by a few short-lived perennial shrubs (Knapp 1996). Cheatgrass has thus effectively—and perhaps irreversibly—transformed the face of the American West.

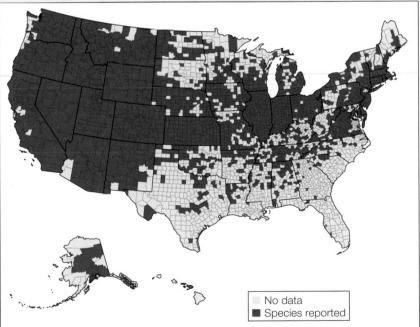

Figure 1 Distribution of cheatgrass in the United States.   Cheatgrass (top left) is now widespread throughout the USA (map, right). It has altered the fire regime in western rangelands, resulting in more frequent fires that favor cheatgrass at the expense of native grasses and shrubs (bottom left).

Source: Cheatgrass photo by Steve Dewey and published under the CC-BY 3.0 License. Cheat grass fire photo courtesy of USDA/NRCS. Map: EDDMapS. 2018. Early Detection & Distribution Mapping System. The University of Georgia - Center for Invasive Species and Ecosystem Health. Available online at http://www.eddmaps.org/.

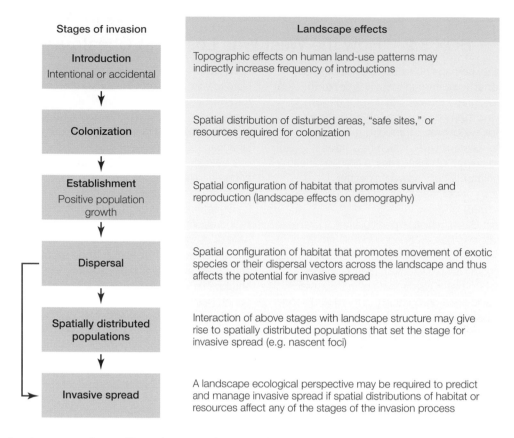

Stages of invasion	Landscape effects
Introduction Intentional or accidental	Topographic effects on human land-use patterns may indirectly increase frequency of introductions
Colonization	Spatial distribution of disturbed areas, "safe sites," or resources required for colonization
Establishment Positive population growth	Spatial configuration of habitat that promotes survival and reproduction (landscape effects on demography)
Dispersal	Spatial configuration of habitat that promotes movement of exotic species or their dispersal vectors across the landscape and thus affects the potential for invasive spread
Spatially distributed populations	Interaction of above stages with landscape structure may give rise to spatially distributed populations that set the stage for invasive spread (e.g. nascent foci)
Invasive spread	A landscape ecological perspective may be required to predict and manage invasive spread if spatial distributions of habitat or resources affect any of the stages of the invasion process

Figure 8.4 The landscape ecology of invasive spread.   The effects of landscape structure on various stages of the invasion process. (After With 2002b.)

development, agricultural land use, or transportation corridors (e.g. Hansen & Clevenger 2005; Cilliers et al. 2008; Gavier-Pizarro et al. 2010).

In the Andes of southern Chile, for example, pastures are a major source of plant invasion within the protected forests of some national parks, which support a great many endemic species (Pauchard & Alaback 2004). Pastures are dominated by agricultural weeds that were imported in livestock feed from Eurasia, and which outcompete native plant species unable to tolerate disturbances caused by forest clearing and cattle grazing. Although pastures are largely confined to low-lying valleys, roadways provide a conduit for the movement of these non-native plants into national parks, especially within secondary forests. Many of these secondary forests had previously been pastures, and thus occur at lower elevations where they are still surrounded by agriculture and other land uses. Thus, secondary forests may suffer greater invasion pressure than other forest types in the parks because of (1) higher rates of introduction from surrounding agricultural areas; (2) more-frequent grazing and more-open canopies that result in higher light levels, which are favored by non-native species; and (3) a milder climate as a consequence of being at a lower elevation, which is conducive

to the establishment of these non-native species that originate from temperate climes within Europe and Asia. Although the effect of elevation and land use on species invasions may often be coupled and therefore difficult to disentangle in many studies, roadways clearly provide an additional avenue for the introduction of non-native species across landscapes.

Introduction is a necessary, but not sufficient, condition for invasive spread. Although human transport and land-use activities may be directly or indirectly responsible for the introduction of many non-native species, the vast majority of introductions do not result in successful colonization and establishment, which we consider next.

STAGES 2 AND 3: COLONIZATION AND ESTABLISHMENT
Whether or not a location is suitable for colonization may well depend on the availability, suitability, distribution, and types of disturbances, habitats, or other resources at the site(s) of introduction (**Figure 8.4**). Much research in invasion ecology has examined the influence of resource and environmental heterogeneity on colonization success, especially for non-native plant species (e.g. Davis 2009). Human-modified landscapes in particular may create disturbed habitats (including

habitat edges) where many non-native species thrive at the expense of native species, which may not tolerate the higher light levels, drier or more exposed conditions, or altered nutrient levels that result (**Box 7.1**). Thus, habitat fragmentation—and its associated edge effects—may enhance the invasibility of landscapes by providing suitable habitat and an environment that favors colonization by non-native species.

Assuming the site of introduction is suitable for colonization, repeated introductions are usually required for successful colonization to occur; this is referred to as **propagule pressure** (or *introduction effort*) and is related to both the number of individuals released at a given time as well as the total number of introductions to that site. Propagule pressure is often the reason why some introduced species are able to gain a toehold (or roothold or finhold) in a new location, whereas others are not (Lockwood et al. 2005). Species (or their propagules) that are released frequently and in large numbers stand a much better chance of becoming established than the one-time, accidental release of a single individual. For example, the repeated flushing of ballast water from countless ships at the various ports within the San Francisco Bay on the western coast of North America has enabled the successful colonization of hundreds of non-native species, making it one of the most heavily invaded estuaries in the world. Although shipping ports are an obvious port-of-entry for many non-native species, propagule pressure may also be influenced by the composition and configuration of the surrounding landscape (Vilà & Ibáñez 2011). As mentioned previously, the degree of agricultural land use, urbanization, and road development have all been found to increase the numbers of non-native species within a landscape, presumably because these land uses provide a continual supply of invading organisms.

Successful colonization does not guarantee that the population will persist; that is, that the non-native species will in fact become established locally. Newly colonized populations are usually small and therefore vulnerable to a host of stochastic demographic and environmental factors that threaten to extinguish the non-native species before it can become established. Establishment requires both survival and successful reproduction, and thus all of the landscape effects on demography that affect population growth rates that we discussed in **Chapter 7** are likely to influence the establishment and persistence of non-native species as well (**Figure 8.4**).

Even if a non-native species becomes established, invasive spread is still not a given. The vast majority of non-native species that become established are not invasive and do not spread beyond the immediate area of introduction. Why some introduced species spread and others do not is one of the most pressing questions in invasive species research. For example, the release of

about two dozen Eurasian tree sparrows (*Passer montanus*) in Saint Louis, Missouri in the central United States in 1870 resulted in the establishment of a small resident population that is still largely confined to this area. By contrast, the closely related European house sparrow (*Passer domesticus*) was released in Brooklyn, New York in the eastern USA in 1852, and has since spread throughout the Americas. Both species are associated with human habitation and thus would be just as likely to benefit from land clearing and urban development. The larger house sparrow is perhaps a bit more aggressive, and thus may simply outcompete its smaller cousin (Barlow & Leckie 2000). This is not an entirely satisfying answer, however, for why the one species has spread so aggressively and the other has not.

STAGES 4 AND 5: DISPERSAL TO NEW AREAS AND THE FORMATION OF SPATIALLY STRUCTURED POPULATIONS

Once established, the non-native species may then disperse and colonize new locations beyond the initial site(s) of introduction. This is the stage at which the non-native species can be said to have spread, although it may not be considered invasive yet. The criteria for identifying when invasive spread has occurred, or is likely to occur, depends on the actual or perceived threat posed by the invasive species in terms of its ability to displace native species or disrupt ecosystem function, the magnitude of societal impacts, rate of spread, and the difficulty or cost of control or eradication. Controlling invasive spread is easier—and less costly—if caught in its earliest stages. For this reason, a cost-benefit analysis may be performed to evaluate the likely bioeconomic risk posed by invasive species and to evaluate the cost-effectiveness of potential monitoring, preventative, or control measures (e.g. Leung et al. 2002; Andersen et al. 2004).

Dispersal and colonization success are clearly influenced by landscape structure, as discussed previously in **Chapter 6**. Hence, our focus at this stage of the invasion process is on the attributes of landscapes that may enhance (or impede) the intrinsic dispersal success of a non-native species following establishment. This is distinguished from the species' initial introduction, which typically involves long-range transport—often mediated by humans—to arrive at new locations. Although repeated introduction to numerous locations (multiple foci of introduction) may result in the establishment of many localized populations, the potential for invasive spread is minimal if the introduced species is not a good disperser on its own.

Many species are naturally capable of long-distance dispersal, either via wind, water, or some animal vector (e.g. birds). Long-range dispersal events—even if rare—may ultimately govern the rate of invasive spread (Higgins & Richardson 1999). As we saw in **Chapter 6**, landscape structure—the spatial arrangement of habitats or resources—may be unimportant where long-range or random dispersal is concerned; only the amount of habitat or resource is important in predicting dispersal success in that case (**Figure 6.27**). Good dispersers are not necessarily good colonizers, however. If habitat suitable for colonization is patchily distributed on the landscape, then long-distance dispersal could be risky, since a large percentage of individuals (or their propagules) may end up in unsuitable areas. If resources are clumped or patchily distributed, then short-range dispersal should increase the likelihood that most individuals or propagules fall within the same local neighborhood, where colonization is most likely to be successful (Lavorel et al. 1995).

As new populations are initiated away from the initial point (or points) of introduction, the distribution of the introduced species becomes spatially structured and may function as a metapopulation, in which local extinctions are offset by immigration from local populations (**Chapter 7**). We should thus anticipate that colonization–extinction dynamics may well occur in a non-native invader, with the added complication that rescue from local extinction may make eradication of unwanted or harmful 'pest' species difficult, if not impossible. Further, some species spread by establishing satellite populations beyond the periphery of the invasion front. These are considered 'nascent foci' and explain why some invading species are able to spread so rapidly (Moody & Mack 1988). As these satellite

Figure 8.5 Invasion of smooth cordgrass (*Spartina alterniflora***).** Smooth cordgrass is an aggressive invader of tidal mudflats along the Pacific coast. It spreads through a pattern of distributed colonies ('nascent foci') that grow and coalesce, eventually forming a monotypic stand that fills in estuaries through a process of sedimentation and accretion, altering hydrology and other ecosystem functions, and threatening habitat for mollusks and the millions of migratory shorebirds and waterfowl that feed upon them.

Source: Aerial photograph courtesy of Thomas W. Forney, Oregon Department of Agriculture.

populations grow, they expand in area and eventually coalesce. The invasive smooth cordgrass (*Spartina alterniflora*), which was initially introduced to the salt marshes of the San Francisco Bay for reclamation purposes, exhibits this pattern of spread (**Figure 8.5**).

STAGE 6: INVASIVE SPREAD   If a non-native species successfully overcomes the various obstacles inherent in each of the preceding stages, the end result may well be invasive spread. Although not all introduced species are invasive and not all invasive species spread, it can be difficult to identify which species are likely to exhibit invasive spread because there is often a latent period or lag phase, from the time the species is introduced (or detected) and when it starts to spread. This is due in part to an initial 'build-up' phase as the population first becomes established and then grows to a size where it is finally detected. Recall, too, that an exponentially growing population will initially show a slow increase in numbers before reaching a point of accelerated growth. Thus, we should anticipate that even if non-native species are capable of exponential growth, they will persist at fairly low numbers for potentially long periods of time (the lag phase) before undergoing explosive growth. For example, the average lag phase of non-native plants introduced to New Zealand was 20–30 years, with some woody species exhibiting lags of nearly 100 years (e.g. scotch broom *Cytisus scoparius*, black elderberry *Sambucus nigra*; Aikio et al. 2010). For invasive tropical plants introduced to the Hawaiian Islands, however, the average lag phase for woody species was 14 years and only 5 years for herbaceous plants (Daehler 2009).

> Today's exotic weeds will become tomorrow's invasive species.
> Ahern et al. (2010)

When is a non-native species likely to become invasive and spread aggressively throughout the landscape, however? The invasion potential of a non-native species is usually assessed in terms of specific life-history attributes that correlate with invasion success (such as high reproductive output or good dispersal abilities) or whether the species has a history of being invasive elsewhere. It has also been suggested that invasive spread might simply be a function of the length of time the species has been present on the landscape ('minimum residence time'), which is consistent with the notion that non-native species may exhibit a lag phase before spreading throughout the landscape. For non-native plants introduced to two distinctly different regions within the USA, Ahern and his colleagues (2010) found that minimum residence time had the greatest effect on the degree of

(A) Michigan

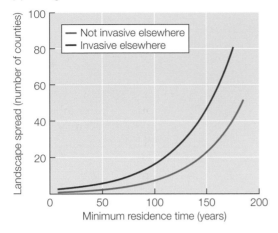

(B) California

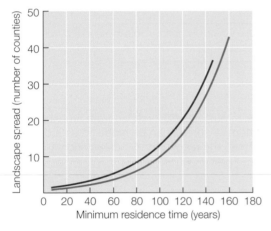

Figure 8.6 Temporal lags in invasive spread. Invasive spread did not occur until many decades after the initial detection of a non-native plant species (the minimum residence time) within two regions of the USA. In Michigan (A), a history of prior invasiveness elsewhere was also a significant predictor of landscape spread.

Source: After Ahern et al. (2010).

landscape spread, although a history of invasiveness elsewhere was also important for predicting landscape spread for non-native plants in Michigan (**Figure 8.6**). Many non-native plants that were introduced into California or Michigan before 1900 are now considered invasive, which is consistent with another regional analysis from Europe that suggested it may take at least 150 years for non-native plants to reach their maximum distribution at broad geographical scales (10^5 km²; Williamson et al. 2009).

Spatial Models of Invasive Spread

Predicting the likelihood, rate, and pattern of invasive spread across landscapes is a research challenge. Given the potential timeframes involved (e.g. 150 years to reach maximum spread), it can also be a difficult one to overcome purely with empirical or experimental

approaches. Thus, spatial models of invasive spread are especially valuable in identifying the landscape scenarios under which invasive spread might occur, as well as providing a means of modeling the potential impact of different management scenarios in controlling invasive spread. We therefore discuss some of the general spatial modeling approaches that have been applied to the problem of invasive spread in this section.

CLASSICAL MODELS OF SPATIAL SPREAD **Reaction-diffusion** (RD) models have enjoyed a long history in the physical and biological sciences. In the context of invasive spread, they were initially used by British ecologist and statistician John Skellam (1951) to model the spread of the North American muskrat (*Ondatra zibethicus*) throughout central Europe, following the escape of five individuals from an estate near Prague in the Czech Republic around 1905 (**Figure 8.7**). Muskrats are prolific breeders, capable of producing several litters of 4–6 young per year. Thus, over the ensuing decades, muskrats spread out from Prague at a rate up to 30 km/yr. Although occasionally hunted for their fur, the muskrat is generally considered a pest species throughout Europe because of its burrowing activities that destabilize banks along waterways, thus threatening levees and dikes.

Let's assume that population spread is similar to a process of diffusion, in which individuals spread out more or less equally in all directions from an initial point of introduction (**Figure 8.8**). For this sort of application of the RD model, the 'reaction' is reproduction or population growth and 'diffusion' pertains to the random movement of individuals. Thus, we are essentially treating populations like an ideal gas (i.e. a large-number system; **Chapter 2**). In other words, we are assuming that for very large populations, the idiosyncratic behavior of individuals (at least in terms of their reproduction and dispersal) can be averaged over the entire population to give rise to certain deterministic properties, such as the average rate at which a population spreads (Andow et al. 1990).

When simple models adequately capture population spread even though complex behaviors are known to be involved, we follow the principle of parsimony; that is, the most appropriate model is that which explains the most with the fewest assumptions. This is not to say that individuals do not use environmental information; it is clear that they do. It rather says that such information [is] often irrelevant to understanding patterns of spread on some scales of interest.

Andow et al. (1990)

To begin, let's imagine that spread occurs only along a single dimension, such as along a river

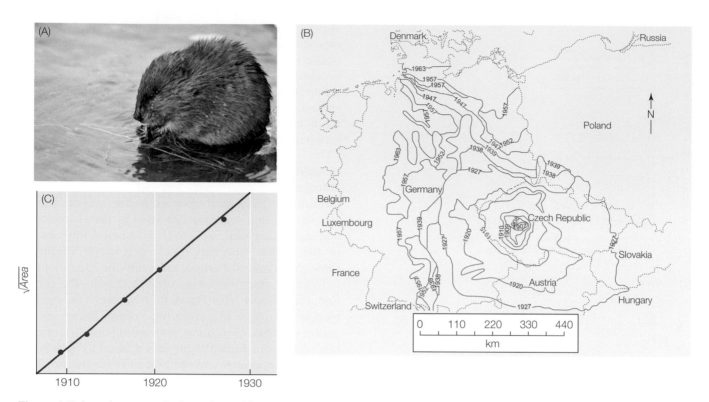

Figure 8.7 Invasive spread of muskrats (*Ondatra zibethicus*) in central Europe. Muskrats (A) escaped from an estate near Prague and spread outward in roughly concentric circles over time (B). This pattern of spread can be modeled as a diffusion process, which predicts that the rate of spread should be constant over time. Thus, the range of the muskrat increases as a linear function of time (C).

Source: After Lockwood et al. (2007).

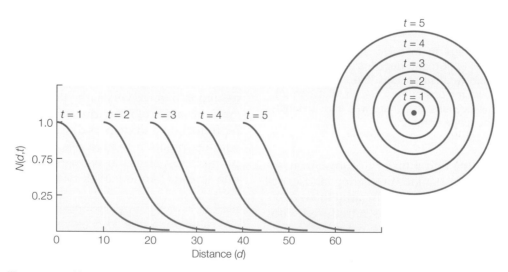

Figure 8.8 Travelling wave of invasive spread. The probability density function [$N(d,t)$] describes the distribution of individuals at various distances (*d*) from the edge of the invasion front at different fixed points in time (*t*). If invasive spread occurs as a diffusion process, then the rate of spread is predicted to be constant, resulting in the movement of the invasion front by a set distance during each time interval, giving the appearance of 'waves.'

Source: After Lockwood et al. (2007).

corridor or coastline, and that the habitat is homogeneous. If we assume that invasive spread is initiated from a single point of introduction, then using partial differential equations, spread is simply a function of the exponential population growth (*rN*) and the

random dispersal of individuals (*D*, the diffusion coefficient, the mean squared displacement of individuals per unit time) along that transect (a single dimension *x*, but movement occurs equally in both directions, so x^2), or

$$\frac{\partial N}{\partial t} = rN + D\left[\frac{\partial^2 N}{\partial x^2}\right].$$ Equation 8.1

If spread occurs in two dimensions (x, y), we can expand Equation 8.1 to

$$\frac{\partial N}{\partial t} = rN + D\left[\frac{\partial^2 N}{\partial x^2} + \frac{\partial^2 N}{\partial y^2}\right].$$ Equation 8.2

The parameters, r and D, can be obtained empirically to provide estimates of population spread for the species of interest. We have previously discussed the intrinsic rate of population increase, r, which can be determined from a life-table analysis (**Chapter 7**). The diffusion coefficient, D, is a measure of the mean-squared displacement per time, and can be estimated using mark-recapture techniques as:

$$D = \frac{(\sum_{n=1}^{M} x / M)^2}{\pi t}$$ Equation 8.3

where M is the total number of marked individuals recaptured, x is the distance from the initial site of capture, and t is the time after initial capture (Shigesada & Kawasaki 1997).

One of the most significant predictions of the RD model is that a species' range should increase linearly with time at a constant rate (i.e. an asymptotic rate of spread; **Figure 8.7**). The rate of population expansion, which is the slope of this relationship, is thus given by:

$$c = 2\sqrt{rD}.$$ Equation 8.4

Since this is the average rate of spread, we can calculate the 95% confidence envelope around this estimate as

$$N(d,t) = 2\sqrt{4Dt}$$ Equation 8.5

which gives the number of individuals within distance d of the initial point of introduction at time t. Basically, this says that 95% of the individuals will be located within a circle of radius $2\sqrt{4Dt}$ at a particular point in time. This equation, describing the distribution of individuals in space assuming a random diffusion process, is an example of a **probability density function** (i.e. a kernel; **Chapter 6**). When plotted, it shows how the density of individuals tapers off as a function of distance from the initial point of introduction (**Figure 8.8**). If we plot this function at different points in time ($t = 1$, $t = 2$, $t = 3$, etc.), we see that the form of the function remains the same but shifts by a constant amount (equal to c) at each time step as the invasion front expands. This produces the characteristic 'travelling wave' of invasive (and disease) spread, in which the rate of spread is constant over time (**Figure 8.8**).

Reaction-diffusion models assume that the environment is homogeneous. Although real landscapes are heterogeneous, the use of RD models at least provides a starting point for quantifying the rate of spread for an invasive species. One consequence of environmental heterogeneity is that the pattern of spread may not be symmetrical. Spread may occur more quickly in certain directions, such as along waterways or roadways, but be impeded along other fronts where populations encounter mountain ranges or large bodies of water. To deal with non-radial spread, one solution might be to divide the range into different sectors and calculate the spread rate separately for each sector. Thus, in the case of the muskrat mentioned previously, the intrinsic rate of population increase, r, was estimated at 0.2–1.1/yr, and the diffusion coefficient, D, ranged from 51 to 230 km²/yr, depending on the range sector in question (Andow et al. 1990). Substitution of these rates into Equation 8.4 yields a predicted spread rate (c) of 6.4–31.8 km/yr, which corresponded reasonably well to observed rates of spread (1–25 km/yr) for muskrats escaping Prague.

'NEO-CLASSICAL' MODELS OF SPATIAL SPREAD As with population growth models (**Chapter 7**), classical models of spatial spread make a number of simplifying assumptions. For example, we have just discussed how rates of invasive spread for a given species (such as the muskrat) may in fact be heterogeneous, reflecting differences in how topography or other landscape features influence dispersal and population growth rates in different parts of the expanding range. We should thus anticipate that relaxing other assumptions inherent in simple RD models will influence estimates of spread rates, and perhaps some fundamental predictions regarding invasive spread as well.

So-called 'neo-classical' spatial spread models address some of these underlying assumptions using various types of analytical approaches (Hastings et al. 2005). The general form of the RD model assumes not only that the environment is homogeneous, but also that all individuals in the population reproduce and disperse pretty much continuously (and at random) throughout their lifetime. We know from our previous study of population growth models that both reproduction and dispersal may be stage- or age-dependent (**Chapter 7**). For example, van den Bosch and colleagues (1992) developed a more detailed modeling approach that included age-structured demography and dispersal-density functions parameterized for the muskrat invasion of Europe, as well as for several introduced bird species (including the aforementioned house sparrow invasion of North America). They found that the classical RD model tended to underestimate observed rates of invasive spread relative to their age-structured model in most cases. In other words, their more detailed model provided a better fit to the observed rates of spread, although the difference between model estimates did not end up being statistically significant. Nevertheless, small differences in estimating spread rates (especially if underestimated) could have large consequences from a management standpoint.

Recall that one of the main predictions of the classical spatial spread model is that the rate of invasion is constant, resulting in a linear expansion of the species' range over time (**Figure 8.7**). Does this prediction hold if we modify assumptions inherent in the RD model, however? Given that many species do not reproduce and disperse continuously throughout their lifetimes, we can break these events into separate stages using **integrodifference equations** (IDE; Kot et al. 1996). The IDE model is composed of a difference equation that describes population growth at each point on the landscape (which, for the sake of simplicity, is assumed here to be one-dimensional and homogeneous) and an integral operator that accounts for the pattern of dispersal (i.e. the dispersal kernel). Thus,

$$N_{t+1}(x) = \int_{-\infty}^{\infty} k(x, y) f[N_t(y)] dy \qquad \text{Equation 8.6}$$

in which the population density at the next time period (N_{t+1}) at a particular location (x) is a function of the number of individuals produced at all other source locations y ($f[N_t(y)]$) that manage to disperse from y to x according to the shape of the dispersal kernel, k. The dynamics of this model are similar to those of the RD model (Equation 8.1) if dispersal distances exhibit a Gaussian (normal) distribution (Hastings et al. 2005).

As mentioned previously, however, many species exhibit occasional long-distance movements, and this is particularly true for non-native species whose movements may be facilitated by human transport. Thus, the shape of the dispersal kernel of such species may have a 'fat tail,' meaning that there are more long-distance dispersal events than expected if dispersal were normally distributed. Interestingly, incorporation of fat-tail dispersal into the IDE model no longer results in the prediction of a constant spread rate; instead, the rate of invasive spread is predicted to *accelerate* over time (Kot et al. 1996). If true, then invasive species will be able to spread farther and faster the longer their spread goes unchecked. Obviously, the potential for invasion to occur as an accelerated wave front is a concern from a management standpoint. Yet again, this emphasizes the importance of early detection and the need for a rapid response in the control of invasive species.

Part of the problem in determining whether invasion speed is likely to accelerate, however, is that data on long-distance dispersal events are difficult to obtain empirically, especially when using mark-recapture techniques. Long-distance dispersal events tend to be rare events, and thus become harder to detect the far-

ther one moves away from the point of initial capture. The tendency to underestimate spread rates for some species has been attributed to a lack of information on these infrequent, long-distance dispersal events. For example, the simple RD model (Equation 8.4) underestimated the observed rate of spread by two orders of magnitude in the case of the cereal leaf beetle (*Oulema melanopus*), a serious pest of wheat and other cereal grains that was first detected in Michigan (USA) in 1962 (Andow et al. 1990). This discrepancy between the predicted and observed rates of spread in the cereal leaf beetle was attributed to its potential for long-distance dispersal, such as by wind or human transport, which was not accounted for in the simple RD model. Fortunately, the availability of molecular techniques is now making it easier to document the occurrence of long-distance dispersal within a population (**Chapter 6 and 9**).

Finally, although simple RD and IDE models may underestimate the rate of invasive spread in fragmented or heterogeneous landscapes, more sophisticated mathematical approaches founded on these types of models have been developed that do incorporate landscape structure or habitat heterogeneity to varying degrees (e.g. as a coupled map lattice using fractal neutral landscapes, Flather & Bevers 2002; or, as a periodic landscape of alternating 'good' and 'bad' patches, Dewhirst & Lutscher 2009). Of particular note, these analytical models have been used to derive quantitative relationships between habitat amount, fragmentation, and the extinction (Flather & Bevers 2002) or invasion (Dewhirst & Lutscher 2009) threshold. If the invasion threshold is defined as the minimum amount of habitat required for population spread, then this threshold is predicted to increase as the landscape becomes more fragmented (i.e. more habitat will be required before the population is able to spread); a species' extinction and invasion threshold would thus be one and the same. The population goes extinct below a certain threshold amount of habitat but is able to spread across the landscape above that threshold. However, if an **Allee effect** occurs, such that the population spreads only above some critical population density, then the invasion threshold actually *decreases* with increasing habitat fragmentation, meaning that less habitat is ultimately needed for population spread in more fragmented landscapes. As we have seen before, high levels of fragmentation initially result in small gaps that can be easily traversed via remaining habitat 'stepping stones.' In fragmented landscapes, more individuals would be able to cross these small gaps to exceed the threshold population density in the neighboring patch, thus facilitating the rate of spread across the landscape. Landscape connectivity is thus important for invasive spread, but invasive spread is ultimately a function of both the dispersal ability and population growth potential of the species.

Landscape Connectivity and the Potential for Invasive Spread

If habitat loss and fragmentation enhance the potential for biological invasions, then it would be useful to have some idea of what critical level of landscape disturbance is likely to facilitate invasive spread, as illustrated above. Empirical information on demographic rates and dispersal are lacking for many species, however, which makes parameterization of population spatial spread models difficult. As a first approximation, we might therefore analyze landscape connectivity, such as by using graph-theoretic or percolation-based approaches (**Chapter 5**), to assess the potential for invasive spread in a given landscape.

Network analysis, based on graph theory, provides one means of assessing connectivity over broad spatial scales and requires minimal information on dispersal to perform. For example, Minor and her colleagues (2009) used a network analysis to assess the connectivity of a fragmented forest landscape in the eastern USA for about 200 plant species, of which nearly a third were non-native species. Network distance, an index of patch connectivity, was used to evaluate whether pairs of sampling locations might be connected via seed dispersal (assumed to be 50 m for these forest plants), and if so, whether locations were within the same or a different forest patch. Network distance thus goes beyond the straight-line distance separating sampling locations, by providing information on the spatial configuration of forest habitat, at least in terms of whether sampling locations are close (<50 m) and within the same patch, close but in different patches separated by a non-forest matrix, or too far apart (>50 m) to be functionally connected via seed dispersal. Locations that fall within the expected seed dispersal range (especially if within the same patch) are expected to be more similar in plant species composition (i.e. to have low **species turnover**) than those that exceed it, which can be evaluated though the use of partial Mantel tests of the species correlation matrix (**Chapter 10**).

Connectivity, as assayed by network distance, influenced species turnover far more than either the straight-line distance or degree to which sites were similar environmentally (in terms of slope, NDVI, soil pH, or forest patch size). This was true for both native and non-native species (**Figure 8.9A**). Connectivity ultimately had a greater effect on non-native plants that were *not* invasive, however (**Figure 8.9B**). Furthermore, connectivity turned out to be more important for non-native species that can spread on their own or via the wind than those that are dispersed by animals (by being ingested or adhering to their fur; **Figure 8.9C**).

Thus, although habitat fragmentation has been thought to facilitate invasive spread, this network analysis suggests that many non-native plant species may face the same sorts of landscape constraints on dispersal that native plants do. In fact, habitat connectivity (a measure of fragmentation) was less important for the spread of invasive plants than non-invasive ones. In other words, invasive plants were invasive precisely because they were not affected by landscape structure (i.e. habitat connectivity). Many of these invasive plants were not restricted to forests and were often found in fields and other open habitats of this landscape. Further, most of these invasive plants (61%) were dispersed by animals, such as white-tailed deer (*Odocoileus virginianus*), which move readily between fields and forests. Thus, forest fragmentation might still play a role in facilitating invasive plant spread in this system through the creation of a more heterogeneous landscape containing open habitats and forest edges that are favored—and easily traversed—by deer.

Neutral landscape models have been used to provide insight into the degree of disturbance that might enable an invasive species to spread across the landscape (i.e. the invasion threshold; With 2002b, 2004). Given simple movement rules related to the dispersal range or gap-crossing abilities of the organism (or its vector), we can assess the potential for invasive spread in a given landscape fairly quickly. In this case, invasive spread is given by the probability of percolation on the landscape, which, as we know from **Chapter 5**, is predicted to occur at some threshold level of disturbance, depending on the scale or pattern of habitat disturbance (**Figure 5.10**). For example, the spread of a non-native species that establishes within disturbed areas of the landscape can occur at lower levels of disturbance when disturbances are large or aggregated in space (thereby creating large patches of disturbed habitat) than if they are small and widely dispersed across the landscape. In the example shown, landscape-wide spread does not occur until 70% of the landscape has been disturbed via small, random disturbances (**Figure 8.10**, top row), but this drops to as little as 30% when disturbances are aggregated on the landscape (**Figure 8.10**, bottom row). This assumes, however, that the species is only able to spread through disturbed habitat that is immediately adjacent (i.e. nearest-neighbor or localized dispersal). If the species can cross even small gaps, then spread is actually facilitated in landscapes with a fine-scale pattern of disturbance (i.e. fragmented) because the species can basically traverse the landscape in a stepping-stone fashion, as mentioned previously. Whether or not an invasive species actually exhibits this sort of threshold response to disturbance, however, ultimately depends on the dispersal capabilities of the species. Species

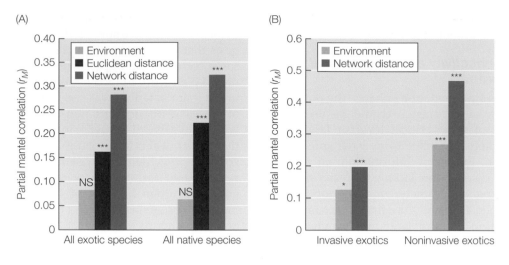

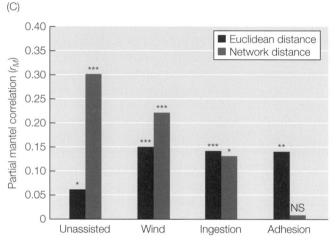

Figure 8.9 Relative effect of habitat connectivity on the spread of non-native versus native forest plants. (A) Network distance, a measure of functional connectivity, was a better predictor of species turnover (degree of similarity in species composition) for both native and non-native species, than either the straight-line (Euclidean) distance or the degree of environmental similarity between locations. (B) The spread of non-invasive plants was affected to a greater degree by habitat connectivity (network distance) than that of invasive plants. (C) Non-native plants that spread by adhering to a passing animal were least likely to be influenced by habitat connectivity (network distance), presumably because these animal vectors, such as deer, moved readily across the landscape. Stars indicate whether the partial Mantel correlation (r_M) is significantly different from 0 (i.e. no correlation): *$P \leq 0.05$; **$P \leq 0.01$; ***$P \leq 0.001$; NS = non-significant.

Source: After Minor et al. (2009).

with extremely good dispersal abilities are less likely to be affected by landscape structure, and thus are unlikely to exhibit invasion thresholds in response to habitat loss and fragmentation.

Dispersal is only part of the invasion equation, however. An invading species must also achieve positive population growth in areas it colonizes. Landscape pattern could thus have different effects on dispersal and demography, leading to contrasting effects on invasive spread. For example, a species with good dispersal or gap-crossing abilities may enjoy higher colonization success on landscapes with a more dispersed (fragmented) pattern of disturbance, which would have a positive effect on spread in fragmented

landscapes. For the same species, however, population establishment and positive growth rates might be harder to attain in small, disturbed patches of fragmented landscapes because populations tend to be small and more vulnerable to stochastic extinction or because of negative edge effects, such as increased competition or predation, on fecundity and survival (**Chapter 7**). Thus, whether fragmentation enhances or impedes invasive spread will depend on the relative effects of landscape structure on different phases of the invasion process (**Figure 8.4**). As a result, there is no single best solution to managing invasive spread that will apply to all species. Nevertheless, dispersal was found to be the most important driver of spread for

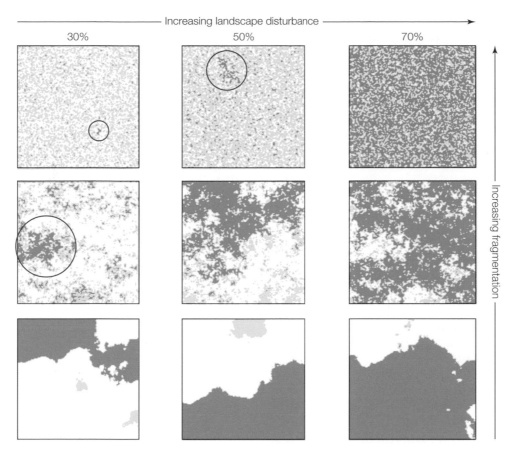

Figure 8.10 Probability of invasive spread as a function of increasing landscape disturbance and fragmentation. Neutral landscape models illustrate how the potential for invasive spread might be assessed in terms of the probability that the species can percolate across the landscape. The largest extent of spread (dark green areas) is highlighted in each landscape, which assumes that the species is only able to move through disturbed habitat (light green areas) that is adjacent (i.e. nearest-neighbor movement).
Source: After With (2002b).

invasive plants in a spatially explicit simulation analysis (Coutts et al. 2011), which suggests that managing dispersal rather than demography may hold the key to containing invasive spread, at least in plants. Still, it is difficult to know which aspect of dispersal should be targeted for management, as this could entail a variety of different strategies, such as disrupting landscape connectivity for the invading species, controlling dispersal vectors, or by removing satellite populations or those that produce high numbers of dispersers or propagules (i.e. source populations). We consider how population source–sink dynamics may facilitate invasive spread next.

Source–Sink Metapopulation Dynamics and Invasive Spread

As mentioned previously in considering the different stages of invasive spread, a non-native species may persist as a metapopulation that is maintained by its own colonization–extinction dynamics. Because not all habitats are optimal for the invading species, we can anticipate that the non-native species will exhibit higher

rates of population growth in some habitats than others, and that these populations might then serve as a source of immigrants or propagules that can colonize new locations or supplement less-successful populations that have already been established in marginal habitats. This is an example of a **source–sink metapopulation**, which we discussed in **Chapter 7**. The same sort of source–sink population dynamics that rescues declining populations from extinction might thus help promote the invasive spread of a non-native species.

To explore the role of source–sink population dynamics in invasive spread, Thomson (2007) studied a Eurasian grass species, the barbed goatgrass (*Aegilops triuncialis*), which is invading the coastal grasslands of California in the western USA. Like many introduced species, goatgrass exhibited a long lag-phase—nearly 70 years—after its initial introduction before it began to spread aggressively. Unlike most grassland invaders in this system, however, it also appears to be more tolerant of the magnesium-rich, nutrient-poor serpentine soils that serve as refugia for many endemic grassland species that are now rare and endangered, including the bay checkerspot butterfly (**Figure 7.14**). Superficially, goatgrass has the appearance of a source–sink metapopulation, occurring as dense stands in core grassland areas and smaller satellite populations along serpentine outcrops. These satellite populations were assumed to be population sinks ($\lambda < 1$), given their occurrence in marginal serpentine habitat, and which therefore were being maintained by immigration from core areas that function as population sources ($\lambda > 1$).

Using mark-recapture techniques to estimate dispersal distances and a matrix population model to estimate population growth rates (**Chapter 7**), Thomson (2007) evaluated whether habitat-specific differences in seed dispersal and population growth rates could give rise to source–sink dynamics and facilitate invasive spread in the barbed goatgrass. As expected, germination rates were lower at the edge of the invasion front within satellite populations on serpentine soils than at the population core (68% vs 89% of seed spikes germinated, respectively). However, plant survivorship did not differ between satellite and core populations (88% vs 81%, respectively), and plants at the periphery were actually larger, perhaps because of lower competition among plants as a result of lower population densities (by an order of magnitude) than in core populations. Larger plants produce more seeds, and thus plant size is a strong correlate of fecundity. Based on these relative differences in demographic rates, population growth rates were estimated to be strongly positive ($\lambda > 1$) in both satellite and core populations, and in fact, were actually higher in satellite populations. Again, this is likely due to higher plant density and competition within the core, as density-removal experiments demonstrated that core populations actually had a higher potential for increase than satellite populations.

Regardless, satellite populations are not functioning as sinks in this system and are not being maintained by immigration from core populations. In fact, there appears to be a net export of seeds *from* satellite to core populations, which again runs counter to expectations if this system were functioning as a source–sink metapopulation. Seeds disperse twice as far along the range edge than at the core, perhaps because the rocky serpentine outcrops have less vegetation and thus seeds can tumble or be blown greater distances before being trapped by vegetation. Although this method of dispersal results in a slow rate of spread (spread is admittedly faster in grazed areas where seed spikes can be dispersed by livestock), management of goatgrass that focuses just on core habitat areas is less likely to be successful than the targeted control of satellite populations. This again reiterates the importance of controlling satellite populations beyond the invasion front, as a means of slowing invasive spread (Moody & Mack 1988).

Species Distribution Models of Invasive Spread

As with species' range shifts in response to climate change, species distribution models can be used to predict the occurrence of a non-native species within a different landscape, especially if we have information on the environmental factors that appear to limit the distribution of the species in its native range (Peterson 2003). Species occurrence (presence/absence or patch occupancy) does not mean that the species is maintaining viable populations at those locations, however. From the standpoint of being able to assess invasion potential and effectively manage invasive spread, it would obviously be advantageous if we could identify those locations where the species is predicted not only to occur, but to thrive. This is especially salient in light of our recent discussion concerning the potential impact of source–sink population dynamics in promoting invasive spread. It would therefore be ideal if we could couple a model of population dynamics with a species distribution model, to identify these high-risk areas on the landscape where an introduced species has the greatest potential to become invasive.

To illustrate the potential of this combined modeling approach, Brown and colleagues (2008) coupled a stage-structured matrix population model with a species distribution model (MaxEnt) for the invasive Malabar plum or rose apple (*Syzygium jambos*) in the Luquillo Mountains of Puerto Rico (**Figure 8.11**). The Malabar plum is an invasive tree that is native to Southeast Asia but was introduced to the West Indies in the first quarter of the 18th century and now occurs throughout Puerto Rico. Birds are believed to be responsible for its localized dispersal, whereas long-distance dispersal may be facilitated by bats or water currents (Brown et al. 2008). Within the Luquillo Mountains, the occurrence of the Malabar plum was best predicted by evapotranspiration rate, rainfall, and the percentage of forest canopy cover in 1936. Apparently, the Malabar plum now occurs most everywhere that is environmentally suitable for it, which is primarily at low elevations along the perimeter of the Luquillo Mountains, where anthropogenic disturbances have historically been greatest (**Figure 8.11A**). In turn, many of these same areas were also identified through independent demographic analysis as being capable of sustaining positive population growth ($\lambda > 1$), although other potential population sources are scattered throughout the region (**Figure 8.11B**). Thus, this coupled modeling approach shows great promise in enabling land managers to map the distribution of habitats that can potentially sustain positive population growth, as

Figure 8.11 Habitat suitability for the invasive Malabar plum (*Syzygium jambos*) in the Luquillo Mountains of Puerto Rico.   The Malabar plum is an invasive tree that is native to Southeast Asia, and is now found throughout the West Indies. The distribution of highly suitable habitat for the Malabar plum is predicted to occur primarily along the northern perimeter of the Luquillo site (A). Many of these same areas are also expected to support viable populations ($\lambda > 1$; B).

Source: Maps after Brown et al. (2008). Photo by B. Navez and published under the CC BY-SA 3.0 License.

a means of identifying targets and setting priorities for invasion control.

Landscape Epidemiology

Disease spread is really a form of biological invasion, but since the discipline of landscape (spatial) epidemiology arose independently, we will address it separately here. The Russian parasitologist Evgeny Pavlovsky is considered to be the founder of landscape epidemiology. Pavlovsky first proposed the concept of **disease nidality**, the idea that diseases basically have an ecological niche (Pavlovsky 1966). The nidality concept, combined with landscape ecology, led to the emergence of landscape epidemiology, the study of the association between pathogens, species, and the landscape or environmental conditions responsible for the incidence and spread of disease (**Figure 8.12**).

The disease nidus can be defined by three factors: (1) climate, since most pathogens have a minimum temperature threshold below which replication, and therefore transmission, cannot occur; precipitation may also be important for pathogen spread, either directly (e.g. a plant fungal disease spread by rain splash to adjacent plants) or indirectly, by influencing the activity or reproductive period of disease vectors (e.g. mosquitoes) or of the hosts themselves; (2) the transmission requirements of the pathogen, in terms of what vectors or host species are required and available to complete its life-cycle; and (3) the habitat associations of pathogens, their vectors, and hosts, which can be mapped and analyzed to identify areas of overlap and thus of potential disease risk (Reisen 2010).

> The few studies that demonstrate how landscape composition...and configuration...influence disease risk or incidence suggest that a true integration of landscape ecology with epidemiology will be fruitful.
>
> Ostfeld et al. (2005)

Landscape epidemiology thus seeks to understand the causes and consequences of spatial heterogeneity that contribute to disease risk (Ostfeld et al. 2005). Disease risk varies in both space and time because various landscape and environmental factors influence the distribution and abundance of vectors, hosts, and

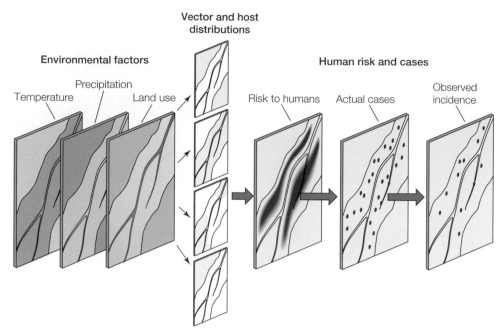

Figure 8.12 A framework for landscape epidemiology.   Landscape and environmental factors (temperature, precipitation) can influence the distribution of vectors and hosts for a particular disease, which influences the likelihood of humans coming into contact with the pathogen, and thus the risk of disease to humans. This variation in disease risk can be mapped (not shown), but will not reflect disease incidence (actual cases), as not all individuals in high-risk areas will be infected. Further, because of differences in surveillance and reporting, the observed incidence of disease in humans is likely to be a subset of actual cases.
Source: After Ostfeld et al. (2005).

pathogens, which in turn affects the likelihood of transmission. We shall turn to the effects of landscape structure on disease risk next.

Landscape Ecology of Disease Spread

Disease spread is influenced by: (1) the dispersal ability of pathogens or their vectors relative to that of the host species, and (2) the spatial dynamics of disease reservoir and host populations. We consider these effects first, before turning our attention to the effects of habitat loss and fragmentation on disease spread.

DISPERSAL OF PATHOGENS, VECTORS, AND HOSTS
Pathogens are spread via several different modes, and many have multiple means by which they can be spread. Plant pathogens in particular exhibit very diverse dispersal mechanisms, from soil-borne pathogens to aerially dispersed fungi and bacteria to vector-borne viruses (Plantegenest et al. 2007). In most cases, the probability of transmission declines dramatically with distance from an infected host; thus, pathogen dispersal tends to be highly localized, requiring some sort of contact between susceptible and infected individuals or vectors, or by coming into contact with an environmental source of pathogens.

Various landscape elements can act as either barriers or conduits for disease spread. For example, large rivers appear to slow the spread of rabies carried by

raccoons (*Procyon lotor*) in the northeastern USA (Smith et al. 2002). Alternatively, rivers can facilitate disease spread, as appears to be the case for the rice yellow mottle virus (RYMV) in Africa (Traore et al. 2005). The RYMV is believed to have originated in the Eastern Arc Mountains of Tanzania and Kenya, and spread westward across the African continent from there. Although chrysomelid leaf beetles are largely responsible for spreading the virus, their dispersal range is ultimately limited by habitat fragmentation and mountain ranges, and thus cannot possibly explain the continent-wide spread of the virus. Instead, RYMV may have spread through wild rice populations growing along major river corridors, such as the Niger and Benoue Rivers, into Central and West Africa. That, coupled with the recent intensification of rice cultivation and production throughout West Africa, may have facilitated the spread of RYMV across the continent.

Water corridors are not the only conduits for disease spread, however. Roads, for example, may also facilitate disease spread, even into remote areas that are otherwise undisturbed by human activities. In the arctic wilderness of northern Alaska, snails infected with parasitic flatworms (trematodes, or 'flukes') were only ever found within 330 m of a major highway that bisected the region (Urban 2006). Mud splattered from vehicles, in addition to water sprayed by road crews for dust control, may have helped 'seed' the roadsides with the snails and their parasites up and down the highway. Snails are not the definitive hosts for trematodes, however, which eventually move into a vertebrate host to complete their life cycle, where they are then capable of causing disease in species such as caribou (*Rangifer tarandus*) and gray wolves (*Canis lupus*). Because roadways are also used as travel corridors by these mammals, they unwittingly assist in spreading their own parasites beyond just the splash zone along road margins.

Another well-documented case of roads facilitating disease spread is that involving the invasive pathogenic fungus (*Phytophthora lateralis*) that causes fatal root rot in Lawson's cypress (also known as the Port Orford cedar, *Chamaecyparis lawsoniana*), a conifer native to the northwestern USA (Jules et al. 2002). The pathogen is water-borne and is carried via free water into the soil, where it can persist indefinitely. Patterns of infection are clearly tied to road crossings at creeks, where mud dislodged from logging trucks, all-terrain vehicles, and automobiles spreads the disease over long distances and into remote areas. As a consequence, the US Forest Service has implemented seasonal road closures during wet periods in an effort to limit its spread.

Just as with invasive species, disease spread can be enhanced by human-mediated transport over long distances. For example, mosquitoes are known carriers of pathogens that cause disease in humans, such as the West Nile virus (WNV), and are frequently transported by humans through travel or commerce. In the case of WNV, however, migratory birds may ultimately have been responsible for its rapid spread across North America (Dusek et al. 2009). First detected in New York in the eastern USA in 1999, WNV had spread westward all the way to the Pacific coast within four years, and by then was found throughout much of Canada and into Mexico and the Caribbean as well. The virus is maintained primarily through a mosquito-bird-mosquito transmission cycle, and the strain introduced to North America is highly virulent, accounting for the deaths of tens of thousands of birds. Unfortunately, the virus can also cause serious disease that is sometimes fatal in humans and livestock, such as horses.

Part of the challenge in studying the dispersal of pathogens directly is that traditional tracking methods, such as mark-recapture techniques, are difficult to apply to pathogens. This may explain why so many landscape epidemiological studies focus instead on animal-vectored diseases. Increasingly, however, molecular markers such as microsatellites are being used to assess patterns of dispersal indirectly, by analyzing the population genetic structure of pathogens and their hosts (Plantegenest et al. 2007; Biek & Real 2010). We will cover landscape genetics more fully in **Chapter 9**.

SPATIAL DYNAMICS OF DISEASE RESERVOIRS AND HOSTS Landscape structure also has the potential to influence disease spread through its effects on the spatial dynamics of reservoirs and hosts (Ostfeld et al. 2005). Reservoirs in particular are important for understanding the emergence and spread of infectious disease. Reservoirs can play a variety of roles in disease dynamics, from simply increasing the total pathogen or inoculant load in the environment, to providing a refuge for pathogens during seasons or years that are unfavorable for disease spread, to providing alternative hosts for pathogens (Plantegenest et al. 2007).

Many pathogens are host generalists, in that they are capable of infecting more than one host species, and thus can be maintained in a species (perhaps at subclinical levels) that can then transmit it to other species, including humans, where it may cause more-serious disease and trigger epidemics. **Disease spillover**, the transmission of disease from a reservoir population to another host population, may drive disease dynamics in many generalist pathogen systems, perhaps more so than transmission within the reservoir population itself (Powell & Mitchell 2004). Thus, the abundance and diversity of alternative hosts can sometimes maintain—and even promote—disease on the landscape. Such spillover effects are rare, however, because the presence of alternative hosts that function as 'dead ends' in the transmission cycle more usually serves to limit disease spread (i.e. other species serve to dilute rather than amplify the risk of disease spread).

Although uncommon, pathogen spillover from other animals has contributed to some serious diseases in humans (**zoonoses**, such as rabies), as well as in our domestic animals and livestock. For example, bison (*Bison bison*) and elk (*Cervus canadensis*) are the primary reservoirs for the pathogen (*Brucella abortus*) that causes brucellosis in cattle and other livestock within the Greater Yellowstone region of North America (Cheville et al. 1998). Similarly, pathogens may spread from wild plant populations to infect our crops, orchards, or nursery stock. For example, Asian soybean rust (*Phakopsora pachyrhizi*), which first emerged in Japan in the early 1900s, is now a serious threat to soybean-producing regions of the world. In North America, soybean rust is maintained seasonally in reservoir populations of wild plants, including kudzu (*Pueraria montana* var. *lobata*), an invasive species that was also introduced from Japan (Fabiszewski et al. 2010). Ironically, then, an invasive plant is now functioning as a reservoir species for an invasive plant pathogen that is capable of spilling over to cause significant losses in an economically important crop (which, it should be noted, is itself an introduced species that also originated in East Asia).

Nor does disease spillover occur in only one direction, from wild to domestic species; pathogen spillover can also occur from domestic to wild species. Indeed, brucellosis in the Greater Yellowstone area was likely first introduced into the native bison population *from* infected cattle in the early 20th century (Meagher & Meyer 1994). Controlling outbreaks of soybean rust within the domesticated host (i.e. soybean) rather than in wild host populations (e.g. kudzu) is likely to be a more successful strategy for disease management and for preventing pathogen spillover into native legumes, several of which are federally listed as threatened or endangered species, in soybean-growing regions of the USA (Fabiszewski et al. 2010). Diseases from companion animals, such as dogs and cats, are also threatening native wildlife, including species of conservation concern. For example, toxoplasmosis is threatening the recovery of the southern sea otter (*Enhydra lutris nereis*) off the California coast, but the origin of the disease appears to have a terrestrial basis. Surface runoff, contaminated with domestic cat feces infected with the parasite (*Toxoplasma gondii*), is believed to be the nonpoint source of this 'pathogen pollution' into the coastal marine environment (Miller et al. 2002).

Because reservoirs and vectors are connected epidemiologically to other species or habitats, their identification and control often represents the key to successful disease management. From a landscape perspective, this entails identifying what landscape features are conducive to the build-up and maintenance of disease reservoirs and vectors, or that increase their frequency of interaction with other host species (hotspots of disease transmission; Bonnell et al. 2010). In some cases, successful disease management might even be achieved through the removal or management of habitat used by disease reservoirs or vectors (habitat-based intervention; Gu et al. 2008). For example, mosquitoes responsible for transmitting malaria and other diseases can be controlled by eliminating or treating stagnant water sources near human dwellings, where ovipositing females can find both breeding habitat and a ready blood-meal in close proximity. There is much interest in the possibility of environmental management of diseases like malaria, especially in regions of the world like Sub-Saharan Africa and Southeast Asia where increasing resistance to antimalarial drugs is a growing problem.

EFFECTS OF HABITAT LOSS AND FRAGMENTATION ON DISEASE SPREAD The transformation of landscapes by humans, coupled with climate change, the introduction of invasive species (including pathogens), and the loss of biological diversity, have created a perfect storm that is giving rise to an increase in emerging infectious diseases worldwide. Of these various agents of global change, habitat loss and fragmentation—particularly through deforestation—is likely contributing most to disease emergence. For example, deforestation

in the Amazon basin is linked to a rise in the incidence of malaria, which results from a pattern of human settlement and land abandonment that ends up creating favorable breeding habitat for the dominant malaria vector in this region (the mosquito, *Anopheles darlingi*; Vittor et al. 2009). Settlers clear the rainforest to carve out farms, create fish ponds, and dig water wells, only to abandon these settlements a few years later when the soils have been exhausted. Abandoned farms transition to secondary forests that shade the former fish ponds, allowing a thick algal mat to grow on the surface, which is then capable of sheltering larval mosquitoes from insectivorous fish, creating a perfect breeding ground for mosquitoes. The prevalence of mosquito larvae was thus correlated with the amount of forest on the landscape: mosquito larvae were 7× more prevalent in landscapes with <20% rainforest remaining than in those with >60% forest cover (Vittor et al. 2009). Not only were mosquito larvae more prevalent in deforested areas, but adult mosquitoes (the actual malaria vector) inflicted more bites on humans than in predominantly forested areas (Vittor et al. 2006). Thus, deforestation may influence disease dynamics by altering the abundance and distribution of disease vectors and susceptible hosts.

Although the previous example demonstrated how habitat loss (deforestation) could lead to an increase in disease prevalence, habitat loss and fragmentation can have positive as well as negative effects on disease dynamics. On the one hand, habitat loss and fragmentation could depress host populations or even lead to a loss of disease vectors or reservoir species, resulting in lowered disease risk. Host density is a major factor driving disease epidemics; if host density falls below some threshold level, then disease may not be able to invade or cannot be sustained within a population owing to lower rates of disease transmission and contact among susceptible and infected individuals (Anderson & May 1979). Thus, disease is expected to be maintained in populations that lie above the **epidemic threshold**, which can be assessed not only in terms of population density, but also in terms of the patterns of contact (connectivity) among individuals that ultimately give rise to a population network capable of facilitating disease spread (Davis et al. 2008; Craft et al. 2009).

Conversely, habitat loss and fragmentation may force individuals to crowd into whatever habitat remains, pushing population densities to unnaturally high levels that increase the risk of disease, especially if individuals are now stressed and their immune systems compromised as a result of crowding, displacement, and diminished food availability. In Uganda, deforestation temporarily resulted in higher densities of the endangered red colobus monkey (*Procolobus rufomitratus*) within the remnant forest, which led to increased nutritional stress and a higher prevalence of parasitic infections within these populations, both of which may have contributed to the eventual decline of the red colobus within forest fragments (Chapman et al. 2006; Gillespie & Chapman 2007). Alternatively, habitat loss and fragmentation could increase disease risk indirectly, by reducing or eliminating the predators that usually prey upon reservoir species, enabling them to build up in the system (Ostfeld & Holt 2004). For example, rodents are reservoirs for many zoonotic diseases, such as Lyme disease, hantavirus pulmonary syndrome, and bubonic plague, but eradication of their vertebrate predators (e.g. foxes, coyotes, birds of prey), either directly through targeted removal programs or indirectly through the removal or degradation of their habitat, may allow rodent populations to increase, thereby increasing the risk of disease spread to humans. Predators may thus protect our health by keeping rodent populations in check.

Although we have previously considered how increasing host diversity can contribute to disease spillover and increase disease risk in some instances, an increase in the number of other host species more usually has a **dilution effect**, which reduces an individual's risk of contracting the disease. Indeed, this effect is intentionally exploited in agroecosystems through intercropping and the use of different genetic cultivars. By planting a mixture of cultivars that differ in their susceptibility to disease, agriculturalists can enhance the diversity and thus heterogeneity of their fields to achieve a dilution effect that may protect against disease (Mundt 2002).

Host diversity can thus provide a strong buffering effect against disease, which is further illustrated by the case of the Lyme disease system in the northeastern USA (Keesing et al. 2010). Lyme disease is transmitted to humans by the bite of an infected blacklegged tick (*Ixodes scapularis*), which acquires the infection during an earlier stage of development by feeding on an infected host, such as the white-footed mouse (*Peromyscus leucopus*). The white-footed mouse is ubiquitous and is the most abundant as well as the most competent host for the Lyme bacterium (*Borrelia burgdorferi*) in these forests. In contrast, most ticks that attempt to feed on North American opossums (*Didelphis virginiana*) are groomed off and killed; the opossum is thus a poor host for the pathogen and can help to reduce the risk of disease transmission in this system. Unfortunately, opossums tend to be absent from heavily degraded or fragmented forests where white-footed mice still thrive. Thus, habitat loss and fragmentation end up amplifying the risk of disease transmission through the loss of a host species that contributed to the buffering capacity of this system against disease.

By increasing the amount of edge, fragmentation can also have a positive effect on disease spread, by increasing the likelihood or rate of transmission between reservoir species and hosts, as evidenced by the various rodent-borne zoonoses we've discussed. To use another example, Sudden Oak Death is a deadly canker disease caused by a non-native fungal pathogen (*Phytophthora ramorum*) that has killed millions of oaks (*Quercus* spp.) and tanoaks (*Lithocarpus densiflora*) along the California and Oregon coasts in recent years (Rizzo & Garbelotto 2003). Proximity to a forest edge (<6 m) is the most important determinant of disease incidence and tree mortality from Sudden Oak Death, probably because trees near edges are more likely to intercept inoculum driven by the wind and rain (Kelly & Meentemeyer 2002).

Despite the potential for this sort of edge-mediated fragmentation effect on disease risk, it appears that the establishment and spread of Sudden Oak Death across the region has been facilitated not by habitat fragmentation per se, but rather by land-use changes that have created denser, more connected woodlands. Decades of fire suppression have led to an expansion of coastal woodlands and an increase in a major foliar host of the pathogen (the California bay laurel, *Umbellularia californica*, in which the pathogen produces only a non-lethal leaf infection), creating a more closed canopy and favorable microclimate that provides the optimal growth conditions for the pathogen, as well as a plentiful reservoir of inoculum (Meentemeyer et al. 2008). Thus, as with many aerially dispersed plant pathogens, the spatial pattern of Sudden Oak Death may be determined more by the ability of the pathogen to become established and spread, than by dispersal per se, which underscores the importance of assessing landscape connectivity in evaluating—and managing—the future spread of Sudden Oak Death (Ellis et al. 2010).

The connectivity of the disease landscape can also be affected by heterogeneity created by host-species diversity. As discussed previously, increased host-species diversity could either contribute to the amplification of disease if the pathogen can infect a wide range of hosts, or conversely, to a dilution effect if most of these species are incompetent hosts that represent 'dead ends' for disease spread. In the case of Sudden Oak Death, the fungal pathogen infects a wide range of hosts, more than 40 of which are native to the coastal forests of California and Oregon (Rizzo et al. 2005). Because species such as the tanoak and bay laurel contribute to increased disease risk, we might imagine that host diversity amplifies disease spread in this particular pathosystem. However, a broad-scale survey of different forest types in the Big Sur region of California revealed that the relationship between host diversity and disease incidence was consistently a negative one

(Haas et al. 2011). Thus, increased host diversity contributes to a dilution effect in the incidence of the fungal pathogen responsible for Sudden Oak Death.

As the examples in this section illustrate, habitat loss and fragmentation—in addition to the attendant loss of biodiversity—can amplify the incidence and spread of disease. Spatial modeling and disease risk mapping, which combine the various pathogen, host, environmental, and landscape factors that give rise to the emergence or prevalence of disease, have become the tools of choice for monitoring and predicting disease outbreaks, as well as planning where targeted control programs are most likely to be effective. In the next sections, we provide an overview of spatial and patch-based models of disease spread, consider the use of landscape connectivity measures for assessing the potential for disease spread, and then conclude with a discussion of the advantages and limitations inherent in disease risk mapping.

Spatial Models of Disease Spread

As with invasive spread, the spread of disease can be modeled as a simple diffusion process. This was first done in the context of modeling the spread of the rabies virus (RABV) in red fox (*Vulpes vulpes*) populations in Europe (Anderson et al. 1981; Murray et al. 1986). Western Europe remained relatively free of rabies until about 1939, when rabies emerged among foxes in Poland and subsequently spread westward at a rate of ~30–60 km/yr (Real & Childs 2006; **Figure 8.13**). In the simplest diffusion model, the fox population is divided into two groups of individuals that are either infected with RABV (*I*) or are susceptible to infection (*S*). Thus, the rate at which these susceptible individuals become infected is influenced by both the disease transmission rate (*r*; how easily the virus passes from individual to individual) and the contact rate between infected and susceptible individuals (the probability that the two types of individuals meet at random, *SI*) as

$$\frac{\partial S}{\partial t} = rSI. \qquad \text{Equation 8.7}$$

Rabies is invariably fatal, and thus infected individuals will die at a rate *d* (*dI*). Foxes are territorial, so the main factor contributing to disease spread in this system is the dispersal of infected individuals across the landscape, which is given by the diffusion coefficient (*D*, km²/yr). Thus, we can model the disease dynamics of rabies in red foxes in terms of the relationship between disease birth and death processes (the infection rate of susceptible individuals minus the death of infected individuals) and the spatial spread of the disease (in one dimension, *x*) as

$$\frac{\partial I}{\partial t} = rSI - dI + D\frac{\partial^2 I}{\partial x^2}. \qquad \text{Equation 8.8}$$

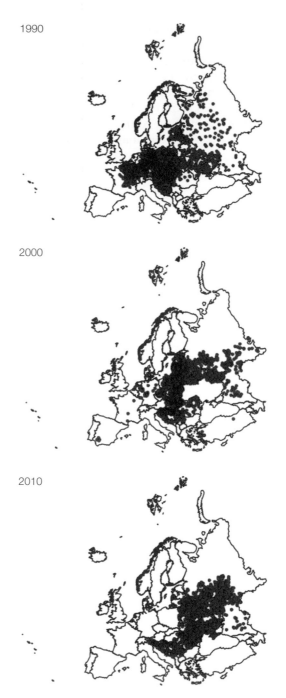

1990

2000

2010

Figure 8.13 Distribution of reported rabies cases in red fox (*Vulpes vulpes*) populations across Europe. Starting at the height of the most recent outbreak in the 1990s, mass vaccination programs have succeeded in pushing the virus south, clearing it from several countries (e.g. France, the Netherlands, Germany, Czech Republic, and Poland) but it is still endemic in many Eastern European countries, especially those along the Black Sea.

Sources: Rabies maps generated by the World Health Organization's Rabies Bulletin database (published online http://www.wildlifeonline.me.uk/rabies.html).

Although more complicated diffusion models have been developed for the spread of rabies in red foxes, this simple model makes a couple of important predictions. First, there is a critical threshold for rabies persistence, in which rabies will die out if the susceptible population declines below $S_c = d/r$. In other words, the disease is expected to die out quickly if it is highly virulent (d is high) relative to the transmission rate (r), because individuals die before the disease can be spread to other individuals. For rabies in the red fox, this critical threshold density may be about 1 fox/2 km², although some estimates from Europe suggest that rabies can spread at much lower densities and that <1 fox/5 km² is needed to stop it from spreading (Anderson et al. 1981).

Second, in areas where densities exceed the critical threshold, rabies will advance with a wave front velocity of

$$c = 2\left[D(rS_0 - d)\right]^{1/2} \qquad \text{Equation 8.9}$$

where S_0 is the density of susceptible individuals prior to the introduction of RABV to the population. Similar to the approach taken with RD models of invasive spread, we can incorporate environmental heterogeneity into this modeling framework by using diffusion coefficients (D) measured within different habitats or in different parts of the species' range to produce more accurate predictions of the rate of disease spread (e.g. Shigesada & Kawasaki 1997).

As with invasive spread, disease spread is also predicted to occur as a travelling wave of constant velocity, c (**Figure 8.14**). This assumes, however, that dispersal occurs as a random walk, in which the move-ment trajectory of individuals consists of a series of adjacent, but randomly oriented, steps (**Figure 6.23**). In the context of disease, this means that transmission occurs via random contact among adjacent hosts. Recall, however, that long-distance dispersal—even if a rare event—may instead produce an accelerating wave front, whose velocity increases in both time and space (Kot et al. 1996). Many important crop diseases, and even some human diseases, are caused by wind-borne or aerially vectored pathogens, and are therefore capable of being dispersed long distances by air currents or birds. We discuss such a case next.

The rate of spread for several major epidemics that have swept through whole continents was analyzed by Mundt and his colleagues (2009a). These epidemics ranged from the historically significant potato late blight, which was responsible for the Irish potato famine in the winter of 1845–46 that contributed to the deaths of a million people (and the forced exodus of millions more), to the modern-day epidemics of WNV and avian influenza ('bird flu') that are capable of causing serious and sometimes fatal disease in humans (**Figure 8.15**, left). All of these epidemics are caused by pathogens that are aerially dispersed, either by wind or birds, and thus can be carried long distances. Dispersal was therefore modeled as an inverse power law, which produces the sort of fat-tailed distribution expected if occasional long-distance dispersal events occur. For diseases capable of long-distance dispersal, the instantaneous velocity (v_t) of the epidemic wavefront is given by

$$v_t = \frac{x_t}{b} \qquad \text{Equation 8.10}$$

where x_t is the position of the wavefront (the distance from the source of infection at time t) and b is the exponent of the inverse power law, which gives the steepness of the disease gradient (Mundt et al. 2009). Thus, the position of the wavefront (x_t) grows exponentially over time as

$$x_t = x_0 \left(\frac{b}{b-1}\right)^t \qquad \text{Equation 8.11}$$

where x_0 is the initial distance of the epidemic front from the origin, which is assumed to be proportional to the transmission rate, r. In practice, b can be estimated from a regression of either the distance from the source ($\ln(x_t)$) against time (**Figure 8.15**, middle) or velocity against distance (**Figure 8.15**, right).

Despite great differences between hosts and mode of transport, all of these epidemics exhibited an accelerated rate of spread with time and distance from the site of the initial outbreak (**Figure 8.15**). Interestingly, the exponents (b) of the power-law dispersal gradients are all close to 2, which means that the distance between the origin and the epidemic front doubled per unit

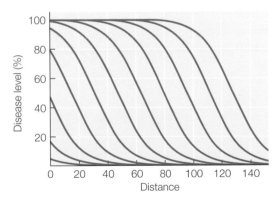

Figure 8.14 Travelling wave for disease spread. Each curve represents the disease gradient (disease incidence as a function of distance) at different (equally spaced) points in time. Thus, the incidence of disease is initially low (<10% at the focus of infection, distance = 0), which increases to almost 20% at the next time step, and reaches 100% infection at the focus by the sixth or seventh time step. By then, the disease has spread outward and there is a high incidence of disease (>50%) some 60–80 distance-units from the initial focus.

Source: After Mundt et al. (2009b).

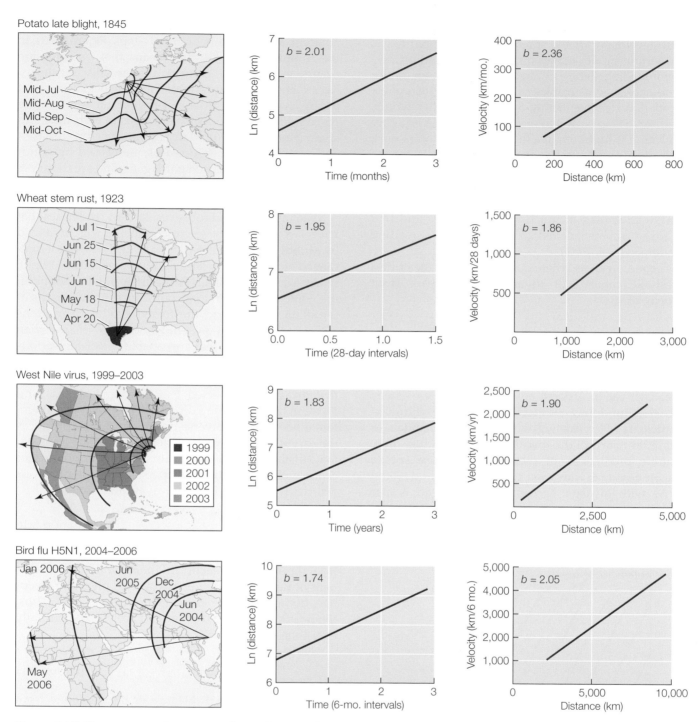

Figure 8.15 Disease spread as an accelerated wave. The continent-wide spread of several epidemics (left) exhibited a linear increase in distance of spread as a function of time (middle) as well as in the velocity of spread with increasing distance from the initial source (right). All of these pathogens are capable of long-distance dispersal and are transported either on air currents (potato late blight, wheat stem rust) or by migratory birds (West Nile virus, bird flu).

Source: After Mundt et al. (2009a).

time, and the slope of the plot of velocity versus distance equaled ~0.5 (i.e. 1/2). It remains to be seen whether other epidemics caused by aerially dispersed pathogens also spread at a rate approximated by an inverse square law. Regardless, accelerating epidemic waves clearly increase the potential for rapid spread over broad areas, and thus present a significant challenge for the management of disease spread.

Metapopulation Models of Disease Spread

The models of disease spread we have discussed thus far assume that the landscape is homogeneous or continuous. If the population is really distributed as a metapopulation, however, how might the movement of infected individuals among populations affect the probability of persistence or extinction risk of the entire metapopulation? To explore this, Hess (1996) extended Levins' classic metapopulation model (Levins 1969; Equation 7.12) to incorporate disease dynamics. Recall that in the basic metapopulation model, the change in patch occupancy is modeled as the difference between patch colonization and extinction rates. Thus, patches are assumed to be in one of two states: the patch has either been colonized (occupied) or the local population has gone extinct (empty).

In the context of a disease that is spreading through a metapopulation, we can further characterize occupied patches in terms of whether the local population has been infected or not. Thus, a patch in our diseased metapopulation can be in one of three states: empty, infected, or susceptible (**Figure 8.16**). Let's assume that the disease is invariably fatal and thus populations that become infected will go extinct. In that case, extinction rates of infected populations (e_i) will be higher than those in susceptible populations (e_s), such that $e_i > e_s$. Let's also assume that individuals move randomly among patches, and that both infected and susceptible individuals have the same rate of movement ($m_i = m_s$, so just m, as in the basic metapopulation model). Finally, we need to account for the transmission rate (r), the probability that a susceptible population becomes infected by an infected immigrant (all individuals in an infected population are assumed to be infected, and thus all immigrants from an infected population are infected). Just as we did not concern ourselves with different population densities among patches in the basic metapopulation model, we will not bother with different infection rates among populations here. Therefore, we can characterize the change in susceptible and infected populations over time as

$$\frac{dS}{dt} = mS(1 - I - S) - Se_s - mrIS$$

and

Equation 8.12

$$\frac{dI}{dt} = mI(1 - I - S) - Ie_i - mrIS$$

where S is the fraction of total patches occupied by susceptible populations and I is the fraction of total patches occupied by infected populations (so, the fraction of patches that are empty = $1 - I - S$). Compare this form to that of the basic metapopulation model (Equation 7.12). The first term in each expression is the rate at which empty patches are recolonized by migrants coming from either susceptible (mS) or infected (mI) populations, respectively (recall that only empty patches can be recolonized). The second term is the rate at which susceptible (Se_s) or infected (Ie_i) populations go extinct, respectively, and the third is the probability that susceptible populations will become infected.

Recall that the basic metapopulation model had a single equilibrium solution for persistence ($\hat{p} = 1 - e/m$). If we rescale by the colonization or movement rate, m, and let $R_s = 1 \; e_s / m$ and $R_I \; 1 - e_i / m$, then substituting in Equation 8.12 above, we have

$$\frac{dS}{d(mt)} = S[R_s - S - (1 + r)I]$$

and

Equation 8.13

$$\frac{dI}{d(mt)} = I[R_I - I - (1 - r)S].$$

The important thing here is that the equilibrium solutions to these two expressions produces four stable equilibria. The stability criteria for each, defining all of the possible disease states in the metapopulation, are as follows:

1. Extinction: no patches of any kind occupied ($S = 0$, $I = 0$)

2. No disease: patch occupancy is given solely by the equilibrium solution for the persistence of susceptible patches ($S = R_s = 1 - e_s/m, I = 0$)

3. Endemic disease: disease persists and is present in some, but not all, populations
$$\left(S = \frac{R_s - R_I - rR_I}{r^2}, \; I = \frac{R_I - R_s + rR_s}{r^2} \right)$$

4. Pandemic disease: all populations are infected and disease persistence is given solely by the equilibrium solution for infected patches ($S = 0, I = R_I = 1 - e_i/m$).

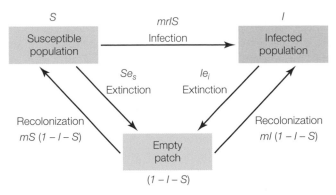

Figure 8.16 Structural diagram for a disease metapopulation model. In this simple disease system, populations can be in one of three states: susceptible (non-infected, S), infected (I), or extinct (empty patch, 1–I–S). The various types of transitions that can occur among these states are indicated by arrows, along with the accompanying term describing that transition in the model (Equation 8.12).

Source: After Hess (1996).

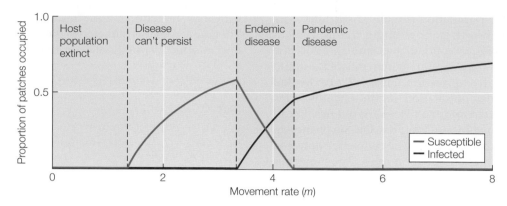

Figure 8.17 Effect of movement rate on disease spread in a metapopulation.   The movement rate (m) between populations can be used as a proxy for disease spread. At low rates of movement ($m = 1–3$), the proportion of occupied patches increases, but remains disease-free. Beyond a certain threshold of exchange among populations ($m = 3$), the proportion of infected populations increases rapidly to become pandemic. Parameters for this scenario: $e_s = 1.4$, $e_i = 2.4$, and $r = 0.5$ (Equation 8.14).
Source: After Hess (1996).

Thus, with some fairly simple modifications to the basic metapopulation model, we can identify the conditions under which a disease will spread (pandemic disease), as a function of the rate at which individuals move among populations (m, also a measure of connectivity) relative to the severity (e_i) and transmission rate (r) of the disease (**Figure 8.17**). If we examine $dI/d(mt)$ in Equation 8.13 for scenarios where disease is present ($I > 0$), then the proportion of infected populations will increase if

$$(1-I-S)+rS > \frac{e_i}{m}.$$ Equation 8.14

So, if infected populations go extinct faster than individuals can spread the disease (a high e_i/m ratio), then the disease will eventually die out. Consider the implications of this for disease spread within a metapopulation. If a disease is easily transmitted (large r), then it can become established at lower movement rates (m) than diseases that are less infectious. Conversely, a disease that spreads easily (high m) but is of mild severity ($e_i \sim e_s$) is unlikely to have much effect on patch occupancy within the metapopulation. Thus, a highly contagious (high r) disease of moderate severity may ultimately pose the greatest threat to metapopulations (Hess 1996).

This metapopulation disease model is basically a patch-structured extension of a more general class of stage-structured or compartmental epidemiological models, in which individuals within a population are either susceptible (S), infected (I), or recovered and now immune (R). The SIR model and its many extensions (e.g. SEIR, to include individuals that have been exposed E, but who are not yet infectious, thereby adding a latent period to disease dynamics) are a staple of disease modeling, and the interested reader is encouraged to explore this further in one of the many texts devoted to the subject (e.g. Keeling & Rohani 2008).

Landscape Connectivity and the Potential for Disease Spread

One of the many challenges of modeling the spread of a disease is that we do not always have good (or perhaps any) demographic information available for the pathogen or host population. All is not lost, however, as we might at least be able to assess the potential risk of spread, based on an analysis of landscape connectivity. Traditionally, the rate at which a population spreads is believed to be influenced more by dispersal (especially long-distance dispersal events) than demography (van den Bosch et al. 1992). Further, as we have just seen, connectivity among host populations can influence the spatial dynamics and persistence of disease on the landscape. In the case of the disease metapopulation model, the degree of connectivity was encapsulated in the movement rate (m), which affected the rate at which individuals (infected or not) moved among populations. High rates of movement signify a highly connected metapopulation, thus increasing the potential for less-infectious diseases to spread widely throughout the population network.

Graph-theoretic or network modeling approaches can be applied to assess landscape connectivity (**Chapter 5**) and the potential for disease or pest spread. Consider the case of managed landscapes, such as plantation forests or agricultural crop fields, which are largely continuous stands of a single species (monocultures). In many parts of the industrialized world, agricultural policies and government incentives actively promote the production of just a few crop types, leading to more homogeneous landscapes that are potentially well-connected and thus susceptible to invasion by pests or disease.

To address this potential, Margosian and her colleagues (2009) adopted a graph-theoretic approach to examine the connectivity of the American agricultural landscape for several economically important crops grown in the USA. More than two-thirds of the cropland in the USA is devoted to the production of just four crops (corn, soybean, wheat, and cotton). Using a 'dropped-edge analysis,' they assessed connectivity at different landscape resistance thresholds, an index based on the acreage of each crop species within adjacent counties (the smallest administrative unit for which crop data were available at a national level). A low resistance threshold corresponded to areas with high crop acreage (high host density) where spread could occur most easily, assuming spread occurs primarily among neighboring counties. Higher resistance thresholds thus indicate that spread is likely to be more difficult because of low host density or larger distances spanning crop-production regions that would have to be traversed by the pathogen or pest. For pests or pathogens capable of long-distance dispersal or transport, however, the agricultural landscape is connected at a national level for crops such as corn (maize) and soybean (**Figure 8.18**). This graph-theoretic analysis of landscape connectivity thus permitted a rapid assessment of the potential risk for disease or pest spread at a national level, which could then be used to identify regions that have a high potential for disease spread. Conversely, such areas might be able to serve as quarantine or containment areas if, for example, crop connectivity can be disrupted through the directed production of alternative crops (increasing heterogeneity of the agricultural landscape) or through the targeted application of pesticides or fungicides.

We have previously explored how percolation theory has inspired the development and application of neutral landscape models, both in terms of assessing landscape connectivity (**Chapter 5**), and the potential for invasive spread in this chapter. It should therefore come as no surprise that similar lattice-based approaches have also been applied to quantify thresholds in disease spread. The threshold for the spread of the ubiquitous fungal plant pathogen, *Rhizoctonia solani*, has been determined experimentally in agar landscapes (Bailey

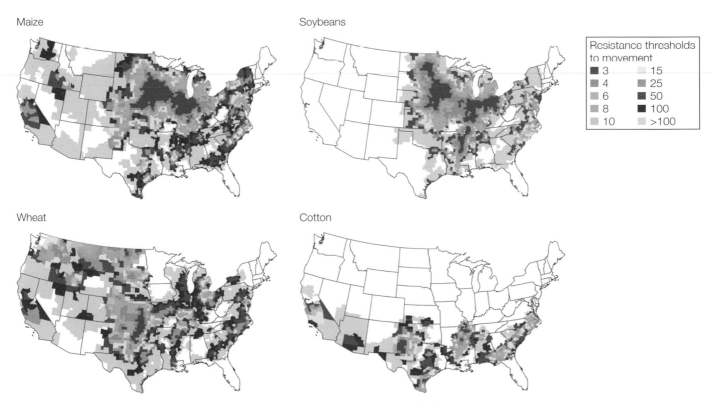

Figure 8.18 Connectivity of the American agricultural landscape to potential disease and pest spread. The predicted extent of spread was mapped for four major crop types based on the ability of pathogens or pests to overcome certain resistance thresholds to movement (low resistance = dark blue, high resistance = gray), which was based on the availability of each crop in the landscape. Spread is especially likely to occur at a regional scale (dark blue areas), but pests or pathogens capable of long-distance movements or transport have the potential to spread across the entire country. The American agricultural landscape is thus well-connected for these crops.

Source: After Margosian et al. (2009).

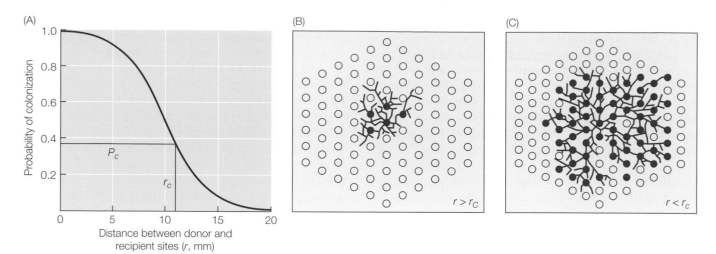

Figure 8.19 Percolation thresholds of disease spread in the fungal plant pathogen, *Rhizoctonia solani.* The threshold distance (r_c) among sites at which successful colonization (infection) of the fungal pathogen can occur (P_c) if spread occurs as a percolation process (A). Actual patterns of spread by the fungal pathogen within lattices in which the inter-site distance is either above (B) or below (C) the threshold distance (r_c). Note that disease spread occurs throughout the agar landscape only when inter-site distances are below the critical threshold (r_c).

Source: After Bailey et al. (2000).

et al. 2000). This fungal plant pathogen spreads through the soil via mycelial growth and expansion of fungal colonies (**Figure 8.19**). Logically, disease spread should occur only if susceptible hosts (i.e. plant seeds or roots) are close enough to enable the pathogen to infect new hosts. As predicted, disease spread (percolation) occurred within experimental arrays in which the inter-site distance among nutrient sites (r) was within the colonization range of the pathogen ($r < r_c$), but not at distances above this threshold ($r > r_c$; **Figure 8.19**). Thus, once again, inter-patch distances may be a critically important component of landscape structure, here influencing infection probabilities as well as the likelihood of disease spread. Intriguingly, this threshold of disease spread (P_c) was modified by resource quality, such that infection could occur over larger distances when nutrient concentrations were high, making spread more likely to occur at lower site (host) density. The potential for disease spread to occur more readily at lower host density in high-quality habitats is not generally appreciated, and represents yet another means by which landscape structure (e.g. differential habitat quality) could potentially influence the pattern and rate of disease spread.

POPULATION CONNECTIVITY AND THE CONNECTIVITY OF THE DISEASE LANDSCAPE Host density is expected to promote disease spread above some abundance threshold based on the basic reproductive number or ratio (R_0), the expected number of new infections following the introduction of a single infectious host into a susceptible population. Recall from **Chapter 7** that in population models, R_0 is the number of offspring pro-

duced by an individual during its lifetime. Thus, in epidemiology, R_0 is the number of secondary infections caused by the initial infected individual during its infectious lifetime (i.e. over the period that it is infectious). The threshold for disease spread is thus defined as $R_0 = 1$; above this epidemic threshold, disease is able to spread throughout the population. Although the basic reproductive ratio is a key concept in epidemiology, and various methods have been devised to calculate it (Heffernan et al. 2005), its usefulness for predicting disease spread in spatially structured populations, or for species that live in social groups, may be more limited. In these systems, the pattern of contact (connectivity) among spatially distributed host populations or social groups is expected to have a large effect on the ability of disease to spread.

For example, the spread of sylvatic plague among great gerbils (*Rhombomys opimus*) in Kazakhstan, which live within family groups in extensive burrow systems, exhibited a clear threshold in which major outbreaks only occurred when at least 33% of the burrows within a region were occupied by gerbils (Davis et al. 2008; **Figure 8.20**). Further, this observed threshold in the occurrence of plague coincided with the predicted percolation threshold of 31% derived from a network-based model of disease spread, but not with the expected R_0 threshold (i.e. $R_0 = 1.5$ rather than the expected threshold $R_0 = 1.0$ at the plague threshold). Thus, a larger abundance threshold (as defined by R_0) may be required before disease can spread in spatially or socially structured populations. This difference is important from the standpoint of disease management, in terms of what targets should

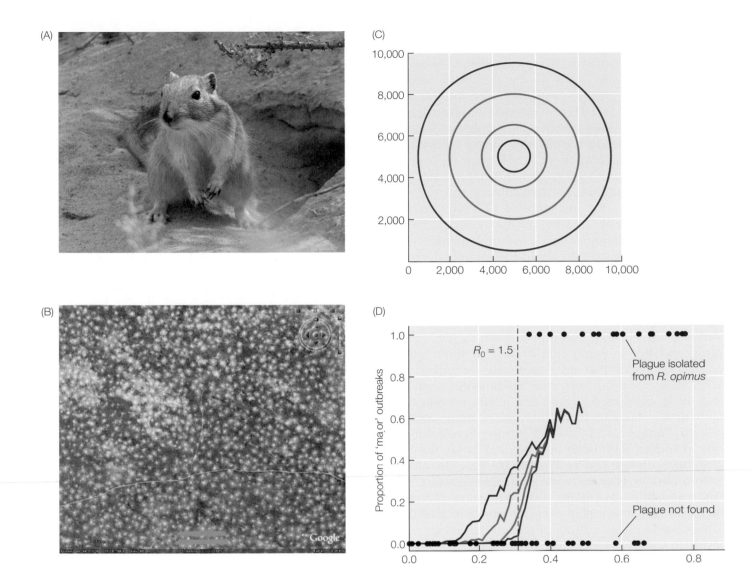

Figure 8.20 Abundance threshold for the spread of plague among great gerbils (*Rhombomys opimus*).   Great gerbils (A) live in family groups within complex burrow systems across arid regions of Central Asia. (B) The white disks that dot the landscape in the satellite image are patches of bare earth, approximately 30 m in diameter, excavated by the gerbils. (C–D) A network-based landscape model was used to predict the theoretical threshold at which sylvatic plague could spread different distances in a random network (red = 750 m, blue = 1.5 km, green = 3 km, purple = 4.5 km). When modeled as a percolation process, disease spread beyond 4.5 km is predicted to occur when 31% of the burrows are occupied (the colored lines in the graph, D, correspond to the colored rings on the hypothetical landscape, C). Plague does not occur when <33% of burrows are occupied, suggesting an abundance threshold for disease spread (dots in lower graph). The predicted threshold for disease spread thus matches the observed threshold for the occurrence of plague in this system. Note, however, that the epidemic threshold (the host density at which disease spreads) does not correspond to $R_0 = 1$ as general epidemiological theory predicts, but rather, to $R_0 = 1.5$. A larger abundance threshold (as defined by R_0) may therefore be necessary before disease can spread in socially structured populations.

Source: A) Photo by Yuriy75 and published under the CC BY-SA 3.0 License. B, C, and D are modified from Davis et al. (2008).

be set for reducing the number of susceptible individuals through wildlife culling or vaccination programs, as fewer individuals may have to be culled or vaccinated to protect the population in spatially or socially structured systems, especially if these measures are directed at populations or groups that have a high degree of connectivity with others (i.e. critical population nodes).

Alternatively, disrupting the social connectivity of a species via culling could have just the opposite effect, leading to an increase in disease incidence. In Great Britain, bovine tuberculosis (bTB) is a major threat not only to the cattle industry, but to the public health as well, given that the disease can be transmitted to humans (although pasteurization kills the bacteria in milk). Although cattle that test positive for exposure to bTB are immediately slaughtered, control of the disease has been complicated by the persistence of the bacterium that causes bTB (*Mycobacterium bovis*) in wildlife populations, such as the European badger (*Meles meles*). Badgers are believed to be the primary wildlife reservoir for bTB, and infected individuals can apparently carry—and spread—the pathogen for years before succumbing to the disease. Badger culling has thus been a major component of bTB-control policy in Great Britain since 1973 (Krebs et al. 1998). Despite the prevalence of cattle disease surveillance and badger-culling programs, the incidence of bTB has actually increased in recent decades (Donnelly et al. 2006). Various culling strategies have been implemented over the years, including the localized culling of badgers in areas where bTB has been detected ('reactive' culling).

Counter to expectations, the localized culling of badgers *increased* the incidence of bTB in neighboring cattle herds by almost 30% (Donnelly et al. 2003, 2006; Vial & Donnelly 2012). European badgers are highly social and live in communal burrow systems (setts) within territories that are vigorously defended. By artificially reducing population densities, culling effectively disrupts the social and territorial behavior of badgers, causing them to wander more widely (>1 km; Pope et al. 2007), which in turn increases their rate of contact with other badgers and cattle, thereby facilitating disease spread. Further, because the culling is localized, it is more widely dispersed across the landscape, thus increasing the proportion of cattle herds that are near culled areas (Donnelly et al. 2006). Thus, 'the localized culling of disease reservoirs may result, in some ecological systems, in social and spatial disturbances that have the potential to contribute to an increased disease risk in and around areas where such operations take place' (Vial & Donnelly 2012, p. 53). Although this would seem to support a more extensive culling program (i.e. over larger areas), this is still costly and problematic for a number of reasons, not the least of which is because the badger is a cultural icon in the UK. Setting culling targets is a challenge, as killing too few badgers will only exacerbate the problem and increase disease incidence, whereas killing too many puts the badger at risk of local extinction, in violation of the Bern Convention on the Conservation of European Wildlife and Natural Habitats[3] (Donnelly &

Woodroffe 2012). Badger culling remains controversial,[4] and vaccination programs of badgers are being implemented as an alternative to culling in some areas (e.g. Wales and Ireland).

Disease Risk Mapping

Disease risk is defined as the probability of infection or exposure to a particular pathogen or infectious disease agent (Ostfeld et al. 2005). **Disease risk mapping** thus involves the combined use of computerized mapping technologies (GIS) and various statistical approaches, including species distribution models, to relate the occurrence and distribution of disease (or its agents) to characteristics of the landscape, or other features of the environment, in order to better understand and predict disease risk (Kitron 1998; Plantegenest et al. 2007). Mapping efforts may be performed at various spatial scales, from an individual locale to an entire continent, and for temporal scales encompassing the duration of a single outbreak to multiyear models. It is worth emphasizing that a disease risk map is in fact a model (a statistical one rather than a mathematical or simulation-based model), and as with any model, the choice of spatial and temporal resolution determines the degree of precision, realism, and general applicability of the risk map (Kitron 2000).

The goal of disease risk mapping is not just a visualization of the spatial variation in disease risk across some geographic extent, but also to identify what landscape or other environmental and societal factors contribute most to that risk (Ostfeld et al. 2005). Ideally, this involves relating spatial information on the distribution of vegetation, land-cover types, land uses, or environmental factors (e.g. temperature, precipitation) to the distribution of the vector, reservoir species, and disease incidence (e.g. reported human cases; **Figure 8.12**). The resulting risk map might then help to identify hotspots of disease transmission, which could be used to guide targeted intervention and control programs involving vaccination, culling, or pesticide applications, as appropriate (Ostfeld et al. 2005). Unfortunately, such fully integrated spatial datasets on all components of the disease system may not be available or would be difficult (or even impossible) to obtain. Further, data are likely to be scarce in the very cases where such models would be most useful in predicting and controlling spread, such as the early stage of a disease outbreak. Thus, assessment of disease risk must necessarily be made on the basis of limited information (LaDeau et al. 2011).

[3]The Council of Europe's Convention on the Conservation of European Wildlife and Natural Habitats (the Bern Convention) is an international legal instrument that was adopted in Bern, Switzerland in 1979 and came into force in 1982. As of 2019, it has 51 contracting parties, including 45 of the 47 member states of the Council of Europe, as well as the European Union and five non-member states (https://www.coe.int/en/web/conventions/full-list/-/conventions/treaty/104/). The aims of the Bern Convention are to conserve wild flora and fauna and their natural habitats, especially where cooperation among several European nations might be required for successful conservation.

[4]Meanwhile, bTB continues to rise in the UK. Badger culls began as a pilot program in 2013 in Somerset and Gloucestershire, and now occur in 21 areas across eight counties (http://www.wildlifetrusts.org/badger-cull). More than 19,000 badgers were culled in 2017 alone.

Disease risk maps are therefore derived most often from either the distribution of a known disease vector (e.g. mosquitoes, ticks) or reservoir species, rather than the actual incidence of the disease. Depending on the disease in question, this may or may not be sufficient. For example, many arthropod vectors are sensitive to climatic factors, such as temperature and precipitation, or have specific habitat requirements (e.g. pools of stagnant water for mosquitoes) that are reasonably easy to map, and thus can be used to assess the probable occurrence of specific vectors and the potential disease risk. However, disease risk is bound to be better correlated with the presence and abundance of *infected* vectors rather than the vector per se (Ostfeld et al. 2005). Many arthropod vectors, such as mosquitoes, are widespread and abundant, and thus simply mapping their distribution may not be sufficient for predicting disease risk without additional information on the pattern of infection within the vector population.

Forecasting disease risk is a growing concern in the face of changing land use, agricultural practices, and climate patterns. The rate of deforestation, in conjunction with the encroachment of human settlements into these new areas, will likely increase contact between people and other species capable of serving as vectors or reservoirs of zoonotic disease. Changing environmental conditions, such as warming temperatures and increased rainfall, coupled with climate-driven shifts in the ranges of disease vectors and/or reservoir species, may well expand the distribution of so-called tropical diseases into new areas (Epstein 2000). Although the role of climate in the recent emergence (or re-emergence) of infectious diseases is a topic of debate (Lafferty 2009; Ostfeld 2009), there is evidence that at least some tropical diseases, such as malaria, cholera, dengue fever, and Rift Valley fever are now appearing (or re-appearing) with increasing regularity in temperate regions of the world, due in large part to recent climatic changes.

For example, leishmaniasis is a zoonotic vector-borne disease of tropical and subtropical regions, which is transmitted to humans from the bite of a sand fly infected with the *Leishmania* parasite. A large ulcer forms at the site of the bite (cutaneous leishmaniasis), but the parasite can also migrate to the spleen, liver, and bone marrow (visceral leishmaniasis), causing death if left untreated. Using a species distribution modeling approach (MaxEnt), González and her colleagues (2010) evaluated how the distribution of the primary disease vector (two species of *Lutzomyia* sand flies) and several reservoir species (four *Neotoma* woodrat species) would likely shift across the North American continent in response to projected climate-change scenarios over the next century. The outlook is not good. Even under the most conservative projection,

leishmaniasis has the potential to spread as far north as southern Canada through the eastern half of the United States, and perhaps farther, limited only by environmental constraints on the distribution of sand flies. This will translate into a doubling of the human exposure risk to leishmaniasis by 2080. This is consistent with the predicted expansion of leishmaniasis in Europe, such as within the Madrid region of Spain, where densities of sand flies (*Phlebotomus*) are expected to triple over the next century in response to climate change (Gálvez et al. 2011).

Disease risk does not necessarily translate into the actual incidence of disease, however. Although spatial data on disease incidence may be used to extrapolate from the current distribution to new geographic areas and into the future, this assumes that incidence and risk are highly correlated (Ostfeld et al. 2005). Correlation is not causation, however. Risk can be underestimated if there is variability in disease reporting across a region, such that disease incidence might be lower than actual cases. Alternatively, mitigative measures can reduce the incidence of disease in high-risk areas, such as through the use of insect sprays to reduce contact with mosquitoes or ticks in high-risk areas of malaria, WNV, or Lyme disease. In other cases, high exposure can lead to the evolution of host immunity, which also reduces the incidence of disease. We consider the evolution of disease systems next.

Evolutionary Landscape Epidemiology

Disease risk may change over time not only in response to changing land use or climate, but also as a result of the evolution of disease systems. Pathogens (or parasites) and their hosts are involved in an ongoing evolutionary arms race, in which hosts may eventually evolve some level of immunity and become less susceptible to infection over time (increased host resistance). In turn, pathogens and parasites may evolve counter-strategies that increase their virulence, ease of spread, or range of hosts that can be infected, perhaps even leading to the emergence of new diseases through the re-assortment of pathogen strains within an intermediate host (another example of disease spillover). For example, many novel strains of influenza capable of causing global pandemics, including the 1918 influenza pandemic that may have killed 30–50 million people worldwide, have their origins within domestic pigs, which serve as 'mixing vessels' for avian, swine, and human influenza strains (Smith et al. 2009). The emerging field of **evolutionary landscape epidemiology** combines landscape epidemiology and evolutionary theory to understand how the ecology and evolution of host-pathogen (or host-parasite) interactions shape the distribution, dynamics, and severity of diseases across heterogeneous landscapes (Deter et al. 2010).

Landscape structure can affect all of the evolutionary processes—genetic drift, gene flow, and selection—that contribute to spatial variation in the genetic structure of pathogens and host populations, and which shape the coevolutionary dynamics of disease systems (**Figure 8.21**). For example, we have previously discussed how landscape structure influences species occurrence and population density within a landscape (**Chapter 7**). Clearly, disease is not an issue if one or more species required for disease transmission does not occur at a particular site, which is why risk mapping might be helpful in identifying potential hotspots, where suitable habitat for pathogen, vectors, and hosts all coincide.

Within these potential disease hotspots, however, the population density of hosts may be an important determinant of disease transmission, not only from the

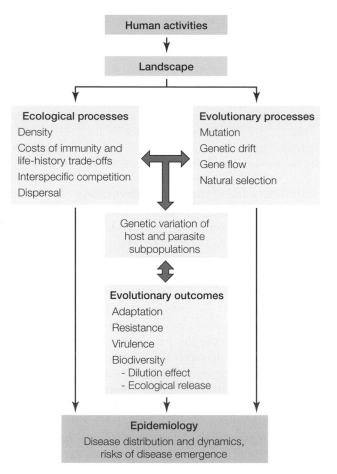

Figure 8.21 Framework for evolutionary landscape epidemiology. This framework emphasizes the role that human activities play in modifying landscape structure, which in turn affects the various ecological and evolutionary processes that influence the genetic structure and adaptive potential of pathogen and host populations in different habitats, all of which shape the distribution, dynamics, and severity of disease across the landscape.

Source: After Deter et al. (2010).

standpoint of evaluating the threshold level at which disease can invade ($R_0 > 1$), but also in terms of the expected level of genetic variation and thus adaptive potential within the population. Population size is inversely related to the rate at which genetic diversity is lost from a population through random **genetic drift**. High genetic diversity within a population may translate into greater variability among individuals in their susceptibility to disease, which increases the likelihood that at least some individuals will survive an outbreak. Furthermore, **inbreeding depression**, another malady of small populations, also leads to a loss of genetic diversity and individual fitness, such as a compromised immune system, making individuals more vulnerable to diseases. For example, American crows (*Corvus brachyrhynchos*) suffer from disease-mediated inbreeding depression, which has been linked to poorer body condition, a weaker innate immune response, and higher disease mortality among inbred birds (Townsend et al. 2010). In addition to a loss of individual fitness, diminished genetic diversity also erodes the evolutionary potential of small populations to adapt to future disease threats, a point we shall return to below.

Dispersal and gene flow are both influenced by landscape structure, creating spatial variation in the dynamics and genetic structure of populations across landscapes (**Chapter 9**). Metapopulation dynamics can strongly influence the ecological and evolutionary processes involved in host-parasite (or host-pathogen) interactions (Burdon & Thrall 1999; Gandon 2002). Clearly, parasites and pathogens may be responsible for disease outbreaks that result in the local extinction of host populations, but the persistence and stability of the host-parasite (or host-pathogen) system are very much dependent on the relative dispersal ability of both species, in terms of the ability of parasites or pathogens to colonize new host populations relative to the ability of hosts to recolonize new habitat patches.

The relative difference in dispersal rates between parasites/pathogens and their hosts may determine the extent to which either species can evolve an adaptive response. As long as dispersal is not so high as to swamp local adaptations, then the species that exhibits the greatest potential for dispersal (whether host or parasite) is the one most likely to exhibit an adaptive response (Gandon 2002). The threshold level of dispersal required for adaptation, however, is modified by the virulence and host specificity of the parasite or pathogen, with greater levels of dispersal being required as virulence and host specificity increase. Gene flow thus provides a source of genetic variability into populations upon which selection can act, thereby promoting local adaptation, as long as it is not too great. Indeed, intermediate levels of connectivity (gene flow that is neither too great nor too little) may be required for the long-term persistence of host-parasite/pathogen populations (Hagenaars et al. 2004).

As we have seen, population and landscape connectivity are important for assessing the potential for disease spread. We do not always know how the disease is being spread, however, given that direct transmission is rarely observed. In that case, landscape genetics (**Chapter 9**) may at least permit us to identify high-risk populations, based on their degree of connectivity to infected populations, as well as landscape characteristics that promote gene flow of the host, from which we can infer the likely pattern of disease spread on the landscape. For example, a landscape genetics approach was used to evaluate the spread of chronic wasting disease among white-tailed deer in the upper midwestern USA (Blanchong et al. 2008). Chronic wasting disease is a fatal prion disease of cervids that is transmitted through direct contact with infected individuals (via saliva) or possibly areas contaminated by the carcasses and excretory wastes of infected animals (Mathiason et al. 2006; Conner et al. 2008). Landscape structure strongly influences the movement and dispersal of white-tailed deer (Long et al. 2005), which in turn should influence gene flow and disease spread. The prevalence of chronic wasting disease was found to be lowest in populations that exhibited the greatest degree of genetic differentiation from the infected population core, from which they were separated by a large river that apparently served as a barrier to dispersal and gene flow, and at least for now, the spread of chronic wasting disease (Blanchong et al. 2008).

Landscape evolutionary epidemiology is clearly at the interface of several, intersecting disciplines, one of which—landscape genetics—will be covered more fully in **Chapter 9**. This section on landscape epidemiology has also demonstrated that we might need to extend our consideration of landscape effects beyond that of single species to consider landscape effects on interacting species. We will consider the effects of landscape structure on other types of species interactions in **Chapter 10**.

Future Directions

The combined threats posed by land-use intensification, habitat loss and fragmentation, and climate change are arguably the most significant—and urgent—environmental challenges we must confront in the coming decades. As this chapter has demonstrated, all of these factors—coupled with global travel and transport—are likely to impact the future distribution of species and their pathogens or parasites, resulting in the extinction of some species, the undesirable spread of others, and, quite possibly, a reshuffling of entire biological communities. Although such broad-scale changes in the distribution of species and their associated biomes have occurred previously in response to global climate cycles, the rate at which such range shifts must now occur to track environmental changes in both climate and landscape are unprecedented. This presents a unique challenge to the management of species that are presently of conservation concern, especially in terms of identifying and managing critical habitat required for the long-term persistence of these species. Basically, such climate-driven changes in species distributions mean that we will necessarily be tracking a moving target, and thus we need to consider how to address this, especially in terms of developing a network of conservation reserves that is flexible and dynamic, rather than fixed and static, in both space and time. Species distribution modeling could thus be useful in assessing the extent to which existing reserves are likely to overlap the future distribution of species in response to climate-driven range shifts. This approach could also be used to develop reserve systems in a climate-change context, in which the location and distribution of reserves are selected on the basis of their future habitat suitability and potential for securing the future distributions of species (Araújo et al. 2004).

Clearly, then, not all population spread is undesirable, but how do we decide which species should be permitted to spread and those whose spread should be controlled? We currently attempt to contain the spread of injurious species or agents of disease, which are typically non-native species or pathogens, and so the target of our control and containment efforts may seem obvious. However, not all non-native species are invasive or injurious. Even native species that shift their ranges in response to climate change may cause the decline or local extinction of other native species, as the mixing of communities produces new ecological interactions that alter competitive and predator–prey/host–parasite relationships. We cannot yet predict with great precision which species—native or otherwise—will increase and spread, what the impact of their arrival in new areas might be, let alone where on the landscape they are most likely to arrive and when. Nor do we yet understand which factors—dispersal, demography, or landscape—are the most important drivers of spread, and thus which should be targeted to manage the spread of an invasive species (Coutts et al. 2011). Further, land-management efforts to contain the spread of an invasive species or disease may be at odds with those directed at preserving viable populations and the spread potential of native species. Designing an optimal strategy to achieve these joint aims will obviously require a fairly comprehensive understanding of landscape effects on population spatial spread for more than just a single species.

The emerging field of landscape epidemiology shows great promise in helping to identify the landscape features that contribute most to disease spread. The hope is that this approach could then be applied toward reducing disease risk through the development of more effective surveillance programs and decision-support systems used to evaluate which control measures are likely to work best, as well as to inform what land-management practices could help control the

incidence and spread of disease (Plantegenest et al. 2007). However, our knowledge of how the landscape interacts with populations of disease vectors, reservoirs, or hosts to influence the emergence and spread of disease is woefully incomplete. Any relationships that link landscape features to these agents of disease must be identified and tested if we are to obtain a mechanistic understanding of disease spread (Ostfeld et al. 2005). In the meantime, approaches such as landscape genetics may be useful for determining patterns of gene flow across the landscape (**Chapter 9**), which might then aid in identifying the origins of a disease as well as the likely migration routes through which the disease could be transmitted among populations.

Chapter Summary Points

1. Species range shifts, biological invasions, and disease spread are all examples of population-based spatial spread processes that can be influenced by landscape structure, and which may be facilitated—directly or indirectly—by humans as a consequence of climate change, land-use intensification, and an increase in global travel and transportation networks.

2. Many species have already begun to shift their ranges toward the poles or toward higher elevations in response to warming temperatures. However, high-latitude and high-elevation species are at greatest risk of range contraction, as warming temperatures may be contributing to the decline and even extinction of these species. Species distribution models are helping to forecast the extent to which species will be able to shift their ranges in response to climate change, as well as where the greatest turnover in species is likely to occur geographically.

3. Under current climate-change scenarios, trees and other species will have to shift their ranges far more quickly than was necessary during previous periods of postglacial warming (e.g. >1000 m/yr), which could be difficult to achieve given the extent of fragmentation and other barriers to migration imposed by the modern Anthropocene landscape. Fortunately, most temperate species appear to have been able to keep pace with warming temperatures thus far.

4. The introduction and spread of non-native species is contributing to the homogenization of biotas on a global scale. Invasive species are capable of completely transforming landscapes, and biological invasions are second only to habitat loss

and fragmentation in terms of the threat posed to native diversity.

5. Landscape structure can influence any or all of the stages that contribute to invasive spread: from the location of the initial introduction(s) of non-native species onto the landscape, to whether a non-native species is able to become established in its new environment, to whether it can successfully disperse and form new populations elsewhere, thereby promoting its spread across the landscape.

6. Landscape structure can have contrasting effects at different stages of the invasion process. For example, habitat fragmentation could simultaneously enhance dispersal success, by creating numerous small areas of disturbance that can be easily traversed or colonized, but adversely affect establishment by limiting reproduction or survival and thus population size. Understanding the potential for these sorts of antagonistic effects on different stages of the invasion process may offer a means of identifying and targeting the stages most amenable to control through landscape management.

7. Identifying which non-native species are likely to become invasive is a challenge in practice, especially since many species exhibit a long latent period or lag phase, during which time the species persists at low population density, perhaps avoiding detection, before it begins to spread. The length of the latent period varies among species and landscapes, from a few years to a few decades or more. In general, the longer the non-native species has been present on the landscape (the minimum residence time), the greater the likelihood that it will spread.

8. Invasive spread has typically been modeled as a reaction-diffusion process, which incorporates both population growth (reaction) and random dispersal (diffusion), and predicts a constant rate of spread over time. Models that allow for occasional long-distance dispersal events, however, predict that invasive spread will accelerate over time. From a management standpoint, it is imperative that the potential for accelerated spread be identified so that the invasive species can be contained early on, while its spread is still relatively slow and when control is therefore most likely to succeed.

9. The potential for landscape-wide spread can be evaluated using network- and percolation-based models, by identifying the critical level of habitat or disturbance where an invasive species or disease might spread, as well as the effect of landscape configuration on promoting (or reducing) the potential for invasive or disease spread. Such spatially explicit modeling approaches may be especially helpful for evaluating and designing mitigative actions to control invasive spread.

10. Source–sink models of invasive spread can help identify potential targets and set priorities for invasion control, based on an assessment of the population's growth potential in different habitat types. While targeting potential source populations is an obvious way to help slow further spread, populations in marginal (sink) habitats should not be ignored. The spread of some species is facilitated through the establishment of satellite populations beyond the invasion front (nascent foci), which may occupy marginal habitat. If so, then targeting these satellite populations may offer the best opportunity for slowing invasive spread.

11. Disease spread is a form of biological invasion. The field of landscape epidemiology investigates how landscape structure, in concert with other environmental factors, influence the distribution and abundance of disease agents across the landscape. Landscape structure can influence disease spread through its simultaneous effects on the spatial dynamics of reservoir, vector, and host populations. Landscape features that promote an increase in disease reservoirs and vectors, or which increase the frequency of interaction between these species and other host species (including humans), can be mapped to identify hotspots of disease risk.

12. Disease risk mapping involves the application of computerized mapping technology (GIS) and statistical approaches, such as species distribution modeling, to link the spatial distribution of reservoir, vector, and host species to various landscape and environmental factors. Disease risk mapping is useful for visualizing the potential disease landscape as well as forecasting disease risk under different scenarios of landscape or climate change, and thus for identifying areas with a high potential for disease outbreaks. Disease risk does not always coincide with the incidence or occurrence of disease, however, and thus disease risk maps should be viewed as predictive models of relative rather than absolute risk of disease or disease exposure.

13. Disease risk may change over time due to the coevolution of host-pathogen/host-parasite interactions. Evolutionary landscape epidemiology combines evolutionary theory and landscape epidemiology to address how landscape structure influences the genetic structure and dynamics of host and pathogen/parasite populations, and thus the potential for new diseases to emerge and spread across landscapes.

Discussion Questions

1. Species have shifted their ranges—sometimes repeatedly—in response to climatic changes in the past, especially following the expansion and contraction of glacial fronts. Why be concerned about recent climate-change effects on species distributions, then? What is so different about the current changes in climate, or the landscapes through which species must now move, that may determine whether species will be able to track environmental changes? What types of species will likely have the greatest difficulty in shifting their ranges in response to environmental changes caused by warming temperatures or altered patterns of precipitation?

2. If many species are already shifting their distributions in response to recent climatic changes, what are the implications of this for the design and establishment of nature reserves and other protected areas? Many national parks in the USA, for example, are located in mountainous or arid regions (e.g. Glacier National Park, Rocky Mountain National Park, Death Valley National Park, Grand Canyon National Park). Consider how climate change is expected to affect these sorts of environments in particular, as you discuss ways in which reserve design might accommodate climate-driven shifts in species' ranges.

376 Essentials of Landscape Ecology

3. The construction of firebreaks—the intentional clearing of vegetation to contain the spread of a fire—is an established practice for managing prescribed burns or controlling wildfires on the landscape ('fighting fire with fire'). To what extent might we be able to adopt similar land-management practices to prevent or control invasive spread? For example, if fragmentation would slow the rate of spread in a forest pest, would intentionally fragmenting the forest be a reasonable means of controlling invasive spread of this species? Consider ways in which landscape management might be used in various types of systems to control the spread of an invasive species.

4. Disease outbreaks and epidemic spread may occur following the introduction of a pathogen that originated on another continent. The long-range transport of exotic diseases is a constant threat in this era of global travel and transport, not only to public health, but also to agriculture, forestry, and native wildlife. In what ways, then, might an understanding of the landscape ecology of invasive spread be helpful in predicting or controlling the spread of exotic diseases?

5. Although disease spread is a type of biological invasion, in what ways does the spread of disease differ from that of an invasive species? In what ways are they similar?

6. Discuss how habitat loss and fragmentation can simultaneously have both positive and negative effects on disease spread. What are the implications of this from the standpoint of managing landscapes to control disease spread? For example, if fragmentation can have both a positive and negative effect on disease spread, how do we decide which aspect of disease spread to control (assuming we can) through strategic land management?

7. Given that gene flow can profoundly affect the adaptive potential of both host and parasite/pathogen populations, how might landscape genetics be used to assess the potential for disease outbreak and spread across a landscape? How might this type of analysis be used to inform land-management strategies and other actions directed at preventing or containing disease spread?

9

Landscape Genetics

Landscape Effects on Gene Flow and Population Genetic Structure

The boundaries of a species' geographic range are dynamic, expanding and contracting in response to broad-scale landscape and climatic changes over time. Populations at the periphery of a species' range are thus often either relictual populations, left behind as ranges contracted, or satellite populations that have become established ahead of an expanding range front. Delineating range boundaries can be a challenge (Fortin et al. 2005), but the genetic signatures of peripheral populations may hold clues as to the nature of the range dynamic (Excoffier et al. 2009; Arenas et al. 2012). For example, range expansions generally lead to a loss of genetic diversity along the expansion axis (Excoffier et al. 2009). Populations at the periphery of a species' range thus tend to have reduced genetic variation (heterozygosity) within populations and increased genetic divergence among populations, compared to populations at the center of the species' distribution.

The collared lizard (*Crotaphytus collaris*) occurs in rocky areas within a variety of habitats throughout its range in the west-central USA. Over the past 100,000 years, the range of the collared lizard has undergone numerous expansions and contractions in response to repeated glacial-interglacial cycles. At the northeastern limits of its current range, the collared lizard is found in a few scattered locations, such as along limestone ridges of the Kansas Flint Hills and among granite outcroppings in forest glades of Missouri's Ozark Mountains. To examine core-periphery effects on population genetic structure, Blevins and her colleagues (2011) studied peripheral populations of collared lizards in the Flint Hills of northeastern Kansas (**Figure 9.1**).

Consistent with expectations, these peripheral populations in the Flint Hills exhibited reduced levels of genetic variation in comparison to populations at the core of the species' range (e.g. Texas; Hutchison 2003). Despite a modest degree of genetic differentiation among populations in the Flint Hills, there is nevertheless evidence of recent gene flow among these scattered populations. Although collared lizards are believed to have expanded their range into this region some 7000 years ago following the last glacial retreat, population divergence is less than expected, probably because the open grasslands and limestone ridges of the Flint Hills facilitate collared lizard dispersal. This is in contrast to peripheral populations in the Missouri glade system, in which forest expansion and fire suppression have effectively isolated populations of collared lizards, leading to a loss of genetic diversity and genetic divergence among populations (Templeton et al. 2001). Landscape history and management context are thus important considerations when evaluating

Essentials of Landscape Ecology. Kimberly A. With, Oxford University Press (2019).
© Kimberly A. With 2019. DOI: 10.1093/oso/9780198838388.001.0001

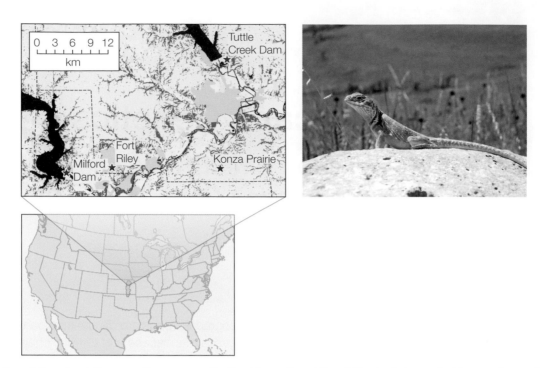

Figure 9.1 Collared lizards at their northern range limit.   The collared lizard (*Crotaphytus collaris*) occupies rocky outcrops throughout its range, including here at the periphery of its geographic distribution in the Flint Hills of northeastern Kansas (USA).
Source: Photo courtesy of Eva Horne. Map after Blevins et al. (2011).

historical as well as contemporary patterns of gene flow that give rise to population genetic structure.

What is Landscape Genetics?

Landscape genetics is a relatively new field that has emerged in the past decade or so at the intersection of population genetics, landscape ecology, and spatial statistics (Manel et al. 2003; Manel & Holderegger 2013). Landscape genetics aims to address how the microevolutionary processes of gene flow, genetic drift, and natural selection interact with environmental heterogeneity to shape patterns of genetic variation across the landscape (Manel et al. 2003; Storfer et al. 2007; Sork & Waits 2010; **Figure 9.2**). Traditionally, population genetics has adopted a simple patch-based view of landscapes, in which discrete populations are assumed to be embedded within an inhospitable matrix. Populations are thus isolated primarily by geographic distance, which gives rise to genetic differentiation among them (**isolation by distance**; Wright 1943). In contrast, landscape genetics makes no a priori assumptions about population structure, and instead seeks to understand how gene flow interacts with landscape structure to give rise to patterns of genetic variation across a range of scales. This focus on how landscape structure influences gene flow and patterns of genetic variation is the key distinction between landscape genetics and population genetics (Holderegger & Wagner 2008).

In particular, landscape genetics considers matrix heterogeneity to be important for evaluating landscape effects on gene flow, and thus populations are expected to be isolated more by the resistance of the intervening matrix to dispersal (**isolation by resistance**) than by distance alone (McRae 2006). Recall from our previous discussion of landscape connectivity in **Chapter 5** that landscapes can be modeled as resistance surfaces, in which different land covers vary in the degree to which they either facilitate or impede movement, and thus gene flow, across the landscape. Landscape genetics is thus expressly concerned with the analysis of functional connectivity. The tools of molecular genetics offer a means by which dispersal success can be evaluated, giving us valuable insights into species' dispersal patterns that would otherwise be difficult or impractical to study directly (Storfer et al. 2010; **Chapter 6**). For example, it might be easier and more meaningful to study plant dispersal via genetic assays of pollen and seed movements, rather than through mark-recapture studies of seeds or by tracking the individual movements of animal pollinators or seed dispersers across the landscape (Sork & Smouse 2006; **Chapter 6**).

Understanding the processes and patterns of gene flow and local adaptation requires a detailed knowledge of how landscape characteristics structure populations.

Manel et al. (2003)

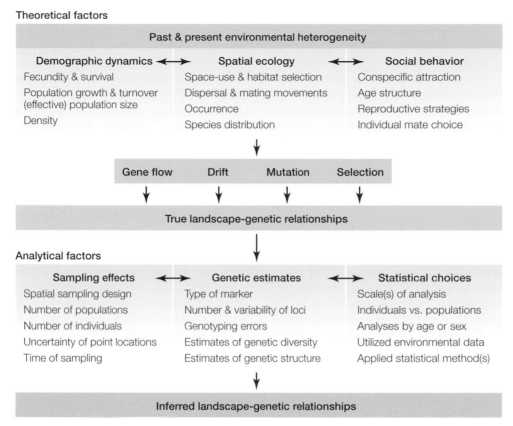

Figure 9.2 Theoretical and analytical framework of landscape genetics.   Landscape structure (past and present environmental heterogeneity) interacts with population processes (demography, spatial ecology, behavior) to influence the four major sources of genetic variation (gene flow, genetic drift, mutation, and natural selection) and give rise to the 'true' landscape-genetic relationships. A variety of analytical factors related to the analysis of genetic variation (sampling design, genetic data, statistical methods) then influences our observations and the inferences we make regarding these landscape-genetic relationships.

Source: After Balkenhol & Landguth (2011).

The molecular tools now available to landscape geneticists permit a fine-scale analysis of genetic structure and contemporary patterns of gene flow, which allow for not only an assessment of population or landscape connectivity, but also the identification of barriers to dispersal and gene flow. Some of these barriers may be obvious ones (a mountain range, a major river, or a dam), whereas others are less so and are thus considered to be 'cryptic barriers' (e.g. climatic or salinity gradients; Storfer et al. 2010). Conversely, a landscape-genetics approach could help to identify corridors or other areas of the landscape that facilitate dispersal and gene flow (Cushman et al. 2009). We would then be able to evaluate the success of efforts aimed at restoring connectivity, such as through the creation of habitat corridors or via the installation of crossing structures placed in strategic locations (e.g. fish ladders or wildlife overpasses). Alternatively, we might be concerned about the potential for disease spread or the spread of genetically modified organisms (GMOs) across a landscape. In that case, identification of landscape corridors or barriers to gene flow could be important for evaluat-

ing risk (as part of a risk-mapping assessment) or for the strategic designation of quarantine areas that could help limit disease spread (Cureton et al. 2006; Real & Biek 2007; Cullingham et al. 2008; **Chapter 8**).

Population genetic structure reflects the successful dispersal and reproduction by individuals over many generations, and thus genetic data are really an amalgamation of behavioral, ecological, and microevolutionary processes operating over a range of scales in both space and time (Anderson et al. 2010; **Figure 9.2**). As such, the study of landscape genetics integrates the various topics we have discussed in the preceding four chapters pertaining to landscape connectivity and the effects of landscape structure on individual movement and dispersal, population distributions and dynamics, and population spatial spread (**Chapters 5–8**). The analysis of spatial genetic variation can aid in delineating (meta)population structure and contribute to a better understanding of species' distributional patterns, especially if that population genetic structure is ultimately influenced by landscape context or whether a given population is at the core or periphery of a species' range (Blevins et al. 2011). Evidence of asymmetrical gene flow among populations can also assist in evaluating source–sink population dynamics, by identifying the population source(s) of immigrants (Andreasen et al. 2012). Understanding how genetic variation is distributed within and among populations is also important for species conservation and management, particularly in terms of identifying which population segments should be protected (e.g. source populations or evolutionary significant units; Manel et al. 2003; Manel & Holderegger 2013).

Spatial variation in genetic structure is the result of both historical and contemporary evolutionary processes, such that population genetic structure may be related more to the historical rather than current landscape configuration (**Figure 9.2**). Unfortunately, the landscape data available to us are usually far more recent than the genetic patterns we seek to explain. Contemporary landscape patterns are thus being used to explain genetic patterns that have potentially evolved over many generations, encompassing time scales measured in decades, centuries, or even, millennia. The study of how the physical structure of the landscape (its geomorphology; **Chapter 3**) might have contributed to genetic divergence among groups of organisms over long time periods has traditionally been the province of **phylogeography** (Manel et al. 2003). Phylogeography combines phylogenetics with biogeography to make inferences about the effects of migration, population expansion, genetic bottlenecks, and vicariance on the evolutionary history of species (Avise 2000). As such, it targets much longer time periods than those studied by landscape genetics, which focuses more on microevolutionary processes (e.g. gene flow and genetic drift) than macroevolutionary ones. Nevertheless, identifying the relative effects of historical versus contemporary landscape effects on microevolutionary processes is a major challenge in landscape genetics, as we'll explore later in this chapter.

To put the field of landscape genetics into context, we begin this chapter with an overview of population genetics, as it is difficult to understand or interpret the results of a landscape-genetics study without a solid understanding of these basic principles. This population-genetics primer provides an overview of the various types of genetic (molecular) markers that are commonly used in both landscape- and population-genetics studies, as well as the basic theory that underlies the definition and estimation of genetic variation within and among populations. This leads to a discussion of how genetic distance is measured, and thus, how gene flow and population connectivity are assayed. With these basic principles under our belt, we then move from population genetics to landscape genetics, by considering the ways in which we can

determine whether population genetic structure exists in our study system. The remainder of the chapter is devoted to applications of landscape genetics, in which we explore the various means by which landscape structure can influence patterns of gene flow and population genetic structure. We conclude this chapter with an overview of evolutionary landscape genetics, by considering the adaptive potential of populations in response to future landscape and climatic changes, as well as some of the emerging frontiers in landscape-genetics research, such as landscape genomics.

A Primer to Population Genetics

In this primer to population genetics, we begin by covering the types of genetic (molecular) markers that are commonly used in population- and landscape-genetic studies. We next explore how genetic variation and divergence are estimated, and then conclude with a section on how gene flow and population connectivity are assessed.

Types of Genetic Markers

Before we can quantify the genetic structure of populations, we must first obtain genetic information from the individuals or populations we wish to study. Many different types of **genetic (molecular) markers** have been used in landscape genetics (at least 18 different types, according to a literature review by Storfer et al. 2010; **Table 9.1**). Historically, **allozymes** were the first markers used in molecular-genetics research but are less commonly used now as more advanced molecular tools have become available. Still, allozymes have been used in some landscape-genetic studies, especially those involving plants (Storfer et al. 2010). Allozymes are the different structural forms of certain enzymes that are produced by changes in the DNA (deoxyribonucleic acid) nucleotide sequence that codes for that enzyme.[1] Only about 30% of the changes that occur at the molecular level (in the DNA) ultimately result in changes to protein structure, however, which means

---

[1] Recall from your general biology class that an enzyme is a type of protein involved in the catalysis of biochemical reactions. Proteins are made up of amino acids, and each amino acid is coded for by a three-base nucleotide sequence (a codon) on the mRNA (messenger ribonucleic acid) molecule, which in turn was transcribed from a strand of DNA. A single base substitution (as by a point mutation) in the nucleotide sequence may result in a different amino acid being used in the synthesis of the protein. For example, the sequence UUC codes for the amino acid, phenylalanine. Substituting G for C in this codon (so, UUG) would instead code for the amino acid, leucine. Although such alterations affect the primary structure of the enzyme (i.e. the linear sequence of amino acids), these changes usually do not affect enzyme function unless they radically alter the shape or binding site of the enzyme. Different structural forms of enzymes thus correspond to different alleles (allozymes) within the population.

TABLE 9.1   Molecular markers commonly used in landscape genetics.

Marker type	Description[1]	Advantages	Disadvantages
Microsatellites (Single Sequence Repeats, SSRs; Simple Sequence Tandem Repeats, SSTRs)	A (usually) non-coding section of nuclear DNA consisting of a very short nucleotide sequence, typically 2–6 base pairs long, that is repeated many times Example: CACACACACACACA is a sequence of two nucleotides (CA) that is repeated 7 times; that is, there are 7 CA repeats	The number of repeats varies widely among individuals, resulting in numerous alleles within a population (i.e. microsatellites are highly polymorphic, which makes them useful as genetic markers) Genetic variation (SSR length) can be ascertained fairly easily in the lab, using either high-resolution gel electrophoresis or an automated DNA sequencer, following amplification of the microsatellite via PCR	Microsatellite primers—the oligonucleotide sequences that bind uniquely to the flanking regions of the microsatellite, thereby permitting its identification, isolation, and amplification via PCR—may not be available for the study species Primers must be designed and synthesized for individual microsatellite loci, and are only transferable among closely related species Typically, only 6–15 microsatellite loci are used in a given study.[1] However, next-generation sequencing technologies are making it easier and cheaper to develop large numbers of microsatellite loci[2]
Amplified Fragment Length Polymorphisms (AFLPs)	A whole-genome approach, in which the genomic DNA is broken into a large number of fragments that vary in length, which are then amplified through a complicated PCR process using AFLP primers	Does not require prior information on the marker's genetic sequence (unlike microsatellites), making the approach ideal for organisms that have not been sequenced (e.g. many plants, fungi, bacteria, and other pathogens) Permits the simultaneous amplification of multiple markers (e.g. tens to hundreds within a single PCR) Produces unique DNA 'fingerprints' that resemble barcodes and depict the presence or absence of individual fragments (bands), as detected by gel electrophoresis or an automated DNA sequencer	Only dominant markers can be studied (these give the bands that make up the barcode), making it impossible to distinguish heterozygotes

(Continued)

TABLE 9.1   Continued

Marker type	Description[1]	Advantages	Disadvantages
Single Nucleotide Polymorphisms (SNPs)	DNA sequences that vary by a single nucleotide among individuals within a population (e.g. ATTCG vs ATTGG) SNPs are usually biallelic (having only two alleles), thus homozygotes have two copies of the same nucleotide (e.g. CC vs GG in the previous example), whereas heterozygotes have a copy of each (CG) They occur within coding as well as non-coding gene regions, and thus contribute neutral as well as adaptive genetic variation within populations	Can sequence many hundreds to thousands of loci across the entire genome Permits the study of neutral as well as adaptive genetic variation within and between populations Possibility of identifying functional genes and regulatory regions that underlie phenotypes Offers much potential for the developing fields of landscape genomics and evolutionary landscape genetics	Requires sequencing many loci within a genome (often hundreds to thousands) across many individuals, which is costly and time-intensive. This will become less of an issue over time as more species' genomes are sequenced and with next-generation sequencing technologies[3] SNPs are prone to ascertainment bias—bias in estimating population parameters—as a result of selecting markers based on their level of polymorphism, using too few individuals to identify SNPs, or when transferring markers between populations[4]

[1]Primary source for information is Holderegger & Wagner (2008), except where noted.
[2]Gardner et al. (2011); Guichoux et al. (2011)
[3]Manel et al. (2010)
[4]Luikart et al. (2003)

that genetic variation may be seriously underestimated using allozymes (Frankham et al. 2010). In addition, enzymes are fragile molecules that begin to break down soon after they form, which necessitates the use of fresh or fresh-frozen samples. Samples, in turn, are usually obtained from the blood, liver, or kidneys of animals, or from the leaves and root tips of plants. In the case of animals at least, this requires some rather invasive or lethal sampling procedures, which might be objectionable for any number of moral, ethical, or legal reasons (e.g. if the animal is a federally listed threatened or endangered species).

Rather than testing for variation in the gene products of different DNA sequences, as in the case of allozymes, we can instead make use of modern molecular techniques to measure variation in the DNA molecule itself. We can obtain nuclear DNA from virtually any type of biological material, including from the hair follicles, feathers, scales, egg shells, saliva, feces, urine, blood, or semen of animals (Frankham et al. 2010). Even preserved specimens in museums can be sampled for DNA, which might be useful in giving us a glimpse into the historical range of genetic variation within a species. We can also obtain genetic samples remotely, such as through the collection of hair or feces left behind by animals, alleviating the need to capture and handle them. Nor do we need to obtain large samples, as even tiny amounts of DNA can be amplified

millions of times through use of the Polymerase Chain Reaction (PCR) in the laboratory.

In addition to nuclear DNA, there is organellar DNA that can be obtained from either the mitochondrion (mtDNA, found in both animal and plant cells) or from chloroplasts (cpDNA, found only in plant and algal cells).[2] Organellar DNA is small and easily isolated, and contains variable regions that make it useful as a molecular marker for tracing patterns of gene flow and the degree of relatedness among individuals (**Table 9.1**). Unlike nuclear DNA, organellar DNA is uniparentally inherited, typically from mother to offspring. Thus, only gene flow associated with the one lineage (either maternal or paternal) can be traced, which may prove limiting in certain applications.

Among the molecular markers used to analyze variation in nuclear DNA, **microsatellites** have been the most popular (Storfer et al. 2010). Microsatellites are short nucleotide sequences consisting of 2–6 base

[2] Recall from your general biology or cell biology course that the mitochondrion is found in eukaryotic cells (i.e. cells that possess a membrane-bound nucleus, such as plant, fungal, and animals cells). The mitochondrion is sometimes called the 'powerhouse of the cell,' because it is the site of energy production via cellular respiration (i.e. the production of ATP, or adenine triphosphate, the energy currency of cells). Chloroplasts contain the photosynthetic machinery of plant and algal cells and are thus involved in the conversion of light energy (sunlight) into chemical energy to synthesize sugars.

pairs (motifs) that are tandemly repeated many times; the number of repeats is highly variable among individuals, giving rise to numerous alleles within the population (i.e. the different sequence lengths correspond to different alleles; **Table 9.1**). Highly variable markers are useful for detecting individual dispersal events and quantifying contemporary patterns of gene flow on the landscape. Microsatellites have a couple of other properties that make them useful in population and landscape genetics: (1) heterozygotes can be distinguished from homozygotes, permitting a more robust estimate of genetic variation within populations; and (2) the majority of microsatellites are selectively neutral, given that they are often located in non-coding regions of the DNA molecule. Neutral genetic markers have typically been used in landscape genetics because any genetic variation observed must then be due to reduced gene flow and genetic drift, both of which are likely to be influenced by landscape structure (e.g. habitat fragmentation), as we shall see in this chapter.

Despite their advantages, microsatellites do have some shortcomings. In order to isolate and amplify a specific DNA segment, such as a microsatellite, it is first necessary to obtain or develop DNA primers, which are short single strands of nucleotides made in the lab, which bind to the flanking regions of the segment of interest; these flanking regions are highly conserved areas of DNA (i.e. they typically do not vary among individuals). Primers tend to be species-specific and thus must be developed for each new species studied, although sometimes primers developed for a closely related species can be used. Designing primers and locating specific microsatellites can thus be a challenge, and there is no guarantee that the specific microsatellites identified will be sufficiently polymorphic (variable) to be useful for a study of landscape genetics. For this reason, most researchers have been limited to using only 8–15 microsatellites in a given study (Storfer et al. 2010). The rise of next-generation sequencing technologies, however, is making it easier and cheaper to develop large numbers of microsatellite loci from partial genome runs for a wide range of organisms (Gardner et al. 2011; Guichoux et al. 2011).

As an alternative to microsatellites, **Amplified Fragment Length Polymorphisms** (**AFLPs**) have been used for organisms that have not been well-studied or have not had parts of their genome sequenced (**Table 9.1**). AFLPs have been widely used in the landscape genetics of plants, for example (Storfer et al. 2010). Unlike microsatellites, this technique does not attempt to isolate a specific DNA sequence, but rather, is a whole-genome approach in which the entire DNA molecule is fragmented into different lengths using restriction enzymes to cut the DNA at specific locations. Each fragment constitutes a molecular marker, whose presence varies

among individuals, and thus this approach enables the amplification of tens to hundreds of markers per individual. The end result resembles a barcode depicting the presence or absence of the different fragments, giving each individual its own unique DNA fingerprint. Although this technique is considered to be fast and efficient, AFLPs are a type of dominant marker (fragments are either present or absent), such that it is impossible to identify heterozygotes (Mueller & Wolfenbarger 1999; Holderegger & Wagner 2008).

The future of landscape genetics, however, belongs to **Single Nucleotide Polymorphisms** (SNPs; Holderegger & Wagner 2008; Manel & Holderegger 2013). A SNP is a single nucleotide substitution of one base (A, T, C, or G) for another within a DNA sequence: **ATTCG** versus **ATTGG**, for example (**Table 9.1**).[3] These substitutions may occur within either a coding or non-coding region of the DNA molecule, and thus contribute to neutral as well as adaptive genetic variation within populations. Next-generation sequencing technologies are contributing to the development of large SNP datasets, in which hundreds or thousands of loci can be assayed per individual (Storfer et al. 2010; Manel & Holderegger 2013). In particular, SNPs have been a real boon to the study of **landscape genomics**, which seeks to identify the environmental or landscape factors responsible for adaptive genetic variation in populations. We will discuss landscape genomics more fully toward the end of this chapter.

In conclusion, many molecular markers are available for estimating genetic variation. No single marker is uniformly best, however, as each has its advantages and disadvantages (**Table 9.1**). Ultimately, the choice of marker will represent a compromise: the needs of the research, in terms of the questions being addressed and the spatiotemporal scale of the genetic processes under study, versus the type and availability of the genetic data or the cost and expertise required to obtain those data (Mueller & Wolfenbarger 1999; Wagner & Fortin 2013).

Estimating Genetic Variation and Divergence

Regardless of the type of genetic marker we use, our dataset will comprise a set of loci that have been genotyped (or sequenced) for all individuals in our sample. These data are then used to obtain allele frequencies, from which we can assess the degree of genetic variation within, or genetic divergence between, the individuals or populations we have sampled. We will

---

[3] A SNP (pronounced 'snip') is a type of polymorphism at a given nucleotide location (which is capable of taking four forms: A, T, C, or G) within a DNA sequence, and which varies among individuals to varying degrees. In practice, the minor allele frequency (i.e. the allele that is least common) should be sufficiently high (e.g. 5–10%) so as to be useful as a marker of genetic variation.

BOX 9.1   Nucleotide Diversity

A major advantage of DNA sequencing is that it permits the direct assessment of genetic variation, in that mutations that occur at the nucleotide level are identified. Heterozygosity at the nucleotide level is assessed by **nucleotide diversity**, which gives the average number of nucleotide differences per site for two homologous DNA sequences selected at random from the population (Nei & Li 1979). Nucleotide diversity (π) can be estimated as

$$\pi = \frac{\sum \pi_{ij}}{n_c},$$

where π_{ij} is the proportion of different nucleotides between the ith and jth sequence, which is summed over all pairwise comparisons and divided by the number of comparisons (n_c; Frankham et al. 2010). For n sequences, there are thus $n(n-1)/2$ comparisons. Nucleotide diversity can be obtained directly from DNA sequence data, or estimated from molecular markers such as AFLP or SNPs (Borowsky 2001).

To illustrate, let's pretend we have a simple DNA sequence of 10 nucleotides that we have obtained from three individuals:

Individual 1: ATGCGTTTTT
Individual 2: ATGGGTTTTT
Individual 3: ATGCGTTTTA

The nucleotide substitutions are underlined. We can see that $\pi_{1,2} = 0.1$ because the two sequences differ at only one of the 10 positions (position 4); $\pi_{1,3} = 0.1$ because these sequences also differ at only one of the 10 positions (position 10); and $\pi_{2,3} = 0.2$ because the sequences differ at two of the 10 positions (positions 4 and 10). Thus, $\pi = (0.1 + 0.1 + 0.2)/[3(2)/2] = 0.4/3 = 0.13$. In other words, there are four differences among 30 base pairs (4/30 = 0.13). We would therefore expect an average of 0.13 nucleotide differences/site between two (randomly chosen) sequences, or equivalently, that sequences will differ by an average of 1.3 nucleotides (0.13 × 10 nucleotides/sequence = 1.3 nucleotides/sequence).

thus consider the various means by which we can estimate genetic variation and divergence in this section.

ALLELIC RICHNESS AND EXPECTED HETEROZYGOSITY
If we are hoping to examine the effects of landscape structure on gene flow and population genetic structure, it obviously makes sense to focus our efforts on markers that are variable among individuals. As a starting point, we can assess what percentage of loci in our sampled population are **polymorphic**, meaning that these loci possess two or more alleles. For example, 80% of the 10 microsatellites used to assess population connectivity in the collared lizard were found to be polymorphic (Blevins et al. 2011; **Figure 9.1**).

For polymorphic loci, we can then calculate the number of alleles per locus and the frequency of those alleles in the population. **Allelic richness** (A, which is sometimes referred to as allelic diversity; Leberg 2002) is the number of alleles per locus, which is averaged over individuals, loci, or populations. In the collared lizard study, for example, average allelic richness varied between 2.7 and 3.6 alleles/locus for the individual populations, with a grand mean of 3.4 alleles/locus over all four populations (Blevins et al. 2011). Because allelic richness is sensitive to sample size (large samples are expected to have more alleles than small samples), rarefaction[4] is typically used

to obtain an unbiased estimate of the expected number of alleles per locus (Petit et al. 1998; Leberg 2002; Kalinowski 2004). For DNA sequence data, we can calculate **nucleotide diversity**, which gives the average number of nucleotide differences at a given site within homologous DNA sequences (**Box 9.1**).

Allele frequencies are the relative proportions of alleles at a given locus (p_i), and are obtained as n_i/N, where n_i is the number of copies of the ith allele and N is the total number of copies across all alleles at that locus. To illustrate, imagine we have sampled 100 diploid individuals (so, $N = 200$ allele copies/locus) and find that there are three alleles at a given locus ($A = 3$) that are distributed as follows: $n_1 = 100$ copies, $n_2 = 80$ copies, and $n_3 = 20$ copies. The frequencies of the three alleles in the population would thus be: $p_1 = 100/200 = 0.5$, $p_2 = 80/200 = 0.4$, and $p_3 = 20/200 = 0.1$. The latter allele is thus relatively uncommon in the population, making up only 10% of allele copies at this locus. In other words, there is only a 10% chance that a randomly selected allele at this locus would be of this particular type.

---

[4] Rarefaction was originally used in ecology to correct for the effects of sample size on species richness, given that the number of species found in a community tends to increase with sampling effort (Hurlbert 1971). Rarefaction curves are generated by repeated resampling of the dataset for different numbers of samples, and

then plotting the average number of species found within a sample of a given size, up to the maximum sample size. For example, if we have surveyed species at 100 sites, how many species do we expect to find—on average—if we visit only two sites, 10 sites, 50 sites, and so on up to a total of 100 sites. Rarefaction curves tend to increase rapidly at first, as the more-common species are encountered even at small sample sizes, but then begin to plateau at larger sample sizes as only the rarest species remain to be sampled. Replace 'species' with 'alleles' in the foregoing explanation, and you'll have an idea of how rarefaction is used to estimate (and correct for) the expected allelic richness for a given sample size.

Intuitively, genetic variation within a population is greatest when all alleles at a locus are equally abundant. We can thus standardize the number of alleles present within a population by their relative frequencies to give the **effective number of alleles** (n_e) as

$$n_e = \frac{1}{\sum p_i^2},$$ Equation 9.1

where p_i is the frequency of each of the i alleles at a particular locus. The effective number of alleles gives the number of alleles that would provide the same level of heterozygosity if all alleles occurred with equal frequency (Frankham et al. 2010). **Heterozygosity** refers to the fraction of individuals in a population of organisms that are heterozygous; that is, that have different alleles at a given locus. Heterozygosity is an important measure of genetic variation and is thus maximized when all alleles at a locus are equally abundant. The **expected heterozygosity** (H_e) at a single locus is obtained as

$$H_e = 1 - \sum_{i=1}^{N} p_i^2$$ Equation 9.2

where N is the total number of alleles at that locus.

To illustrate, let's once again imagine a locus with three alleles ($A = 3$). If each allele has the same frequency ($p_1 = p_2 = p_3 = 0.33$), then the expected heterozygosity is

$$H_e = 1 - [(0.33)^2 + (0.33)^2 + (0.33)^2] = 1 - 0.33 = 0.67.$$

In other words, 67% of the population is expected to be heterozygous when all three alleles are equally abundant. As a check, the effective number of alleles is $n_e = 1/0.33 = 3.03$ (Equation 9.1), which is basically the same as the number of alleles known to occur at this locus ($A = 3$). Again, this is the maximum expected heterogeneity for three, equally abundant alleles.

Now, let's imagine another population in which allele frequencies at this locus are not equal: $p_1 = 0.85$, $p_2 = 0.1$, and $p_3 = 0.05$. The expected heterozygosity (H_e) is

$$H_e = 1 - [(0.85)^2 + (0.1)^2 + (0.05)^2] = 1 - 0.74 = 0.26.$$

Only 26% of the population is expected to be heterozygous at this locus, which is 2.5× less than our previous example. The effective number of alleles within the population is $n_e = 1.4$, which is about half the number of alleles known to occur at this locus ($A = 3$). In effect, then, the level of heterozygosity observed at this locus is what we would expect if only 1.4 alleles were present and occurred with equal frequency.

ESTIMATING EXPECTED GENOTYPE FREQUENCIES WITHIN POPULATIONS

Once we have information on allele frequencies, we can estimate the expected genotype frequencies within the population. To illustrate, let's focus on a single locus that has two alleles: A_1 and A_2. Because we are working with a diploid organism

here, there are three possible genotypes: A_1A_1, A_1A_2, and A_2A_2. Each genotype is assumed to represent the random and independent union of two gametes, each carrying an allele for this locus. Thus, the expected frequency of each genotype within the population is given by the product of their allele frequencies. Let **p** be the frequency of A_1 and **q** be the frequency of A_2. If **p** = 0.4 and **q** = 0.6, then the expected genotype frequency of A_1A_1 is 0.16 [**p** × **p** = **p²** = $(0.4)^2$ = 0.16], whereas the expected genotype frequency of A_2A_2 is 0.36 [**q** × **q** = **q²** = $(0.6)^2$ = 0.36]. Because there are only three genotypes possible and genotype frequencies must sum to 1, we can deduce that the expected proportion of heterozygotes (A_1A_2) is 0.48 (1−0.16−0.36 = 0.48). That is, 48% of the individuals in this population are expected to be heterozygous, given the stated allele frequencies.

The expected genotype frequencies within this population can thus be summarized as

$$p^2 + 2pq + q^2 = 1.$$ Equation 9.3

Notice that the expected proportion of heterozygotes in the population is not **pq**, but rather, **2pq**. That is because there are two ways in which heterozygotes may be produced (A_1 comes from the mother and A_2 from the father, or vice versa), as illustrated by the classic Punnett Square, which gives all of the possible combinations between alleles for this locus (i.e. all of the possible fertilization events between parental gametes):

♀ \ ♂	A_1	A_2
	p	q
A_1 p	A_1A_1 pp	A_1A_2 pq
A_2 q	A_1A_2 pq	A_2A_2 qq

Equation 9.3 is known as the **Hardy–Weinberg equation**.[5] For a population that is assumed to be in equilibrium (more on that in a moment), we can use the observed frequencies of the two alleles (**p** and **q**) to calculate the expected genotype frequencies within our population. Furthermore, we can estimate the expected heterozygosity (H_e), which is simply the

---

[5] The Hardy–Weinberg equation can be applied to three or more alleles through a binomial expansion. For example, if there are three alleles for a particular trait (**p**, **q**, and **r**), the equation would be of the form: $(p + q + r)^2 = p^2 + 2pq + 2pr + q^2 + 2qr + r^2 = 1$. Note that there are now six possible genotypes within the population. Because the expected frequency of heterozygotes is $1 - (p^2 + q^2 + r^2)$, which is simply 1 minus the proportion of all homozygotes in the population, it is easier just to use Equation 9.2 to estimate the expected heterozygosity of polymorphic loci.

proportion of the population that is not homozygous (i.e. neither $\mathbf{p}^2$ nor $\mathbf{q}^2$). If we rearrange Equation 9.3, we can see that this is simply a restatement of Equation 9.2:

$$H_e = 1 - \sum_{i=1}^{2} p_i^2 = 1 - \left(\mathbf{p}^2 + \mathbf{q}^2\right) = 2\mathbf{pq}. \qquad \text{Equation 9.4}$$

Expected heterozygosity (H_e) is thus given by the expected proportion of heterozygotes within a population: $H_e = 2\mathbf{pq}$.

What is meant here by a population in equilibrium? In population genetics, we are referring specifically to a population that is in **Hardy–Weinberg equilibrium (HWE)**, which means all allele and genotype frequencies remain constant, such that the level of heterozygosity does not change over time. In other words, if $\mathbf{p} = 0.3$, it will persist at this frequency from one generation to the next. In the real word, there are a number of forces that can cause allele and genotype frequencies to change over time, resulting in deviations from the expected Hardy–Weinberg proportions. These forces include: (1) mutations, which introduce new alleles to the population, thus changing allele frequencies, (2) non-random mating between individuals, such as a higher probability of mating with close relatives (inbreeding), which changes genotype frequencies (leading to increased homozygosity in the case of inbreeding), (3) natural selection, either for or against certain traits that affect individual fitness (reproduction and/or survival), which alters allele frequencies over time, (4) gene flow, the movement of alleles into or out of populations, and (5) genetic drift, random fluctuations in allele frequencies that can lead to a loss of alleles in small populations. Thus, the **observed heterozygosity** (H_o)[6] within a population may deviate from that expected (H_e) under HWE.

The effects of genetic drift and gene flow on heterozygosity are of particular interest in landscape genetics, given the usual focus of these studies on selectively neutral markers. **Genetic drift** refers to changes in the genetic composition of populations as a result of random fluctuations in allele frequencies over time. The effects of genetic drift are especially pronounced in small populations, where the frequency of certain alleles may 'drift' to the extent that they either become lost ($p = 0$) or 'fixed' ($p = 1$) within the population. This loss of heterozygosity due to genetic drift can be offset by **gene flow**, the movement of new alleles into the population as a result of migration.[7]

Very little migration is actually required to offset genetic drift in small populations, however, perhaps as few as one migrant/generation (Wright 1931; Slatkin 1987; but see Vucetich & Waite 2000). The degree to which genetic drift is being offset by gene flow enables us to determine whether and how fast heterozygosity is being lost, as we explore next.

ASSESSING THE EFFECT OF GENETIC DRIFT ON THE LOSS OF HETEROZYGOSITY   The probability that a particular allele will eventually become fixed (or alternatively, lost) via genetic drift is given by its initial frequency. For example, if there are initially two, equally abundant alleles at a given locus ($\mathbf{p} = \mathbf{q} = 0.50$), each has a 50% chance of becoming fixed within the population. If the one allele has an initial frequency $\mathbf{p} = 0.9$, there is a 90% chance that this allele will become fixed (although, there is still a 10% chance that the other allele will become fixed instead). Rare alleles are thus more likely to be lost via genetic drift.

We can also estimate the probability that an allele will become fixed for a given population size, or rather, a given **effective population size** (N_e). The effective population size is the number of individuals that would lose heterozygosity at a rate equivalent to that of an ideal population (i.e. a population in HWE, in which mating is random and no genetic drift occurs; Wright 1931).[8] If we assume a diploid population, then the total number of allele copies is $2N_e$ ($N_e \times 2$ alleles = $2N_e$). Thus, the probability of fixation (F) can be obtained as

$$F = \frac{1}{2N_e}. \qquad \text{Equation 9.5}$$

In other words, the probability that a given allele will become fixed is inversely related to the effective population size (N_e); the probability of fixation decreases as the effective population size increases. Another way of looking at F is that this provides the rate at which heterozygosity is being lost per generation, such as through genetic drift.[9] If $N_e = 100$, then heterozygosity

---

[6] Observed heterozygosity (H_o) is simply the number of heterozygotes observed in a population, divided by the total number of individuals sampled.

[7] Population geneticists refer to this as 'migration' whereas ecologists would call this 'immigration' or simply 'dispersal.' As discussed in **Chapter 6**, different disciplines have adopted different terms to describe the movements of organisms and their

propagules. Thus, it is important to be aware of the context and not confuse the terminology (i.e. migration does not refer here to the seasonal movements of animals).

[8] Among the various factors that affect N_e are fluctuating population sizes (where populations vary greatly in size from one year to the next); an unequal breeding sex ratio; overlapping generations; variation in offspring production; and the spatial distribution of populations, which is a function of both density (δ) and the variance in dispersal distances (σ^2) as $N_e = 4\pi\sigma^2\delta$. Thus, N_e is usually less than N, the surveyed size of the population.

[9] For loci in organellar DNA (mtDNA and cpDNA), which are haploid and maternally inherited (usually), N_e is approximately half that for nuclear DNA: $N_e = N_{ef}/2$, where N_{ef} is the effective number of females in the population (assuming this is maternally inherited). Substituting $N_{ef}/2$ for N_e in Equation 9.5, we find that maternally transmitted markers lose expected heterozygosity at a

is expected to be lost at a rate of 0.005/generation (i.e. 0.5% of heterozygosity is lost per generation). If $N_e = 10$, however, heterozygosity is expected to be lost at a rate of 0.05/generation (5%/generation), which is a 10-fold increase.

We can focus instead on how much heterozygosity is expected to remain within the population each generation, which is simply

$$1 - F = \left(1 - \frac{1}{2N_e}\right),\qquad \text{Equation 9.6}$$

given that the fraction of heterozygosity that remains is just the difference between the total and whatever fraction is lost (i.e. the fraction remaining and the fraction lost must sum to 1). Thus, to use our example above for $N_e = 10$, if we lose heterozygosity at a rate of 5%/generation, we would still have 95% remaining, which doesn't seem so bad. Like compounded interest, however, the loss of heterozygosity accumulates over time. We can estimate the proportion of the initial heterozygosity remaining after t generations (F_t) as

$$1 - F_t = \left(1 - \frac{1}{2N_e}\right)^t.\qquad \text{Equation 9.7}$$

Thus, after 10 generations, we would expect to have only about 60% [$(0.95)^{10} = 0.599$] of the initial heterozygosity remaining in this population (again, if $N_e = 10$). The total amount of heterozygosity lost in this timeframe can then be obtained through subtraction as $1 - 0.60 = 0.40$. In other words, we expect to lose about 40% of the initial heterozygosity within 10 generations. More generally, we can write the expected loss in heterozygosity over time (F_t) as

$$F_t = 1 - \left(1 - \frac{1}{2N_e}\right)^t.\qquad \text{Equation 9.8}$$

If heterozygosity is being lost in a population due to genetic drift, then we expect the observed heterozygosity to be less than that expected under HWE ($H_o < H_e$). The heterozygosity observed (H_o) is the product of the expected heterozygosity ($H_e = 2\mathbf{pq}$) and whatever proportion of heterozygosity still remains in the population ($1 - 1/2N_e$). For example, if $\mathbf{p} = 0.3$ and $\mathbf{q} = 0.7$, then $H_e = 2(0.3)(0.7) = 0.42$. If the effective population size (N_e) is only 10 individuals, however, then heterozygosity is reduced at a rate of 0.05/generation, which means that 95% of the expected heterozygosity ($H_e = 0.42$)

remains. In that case, $H_o = 0.42(0.95) = 0.40$. If we compare the ratio of H_o to H_e, we get $0.40/0.42 = 0.95$, which is simply the proportion of heterozygosity remaining ($1 - 1/2N_e$) in each generation! So,

$$\frac{H_o}{H_e} = \frac{2\mathbf{pq}\left(1 - \frac{1}{2N_e}\right)}{2\mathbf{pq}} = 1 - \frac{1}{2N_e},$$

which gives us Equation 9.6 once again. Thus, we can rewrite this expression as

$$\frac{H_o}{H_e} = 1 - F,$$

which upon rearranging becomes

$$F = 1 - \frac{H_o}{H_e}.\qquad \text{Equation 9.9}$$

This is an example of an **F-statistic**, which was first introduced by the American geneticist Sewall Wright (Wright 1943). As we shall see in the next section, the degree to which the observed heterozygosity deviates from that expected is the basis for evaluating population genetic structure; that is, how heterozygosity is distributed within versus between populations.

MEASURING POPULATION GENETIC STRUCTURE USING F-STATISTICS Population subdivision is expected to occur as a consequence of natural and anthropogenic barriers to gene flow. If two populations have become completely isolated from one another (i.e. no gene flow), then genetic drift may lead to fixation within these populations, such that $H_o = 0$ and thus $F = 1$ (Equation 9.9). Population subdivision is expected to reduce observed heterozygosity (H_o) relative to that expected under HWE (H_e). Heterozygosity might still be maintained *between* populations, however, if different alleles have been fixed in each population. We thus need to understand how the observed heterozygosity is partitioned within versus between populations in relation to the level of heterozygosity expected for the entire population if it were not subdivided (i.e. if the entire population or metapopulation were in HWE). If heterozygosity between populations has been reduced, then overall heterozygosity is lower as well, even if the individual populations are in HWE. The overall reduction in heterozygosity as a consequence of population subdivision is referred to as the **Wahlund effect**.

We can explore whether population genetic structure occurs in our study system by calculating three F-statistics. Basically, we can visualize a population of populations (i.e. a metapopulation; **Chapter 7**), in which heterozygosity is examined at three levels: within individual populations, between individual populations,

rate of $1/N_{ef}$. Loss of heterozygosity due to genetic drift is thus expected to be faster for loci in organellar DNA than for diploid loci. How much faster? If we assume that $N_e = 10$ (5 males and 5 females), then the rate of loss for diploid loci in this example is $1/2N_e = 1/20$, whereas for mtDNA or cpDNA, the rate of loss is $1/N_{ef} = 1/5$. Thus, loci found in organellar DNA are expected to lose heterozygosity four times faster than that of diploid loci.

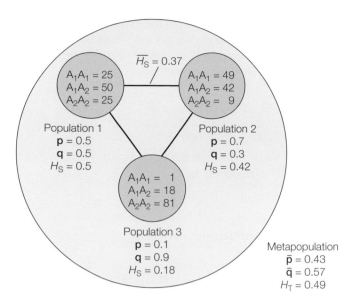

Figure 9.3 Evaluation of population genetic structure using F_{ST}. In this example, 100 individuals were genotyped at a biallelic locus in each population, where **p** is the frequency of the A_1 allele and **q** is the frequency of the A_2 allele. Population structure is determined by comparing the average heterozygosity among all populations ($\overline{H_S}$) to the total heterozygosity expected across the metapopulation as a whole (H_T) as $F_{ST} = 1 - (\overline{H_S} / H_T)$. In this case, $F_{ST} = 0.24$, which in practice is considered a 'great' degree of genetic differentiation among populations (cf **Table 9.2**).

and for the entire collection of populations (the metapopulation; **Figure 9.3**).[10] At the level of the individual population, we can examine how the observed mean heterozygosity within a given population (i.e. the observed proportion of heterozygotes in each population, averaged over all populations; $\overline{H_I}$)[11] compares to that expected for the average population (the expected proportion of heterozygotes within each population ($2\mathbf{p}_i\mathbf{q}_i$), averaged over all n populations; so,

$$\overline{H_S} = \left(\frac{\sum_{i=1}^{n} 2\mathbf{p}_i\mathbf{q}_i}{n} \right) \text{ as}$$

$$F_{IS} = \frac{H_e - \overline{H_o}}{H_e} = \frac{\overline{H_S} - \overline{H_I}}{\overline{H_S}} = 1 - \frac{\overline{H_I}}{\overline{H_S}}. \qquad \text{Equation 9.10}$$

Notice that this is a variant of Equation 9.9. If individuals are breeding at random within populations, then $\overline{H_I} = \overline{H_S}$ and $F_{IS} = 0$. For this reason, F_{IS} is sometimes referred to as the inbreeding coefficient, for it measures the extent of non-random mating within populations (Wright 1969).[12] If $F_{IS} = 1$, then all individuals within the population are homozygous (fixation has occurred), whereas if $F_{IS} = -1$, all individuals are heterozygous.

To determine the degree of genetic differentiation between populations, we can examine the extent to which the average expected heterozygosity within populations ($\overline{H_S}$) deviates from that expected for the entire metapopulation ($H_T = 2\overline{\mathbf{p}}\overline{\mathbf{q}}$) as

$$F_{ST} = \frac{H_T - \overline{H_S}}{H_T} = 1 - \frac{\overline{H_S}}{H_T}. \qquad \text{Equation 9.11}$$

This is the F-statistic of greatest interest in population and evolutionary genetics (Holsinger & Weir 2009). More than half of all studies in landscape genetics have used F_{ST} or one of its analogues (Storfer et al. 2010).[13] Populations are not differentiated when $F_{ST} = 0$ (the population is in HWE), and are completely differentiated when $F_{ST} = 1$, in which a different allele becomes fixed (at random) within each population.

To illustrate, let's imagine a metapopulation consisting of three populations, each of which has a different frequency for the A_1 allele at a biallelic locus (**Figure 9.3**). The expected heterozygosity ($H_S = 2\mathbf{pq}$) thus varies among these three populations, resulting in an average expected heterozygosity ($\overline{H_S}$) of 0.37. Across the entire metapopulation, the expected frequency of the A_1 allele ($\overline{\mathbf{p}}$) is 0.43, the expected frequency of the A_2 allele ($\overline{\mathbf{q}}$) is thus 0.57, and the expected heterozygosity for the metapopulation as

---

[10] In genetics, these individual 'populations' are more properly referred to as **demes** (or subpopulations), a group of interbreeding individuals. Demes may be separated from other demes within the larger population. A population is thus a collection of demes and may be separated from other populations within a larger area. It is important to be aware of the differences in terminology between fields, given that the same term (population) can have different connotations depending on whether one is an ecologist or a geneticist. Because of the similarity to the metapopulation structure discussed in **Chapter 7**, however, I refer to demes as populations here in this chapter for consistency, and to reinforce the idea that populations are potentially linked via dispersal and gene flow. The important point is that there is the potential for hierarchically nested structure in the distribution of genetic variation, whatever we choose to call those lower-level groups (demes, subpopulations, or populations).

[11] For genetic markers, H_I is best calculated as the average observed heterozygosity across all loci. Individual loci are likely to have very different allelic frequencies (and thus, F-statistics), such that a more robust estimate of genetic differentiation can be obtained by averaging over many loci.

[12] The theoretical effects of genetic drift and inbreeding on heterozygosity are mathematically equivalent. F-statistics are therefore known more generally as 'fixation indices,' since a complete loss of heterozygosity represents fixation ($F = 1$), regardless of whether it comes about through inbreeding or genetic drift.

[13] The form of the F_{ST}-statistic given in this chapter is really that proposed by Nei (1973) for multiple alleles, which he termed G_{ST}. Since this is basically equivalent to F_{ST}, however, we will just refer to it as such to simplify our discussion. It is worth noting, however, that hypervariable markers such as microsatellites have maximum F_{ST} (or G_{ST}) values far less than 1 (Meirmans & Hedrick 2011). A standardized form of the G_{ST}-statistic has thus been proposed (G'_{ST}; Hedrick 2005). Alternatively, the R_{ST}-statistic can be used for microsatellite data, which examines the variance in allele size (S) assessed at the level of the subpopulation (S_P) relative to that for the entire population (S_T) as $R_{ST} = 1 - S_P/S_T$ (Slatkin 1995). Although similar in form to F_{ST}, the R_{ST}-statistic assumes a stepwise mutation model (as opposed to the infinite alleles model of F_{ST}), and thus provides a less biased estimate of genetic divergence for microsatellite loci. The use and interpretation of F_{ST} and its various analogs are a frequent topic of debate in the literature (e.g. Jost 2008; Heller & Siegismund 2009; Whitlock 2011; Meirmans & Hedrick 2011).

a whole ($H_T = 2\overline{p}\overline{q}$) is 0.49. The degree of genetic differentiation among these populations (F_{ST}) is thus

$$F_{ST} = 1 - \frac{\overline{H_S}}{H_T} = 1 - \frac{0.37}{0.49} = 1 - 0.76 = 0.24.$$

This indicates that 24% of the total genetic variation is distributed among populations, which means that 76% of the variation occurs within populations. Although it may seem that a relatively small fraction of total genetic variation is distributed among populations, this is actually considered a 'great' degree of genetic differentiation in practice (Wright 1978; **Table 9.2**).

Finally, we can compare the average heterozygosity observed within individual populations ($\overline{H_I}$) to that expected within the entire metapopulation (H_T) as

$$F_{IT} = \frac{H_e - \overline{H_o}}{H_e} = \frac{H_T - \overline{H_I}}{H_T} = 1 - \frac{\overline{H_I}}{H_T}. \quad \text{Equation 9.12}$$

In the extreme case, where all heterozygosity has been lost (the same allele has been fixed in all populations), then $\overline{H_I} = 0$ and $F_{IT} = 1.0$. F_{IT} is thus sometimes called the overall fixation index.

To reinforce the connection among these F-statistics, all three indices are related as

$$(1 - F_{IT}) = (1 - F_{IS})(1 - F_{ST}).$$

If we substitute the appropriate equivalencies given in the equations above (Equations 9.10–9.12), this expression simplifies to

$$\frac{\overline{H_I}}{H_T} = \frac{\overline{H_I}}{\overline{H_S}} \times \frac{\overline{H_S}}{H_T}.$$

Again, the extent to which the observed mean genetic variation within individual populations matches that expected for a metapopulation in HWE ($\overline{H_I}/H_T = H_o/H_e$) depends on how that variation is partitioned within ($\overline{H_I}/\overline{H_S}$) versus between ($\overline{H_S}/H_T$) populations.

MEASURES OF GENETIC DISTANCE   The fixation index, F_{ST}, can also be interpreted as a measure of **genetic distance** that assesses the degree of genetic differentiation between two populations. **Pairwise F_{ST}** values can thus be calculated between all possible pairs of populations (or sites) to quantify the genetic distance between each pair. Low values ($F_{ST} \to 0$) indicate that populations are genetically similar, whereas high values ($F_{ST} \to 1$) indicate that the two populations have diverged genetically. In the collared lizard study, for example, pairwise F_{ST} values ranged from 0.05 to 0.13 ($\overline{F_{ST}} = 0.08$), which suggests a 'moderate' degree of genetic differentiation between populations (Blevins et al. 2011; **Tables 9.2 and 9.3**).

To illustrate the calculation of pairwise F_{ST} values, let's compute these for each of the three pairs of populations depicted in **Figure 9.3**. Note that ($\overline{H_S}$) and H_T are now being calculated just over the two populations being compared (e.g. $H_T = 2\overline{p}\overline{q} = 2(0.6)(0.4) = 0.48$ for Populations 1 and 2). Thus,

$$\text{Populations 1 and 2: } F_{ST} = 1 - \frac{\overline{H_S}}{H_T} = 1 - \frac{0.46}{0.48}$$
$$= 1 - 0.96 = 0.04$$

$$\text{Populations 1 and 3: } F_{ST} = 1 - \frac{\overline{H_S}}{H_T} = 1 - \frac{0.34}{0.42}$$
$$= 1 - 0.81 = 0.19$$

$$\text{Populations 2 and 3: } F_{ST} = 1 - \frac{\overline{H_S}}{H_T} = 1 - \frac{0.30}{0.48}$$
$$= 1 - 0.63 = 0.37$$

There is a high degree of genetic similarity between Populations 1 and 2, a great degree of genetic differentiation between Populations 1 and 3, and a very great degree of genetic differentiation between Populations 2 and 3 (**Table 9.2**). Genetic distance is thus greatest between Populations 2 and 3, which suggests that Population 3 is somewhat isolated from the other populations (so, genetic drift > gene flow), especially since the genetic distance between Populations 1 and 3 is also considered to be great.

There are other measures of genetic distance that are not based on F_{ST}, however. For example, Nei's genetic distance (D_n; Nei 1972) is commonly used in population genetics (Frankham et al. 2010), as well as in some landscape-genetic studies (Storfer et al. 2010). Nei's genetic distance (D_n) is based on an index of genetic similarity (I_n) between two populations which is obtained as

$$I_n = \frac{\sum_{i-1}^{m}(p_{ix}p_{iy})}{\left[\left(\sum_{i-1}^{m}p_{ix}^2\right)\left(\sum_{i-1}^{m}p_{iy}^2\right)\right]^{0.5}}, \quad \text{Equation 9.13}$$

where p_{ix} is the frequency of allele i in population x, p_{iy} is the frequency of allele i in population y, and m is the

TABLE 9.2   Assessment of the degree of genetic differentiation among populations, based on F_{ST} values.

F_{ST}-statistic range	Assessment
0.0–0.05	Little genetic differentiation among populations
0.05–0.15	Moderate genetic differentiation among populations
0.15–0.25	Great genetic differentiation among populations
>0.25	Very great genetic differentiation among populations

Source: After Wright (1978).

total number of alleles at a locus. Using I_n from Equation 9.13, Nei's genetic distance (D_n) is thus

$$D_n = -\ln(I_n).$$ Equation 9.14

If allele frequencies are the same in both populations ($p_{ix} = p_{iy}$), then $I_n = 1.0$ and $D_n = 0$. Conversely, if the populations share no alleles in common, then $I_n = 0$ and $D_n \to \infty$.

Nor is F_{ST} the only way that population genetic structure can be assessed (Holsinger & Weir 2009). For example, statistical methods similar to an Analysis of Variance (ANOVA) have been used to evaluate population genetic structure, by partitioning the variance in genetic variation into within- versus between-population components (Weir & Cockerham 1984). An Analysis of Molecular Variance (AMOVA) can also make direct use of molecular data (i.e. the presence or absence of particular molecular markers among individuals) rather than allele frequencies to assess population genetic structure (Excoffier et al. 1992). According to one survey, AMOVA was used in about a quarter of landscape-genetic studies (Storfer et al. 2010). For example, an AMOVA found that 86% of the total genetic variation occurred within rather than among collared lizard populations in the Flint Hills (Blevins et al. 2011; Table 9.3). We'll discuss some other ways of identifying population genetic structure a bit later in this chapter.

Measuring Gene Flow and Population Connectivity

We can use genetic distances to make inferences about gene flow; that is, the degree of population connectivity. Recall that genetic drift is greater in small populations and can be offset by gene flow (i.e. migration). Thus, we can determine what level of migration would be needed to offset genetic differentiation that would

TABLE 9.3 Symmetrical distance matrices depicting the pairwise Euclidean distances (km; top diagonal) and genetic distances ($F_{ST}/(1 - F_{ST})$; bottom diagonal) among four collared lizard populations in northeastern Kansas, USA (cf Figure 1).

	Konza Prairie	Fort Riley	Milford Dam	Tuttle Creek Dam
Konza Prairie	0	17.1	20.2	17.5
Fort Riley	0.06	0	13.0	22.0
Milford Dam	0.05	0.05	0	33.8
Tuttle Creek Dam	0.09	0.09	0.13	0

Source: After Blevins et al. (2011).

otherwise emerge in small populations via genetic drift; this is the **drift–migration equilibrium**. Using F_{ST} as our measure of genetic distance, and assuming that all population sizes and migration rates between them are the same, we have

$$F_{ST} = \frac{1}{4N_e m + 1}$$ Equation 9.15

where m is the rate of migration per generation. The number of migrants per generation (the effective migration rate) is thus $N_e m$. As you'll recall from our earlier discussion, it has been suggested that only 1–2 migrants/generation are needed to offset genetic drift (Wright 1931; Slatkin 1987). If we assume that there is only 1 migrant/generation ($N_e m = 1$), then the expected distance between populations is

$$F_{ST} = \frac{1}{4(1)+1} = \frac{1}{5} = 0.2.$$

If the effective migration rate is reduced to 1 migrant/10 generations ($N_e m = 0.1$), however, then $F_{ST} = 0.71$. Thus, a reduction in migration rate leads to greater population differentiation ($F_{ST} \to 1.0$), consistent with our expectations.

We can rearrange Equation 9.15 to infer the level of gene flow from a given F_{ST} as

$$N_e m = \frac{1 - F_{ST}}{4F_{ST}}.$$ Equation 9.16

For example, if the average F_{ST} among collared lizard populations was 0.08 (Blevins et al. 2011), we can substitute this into Equation 9.16, which gives

$$N_e m = \frac{1 - 0.08}{4(0.08)} = \frac{0.92}{0.32} = 2.88,$$

or approximately 3 migrants/generation.

Using a completely different approach, involving maximum likelihood estimation based on coalescence theory[14] (Beerli & Felsenstein 2001), Blevins et al. (2011) had estimated the average rate of migration among collared lizard populations to be 1.56 migrants/generation, which is roughly half the migration rate obtained from Equation 9.16. Much like the classic metapopulation model (**Chapter 7**), F_{ST} is derived from a patch-based (island) model, in which all populations are assumed to be the same size with equal rates of migration occurring

[14] Coalescence theory treats genetic drift as a retrospective process, rather than as a prospective process (i.e. how much heterozygosity is predicted to be lost over time) as we've been discussing in this chapter. Basically, coalescence theory works backward to deduce how the observed genetic variation might have arisen based on how far back in time one would have to go to find a common ancestor with the same allele or genotype (e.g. time to coalescence). From that standpoint, we can think of F_{ST} as a measure of the difference in coalescence times within populations relative to the coalescence times of all individuals in the sample (Slatkin 1991).

among populations, which is unlikely to apply in most real-world applications (Whitlock & McCauley 1999). These assumptions can be relaxed with the sort of maximum likelihood approach adopted by Blevins et al. (2011), and this approach should therefore provide a more robust estimate of the effective migration rate than Equation 9.16. Regardless of the approach used here, however, the migration rate among collared lizard populations is still greater than 1–2 migrants/generation, which may explain why only a modest degree of population differentiation was observed among these populations (**Table 9.3**).

ISOLATION BY DISTANCE   Because of limits to dispersal, populations in close proximity should experience more gene flow and thus have greater genetic similarity than populations that are farther apart. Indeed, this can be viewed as the population-genetics corollary to Tobler's first law of geography ('near things are more related than distant things'). As we have seen, gene flow (migration) and F_{ST} are expected to be inversely related (Equations 9.15 and 9.16). Thus, if gene flow decreases with distance, then F_{ST} should increase with distance ($F_{ST} > 0$). That is, genetic distance (F_{ST}) should increase with geographic distance, and thus we expect populations to exhibit genetic **isolation by distance** (Wright 1943).

To test for isolation by distance, we first calculate F_{ST} for all pairs of populations (pairwise F_{ST} values), as well as the pairwise geographic distances between them. Although we could simply regress F_{ST} on distance, the relationship between the two is not a linear one mathematically, and thus we need to transform both variables to make the relationship linear. A commonly used approach suggested by Rousset (1997) adjusts F_{ST} by $1 − F_{ST}$, and is sometimes referred to as the **linearized F_{ST}**:

$$\frac{F_{ST}}{1 - F_{ST}}. \qquad \text{Equation 9.17}$$

The expected relationship between the linearized F_{ST} and the logarithm of geographic distance is linear, and so we can now use linear regression to determine whether there is genetic isolation by distance (i.e. whether the slope of the line is positive, $\beta > 0$). Testing for the significance of this relationship (e.g. $\beta > 0$ at $P < 0.05$) is not recommended, however, given that our pairwise data are not independent, thus violating one of the assumptions of this parametric test. It would instead be preferable to use a non-parametric test, such as the Mantel test described next.

Mantel test. The **Mantel test** (Mantel 1967) has been widely used in both population and landscape genetics to test for isolation by distance (Storfer et al. 2010; Legendre & Fortin 2010). The Mantel test operates on two distance matrices, which are symmetrical.[15] The pairwise genetic distances between populations (the linearized F_{ST} values) are in one matrix (**A**), and the corresponding pairwise geographic (the straight-line or Euclidean) distances between those populations are in the second matrix (**B**). The Mantel statistic (Z_M) is then obtained as the sum of the cross-products of these two distance matrices ($Z_{AB} = \mathbf{A} \cdot \mathbf{B}$). A standardized Mantel statistic (r_M) is also usually computed, as this can be interpreted like a Pearson's correlation coefficient ($-1 < r_M < 1$), although it is not directly comparable, given that the Mantel correlation is based on the cross-products of the two distance matrices and not the actual data themselves (Fortin & Dale 2005). Randomization procedures are then used to test the significance of the observed r_M [Prob(r_M) < 0.05].

By way of example, let's perform a Mantel test of isolation by distance among the collared lizard populations we've been discussing in this chapter (**Table 9.3**). Does genetic distance increase with geographic distance among these populations? At first glance, it would appear that the populations that are farthest apart (Tuttle Creek Dam vs Milford Dam or Fort Riley) do indeed have the greatest genetic distances (linearized $F_{ST} = 0.09 - 0.13$), whereas the two closest populations (Milford Dam and Fort Riley) have the lowest genetic distance (linearized $F_{ST} = 0.05$). Although the Mantel test statistic suggests a high correlation between genetic and geographic distances, this correlation was not significant, possibly because of the low power associated with only four sites [$r_M = 0.85$, Prob(r_M) = 0.08]. Nevertheless, some populations that are far apart have genetic distances similar to those that are much closer (e.g. compare the genetic distance between populations at Milford Dam and the more distant Konza Prairie, to that between Milford Dam and nearby Fort Riley; **Table 9.3**). There thus appears to be a sufficient amount of gene flow occurring among these populations, consistent with our previous assessment that genetic variability is much greater within than between populations in this region.

Looking ahead to possible landscape-genetic applications, we might be interested in assessing the correlation between genetic distance and some other landscape feature, such as elevation. Given that many environmental gradients are influenced by elevation,

---

[15] The matrix is said to be symmetrical because the distance from **a** to **b** is the same as from **b** to **a**, where **a** and **b** are each matrix elements. In other words, the genetic distance between population **a** and population **b** is the same as that between population **b** and population **a**. Note that distance in the context of a 'distance matrix' is not just geographic or linear (Euclidean) distance (although it might be), but rather is being applied in the more general sense of dissimilarity or the degree of difference between two locations (e.g. the genetic distance between two populations).

and elevation by definition represents a change in (vertical) distance, we might wonder whether genetic distance is really related more to the environmental gradient (or its proxy, elevation) than distance per se. For example, populations separated by 1 km could exhibit greater genetic differentiation when distance is being assayed along an elevational gradient than as a simple linear distance. In that case, we would like to be able to factor out the effects of linear distance in order to evaluate whether elevation is still correlated with genetic distance. To do this, we could perform a **partial Mantel test**, which permits a comparison of two distance matrices while controlling for the effects of a third (Smouse et al. 1986; Fortin & Dale 2005). In our example here, we would thus be comparing the genetic distance matrix (**A**) to a matrix of the elevational differences among our sites (**B**), while controlling for the effects of the linear distance between sites (**C**). The partial Mantel test statistic ($r_{AB.C}$) is then calculated by constructing a matrix of residuals, **A′**, of the regression between **A** and **C**, and a matrix of residuals, **B′**, of the regression between **B** and **C**. The two residual matrices, **A′** and **B′**, are then compared using the standard Mantel test (Fortin & Dale 2005).

Despite the widespread use of the Mantel and partial Mantel tests in population and landscape genetics, there are some known statistical issues that may limit their performance and ability to detect spatial relationships between genetic data and geographic distances (Raufaste & Rousset 2001; Balkenhol et al. 2009a; Guillot & Rousset 2013). Other approaches, such as spatial regression, causal modeling, or multivariate ordination techniques (to be discussed later), might thus be considered in cases where more complex (multivariate) spatial relationships are suspected to influence gene flow and genetic structure (Balkenhol et al. 2009a; Cushman & Landguth 2010; Legendre & Fortin 2010). In particular, Mantel and partial Mantel tests are not viewed as appropriate when spatial autocorrelation is present (Legendre & Fortin 2010; Guillot & Rousset 2013), and yet, that is precisely what isolation by distance predicts: that genetic data are autocorrelated over some distance. We discuss the analysis of spatial autocorrelation in genetic data next.

Spatial autocorrelation. Given that **spatial autocorrelation** in genetic data is expected to occur over some range of distances, we could use autocorrelation analysis to uncover the scale(s) at which gene flow becomes disrupted and population structure emerges. As might be expected, autocorrelation analysis is often performed as part of a landscape-genetics study (Storfer et al. 2010). To continue with our lizard case study, we can examine the application of spatial autocorrelation among collared lizards in one of these populations (Konza Prairie) for which Blevins et al. (2011) had obtained information on

the pairwise geographic and genetic distances among a large number of individuals at the site.[16] This scale of analysis thus has the potential to reveal finer details of dispersal and gene flow than our previous analysis of isolation by distance among populations.

Genetic distances among individuals were significantly autocorrelated at distances <1 km (**Figure 9.4**), which is consistent with a mark-recapture study that found that collared lizards generally moved only 100–300 m, with none moving >1 km (Blevins et al. 2011). In addition, genetic distances among individuals became negatively autocorrelated at 3–4 km, which suggests that this might be the scale at which gene flow becomes disrupted. However, this is far less than the distances separating the various collared lizard populations, which as you'll recall, showed little evidence for isolation by distance and only modest genetic differentiation (**Table 9.3**). Thus, it can be difficult to reconcile individual movements with the apparent degree of migration and gene flow that occurs among populations, as these entail different processes operating at different scales. Gene flow involves both successful dispersal and reproduction, and is thus measured on a generational time scale. Recall that it may take only 1–2 migrants per generation to offset genetic drift and reduce genetic differentiation among populations. As we discussed in **Chapter 6**, it is precisely these sorts of infrequent, long-distance dispersal events that are the most difficult to detect in individual movement studies, which is why molecular-genetic approaches have

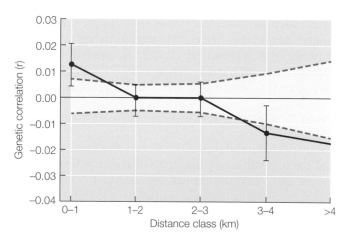

Figure 9.4 Evaluation of population genetic structure using spatial autocorrelation. Genetic correlation (r) among individual collared lizards (*Crotaphytus collaris*) as a function of distance at Konza Prairie (cf **Figure 9.1**), based on an analysis of pairwise genetic distances.

Source: After Blevins et al. (2011).

[16] Genetic distances (such as F_{ST}) can be calculated at an individual level by assessing the degree of heterozygosity within versus between individuals, based on their multilocus genotypes (i.e. the proportion of loci within individuals that are heterozygous).

become invaluable for uncovering evidence of migration among populations.

From Population Genetics to Landscape Genetics

Most of the population-genetic measures we have discussed thus far assume that we are working with populations that can be identified a priori (Manel et al. 2003). For example, in order to calculate a pairwise F_{ST}-statistic between two populations, we must first decide what constitutes a population. Admittedly, our definition of populations is often subjective, perhaps based on the fact that we have sampled individuals at different locations. This may or may not reflect the scale(s) at which gene flow and population genetic structure actually occur, however.

Instead of defining populations a priori, we can let our genetic data inform where those population boundaries lie (Manel et al. 2003). By obtaining genetic information from many individuals, whose exact geographic locations are known, we can apply various statistical approaches to determine what (if any) spatial genetic structure occurs in the system. Ideally, then, individuals represent the unit of analysis in landscape genetics, rather than populations. This is a key distinction between population genetics and landscape genetics. Although populations or habitat patches (nodes) can also be used as the unit of analysis in landscape genetics, a focus on individuals permits a finer scale of analysis that is less biased by our own preconceived notions of how the system is structured genetically. We thus begin with a discussion of Bayesian genetic clustering methods, before turning our attention to multivariate ordination techniques for detecting population genetic structure a posteriori.

Bayesian Genetic Clustering

Bayesian assignment (clustering) methods are currently the most widely used approach for delineating population genetic structure a posteriori (Beaumont & Rannala 2004; Storfer et al. 2010). With **Bayesian genetic clustering**, we assume that our samples have been drawn from some unknown number (K) of populations (clusters), which can be characterized in terms of their allele frequencies (p) for various loci (Pritchard et al. 2000). As the name implies, Bayesian clustering uses the methods of Bayesian statistical inference[17] to derive a posterior distribution; that is, the probable values of both p and K given our genetic data (X). The posterior distribution is

the product of two functions: (1) the likelihood function, which is a conditional probability distribution that specifies how well the genetic data, X, appear to conform to our hypothesized values of p and K based on some model or theoretical distribution; and 2) the prior distribution, a probability distribution that reflects our initial assumptions about how p and K might be distributed. To put it more succinctly, in mathematical terms:

$$\mathrm{Prob}(p,K|X) \propto \mathrm{Prob}(X|p,K)\mathrm{Prob}(p,K). \text{ Equation 9.18}$$

The posterior distribution is typically approximated using methods such as Markov Chain Monte Carlo (MCMC)[18] simulations, which generate a probability distribution from which statistical inferences about p and K can be made (Pritchard et al. 2000; Beaumont & Rannala 2004). Basically, MCMC starts with an arbitrary configuration of parameter values and then updates one or more of these parameters to new values in an iterative fashion, based on the observed data and the current values of the other parameters (Falush et al. 2007). After many such iterations, the probability distribution should eventually reach an equilibrium that converges on the posterior distribution.[19]

The first computer program developed for implementing Bayesian genetic clustering was STRUCTURE (Pritchard et al. 2000), and this is still the most commonly used program for genetic cluster analysis (François & Durand 2010). We'll therefore focus on two modeling approaches that are commonly implemented within STRUCTURE: (1) models with admixture, and (2) models without admixture (François & Durand 2010). Regardless of which clustering model or program is used, the likelihood function is always the same, in that each population in our sample is hypothesized to be in HWE and to exhibit linkage equilibrium[20] (François &

---

[17] Bayesian inference employs Bayes' rule, which in simple terms can be stated as: Posterior Distribution ∝ Likelihood Ratio × Prior Distribution. As you can see, this is just a more generic statement of Equation 9.18.

[18] A Markov Chain is a stochastic model that describes a set of random variables or a sequence of events, in which the probability of each depends only on the state of the variable or event that immediately precedes it; in other words, a Markov Chain is basically a random walk in one dimension (**Chapter 6**). Monte Carlo methods refer to a broad class of computational algorithms, which rely on repeated sampling or simulations to obtain an unknown probability distribution. Thus, the Markov Chain Monte Carlo methods employed in Bayesian clustering programs combine both of these approaches to approximate the posterior probability distribution.

[19] It is common practice to allow for an initial 'burn-in' period, where the MCMC simulation is allowed to run, but these initial iterations are discarded so as to guard against the inclusion of transient values that are ultimately unrepresentative of the equilibrium distribution (i.e. the posterior distribution upon which the run ultimately converges). For example, Blevins et al. (2011) used an initial burn-in of 100,000 iterations that was then followed by 1,000,000 iterations for each run (30 runs total).

[20] Linkage disequilibrium occurs when two or more loci are linked (perhaps because they occur close together on the same chromosome), such that alleles at these loci exhibit a non-random association, meaning that allele frequencies tend to depart from that expected under HWE. It is therefore advisable to first test genetic

Durand 2010). Thus, it is the prior distribution (our assumptions about how p and K are distributed) that differs between these two approaches, and is the main source of difference among the various clustering programs that are now widely available (e.g. STRUCTURE, Pritchard et al. 2000; GENELAND, Guillot et al. 2005; TESS, Chen et al. 2007; BAPS, Corander et al. 2008).

In models without admixture, we assume that our samples were obtained from a mixture of K populations that are diverging; that is, all individuals originated from the same parental population but no mixing is now occurring among populations (i.e. gene flow is limited among populations; **Figure 9.5**). Given that individuals from the same population are likely to have greater genetic similarity than those from different populations, we can assign individuals probabilistically to one of the genetic clusters. If we define populations according to Hardy–Weinberg and linkage equilibria criteria, we can then use evidence of linkage disequilibrium and departures from the HWE, such as a reduction in the expected frequency of

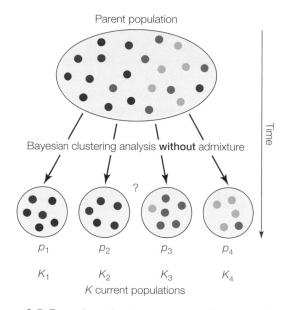

Figure 9.5 **Bayesian cluster analysis without admixture.** Bayesian cluster analysis without admixture assumes that individual populations have diverged from a parent population over time. Barriers to gene flow, followed by genetic drift, are assumed to have given rise to genetic structure that can be identified as a result of differences in allelic frequencies (p) among the K populations (i.e. via the Wahlund effect), where K may be unknown. Because individuals that share a particular genotype are more likely to be from the same K population, individuals can be assigned probabilistically to a population based on their genotypes. Population structure (K) is then inferred by assessing how well the model fits the genetic data for different values of K (e.g. K = 1, 2, 3, or 4).

data for departures from HWE and linkage disequilibrium, using computer programs such as GENEPOP (Raymond & Rousset 1995).

heterozygotes, to identify whether population structure exists and thus how many populations are present (i.e. we can estimate K). Bayesian clustering without admixture thus capitalizes on the Wahlund effect, in which population structure is expected to emerge over time in the absence of gene flow due to genetic drift (François & Durand 2010).

The second clustering approach includes admixture, and is the more robust and thus commonly used of the two models (François & Durand 2010). In this model, we assume that individuals come from an unknown number of K parent populations, which may or may not be present today, but among which gene flow (interbreeding) occurred at some point in the past, such that our current sample contains individuals of mixed ancestry, in that their genotypes represent a mixture of alleles derived from the K populations (**Figure 9.6**).[21] For each individual in the sample, we thus want to estimate the proportion of the individual's multilocus genotype that originated from each of the putative K populations (the admixture proportion or ancestry coefficient), which is given in a Q matrix of individuals and the various K populations (Pritchard et al. 2000). By varying K, we can examine the effect this has the distribution of the ancestry estimates (i.e. the admixture proportions) for individuals in our sample. For example, if we assume that there are at least two populations ($K = 2$), which can be identified on the basis of their different allele frequencies, then we can assign individuals to one or the other population based on the proportion of their multilocus genotype that is shared with each. Alternatively, it may be that we get a better fit to our data if we assume that there are three, four, or more populations. The number of populations, K, is thus inferred by maximizing the posterior probability of K and assessing the log-likelihood of the model fit to the data (i.e. $\ln[\text{Prob}(X\,|\,K)]$; Pritchard et al. 2000). In addition, the rate of change in the likelihood as a function of K (ΔK) is often used (Evanno et al. 2005).

To illustrate, we'll shift gears and consider a new case study, given that a Bayesian clustering analysis of the collared lizards we've been discussing found little evidence of population structure, as might be expected given that gene flow was apparently sufficient to offset genetic drift (1.56 migrants/generation), resulting in a low degree of genetic differentiation among populations (based on the results of an AMOVA; Blevins et al. 2011). We'll thus turn our attention to a consideration of the genetic structure of wild canid populations in Ontario, Canada (Rutledge et al. 2010; **Figure 9.7**). In this region of Ontario, there are actually three species

[21] Other models may be more appropriate if gene flow is directional (from one population to another), or if genetic drift is likely to have occurred since the admixture event (e.g. Long 1991).

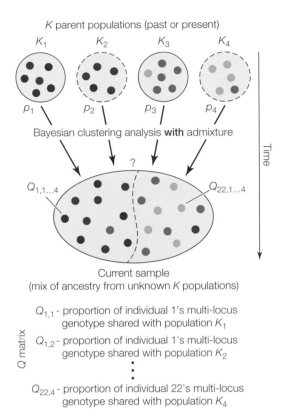

K parent populations (past or present)

K_1 K_2 K_3 K_4

p_1 p_2 p_3 p_4

Bayesian clustering analysis **with** admixture

Time

?

$Q_{1,1...4}$ $Q_{22,1...4}$

Current sample
(mix of ancestry from unknown *K* populations)

Q matrix

$Q_{1,1}$ - proportion of individual 1's multi-locus
genotype shared with population K_1

$Q_{1,2}$ - proportion of individual 1's multi-locus
genotype shared with population K_2

$Q_{22,4}$ - proportion of individual 22's multi-locus
genotype shared with population K_4

Figure 9.6 Bayesian cluster analysis with admixture. Bayesian cluster analysis with admixture assumes that individuals are derived from an unknown number of parent populations, some of which may no longer be extant (denoted here by the dotted lines around populations K_2 and K_4), among which gene flow occurs (or did at some point in the past). Individual genotypes thus reflect a mixture of alleles that have originated from one or more of these populations, assuming no genetic drift has occurred. The proportion of an individual's multilocus genotype shared with each of the putative *K* populations is given by the admixture proportion (otherwise known as the ancestry coefficient) for each individual in a *Q* matrix. Population structure (*K*) is then inferred by assessing how well the model fits the data for different values of *K* (e.g. *K* = 1, 2, 3, or 4).

of canids that have been interbreeding extensively, creating a 'hybrid swarm' among gray wolves (*Canis lupus*), eastern wolves (*Canis lycaon*), and coyotes (*Canis latrans*). Species boundaries have thus become blurred, resulting in a genetic gradient of admixed individuals that runs from southern Ontario along the Frontenac Axis (FRAX) where coyotes predominate, to northeastern Ontario (NEON) where gray wolves are most prevalent. Although gray wolves and coyotes do not normally interbreed, introgression between the two species may occur within this region owing to the presence of eastern wolves, which are intermediate in both their size and location along this gradient (Algonquin Provincial Park, APP), and thus may serve as a 'genetic bridge' between gray wolves and coyotes.

To test this, Rutledge and her colleagues (2010) conducted a Bayesian cluster analysis with admixture for over 200 individuals that were sampled along this gradient (using microsatellite data and STRUCTURE; **Figure 9.7**). The results of the cluster analysis revealed not only the existence of three distinct population clusters (*K* = 3, **Figure 9.7A**), but also reinforced the regional affiliations of these different species-types (i.e. coyotes to the south, eastern wolves in the middle, and gray wolves to the north; **Figure 9.7B**). Nevertheless, the individual assignment tests uncovered a high degree of admixture (*Q*), with hybrid individuals found throughout the gradient (**Figure 9.7B**). Hybridization appears to be occurring primarily between coyotes and eastern wolves, and between eastern wolves and gray wolves. Thus, eastern wolves appear to be acting as a genetic bridge between coyotes and gray wolves, as evidenced by the small amount of coyote DNA that appears in the genetic profile of some gray wolf/eastern wolf hybrids (i.e. red lines are present within the blue/green lines of the 'NEON' cluster made up of predominantly gray wolves in **Figure 9.7B**).

Given the pronounced gradient in the genetic composition of these canid populations, Rutledge and her colleagues (2010) additionally performed a **spatial Bayesian cluster analysis** (using GENELAND;[22] Guillot et al. 2005, 2008). In spatial Bayesian clustering, information on the geographic coordinates of individuals is used to infer not only the number of populations (*K*), but also to map the location of genetic discontinuities between those populations (i.e. the population boundaries). The prior distribution assumes that individuals can be grouped on the basis of their multilocus genotypes into discrete groups (i.e. random mating populations), and uses tessellation[23] to help define those groups (Guillot et al. 2005; François & Durand 2010). In the case of the Ontario canid populations, the spatial Bayesian cluster analysis agreed with the non-spatial analysis (STRUCTURE) in finding evidence for three distinct population clusters across the region. In addition, the spatial clustering analysis revealed a very sharp genetic discontinuity between the FRAX (predominantly coyote) and APP (predominantly eastern wolf) populations. This is not an absolute boundary, for we have seen how hybridization is occurring among

[22] Unlike STRUCTURE, the cluster model in GENELAND does not consider admixture (François & Durand 2010).

[23] Tessellation or 'tiling' involves breaking a plane into multiple polygons based on the proximity of neighboring points (e.g. nearest neighbors; Fortin & Dale 2005). The spatially explicit clustering models implemented within GENELAND, TESS, and BAPS all use Voronoi tessellation to do this, but in slightly different ways to define neighborhoods of related points (François & Durand 2010). Regardless of how Voronoi tessellation is performed, the end result is that each polygon (population) consists of points (individuals) that are closer to each other (genetically) than to any other points.

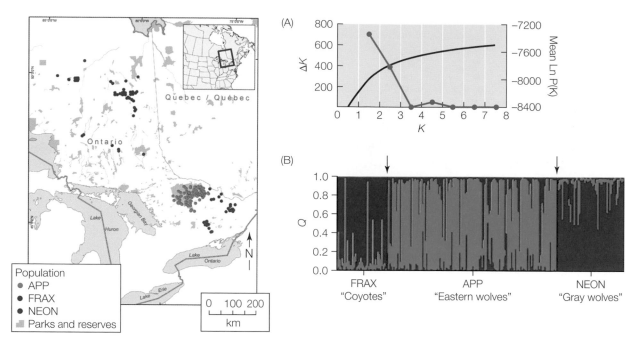

Figure 9.7 Bayesian cluster analysis of wild canid populations in southern Ontario. Three species of canids (coyotes, *Canis latrans*; eastern wolves, *Canis lycaon*; gray wolves, *Canis lupus*) hybridize along a gradient extending from southern Ontario along the Frontenac Axis (FRAX), through the Algonquin Provincial Park (APP), and into northeastern Ontario (NEON; left). Nevertheless, Bayesian cluster analysis with admixture (cf **Figure 9.6**) revealed the existence of three distinct population clusters, as demonstrated by the concordance of the log-likelihood and ΔK values at $K = 3$ (A). However, individual assignments within program STRUCTURE still provide evidence for hybridization among species in this region (as given by the admixture proportion, *Q*; B). Each colored line corresponds to the genetic profile of a single individual, and thus hybrid individuals are denoted by lines consisting of >1 color (red = coyote; green = eastern wolf; blue = gray wolf). The proportion of the individual's genotype from each species is given by the relative length of the colored line; thus, a coyote/eastern wolf hybrid would have a line that is some combination of red and green. Arrows mark the boundaries of the regions from which these individual samples were obtained.

Source: After Rutledge et al. (2010).

coyotes and eastern wolves (**Figure 9.7B**). Nevertheless, coyotes are still 'more different' genetically from eastern wolves, than eastern wolves are from gray wolves.

Multivariate Ordination Techniques

Bayesian clustering programs, such as STRUCTURE and GENELAND, are useful tools for delineating population structure from multilocus genetic data. Recall, however, that Bayesian clustering models all invoke the same likelihood function: populations are assumed to be in HWE and linkage equilibrium is assumed to exist among loci. If these assumptions are not met, all is not lost, for we can use other statistical methods that are not dependent on genetic equilibrium assumptions, such as multivariate ordination techniques. These methods may also be used to corroborate inferences from a Bayesian cluster analysis, given that they make fewer assumptions regarding the underlying genetic data structure, and thus can provide a complementary analysis (e.g. Rutledge et al. 2010).

In landscape genetics, a variety of ordination methods have been used (Storfer et al. 2010), including

principal components analysis (PCA), principal coordinates analysis (PCoA), and non-metric multidimensional scaling (NMDS). To illustrate each of these approaches, we'll begin with an example of PCA and revisit the cluster analysis of the wild canid populations in Ontario (Rutledge et al. 2010). In addition to Bayesian cluster analysis, the researchers used PCA to explore the clustering of individuals based on their multilocus genotypes. Basically, PCA attempts to fit a multidimensional ellipsoid to the data, where each axis of the ellipsoid corresponds to one of the principal components. Each principal component represents a linear transformation of the original data matrix, and is rotated such that the largest axis (the first principal component) explains the greatest amount of variation in the dataset, the next-largest axis (the second principal component) is rotated to be orthogonal to the first and explains the second-greatest amount of variation, and so forth.[24] In the case of the *Canis* genotypes, the first two

[24] Principal components analysis employs an eigenanalysis of either the correlation or covariance matrix of the dataset to obtain the

principal components clearly differentiated three clusters that corresponded well with the regional populations (NEON, APP, and FRAX) and the results of the Bayesian cluster analyses (Rutledge et al. 2010; **Figure 9.8**).

Going back to the collared lizard study, Blevins et al. (2011) applied PCoA to search for evidence of genetic structure among the four sites they surveyed (**Figure 9.1**). Principal coordinates analysis is a distance-based ordination method that is applied to a distance (dissimilarity) matrix, such as pairwise geographic or genetic distances. Much like PCA, the result is a set of orthogonal axes (the principal coordinates) whose magnitudes (eigenvalues) indicate the relative amount of variation captured by each axis.[25] In the case of collared lizards, the first three coordinates explained 59% of the variation in the dataset. A scatterplot of the first two coordinates (which make up most of that 59% total variation explained) showed no clear pattern in the distribution of individual genotypes sampled at different locations. Consistent with previous analyses, then, the PCoA revealed little evidence of genetic structure among collared lizards sampled at these four sites in the northern Flint Hills.

As a final illustration of how ordination techniques can be applied to assess population genetic structure, we consider a study of genetic diversity in golden-cheeked warblers (*Setophaga chrysoparia*), a federally listed endangered species that breeds only in the Ashe juniper (*Juniperus ashei*)-deciduous oak (*Quercus* spp.) woodlands of Texas, USA (**Figure 9.9**). In this study, Lindsay and her colleagues (2008) used NMDS to view the genetic relationships among warblers sampled at different locations across a forest fragmentation gradient. Much like PCoA, NMDS allows visualization of the relationships among data points (allele frequencies in this case) based on a distance matrix, but the goal of NMDS is to minimize the 'stress' among the data points. Stress is a measure of the correspondence (or lack thereof) between the original distances (genetic distances here, derived from microsatellite data) and the ordination.[26] A stress of 0% would thus represent a per-

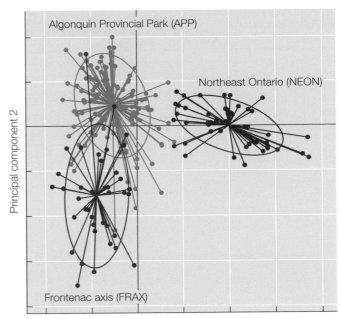

Figure 9.8 Multivariate ordination techniques and population genetic structure. Principal components analysis (PCA) was used to corroborate the inferences of Bayesian cluster analyses for wild canid populations in Ontario (**Figures 9.7**). The first two principal components were able to differentiate individuals associated with the three regional populations on the basis of their multilocus genotypes (12 microsatellite loci). The APP population is intermediate between the FRAX and NEON populations, and shows greater overlap with the FRAX population owing to the greater degree of hybridization between eastern wolves (APP) and coyotes (FRAX). The ovals represent the 95% inertia ellipses used to delineate the different population clusters.

Source: After Rutledge et al. (2010).

---

eigenvectors (the principal components, PCs) and their associated eigenvalues (the amount of variation 'extracted' by a particular eigenvector). Principal components are ranked by their eigenvalues. The greatest amount of variation is projected on the first axis (PC1), the maximum variation uncorrelated with the first axis is projected on the second axis (PC2), the maximum variation uncorrelated with the first and second axis is projected on the third axis (PC3), and so on. The usual practice is to graph and interpret just those principal components with eigenvalues ≥ 1. Each principal component is a linear function of the data, and thus we can examine the factor loadings to determine which variable(s) best explain a given principal component.

[25] If the distance matrix contains the pairwise geographic (Euclidean) distances, then PCoA and PCA are equivalent.

[26] Unlike PCA and PCoA, the axes of the NMDS plot do not relate to the amount of variation explained and are not interpretable with

fect correspondence between genetic distances and the ordination (Lindsay et al. 2008). In this analysis, the stress was 3.7%, which represents an 'excellent depiction' of the relationships among the original data. From this depiction of genetic differences among sampling locations, it is readily apparent that the northern-most site (#7) is genetically distinct from the other sites (**Figure 9.9**). This site is also the most isolated, not only in terms of its geographic distance at the periphery of the breeding range, but also in terms of its effective isolation, given that it is now largely surrounded by agricultural land (i.e. isolation by resistance). Perhaps surprisingly, given that this species is a long-distance migrant that overwinters in Central America, the loss and fragmentation of their breeding habitat appears to have disrupted dispersal and gene flow among golden-cheeked warblers. These warblers are fairly philopatric, and usually nest

respect to the relative contributions of individual loci to any genetic differences observed among data points.

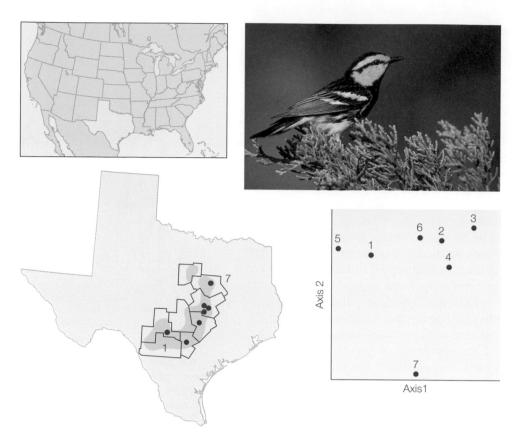

Figure 9.9 Distance-based ordination and population genetic structure. Distance-based ordination methods, such as non-metric multidimensional scaling (NMDS), reveal population genetic structure in the endangered golden-cheeked warbler (*Setophaga chrysoparia*), which has a very small breeding range (orange area) located entirely within the state of Texas, USA (left). A NMDS analysis of genetic similarity among warblers sampled along a fragmentation gradient revealed that one population appeared to be genetically distinct (#7; bottom right). This population is more isolated from other the others, separated by vast stretches of agricultural land, and breeding habitat is more fragmented within this part of the bird's range.

Source: After Lindsay et al. (2008). Photo by Steve Maslowski/USFWS.

within 4 km of their natal territory (i.e. where they were hatched). Thus, barriers that do not ordinarily restrict movement during migration, may nevertheless do so at other times, such as when individuals are searching for suitable habitat in which to breed.

The first step in a landscape-genetics study is to locate genetic discontinuities among populations (**Figure 9.2**). Now that we have explored a variety of means for determining population genetic structure, we can move on to stage two, which is to uncover what landscape or environmental features might have contributed to that genetic structure or influenced patterns of gene flow. This is where landscape genetics begins to diverge from population genetics and come into its own, through its explicit emphasis on testing the relative influence of landscape and environmental features on gene flow and population genetic structure (Storfer et al. 2007). We will thus explore these applications of landscape genetics in the remainder of the chapter.

The key distinction between landscape genetics and traditional population genetics studies is the incorporation of explicit tests of landscape heterogeneity on gene flow and genetic variation within and among populations.

Storfer et al. (2010)

Landscape Genetics

Some of the major research questions currently being addressed within the field of landscape genetics focus on the following six areas of inquiry: (1) uncovering the landscape correlates of population genetic structure; (2) assessing functional landscape connectivity; (3) identification of movement corridors and barriers to gene flow; (4) characterization of source–sink population dynamics; (5) determining the relative influence of current versus past landscape structure on population genetic structure; and (6) evaluating the potential for adaptive responses to changing landscapes in the face of future climate change (Storfer et al. 2007, 2010). We'll consider each of these topics in turn, along with a more explicit consideration of the relative effects of habitat area versus fragmentation on population genetic structure.

Landscape Correlates of Population Genetic Structure

One obvious way to explore the potential relationship between landscape pattern and genetic processes is to examine the degree of correlation between specific landscape features or metrics (**Chapter 4**) and population genetic structure. Different elements of landscape structure, such as elevation, slope, topographic relief, degree of habitat isolation, or the percentage of certain landcover or land-use types, may be variously correlated— whether positively or negatively—with genetic distances among sites (e.g. pairwise F_{ST} values). For example, Lindsay et al. (2008) used Mantel tests to evaluate individually the correlation between various landscape measures and the pairwise genetic distances between sampling locations for golden-cheeked warblers. Pairwise genetic distances were most correlated with the percentage of agricultural land, even more so when the influence of geographic distance was removed from the analysis ($r_M = 0.72$, $P < 0.001$). Thus, golden-cheeked warblers may be averse to dispersing across open habitats in search of suitable breeding habitat, which could then lead to genetic divergence of populations separated by wide swaths of agricultural land (e.g. site #7, **Figure 9.9**).

Nor are such relationships between landscape structure and genetic structure limited to just terrestrial species. For riverine fishes, gene flow tends to be restricted by the hierarchical nature of the drainage network (**Figure 2.11**) and can be further constrained by temporal changes in hydrological connectivity across the network (**Figures 3.24 and 3.28**). Thus, both landscape and riverscape variables, such as elevation, the slope or stream gradient, river network structure, stream distance, variability in streamflow, temperature, or the distribution and variety of riverine habitats, have all been linked to genetic patterns in freshwater fish (e.g. Storfer et al. 2010). For example, genetic diversity (allelic richness) in the endangered Macquarie perch (*Macquaria australasica*), which inhabits upland streams in the Murray-Darling river basin of southeastern Australia, was best explained by a single landscape variable, the stream gradient or river slope (i.e. the change in elevation within a 10 km range spanning each sampling location; Faulks et al. 2011). As river slope appears to be correlated with the availability of the species' riffle habitat, it stands to reason that genetic diversity of this habitat specialist should decline with a decrease in its preferred habitat. The

genetic differentiation between populations (pairwise F_{ST}), however, was positively correlated (r_M) with both distance from the river source (i.e. the headwaters) and the number of anthropogenic barriers (e.g. weirs, dams, or impoundments) between sites. These barriers still explained a significant degree of genetic differentiation among populations even after distance was factored out in a partial Mantel test. Thus, the population genetic structure of the Macquarie perch reflects not only broader-scale divisions among drainage basins and major catchments within these basins, but also the effects of major anthropogenic structures such as dams that fragment the river and create barriers to dispersal and gene flow. We'll discuss more fully the effect of landscape barriers on gene flow in a moment, but first we'll explore how landscape genetics can contribute to an analysis of the functional connectivity of landscapes.

Functional Connectivity

The most common objective of landscape genetic studies is to identify the landscape or environmental features that facilitate or constrain genetic connectivity.

Sork & Waits (2010)

Earlier in the book, we touched upon how genetic data could be used to assess functional landscape connectivity (**Chapter 5**). Recall that functional connectivity incorporates the interaction between an ecological process or organism and landscape structure. Thus, we can determine the degree to which landscapes facilitate the movement of organisms (or their propagules) by relating gene-flow patterns to landscape structure. Landscape genetics is a real boon to the study of functional connectivity, particularly in species whose movements or dispersal are rarely observed, in which individual marking and tracking are virtually impossible or just impractical (e.g. insects, pelagic larvae of marine invertebrates, plant pollen, or seed dispersal; Holderegger & Wagner 2008; **Chapter 6**). Even for those species whose movements can be observed or tracked, however, evidence of successful movement across the landscape is not evidence of success, in terms of the longer-term survival and reproduction of those individuals. Thus, we can amend the definition for functional connectivity given in **Chapter 5** (Calabrese & Fagan 2004) to highlight that genetic data can provide better evidence of actual landscape connectivity, by demonstrating that both successful dispersal and reproduction have occurred. Measures of spatial genetic structure are therefore commonly used as measures of functional connectivity (Spear et al. 2010).

As we've discussed in this chapter, population genetics has traditionally adopted an island or meta-population perspective in describing the genetic relationships among populations. In that context, the intervening matrix is ignored and thus populations are isolated only by distance. In landscape genetics, however, the relative permeability of the matrix is seen as a major determinant of gene flow across the landscape, and thus for understanding the genetic relationships among individuals or populations (Holderegger & Wagner 2008). It is for this reason that landscape genetics has been credited with contributing to a paradigmatic shift from 'the study of gene flow in a purely theoretical space characterized by geographical distances only, to the study of gene flow in heterogeneous and fragmented landscapes' (Manel & Holderegger 2013: 615). That is, landscape genetics has contributed to the study of genetic **isolation by resistance**.

Incorporation of the matrix is a discriminating difference between landscape genetics and population genetics.

Holderegger & Wagner (2008)

We discussed the characterization of landscapes as resistance surfaces in **Chapter 5**. Recall that land-cover types that offer a high resistance to movement are expected to impede dispersal and gene flow (low connectivity), whereas low-resistance land covers should facilitate dispersal and gene flow (high connectivity). Regardless of the approach taken (see below), this entails making certain assumptions about how different land covers and other elements of the landscape are expected to influence dispersal, and thus, gene flow. As discussed in **Chapter 5**, we can use genetic data to make inferences about landscape resistance, although this still provides only an indirect rather than direct measure of resistance (Zeller et al. 2012) and may reflect the historical rather than contemporary effects of landscape structure on gene flow (Landguth et al. 2010). The biggest challenge in modeling resistance surfaces is assigning resistance values to different landscape features, which are rarely known a priori. Thus, resistance surfaces are best viewed as 'hypothesized relationships' between landscape structure and movement or gene flow (Spear et al. 2010). Nevertheless, the varied approaches that have been applied to the problem of modeling resistance surfaces offer the best examples to date of how we can link genetic processes to landscape patterns and evaluate the functional connectivity of landscapes. Such approaches include causal modeling (Cushman et al. 2006), graph theory (Garroway et al. 2008), circuit theory (McRae 2006; McRae & Beier 2007), and gravity models (Murphy et al. 2010). We consider each of these approaches in turn.

CAUSAL MODELING   As the name implies, causal modeling seeks to understand the causal relationships (also known as structural relationships or paths)[27] among a group of variables that are hypothesized to be responsible—either directly or indirectly—for a particular phenomenon. In practice, this approach does not identify causation so much as it identifies correlation, which then enables us to rule out some alternative hypotheses for the observed patterns. For example, Cushman and his colleagues (2006) hypothesized that the observed genetic connectivity among American black bears (*Ursus americanus*) within a national forest in Idaho was produced by some combination of effects due to isolation by distance, landscape-resistance gradients, and/or the presence of geographic barriers, such as a major river valley, on gene flow (**Figure 9.10**). Given that they did not know a priori which of these factors best explained genetic structure, Cushman et al. (2006) tested all possible combinations of factors using a combination of simple and partial Mantel tests to assess the degree of support for these alternative models (i.e. a total of seven causal models: three single-factor models, three two-factor models, and one three-factor model; **Figure 9.10**).

Because landscape resistance was hypothesized to influence gene flow, Cushman et al. (2006) additionally produced more than 100 landscape-resistance surfaces that encompassed all possible combinations of the hypothesized resistances of elevation, slope, roads, and non-forest cover to gene flow (e.g. roads had either a high or low resistance to gene flow). Although these four factors are known to influence black bear movements, it was not known a priori which of these would most strongly influence patterns of gene flow, hence the need to develop and test a large set of hypothetical resistance surfaces. From these surfaces, the authors then derived cost matrices that gave the least-cost distances (**Chapter 5**) between each sampled bear and every other bear across the hypothesized landscape-resistance map. Mantel tests were used to assess the degree of correlation between each of the hypothetical least-cost distance matrices (a total of 108 resistance surfaces in all) and the genetic-distance matrix of the individual black bears ($n = 146$; genetic distance was assayed as the percentage dissimilarity among individuals based on a nine-locus genotype derived from an analysis of microsatellites). A significant Mantel correlation (r_M) between the two matrices would thus support the hypothesis of genetic isolation by resistance (**Figure 9.10**).

---

[27] Causal modeling includes path analysis, which in turn may include a variety of multivariate approaches such as multiple regression, factor analysis, discriminant analysis, and multivariate analysis of variance or covariance; all of these fit under the general umbrella of structural equation modeling (Wright 1921).

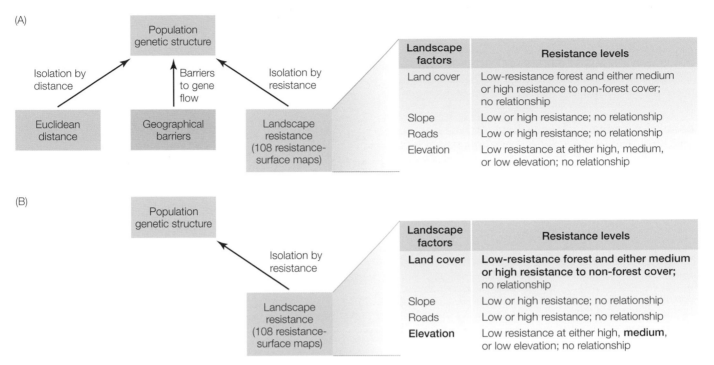

Figure 9.10 Causal modeling of factors affecting genetic structure. Causal modeling was used to investigate which factors influenced genetic structure in American black bears (*Ursus americanus*). A set of models was created that explored all possible combinations of the effect of Euclidean distance, a geographical barrier (a large river valley), and landscape resistance on gene flow (A). Landscape resistance included an additional 108 models (resistance surfaces) involving all possible combinations of hypothesized resistances of various landscape factors (non-forest land covers, slope, roads, and elevation) on gene flow (table A). In the end, the genetic structure of this black bear population was most consistent with the hypothesis of isolation by resistance (B). Specifically, high forest cover at intermediate elevations appeared to facilitate gene flow among black bears in this landscape (boldfaced terms in table B).

Source: After Cushman et al. (2006).

Of the seven causal models, the model in which gene flow was influenced solely by landscape resistance was best supported by the data (i.e. the single-factor model that contained only the effects of landscape resistance on gene flow). That still leaves some 108 hypotheses as to how landscape structure might influence gene flow, however. After the effects of distance were removed with a partial Mantel test, the candidate model set was reduced to 10 significant partial models, in which the best-supported model was associated with gradients in non-forested land cover and elevation. Specifically, gene flow in this black bear population appears to be facilitated by contiguous forest cover at middle elevations (Cushman et al. 2006).

Correlation is not causation, however. As discussed previously, the use of Mantel tests in landscape genetics is not without controversy, as these can lead to spurious correlations and incorrect inferences about the relationship between landscape pattern and genetic processes (Balkenhol et al. 2009a; Cushman & Landguth 2010). Nevertheless, a causal modeling framework using partial Mantel tests and multi-model inference,

where the relative support among multiple competing models is evaluated statistically (e.g. Cushman et al. 2006), appears reasonably robust to the sorts of spurious correlations that can plague simple Mantel tests (Cushman & Landguth 2010).

GRAPH THEORY We first discussed graph-theoretic approaches for assessing landscape connectivity in **Chapter 5**. To review, a **graph** is a spatial representation of the landscape as a network of nodes (individuals, populations, or sites) that are actually or potentially connected by dispersal or gene flow, as denoted by lines (edges) between the nodes (**Figure 5.12**). Edges represent the functional linkages among nodes; they can be weighted so as to reflect the resistance of the intervening matrix to dispersal or gene flow, making graph theory a useful application for landscape genetics.

To illustrate, we'll consider a study that examined the population genetic structure of fishers (*Martes pennanti*), a large weasel of the boreal forest that is trapped for its fur, among three-dozen locations in the Great Lakes region of Ontario (Garroway et al. 2008; **Figure 9.11**). The pairwise genetic distances among populations was

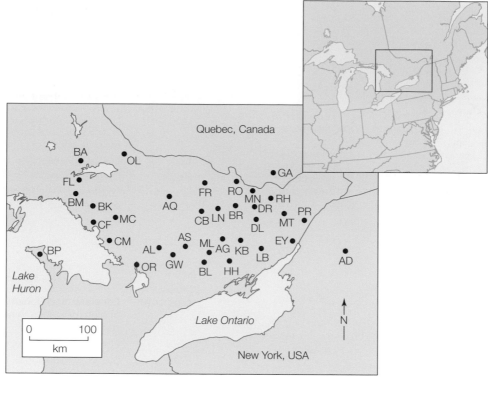

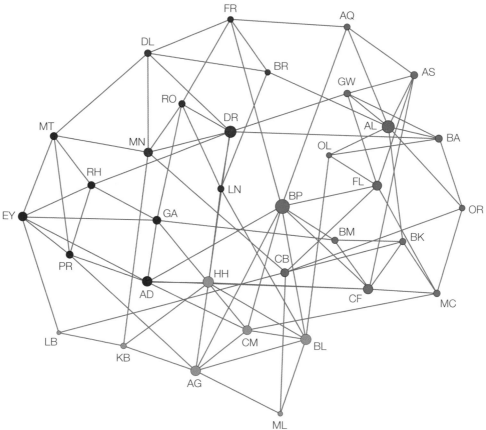

Figure 9.11 Graph-theoretic analysis of genetic connectivity. A graph-theoretic analysis of genetic connectivity among fisher (*Martes pennanti*) populations in the Great Lakes region of Ontario, Canada. Genetic samples were obtained from 34 locations (map inset). The genetic relationships among these fisher populations is represented as a graph, in which the locations are indicated by colored nodes. Node size is related to the degree of connectivity (large nodes are more highly connected) and edge length is related to the genetic distance between populations (short edges correspond to populations that are genetically similar). Node colors identify clusters of populations that share the greatest genetic similarity.

Source: Adapted from Colin J. Garroway, Jeff Bowman, Denis Carr, et al. (2008).

based on an analysis of 16 microsatellite loci from 20 or so fishers in each population (722 fishers in all). Based on a graph analysis of this population-genetic network, populations were more closely related and clustered genetically than would be expected if populations were connected at random (i.e. this network exhibited small-world properties; **Box 5.1**). Fishers are highly territorial and philopatric, and thus a high degree of genetic clustering is to be expected, especially if long-distance dispersal has high fitness costs. Higher clustering also leads to greater redundancy in the linkages among these landscape populations (nodes), which enhances the robustness of the network to perturbations, such as the targeted removal of individuals or populations during trapping season. Thus, current harvest levels are not expected to affect genetic connectivity or induce genetic differentiation among these fisher populations. Moreover, the authors found that two measures of node centrality (degree and eigenvector centrality; **Table 5.1**) were negatively related to both the proportion of immigrants from a landscape node and the snow depth in that landscape (Garroway et al. 2008). Well-connected nodes (e.g. BP, AL, and DR in **Figure 9.11**) thus appear to be functioning as population sources (**Chapter 7**), in that they have greater habitat suitability (i.e. less snow cover) and produce a greater proportion of migrants than less-connected nodes. Although the overall network may be sufficiently robust to perturbations, these highly connected 'central nodes' are clearly important for maintaining gene flow and connectivity in this population network.

CIRCUIT THEORY Circuit theory was also introduced as a method for assessing landscape connectivity in **Chapter 5**. As you'll recall, circuit theory similarly depicts the landscape as a graph, in which nodes (individuals, populations, or sites) are separated by a matrix of varying resistances to dispersal and gene flow. Unlike graph theory, however, the calculation of connectivity (the resistance or effective distance) is based not just on a single optimal pathway, but also takes into consideration the availability of alternative pathways (i.e. redundancy; **Figure 5.15**).

Although we saw an application of circuit theory for landscape genetics in **Chapter 5**, we'll consider another study here that examines the linkage between finer-scale landscape features (<10 km) and the population genetic structure of wood frogs (*Lithobates sylvaticus*) at a regional scale (10–100 km) in northeastern North America (Lee-Yaw et al. 2009). Pairwise genetic distances (linearized F_{ST} values) were obtained among 16 populations throughout this region, based on an analysis of multilocus genotype data (six microsatellites and 288 frogs total). The authors then used CIRCUITSCAPE (McRae 2006; McRae et al. 2013) to create a number of different landscape-resistance surfaces with which to test differ-

ent hypotheses about how major water barriers and other landscape features appear to be influencing gene flow among wood frog populations in this region.

Models that incorporated barriers or landscape resistance were more strongly correlated with population genetic structure than a hypothetical landscape of uniform resistance, in which populations were only isolated by distance ($r_M \approx 0.9$ vs $r_M \approx 0.3$, respectively). Much of this effect (91%), however, was apparently due to the presence of large water bodies, which function as absolute barriers to dispersal and gene flow in wood frogs. For example, the St Lawrence River is a major barrier that structures the genetics of populations for a subset of sites located on either side of the river ($r_M = 0.69$ vs $r_M = 0.16$ for the uniform landscape). Conversely, water barriers were not important for another subset of sites that were located in an area without any major bodies of water. Even for these sites, however, populations were primarily isolated by distance rather than by the resistance of the intervening landscape matrix. Although features such as land cover, slope, and moisture levels are all known to influence wood frog dispersal within landscapes, gene flow in wood frogs is ultimately limited more by broad-scale geographic barriers, such as large water bodies. In retrospect, this is perhaps not surprising given the widespread distribution of this species. Regional patterns of genetic structure in wood frogs are more likely to reflect the effects of similarly broad-scale landscape and climatic features that have influenced this species' pattern of colonization and range dynamics over time, such as in response to glacial-interglacial cycles. We are reminded once again that different processes—and thus different scales—are at play here, even though dispersal, colonization, and gene flow are all obviously linked to some degree.

We'll return to the identification of barriers to gene flow a bit later in this section. Next, however, we consider one last class of models that have been used to evaluate genetic connectivity in heterogeneous landscapes.

GRAVITY MODELS As the name implies, gravity models are derived from Newton's law of universal gravitation, which states that the strength of interaction between two 'bodies' is proportional to the product of their size and inversely proportional to the squared distance between them. In other words, big near things should interact to a greater degree than small far things. Gravity models have been used by urban geographers to study movement along trade and transportation networks, and thus it is but a small step to adapt these sorts of models to deal with landscapes and genetic networks of interacting populations (nodes), especially given that populations tend to vary in their size and are connected to varying degrees by dispersal and gene flow (e.g. source–sink dynamics, **Chapter 7**). Thus, we expect genetic connectivity to be influenced not only by the intervening landscape (geographic distance or landscape resistance), but also by the properties of the sites or populations themselves (e.g. whether populations are sources or sinks). Gravity models offer a way to examine both in one framework.[28]

To illustrate, Murphy and her colleagues (2010) applied gravity models to study the landscape genetics of Columbia spotted frogs (*Rana luteiventris*), a pond-breeding amphibian that can be found at high elevations in western North America. As we have seen from the previous example, frogs tend to be limited in their dispersal abilities, in terms of both distance travelled and landscape resistance. The Bighorn Crags area of the Salmon River Mountains (Idaho, USA) is noted for its rugged topography, consisting of high granitic peaks (>3000 m) and ridgelines, which should function as barriers to frog dispersal. At such high elevations, the growing season is short, which limits primary productivity within ponds and lakes. This in turn limits the length of the breeding period and reproductive output of frogs. In addition, the introduction of non-native, predatory trout into some of the deeper alpine ponds and lakes in this region pose a significant threat to spotted frogs. Thus, there was good reason to expect a priori that spotted frogs in this area would exhibit source–sink dynamics. Further, given their physiological limitations, spotted frogs were hypothesized to disperse primarily through wetter areas (meadows) than forests, and to travel farther in areas of low topographic relief, especially in large catchments (= wetter areas) with a sufficiently long growing season (a function of elevation and the length of the frost-free period) to permit long-range dispersal. Thus, gene flow in spotted frogs was expected to be influenced by all three parameters that go into a gravity model: the distance between sites; local-site variables: elevation and water depth, as proxies for productivity and the presence of predatory fish, respectively; and between-site variables: resistance of the landscape to gene flow, based on the ratio of meadow to forest habitats, the degree of topographic relief, and length of the frost-free period.

The authors constructed two-dozen candidate models made up of various combinations of these three types of parameters (distance, local-site variables, between-site variables) for both unconstrained and constrained gravity models. Unconstrained gravity models do not distribute flows based on whether a site is an origin or a destination (e.g. a source or a sink), whereas singly constrained models incorporate local-site and

between-landscape effects on either production (origin sites and the landscape processes influencing flows from a site, i.e. population sources) or attraction (destination sites and the landscape factors influencing flows arriving at a site, i.e. population sinks). The best-fitting model (evaluated using AIC) was then compared to the observed genetic distances (pairwise F_{ST} between sites, based on an analysis of eight microsatellite loci from >400 frogs in 37 ponds distributed among eight drainage basins) to see how well this explained patterns of gene flow among spotted frog populations (i.e. functional connectivity).

Although frog populations were clearly isolated by distance, given that steep ridges separated drainage basins and served as major barriers to dispersal, the gravity models that included information on the local site as well as the landscape resistance between sites were better supported than the model with distance alone (Murphy et al. 2010). The best-supported gravity models (whether constrained or unconstrained) all included distance, site productivity (water depth), and landscape resistance (degree of topographic relief and length of the frost-free period); interestingly, habitat permeability (the ratio of meadow to forest habitat) did not influence landscape resistance to gene flow. Moreover, the singly constrained gravity models, which incorporated the relative production or attraction strength of local sites, generally had higher support than unconstrained models. Thus, the authors concluded from this, in conjunction with evidence of a genetic bottleneck at sites that purportedly contained fish, that the spotted frog exists as a source-sink metapopulation within this region.

We'll discuss more fully how landscape genetics can be used to detect source–sink dynamics in a bit. Before we leave the topic of functional connectivity, however, we should take a closer look at how we can use landscape genetics to detect movement corridors and barriers to gene flow, as well as the relative effects of habitat loss versus fragmentation on population genetic structure. We therefore consider these applications in the following two sections.

Identification of Movement Corridors and Barriers to Gene Flow

Inevitably, the study of landscape effects on genetic structure, particularly in terms of assessing genetic connectivity among populations (functional connectivity), leads to the identification of landscape features that serve either as movement corridors or—more likely—barriers to gene flow. Identifying barriers to gene flow is a major focus in landscape genetics because natural and anthropogenic barriers so clearly disrupt functional connectivity among populations, and because the construction of human transportation and water-management infrastructure (roads, dams, canals) create barriers to dispersal and gene flow for many species (Storfer et al. 2010; Sork & Waits 2010). For example, roads may fragment populations by creating a barrier to individual movement and dispersal, either as a result of behavioral avoidance (the organism perceives the roadway to be a hard edge; **Figure 6.10**) or increased mortality (wildlife–traffic collisions), resulting in a number of genetic consequences involving reduced gene flow (decreased genetic connectivity), decreased genetic diversity, and increased genetic structure among populations (Balkenhol & Waits 2009; **Figure 9.12**). Dams and weirs have similar effects in fragmenting and disrupting gene flow among fish populations in riverine networks (e.g. Faulks et al. 2011; Horreo et al. 2011). In other cases, barriers to gene flow are natural ones, such as major rivers, mountain peaks or ridgelines, as highlighted in the frog case studies discussed previously (Murphy et al. 2010; Lee-Yaw et al. 2011). We next discuss various approaches that have been used to identify movement corridors and barriers to gene flow, starting with the latter.

BARRIERS TO GENE FLOW   Barriers to gene flow can be detected or inferred in a variety of ways (Holderegger & Wagner 2008; Safner et al. 2011; Blair et al. 2012). We'll consider three approaches here, based on genetic-distance measures, Bayesian genetic clustering and assignment tests, and parentage analysis.

Genetic-distance measures. As discussed previously, genetic distance can be calculated at an individual or population level to give a matrix of pairwise genetic distances, which can then be correlated with pairwise effective distances (the resistance or least-cost distances) using various approaches, such as Mantel tests (e.g. Cushman et al. 2006) or circuit theory (McRae et al. 2013). For example, specific landscape features that are hypothesized to act as barriers to gene flow, such as a road, dam, or large water body, are assigned maximal resistance values on a landscape map, such that the effective distance between individuals or populations is far greater than suggested by their geographic distance. A significant correlation (r_M) between a 'barrier model' of landscape resistance and genetic

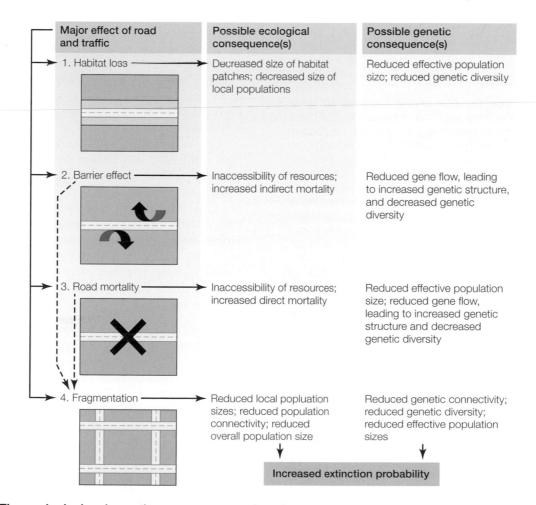

Figure 9.12 The ecological and genetic consequences of roads on populations.
Source: After Balkenhol & Waits (2009).

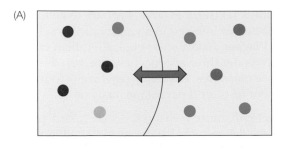

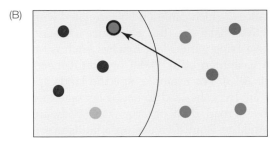

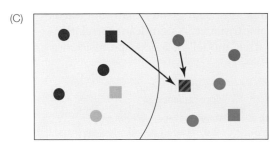

Figure 9.13 Methods for identifying barriers to gene flow. In each of the three landscape scenarios, the black line represents a putative barrier (e.g. a river), and the colored symbols denote different genotypes of individuals sampled on either side of this barrier. (A) Genetic distances between all sampled individuals can be analyzed with respect to the barrier. In this case, the largest genetic differences are found among individuals located on either side of the barrier (denoted by the double-headed arrow). (B) Assignment tests can be used to assess recent gene flow between populations. Here, one individual (the heavily outlined green circle) is likely an immigrant from the population located on the other side of the barrier (population origin indicated by arrow). (C) Current gene flow is assessed through a parentage analysis of offspring (squares). In the example given, the focal offspring (hatched square) is the product of parents located on either side of the barrier (indicated by arrows).

Source: Adapted from Holderegger & Wagner (2008).

distances would suggest that this landscape feature represents a barrier to gene flow (**Figure 9.13A**).

Bayesian genetic clustering and assignment tests. Alternatively, we could compare the boundaries of genetic clusters identified using Bayesian clustering approaches (**Figures 9.5 and 9.6**) to the location of specific barriers (e.g. a river or road). If the genetic boundary coincides with the hypothesized barrier (**Figure 9.13B**), we might then infer that this feature functions as a barrier to gene flow. For example, two

genetically distinct clusters of raccoons (*Procyon lotor*), which were identified through a spatial Bayesian cluster analysis (based on 10 microsatellite loci), were very clearly divided by the Niagara River that runs north from Lake Erie to Lake Ontario along the US–Canadian border (Cullingham et al. 2009; **Figure 9.14**). This was supported by individual assignment tests, which assess the probability that an individual belongs to the cluster from which it was sampled, based on certain criteria (e.g. the ancestry or membership coefficients of the Q matrix; **Figure 9.6**). In this study, the authors used an approach that maximized the average pairwise similarity among Q matrices generated by a replicated series of cluster analyses on the genetic data (using the program CLUMPP; Jakobsson & Rosenberg 2007). Migrants were identified as those individuals whose sampling location fell within the mapped area of one cluster, but whose average assignment value indicated a high likelihood of membership (>0.9) in the other cluster. Based on this criterion, only a handful of migrants were identified (**Figure 9.14**). Thus, the Niagara River appears to be a barrier to raccoon dispersal and gene flow, and as such, may also serve as a natural barrier to disease spread: the rabies virus is prevalent among raccoon populations east of the Niagara River in New York, but not west of the river in Ontario. Landscape genetics can therefore contribute to an understanding of how landscape features influence the potential for disease spread (i.e. landscape epidemiology; **Chapter 8**).

Unlike genetic-distance measures, in which gene flow is inferred, assignment tests provide direct evidence of recent migration events. These events are considered 'recent' (i.e. first-generation migrants) because all evidence of migration (dispersal) will be erased within one or more generations once the individual begins breeding with individuals from the local population (Holderegger & Wagner 2008). Contemporary patterns of gene flow can also be assayed using parentage analysis, an approach we consider next.

Parentage analysis. Current gene-flow patterns can be assayed directly through an analysis of parentage, in which the likely origins of an individual are identified based on an evaluation of all possible parents on the landscape (Jones et al. 2010; **Figure 9.13C**). In general, parentage analysis works by either excluding individuals as parents based on the lack of a shared genotype with offspring, or by assessing the likelihood of parentage using maximum likelihood techniques (e.g. program CERVUS; Kalinowski et al. 2007) or Bayesian approaches (Jones et al. 2010). Parentage analysis has been widely used in plants, especially in trees (Sork & Smouse 2006; Holderegger et al. 2010), although it can obviously be used in animal studies as well.

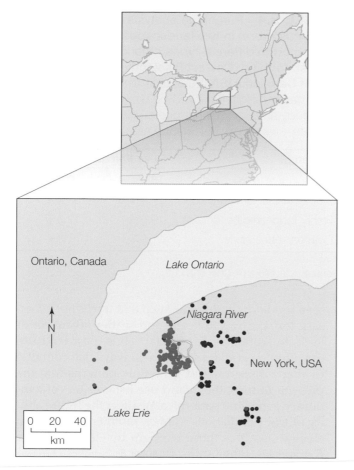

Figure 9.14 Identifying barriers to gene flow using spatial Bayesian clustering.   Genetic samples from individual raccoons (*Procyon lotor*) were assigned to one of two clusters (identified using TESS), which were separated by the Niagara River. Immigrants are identified as those individuals whose average assignment values correspond to a genetic cluster different from the one where they were found (e.g. blue symbols in New York indicate first-generation migrants from Ontario). The Niagara River is thus a major barrier to dispersal and gene flow, as well as the spread of rabies. The disease is found among raccoons to the east of the river in New York, but not to the west in Ontario.

Source: After Cullingham et al. (2009).

The study of gene flow in plants is complicated by the fact that this typically involves the differential movements of two types of propagules: haploid pollen (male gametes) and diploid (or polyploid) seeds, each of which may be dispersed by entirely different processes within the same plant (e.g. a wind-pollinated plant with animal-dispersed seeds). Pollen flow is determined via a paternity analysis, in which the pollen sources (fathers) responsible for fertilizing the seeds (offspring) of known plants (mothers) are identified from their multilocus genotypes (i.e. obtained from an analysis of microsatellites). In addition to identifying the location of pollen sources, paternity analysis can also help to estimate the pollen dispersal

neighborhood or kernel (the distribution of dispersal distances) and the proportion of pollen coming from outside the area (i.e. immigration; Sork & Smouse 2006). Pollen flow alone does not create spatial genetic structure, however. Seed-mediated gene flow can be determined via a maternity analysis, using either nuclear DNA (leaf tissue) or maternally inherited seed-coat tissue or cpDNA (Sork & Smouse 2006) to determine the likely mothers of dispersed seeds or seedlings (Holderegger et al. 2010). Maternity analysis can thus provide an estimate of the seed dispersal kernel as well as the amount of immigration into the study area (i.e. if seeds are not simply deposited within the 'seed shadow' of the mother).

We have learned a great deal about pollen movement and a little about seed movement across extended landscapes, but we know little about the combined effects of the two processes…Beyond the interaction of pollen and seed flow is the interaction of pollen and seed flow with the landscape itself. So far, most of the work has neglected the landscape.

Sork & Smouse (2006)

In plants, as in other sessile organisms, the probability of parentage is generally a function of distance, which may differ for pollen versus seeds, depending on how these propagules are dispersed (wind vs animal; Ashley 2010). Parentage analysis has revealed that gene flow in plants can occur over far greater distances than evident from ecological studies of plant dispersal (**Chapter 6**), although landscape resistance to gene flow has rarely been considered beyond the study of fragmentation effects on seed or pollen flow (Sork & Smouse 2006; Holderegger et al. 2010). The intervening matrix can serve as a barrier to gene flow, with some types of habitat or land use being more—or less—conducive to the dispersal of seeds and pollen, depending on their mode of dispersal or that of their vectors. For example, small habitat fragments tend to receive a significant fraction of pollen and seeds from outside the patch, often from hundreds of meters away, suggesting a far greater connectivity than might otherwise be suspected given the intervening distance (Sork & Smouse 2006). In the case of wind-dispersed species, the clearing of habitat in the intervening matrix might actually enhance dispersal within fragmented landscapes, giving rise to the high levels of immigration and gene flow observed within these fragments. The situation is admittedly more complicated for species whose seeds or pollen are dispersed by animals, as it is the animal agents themselves that are influenced by landscape structure, and not the plant propagules per se (Holderegger et al. 2010).

The differing effects of landscape structure on the movements of seeds versus pollen are illustrated in the case of two red oak (*Quercus rubra*) populations in the Piedmont region of North Carolina, USA (Moran & Clark 2012). Oaks have large seeds (acorns) that are dispersed by scatter-hoarding rodents or birds, such as gray squirrels (*Sciurus carolinensis*) and blue jays (*Cyanocitta cristata*). Long-distance dispersal of pollen can occur via wind, although effective dispersal distances may be limited in dense stands owing to the swamping effect of pollen produced by neighboring trees. One oak population (Duke Forest, 12 ha) occurred in a landscape that had been virtually cleared of trees and was farmed until the early 1900s, whereas the other (Coweeta, 7.5 ha) was located in a landscape that had only been selectively logged in the early 1900s. As a consequence of these different land-use histories, the densities of adult oaks and oak seedlings were about three times greater at the Coweeta site, which produced more than twice as many acorns per hectare as Duke Forest.

Consistent with previous studies of plant gene flow in fragmented or human-disturbed landscapes, the vast majority of oak seedlings at both sites had at least one parent coming from outside the study area (Coweeta: 77%; Duke Forest: 84%; **Table 9.4**). Although 60% of seedlings at both sites were 'fathered' by a tree outside the study area (father outside or both parents outside), most seedlings at Coweeta (74%) were produced by a maternal tree on-site, whereas fewer than half (43%) were at Duke Forest (mother inside or both parents inside). Pollen flow thus seems to have about equal importance at these two sites, whereas seed dispersal is playing a much greater role in the colonization and genetic diversity of red oaks in Duke Forest: nearly 57% of seedlings from Duke Forest were produced from acorns that originated outside the study area, whereas only 26% of seedlings at the Coweeta site were from acorns produced elsewhere (mother outside or both parents outside; **Table 9.4**). This was also reflected in the estimated dispersal distances for pollen versus seeds at the two sites. Although pollen dispersal distances were about the same (Duke Forest: 178 m; Coweeta: 146 m), seed dispersal distances were eight times greater in Duke Forest (125 m) than in Coweeta (15 m). It appears that most acorns sprout where they fall at Coweeta (i.e. very near to the maternal tree), whereas most of the seedlings at Duke Forest originate from animal-dispersed seeds (Moran & Clark 2012). Given that squirrel density did not differ much between sites, the authors posited that the greater tree density and acorn production at Coweeta may have saturated

TABLE 9.4 Parentage analysis of red oaks (*Quercus rubra*) in two landscapes previously subjected to different types of land use (Duke Forest: farming; Coweeta: selective logging) in North Carolina, USA. Parentage assesses whether both parents of an oak seedling were located within the study area, or whether one or the other parent—or both—were from outside the area.

Parentage	Duke Forest	Coweeta
Both parents inside	16.0%	22.9%
Mother inside, father outside	27.4%	51.4%
Father inside, mother outside	19.6%	17.9%
Both parents outside	37.0%	7.8%

Source: After Moran & Clark (2012).

seed predators at this site, such that it was not necessary for squirrels to horde and scatter acorns, or at least, to scatter them very far. Thus, past land use can alter forest stand structure in ways that continue to shape dispersal and gene-flow patterns into the present. In turn, these land-use legacies on contemporary gene flow patterns may affect the future ability of tree populations to track rapid environmental changes, such as those wrought by climate change (i.e. species' range shifts; **Chapter 8**).

TIME LAGS IN THE POPULATION-GENETIC RESPONSE TO LANDSCAPE BARRIERS   We have now discussed three ways for identifying landscape barriers to gene flow. Regardless of how we do this, we should remain mindful of the fact that *when* we look is as important as *how* we look for barrier effects on gene flow. Our ability to detect barriers to gene flow is very much dependent on the nature of the time lag involved (Landguth et al. 2010). How soon can we expect to see genetic divergence among populations once gene flow has been disrupted by a barrier? Conversely, how long will the signal of some past barrier, such as a glacial lobe that isolated populations during the last glacial maximum, persist and remain detectable? Using a spatially explicit simulation model, Landguth and her colleagues (2010) found that genetic divergence between populations on either side of a new barrier could be detected within 15 generations, depending on the generation time of the organism in question, which is a pretty rapid response. Although that may sound like good news in terms of our ability to detect landscape barriers to gene flow, it also means that such barriers begin to affect the genetic structure of populations almost immediately, which may not be such good news from the standpoint of the species, especially if this leads to a loss of heterozygosity.

Given that species conservation and management is primarily concerned with recent or future changes, if landscape genetics is to be used in this context then there is an urgent need to rigorously assess both the effects that legacies of past landscape change have on observed genetic patterns and the speed at which these genetic patterns change in response to alterations to existing landscapes.   Landguth et al. (2010)

The degree of lag, in terms of the number of generations before significant genetic divergence is observed, may depend on a species' dispersal ability, however. For example, the genetic consequences of a landscape barrier may not be evident for many generations (>100) for species that have limited dispersal abilities (Landguth et al. 2010). In the case of such long time lags, we would likely conclude (albeit incorrectly) that the putative barrier is not a concern, which could have serious repercussions for the management or conservation of sensitive species. The flip side of that, however, is that the ghosts of landscapes past (**Chapter 3**)—the historical barriers to gene flow from more than 100 generations ago—will continue to haunt the genetic structure of populations present. In that case, populations with relatively restricted dispersal abilities may retain a genetic signal from a past landscape barrier for tens to hundreds of generations (Landguth et al. 2010). Obviously, we will want to disentangle the effects of past landscape structure from current landscape structure on patterns of gene flow, a topic we will return to a bit later in this section.

MOVEMENT CORRIDORS   Landscape genetics also holds great promise for identifying movement corridors, that is, areas of the landscape that facilitate movement and thus gene flow. Movement corridors may be obvious features of the landscape, such as riparian areas or mountain valleys, but can also arise from the juxtaposition of different habitats or land uses, and thus are not always obvious (Gustafson & Gardner 1996). In **Chapter 5**, we discussed the use of a genetically based landscape-resistance model that was used to identify three major movement routes for black bears (*Ursus americanus*) in the northern Rockies (Cushman et al. 2009). We'll consider a different case study here, but this one also uses a least-cost path analysis of landscape resistance to gene flow in order to uncover dispersal routes among breeding populations of the federally endangered California tiger salamander (*Ambystoma californiense*).

Like most amphibians, California tiger salamanders breed in ponds, where they spend the first 3–6 months of their life as aquatic larvae. Once they mature, juveniles disperse over land for 1 km or more, returning as adults in subsequent years to breed in their natal pond, or occasionally, in a different pond. Adult salamanders are typically associated with grassland habitats, where they spend much of their time underground in burrows created by small mammals. To determine the likely movement routes taken by adults to breeding ponds, Wang and his colleagues (2009) obtained genetic information (13 microsatellites) from larval salamanders sampled from a collection of 16 ponds in central California. They then calculated least-cost distances between ponds, based on hypothetical costs assigned to different habitat types through which salamanders would have to disperse, which resulted in >24,000 analyses to address all of the possible combinations. They then compared these least-cost distances to those predicted by the gene-flow estimates derived from genetic assignment tests (i.e. the presence of larvae with immigrant ancestry), by assuming that the rate of gene flow between two populations is inversely proportional to the cost of moving between them. Much to their surprise, given the association of adult salamanders with grasslands, movement through grassland habitats appeared to be twice as costly as movement through chaparral, although not nearly as costly as movement through oak woodlands (**Figure 9.15**). This underscores that animals may move through very different sorts of habitat from the one(s) in which they reside or breed, which must be considered in any habitat-management plans for a species, especially one of conservation concern.

Habitat Area versus Fragmentation Effects

Isolation by resistance proves that landscape structure can have a major influence on gene-flow patterns, and thus, population genetic structure. As we have seen, landscape resistance is a product of how species interact with the composition and configuration of different habitat types or land covers on the landscape, as reflected in the cumulative movement costs of individuals during dispersal. Which aspect of landscape resistance has the greater effect on the genetic connectivity of populations, however? If habitat fragmentation disrupts dispersal and gene flow, then perhaps the configuration of habitats is the most important determinant of genetic connectivity. However, population genetic structure is also influenced by the distribution and dynamics of populations, in which case, habitat area generally has a larger influence on patch occupancy and population persistence (**Chapter 7**). From a landscape- or conservation-genetics standpoint, it would be useful to know whether the genetic connectivity of a species is best maintained (or restored) by increasing the amount and/or connectivity of low-resistance land covers, or through a reduction in land covers that are highly resistant to movement, or both.

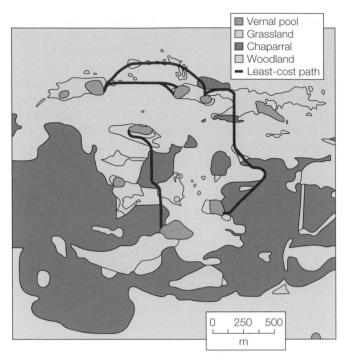

Figure 9.15 Least-cost path analysis of gene-flow patterns. Movement corridors among breeding ponds for California tiger salamanders (*Ambystoma californiense*) identified using a least-cost path analysis of gene-flow patterns. The least-cost paths are highlighted in red. Although adults reside in grasslands, dispersal through grassland habitats was twice as costly (and movements through woodlands five times as costly) as dispersal through chaparral.

Source: After Wang et al. (2009).

To explore the relative effects of habitat area and fragmentation on population genetic structure, Bruggeman and his colleagues (2010) used an individual-based, spatially explicit population model (**Chapter 7**) developed for the red-cockaded woodpecker (*Picoides borealis*) and applied this to a series of simulated landscapes in which the amount and configuration of habitat could be adjusted independently of one another (**Figure 9.16**). The red-cockaded woodpecker is a habitat specialist of old-growth (>60 years) pine forests in the southeastern USA. Forestry operations and other activities have contributed to the widespread loss and fragmentation of mature forests throughout the woodpecker's historical range, such that it is now federally listed as an endangered species. The red-cockaded woodpecker lives in small family groups and is a cooperative breeder, in which offspring from previous breeding seasons (usually male offspring) serve as 'helpers' in raising subsequent broods. Thus, the potential for inbreeding depression is high in such a species, especially if populations are being increasingly isolated through habitat loss and fragmentation.

The woodpecker population model included both demographic and genetic processes, as well as demographic-genetic feedbacks (i.e. the potential for inbreeding). Dispersal was modeled as a correlated random walk (**Chapter 6**) that incorporated the habitat preferences and known movement behaviors of this species. (e.g. females avoid gaps >600 m wide). In addition, survivorship was adjusted by the amount of time individuals spent in forested versus non-forested areas, by assuming a greater predation risk in the latter habitat. Thus, landscape structure influenced both dispersal and demography (survival) in this model. By keeping the amount of high-quality habitat constant and varying the degree of spatial contagion (**Figure 9.16**), the authors could partition out the relative contribution of fragmentation on population size (number of potential breeding groups), the genetically effective population size (N_e), and the proportion of total genetic variation due to population subdivision (F_{ST}).

Nearly all (99%) of the variance in the number of potential breeding groups could be explained by the amount of habitat on the landscape. Conversely, most (~80%) of the variation in N_e and F_{ST} was accounted for by habitat fragmentation. Thus, analyzing only the demographic response of this model would lead one to conclude that habitat fragmentation did not affect population viability in red-cockaded woodpeckers, whereas an analysis of the genetic response (N_e and F_{ST}) suggested just the opposite. In their conclusion, Bruggeman et al. (2010: 3688) emphasized the importance and implications of this finding for conservation:

> *This finding is a strong argument against the viewpoint that spatial configuration makes little or no difference for population viability compared with the total amount of suitable habitat in the landscape....***The conservation implication is that increases in habitat adjacency could more effectively ameliorate the erosion of genetic variation than increases in habitat area alone**. *Further, our results suggest that* **conservation planning based solely on demographic criteria could lead to fragmented landscapes in which populations expected to be demographically stable maintain a much smaller effective population size than would be possible if landscape and genetic factors were recognized**. (emphasis added)

Admittedly, red-cockaded woodpeckers are unusual in that they are highly philopatric (at least in the case of males), live in small family groups year-round, and are cooperative breeders; thus, they might just be especially sensitive to the genetic consequences of habitat fragmentation. So, lest we be concerned over the generality of this finding, Cushman and his colleagues (2012)

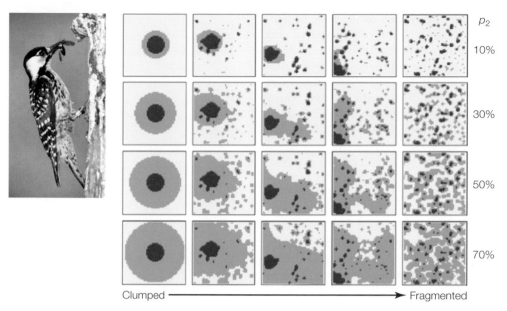

Figure 9.16 Relative effects of habitat amount versus fragmentation on population genetic structure. An individual-based, spatially explicit population model for red-cockaded woodpeckers (*Picoides borealis*, left) was applied to simulated landscapes (right) in which the amount and fragmentation of three habitat types were adjusted independently. High-quality habitat (mature pine forest; dark green) was fixed at 7%, whereas the amount of lower-quality second-growth forest (p_2; green) was varied between 10% and 70%. The light-green areas are non-forested habitat.

Source: Photo by James Hanula, USDA Forest Service. Figure from Bruggeman et al. (2010).

implemented a more general individual-based, spatially explicit population model on fractal neutral landscapes (**Figure 5.7**), with a similar goal of separating the effects of habitat area (p) and fragmentation (H) on genetic differentiation. In addition, the authors varied the resistance of the intervening matrix to movement; that is, the landscape was modeled as a resistance surface, such that individuals accrued a cost during dispersal. In the end, they likewise found that habitat configuration was more important than habitat area in driving genetic differentiation. Several habitat configuration metrics (**Chapter 4**) in particular—correlation length, patch cohesion, and the aggregation index—were found to be the single-best predictors of genetic differentiation, even after the effect of habitat area was removed.

Although these two simulation-based studies are not sufficient to conclude that habitat fragmentation is always more important than habitat area for understanding or predicting genetic differentiation, they do at least underscore that different aspects of landscape structure can have different, possibly contrasting effects depending on the process or criterion (e.g. demographic vs genetic) in question.

Source–Sink Population Dynamics

Given that patterns of gene flow provide evidence of successful dispersal and reproduction, genetic techniques are increasingly being used to identify source–

sink population dynamics (**Chapter 7**). For example, we previously discussed an application of gravity models for detecting source–sink dynamics in a spotted frog metapopulation (Murphy et al. 2010). Here, we consider another approach involving Bayesian clustering and assignment tests which, as we discussed previously in the raccoon case study (Cullingham et al. 2009), can be used to uncover evidence of recent (first-generation) migration events. In particular, assignment tests can aid in evaluating whether asymmetrical flows occur among populations (from sources to sinks), as well as the likely origin of immigrants (population sources), as illustrated by the following case study of source–sink dynamics in mountain lions (*Puma concolor*).

The mountain lion or cougar is a large predator whose range formerly encompassed much of the western hemisphere. Although it has been extirpated from the eastern half of the USA,[29] it still ranges throughout the western mountain states where its hunting is permitted in all but the state of California. Variation in hunting pressure is thus a major source of differential

---

[29] Occasional sightings of mountain lions are still reported from as far east as Connecticut in the eastern USA. In 2011, genetic tests revealed that one mountain lion that had been hit and killed by a vehicle in Connecticut had likely originated from the Black Hills of South Dakota (some 1500 miles or 2400 km away), which certainly attests to this species' potential for long-range movements. A small relictual population also resides in southern Florida and is considered a unique subspecies of mountain lion (*P. concolor coryi*, the Florida panther) that is protected under the US Endangered Species Act.

mortality among populations, which could drive source–sink dynamics. Using a combination of non-spatial and spatial Bayesian clustering analyses, Andreasen and her colleagues (2012) evaluated the population genetic structure and migration rates of mountain lions throughout the Basin and Range region of Nevada and the Sierra Nevada mountain range of California (13 microsatellites from 739 individuals in total; **Figure 9.17**). Bayesian cluster analyses (using both STRUCTURE and TESS) revealed the existence of five genetically distinct populations: one population in the Sierra Nevada and the other four in Nevada. Immigration among these five populations was found to be asymmetrical, with mean migration rates ranging from almost no migration between populations (from the western Sierra Nevada population to the northern Nevada population) to a high of 40% from the southern to eastern populations in Nevada (**Figure 9.17**).

Counter to expectations, the western population in the Sierra Nevada range of California, where hunting of mountain lions is prohibited, was not a net exporter of dispersers (i.e. a putative population source). Rather, the southern Nevada population proved to be the largest net exporter, thanks to the presence of several large refuges within this region where mountain lions are not in fact hunted. Source populations can occur in areas with less hunting pressure, therefore. Nevertheless, we should not infer from this sort of analysis that the Sierra Nevada population is an absolute population sink, based as it is only on evidence of migration among these particular populations. In other words, we can only assert here whether a population is a *relative* source or sink. When evaluated in a different context, the Sierra Nevada population may well be a net source of migrants to other mountain lion populations to the north and west of the Sierras. East of the Sierras, the presence of several large desert playas (the Lahontan Basin) may limit dispersal between the Sierra Nevada and the Great Basin populations of mountain lions. In other words, the dry remains of an ancient lakebed[30] may now be serving as a barrier to gene flow among modern-day populations of mountain lions.

As this last point illustrates, teasing apart the effects of current versus historical landscape effects on gene flow and population genetic structure can present a challenge, which is why we will consider this topic next.

[30] The Lahontan Basin is an endorheic basin (a closed drainage basin that does not flow to a river or ocean) that once contained a very large lake, Lake Lahontan (22,000 km² = 8,500 mi²), during the mid-to-late Pleistocene. For comparison, Lake Lahontan was a bit bigger than present-day Lake Ontario, one of the Great Lakes that span the US-Canadian border (cf **Figures 9.7 and 9.11**).

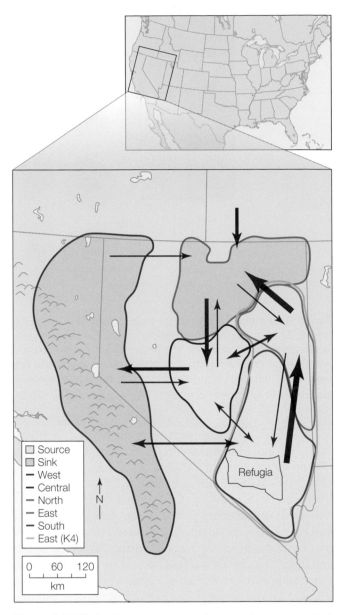

Figure 9.17 Using landscape genetics to infer source–sink population dynamics. Bayesian genetic clustering analysis and assignment tests were used to identify source and sink populations for mountain lions (*Puma concolor*) in the western USA (the Sierra Nevada mountain range in California and the Great Basin mountain ranges of Nevada). There are five genetically distinct subpopulations occurring within these two regions (West, Central, North, East, and South), with the two populations in the South and East possibly representing a single population according to some models [i.e. the East (K4) population, if *K* = 4]. Arrows indicate the direction and rate (thickness) of recent migrations, showing asymmetries in dispersal among populations. Source populations are net exporters of individuals while sink populations are net importers.

Source: Adapted from Andreasen et al. (2012).

Current versus Historical Landscape Effects on Population Genetic Structure

As we have seen, landscape genetics typically studies how the current landscape influences contemporary patterns of gene flow and population genetic structure. Indeed, the finer spatiotemporal resolution of landscape-genetic studies has been viewed as one of its defining characteristics that distinguish it from phylogenetics (Manel et al. 2003). Phylogenetics focuses on the geographic and historical factors that have shaped the evolutionary relationships (phylogenies) among organisms, and as such, considers genetic patterns at much broader scales in space and time than landscape genetics. Recall, however, that the ghosts of landscapes past may haunt the landscape genetics of the present (Landguth et al. 2010). The genetic structure of populations today may thus carry the signature of these historical landscapes, as well as that of the contemporary landscape.

One way to disentangle the relative effects of historical versus contemporary landscapes on population genetic structure is to examine the correlation between genetic distance and some aspect of landscape structure that has been assayed at different points in time. For example, we have seen how habitat loss and fragmentation can disrupt genetic connectivity among populations, through a combination of effects on dispersal and demography, and thus, gene flow. However, the genetic consequences of habitat loss, fragmentation, or other landscape barriers may not be evident for many generations (>100) in species that have limited dispersal abilities (Landguth et al. 2010). Historical landscape factors may thus play a larger role in determining patterns of genetic variation in such species.

To illustrate, we once again consider a study on wood frogs (*L. sylvaticus*), this time from southeastern Michigan, USA (Zellmer & Knowles 2009). Recall from our previous discussion that geographic barriers, such as large water bodies, and climate change have likely contributed to the genetic structure of wood frogs at a broad region-wide scale in the northeastern USA, presumably as a consequence of species' range shifts in response to glacial-interglacial cycles (Lee-Yaw et al. 2009). Given that wood frogs are believed to have recolonized southern Michigan some 10,000 years ago, following the last glacial maximum, it seems unlikely that the population genetic structure of wood frogs today would still bear the imprint of effects dating from the last ice age, at least when examined at the scale of individual landscapes. Instead, more recent landscape changes would seem far more likely to affect population genetic structure. Like much of the eastern USA, southern Michigan has undergone a dramatic transformation over the past 200 years, following

European settlement and clearing of the native forest for agriculture and other land uses. Amphibians are generally sensitive to habitat loss and fragmentation (Cushman 2006), especially in this case where forests are used for both foraging and dispersal among breeding ponds (they are not called wood frogs for nothing!). If wood frogs exhibit a lagged response to habitat loss and fragmentation (Landguth et al. 2010), however, then it is possible that the genetic structure of wood frog populations better reflects the structure of the pre-settlement landscape than that of the contemporary landscape.

To test this, Zellmer and Knowles (2009) evaluated the relationship between the pairwise genetic (F_{ST}) and geographic distances among breeding populations of wood frogs (9 microsatellites obtained from >1000 tadpoles across 51 ponds), as well as to the resistance distances (calculated in CIRCUITSCAPE) obtained from landscape maps representing three different time periods: a reconstructed landscape prior to European settlement (early 1800s); a 1978 landscape obtained from satellite imagery; and a 2001 landscape obtained from the National Land-Cover Database for the USA (Homer et al. 2004). Although there were significant isolation-by-distance and isolation-by-resistance effects, partial Mantel tests that controlled for geographic distance indicated that landscape structure (isolation by resistance) ultimately explained a greater amount of genetic variation among populations (Zellmer & Knowles 2009). Given that landscape structure from each time period was significantly correlated with genetic structure, the authors also assessed the effects of each time period independently of the other time periods using partial Mantel tests. When the effects of the contemporary landscape were removed, the structure of the historical (pre-European settlement) landscape was no longer correlated with the observed genetic structure of wood frog populations. Thus, the current landscape appears to have exerted a greater influence than the past landscape on the population genetic structure of wood frogs in this region (Zellmer & Knowles 2009).

That is not always the case, however. Sometimes we do have to look to the past to understand the present. To give a different example, we'll consider a study that examined how climate, geography, and the distribution of suitable habitat over time have shaped patterns of genetic variation in the dotted-line robust slider (*Lerista lineopunctulata*), a type of skink found along the coastal sand plains and dunes of Western Australia (He et al. 2013). Specialized for burrowing in the sand, the robust slider is more snake than lizard, possessing only a tiny set of vestigial hind limbs. Sea-level changes during previous glacial cycles have caused repeated shifts in the distribution of the robust slider's coastal sand habitat, such that genetic divergence among present-day populations may well reflect these sorts of historical effects on its range dynamics. Alternatively, genetic variation among populations may reflect the effects of the current habitat distribution on contemporary patterns of gene flow.

To tease apart current from historical effects on population genetic structure, the authors combined ecological niche modeling, which was used to estimate habitat suitability and predict the species' distribution under both current and past climate conditions (using MaxEnt; **Chapter 7**), with spatially explicit demographic and genetic-coalescence models to generate patterns of genetic variation (what they refer to as an integrated distributional, demographic, and coalescent—or iDDC—modeling approach; He et al. 2013). That is, the authors used species distribution modeling to map the predicted variation in habitat suitability across space and time; these habitat-suitability estimates in turn informed a spatially explicit demographic model, whose parameters (local population sizes and migration rates, à la source-sink population dynamics) were then used in coalescent simulations.[31] For the demographic simulations, the authors tested three hypotheses related to gene flow: (1) isolation by distance, in which genetic variation reflects only the current configuration of habitat; (2) a contemporary landscape model, in which genetic variation is hypothesized to be related to habitat configuration as well as differences in the suitability of the current environment (i.e. habitat heterogeneity), through influences on local population sizes and the degree of gene flow among them (isolation by resistance); and (3) a dynamic landscape model, in which genetic variation reflects the shifting distribution of the species over time in response to environmental changes since the last glacial maximum. Model-based inference was then used to compare the performance of these different models against the observed pattern of genetic variation among robust slider populations.

Of these three hypotheses, the dynamic landscape model, in which the distribution of the dotted-line robust slider tracked climate-induced shifts in its habitat over time, best explained the observed patterns of genetic variation in this species over the other two hypotheses that considered only the present habitat configuration. This finding is especially noteworthy because more-traditional tests of the correlation between geographic distances, resistance distances, and genetic distances (based on Mantel tests) had found an overall effect of isolation by distance, even after controlling for climate and habitat suitability (using partial Mantel tests).

Correlation is not causation, however. Different processes can give rise to similar patterns, and different models can generate similar predictions. Such confounding presents a real concern for landscape genetics, as this can obviously undermine our ability to link pattern and process, especially if we fail to consider any other explanation or hypotheses for the observed pattern.

In the case of the dotted-line robust slider, isolation by distance seemed an unlikely explanation for how genetic structure could have emerged among populations, given their limited dispersal ability and the present-day distribution of their coastal sand habitat. There were some obvious incongruities in the dataset. For example, it is difficult to envision how two populations that appear close in terms of geographic distance, but which are in fact separated by ocean given their relative positions along the convoluted coastline of Western Australia, could have obtained such a high degree of genetic similarity, given that dispersal—and thus gene flow—should be impossible between these populations. Similarly, although isolation by resistance provided a somewhat better statistical fit to the data, this model also did not provide a satisfactory explanation for the observed pattern of genetic variation among populations: populations now isolated by ocean or other unsuitable habitats still should not exhibit such a high degree of genetic similarity.

Thus, it was only in the context of dynamic landscape change, in which the species' distribution is assumed to track climate-driven shifts in the distribution of its habitat, that the observed pattern of genetic variation made sense: the coastline and its associated sand habitat have retreated in response to rising sea levels, such that formerly continuous populations are now separated by inhospitable habitat, but still share a common genetic ancestry. Differences among populations elsewhere along the coast were thus more likely a consequence of historical processes that affected colonization dynamics and contributed to this species' shifting distribution over time, rather than more recent effects of isolation or landscape structure on gene flow. In other words, just because geographic isolation or landscape resistance *can* give rise to genetic differentiation among populations doesn't mean that they have necessarily done so, even though the relationship is statistically significant. As shown in this study, genetic divergence among populations can come about through entirely different processes, such as through colonization–extinction dynamics that underlie species' range shifts in response to climate change (He et al. 2013; **Chapter 8**).

Genetic divergence among populations can thus come about because of isolation by distance (when genetic drift exceeds gene flow), through the effects of landscape structure on gene-flow patterns (isolation

---

[31] Recall from footnote 14, that the coalescent approach analyzes genealogies and works backward to deduce how the observed genetic variation might have arisen.

by resistance), or through the combined effects of geographic barriers and other landscape features— past or present—on colonization–extinction dynamics (species' range dynamics). There is also the possibility that populations diverge because of adaptive responses to fine-scale heterogeneity (microgeographic adaptation). We therefore consider this more fully in the next section on evolutionary landscape genetics, which studies how migration and population structure affect evolutionary processes in heterogeneous landscapes (Petren 2013).

Evolutionary landscape genetics will ultimately help to unify ecology and evolution by connecting metapopulation theory with classical theoretical population genetics.

Petren (2013)

Evolutionary Landscape Genetics

Throughout this chapter, our discussion has focused entirely on neutral genetic variation; that is, genetic variation that has no direct adaptive value and which is therefore not acted upon by natural selection (Holderegger et al. 2006). To a large extent, this is simply a reflection of the prevailing view in population genetics, in which gene flow and genetic drift are seen as largely responsible for population genetic structure (Orsini et al. 2013). This 'neutralist view' holds that genetic divergence occurs among populations through the combined effects of genetic drift and isolation, either through distance or because of landscape resistance to gene flow that ultimately disrupts the drift-migration equilibrium. Thus, genetic divergence among populations is believed to come about primarily through a process of dispersal limitation, where neutral genetic variation is concerned (Orsini et al. 2013).

To what extent might natural selection drive genetic differences among populations, however? If selective pressures vary among habitats or as a function of different environmental conditions across the landscape, might this be an important mechanism for understanding population genetic structure? In the past, natural selection has not been considered very important for understanding genetic divergence at the population or metapopulation scales, which are ultimately defined in terms of the size of a species' dispersal neighborhood (or dispersal kernel; **Figure 9.18**). This is because, according to theory, the strength of selection would have to be very strong at a local or 'microgeographic'[32]

scale in order to offset the homogenizing effects of dispersal and gene flow within that neighborhood (Richardson et al. 2014). In other words, genetic divergence was assumed to occur only when the selection differential between habitats (s) greatly exceeded the **effective gene flow** between populations (m) in these habitats (i.e. through a disruption of the selection-migration equilibrium; Wright 1931). Although gene flow involves the exchange of genetic information among populations as a consequence of the successful dispersal and reproduction of migrants, effective gene flow is defined as the 'movement and establishment of novel genes that are not currently present in the recipient population' (Richardson et al. 2014).

Microgeographic adaptation is particularly interesting because it occurs despite a high potential for mixing within a dispersal neighborhood, making it unlikely that neutral processes can generate appreciable variation at this scale.

Richardson et al. (2014)

Until recently, such strong selection at a microgeographic scale was not thought to be sufficiently commonplace to merit consideration as a driver of population genetic structure. Any adaptive traits that arose within a population would be quickly swamped by dispersal and gene flow. In fact, this is why most landscape-genetic studies focus on neutral genetic variation in the first place, owing to the greater likelihood of detecting unique alleles that have entered into the population through migration. However, strong selection at microgeographic scales may be more common than previously thought (Hereford 2009). For example, spotted salamanders (*Ambystoma maculatum*) were found to vary both genetically and phenotypically among populations as a function of differential predation risk within breeding ponds in southern Connecticut, USA (Richardson & Urban 2013). Some of these ponds contain larval marbled salamanders (*A. opacum*; **Figure 5.14A**), which are voracious predators on the smaller larvae of spotted salamanders. Subsequently, strong predation pressure within these ponds has apparently selected for spotted salamanders that forage more rapidly and thus grow quicker, enabling them to achieve a 'size refuge' from predation faster than those from ponds that lack marbled salamanders. That these traits

---

[32] The term 'microgeographic' was coined to distinguish finer scales of adaptation or population genetic divergence from the macrogeo-

graphic scales that are more typically studied by evolutionary biologists. Although Richardson et al. (2014) sought to distinguish microgeographic adaptation (adaptation scaled relative to the dispersal neighborhood of the organism) from local adaptation (a looser term not standardized by dispersal scale), we will use the terms interchangeably and note that what is considered a 'local adaptation' should obviously be scaled relative to the organism.

Figure 9.18 Microgeographic adaptation.   The microgeographic scale of adaptation is defined as adaptive responses that occur as a function of environmental heterogeneity within two standard deviations (2σ) of a species' dispersal neighborhood (i.e. the dispersal kernel). In the illustration, two different habitats result in adaptive divergence in moth phenotypes (coloration) in spite of the high potential for gene flow between populations (i.e. both populations lie within the same dispersal neighborhood).
Source: From Richardson et al. (2014).

have a genetic basis was demonstrated through **common garden experiments** (**Box 9.2**). Importantly, this adaptive divergence in foraging traits among spotted salamander populations occurs in spite of gene flow among ponds, all of which are located well within the usual dispersal range of this species. Thus, spotted salamanders in this system exhibit adaptive divergence at a microgeographic scale; that is, they exhibit **microgeographic adaptation** (**Figure 9.18**).

Beyond Strong Selection: Other Mechanisms of Microgeographic Divergence

Strong natural selection is not the only mechanism that can give rise to microgeographic divergence and adaptation, however. Any mechanism that reduces the flow of maladapted genes relative to that expected based on the organism's dispersal range can lead to microgeographic adaptation. Other such mechanisms include adaptive landscape barriers, spatially autocorrelated selection, and adaptive dispersal coupled with habitat selection (Richardson et al. 2014). We consider each of these in turn.

ADAPTIVE LANDSCAPE BARRIERS   As we have seen in this chapter, landscape barriers, such as roads, rivers and dams, can restrict gene flow between populations, but so

too can other features of the landscape, such as habitats or environments that limit dispersal and gene flow. If a landscape feature limits the flow of traits that would be maladaptive in a different habitat or environmental context, it provides an 'adaptive barrier' that, should this occur within the dispersal neighborhood of a species, can lead to microgeographic divergence and adaptation. Such appears to have occurred in the case of sockeye salmon (*Oncorhynchus nerka*) in the Copper River basin of Alaska, USA (Ackerman et al. 2013).

The Copper River basin encompasses a huge (~64,000 km²) area,[33] and as a consequence, drains two completely different types of landscapes: the upper Copper River and its tributaries drain large areas of tundra and are relatively low gradient, whereas the middle and lower portions of the Copper River drain a forested, mountainous landscape and are thus higher in gradient and turbidity. The authors used SNPs (**Table 9.1**), which exhibit both neutral and putative adaptive variation, to evaluate microgeographic divergence among sockeye salmon in the Copper River (44 SNPs from 4100 individuals across 28 populations). Their analysis uncovered the existence of an adaptive barrier (identified by an abrupt difference in allele

---

[33] For comparison, the Copper River basin is bigger than 10 out of the 50 states in the USA.

BOX 9.2   Using Common Garden Experiments to Study Adaptive Variation among Populations in Different Environments

Individuals from different populations may differ phenotypically in their morphology, physiology, or behavior. If these populations are located in different types of environments, we might wonder about the adaptive value of these phenotypic differences and posit a genetic basis for these traits. Our suppositions might be further reinforced if we also find evidence of genetic variation among these same populations. Nonetheless, we still have not demonstrated that the genetic differences we observe are responsible for those particular traits, that is, that the genetic variation we have observed is adaptive. Thus, we are still left to speculate as to whether or how much of the phenotypic variation observed among populations is due to genetic differences, as opposed to environmental differences.

One way to address this is to conduct a **common garden experiment**. In a common garden experiment,

(A)

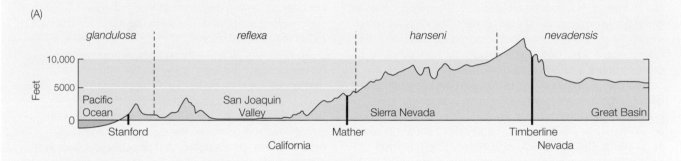

(B)

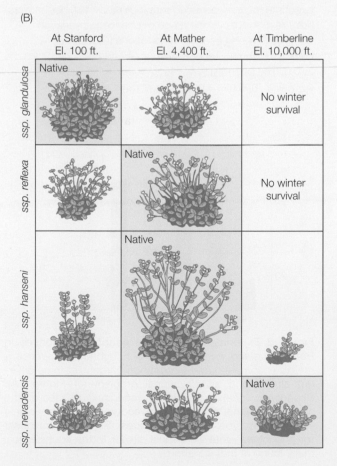

Figure 1 Phenotypic variation of sticky cinquefoil (*Potentilla glandulosa*).   (A) Different ecotypes (subspecies) vary in growth form along an elevational gradient across central California, USA. The locations of the three sites where common garden experiments were performed are also indicated (Stanford, Mather, and Timberline). (B) Results of the common garden experiments revealed that both environment (rows) and genetics (columns) contribute to phenotypic variation in this species.

Source: After Clausen et al. (1940).

(Continued)

BOX 9.2 Continued

individuals from different populations are grown or raised in a common environment (i.e. in the field or lab). If the phenotypic differences among these populations persist in spite of being reared in the same environment, then these differences are assumed to have a genetic basis. If, however, individuals from these different populations exhibit similar phenotypes when grown in a common environment, then we would conclude that the species merely exhibits phenotypic plasticity in response to different environmental conditions. For example, plants may grow taller in certain areas simply because they receive more rainfall; shorter plants from drier locations are capable of growing just as tall when grown in a common environment that receives more rainfall. Thus, the phenotypic differences among populations in this case are not likely to be genetic, but are simply a product of the environment.

The use of common garden experiments to distinguish genetic from environmental effects on phenotypic variation was famously demonstrated by Jens Clausen, David Keck, and William Hiesey during the 1930s and 1940s (Clausen et al. 1940). Clausen, Keck, and Hiesey (CKH) studied ecotypic variation in native plant species along an elevational gradient, encompassing a wide range of climatic conditions, across the whole of central California, USA (**Figure 1**). Their pioneering studies into the role of natural selection in driving ecotypic differences among plant populations (i.e. that these differences were adaptive and had a genetic basis) were clearly ahead of their time, given that the modern synthesis had not yet been fully realized in evolutionary biology (Núñez-Farfán & Schlichting 2001).

Among their many and varied contributions, CKH's experimental work on sticky cinquefoil (*Potentilla glandulosa*) nicely demonstrated how ecotypic variation among regional populations (subspecies) can result from the interplay between genetic differentiation and the environment. Sticky cinquefoil exhibits different growth forms along this elevational gradient (**Figure 1A**), with plants native to higher elevations (*P. g. nevadensis*) considerably smaller than those native to lower elevations (*P. g. glandulosa*; **Figure 1B**). Plants from each of the four subspecies were grown by CKH in experimental gardens at three sites along this elevational gradient: Stanford (30 m elevation), Mather (1400 m), and Timberline (3050 m). Phenotypic variation in sticky cinquefoil apparently does have an environmental component, as evidenced by the differences in growth and survivorship across sites for each subspecies (i.e. compare differences in growth form across a row in **Figure 1B**). For example, plants from the *hanseni* population vary greatly in size depending on where they are grown. Thus, growth form in this species is phenotypically plastic, in that it is influenced by the environmental conditions in which the plant grows.

Still, there are also differences in growth form among subspecies grown at each site (compare differences within each column of **Figure 1B**). Given the common environment at each site, it seems likely that there is also a genetic basis to the phenotypic differences observed in this species. Most subspecies seem to do best when grown at a site that falls within their native range. For example, plants from the *hanseni* population achieved maximal growth when grown at the Mather site, which is within the native range of this subspecies (**Figure 1A**). We might also consider whether the fact that only plants from the *hanseni* and *nevadensis* populations survive the winter at the Timberline site reflects some sort of adaptive response to the harsh conditions found at higher elevations.

To determine whether these sorts of phenotypic differences have a genetic basis and whether these differences are in fact adaptive, Clausen and Hiesey (1958) additionally performed a large number of breeding experiments in which they examined crosses between subspecies and quantified the expression of various morphological, reproductive, and physiological traits in subsequent generations. Based on these breeding experiments, they deduced that more than 100 genes were likely responsible for the range of phenotypic variation observed among ecotypic populations (Clausen & Hiesey 1958; see Table 1 in Núñez-Farfán & Schlichting 2001). The phenotypic variation observed in sticky cinquefoil thus has a genetic basis, but is it adaptive?

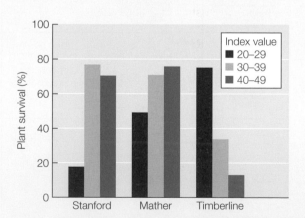

Figure 2 Experimental study of adaptive responses in *Potentilla glandulosa*. Survival of F$_2$ hybrids (*nevadensis* × *reflexa* ecotypes) grown in common gardens at three different sites along an elevational gradient (cf **Figure 1A**). Phenotypic variation of hybrids was indexed based on the similarity to either the *nevadensis* ecotype (low index values) or *reflexa* ecotype (high index values). Survival was highest for phenotypes that were most similar to the ecotype normally found at that elevation.

Source: After Núñez-Farfán, & Schlichting (2001).

To test this, Clausen and Hiesey (1958) developed a second-generation hybrid population (F_2) by self-pollinating the F_1 progeny obtained from crosses between *reflexa* and *nevadensis* plants.[33] By scoring phenotypic differences among the hybrids, they derived an index of similarity to characterize the range of variation evident within populations (low index values = greatest similarity to the *nevadensis* ecotype; high index values = greatest similarity to *reflexa* ecotype). These hybrid plants were then grown in common gardens at each of the three sites. The phenotypes with the highest survival at each site were those that most closely matched the ecotype normally found at that elevation (**Figure 2**). That is, plants most closely resembling the *nevadensis* ecotype (low index values) had the highest survival at the high-elevation

Timberline site, whereas those most similar to the *reflexa* ecotype (high index values) did best at the lower-elevation sites. Thus, these ecotypic differences between populations do indeed represent adaptive responses to the environment (i.e. they are the result of natural selection).

Although this work was done 60 years ago, it still resonates today and represents a 'keystone for the study of natural selection in the wild' (Núñez-Farfán & Schlichting 2005: 118):

> *The efforts of CKH to demonstrate the action of natural selection as the process responsible for adaptive physiological differences between climatic races of Potentilla glandulosa were ahead of their time. The approaches they took convincingly supported the view that population differentiation (and local adaptation) is the product of natural selection in P. glandulosa, and their inquiries into reaction norm evolution through the use of cloned genotypes were pioneering.*

[33] A second-generation filial (F_2) population was created because such a population includes individuals representing both parental types as well as a range of recombinations. Thus, a F_2 population maximizes the range of phenotypic and genetic variation upon which natural selection can act.

---

frequencies at two outlier loci[34] associated with the major histocompatibility complex[35]), not between the upper and middle/lower Copper River as expected, but within the Klutina tributary of the middle Copper River.

The Klutina River is fed by glacial meltwater from the Chugach Mountains, and flows through a natural lake (Lake Klutina) before joining the Copper River some 100 km downstream. The adaptive barrier occurred between sockeye populations that spawned in the upper Klutina River above the lake and those that spawned in the mainstem below Klutina Lake. Furthermore, this adaptive barrier coincides with the distribution of two different sockeye ecotypes:[36] a lake ecotype (found in those populations spawning above

Klutina Lake), which spend half their life in the lake before migrating to the ocean to complete development; and a sea/river ecotype (found in the population that spawns in the mainstem below Klutina Lake) that spends less time maturing in freshwater before migrating to the sea. Differences in the spawning and rearing habitats experienced by sockeye salmon in the upper versus lower Klutina River appear to have contributed to the microgeographic adaptation observed between ecotypes in this system (Ackerman et al. 2013).[37] Thus, Klutina Lake effectively functions as an adaptive landscape barrier in that it appears to limit maladaptive gene flow between populations, even though salmon are able to migrate freely throughout the entire Klutina river-lake system.

SPATIALLY AUTOCORRELATED SELECTION As illustrated by the previous example, environments with different selective pressures are not usually distributed at random throughout the landscape, but instead exhibit a clumped (patchy) or clinal (gradient) distribution. That is, the selective environment tends to exhibit positive spatial autocorrelation, which may help to give rise to—and subsequently reinforce—microgeographic adaptations (Richardson et al. 2014). For example, if different adaptive environments are patchily distributed or arrayed along a gradient, and most

[34] Outlier loci are genetic markers that exhibit locus-specific behavior or patterns of variation that are extremely divergent from the rest of the genome; they are therefore usually omitted from analysis in studies of neutral genetic variation. However, this divergent behavior may be the result of divergent selection and might therefore be associated with different adaptations, which makes these loci useful for identifying putative adaptive divergence among populations (Luikart et al. 2003).

[35] The major histocompatibility complex (MHC) refers to a group of genes that code for cell-surface proteins that bind to pathogens and thus aid in their recognition and eventual elimination or neutralization by the immune system. The MHC is both polygenic, in that it is encoded by several different genes, and polymorphic, having multiple alleles at each locus. The highly variable nature of MHCs, coupled with their presumed adaptive importance, explains the appeal of these loci for evolutionary landscape genetics.

[36] Ecotypes are genetically distinct populations of a species that are adapted to specific environments. Ecotypes exhibit phenotypic differences in their morphology, physiology, or life-history traits in response to environmental heterogeneity.

[37] It should be noted that these different ecotypes of sockeye salmon are not just found in the Klutina River but occur throughout the range of the species (Wood 2007).

dispersal occurs within a local neighborhood, then a higher fraction of migrants should possess the necessary adaptations for a given environment.

Some of the best examples of microgeographic adaptation have been identified in adjacent populations along environmental gradients or ecotonal boundaries. The evolution of metal tolerance in sweet vernal grass (*Anthoxanthum odoratum*) growing on mine tailings in Flintshire, North Wales (UK) is a classic illustration of how divergent selection can occur over very short distances (Antonovics 1971).[38] Sweet vernal grass grows in pastures adjacent to the mine tailings, but ordinarily does not possess a tolerance for heavy metals (e.g. pasture plants die when grown in metal-contaminated soil). Thus, metal tolerance is not a trait inherent to this species but has evolved in situ. The adaptive divergence between metal-tolerant and -intolerant populations is all the more remarkable given the short distances separating populations along this cline (they diverge rather sharply over a distance of a few meters) and because grass is wind pollinated, meaning gene flow should not be limited (**Figure 9.19**).

Effective gene flow between populations apparently *is* limited, however. Metal tolerance is not the only difference between these adjacent populations. Sweet vernal grass on the mine tailings flowers about a week earlier than those in the pasture (**Figure 9.19**). Asynchrony in flowering results in strong assortative mating and thus reproductive isolation between metal-tolerant and -intolerant populations. These differences in flowering time have persisted for more than 40 years and might have evolved due to selection against gene flow between habitats (Antonovics 2006). The steep cline observed here between metal-tolerant and -intolerant populations is thus maintained by divergent selection and reproductive isolation at the mine-pasture boundary; that is, by selection that is spatially autocorrelated and reinforced by pre-reproductive isolating mechanisms.

ADAPTIVE DISPERSAL AND HABITAT SELECTION　Although theoretical population genetics may treat all dispersers equally, by assuming that dispersers comprise a random sample of the larger population or gene pool, clearly not all dispersers are created equal (Richardson et al. 2014). Individuals may differ in their propensity for dispersal as well as in their habitat preferences, both of which influence patterns of settlement and distribution

---

[38] Antonovics (1971: 592) viewed this as an example of sympatric divergence (divergence in the absence of geographical isolation), and posited that this could provide a mechanism for sympatric speciation (i.e. speciation in the absence of geographic barriers to gene flow). A heretical idea at the time, controversy regarding the when and how of sympatric speciation continues to this day (Jiggins 2006; Nosil 2008).

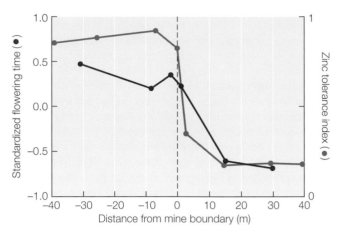

Figure 9.19 Spatially autocorrelated selection can drive microgeographic adaptation along gradients or ecotonal boundaries. Sweet vernal grass (*Anthoxanthum odoratum*) is wind-dispersed and thus has the potential for high gene flow. Nevertheless, it exhibits a sharp cline in metal tolerance, an adaptive trait that permits it to grow on the contaminated soils of mine tailings. Flowering time also differs between populations, which contributes to reproductive isolation and helps to maintain adaptive divergence in the face of high gene flow.

Source: After Antonovics (1971, 2006).

on the landscape (**Chapters 6 and 7**). Further, if these differences have a genetic basis (genotype-dependent dispersal), genetic differentiation can arise quickly between populations, especially if gene flow is directed and adaptive (Bolnick & Otto 2013). In that case, population genetic structure is actually *facilitated* by gene flow, rather than impeded by it. Microgeographic adaptation can thus occur in spite of high gene flow when dispersal is adaptive; that is, when individuals selectively disperse and settle within habitats that confer greater fitness. Adaptive dispersal and habitat selection therefore afford great potential for microgeographic adaptation, given that individuals are necessarily operating within the scale of the species' dispersal range as they search for suitable habitat in which to settle on the landscape (Richardson et al. 2014; **Figures 6.2 and 9.18**).

Strong habitat preferences were found to play a role in the adaptive divergence between lake and stream populations of the three-spine stickleback (*Gasterosteus aculeatus*) on Vancouver Island in British Columbia, Canada. Stream and lake populations of stickleback are morphologically and genetically distinct: stream sticklebacks tend to be larger, have deeper bodies, and shorter dorsal spines than lake sticklebacks. Common garden experiments (**Box 9.2**) have previously shown that these morphological differences have a genetic basis. So, to evaluate the degree of habitat preference among sticklebacks, Bolnick and his colleagues (2009) used a mark-transplant-recapture design, in which an equal number of fish from both the lake and stream were marked (about

1400 in all), and then released into a channel a short distance above the lake so as to give the fish an equal opportunity to move either upstream or down into the lake.

Sticklebacks apparently have strong site fidelity. Almost 90% of recaptured fish ($n = 252$) returned to their 'native' habitat; that is, stream fish returned to the stream whereas lake fish returned to the lake. The most compelling evidence for phenotype-dependent dispersal and habitat preferences actually came from the fish that didn't return to their native habitat, however. The fish that dispersed between habitats looked more like the residents found in their new habitat: stream fish that dispersed to the lake had a more lake-like morphology than stream fish that returned to the stream, and lake fish that dispersed to the stream had a more stream-like morphology than lake fish that returned to the lake. How they accomplish this matching of their phenotype to the habitat is unknown. Regardless, the strong habitat preferences exhibited by sticklebacks pose a substantial barrier to gene flow between lake and stream populations. Bolnick et al. (2009) estimated that dispersal between these populations has been reduced by more than 76% given their proximity and the dispersal range of stickleback. Such a major reduction in dispersal, coupled with strong site fidelity, would greatly facilitate adaptive divergence between populations in response to different selective pressures within these habitats (lake vs stream). Increased genetic and phenotypic divergence could in turn lead to stronger habitat preferences, resulting in a positive feedback loop that further drives increased reproductive isolation between populations (Bolnick et al. 2009).

As you read through the foregoing examples, it perhaps did not go unnoticed that these mechanisms overlap or could even act in concert to give rise to microgeographic divergence. Adaptive dispersal and strong habitat preferences may also contribute to ecotypic differences in sockeye salmon within the Klutina River, just as they do in sticklebacks. For both types of fish, these different selection environments (stream vs lakes) clearly exhibit spatial autocorrelation, which, as we saw, drives adaptive divergence in sweet vernal grass. Thus, these mechanisms are not mutually exclusive and are reinforced by various other types of evolutionary mechanisms (e.g. pre- or post-reproductive isolating mechanisms, selection against migrants, evolutionary monopolization effects[39]), all of which help to drive local adaptations at a microgeographic scale (Richardson et al. 2014).

[39] Evolutionary monopolization occurs when the initial colonists of a new environment evolve local adaptations that enable the founding population to grow rapidly and gain a numerical advantage that helps to resist further immigration and establishment by non-adapted migrants; in other words, it is an evolution-mediated priority effect (De Meester et al. 2002) that can lead to 'isolation by colonization' (Orsini et al. 2013).

Importantly, these mechanisms all describe interactions between evolutionary processes with fine-scale heterogeneity in the selection environment, leading to a reduction in effective gene flow, which simultaneously helps to give rise to, and to reinforce, local adaptations. Thus, a pattern of 'isolation by adaptation' (Nosil et al. 2008) can lead to a pattern of 'isolation by environment' (Orsini et al. 2013; Wang & Bradburd 2014). This is significant, for it runs counter to the traditional view in evolutionary biology, by which geographic isolation (at continent-wide scales) leads to isolation by adaptation (e.g. allopatric speciation). Instead, there is now increasing support for adaptive divergence—and possibly speciation—in the absence of major geographic barriers to gene flow and at finer scales than previously thought possible, which brings this sort of evolutionary research well within the domain of landscape ecology.

While exploring the distribution of neutral genetic variation can definitely inform us about the patterns and processes that limit gene flow, landscape genetics has yet to develop a framework to understand how landscape features contribute to the distribution of adaptive genetic variation. **Taking a landscape perspective could have huge implications for evolutionary biology.** (emphasis added)   Lowry (2010)

Central to this emerging field of evolutionary landscape genetics is the role of adaptive versus neutral genetic variation in contributing to patterns of genetic divergence. This is the purview of landscape genomics, which is presented next.

Toward a Landscape-Genomic Approach to the Study of Adaptive Genetic Variation

Adaptation at a local or microgeographic scale typically results not just in a single trait change, but in many phenotypic changes among populations in different habitats. For example, different ecotypic populations of sticky cinquefoil, sockeye salmon, sweet vernal grass, and three-spine sticklebacks all varied in a number of morphological, behavioral, or physiological traits between habitats. If we assume that these trait changes have a genetic basis, consider the sheer enormity of trying to identify which genes are ultimately responsible for each, especially if the trait is polygenic and coded for by more than one gene at different loci on different chromosomes. Recall that Clausen and Hiesey (1958) estimated that more than 100 genes were responsible for some 20 traits they measured among sticky cinquefoil populations across an elevational gradient in central California, USA (**Box 9.2**).

There are a number of ways we can go about decoding the genetic basis for phenotypic differences among populations. We discuss each of these approaches in turn.

THE CANDIDATE-GENE APPROACH   That certain traits have a genetic basis and are adaptive can be illustrated through experimental breeding and common garden experiments, as we've discussed (**Box 9.2**). Nevertheless, linking specific traits to the particular gene(s) responsible for those traits remains a challenge. The 'gene-first' or **candidate-gene approach** requires a priori information on the gene responsible for a given trait, and then uses the tools of population genetics to determine how allele frequencies for that gene vary among ecotypic populations. This is referred to as a 'bottom-up approach,' in that the starting point of the investigation is the gene, and the goal is to link variants of the gene (genotypes) to phenotypes within different environments (Barrett & Hoekstra 2011; **Figure 9.20**).

For example, the three-spine stickleback is actually a marine fish that has repeatedly colonized freshwater lakes and streams throughout the northern hemisphere following the last glacial retreat. Remarkably, sticklebacks have shown consistent patterns of convergent adaptation within freshwater environments, namely in the loss of lateral plating or 'armor' (Schluter et al. 2010). Armor plating is thought to provide some measure of protection against predation. Marine sticklebacks are completely plated, so the repeated loss of this armor in freshwater populations suggests that different selective pressures are at work in these two environments. Indeed, fish are the main predators of sticklebacks in marine environments, whereas predatory insects are the main threat to juvenile sticklebacks in freshwater. Young fish that lack armor grow more rapidly, which gives them a growth advantage over plated fish in freshwater habitats, by enabling them to achieve a size refuge from predatory insects more quickly (Barrett et al. 2008). Armor plating in sticklebacks is known to be coded for by the *Ectodysplasin-A* (*Eda*) gene, and this knowledge makes it possible to link phenotypic variation to genotypic variation within populations, even though this is not a perfect one-to-one correspondence (plating is not a discrete trait, as we'll learn in a moment). The genetic variant for the freshwater trait (reduced plating) exists naturally at low frequency (~1%) within marine populations, which explains why the loss of armor occurs with such regularity and fairly soon after sticklebacks colonize freshwater habitats.

To quantify selection strength in freshwater environments, Le Rouzic and colleagues (2011) analyzed a population of sticklebacks whose phenotypic variation has been tracked for 20 years within a small experimental pond, in which predation pressure was minimal (no predatory fish and few macroinvertebrates were present). Although equal numbers of armored and unarmored fish were initially released, almost all of the fish that survived the first year were armored, which introduced an unintended bottleneck effect into the experiment. As it turned

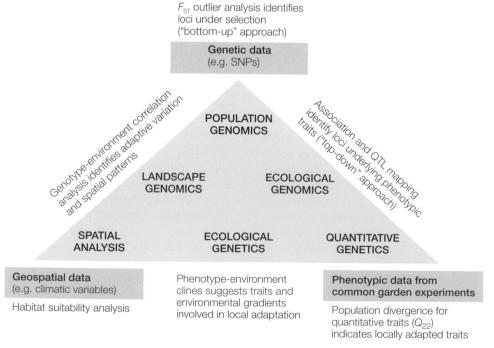

Figure 9.20 Approaches to the study of adaptive genetic variation.
Source: After Sork et al. (2013).

out, this chance event made the experimental results even more compelling. Despite making up only 1% of the population after the bottleneck, the proportion of unarmored fish steadily increased—and the proportion of completely armored fish decreased—over the ensuing decades, such that the experiment ended with nearly equal percentages of both ecotypes (the unarmored fish were a bit more abundant). Based on a genotypic analysis of the last stickleback generation, the authors found a strong fitness advantage (92%) for fish that carried the 'freshwater' genotype (i.e. homozygous recessive) over the completely armored 'marine' genotype (homozygous dominant). That a reduction in armor plating still appears to be under strong selection in this freshwater environment despite limited predation pressure suggests that this trait may not be the sole—or even the main—target of selection, just an obvious one. The *Eda* gene may well be influencing a number of other traits (pleiotropy) that are necessary for adapting to a freshwater environment and which are likewise under selection. Once again, it is highly unlikely that a single trait is the sole target of natural selection.

THE QUANTITATIVE TRAIT LOCUS APPROACH   While the candidate-gene approach can be extremely powerful for studying the genetic architecture of complex traits, it is ultimately limited by a lack of a priori information on candidate genes for most traits and species, unless one is dealing with a model organism that has been well studied, such as the three-spine stickleback. Another approach is thus to analyze **quantitative trait loci** (QTL). Quantitative traits are those that exhibit continuous variation, such as plant height, as opposed to discrete traits that have just two or only a few character values (e.g. short vs tall). For example, *Eda* is actually a quantitative trait locus, for the range of lateral plating in sticklebacks spans a continuum from complete plating to reduced or limited plating (Colosimo et al. 2005). Quantitative traits tend to be polygenic (controlled by two or more genes, often on different chromosomes) and exhibit multifactorial inheritance that includes interactions with the environment, which again underscores the difficulty in trying to decode the genetic architecture of these complex traits. Of course, not all quantitative traits are necessarily under selection; some are effectively neutral (Holderegger et al. 2006). The goal of a QTL analysis is thus to identify regions of the genome (loci) that are responsible for observed phenotypic differences, via a statistical analysis of the genetic differences (assayed using molecular markers such as AFLPs or SNPs) between ecotypic populations (Miles & Wayne 2008). The analysis of QTLs thus represents a 'top-down approach,' in that we are starting at the genome level and are attempting to drill down to find loci that correlate with phenotypic variation for locally adaptive traits, and which therefore may be under selection (Barrett & Hoekstra 2011; **Figure 9.20**).

To illustrate, a QTL analysis was used in conjunction with a reciprocal transplant experiment to study adaptive divergence in salt tolerance between coastal and inland populations of the yellow monkeyflower (*Mimulus guttatus*). The yellow monkeyflower occurs in a wide variety of wet or mesic habitats throughout western North America, and thus exhibits a great deal of variation in both its growth form and its life-history traits throughout its range. Coastal populations are continually inundated by salt spray from the ocean, which coats their leaves with salt and increases salinity levels of the soils in which they grow. High salt concentrations are generally toxic to plants, but coastal monkeyflowers have apparently developed physiological mechanisms for dealing with higher salt levels in their environment. Such mechanisms are evidently lacking from inland monkeyflowers, given that these plants do not fare very well when they are transplanted to the coast (e.g. they experience severe leaf necrosis). The physiological and genetic basis for this difference in salt tolerance was not known, however.

Lowry and his colleagues (2009) thus performed a series of experimental assays to determine which physiological mechanisms were responsible for salt tolerance in monkeyflowers. They found that, relative to inland plants, coastal monkeyflowers can withstand far greater leaf sodium concentrations (an advantage, given that their leaves are often coated with salt spray) and their shoots also have higher salt-tolerance levels (another advantage, given that they grow in soils with high salinity levels). The authors then used QTL mapping[40] to uncover the specific loci involved in salt tolerance and leaf sodium concentrations (three QTLs were found to contribute to salt tolerance and two QTLs were found to contribute to leaf sodium concentrations; notice that more than one locus is involved in each trait and that more than one trait may give rise to overall salt tolerance). Finally, they evaluated whether these QTLs could contribute to adaptive divergence in salt tolerance between populations, by examining the fitness of each ecotype in a reciprocal transplant experiment in the field. All three QTLs linked to higher salt tolerance within shoots were positively correlated with fitness at the coastal site (where seed production was ~2.5× greater in coastal than inland plants), but interestingly, did not influence fitness at the inland site (both ecotypes had the same fitness). Thus, while salt tolerance is advantageous in coastal habitats, it does not appear to have any negative consequences for coastal monkey-

---

[40] QTL mapping is a rather involved process, and thus the interested reader is referred to Miles & Wayne (2008) for an overview of the approach.

flowers that are grown elsewhere. Traditionally, adaptation has been viewed as a trade-off between positive selection for certain traits in a given environment, and negative selection against these same traits everywhere else they do not have adaptive value (i.e. they carry a fitness cost elsewhere). The results of this study thus raise the interesting question of whether positive selection alone can give rise to local adaptations.

THE GENOMICS APPROACH   As we've just seen in the case study on monkeyflowers, the search for QTLs involves scanning and mapping large regions of the genome, followed by a good deal of experimental work to reduce thousands of putative genes down to a few dozen (or more) candidates, which then must be validated to determine whether these are in fact responsible for adaptive variation in the trait(s) of interest. Thanks to the genomics revolution, such genome scans are becoming increasingly common, efficient, and cost-effective, and have spawned new fields of inquiry at the interface of existing disciplines. Thus, the emerging field of **population genomics**[41] uses genome-wide sampling to gain a better understanding of the roles played by different evolutionary processes (selection, genetic drift, gene flow) in giving rise to both neutral and adaptive genetic variation within populations (Luikart et al. 2003; Lowry 2010; **Figure 9.20**). More precisely, population genomics seeks to distinguish the evolutionary effects that influence specific loci (locus-specific effects related to natural selection) from the evolutionary effects that influence the entire genome (genome-wide effects related to genetic drift and gene flow). Loci under selection exhibit patterns of variation that differ significantly from the rest of the genome (i.e. at neutral loci), and thus identification of these outlier loci is a crucial step in the population-genomic approach for detecting adaptive genes (Black et al. 2001; Luikart et al. 2003; **Figure 9.20**). Recall that identification of outlier loci enabled Ackerman et al. (2013) to identify an adaptive landscape barrier (a lake) separating two ecotypes of sockeye salmon within the Klutina River of Alaska.

Building upon population genomics, theoretical population genetics, and spatial analysis, **landscape genomics** studies the spatial distribution of adaptive genetic variation by combining the tools of the population genomicist with those of the landscape ecologist (Joost et al. 2007, 2013; **Figure 9.20**). The main goal of landscape genomics is to examine the relationship between genomic and environmental data to identify loci that have adaptive significance (Manel et al. 2010). Rather than search for outlier loci, however, landscape genomics attempts to identify loci that are strongly correlated or associated with environmental factors, based on the premise that these sorts of loci are most likely under the influence of natural selection and therefore should represent adaptive genetic variation (Joost et al. 2007). Environmental association analysis has been performed using a variety of statistical approaches (Joost et al. 2013), but we will highlight correlation-based methods here.

Correlation-based methods, such as logistic regression, have generally been found to perform better than outlier-detection methods in discovering loci under selection within heterogeneous landscapes (De Mita et al. 2013). Because this approach is based on uncovering correlations between loci and environmental variables, however, it does run the risk of returning 'false positives,' that is, loci that are correlated with environmental variables but which are not in fact a product of natural selection (i.e. adaptive). For example, many environmental and climatic variables, such as temperature or precipitation, exhibit geographical variation. The genotypic frequencies of certain loci may likewise vary geographically, giving the appearance of a very strong genotype-environment correlation, which would lead us to conclude that these loci are under selection and are responsible for the adaptive traits evidenced by populations in these different environments. However, as we've discussed previously in this chapter, it is entirely possible for genetic variation to arise in neutral loci either as a function of distance (isolation by distance) or because of environmental heterogeneity (isolation by resistance), resulting in population genetic structure that is the result of genetic drift and gene flow, rather than natural selection. Thus, neutral genetic structure can complicate the search for adaptive loci by obscuring the signal between genetic variation and the environment, especially when geographic variation is correlated with both (Holderegger et al. 2006). Ideally, then, the approach we adopt in our search for adaptive loci will also control for neutral genetic variation (i.e. population structure).

To illustrate the landscape-genomic approach to detecting adaptive loci, we consider a case study involving loblolly pine (*Pinus taeda*), an ecologically and economically important tree species in the southeastern USA. Like most tree species, the distribution of loblolly pine is limited by climate: low temperatures set the northern range limit for this species, but the western edge of its distribution is limited by inadequate rainfall. Water availability is a major climatic stressor for conifers and is thus a major environmental driver of local adaptation. Traits related to drought tolerance have a genetic basis in many conifers, including

---

[41] Just to be clear, population genomics differs from population genetics in that the latter typically focuses on individual genes in isolation (e.g. the distribution and change in allele frequencies within or among populations) whereas genomics considers the interaction among a great many genes (i.e. the entire genome) and the environment.

loblolly pine, and variation in these traits has been shown to be adaptive (González-Martínez et al. 2006). Given the precipitation gradient across the range of the loblolly pine, we might therefore expect to see selection for increased water-use efficiency and other physiological traits that increase drought tolerance in the drier, western parts of its range (Eckert et al. 2010a).

Using environmental association analysis, Eckert and his colleagues (2010b) searched for correlations between climate variables and adaptive genetic variation (assayed using 1730 SNPs from 682 trees sampled at 54 sites) across the geographical range of the loblolly pine. Because of the potential confounding of geographic (neutral genetic) variation with climatic or environmental (adaptive genetic) variation, Eckert et al. (2010b) also accounted for population structure in their analysis. To do this, they adopted a correlation-based modeling approach founded on Bayesian inference, which they refer to as a 'Bayesian geographical analysis' (Coop et al. 2010; Eckert et al. 2010b). In this approach, a null model is constructed in which the prior distribution is based on the observed pattern of covariance[42] in allele frequencies among populations (sampling sites) for a set of markers (SNPs). Populations that covary in the degree to which their allele frequencies deviate (drift) from the global allele frequencies for individual loci are thus assumed to possess shared ancestry and gene flow (i.e. population structure). This null model is then tested against a model in which the climate variable has a linear effect on allele frequencies at individual SNPs. If the correlation between allele frequencies at a particular SNP locus and the climate variable is greater than expected, based on the null model, then we would conclude that this locus is likely under natural selection and is potentially contributing to local adaptation.

The analysis identified four-dozen SNPs that were strongly correlated with climatic variables related to temperature, precipitation, and aridity (an index of precipitation to potential evapotranspiration), among other factors. Many of these potentially adaptive SNPs are related to different physiological mechanisms by which plants respond to abiotic stress, including genes involved in oxidative stress, protein degradation, and sugar metabolism. Increased protein degradation has been observed in plants as a way to eliminate damaged proteins or to mobilize nitrogen in response to

environmental stress, and the accumulation of sugars represents a common means by which plants and animals deal with osmotic stress (Eckert et al. 2010b). Many of the strongly supported SNPs were thus located in genes that likely play a role in adaptations to warmer, drier conditions, consistent with the expectation of diversifying selection among loblolly pine populations along this climatic gradient. For example, a SNP on a locus that encodes a potassium/hydrogen antiporter, a class of genes associated with drought and salt tolerance, was fixed in the western region of the loblolly pine's range, which experiences higher summer and fall aridity (**Figure 9.21**). Thus, a landscape genomics approach facilitated the identification of putative adaptive loci, by directly assessing the link between genetic variation and environmental variation.

As intensive as this sort of genomic analysis is, it still represents only a first step in understanding the molecular basis of adaptive genetic variation (Eckert et al. 2010b). Identifying putative adaptive loci is only part of the challenge, for we still have a very limited understanding of the encoded genes and their functions for all but a few model organisms. Therefore, a pluralistic approach is recommended (Manel et al. 2010). The various methods mentioned above all have different sensitivities in their abilities to detect loci under selection, partly because of differences in the genomic architecture of adaptive traits and partly because of differences across loci in the timing and strength of selection, as well as overall differences in the background level of differentiation across loci (i.e. in neutral population structure). Combined approaches that utilize information on population phenotypes, genotypes, and environments will thus be 'the most informative for understanding the genetic and genomic basis of local adaptation, the environmental characteristics that drive divergent selection, and the phenotypic traits that confer fitness to those environments' (Sork et al. 2013: 904).

Adaptive Responses to Future Climate and Landscape Changes

In the latter half of this chapter, we have studied contemporary versus historical landscape effects on population genetic structure, as well as the potential for adaptive genetic responses to different environmental conditions, all of which are assumed to have occurred within the recent past (e.g. since the last glacial maximum or in the centuries or decades following human alteration of the landscape). Looking ahead, the future is likely to be marked by unusually rapid and widespread environmental changes due to climate change, the ongoing exploitation and conversion of landscapes by humans, and the biological invasion of those landscapes by non-native species and pathogens. Indeed, such changes are already taking place and are

---

[42] Recall from your basic statistics course that covariance provides a measure of the strength of correlation between two (or more) sets of random variables. If the variables change together (greater values of one variable correspond to greater values of the second variable), then the covariance is positive. Alternatively, if greater values of one variable correspond to smaller values of the second variable, then the covariance is negative. The sign of the covariance therefore shows how the linear relationship between two variables changes together, that is, how they covary.

(A) Environmental variation

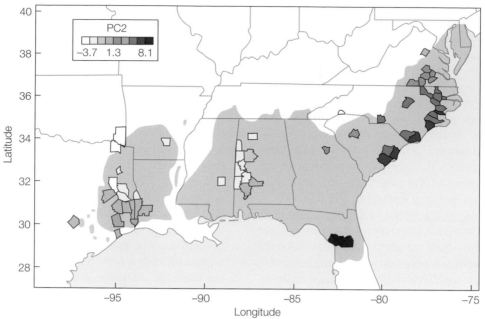

(B) Environmental associations

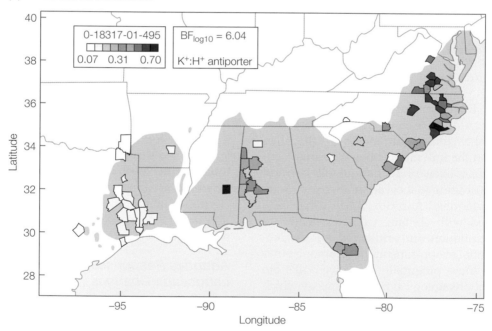

Figure 9.21 Top-down approach to the study of adaptive genetic variation. The top-down approach attempts to identify adaptive genetic variation from a large set of random markers (thousands of SNPs, in this case) by searching for significant relationships between environmental variation and genetic variation. For the loblolly pine (*Pinus taeda*), a composite variable (principal component, PC2) that explained a gradient in temperature and precipitation across the species' range in the southeastern USA (A) was found to be associated with a SNP that encodes a potassium/hydrogen antiporter (SNP 0-18317-01-495), a class of genes associated with drought and salt tolerance (B). This trait was basically fixed in the western part of the tree's range, which is hotter and drier than the eastern portion of its range.

Source: After Eckert et al. (2010b).

altering the selection environment for many organisms (Parmesan 2006). We might thus wonder whether plants and animals will be able to adapt quickly enough, via the sorts of contemporary evolutionary processes we have been discussing in this chapter.

This question was the focus of a study on the adaptive potential of valley oaks (*Quercus lobata*), an endemic species native to the Coast Range and Sierra Nevada foothills that encircle California's Central Valley (USA). Trees, like any species, can survive environmental change either by tolerating the new conditions (via phenotypic plasticity), shifting their ranges to track environmental change (**Chapter 8**), or by adapting. Sork and her colleagues (2010) took a multipronged approach to investigate which of these options would likely apply to valley oaks in response to future climate-change projections. The investigators first sought to characterize historical patterns of colonization related to past range shifts, by analyzing neutral variation in chloroplast DNA (cpDNA) to track seed migration (recall that cpDNA is maternally inherited). Population connectivity was then assayed across the species' range using a graph-theoretic approach, which revealed evidence of extensive gene flow, even across portions of the Central Valley where riparian corridors likely helped to facilitate seed dispersal. Since pollen flow occurs over an even greater distance than seed dispersal, it seems unlikely that valley oaks experienced significant barriers to gene flow in the past. Thus, the observed genetic structure among regional populations is unlikely to be the result of past dispersal limitation (i.e. a product of genetic drift), which leaves open the possibility of adaptive divergence.

To address that possibility, the authors linked neutral genetic variation (microsatellites) with climatic variation across the species' current range. Although we do not ordinarily expect neutral loci to exhibit a strong relationship with environmental or climatic variables (Holderegger et al. 2006), these might nevertheless correlate with adaptive potential if a large degree of genetic variation is examined (as it was here, in which genetic variation was treated as a multivariate response variable), if neutral and adaptive loci are linked to some degree, and we assume that natural selection is operating on the genome as a whole rather than on individual loci.[43] In the case of valley oaks, neutral genetic information appeared to be a reasonable surrogate for adaptive genetic information: multilocus genetic structure showed a strong association with climatic gradients across the species' range, which held even when the authors partitioned out the effect of geographic location. Populations in different portions

of the species' range thus experience different climatic conditions, which might then have contributed to divergent selection among regional populations and in turn could influence the response of local populations to future climate change (Sork et al. 2010).

To evaluate whether valley oaks might be able to shift their distribution in response to future climate change, Sork and colleagues (2010) fitted species distribution models with data from different climate-change scenarios (using MaxEnt; **Chapter 7**). These climate-based distribution models similarly exhibited regional variation, such that models parameterized for one region had low accuracy in predicting the occurrence of oaks in another region. Thus, valley oak populations are expected to experience regional differences in how climate change will influence their distribution. For example, populations along the central coast are predicted to be especially vulnerable because the projected climatic shifts in that region lie outside the migration potential of the species (i.e. 6–8 km over a period of 60–80 years).

It therefore seems unlikely that many valley oak populations will be able to track rapid environmental change via range shifts under the current climate-change projections. Nor does it seem likely that populations will be able to evolve adaptive responses quickly enough in the face of such rapid climate change. Despite the observed relationship between genetic and climatic variation, which suggests that adaptive divergence among regional populations may have arisen in response to past climate change, an adaptive response to future climate change now seems unlikely given the rate of change relative to the demographic response rate of this species (valley oaks can live for 100–200+ years). That leaves the possibility that some local populations might at least have sufficient phenotypic plasticity to tolerate climate change. Of the possible responses to rapid climate change, this is perhaps the most likely, given the high degree of genetic variation within populations (a legacy of historical gene flow as discussed previously), and because some regional populations already experience highly variable climatic conditions for which they have presumably adapted (i.e. the potential for phenotypic plasticity is itself an adaptive trait; Sork et al. 2013).

While this study nicely illustrates how a variety of different approaches can be used to infer the adaptive potential of species to future climate or landscape change, it would obviously be ideal if this could be assessed more directly, such as by using candidate genes or a landscape-genomics approach. We therefore conclude this section with a research agenda for investigating the adaptive potential of species to future environmental change.

---

[43] Recall that even putative adaptive loci, such as candidate genes, may not exhibit strong correlations with the environmental variables of interest either.

AN AGENDA FOR THE FUTURE   In considering the potential for adaptive responses to future environmental change, we need to distinguish adaptive evolutionary change from short-term phenotypic changes (i.e. phenotypic plasticity; **Table 9.5**). As we have seen, many species can exhibit shifts in phenotypic traits in response to local environmental conditions. Although phenotypic plasticity can provide an initial response to environmental change that can eventually lead to adaptation, we should not assume that all phenotypic changes are necessarily reflective of (or will lead to) adaptive genetic shifts. This underscores the difficulty of trying to forecast longer-term adaptive responses to environmental change. If the observed phenotypic changes are entirely plastic, then the ability of populations to respond to further climate or landscape change will be seriously curtailed once the degree of environmental change exceeds the limits of that plasticity (Hansen

TABLE 9.5 A framework for studying adaptive responses to environmental change, based on six criteria needed to establish that the observed trait changes have a genetic basis and are adaptive.

Criterion	Possible approaches for documenting criterion	
	Molecular markers	Quantitative traits
1. Demonstrate that genetic variation exists in the population	Identify or analyze candidate loci previously demonstrated to be under selection	For putative traits under selection, use common garden or reciprocal transplant experiments to quantify genetic vs environmental components of trait variation
2. Link genetic variation to a specific environmental stressor (selection agent)	Analyze candidate loci relevant to environmental stress (e.g. MHC genes) Test for association between candidate gene variants and environmental variables associated with selection	Demonstrate that variation in the trait covaries with clinal or microgeographic variation in environmental conditions Demonstrate that variation in this trait is important for fitness
3. Test genetic changes over time	Begin a continuous sampling program to track genetic changes from here on out Perform a retrospective analysis, using genetic material from archived herbarium or museum specimens	Begin a continuous sampling program; pedigree analysis may also be helpful in tracking genetic variation within families over time Conduct 'resurrection studies,' in which trait changes are compared between contemporary and resurrected individuals (e.g. from long-dormant seeds or eggs)
4. Confirm that selection has occurred	Conduct statistical tests (e.g. outlier tests; landscape genomics and genetic association tests) Hypothesis testing (e.g. expected trait changes are consistent with a certain type of environmental change)	Conduct statistical or empirical tests (e.g. trait has changed in a predictable way along an environmental gradient)
5. Link the observed genetic change(s) to selection associated with a particular environmental factor	Assess or test that the observed adaptive change coincides with the specific environmental change. If possible, obtain information on many environmental parameters to rule out confounding variables	Assess or test that the observed adaptive change coincides with the specific environmental change. If possible, obtain information on many environmental parameters to rule out confounding variables
6. Rule out the possibility of replacement (i.e. that a better adapted population replaced the original population)	Demonstrate genetic continuity over time using molecular markers (e.g. by estimating temporal differentiation) Obvious if immigration from other populations is unlikely (i.e. population is isolated)	Demonstrate genetic continuity over time using molecular markers (e.g. by estimating temporal differentiation) Obvious if immigration from other populations is unlikely (i.e. population is isolated)

Source: After Hansen et al. (2012).

et al. 2012). Distinguishing actual adaptive change from phenotypic plasticity is admittedly a challenge in practice (**Table 9.5**), especially since most phenotypic traits are polygenic and genes can interact epistatically, as we have seen. Nevertheless, the approaches we have discussed previously, such as common garden experiments (**Box 9.2**) and reciprocal transplants in the field, can facilitate this task, by permitting the decomposition of quantitative trait variation into its constituent genetic and environmental components (Hansen et al. 2012).

Of course, the study of phenotypic trait variation represents only one approach that we can use to identify or monitor adaptive responses to environmental change. Another approach is to study adaptive genetic variation directly through the use of molecular markers (**Table 9.5**), such as those known to encode ecologically important genes (e.g. the *Eda* gene that leads to armor loss in freshwater sticklebacks; the QTLs that contribute to salt tolerance in coastal yellow monkeyflowers; the outlier loci that were part of the MHC in sockeye salmon). As we explored previously in this section on evolutionary landscape genetics, associating genetic variation at particular loci with phenotypic and environmental variation is no small undertaking, especially given that most adaptive traits have a complex genetic architecture. Thus, failure to demonstrate genetic change at a specific candidate locus does not rule out that change has occurred at other loci affecting the same trait (Hansen et al. 2012). This is where genome-wide sequencing has become a real boon in the study of adaptive responses, by identifying other loci under selection from the same environmental stressors.

Landscape genomics may have particular utility in this regard, not only in terms of identifying adaptive genetic loci, but also in terms of assessing how populations might respond to future environmental change (**Table 9.5**). For example, as the valley oak study illustrated, tree populations are expected to be particularly vulnerable to rapid climate change because their long generation times and immobility (once seedlings are established) limit how quickly they can respond to environmental change (Sork et al. 2013). As we saw in the case study of the loblolly pine (Eckert et al. 2010a,b), landscape-genomic approaches can greatly facilitate our understanding of how selection shapes local adaptations, while controlling for the potentially confounding historical effects of gene flow and genetic drift. The growing availability of genomic resources that have been developed specifically for trees and other species will increasingly reveal which loci are associated with adaptive traits, and thus whether current populations possess the necessary genetic diversity to adapt to new climatic conditions. Landscape genomics could thus help to identify geographic regions where populations lack the capacity for adapting to rapid environmental change and might then be used to locate suitable donor populations that could serve as sources, should it become necessary or desirable to repopulate these areas in response to future climate change (Sork et al. 2013).

Future Directions

Landscape genetics is more than just spatial genetics. It is the interactions between genetics (or the processes that population genetics can trace) and the landscape in which we are interested. Real landscape genetic studies have to include the quantitative and qualitative characteristics of the landscape studied, such as the types, size and spatial arrangement of potential barriers to migration. Then, and only then, will landscape genetics hold the great expectations that it presently evokes.   Holderegger et al. (2006)

In the decade or so since its introduction, landscape genetics has grown exponentially, spawning numerous review articles, several special issues in a variety of journals (2006 in *Landscape Ecology*; 2010 in *Molecular Ecology*; 2013 in *Conservation Genetics*), and a recently published edited volume (Balkenhol et al. 2015), all of which seek to establish the field, summarize its contributions, and chart its future (Balkenhol et al. 2009b; Sork & Waits 2010; Storfer et al. 2010; Manel & Holderegger 2013; Wagner & Fortin 2013; Bolliger et al. 2014). To a large degree, this growth has been spurred by rapidly advancing molecular technologies that have now made routine what was once unimaginable when landscape ecology first emerged in the mid-1980s. Landscape genetics is still very much a frontier science, however, with tremendous growth potential, especially in the area of evolutionary landscape genetics. Thus, landscape genetics should prove increasingly relevant for addressing a number of global challenges that now confront landscape ecology, particularly in regards to managing genetic resources and species distributional shifts in response to rapid climate change. This is critically important not only for the successful management or conservation of species, but also for safeguarding our own future resource needs, especially those involving agriculture, forestry, and fisheries.

To meet these challenges, landscape genetics needs to move forward on a number of different fronts. Although its primary focus to date has been on landscape effects on gene flow and spatial genetic variation, landscape genetics should now be prepared to address the relative roles of landscape structure versus other population processes in giving rise to population genetic structure (i.e. the theoretical factors, such as demographic dynamics and social behavior in **Figure 9.2**). That is, we need to understand not just whether landscape composition or configuration is the more important predictor of genetic structure, but how large a role the landscape plays in the first place, compared to other processes that likewise contribute to genetic variation, especially given that different processes can give rise to similar patterns (Balkenhol & Landguth 2011; Bolliger et al. 2014). Landscape management alone will not be sufficient to preserve gene flow and genetic connectivity if we ignore how these other processes contribute to population genetic structure.

Along these lines, analytical and statistical methods for evaluating landscape effects on genetic variation must move beyond simple correlative tests (Mantel or partial Mantel tests of isolation by distance or isolation by resistance), and instead consider alternative hypotheses that more fully explore the link between landscape pattern and genetic processes (Sork & Waits 2010; Manel & Holderegger 2013; Bolliger et al. 2014). We have seen in this chapter some examples of how this might be accomplished, such as through the use of multivariate analyses or multimodel inference using Bayesian approaches to explore more complex relationships between landscape heterogeneity and genetic variation (e.g. Cushman et al. 2006; Balkenhol et al. 2009a; Legendre & Fortin 2010). Similarly, individual-based simulation models provide a means of increasing both the spatial and biological realism of landscape-genetic models, which might then permit a more complete evaluation of landscape and demographic effects on genetic variation, and could be used to develop a better understanding of landscape effects on selection and adaptive genetic variation (Epperson et al. 2010; Balkenhol & Landguth 2011).

The recent emphasis in landscape genetics on adaptive genetic variation represents an important step forward, one that has been greatly facilitated by the genomics revolution. As we have seen in this chapter, landscape genomics is an important tool for understanding adaptive genetic variation (Joost et al. 2007, 2013; Manel & Holderegger 2013). With the rise of next-generation sequencing technologies, it is now feasible to monitor genomes rather than genes (Hansen et al. 2012). The monitoring of genomes can provide unprecedented information about the genes involved in adaptive responses to specific environmental changes, which is a boon not only to the study of evolutionary landscape genetics, but also to forecasting adaptive responses to future landscape and climatic changes. In the future, we will need to go beyond the mere identification of genomic regions under selection to uncover the ecological function of adaptive genetic variation (functional genomics), if we hope to forecast the adaptive potential of populations to future environmental change with any great accuracy (Bolliger et al. 2014).

Sometimes we need to look to the past to understand the present and predict the future, especially if past landscape legacies still exert an influence on current evolutionary trajectories and thus how populations might respond to future climate and landscape changes. Parsing out the historical versus contemporary effects of landscape structure on population genetic structure is another critical research need in landscape genetics (Sork & Waits 2010; Bolliger et al. 2014). Increasingly, the use of molecular markers coupled with the development of new sequencing technologies (e.g. parallel sequencing of genome-enabled markers) is blurring the conventional disciplinary distinction between landscape genetics and phylogenetics, which opens up the possibility of exploring the joint effects of current and historical landscape effects on population genetic structure (Thomson et al. 2010; He et al. 2013). Historical samples can also be used for retrospective monitoring of molecular markers over decadal time scales, and we may soon be able to extend the time scale by centuries or even millennia through the analysis of ancient DNA, thanks to next-generation sequencing technologies (Hansen et al. 2012).

Regardless of what the future may hold for landscape genetics, it is important to maintain perspective and not confuse the science with its tools. To quote Manel and her colleagues (2003: 195): 'landscape genetics is not an end in itself. The techniques used in this field will describe spatial genetic patterns, **but more importantly will lead to exploration of the processes that caused the patterns**' (emphasis added). The search for pattern-process linkages should thus continue.

Landscape genetics can provide valuable insights into how the evolutionary processes of gene flow, genetic drift, and natural selection have been influenced by landscape-scale processes to shape current patterns of genetic variation. **Understanding these processes will allow us to predict the response of current populations to anthropogenic forces, such as climate change, human population growth, habitat destruction, and fragmentation.** (emphasis added)

Sork & Waits (2010)

Chapter Summary Points

1. Landscape genetics is a relatively new field that has emerged at the interface of population genetics, landscape ecology, and spatial statistics. It addresses how microevolutionary processes such as gene flow, genetic drift, and natural selection interact with environmental heterogeneity to shape patterns of genetic variation across landscapes.

2. The key distinction between population genetics and landscape genetics is that the latter explicitly considers how landscape structure influences gene flow and genetic variation. Individuals or populations that are closer geographically are expected to have greater genetic similarity than those that are farther apart. This forms the basis of the isolation-by-distance hypothesis that underlies much of spatial population genetics. Landscape genetics, by contrast, considers matrix heterogeneity to be important for evaluating patterns of gene flow and genetic similarity. Individuals or populations might thus be isolated more by the resistance of the intervening matrix to dispersal than simply by geographic distance. The hypothesis of isolation by resistance is a major theme in landscape genetics.

3. Another important distinction between population genetics and landscape genetics pertains to the operational unit of analysis. Ideally, the individual is the unit of analysis in landscape genetics, whereas most applications in population genetics require that we define populations a priori. Although populations or habitat patches (nodes) can also be used as the unit of analysis in landscape genetics, a focus on individuals permits a finer scale of analysis that is less biased by our preconceived notions of how the system is structured genetically. Delineating population genetic structure is thus frequently the first stage in a landscape-genetics study.

4. Major research themes in landscape genetics include: uncovering the landscape correlates of population genetic structure; assessing functional landscape connectivity; identification of movement corridors and barriers to gene flow; characterization of source–sink population dynamics; determining the relative influence of current versus past landscape structure on population genetic structure; and, evaluating the potential for adaptive responses to changing landscapes in the face of future climatic and landscape changes.

5. Habitat fragmentation may affect genetic and demographic processes differently, such that an assessment of the relative effects of habitat amount versus fragmentation on populations may depend on whether we adopt a demographic or genetic perspective. For example, habitat amount may have a greater effect on breeding population size and population growth rates, whereas habitat fragmentation may have a greater effect on effective populations sizes and gene flow. Very different conclusions might thus be reached regarding the relative importance of habitat amount versus fragmentation on population viability.

6. Although landscape genetics is primarily concerned with how the current landscape influences contemporary patterns of gene flow and genetic variation, there is still the possibility that the genetic structure of populations today carries the signature of past landscapes. For example, population genetic structure may exhibit a lagged response, such that the genetic consequences of landscape barriers to gene flow may not be evident for many generations (especially if the species has limited dispersal abilities), but then may persist for many more generations, even when the barrier is no longer present. The genetic structure of contemporary populations may therefore bear the signature of past landscape effects on gene flow. Disentangling the relative effects of historical versus contemporary landscapes on population genetic structure is thus a major research challenge.

7. Evolutionary landscape genetics seeks to understand how natural selection contributes to patterns of genetic variation across heterogeneous landscapes. This is in contrast to the traditional neutralist view in population and landscape genetics, in which genetic divergence is believed to come about primarily through a process of dispersal limitation.

8. Natural selection traditionally has not been considered important for understanding genetic divergence at population or metapopulation scales, because it was thought that very strong selection pressure would be necessary to offset the homogenizing effects of dispersal and gene flow at these scales. However, strong selection at such 'microgeographic scales,' which lie within the dispersal neighborhood of a species, may be more common than previously thought.

9. Nor is strong selection the only mechanism by which microgeographic divergence can occur. Any mechanism that reduces the flow of

maladapted genes relative to that expected based on a species' dispersal neighborhood can lead to microgeographic adaptation. Other mechanisms therefore include adaptive landscape barriers, spatially autocorrelated selection, and adaptive dispersal coupled with habitat selection. These mechanisms may act singly or in concert to create adaptive genetic variation at microgeographic scales.

10. Landscape genomics builds upon theoretical population genetics, population genomics, and spatial analysis to study the spatial distribution of adaptive genetic variation. It seeks to identify loci that are strongly correlated or associated with environmental or landscape factors, under the assumption that these sorts of loci are most likely to be under the influence of natural selection and should therefore represent adaptive genetic variation. Landscape genomics thus shows great promise for evaluating the adaptive potential of species to future climatic and landscape changes.

Discussion Questions

1. Should genetic variation be assessed at the level of the individual, the population, the metapopulation, or the habitat patch (or node)? To what extent might this depend on the species, question, process, or scale of the study?

2. Given that gene flow occurs through a process of dispersal, how or why does gene flow differ from dispersal? Are the scale(s) of these processes similar? If landscape structure affects dispersal, will it also affect gene flow? How about the converse?

3. Landscape-genetic approaches are increasingly being used to assess the functional connectivity of landscapes. Is an assessment of landscape connectivity that is based on genetic data inherently better than one based on movement or dispersal? Or does an assessment of functional connectivity founded on genetic processes measure something fundamentally different from one based on movement or dispersal?

4. What are the implications for land management or conservation if different sorts of landscape features are found to impede gene flow for males versus females, or for pollen versus seeds? How do we decide which is the more important component of overall gene flow in that species? What if a landscape barrier for one species is a movement corridor for another? Which takes precedence in a comprehensive land-management plan?

5. In thinking about the landscape where your research is based (or one suggested by your instructor), how would you construct a resistance surface for a given species within that landscape? In other words, what costs would you assign to the different land covers or land uses within this landscape, in terms of how you expect those features to influence dispersal and gene flow for the focal species? What landscape features are likely to serve as barriers to gene flow? Which might function as movement corridors?

6. Explain how or why our assessment of the relative effects of habitat area versus fragmentation (habitat configuration) on population viability may vary depending on whether we use demographic or genetic criteria. What are some advantages of using genetic criteria over demographic criteria? What are the disadvantages? Should population viability analysis be based primarily on demographic or genetic criteria, therefore?

7. Is evidence of asymmetrical gene flow between populations sufficient for identifying source–sink population dynamics? Explain why or why not. If not, what else do we need to know?

8. Many extant populations have experienced dramatic shifts in their ranges in response to past climatic changes. Given that population genetic structure may bear the legacy of historical climatic and landscape effects, to what extent can we use retrospective analysis to gain insight into how populations will likely respond to future environmental change? In other words, can we extrapolate from how populations have responded to past climatic and landscape changes to predict how they will respond in the future? Why or why not?

9. Most landscape-genetic studies to date have used neutral genetic markers (e.g. microsatellites) for assessing gene flow and population genetic structure. Increasingly, however, the focus is shifting to the assessment of genetic variation that is adaptive (e.g. SNPs). Do you think next-generation sequencing technologies will soon render the study of neutral genetic variation passé in landscape genetics? What sorts of questions (if any) might be better addressed with neutral genetic variation than with adaptive genetic variation?

10. Gene flow is seen as a mechanism that homogenizes genetic variation among populations, and yet, the spread of locally adaptive traits across the landscape is an important means by which species might be able to track changing environmental conditions. How does landscape structure differentially affect gene flow versus natural selection? Which do you think is likely to be more important in giving rise to population genetic structure: a disruption of gene flow or local adaptation? In other words, does landscape structure ultimately have a greater effect on neutral or adaptive genetic variation?

Landscape Effects on Community Structure and Dynamics

As the pace of landscape change accelerates due to increased human land use and rapid climate change, ecologists and resource managers have the daunting task of trying to anticipate the consequences of future environmental changes on today's ecological communities. The hope is that such foresight will lead to better conservation and management policies capable of meeting our current and future resource needs, while preserving biodiversity and the goods and services that ecosystems provide. Short of a crystal ball, however, the only way to do this is through ecological forecasting, using modeling approaches such as species distribution models (**Chapter 7**). Recall that these models can couple information on the current landscape (e.g. land covers, vegetation types, land uses) with a climate-change model to forecast how future landscape conditions are likely to influence the distribution of species, given their current environmental or habitat requirements.

To illustrate, consider how species distributions are likely to change in response to the climatic and landscape changes that are already underway in western North America (**Chapter 3**). Like much of the western USA, California has suffered from severe drought in recent years, which has left nearly a billion trees dead or dying in its wake (Asner et al. 2016). Because of its large size and topographic diversity, the landscapes of California will not be impacted by drought and other climatic changes to the same degree. Nevertheless, bird species richness is expected to decrease across all major vegetation types throughout much of the state by 2070 (Wiens et al. 2009; **Figure 10.1A, B**). Furthermore, the composition of these bird assemblages is also predicted to change, especially in areas with steep environmental gradients, such as within the Sierra Nevada mountain range (**Figure 10.1C**). Subsequently, species that are not currently found together may co-occur in the future if their distributions shift, causing them to overlap, whereas species that do co-occur at present may not be found together in the future. Such changes would likely lead to a restructuring of present-day bird communities, creating new assemblages that may have no counterpart in today's landscapes (Wiens et al. 2009).

Whether such changes will come to pass is less certain, as much depends on whether species are able to shift their distributions quickly enough to track rapid environmental change (**Chapter 8**). Species distribution models typically assume that species will shift their distributions independently of one another, but the change in the abundance and occurrence of individual species can greatly alter competitive interactions, predator–prey relationships, host–parasite relationships, or the ecology of plant-pollination systems. Even if the loss of individual bird species

Essentials of Landscape Ecology. Kimberly A. With, Oxford University Press (2019).
© Kimberly A. With 2019. DOI: 10.1093/oso/9780198838388.001.0001

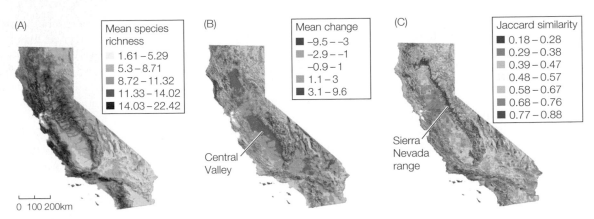

Figure 10.1 Forecasting climate-mediated changes in bird assemblages.   Environmental niche models developed individually for 60 bird species reveal that the current mean species richness of most bird assemblages (A) is expected to decline over much of the state of California (USA) by 2070 in response to climate change (B). The greatest decreases in bird species richness are projected to occur in the northern Central Valley and desert regions of the state (dark red areas, B), with the greatest projected increases occurring along the north coast and mountainous areas (dark blue areas, B). As a result, the composition of these bird communities (as assayed by the Jaccard similarity index) will likely be very different from those found in today's landscapes, especially within the Sierra Nevada mountain range (dark red area; C).

Source: After Wiens et al. (2009).

or the reshuffling of entire bird assemblages doesn't strike you as particularly troubling, consider how the loss of native pollinators and natural enemies (predators and parasites) as a result of landscape change (Klein et al. 2007; Tscharntke et al. 2007; Chaplin-Kramer et al. 2011), along with the climate-mediated shift in the distribution of crop pests and pathogens (Bebber et al. 2013), would have devastating consequences for agricultural production and food security, especially in places like California, which is the largest producer and exporter of agricultural commodities in the USA. Species distribution models largely ignore the role that biotic interactions play in shaping species' distributions, even though the ecological impacts of some of these interactions can outweigh the effects of climate and landscape change (e.g. Araújo & Luoto 2007; Suttle et al. 2007; Wiens et al. 2009). Clearly, we need a better understanding of how species interactions are influenced by landscape structure and dynamics, especially as climate-mediated disturbances, such as drought and fire, are already altering communities in the landscapes of western North America, and elsewhere.

A Landscape Perspective on the Structure and Dynamics of Communities

Ecological communities are made up of species that interact to varying degrees within the same geographical area. By definition, then, communities exist within a landscape context. The species present within a community are a subset of the larger regional or biogeographic species pool, in which climate, productivity, disturbance history, dispersal, colonization–extinction dynamics, and evolutionary processes, along with human land use and species introductions or extirpations, all play a role in shaping the numbers and distribution of species, albeit at very different scales in space and time. To quote Dray and colleagues (2012: 257):

> *Species spatial distributions are the result of population demography, behavioral traits, and species interactions in spatially heterogeneous environmental conditions. Hence the composition of species assemblages is an integrative response variable, and its variability can be explained by the complex interplay among several structuring factors.* **The thorough analysis of spatial variation in species assemblages may help infer processes shaping ecological communities**. (emphasis added)

Not only do ecological communities occur within a landscape context, but the study of how the composition of species assemblages vary across landscapes can provide insight into the processes that shape ecological communities. The study of the reciprocal effects of spatial pattern on ecological processes is what landscape ecology is all about (**Chapter 1**), and thus adopting a landscape perspective can contribute to a better understanding of the structure and dynamics of ecological communities.

There are two ways we might approach the study of landscape effects on community structure and dynamics. We could adopt a 'bottom-up' reductionist approach that builds upon the previous four chapters and extends our understanding of how landscape structure influences individual and population processes to an entire assemblage of multiple, potentially interacting species. Extrapolating from the species to community level might prove difficult, however, if ecological communities are more than the sum of their species (an 'integrative response variable') that comes about because of the 'complex interplay of several structuring factors' (e.g. Cohen et al. 2009; Dray et al. 2012; Clark et al. 2014). In that case, we might instead adopt a 'top-down' approach (*sensu* Whittaker et al. 2001) that examines how landscape structure influences the statistical properties of communities, and search for general patterns in the distribution of diversity at broader scales, what is commonly referred to as **macroecology** (Brown 1995; Gaston & Blackburn 2000; Keith et al. 2012).

Ultimately, we will need to do both. As you may recall from our discussion of hierarchy theory earlier in the text (**Chapter 2**), an ecological system can be studied as a hierarchically nested structure comprising three (or more) levels (a triadic structure; **Figure 2.10**). According to hierarchy theory, interactions among lower-level subunits (different species, in this case) give rise to the focal level (e.g. the community). Lower-level mechanisms and processes are thus responsible for contributing to the structure and dynamics of the focal level (i.e. these processes exert bottom-up controls). The upper level (the landscape or biogeographic region in which the community is embedded) not only provides the broader context for the focal system, but also exerts top-down controls on the structure and dynamics of that system (e.g. reflected in the different regional species pools and evolutionary histories of different biota). Thus, while an understanding of colonization–extinction dynamics or predator–prey relationships may help us explain the relative level of diversity found within a given community, we still need to know something about the landscape in which that community occurs, in terms of its disturbance history or the type and intensity of land use, to put our mechanistic understanding of species interactions in the proper context. Otherwise, our explanations of species diversity patterns may be off-base, especially if the landscape modifies species interactions and there are feedbacks between species and the landscape, as we shall see later in this chapter and as previously discussed in **Chapter 2** in the context of trophic cascades and invasional meltdowns (examples of a disrupted negative feedback and of a positive feedback, respectively).

In this chapter, we thus begin with a top-down approach, by examining statistical measures of community structure and the different scales at which species diversity can be assessed. We then consider how species diversity can be partitioned spatially, which gives us a way of analyzing multiscale patterns of diversity. From there, we shift to a discussion of species diversity patterns, where we begin our quest for under-standing the drivers and underlying mechanisms responsible for these patterns, starting at broad biogeographic scales with latitudinal diversity gradients. We then segue to a discussion of elevational diversity gradients, which contribute a great deal to the overall diversity of vegetation types and species found within landscapes. Although elevational diversity gradients have long been thought to mirror latitudinal diversity gradients, there are some important differences as to how species richness (the number of different species) varies as a function of elevation versus latitude, as we shall see.

Regardless of whether we are talking about elevational or latitudinal gradients, however, we generally find that—all else being equal—larger areas contain more species. The species–area relationship is one of the most ubiquitous scaling relationships in ecology, and can be applied to evaluate the rate at which species richness increases (or decreases) as a function of increasing (or decreasing) habitat area. The species–area relationship has thus been one of the cornerstones for predicting the effects of habitat loss on species diversity. Because habitat loss is often accompanied by habitat fragmentation, we can additionally consider the effects of habitat isolation on colonization–extinction dynamics by applying the theory of island biogeography, another cornerstone in the study and conservation of species diversity within fragmented landscapes. Next, we shift gears and adopt more of a bottom-up approach, by considering how landscape structure influences various types of species interactions, from competitive interactions to predator–prey and host–parasitoid relationships to plant–pollinator interactions. We then conclude this chapter with a discussion of how the interactions of species among different communities contributes to the higher-order structure and dynamics of the **metacommunity**.

Measures of Community Structure

Although **biodiversity** represents the sum total of all biological diversity, from genes to ecosystems, it is most often equated with species diversity. Besides providing a means of characterizing community structure, measures of species diversity are important for documenting the loss of biodiversity due to landscape and climate change, which has implications for biological conservation and the maintenance of ecosystem function, including the goods and services they provide (Pereira et al. 2010; **Chapter 11**). The indices used to characterize community structure are essentially the same we used to describe landscape composition (**Chapter 4**), as those landscape metrics were simply adapted from measures of species diversity. The notation differs a bit, however, and thus we'll provide a quick review here, as well as discuss some additional measures and issues pertaining to the characterization of community structure.

Species Richness

The most fundamental measure of community structure is given by **species richness** (S), which is simply the number of species present at a given location. Species richness is sensitive to sample size or the amount of time spent sampling (sampling effort), however. The more we sample, the more likely we are to encounter new species in our samples. Indeed, we can see this graphically by plotting the cumulative number of species as a function of sampling effort (the number of samples or individuals collected in our samples) to produce **species-accumulation curves** (what are sometimes called 'collector's curves'; Colwell & Coddington 1994).

The rate at which species are added as a function of sampling effort provides some information about the relative abundance and distribution of species in the community (Magurran 2004). Because of this, however, species-accumulation curves are sensitive to the order in which species are added or encountered in our samples. This can be overcome by using procedures such as rarefaction. **Rarefaction** techniques are used to standardize sampling effort, which can then facilitate comparisons among locations or studies that differ in the number of individuals or sites sampled (Gotelli & Colwell 2001). Rarefaction curves are produced by repeatedly resampling the pool of individuals or samples collected, and then plotting the average number of species expected if we had collected only 1, 2, 3,...N individuals or samples.[1]

To illustrate why standardization of sampling effort might be necessary for making comparisons, consider the problem of trying to evaluate the effect of habitat loss on species richness. All else being equal, landscapes with a greater amount of habitat are expected to have more species than landscapes with less habitat. This is the basis of the species–area relationship, to be discussed later. Because of this relationship, however, it is not clear whether the number of species found in a landscape with only 10% habitat is what we would find if we obtained a random sample of a similar size from, say, a landscape with 80% habitat. Because of negative edge effects (**Box 7.1**; **Figure 7.5**), we might expect that a landscape with only 10% habitat would have far fewer species than a similar-sized area within an 80% landscape, for example.

To address this sort of issue, With and Pavuk (2012) calculated rarefaction curves for arthropod communities within a set of experimental landscapes that varied in the amount and fragmentation of habitat (**Figure 5.11**). Because 10% of the available habitat was randomly sampled in each landscape, sampling effort was greater in landscapes with 80% habitat than in 10% landscapes (i.e. a greater number of habitat cells were sampled in 80% landscapes than in 10% landscapes). Unsurprisingly, 80% landscapes were found to have 3.5 times more species than landscapes with only 10% habitat (**Table 10.1**). However, rarefaction curves revealed that, for the most part, landscapes with different amounts of habitat gained species at approximately the same rate, when adjusted by the number of individuals obtained within these samples (more samples = more individuals = more species; **Figure 10.2**). This suggests that arthropod communities in this agriculturally dominated system form via a process of random assembly: species found within a landscape are essentially a random draw from the larger species pool. This does not mean that edge effects are unimportant, as they could still influence the number and composition of species at a local patch scale, but overall, landscapes with less habitat appear to be sampling the regional species pool less thoroughly than landscapes with more habitat.

Species Diversity

Species diversity is a function of both species richness (S) and the relative abundance (p_i) of those species in the community. Many measures have been proposed for the measurement and estimation of species diversity (reviewed in Magurran 2004). The most commonly

[1] The *vegan* package in the statistical programing language R provides a suite of tools for the analysis of ecological communities, including rarefaction and many of the other diversity measures, similarity measures, and analyses discussed in this chapter (Oksanen et al. 2019). In addition, the *BiodiversityR* package provides a nice Graphical User Interface (GUI) and utility functions (most of which are based on *vegan*) for diversity and community analysis (Kindt 2019).

TABLE 10.1 Comparison of diversity measures for arthropod communities within experimental landscapes with different habitat amounts (red clover, *Trifolium pratense*; Figure 5.11).

Diversity measure	10% habitat	80% habitat
Alpha diversity (α)		
Mean species richness (S)	37	128
Shannon diversity index (H')	2.85	3.49
Evenness (E)	0.79	0.72
Effective species richness (S_e)	17.3	32.7
Effective evenness (E')	0.47	0.26
Gamma diversity (γ)	270 species	
Beta diversity (β)		
Multiplicative β ($\beta = \gamma/\alpha$)	7.3	2.1
Additive β ($\beta = \gamma - \alpha$)	233	142

Source: With & Pavuk (2012).

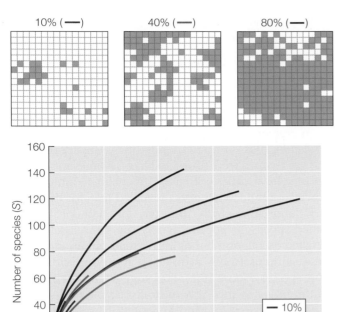

Figure 10.2 Rarefaction curves. Arthropods were sampled within experimental landscapes plots that varied in the amount and fragmentation of habitat (Figure 5.11). Because only 10% of the habitat cells were sampled in each landscape plot, sampling effort was greater in landscapes having more habitat, which could bias comparisons of habitat-area effects on species richness and diversity. To correct for this, rarefaction curves were calculated by repeatedly resampling the individuals collected and plotting the average number of species expected for a given sample size (number of individuals, N). The overlapping rarefaction curves indicate that landscapes with different amounts of habitat gained species at approximately the same rate. Arthropod assemblages in this system likely form via a process of random assembly, in which landscapes with less habitat are simply sampling the regional source pool less effectively than those with more habitat.

Source: After With & Pavuk (2012).

used measure of species diversity is the **Shannon diversity index (*H'*)**:

$$H' = -\sum_{i=1}^{S} p_i \ln(p_i). \qquad \text{Equation 10.1}$$

All else being equal, a community with species having the same relative abundances (an even distribution) would be considered more diverse than one in which a single species dominates. Although the Shannon diversity index is unbounded, observed values for ecological communities typically fall within the range $1.5 \leq H' \leq 3.5$ (Magurran 2004).

The maximum diversity (H'_{max}) that can be attained for a community with S equally abundant species is given by $H'_{max} = \ln(S)$. An **evenness index (*E*)** can be calculated as

$$E = H' / H'_{max}, \qquad \text{Equation 10.2}$$

which gives the proportion of maximum diversity attained by a community with S species. It is thus a bounded index ($0 < E \leq 1$), unlike H'. Furthermore, since we are standardizing H' by the maximum diversity possible for a given number of species, we can more readily make comparisons among communities or locations that differ in species richness.

To illustrate, let's compare diversity indices for arthropod communities within the experimental landscape plots mentioned previously (With & Pavuk 2012; **Table 10.1**). Recall that landscapes with 80% habitat had 3.5 times more species than 10% landscapes. The higher richness of the 80% landscapes also translated into a higher Shannon diversity index ($H' = 3.49$) relative to the 10% landscapes ($H' = 2.85$). However, evenness was marginally greater in the 10% landscapes ($E = 0.79$) than in the 80% landscapes ($E = 0.72$). These differences are not very great, however, especially given the great disparity in species richness between landscapes.

What do these values actually mean, however? There is a 20% difference between the H' values for arthropod communities in landscapes with 10% versus 80% habitat, so does that mean that the 80% landscape is 20% more diverse? No, it does not. Although a H' is an *index* of diversity, it is not an actual measure of diversity (Jost 2006). Consider this analogy made by Jost (2006: 363):

> *The radius of a sphere is an index of its volume but is not itself the volume, and using the radius in place of the volume in engineering equations will give dangerously misleading results. This is what biologists have done with diversity indices.*

A better measure of species diversity is given by the **effective species richness** (MacArthur 1965; Hill 1973a; Jost 2006). The effective species richness (S_e) gives the level of species richness expected if all species were equally abundant, and is obtained as

$$S_e = e^{H'}, \qquad \text{Equation 10.3}$$

in which H' is the Shannon diversity index calculated in the same manner as before (Equation 10.1) and e is the base of the natural logarithm (~2.718). If we apply this to the diversity indices obtained from the aforementioned experimental landscape study, we find that 80% landscapes have an effective species richness, $S_e = e^{3.49} = 32.7$, whereas 10% landscapes have an effective species richness, $S_e = e^{2.85} = 17.3$ (**Table 10.1**). Although 80% landscapes averaged 128 species, their level of diversity is what we would expect if these communities had only about 33 equally abundant species. That is, these communities are far less diverse than suggested by their level of richness. Nevertheless, we can now state that 80% landscapes are twice as diverse as 10% communities,[2] something we could not say previously using only H'.

What about the fact that communities in 10% landscapes were more even than those in 80% landscapes? Recall that the evenness index (E) is obtained as a ratio of the observed diversity to that expected if all species had equal abundance. We can thus calculate an evenness index (E') based on effective species richness (S_e) as

$$E' = e^{H'} / S = S_e / S. \qquad \text{Equation 10.4}$$

Thus, the effective evenness for the 80% landscapes is $E' = 32.7/128 = 0.26$, whereas evenness for the 10% landscapes is $E' = 17.3/37 = 0.47$ (**Table 10.1**). In other words, 10% landscapes are about half as diverse as they could be, given their level of richness, whereas 80% landscapes are only a quarter as diverse as expected, given the large number of species they contain. Although this supports our initial finding that 10% landscapes have more even communities than 80% landscapes, the degree of difference between these landscapes is more evident now.

Although evenness is a component of diversity, it too is an index, and by itself does not indicate which community is more diverse. We still need information on species richness (S) to make that determination. Given that the Shannon diversity index (H') subsumes information on both species richness and evenness, we can decompose H' into its constituent parts as

$$H' = \ln(S) + \ln(E'). \qquad \text{Equation 10.5}$$

Thus, for the 80% landscape we have

$$H' = \ln(128) + \ln(0.26) = 3.50,$$

and for the 10% landscape,

$$H' = \ln(37) + \ln(0.47) = 2.86.$$

You'll note that these are essentially the same values for H' obtained previously using Equation 10.1. Although this may appear redundant, it gives us a way of evaluating the relative contributions of richness and evenness to this index of diversity, since different combinations of the two can give rise to the same H' value. We can thus determine at a glance that while 80% landscapes have 3.5 times the richness, they only have half the effective evenness of 10% landscapes. This could be important when evaluating how land-use change affects different aspects of diversity (richness vs evenness), for example.

Spatial Partitioning of Diversity

Species diversity can also be partitioned into different spatial components, such as the level of diversity

found within versus between communities within a landscape or along an environmental gradient. Up to now, we have been considering diversity within individual communities. Species diversity within a given community, location, or sample is referred to as **alpha diversity** (α diversity); it is therefore a measure of local diversity.

By contrast, the pooled diversity across a group of communities, locations, or samples is called **gamma diversity** (γ diversity). Gamma diversity is a measure of total species richness across the entire landscape or region. As such, it is assayed using the various diversity indices discussed previously (Equations 10.1–10.5), only now applied to a larger area or collection of samples. For example, a total of 270 arthropod species were found within that experimental landscape system mentioned previously (With & Pavuk 2012; **Figure 5.11**; **Table 10.1**). Gamma diversity is thus 2–7 times greater than the average alpha diversity reported for landscape plots having either 10% or 80% habitat (37 and 128 species, respectively). Given that gamma diversity is greater than alpha diversity, there are clearly more species in this system than can be found in any one landscape plot.[3] That is, species richness in a given landscape is a subsample of species that occur within the regional species pool. Although communities within landscapes with the same amount of habitat might be expected to exhibit some degree of similarity, especially with respect to the more common species, they nevertheless differ as to which particular species—out of 270 total species—they contain (With & Pavuk 2012).

The degree to which species composition differs or changes among communities is assayed by **beta diversity** (β diversity). Although this definition may appear straightforward, beta diversity has been defined in different ways, which in turn has led to a great many ways for assessing it (Vellend 2001; Tuomisto 2010a,b; Anderson et al. 2011; Legendre & De Cáceres 2013). A handy distinction among β diversity concepts was made by Anderson and colleagues (2011), who recognized two major types of beta diversity: (1) the degree or rate of species turnover between communities, especially along an environmental gradient, and (2) variation in community structure among samples within a given area. We will discuss each of these types of beta diversity in turn.

---

[2] Note that 80% landscapes still have 3.5 times the richness, but only 2 times the diversity, of 10% landscapes. That is because diversity includes information on both species richness *and* evenness.

[3] Although gamma diversity is defined to be greater than alpha diversity, there are some diversity measures (e.g. Fisher's α) in which the summed diversity of a set of samples may be less than the mean of those samples (i.e. $\gamma < \bar{\alpha}$; Lande 1996). The measures of diversity we've discussed in this chapter, such as species richness (S) and the Shannon index (H'), all have the desirable property of *concavity* (Lande 1996), in which the total diversity in a pooled set of samples exceeds (or equals) the average diversity within those samples (i.e. $\gamma \geq \bar{\alpha}$). Concavity is what permits us to perform an additive partitioning of diversity measures, as we discuss a bit later.

SPECIES TURNOVER BETWEEN COMMUNITIES   For the first type of beta diversity, the overarching objective is to characterize species turnover between pairs of communities (samples or locations; **Figure 10.3**). This is commonly done by calculating the degree of similarity (or its inverse, dissimilarity) between the two communities. The simplest index for evaluating pairwise similarity is the **Jaccard coefficient** (Jaccard 1912):

$$C_J = a / (a+b+c), \qquad \text{Equation 10.6}$$

in which a is the number of species found in both communities, b is the number of species found only in the first community, and c is the number of species found only in the second community. A related measure is the **Sørensen coefficient** (Sørensen 1948). In this index, greater weight is given to species found in both communities:

$$C_S = 2a / (2a+b+c). \qquad \text{Equation 10.7}$$

In either case, communities that have all species in common would have $C = 1$ (i.e. complete similarity).[4] Dissimilarity (β diversity) is thus obtained as $1 - C$, which is often multiplied by 100 so as to express this as a percentage. For example, two communities that have 40% of their species in common ($C = 0.4$) would have 60% dissimilarity.

The Jaccard and Sørenson coefficients are both based on the incidence of species (i.e. presence-absence data). They thus treat rare and common species equivalently, which may not be ideal for comparing the similarity of two communities if one contains numerous rare species. Various abundance-based similarity indices have therefore been developed, including extensions to the traditional Jaccard and Sørenson indices (e.g. Chao et al. 2005, 2006). One of the most popular abundance-based indices is the **Bray–Curtis dissimilarity coefficient** (Bray & Curtis 1957), which incorporates the relative difference in species abundances between two communities, locations, or samples as

$$C_{BC} = \sum_{i=1}^{S} |n_{i1} - n_{i2}| / (N_1 + N_2), \qquad \text{Equation 10.8}$$

where n_{i1} is the abundance of species i in the first community, n_{i2} is the abundance of species i in the second community, N_1 is the total abundance of all species in the first community, and N_2 is the total abundance of all species in the second community. Given that this is calculated in terms of relative differences, a value of $C_{BC} = 1$ means that the two communities share no spe-

cies in common (complete dissimilarity).[5] For example, the greatest dissimilarity between arthropod communities in that experimental landscape system we've been discussing (With & Pavuk 2012) was between landscape plots with 10% habitat and those with either 60% or 80% habitat (C_{BC} = 82% dissimilarity; **Table 10.2A**). In general, community similarity was highest among landscapes having abundant habitat ($\geq$ 50% habitat), which had 50–60% of their species in common ($1 - C_{BC}$; **Table 10.2A**).

Calculating an index of similarity (or dissimilarity) is only the first step in analyzing species turnover or differences in community composition, however. If the goal is to relate species turnover to changes along a geographic or environmental gradient, then the percent dissimilarity between adjacent pairs of communities (or locations) can be analyzed with respect to distance or whatever environmental factors are being used to characterize that gradient (e.g. precipitation or temperature) using linear or non-linear regression

TABLE 10.2   Community similarity among arthropod communities in experimental landscapes that differed in the amount of habitat cover (red clover, *Trifolium pratense*; Figure 5.11). (A) Mean pairwise dissimilarity between communities based on the Bray–Curtis dissimilarity coefficient (C_{BC}). (B) Analysis of similarity (ANOSIM) based on the Bray–Curtis dissimilarity coefficient (C_{BC}). Statistically significant differences ($P < 0.05$) for the ANOSIM are indicated in boldface.

Habitat amount (%)	Habitat amount (%)				
	10	20	40	50	60
(A) Percent dissimilarity (C_{BC})					
20	59.8				
40	74.9	51.5			
50	74.5	45.3	41.6		
60	82.3	58.6	46.4	41.4	
80	82.1	61.1	50.0	42.5	41.2
(B) Analysis of similarity (*R* statistic)					
20	**0.47**				
40	**0.75**	**0.41**			
50	**0.75**	**0.65**	-0.29		
60	**0.98**	**1.00**	-0.05	**0.70**	
80	**0.98**	**0.98**	0.00	**0.44**	0.11

Source: With & Pavuk (2012).

---

[4] An alternate formulation of C_J and C_S involves the total number of species in each sample, rather than the number of unique species. In that case, $C_J = S_{1,2}/(S_1 + S_2 - S_{1,2})$ and $C_S = 2S_{1,2}/(S_1 + S_2)$, respectively, where $S_{1,2}$ = number of species common to both samples, S_1 = number of species in the first sample, and S_2 = number of species in the second sample.

[5] It should be noted that $C_{BC} = C_S$ if species presence-absence data (1 or 0, respectively) are used rather than abundance data. The Bray–Curtis dissimilarity coefficient can therefore be considered a modified form of the Sørensen index (Magurran 2004).

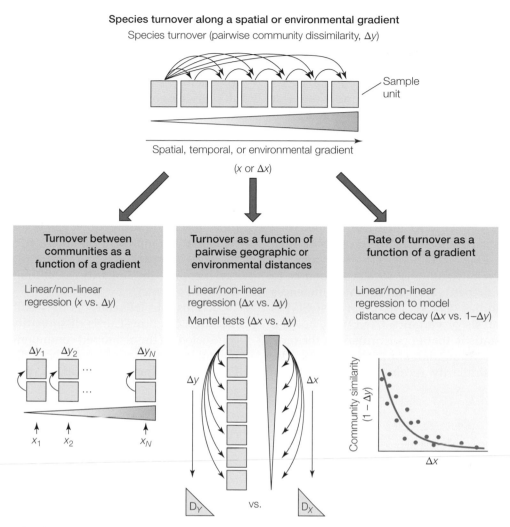

Figure 10.3 Assessing beta diversity as species turnover between communities.
Source: After Anderson et al. (2011).

(Tuomisto 2010b; Anderson et al. 2011; **Figure 10.3**). Given that we expect community similarity to decrease as a function of distance, we could use linear or non-linear regression to obtain the rate of species turnover as the slope of a distance-decay model (Tuomisto 2010b; Anderson et al. 2011). Or, if we calculate pairwise dissimilarities between all communities or locations, we could use Mantel tests (r_M) to compare the community dissimilarity matrix (which is a type of difference matrix) to a pairwise environmental difference or geographic distance matrix (Tuomisto 2010b; Anderson et al. 2011; **Figure 10.3**). Recall from our discussion in **Chapter 9** that Mantel tests are used to test for linear relationships between two sets of distance (or difference) matrices.

This approach was taken by Grenouillet and his colleagues (2008), who examined the similarity of stream assemblages for three taxonomic groups (stream fish, benthic macroinvertebrates, and diatoms) in relation to environmental conditions and geographic distance

along the Viaur River in southwestern France. A community dissimilarity matrix was constructed for each group, based on the Bray–Curtis dissimilarity coefficient (C_{BC}) between all pairs of sites ($n = 13$ sites sampled along the river, so 78 site-pairs). Although significant correlations (r_M) were found between the stream assemblages and both environmental and geographic distance, partial Mantel tests (which explore the relationship between two distance matrices while controlling for the effects of a third) revealed that the correlations between the stream assemblages and environmental distance were not in fact significant once the effects of geographic distance were removed (Grenouillet et al. 2008). Thus, species turnover along the longitudinal gradient of the stream is influenced more by distance effects than by environmental conditions per se. Remarkably, all three groups exhibited a similar response to distance along this stream gradient, with species assemblages transitioning from a positive to negative autocorrelation somewhere between 23 and 54 km, which is about a third of this river's total length (**Figure 10.4**).

VARIATION IN COMMUNITY COMPOSITION For the second type of beta diversity, the primary goal is to measure the variation among communities, and either test for the significance of that variation, relate it to a set of spatial or environmental factors, or partition it among a hierarchically nested set of spatial scales. As with the first type of beta diversity, there are numerous ways this can be done (Tuomisto 2010a; Anderson et al. 2011; Legendre & De Cáceres 2013).

To begin, we could use the indices discussed previously (C_J, C_S, or C_{BC}) to measure all pairwise similarities between communities in our sampling area (**Figure 10.5**, left). One could then perform an analysis of similarity (ANOSIM) on the similarity matrix to evaluate whether there are significant differences between communities (Clarke 1993). An ANOSIM is similar to an analysis of variance (ANOVA), but is a non-parametric statistic that is performed on the rank similarities of the triangular similarity matrix. The highest pairwise similarity receives a rank of 1, the next-highest similarity receives a rank of 2, and so on. The test statistic is given by

$$R = \left(\overline{r_B} - \overline{r_w}\right) / \left(n(n-1)/4\right), \qquad \text{Equation 10.9}$$

in which $\overline{r_B}$ is the average rank similarity arising from all pairwise comparisons between sites, $\overline{r_w}$ is the average of all rank similarities within sites, and n is the total number of samples (communities) under consideration. The test statistic R is bounded between $-1 \leq R \leq 1$, in which $R = 1$ indicates that all of the variation in similarity is found within communities (i.e. communities all belong to the same group), whereas $R = -1$ means that all of the variation occurs between communities (each community belongs to a different group). Note that if a matrix of dissimilarity is used (e.g. C_{BC}), then the interpretation is reversed: $R = 1$ would indicate complete dissimilarity (i.e. communities belong to different groups).

To illustrate, let's consider the results of an ANOSIM based on the Bray–Curtis dissimilarity index that was conducted on the arthropod communities in that experimental landscape study we've been discussing (With & Pavuk 2012; **Figure 5.11**). Overall, communities within a

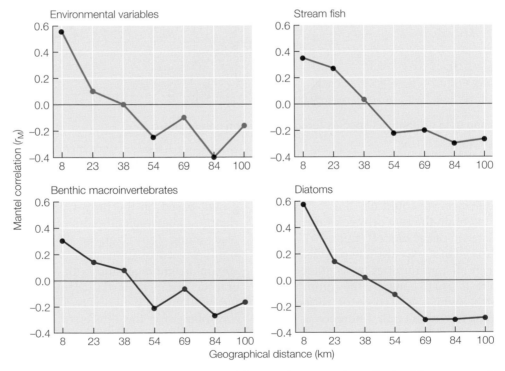

Figure 10.4 Assessing community similarity along a spatial or environmental gradient. Stream assemblages in the Viaur River of southwestern France become increasingly dissimilar with increasing distance, based on a Mantel test of pairwise Bray–Curtis dissimilarity coefficients (C_{BC}) among sites for each of three different taxonomic groups. Although environmental variables also vary along this stream gradient (top left), a partial Mantel test (not shown) revealed that the correlations between the stream assemblages and environmental distance were not significant once the effects of geographic distance were removed. All three groups had a similar response to distance along this stream gradient, exhibiting a significant turnover in community composition somewhere between 23 and 54 km (black symbols indicate significant correlations between the pairwise community dissimilarity coefficients and geographic distance).

Source: After Grenouillet et al. (2008).

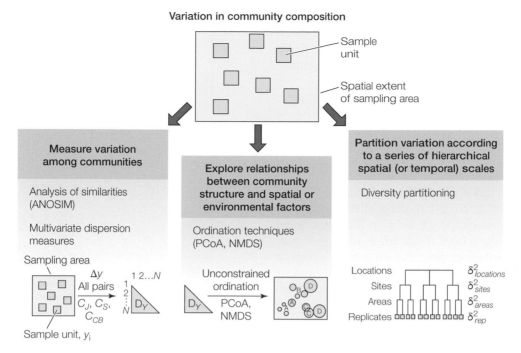

Figure 10.5 Assessing beta diversity as variation in community composition.
Source: After Anderson et al. (2011).

given landscape type—defined in terms of the amount of habitat—were more similar than expected by chance, and thus community composition differed significantly among landscape plots having different amounts of habitat. The greatest difference between groups was observed between communities in 10% or 20% landscapes and those in 60% or 80% landscapes ($R = 0.98 - 1.00$; **Table 10.2B**). Recall that communities within 10% landscapes were previously shown to be the least similar to those in 60–80% landscapes on the basis of the Bray–Curtis dissimilarity index (**Table 10.2A**).

Conversely, we could assess beta diversity in terms of how far each community lies from the group centroid in multivariate space; that is, we can use multivariate dispersion (variance) as a measure of beta diversity (Anderson et al. 2006; Anderson et al. 2011; **Figure 10.5**, left). In this context, the group centroid is defined in the principal coordinate space of whatever dissimilarity measure we are using (e.g. $1 - C_J$, $1 - C_S$, C_{BC}), such that beta diversity is measured as the average distance (dissimilarity) of an individual location or community from the group centroid. Intuitively, communities that are similar would all lie pretty close to the group centroid, resulting in low multivariate dispersion (i.e. low beta diversity). Differences in beta diversity among habitats or landscapes can then be evaluated using a multivariate test for homogeneity in dispersions (Anderson et al. 2006).[6]

For example, Anderson and her colleagues (2006) applied various measures of multivariate dispersion to test for differences among invertebrate assemblages inhabiting the holdfasts of kelp (*Ecklonia radiata*) that grow along the New Zealand coast. Holdfasts are root-like structures that anchor the kelp to the ocean floor, and which provide shelter and habitat for a variety of invertebrates, including arthropods, annelids (segmented worms), mollusks, and bryozoans (moss animals). Interestingly, tests based on C_J or C_S, which emphasize the compositional differences between communities, revealed that small holdfasts had significantly greater beta diversity (multivariate dispersion) than large holdfasts, whereas a dissimilarity measure that primarily emphasized relative differences in species' abundances found the opposite (i.e. greater variation among large holdfasts than small ones).

Our assessment of beta diversity may therefore depend on which measure we use, and thus, which aspect of diversity is primarily being assayed by that measure: variation in species composition or variation in species' abundances. This raises the possibility that a given environmental factor could differentially influence species composition versus abundance, thereby complicating the analysis of how environmental heterogeneity influences beta diversity. Given that more than a dozen dissimilarity coefficients have been

[6] This is analogous to the Levene's test of homogeneity of variances, which is often performed prior to certain statistical tests (e.g. ANOVA) to verify the standard assumption that variances are equal across the samples or groups being compared.

developed, which vary with respect to more than a dozen properties, some careful thought should obviously be given as to whether the selected measure will be up to the challenge (Anderson et al. 2011; Legendre & De Cáceres 2013).

Alternatively, multivariate ordination techniques, such as principal coordinates analysis (PCoA) or non-metric multidimensional scaling (NMDS), can also be used to explore relationships among communities and identify how communities are grouped, whether in space or with respect to some other environmental factor (Clarke 1993; Anderson et al. 2011; **Figure 10.5**, middle). Recall that we discussed these ordination techniques previously in **Chapter 9** on landscape genetics, where these methods were used to uncover population genetic structure given a matrix of pairwise genetic distances. In the present context, we can use these methods to uncover structure within a matrix of pairwise community similarities (or dissimilarities).

For example, NMDS was used to compare the effects of an ecosystem engineer, the North American beaver (*Castor canadensis*), on plant community composition in riparian areas of the central Adirondacks of New York, USA (Wright et al. 2002). Recall from **Chapter 3** that beavers create ponds that eventually fill in or are abandoned and undergo succession, forming meadows or forested wetlands that may persist for decades (**Figure 3.34**). As ecosystem engineers, beavers increase landscape heterogeneity, but might they also increase plant diversity within habitats? To address this question, Wright and colleagues (2002) compared the diversity of herbaceous plants within three types of riparian habitats: wet meadows and shrubby swamps created as a result of beaver activity, and forested areas that had no history of beaver activity for at least 60 years. Although the beaver-modified and forested habitats did not differ in species richness, the composition of these habitats was very different: there was very little overlap of species between beaver-modified and non-modified sites (only 17% similarity), and these forested communities were distinct from the beaver-modified habitats, which had 58% of their species in common (**Figure 10.6**). Clearly, then, beavers are capable of influencing diversity across a range of scales, from the habitat patch up to the scale of the landscape mosaic.

The last approach that we will discuss involves **diversity partitioning**, which has become a popular way of analyzing multiscale patterns of diversity (Veech et al. 2002; Crist et al. 2003; Veech & Crist 2010a,b; **Figure 10.5**, right). As we've discussed, beta diversity links alpha and gamma diversities (and thus local and regional diversities), but exactly how it does so is a matter of debate (Ellison 2010). Although alpha and

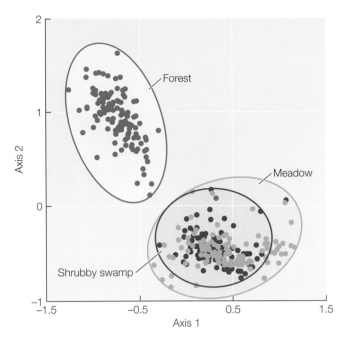

Figure 10.6 Beavers alter the community composition of riparian habitats. Non-metric multidimensional scaling (NMDS) was used to evaluate differences in the species composition of plant communities from beaver-modified habitats (shrubby swamps and meadows) versus riparian forests where beavers had been extirpated for >60 years. Although all three habitats had a similar number of species, they clearly differed in terms of what species made up those communities. Beaver-modified habitats exhibited a great deal of overlap, having most (58%) of their species in common, whereas there was little overlap in community composition between these habitats and the riparian forests that lacked beavers. Beavers are thus capable of modifying diversity across a range of scales, from the habitat patch up to the landscape mosaic (**Figure 3.34**).

Source: After Wright et al. (2002).

gamma diversities can be measured directly (e.g. as species richness), beta diversity is a derived quantity.[7] As originally proposed by Whittaker (1960), gamma diversity was viewed as the product of alpha and beta diversity: $\gamma = \alpha \times \beta$. In that case, we can derive beta

---

[7] It has also been argued that alpha and beta diversities could be independent of one another, such that gamma diversity is the derived quantity (Jost 2007, 2010). If neither gamma diversity nor community size (N, the number of individuals) is fixed, such that these can vary with random draws of alpha and beta values, then it is theoretically possible to achieve independence of alpha and beta diversity. In practice, however, both gamma and N are fixed in real community data sets, such that beta can only be determined by first obtaining alpha and gamma, both of which are measured directly (Veech & Crist 2010b). For example, Kraft and his colleagues (2011) demonstrated in their analysis of woody plant diversity along latitudinal and elevational gradients how alpha diversity and all common measures of beta diversity depend on both gamma and N. Interestingly, although beta diversity (species turnover) is typically higher in the tropics and at low elevations, these latitudinal and elevational differences disappeared once the variation in gamma diversity was removed (Kraft et al. 2011).

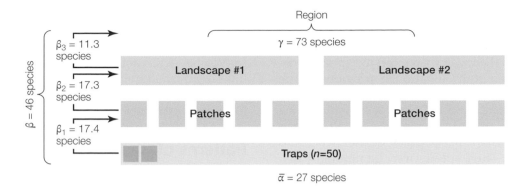

Figure 10.7 Hierarchical sampling and diversity partitioning. Seventy-three species of fruit-feeding butterflies occur within a region of the Atlantic Forest in Brazil, but only a third of these species are generally found at any given site ($\bar{\alpha}$ = 27 species). By partitioning beta diversity (β) among a series of hierarchically nested samples (fruit-baited traps nested within forest patches nested within landscapes), it becomes apparent that much of the variation in species richness among scales (species turnover) is to be found among traps within forest patches (β_1 = 17.4 species) and among patches within landscapes (β_2 = 17.3 species).
Source: After Ribeiro et al. (2008).

diversity as $\beta = \gamma/\alpha$. For example, if the average community from a landscape with 80% habitat has $\alpha = 128$ species and $\gamma = 270$ species, then $\beta = 270/128 = 2.1$ (**Table 10.1**). Gamma diversity is thus twice as much as alpha diversity, which means that the average 80% landscape only contains about half of the species present in this system (or alternatively lacks half of the species found across the entire system).

Conversely, it has been suggested that gamma diversity should be an additive property of alpha and beta diversities: $\gamma = \alpha + \beta$ (MacArthur et al. 1966; Lande 1996; Veech et al. 2002). In that case, beta diversity is derived as $\beta = \gamma - \alpha$. Thus, using our previous example, $\beta = 270–128 = 142$ species (**Table 10.1**). That is, an average of 142 species are added as we move from the scale of the individual landscape to that of the entire region (a collection of landscapes). Although diversity partitioning can be performed in different ways and using different measures of diversity, the overarching goal is largely the same: to evaluate how diversity might be spatially structured across a range of scales, such as among a series of hierarchically nested samples (**Figure 10.5**, right).

To illustrate, we'll consider a study that employed an additive partitioning of diversity measures (both S and H') among hierarchically nested samples of fruit-feeding butterflies (Nymphalidae) in the Atlantic Forest of Brazil (Ribeiro et al. 2008). Five baited traps were surveyed within each of ten forest patches located within two fragmented landscapes (**Figure 10.7**). Over the survey period, a total of 73 species of butterflies were captured across the entire study region ($\gamma = 73$ species). At a local scale, however, individual traps averaged 27 species ($\bar{\alpha} = 27.1$ species, $n = 50$ traps), or only about a third of the regional diversity, which is significantly less than expected from sample-based

randomization tests (see Crist et al. 2003 for how this is done). Nevertheless, an average of 46 species is gained by scaling up from the local to regional scale ($\beta = 73–27 = 46$ species).

Beta diversity can be partitioned more finely among these scales, however, by calculating the difference in mean richness between each level in the sampling hierarchy. Thus, the variation in species richness between traps is obtained as the difference between the mean patch richness and the mean trap richness ($\beta_1 = 17.4$ species), the variation among fragments as the difference between the mean patch richness and the mean landscape richness ($\beta_2 = 17.3$ species), and the variation between landscapes as the difference between the regional richness and the mean landscape richness ($\beta_3 = 11.3$). Note that the sum of these three betas equals the overall beta diversity (i.e. partitioning is additive, so $\beta = \beta_1 + \beta_2 + \beta_3$).[8] As a result, we can now determine that the greatest variation in species composition occurs among traps within patches, and among patches within landscapes (**Figure 10.7**). Based on randomization tests, beta diversity was higher than expected among traps within forest fragments, but not among fragments within a landscape (Ribeiro et al. 2008). Thus, butterflies are not randomly distributed within forest fragments, but the butterfly community within a given forest fragment is a random sample of the regional species pool.

In addition to exploring patterns of species richness, Ribeiro and colleagues (2008) also partitioned the Shannon diversity index (H') among scales, and found that most of the variation in beta diversity (86.4%)

[8] Diversity partitioning can also be multiplicative: $\gamma = \alpha \times \beta_1 \times \beta_2 \times \beta_3$ (Veech & Crist 2010a). In the present example, the multiplicative partitioning of gamma diversity (γ) is: $27.1 \times 1.64 \times 1.39 \times 1.18 = 73$ species.

occurred at the local, within-trap scale ($a_{H'}$ = 2.67), where the observed level of diversity was significantly less than expected, based on randomization tests. This is consistent with the lower than expected species richness found within traps mentioned previously. The level of diversity found among fragments was significantly greater than expected, however, which runs counter to the results for species richness. Recall that the Shannon diversity index gives greater weight to common species. Thus, species may be randomly distributed among fragments, but forest patches apparently differ in their relative abundances of the most common species. Similarly, the level of diversity (H') between landscapes was not significantly different from that expected, even though species richness had been greater than expected, suggesting that the dominant species are the same in both landscapes.

Differences in the diversity of butterfly assemblages among these landscapes are thus driven by the number of rare species they contain (Ribeiro et al. 2008). The authors suggest that this is consistent with the **landscape-divergence hypothesis** (Laurance et al. 2007), in which the composition of species assemblages within fragments of a given landscape will tend to have similar dynamics and trajectories of change, but which differ from those of other landscapes, causing species assemblages in different landscapes to diverge over time. Again, this divergence may only occur with respect to rare or uncommon species, as communities in different landscapes can still all contain the same set of 'core species,' which tend to be the most common or widespread species, habitat generalists, and/or species that are tolerant of disturbances.

Regardless of whether beta is partitioned or not, consideration of both alpha and beta diversity is necessary for evaluating how species diversity is spatially structured across landscapes. Although there are myriad ways in which we can do this, as this section of the chapter demonstrates, all of these measures and approaches ultimately assume that species are equal, in that only the presence or relative abundance of a species is used to determine its contribution to diversity. Yet, we know that some species are more equal than others.[9] The concept of a keystone species or ecosystem engineer highlights the disproportionate effect that some species can have on ecosystems and landscapes, impacts that far exceed their own individual size or relative abundance in the community (recall that the beaver is an example of both a keystone species and an ecosystem engineer; **Chapter 3**). **Functional diversity**, which assesses the range of functional traits represented by species within a community, is another component of biodiversity we might therefore consider (Díaz and Cabido 2001). In assessing functional diversity, the focus is on what species 'do' within ecosystems (i.e. how they assimilate energy or utilize resources), rather than on the numbers of species per se (Petchey & Gaston 2006). Functional diversity thus links organismal traits and community composition to ecosystem processes, which in turn has consequences for ecosystem functioning and resilience (**Chapter 11**).

Patterns of Species Diversity

Even the most casual observer is aware that species diversity is not distributed uniformly across the Earth's surface. The wet tropics are teeming with life, whereas far fewer species inhabit the frigid poles or arid deserts of our planet. The study of species diversity patterns at such broad geographic or global scales has historically been the province of **biogeography**, and more recently, the subfield of ecology known as **macroecology** (Brown 1995; Gaston & Blackburn 2000). However, these broader biogeographic or macroecological patterns provide a basis for understanding spatial patterns in species diversity at whatever scale they occur, whether in terms of providing a broader geographic or historical context for understanding local community patterns, or because similar mechanisms influence diversity across a range of scales.

It should come as no surprise that 'scale matters' for understanding patterns of species diversity (Willis & Whittaker 2002; Willig et al. 2003). Indeed, we first encountered this in **Chapter 2,** where a space-time diagram was used to depict the scaling domains of various types of disturbances and ecological processes that shape terrestrial vegetation patterns from local to global scales (**Figure 2.2**). In a similar vein, we can use a space-time diagram to illustrate more generally the mechanisms that give rise to species diversity across a range of spatial and temporal scales (**Figure 10.8**). For example, the fundamental differences between biogeographic realms, such as between the Nearctic and Neotropics, are largely the result of planetary surface processes, such as plate tectonics and sea-level changes that have occurred over tens to hundreds of millions of years. At a continental or regional scale, however, mountain-building episodes, glacial-interglacial cycles, and variation in the availability of water and energy contribute to patterns of species richness such as the **latitudinal diversity gradient**, which has been shaped over tens of thousands to millions of years. Topographic variation additionally exerts a major influence on soil types and other edaphic and climatic factors that shape species distributions at traditionally defined 'landscape scales' (**Chapter 3**; **Figure 10.8**). Topographic variation is thus an important driver of spatial diversity patterns within landscapes, especially as it can give rise to elevational gradients in species richness. Landscape disturbance dynamics—whether abiotic,

---

[9] With apologies to George Orwell, author of *Animal Farm* (1945).

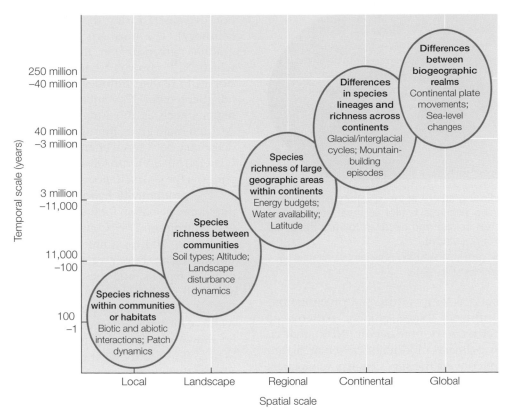

Figure 10.8 Space-time diagram of factors influencing species diversity.
Source: After Willis & Whittaker (2002).

biotic or anthropogenic—can have an overwhelming influence on the levels of diversity found in a given landscape or site, as discussed in **Chapter 3**. Finally, the interactions between species and their environment, including the spatial distribution of resources, affects the structure and dynamics of communities at a local site scale (**Figure 10.8**).

We thus begin with a discussion of latitudinal gradients in species richness, before moving to discuss the similarities to **elevational diversity gradients**, which are a major contributor to patterns of species diversity within landscapes. We will also cover the species–area relationship and the theory of island biogeography in this section, given that both are used to evaluate patterns of species diversity within landscapes, and have long provided a theoretical foundation and research paradigm for the study and management of habitat loss and fragmentation effects on species diversity.

Latitudinal Gradients in Species Richness

Why are there so many species in the tropics compared to polar regions? The latitudinal gradient in species richness, in which species richness increases from the poles to the equator, is one of the most pervasive and best-known biogeographic patterns of species diversity (**Figure 10.9**). It was remarked upon some 300 years ago by European naturalist-explorers and has

been found to occur within a wide range of taxonomic groups—from microbes to vertebrates—across the terrestrial, freshwater, and marine realms (Hillebrand 2004; Gaston 2007; Brown 2014). One would think that 'an almost universal pattern must have some common explanation' (Rohde 1992: 515), and yet, there is little consensus as to why diversity is greater in the tropics. At last count, more than 30 hypotheses have been proposed to explain the latitudinal diversity gradient (Willig et al. 2003; Lomolino et al. 2010; Brown 2014). We will cover a few of these hypotheses here.

To begin, we should note that latitude per se is not the actual driver of the latitudinal diversity gradient. Rather, latitude is correlated with a number of climatic and environmental factors that do contribute to species diversity (Gaston 2000; Hawkins et al. 2003). For example, much of the latitudinal gradient in species richness can undoubtedly be explained by the latitudinal gradient in ecosystem productivity (Hawkins et al. 2003; Brown 2014), which is greatest in the tropical regions of the world owing to higher annual insolation (the amount of solar radiation or 'energy' that strikes the Earth's surface) at the equator relative to the poles.[10] The annual amount of insolation influences

[10] The sun is the ultimate source of energy for nearly all life on the planet, but the amount of solar radiation reaching the Earth's surface varies by latitude because the planet is spherical and orbits

Figure 10.9 Latitudinal gradient in the diversity of terrestrial vertebrates. Species richness is highest near the equator (dark red and warmer colors) and declines toward the poles (blue end of the color spectrum).

Source: Mannion et al. (2014). Published under the CC-BY License. Figure created from Jenkins et al. (2013).

climatic factors, such as temperature and precipitation, which in turn influence biogeographic patterns of vegetation and net primary production (NPP), that is, the rate at which primary producers (vascular plants and aquatic algae) convert the sun's energy to biomass that can then be made available to other trophic levels.[11] Since March 2000, data obtained from the Moderate Resolution Imaging Spectroradiometer (MODIS) aboard NASA's Terra and Aqua satellites have been used to create weekly estimates of terrestrial NPP across the entire Earth's surface (**Chapter 4**). The latitudinal gradient in productivity is especially evident when comparing the seasonal changes in NPP that occur in the northern hemisphere: NPP is highest during the months of greatest insolation (June–August = summer), but then gradually declines as solar insola-

tion decreases during the winter months (December–February). The northern hemisphere thus exhibits a gradual 'green-up' in the spring (March–May), in which NPP appears to advance northward from the tropics to higher latitudes, only to retreat again in the fall (September–November).[12] By contrast, equatorial regions such as the Amazon Basin and Sub-Saharan Africa remain green throughout the year, although NPP does vary between wet and dry seasons in these areas (visit the NASA Earth Observations website to see these patterns for yourself: neo.sci.gsfc.nasa.gov).

The **productivity hypothesis** has thus long been a popular explanation for the latitudinal diversity gradient (e.g. Hutchinson 1959; Connell & Orias 1964; Pianka 1966; see Brown 2014 for a more recent review). To reiterate, tropical regions have greater productivity than temperate or polar regions, and thus a greater number of species across all trophic levels can be supported. Envision the flow of energy through trophic levels as a pyramid: the width of the base is the amount of energy made available by primary producers (e.g. NPP), only some of which (~10%) flows to the next trophic level of primary consumers (herbivores), a fraction of whose energy supports the next level of secondary consumers or predators, and so on. Less energy is available at each subsequent trophic level because some is used to support organisms in the previous level(s) and the transfer of energy between levels is inefficient. Given that NPP in terrestrial environments is limited primarily by climatic factors, such as precipitation and temperature, which in turn are influenced by annual insolation (i.e. energy from the sun), the productivity hypothesis can be viewed as a corollary of the **energy hypothesis**, which claims that the availability of energy is really what drives the latitudinal gradient in species richness (Currie 1991; Hawkins et al. 2003). Indeed, the **metabolic theory of ecology** views metabolism—the uptake and processing of energy by organisms—as a fundamental biological constraint that controls ecological processes at every level of organization, from individuals to ecosystems, and therefore influences rates of biomass production (NPP), trophic dynamics, and patterns of species diversity (Brown et al. 2004; Brown 2014).

The tropics are also more benign from a physiological standpoint, which again could account for the greater diversity found there. Higher latitudes, especially in the northern hemisphere, have undergone repeated cycles of glaciation, causing large regions of affected continents to become uninhabitable for extended periods of time.[13] The tropics have thus been more stable

the sun at a slight tilt (23.5º). Thus, the 'tropics' are defined as the zone around the equator where the sun's irradiance is perpendicular to the Earth's surface at least once during the year, which is delimited by the Tropic of Cancer in the northern hemisphere (23.5º N) and the Tropic of Capricorn in the southern hemisphere (23.5º S). The polar regions receive far less insolation than the tropics because the sun's rays strike these areas of the Earth's surface more obliquely, such that the amount of solar radiation received is spread over a larger area (= less energy per unit area). Furthermore, polar regions receive less insolation annually, given that the poles are tilted away from the sun for half the year and thus receive little or no sunlight at that time. The Arctic Circle delimits the northern polar region (66.5º N, which is located approximately 23.5º south of the North Pole), and the Antarctic Circle delimits the southern polar region (66.5º S, which is located approximately 23.5º north of the South Pole). The temperate zone thus lies at the 'mid-latitudes' between these polar and tropical zones.

[11] Recall from your general biology course that photosynthesis involves the 'fixation' (conversion) of carbon obtained from CO_2 in the atmosphere to produce carbohydrates (e.g. glucose), using water and light energy. These carbohydrates are then broken down during cellular respiration to produce chemical energy needed to power the cellular functions necessary for life. Net primary production (NPP) is thus the difference between the total amount of carbon that is fixed by primary producers (gross primary production, GPP) and the amount of carbon used by primary producers to meet their own metabolic needs (cellular respiration).

[12] The focus here on the northern hemisphere is simply because most of the Earth's land area is currently concentrated in the northern hemisphere, and thus the seasonal changes in terrestrial NPP are more evident there than in the southern hemisphere.

[13] There have been five major 'ice ages' over the past 2.4 billion years, which appear to be driven by cyclic changes in the Earth's orbit (Milankovitch cycles, which affect the eccentricity or shape of

climatically, providing favorable environmental conditions for tropical species throughout much of their evolutionary history (the **climatic stability hypothesis**; Pianka 1966).[14] As a result, tropical species tend to have narrower environmental tolerances than species that occur at higher latitudes (Janzen 1967; Stevens 1989). Their ecological niches are also smaller, as a consequence of greater specialization and a finer subdivision of habitat and other resources in the tropics, which allows more species to coexist (i.e. there is greater 'niche packing;' Pianka 1966; MacArthur & Levins 1967). Conversely, the degree of individual niche variation might actually be greater in the tropics; that is, species as a whole might have narrower niche breadths, but there is greater variability among individuals within a population along that niche axis, relative to species at higher latitudes (Araújo & Costa-Pereira 2013). If ecological diversity within species (intraspecific ecological diversity) also exhibits a latitudinal gradient, then this too could provide a source of variation that leads to greater speciation in the tropics. Further, if most lineages originated in the tropics, then climatic conditions may have constrained the geographical range expansion of many tropical species into more temperate areas, as only those species with broad environmental tolerances, perhaps owing to greater intraspecific niche variation, could expand their ranges northward as the continental glaciers retreated. The northern limits of many species' distributions today are constrained by their cold tolerance, for example, consistent with the idea that many tropical species have been unable to expand their ranges during the last interglacial owing to **niche conservatism** (i.e. they conserved their narrow environmental tolerances over evolutionary time; Wiens & Graham 2005; Wiens et al. 2010).

From an evolutionary standpoint, there may be more species in the tropics because of higher rates of speciation. Indeed, it has been suggested that the effect of higher temperatures on various physiological processes causes faster rates of mutation as well as a more rapid response to selection pressures (e.g. shorter generation times), resulting in faster evolutionary speeds, and hence, higher speciation rates in the tropics (the **effective evolutionary time hypothesis**; Rohde 1992; Gillman & Wright 2014; Puurtinen et al. 2016). Selection pressures induced by host or prey defenses, competition, predation, or parasitism are also expected to be greater in the tropics (Dobzhansky 1950; Pianka 1966). As diversity increases, so too do the interactions among species, which can result in a 'evolutionary arms race,' in which species continually evolve (or co-evolve) in response to their hosts, prey, competitors, predators, or parasites so as to maintain their own evolutionary fitness (the **'Red Queen' hypothesis**; Van Valen 1973).[15] In short, diversity begets diversity. This is also consistent with metabolic theory, in which the relatively high temperatures in the tropics generate and maintain high levels of diversity because of higher rates of biological activity leading to more—and faster—ecological interactions, which in turn increase rates of coevolution. In other words, 'the Red Queen runs faster when she is hot' (Brown 2014: 8).

The latitudinal gradient of biodiversity is so ancient and pervasive because the relationship of the Earth to the sun and the variation in solar energy input creates a gradient of environmental temperature. Temperature affects the rate of metabolism and all biological activity, including the rates of ecological interactions and coevolution. 'Diversity begets diversity' in the tropics, because 'the Red Queen runs faster when she is hot'.

Brown (2014)

Aside from overheated Red Queens and higher speciation rates, the tropics could have so many species because many groups of organisms originated there (the tropics as the **'cradle of life' hypothesis**; Stebbins 1974), and thus have been able to diversify

the Earth's orbit, its axial tilt, and its precession or 'wobble' as it spins on its axis). We are currently in the fifth of these major ice ages, known as the Quaternary glaciation period, which began about 2.58 million years ago. An ice age is defined by the presence of at least one perennial ice sheet (e.g. the Greenland and Antarctica ice sheets at present). Within a period of glaciation, there are multiple glacial and interglacial cycles during which ice sheets repeatedly advance and retreat. The most recent glacial maximum was reached during the Pleistocene about 24,500 years ago, when vast ice sheets covered northern Europe, Canada, and the northern third of the USA in the northern hemisphere, and the southern third of Chile and parts of neighboring Argentina in the southern hemisphere. We are currently in a warm interglacial period that began about 11–12,000 years ago (i.e. the start of the Holocene). Even so, we have experienced an extraordinarily rapid warming of global temperatures in recent decades, during what is now being viewed as a new geological epoch, the Anthropocene.

[14] Although the tropics were not buried under thick sheets of ice, climatic changes related to glacial-interglacial cycles also affected tropical regions, such as the Amazon Basin. During the last glacial maximum in the late Pleistocene, global temperatures dropped an estimated 5–6° C, which led to a drying and fragmentation of rainforests in Amazonia and may have created refugia that contributed to speciation in some groups (e.g. birds; Haffer 1969; Bonaccorso et al. 2006).

[15] The Red Queen Hypothesis was coined by American evolutionary biologist Leigh Van Valen, who took his inspiration from Lewis Carroll's *Through the Looking-Glass and What Alice Found There* (1871). Upon meeting in Looking-Glass Land, the Red Queen takes Alice by the hand and encourages her to run, faster and faster, only to find upon stopping that they are still in the same place. In response to Alice's confusion, the Red Queen explains the strange physics of her world (where everything is a mirror image to Alice's world): 'Now, here, you see, it takes all the running you can do, to keep in the same place.' In the context of an evolutionary arms race, species have to 'run' or evolve in response to the adaptive responses of other species so as to remain extant (i.e. 'to keep in the same place'). The faster rabbits run to escape their predators, the faster their predators need to be able to run in order to catch them.

over a longer period of time (the **time-for-speciation hypothesis**; Pianka 1966; Wiens & Donoghue 2004). Tropical regions once had a much greater geographical extent than they do today; the temperate zone began increasing in size around 30–40 million years ago. Larger geographic areas might be expected to have more species regardless of where they occur (the **geographic area hypothesis**; Blackburn & Gaston 1997), but if much of the world was once tropical, and for longer than it has been mostly temperate, then it stands to reason that more of the lineages present today would have originated in the tropics than in the temperate zone (Wiens & Donoghue 2004).

Diversification in some tropical regions, such as the Amazon Basin, may have been further enhanced by the dramatic landscape changes that followed the uplifting of the Andes Mountains, which includes the formation of the Amazon River system some 7–11 million years ago (Hoorn et al. 2010). Isolation of formerly continuous populations, whether by mountain ranges or major rivers, can lead to the formation of new species via allopatric speciation (**vicariance biogeography;** Wiley 1988). Diversification in at least some Amazonian bird groups appears to have occurred more recently, however (i.e. within the last 3 million years; Ribas et al. 2012; Smith et al. 2014). In that case, speciation may have been driven more by the existence of various landscape barriers to dispersal and gene flow (ecologically mediated vicariance; Smith et al. 2014; **Chapter 9**). Lineages with lower dispersal ability that have persisted sufficiently long on the landscape are more likely to diversify and have higher rates of speciation than those with good dispersal abilities. If true, then the main effect of the Andean orogeny was to create a topographically and edaphically complex, heterogeneous landscape in which subsequent species diversification occurred. Species diversity is generally expected to be greater in more heterogeneous landscapes, as well as in more structurally complex habitats such as tropical rainforests (the **habitat heterogeneity hypothesis**; MacArthur & MacArthur 1961; Pianka 1966; Tews et al. 2004). Once again, diversity begets diversity. Habitat heterogeneity by itself is not sufficient to explain the latitudinal diversity gradient, however, as strong latitudinal gradients in species richness can also be found among grasslands, wetlands, and other less heterogeneous or structurally complex habitats, such as within pelagic areas of the ocean (Hillebrand 2004).

Speciation is only part of the biodiversity equation, however. Regardless of where various groups originated, the tropics may simply have accumulated more species over time owing to lower rates of extinction (the tropics as **'museums of biodiversity' hypothesis**; Stebbins 1974; Gaston & Blackburn 1996; Chown & Gaston 2000). Of course, the tropics could be both cradles *and* museums of

biodiversity (high speciation rates coupled with low extinction rates; Mittelbach et al. 2007), as well as an engine that drives biodiversity elsewhere (the **'out of the tropics' hypothesis**, in which lineages originate in the tropics and then spread outward from there; Jablonski et al. 2006).[16] Ultimately, the generation and maintenance of diversity depends on species dynamics: how rates of colonization, speciation, and extinction vary with the number and interactions between species, which in turn drive coevolutionary processes (Brown 2014). In birds and mammals, for example, species turnover appears to have been greater at higher latitudes than in the tropics, owing to higher rates of speciation and extinction in temperate areas in the recent past (within the last million years or so), and which therefore may contribute to the observed latitudinal diversity gradient in these groups (Weir & Schluter 2007).

So, why are there so many species in the tropics? Despite a multitude of hypotheses, we have yet to arrive at a definitive answer. Some of these hypotheses offer different levels of explanation for essentially the same mechanism, and thus are not completely independent (Gaston 2000). It also seems likely that more than one mechanism could be responsible, especially when considering different groups of organisms, or even different scales in space and time (Hillebrand 2004). For example, the current latitudinal diversity gradient, in which diversity peaks in the tropics, has persisted for at least 30 million years, which is certainly a very long time from our perspective (Mannion et al. 2014). Yet, at other times during Earth's history, species richness appears to have peaked at temperate latitudes or exhibited no peak at all (i.e. a flattened gradient). Thus, the latitudinal diversity gradient we know today appears to occur during especially cold periods (recall that we are still technically in an ice age) that produce steep climatic gradients ('icehouse worlds'), whereas a temperate peak or flattened diversity gradient is characteristic of warmer periods in which conditions are more equitable worldwide ('greenhouse worlds'; Mannion et al 2014).

---

[16] Our own species (*Homo sapiens*) also originated in the tropics (i.e. East Africa) some 200–250,000 years ago, spreading out from there in multiple waves to populate Eurasia and Australasia beginning as early as 120,000 years ago and as recently as 50–60,000 years ago ('the Out of Africa' hypothesis; Groucutt et al. 2015). Interestingly, the expansion of anatomically modern humans into Europe and Asia may have been facilitated by the acquisition of beneficial traits through interbreeding (admixture) with other human species or subspecies that had previously colonized these areas, such as Neanderthals (*H. neanderthalensis*) and Denisovans (*H. sapiens* ssp. *Denisova*), enabling them to more quickly adapt to non-tropical environments. Humans today bear the genetic legacy of our interbreeding ancestors, as evidenced by the retention of Neanderthal alleles for the integument, immune function, and metabolism in people of non-African descent (Simonti et al. 2016). Such traits were presumably adaptive for dealing with the new diseases, lower insolation, and colder climates found at higher latitudes.

Although this may seem to point to the role of climate as the primary driver of latitudinal diversity gradients, that still leaves a lot of room for exploring how these patterns of diversity have emerged within the past 30 million years, especially in terms of what this might presage in the face of a rapidly warming planet.

Elevational Gradients in Species Richness

Mountains cover about a quarter of the Earth's land surface (Körner 2007), and thus are a distinctive and prominent feature of many landscapes (**Chapter 3**). Besides increasing topographic diversity, mountains also create a wide range of climatic conditions within a fairly localized area. As one ascends a mountain, the drop in temperature is perhaps the most noticeable change, with temperatures falling about 6.4°C for every 1000 meter gain in elevation. As a result, precipitation may fall as snow at sufficiently high elevations, with some mountains sporting snow-capped peaks year-round.[17] Otherwise, precipitation does not exhibit a predictable trend with increasing elevation; precipitation may either increase or decrease with elevation, or even peak at mid-elevations, depending on the region (Körner 2007). Slope aspect further contributes to climatic variability in montane landscapes, as south-facing slopes receive more insolation and are therefore warmer and drier than north-facing slopes. Finally, mountains can interact with prevailing winds and influence precipitation patterns on a region-wide and even continental scale. Mountain ranges contribute to the **rain shadow effect**, in which warm, moist air cools and condenses as it rises up and over the mountain, causing more precipitation to fall on the windward side than on the leeward side (the 'shadow') of the mountain. Because of the rain shadow effect, vegetation zones found at particular elevations may vary depending on which side of the mountain or range they occur.

[Mountains] differ in age, size, historical stability, climate regimes and topographic complexity making them excellent natural laboratories to determine how environmental, geographic and biotic factors interact to promote diversification and maintenance of mountain flora and fauna.   Graham et al. (2014)

Vegetation zones thus provide the clearest example of an elevational diversity gradient, in which different types of vegetation form a seemingly ordered sequence up a mountain's flanks. Indeed, it was the study of elevational zonation in vegetation communities that led German geographer Carl Troll to first propose the ecological study of landscapes (Troll 1939), which as you'll recall, was the genesis for the modern field of landscape ecology (**Chapter 1**). That mountains exhibit elevational zonation in vegetation had long been recognized, however, at least since the pioneering work of German geographer and naturalist Alexander von Humboldt, the 'father of phytogeography,' in the early 19th century. Later that same century, the American naturalist, Clinton Hart Merriam, developed the **life-zone concept** to describe the elevational zonation of plant communities in western North America (Merriam 1890; **Figure 3.2**). Merriam drew parallels between the elevational and latitudinal zonation he had observed among major vegetation types in western North America, which is readily apparent in the names he assigned to these life zones. For example, spruce-fir forests similar to the high-latitude boreal forests of Canada can be found at high elevations in the mountains of northern Arizona, such as the San Francisco Peaks (**Figures 3.1 and 3.2**), and thus Merriam called this vegetation zone the 'Canadian Life Zone.' From a historical standpoint, it is also worth noting that many of the now-classic studies in ecology concerning the distribution and turnover of species along environmental gradients (i.e. β diversity) were performed along elevational gradients, most notably in the seminal work of American plant ecologist Robert Whittaker, who quantified how vegetation changed along a topographic moisture gradient in the Great Smoky Mountains (Tennessee, USA) and in the Siskiyou Mountains of the Pacific Northwest (Whittaker 1956, 1960).

Temperature and precipitation are clearly important for understanding patterns of vegetation, which is why vegetation zones are delineated in terms of moisture gradients (xeric to mesic) or humidity provinces (**Figure 3.3**). Given that temperature in particular is influenced by elevation (Körner 2007) and temperature influences primary production, a major driver of the latitudinal gradient in species richness, can we expect a similar diversity gradient with respect to elevation? That is, does species richness decrease with increasing elevation, in much the same way that diversity decreases with increasing latitude? There are far fewer species that live above the snow line at the top of a mountain than at lower elevations, for example, just as there are fewer species that live at the poles than at the equator.

Elevational diversity has long been thought to mirror the latitudinal diversity gradient (Rahbek 1995). The life-zone concept is an obvious example of this. However, the relationship between species richness and elevation is complicated by latitude and other

---

[17] The 'snow line' is the elevation at which snow and ice are retained throughout the year, which is influenced by latitude and other factors, such as slope aspect and proximity to a coastline and the moderating influences of the ocean. To illustrate, the snow line near the equator occurs at about 5500 m on Mount Kilimanjaro in Tanzania (3° S), but farther south the snow line drops to 2500 m on the North Island of New Zealand (37° S), which then falls to 1600 m on New Zealand's South Island (43° S), and reaches its lowest point—sea level—in Antarctica (snow level = 0 m, 70° S).

'regional peculiarities' (Körner 2007). Although temperatures vary with elevation in the tropics (albeit to a lesser degree than at higher latitudes), annual temperatures are still fairly constant at a given elevation, whereas a montane site at higher latitudes in the temperate zone experiences large seasonal fluctuations in temperature. As a result, there is almost no overlap between the temperature ranges of high and low elevations in the tropics, whereas temperatures are more likely to overlap between the higher and lower elevations of temperate mountains. Tropical mountains may thus pose a greater physiological barrier to dispersal than temperate mountains of the same height, such that 'mountain passes are higher in the tropics' as American evolutionary ecologist Daniel Janzen explained in his classic paper (1967: 243):

> ...mountains are higher in the tropics figuratively speaking; they are harder to get over because, for a given elevational separation, the probability is lower in the tropics that a given temperature found at the higher elevation will fall within the temperature regime of the lower elevation than is the case in the temperate area.

Recall that species are expected to have narrower environmental tolerances in the tropics, which means that they would have a more difficult time dispersing or expanding their range if they are constrained to a particular temperature range, and hence, elevation. Thus, if mountains are more likely to represent dispersal barriers to tropical species than to temperate species, this could lead to greater isolation and diversification in tropical montane species. Still, the tropics in general are expected to support higher diversity, so to what extent is diversity enhanced by elevational gradients?

In birds, at least, latitudinal and elevational gradients both appear to be driven by climatic factors, just over different spatial scales (Ruggiero & Hawkins 2008). Mountains create elevational climate gradients, but they also increase topographic complexity, which can also increase isolation and lead to greater diversification among species in montane landscapes. To tease these factors apart, Ruggiero and Hawkins (2008) used causal models (path or structural equation models), which we discussed previously in the context of landscape genetics (**Chapter 9**), to evaluate the relative effects of topography versus climate on bird species richness across the Americas. After controlling for the latitudinal climate gradient that underlies the latitudinal gradient in bird diversity, they found that elevational climate had a greater effect than topographic complexity on species richness, but in different ways for montane versus lowland species: lowland species richness decreases, but montane species richness increases, with an increasing range in temperature across elevations. That is, montane bird diversity increases sharply in response to the steep climatic gradients typical of mountains, whereas lowland bird diversity varies gradually as a result of a broader and shallower climate gradient. Climatic gradients rather than topographic complexity per se drive bird diversity. Thus, mountains appear to have more bird species for the same reason that the tropics do: birds are responding to climatic gradients, albeit at a more local scale than for latitudinal gradients.

Even if species richness is influenced by elevational climate gradients, we still have not answered the question of whether diversity generally declines with increasing elevation, as it does with increasing latitude. A review of the literature suggests that it often does, although the pattern of decline is not always a linear one (Rahbek 1995, 2005). Instead, elevational patterns of species richness often exhibit a 'hump-shaped' distribution, in which the number of species peaks at mid-elevations (**Figure 10.10**). The observed elevational diversity pattern is highly sensitive to scale, however. A linear decline with increasing elevation is more likely to be observed when the sampled elevational gradient is too shallow or has been truncated (e.g. samples span <2000 m in elevation), whereas a hump-shaped diversity distribution is typically found for elevational gradients encompassing a larger spatial extent (Rahbek 2005; Nogués-Bravo et al. 2008). A similar pattern is found when analyzing species diversity patterns at different scales (grain sizes), with hump-shaped distributions more likely to be found at finer sampling resolutions or scales of analysis (e.g. <0.1–1 km^2; Rahbek 2005). Thus, insufficient sampling along an elevational gradient, either by covering too narrow an elevational range or adopting too coarse a scale of sampling or analysis, can miss the 'hump' in the distribution and give the appearance of a linear decline in species richness with increasing elevation (Nogués-Bravo et al. 2008; **Figure 10.10**).

As a result of this scale sensitivity, altitudinal gradients appear to be an excellent choice to study the effects of scale on patterns of species richness.   Rahbek (2005)

Besides sampling effort, geographic **area effects** might also influence the appearance of the elevational diversity gradient. Depending on the steepness of the terrain, the vegetation and climatic zones at higher elevations tend to occupy a more narrow or smaller areal 'band' than those at lower elevations (e.g. envision a mountain as a pyramid, in which the base is much wider and larger overall than the area at the peak). Subsequently, species richness could be inflated at low elevations, giving rise to the appearance of a

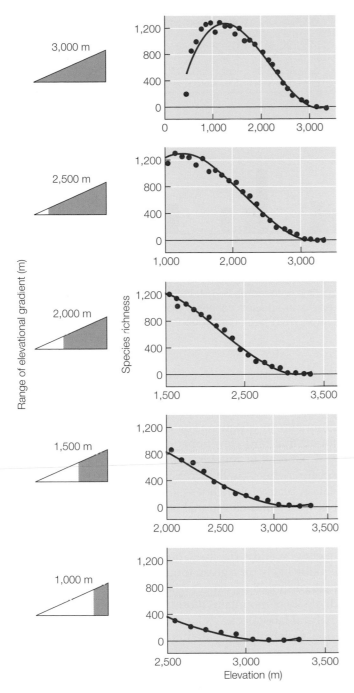

Figure 10.10 Effect of spatial extent on the appearance of the elevational diversity gradient. Plant species richness in the central Pyrenees of Spain exhibits a hump-shaped distribution when the entire elevational range is sampled (i.e. over 3,000 m; top). The distribution becomes increasingly linear as the range is truncated to exclude more of the low-elevation samples, however (white areas of the triangular gradients depict the range of samples being excluded). In this system, changing the spatial grain of analysis had little effect on the resulting diversity pattern (results for the 1 km² grain size are shown here).

Source: After Nogués-Bravo et al. (2008).

linear decline with increasing elevation, simply because larger areas support more species than smaller areas (see ***Species–Area Relationships***).[18] After correcting for area effects, elevational diversity gradients often reveal a hump-shaped distribution in species richness with increasing elevation (Rahbek 2005).

Why the elevational diversity gradient might be humped-shaped rather than linear is another question, and one that does not yet have a good answer (e.g. Graham et al. 2014). It could be that mid-elevations, which by definition are located in the middle of an environmental gradient, tend to provide the optimum conditions in terms of temperature, precipitation, and productivity for the greatest number of species in a given region (a manifestation of the Goldilocks principle,[19] in which conditions are 'just right' for most species, being neither too hot or dry nor too cold or wet). For example, the relationship between plant diversity and productivity was first proposed by British plant ecologist J. Philip Grime to exhibit a hump-shaped distribution, in which plant diversity peaks at intermediate levels of productivity (Grime 1973b). In less-productive ecosystems, species richness is limited by environmental stresses, such as insufficient water, which few species can tolerate. At the other extreme, competitive exclusion by a few highly competitive species limits diversity in highly productive ecosystems. Thus, plant species diversity should be maximized at intermediate levels of productivity.[20] Although not specific to elevational gradients per se, the hypothesized hump-backed model for the productivity-diversity relationship might still apply if there is an

[18] A similar suggestion has also been made regarding latitudinal diversity gradients. The spherical shape of the Earth necessarily produces latitudinal bands that differ in circumference: the circumference at the equator is about 40,000 km, whereas the circumference at the polar circles is 17,662 km. The polar regions (between the poles and the polar circle) thus encompass about 8% of the Earth's surface area, whereas the tropics (the region between the Tropics of Cancer and Capricorn) cover about 40% of the Earth's surface area. All else being equal, larger areas should have more species than smaller areas.

[19] The reference comes from a children's fairy tale, 'Goldilocks and the Three Bears,' in which a little girl (Goldilocks) enters the forest home of a bear family while they are away. She makes herself at home, eating their food and sitting on their furniture, and eventually, falls asleep on the bed that is neither too hard nor too soft, but 'just right.' The bears are understandably outraged to find an intruder in their bedroom, but Goldilocks manages to escape unharmed by jumping out a window and running away.

[20] We have discussed this previously in the context of the Intermediate Disturbance Hypothesis (**Chapter 3**). As with most general hypotheses in ecology, however, there has been a good deal of debate in the literature as to whether the productivity-diversity relationship generally exhibits a unimodal or hump-shaped distribution (e.g. Adler et al. 2011; Fridley et al. 2012; Fraser et al. 2015). Detection of the 'hump' may once again depend on sampling effort and methodology, the spatial grain and extent of the analyses, the range of site productivities sampled, how productivity is measured, and the type of plant community considered (Fraser et al. 2015).

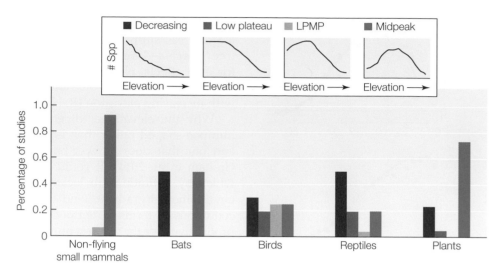

Figure 10.11 Patterns of elevational diversity.   Global analyses of elevational gradients for plants and different vertebrate groups revealed four general patterns, ranging from a monotonic decline with increasing elevation, to low-elevation plateaus, to low-plateau mid-elevational peaks (LPMP), to mid-elevational peaks in species richness. Some patterns predominate in certain taxonomic groups (e.g. non-flying small mammals exhibit a strong mid-peak distribution), whereas others show all four patterns in roughly equal proportions (e.g. birds).
Source: After McCain & Grytnes (2010).

elevational gradient in climatic conditions that causes productivity—and hence diversity—to be greater at intermediate elevations.

Conversely, spatial constraints imposed by the base and summit of a mountain may limit the distribution of species, such that species having large and intermediate-sized ranges will necessarily overlap in the center of the gradient, creating a peak in species richness at mid-elevations. This **mid-domain effect** demonstrates how a peak in species richness can arise even in the absence of a climatic or environmental gradient, simply as a random association between the size and placement of species' ranges along an elevational (or latitudinal) gradient (i.e. this is a null model of species' distributions; Colwell and Hurtt 1994; Colwell et al 2004).[21] Although there has been some support for a mid-domain effect on the elevational diversity of plants, ants, and small mammals (Sanders 2002; Grytnes & Vetaas 2002; McCain 2004), the effect may be attenuated by climate, area effects, and scale (e.g. whether diversity is being assessed at a local or regional scale; McCain 2005). For example, peaks in the gamma diversity (γ) of small mammals were generally shifted toward the lower third of the elevational gradient, which suggests that area effects may constrain species richness at a regional scale. By contrast, peaks in alpha diversity (α) were shifted toward higher elevations above the mountain midpoint, which may indicate that climatic constraints have a greater effect on species richness at a local site scale (McCain 2005).

Then again, different types of elevational diversity patterns may emerge, depending on the specifics of the mountain in question. For example, a global analysis found that bird elevational diversity follows four general patterns in roughly equal measure: (1) decreasing species richness with increasing elevation; (2) a low-elevation plateau in richness; (3) a low-elevation plateau, but with a mid-elevational peak (LPMP); and (4) a mid-elevational peak in richness (McCain 2009; **Figure 10.11**). The fact that these four patterns occur with equal frequency worldwide does not mean that they are all equally likely for a given mountain, however. But which pattern should we expect where? The sheer number and diversity of mountains worldwide makes them excellent 'natural laboratories' (Graham et al. 2014) that can provide a globally replicated and therefore 'powerful test system' for evaluating among the various drivers of elevational diversity patterns (McCain 2009).

Of the various hypotheses proposed (e.g. climate, area effects, mid-domain effects), it appears that elevational gradients in bird diversity are best explained by an **elevational climate model** that incorporates a linearly decreasing temperature gradient (recall that temperature decreases fairly reliably with increasing elevation) and a unimodal water availability gradient (McCain 2007; **Figure 10.12**). According to this model, species richness should peak at mid-elevations in arid and/or temperate mountains because diversity is limited by water availability at low elevations and by extreme cold at high elevations. In contrast, species richness should either decrease with increasing elevation or exhibit a low-elevation plateau in humid and/

[21] The mid-domain effect is also one of the 30+ hypotheses that have been proposed to explain latitudinal diversity gradients.

or tropical mountains, where precipitation is plentiful and diversity is constrained more by the temperature gradient (**Figure 10.12**). Consistent with expectations, mid-elevational peaks in bird diversity were typically found in arid regions (e.g. southwestern USA), whereas all of the decreasing diversity gradients and almost all of the low-elevation plateaus in species richness were found in humid regions like the tropics (McCain 2009). Furthermore, these elevational diversity patterns were seen across different types of bird assemblages (e.g. breeding birds, forest birds, all birds combined), and within and among different biogeographical regions (McCain 2009).

We thus have come full circle in our discussion of elevational diversity gradients. Elevational gradients in climate impose physiological or energetic constraints that set boundaries on the elevational ranges of species, especially if these same conditions also contribute to resource limitations (Parmesan 2006; McCain 2009). The optimum conditions for species diversity

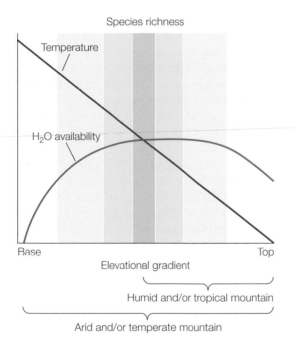

Figure 10.12 Elevational climate model. Patterns of elevational diversity in some groups, such as birds and bats, are best explained by an elevational climate model that incorporates a linearly decreasing temperature gradient in conjunction with a unimodal water availability gradient. Species richness is expected to peak at mid-elevations in arid and/or temperate mountains because diversity is limited by water availability at low elevations and by extreme cold temperatures at high elevations. Conversely, species richness should either decrease or exhibit a low-elevation plateau with increasing elevation in humid and/or tropical mountains where precipitation is plentiful and diversity is constrained more by the temperature gradient. Elevational diversity gradients thus exhibit different patterns depending on the specific climatic or biogeographic region where the mountain occurs.

Source: After McCain (2007).

thus vary as a consequence of the interaction between climate and elevation, which has important implications for the ability of montane species to respond to climate change. As global temperatures increase, the optimum climatic conditions for many montane species are shifting upwards, which means that species will need to shift their ranges to higher elevations, if they can (Parmesan 2006; **Chapter 8**). In the mountains of Western Europe, for example, the optimum elevation for forest plant species (defined as the elevation within a species' range where it has the highest probability of occurrence) has shifted upward by 29 m/decade over the past 30 years (Lenoir et al. 2008). Significantly, this upward shift in species distributions coincides with a temperature shift in the region: mean annual temperatures have consistently been 'above normal' since 1986. Subsequently, more than two-thirds of plant species (118/171) have shifted their optima upward, whereas only one-third (53/171) shifted their optima downward (Lenoir et al. 2008).

Species are thus not shifting their distributions in unison, as intact assemblages. There is wide variability in the magnitude of optimum elevation shifts, which could end up disrupting biotic interactions (e.g. plant-pollinator interactions) and the ecological networks in which these species are embedded (Lenoir et al. 2008). In addition, there are geometric limits to how far upwards a montane species can shift its range before it reaches the summit. Montane species are thus at an increased risk of extinction from a warming climate, especially in the tropics where species have narrower elevational ranges and environmental tolerances, and where climate change is likely to produce novel climates rather than just upward shifts in current climatic conditions (Williams et al. 2007). A high proportion of tropical species may thus face **range-shift gaps** between their current and projected elevational ranges, which would exacerbate the projected loss of diversity in the wet tropics from global warming (Colwell et al. 2008).

Species–Area Relationships

Area effects are one of the hypotheses that have been proposed to explain gradients in elevational and latitudinal diversity. All else being equal, larger areas should have more species than smaller areas. Larger areas tend to have greater habitat diversity, which can therefore host a greater variety of species than smaller areas (the habitat heterogeneity hypothesis; Williams 1964). Population sizes are also expected to be larger, which helps to reduce species' extinction risk, thereby contributing to greater diversity in large versus small areas (the area per se hypothesis; Simberloff 1976). Conversely, large areas are more likely to receive more colonists than small areas, leading to a greater accumulation of species (the passive sampling hypothesis; Connor & McCoy 1979).

Species–area effects are not limited solely to broad-scale diversity gradients, however. The **species–area relationship** (SAR) is arguably the most ubiquitous scaling relationship in ecology, having been called a 'general rule' and even a 'law' of ecology (Lawton 1999; Lomolino 2000). The first empirical species–area curve is attributed to the renowned British botanist and plant geographer, Hewett C. Watson. In 1859, Watson illustrated how the number of plant species increased as a function of the size of the area surveyed, from the finest scale of a small plot within the county of Surrey to the whole of Great Britain (Rosenzweig 1995: 9).

Some 60 years later, Swedish ecologist Olof Arrhenius published a mathematical description of a SAR, in which plant species of the Stockholm archipelago were found to increase as a power of island size (Arrhenius 1921). Recall from **Chapter 2** that a power-law function has the general form $y = ax^k$ (Equation 2.1; **Figure 2.13**). Thus, if species richness (S) increases as a power of island area (A), we can represent the SAR as

$$S = cA^z ,\qquad \text{Equation 10.10}$$

where c is the number of species when $A = 1$, and z is the scaling exponent that specifies the rate at which species richness increases as a function of area. Because c and z are obtained by fitting an empirical species–area curve, a common practice is to take the logarithmic or natural log form of Equation 10.10 so as to linearize this relationship (**Figure 10.13C**), thereby making it easier to fit using ordinary least-squares regression. For example, the logarithmic form of Equation 10.10 is:

$$\log_{10} S = \log_{10} c + z \log_{10} A.$$

You'll note that this is the equation of a line: $y = mx + b$, where m is the slope and b is the y-intercept. Thus, z is the slope of the linearized species–area relationship, whereas c is its y-intercept (**Figure 10.13C**).[22] The slope of the species–area relationship (z) varies among biomes and taxonomic groups, but generally lies within the domain of $0.1 \le z \le 0.4$ (Rosenzweig 1995; Durrett & Levin 1996). For example, the scaling of plant species richness has been shown to vary among different biomes, from $z = 0.11$ within deserts and xeric shrublands to $z = 0.33$ in the tropical rainforests of Central America (Kier et al. 2005).

Although the power function (Equation 10.10) is the most commonly employed SAR model, it is by no means the only model. At least 20 different functions have been proposed for modeling SARs, which vary in their general form and complexity (having from two to four parameters), as well as in their theoretical origins and justification (Dengler 2009; Tjørve 2009; Williams et al. 2009; Triantis et al. 2012). One of the main points of contention is whether the SAR possesses an asymptote, and thus what functional form the species–area curve actually takes. For example, the power function (Equation 10.10) increases without bound, and thus the species–area curve based on this function is convex upward without an asymptote. Some have argued that the species–area curve must have an upper asymptote, corresponding to the size of the regional or biogeographic species pool, and suggest that the SAR is instead sigmoidal in shape (Lomolino 2000), leading others to point out that even if there is an upper limit to the number of species in a given area, the species–area curve need not approach this limit asymptotically (Williamson et al. 2001). Mathematical functions that lack asymptotes, such as the power function, can still have limits. For example, the SAR is necessarily limited at both ends of the curve, whatever its form: neither species nor area can be a negative number ($S \ge 0$ for all $A > 0$), and both area and species richness are finite at some scale (e.g. species are limited to this one planet, as far as we know).

TYPES OF SPECIES–AREA RELATIONSHIPS Ultimately what the debate over the form of the SAR reveals is that there is more than one type of SAR. Different sampling designs give rise to different types of SARs (Scheiner 2003; Whittaker & Triantis 2012). Some sampling schemes are more efficient than others in a particular setting (Scheiner et al. 2011), but because we can sample area in different ways, there are different ways we can construct species–area curves.

Whilst the increase in species richness with area is general, it is not universal to all data sets and circumstances. What form the relationship takes in particular circumstances is of key concern both for what it reveals of the factors controlling diversity pattern and for the predictive value of [species–area relationships] in relation to the biodiversity consequences of anthropogenic change (habitat alteration and destruction, climate change, etc.).
Whittaker & Triantis (2012)

For example, we could adopt a nested sampling design, in which we count the total number of species encountered within areas of progressively larger size (**Figure 10.13A**). Each data point is thus a single measure of species richness for that particular area. Conversely,

[22] If we look at the power-law form of the SAR (Equation 10.10), it is apparent that c is not really an intercept, but rather, the 'initial slope' of the relationship between S and A^z (Rosenzweig 1995; Lomolino 2001). That is, c helps determine how quickly the empirical species–area curve rises as a function of increasing area. For example, species–area curves from different regions could all have the same slope (z values), but very different c values depending on their initial species richness (e.g. areas of high productivity will have higher initial richness, and thus a higher c value, than areas of low productivity). Thus, both z and c are needed to describe (and compare) species–area curves (Rosenzweig 1995).

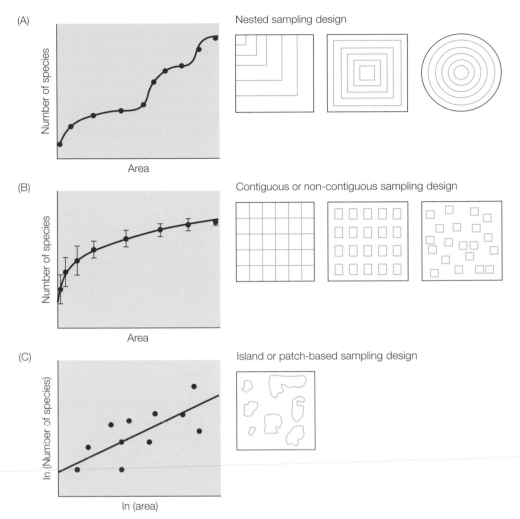

(A)   Nested sampling design

Number of species

Area

(B)   Contiguous or non-contiguous sampling design

Number of species

Area

(C)   Island or patch-based sampling design

ln (Number of species)

ln (area)

Figure 10.13 Different types of species–area relationships. Species–area curves can be generated in different ways, depending on the sampling design. (A) Nested sampling designs produce species-accumulation curves that increase as an irregular staircase, depending on where a habitat boundary is crossed, and each data point represents the accumulated richness across the nested subset of areas up to that point. (B) Contiguous or non-contiguous sampling designs, such as a grid-based survey using quadrats that are either adjacent or not, produce a smoothly increasing species-accumulation curve, in which each point is the mean richness of samples that are aggregated to encompass increasingly larger areas. (C) Island or patch-based sampling designs, in which each sampling point is the richness of the entire patch or island area.

Source: After Scheiner (2003).

we may have sampled species richness at some fixed grain size (e.g. within quadrats or study plots), using either a contiguous or non-contiguous sampling design, such that we can aggregate samples over an increasing range of scales to obtain a mean species richness for these different-sized areas (**Figure 10.13B**).[23] Regardless

[23] Scheiner (2003) further subdivides SARs derived from contiguous and non-contiguous sampling schemes into two types, depending on whether only adjacent or neighboring samples within a given-sized area are being averaged together (as in a moving window analysis; **Figure 4.27**), or whether all samples within a given distance (i.e. all possible combinations among samples within that range) are averaged to construct the SAR. These are referred to as 'Type A' and 'Type B,' respectively, by Scheiner (2003). Type A species–area curves are spatially explicit because their construction is constrained by the spatial arrangement of the sampling units (i.e. to the adjacent or neighboring samples), whereas Type B curves are not (Scheiner et al. 2011).

of the specific sampling scheme (nested, contiguous, or non-contiguous), these SARs are all constructed as the cumulative number of species across samples that have been aggregated to produce different-sized areas. These SAR types are thus referred to as aggregate species richness relationships (Scheiner et al. 2011) or species accumulation curves (Whittaker & Triantis 2012). These differ from SARs that are generated as the total number of species surveyed within discrete patches that vary in size (e.g. oceanic islands or habitat fragments), which are referred to as independent species richness relationships (Scheiner et al. 2011) or island species–area relationships (Whittaker & Triantis 2012; **Figure 10.13C**). The different SAR types thus vary primarily in their **focus**, which is defined as the cumulative area or inference space represented by each data point (e.g. whether each point in the species–area curve represents a single value or a mean value; Scheiner et al. 2000; Scheiner 2003). This has implications for the estimation and interpretation of diversity measures, as we'll discuss in a moment.

Different SAR types also vary in the spatial scales encompassed by the analysis (Scheiner 2003; Scheiner et al. 2011). For all types, the spatial grain is either the size of the sampling unit used to measure species richness (e.g. the dimensions of the sampling quadrat or study plot) or the standardized unit to which all data are adjusted prior to analysis. All data should obviously be adjusted to a common grain size prior to making comparisons among sites or studies (**Chapter 2**). For example, if one study has a sampling grain of 1 m² but another used 10 m², then the former dataset should be adjusted by aggregating samples to give the same 10 m² resolution for both studies. The different types of SARs therefore differ in how spatial extent, the broadest scale encompassing all of the sampling units, is defined. Depending on the type of SAR, this is either the largest area or cumulative area sampled (for nested and contiguous sampling designs) or the maximum distance among sampling units (for non-contiguous sampling designs and island species–area curves; Scheiner et al. 2011). A failure to recognize differences in the spatial grain and extent encompassed by different types of SARs can compromise our ability to make comparisons among studies, and could lead to incorrect conclusions about what aspects of diversity these different SARs measure or can be used to estimate, as well as what processes might be responsible for the observed patterns of diversity in the first place.

IMPLICATIONS FOR THE ESTIMATION AND INTERPRETATION OF DIVERSITY One of the reasons the type of SAR matters is that it affects how different aspects of diversity (α, β, and γ diversity) are measured and interpreted (Tuomisto 2010b; Scheiner et al. 2011). Although all SARs measure species richness and how it changes across spatial scales, different sampling designs estimate different components of diversity or do so in different ways.

For example, SARs constructed from a nested sampling design have only a single measure of richness (S) for each unit of area (A), but is this a measure of the local (site) diversity (α) or of the total diversity (γ) in that area? It has been argued that since the size of the sampling unit (i.e. the grain) is the same as the focus of the analysis in this case (grain = focus), nested SARs are really measuring how total species richness (γ) increases as one increases the size of the area being surveyed (Scheiner et al. 2011). In other words, α diversity is not being measured, and therefore cannot be estimated from a SAR based on a nested sampling design (**Figure 10.13A**). A similar argument can be made for

island species–area curves, in which each data point represents a single measure of species richness for that particular island or patch area. As a result, it has been argued that island SARs provide a measure of how γ diversity scales with area (Scheiner et al. 2011; **Figure 10.13C**).[24] Other investigators, however, define γ diversity as the total species richness observed over the entire collection of islands (or patches), and then estimate α diversity by obtaining an average S across all islands or patches regardless of size (e.g. Crist & Veech 2006).

For contiguous and non-contiguous sampling designs, each data point represents the mean species richness for a given area (**Figure 10.13B**), but again, are we estimating the mean local site diversity ($\bar{\alpha}$) or the mean total diversity ($\bar{\gamma}$) in that area? In this case, it depends on how we calculate the mean, in terms of whether we base this only on adjacent or neighboring sampling units within a specified area (a spatially explicit or 'Type A' curve), or whether we derive a mean value based on a collection of samples within some specified area or distance (non-spatially explicit or 'Type B' curves; Scheiner et al. 2011). Contiguous and non-contiguous sampling designs that are spatially explicit (Type A curves) are estimating how $\bar{\alpha}$ diversity increases with sampling scale (i.e. increasing grain size; Tuomisto 2010b), whereas their non-spatially explicit counterparts (Type B curves) are estimating how $\bar{\gamma}$ diversity increases with the size of the area being surveyed (i.e. involving an increase in focus, the area over which sampling units are aggregated; Scheiner et al. 2011). Thus, to quote Scheiner and colleagues (2011: 202): 'The conceptual difference between α- and γ-diversity is thus tied to the relationship between the area of the grain and the focus. If the grain is less than the focus, average species richness is an estimate of α-diversity; if the grain equals the focus, average richness is an estimate of γ-diversity.'

That brings us to whether or how β diversity can be estimated from the SAR. One of the greatest misconceptions is that the slope (z) of the SAR is a measure of β diversity. Although the slope can be used to estimate components of β diversity (e.g. species turnover), it is not actually an estimate of β diversity (Tuomisto 2010b; Scheiner et al. 2011). To understand why, consider that the slope gives the rate of change in either α or γ diversity as a function of area, depending on the type of SAR. Although β diversity has been defined in different ways (**Figures 10.3 and 10.5**), one way to think about it is as the effective number of sampling units (communities) necessary to give the total richness found in an

---

[24] Unlike nested sampling designs, islands are independent sampling units, which may or may not be drawing from the same regional species pool (e.g. if islands or patches are very far apart). The resultant SAR may or may not describe how total diversity (γ) changes with an increase in area, therefore (Scheiner et al. 2011).

area (ɣ diversity), given that each sample consists of a certain number of species (local richness, α) with a mean diversity equal to $\bar{\alpha}$ (Tuomisto 2010a). Thus, one way to estimate β diversity is as a ratio of ɣ and $\bar{\alpha}$ diversities, $\beta = \gamma/\bar{\alpha}$, which as you'll recall, derives from Whittaker's (1960) original suggestion that ɣ diversity is a product of both α and β diversities.

Alternatively, diversity can be partitioned additively into its different components (Crist & Veech 2006). To illustrate the additive partitioning of the SAR, Crist and Veech (2006) first defined α diversity as the mean richness observed within habitat patches, averaged over all patches regardless of size ($\bar{\alpha}$). They defined ɣ diversity is the total species richness found across all patches. The β component of diversity is then determined by subtraction as $\beta = \gamma - \bar{\alpha}$. Because species richness is expected to vary as a function of area, however, it is possible to assess how much of the total β diversity is due to the effect of area (β_{area}) and how much is due to species 'replacement' ($\beta_{replace}$); that is, differences in species composition among patches (species turnover). The area effect is estimated as the mean deviation between the species richness of the largest patch (S_{max}) and each of the n smaller patches (S_i) as:

$$\beta_{area} = \frac{1}{n}\sum_{i=1}^{n}\left(S_{max} - S_i\right). \qquad \text{Equation 10.11}$$

Thus, the additive partition of the SAR can be represented as:

$$\gamma = \bar{\alpha} + \beta_{area} + \beta_{replace}. \qquad \text{Equation 10.12}$$

Using a power-law function (Equation 10.10) to model the SAR for 179 moth species (ɣ) inhabiting forest fragments within an agricultural landscape, Crist and Veech (2006) demonstrated how β diversity can be calculated via additive partitioning (**Figure 10.14**). The average species richness across all fragments ($\bar{\alpha}$) was 66.7 species, which is 37% of the total species richness (ɣ). Using Equation 10.11, they estimated β_{area} to be 23.8 species, which is only 13% of the total richness. Rearranging Equation 10.12, we can solve for $\beta_{replace}$: $\beta_{replace} = \gamma - \bar{\alpha} - \beta_{area} = 179 - 66.7 - 23.8 = 88.5$ species, or about 50% of the total richness (**Figure 10.14**).

Thus, even though species richness varied predictably with area (i.e. the power-law SAR explained 73% of the variation in species richness among fragments), forest fragments have very different moth assemblages (a large $\beta_{replace}$) due to factors other than area. In this case, it appears that differences in the tree species composition of forest fragments (habitat heterogeneity) may be responsible for the observed differences in moth assemblages among fragments. Importantly, the observed slope of the SAR ($z = 0.146$) would have underestimated the total β diversity for moths in these forest fragments by 79%, assuming an additive

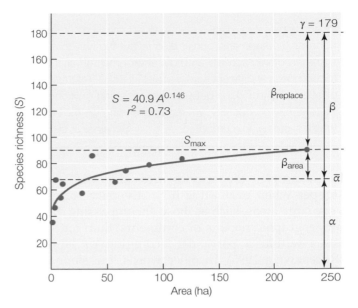

Figure 10.14 Additive partitioning of the species–area relationship. Components of diversity can be estimated from the SAR through additive partitioning. A total of 179 moth species (ɣ diversity) were found within forest fragments in an agricultural landscape, with fragments averaging 66.7 species ($\bar{\alpha}$ diversity). Although area explained 73% of the variation in species richness ($r^2 = 0.73$), only 13% of the total β diversity is due to area effects ($\beta_{area} = 23.8$ species). Thus, about half of the total richness is due to effects other than area ($\beta_{replace} = 88.5$ species), most likely to differences in tree species composition among forest fragments (i.e. habitat heterogeneity).

Source: After Crist & Veech (2006).

definition of β diversity (and multiplicative partitioning would have underestimated total β diversity by 49%; Crist & Veech 2006). Put another way, the slope of the species–area curve would have to be considerably larger ($z = 0.578$) to give the observed level of β diversity in this system. To reiterate, the slope (z) of the SAR does not provide a direct measure of β diversity.

SCALE-DEPENDENCE OF THE SPECIES–AREA RELATIONSHIP   The SAR is often extrapolated across scales to predict the relationship between species richness and habitat area in different spatial contexts (e.g. extrapolating across scales from study plots to biomes; Harte et al. 2009). This assumes that the slope of the relationship (z) is constant across the range of scales considered or to which it is being applied. The slope of the SAR is known to be scale-dependent, however, exhibiting different scaling domains that presumably reflect different processes or constraints on how species richness varies as a function of area (Palmer & White 1994; Rosenzweig 1995; Lomolino 2000; Turner & Tjørve 2005).[25]

[25] Although Rosenzweig (1995: 21–22) claimed that z is not scale-dependent, he was referring to the fact that z values do not

In addition to providing a standardized measure of species richness, species–area curves also reflect the way that diversity is structured spatially and how environmental variables affect richness at different spatial scales. If scale 'matters', then observed relationships between richness and environmental factors, say productivity, will vary depending on the scale at which systems are compared.   Scheiner et al. (2000)

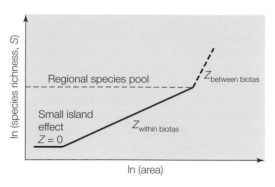

Figure 10.15 Scale-dependence of the species–area relationship. The SAR could exhibit different domains of scale, reflecting shifts in the processes responsible for the relationship between species and area. For example, the SAR may not hold at finer scales encompassing small habitat patches, where species richness is more a function of stochastic factors or specific environmental conditions unrelated to area (the small island effect). Typically, the SAR is calculated within a particular habitat type or region, where species' colonization–extinction dynamics are influenced by area effects ($z_{\text{within biotas}}$). Beyond the regional scale, different evolutionary histories give rise to different biota, such that species richness increases rapidly with further increases in area ($z_{\text{between biotas}} > z_{\text{within biotas}}$).

For example, the species–area slope is typically steeper when evaluating how diversity scales over very large areas, such as those encompassing different biogeographic regions or continents, than for smaller areas within a particular ecoregion or landscape (Rosenzweig 1995). In other words, the slope of the SAR between biotas is much steeper than within biotas ($z_{\text{between biotas}} > z_{\text{within biotas}}$). Ordinarily, we would expect species richness to increase linearly with area up to some point, such as that encompassing the regional species pool (**Figure 10.15**). If we extend the SAR to encompass regions with very different evolutionary histories, however, then entire biotas are being added with further increases in area, which is why z (the rate at which species richness increases as a function of area) is expected to be steeper in this domain, as large numbers of species are added per increase in unit area (Lomolino 2001; **Figure 10.15**). Conversely, at the other extreme, species richness may vary independently of area when very small areas are considered, which is referred to as the **small island effect** (MacArthur & Wilson 1967; Lomolino 2001; **Figure 10.15**). In this domain, the establishment of species on small islands or habitat fragments is more tenuous and likely to depend on stochastic factors or the specific location and environmental conditions of that island or fragment, which are unrelated to its size (Triantis et al. 2006).

Even within biomes, however, the slope of the SAR may be scale-dependent. This was demonstrated by Palmer and White (1994) in an analysis of SARs for vascular plants within a heterogeneous forest landscape. Using a nested sampling design, variation in the spatial grain of the analysis was achieved through different quadrat sizes (linear dimensions = 0.125, 0.25, 0.5 … 16 m) within plots, whereas spatial extent varied in terms

of the distance between plots being compared (e.g. the number of new species encountered in a plot that is spaced 16, 32, 48, 64, 128 … 256 m away from the first). Although both grain and extent influenced species–area curves in this system, increasing the grain size had a far greater effect on the relationship between species richness and area than increasing the spatial extent. That is, the number of new species increased with increasing grain size, but not so much with increasing distance, except at large spatial extents. However, unlike the sharp thresholds or 'breakpoints' between scaling domains for the idealized multiscale SAR (**Figure 10.15**), the transition between scales occurred more gradually in this system, perhaps reflecting a more continuous variation in environmental heterogeneity and species distribution patterns (i.e. heterogeneity and species were not so patchily distributed). In any case, the fact that the SAR can exhibit scale-dependence even within the same landscape means that there may not be a single species–area curve for a given location (Palmer & White 1994). If true, this would obviously complicate efforts to extrapolate SARs across landscapes, even within the same region or for the same vegetation type.

If scale dependency in the SAR is the norm, then how do we determine which scale or range of scales to use when attempting to evaluate how patterns of diversity compare between sites? Should we define a standardized area or range of scales when constructing SARs, so as to permit comparisons among sites or studies? Finding an optimum scale range will require a great deal of work,

depend on the units of measurement used in the analysis (e.g. whether area is measured as km² or mi² or we use $\log_{10}$ versus the natural log of area), whereas we must rescale c to ensure that these values are in equivalent units when comparing different species–area curves (e.g. so that we can conclude that an area with a greater c value really does have higher richness than an equivalent area with a lower c value). In our case, we are using the term 'scale dependence' in reference to how z changes across a range of spatial scales (across different scaling domains), which Rosenzweig (1995) also discussed when highlighting how the slope of the SAR changes, depending on whether species richness is assayed 'within biotas' versus 'between biotas.'

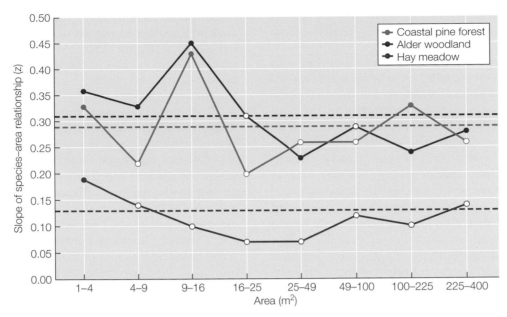

Figure 10.16 Species–area relationships may exhibit scale-dependence even within landscapes.   Different scaling relationships (z) emerge across different ranges of scale for three vegetation types found on the Russian Curonian Spit along the southeastern Baltic Sea. The relationship between species and area is significant only for certain scaling ranges (filled symbols: $P \leq 0.05$). However, a better fit is obtained when the SAR is obtained across this entire range of scales (dashed horizontal lines give the z value for 1–400 m²; all $R^2 \geq 0.88$ and $P \leq 0.001$).

Source: After Dolnik & Breuer (2008).

especially as this is likely to be specific to a particular vegetation type or region. For example, Dolnik and Breuer (2008) compared the slopes (z) of SARs calculated for 14 different scaling ranges in each of three vegetation types found on the Russian Curonian Spit, which lies along the southeastern coast of the Baltic Sea: coastal pine forests (*Empetro nigri-Pinetum sylvestris*), alder woodlands (*Carici elongatae-Alnetum glutinosae*), and hay meadows (pasture). In each vegetation type, they recorded the number of vascular plants, bryophytes, and lichen species using a nested sampling design consisting of 13 nested square plots that varied in size from 0.065 to 900 m². These quadrat sizes encompass some of the most commonly used in plant community studies, including those used in global monitoring programs.

Although z values were higher in the two forest types than in the hay meadows, there was a great degree of variability in z values among the different scaling ranges, especially at the finer scales where forest species were more likely to be absent and where quadrats with unusually low or high numbers of species ('outliers') had a greater effect on the mean richness at that scale (**Figure 10.16**). Area was therefore not always a strong or significant predictor of species richness within these scale ranges (as assayed by the R^2 and P values of the species–area regression; **Figure 10.16**). A better and more consistent fit was obtained when the SAR was obtained over a greater range of scales (e.g. 1–400 m², where $P < 0.001$ and R^2

$= 0.88 – 0.92$ depending on vegetation type). The authors therefore concluded that a 1–400 m² scale range was best for comparing forest and grassland vegetation types in this landscape, but cautioned that a standardized scale range that could be applied to all plant communities should obviously not be based on this one study. Instead, they advocate for establishing a standardized reference system for surveying plant communities across a nested range of scales, as with the Whittaker or modified-Whittaker nested plot designs, which encompass a range of scales from 1 to 1000 m² (Shmida 1984; Stohlgren et al. 1995).

Finally, any landscape disturbance that alters the distribution and abundance of species can potentially have scale-dependent effects on the SAR. For example, Powell and her colleagues (2013) found that invasive plants had scale-dependent effects on the SARs of understory plants within forest communities. Invaded sites had lower intercepts (c) but steeper slopes (z) than non-invaded sites. Importantly, this trend was consistent among three different forest types, each of which had been invaded by a different species of non-native plant: the cerulean flax lily (*Dianella ensifolia*) in the hardwood hammock forests of Florida, the Amur honeysuckle (*Lonicera maackii*) in oak-hickory forests of eastern Missouri, and the fire tree (*Morella faya*) in the tropical mesic forests of Hawai'i. In each case, the plant invader greatly reduced plant species richness at fine spatial scales (≤25 m²), but this effect largely disappeared at broader spatial scales (500 m²) owing to the steeper species-accumulation curves (higher z values) of these invaded sites.

Invasion substantially reduced plant abundances in these plots (by 65–91%), which is why far fewer species were encountered at local scales within invaded plots. Paradoxically, species that were common in the non-invaded plots appeared to be more strongly affected by invasion than rare species. The authors speculate that this may result from a high degree of niche overlap between the common species and the invader, and/or because rare species just happened to possess certain life-history traits (e.g. shade tolerance) that permitted them to maintain their (low) level of abundance within these systems despite invasion (Powell et al. 2013). This underscores an important point, however: invasion impacts will vary in space and time owing to different species traits, including those of the invader, as well as a host of other factors such as invasion severity, time since invasion, the nature of species interactions, and so forth (Stohlgren & Rejmánek 2014). Thus, we should not assume that all invasive plants (or that all types of disturbances) will necessarily have these same sorts of scale-dependent effects on the SAR, but should consider the extent to which they could.

SHOULD PATCHES OR LANDSCAPES BE THE FOCUS OF SPECIES–AREA RELATIONSHIPS? Besides varying in spatial grain and extent, SARs can also differ in their focal scale, as discussed. Although this is generally taken to mean the cumulative area or inference space of each data point (e.g. Scheiner 2003; Scheiner et al. 2011), the focus of a SAR can refer to what type of spatial units form the basis of the analysis, such as whether the SAR is based on the areas of individual habitat patches or on the habitat areas of entire landscapes (Turner & Tjørve 2005; Rybicki & Hanski 2013; With 2016). Landscapes are made up of patches of varying sizes, but it is unclear to what extent we can extrapolate from a patch-based SAR to predict the effects of habitat loss and fragmentation on species richness at a landscape scale, for example. This is a salient point, given that most studies that examine the effects of habitat loss and fragmentation typically adopt a patch-based approach, where habitat fragments of different sizes are the focus, rather than landscapes with different habitat amounts (McGarigal & Cushman 2002; Fahrig 2003). Habitat loss and fragmentation are really a landscape-wide phenomenon, however. Although fragmented landscapes have more small patches and a greater amount of edge habitat than clumped landscapes (**Table 4.7; Figure 7.2**), the configuration of those patches can influence species dynamics beyond simple patch-area effects. For example, patches that are closer together (a clumped habitat

distribution) could help mitigate extinction risk, enabling species to persist across a greater range of habitat loss (**Figure 7.16**). As a result, we expect a greater loss of species in fragmented landscapes for a given degree of habitat loss ($z_{\text{fragmented}} > z_{\text{clumped}}$; Rosenzweig 1995).

To examine the congruence of patch-based and landscape-based SARs, With (2016) compared species–area curves obtained from a survey of arthropods within that experimental landscape system mentioned previously, where the amount and fragmentation of habitat varied among landscape plots (**Figures 5.11 and 10.2**). The patch-based SAR examined how species richness within individual patches (S_p) varied as a function of habitat patch area (A_p) across all landscapes, whereas the landscape-based SAR examined how species richness within individual landscape plots (S_L) varied as a function of the total habitat area in each landscape (A_L). The two types of SARs thus differ in their focus: patch-based SARs represent the conventional approach in which the relationship between species richness and patch area is of interest (each data point on the curve represents a patch; **Figure 10.13C**), whereas landscape-based SARs are focused on the relationship between species richness and the total habitat area within individual landscapes (each data point on the curve represents a landscape, that is, a collection of patches).

Patch-based and landscape-based SARs were not congruent in this experimental landscape system. Patches gained species at a faster rate than did landscapes ($z_p = 0.37$ vs $z_L = 0.26$), which contributed to domains of incongruity between the two types of curves (**Figure 10.17**). Below about 30% habitat, landscape richness (S_L) exceeded patch richness (S_p) for a given habitat area (i.e. where $A_L = A_p \leq 0.30$). In this domain, species richness was greater for a collection of patches (a landscape) than it was for a single patch of the same total habitat area (i.e. the size of some patches could match or exceed the total habitat area of some landscapes, especially those with low habitat amounts). Thus, information obtained at a patch scale would tend to underestimate species richness at the landscape scale. In this system, species richness was actually greater in fragmented than in clumped landscapes with <30% habitat (With 2016), probably because habitat patches were passively sampling the larger species pool, and species turnover (β diversity) was especially high among small patches, which are more numerous in fragmented landscapes (With & Pavuk 2012). Landscape configuration (whether the habitat distribution was clumped or fragmented) was thus important for evaluating the SAR in this domain.

At the other extreme, patch richness exceeded landscape richness when habitat was very abundant ($S_p > S_L$ when $A_L = A_p \geq 0.6$). In this domain, most landscapes

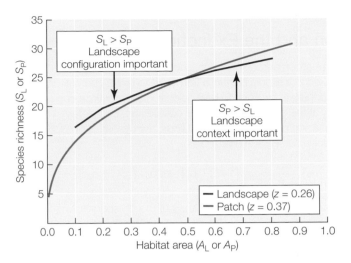

Figure 10.17 Species–area relationships based on patches versus landscapes are not congruent. A survey of arthropods in an experimental landscape system (**Figure 5.11**) revealed that patches gained species (S_P) at a faster rate than did landscapes (S_L) with increasing habitat area (patch size, A_P, or total habitat amount, A_L). Species richness was greater in landscapes than in patches at low habitat levels ($\leq$30% habitat), but greater in patches than in landscapes at high habitat levels ($\geq$60% habitat). Landscape configuration, in terms of whether the landscape was clumped or fragmented, was important for understanding species–area relationships when habitat was limiting, whereas landscape context, whether the largest habitat patch was in a fragmented or clumped landscape, was more important when habitat was abundant.

Source: After With (2016).

having >60% habitat were dominated by a single large patch, and yet landscapes were not simply functioning as large habitat patches. Although the size of the largest patch did not differ between fragmented and clumped landscapes with >60% habitat, the largest patch in fragmented landscapes still had twice as much edge as the largest patch in clumped landscapes. If negative edge effects are responsible for a reduction in species richness at a patch scale (**Boxes 7.1 and 7.2**), then this could result in fewer species within fragmented landscapes. Further, the mean patch size was lower in these fragmented landscapes with abundant habitat, owing to a greater number of small patches, which contributed to a lower mean patch richness as well (With 2016). Thus, landscape context—whether the largest habitat patch was in a fragmented or clumped landscape—had an influence on the SAR in this domain. It would seem that landscapes are not simply the sum of their patches, in which case we should probably reconsider using patch-based SARs to predict the landscape-scale effects of habitat loss and fragmentation on biodiversity.

Admittedly, the difference here between patch-based and landscape-based SARs may appear rather

small, a difference of a few species at most (**Figure 10.17**). Given that this experimental system was established in an agriculturally dominated system, with many widespread and generalist species, dispersal limitation is likely not an issue (With & Pavuk 2011, 2012). Most arthropod species were able to colonize the available habitat, such that community assemblages appeared to be a random sample of the regional species pool. We might therefore expect greater differences between patch-based and landscape-based SARs in systems where dispersal limitation is more of an issue, such as in systems with many endemic species or habitat specialists. Put another way, habitat specialists or endemic species are generally the first species to go extinct as a result of habitat loss, whereas more widespread or generalist species are able to maintain viable populations (at least for a while), especially if they are being supported by immigration from elsewhere (i.e. source–sink landscape dynamics; With et al. 2006). Even habitat generalists or widespread species may go extinct in time, however, if the degree of habitat loss is severe enough and species exhibit a lagged response to landscape change (e.g. an **extinction debt**, Tilman et al. 1994; **Box 10.1**).

For these reasons, the traditional SAR can overestimate the numbers of species expected to go extinct as a function of habitat loss (Kinzig & Harte 2000; He & Hubbell 2011). This is especially true when SARs are derived as species-accumulation curves, giving the rate at which species richness increases as a function of increasing habitat area, which is usually not the same rate at which species are lost as a function of decreasing habitat area (i.e. habitat loss). The distinction is an important one, given that the most common method for predicting the consequences of habitat loss on biodiversity is to take the species–area accumulation curve and extrapolate backwards to estimate the number of species expected in smaller habitat areas. Species are typically encountered in landscapes with far less habitat than strictly needed to support viable populations, however, which means that endemic species in particular will ultimately require a greater amount of habitat to persist than predicted by the traditional SAR (He & Hubbell 2011). It has therefore been suggested that the predicted species losses from habitat destruction be based on an **endemics–area relationship** (Harte & Kinzig 1997; Kinzig & Harte 2000).

THE ENDEMICS–AREA RELATIONSHIP The endemics–area relationship is related to the traditional SAR as follows. First, let's rewrite the equation for the traditional SAR (Equation 10.10) to emphasize the number of species found (S) in a given area of size A relative to the total number of species (S_0) found in the landscape being surveyed (A_0):

BOX 10.1   Habitat Destruction and the Extinction Debt

Habitat destruction is the major cause of species extinctions worldwide (Hanski 2011). Although species have different extinction thresholds, in terms of the amount of habitat loss they can withstand (**Figures 7.15 and 7.16**), most species do not go extinct immediately once their extinction threshold has been exceeded. Recall that rapid landscape change can lead to a decoupling of population and landscape dynamics, such that populations exhibit a lagged response to habitat loss (**Chapter 7**). In that case, population declines could play out for years until the species inexorably goes extinct (e.g. Schrott et al. 2005). Thus, the consequences of habitat loss on a community may not be immediately apparent, creating an **extinction debt** that continues to be paid off long after the initial disturbance occurred.[28]

Species are not living in a vacuum, however. Other species could exacerbate the effects of habitat destruction, causing species to go extinct sooner than expected given their individual extinction thresholds. Imagine a forest community, in which all tree species are competing for the same limiting resources, such as light, water, or nutrients. Some species are better competitors for these resources, perhaps because they grow faster and can shade out other species or because they have deeper root systems and can obtain water and nutrients that lie below the rooting zone of other species. Species' life-history traits often involve trade-offs, however. Species that are good competitors may not be very good at colonizing new areas, for example (Tilman 1994). Conversely, other species may have superior dispersal abilities and are good colonizers, but are unable to persist very long once species that are better competitors arrive.

Over time, we might expect that good competitors will come to dominate the landscape, given that their greater abundance should make stochastic extinction less likely. We might also expect that these dominant competitors will drive many of the other species extinct, especially when good competitors are very abundant on the landscape. Under a scenario of habitat loss, however, we could be wrong on both counts.

In a pair of now-classic papers, David Tilman and his colleagues extended a metapopulation model (Equation 7.12) to predict the consequences of habitat destruction on the coexistence of multiple species competing for a single limiting resource (Tilman 1994; Tilman et al. 1994). Species are assumed to possess the aforementioned trade-off between

competition and colonization, such that species can be ranked in a simple hierarchy from best to worst competitor (and thus, from worst to best colonizer). Competition occurs only at the site level, with better competitors completely displacing poorer competitors. Thus, the dynamics of each species can be summarized as

$$\frac{dp_i}{dt} = c_i p_i \left(1 - \sum_{j=1}^{i} p_j\right) - m_i p_i - \left(\sum_{j=1}^{i-1} c_j p_j p_i\right),$$

where p_i is the patch occupancy of the ith species, which is a function of its species-specific probability of colonization (the first term), local extinction (mortality) rates (the second term), and its probability of competitive displacement by other species (the third term). Habitat destruction would reduce the proportion of sites available for colonization, and thus Tilman et al. (1994) modified the equation above by adding a single parameter (D) to the colonization term as

$$\frac{dp_i}{dt} = c_i p_i \left(1 - D - \sum_{j=1}^{i} p_j\right) - m_i p_i - \left(\sum_{j=1}^{i-1} c_j p_j p_i\right), \quad \text{Equation 1}$$

where D is the proportion of sites permanently destroyed by habitat destruction. For the best competitor (species 1), its abundance at equilibrium (Equation 7.13) is thus

$$\hat{p}_1 = 1 - D_1 - \frac{m_1}{c_1}.$$

Rearranging this, we can define an extinction threshold for species 1 (i.e. when $\hat{p}_1 = 0$) as

$$D_1 \geq 1 - \frac{m_1}{c_1};$$

that is, species 1 will go extinct when the amount of habitat destroyed equals or exceeds its initial abundance (patch occupancy) on the landscape (i.e. before habitat destruction; $\hat{p}_1 = 1 - \frac{m_1}{c_1}$ when $D = 0$).

For the other species, their initial abundances are assumed to form a geometric series of decreasing abundance with respect to species 1, such that species are predicted to go extinct in order, from the best to worst competitor as habitat destruction increases (i.e. $D_1 < D_2 < D_3 \ldots D_n$). Thus, the most abundant species (species 1) is also the most vulnerable to habitat destruction, which seems counterintuitive. Recall that the most abundant species is the best competitor, which means that it is also the worst colonizer. Although habitat destruction lowers the effective colonization rate of all species (Equation 1), it has the greatest impact on species with low colonization rates (c_i), which in this case are the better competitors (Tilman et al. 1994).

To illustrate, Tilman and his colleagues (1994) used this metapopulation model to simulate the dynamics of 20

[28] The extinction debt is similar to the concept of **relaxation time** in island biogeography (Diamond 1972). Relaxation is the amount of time it takes for island biota to return to an equilibrium number of species (**Figure 10.18**) following a perturbation, such as the contraction of island areas or the severing of land bridges due to rising sea levels during interglacial cycles. In his 1972 paper, Jared Diamond also considered the implications of faunal relaxation in tropical rainforests undergoing deforestation and how this should factor into the design of nature reserves.

(A)

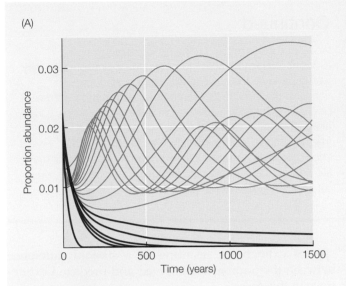

(B)

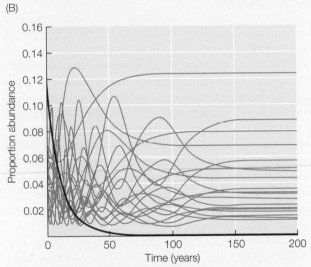

Figure 1 The extinction debt. Species exhibit a lagged response to habitat loss (30% initial habitat destroyed at Time = 0), going extinct many decades and even centuries later. Assuming a trade-off between competitive ability and colonization ability, the best competitors (= worst colonizers) are the first to go extinct. More species go extinct in landscapes where species are initially rare (3% site occupancy, A) than where they are initially more abundant (20% site occupancy, B).

Source: After Tilman et al. (1994).

hypothetical tree species following a bout of habitat destruction, in which a third of the landscape was deforested. In the first scenario, the best competitor occupied only 3% of sites ($q = 0.03$), which was intended to mimic a tropical forest where species richness is high, but the relative abundances of species are low. In the second scenario, the best competitor occupied 20% of sites ($q = 0.2$), which is more akin to a temperate forest where one or a few species tend to dominate. In both scenarios, the most

abundant species was the first species to go extinct; it was also the only species to go extinct in the temperate forest scenario (**Figure 1B**). By contrast, a half-dozen other species also went extinct in the tropical forest scenario (**Figure 1A**), in which initial abundances for all species were considerably lower than those in the temperate forest scenario. Importantly, species did not go extinct immediately following the initial bout of habitat destruction. Instead, species declined and gradually went extinct some 40–400 (or more) years later.

Although the original formulation of the extinction debt was based on an assumed trade-off between the competitive dominance and dispersal abilities of species within a community, such trade-offs are not necessary for extinction debts to occur. Nor is habitat loss the only type of disturbance that can give rise to an extinction debt. The extinction debt can apply to a wide range of disturbances that decrease the amount or quality of available habitat, such as those resulting from climate change or invasion by a non-native species (e.g. Dullinger et al. 2012). The extinction debt thus represents a major challenge for conservation biology (Kuussaari et al. 2009). If the extinction debt is large, then the number of species that are effectively endangered by habitat loss and fragmentation is likely to be underestimated. But how can we estimate the extinction debt, if it takes years, decades, or even centuries to pay out? Such long time-series data are unlikely to be available for most communities or landscapes. Although the extinction debt can be estimated in various ways, the most common approach is to determine if past landscape patterns better explain current species richness than the contemporary landscape pattern (Kuussaari et al. 2009). If they do, then this is interpreted as evidence of an extinction debt.

Ultimately, the probability and magnitude of an extinction debt depend on the life-history traits of the species, the amount and configuration of habitat on the landscape, the time since the habitat was altered, and the severity of that alteration (Kuussaari et al. 2009). For example, the extinction debt is expected to be particularly great in fragmented landscapes where the amount of habitat has been reduced below most species' extinction thresholds (Hanski & Ovaskainen 2002). Conversely, the extinction debt may be absent in landscapes in which the disturbance occurred a long time ago (the extinction debt has since been paid off), or in which the extinction debt was paid off very quickly owing to the widespread extent and severity of the disturbance.

In an example of the first case, forest plants were found to carry an extinction debt that was created more than a century ago following the destruction and fragmentation of forest cover in one landscape, but not within a similar landscape that had been deforested to its current state 1000 years ago (both landscapes now have <10% forest cover; Vellend et al. 2006). Furthermore, the extinction debt was most evident in plant species that had low

(Continued)

BOX 10.1 Continued

colonization–extinction rates (species with 'slow' metapopulation dynamics), which are precisely the types of species we would expect to show a lagged response to landscape change.

In the second case, the extinction debt might be paid off fairly quickly if the remaining habitat patches are small and isolated. In European grasslands, for example, the extinction debt appears to have been paid off in landscapes with <10% grassland remaining, based on the fact that plant species richness is better predicted by the current landscape pattern than by the historical landscape (Cousins 2009). Our ability to detect the extinction debt may well depend on the scale at which we assess species richness, however. In this same grassland system, gamma diversity (γ diversity) was found to degrade more slowly than alpha diversity (α diversity) following habitat loss and fragmentation (Cousins & Vanhoenacker 2011). Thus, we should consider the extent to which an extinction debt may still persist at the broader landscape or regional scale, even if it appears to have been settled at a local scale.

$$S(A) = (A/A_0)^z S_0. \qquad \text{Equation 10.11}$$

For the endemics–area relationship, we will be focusing on the number of species that are found only in the area of size A and nowhere else within the larger landscape, A_0 (i.e. the endemic species, E):

$$E(A) = (A/A_0)^{z'} S_0. \qquad \text{Equation 10.12}$$

The exponent z' for the endemics–area relationship (Equation 10.12) is related to z of the SAR (Equation 10.11) as

$$z' = -\ln(1 - 0.5^z)/\ln(2) \qquad \text{Equation 10.13}$$

(see Harte & Kinzig 1997 for the derivation of z'). Thus, if $z = 0.2$, then $z' = 2.95$.

To illustrate the differences between predictions of the endemics–area relationship and the traditional SAR, Kinzig and Harte (2000) used the example of avian extinctions in the northeastern USA. There have been far fewer bird extinctions than expected given the degree of forest clearing following European settlement of this region in the 1600s (**Chapter 3**). From an initial total of 160 bird species (S_0), a 50% loss of forest ($A/A_0 = 0.5$) should have resulted in a loss of about 24–25 species, assuming $z = 0.25$, as is often done for the purposes of estimating species losses in fragmented landscapes[26] (Equation 10.11):

$$S(A) = (0.5)^{0.25} 160 = 134.5,$$

where the number of species lost is calculated as $S_{lost} = S_0 - S(A) = 160 - 134.5 \approx 24-25$ species. Instead, only four species have gone extinct in this region (Pimm & Askins 1995).[27] Again, the SAR tends to overestimate species extinctions, resulting in a six-fold difference between the number of observed and predicted extinctions in this case.

However, only 28 of these 160 bird species are endemic to the northeastern forests of the USA (Pimm & Askins 1995). Note that the use of the endemics–area relationship (Equation 10.12) requires that we include the total landscape area that encompasses all forest birds (most of whom have ranges extending well beyond the northeastern USA, given that so few species are endemic to the region), as well as any other birds within that larger landscape area (i.e. non-forest endemics). We thus need to expand our measure of A_0 and S_0. As a very rough estimate, we can assume that if 28 species are endemic to the forested half of the northeastern USA, then another 28 species would be endemic to the non-forested half, giving a total richness of $160 + 28 = 188$ bird species for the entire landscape area (i.e. $S_0 = 188$; see Kinzig & Harte 2000 for a more precise way of estimating this, which gave a total of 182 species). Thus, if northeastern forest made up 53% of this enlarged landscape area, A_0 (Pimm & Askins 1995), and half of that was cleared (A_{lost}), then only 26.5% of the forested habitat within the larger landscape (A_0) was ultimately lost ($0.53 \times 0.5 = 0.265$). Assuming a pre-clearing relationship between forest bird species and area of $z = 0.2$ (Pimm & Askins 1995; note that the pre-clearing z is expected to be lower than the z value used previously to predict species losses in fragmented landscapes), the corresponding z' for the endemics–area relationship would be 2.95 (Equation 10.13). Thus, the expected number of species lost (Equation 10.12) would be

---

[26] The slope value $z = 0.25$ stems from MacArthur & Wilson's (1967) theory of island biogeography, and was subsequently found in many empirical species–area relationships derived from oceanic islands (Connor & McCoy 1979; Rosenzweig 1995).

[27] According to Pimm & Askins (1995), the four forest bird species that went extinct in eastern North America are the passenger pigeon (*Ectopistes migratorius*), Carolina parakeet (*Conuropsis carolinensis*), ivory-billed woodpecker (*Campephilus principalis*), and Bachman's warbler (*Vermivora bachmanii*). Although the last two species are still listed by the IUCN as critically endangered, both are probably extinct given that there have been no confirmed sightings of either species in some 55-75 years (a supposed 2004 sighting and blurry video of an ivory-billed woodpecker in Arkansas was met with much skepticism by some in the ornithological community, especially as subsequent searches have failed to uncover unequivocal evidence of the bird).

$$S_{\text{lost}} = \left(A_{\text{lost}} / A_0\right)^{z'} S_0 = (0.265)^{2.95} \, 188 = 3.73 \text{ species,}$$

Equation 10.14

which agrees quite well with the actual number of bird species that actually went extinct (i.e. 4 species; Kinzig & Harte 2000).

Although endemic species are expected to go extinct immediately upon the destruction of their habitat, other species may go extinct later if they exhibit a lagged response to habitat loss and fragmentation, creating an **extinction debt** that is paid out over the ensuing years or decades (Tilman et al. 1994; Schrott et al. 2005; **Box 10.1**). Might the extinction debt thus explain the discrepancy between the predictions of the SAR and the actual number of extinctions observed on the landscape? To evaluate this, Rybicki and Hanski (2013) explored the relationship between habitat area and the remaining species on the landscape, that is, the non-endemics that survived the initial loss of habitat. Using a dynamic spatially explicit simulation model, Rybicki and Hanski (2013) found that many non-endemic species also go extinct soon after habitat loss, especially when a great deal of habitat is destroyed (e.g. 80–90%). In that case, the traditional SAR gave a better prediction of short-term extinctions than the species–area relationship based only on the fraction of non-endemic species that remained following habitat loss (the remaining species–area relationship). However, the traditional SAR greatly underestimated the number of species going extinct in highly fragmented landscapes (Rybicki & Hanski 2013; Hanski et al. 2013).

Thus, the traditional SAR does not always overestimate species extinctions from habitat loss, especially when habitat loss is great and the habitat is also being fragmented. Indeed, the greater concern is the extent to which it might underestimate extinctions in landscapes undergoing rapid habitat loss and fragmentation, as this is when lagged responses to landscape change and extinction debts are most likely to occur. Even though predictions of species extinctions in the face of habitat loss and fragmentation should probably be based on landscape-based SARs rather than patch-based (island) SARs (Hanski et al. 2013; With 2016), patch-based or island SARs have long been used to predict the loss of biodiversity due to habitat loss and fragmentation, which has been reinforced by the application of island biogeographic theory to fragmented landscapes, as we discuss next.

Island Biogeography and the Habitat Fragmentation Paradigm

Although the empirical relationship between species and area was described mathematically early on, it was not actually predicted on theoretical grounds until

Robert MacArthur and Edward O. Wilson published their seminal work on the **theory of island biogeography** in the 1960s (MacArthur & Wilson 1963, 1967). As with most general theories in ecology, the theory of island biogeography is elegant in its simplicity: the equilibrium number of species ($\hat{S}$) found on an island is predicted to represent a balance between species' immigration (colonization) and extinction rates (**Figure 10.18**). The model makes certain assumptions, namely that species only arrive on an island via immigration from the mainland,[28] such that the number of island species (S) represents a subset of species found on that mainland (the mainland source pool, S_M). The number of species present on an island therefore lies somewhere within the domain of $0 \leq S \leq S_M$, as determined by the relative immigration and extinction rates for the island.

Let's explore this graphically. The model predicts that as the number of island species increases, the overall immigration rate (I) decreases whereas the overall extinction rate (E) increases (**Figure 10.18**). The overall immigration rate (I) to the island is expected to decrease as the number of species on the island approaches the size of the mainland source pool ($I \to 0$ as $S \to S_M$; **Figure 10.18**). If all mainland species colonize the island, then there would be no new species left to immigrate ($I = 0$ when $S = S_M$). Not all mainland species will colonize the island, however. Although very good dispersers are likely to colonize the island quickly (e.g. the 'supertramps;' Diamond 1974), the rate at which new species continue to arrive begins to taper off at some point, as fewer and fewer of the remaining mainland species are able or likely to ever immigrate (i.e. very poor dispersers may never colonize the island). The more species there are on the island, however, the more species there are to go extinct, especially if species interact and contribute to each other's extinction (e.g. through increased competition for limited resources). The overall extinction rate of the island (E) is thus expected to increase with increasing species richness on the island. Thus, the point at which these two curves intersect graphically is where the immigration rate balances the extinction rate (= equilibrium), which then gives the expected number of species on that island at equilibrium ($\hat{S}$).

[28] Because the equilibrium model of island biogeography emphasizes processes that supposedly take place on ecological time scales (i.e. colonization and extinction), it does not include in situ speciation rates (e.g. adaptive radiation) as a process by which new species appear on islands because this is assumed to take place over longer, evolutionary time scales. This is debatable, since immigration to remote islands may occur over what we traditionally consider evolutionary time scales, and evolution has been documented to occur within ecological time scales. Thus, efforts are now underway to link both ecological and evolutionary dynamics into a more unified theory of island biogeography (Heaney 2007; Whittaker et al. 2008; Chen & He 2009).

(A)

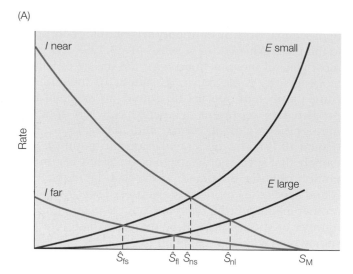

(B)

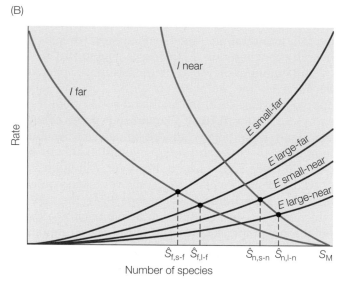

Figure 10.18 The equilibrium theory of island biogeography and the rescue effect. (A) The equilibrium number of species ($\hat{S}$) on an ocean island is the balance between immigration (I) and extinction (E) rates. Immigration rates are expected to be higher for islands near to the mainland source, which contains all of the species that might possibly be found on the island (S_M), than for islands that are farther away ($I_{near} > I_{far}$). Extinction rates are expected to be greater on small islands than on large islands ($E_{small} > E_{large}$). The combination of island size and distance from the mainland (isolation) thus predicts different equilibrium numbers of species for each island type (e.g. $\hat{S}_{n,l}$ vs $\hat{S}_{f,l}$). (B) Species may be rescued from extinction if there is immigration from the mainland, such that species on near islands—regardless of size—will have lower extinction rates than those farther away. Because of this rescue effect, small islands that are near could have lower extinction rates than islands that are large but farther from the mainland ($E_{large-far} > E_{small-near}$).

Source: (A) After MacArthur & Wilson (1967), (B) After Brown & Kodric-Brown (1977).

Immigration rates are not just a function of species' dispersal abilities, but are also likely to depend on the proximity of the island to the mainland. More species should arrive on an island that is close to the mainland than on one that is farther away ($I_{near} > I_{far}$; **Figure 10.18A**). All else being equal, then, the equilibrium number of species should be greater on near islands than on far islands ($\hat{S}_n > \hat{S}_f$). And, as we know from our previous discussion of SARs, large areas have more species than small areas. One reason for this is that large areas can accommodate larger population sizes, which generally have lower extinction rates compared to small populations in small areas. Graphically, we can depict this as a higher overall rate of extinction on small versus large islands ($E_{small} > E_{large}$; **Figure 10.18A**). All else being equal, the equilibrium number of species should be greater on large islands than on small islands ($\hat{S}_l > \hat{S}_s$), consistent with our expectations from the SAR. If we combine the area effects of island size with those of island distance from the mainland (isolation effects), we can predict species richness for different island configurations. Although it might seem obvious that large islands that are near to the mainland should have more species than ones that are small and far away ($\hat{S}_{n,l} > \hat{S}_{s,f}$), we can now determine whether species richness should be greater on small near-shore islands than on those that are large but farther from the mainland. Ultimately, that will depend on how isolation and area effects differentially influence rates of immigration and extinction, respectively (**Figure 10.18A**).

Proximity to a mainland source could also influence extinction rates through the **rescue effect** (Brown & Kodric-Brown 1977). Islands nearer to the mainland—regardless of size—have higher immigration rates than islands that are farther away. As a result, populations on near-shore islands may be 'rescued' from extinction by a continual influx of immigrants, such that extinction rates are lower for near islands than those farther from the mainland (**Figure 10.18B**). Thus, it is possible that small islands that are near could have lower extinction rates than large islands that are farther away ($E_{large-far} > E_{small-near}$; **Figure 10.18B**). This is analogous to our previous discussion of population source–sink dynamics and the structure of certain metapopulation models (especially mainland-island metapopulations) in **Chapter 7**, where small populations in marginal habitats (sinks) are subsidized by immigration from larger and/or closer source populations, thereby rescuing them from certain extinction (**Figures 7.13 and 7.14**). That similarity is due in no small part to the application of principles from island biogeographic theory to patchy populations in fragmented landscapes.

THE APPLICATION OF ISLAND BIOGEOGRAPHY TO FRAGMENTED LANDSCAPES The parallels between oceanic islands and habitat fragments were certainly not lost on MacArthur and Wilson. Recall that the first figure in their monograph on the theory of island biogeography was that time-series depicting the loss and fragmentation of forest in the Cadiz Township of Wisconsin following European settlement (**Figure 3.6**). They thus anticipated the application of island biogeographic theory to fragmented landscapes: 'The same principles apply, and will apply to an accelerating extent in the future, to formerly continuous natural habitats now being broken up by the encroachment of civilization' (MacArthur & Wilson 1967). Little wonder, then, that island biogeographic principles, especially those pertaining to the size and isolation of habitat fragments, have dominated the study of fragmented landscapes ever since. This has been especially evident in the design of fragmentation experiments, in which the size and distances between patches are manipulated to create fragmented landscape patterns, so as to test hypotheses about how differences in habitat area and isolation influence patterns of species diversity (Debinski & Holt 2000; McGarigal & Cushman 2002). The size and isolation of habitat fragments are also among the most commonly used metrics to characterize the degree to which landscapes have become fragmented, and thus to evaluate the effects of habitat fragmentation on biodiversity (McGarigal & Cushman 2002; Fahrig 2003; Ewers & Didham 2006).

Following from the predictions of island biogeographic theory, the number of species within habitat fragments should decline as a function of decreasing patch size and increasing patch isolation. Although these general expectations are met in some studies, they are not met in others (Fahrig 2003; Ewers & Didham 2006; Collinge 2009). Understanding how or why different types of species respond differently to the different patch-scale consequences of habitat loss and fragmentation has been a major theme in fragmentation research, generating hundreds—if not thousands—of research papers in the decades since the publication of MacArthur and Wilson's (1967) monograph on the theory of island biogeography.[29] Although the relationship between species diversity and patch area or isolation represents a community-level response to habitat loss, the exact nature of that response ultimately depends on what type of community or taxonomic group is being studied, the composition and relative abundances of species in that community, and

the specific traits of those species, such as body size, habitat specificity, trophic position, and dispersal ability (Ewers & Didham 2006; Öckinger et al. 2010; Franzén et al. 2012; van Noordwijk et al. 2015). We will thus consider the trait-based sensitivity of species to patch-area and isolation effects in the next couple of sections, before returning to our discussion on the broader application of island biogeographic theory to communities in fragmented landscapes.

Patch-area effects. The species most susceptible to a decrease in patch area tend to be those that are:

- large
- rare
- habitat specialists
- top consumers (e.g. predators)
- sedentary, or conversely, species that are very mobile.

Recall from our discussion of allometric scaling relationships in **Chapter 6** that large-bodied species have larger home range sizes or minimum area requirements than small-bodied organisms. The area requirements of large species are thus unlikely to be fulfilled by small habitat patches (**Figure 10.19C**). As a result of their large home range or territory sizes, large species also tend to occur at low density and thus have smaller population sizes (Peters 1983). Rarity increases extinction risk (**Figure 10.19E**), and thus rare species, which by definition have small populations and/or low patch occupancy, are especially vulnerable to even small losses of habitat that further reduce patch sizes or the availability of those patches (**Figure 7.15; Chapter 7**). Rare species should thus be lost at a greater rate than common species as habitat area declines ($z_{rare} > z_{common}$). For example, the slopes of the SAR (z) for rare species were at least three times larger than those for common species across three different trophic groups comprising plants, insect herbivores (leaf miners), and their insect parasitoids within dry-forest fragments of the Chaco Serrano in central Argentina (Cagnolo et al. 2009).

Habitat specialists are more likely to be affected by a reduction in their requisite habitat than habitat generalists (**Figure 10.19D**), as the latter can potentially make use of other habitats in the matrix via landscape supplementation (**Figure 7.3**). For example, habitat patch area was found to be the most important predictor of butterfly diversity in calcareous grasslands of Lower Saxony, Germany, but habitat specialists exhibited greater area sensitivity than habitat generalists (Krauss et al. 2003). This was demonstrated by a steeper SAR for butterfly species considered to be endemic to calcareous grasslands (i.e. habitat specialists) than for habitat generalists ($z_{specialists} > z_{generalists}$; Krauss et al. 2003). Thus, not only was patch species richness lower for habitat specialists than for habitat generalists, but more

---

[29] A Web of Science™ search performed in January 2017 using the keywords 'habitat fragmentation + patch area + patch isolation' returned ~550 citations, but other searches such as 'habitat fragmentation + patch' returned >4500 citations; 'habitat fragmentation + patch area' or 'habitat fragmentation + isolation' each produced >2700 citations, whereas 'habitat fragmentation + patch isolation' returned ~900 citations.

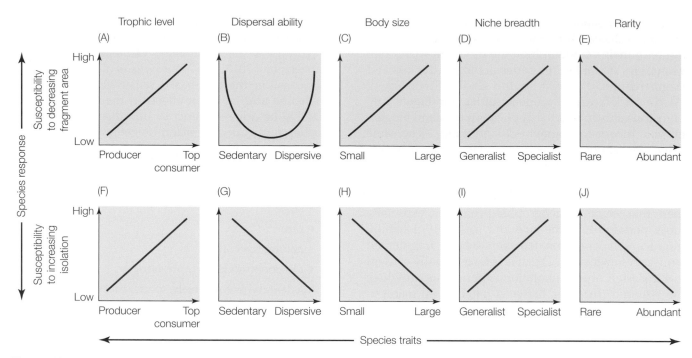

Figure 10.19 Species' responses to patch area and isolation are predicted to vary as a function of species-specific traits.
Source: After Ewers & Didham (2006).

specialist species are expected to be lost as a result of decreasing habitat area in this system.

Another type of specialization that can affect SARs involves trophic specialization, the degree to which species are constrained to feed only on species at lower trophic levels (**Figure 10.19A**). At the extreme, a predator that specializes on only one type of prey would be considered to be a highly specialized consumer, for example. Within a trophic level, specialists are expected to exhibit a higher sensitivity to patch-area effects than generalists. For example, the species–area slope was three times greater for specialist parasitoids than for generalist parasitoids in the aforementioned study in the Chaco Serrano (Cagnolo et al. 2009). Across levels of a food chain, trophic specialization results in a compounding of spatial effects, resulting in a steeper species–area slope for higher trophic levels than for lower trophic levels (e.g. $z_{predators} > z_{herbivores}$; Holt et al. 1999). This occurs because specialists of high trophic rank (e.g. top predators) are constrained in their distribution not only by processes that operate directly on their own dynamics, but also by spatial constraints on the distribution and dynamics of the lower-ranking species upon which they depend (Holt et al. 1999). In addition, species of higher trophic rank (predators or parasitoids) typically have larger resource utilization scales than species at lower trophic levels (**Chapter 6**); they also tend to be larger bodied, have larger minimum area requirements, and occur at lower population density, all of which increase a spe-

cies' susceptibility to declines in habitat area, as mentioned previously. We therefore find that while the loss of species with decreasing habitat area was greater for specialists in the Chaco Serrano, the rate of loss was more pronounced for parasitoid specialists than for leaf miner specialists, for example (Cagnolo et al. 2009).

Is specialization therefore more important than other traits for evaluating a species' sensitivity to declining patch area? In a global analysis of terrestrial vertebrates, Keinath and colleagues (2017) found that while patch area was the single most important predictor of a species' occurrence within habitat fragments,[30] habitat specialists, carnivores, and larger species were all less likely to be present than other types of species. In terms of the sensitivity of these species to patch area, however, habitat relationships played a greater role than other species' traits (Keinath et al. 2017). Habitat specialists and species that occurred in forests or shrublands were more sensitive to changes in patch area than were habitat generalists or grassland species (note that sensitivity to patch area was defined in this study as a large proportional change in a species' probability of occurrence across a range of patch sizes). Habitat specialization affected both the rarity (probability of occurrence) and sensitivity of a species to habitat loss and fragmentation (Keinath et al. 2017). Thus, for terrestrial vertebrates at

---

[30] Patch isolation was not considered in this analysis, which nevertheless incorporated other landscape metrics indicative of habitat fragmentation, such as number of patches.

least, habitat relationships were ultimately more important than life-history traits for predicting species' responses to habitat loss and fragmentation, with forest specialists exhibiting the greatest sensitivity of all (Keinath et al. 2017).

A reduction in patch area is also expected to have a greater effect on species that are sedentary or poor dispersers than those capable of dispersing to another patch (**Figure 10.19B**) to escape deteriorating conditions due to crowding, resource limitation, and increased competition or predation (**Chapter 6**). Recall that species may be particularly averse to crossing patch boundaries, which are more likely to be encountered in a small habitat patch, if the boundary is perceived to be a hard edge (**Figure 6.10**), and that species with low mobility are more likely to perceive patch boundaries as hard edges (**Figure 6.8**). Thus, reduced movement out of (or into) small patches could exacerbate extinction risk for species that are poor dispersers, especially if their rates of colonization are insufficient to offset local extinctions, thereby reducing their persistence on the landscape (**Chapter 7**).

At the other extreme, highly mobile species might also be sensitive to decreasing patch size if this ends up increasing their rate of emigration from small patches (**Figure 10.19B**). Species with high mobility are more likely to perceive the boundary as permeable (**Figure 6.8**) and/or encounter the boundary with greater frequency, leading to higher emigration rates in small versus large patches (**Chapter 6**). For example, emigration and immigration rates for the bog fritillary butterfly (*Boloria* [*Proclossiana*] *eunomia*), a habitat specialist, were about 2.5× greater in very small patches (<0.2 ha) than in very large patches (>3.0 ha) within a fragmented Belgian landscape (Mennechez et al. 2003). Increased emigration rates from small patches might well convert these to population sinks (**Figure 7.12**), and force more individuals into the matrix where they could then suffer higher mortality, leading to a reduction in dispersal or colonization success even for highly mobile species.

Dispersal ability is a relative concept, however. Species with intermediate dispersal abilities (not too sedentary but not too mobile) relative to the scale of habitat loss and fragmentation might thus exhibit the least sensitivity to a reduction in patch area (explaining the 'U-shaped' curve in **Figure 10.19B**), perhaps because these species are best able to balance the costs and benefits of dispersal (yet another example of the Goldilocks principle). These species are able to escape deteriorating conditions in small patches as needed and can incorporate nearby patches to fulfil their habitat requirements or resource needs, but are not so mobile or dispersive as to increase their likelihood of emigrating from patches, thereby risking higher mortality in the matrix.

Patch-isolation effects. The species most susceptible to increasing patch isolation tend to be those that are:

- small
- rare
- habitat specialists
- top consumers (e.g. predators)
- poor dispersers.

It should be obvious that many of the traits that heightened a species' sensitivity to a reduction in patch area also increase a species' sensitivity to patch isolation, which highlights the potential confounding between patch-area and isolation effects. We'll return to this issue in a bit, but for now, note that rarity, habitat specialization, and trophic position are all traits that have been identified as contributing to patch-isolation effects in various groups of organisms (Ewers & Didham 2006; **Figure 10.19F, I, and J**).

To illustrate how these three traits enhance sensitivity to patch isolation, we'll consider an experimental study conducted in a fragmented agricultural landscape in southwestern Germany (Kruess & Tscharntke 1994). Small patches (1.2 m²) of clover were established at varying distances (100–500 m) from the nearest meadow, which also contained clover. Consistent with expectations, the effect of patch isolation varied among trophic groups, with parasitoids being more strongly affected by patch isolation than their herbivorous insect hosts. Parasitoid species richness in the experimental clover patches decreased with increasing distance to the nearest meadow, such that isolated patches had only 17–50% of the parasitoid species found in the larger meadows (S_{meadow} = 8–12 species). As a result, parasitism rates also decreased with increasing patch isolation: the most isolated patches had rates of parasitism that were only 30% of the parasitism rates observed in meadows for some insect groups (e.g. stem-boring weevils; Kruess & Tscharntke 1994). In general, the species of herbivores and parasitoids that failed to colonize isolated clover patches occurred as populations that were both small and highly variable. Parasitoids are thus expected to be especially sensitive to patch-isolation effects given that their populations tend to be smaller and more variable than those of their herbivorous hosts. All but one of the parasitoids in this system were also specialists on clover or clover herbivores, and the effects of patch isolation are expected to be especially pronounced for habitat specialists at higher trophic levels.

Even among herbivores, however, isolation reduced the abundance of some species (stem-boring weevils) more than others (seed-feeding weevils; Kruess & Tscharntke 1994). This was attributed to differences in their flight capabilities and dispersal ability, given that there was a higher proportion of short-winged females among stem-boring weevils. By contrast, the most abundant species of seed-feeding weevils always have large wings, and thus should be better able to disperse and colonize patchily distributed habitats in a fragmented landscape such as this one.

Not unexpectedly, then, poor dispersers should be more sensitive to patch isolation than good dispersers (**Figure 10.19G**). Greater patch isolation could also increase predation risk for those individuals that do attempt to move through the matrix. For example, female root or tundra voles (*Microtus oeconomus*) experienced very high rates of avian predation outside of vegetated patches within an experimentally fragmented grassland in Norway (Andreassen & Ims 1998). Seemingly in recognition of this threat, females were 2.5× more likely to move between grassland fragments when inter-patch distances were short (1.5–3 m) than when patches were farther apart (7.5–15 m), where their exposure to predators presumably would have been greater (Andreassen & Ims 1998).

Recall, too, from our discussion of allometric scaling relationships in **Chapter 6** that small-bodied species generally do not move as far as large-bodied organisms, and therefore operate over a smaller spatial extent than large-bodied organisms (**Figure 6.4**). As such, small-bodied species are less likely to be able to incorporate neighboring patches to fulfill their minimum area requirements if these patches exceed the spatial extent of their space use. In the case of the aforementioned root vole, which is a small mammal (50 g), females were more likely to incorporate neighboring patches into their home range when these were close together (<3 m) than when patches were farther apart (Andreassen & Ims 1998). Small-bodied species are thus more likely to experience patch-isolation effects than large-bodied species (**Figure 10.19H**).

Further evidence of the negative effects of patch isolation on species is given by the ability of habitat corridors to mitigate those effects, by increasing the movement and dispersal of species among habitat fragments, thereby maintaining or increasing species richness within those fragments relative to unconnected fragments. For example, habitat corridors increased plant species richness within an experimentally fragmented landscape in the longleaf pine forest of the southeastern USA (Damschen et al. 2006; **Figure 6.13**). In the case of plants, corridors can increase native species richness by facilitating dispersal and enhancing colonization success through increased seed deposition, by increasing within-patch recruitment through enhanced pollen movement, and by altering the movements and foraging behavior of seed predators that could benefit species that might otherwise be excluded by seedling competition (Damschen

et al. 2006). Damschen and her colleagues (2006) found that plant species richness was greater in connected patches than in unconnected patches, and furthermore, that this difference increased over time. Over a five-year period, connected patches contained 20% more plant species than unconnected patches, the vast majority of which were species native to longleaf pine forests, including more than two dozen uncommon or rare species (Damschen et al. 2006).

Importantly, habitat corridors did not promote the spread of non-native or invasive plant species in this system. The potential for habitat corridors to enhance the invasive spread of non-native species that might decrease native species diversity is one of the concerns typically raised in regards to the potential value or efficacy of habitat corridors (Procheş et al. 2005; **Chapters 5 and 8**). For example, non-native plants that have attractive flowers or fruits could hijack the generalist pollinators and seed dispersers of native plant species (Richardson et al. 2000), resulting in an increase in seed production and deposition of the invasive species, such that it eventually outcompetes the native species (Lockwood et al. 2005). Although a recent meta-analysis did not find that habitat corridors generally increased the risk of invasive spread or have other undesirable effects on biodiversity (Haddad et al. 2014), these results should not be taken to mean that they never do. Corridors can indeed increase invasion by some non-native species. For example, corridors in the above-mentioned experimentally fragmented forest (**Figure 6.13**) facilitated the spread of the non-native fire ant (*Solenopsis invicta*), which led to decreased diversity of native ant species within these habitat fragments (Resasco et al. 2014). As with other species' responses to landscape structure, then, we must consider the specific group or species-specific trait when assessing corridor function and utility.

Finally, connected habitats may do better at maintaining their full complement of species following a disturbance than isolated habitats. To address this, Shackelford and her colleagues (2017) analyzed the response of more than a dozen plant and animal communities to disturbance within a wide range of ecosystems from around the world. The nature of the disturbance was as diverse as the ecosystem in which it occurred: grazing in grasslands, clearcutting of forests, hurricanes and storm surges in coastal wetlands, and bleaching of coral seascapes. Nevertheless, the strongest predictor of community response to disturbance was whether or not the habitat had become isolated from the surrounding landscape or seascape by the disturbance event. Connected communities were significantly more similar to control sites following a disturbance than were isolated communities (Shackelford et al. 2017). Indeed, connectivity was far more important to community recovery than the relative intensity of the disturbance, the amount of time that had elapsed since the disturbance, or even the size of the regional species pool. If isolated communities are less likely to recover to their pre-disturbed state, then isolation is yet another factor that must be considered when evaluating the potential resilience of landscapes and ecological communities to disturbance (**Chapter 3**).

Which has the greater effect on species diversity: patch area or isolation? The fact that similar species traits are sensitive to both decreasing patch area and increasing patch isolation illustrates the inherent challenge of trying to disentangle the two factors, in terms of which has the greater effect on species diversity. Such efforts are not helped by the fact that patch area and isolation are themselves often confounded or not controlled for in most patch-based fragmentation studies (McGarigal & Cushman 2002). Perhaps as a result, patch area and isolation are often poor predictors of species' distribution patterns (Prugh et al. 2008). Of the two, patch area was generally a better predictor than patch isolation, although this depended on which measure of patch isolation was used (**Figure 5.3**). Area and isolation had about equal effects on patch occupancy when a measure that incorporated distance to the nearest occupied patch was used, for example (Prugh et al. 2008). Species' responses to patch area and isolation were also strongly affected by the intervening matrix, with patch-area effects being greater in landscapes fragmented by human activities (agricultural, forestry, urbanization) and patch-isolation effects being greater in clearcut forests (Prugh et al. 2008). This underscores an important limitation to applying island biogeography to fragmented landscapes: habitat fragments are not islands.

NO HABITAT FRAGMENT IS AN ISLAND: THE IMPORTANCE OF THE LANDSCAPE MATRIX Unlike oceanic islands, habitat fragments are not embedded in a uniformly hostile matrix (Wiens 1995b; Haila 2002). Rather, the matrix is a mosaic of different land covers and uses that differ in their permeability and suitability to species (**Chapters 5, 6, and 9**). Recall that whether or how a species moves through the matrix is determined not only by its dispersal or gap-crossing abilities, but also by the composition and configuration of matrix habitats (**Figure 6.5**). The differential movement responses by a species to habitats within the matrix contributes to the effective isolation of habitat patches, such that some patches may be farther away than they otherwise appear based on distance alone (**Figure 5.4**). Conversely, the presence of movement corridors, which are not always discrete features on the landscape (i.e. habitat corridors) but can arise from the juxtaposition of

certain habitat types preferred by the species for movement, can increase the degree to which individuals move from patch-to-patch despite their apparent isolation (**Figures 6.28 and 9.15**). The matrix can also influence species' perceptions of patch boundaries (hard vs soft edges), which might then affect whether individuals leave or enter patches (edge permeability; **Figure 6.10**), possibly creating a population differential in abundance, survival, or reproductive rates on either side of the boundary (negative or positive edge effects; **Figure 7.5**). Negative edge effects can extend many tens to hundreds of meters into habitat fragments, reducing their core area and thus the effective size of patches (**Box 7.1**). Habitat edges can also change the frequency or intensity of interactions between species, such as increasing the numbers or activity of generalist predators along forest edges, which can then exert higher predation pressure on, for example, songbirds nesting within those forest fragments (**Box 7.2**).

The landscape matrix can thus have a strong influence on habitat fragments that may be just as important—if not more so—than the ecological processes that occur within those fragments (Wiens 1995b; Kupfer et al. 2006). Patch area and isolation can both be modified by the matrix, such that landscape context is likely to be important for interpreting patch-area and isolation effects on species diversity. Landscape context may thus partly explain why species diversity is not always well predicted by patch area or isolation measures alone. For example, fragment size was the most important factor affecting the species richness of bees, wasps, and their natural enemies within fragmented orchard meadows in Lower Saxony, Germany (Steffan-Dewenter 2003). However, the species richness of natural enemies (insect predators and parasitoids) also increased in response to greater habitat diversity in the surrounding matrix (i.e. within a 250 m radius of the focal patch). In the case of natural enemies, then, it appears that a mix of habitat types may increase either the amount of available habitat (landscape complementation) or the variety of resources (landscape supplementation) that can support or be utilized by these species.

TOWARD A LANDSCAPE PERSPECTIVE ON FRAGMENTATION EFFECTS Although they may be among the most frequently used metrics to characterize habitat fragmentation, patch area and isolation are not always good measures of fragmentation per se (Fahrig 2003; **Chapter 4**). Many of the supposed consequences of habitat fragmentation, such as a decrease in mean patch size or the increased isolation of patches, may be due entirely to the effects of habitat loss rather than fragmentation per se (**Figure 7.2**). Fragmentation research in landscape ecology is therefore moving away from a purely patch-based, island biogeographic approach to embrace more of a landscape perspective. In this context, a landscape perspective entails more than just a study of how the landscape matrix modifies patch-area and isolation effects, although that is certainly one component of it (Laurance 2008). More generally, the landscape approach to the study of fragmentation effects involves a focus on whole landscapes rather than individual patches (McGarigal & Cushman 2002; Fahrig 2003). We thus turn our attention to the broader landscape effects of habitat loss and fragmentation on species diversity.

Effects of Habitat Loss and Fragmentation on Species Diversity

Overall, it seems that, although we have learned a great deal about patch-level processes related to island biogeography, we have yet to address many of the most interesting and relevant questions about fragmentation at the landscape level.... fragmentation is a landscape-level process and ultimately can be understood only by investigating landscapes, not patches. McGarigal & Cushman (2002)

Habitat loss and fragmentation are landscape-wide phenomena, which means that the study of fragmentation effects on species diversity should ideally be performed at a landscape rather than patch scale (McGarigal & Cushman 2002; Fahrig 2003). Trying to infer fragmentation effects on species diversity from patch-scale studies has been likened to inferring metapopulation dynamics from the study of individual populations (McGarigal & Cushman 2002). Although population dynamics are important for understanding whether a given population can serve as a source or sink, and thus whether the population is likely to be a net exporter or importer of individuals, the probable persistence of a species on the landscape can only be assessed by understanding the source–sink dynamics of those interacting populations, that is, at the scale of the entire metapopulation (Chapter 7). Thus, the study of patch-scale properties, such as their size or relative isolation, is important for understanding the mechanisms that give rise to fragmentation effects, but should not themselves be confused with fragmentation effects, which are usually better addressed at the landscape scale (i.e. as a collection of patches).

Because landscapes often become fragmented as a result of habitat loss, the two processes are inextricably linked and thus their effects tend to be confounded in most studies (Ewers & Didham 2006). Trying to disentangle the effects of habitat fragmentation from those

due to habitat loss has been a challenge in practice. This can be accomplished experimentally, as in that experimental landscape system in which both the amount and fragmentation of habitat were adjusted individually at the scale of the entire landscape (With & Pavuk 2011, 2012; With 2016; Figure 5.11), or can be teased apart statistically through the study of species' responses to landscapes that vary in the amount and configuration of habitat.

A nice illustration of the latter approach is given by Trzcinski and his colleagues (1999), who tested the individual effects of forest cover versus fragmentation on the occurrence of 31 forest-breeding bird species in landscapes throughout southern and eastern Ontario, Canada. The amount of forest cover in these landscapes (100 km^2) varied from 2.5% to 55.8% as a result of forest conversion to agriculture. An index of fragmentation was obtained as a composite measure derived from the mean forest patch size, number of patches, and total forest edge using principal components analysis. The first principal component (PC1) explained 52% of the variation among landscapes, but was highly correlated with the percentage of forest cover. The second principal component (PC2) explained 42% of the variation, but was not significantly correlated with forest cover ($r = 0.18$). The second PC loaded most heavily on the number of patches and the total amount of edge, both of which are fairly unambiguous measures of fragmentation (Table 4.6; Figure 7.2). Nevertheless, the authors went a step further and removed this small degree of correlation between forest cover and PC2 by taking the residuals from a linear regression between the two as their measure of forest fragmentation.[31]

The presence of most forest-breeding birds (80%) was strongly related to the amount of forest cover on the landscape, whereas the effects of forest fragmentation were generally weak and highly variable among species. The lone exception was the hermit thrush (*Catharus guttatus*), a forest-interior species that was negatively affected by fragmentation, and whose presence was determined more by the degree of forest fragmentation than forest cover. Although it has been

[31] Although commonly used, residual regression may not be the best way to disentangle the effects of habitat amount and fragmentation (Koper et al. 2007). Using residuals as a fragmentation index tends to make the effect of habitat amount appear stronger and more influential, whereas using residuals as an index of habitat amount makes the effect of fragmentation appear more influential (Koper et al. 2007). Multiple regression was found to perform as well as, or better than, other methods, including residual regression, such that standardized partial regression coefficients were deemed useful predictors of effect strength, even when indices of habitat amount and fragmentation are correlated (Smith et al. 2009). Of course, this assumes that habitat loss and fragmentation are independent effects that can be separated out, as opposed to representing interdependent effects that should not be treated as orthogonal variables (Didham et al. 2012).

suggested that the effect of fragmentation may only occur at low levels of habitat (e.g. <20–30% habitat; Andrén 1994; Fahrig 1997), no significant interaction between forest cover and fragmentation was found for most of these species, except in the case of the hermit thrush and black-billed cuckoo (*Coccyzus erythropthalmus*). From a community standpoint, then, habitat amount had a greater effect than fragmentation on the distribution of these forest-breeding birds (Trzcinski et al. 1999).

In general, habitat amount is typically found to have large negative effects on biodiversity, whereas the effects of fragmentation (habitat configuration) tend to be weaker and more equivocal (Fahrig 2003). This has led to the **habitat-amount hypothesis**, in which species richness at a given site is predicted to be related to the amount of habitat in the surrounding landscape, rather than to the configuration of that habitat (Fahrig 2013). We thus discuss the habitat-

amount hypothesis, its predictions, and implications more fully in the next section.

Habitat-Amount Hypothesis

The premise behind the habitat-amount hypothesis is this: species richness at a given sampling site should increase as a function not of local patch size, but of the total amount of habitat in the surrounding landscape (**Figure 10.20**). That is, patch size and isolation are important only to the extent that both factors influence the total amount of habitat in the surrounding landscape (Fahrig 2013). This of course assumes that community dynamics are primarily influenced by habitat just within the surrounding landscape, and that we can define the landscape at a scale appropriate to these dynamics (e.g. Jackson & Fahrig 2012). This approach thus has much in common with neighborhood measures of patch connectivity, in which the amount of habitat within some radius of the patch defined by the species'

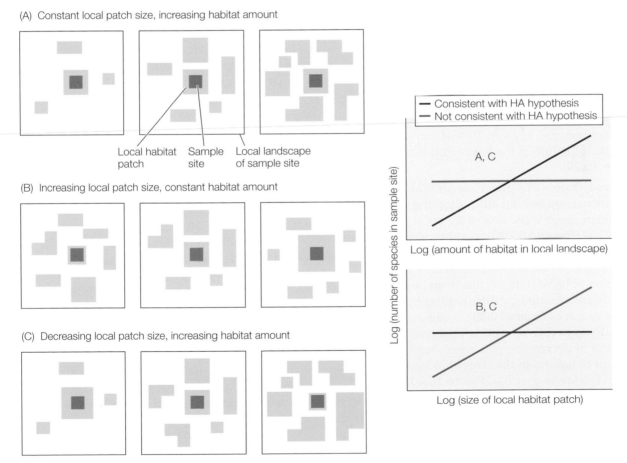

Figure 10.20 Predictions of the habitat-amount hypothesis. For a given sample site within a habitat patch, the habitat-amount hypothesis predicts that species richness will increase as a function of increasing habitat area in the surrounding landscape (A), regardless of variation in the size of the local patch (B and C). Thus, increasing the local patch size would not lead to increased site-based richness when habitat amount in the surrounding landscape remains constant (B). Similarly, site-based richness should increase with increasing habitat amount in the surrounding landscape, even if the size of the local patch size decreases (C).

Source: After Fahrig (2013).

dispersal range (the buffer size) is considered to be accessible and a source of potential immigrants (**Figure 5.3B**; **Chapter 5**). The difference here, however, is that the individual sampling location within the habitat patch, rather than the habitat patch itself, is the focus.

The habitat-amount hypothesis makes several predictions (Fahrig 2013):

1. Holding patch size constant, the number of species at a sampling site within that patch should increase as the amount of habitat in the surrounding landscape increases (**Figure 10.20A**). This contrasts with the usual assumption under the island-biogeographic/habitat fragmentation paradigm, in which species richness is expected to be similar among patches of the same size, regardless of the amount of habitat in the surrounding landscape (the horizontal blue line in the upper graph of **Figure 10.20**).

2. Site-based richness should not vary among patches of different sizes if the overall amount of habitat in the landscape is the same (**Figure 10.20B**). That is, a site within a small patch would be expected to have the same richness as a site within a large patch, so long as both patches come from landscapes having the same total habitat amount (the horizontal red line in the lower graph of **Figure 10.20**). Under the patch-based fragmentation paradigm, we would expect to see species richness increase as a function of increasing patch size (the blue line in the lower graph of **Figure 10.20**).

3. Species richness at a given site should increase with increasing habitat amount in the surrounding landscape, even if the size of the local patch decreases (**Figure 10.20C**). In other words, species richness should be greater in sites located within small patches that are surrounded by large amounts of habitat, than sites from large patches that are surrounded by little habitat (red line in upper graph of **Figure 10.20**). Again, we would typically expect to see species richness decline as a function of decreasing patch size, regardless of the amount of habitat in the larger landscape (the blue line in the lower graph of **Figure 10.20**).

The habitat-amount hypothesis thus has some important implications about the scale at which the surrounding habitat is likely to influence local community structure and dynamics. Indeed, the distinction between the patch and landscape scale would become irrelevant, as all habitat within the landscape, including the local patch, would influence species richness at a given sampling site (Fahrig 2013). In addition, it would obviously simplify matters if we could replace patch area and isolation with a single landscape measure (i.e. habitat amount). No longer would we need to worry about disentangling the relative effects of patch area and isolation on species richness, for both would contribute to the overall amount of habitat in the landscape, and thus, to the habitat-amount effect on species richness. The corollary of all this, however, is that the configuration of habitat in the landscape would then have little or no effect on species richness at a given site (Fahrig 2013). The habitat-amount hypothesis thus provides a basis for evaluating whether the spatial configuration of habitat adds significant explanatory power beyond habitat amount alone. From a statistical modeling standpoint, the most parsimonious explanation—the model with the fewest parameters—is generally preferred.

To demonstrate how the habitat-amount hypothesis can be tested, we'll consider a study by Evju and Sverdrup-Thygeson (2016), who evaluated plant species richness within equal-sized quadrats (0.25 m²) in dry calcareous grasslands in southern Norway. Calcareous grasslands naturally occur as discrete patches, and thus habitat patches could be selected to encompass a range of patch sizes and relative distances to neighboring patches, giving all possible combinations of patch types (e.g. small or large isolated patches vs small or large patches that were less isolated). The investigators quantified the amount of habitat within radii of different distances from the focal patch (0.5 km, 1 km, 2 km...7 km) to determine the most appropriate **scale of effect**, the spatial scale over which the amount of habitat in the surrounding landscape had the largest effect on site-based richness (Holland et al. 2004; Jackson & Fahrig 2012). In this case, the amount of habitat within 3 km of the focal patch was ultimately found to have the greatest effect on plant species richness within sampling sites.

Once the total amount of habitat at the 3 km landscape scale was taken into account, however, the size of the local patch was still found to have a significant effect on species richness: site-based richness increased more rapidly with increasing habitat amount when the focal patch was large than when it was small (Evju & Sverdrup-Thygeson 2016). This is contrary to the predictions of the habitat-amount hypothesis, in which only the amount of habitat in the surrounding landscape should matter and not the size of the local patch that contributes to that total. The authors therefore concluded that 'the spatial configuration of habitat matters' for plant species richness in these calcareous grasslands (Evju & Sverdrup-Tygeson 2016).

This, of course, assumes that patch size or isolation can be viewed as measures of spatial configuration, which harkens back to our earlier discussion regarding the application of island biogeography to fragmented landscapes. As we've discussed, patch area and isolation are not always good measures of fragmentation, especially if habitat loss is primarily responsible (Fahrig 2003; **Figure 7.2**). Nevertheless, the spatial configuration of naturally fragmented systems is typically characterized in terms of patch size and isolation, especially when the habitat covers only a small percentage of the study region, as with the dry calcareous grasslands in the above study (which occupy <1% of the landscape). Still, we seem to be back to where we started, trying to tease apart the effects of habitat area from fragmentation per se, and focusing on patch-scale rather than landscape-scale effects on species richness. Rest assured that we are not simply going around in circles, but rather are exploring the interdependence of these concepts, as we consider in greater detail next.

Interdependence of Habitat Loss and Fragmentation

Habitat loss and fragmentation may be landscape-wide phenomena, but they clearly have patch-scale consequences. Many of the supposed measures of habitat fragmentation, such as patch area and isolation, are dependent on the amount of habitat remaining on the landscape. This in turn contributes to the inevitable confounding of habitat area and fragmentation effects, which is reflected in the broad overlap in the variance that is potentially attributable to habitat loss versus fragmentation per se (Ewers & Didham 2006; Didham et al. 2012). Whether we end up attributing this intercorrelated variance to habitat loss or fragmentation, however, is likely to depend on our perspective, in terms of whether we adopt a patch-based or landscape perspective.

From a patch-based or island-biogeographic perspective, the interdependence of habitat loss and fragmentation at the landscape scale is typically ignored, with the result that all the intercorrelated variance tends to be attributed to the effects of habitat fragmentation (Didham et al. 2012). Conversely from the landscape perspective, in which habitat loss and fragmentation are treated as independent effects, most of the intercorrelated variance is inevitably attributed to habitat loss (Fahrig 2003). If the effects of habitat loss and fragmentation are interdependent, however, then the intercorrelated variance cannot be directly attributable to either factor alone (Didham et al. 2012). To remedy this, Didham and his colleagues (2012) suggest a third approach which integrates both patch-based and landscape perspectives, using hierarchical causal models (i.e. structural equation modeling or path analysis) to explore the direct and indirect relationships by which the amount and spatial arrangement of habitat can affect species' responses to landscape change (**Figure 10.21**).

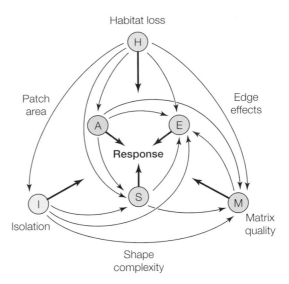

Figure 10.21   A hierarchical causal model of the direct and indirect effects of habitat loss and fragmentation. The outer variables are landscape factors that influence the amount, configuration, and/or quality of habitat in which the species occurs: degree of habitat loss (H), habitat connectivity (isolation, I), and matrix quality (M). Each of these landscape factors has an influence on various patch-scale attributes, such as patch area (A), patch shape complexity (S), and patch edge (E). Each pathway is a hypothesis as to how the landscape or patch-scale factor influences—whether directly or indirectly—the measured response variable. As such, not all of these pathways will be relevant to every response variable or species. Arrows indicate the direction of causal inference, which will also vary depending on the type of response and species in question. The community response thus emerges based on the way these spatial factors influence individual species' responses, as well as the strength of interactions among these species (not shown). Even if a species did not respond directly to habitat loss and/or fragmentation, there could still be indirect effects related to how these spatial factors influence other species in ways that alter the strength of species interactions (e.g. increased competition or predation).

Source: After Raphael K. Didham, Valerie Kapos, and Robert M. Ewers (2011).

[A] hierarchical causal model would help resolve the striking paradox between the landscape-biased conclusions that habitat fragmentation has negligible effects on biodiversity after habitat amount is taken into account, versus the conclusions of many thousands of patch-based studies showing strong ecological effects of patch area, matrix hostility, edge effects and so on. The most parsimonious explanation for this apparent paradox is that the effects of habitat loss are mediated in large part by changing spatial arrangement of habitat – that is, habitat loss acts via the change in habitat arrangement, not *independently* of it.   Didham et al. (2012)

Given that habitat fragmentation is defined as a process by which habitat loss leads to a greater number of smaller patches that are isolated from each other by a matrix of dissimilar habitat(s), we could represent the direct and indirect causal relationships hierarchically, as shown in **Figure 10.21**. The outermost variables describe the landscape context, which encompasses the amount (or loss) of habitat (H), matrix quality (M), and the degree of habitat connectivity, or its converse, isolation (I). All of these landscape factors exert an influence on patch attributes, which are represented by the innermost variables in the hierarchical causal model. For example, patches can be defined in terms of their area (A), shape complexity (S), and amount of edge (E). Each path in the hierarchical causal model thus represents a hypothesis as to how these patch and landscape variables are expected to influence species' responses to habitat loss and fragmentation via one or more mechanisms, either directly or indirectly (**Chapters 6–9**). For example, habitat loss results in smaller patches that contribute to greater edge effects, such as increased predation or parasitism rates, leading to decreased nesting success for songbirds breeding within habitat fragments (**Box 7.2**).

Species will obviously differ in their response to these various spatial measures of habitat loss and fragmentation. Not all of the potential paths depicted in **Figure 10.21** will necessarily apply to every species, therefore. Nor is the strength or direction of causality always the same, as this will likely differ among species and even for the same species in different landscape contexts. For example, some species have a positive rather than negative response to habitat edge, and the type of edge response may vary even within a given species depending on the quality of the matrix habitat (**Figure 7.5**). Within a species, different demographic processes (the BIDE factors) are influenced to varying degrees by the amount and spatial arrangement of habitat (**Chapters 6 and 7**), perhaps even in contrasting ways, resulting in different causal relationships that ultimately drive population responses to habitat loss and fragmentation. These direct and indirect effects of habitat loss and fragmentation on individual species' abundances are then reflected in the community-level response to landscape change.

Species do not exist in a vacuum, however. They are interacting with each other, as well as with the landscape. Thus, even if habitat fragmentation does not have a direct effect on a species, the spatial structure of the landscape may still have indirect effects, by altering the interaction strength among species (Didham et al. 2012). For example, habitat loss and fragmentation may facilitate the spread of an invasive, competitively dominant species that then has a suppressive effect on populations of native species

(**Chapter 8**). Fragmentation would thus be having an indirect effect on these native species, which is mediated though a change in interaction strength among species (i.e. competition has increased as a result of the invasive spread of a competitively dominant species). Because of the potential for these sorts of indirect effects of habitat loss and fragmentation on species, we'll turn our attention next to the ways in which landscape structure might affect different types of species interactions.

Landscape Effects on Species Interactions

Species diversity represents the coexistence of multiple interacting species within a given area. Up to now, our focus has largely been directed at the effects of landscape structure on the patterns and scaling of species diversity, from the local site or patch scale to the broader landscape or regional scale. We are therefore ready to shift our focus to the spatial dynamics of interacting species, which help give rise to patterns of diversity. In this section, we thus consider how landscape structure influences species interactions, such as those that affect competitive coexistence, predator–prey and host–parasitoid dynamics, and plant–pollinator interactions.

Competitive Coexistence

Interspecific competition has long been thought to play a major role in the level of diversity attained by ecological communities. The maxim that 'complete competitors cannot coexist' (Hardin 1960) neatly summarizes this view: species that are too similar ecologically will be unable to coexist if they are competing for the same limiting resource (e.g. food, space, light, water, nutrients), such that all but the best competitor will eventually be excluded from the community. This **competitive exclusion principle** is generally attributed to the Russian biologist, Georgiĭ F. Gause, who famously demonstrated that two species of *Paramecium*, that ubiquitous ciliated protozoan of general biology labs, were unable to coexist for very long when mixed together in the same culture. The species with the faster growth rate (*P. aurelia*) inevitably outcompeted the other species (*P. caudatum*) as their bacterial food source became limiting (Gause 1934). Although Gause generally gets credit for the competitive exclusion principle (it is sometimes even called Gause's Principle or Gause's Law), the potential for competitive exclusion had been predicted previously by the two-species competition models published independently (and nearly simultaneously) by mathematicians Alfred Lotka and Vito Volterra (Lotka 1925; Volterra

1926),[32] and the general idea appeared even earlier in the writings of Joseph Grinnell (who coined the term 'niche') and Charles Darwin (Hardin 1960; Palmer 1994).

Through the lens of competitive exclusion, community ecologists began attributing the coexistence of species within communities to **niche partitioning**, in which ecologically similar species use a shared resource in different ways, thereby limiting the degree of niche overlap—and hence competition—among them (Holt 2001). Competition for resources can be direct if individuals interfere with the ability of others to acquire resources via aggressive behaviors such as overgrowth or territoriality (**interference competition**), or indirect if individual resource use depletes the amount available to others (**exploitation competition**). For example, species may avoid competition by foraging in different areas, in different ways, or at different times, which then prevents them from overlapping too greatly in resource use, whether spatially or temporally.

The textbook example of niche partitioning is Robert MacArthur's classic study on the foraging ecology of wood-warblers (Parulidae) in the spruce-fir forests of the northeastern USA (MacArthur 1958). Despite the fact that all five species of wood-warbler that MacArthur studied are insectivorous and of similar size and bill morphology, they all co-occur in the same forest stands and even forage and nest within the same types of trees (albeit, not necessarily all at once in the same tree). MacArthur found that despite some overlap between species, each tended to forage at a different height and within different 'zones' of the trees (e.g. the outer foliage vs inner branches), and even used different foraging behaviors to obtain prey (e.g. gleaning insects from foliage vs flycatching). Such differences might well enable these otherwise similar species to partition a common food resource (arboreal arthropods), by subdividing their habitat spatially (different foraging zones) as well as temporally (the peak nesting period—a time of greatest food demand—also varied among species), thereby permitting them to coexist in the same forest. By reducing the degree of niche overlap among species, resource partitioning results in niche differentiation among species, thereby allowing for greater 'species packing' within communities. From that standpoint, then, niche differentiation helps

maintain species diversity by limiting competitive exclusion.

Following MacArthur's lead, ecologists soon found that most species differed in some aspect of their foraging ecology or habitat use, such that even subtle differences were being ascribed to niche partitioning and taken as evidence that interspecific competition was structuring communities, whether now or at some point in the past (i.e. the ghost of competition past; Connell 1980). Besides conflating pattern with process, competitive exclusion was ultimately deemed too restrictive and insufficient a mechanism to be the primary explanation for patterns of diversity (Mouquet et al. 2005): too restrictive because it is concerned only with the effect of a single limiting resource on species diversity at the local community scale (e.g. within a homogeneous patch), and insufficient because it could not explain the unexpectedly high levels of diversity found in systems with a limited number of resources (e.g. the paradox of the plankton; Hutchinson 1961). Indeed, there are a number of conditions required for competitive exclusion (Palmer 1994): species must be limited by a single resource; rare or competitively inferior species cannot possess an advantage (e.g. no competitive reversals or trade-offs between competitive ability and dispersal ability); the community must be closed, with no immigration (i.e. no rescue effect); there must be sufficient time and opportunity for species to compete; and, the habitat is homogeneous and static (i.e. no disturbances). The corollary to all this, then, is that violation of one or more of these conditions should increase the number of species that can potentially coexist in an area (Palmer 1994).

Given that landscapes are patchy, heterogeneous, and dynamic (**Chapter 3**), we can anticipate that one or more of these factors will help to enhance competitive coexistence among species. Thus, even if species do not co-occur at a local patch or community scale, coexistence might still be possible at the broader landscape scale. Although there are several ways by which competitive coexistence can be enhanced at a landscape scale (Hanski 1995; Chesson 2000a; Amarasekare 2003), we will focus here on two of the most commonly studied mechanisms: (1) habitat heterogeneity and competitive reversals and (2) competition-colonization trade-offs (Holt 2001; Kneitel & Chase 2004; Hoopes et al. 2005).

---

[32] Indeed, Gause's experiments with *Paramecium* were apparently inspired by the theoretical predictions of the Lotka–Volterra equations: '[C]an two species exist together for a long time in a microcosm, or will one species be displaced by the other entirely? This question has already been investigated theoretically by Haldane [1924], Volterra [1926] and Lotka [1932]. It appears that the properties of the corresponding equation of the struggle for existence are such that if one species has any advantage over the other it will inevitably drive it out completely...' (Gause 1934: 98).

HABITAT HETEROGENEITY AND COMPETITIVE REVERSALS If the landscape consists of many habitat types and each species specializes in a different habitat, then coexistence can be achieved at a landscape scale if species are at a competitive advantage in their preferred habitat (Holt 2001; Hoopes et al. 2005). In other words, the ranking of species in the competitive hierarchy varies by habitat: species that are superior competitors in one habitat type are inferior competitors in others (a **competitive reversal**). From that standpoint, then, there could be as many species as there are habitat types in the landscape, so long as each species is able to persist within its preferred habitat (Hoopes et al. 2005). Note that this is not so different from classical niche partitioning that permits coexistence at a local scale. At the landscape scale, however, competitive coexistence is occurring through the partitioning of different habitats among species (**spatial niche differentiation**), rather than through the partitioning of a single limiting resource.

If species differ in their competitive dominance among habitats, this can give rise to a **spatial storage effect**, in which each species' population dynamics are essentially 'buffered' against spatial variation in population growth rates across a heterogeneous landscape (Chesson 2000b). Under a spatial storage effect, decreasing population growth rates in unfavorable habitats, where species are inferior competitors, would be offset by positive population growth rates in favorable habitats, where they are superior competitors.[33] In effect, spatial storage effects enable species coexistence through source–sink dynamics. Competitive exclusion of inferior competitors within unfavorable habitats (sinks) is prevented by dispersal from favorable habitats where they are superior competitors (sources).

For competitive coexistence to occur via source-sink dynamics, there must be some sort of negative feedback mechanism that limits population growth in favorable habitats, else a superior competitor will spread and eventually exclude other species within the landscape. Given that a species' competitive advantage will typically lead to higher population densities in favorable habitat (sources), species will experience increasingly high levels of *intra*specific competition in that habitat as population size increases (i.e. stronger competition among individuals of the same species as opposed to individuals of different species), which acts to limit further population growth in source habitats. In unfavorable habitats where they are inferior competitors (sinks), species are mainly limited by interspecific competition in sink habitats, resulting in low population densities that may be supplemented by immigration from source habitats. Such differences in the strength of competitive interactions and population densities among habitats mean that species experience lower competition on average than if all populations had the same mean density and experienced the same mean level of competition (Hoopes et al. 2005). Spatial storage effects thus enhance competitive coexistence by limiting the population growth rates of dominant species (i.e. in less favorable habitats), while providing spatial refugia for otherwise inferior competitors to persist on the landscape (Sears & Chesson 2007).

Habitat degradation as a result of climate change or land-use practices can actually cause competitive reversals, which in some cases can facilitate invasion by non-native species (Didham et al. 2007). For example, the European earthworm *Aporrectodea trapezoides* has been introduced all over the world, either inadvertently with the spread of European agriculture or intentionally for use as fishing bait (Fernández et al. 2010). In California, it has displaced the native earthworm *Argilophilus marmoratus* in disturbed grasslands (fertilized pastures grazed by livestock) owing to its faster growth rates and earlier onset of reproduction (Winsome et al. 2006). In undisturbed native grasslands, which are less productive, the invasive earthworm is unable to acquire enough resources to maintain its rapid growth and reach reproductive maturity. Thus, although the native earthworm is the weaker competitor of the two, a competitive reversal occurs in less productive grasslands, where the native species can exert a negative effect on the invasive species that exacerbates the role of resource limitation in preventing spread of the non-native earthworm into undisturbed native grasslands (Winsome et al. 2006). Invasive spread may depend not only on the abundance or population growth rates of the invader (**Chapter 8**), but also on the degree to which habitat degradation alters competitive interactions between invasive and native species (Didham et al. 2007).

Then again, the competitive exclusion of a native species by an invasive species is expected to be a slow process at the landscape or regional scale (Yackulic 2017). Although competition may drive declines in the local distribution and abundances of species over relatively short time periods (years to decades), competitive exclusion across a region can take many decades or centuries. Besides outcompeting native species, invasive species can alter landscape structure in ways that negatively affect the metapopulation persistence of native species, thereby contributing to an **extinction debt** in which native species may persist for decades (or even centuries) in spatial refugia before going extinct (**Box 10.1**). This appears to be the case for native plants in the serpentine grasslands[34] of California (USA), which are

---

[33] In the analogous temporal case, from which the term 'storage effect' was derived, seed banks or long-lived adults 'store' the effects of favorable years, which buffer the effects of bad years when populations decline (Chesson 1994). In spatial storage effects, dispersal across space (source–sink dynamics) has the same effect as long-lived life stages in the temporal storage effect (dispersal through time; Hoopes et al. 2005).

[34] Recall that serpentine grasslands form on serpentine soils, which are dry nutrient-poor soils that contain magnesium and heavy metals toxic to most plants not adapted to these soils. A unique flora is therefore associated with serpentine soils, including many native plant species that are now rare and endangered.

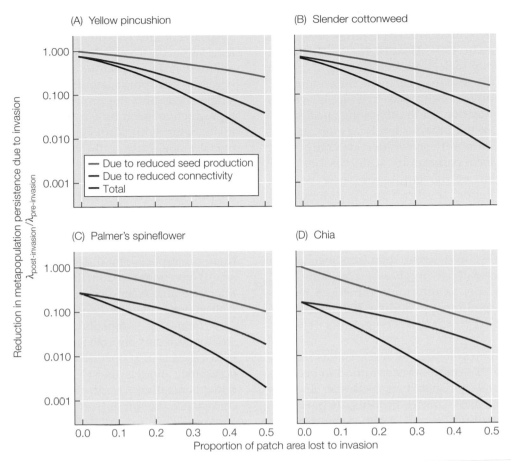

Figure 10.22 Non-native grasses outcompete native annuals in serpentine grasslands. Invasive grasses reduce metapopulation persistence in several annual plant species ($\lambda_{post\text{-}invasion}/\lambda_{pre\text{-}invasion}$ < 1; black lines). The impact of invasive grasses on annual plants can be partitioned into the multiplicative effects of reduced seed production and reduced habitat connectivity. Invasive grasses decrease the size of native habitat patches, thereby reducing the population size and seed production of native plants (blue lines), which has consequences for their colonization–extinction dynamics and results in decreased metapopulation persistence. Furthermore, invasive grasses outcompete native bunchgrasses in the matrix, which negatively affects dispersal of native annuals and thus reduces habitat connectivity (red lines), causing further reductions in metapopulation persistence. The relative effects are depicted here for several native plants: A) yellow pincushion (*Chaenactis galibriuscula*), B) slender cottonweed (*Micropus californicus*), C) Palmer's spineflower (*Chorizanthe palmerii*), and D) Chia (*Salvia columberiae*).
Source: After Gilbert & Levine (2013).

being invaded by European grasses (Gilbert & Levine 2013). Not only have invasive European grasses reduced the availability of habitat for native annuals, but they have also decreased the permeability of the intervening matrix to dispersal by these native species. By reducing the availability of habitat, invasion leads to smaller population sizes and reduced seed production by native species within serpentine patches, which in turn alters their colonization–extinction dynamics and leads to decreased patch occupancy, thereby reducing metapopulation persistence across the landscape (**Figure 10.22**). By replacing the native bunchgrass species that formerly occurred in the matrix, European grasses have also altered the connectivity of the landscape for native annuals, by significantly limiting their ability to colonize patches only a few meters away. Reduced connectivity ultimately had a greater effect than reduced seed production on metapopulation persistence in invaded landscapes for all of the native annuals examined in this study (**Figure 10.22**). Even under low levels of invasion, most species that occupy <10% of habitat patches may enter an extinction debt, which could take hundreds of years to play out (Gilbert & Levine 2013). The appearance of a stable coexistence between native and invasive species at the landscape scale might therefore be illusory: just a long interlude during the slow march toward competitive exclusion.

COMPETITION–COLONIZATION TRADE-OFFS Even if all habitat patches are of the same type, coexistence at the landscape scale might still be possible if there is a trade-off between a species' competitive and colonization abilities (Levins & Culver 1971; Hastings 1980; Nee & May 1992; Tilman 1994; **Box 10.1**). Recall that at the local community scale, trade-offs among species in the use of a limiting resource (i.e. niche differentiation) enhances competitive coexistence. In a similar vein, a competition-colonization trade-off among species permits coexistence at the broader landscape or regional scale (the metacommunity scale), despite competitive exclusion at the local community scale. Species coexistence is possible at the landscape scale because competitively inferior species have a colonization advantage: they can colonize new patches before the better competitors arrive, such that inferior competitors exist as 'fugitive species' on the landscape (Horn & MacArthur 1972). In essence, then, this mechanism of competitive coexistence involves a 'constant chase' across the landscape, as inferior competitors repeatedly escape competition by finding habitat patches not yet colonized by superior competitors (Hoopes et al. 2005).

Most competition-colonization research has focused on plants, where this trade-off is assumed to occur in the negative correlation between seed size and the number of seeds a plant species produces (e.g. Harper et al. 1970; Turnbull et al. 1999; Jakobsson & Eriksson 2000; Levine & Rees 2002). For the same reproductive effort, species could produce many small seeds or a few large ones. A species' competitive ability is thought to be enhanced through the production of large seeds, which result in higher seed and seedling survival, and thus are more likely to become established at a site. Colonization ability, by contrast, might be facilitated by the production of many small seeds, which are lighter and more easily dispersed, contributing to a greater 'propagule rain' that enhances colonization success. The trade-off between seed size and seed number is thus often used as a proxy for the competition–colonization trade-off, even though empirical support for such a trade-off has been equivocal. In many cases, field research fails to find evidence of a competition-colonization trade-off either because the trade-off is an imperfect one (Banks 1997) or because dispersal rather than colonization is being assayed in these studies (Cadotte et al. 2006). Recall that dispersal describes the movement of individuals or propagules (**Chapter 6**), whereas colonization also includes the ability of populations to become established, which entails both successful dispersal and positive population growth rates in the new site (**Chapters 7 and 8**).

This distinction was demonstrated in a comparison of 15 wind-dispersed flowering plants (Asteraceae) in southeastern Sweden by Jakobsson and Eriksson (2003), who evaluated each species' dispersal and colonization abilities relative to its competitive abilities. Dispersal ability was measured at the individual seed level as seedfall velocity (m/s), whereas colonization ability was assayed at the 'brood level' using a measure that incorporated both the dispersal abilities of single seeds and the number of seeds produced per reproductive event by a species; the latter should correlate better with plant colonization success than dispersal alone. Despite the observed trade-off between seed size and seed numbers among this group of species, Jakobsson and Eriksson (2003) ultimately found support for a competition-*dispersal* trade-off, but not for a competition-colonization trade-off. Thus, the trade-off between seed size and seed number may not always be a good proxy for a competition-colonization trade-off, especially if differences in seed size (large vs small) have more to do with a trade-off between stress tolerance and fecundity than competition and colonization per se (Muller-Landau 2010).

Whether a competition-colonization trade-off is strictly necessary for coexistence in spatially structured landscapes is also open to debate. The classical patch-occupancy models that demonstrate how a competition-colonization trade-off can promote species coexistence (Levins & Culver 1971; Hastings 1980; Nee & May 1992; Tilman 1994; **Box 10.1**) make several simplifying assumptions, chief among them being that competition is asymmetric and not preemptive. In other words, species' competitive abilities can be ranked with respect to a strict dominance hierarchy, and a better competitor is always able to displace competitively inferior species from patches, but not vice versa. Unsurprisingly, relaxing these and other assumptions affects whether a competition-colonization trade-off can promote species coexistence, as in some cases it clearly would not (Amarasekare 2003).

For example, a superior competitor may be unable to displace an inferior competitor once the latter has become established at a site, especially if propagules or other early life stages are physically unable to supplant larger or later life stages (e.g. a seed cannot displace a tree, regardless of each species' relative competitive abilities). If competition for newly available sites is basically a lottery among propagules or juveniles,[35] then a competition-colonization trade-off no longer

---

[35] Originally developed to explain the high diversity of coral-reef fishes, the lottery hypothesis (Sale 1977) posits that high diversity can be maintained through random recruitment of available sites (free space) so long as species all have equal colonization abilities and are unable to be displaced once they have colonized a site (i.e. there is a priority effect). In the present context, where competition for space is preemptive and available sites are colonized by whichever species arrives first (akin to winning the lottery), the best colonizers—whether because of their superior dispersal abilities or greater fecundity—will generally have an advantage and will eventually replace poorer colonizers over time.

guarantees coexistence because the best colonizers (i.e. inferior competitors) will win the race for space (Yu & Wilson 2001). In that case, species coexistence might still be possible if there is spatial variation in the distribution of habitat, owing to differences in patch size or habitat density.[36] Superior competitors have the advantage in large patches of intact habitat (high habitat density) because of their suppressive effects on the colonization rates of inferior competitors. For example, superior competitors might outcompete other species for pollinators or seed dispersers, thereby reducing the fecundity and dispersal abilities of inferior competitors, which effectively negates their main advantage (i.e. superior colonization ability). Conversely, inferior competitors have the advantage over superior competitors when habitat patches are small and fragmented (low habitat density) owing to their superior colonization abilities. Thus, a mix of patch sizes or densities across a landscape would ensure that each type of species 'does best' somewhere, thereby permitting species coexistence at the landscape scale (Yu & Wilson 2001). One implication of all this is that the pattern of habitat loss (fragmentation) could be just as important as the amount of habitat loss for understanding community responses to habitat loss and fragmentation. Although this sort of patch-based model had previously predicted that habitat loss would lead to the sequential extinction of superior competitors before competitively inferior ones (Tilman et al. 1994; **Box 10.1**), relaxing the model assumptions to include preemptive competition illustrates how the pattern of habitat loss could tip the balance toward one type of species or another (good competitors vs good colonizers), depending on the resulting patch size and/or habitat density within the landscape (Yu & Wilson 2001).

Spatially explicit simulation models have demonstrated how coexistence can be maintained in patchy landscapes even in the absence of a competition-colonization trade-off. For example, constraining dispersal and competition to a local neighborhood obviated the need for such a trade-off in the first place; that is, inferior competitors no longer needed a colonization advantage to coexist with a superior competitor (Higgins & Cain 2002). In that case, coexistence may be promoted by the creation of refuges from competition: restricted dispersal leads to greater spatial aggregation, such that populations experience more intraspecific than interspecific competition. Furthermore,

if displacement by a superior competitor is not instantaneous, which is certainly true in plants, then inferior competitors may also experience a brief temporal refuge, affording them sufficient time to reproduce and disperse seeds into a patch where their competitors are absent.

This is similar to the findings of a simulation model of competition between two annual plant species in hierarchically structured landscapes (hierarchical neutral landscape models; **Figure 5.9**), in which both spatial and temporal storage effects promoted species coexistence (Lavorel et al. 1994). For species that differed only in their dispersal abilities, such that competition for space was a lottery, the species with the shorter dispersal range would always win out over the better disperser, with competitive exclusion occurring most rapidly when habitat was limiting and sites were aggregated. Here, short dispersal leads to propagule clumping, such that the poorest disperser is actually the best colonizer (illustrating yet again why dispersal ability does not equate to colonization ability). Temporal storage effects were also introduced by allowing species to differ in dormancy type, in which some fraction of seeds became dormant each year (creating a seed bank) and then either germinated at some constant rate (a risk-spreading strategy) or in response to a disturbance that created vacant sites for colonization (a disturbance-broken strategy). Although dormancy allows for seeds to persist for a time at any site the species reaches by dispersal, its primary effect was simply to delay competitive interactions that emerged as a result of spatial storage effects. Thus, coexistence occurred at the landscape scale only if the two species had similar dispersal abilities. Note that in this case, habitat patchiness alone does not guarantee coexistence because the environment is equally suitable for both species. Rather, it is the resulting pattern of spatial segregation that emerges between the two species, in which intraspecific competition is increased relative to interspecific competition, that promotes coexistence.

It has been suggested that trade-offs between traits, such as competition and colonization ability, can contribute to competitive coexistence only if there is some sort of negative feedback mechanism that limits a superior competitor's population growth in favorable habitats (sources) or when exploiting its preferred resource, else it will come to dominate the landscape (Kneitel & Chase 2004). As we have seen previously for spatial storage effects and source–sink dynamics, increasing levels of intraspecific competition within preferred habitat limit further population growth of superior competitors. Species are not just limited by habitat or resources, however. Natural enemies, such as predators or parasites, can also limit populations of their prey or host species. For example, two prey

---

[36] However, trade-offs in the real world are 'messy' (Banks 1997), and most communities will therefore lie somewhere in between the extremes of pure dominance (displacement) competition and fully preemptive competition depicted here. In fact, preemption could theoretically increase species coexistence if the competition-colonization trade-off is only of moderate intensity and the community is not strongly limited by colonization (Calcagno et al. 2006).

species that compete for the same limiting resource may still be able to coexist if the superior competitor is more strongly limited by predation (**apparent competition**; Holt et al. 1994). In other words, competitive coexistence can also occur if there is a trade-off between prey competitive ability and susceptibility to predation. Because the type and abundance of predators, as well as a species' susceptibility to predation, can all vary among habitats, we consider next the effects of landscape structure on predator–prey dynamics.

Predator–Prey Dynamics

Prey species must also make a trade-off while foraging: they must balance their intake of food (energy) against the risk of themselves being eaten by a predator. The ideal habitat or resource patch would therefore enable prey species to maximize their rate of energy intake while minimizing their predation risk (Moody et al. 1996). The distribution of predators relative to their prey can thus be thought of as a behavioral space race: predators prefer areas where prey are more abundant, whereas prey tend to avoid areas where predators are most abundant (Sih 2005).

In the absence of predators, prey should concentrate within areas of the landscape where resources are most abundant (i.e. high-quality patches or habitats). As prey densities increase within high-quality patches, however, competition for resources—whether from conspecifics or other species—will also increase. At some point, a forager might therefore do better to select lower-quality habitat patches to achieve a certain food intake rate and meet its energetic needs (i.e. an ideal free distribution; **Figure 6.14**). In that case, the distribution of prey species would mirror the availability of resources among patches, being highest in high-quality patches but occurring at lower abundance in lower quality habitats (Moody et al. 1996). If predators are attracted to areas of high prey abundance, however, then habitat quality must also be assessed in terms of predation risk. In that case, prey species are unlikely to match the distribution of resources (e.g. prey densities are lower than expected in high-quality patches). Conversely, their use of high-risk patches may depend on how much they stand to gain energetically. For example, a high reward might offset predation risk (Moody et al. 1996). The extent to which predators and their prey overlap in their distributions will obviously influence the strength of predator–prey interactions, and therefore has important consequences for community structure and dynamics (Sih 2005). We thus consider the scale of spatial overlap between predators and prey next.

THE SCALE OF SPATIAL OVERLAP BETWEEN PREDATORS AND PREY Different constraints or mechanisms are likely to influence the distribution of predators rela-

tive to their prey and at different scales. Predators tend to be more mobile than their prey, and are thus more likely to match the distribution of their prey when spatial overlap is assessed at a broad scale. In addition, prey may reduce their activity to avoid detection in high-risk areas, thereby reinforcing this positive association between predators and prey at broader scales. Conversely, prey should track their resource distribution at a finer scale relative to that of the predator, given their trophic position and lower mobility. Furthermore, interference competition may be stronger among predators than prey, such that individual predators are unlikely to share the same patch, preventing predators from precisely tracking prey at those finer scales. As such, there may be no spatial overlap between predators and prey when their joint distribution is assessed at finer scales of patchiness. Alternatively, the presence of predators within patches could drive prey out, creating a negative relationship between predators and prey at the patch scale. Because of these contrasting responses by predators and prey at different scales, predators may ultimately exhibit little to no spatial overlap with prey at fine scales, but exhibit a high degree of spatial overlap at broader scales.

With fully interacting predators and prey, the spatial dynamics should be a balance of the prey's attempts to be in areas with low predation risk and the predator's attempts to be in areas with more prey. How these preferences play out over space and time have large population and community consequences in setting species distributions, interaction rates and ultimately, species abundances.   Hammond et al. (2012)

To test this hypothesis, Hammond and his colleagues (2012) explored the degree of spatial overlap between dragonfly larvae, which are voracious predators in many aquatic systems, and their tadpole prey, at two nested scales of patchiness within experimental tanks (aquatic landscapes). At the broadest scale, resource patchiness was created by dividing the tank into a 'high-resource area' (containing 80% of the total resource abundance) and a 'low-resource area' (containing 20% of the total resource abundance). Within each of these areas, a finer scale of resource patchiness was created by placing different-sized patches (large vs small algal discs, which provided a food source for the tadpoles) at opposite ends, with patch sizes being smaller overall in the low-resource area than in the high-resource area. Space-use patterns were assessed for predators and prey when each species was alone versus when both were together in the tank.

At the broadest scale, prey exhibited a clear preference for high-resource areas when predators were absent, whereas predators showed no preference for either area when prey were absent. When prey were present, however, predators exhibited a strong preference for the high-resource areas, presumably because their prey were concentrated in those areas. Predators thus shifted their distribution in the presence of prey, resulting in a high degree of spatial overlap between predators and their prey, at least initially. Prey eventually started moving out of these high-resource areas, presumably because of higher predation pressure, and thus became more uniformly distributed over time.

At the finer patch scale within resource areas, prey always exhibited a preference for the high-resource patch (i.e. the larger of the two patches) regardless of whether predators were present, but this preference for larger patches was especially pronounced in high-resource areas. Predators, by contrast, were uniformly distributed with respect to the different-sized patches regardless of the presence or distribution of their prey. Consistent with expectations, then, predators and prey exhibited a high degree of spatial overlap only when assessed at the broader scale of resource patchiness, whereas only prey responded to the finer-scale patchiness of the resource distribution (Hammond et al. 2012). The spatial overlap between predators and prey thus depends on the scale at which it is examined.

EFFECT OF SPATIAL REFUGIA ON THE STABILITY OF PREDATOR–PREY DYNAMICS

Given that resources or habitat—and thus prey—tend to be patchily distributed at some scale across the landscape, how might this influence predator–prey interactions? Does it have a stabilizing or destabilizing effect on predator–prey dynamics? The classic Lotka–Volterra predator–prey model, which predicts reciprocal oscillations between predator and prey populations, assumes that the environment is homogeneous. In the lab and field, however, it has been difficult to find evidence of such regular cycling between predators and their prey.[37] For example, Gause (1934), who also studied predator–prey interactions in his experiments with ciliated protozoans, found that a predator (*Didinium nasutum*) invariably drove its *Paramecium* prey (*P. caudatum*) extinct in a homogeneous environment, unless the prey was given a refuge (a substrate in which they could hide from the predator). Even then, the predator went extinct unless assisted immigration occurred (Gause supplemented the predator population by adding new individuals every few days). Recall from our discussion of Huffaker's classic mite experiment in

Chapter 3 that the predator and its prey could only coexist when spatial refugia were added and prey could out-disperse their predators (Huffaker 1958; Huffaker et al. 1963). From this we might conclude that spatial heterogeneity or patchiness has a stabilizing effect on predator–prey interactions, so long as each species is able to persist through asynchronous colonization–extinction dynamics; that is, as a predator–prey metapopulation (Murdoch et al. 1992).

If patchiness interferes with the search efficiency of predators, however, it can have a destabilizing effect on predator–prey dynamics. In a now-classic experiment, Kareiva (1987) compared population densities of the predatory seven-spotted lady beetle (*Coccinella septempunctata*) and their aphid prey (*Uroleucon nigrotuberculatum*) in continuous versus patchy arrays of goldenrod (*Solidago canadensis*). Although both species were more abundant in the patchy arrays, aphids were up to ten times more abundant, and were three times more likely to achieve an 'outbreak' (defined as aphid densities >100 individuals/stem), in the patchy versus continuous goldenrod arrays. Because aphid colonization and population growth rates were the same in both treatments (the latter assessed when predators were absent), Kareiva hypothesized that patchiness was not having a direct effect on aphids, but rather, an indirect effect on the interaction between predators and their prey. Mathematical modeling subsequently demonstrated that the ability of lady beetles to control aphid populations was dependent on how rapidly predators could aggregate in response to aphid clusters (Kareiva and Odell 1987). This was confirmed experimentally by creating incipient aphid outbreaks in both types of arrays, and observing that lady beetles were twice as successful at eliminating aphid colonies within continuous versus patchy habitat arrays. Thus, it appears that habitat patchiness promotes aphid outbreaks in this system because it interferes with the ability of coccinellid predators to aggregate in areas of high prey density.

Consider the implications of a patchy or fragmented habitat distribution for the biological control (**biocontrol**) of agricultural and horticultural pests like aphids. Biocontrol is most successful when prey are unable to find refuge from their natural enemies (Hawkins et al. 1993). If habitat patchiness disrupts the search efficiency and aggregative response of natural enemies, then pest populations can build up within habitat patches and potentially spread to other areas, thereby increasing the potential for pest outbreaks across the broader landscape. Is the ability to track the distribution of prey populations at a landscape scale the key to successful biological control, then? To address this, With and her colleagues (2002) compared the distributions of two coccinellids—the native

[37] One of the best-known examples of regular cycling between predators and prey is that between snowshoe hares (*Lepus americanus*) and Canada lynxes (*Lynx canadensis*), which was determined from an analysis of 100 years of pelt records from the Hudson Bay Company in Canada (Elton & Nicholson 1942).

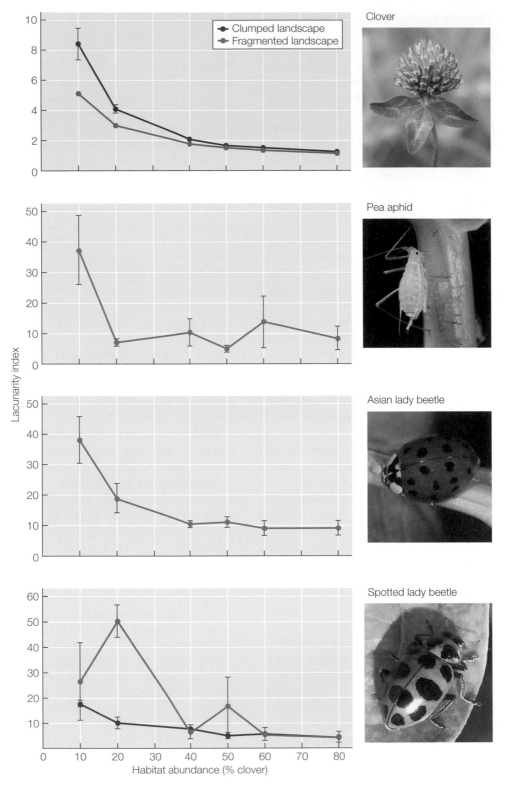

Figure 10.23 Biocontrol thresholds. A predatory beetle (Asian lady beetle, *Harmonia axyridis*) that was introduced as a biocontrol agent of agricultural pests like the pea aphid (*Acyrthosiphon pisum*), was able to track the distribution of its aphid prey even as they became scarce and more widely dispersed at low habitat levels (<20% clover). A native predator (spotted lady beetle, *Coleomegilla maculata*) did not exhibit a similar congruence with aphid distributions, especially in fragmented land-scapes (blue line). Data points represent the mean lacunarity index (±1 SE) for each species' distribution within landscape plots having either a clumped or fragmented distribution of clover for a given level of habitat (**Figure 5.11**). Data were pooled between landscape types if not significant (i.e. pea aphids, Asian lady beetle).

Source: After With et al. (2002).

spotted lady beetle (*Coleomegilla maculata*) and the Asian lady beetle (*Harmonia axyridis*), which has been introduced all over the world as a biocontrol agent—in relation to their aphid prey (*Acyrthosiphon pisum*) in experimental landscapes that differed in the amount and fragmentation of red clover (**Figure 5.11**). Perhaps unsurprisingly, the distribution of aphids mirrored that of their clover host plant, even exhibiting a similar threshold below 20% habitat where they became scarcer and more patchily distributed across the landscape (**Figure 10.23**).[38] Interestingly, the distribution of the biocontrol agent—but not the native coccinellid—also tracked this threshold in aphid distribution (**Figure 10.23**).

Although habitat fragmentation per se did not affect the distribution of aphids or the biocontrol agent, it did influence the distribution of the native predator, which occupied 57% fewer clover cells within fragmented than clumped landscapes (With et al. 2002). The greater effect of fragmentation on the native coccinellid (*C. maculata*) could be due to the fact that they operate at a finer spatial scale than the biocontrol agent (*H. ayxridis*): *C. maculata* moved 1.5× faster among individual clover stems within habitat cells, but *H. axyridis* moved 2.2× faster among clover cells, largely because it preferred to fly among clover plants rather than crawl like *C. maculata*, and thus *H. axyridis* travelled 2.5× farther across these landscape plots than *C. maculata* (With et al. 2002). Furthermore, the biocontrol agent was nearly 3× more likely to travel between landscape plots than the native coccinellid. Given its greater mobility, the biocontrol agent should thus be more effective at tracking the spatiotemporal dynamics of aphids within and among landscape plots in this experimental system.

High predator efficiency in locating and controlling pest populations when they occur at low density and/or patch occupancy is a desirable trait for a biological control agent. Indeed, *H. axyridis* was more successful than *C. maculata* in locating aphids even when they were scarce and patchily distributed on the landscape (i.e. plots with ≤20% clover habitat). In these landscapes, the biocontrol agent foraged within clover cells that had 2.5–3× more aphids than those in which *C. maculata* was found (With et al. 2002). Successful biological control may thus depend on the ability of introduced predators to track prey distributions at the broader landscape or regional scale, especially if thresholds occur in the distribution of prey at low

habitat levels that would otherwise make it difficult for other (native) predators to locate and aggregate in response to a scarce and patchily distributed prey (biocontrol thresholds; With et al. 2002). It should be noted that high mobility is a characteristic of many biocontrol agents (Murdoch & Briggs 1996). In the absence of a highly mobile predator or biocontrol agent, then, such thresholds at low habitat levels might create spatial refugia for prey, which could well destabilize predator–prey interactions and lead to pest outbreaks (Hawkins et al. 1993).

EFFECTS OF HABITAT LOSS AND FRAGMENTATION ON PREDATOR–PREY INTERACTIONS As the above examples demonstrate, the amount and/or spatial arrangement (fragmentation) of habitat can alter interactions between predators and their prey. The magnitude of the effect ultimately depends on the type of predator and/or prey species, and thus the nature of the predator–prey interaction (Ryall & Fahrig 2006). For example, a predator that is a specialist, which feeds on only one type of prey, is restricted to the same habitat as its prey. The loss of prey habitat is thus the loss of predator habitat. Even so, habitat loss is expected to have a greater negative effect on specialist predators than on their prey (**Figure 10.24A**). Not only do predators generally have larger area requirements and require more habitat to persist than their prey, but specialist predators are also strongly limited by prey abundance, which is likewise declining with habitat loss (Prey response 1 in **Figure 10.24A**). Alternatively, a reduction in predator numbers might cause prey populations to increase at low-to-moderate levels of habitat loss (Prey response 2 in **Figure 10.24A**). This is expected to occur when the cost of predation is greater than the cost of habitat loss on prey populations (Ryall & Fahrig 2006). Nevertheless, prey populations will eventually decline once some threshold level of habitat loss is reached and prey begin to overexploit their resources (Prey response 2 in **Figure 10.24A**).

An omnivorous predator is one that consumes a variety of prey species, and thus is not limited to occurring only in the prey's habitat but can also persist in the matrix, albeit with lower survivorship than in prey habitat. Omnivores are considered to be 'matrix tolerant,' and so should be less vulnerable to habitat loss than specialist predators (Ryall & Fahrig 2006; **Figure 10.24B**). As a consequence, prey are expected to exhibit greater population declines in habitat that also contain omnivorous predators (**Figure 10.24B**). Generalist predators also consume many prey species but live primarily in the matrix (they are 'matrix-based'), and are therefore not limited by the abundance of the focal prey species that is experiencing habitat loss. They might therefore be expected to benefit from prey habitat loss if this increases the matrix and provides

---

[38] Aphids did not occur in every clover cell, however, which is why their distribution had a higher lacunarity index than that of the clover distribution (**Figure 10.23**). Recall that lacunarity measures the 'gap structure' of a spatial pattern or landscape, with higher values indicating larger and more variable gap sizes between occupied cells (i.e. the distribution is 'gappier' or clumped).

(A)

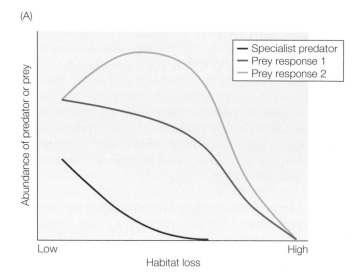

(B)

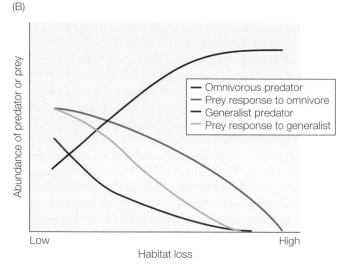

Figure 10.24 The predicted effects of habitat loss on different types of predators and their prey. (A) Specialist predators, which occur in the same habitat as their prey, are negatively affected by habitat loss to a greater degree than their prey. Although prey populations might also be expected to decline as their habitat is lost (Prey response 1), some prey species might initially increase in abundance (Prey response 2) as predator populations decline. (B) Omnivores, which are tolerant of the matrix, are less sensitive to habitat loss than specialist predators, but still more sensitive than prey. Conversely, generalist predators, which occur in the matrix, increase in abundance as prey habitat is lost and more matrix habitat is created. Habitat loss, coupled with an increasing abundance of generalist predators, thus causes the greatest decline in prey populations.

Source: After Ryall & Fahrig (2006).

them with additional resources. Unfortunately for the focal prey species, however, the build-up of generalist predators in the matrix increases the likelihood of spillover into the prey's shrinking habitat, resulting in increased predation that exacerbates the effects of habitat loss on prey population declines (**Figure 10.24B**).

Habitat fragmentation—the pattern of habitat loss—can either stabilize or destabilize predator–prey relationships depending on how patch structure affects the colonization–extinction dynamics of predators relative to their prey, which again depends on the type of predator and thus the nature of the predator–prey interaction. For a specialist predator, fragmentation per se is predicted to stabilize predator–prey interactions when: (1) prey are more mobile than predators; (2) local dispersal enhances the patchiness of predators and prey (i.e. prey are locally common in some patches, but scarce in others, making them more difficult to find); (3) prey have higher rates of emigration from patches when specialist predators are present (i.e. prey exhibit anti-predator behaviors); and, perhaps counterintuitively, (4) prey populations have lower extinction rates when predators are present, because as predators reduce prey numbers they simultaneously reduce the potential for prey to overexploit their resources (Ryall & Fahrig 2006). Conversely, habitat fragmentation is predicted to destabilize the interactions between a specialist predator and its prey when: (1) prey are more restricted in their movements than their predators and (2) predators increase the extinction rate of prey populations (Ryall & Fahrig 2006). For specialist predators, then, the relative movement and dispersal abilities of their prey, the impact of prey on their resources, and the impact of predation on prey populations all strongly affect the predicted effects of habitat fragmentation on predator–prey interactions (Ryall & Fahrig 2006).

For generalist predators, which live in the matrix, fragmentation of the prey's habitat is expected to have a positive effect on the generalist predator, but a negative effect on prey populations, owing to increased predation (Ryall & Fahrig 2006). Recall that habitat fragmentation increases the amount of edge beyond that due to habitat loss alone (compare the total edge (TE) for the two landscapes depicted in **Figure 4.30** versus **Figure 4.31**, in which fragmentation is greater in the former). Fragmentation can thus lead to an even greater spillover of generalist predators from the matrix into the prey's habitat. For example, predation rates of forest-breeding songbird nests are higher along forest edges, especially in agriculturally dominated landscapes, owing to an increase in predation pressure from generalist predators (Chalfoun et al. 2002a,b; **Box 7.2**). Although increased predation along habitat edges is usually viewed as a negative, such spillover effects might prove beneficial in other contexts. For example, insect predators and other natural enemies associated with natural habitats can help control pest outbreaks in agricultural fields. Generalist insect predators associated with the non-crop matrix often invade crop fields earlier in the season than specialists;

that is, they arrive when pest densities are still low and therefore easier to control (Tscharntke et al. 2005b). The early colonization of fields and suppression of pest populations by generalist predators may thus be key to the natural control of insect pests in agricultural systems, by helping to prevent serious outbreaks later in the season (Tscharntke et al. 2005b).

From a conservation standpoint, the coexistence of predator and prey species is essential for the maintenance of biodiversity. By disrupting the connectivity of landscapes, habitat loss and fragmentation may have disproportionate effects on predators, which can then have cascading effects on prey and other species in the community (i.e. trophic cascades). By coupling an individual (agent) based model with a landscape graph (**Figure 5.12**), Baggio and his colleagues (2011) explored how altering connectivity among habitat patches (nodes) affected predator–prey population dynamics. While overall network connectivity was important to the persistence of both predators and prey, prey populations tended to reach their highest abundance at lower node-centrality levels (a measure of how well an individual node is connected to other nodes; **Table 5.1**) compared to predators. Lower patch connectivity helps to create spatial refugia for prey because increased fragmentation interferes more with the predator's movements and search efficiency. Nevertheless, both predator and prey benefit from living in centrally located, well-connected habitat patches.

How well-connected, therefore, must habitat patches be to promote the coexistence of predators and their prey? If patches are too well connected, predators can move freely and extirpate prey populations wherever they occur, thereby driving themselves extinct in the process. Conversely, if patches are isolated, then prey populations go extinct via overexploitation of their resources or because of predation (if predators are also present in the patch), whereas predator populations go extinct when they deplete the prey population in the habitat patch. Thus, the level of connectivity needed to promote predator–prey coexistence likely lies at some intermediate level of connectivity, which must be assessed in terms of how the relative dispersal of predators and prey are affected by the patch structure of the landscape (Salau et al. 2012; **Chapter 5**). As with single-species metapopulation dynamics (**Chapter 7**), the persistence of predator–prey metapopulations may likewise occur within the Goldilocks Zone, where landscape connectivity is neither too great nor too little, but just right.

Parasite–Host and Host–Parasitoid Dynamics

Predators are not the only type of natural enemy capable of regulating the populations of the species they eat; parasites and parasitoids can also regulate the populations of their host species. Parasitoids are insects (usually wasps or flies) in which the female lays one or more eggs in, on, or near the body of another insect species (the host), and so their larvae are parasitic. We will therefore consider parasitoids and parasites together here.

Parasitism is sometimes considered a special case of predation, in that predators and parasites both feed on other organisms (the parasite-as-predator analogy; Raffel et al. 2008). From a consumer-resource standpoint, there are clearly similarities between predator–prey and parasite–host/host–parasitoid interactions: both types of interactions are dependent on the abundance and distribution of the interacting species, and the outcome of those interactions in turn determines the relative abundances of the interacting species and thus their ability to coexist (Hall et al. 2008). Such similarities are also reflected in the theoretical foundations that underlie the modeling of predator–prey and parasite–host/host–parasitoid interactions, despite some differences in the biological and mathematical forms of the models (Royama 1971). For example, Anderson & May (1978) drew upon the predator–prey literature to develop simple models of parasite–host interactions (Raffel et al. 2008), and much of the current theory on host–parasitoid population dynamics is based on the Nicholson–Bailey model (Briggs 2009). Developed in the 1930s by Australian entomologist Alexander John Nicholson and physicist Victor A. Bailey, the model describes the dynamics of a coupled host-parasitoid system in which the population of a single host species is attacked by a single parasitoid that randomly searches for hosts within a homogeneous environment (Nicholson & Bailey 1935). The model is thus closely related to the Lotka–Volterra predator–prey model, except that the Nicholson–Bailey model uses discrete-time difference equations rather than the continuous-time differential equations of the Lotka–Volterra model. Such similarities between predator–prey and parasite–host/host–parasitoid interactions have thus led to calls for the conceptual and theoretical unification of these fields, perhaps under the umbrella of 'natural enemies ecology' (Hall et al. 2008; Raffel et al. 2008).

Similar does not mean same, however. The main difference between parasites and predators is that predators kill and consume multiple prey individuals over their lifetimes, whereas parasites infect only a single host during each of their life stages and do not usually kill their hosts outright (Lafferty & Kuris 2002).[39]

---

[39] There are many exceptions, however. Parasitoids do kill their hosts, and predators do not always kill theirs. For example, micropredators include species like mosquitoes, the females of which obtain blood meals from multiple 'prey,' which they do not kill, at least not directly. Micropredators can be effective vectors for parasites and

Indeed, it is advantageous for parasites *not* to kill their hosts immediately, as a long period of infection provides more opportunities to transmit the parasite to other hosts (**Chapter 8**). Another important distinction between predators and parasites is that parasites are much smaller than their hosts, whereas predators are typically larger than their prey (micropredators and social predators, like wolves and lions, being exceptions). As a result, parasites and predators may perceive their environment and resources at very different scales, resulting in large variations in their respective dynamics, especially in the size of those oscillations (Hall et al. 2008). There are thus major quantitative differences that underlie predator–prey and parasite–host interactions. Still, if the main quantitative difference between these interactions revolves around scale, and body size in particular, then perhaps scaling relationships (**Chapter 2**) could be used to develop a unified model of natural-enemy interactions after all (Hall et al. 2008).

EFFECT OF LANDSCAPE STRUCTURE ON PARASITISM RATES If parasite–host/host–parasitoid interactions are qualitatively similar to predator–prey interactions, then we already have some insights into how landscape structure might influence these interactions, given our previous discussion regarding landscape effects on predator–prey interactions. For example, forest-nesting songbirds experience increased rates of brood parasitism by brown-headed cowbirds (*Molothrus ater*), in addition to higher nest predation rates from generalist predators, along forest edges in fragmented agriculturally dominated landscapes in the eastern USA (Robinson et al. 1995; Donovan et al. 1997; **Box 7.2**). Conversely, forest fragmentation may lead to a reduction in parasitism if it interferes with the search efficiency and aggregative response of a given parasite or parasitoid, just as we saw in the case of some predator–prey systems (e.g. Kareiva 1987).

For example, the forest tent caterpillar (*Malacosoma disstria*) is a major defoliator of hardwood trees in North America. The duration of forest tent caterpillar outbreaks in aspen-dominated (*Populus tremuloides*) forests in Canada has been linked to the degree of forest fragmentation, which was assessed in terms of forest edge density (km forest edge/km²; Roland 1993). The duration of tent caterpillar outbreaks increased by ~1 year for every unit increase in forest edge density.

other disease-causing organisms (e.g. mosquitoes can carry and spread the *Plasmodium* parasite that causes malaria), and as such, might also be considered ectoparasites (i.e. feeding on the outside of organisms) that may indirectly lead to the death of their host if they manage to infect them with a vector-borne disease (e.g. mosquitoes transmit malaria, dengue, yellow fever, and West Nile Virus, among other diseases). Life is incredibly diverse, and thus often defies our attempts at making these sorts of tidy generalizations.

Forest fragmentation may increase forest tent caterpillar populations in two ways: (1) directly, if female moths preferentially lay their eggs along forest edges, as many forest Lepidoptera are known to do, and the warmer temperatures associated with forest edges (**Box 7.1**) result in more rapid development of tent caterpillar larvae and pupae; and (2) indirectly, by interfering with the dispersal and search behavior of their natural enemies, such as parasitoids (Roland 1993). Indeed, rates of parasitism by three major parasitoids of tent caterpillars were all found to be lower along forest edges and in areas of greatest forest fragmentation (Roland & Taylor 1997).

As with predators, the response of parasitic species to habitat loss and fragmentation will depend on the scale(s) at which species interact with landscape structure. For example, the various fly species that parasitize forest tent caterpillars in aspen-dominated forests respond to landscape structure at very different scales, consistent with their size differences: parasitism by the largest species (*Leschenaultia exul*; Tachinidae) was most strongly correlated with forest structure at broad scales (850 m), whereas the smallest parasitoid (*Carcelia malacosomae*; Tachinidae), which is half the size of the largest parasitoid, responded to forest structure at a much finer spatial scale (53 m; Roland & Taylor 1997). A two-fold difference in body size thus amounts to a 16-fold difference in how these species are scaling the landscape. Interestingly, the smallest parasitoid (*C. malacosomae*) was the only species of the four studied that exhibited a higher rate of parasitism along forest edges and in fragmented forest areas compared to intact forest. For this species, the forest itself may act as a barrier to movement, but since it only responds to landscape structure at finer spatial scales, it could simply be unable to play a major role in suppressing tent caterpillar outbreaks given the scale of fragmentation in this forested landscape.

Just as we should not assume that all parasitoids respond similarly to habitat loss and fragmentation in a given landscape, we likewise should not assume that a given species of parasitoid will respond similarly to habitat loss and fragmentation in different landscape contexts. A case in point: subsequent research in another aspen forest elsewhere in Canada failed to find support for the 'parasitoid movement hypothesis' (i.e. that fragmentation results in longer tent caterpillar outbreaks because clearings interfere with parasitoid movements, thereby disrupting their ability to aggregate in response to elevated host densities in forest fragments; Roth et al. 2010). Although two of the three parasitoid species were the same as in the Roland and Taylor (1997) study, including the small parasitoid (*C. malacosomae*) that was found to exhibit a positive response to forest edges and fragmentation, Roth and his colleagues (2010) now found no difference in parasitism rates between forest fragments and continuous forest tracts for either species. Instead, they found that a third parasitoid species (*Lespesia frenchii*; Tachinidae) exhibited elevated rates of parasitism in forest fragments and along forest edges, which they posited might be due to parasitoids getting 'stuck' within forest fragments if they are unwilling to cross the forest edge, resulting in repeated encounters with hosts, and thus elevated rates of parasitism in forest fragments (Roth et al. 2010). Regardless, none of these species showed patterns of parasitism in fragmented landscapes that were consistent with the expectations of the parasitoid movement hypothesis. To add another wrinkle, the duration of forest tent caterpillar outbreaks in the hardwood forests of the northeastern USA, which are dominated by sugar maple (*Acer rubrum*), was found to be *shorter* in more fragmented landscapes (Wood et al. 2010). Clearly, landscape context matters for understanding the interaction between parasitoids and their hosts, an issue we'll return to in a bit.

EFFECT OF LANDSCAPE HETEROGENEITY ON PARASITISM RATES Parasitism rates may also be affected by landscape composition or the degree of heterogeneity within the landscape matrix (i.e. landscape complexity; Chaplin-Kramer et al. 2011). As mentioned previously, natural or seminatural habitats can serve as a source of parasitoids and other natural enemies, which may then spill over and provide pest-control services in nearby agricultural fields (Tscharntke et al. 2005b). Natural or seminatural habitats may benefit parasitoids in at least three ways: (1) natural habitats are less disturbed than agricultural fields and thus provide overwintering refuge, enabling parasitoid populations to build up over

time; (2) natural habitats contribute to greater landscape heterogeneity, which in turn leads to a greater diversity of host plants and thus insect hosts, which may enhance parasitoid populations; and (3) a more heterogeneous landscape may include natural habitats that contain a greater variety of flowering plants, thereby providing more pollen and/or nectar resources that can serve as a food source for adult parasitoids (Thies et al. 2003).

As an example, parasitism rates of the pollen beetle (*Meligethes aeneus*), a major crop pest of oilseed rape (*Brassica napus*), increased in agricultural landscapes that had more seminatural habitat (e.g. fallow fields, grasslands, and hedgerows) in Lower Saxony, Germany (Thies et al. 2003; **Figure 10.25**). The relationship was only significant at scales of 1–2 km, however, with the maximum scale of effect occurring within a radius of 1.5 km around the focal crop patch. Given that these are agriculturally dominated landscapes, a greater amount of seminatural habitat (non-crop areas) signifies greater landscape heterogeneity, which in turn is correlated with higher rates of parasitism—and thus natural biological control—within agricultural fields.

How much natural or seminatural habitat might therefore be necessary to achieve successful biological control of crop pests like the pollen beetle? For the agricultural landscapes of Lower Saxony, parasitism rates fell below the 32–36% threshold generally considered necessary for successful biological control (Hawkins & Cornell 1994) when landscapes comprised <20% seminatural habitat (Thies et al. 2003; **Figure 10.25**). We might therefore conclude that natural or seminatural habitats should make up at least 20% of the landscape within a 1.5 km radius of agricultural fields. Whether this will be sufficient elsewhere, however, depends on how host–parasitoid interactions are

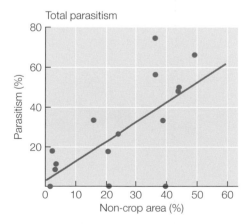
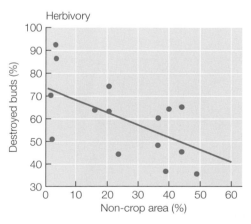

Figure 10.25 Effect of landscape heterogeneity on host–parasitoid interactions. Increasing landscape heterogeneity, as assayed by the amount of seminatural habitat (% non-crop area) in the surrounding landscape, results in greater parasitism rates and decreased rates of herbivory by pollen beetles (*Meligethes aeneus*) on oilseed rape (*Brassica napus*) in 15 landscapes in Lower Saxony, Germany. The relationship shown here is for the maximum scale of effect, which was observed within a 1.5 km radius around the focal crop patch.

Source: After Thies et al. (2003).

influenced by the amount and configuration of habitat in different landscape contexts, as we discuss next.

EFFECT OF LANDSCAPE CONTEXT ON PARASITISM RATES   Landscape context can have a significant effect on host–parasitoid interactions, owing to differences in the intervening land-use or habitat matrix. Recall that the matrix can influence a species' perception of edge permeability (**Figures 6.8–6.10**), which might then alter movement rates into or out of patches, creating a population density gradient across the patch boundary (**Figure 7.5**). Although we have previously discussed how different species of parasitoids might respond— whether positively or negatively—to habitat edges, we now consider whether parasitism rates by an individual species are altered in different landscape contexts.

To illustrate, we'll consider a case study involving the minute parasitic wasp *Anagrus columbi* and its planthopper host (*Prokelisia crocea*), which inhabit prairie cordgrass (*Spartina pectinata*) patches within wet tall-grass prairies of the northern Great Plains, USA

(Cronin 2003a; Reeve & Cronin 2010; **Figure 10.26**). Prairie cordgrass patches are embedded within a matrix consisting of either native grasses, smooth brome (*Bromus inermis*), or mudflats (**Figure 10.26**). The distribution of *A. columbi* within these cordgrass patches is strongly matrix-dependent. Parasitoid densities exhibited a sharply negative response to cordgrass-mudflat edges, declining by 59% from the patch interior to the edge (Cronin 2003a; **Figure 10.27A**). In comparison, the decline in parasitoid density from the patch interior to the edge was far more gradual in cordgrass patches surrounded by native grass, whereas parasitoid densities were uniformly low in cordgrass patches surrounded by brome (**Figure 10.27A**). The absence of an edge effect on parasitoid densities in the latter case demonstrates how structurally similar brome is to cordgrass.

Paralleling these edge effects on parasitoid density, parasitism rates also exhibited a negative edge effect, in which parasitism on planthopper eggs averaged 48% higher in the patch interior than along the edge (Cronin 2003a; **Figure 10.27B**). Put another way, half as

Figure 10.26 Host–parasitoid interactions in different landscape contexts.   In the northern Great Plains (USA), patches of prairie cordgrass (*Spartina pectinata*) are embedded in a variable matrix consisting of either native grasses, invasive brome (*Bromus inermis*, A), or mudflats (B). Prairie cordgrass is the primary host plant of the planthopper, *Prokelisia crocea* (C), whose eggs in turn are the primary host of the minute parasitoid, *Anagrus columbi* (D).
Source: Photos courtesy of James T. Cronin.

(A)

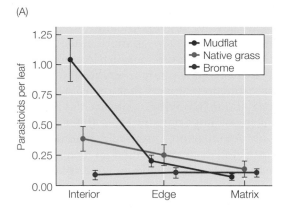

(B)

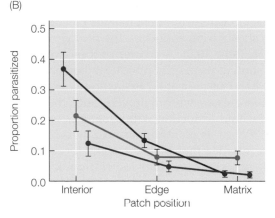

Figure 10.27 Effect of landscape context on host–parasitoid interactions.   In the northern Great Plains (USA), patches of prairie cordgrass (*Spartina pectinata*) are embedded in a variable matrix consisting of either native grasses, invasive brome (*Bromus inermis*), or mudflat (cf. **Figure 10.26**). (A) Although densities of the parasitoid (*Anagrus columbi*) tend to be higher in the interior of cordgrass patches than at the patch edge or out in the matrix (3 m from the patch edge), densities are higher, and the decline across the patch boundary steeper, in cordgrass patches surrounded by mudflats. (B) The proportion of planthopper (*Prokelisia crocea*) eggs parasitized also exhibit a decline from patch interiors out into the matrix, but parasitism rates are likewise greater and steeper in cordgrass patches surrounded by mudflats than other matrix types.

Source: After Cronin (2003a).

many hosts were parasitized per female parasitoid along the patch edge than in the interior. Although this was generally true across all matrix types, parasitism rates declined more steeply across cordgrass-mudflat edges than for other edge types, consistent with the trends in parasitoid density (**Figure 10.27B**). It is worth noting that these negative edge effects on parasitoid density and parasitism rates occurred even though planthoppers tend to become concentrated along the edges of cordgrass patches in mudflats (i.e. planthoppers exhibit a positive edge response across the cordgrass-mudflat boundary; Haynes & Cronin 2006). Nevertheless, the density of planthopper eggs (the

actual host of this egg parasitoid) did not vary within or among cordgrass patches in different matrix contexts, and so cannot explain the observed gradient in parasitoid densities and parasitism rates (Reeve & Cronin 2010).

Because of this strong gradient in parasitoid density and parasitism rates across the cordgrass-mudflat boundary, we might hypothesize that this represents a hard edge (**Figure 6.9**) that is actively avoided by parasitoids. This does not appear to be the case, however. A subsequent analysis of parasitoid movements (based on mark-recapture experiments) found that parasitoids released either in the center of a cordgrass patch or out in the matrix had a strong tendency to move toward the edge regardless of matrix type (i.e. whether the patch was in brome or mudflats; Reeve & Cronin 2010). Although parasitoids are not actively avoiding edges, they do have a greater tendency to move toward the patch center when released at a cordgrass-mudflat edge, and that tendency alone could explain the gradient in parasitoid density and parasitism rates observed within patches surrounded by mudflats (**Figure 10.27**). Thus, to quote Reeve and Cronin (2010: 489): 'although patterns of abundance and distribution can provide clues regarding movement behavior, there is no substitute for explicit movement experiments.' In other words, evidence of a negative edge effect should not be interpreted as evidence of edge avoidance. Yet again, we should exercise care when attempting to infer process from pattern.

LANDSCAPE EFFECTS ON THE COLONIZATION–EXTINCTION DYNAMICS AND STABILITY OF HOST–PARASITOID INTERACTIONS   Ultimately, the stability of host–parasitoid dynamics depends on how tightly coupled parasitoid and host populations are at the scale of the entire landscape. Recall that metapopulation persistence is possible because of a stochastic balance between extinction and colonization rates. As we saw in the case of predator–prey metapopulations, however, parasitoids and host populations cannot be too tightly coupled or else they both go extinct (parasitoids are too efficient and eradicate hosts), whereas an insufficient degree of connectivity will lead to decoupled dynamics, again resulting in the extinction of both species (parasitoids are unable to locate hosts, and hosts eventually overexploit their resources), or at the very least, protracted pest outbreaks if pest numbers are no longer being controlled by the parasitoid (e.g. Roland & Taylor 1997). Parasitoids, like predators, tend to be more mobile than their prey, and thus, their colonization success is expected to be less affected by patch area and isolation effects. Conversely, species at higher trophic levels, like parasitoids, are more prone to extinction than their hosts, given that they occur at lower densities, have narrower resource requirements, and are dependent on the distribution and abundance of their hosts (Kruess & Tscharntke 1994; Thies et al. 2003). Landscape structure is thus expected to have contrasting effects on colonization versus extinction for parasitoid populations relative to their hosts, which underscores the challenge of predicting the effects of landscape change on the stability of host–parasitoid dynamics.

To illustrate, we'll return to that prairie cordgrass system we've been discussing (**Figure 10.26**), and examine what factors affect the colonization–extinction dynamics of the minute egg parasitoid (*A. columbi*) and its planthopper host (*P. crocea*). Planthoppers occupy only about 77% of cordgrass patches at a time (Cronin 2003b). Of the patches that were occupied, planthoppers went extinct in about a quarter of these (23%) each generation, with extinction risk increasing as a function of decreasing patch size (Cronin 2004; **Table 10.3**). Extinction risk was also affected by landscape context, in that the likelihood of planthopper extinction decreased as the proportion of mudflat increased in the matrix (**Table 10.3**). Given that planthoppers are reluctant to cross cordgrass-mudflat edges, a predominantly mudflat matrix effectively

reduces planthopper emigration losses, resulting in higher egg densities and larger population sizes—both of which lower extinction risk—in mudflat-embedded patches (Haynes & Cronin 2003; Cronin 2004). Interestingly, parasitism was not a significant driver of planthopper extinctions, given that parasitism rates averaged only about 16% within patches. Patch isolation was the most important factor influencing the likelihood of colonization, although planthoppers apparently are not dispersal-limited in this system: more than half (53%) of the patches in which planthoppers went extinct were recolonized one generation later, and almost 90% of vacant patches were colonized within three generations (Cronin 2004; **Table 10.3**).

As for the parasitoid, about half of the patches occupied by *A. columbi* went extinct by the next generation (Cronin 2004; **Table 10.3**). A quarter of these were due to the extinction of its planthopper host, however, and so the likelihood of parasitoid extinction was around 39% once those cases were removed. Still, the likelihood of parasitoid extinction within patches was 1.7× higher than that of its planthopper host, which is consistent with the idea that higher trophic levels are at greater risk of extinction than lower trophic levels.

TABLE 10.3 Summary of landscape effects on the colonization and extinction rates (proportion of patches/generation) in a host–parasitoid system, involving the planthopper (*Prokelisia crocea*) and its primary egg parasitoid (*Anagrus columbi*), over a 3.5 year period (7 generations) within patches of native prairie cordgrass (*Spartina pectinata*) in the northern Great Plains, USA. A 'Negative' relationship indicates that the rate decreased in response to an increase in the specified patch attribute, whereas a 'Positive' relationship indicates that the rate increased with an increase in the specified attribute. Only significant effects are highlighted here.

Attributes affecting colonization or extinction rates	Planthopper	Parasitoid
Patch occupancy (% cordgrass patches)	77%	48%
Extinction rate (% occupied patches)	23%	52% (all patches) 39% (excluding patches in which hosts went extinct)
Patch parasitoid density		Negative
Patch planthopper density		Negative
Patch size	Negative	
Landscape context: Increasing proportion of mudflat matrix	Negative	Positive
Spatially correlated?	No	Yes, spatially correlated extinctions over short distances (≤25 m)
Colonization rate (% vacant patches)	53% (1 generation) 88% (3 generations)	33% (1 generation) 80% (3 generations)
Patch planthopper density		Positive
Patch isolation	Negative	Negative
Spatially correlated?	No	Yes, but no obvious relationship to distance

Source: Cronin (2004).

Parasitoid extinctions were also spatially correlated: patches in close proximity (≤25 m) had a higher likelihood of extinction than those that were farther apart (200–400 m; Cronin 2004). Patches that are close together are more likely to be embedded in the same matrix type, such as mudflat, and so populations should experience similar matrix effects (**Table 10.3**). In the case of the parasitoid, the likelihood of extinction increased with an increasing proportion of mudflat in the matrix, which is the opposite of planthoppers (Cronin 2004; **Table 10.3**). The lack of a strong rescue effect may be responsible for higher parasitoid extinction rates in mudflat-embedded patches, given that patch colonization rates were six times lower in patches surrounded by mudflats that those in brome (Cronin 2003a). Defying conventional wisdom, parasitoids also appear to be more dispersal limited than their hosts in this system: only a third of patches in which A. columbi went extinct were recolonized the following generation, although nearly 80% of vacant patches were colonized within three generations (Cronin 2004).

The colonization–extinction dynamics of planthoppers and their primary egg parasitoid is thus influenced by different landscape factors, and as such, is likely to be differentially affected by the loss and fragmentation of their cordgrass habitat caused by the invasive spread of smooth brome (*Bromus inermis*). Brome not only displaces the native cordgrass, but is also altering the matrix in which cordgrass patches are embedded. Parasitoids in general are expected to be more sensitive to the effects of habitat loss and fragmentation (Kruess & Tscharntke 1994), but in this prairie cordgrass system, the host is also likely to suffer an increased risk of extinction as brome comes to dominate the matrix. This was demonstrated experimentally through the creation of a replicated series of cordgrass patch networks that were embedded in either a mudflat or brome matrix (Cronin 2007). Planthoppers and parasitoids in the brome-embedded patches had densities that averaged 50% lower, and a patch extinction rate that was 4–5× higher, than patches in mudflats. At the scale of the whole network (the metapopulation scale), the complete extinction of parasitoids and their planthopper hosts occurred within 4–5 generations (2–2.5 years; Cronin 2007). Importantly, no host–parasitoid extinctions occurred in the mudflat landscapes. Extinction was so much higher in the brome-embedded patches largely because cordgrass-brome boundaries are permeable to planthoppers (and parasitoids, to a lesser extent), such that emigration losses greatly exceeded gains from reproduction and immigration. Thus, the alteration of the landscape matrix from mudflat to brome is sufficient by itself to convert cordgrass patches into extinction-prone 'population sieves' (Thomas & Kunin 1999)

capable of destabilizing host–parasitoid dynamics and causing the extinction of both species (Cronin 2007).

Of course, not all species interactions are antagonistic, causing harm to one or more of the interacting species, as in the case of competitive, predator–prey, and host–parasitoid interactions. Some interactions between species are mutually beneficial, such as between animal pollinators and the plant species that depend upon their pollination services. We thus conclude this section by considering the effect of landscape structure on plant–pollinator interactions, which falls within the domain of **pollination ecology**. Pollination ecology studies the reciprocal interactions between flowering plants and pollinators across a range of scales, from the effectiveness of a single pollinator for a given plant species, to the interactions among floral visitors and plants within a community, to regional and biogeographic patterns in the distribution of pollinators (Rafferty 2013).

Plant–Pollinator Interactions

Pollinators have experienced widespread declines in recent decades, which threaten not only native plant diversity but also our own food security (Steffan-Dewenter et al. 2005; Potts et al. 2010). For example, 75% of the crops consumed directly by humans for food (i.e. as fruits, vegetables, seeds, or nuts) are pollinated by animals (Klein et al. 2007). [40] Among animal pollinators, bees are the most important in terms of crop pollination services. Of the 57 insect and vertebrate species known to pollinate crops grown for direct human consumption, 82% of these species are bees (Klein et al. 2007). Honey bees (*Apis mellifera*) in particular are managed throughout the world to enhance agricultural production, but native wild bee species also provide significant crop pollination services (Winfree et al. 2008). Among the world's leading crops, 42% were pollinated by at least one wild bee species (Klein et al. 2007). The global economic value of insect pollination services has thus been estimated to be

[40] Although a majority of the world's food crops rely on animal pollinators, most of the world's agricultural output (60% of total production) does not in fact depend upon animal pollination services, given that grains and cereal crops (e.g. maize, rice, and wheat) are among the top staple foods produced worldwide and these are either wind-pollinated or passively self-pollinated (Klein et al. 2007). Still, this means that more than a third of the global agricultural output *is* dependent on animal pollination services, which is by no means trivial, especially given the ever-increasing demands placed upon agricultural production systems to feed a growing human population that is projected to top 11 billion by the end of the century. In addition, even some crops that are primarily wind- or self-pollinated, like coffee (e.g. *Coffea arabica*), achieve much higher yields when they are cross-pollinated by animal pollinators like native bee species (Roubik 2002; Klein et al. 2003). The decline of native and managed bee populations should thus concern anyone for whom coffee is a staple (e.g. the author of this textbook).

around $179 billion per annum (Gallai et al. 2009). In addition, most wild plant species (80%) are dependent on insect pollination for fruit and seed set, and most plant populations (62–73%) experience pollination limitation (Ashman et al. 2004; Aguilar et al. 2006; Potts et al. 2010). A loss of pollination services thus has considerable ecological as well as economic consequences.

Although wild and managed honey bees have been in decline for many decades, the massive die-offs of managed honey bee stocks in the USA, Europe, China and elsewhere in recent years due to 'colony collapse disorder' is worrisome (Stokstad 2007; Kluser et al. 2010).[41] In the USA, honey bees are responsible for pollinating 95 kinds of fruits and nuts, along with crops like soybeans (Stokstad 2007; Kluser et al. 2010), providing about $15 billion annually in crop pollination services (Calderone 2012). The honey bee is not native to North America, however, so if honey bee stocks continue to decline, will native bee species be able to replace their crop pollination services? In other words, can native bees provide insurance against ongoing honey bee losses (Winfree et al. 2007)?

Native bee species will be unable to fill the void left by honey bees if the landscape is devoid of native bees. Of the various drivers contributing to the global decline of bees and other pollinators, human land use that causes the loss, fragmentation, and degradation of natural and seminatural habitats is believed to pose the biggest threat (Rathcke & Jules 1993; Steffan-Dewenter & Westphal 2008; Potts et al. 2010). For example, a meta-analysis found that while anthropogenic disturbances (e.g. grazing, fire, logging, agriculture) negatively affected bee richness and abundance, much of this effect was ultimately due to the loss of bee habitat (Winfree et al. 2009). As a group, bees are known to be sensitive to habitat-area and -isolation effects (Steffan-Dewenter et al. 2006; Ricketts et al. 2008). Habitat loss and fragmentation have thus been cited as one of the greatest threats to native bee diversity (Brown & Paxton 2009). We therefore consider the expected effects of habitat loss and fragmentation on plant–pollinator interactions, especially for native bees, in the next section. Although other insect groups (e.g. butterflies and beetles) and some vertebrates (hummingbirds and bats) are important

pollinators in many ecosystems, much of the landscape research on pollinators has admittedly been conducted in agroecosystems, and thus on the potential of native bees to provide crop pollination services.

EFFECTS OF HABITAT LOSS AND FRAGMENTATION ON PLANT–POLLINATOR INTERACTIONS   Although plant-pollinator systems vary in their degree of specialization, we begin by assuming a highly specialized relationship between one pollinator species and a single plant species, as this type of plant–pollinator interaction is most likely to be vulnerable to habitat loss and fragmentation (Rathcke & Jules 1993; Steffan-Dewenter et al. 2006), before expanding our focus to consider the broader effects of habitat loss and fragmentation on pollinator communities.

For both pollinators and plants, specialization for mutualistic partners appears to be a key characteristic that increases their risk of local extinction in fragments.

Rathcke & Jules (1993)

Habitat loss and fragmentation can influence pollination success in at least three ways, by altering (1) plant density, (2) pollinator density, and/or (3) pollinator movements (Hadley & Betts 2012; **Figure 10.28**). Habitat loss is known to have strong negative effects on the abundance and distribution of both plants and pollinators (e.g. Vellend 2003; Vellend et al. 2006; Winfree et al. 2009; **Figure 10.28**). A decline in plant density directly reduces pollination rates, since fewer plants means less pollen is produced, which in turn means less pollen is available to be transferred among plants by pollinators (**Figure 10.28**). In addition, a decline in plant density could have adverse effects on pollinator densities, owing to a reduction in the availability of floral resources (pollen, nectar) and thus in the frequency of floral visits by pollinators. Lower plant densities may also alter the movement or foraging behavior of pollinators, requiring them to travel farther afield or to switch to other plant species (Smithson & McNair 1997; Kremen et al. 2007; **Figure 10.28**). If pollinators switch from a specialist to generalist foraging strategy, then pollination success will likely decline as well, especially if the pollinator visits multiple plant species within a foraging bout and ends up delivering a mixture of pollen that is largely incompatible with its usual plant partner. The combined effects of habitat loss on plant density, pollinator density, and pollinator movement thus contribute to lower pollination success, creating a positive

---

[41] The factors responsible for colony collapse disorder, which refers to the unexplained abandonment of hives by bees (specifically, the female workers), are unknown and likely multifaceted. Some of the factors that have been proposed are: ectoparasites (especially the *Varroa destructor* mite); pathogens (fungi, viruses, and the *Nosema* spp. gut parasite); poor hive management (use of antibiotics and long-distance transport of bee hives to agricultural fields across the country); habitat loss; climate change; and the use of neonicotinoids and other agricultural pesticides (Oldroyd 2007; Potts et al. 2010; Tsvetkov et al. 2017).

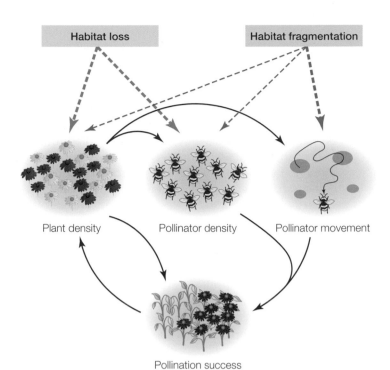

Plant density   Pollinator density   Pollinator movement

Pollination success

Figure 10.28 The direct and indirect effects of habitat loss and fragmentation on pollination success. Habitat loss is expected to have strong negative effects on plant and pollinator densities. A reduction in plant density reduces pollination success simply because less pollen is available, but also because fewer pollinators are then attracted to declining floral resources within the habitat patch. Habitat loss can thus directly and indirectly alter the movement behavior of pollinators, resulting in less pollen being transferred. Although habitat fragmentation (the spatial arrangement of patches) may also negatively affect plant and pollinator densities, it is expected to have the greatest effect on pollinator movement and foraging behavior.

Source: After Hadley & Betts (2012).

feedback loop that contributes to further declines in plant density (Hadley & Betts 2012; **Figure 10.28**).

Although habitat fragmentation may also have negative effects on the densities of plants and their pollinators, these effects are expected to be weaker than those due to habitat loss alone (Hadley & Betts 2012). Habitat fragmentation is thus expected to exert its strongest effect on the movement of pollinators (**Figure 10.28**). By restricting pollinator movements, fragmentation could reduce gene flow among plant populations, which in turn may cause a reduction in reproductive success through inbreeding depression, thereby exacerbating plant population declines within habitat remnants (Rathcke & Jules 1993; Lennartsson 2002; **Chapter 9**). The effects of habitat fragmentation on pollinator movements are usually assessed indirectly, in terms of quantifying floral visitation rates, pollen transfer rates, or seed set of flowering plants at increasing distances from patches of natural or seminatural habitat (Hadley & Betts 2012).[42] Although patch distance or isolation may not capture the effects of habitat fragmentation so much as habitat loss, it has nevertheless been one of the most widely used measures for assessing fragmentation effects on plant–pollinator interactions (Hadley & Betts 2012).

Patch isolation is expected to decrease the richness and abundance of the pollinator community, thereby limiting the number of available pollinators, which in turn would reduce pollen transfer among plant populations, causing a reduced seed set. For example, the number of wild bees that visited mustard (*Sinapis arvensis*) flowers decreased with distance to the nearest calcareous grassland patch in an agricultural landscape near Göttingen, Germany (Steffan-Dewenter & Tscharntke 1999; **Figure 10.29A**). This in turn appears to have resulted in decreased seed set (the number of seeds produced per mustard plant) in more isolated patches: mean seed set was reduced by 50% for mustard plants located 1 km from a grassland source of pollinators (**Figure 10.29B**). Interestingly, this is generally consistent with the overall distance at which species richness and floral visitation rates were found to be halved (at 1.5 km and 668 m, respectively) in a meta-analysis that investigated patch-isolation effects on wild pollinators in agricultural landscapes (Ricketts et al. 2008).[43]

As with other types of species' responses to landscape structure, not all pollinators are affected to the same degree, or even in the same way, by patch isolation. In contrast to wild bees, floral visitation rates by hoverflies (Diptera: Syrphidae) did not change with increasing patch isolation in the aforementioned agricultural landscape in Germany (Steffan-Dewenter & Tscharntke 1999). Indeed, the abundance of hoverflies was found to *increase* with distance from seminatural habitat in some agricultural landscapes (Jauker et al.

---

[42] Fragmentation effects on pollinator movements can also be inferred from the analysis of plant gene flow and population genetic structure, as we discussed in **Chapter 9**.

[43] However, the effect of patch isolation on fruit or seed set was much weaker than patch-isolation effects on pollinator richness and floral visitation rates in this meta-analysis (Ricketts et al. 2008).

(A)

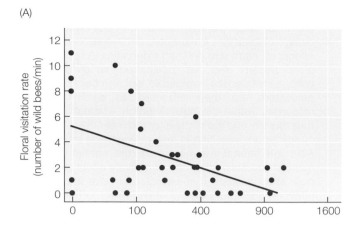

(B)

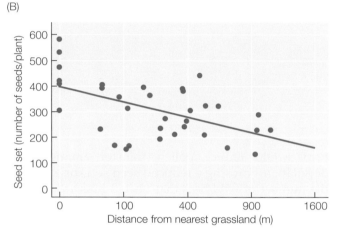

Figure 10.29 Effect of habitat isolation on pollinator visitation rates and pollination success. Floral visitation rates by wild bees (A) and seed set (B) within mustard plants (*Sinapis arvensis*) both decreased with distance to the nearest calcareous grassland in an agricultural landscape near Göttingen, Germany.

Source: After Steffan-Dewenter & Tscharntke (1999).

2009). The contrasting responses of bees and hoverflies to patch isolation might be explained by the differences in their resource requirements and foraging behaviors. Most bees consume nectar and pollen, but adults also need to collect these floral resources to feed their larval brood, which are located in a nest or hive. Bees are thus central-place foragers, and so may be more constrained to forage near flower-rich natural and seminatural habitats. By contrast, only adult hoverflies feed on pollen and nectar, whereas the offspring of some species are predators on aphids, a major crop pest. Thus, hoverflies may actually benefit from having some agricultural habitat in the surrounding landscape if their predatory larvae preferentially feed on aphids, given that adults must travel away from flower-rich patches of natural or seminatural habitats in order to oviposit within aphid-rich crop fields. Because of their affinity for both seminatural and agricultural habitats, hoverflies provide twice the ecosystem service: they contrib-

ute to the biological control of crop pests while also helping to maintain pollination services in agricultural landscapes that are otherwise unsuitable for more specialized or less-mobile bee species (Jauker et al. 2009).

The effects of habitat loss and fragmentation can also be expected to vary among different types of species within a pollinator group. For example, the strongest habitat-area effects on the richness and abundance of bees in central Germany were observed for species that are non-social (solitary bees), brood-parasitic, oligolectic (i.e. they collect pollen from only one plant species or genus), and/or nest above ground in pithy plant stems or in tree cavities (Steffan-Dewenter et al. 2006). Recall that habitat-area effects tend to be greater for specialists than for generalists, which might explain the greater sensitivity of brood-parasitic bees ('cuckoo bees') and oligolectic species to habitat loss, given that the former represent a high degree of trophic specialization (compared to bees that collect pollen to provision their broods) and the latter are by definition resource specialists. Bee species that nest above ground in tree cavities have requirements for a limiting resource, and so could also be viewed as specialized in terms of their resource requirements, especially if trees become even more limiting due to habitat loss (Steffan-Dewenter et al. 2006). Note that habitat loss may have less of an effect on social bees in this region because this group is dominated by bumblebees (*Bombus* spp.), which have large foraging ranges and generalized nesting and resource requirements (Westphal et al. 2003, 2006), traits that ought to make them less sensitive to the loss of natural or seminatural habitats.

The apparent effects of habitat loss and fragmentation on different pollinator groups are likely to depend on the landscape or regional context, and thus on the composition of pollinator communities within those different regions. As mentioned above, social bees were generally found to be less sensitive than solitary bee species to habitat loss and fragmentation (Steffan-Dewenter et al. 2006). However, a global meta-analysis of landscape effects on crop pollination services appeared to find just the opposite, that social bees are more sensitive to habitat loss and fragmentation than are solitary bees (Ricketts et al. 2008). This seeming contradiction can be explained by considering the different regional context of studies included in the meta-analysis, which encompassed tropical as well as temperate regions. Floral visitation rates by bees were found to decline more steeply with patch isolation in tropical than temperate regions, such that patch isolation had a greater overall effect in the tropics than in temperate regions: floral visitation rates declined by half at a distance of 589 m in the tropics versus 1.31 km in temperate regions (Ricketts et al. 2008). In the tropical regions included in this meta-analysis, bee communities are dominated by social bee species, most of which are meliponines ('stingless bees') that tend to be short-range foragers and cavity nesters (Michener 2000). Thus, traits of social bees in the tropics differ from those in temperate regions, which again, tend to be dominated by bumblebees that are long-range foragers and have generalized nesting and habitat requirements.

However, this does not mean that more generalized species, like bumblebees, are completely unaffected by landscape structure. Different species may still respond to the abundance of habitat or floral resources at different scales (Steffan-Dewenter et al. 2002; **Figure 7.7**). For example, the densities of four species of bumblebees were related to the availability of mass-flowering crops (e.g. oilseed rape) within agricultural landscapes of the Lower Saxony region in Germany, but at different scales that correlated with their foraging ranges, which in turn were related to body-size differences among species (Westphal et al. 2006). Thus, the two largest bumblebee species (*Bombus lapidarius* and *B. terrestris*) had the largest foraging ranges (2.75–3 km), whereas the smallest bumblebee (*B. pratorum*) had the smallest foraging range at 250 m; the foraging range of the intermediate-sized species (*B. pascuorum*) was in between these other species (1 km; Westphal et al. 2006).

Such scale-dependent responses by species to landscape structure underscores the importance of understanding the effects of land management at both a local field and landscape scale, especially when making land-use decisions to maintain or bolster crop pollination services in agroecosystems (Tscharntke et al. 2005a). For example, an analysis of local versus landscape effects on wild bee pollinators, representing more than three-dozen agroecosystems from around the world, found that bee abundance and richness were generally higher in organic and diversified fields, as well as in landscapes that had more high-quality habitats surrounding crop fields (Kennedy et al. 2013). From a management standpoint, these results suggest that the negative effects of agricultural intensification might be addressed at either a local or landscape scale. For example, bee richness and abundance increased by an average of 37% for each additional 10% increase in the amount of high-quality bee habitat in the surrounding landscape (Kennedy et al. 2013). Increasing the amount of natural or seminatural habitat at a landscape scale may be beyond the capacity of most individual producers, however. In that case, switching from conventional farming to organic farming could lead to an average increase in bee abundance of 74% and an average increase in bee richness of 50%, whereas increasing crop diversity within fields could lead to an average 76% increase in bee abundance (Kennedy et al. 2013). Put another way, the composition of the surrounding landscape becomes less important if most producers diversify their crops and/or switch to organic-farming practices. In that case, the distinction between local farm management and landscape effects begin to blur, as the agricultural landscape becomes more a 'multifunctional matrix' that sustains both crop productivity and crop pollination services (Kennedy et al. 2013).

[T]he interactions between local and landscape factors suggest that the local benefits of a diversity of crops or natural vegetation and organic management could transcend an individual field or farm because the improved quality of habitats on one field can provide benefits to adjacent or nearby fields... In this way, the distinction between local farm management and landscape effects blur. As a result, the agricultural landscape becomes more of a multifunctional matrix that sustains both crop productivity and natural capital rather than being a single purpose landscape with limited biodiversity value.   Kennedy et al. (2013)

Interestingly, this analysis found that wild bees are impacted more by the amount of high-quality habitat in the surrounding landscape rather than by the configuration of those habitats (Kennedy et al. 2013). That is, habitat loss rather than fragmentation per se appears to be the more important driver of wild bee declines (Potts et al. 2010; Hadley & Betts 2012). If so, then we might expect bees to tolerate fragmentation so long as sufficient high-quality habitat remains within their respective foraging ranges. We therefore address the habitat-area requirements of pollination services next.

THE HABITAT-AREA REQUIREMENTS OF POLLINATION SERVICES Although domesticated honey bees are widely used to enhance pollination services to crops, native bee species may ultimately prove crucial for the maintenance of pollination services within agroecosystems, by providing insurance against the extensive die-offs of honey bee populations that are occurring in North America and elsewhere (Winfree et al. 2007). As we have seen, the abundance and richness of native bee species depend on the amount and (to a lesser extent) the configuration of natural and seminatural habitats in the agricultural landscape. The loss and fragmentation of such habitats is thus expected to lead to a loss of native bees, and presumably, the crop pollination services they provide. How much habitat is needed to preserve pollination services by native bees, and at what scale(s)?

In an agricultural area of California (USA), Kremen and her colleagues (2004) found that pollination services provided by native bees depended strongly on the proportion of natural and seminatural habitats located within 1–2.5 km of the farm site. In other words, if farmers had to rely solely on native bees for pollination, then natural or seminatural habitats would need to make up at least 30–40% of the surrounding landscape (i.e. within 1–2.5 km of their fields). Interestingly, the area requirements of pollination services in this landscape are consistent with the maximal foraging distance of one of the largest bee species in this system (i.e. 2.2 km; Kremen et al. 2004).

Similarly, the results of a spatially explicit simulation model of plant–pollinator dynamics revealed that extinction thresholds for wild plants and their pollinators occurred when 50–60% of the original habitat had been converted to cropland (Keitt 2009). That is, the plant-pollination system collapsed when less than 40–50% habitat remained on the landscape. In this case, the collapse was forestalled (i.e. the threshold shifted from 50% to 60% habitat loss) if the pollinator was a habitat generalist capable of nesting in cultivated areas. However, if pollinators are generalist foragers that can feed on crop pollen, then the extinction threshold for pollinators can be extended beyond 90% habitat loss, or may disappear entirely if pollinators can nest in cultivated areas (Keitt 2009). Some wild bees may indeed benefit from agriculture, such as ground-nesting bees that use disturbed areas for nesting or bees that can forage in pollen-rich crop fields, such as oilseed rape (Westphal et al. 2003; Westphal et al. 2006).

Keitt (2009) additionally sought to identify the properties of landscapes that produced the greatest return in crop pollination services. His simulation results suggest that both landscape pattern and the scale of fragmentation can have large effects on crop pollination services in agroecosystems. As discussed previously, crop pollination services can be maximized by having enough natural and seminatural habitats (or functionally connected habitat patches of sufficient size) to support pollinator populations, while ensuring that agricultural fields are sufficiently close to natural habitats so as to capitalize on the spillover of pollinators into crop fields. Thus, management should ideally strive to match the scale of habitat availability, in terms of the size and configuration of habitat patches, to the scales of foraging and dispersal by the primary pollinators. In the context of the simulation model, for example, this optimal landscape configuration occurred when the size of remnant habitat patches was equal to half the mean foraging or dispersal distance of pollinators, and the spacing between remnant patches was equal to the mean foraging or dispersal distance (Keitt 2009). This accords well with empirical studies of patch-area and isolation effects on bee abundance and richness, in which the amount of natural and seminatural habitats within the foraging range of bees had

a significant effect on floral visitation rates (e.g. Kremen et al. 2004).

EFFECT OF HABITAT CORRIDORS ON POLLINATORS AND POLLINATION SERVICES Because pollination services are so critically dependent on the amount of habitat within the foraging or dispersal range of pollinators, we could enhance pollination services either by protecting or restoring more habitat within the landscape, which is not always an option in production landscapes, or possibly by increasing the connectivity of remaining habitat via habitat corridors (e.g. hedgerows along farm fields). If habitat corridors enhance the movement of pollinators between habitat patches, especially of less-mobile species, then we should see increased pollen transfer among plant populations that are otherwise isolated by distance or by a matrix that is largely resistant to pollinator movements (**Chapters 5 and 9**). To test this 'transfer corridor hypothesis,' Townsend and Levey (2005) studied the movement of fluorescent powder (a stand-in for pollen) among flowering plants by insect pollinators (bees and wasps) in an experimentally fragmented longleaf-pine forest (**Figure 6.13**). They found that corridors did indeed facilitate pollen flow between neighboring patches: flowers within patches connected by corridors were twice as likely to receive pollen than were flowers in isolated patches (Townsend & Levey 2005). As the flowers studied were only pollinated by bees or wasps, which are unlikely to move through forest, we can be reasonably certain that these differences in pollen flow represent the effect of corridors on pollinator movements.

More direct evidence of corridor effects on pollinator movement can be had by observing the individual movement responses of pollinators along corridors. For example, Cranmer and her colleagues (2012) studied bumblebee movements along hedgerows within an agricultural landscape in Northamptonshire, England. In these intensively managed landscapes, hedgerows are frequently the only connections left among fragments of seminatural habitat. By observing bumblebee movements along natural hedgerows and within experimental arrays of patches (potted flowers) connected by artificial corridors (essentially drift fences created with swaths of fabric), Cranmer and colleagues demonstrated that bees were preferentially orienting their movements toward corridors, and furthermore, were moving along them to access floral resources on the landscape. As a result, floral patches connected by more hedgerows received more pollinators, causing more pollen to be deposited per flower, contributing to higher seed set (Cranmer et al. 2012). Hedgerow corridors thus enhanced plant population connectivity and pollination flow in this landscape.

If habitat corridors are effective in facilitating the movements of pollinators, then corridors should be able to enhance or restore pollination services in fragmented landscapes. To address this, Kormann and his colleagues (2016) studied the movement of hummingbirds along forest corridors to determine whether the presence of corridors increased floral visitation rates, pollen movement, and pollination success within Costa Rican forest fragments. Like hedgerows in the agricultural landscapes of Western Europe, treed fencerows ('living fences') and narrow riparian buffers often provide the only forested structures left connecting forest remnants in many agricultural landscapes in the tropics. Forest corridors significantly increased floral visitation rates by four hummingbird species that are forest specialists, but corridors had no effect on the frequency of floral visits by two habitat generalists (Kormann et al. 2016). Again, habitat specialists are expected to be more sensitive than habitat generalists to habitat loss and fragmentation, and so are more likely to depend on habitat corridors to facilitate movement through a fragmented landscape.

The benefits of corridors on the movements of forest-dependent hummingbirds translated into increased pollen flow: pollen transfer among artificial flowers was severely limited within pastures (the non-forest matrix), but forest corridors helped to mitigate this effect by increasing pollen transfer, although the flow of pollen was still greatest in unfragmented forest (Kormann et al. 2016). The ability of corridors to promote pollen flow is not unlimited, however. Pollen flow decreased with increasing distance from a pollen source, becoming negligible at about 100 m. Corridors thus may not be able to function as conduits for pollen (gene) flow among forest plant populations if inter-patch distances exceed this limit. That corridors can be effective at increasing pollination rates, however, was evidenced by the virtual absence of pollination in isolated forest fragments. Not coincidentally, corridors also increased the likelihood of patch occupancy by forest-dependent hummingbirds, such as the green hermit (*Phaethornis guy*), which was otherwise absent from isolated patches (Kormann et al. 2016). Thus, habitat corridors have the potential to promote functional connectivity by helping to maintain pollinator diversity—and the pollination services they provide—within fragmented landscapes.

LANDSCAPE EFFECTS ON POLLINATOR DIVERSITY AND THE STABILITY OF POLLINATION SERVICES Because a diverse and abundant pollinator community (= high evenness) should increase the stability of pollination services in natural and agricultural systems (Garibaldi et al. 2011), understanding how habitat loss and fragmentation affect the evenness of pollinator communities is ultimately important for the maintenance and management of this critical ecosystem function. A pollinator community dominated by a few generalist

species is unlikely to guarantee the pollination of flowering plants in isolated forest fragments, as we saw in the previous example.

Habitat area and connectivity may have contrasting effects on the evenness of pollinator communities, however. In Europe, for example, Marini and his colleagues (2014) found that a reduction in habitat area increased the evenness of wild bee and butterfly communities, whereas a loss of habitat connectivity had the opposite effect. Pollinator communities in small fragments were mostly composed of mobile and/or generalist species, and so higher evenness in these patches may result from frequent inter-patch movements by highly mobile species, especially as many of these are also generalists and can use resources in the matrix. In small fragments, dispersal processes thus appear to be more important than local recruitment processes in determining the relative abundance of species; a large exchange of highly mobile species reduces the likelihood that a few species will become numerically dominant in the community, resulting in higher evenness (Marini et al. 2014).

This may explain why increasing connectivity also promotes greater evenness within habitat fragments: isolated patches tend to be dominated by a few highly mobile and/or generalist species, resulting in low evenness, whereas connectivity facilitates the dispersal of more sedentary and/or specialist species and so helps to increase diversity and evenness, especially among species that are neither very common nor too rare (Marini et al. 2014). So, although habitat area and connectivity often decrease simultaneously, the resulting composition of the pollinator community will depend on the relative strength of local recruitment versus dispersal processes, which in turn have different consequences for species richness versus evenness.

Does a diverse pollinator community beget greater stability of pollination services, however? In agricultural landscapes, for example, pollinator richness and flower visitation rates both decline with distance from natural and seminatural areas (Ricketts et al. 2008; Garibaldi et al. 2011). If we define the stability of crop pollination services as the inverse of variability (e.g. the inverse of the coefficient of variation, CV^{-1}), then we can quantify how pollination services vary both spatially (among sites) as well as temporally (among days or weeks within the flowering season). Using a large dataset representing different biomes, crop species, and pollinator communities, Garibaldi and his colleagues (2011) found that the spatial stability of pollination services within crop fields located 1 km from a natural area declined by up to 25%, whereas the temporal stability of crop pollination services declined by up to 34%, depending on whether pollinator richness, floral visitation rate, or fruit set was assayed.

Importantly, these results held even when honey bees were abundant, which again underscores the importance of managing agricultural landscapes so as to preserve or restore natural habitats, which in turn can help to enhance the abundance and diversity of native pollinators, thereby ensuring greater reliability of crop pollination services (Garibaldi et al. 2011).

Given that global declines in pollinators have largely been driven by agricultural intensification, it might seem counterproductive to argue for the conservation of native pollinators and natural habitats *within* agricultural landscapes. As this section illustrates, however, a significant percentage of our agricultural production is dependent on the crop pollination services provided by native bees and other insects, which in turn requires protecting or restoring at least some natural or seminatural habitat in the surrounding landscape. **Agri-environment schemes** (AES) have therefore been implemented in many countries to provide incentives to farmers and landowners to manage agricultural landscapes for the conservation of native species and ecosystem services. In Europe, AES have become the main tool to conserve biodiversity, thanks to the European Union's Common Agricultural Policy (Batáry et al. 2015). In fact, AES are a major source of conservation funding—as well as the highest conservation expenditure—in the EU (Batáry et al. 2015). Recall that Europe has a very long history of human occupation and intensive land use, which in some areas goes back millennia to the dawn of agriculture itself (**Chapter 3**). Thus, to quote Batáry and his colleagues (2015: 1007):

> *Although agricultural intensification is generally considered the most important driver of global terrestrial biodiversity loss, through habitat loss, habitat fragmentation, and habitat conversion…in Europe[,] agriculture itself has long been understood as part of the solution. Much of current European nature conservation aims to halt the on-going loss of farmland biodiversity, evolved during millennia of extensive management…and abandonment of agriculture is generally seen as a threat to biodiversity…In this sense, Europe is different from other continents, particularly the Americas, where areas of high biodiversity interest are rarely in use for commercial production of food, and agricultural practices are not prominent in conservation strategies.*

In conclusion, the maintenance and stability of crop pollination services, as well as biocontrol services, are dependent on the spillover of native pollinators and natural enemies from nearby habitats into crop fields. This in turn implies that these neighboring communities and ecosystems are—or can be—linked by the movement of species between them. The linkage of communities by multiple, potentially interacting species is referred to as a **metacommunity**, whereas the linkage of interacting ecosystems by the flux of organisms or materials (matter and energy) is referred to as a **metaecosystem** (Loreau et al. 2003; Massol et al. 2011). We'll hold off on discussing metaecosystems and the management of multifunctional landscapes until the next chapter, but will conclude this chapter with a discussion of the different ways in which dispersal and landscape heterogeneity can give rise to the structure and dynamics of metacommunities.

Metacommunity Structure and Dynamics

Just as populations linked by dispersal require a larger spatial perspective—the metapopulation—to assess a species' distribution and extinction risk (**Chapter 7**), so too do communities linked by dispersal require a broader perspective in order to make sense of patterns of diversity and community dynamics at the landscape or regional scale. The **metacommunity** is thus a logical extension of the metapopulation concept (Resetarits et al. 2005). As such, it provides a useful way to think about linkages between different spatial scales in community ecology, which historically was concerned only with the local community scale (Leibold et al. 2004; Logue et al. 2011).

This idea of processes at different scales being important and interacting to affect local community composition and diversity embodies the core of metacommunity theory.

Logue et al. (2011)

A metacommunity is defined as a set of local communities that are linked by the dispersal of multiple, potentially interacting species (Wilson 1992; Leibold et al. 2004). The structure of a metacommunity is determined by the spatial structure of the landscape (its composition and spatial configuration), the composition of communities that make up the metacommunity, and the extent of dispersal among interacting species. **Metacommunity dynamics** refer either to the spatial dynamics or to the regional properties (resulting from the spatial dynamics) of communities occupying two or more interconnected patches (Holyoak et al. 2005: 11). In a metacommunity, the dynamics of individual species are likely to be altered by the dispersal and interactions among species occupying different habitat patches, as opposed to the local community dynamics of interacting species within the same habitat patch.

Four theoretical frameworks have been proposed to explain the structure and dynamics of metacommunities:

(1) the patch-dynamic perspective, (2) the species-sorting perspective, (3) the mass-effects perspective, and (4) the neutral-model perspective (Leibold et al. 2004; Cottenie 2005; Holyoak et al. 2005). Each framework focuses on different mechanisms to explain species coexistence within a metacommunity, and predicts changes in local community composition based on some combination of environmental heterogeneity, species niche characteristics, and dispersal (spatial) processes (Cottenie 2005; Holyoak et al. 2005; Logue et al. 2011; **Table 10.4**). We consider each of these four perspectives in turn.

1) **Patch-dynamic perspective.** The patch-dynamic perspective basically extends metapopulation theory (**Chapter 7**) to more than two species (Holyoak et al. 2005). As with metapopulation theory, this perspective assumes that the landscape is patchy and that all patches are of the same type (a homogeneous patchy landscape; **Figure 7.13**). Species extinctions occur stochastically (as in standard metapopulation models) as well as deterministically (as in metapopulation models for two interacting species; Hoopes et al. 2005). Colonization offsets local extinctions, and it is in their dispersal abilities that species differ and which ultimately permits their regional coexistence. For species coexistence to occur at the landscape or regional scale, dispersal rates must be sufficiently limited so that the dominant species cannot drive other species (e.g. its competitors or prey) to regional extinction. For example, regional coexistence is possible if there is a trade-off between the colonization and competitive abilities of species, as discussed previously. However, the addition of competitors or predators capable of causing the local extinctions of other species reduces the conditions under which regional coexistence is possible (Holyoak et al. 2005). Recall that in predator–prey metapopulations, for example, prey must disperse faster than predators and colonize patches faster than they are driven extinct. The regional persistence of both predators and their prey species may thus be possible only for intermediate rates of dispersal or levels of connectivity (e.g. Salau et al. 2012).

2) **Species-sorting perspective.** Much as niche differentiation is assumed to enhance competitive coexistence among species within communities, the species-sorting perspective assumes that species are able to coexist regionally because of niche differentiation in their use of habitats (Holyoak et al. 2005; Logue et al. 2011). Recall that we discussed spatial niche differentiation previously, in the context of how coexistence might occur as a result of competitive reversals in a heterogeneous landscape (Holt 2001; Hoopes et al. 2005). Coexistence

is possible at a landscape or regional scale if each species is at a competitive advantage in their preferred habitat. The species-sorting perspective thus assumes that dispersal is sufficient to enable species to colonize habitats (i.e. dispersal is not limiting), but differences in their ability to persist in different environments is what ultimately determines their regional coexistence (i.e. species are 'sorted' among different habitats).

3) **Mass-effects perspective.** The mass-effects perspective (Shmida & Wilson 1985) is basically a multispecies version of population source–sink dynamics (**Chapter 7**) involving rescue effects (Holyoak et al. 2005). Low densities (mass) in sink habitats are offset by dispersal from source habitats. Dispersal is a double-edged sword, however: immigration can enhance local population densities beyond what might be expected if the community were closed (rescuing those species from extinction), whereas emigration can enhance the rate of loss (and thus extinction) from local communities (Holyoak et al. 2005). Although mass effects can occur in the absence of habitat heterogeneity, they are more predictable when the landscape is heterogeneous. Assuming the landscape is heterogeneous, the main difference between the mass-effects and species-sorting perspectives is that local extinctions are primarily due to differences in the ability of species to persist in different habitats in the species-sorting perspective, as opposed to differences in flux rates among populations due to asymmetrical dispersal (from sources to sinks) in the case of the mass-effects perspective (Holyoak et al. 2005).

4) **Neutral-model perspective.** As we have seen, most models of community and metacommunity structure focus on how species differ, in terms of their niche relationships with the local environment and/or in their ability to disperse and avoid local extinction (Holyoak et al. 2005). The neutral-model perspective, by contrast, assumes ecological equivalence among species (Hubbell 2001). By that view, local community composition is not driven by differences among species in their competitive or dispersal abilities, but instead is a stochastic process of assembly in which immigration and speciation offset local extinctions (Cottenie 2005). Hubbell's (2001) neutral model explores how speciation counteracts the extinction process due to 'ecological drift,' the gradual loss of species over time as a result of stochastic processes. Just as genetic drift describes the loss of genetic diversity over time, in which alleles exhibit a random walk toward fixation (**Chapter 9**), so too ecological drift describes the loss of species diversity due to a random walk of species toward extinction.

TABLE 10.4   Comparison of characteristics among the four metacommunity perspectives.

Characteristic	Metacommunity Perspectives			
	Patch dynamics	Species sorting	Mass effects	Neutral models
Habitat heterogeneity?	No	Yes	Yes	No
Degree of inter-patch dispersal	Low	Sufficiently high to permit colonization of suitable habitats, but lower than for mass effects	High and asymmetrical (from sources to sinks)	Limited
Niche differences?	Maybe; trade-offs in competitive and dispersal abilities required in the case of competing species	Yes; species differ in their ability to persist in different habitats	Yes; species differ in their ability to persist in different habitats	No
Spatial dynamics	Asynchronous	Not specified	Synchronous, because of high inter-patch dispersal	Not specified, but at least some asynchrony
Local communities in equilibrium?	No, because of low inter-patch dispersal	Yes	No, because of high inter-patch dispersal, but can reach a new equilibrium if dispersal changes	No, because of ecological drift
Local vs regional species composition	Local: varies through time Regional: more constant	Local and regional both constant	Local and regional both constant, assuming inter-patch dispersal is constant	Local and regional both vary through time

Source: Holyoak et al. (2005).

The neutral model thus provides a null hypothesis of what metacommunity patterns occur in the absence of the sorts of species or habitat differences featured in the other metacommunity perspectives (Holyoak et al. 2005). In the case of the neutral model, dispersal (immigration) helps to maintain regional diversity by balancing local extinctions. Dispersal cannot be too great, however, as otherwise local communities will all come to have the same set of species, causing regional diversity to decline. At the other extreme, an absence of immigration would preclude metacommunity structure and dynamics altogether. Once again, some intermediate level of dispersal or immigration is likely necessary to achieve regional coexistence.

From an empirical standpoint, the relative importance of environmental factors (habitat heterogeneity) versus spatial dispersal processes can help distinguish among the different metacommunity types (Cottenie 2005; Table 10.4). For example, Cottenie (2005) found that the structure of most metacommunities (44% of 158 data-

sets) was consistent with the species-sorting model, in that habitat heterogeneity and associated species-sorting dynamics were most important. This does not mean that spatial dispersal processes were wholly unimportant, however, as dispersal must be sufficient to enable species to colonize sites with appropriate environmental conditions (**Table 10.4**). In fact, nearly a third (29%) of metacommunities were structured equally by environmental and spatial variables, thus exhibiting characteristics of both the species-sorting and mixed-effects models. Only 8% of metacommunities in this survey best fit the neutral or patch-dynamics models.[44]

In addition, the diversity in metacommunity types could be partly explained by spatial scale, habitat type, and dispersal type, with dispersal type (passive vs

[44] Although these latter two models invoke very different processes (e.g. a competitive-colonization trade-off in the case of the patch-dynamics model), the analytical methods and datasets used by Cottenie (2005) did not permit a separation between the two models. Regardless, these two models together fit only a small percentage (8%) of the metacommunities in the survey.

active dispersal) having the greatest effect (Cottenie 2005). Given that dispersal type is associated with dispersal rates, it makes sense that this would be an important determinant of metacommunity type (**Table 10.4**). Passive dispersal, such as by freshwater plankton and terrestrial plants, resulted in species compositions that tracked environmental heterogeneity across all scales. By contrast, active dispersers in some habitats (namely estuarine, stream, and terrestrial systems) exhibited species-sorting and neutral-model dynamics at spatial scales up to 948 km. Thus, environmental processes were most important for metacommunity structure for passive dispersers and active dispersers in some habitat types (lakes and marine habitats), whereas pure spatial dynamics (dispersal limitation) becomes increasingly important for active dispersers in other habitats, especially at broader spatial scales (>948 km; Cottenie 2005).

Spatial dispersal processes should thus be part of every study in (community) ecology, and should be explicitly modeled in the analyses.   Cottenie (2005)

Although these models make different predictions about metacommunity structure and dynamics, most metacommunities are likely to exhibit characteristics of more than one type, especially given that different processes can generate similar patterns (Chase et al. 2005; **Table 10.4**). At this point, it is more informative to evaluate the relative importance of these various processes on metacommunity structure and dynamics, rather than worry about whether the observed metacommunity fits one model better than another (Chase et al. 2005). Most experimental and observational studies of metacommunities have focused on testing predictions related to patterns of abundance and local diversity, but ultimately the goal is to tie the structure and dynamics of metacommunities to ecosystem function and the provisioning of critical ecosystem services, such as biological control and pollination services (Logue et al. 2011). As human land use continues to intensify, the management of landscapes to preserve ecosystem services is a growing concern in the Anthropocene, as we explore in the next—and final—chapter of this text (**Chapter 11**).

Future Directions

We end this chapter the way it began: by looking to the future. Tools for predicting the future state of communities and ecosystems are urgently needed if we are to confront today's global environmental challenges and develop better conservation, land-use, and management strategies for dealing with the sorts of climatic and land-use changes that are projected to occur in the coming decades (Petchey et al. 2015). Although the ability to make accurate predictions is viewed as one of the hallmarks of a successful science (Evans et al. 2012), the sheer complexity and inherent variability of ecological communities and ecosystems would seem to preclude reliable forecasts, at least beyond a certain distance or time period (the **forecast horizon**;[45] Petchey et al. 2015). To develop reliable forecasts, ecologists need to know what properties and components of ecological systems are forecastable, along with the accuracy and uncertainty associated with those forecasts (the **forecast proficiency**; Petchey et al. 2015). In other words, our forecast proficiency will depend on what we are trying to predict and with what degree of accuracy or precision. For example, certain predictions at the community or ecosystem level might still be

possible even when predictions at the population level are not (Petchey et al. 2015). A case in point: population abundances of individual species may exhibit high inter-annual variability (low predictability) in highly variable environments like grasslands, resulting in increased competition during drought or 'lean' years ('ecological crunches;' *sensu* Wiens 1977), but community biomass and ecosystem productivity might nevertheless remain fairly constant (high predictability) owing to compensatory increases in those species unaffected by competition (Tilman 1996). Successful ecological forecasting thus demonstrates the advances we've made in understanding how ecological systems work (ecology is a successful science!), while also revealing the gaps in our understanding, the challenges we still need to overcome, and the limits to our predictive powers (Coreau et al. 2009).

Recent quantitative and methodological advances have increased our understanding of the structure and determinants of ecological communities, which in turn, should help to improve the reliability of ecological forecasts aimed at predicting community responses to environmental change. For example, improvements to our ecological forecast horizons will be made as we continue to gather data with better spatial coverage, at finer resolutions, and over longer temporal extents (Petchey et al. 2015). This should help to improve projections by the next generation of species distribution

---

[45] The ecological forecast horizon is the distance in space or time at which forecast proficiency (measured in terms of accuracy and/or precision) falls below the forecast proficiency threshold, the value at which forecasts are no longer considered useful (i.e. forecasts that are inaccurate and/or have a high degree of uncertainty; Petchey et al. 2015).

models, for example, which are moving beyond 'stacked' single-species models to adopt a more integrative community approach. This may be accomplished either by considering the joint distribution of all species together (Clark et al. 2014), or by using mechanistic, process-based models that incorporate greater structural realism, such as information on species' dispersal abilities and interspecific interactions, and how these processes are likely to be influenced by changing landscape structure (Singer et al. 2016).

Next generation species distribution models can and should take into account not only the (abiotic) landscape change correlated to the realized species niches, but explicitly the (biotic) population and community processes that are involved in the response to the abiotic change.   Singer et al. (2016)

Regardless of whether our objective is to improve forecasts of community change or simply to better understand the structure and determinants of ecological communities, we can expect to make progress through the continued study of how spatial structure influences community structure and dynamics across a range of scales. For example, recent methodological developments have made the multivariate analysis of species assemblages more spatially explicit, including the ability to analyze community spatial structure at multiple scales (Dray et al. 2012). Although spatialized multivariate techniques are still very much at the cutting edge of community analysis, such methods may be used for detecting and characterizing spatial patterns in community structure; determining whether spatial variation in community composition can be explained by environmental or landscape factors; and identifying characteristic scales of community spatial structures, such as the scales at which variation in community composition is well modeled by environmental factors, and whether there are other scales at which spatial structures are observed but not explained by environmental descriptors (Dray et al. 2012). This fits well with recent advances in quantifying species turnover (β diversity) along spatial or temporal gradients, as well as analyzing spatial variation in community composition (**Figures 10.3 and 10.5**; Anderson et al. 2011), including the hierarchical partitioning of diversity among spatial or temporal scales (**Figures 10.7 and 10.14**; Veech et al. 2010a,b).

Understanding how diversity scales and the form taken by the species–area relationship (e.g. **Figures 10.15 and 10.16**) continue to be important areas of research that could act as 'major stepping stones for [developing] a deeper understanding of both species–area relationships and broad-scale variation in species

richness generally' (Whittaker & Triantis 2012: 625). We still have much to learn about the ecological and evolutionary determinants of broad-scale patterns of diversity, such as those that underlie elevational diversity gradients (**Figure 10.11**; McCain & Grytnes 2010). Mountains enhance the range of climatic and environmental conditions—and thus species diversity—that may be found within a landscape, and thus can serve as natural laboratories for studying the mechanisms underlying spatial patterns of biodiversity (Graham et al. 2014).

As human land use intensifies, research on the effects of habitat loss and fragmentation will remain a research priority. The study of fragmentation effects on species and their interactions is still largely patch-based, given that much of this work to date has focused on the effects of patch area or isolation (e.g. distance from the nearest natural habitat patch). To be sure, the study of patch-area and isolation effects is still important for understanding the mechanisms that contribute to fragmentation effects (e.g. **Figure 10.19**), but should not themselves be confused with fragmentation effects, which are usually better addressed at the landscape scale (McGarigal & Cushman 2002). Fragmentation research in landscape ecology is thus moving away from a purely patch-based approach to more of a landscape perspective, in which entire landscapes that vary in the amount and configuration of habitat are the focus of study. This is evidenced in the design of fragmentation experiments that manipulate the distribution of habitat at a landscape-wide scale in order to test the independent effects of habitat amount versus fragmentation (**Figures 5.11 and 10.23**; With & Pavuk 2011, 2012). Future research should thus continue to investigate how habitat loss and fragmentation interact, and whether it is the amount or the configuration of habitat that is ultimately important for predicting community responses to landscape change. This will require more rigorous testing of the habitat-amount hypothesis (**Figure 10.20**; Fahrig 2013), as well as empirical tests of the direct and indirect causal relationships that are hypothesized to underlie species' responses to habitat loss and fragmentation at different scales (**Figure 10.21**; Ruffell et al. 2016).

Studies designed to disentangle the independent (or interdependent) effects of habitat loss and fragmentation are also essential for gaining insight into landscape-mediated declines in the ecosystem services provided by natural enemies and pollinators, as well as for evaluating AES schemes and other conservation strategies that seek to restore or optimize these services within landscapes (Hadley & Betts 2012; Cronin & Reeve 2014). Continued advances in understanding how landscape or spatial structure influence species interactions will require a more spatially explicit and integrative research approach. Ideally, this would combine experiments on movement behavior, manipulation of different aspects of landscape structure, and spatially explicit models (Cronin & Reeve 2014). Studies of predator–prey/host–parasitoid interactions in different landscape contexts are still relatively uncommon, and spatially explicit models that explore the interplay between spatial heterogeneity and realistic aspects of the movement of both species are currently unavailable (Cronin & Reeve 2014). The dearth of quality movement and dispersal data represents the biggest hurdle to this approach, but as new methods for marking and tracking the movements of species become available (**Chapter 6**), it will be increasingly feasible to integrate these movement data with spatially explicit models and to test model predictions experimentally. Currently, the lack of manipulative studies that provide a causal link between changes in landscape structure, habitat connectivity, and the local and regional population dynamics of interacting species remains a key impediment to advancement in these fields. For example, Hadley and Betts (2012) suggest that research on how the matrix influences pollinator movements could contribute to a better assessment of functional connectivity and help advance the field of landscape pollination ecology. We thus need to look beyond just the amount or configuration of habitats to consider how other aspects of landscape structure, such as the composition of the matrix, might influence species interactions.

A related concern is identifying the scale(s) at which species interact with landscape structure. The spatial scales at which different species move and interact with the landscape are almost always different. For example, 92% of the studies in one survey found that prey species and their predator(s) differed significantly in the scales at which they disperse or respond to spatial structure (Cronin & Reeve 2014). However, most research on predator–prey/host–parasitoid interactions assume that they operate on comparable spatial scales. Compounding the problem is the fact that most studies measure natural enemy responses to landscape structure, but far fewer also assess pest responses, let alone measure the effect that natural enemies are having on reducing pest populations in those landscapes (Chaplin-Kramer et al. 2011). Studies or experiments that quantify the ability of natural enemies to suppress pest populations and/or reduce plant damage, coupled with spatially explicit models of predator–prey/host–parasitoid population dynamics, would not only enhance our understanding of the landscape conditions under which biological control can be achieved, but might also inform release strategies of natural enemies across the landscape so as to enhance the success of biological control programs (Barbosa 1998; Chaplin-Kramer et al. 2011).

Finally, with information about the movement behavior of different interacting species, we can expand our perspective to investigate landscape effects on metacommunity structure and dynamics. At present, our understanding of metacommunity dynamics is predominantly theoretical in nature, although increasing interest has led to a growing number of empirical studies that address different aspects of metacommunity theory, such as testing assumptions from its four paradigms experimentally or by assessing local community assembly within its theoretical framework empirically (Logue et al. 2011). Much of metacommunity theory is founded on competitive interactions, and so should be extended to include other types of species interactions (predator–prey and plant–pollinator interactions), which also have a clear spatial component. Thus, metacommunity ecology would benefit from investigating the effect of more complex spatial configurations on different types of community organizations. For example, different trophic guilds have different levels of mobility and distributions, which again underscores the need for empirical data on dispersal rates. In one review, only 5 of 74 empirical metacommunity studies (7%) had measured dispersal rates (Logue et al. 2011). In all other cases, dispersal information was inferred indirectly (e.g. via spatial distance, spatial variability in species abundance and composition, connectivity, or isolation). Only a few studies addressed the functional consequences of altered metacommunity structure, such as in productivity or consumption rates, although this is addressed more by metaecosystem models, as we discuss in the next chapter.

Chapter Summary Points

1. Landscape ecology can contribute to the study of community structure and dynamics through its emphasis on the causes and consequences of spatial variation in the distribution of diversity across a range of scales.

2. Species diversity can be partitioned into different spatial components. Alpha diversity (α diversity) is the richness or diversity of species that occurs within a given community, location, or sample. Gamma diversity (γ diversity) is the total species richness found across all communities, locations, or samples, and thus provides a measure of landscape or regional diversity. The degree to which species composition changes or differs

among communities, locations, or samples within that landscape or region is assayed by beta diversity (β diversity).

3. Beta diversity (β diversity) can be assayed in many different ways, but is generally characterized as either (a) the degree of species turnover between communities, by calculating the degree of similarity (or dissimilarity) between pairs of sites or samples; or (b) the spatial variation in community composition using measures of multivariate dispersion or multivariate ordination techniques, where the goal is to relate that spatial variation in community composition to various environmental or landscape factors.

4. Multiscale patterns of diversity can also be uncovered through diversity partitioning. Multiplicative partitioning assumes that γ diversity is the product of α and β diversities ($\gamma = \alpha \times \beta$), in which case, β can be derived as $\beta = \gamma / \alpha$. Conversely, if γ diversity is an additive property of α and β diversities ($\gamma = \alpha + \beta$), then $\beta = \gamma - \alpha$. Additive partitioning is often preferred because it expresses β diversity in terms of the number of species, the same units as γ and α diversities, rather than as a proportion (i.e. the effective number of communities or samples needed to attain a particular γ diversity).

5. The processes responsible for patterns of diversity operate across a wide range of spatial and temporal scales: from the local site scale, in which interactions between species and the environment influence habitat suitability and site occupancy and which take place on a time scale of years to decades; to landscape and regional scales where landscape disturbance dynamics, environmental heterogeneity, and ecosystem productivity interact with species' colonization–extinction dynamics to influence the spatial distribution and persistence of species over centuries and millennia; all the way up to the broader continental and global scale, where climate cycles, geomorphic processes, and plate tectonics interact with evolutionary processes to shape different biogeographic realms over tens to hundreds of millions of years.

6. The latitudinal diversity gradient, in which species richness increases from the poles toward the equator, is one of the best-known biogeographic patterns of species diversity, having been found within a wide range of taxonomic groups, from microbes to vertebrates, and across the terrestrial, freshwater, and marine realms. More than 30 hypotheses have been proposed to explain this pattern, and although no consensus has yet emerged, the latitudinal diversity gradient is likely driven by the latitudinal gradient in productivity and/or energy, which similarly increases toward the equator.

7. Species diversity also exhibits an elevational gradient, in which the number of species decreases as one moves up in elevation from the lowlands to the summit of a mountain. The elevational diversity gradient is thought to be due to interactions between temperature and precipitation, which then impose physiological or energetic constraints on the elevational ranges of species. Diversity does not decline monotonically with elevation in all groups, however. In some groups, diversity peaks at middle elevations. Regardless of the exact elevational diversity pattern, mountains greatly add to the topographic, climatic, vegetation, and species diversity within a landscape.

8. The species–area relationship (SAR) is perhaps the most ubiquitous scaling relationship in ecology, to the extent that it has been called one of its few laws. Simply stated, species richness (S) increases with the size of the area (A) surveyed, usually as a power-law of the form $S = cA^z$. The slope of the species–area relationship (z) gives the rate at which species richness increases per unit area.

9. There are several different types of SARs, depending on how area is surveyed, and which therefore have different implications for the estimation and interpretation of diversity measures (α, β, and γ diversity). Regardless of whether the SAR ultimately depicts how α or γ diversity changes with spatial scale, the slope of the relationship (z) is not a measure of β diversity, despite previous claims in the literature to the contrary.

10. The slope of the SAR (z) may be scale-dependent, exhibiting different scaling domains when a large range of areas is considered. For example, the slope of the SAR tends to be steeper when considering how diversity scales over very large areas encompassing different biogeographic regions or continents (between biotas) than when evaluating how diversity scales over a range of smaller areas within a particular landscape (i.e. within biotas; $z_{\text{between biotas}} > z_{\text{within biotas}}$). At the other extreme, species richness may vary independently of area when very small areas are considered owing to the 'small island effect,' in which the particular location, environmental conditions, or stochastic factors unrelated to area have a greater effect on species occurrences. Because of the potential for scale-dependence, z should not be extrapolated blindly across scales to make predictions about habitat-area effects on species richness.

11. Although the SAR is often used to predict the effects of habitat loss and fragmentation on species richness, the traditional SAR can overestimate the numbers of species expected to go extinct when species exhibit a delayed response to habitat loss, incurring an extinction debt. The rate at which species richness increases as a function of increasing habitat area is usually not the same rate at which species are lost as a function of decreasing habitat area (habitat loss). For this reason, it has been suggested that the SAR be based only on endemic species, which should be most sensitive to the loss of habitat (the endemics-area relationship).

12. Because habitat loss and fragmentation occur at a landscape-wide scale, SARs should be based on the habitat area of entire landscapes rather than the areas of individual patches within a landscape, especially if patch-based and landscape-based SARs are not congruent.

13. The theory of island biogeography has long provided the theoretical underpinning for predicting the loss of species in landscapes undergoing habitat loss and fragmentation. According to the theory of island biogeography, the equilibrium number of species within a habitat fragment represents a balance between immigration (colonization) and extinction rates. Immigration rates are primarily influenced by the relative isolation of the patch from a habitat source. Extinction rates are primarily influenced by the size of the habitat fragment. All else being equal, large patches that are close together will have more species than those that are small and far apart.

14. Species vary in their sensitivity to patch-area and isolation effects. Species sensitive to patch-area effects tend to be those that are large-bodied, rare, habitat specialists, top consumers, and/or sedentary, or conversely, so highly mobile as to cause emigration losses to the matrix. The types of species most susceptible to increasing patch isolation tend to be those that are small-bodied, rare, habitat specialists, top consumers, and/or poor dispersers. The fact that many of the same traits increase a species' sensitivity to decreasing patch area and increasing isolation highlights the difficulty of disentangling these relative effects in determining whether habitat loss or fragmentation has the greater effect on species richness.

15. Habitat loss generally has large negative effects on species richness, whereas the effects of fragmentation (habitat configuration) tend to be weaker and more equivocal. This has led to the habitat-amount hypothesis, which predicts that local species richness within a patch is more related to the amount of habitat in the surrounding landscape than to patch size or habitat configuration.

16. Then again, if habitat loss and fragmentation are really *inter*dependent rather than independent processes, it may not be possible to tease apart their relative effects on species diversity. Further, habitat loss and fragmentation are landscape-scale processes that clearly have patch-scale effects. Hierarchical causal models could help to reconcile these patch-based and landscape perspectives, via a path analysis of all the hypothesized causal relationships as to how changes in various landscape and patch-based measures are expected to influence species' responses to habitat loss and fragmentation.

17. Even if the fragmentation or spatial configuration of habitat does not have a direct effect on a species, it could still have indirect effects, such as by altering the interaction strength between species. Spatial structure can either stabilize or destabilize species interactions, depending on how the interacting species are affected by the patchiness or heterogeneity of the landscape, which in turn affects species coexistence and diversity at the scale of the landscape.

18. Competitive coexistence at a local community scale has long been attributed to niche partitioning, in which ecologically similar species utilize a shared resource in different ways, in different areas, or at different times so as to reduce competition ('complete competitors cannot coexist'). At a landscape scale, however, coexistence between similar species can occur via competitive reversals in different habitats (each species is at a competitive advantage in one habitat but an inferior competitor in all others), or if there is a trade-off between dispersal and competitive abilities (the best competitors are the worst dispersers) such that inferior competitors are able to persist by colonizing habitat patches not yet occupied by superior competitors.

19. Predators and prey can be viewed as engaging in a behavioral space race: predators prefer areas where their prey are most abundant, whereas prey tend to avoid areas where predators are most abundant. The degree of spatial overlap between predators and prey influences the strength and stability of predator–prey interactions, and thus the likelihood of species coexistence at the landscape scale.

20. Spatial heterogeneity or patchiness can create refugia that help stabilize predator–prey interactions if predators would otherwise overexploit

and extirpate their prey. Patchiness, such as that caused by habitat fragmentation, can have a destabilizing effect on predator–prey interactions if it interferes with the search efficiency of predators and permits prey populations to build-up and overexploit their resources.

21. Habitat loss and fragmentation are expected to have the greatest effect on specialist predators (a predator that feeds on only one type of prey): not only do predators generally have larger area requirements than their prey, but specialist predators are also strongly limited by the abundance of their prey, which are also declining in response to habitat loss and fragmentation. Conversely, generalist predators, which consume many types of prey and are associated with the matrix, may benefit from habitat loss if this increases matrix habitat and provides them with additional resources. Unfortunately for the prey species in the habitat being lost, however, the build-up of generalist predators in the matrix can cause them to spill over into the prey's shrinking habitat, resulting in increased predation that exacerbates prey population declines caused by the loss and fragmentation of their habitat.

22. The effects of landscape structure and heterogeneity on parasite–host/host–parasitoid dynamics are similar to those for predator–prey dynamics. Matrix heterogeneity can increase parasitoid diversity and enhance biological control of crop pests if parasitoids spill over from the matrix into agricultural fields. Parasitoids respond in species-specific ways to habitat edges, which makes them differentially sensitive to habitat loss and fragmentation, as well as to changes in the land-use matrix (landscape context). This in turn can destabilize host–parasitoid interactions and cause pest outbreaks if the altered landscape structure interferes with the search behavior or habitat use of parasitoids and prevents them from aggregating in response to their hosts.

23. The loss, fragmentation, and degradation of natural and seminatural habitats have contributed to global declines in bees and other native pollinators. Habitat loss and fragmentation can affect plant–pollinator interactions in at least three ways, by altering plant density, the density of pollinators, and the foraging or dispersal behavior of pollinators. Habitat loss is expected to have a greater effect on the densities of plants and pollinators, whereas habitat fragmentation is expected to have the greatest effect on pollinator

movement. The combined effects of habitat loss and fragmentation result in less pollen transfer among plant populations, which in turn can lead to reduced seed set and crop yields. As with other types of species' responses to landscape structure, however, not all pollinators are affected to the same degree, or even in the same way, by habitat loss and fragmentation.

24. Pollination services are critically dependent on having a sufficient amount of natural or seminatural habitat capable of supporting a diverse pollinator community that is also within the foraging or dispersal range of the pollinators. Although habitat amount is generally more important than fragmentation per se for the maintenance and stability of pollination services, corridors can nevertheless help to promote the movement of pollinators in some landscapes, thereby increasing pollen flow among isolated plant populations.

25. A metacommunity is a set of local communities linked by the dispersal of multiple, potentially interacting species. The composition of local communities is predicted to depend on some combination of environmental heterogeneity, species niche characteristics, and dispersal (spatial) processes.

26. Four conceptual frameworks have been developed to describe the structure and dynamics of metacommunities. The patch-dynamics perspective assumes that the habitat is patchy but homogeneous and that species differ in their dispersal abilities, which subsequently affects their colonization–extinction dynamics, and thus their ability to persist at the regional scale. In contrast, the species-sorting perspective assumes a heterogeneous landscape in which all species can disperse freely but are able to coexist regionally because species differ in their habitat niche requirements. The mass-effects perspective invokes source–sink dynamics to explain regional coexistence: species differ in their flux rates among habitats owing to dispersal from sources to sinks. Finally, the neutral-model perspective posits that species are ecologically equivalent, and thus do not exhibit niche differences or differences in their dispersal or flux rates. Speciation and immigration rates offset species extinctions, which occur as a random walk through time (ecological drift). The neutral model thus serves as a null hypothesis for evaluating what patterns are expected in the absence of the sorts of processes featured in these other models of metacommunity structure and dynamics.

Discussion Questions

1. For your study area, or an example provided by your instructor, how would you approach the measurement of beta diversity (β diversity) in this system? Does the characterization of β diversity—in terms of whether this is assayed as species turnover along an environmental or spatial gradient or as variation in community composition—depend on the type of landscape, for example? In other words, do certain types of landscapes or habitat distributions lend themselves to particular definitions of β diversity?

2. How would you devise a hierarchical sampling scheme to partition species diversity in your study system (**Figure 10.7**)? Do you expect species turnover to be greater among certain scales than others? What explanation can you offer for why differences in species turnover might occur at certain scales in this system?

3. What sampling design (**Figure 10.13**) would you advise for exploring how species diversity scales in your study area? Are particular study designs better suited for certain types of spatial distributions than others? Explain.

4. Species–area curves depict the scaling of species richness in space, but how might species richness scale over time? For a given area, how would you expect species richness to vary as a function of time, such as the length of study or the number of times the site is surveyed? How might time interact with space to influence species–area curves? For example, if highly diverse systems such as the tropics have a steep species–area relationship, do they also have a steep species–time relationship (a positive time-by-area interaction)? Or, is the time-by-area interaction more likely to be negative (e.g. a steep species–area relationship but a shallow species–time relationship)? Explain your reasoning.

5. Given your understanding of latitudinal and elevational gradients in species richness, how would you expect the values of c and z in the species–area relationship (Equation 10.10) to vary with increasing latitude or elevation? For example, do you expect c and z values to increase, decrease, or stay the same, as you move from the tropics to the poles or from the lowlands up to a mountain's summit? Do these two parameter values change independently of one another, or are they linked in some fashion? Explain.

6. Do you expect the species you study to be influenced by patch area, patch isolation, or both? What traits do they possess that might make them differentially susceptible to patch-area versus patch-isolation effects? Based on their individual responses to decreasing patch area and/or increasing patch isolation, can we infer how these species are likely to respond to habitat loss and fragmentation? Why or why not?

7. Is habitat amount or habitat fragmentation (the spatial configuration of habitat) more important for understanding patterns of diversity in your study system? How would you test which has the greater effect on species diversity?

8. Diagram the expected direct and indirect effects of habitat loss and fragmentation on the population responses of one or more species in your study system, using a hierarchical causal modeling approach (**Figure 10.21**). Which population responses are likely to be most affected by habitat loss and/or fragmentation? Are most of the effects of habitat loss and fragmentation on these species likely to be direct or indirect?

9. Identify two species that interact in your study system (e.g. as competitors, predator–prey, host–parasitoid, or plant–pollinator). How is the interaction strength of these two species likely to be affected by habitat loss and/or fragmentation? How might these landscape-induced changes in interaction strength affect the population responses of these interacting species to habitat loss and fragmentation (e.g. as diagrammed in response to the previous question)?

10. Explain how spatial structure can either stabilize or destabilize species interactions, and thus affect the likelihood of species coexistence and overall diversity in a landscape. Under what conditions is spatial structure stabilizing? Under what conditions is spatial structure destabilizing? How might this information be used to manage landscapes for a species of conservation concern? To prevent the spread of a forest or crop pest, an invasive species, or a zoonotic disease? To conserve high native diversity?

Landscape Effects on Ecosystem Structure and Function

11

The tropical rainforests of the Amazon basin are highly productive ecosystems. They are lush, verdant, and teeming with life, and thus we might naturally assume that all this biomass is the product of growing conditions that are highly conducive to plant growth, which they are, at least climatically. More puzzling, however, is how the thin, nutrient-poor soils of the Amazon forests can possibly sustain such high levels of productivity. The same climatic conditions that promote vigorous plant growth have also led to the rapid weathering of soils that have all but leached them of nutrients. Nutrient cycling is apparently very efficient in tropical forests, but even so, most nutrients are locked up in the living tissues of plants and other organisms, and thus are unavailable for uptake (Vitousek & Stanford 1986).

Half a world away, in the harsh deserts of northern Africa, we go from one extreme to the other. Hot, arid, and sparsely vegetated, the Sahara Desert could hardly be more different from the tropical rainforests of the Amazon basin. Yet, improbably, they are connected. The Sahara was not always a desert. Some 15,000 years ago, the region was much wetter and supported a grassland savanna. Lake Mega-Chad was then the largest freshwater lake in the world, about the size of the Caspian Sea today. As the climate changed and became more arid, Lake Mega-Chad began to dry and shrink to its current size, Lake Chad, leaving behind lakebed sediments in what is now called the Bodélé Depression (**Figure 11.1**). Strong winds over this depression create massive dust storms, capable of carrying and depositing its mineral-rich dust many hundreds to thousands of kilometers away, over the Atlantic Ocean and onto the Amazon basin (Koren et al. 2006). It would thus appear that the productivity of the tropical rainforests in the Amazon basin is being subsidized by the past productivity of an ancient lake in Africa (Bristow et al. 2010). Understanding the extent of such spatial subsidies—the linkage among ecosystems in space and time—is obviously important for evaluating their productivity, structure, and function. We thus turn our attention to the effects of landscape structure and connectivity on ecosystem processes in this final chapter.

Why is a Landscape Ecology of Ecosystems Needed?

Landscapes and ecosystems are inextricably linked, to the point that the two terms are often used interchangeably. One could just as easily refer to a forest landscape as a forest ecosystem, for example. Since landscapes and ecosystems both supply the goods and services upon which human societies and economies ultimately depend

Essentials of Landscape Ecology. Kimberly A. With, Oxford University Press (2019).
© Kimberly A. With 2019. DOI: 10.1093/oso/9780198838388.001.0001

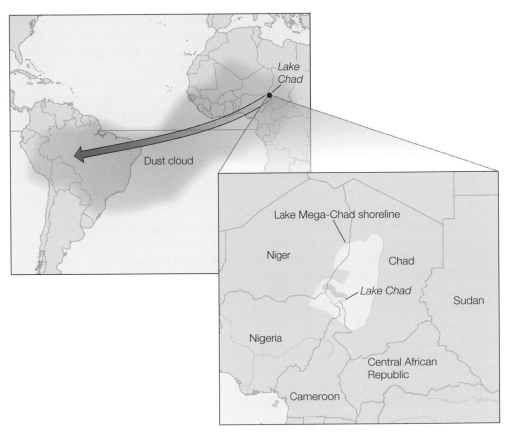

Figure 11.1 Connections among ecosystems can occur at a global scale. The rainforests of the Amazon basin are fertilized by nutrient-rich dust from the Sahara Desert, including the Bodélé Depression, the remains of an ancient lakebed of what was once the largest lake in the world, Lake Mega-Chad.

(Daily 1997; Termorshuizen & Opdam 2009), this blurring of concepts is perhaps understandable. Landscapes are not ecosystems, however, nor are they equivalent in form or function.

By now, you should have a fairly good idea of what a landscape is and how landscape structure is assessed. We'll hold off on discussing landscape function and the multifunctionality of landscapes until a bit later in this chapter. What, therefore, is an ecosystem and how does it differ from a landscape? An **ecosystem** consists of an ecological community and its abiotic environment, which are linked via the fluxes of matter and energy. The ecosystem concept thus emphasizes the transfer of materials, nutrients, and energy among organisms, and between organisms and their environment, which is why the ecosystem has been considered to be the most basic or fundamental unit of ecology by some (e.g. Tansley 1935; Odum 1953, 1964; Willis 1997). **Ecosystem structure** refers both to the diversity of organisms that make up the ecological community (i.e. the microbial, plant, and/or animal communities) and to the physical features of that system. **Ecosystem function**, by contrast, might variously refer to the processes that regulate the fluxes of matter and energy

within ecosystems, the extent to which those processes are influenced by species (the functional role of species within ecosystems), the functioning of the entire ecosystem, or the services provided by ecosystems (Jax 2005). Both structural and functional measures are needed to assess the status or 'health' of an ecosystem, as well as their continued potential to provide goods and services, such as biocontrol and pollination services (**Chapter 10**) or water for drinking and irrigation (Palmer & Febria 2012).

Ecosystem structure refers to attributes that can be evaluated with point-in-time measurements and that are assumed to reflect the existing status or condition of an ecosystem. Structural attributes may be easy to measure, but they do not capture the dynamic properties of an ecosystem that represent its actual performance. Functional measurements, on the other hand, attempt to capture system dynamics through repeated measurements that quantify key biophysical processes.   Palmer & Febria (2012)

For example, the ecosystem structure of rivers and streams can be described by their channel morphology, type of streambed substrates, water characteristics (physical, chemical, and/or microbial), and biological diversity (Sandin & Solimini 2009; Palmer & Febria 2012). Note that some of these same structural attributes, such as channel morphology and streambed substrates, can also be used to describe the structure of the riverine landscape (Allan 2004; **Figures 2.11 and 3.7**). Landscape structure can thus contribute to ecosystem structure, and may even influence aspects of that structure (e.g. water quality, community composition). In contrast, the functioning of a river or stream ecosystem is typically assayed in terms of various processes related to its metabolism (the balance between the energy created by primary production and that used by organisms), decomposition rate of organic matter (the rate at which fine or coarse particulate organic matter is broken down by microbial decomposers and animal detritivores), and nutrient cycling or flux rates (Sandin & Solimini 2009; Palmer & Febria 2012; von Schiller et al. 2017). Which combination of structural and functional measures should be used to evaluate river and stream ecosystems is a matter of debate, however. The structural and functional components of ecosystems interact in varied and complex ways, with the result that measures of river and stream functioning may vary widely and exhibit both scale and landscape dependence (Allan 2004; Palmer & Febria 2012; von Schiller et al. 2017).

Understanding how landscape structure influences ecosystem processes is a major focus of **landscape ecosystem ecology** (Massol et al. 2011). Traditionally, landscapes were viewed as encompassing a broader spatial extent than that studied in ecosystem ecology (Forman & Godron 1986; Turner & Chapin 2005; **Figure 2.9**). Thus, the initial 'landscape approach' to ecosystem research simply put the study of ecosystem processes within a broader landscape context, by comparing ecosystem measures between different locations on the landscape (e.g. the effect of topographic position on net primary production, which will be covered in the next section). Subsequently, it became clear that ecosystem processes are spatially and temporally variable across the landscape, reflecting differences in topography, soils, vegetation, and the action of species, including human land-use activities (White & Brown 2005; Shen et al. 2011). A landscape ecology of ecosystems has thus emerged that is concerned with uncovering the nature of the observed heterogeneity in ecosystem function, as well as the interactions among system components (e.g. habitat patches or nutrient pools) that influence the structure and dynamics of ecosystems (Lovett et al. 2005). In turn, landscape heterogeneity may be generated or sustained by the redistribution of nutrients or organisms among patches both within and among ecosystems. Indeed, the landscape has been defined in some contexts as a system of interacting ecosystems (Forman & Godron 1986; Lovett et al. 2005), which again highlights the interconnected nature of these systems.

Thus, landscape and ecosystems *are* linked, which perhaps excuses their sometimes-synonymous usage. But are landscapes really a higher-order system composed of interacting ecosystems, or are they simply the spatial context in which ecosystem processes play out? Perhaps they are both? Is a landscape ecology of ecosystems simply ecosystem function assessed at a broader scale, so as to encompass the effect of environmental heterogeneity on ecosystem processes? Or does landscape structure modify ecosystem processes in ways that cannot be understood simply by scaling up ecosystem functions? If landscapes influence ecosystem processes, and ecosystem processes modify the landscapes in which they occur, then how does ecosystem function relate to landscape function? What is the function of a landscape and how does it compare to ecosystem function?

In this chapter, we begin with an overview of how landscape context influences ecosystem processes. Next, we consider the linkages among systems (**metaecosystems**) and how spatial subsidies are important for understanding ecosystem function in a landscape context. We then review the twin concepts of ecosystem and landscape function and explore under what conditions landscape function becomes disrupted. Finally, we conclude by considering the implications and challenges of managing for functional—and multifunctional—landscapes (i.e. landscape sustainability).

Ecosystem Processes in a Landscape Context

Four major processes are considered essential to all ecosystems: (1) the water or hydrological cycle, (2) biogeochemical or nutrient cycles (e.g. the nitrogen or carbon cycle), (3) trophic or food-web dynamics, and (4) energy flows among trophic levels.[1] All of these processes interact and are influenced by one another. For example, the water and carbon cycles are intimately linked: precipitation strongly regulates plant growth and the productivity of ecosystems (net primary production) because water is used both for transpiration (the movement of water and nutrients from the soil

---

[1] Notice that nutrients cycle, but energy flows. Energy cannot be recycled and thus must be continuously replenished, especially given that only about 10% of the energy at each trophic level gets transferred to the next. Most of the energy at a given trophic level goes to maintaining individual homeostasis (e.g. cellular respiration) rather than to the production and storage of biomass that can be consumed by individuals at the next level. The flow of energy through an ecosystem can thus be depicted as a pyramid, being greatest at the base (primary producers) and tapering off by a factor of 10 with each successive trophic level.

up through the plant) and during photosynthesis (the fixation of atmospheric carbon, in the form of CO_2, to produce organic compounds).[2] In turn, the evaporative water loss from vegetation (evapotranspiration) contributes to atmospheric water vapor that condenses and falls as precipitation, completing the water cycle. Precipitation also affects soil biogeochemical processes and decomposition rates, such as the breakdown of soil organic carbon by microorganisms, which then respire carbon back into the atmosphere (as CO_2 during the process of cellular respiration), thereby completing the carbon cycle.

Landscape structure, including human land uses, can affect any or all of these ecosystem processes. For example, we discussed earlier in the text how surface water flows are both affected by landscape structure and simultaneously contribute to the formation and evolution of landscapes, due to the interplay of erosional and depositional processes (**Figures 3.14 and 3.17**). As part of the water cycle, precipitation and surface runoff (glacial meltwaters, snowmelt, streamflow) are major ways in which minerals, including carbon, nitrogen, phosphorus, and sulfur, are cycled from land to water. Recall, too, that periodic flooding—a component of the natural flow regime—has the potential to mobilize large amounts of material (sediments, nutrients, organisms) across the landscape and deposit these elsewhere, contributing to unique habitats and landscapes (e.g. riparian habitats, floodplain landscapes). That is why rivers have been called 'the skeleton and the circulatory system of Earth's landscapes' (Perron et al. 2012), because they both contribute to landscape structure and serve as transportation networks that carry and deliver materials throughout the landscape. Thus, human alteration of hydrological flows, such as through the construction of dams, has profoundly altered not only the structure, dynamics, and connectivity of these riverine systems, but also affects the transport of sediments and nutrients downstream (the **nutrient spiraling concept**; Newbold et al. 1981; Ensign & Doyle 2006), which in turn alters channel morphology, riparian structure, species richness, and ultimately, the productivity and food-web structure of downstream ecosystems (Gomi et al. 2002). In addition, nutrient loading from the surrounding landscape, such as from agricultural or urban runoff, may affect the retention and export of organic matter and nutrients (nitrogen, phosphorus) within these watersheds, which can impair water quality and alter ecosystem structure and function downstream (e.g. Bernot et al. 2006; Hobbie et al. 2017).

Because we have discussed landscape and land-use effects on surface water flows previously (**Chapter 3**), much of our focus in this section of the chapter will center on how landscape structure influences biogeochemical or nutrient cycles, which has implications for food-web dynamics and energy flows within and between ecosystems. Nutrient cycles are key ecosystem processes, and thus have been a major focus of research in ecosystem ecology. Nutrient cycles are referred to as biogeochemical cycles because the element in question (carbon, nitrogen, phosphorus) is cycling between an inorganic pool (rocks, soils) and an organic pool (plants, animals, microbes), often involving direct or indirect exchanges with the atmosphere and/or water. For example, mention was made previously of the **carbon cycle**, by which plants and algae take CO_2 from the atmosphere to create carbohydrates and biomass via photosynthesis, but CO_2 is returned to the atmosphere via cellular respiration (plant and algal cells respire too!) or by the combustion of that biomass (i.e. when forests and grasslands burn as a result of natural or anthropogenic fires). The decomposition of plants and animals contributes to carbon stores found in the soil, which, given enough time and pressure, form fossil fuels that can be burned and contribute to the addition of CO_2 to the atmosphere (hence CO_2 is the primary greenhouse gas emitted by human activities). Carbon dioxide also dissolves in water, where it is used in the production of shells (as $CaCO_3$) for mollusks, foraminifera (marine protists), and corals, whose skeletal remains contribute to the formation of limestone via sedimentation. Carbon moves through these various pools (biotic, atmosphere, water, soils/rock) at different rates, and can accumulate and be stored for long periods in reservoirs called sinks. Oceans, soils, and forests are all major carbon sinks, for example.

Nutrient cycling and availability are controlled by interactions among factors that operate across a range of spatial and temporal scales. At a regional scale, climate and geomorphology influence soil development and the formation of different soil types (**Chapter 3**; **Figure 3.18**). In terrestrial ecosystems, the rates at which nutrients are exchanged among pools and accumulate are primarily governed by soil properties and climatic factors. Within a landscape, topographic position (whether a site is at the top or bottom of a hill) and human land-use or land-management practices are the primary factors affecting soil properties, and which therefore contribute most to spatial variation in organic matter pools and nutrient availability. Temporally,

---

[2] Recall from your basic biology course that photosynthesizers use carbon dioxide (CO_2) and water (H_2O) in the presence of light to make carbohydrates ($C_6H_{12}O_6$), with oxygen (O_2) produced as a byproduct: $6CO_2 + 6H_2O \rightarrow C_6H_{12}O_6 + 6O_2$. By contrast, the process of cellular respiration (metabolism) breaks down carbohydrates in the presence of oxygen to produce chemical energy the cell can use to power its various processes; carbon dioxide and water are a byproduct of those reactions. Thus, the chemical equation for cellular respiration is the reverse of that for photosynthesis: $C_6H_{12}O_6 + O_2 \rightarrow 6CO_2 + 6H_2O$. The symmetry of these two processes represents another cycle, one that supports nearly all life on the planet.

nutrient dynamics may be linked to seasonal or annual variations in temperature and precipitation, which can have both broad-scale landscape and local-scale site effects on nutrient availability (Burke 1989; Briggs & Knapp 1995; Hook & Burke 2000). The result is a spatially variable and dynamic resource landscape, in which some areas may be in nutrient debt or absorb more of a certain nutrient than they release (sinks), whereas other areas will have a surplus or release more of a certain element than they absorb (sources). To develop a deeper understanding of how landscapes can influence ecosystem processes, we consider topographic and land-management effects on nutrient availability and ecosystem productivity next, before addressing the effects of land-use change on nutrient dynamics in the following section.

Topographic and Land-Management Effects on Nutrient Availability and Ecosystem Productivity

Topographic variation (relief) is a major contributor to landscape structure and heterogeneity, as we've discussed in previous chapters regarding landscape formation (**Chapter 3**) and elevational gradients in productivity, vegetation zones, and species diversity (**Chapter 10**). Even less vertiginous landscapes can have these sorts of effects, however, albeit over a shorter range of distances and elevation. In hilly landscapes, topographic or landscape position exerts strong controls on the hydrologic and geomorphic processes that influence patterns of soil development, biogeochemistry, and vegetation structure along hillslopes. For example, a **catena** describes the sequence of soils that occur in a fairly regular and predictable fashion along the length of a hillslope (i.e. from top to bottom). Formalized by the British soil scientist, Geoffrey Milne,[3] the **catena concept** provides a useful framework for understanding the processes by which soils form and are expected to vary across hilly landscapes:

> The soil differences are brought about by differences of drainage conditions, combined with some differential re-assortment of eroded material and the accumulation at lower levels of soil constituents chemically leached from higher up the slope (Milne 1935, p. 346).

According to the catena concept, biogeochemical differences between upland and lowland sites are assumed to reflect the redistribution of soil constituents (mineral materials, organic matter) as a result of hillslope hydrology and erosional/depositional processes. Consequently, soils at the bottom of a hill (the toeslope) tend to be

deeper, heavier in clay content, richer in organic matter, and possess a higher soil moisture content than those on the ridgetop or steeper hillside (e.g. Schimel et al. 1991). Given these sorts of differences in soils and nutrient availability, we should expect that plant diversity and biomass (net primary production) will likewise vary along a topoedaphic gradient (Schimel et al. 1991; Hook & Burke 2000). How productivity and species composition actually vary, however, will depend on how topography interacts with other factors, such as different disturbances or land-management practices (e.g. fire, grazing).

To illustrate, we'll consider a case study involving topoedaphic effects on aboveground net primary production (ANPP)[4] and plant species diversity in the tallgrass prairie of the Flint Hills in eastern Kansas (USA), near the eastern edge of the Great Plains. This landscape was formed by the differential erosion of Permian-age limestone and shale layers, resulting in rolling hills and flat-topped ridges separated by wide valleys. Ridges are capped with shallow, rocky soils (comprising chert or 'flint,' which gives the region its name), whereas lowlands have deep, permeable soils (**Figure 11.2**). The structure, productivity, and diversity of the tallgrass prairie are governed by the interaction between climate (precipitation), fire, and grazing (Knapp et al. 1998; **Figure 3.35**). Fire in particular is necessary for the maintenance of tallgrass prairie, which otherwise would succumb to woody invasion and become a savanna or woodland (a system state change we'll discuss later in this chapter). Fire is also used as a management tool throughout the Flint Hills to increase the productivity and quality of the native rangeland for cattle.

Because tallgrass prairie is a fire-adapted system, fires increase ANPP due to a combination of effects related to the release of readily available nitrogen and phosphorus from plant material, increased nitrogen mineralization rates, enhanced nitrogen fixation, warmer soils, and less shading from litter or standing biomass (Knapp & Seastedt 1986; Knapp et al. 1993; Turner et al. 1997; **Box 11.1**). Most of these effects occur soon after a fire (Ojima et al. 1994a). Too much of a good thing is not a good thing, however. Annual burning, which is often performed on cattle-grazed rangeland, may decrease soil organic matter and contribute to large losses of nitrogen through volatilization and immobilization (Benning & Seastedt 1994; **Box 11.1**). Over the long term, repeated annual burning can cause a significant reduc-

---

[3] The general idea had been outlined a few years earlier by the British soil chemist, W. S. Martin (Grunwald 2005). Geoffrey Milne coined the term *catena*, however, and expanded upon (and published about) the concept.

[4] Although the NPP of an ecosystem comprises both aboveground and belowground components, ANPP is more typically assayed because it is easier to measure (e.g. as peak standing biomass) and is assumed to correlate with NPP. Given that much of the biomass in tallgrass prairie lies belowground in the deep root systems of the predominant grass species, however, the interactive effects of climate, fire, and grazing can have very different effects on BNPP than on ANPP (e.g. Johnson & Matchett 2001).

BOX 11.1 Importance of the nitrogen cycle in terrestrial ecosystems

The availability of nitrogen is important for evaluating productivity within ecosystems because nitrogen is needed to make chlorophyll, the pigment used in photosynthesis, which is responsible for primary production in almost all ecosystems, including aquatic and marine ones.[10] Furthermore, nitrogen is required for the synthesis of nucleic

acids (i.e. deoxyribonucleic acid or DNA, and ribonucleic acid or RNA) and the proteins they encode, which are the building blocks of all organisms. Although nitrogen is the most abundant element in Earth's atmosphere (78% by volume), it occurs in a form (nitrogen gas, N_2) that is not readily accessible to plants and animals, which is why it is

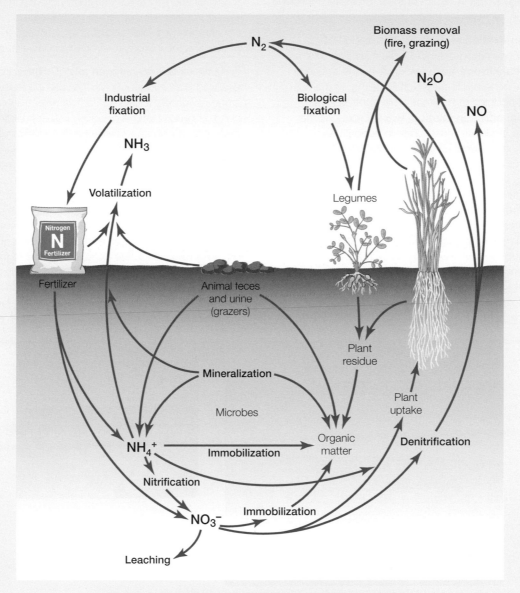

Figure 1 The nitrogen cycle in a terrestrial ecosystem.

[10]Chemosynthetic bacteria around hydrothermal vents and cold-water methane seeps are responsible for primary production in the deep sea. In these systems, bacteria manufacture carbohydrates from carbon dioxide (CO_2) or methane (CH_4) using the oxidation of hydrogen sulfide (H_2S) or hydrogen gas (H_2) as an energy source, rather than sunlight.

limiting in most ecosystems. Nitrogen is also contained in the tissues of dead animals and plants, but again, is not in a form that is accessible for plant uptake in the soil.

Nitrogen in the atmosphere or in organic matter must thus be converted into nitrogen compounds that can be

(Continued)

BOX 11.1   Continued

used by plants, such as ammonium (NH_4^+) and nitrate (NO_3^-), which is primarily accomplished by soil microbes in terrestrial ecosystems. Basically, the processes of nitrogen fixation, mineralization, and nitrification increase the availability of nitrogen to plants, whereas the processes of volatilization, immobilization, and denitrification result in the loss of nitrogen to plants. Each of these processes involved in the nitrogen cycle is described more fully, as follows:

- **Nitrogen fixation** involves the conversion of atmospheric nitrogen (N_2) to a plant-available form. **Biological nitrogen fixation** involves the action of nitrogen-fixing bacteria, which live either in the soil or in the root nodules of legumes. This contrasts with **industrial nitrogen fixation**, which produces the synthetic fertilizers widely used in agricultural production to boost crop yields, but which are applied to the soil surface and can therefore run off. Synthetic fertilizers are thus a major contributor to non-point-source pollution in aquatic systems.
- **Volatilization** involves the loss of nitrogen due to the conversion of NH_4^+ (largely from the urine or feces of animals) to ammonia gas (NH_3), which is then released to the atmosphere. Volatilization increases under conditions that favor evaporation (e.g. during a fire or under hot, dry, and windy conditions).
- **Mineralization** is the process by which soil microbes convert organic nitrogen to NH_4^+, an inorganic form of

nitrogen that can be used by plants; NH_4^+ can be further converted by soil microbes to NO_3^-, the form of nitrogen most available to plants, through the process of **nitrification**. Mineralization and nitrification rates occur most rapidly when soil is warm, moist, and well aerated.
- **Immobilization** is the reverse of mineralization and nitrification, in which inorganic nitrogen in the soil (NH_4^+ and/or NO_3^-) is converted by soil microbes into organic nitrogen (e.g. biomass of soil microbes), thereby limiting the inorganic nitrogen available to plants.

Finally, the **carbon-to-nitrogen ratio (C:N)** is important because it is an indicator of nitrogen use efficiency and soil nitrogen availability for plants. For example, plant materials having a high C:N ratio (e.g. woody tissues) decompose more slowly than those with a low C:N ratio (e.g. grass litter). Microbes responsible for this decomposition need a C:N ratio of about 25:1 to meet their own growth requirements (Chapin et al. 2011). If the C:N ratio in the plant material or soil organic matter is too high, microbes must import nitrogen and will start immobilizing inorganic nitrogen in the soil, thereby limiting nitrogen available for uptake by plants. At low C:N ratios, the amount of nitrogen exceeds microbial growth requirements, and thus microbes excrete excess nitrogen into the soil (nitrogen mineralization), where it is then available for plant use. The dividing line between nitrogen immobilization and mineralization, and thus nitrogen availability for plants, therefore occurs around a C:N ratio of about 25:1 (Chapin et al. 2011).

Figure 11.2 Topographic position influences ecosystem processes.   In the tallgrass prairie of the Flint Hills of Kansas (USA), soils are thin and rocky along ridgetops (inset) but deeper and more permeable in low-lying areas at the bottom of hills (the toeslope). As a result, grassland productivity is greater in lowland sites than in upland sites (cf. **Figure 11.3A**).

Source: Photos by Eva Horne.

tion in soil organic nitrogen, lower microbial biomass, lower nitrogen availability, and higher C:N ratios in soil organic matter (Ojima et al. 1994a; Turner et al. 1997; **Box 11.1**). As a result, nitrogen limitation has long been thought to be a general characteristic of the tallgrass

prairie ecosystem (Seastedt et al. 1991; Ojima et al. 1994a). That nitrogen isn't always limiting, however, is due to a complex interplay between the effects of precipitation, fire, and topography on productivity and plant species composition.

Topographic variation can interact with precipitation and fire to influence patterns of productivity and diversity across the tallgrass prairie landscape. The tallgrass prairie is a mesic grassland system, and so ANPP is generally less limited by water than in other grasslands, except during droughts (Briggs & Knapp 1995). Even so, the amount of precipitation within a year (whether in total or just during the growing season) was found to have a strong, significant effect on ANPP in annually burned uplands, a moderate effect in annually burned lowlands, but no effect in unburned sites, except for a weak relationship between ANPP and growing-season precipitation on unburned uplands (Briggs & Knapp 1995; **Figure 11.3B**). This makes sense, given that upland sites are more water-limited than lowland sites (i.e. soil moisture levels in the uplands are lower to depths of 1 m), especially if burned, and so behave more like an arid grassland system (Briggs & Knapp 1995; **Figure 11.3A**). Indeed, the stature and species composition of upland sites resemble that of the

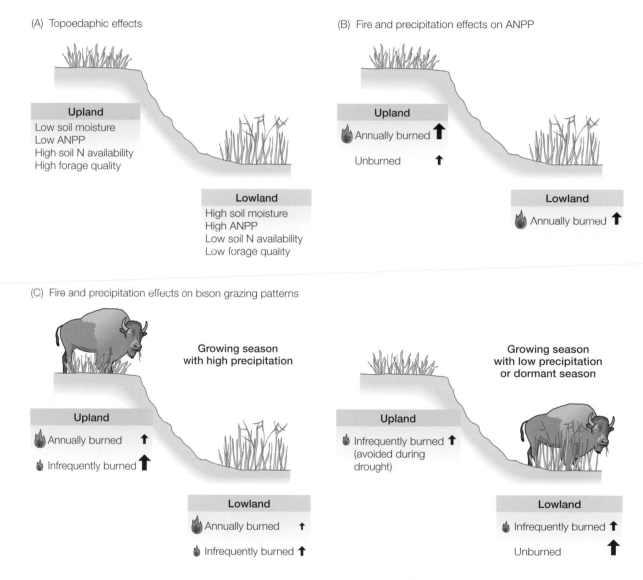

Figure 11.3 Topographic variation and land-management effects on ecosystem processes in the tallgrass prairie. (A) Effect of topographic position on factors related to aboveground net primary production (ANPP) and forage quality (as assayed by foliar nitrogen or protein content); (B) Topographic position mediates the interactive effects of fire and precipitation on grassland productivity (ANPP), such that topographic position has the greatest effect on ANPP of annually burned upland sites, but no effect on unburned lowland sites, which are less water-limited; (C) Grazing patterns of bison (*Bison bison*) switch from uplands to lowlands in response to forage availability, which varies seasonally (whether grasses are actively growing or senescent) and in response to inter-annual variation in precipitation (especially if it is a drought year) and fire frequency. Bison, like many grazers, prefer recently burned grasslands, and forage mostly on upland sites during the growing season in years of sufficient rainfall, where forage quality is greatest. Size of arrows is related to effect size (B) or selection strength (C). After: (A) Briggs & Knapp (1995), Blair (1997), Turner et al. (1997); (B) Briggs & Knapp (1995); (C) Raynor et al. (2017).

shortgrass prairies that occur along the western edge of the Great Plains, in the rain shadow of the Rocky Mountains (Gibson & Hulbert 1987).

Although much of the total ANPP in tallgrass prairie is due to grass production (Briggs & Knapp 1995), most of the plant diversity is due to the richness of its forb species (Freeman & Hulbert 1985). Forb species richness has been reported to be greater on upland than lowland sites (Abrams & Hulbert 1987; Gibson & Hulbert 1987), but fire and grazing ultimately appear to be more important to forb diversity than topoedaphic effects per se (Collins & Calabrese 2012). Unlike the dominant grasses in this system, forb ANPP does not appear to be directly affected by precipitation or topography (Briggs & Knapp 1995). Given that forb ANPP is negatively affected by grass ANPP, however, conditions that favor grass production, such as frequent fires and topoedaphic effects, might well reduce forb production (Briggs & Knapp 1995). Indeed, grasses appear to outperform forbs under frequent fire conditions not only because they are better adapted to the direct effects of burning, but also because they grow better under the low-nitrogen regimes created by frequent fires (i.e. the dominant grasses have higher nitrogen use efficiency; Seastedt et al. 1991; Ojima et al. 1994a; Briggs & Knapp 2001).

This helps explain why annually burned sites have higher productivity than expected, despite lower nitrogen availability. What about the effect of topographic position on productivity, then? Upland sites have lower productivity than lowland sites, and yet soil nitrogen availability (as assayed by nitrogen mineralization rates) was found to be *greater* in upland sites (Turner et al. 1997; **Figure 11.3A**). These patterns are the opposite of what we would expect if nitrogen primarily limits productivity in this system. Again, compensatory mechanisms, such as shifts in carbon allocation patterns or increased nitrogen use efficiency by grass species, along with changes in plant community composition, may offset nitrogen losses on upland sites, thereby helping to conserve soil nutrient pools even in the face of frequent fire (e.g. Turner et al. 1997; Wilcox et al. 2016). Consequently, productivity in the tallgrass prairie may be less constrained by nitrogen limitation than previously believed (e.g. Seastedt et al. 1991; Ojima et al. 1994a). In any case, different factors appear to control soil nitrogen availability versus plant productivity in tallgrass prairie, which underscores the difficulty of unraveling the various controls on ecosystem processes in this—or any—system.

To complicate matters further, grazing interacts with fire and precipitation to affect productivity and nutrient cycling in the tallgrass prairie. The tallgrass prairies of the Great Plains were historically grazed by native ungulates, such as the American bison (*Bison bison*), which once numbered in the tens of millions but were hunted nearly to extinction by the late 19th century. Today, there are fewer than 500,000 bison in North America, scattered among protected areas and private ranches, and their ecological role as a keystone herbivore has now largely been replaced by cattle (Knapp et al. 1999; Allred et al. 2011a; **Figure 3.35**). For example, the Flint Hills supports one of the largest beef cattle production regions in the USA (USDA 2014). Nevertheless, we do have some insights into how and where bison forage, and thus can assess the critical role they play (or once did) in the cycling and redistribution of nitrogen in the tallgrass prairie. Bison (and cattle) preferentially graze burned areas of the prairie, and so fire and grazing are linked disturbances, which interact to create a shifting disturbance mosaic that enhances heterogeneity across a range of scales (Hobbs et al. 1991; Fuhlendorf & Engle 2004; Fuhlendorf et al. 2009; Allred et al. 2011b; **Chapter 3**). By preferentially foraging in burned areas, grazers reduce grass biomass and decrease fire frequency in those areas, whereas other areas untouched by fire (and grazers) will accumulate grass fuels that will burn when the next fire sweeps through the prairie, causing grazers to shift their foraging preferences to these recently burned sites; the landscape thus consists of a shifting mosaic of burned and unburned patches that are being differentially grazed (i.e. fire-driven grazing or pyric herbivory; Fuhlendorf et al. 2009).

Despite the coupling of fire and grazing in this system, the two disturbances have different effects on ANPP and nutrient availability. Although both fire and grazing remove aboveground biomass, fire does it all at once, resulting in large combustion losses of nitrogen, whereas grazing removes live biomass periodically throughout the growing season (Turner et al. 1997). Fire stimulates the aboveground productivity and nutritional quality of grasses, however, which explains why grazers are attracted to recently burned grassland, as these areas contain higher quality forage. Grazing can similarly stimulate increased aboveground production in grasses, but as grazers also excrete nitrogen in their urine and feces, this can enhance aboveground productivity and reduce nitrogen losses caused by burning (by half, in some cases), thereby conserving nitrogen stocks that would otherwise be lost from burned prairie that is not grazed (Hobbs et al. 1991; Allred et al. 2011b; **Box 11.1**). Thus, grazing has been shown to accelerate the cycling of nitrogen and to enhance soil nitrogen availability, even in annually burned tallgrass prairie (Johnson & Matchett 2001).

Given that the effects of precipitation and fire on grass production are dependent on topographic position in tallgrass prairie, can we expect grazers to exhibit similar responses to these topoedaphic gradients?

Topographic position has been shown to influence bison foraging and space-use patterns, but how it does so depends on precipitation, fire frequency, and the time of year. During the growing season, bison strongly favor upland sites in recently burned areas, especially if the grassland has not been burned for a few years (e.g. in the past 2–4 years; Raynor et al. 2017; **Figure 11.3C**). Although grass productivity is greater in lowland areas, forage quality (based on grass nitrogen content) is ultimately higher in upland sites (Blair 1997; **Figure 11.3A**). Even in burned grasslands, however, bison selected patches that contained lower grass biomass and higher foliar protein than nearby, available patches (Raynor et al. 2016). Bison return repeatedly to graze in these preferred patches (grazing lawns), which helps to maintain grasses in a tender, nutritious state throughout the growing season. During the winter when grasses are dormant, or during drought years when forage production is poor, bison switch to foraging in lowland areas, which offer low-quality forage but in sufficient quantity to meet their energy requirements (**Figure 11.3C**). Thus, winters and drought reduce the use of recently burned, upland areas by bison in tallgrass prairie.

The interaction between topoedaphic effects, precipitation, fire, and grazing is what contributes to such a heterogeneous and dynamic landscape in the tallgrass prairie ecosystem. The tallgrass prairie is by no means unique in this regard, however. Topographic or landscape position has been shown to have an important effect on ecosystem processes in a wide range of terrestrial ecosystems, including arctic tundra (Stewart et al. 2014), desert scrub (Schade et al. 2003), arid grasslands (Hook & Burke 2000), temperate hardwood forests (Johnson et al. 2000), coniferous forests (Swetnam et al. 2017), and tropical lowland forests (Weintraub et al. 2015). Much as research on elevational gradients has revealed how environmental, geographic, and biotic factors interact to influence species diversity within landscapes (**Chapter 10**), so too has the study of topographic gradients been instrumental in uncovering the biotic and abiotic drivers of various ecosystem processes that influence nutrient cycling, soil nutrient availability, and ecosystem productivity across landscapes.

Effects of Land-Cover Change on Nutrient Dynamics

Human land use can profoundly alter the structure and functioning of ecosystems, producing enduring land-use legacies that may persist for many decades or centuries, even after that land use is discontinued and the system is allowed to recover on its own ('passive restoration' via secondary succession) or is actively restored (**Chapter 3**). The effects of disturbance and successional dynamics on ecosystem processes have been a major area of research in ecosystem ecology for decades (e.g. Odum 1960, 1969; Paul 1984). Thus, our interest here lies in understanding how land-cover change, particularly that driven by human land use, affects nutrient dynamics, especially in regards to carbon storage.

Land uses that cause land-cover change, as when agricultural expansion results in the conversion of forests and grasslands to cropland, cause significant carbon losses that currently account for about 12% of all anthropogenic carbon emissions (Houghton et al. 2012).[5] As we've discussed previously in regards to the carbon cycle, human activities like deforestation have greatly increased greenhouse gas emissions, of which CO_2 is a major component, and demonstrates why land-use/land-cover change is so important for reconciling the global carbon budget (Houghton 2007). Most of the carbon in terrestrial ecosystems is actually stored in the soil rather than in the vegetation, however. Most soils are thousands of years old, and have been slowly accumulating carbon over the millennia; the global average is on the order of 0.1 Pg C/year (Amundson 2001; Houghton 2007).[6] As a result, soil typically functions as a carbon sink, in that it stores (sequesters) more carbon than it releases. Soil is therefore the largest terrestrial pool of carbon, being larger than the amount of carbon stored in terrestrial vegetation and the atmosphere combined (Amundson 2001; Houghton 2007). The loss of carbon from soil can therefore have a significant effect on atmospheric CO_2 levels, and hence on the global climate.

The amount of organic carbon stored in the soil represents a balance between carbon inputs (the amount and quality of the soil organic matter) and carbon outputs, specifically the CO_2 produced by soil organisms during decomposition of soil organic matter (Post & Kwon 2000). The same factors that influence plant production (the primary source of soil organic matter) and decomposition rates, namely soil type and climate, also influence the amount of soil organic carbon (SOC). Precipitation primarily increases plant production (increasing carbon inputs), whereas temperature primarily increases decomposition rates (increasing carbon outputs); soil texture plays an important role in the turnover rates of different carbon compounds, with carbon outputs decreasing as the clay content of the soil increases. Thus, the total organic carbon content of soils tends to increase with precipitation and the clay

---

[5] Prior to the 1980s, land-cover change accounted for about 33% of total anthropogenic carbon emissions (Houghton 1999). This declined to 20% in the 1980s and 1990s, down to its current rate of about 12% in the 2000s, because of increasing fossil fuel emissions, which now account for most of our total carbon emissions (Houghton et al. 2012).

[6] 1 Pg = 1 Petagram = 10^{15} g = 10^9 metric tons

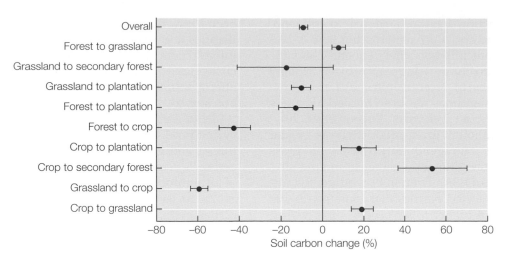

Figure 11.4 Effects of land-cover change on soil carbon stocks. The conversion of forests and grasslands to cropland reduces soil carbon stocks, whereas the conversion of cropland to grassland or forest increases soil carbon stocks. Globally, the net effect of land-cover change is an overall reduction in soil carbon stocks.
Source: After Guo & Gifford (2002).

content of soils, and to decrease with temperature, across a diverse array of soils and vegetation types (Jobbágy & Jackson 2000). Perhaps unsurprisingly, then, the biomes having the highest global storage of SOC are tropical evergreen forests (158 Pg C) and tropical grasslands/savannas (146 Pg C).

Soil organic carbon is not distributed uniformly throughout the soil profile, however. Soil exhibits considerable heterogeneity vertically as well as horizontally across landscapes. Most of the total organic carbon distributed across a typical 3-meter-deep soil profile occurs within a meter of the surface (64%), and almost half of that (42%) is found within the first 20 cm (i.e. in the topsoil; Jobbágy & Jackson 2000). How SOC is distributed across the soil profile varies by vegetation type and climate, however. For example, the percentage of SOC in the upper 20 cm of the first soil meter varied from 29% in cold arid shrublands to 57% in cold humid forests and, for a given climate, was always deepest in shrublands, intermediate in grasslands, and shallowest in forests (Jobbágy & Jackson 2000).

Vegetation thus appears to regulate the amount and distribution of SOC within the soil profile, owing to differences in the amount and decomposability of litterfall inputs, as well as differences in the allocation to aboveground versus belowground biomass among different plant groups. For example, temperate forests have a fairly high aboveground allocation relative to most temperate grasslands, and thus have shallower SOC profiles than many temperate grasslands. Conversely, the relatively deep root distributions of shrubs may lead to soil carbon profiles that are deeper than those in arid grasslands (Jobbágy & Jackson 2000). This is important because our assessment of the effects of land-use change on soil carbon stocks will depend on

the depth we sample relative to the distribution of carbon in the soil profile. The distribution of SOC within the soil profile also gives us an indication of the potential magnitude of carbon loss and how quickly those carbon stocks are likely to be depleted by different types of land-use or land-cover changes.

Land uses that result in land-cover change significantly alter the amount of organic carbon stored in the soil. Over the past 150 years or so, soils have lost 40–90 Pg carbon globally through cultivation and disturbance, with current rates of carbon loss due to land-use change estimated at about 1.14 Pg C/year, mostly as a consequence of deforestation in the tropics (Schimel 1995; Houghton 1999; Smith 2008; Houghton et al. 2012). The net effect of land-use change has been to reduce soil carbon stocks by 9% overall (Guo & Gifford 2002; **Figure 11.4**), amounting to a loss of 2.52 Mg C/ha, which is equivalent to a rate of loss of 0.39 Mg C/ha/yr (Deng et al. 2016).[7]

The greatest loss of SOC occurs when forests or grasslands are converted to cropland, presumably because of the great reduction in organic matter inputs to the system (perennial natural vegetation is removed and replaced with annual crops, which are harvested), the faster decomposition rates of crop residues compared to other vegetation types, and tillage effects that break up and homogenize the soil, exposing more of it to decomposers (Post & Kwon 2000). The net effect is to hasten the turnover rates of various carbon compounds in the soil, which, coupled with the warmer soil temperatures of cultivated fields, lead to increased soil respiration rates (i.e. increased release of CO_2 produced by heterotrophic soil microbes; Post & Kwon 2000). Thus, soil carbon stocks have declined by an average

[7] 1 Mg = 1 Megagram = 10^6 g = 1 metric ton

of 42% when native forest is converted to cropland, and by 59% when grassland is converted to cropland (Guo & Gifford 2002; **Figure 11.4**). Although the magnitude of soil carbon loss is greater in grassland converted to cropland, soil carbon is actually lost at a faster rate when forest is converted to cropland (1.74 Mg C/ha/year vs 0.89 Mg C/ha/year for cropland; Deng et al. 2016).

Fortunately, from an ecosystem management, ecological restoration, and carbon mitigation standpoint, the conversion of cropland back to grassland or forest can reverse the loss of SOC stocks. Large-scale agricultural abandonment has occurred in some parts of the world, including North America, Europe, and China, due to socioeconomic changes, increasing urbanization and globalization, shifts in agricultural policies, and/or programs providing market-based incentives to restore and recover ecological functioning on degraded cropland (e.g. to control soil erosion; **Chapter 3**). Ecological restoration can be an effective means of increasing soil carbon stocks in former cropland. Soil carbon stocks increase by an average 19% when cropland is restored to grassland, but by 53% when former cropland is restored to forest (Guo & Gifford 2002; **Figure 11.4**). In the latter case, this magnitude of carbon increase might well suffice to recover the carbon lost when forests were cleared for cropland (when 42% of soil carbon stocks are lost), although this will depend on initial carbon stores, how badly those stores were depleted, and how many years it has been since the restoration occurred (i.e. SOC tends to accumulate over time). Nevertheless, these results prove that cropland conversion can increase soil carbon stores, and thus presents a very real opportunity for mitigating carbon emissions by increasing soil carbon sequestration rates in terrestrial ecosystems (i.e. by increasing the sink potential of the soil).

Adoption of a restorative land use and recommended management practices...on agricultural soils can reduce the rate of enrichment of atmospheric CO_2 while having positive impacts on food security, agro-industries, water quality and the environment. A considerable part of the depleted SOC pool can be restored through conversion of marginal lands into restorative land uses, adoption of conservation tillage with cover crops and crop residue mulch, nutrient cycling including the use of compost and manure, and other systems of sustainable management of soil and water resources...The soil [carbon] sequestration is a truly win-win strategy.   Lal (2004)

IMPLICATIONS FOR THE ECOLOGICAL RESTORATION OF DEGRADED LANDSCAPES Halting agricultural expansion and other land uses that lead to land-cover change would be the most obvious and effective means

of reducing soil carbon losses, but this is not wholly realistic in view of a growing human population that is projected to exceed 11 billion by 2100, and which will likely require more land clearing to meet our increasing food demands (Smith 2008). Agricultural intensification, which entails making current agricultural lands more productive, might at least help to reduce the amount of land cleared for agriculture, especially if this is done strategically, by targeting high-yielding technologies within nations that are currently under-yielding (Tilman et al. 2011; **Chapter 3**). The adoption of management practices that increase carbon inputs to the soil (e.g. improved residue and fertilizer management) or that reduce carbon losses (e.g. no tillage, reduced or conservation tillage, reduced residue removal) would also help to maintain or increase SOC levels on agricultural lands (Smith 2008). We should thus consider soil carbon sequestration to be part of a broader sustainable development framework (Smith 2008). Although soil carbon sequestration will not be sufficient by itself to mitigate all anthropogenic carbon emissions, it could nevertheless help buy us time until longer-term solutions are developed and/or widely adopted (e.g. alternative energy solutions that reduce our dependence on fossil fuels; Lal 2004).

To illustrate the potential that ecological restoration can have for recovering soil carbon stocks, we can point to the massive effort undertaken by China's Grain-for-Green Program (GFGP), which has been called the 'largest ecological restoration and rural development program in the world' (Delang & Yuan 2015). We touched upon this program briefly in **Chapter 3** in regard to how reforestation in some parts of the world, such as China, have been successful in reversing deforestation trends. The GFGP, which was initiated by the Chinese government in 1999 for the primary purpose of controlling flooding and soil erosion caused by rampant deforestation, provides cash incentives to rural farmers to restore natural vegetation (forest, shrub, or grassland) on a targeted 32 million ha of degraded cropland (Delang & Yuan 2015). For comparison, that is an area a bit larger in size than Vietnam or Norway.

Overall, the GFGP has proved beneficial for recovering soil carbon stocks on cropland converted to natural vegetation; the average rate of carbon sequestration across all converted croplands is estimated to be 0.75 Mg C/ha/yr (Deng et al. 2014). Although soil carbon accumulates over time following cropland conversion, it does not increase at a constant rate, nor does it begin increasing immediately, especially when cropland is planted to forest. Indeed, soil carbon stocks actually *decreased* in the first five years following forest planting (Deng et al. 2014). This initial reduction in soil carbon is likely due to decreased carbon inputs as a result of lower production and quality of litterfall during the early years of forest establishment, coupled with greater carbon losses caused by increased soil disturbance and erosion during the initial preparation of the site prior to tree planting. Thus, it takes an average of 5 years after cropland conversion before SOC starts to accumulate within coniferous forests, and an average of 10 years before SOC begins to accumulate within broadleaf forests. The rate of carbon sequestration in the soil peaks at around 6–10 years in coniferous forests, and at 11–30 years in broadleaf forests, at which point soil carbon stocks decrease again before stabilizing some 40 years after cropland conversion (Deng et al. 2014). Although the magnitude and dynamics of soil carbon stocks may differ among forest types, and between forest and other vegetation types, these results highlight that recovery will not be quick, and in fact, things may get worse before they get better (e.g. soil carbon stocks may initially decline before increasing and stabilizing). We should thus anticipate that soil carbon stocks will exhibit a lagged response to restoration, especially when attempting to establish forest on former cropland.

EVALUATING RESTORATION SUCCESS IN THE FACE OF LAND-USE LEGACIES   Past agricultural land use tends to leave an enduring legacy on the availability and distribution of soil nutrients, which may persist for decades, centuries, or even millennia after farmland abandonment, and could thus alter soil nutrient pools and the successional trajectories taken by terrestrial ecosystems (**Chapter 3**).

For example, forest soils in northeastern France still bear the imprint of their past agricultural use during Roman times, some 2000 years ago (Dupouey et al. 2002). For 200 years, Roman farmers grew cereal crops like wheat and barley on cleared fields that they fertilized regularly using ashes, animal, or green manures (i.e. cover crops like clover or alfalfa). Cultivation helped to crush and distribute limestone gravels throughout the soil, which contributed to a higher pH that may have favored the breakdown and accumulation of soil organic matter relative to undisturbed forest areas. The application of manures for fertilizer appears to have decreased the soil C:N ratio, which in turn favors nitrogen mineralization, increasing nitrogen availability for plants (Dupouey et al. 2002; **Box 11.1**). Today, a quarter of understory plant species are found only in areas of the forest where Roman farmers once tilled their fields. These species are typically ruderal or pioneer species that have a high demand for nitrogen, such as the common dandelion (*Taraxacum officinale*), bitter cress (*Cardamine pratensis*), and cypress spurge (*Euphorbia cyparissias*), and which are more usually associated with meadows (Dupouey et al. 2002). Conversely, 13% of plant species are found only in undisturbed areas of the forest that were not cultivated. So, to emphasize,

ancient farming practices have induced gradients in soil nutrient availability and plant diversity that are *still* measurable some 2000 years after these fields were abandoned and returned to forest. In view of the high level of disturbance created by modern agricultural practices, it seems unlikely that restoration of recently abandoned agricultural lands can ever achieve a level of recovery that matches the soil properties and diversity of undisturbed areas (Dupouey et al. 2002).

Then again, that level of recovery—a return to the pre-disturbed state—is hardly necessary for ecological restoration to be successful. The success of ecological restoration is typically evaluated in ecological terms, particularly in regards to the restoration of vegetation structure, species diversity and abundances, and, increasingly, ecological processes (Wortley et al. 2013). Of the restoration projects that have assayed ecological processes, most measured some aspect of nutrient cycling, soil structure, or carbon storage (Wortley et al. 2013). This focus on the restoration of ecosystem function is viewed as a positive development in the field of restoration ecology, in that 'an increased understanding of how restoration affects processes such as nutrient cycling, pollination, and erosion control is critical for the long-term persistence and stability of the projects, **as well as understanding the role of restoration in the landscape context**' (Wortley et al. 2013; emphasis added).

The Metaecosystem: Interacting Ecosystems in a Landscape Context

Given that ecosystems are embedded in a broader landscape context, they are likely to receive inputs from other systems within that landscape, depending on their relative position and extent to which they are linked to those systems by the flow of water, materials, nutrients, or organisms. In other words, ecosystems are open systems, which means that the productivity of a given ecosystem may be subsidized by inputs originating from elsewhere outside the system (allochthonous inputs or **spatial subsidies**). The corollary, however, is that the productivity and the structure or dynamics of food webs may not be satisfactorily explained or predicted simply by examining the ecosystem in isolation (i.e. as a closed system).

A **metaecosystem** is thus a set of interacting ecosystems connected by the flow of energy, materials, and organisms (Loreau et al. 2003). Recall that landscapes have sometimes been defined as systems of interacting ecosystems (Forman & Godron 1986). While landscape ecosystem ecology considers the spatial flows of materials or nutrients among systems, metaecosystem ecology additionally seeks to understand how these flows are mediated by the movement of particular types of organisms—such as frugivores or migratory species—among systems, or how the movement of such species can regulate primary production in other systems as a result of trophic cascades (Polis et al. 1997; Loreau et al. 2003; Knight et al. 2005; Massol et al. 2011; Semmens et al. 2011). Although resource subsidies have long been known to influence the structure and dynamics of ecosystems, the study of subsidies has generally been approached from two different viewpoints: from the standpoint of food-web dynamics, in terms of how subsidies modify animal population dynamics, trophic pathways, and interaction strengths among species; or from an ecosystem perspective, which studies the effects of subsidies on energy budgets and the flow of carbon or nutrients (Marcarelli et al. 2011). The meta-ecosystem concept thus attempts to combine these two perspectives, by studying how spatial subsidies among ecosystems are mediated by the movements of organisms representing different trophic levels. We begin by discussing how spatial subsidies among systems can influence ecosystem structure and dynamics before turning our attention to the sorts of species involved in the transfer of materials or energy (**mobile-link species**) and how the differential flows among systems can generate source–sink metaecosystem dynamics.

Spatial Subsidies

The study of spatial subsidies—how the productivity and trophic dynamics of one ecosystem are supported by the productivity of another—reflects a fairly recent integration of concepts from landscape and food-web ecology (Polis et al. 1997), but the importance of terrestrial resource subsidies to the structure and function of ecosystems has long been studied by ecosystem ecologists, particularly for freshwater systems (Minshall 1967; Cummins 1974; Likens & Bormann 1974). Recall that under the river continuum concept, allochthonous (external) inputs from the surrounding landscape may exceed autochthonous (internal) inputs along certain lengths of the river, depending on the amount of shade cover relative to the river's width, for example (**Figure 5.18**). This difference in the relative importance of resource subsidies along the river's length is then reflected in differences in trophic structure, such as switching from communities dominated by primary producers and herbivores to ones dominated by collectors and predators downstream, which can utilize the energy inputs from further upstream (Vannote et al. 1980).

Allochthonous inputs not only subsidize the productivity of aquatic ecosystems, but can also influence the strength of trophic interactions. Under the **subsidy hypothesis**, ecosystems receiving a larger fraction of their resource inputs from allochthonous sources are expected to experience stronger trophic cascades (Leroux

& Loreau 2008). Recall that a trophic cascade is a top-down form of ecosystem regulation, in which a higher trophic level (e.g. a predator) controls primary production via suppressive effects on the abundance or behavior of an intermediate trophic level (e.g. an herbivore). One way we can evaluate the strength of a trophic cascade is to quantify the relative effect that predators have on plant biomass when they are present versus when they are absent (Shurin et al. 2002). Using this approach, Leroux & Loreau (2008) demonstrated mathematically that ecosystems that were more heavily subsidized should indeed experience stronger trophic cascades. This may help to explain why trophic cascades are stronger in aquatic ecosystems than in terrestrial ecosystems (Shurin et al. 2002), given that aquatic systems receive more allochthonous inputs by virtue of their location on the landscape (i.e. in low-lying areas).

> Although subsidies flow into and out of all ecosystems, inputs to lakes and streams tend to be much larger than their reciprocal flows to adjacent terrestrial ecosystems because of the position of lakes and streams at convex locations in the landscape, and because they are linked through watersheds by the downhill flow of water.   Marcarelli et al. (2011)

Most aquatic ecosystems rely on terrestrial subsidies for their productivity. Because water flows downhill, terrestrial inputs into lakes and rivers tend to be greater than flows in the opposite direction (Marcarelli et al. 2011; Bartels et al. 2012). Different types of land uses can also affect surface and groundwater flows across the landscape, potentially increasing the load of sediments, organic nutrients, inorganic chemicals, heavy metals, and disease-causing pathogens that are ultimately discharged into these systems, especially in the form of stormwater, agricultural, and industrial wastewater runoff (**Chapter 3**). Human-induced eutrophication (**cultural eutrophication**) is caused by an increased flow of otherwise limiting nutrients, such as nitrogen and phosphorus, into water bodies as a result of point and non-point sources of pollution, such as from sewage or fertilizers in agricultural runoff (Carpenter et al. 1998). If excessive nutrient loading increases the amount of internal (autochthonous) primary production, then these aquatic food webs may no longer be supported principally by terrestrial subsidies, which could alter the strength of trophic interactions and radically alter nutrient dynamics and ecosystem functioning.

To illustrate, Carpenter and his colleagues (2005) tested whether nutrient enrichment diminished the importance of terrestrial subsidies among several small lakes in the upper Great Lakes region of the USA. By first adding dissolved inorganic ^{13}C to the lakes (in the form of NaH^{13}CO$_3$), the investigators were able to label and trace the flow of carbon through the ecosystem, enabling them to tease apart the relative contributions of in-lake primary production (autochthony) versus terrestrial inputs (allochthony) to these aquatic food webs.[8] Allochthonous organic carbon represented a substantial subsidy to the food webs of these lakes. Carbon flow to herbivorous zooplankton was 22–75% allochthonous, whereas fish allochthony was even higher (up to 80% allochthonous carbon) as a result of a diet rich in terrestrial prey (e.g. insects) or zoobenthic prey of terrestrial origin (Carpenter et al. 2005). When one of the lakes was experimentally enriched with liquid fertilizer, simulating cultural eutrophication, the allochthony of both zooplankton and fish declined, with zooplankton now supported almost entirely by within-lake primary production (0–12% allochthony) and fish allochthony declining by one-third to one-half. Consistent with predictions, nutrient enrichment reduced the importance of terrestrial subsidies to these aquatic food webs, in essence decoupling the lake from its watershed (Carpenter et al. 2005).

Tremendous spatial heterogeneity exists, and natural systems are open: nutrients, organic matter, and organisms move among habitats. Ecological dynamics are rarely contained within the boundaries of the area selected for study, and factors outside a focal system may exert substantial effects on the patterns and dynamics observed in the system. In many cases, between-habitat influences are relatively important as compared to internal, within-habitat effects . . . **we stress that the flow of materials and organisms among habitats is often a key feature of population dynamics, energetics, and the structure of food webs and communities.** (emphasis added)   Polis & Hurd (1996)

Although freshwater systems tend to be more subsidized than terrestrial systems (Bartels et al. 2012), resource subsidies may also occur from freshwater or marine systems to terrestrial ones (Richardson & Sato 2015). This was illustrated by the classic work of American ecologist Gary Polis and his colleagues (Polis & Hurd 1995, 1996; Anderson & Polis 1999; Polis et al.

2004) on marine subsidies to desert islands within the Gulf of California. Being deserts, these islands receive precious little rainfall (averaging only 59 mm annually) and thus have intrinsically low levels of net primary productivity (115 gC/m^2/y), as evidenced by the sparse vegetation cover on most islands (<10%; Polis & Hurd 1996). Despite this, these desert islands support a far richer biota than can be explained based on terrestrial productivity alone.

In this case, no desert is an island, or rather, no island is truly a desert. The warm waters of the Gulf of California are extraordinarily productive, with primary productivity levels nearly five times greater than that of the islands. Such high primary productivity in turn supports very high numbers of squid, fish, seabirds, and marine mammals. As any beachcomber knows, a great deal of marine debris gets washed ashore during high tides, and this biological wrack—in the form of seaweed and dead sea creatures—provides an important source of nutrient inputs to the otherwise desolate islands in the Gulf of California. In fact, marine inputs exceeded terrestrial primary productivity for most islands, but smaller islands were subsidized to a greater degree than larger islands (**Figure 11.5**). Small islands have proportionately more shoreline (perimeter) relative to their area, and thus receive more marine wrack per unit area than larger islands (an example of a positive edge effect). Nutrient inputs from the sea are thus a major top-down factor that subsidizes terres-

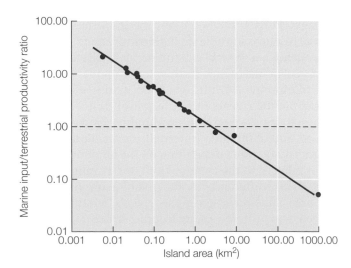

Figure 11.5 Magnitude of spatial subsidies as a function of island area. The magnitude of the spatial subsidy is given by the ratio of marine inputs (MI) to the amount of terrestrial productivity (TP) on 19 desert islands in the Gulf of California. All but three of the islands lie above a MI/TP Ratio = 1, indicating that their marine subsidies far exceed their own terrestrial productivity. Small islands (<1 km^2) are more heavily subsidized than large islands.

Source: After Polis & Hurd (1996).

---

[8] Stable isotopes, such as ^{13}C, have been used in food-web studies to trace the flow of nutrients through the ecosystem (e.g. from primary producers to consumers; Middleburg 2014). For example, if ^{13}C is pumped into a lake so as to become the most prevalent isotope of carbon, it will be preferentially used during photosynthesis (^{13}CO$_2$) by primary producers (e.g. phytoplankton and periphyton), which in turn are incorporated into the biomass of consumers, creating a distinct carbon signature that makes it possible to separate out what fraction of total carbon (^{12}C + ^{13}C) is due to internal primary production (autochthony) versus terrestrial carbon inputs (which are mostly ^{12}C).

trial consumer communities—directly and indirectly—on these desert islands (Polis & Hurd 1995, 1996).

Small island size has another advantage: seabirds frequently nest and rest on small islands because their predators are usually absent (recall from **Chapter 10** that higher trophic levels are usually more affected by patch—or island—size than their prey). Nesting seabirds provide a second avenue for marine inputs onto islands, through their guano and other detritus (feathers, egg shells, dead chicks, fish carcasses). Indeed, seabirds have even been called 'biological pumps' between marine and terrestrial ecosystems (Otero et al. 2018). As seabird guano is high in both nitrogen and phosphorus, it is a highly effective fertilizer that enriches the nutrient-poor soils of these desert islands, leading to a 14-fold increase in terrestrial productivity compared to islands without seabirds (Polis et al. 1997; Anderson & Polis 1999). Plant quality also increases. Nutrient concentrations in a variety of plants (*Opuntia* cactus; saltbush, *Atriplex barclayana*; various annuals) were found to be 1.6–2.4× greater on islands with nesting seabird colonies than those without them. Further, an analysis of stable isotope ratios in *Atriplex* showed that shrubs from islands with seabird colonies were enriched in ^{15}N, reflecting the high ^{15}N signatures of the guano-based soils. Thus, the seabird-mediated transport of marine nutrients, such as nitrogen and phosphorus, can also act as a major bottom-up factor stimulating primary productivity in this desert island system. For example, insects are 2.8× more abundant on these guano-enriched islands, which in turn stimulates large populations of higher-level consumers, such as spiders and lizards, which are 4–5× more abundant on islands with seabirds than those without (Polis & Hurd 1996; Polis et al. 1997).

Desert islands are not the only terrestrial ecosystems subsidized by inputs from marine or aquatic ecosystems. During the late summer and fall, millions of Pacific salmon (*Oncorhynchus* spp.) return to their natal freshwater streams in northwestern North America to spawn, sometimes travelling up to 1000 km upriver (Gende et al. 2002). It's a bittersweet homecoming, as the fish will die soon after spawning, assuming they survive the journey upriver. Although born of freshwaters, salmon spend most of their lives at sea, where they obtain most of their body mass (90%) and the reserves they'll need for their return migration to their spawning grounds. Salmon thus enrich the freshwater systems they enter with marine-derived nutrients in a number of ways, via excretion of nitrogenous wastes, the decomposition of their bodies after death, and the direct consumption of their eggs and flesh by a host of aquatic consumers (Gende et al. 2002; Naiman et al. 2002). Furthermore, salmon subsidize the productivity of terrestrial ecosystems in the surrounding landscape

(Ben-David et al. 1998; Willson et al. 1998; Gende et al. 2002). Predators and scavengers that feed on salmon help to distribute marine-derived nutrients widely across the landscape (Gende et al. 2002; Adams et al. 2017). For example, brown bears (*Ursus arctos*) feeding on salmon along rivers on the Kenai Peninsula of Alaska were found to deposit marine-derived nitrogen (which has a higher ^{15}N signature) some 500 m or more into the surrounding boreal forest (Hilderbrand et al. 1999). Most of this salmon-derived nitrogen is excreted by bears in their urine (96%), where it is then readily available for uptake by plants (**Box 11.1**). Nitrogen is frequently a limiting nutrient in northern forests, and so, absent nitrogen fixation (e.g. by alders, *Alnus crispa*), riparian vegetation may derive up to 26% of their foliar nitrogen from salmon (Hilderbrand et al. 1999; Helfield & Naiman 2002). The streams and rivers in northwestern North America thus serve as conduits for the input of marine-derived nutrients to freshwater and terrestrial systems via migrating salmon (Gende et al. 2002). Because salmon subsidies are so important to the productivity of many freshwater and terrestrial systems in this region, the loss of salmon runs—whether through overfishing, dams that prevent their migration, pollution, or the loss and degradation of spawning habitat—is likely to have a major effect on nutrient budgets, trophic interactions, and food-web dynamics in ecosystems far removed from the ocean.

Such **cross-boundary subsidies** between ecosystems reflect the interdependence between aquatic and terrestrial food webs (Nakano & Murakami 2001; Richardson & Sato 2015). Pacific salmon (or at least the nutrients contained in their bodies) cross many boundaries between oceanic, freshwater, and terrestrial ecosystems. Species like salmon that connect ecosystems by transferring nutrients as they move between them are an example of a mobile-link species, which we discuss next.

Mobile-Link Species

Highly mobile species that move between ecosystems have the potential to function as 'mobile links' that connect these systems through the provisioning or support of essential ecosystem functions (Lundberg & Moberg 2003). For example, bird or bat species that provide critical ecosystem services, such as crop-pest control, seed dispersal, and pollination services (Wenny et al. 2011; Boyles et al. 2011), are canonical examples of mobile-link species, and the services they provide have therefore been termed 'mobile agent-based ecosystem services' (Kremen et al. 2007). We can categorize mobile-link species according to the type of function they perform (Lundberg & Moberg 2003):

1) **Resource linkers** transport organic material and nutrients among ecosystems. Examples of resource linkers already discussed in this chapter include seabirds that deposit their nutrient-rich guano on desert islands, and the delivery of marine-derived nutrients by Pacific salmon and brown bears to freshwater and terrestrial ecosystems, respectively.

2) **Genetic linkers** carry genetic information among ecosystems (e.g. seed dispersers and pollinators; **Chapters 6, 9, and 10**).

3) **Process linkers** connect ecosystems by providing or supporting an essential process (e.g. pest-control services; **Chapter 10**), including some types of biotic disturbances like grazing (**Chapter 3**).

These categories are not mutually exclusive, as a species might fulfill more than one type of function. For example, large herbivores could serve as process linkers through their grazing or browsing effects on ecosystems, but could also be genetic linkers if they disperse seeds between these systems, either via their feces or adhered to their fur (Vellend et al. 2003; Couvreur et al. 2004).

Salmon link oceans to lakes and streams via their migrations and link streams to lakes via their nest-digging and spawning activities.   Moore et al. (2007)

To use an example from before, Pacific salmon are resource linkers between marine and freshwater systems, but they might also qualify as process linkers, given that they extensively modify their freshwater spawning habitat while digging their rather large nests (up to 17 m² and 35 cm deep, depending on individual size, species, and location; Moore et al. 2004). Because salmon typically spawn at high densities, the disturbance created by nest-digging salmon can have a significant effect on community structure and ecosystem processes within these freshwater systems (Moore et al. 2004). The act of digging into the streambed increases the suspension of nutrient-rich sediments and salmon eggs (a form of bioturbation), resulting in substantial sediment and nutrient exports to estuaries and lakes downstream (Moore et al. 2007). For example, the spawning activities of sockeye salmon (*Oncorhynchus nerka*) in the streams of southwestern Alaska result in an export of phosphorus equivalent to 60% of all phosphorus imported in their bodies, and in some streams, salmon actually export more phosphorus than they import (Moore et al. 2007). Again, the link between ecosystems may result in reciprocal subsidies (Nakano & Murakami 2001; Bartels et al. 2012), which can vary in magnitude and direction, in addition to being spa-

tially and temporally variable (e.g. varying between seasons or years). In the case of Pacific salmon, for example, there is a large seasonal pulse of marine-derived nutrients to freshwater and terrestrial systems that coincides with the spawning season, but there is also variation among streams and years in the export of nutrients, such as phosphorus, as a result of their spawning activities, owing to differences in total summer precipitation and stream discharge (Moore et al. 2007).

IMPLICATIONS FOR ECOSYSTEM CONNECTIVITY AND RESILIENCE By maintaining functional connectivity among different habitats or ecosystems, mobile-link species not only influence the structure and functioning of these systems, but also their resilience; that is, their ability to recover following a disturbance (Lundberg & Moberg 2003; **Chapters 3 and 5**). For example, frugivorous birds tend to be highly mobile and can facilitate forest recovery after fire or logging through the 'perch effect,' whereby they deposit the seeds of distant forests below perches within the disturbed area (e.g. below snags or standing burned trees; Cavallero et al. 2013). Of course, not all frugivorous birds are equally good at dispersing seeds between habitats, which is why maintaining **functional heterogeneity**—the degree of variability in the functional capabilities of species—is so important for ecosystem resilience (Walker 1992, 1995; Hooper et al. 2005).[9] Functional heterogeneity can enhance ecosystem resilience through a combination of **functional redundancy**, in which multiple species perform the same function within an ecosystem, and **response diversity**, in which species that contribute to the same ecosystem function respond differently to disturbance (i.e. species complementarity; Elmqvist et al. 2003).

In the case of frugivorous birds, for example, species have been found to differ in their propensity to deposit seeds across an agricultural landscape, largely because of differences in habitat use (species differed in their relative use of the agricultural matrix), the types of perches they are willing to use (isolated trees vs artificial structures like transmission towers), and how far out from the forest edge they are willing to travel (González-Varo et al. 2017). Still, because species overlap in their spatial pattern of seed dispersal, the end result is a nearly uniform 'seed rain' from the forest edge out into the agricultural matrix. That is, an evenly distributed seed-rain pattern comes about because of unevenly distributed frugivore contributions to seed

---

[9] Note that functional heterogeneity, as used here, has a different meaning than when used in the context of quantifying species' responses to landscape pattern (i.e. environmental heterogeneity that 'matters' to the species in question, as opposed to measured or structural heterogeneity; **Chapter 4**).

dispersal between different habitats (González-Varo et al. 2017). Because these species operate unevenly as mobile links (high functional heterogeneity), the loss of any one frugivore would have only a small effect on these birds' overall seed-dispersal function, such that the process of forest recovery would be unaffected. Indeed, others have argued that the functional heterogeneity of the entire plant-frugivore assemblage (i.e. bird responses to the fruiting patterns of different tree species within and between years) is what enhances the resilience of many such forest ecosystems to anthropogenic habitat loss and fragmentation (García et al. 2013).

High mobility does not guarantee, however, that mobile-link species will always be able to maintain ecosystem connectivity in the face of ongoing habitat loss or ecosystem degradation. Migratory species, for example, establish functional linkages between distant ecosystems, which may be located on different continents, or even in different hemispheres (migratory connectivity; **Chapter 6**). Different habitats or locations within the range of the migratory species are bound to differ in their ability to support viable populations of that species, however, resulting in differences between the migratory services these locations receive versus the migration support they provide to other locations (Semmens et al. 2011). In other words, spatial mismatches may exist between the habitats or locations that support high population viability of migratory species (high migration support), and the areas where these species provide the most ecosystem services (Semmens et al. 2011). Clearly, the potential for spatial mismatches creates a management challenge, as it may be necessary to protect habitat in other parts of a species' range in order to ensure the continued provisioning of ecosystem services within a given landscape. This, then, sets the stage for source–sink metaecosystem dynamics, which we discuss next.

Source–Sink Metaecosystem Dynamics

Although we tend to think of certain ecosystems as inherently productive and therefore functioning as nutrient sources, while other less-productive ecosystems function as nutrient sinks, the source–sink dynamics of these systems are ultimately determined by the nature of the spatial flows between them. As we've discussed in this chapter, less productive ecosystems (nutrient sinks) may be subsidized by more productive ecosystems (nutrient sources), which sets up a source-sink dynamic between these ecosystems. However, we've also discussed how flows between ecosystems can be reciprocal, with the strength and direction of ecosystem fluxes changing in response to external perturbations (whether natural or anthropogenic), seasonal or annual variation in environmental conditions, or by the number and types of mobile-link species

present. In addition, it appears that the balance between different types of spatial flows, such as that between inorganic nutrient flows (direct nutrient inputs) and organic matter flows (indirect nutrient inputs), can also determine the net direction of nutrient flows between ecosystems with contrasting productivities (Gravel et al. 2010). That is, spatial flows could theoretically reverse the source–sink status of a given ecosystem, turning a sink into a source, or a source into a sink.

To illustrate, let's consider two ecosystems or habitats that vary in their environmental conditions, such that a primary producer (e.g. a plant species) is capable of maintaining a viable population only within one of these locations. Ordinarily, we would consider that location to be a source (**Chapter 7**). Higher productivity in the source habitat thus results in increased plant biomass, which increases organic matter flows through other trophic levels in that ecosystem (i.e. indirect nutrient flows). However, high primary production might also deplete the availability of limiting nutrients within this ecosystem, to the point that nutrient concentrations are actually higher in the other, less productive ecosystem. In other words, the less productive ecosystem has the potential to serve as a nutrient source if these two ecosystems are coupled via the direct flow of nutrients (e.g. leaching and water runoff). Furthermore, detritus and other organic matter inputs, such as from the dead bodies of consumers, could cause an indirect flow of nutrients from the system with higher productivity to the one with lower productivity (i.e. from an organic-matter source to sink). The recycling of these imported nutrients within the low-productivity system (e.g. via breakdown of detritus by soil microbes and other detritivores) might then result in some fraction being exported back to the high-productivity system (from nutrient source to sink). In that case, the productive ecosystem would basically be subsidizing its own productivity. Whether it does or not, however, depends on the relative strength of these direct and indirect spatial flows between ecosystems, in terms of whether an ecosystem is functioning as an overall source (a net exporter) or sink (a net importer) of nutrients.

These spatial fluxes of nutrients can even become the driver of habitat suitability. For example, the coupling of ecosystems could enrich a sink habitat to the point where it could facilitate the establishment of a producer population (i.e. the sink becomes an organic matter source). This complicates our ability to assess habitat quality based on positive population growth rates, at least for primary producers (**Chapter 7**). A low-productivity habitat subsidized by nutrient inputs from surrounding productive habitats would be assessed as a source, while a similar habitat surrounded by less productive ones would be assessed as a sink (Gravel et al. 2010). The composition and configuration of habitats (ecosystems) on the landscape is thus expected to influence source–sink metaecosystem dynamics.

The composition and trophic structure of these coupled ecosystems are also important for understanding source–sink metaecosystem dynamics (Gravel et al. 2010). For example, the replacement of a productive primary producer within the source ecosystem by a less productive one would ultimately be detrimental to the sink ecosystem if indirect organic matter flows are important. Similarly, the introduction of an herbivore to the source ecosystem might result in a spatial trophic cascade (Schmitz 2008), owing to a reduction of plant biomass and thus nutrient uptake in that ecosystem, which then modifies the flux of nutrients between the two ecosystems (Gravel et al. 2010). Even if the introduced herbivore does not affect the net direction of nutrient flows, it can still affect the primary productivity of the recipient ecosystem. Thus, the loss or replacement of important trophic species in one ecosystem can potentially alter the structure and functioning of another, even though species within these different ecosystems are not interacting directly.

From Ecosystem Function to Landscape Function

Ecosystems and landscapes both support a variety of critical ecological functions that in turn provide the goods and services upon which humans—and other species—depend. **Ecosystem function** is generally defined in terms of the condition or 'performance' of the system, its capacity to produce, regulate, or maintain one or more services, especially in relation to human needs (Millennium Ecosystem Assessment 2003). For example, four classes of ecosystem services related to ecosystem function have been recognized by the Millennium Ecosystem Assessment (**Table 11.1A**):

1. **Provisioning services**: products or goods obtained from ecosystems, such as food, fiber, fuel, genetic resources, or pharmaceuticals
2. **Regulating services**: benefits obtained through the regulation of ecosystem processes involving climate, disease, or water
3. **Cultural services**: non-material benefits obtained from the ecosystem, such as opportunities for recreation, education or ecotourism, or sites that have aesthetic, cultural, spiritual, or religious significance
4. **Supporting services**: necessary for the production of all other services, such as soil formation, productivity, and nutrient cycling

Ecosystem function is thus often defined in utilitarian terms, especially in relation to the economic valuation of ecosystem services for human benefit (Costanza et al. 1997). Because ecosystems provide so many different types of services, they are considered to be **multifunctional**.

Landscapes are also multifunctional. This notion clearly underlies the multiple-use concept of public lands in the USA (e.g. US Forest Service lands) and elsewhere in the world, which are managed for different—sometimes competing—purposes, including natural resource and energy extraction, livestock grazing, recreation, and the conservation of biological diversity. In Europe, for example, multifunctionality is at the center of agricultural policy reform (i.e. Common Agricultural and Rural Policy of Europe, CARPE), reflecting a shift away from valuation of land use solely in terms of agricultural or commodity production (e.g. crop yields) toward an emphasis on 'joint production,' in which non-commodity outputs (erosion control, wildlife habitat, tourism, and recreation) are also valued (Vejre et al. 2007). Landscapes are thus multifunctional 'through their simultaneous support of habitat, productivity, regulatory, social and economic functions' (Mander et al. 2007).

Landscape function has been defined on the basis of four categories related to provisioning, regulation, cultural/amenity, and habitat functions (**Table 11.1B**, de Groot & Hein 2007). The similarities here to ecosystem function (**Table 11.1A**) are no coincidence; this table was initially presented as a typology of ecosystem function that was then used to illustrate various indicators of landscape function for rural agricultural landscapes (de Groot & Hein 2007). Although this example further illustrates the interchangeability of ecosystem and landscape concepts, there is an important distinction here in terms of the emphasis placed on the *habitat function* of landscapes for the maintenance of biological and evolutionary processes, and the importance of human land use and other spatial properties of landscapes in defining or controlling ecological functions. Thus, this emphasis on how spatial properties of the landscape interact with ecological and biophysical processes is what uniquely defines landscape function, thereby distinguishing landscape function from ecosystem function. We expand upon this in the next section.

Landscape Function and 'Dysfunctional' Landscapes

Various definitions of landscape function or multifunctionality have been debated, reflecting the varied perspectives of the participants in these debates as to whether one is a scientist, politician, or economist (Hagedorn 2007). The basic concept of 'joint production,'

TABLE 11.1 **Comparison of landscape and ecosystem functions.** Relevant landscape ecological principles are highlighted.

Function	Description	Examples*
Ecosystems[1]		
Provisioning	Goods and services	Food, fiber, fuel, pharmaceuticals, genetic resources, water
Regulation	Direct benefits obtained from ecosystem processes (e.g. biogeochemical and water cycles)	Climate regulation, erosion control, regulation of air and water quality, pollination, biological control, disease regulation
Cultural/ Amenity	Aesthetic, recreational, spiritual, inspirational, scientific, or educational values	Scenery, ecotourism, heritage/cultural sites, spiritual/ religious sites, cultural expressions (folklore, art), research and education (nature centers, biological research stations)
Supporting	Services necessary for the production of all other functions	Soil formation and retention, photosynthesis, primary production, production of atmospheric oxygen, nutrient and water cycling, ***provisioning of habitat***
Landscapes[2]		
Provisioning	Divided into *production* and *carrier* functions that reflect the resources (goods and services) and the ***human manipulation or use of space*** to obtain those resources	Production: Food, raw materials, pharmaceuticals Carrier: cultivation (agriculture, aquaculture, plantations), resource extraction (ore mining, fossil fuels), energy conversion (wind, solar), transportation networks (including waterways)
Regulation	Direct benefits from ecosystem processes (e.g. biogeochemical and water cycles), ***which often have an important spatial (connectivity) aspect***	Climate regulation, maintenance of soil fertility, erosion control, air and water quality, water regulation, pollination, population/biological control (pests and diseases)
Cultural/ Amenity	***Landscape properties*** that have aesthetic, recreational, inspirational, spiritual, scientific or educational value	Scenery, ecotourism, heritage/cultural sites, spiritual/religious sites, cultural expressions, research and education
Habitat	*Importance of ecosystems and landscapes for the maintenance of biodiversity and evolutionary processes*	Persistence of rare or endemic species, presence of refugia or breeding habitat on the landscape

*Categories overlap and thus services or benefits may have more than one function (e.g. water has both provisioning and regulatory functions, as well as supporting and possibly cultural functions)
[1]Millenium Ecosystem Assessment (2003)
[2]de Groot & Hein (2007)

in which landscapes should simultaneously support the production of commodities as well as non-commodities, appears to be the prevailing theme shared by these different perspectives, at least conceptually. Landscape function, however, more properly refers to the interactions among the spatial elements of a landscape resulting from the flows of energy, materials, and species (Forman & Godron 1986: 595). In other words, it is the interaction between pattern (spatial elements) and processes (flows of energy, materials, and species) that ultimately gives rise to landscape function. This is, you'll recall, one of the core tenets of landscape ecology, whose disciplinary focus is the study of how spatial pattern affects ecological process (Turner 1989;

Chapter 1). Landscape ecology might thus be said to be expressly concerned with the study and management of landscape function.

To gain a better understanding of landscape function, it is useful to consider the process by which landscape function becomes disrupted or altered (i.e. **landscape dysfunction**). Although a loss or alteration of landscape function may be triggered by various types of stressors and vary in its consequences, the following case study illustrates many of the key features by which landscape function can become disrupted.

The term 'landscape dysfunction' was initially applied to the effects of overgrazing on the semi-arid rangelands of Australia (Ludwig & Tongway 1997). These rangelands encompass a variety of grassland types that naturally exhibit a high degree of patchiness across a range of scales, such as grassland savannas in which mulga (*Acacia anuera*) groves are interspersed with smaller patches of shrubs, grass clumps, soil hummocks, and individual grass tussocks at the finest scales. Patches of vegetation effectively act as 'reserves,' intercepting and trapping the flow of water, soil sediments, nutrients, litter, and seeds that are driven across the landscape either by the wind or by runoff following a rainfall event. These naturally patchy landscapes are considered to be conserving or highly functional systems, because they efficiently capture, retain, and recycle water and nutrients, thereby reducing runoff and erosion (Ludwig et al. 2005). As a consequence, functional landscapes in this system exhibit production pulses in response to the sporadic rainfall of the region, creating positive feedbacks that result in the further accumulation of biomass, seeds, and organic matter within patches (**Figure 11.6A**). In contrast, landscapes in which the vegetation has been denuded, such as through chronic overgrazing, especially during a prolonged dry period, have few patches to intercept water and nutrients, most of which are then lost from the system. These denuded landscapes are thus considered to be 'leaky' or 'dysfunctional' (**Figure 11.6B**). Leaky landscapes are less productive because production pulses no longer occur in response to rainfall events, and are therefore less profitable and sustainable economically (Ludwig et al. 2005). The implication here, then, is that preserving landscape function is necessary for sustainability.

This case study illustrates several important concepts regarding landscape function: (1) the disruption or alteration of landscape drivers, such as the disturbance regime (grazing, in this case) can precipitate (2) changes in landscape structure such as in the size and density of patches, leading to (3) alteration of flows across the landscape, and subsequently, (4) a disruption in pattern-process linkages (switching from positive to negative feedbacks, or vice versa) that ultimately give rise to (5) a loss of landscape function (lower productivity owing to the excessive 'leakage' of water and nutrients from the system). In its most serious form, a disruption of landscape function can manifest as a system state change, such as desertification, that may well be irreversible. We consider these sorts of changes that can be assayed through a study of landscape structure and function next.

Assessing and Monitoring Landscape Function

Landscape function can be assessed in several different ways (**Table 11.1B**). The drivers of landscape change leading to a loss or alteration of landscape function, as well as the specific trajectory taken, are expected to vary depending upon the landscape in question. Subsequently, not all landscapes will lose function in the same way, and a landscape may lose certain functions while maintaining others (again, landscapes are multifunctional). Thus, a landscape is not inherently functional or not. Nevertheless, a few early warning signs and less-subtle indicators of compromised landscape function are outlined here.

ALTERATION OF LANDSCAPE STRUCTURE As we discussed in **Chapter 3**, the patch structure of a landscape sometimes gives clues to the formative processes that shaped it. Processes that shape landscapes operate

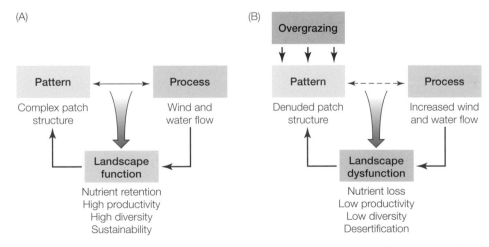

Figure 11.6 Disruption of landscape function.   (A) For mulga grasslands in eastern Australia, the interaction between spatial patterns and various ecological processes give rise to positive feedback mechanisms (local facilitation) that help to sustain an array of landscape functions (e.g. nutrient retention), leading to a productive and sustainable system. (B) Overgrazing, especially when coupled with drought, disrupts pattern-process linkages (denoted by the dashed line), leading to feedbacks that erode landscape function and may result in desertification, a system state change.

Source: Based on work by Ludwig & Tongway (1997); Ludwig et al. (2005).

across a wide range of spatial and temporal scales, but may nevertheless leave a characteristic signature that can be discerned by examining the nature of patchiness and the scale(s) at which it emerges on the landscape. Although process cannot always be inferred from pattern because different processes may give rise to similar patterns (Cale et al. 1989), an analysis of landscape pattern offers a useful diagnostic tool for evaluating the effects (or presumed effects) of a given process on landscape structure, especially for documenting the consequences of human land-use and other landscape changes over time (Krummel et al. 1987). Thus, dramatic changes in landscape structure can give an early warning sign that critical processes have been altered or that the disturbance regime has been disrupted, which may interfere with landscape function.

Numerous patch-based metrics have been developed to describe landscape pattern, involving the number, size, shape, type, distribution, or proximity of patches on the landscape (Gustafson 1998; **Chapter 4**). Anthropogenic disturbances that contribute to habitat loss and fragmentation typically result in a reduction in the number and size of remaining patches, an increase in the distance (or gap size) between patches, an increase in the amount of edge habitat, and a change in the surrounding land-use matrix. Such changes in patch structure could decrease the amount of suitable habitat for native species, increase the impact of negative edge effects (such as increased predation or competition), and decrease dispersal or colonization success, all of which could lead to increased extinction risk and thus lower diversity on fragmented landscapes (**Chapters**

6–10). Given the habitat function of landscapes (**Table 11.1B**), the diminished potential of fragmented landscapes to support viable populations and thus maintain a full complement of biological diversity is often correlated with a decline in function related to the stability or resilience of the system to withstand future disturbances (Peterson et al. 1998; Gunderson 2000).

At the extreme, altered patch structure may reflect a landscape in transition, a system for which state change is perhaps imminent. For example, Kéfi and her colleagues (2007) documented that vegetation patterns of arid rangelands in the Mediterranean naturally exhibited a power-law distribution of patch sizes, in which landscapes have many small but few very large patches (**Figure 11.7**). Recall that power laws are considered the fingerprint of self-organized systems (Solé 2007; **Box 2.1**), in which localized interactions among system components give rise to higher-order patterns and system behavior. Vegetation patches within these arid Mediterranean systems, as with the arid rangelands in Australia, retain water and create a local environment that enhances the survival and establishment of other plants. Positive feedbacks thus permit denuded areas to be re-vegetated by neighboring patches (local facilitation). Under high grazing pressure, however, the rate of degradation exceeds the capacity of the system to recover; local interactions are swamped and vegetative spread is sharply reduced. Overgrazed landscapes thus exhibit a truncated power-law distribution owing to a dearth of large patches. Desertification may occur suddenly under continuous stress resulting from prolonged drought and overgrazing, producing a system state change that may well be irreversible (Solé 2007).

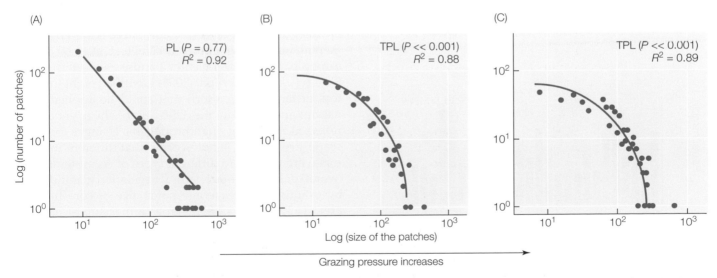

Figure 11.7 Early warning of system state changes: altered patch-size distributions.   (A) The distribution of patch sizes within arid Mediterranean grasslands (Spain) initially exhibited a power-law (fractal) distribution (PL). (B, C) The patch-size distribution became truncated (TPL) in landscapes subjected to increasing levels of overgrazing, owing to the loss of large patches. Thus, the altered patch-size distribution may signify that desertification is imminent.

Source: After Kéfi et al. (2007).

Other measures of landscape structure may also indicate the potential for altered landscape function. Ludwig and his colleagues (2007) derived a 'leakiness index' of landscape function, in which they quantified the potential for the landscape to lose water, nutrients, or soil sediments in relation to the amount and configuration of vegetation on the landscape. Their approach is based on calculating the progressive flow of water or materials among sites (e.g. pixels of a remotely sensed image), as a function of both the amount of vegetation cover and position (elevation) relative to neighboring sites. For example, sites with high vegetation cover should do a much better job of intercepting flows and retaining water or nutrients than those with less vegetation, affecting the degree to which sites lower in elevation will receive water or material runoff. In the semi-arid grasslands of Australia, for example, the leakiness index exhibited a negative relationship with vegetation cover in recovering rangelands, which had previously been degraded as a result of overgrazing (**Figure 11.8A**). The recovery of landscape function could be quantified by assessing how the rangeland's potential to retain water and nutrients improved as vegetation increased and the landscape became less leaky. Once restored, this

rangeland was then able to preserve landscape function even during years of low rainfall (**Figure 11.8B**).

Landscape heterogeneity is another dimension of landscape structure whose assessment may likewise be relevant for evaluating landscape function (e.g. landscape diversity index; **Chapter 4**). For example, a decrease in landscape heterogeneity resulting from the loss of particular habitats—representing entire biological communities or ecosystems—may disrupt nutrient source–sink dynamics and spatial subsidies across the landscape. Diversity typically confers greater system resilience by enhancing the capacity of the system to recover after disturbances. Thus, maintaining habitat heterogeneity might be viewed as important for preserving landscape function in much the same way that maintaining species diversity is important for preserving ecosystem function (e.g. functional heterogeneity). This involves not only protecting or restoring a variety of landscape components (different habitat or patch types), but also the connections or interactions among those components. Landscapes are complex adaptive systems, in which the interactions and positive feedbacks among a diverse array of components (patches, species, nutrient pools) simultaneously give rise to system structure (aggregation and self-organization on multiple scales) and ecological function (Ryan et al. 2007). A loss of heterogeneity may thus lead to a loss of landscape function, as no single component of the landscape (e.g. a particular plant functional type) can possibly interact with all other components or function under all possible environmental conditions, which is especially salient given the immediacy of climate change.

Dramatic changes in landscape structure may thus presage disruptions to landscape function, which is why quantifying landscape change has become such an integral component of many large-scale environmental monitoring efforts (O'Neill et al. 1997). For example, the Heinz's Center Landscape Pattern Task Group (Christensen et al. 2008) identified eight indicators of landscape pattern that could be applied at a nationwide scale in the USA, and which generally reflected the extent of anthropogenic changes to landscape structure. It is noteworthy that different indicators were tailored to different types of ecosystems: the overall pattern of intact natural vegetation was indexed by examining the average patch size of core habitat areas (>240 acres = >97.1 ha) as a measure of habitat loss and fragmentation (a potential indicator of a loss of habitat functions), whereas the proximity of cropland to residences was used as a landscape indictor in agroecosystems, reflecting the threat from increasing human development that may lead to the permanent removal of high-quality farmland from production.

Importantly, these system-specific indicators demonstrate that no single landscape metric will suffice to

(A)

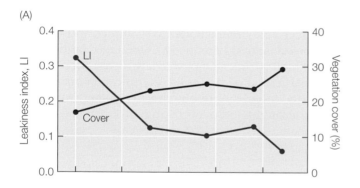

(B)

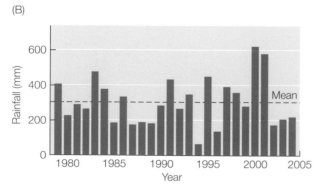

Figure 11.8 Assessing landscape function: the landscape leakiness index. (A) The leakiness index is negatively related to vegetation cover in an Australian rangeland. As the rangeland is allowed to recover, landscape function is restored in that the landscape becomes less 'leaky', even during dry periods (B).

Source: After Ludwig et al. (2007).

gauge a potential loss of landscape function in all cases. With the widespread availability of remotely sensed imagery and landscape analytical software (**Chapter 4**), such broad-scale assessments of landscape structure are a reasonably efficient and inexpensive means for assessing landscape function. Nevertheless, these indicators merely indicate a *potential* loss of function, and should not be viewed as the sole criterion or diagnostic for evaluating landscape function. Simply monitoring changes in landscape structure is no substitute for evaluating what agents are driving landscape change or for understanding the linkage between altered landscape structure (pattern) and ecological processes (e.g. **Figure 11.6**). We therefore discuss this in the context of landscape connectivity next.

ALTERATION OF LANDSCAPE CONNECTIVITY Landscape connectivity is considered critical for maintaining many important ecological processes, such as dispersal and gene flow among populations, which enhance population viability and species persistence on the landscape (Taylor et al. 1993; Bélisle 2005; Kindlmann & Burel 2008; **Chapters 5 and 9**). Recall that the disruption (or emergence) of landscape connectivity may occur as a threshold response to habitat loss and fragmentation (Gardner et al. 1987; With 1997). That is, small changes in the amount or distribution of habitat at some critical point can result in a sudden, system-wide change in connectivity. The exact point at which this occurs depends upon the scale or distance across which flows can occur among patches. The disruption of landscape connectivity thus represents an uncoupling of landscape pattern and some ecological process, which might well lead to a disruption or loss of landscape function. If this occurs as a threshold response to environmental change (whether natural or anthropogenic), then it is imperative that we anticipate when and where these thresholds are likely to occur, as it will be far more difficult (and expensive) to try to recover or restore landscape function once the threshold has been exceeded (Groffman et al. 2006).

As a consequence, landscape connectivity is generally viewed as something that must be conserved or even enhanced (Taylor et al. 2006). Not all connectivity is necessarily beneficial or desirable, however. For example, the transformation of landscapes by human land use may lead to increased connectivity of whatever new land cover comes to dominate the landscape, as in the case of agricultural landscapes where policy or economic incentives favor the production of vast acreages of only one or a few crop types. As a case in point, the American agricultural landscape exhibits connectivity at a nationwide scale for certain crops (maize, soybean), which increases the spread potential of any number of economically devastating crop pests

or disease (Margosian et al. 2009; **Figure 8.18**). Beyond the spread of disease, we have also discussed how connectivity may enhance the spread of invasive species (With 2002b; **Chapter 8**). Further, as discussed in the previous section, connectivity (of bare-ground areas) may contribute to leaky landscapes if it leads to the excessive flow of water, materials, or nutrients from the landscape (Ludwig et al. 2005). Indeed, the various forms of desertification and land degradation that plague arid landscapes—from the loss of vegetation as a result of intensive agricultural and grazing practices, to shrubby encroachment of grasslands, to the invasion of shrublands by exotic grasses—might all stem from changes in connectivity that increase the length of connected pathways that serve as conduits for the movement of fire, water, or soil resources borne by water or wind (Okin et al. 2008).

A loss of heterogeneity can also contribute to increased landscape connectivity, which might then be able to promote the spread of fire or other unwanted disturbances across a much broader spatial extent. For example, the historical, pre-settlement landscape of southern Wisconsin (USA) exhibited an intermediate degree of connectivity that was likely a consequence of fire operating on—and maintaining—a heterogeneous landscape comprising prairies, savannas, and different forest types (Bolliger et al. 2003). These different vegetation types had different sensitivities to and thus frequencies of fire disturbance, but because of the initial heterogeneity, fire was unlikely to spread across the entire landscape, despite the high degree of contagion within individual habitat types. A power-law distribution was evident in these landscapes, leading the researchers to propose that these landscapes were self-organized (**Box 2.1**) and that self-organized criticality promotes intermediate connectivity in landscapes (Bolliger et al. 2003).

However, power-law distributions may also result from positive feedbacks (such as from local-scale facilitation) even in the absence of the sort of threshold behavior commonly associated with criticality (Scanlon et al. 2007); that is, systems need not be self-organized to a critical state to exhibit power-law scaling in some property. For example, we discussed previously how altering the natural disturbance regime (e.g. by overgrazing) disrupted positive feedbacks among vegetation patches in a dryland system, resulting in truncated power-law distributions in the patch structure of these landscapes (Kéfi et al. 2007). This in turn may lead to overconnected landscapes (in terms of bare-ground areas) that cause excessive loss of water and nutrients (a leaky landscape; Ludwig et al. 2005, 2007). In this case, a hyper-connected landscape would lead to a loss of system resilience (the ability of vegetation to recover from a drought) and, if severe, could cause a system state change (e.g. desertification). We consider system state changes next.

SYSTEM STATE CHANGES   A system state change is a type of critical threshold or transition, in which small changes near the threshold cause the system to flip suddenly to a different state (Scheffer et al. 2001; Folke et al. 2004; Groffman et al. 2006; Scheffer et al. 2009; **Figure 11.9**). In dynamical systems models, such critical thresholds between system states correspond to 'bifurcation points,' where the ecosystem response curve 'folds' back on itself (Scheffer et al. 2001, 2009; **Figure 11.9B**). This means that, over a certain range of conditions, the system could exist in one of two states (stable states or 'basins of attraction'). The system thus has three equilibria: two stable ones corresponding to each of the alternative states, and an unstable one in the domain between the two bifurcation points (P_1 and P_2 in **Figure 11.9B**). For each of the stable states (the upper or lower portions of the response curve), small perturbations to the system cause a rapid return to the equilibrium state; small changes in environmental conditions would thus appear to have minimal effect on the system (i.e. the system is resilient to environmental change in this domain). In the unstable domain between the bifurcation points (the dashed middle section of the response curve), small perturbations cause the system

to move away from this part of the curve rather than returning. Small changes in this domain near a bifurcation point could thus cause an abrupt transition between the two system states, seemingly without warning (i.e. 'ecological surprises;' **Chapter 3**).

For example, increased forcing of a semi-arid grassland, due to chronic overgrazing and protracted drought, could push the system from a grassland (a stable state) to a desert (another stable state; **Figure 11.9B**). Note that this is similar to the critical landscape thresholds we've discussed previously in this textbook (e.g. critical thresholds in landscape connectivity; **Chapter 5**). Small perturbations (loss of grass cover) near the threshold can cause a disruption in grassland connectivity (**Figure 11.9A**). Whether this results in a system state change (desertification), however, depends on how a disruption of landscape connectivity affects the various flows and ecosystem processes that contribute to grassland resilience. If decreased vegetation connectivity leads to increased soil erosion or loss of nutrients (i.e. a change in function from a conserving landscape to a leaky landscape; Ludwig et al. 2005, 2007), then the altered patch structure and connectivity of the landscape might well lead to a 'catastrophic' state change, which in some cases, may be effectively irreversible on human time scales (Scheffer et al. 2001; Folke et al. 2004).

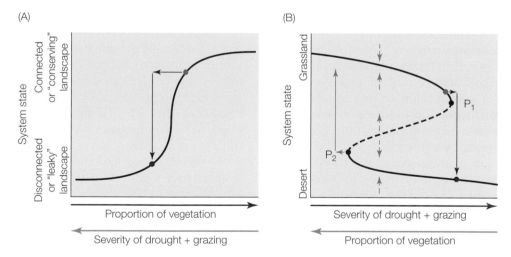

Figure 11.9 Critical thresholds and system state changes: the desertification of an arid grassland.   (A) Grazing and drought may reduce grassland cover, causing a sudden loss of landscape connectivity past a critical threshold in the amount of vegetation. If landscape connectivity influences ecosystem flows (e.g. water, nutrients, organic matter, or seeds), then this threshold might also signify the transition from a conserving landscape to a leaky landscape (cf. **Figure 11.8**), which could precipitate a system state change (desertification). (B) The state change from grassland to desert is predicted to occur at a critical threshold in drought and grazing severity (P1). Small changes near this point (a bifurcation point) could cause an abrupt change from grassland to desert. Note that a return to a grassland state is theoretically possible, but will require the amelioration of drought and a reduction in grazing pressure to levels far below those that caused the system state change in the first place (the critical threshold from desert back to grassland occurs at P2, and not at P1). The system thus has three equilibria: two stable ones associated with the alternative states (grassland or desert), where small changes along each trajectory result in a return to that state (indicated by the small arrows along the upper and lower portions of the response curve), and a third unstable one (the dashed portion of the response curve) in which any change causes the system to move away from that point to one or the other of the stable states.

Source: After Scheffer et al. (2009).

Assuming that a system state change is reversible, however, recovery or restoration of that former state will likely require far more than just a return to the environmental conditions that existed prior to the threshold (Scheffer et al. 2001). For example, if overgrazing and drought contribute to the desertification of grassland at a particular threshold of grazing and drought severity (P_1), it might be necessary to alleviate the drought and reduce grazing pressure to much lower levels (P_2) than those that caused the state change in the first place (**Figure 11.9B**). In other words, desertification and grassland recovery may occur at very different thresholds in critical conditions. When the shift from one system state to another occurs at different critical thresholds, this is referred to as **hysteresis** (Scheffer et al. 2001). The road to system recovery is thus different from the one that led to its degradation in the first place.

System state changes are not unique to arid or semi-arid grasslands, although they are perhaps best illustrated by these sorts of systems because they are water-limited and so might already be very near to the critical transition threshold if pushed further by one or more exogenous drivers (overgrazing and/or drought). Although these systems are particularly vulnerable to state shifts, more mesic systems are not immune. System state changes have been documented in a wide variety of systems, including temperate and tropical lakes, coral reefs, kelp forests, and in temperate and tropical forests (Folke et al. 2004). System shifts might even occur more rapidly in mesic than arid systems because of greater resource abundance and production potential (Briggs et al. 2005).

For example, mesic grasslands such as the tallgrass prairies of North America are rapidly undergoing a transition to a savanna state, owing to an alteration of the natural fire regime that historically shaped this system and due to increased grassland fragmentation. Fire favors the warm-season grasses that characterize the tallgrass prairie and helps to suppress woody species (e.g. rough-leaf dogwood, *Cornus drummondii*; eastern red-cedar, *Juniperus virginiana*) that would otherwise invade (an example of a negative feedback system). Fire suppression thus results in the dramatic transformation of these grasslands into savanna grasslands or mixed shrublands in fewer than 20 years, with the complete conversion to closed-canopy red-cedar forests in as little as 40 years (Briggs et al. 2002, 2005). Even if fire is eventually returned to the system, the large shrub 'islands' that have formed in the interim do not burn, and can thus provide safe sites for trees, thus further promoting woody invasion and the rapid transition of grassland to savanna (i.e. a reversal in process constraints and a shift from negative to positive feedback; **Figure 11.10**).

Figure 11.10 System state change in a mesic grassland. The tallgrass prairie of North America is maintained by a combination of periodic grazing and fire that give rise to a diverse grassland community. Infrequent fire causes woody vegetation to become established, causing a shift to a savanna state that may be irreversible even when fire is returned to the system. (A) An annually burned watershed at the Konza Prairie Biological Station (KPBS; Kansas, USA), which is clearly dominated by grasses. (B) A grassland site at the KBPS that is burned every four years (including the year this photo was taken) has numerous 'shrub islands' established throughout the watershed. Despite some charred stems, the shrubs (*Cornus drummondii*) are vigorously leafing out and flowering in the weeks following the spring burn.
Source: (A) Melinda Smith, (B) John Briggs.

Nutrient dynamics are also affected, such that these systems are altered functionally as well as structurally (Norris et al. 2007; McKinley & Blair 2008). Although soil nitrogen availability did not vary between grassland sites and those invaded by red-cedar, the mean aboveground plant productivity of red-cedar forests was 2.5× greater than that of comparable grassland sites, resulting in an increased nitrogen use efficiency in red-cedar forests that was more than double that of the grasslands they replaced (Norris et al. 2007). Conversely, soil carbon exhibits slower turnover in forests compared to grasslands (McKinley & Blair 2008). The end result is the rapid accrual and storage of aboveground stocks of nitrogen and carbon in red-cedar forests compared to grassland, and a shift in the mean C:N ratios in aboveground biomass from about 54 in grasslands to about 126 in red-cedar forests (McKinley & Blair 2008; **Box 11.1**). Thus, the transition of tallgrass prairie ecosystems to savanna or woodland represents a system state change that is no less dramatic or important than the desertification of arid grasslands, and may be irreversible without significant interventions (e.g. mechanical removal of shrubs and trees) that may be too costly or impractical to implement on a broad landscape or regional scale.

Given that many types of system state changes are deemed undesirable, owing to a loss of ecosystem services and landscape functions provided by the original system, it would be ideal from an ecosystem or land management standpoint to be able to predict when such state changes are imminent, in the hopes that preemptive action might then be taken to avoid a catastrophic shift in system state. How to determine whether a system state change is imminent, however? It is notoriously difficult to predict when such changes might occur, because the system may show little change before the threshold or tipping point is reached (Scheffer et al. 2009). Nevertheless, early warning signals tend to arise as systems approach the bifurcation point, which may be reflected by changes in the spatial properties of the landscape or in its rate of recovery following a disturbance (Dakos et al. 2012; Kéfi et al. 2014). We have discussed previously in this chapter how changes in landscape structure and connectivity can alter ecological flows and landscape function, and thus might provide clues to an impending state change. We'll thus focus here on how changes in the rate of recovery, what is referred to as **critical slowing down**, might presage a system state change.

In critical slowing down, the closer the system is to the bifurcation or tipping point, the longer it takes to recover from small perturbations (Kéfi et al. 2014). The phenomenon of critical slowing down leads to three possible early warning signals in the dynamics of the system: (1) slower recovery from perturbations, (2) increased temporal autocorrelation, and/or (3) increased variance in the pattern of fluctuations (Scheffer et al. 2009). Because critical slowing down causes the system to exhibit a slower response to disturbance (i.e. a lagged response), the system will appear very similar to its previous state, causing an increase in the 'short-term memory' of the system (a higher degree of temporal autocorrelation at short time-lag intervals; e.g. **Figure 3.9**) near the bifurcation point (Dakos et al. 2012). Slow

return rates back to a stable state can also make the system drift widely around the stable state, especially if the impacts of disturbance do not decay, and their accumulating effects increase the variance of the state variable (Scheffer et al. 2009; Dakos et al. 2012). Conversely, it has been argued that critical slowing down could reduce the ability of the system to track environmental fluctuations, thereby leading to a reduction in variance (Scheffer et al. 2009). Nevertheless, rising variance near the threshold is generally expected to occur in most ecological systems (Brock et al. 2006; Scheffer et al. 2009).

For example, we've discussed previously how human land-use activities can cause increased nutrient inputs into lakes, which could result in eutrophication (another type of system state change; Carpenter et al. 1998). Using a simple model parameterized with lake phosphorus data obtained over many years from Lake Mendota in Wisconsin (USA), Carpenter and Brock (2006) simulated nutrient dynamics under different conditions to evaluate whether lake phosphorus levels exhibited increased variability as the system approached the eutrophication threshold. As phosphorus loading increased (as indexed by c, the input coefficient), the total phosphorus density of the lake water (g/m^2) increased linearly until a critical input level was reached ($c = 0.0025$), at which point phosphorus density suddenly exhibited a very large 'bump,' consistent with the transition from an oligotrophic to eutrophic lake (i.e. a system state change; **Figure 11.11A**). Interestingly, the variance of phosphorus density (as indexed by the standard deviation, SD) also exhibited a sudden increase near the threshold, and in fact, preceded it slightly ($c = 0.0021$; **Figure 11.11B**). Thus, increased variability in total phosphorus density of lake water could well provide early warning of impending eutrophication, giving ecosystem managers an opportunity to take action to prevent this (Carpenter & Brock 2006).

Because of hysteresis, we know that it will be far more difficult to recover or restore the original system state once the ecosystem has been pushed past the critical transition threshold (e.g. the eutrophication or desertification threshold; P_1 in **Figure 11.9B**). The 'restoration threshold' represents a different bifurcation point corresponding to a different set of conditions (P_2 in **Figure 11.9B**) from those that caused the system state change in the first place. Nevertheless, this suggests that the transition from a degraded to a healthy ecosystem may also occur suddenly, in which case, it would be useful to have some sort of indicator to assess the efficacy of our restoration efforts in achieving a 'reverse transition.' Might indicators related to critical slowing down or rising variability also be useful in assessing whether a restoration threshold is imminent?

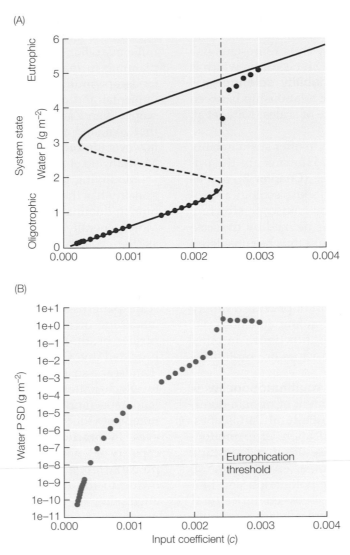

Figure 11.11 Early warning of system state changes: increased variability. (A) The density of phosphorus (P) in the water of a freshwater lake is expected to increase with increased nutrient loading (input coefficient, *c*), but exhibits an abrupt increase at a critical point (the eutrophication threshold) when the system shifts from an oligotrophic to a eutrophic lake. (B) The variability in water phosphorus density (as indexed by the standard deviation, SD) also exhibits a sharp increase near the eutrophication threshold, but importantly, occurs slightly before that transition threshold is reached. Rising variance might thus provide an early warning signal of an impending system state change.

Source: After Carpenter & Brock (2006).

In assessing the restoration of a dryland ecosystem in China, Chen and his colleagues (2018) found that the temporal dynamics of grass cover exhibited increased variability as the system transitioned from a barren to vegetated state. Thus, rising variability in a state variable might also prove useful as an indicator of a reverse transition.

Managing for Landscape Multifunctionality and Sustainability

Sustainability or sustainable development is a necessity, not a choice.   Wu (2013)

By assessing and monitoring landscape function, we can hopefully avoid catastrophic state changes that would result in a loss of productivity or of critical ecological processes, thereby compromising the ability of landscapes to provide the various goods and services that are essential to our own physical, environmental, and socioeconomic well-being. That is in fact the aim of **sustainable development**, which strives to meet human development goals while sustaining the ability of natural systems to provide the natural resources and ecosystem services upon which the economy and society depend, all without compromising the ability of future generations to meet their own needs (World

Commission on Environment and Development 1987; National Research Council 1999; Millennium Ecosystem Assessment 2005). Sustainability thus represents one of the greatest challenges of the 21st century, which is why a new discipline—**sustainability science**—has emerged to address the dynamic relationship between nature and society across a range of scales (Kates et al. 2001; Kates 2011; Wu 2013).

Landscape ecology has much to offer sustainability science (Wu 2006; Wiens 2013). Humans are the primary driver of landscape change (**Chapter 3**), and as we've seen throughout this textbook, ecological processes are profoundly affected by landscape structure and dynamics. The landscape, as defined by the scale of human land-use activities, might thus be considered *the* basic spatial unit or operational scale for managing and achieving sustainability (Wu 2006; 2013). Land use is the key activity that determines to what degree landscapes can simultaneously support or provide a variety of environmental, social, and economic functions (i.e. ecosystem services or landscape functions); in other words, the extent to which landscapes are multifunctional (Mander et al. 2007).

The concept of **landscape multifunctionality** is typically applied to the management of human-dominated landscapes, such as agricultural landscapes. Because of agricultural intensification, crop production is maximized at the expense of other landscape functions, such as carbon storage, erosion control, wildlife habitat, biodiversity, or pollination and biocontrol services. Thus, a balance between agricultural production and biodiversity is typically achieved only through **land sparing**, in which homogeneous areas of farmland are intensively managed so as to maximize yields, while separate natural areas or reserves are set aside elsewhere for the purposes of providing wildlife habitat and protecting biodiversity (Green et al. 2005; Fischer et al. 2008).

Conversely, the joint production of both agricultural commodities and biodiversity might be achieved through **land sharing**, by integrating production and conservation within the same heterogeneous landscape; that is, by managing for multifunctional landscapes (Fischer et al. 2008). For example, the adoption of wildlife-friendly farming practices, such as retaining or restoring habitat elements adjacent to farm fields (e.g. scattered trees, hedgerows, grass, or flower margins), could provide habitat and resources for birds, pollinators, and/or natural enemies of crop pests. Not only do wildlife-friendly farming practices help to maintain biodiversity within agricultural landscapes, but the ecosystem services provided by those species might even help to increase crop yields (i.e. ecological intensification; Pywell et al. 2015). Managing for landscape multifunctionality is thus common in Europe and in other regions of the world that have a long history of agricultural land use (e.g. on the order of centuries to millennia). For example, the stated goal of landscape planning in many European countries is to increase landscape multifunctionality, by enhancing the environmental and cultural functions of production landscapes (van Zanten et al. 2014). How best to achieve that goal, and at what scale, is still an open question, however (de Groot & Hein 2007; Lovell & Johnston 2009; Mastrangelo et al. 2014; Stürck & Verburg 2017).

Modifying landscapes to increase their multifunctionality has the potential to enhance sustainability in human-dominated landscapes (Stürck & Verburg 2017). More generally, **landscape sustainability** is defined as 'the capacity of a landscape to maintain its basic structure and to provide ecosystem services in a changing world of environmental, economic, and social conditions' (Wu 2012). The focus here is squarely on landscape patterns and processes, ecosystem services, and landscape or spatial resilience, which is reinforced further by the definition of landscape sustainability provided by Cumming and his colleagues (2013): 'Landscape sustainability can be viewed as the degree to which patterns and processes occurring within a landscape (and their interactions) can persist indefinitely into the future.' Indefinitely is a long time, however. In practice, our ability to predict the future capacity of landscapes to provide ecosystem services is probably limited to a few decades or a century at most, owing to the inherent uncertainties associated with attempting to make long-range forecasts about the dynamics of complex systems (Wu 2013; **Chapter 2**). In the present context, the landscape can thus be considered a complex human-environment system (Wu 2013).

The goal of landscape sustainability is thus to provide landscape-specific ecosystem services that are essential for maintaining and improving human well-being in the face of environmental and socioeconomic changes (Wu 2013). **Landscape resilience** is a key component of landscape sustainability. Landscape resilience has been defined as 'the capacity of a system to tolerate disturbance without shifting to a qualitatively different state' (i.e. the ability to resist a system state change), which then affects the capacity of that landscape to provide ecosystem services for current and future generations (i.e. landscape sustainability; Turner et al. 2013, p 1082). Landscape resilience is thus seen as essential for sustaining future ecosystem services, which means that we should seek to strengthen the capacity of landscapes to recover from disturbances if we are to achieve our sustainability goals. For example, we discussed in the previous section of this chapter how landscape heterogeneity is important for evaluating landscape function, given that spatial heterogeneity can enhance the ability of the system to recover after a disturbance; that is by contributing to greater landscape resilience. In that case, spatial heterogeneity might very well be the key to landscape sustainability (Wiens 2013).

The role of landscape heterogeneity in buffering environmental variation, enhancing the resilience of ecosystems, and forestalling transitions past thresholds to alternative states may be key to resource and landscape sustainability.　　Wiens (2013)

To illustrate, Turner and her colleagues (2013) investigated the importance of spatial heterogeneity for the resilience of forested landscapes in the face of changing climate, disturbance regimes, and land use. Recall from our discussion in **Chapter 3** that large, frequent forest fires are becoming the 'new normal' in western North America. The frequency of large and unusually severe forest fires has increased in the last quarter-century, largely due to climatic changes that have produced warmer and drier conditions (Romme et al. 2011). Climate is a major top-down driver of fire activity, but humans can also exert top-down effects on fire regimes through our land use and land-management decisions, which can interact with climate and exacerbate those effects on the fire regime. For example, livestock grazing and fire-suppression policies have contributed to a growing fire deficit in many parts of this region, even as successional and climatic changes have increased the fire potential of forest landscapes (Marlon et al. 2012). In the Greater Yellowstone Ecosystem, for instance, the increased frequency of large, severe fires is predicted to shorten the fire-return interval (from 300 years to 30 years) and compromise forest regeneration, which will likely cause a shift from forest to savanna or shrubland (a system state change) and a concomitant loss of forest-based ecosystem services (e.g. carbon storage, timber, habitat for forest species; Westerling et al. 2011; Turner et al. 2013).

Spatial heterogeneity is important to the resilience of forest landscapes because it enhances their ability to recover after fire as well as to continue to provide a variety of forest-based ecosystem services. Even the most severe large fires create a heterogeneous disturbance mosaic, in which unburned patches and surviving 'legacy' trees serve as seed sources for forest regeneration (**Figure 3.31**; Turner et al. 2013). Although almost any part of the landscape is likely to burn under extreme weather conditions (i.e. dry and windy), spatial heterogeneity in the distribution of fuels owing to topographic variation, the presence of natural fire breaks (e.g. rivers), or even forest-management practices like tree thinning, can reduce the spread of fire under more moderate burning conditions (Turner et al. 2013; **Chapter 3**). Uneven-aged forest management creates spatial heterogeneity within stands, contributing to a more complex pattern of fuels that can influence the behavior and severity of future fires, especially if an uneven distribution of suitable tree hosts also halts the spread of forest insect pests and pathogens that might otherwise kill trees and increase fuel loads (Turner et al. 2013). Spatial heterogeneity also plays an especially important role in the availability of wildlife habitat, given that the composition and configuration of different habitats on the landscape will determine the occurrence, abundance, and population viability of many forest species (Turner et al. 2013; **Chapter 7**). Furthermore, as we have discussed in the second half of this textbook, the connectivity of populations, communities, and ecosystems are all affected by spatial heterogeneity (**Chapters 7–11**). Maintaining ecological flows and functional linkages among these various types of metasystems (**connectivity conservation**; Crooks & Sanjayan 2006) is essential not just for the resilience of those systems, but ultimately, for landscape resilience and sustainability.

Future Directions

Developing a true landscape ecology of ecosystems will entail more than simply understanding how landscape position, spatial heterogeneity, and current or past land uses influence ecosystem structure and function, although those are obviously important contributions to the emerging field of landscape ecosystem ecology (Massol et al. 2011). As we've discussed, the composition and configuration of habitats, land covers, or land uses within a landscape are expected to influence the productivity, trophic structure, and food-web dynamics of different ecosystems. However, it's understanding how these different ecosystems interact through the

spatial flows of nutrients and organic matter, whether directly or indirectly, across that landscape that brings a new synergy to the field. From that perspective, the metaecosystem framework has provided a conceptual and theoretical basis for understanding the spatial dynamics of functionally connected ecosystems (e.g. Loreau et al. 2003; Massol et al. 2011; Guichard 2017). Understanding how spatial and temporal variation in spatial flows and ecosystem productivity contribute to the source–sink dynamics of ecosystems is important not only for advancing landscape ecosystem ecology, but ultimately, for ensuring the continued provisioning of landscape-specific ecosystem services, and thus, for the management of landscape sustainability. A major focus of metaecosystem research is thus on uncovering the nature of the functional linkages among coupled ecosystems, especially in regards to quantifying the magnitude and direction of spatial subsidies as well as the role played by mobile-link species in connecting these systems. For not only is connectivity vital to the structure and functioning of these coupled systems, but preserving functional heterogeneity can also confer greater landscape resilience, and thus enhance sustainability.

Because landscapes and ecosystems provide the goods and services upon which human societies and economies depend, there is an ever-present need to monitor and manage landscape-specific ecosystem services. Landscape ecology can clearly contribute to these efforts, given that it has developed numerous methods for quantifying changes in the spatial structure, heterogeneity, and connectivity of landscapes, and is expressly concerned with uncovering the effects of spatial pattern on ecological processes. The alteration of landscape structure or connectivity can modify the strength and direction of ecosystem flows, which might then threaten the provisioning or support of critical ecosystem services. Indeed, dramatic changes in the spatial properties or dynamics of landscapes, such as in the patch-size distribution of vegetation or the recovery rate of a state variable following a disturbance, have been proposed as early warning signs of an ecosystem in transition, one that is perhaps on the verge of a catastrophic state shift (Dakos et al. 2012; Kéfi et al. 2014). Unfortunately, empirical validation of these spatial and temporal measures is hampered by the dearth of long-term datasets that track the system through a range of environmental conditions. Nevertheless, research is urgently needed to determine which measures can best serve as leading indicators of degraded landscape function and impending state change within different types of landscapes, especially as many systems may soon be reaching a tipping point due to changes in environmental conditions, disturbance regimes, and land use, all of which are being exacerbated by climate change.

Because of its focus, landscape ecology is uniquely positioned to provide the scientific basis for sustainable landscape development and to inform land-change policies related to land use, urbanization, and climate change (Termorshuizen & Opdam 2009; Mayer et al. 2016). For example, managing for landscape multifunctionality through land-sharing programs that promote wildlife-friendly farming practices and urban green spaces, or the restoration of habitat to recover certain regulatory functions (e.g. erosion control, flood protection, pollination services), have the potential to enhance sustainability in human-dominated landscapes, such as farmlands and cities. True, decisions as to which ecosystem services or attributes of landscapes are to be sustained, and whether or how the trade-offs among competing needs and goals can be balanced, are ultimately societal, not scientific, decisions (Wiens 2013). Still, science can—and should—contribute to those decisions, and landscape ecology is arguably in the best position to provide that scientific basis for sustainable development and managing for landscape multifunctionality (Termorshuizen & Opdam 2009). As landscape ecologists, we will need to marshal all of our collective energies and expertise to understand how land use, habitat loss and fragmentation, climate change, and altered disturbance regimes are affecting the structure, function, and dynamics of landscapes, in ways that compromise ecological resilience and undermine sustainability. Sustainability is our most herculean challenge, for as Wu (2013: 1020) has observed:

> Sustainability is more of a process than a state. Thus, landscape sustainability is a constantly evolving goal. We cannot predict it precisely; we cannot fix it permanently; but we must, and we can, make our landscapes sustainable by continuously improving the human–environment relationship based on what we know and what we are learning.   (emphasis added)

Chapter Summary Points

1. Ecosystem ecology studies the structure and functioning of ecosystems, especially how matter, nutrients, and energy are distributed and transferred among organisms, and between organisms and their environment. The major processes essential to all ecosystems are the water (hydrological) cycle, nutrient (biogeochemical) cycles, food-web (trophic) dynamics, and energy flows among trophic levels. Landscape structure—including human land use—can affect any or all of these processes in a given ecosystem. Understanding how landscape structure influences ecosystem processes is thus a major focus of landscape ecosystem ecology.

2. In terrestrial ecosystems, the interaction between climate (precipitation, temperature) and soil properties (organic matter content, mineral composition, permeability) governs many important ecosystem processes, such as the rate of decomposition and nutrient turnover in the soil, thereby affecting the abundance and bioavailability of chemical elements essential to all lifeforms (e.g. nitrogen, carbon). Factors affecting nutrient

cycling operate over a range of scales in space and time. The result is a spatially heterogeneous and dynamic resource landscape, in which some areas act as nutrient sinks (absorbing more of a certain element than they release) whereas others are nutrient sources (releasing more of a certain element than they absorb).

3. Topographic position, human land use, and land-management practices are the primary factors affecting soil properties in most terrestrial landscapes, and therefore have a major influence on the distribution and availability of soil nutrients across the landscape.

4. The study of topographic gradients has been instrumental in uncovering the biotic and abiotic drivers of various ecosystem processes that influence nutrient cycling, soil nutrient availability, and ecosystem productivity across landscapes. Topographic position has been found to affect ecosystem processes in a diverse array of terrestrial systems, including arctic tundra, desert scrub, arid and mesic grasslands, temperate hardwood forests, coniferous forests, and tropical lowland forests.

5. Agricultural conversion of forests and grasslands causes significant carbon losses that account for about 12% of all anthropogenic carbon emissions. Most of the carbon is stored in the soil rather than in the vegetation, however, making soil the largest terrestrial pool of carbon, a sum larger than the amount of carbon stored in the atmosphere and terrestrial vegetation combined. Soil carbon losses can therefore have a significant effect on atmospheric CO_2 levels, and thus, on the global climate.

6. Land uses that cause land-cover change, such as the agricultural conversion of forests and grasslands, can significantly alter the amount of carbon stored in the soil. The loss of soil organic carbon due to land-use change has been estimated to be around 1.14 Pg C/year, much of which is due to deforestation in the tropics. Land-use change has reduced soil carbon stocks by 9% overall.

7. The agricultural conversion of forests and grasslands causes the greatest loss of soil organic carbon. Soil carbon stocks decline by an average of 42% when native forest is converted to cropland, and by 59% when grassland is converted to cropland. Although the magnitude of soil carbon loss is greater in grassland converted to cropland, soil carbon is lost nearly twice as fast when forest is converted to cropland.

8. Ecological restoration can be an effective means of increasing soil carbon stocks in former cropland.

Soil carbon stocks increased by an average 19% when cropland is restored to grassland, but by 43% when former cropland is restored to forest. The latter figure might even be sufficient to offset losses incurred during the initial forest clearing for agricultural land use. The tremendous potential for ecological restoration to recover soil carbon stocks on former cropland is evidenced by China's 'Grain-for-Green Program,' in which some 32 million ha of degraded or highly erodible farmland is being converted back to forests and grasslands.

9. Past agricultural land use can leave an enduring legacy on the availability and distribution of soil nutrients. Land-use legacy effects on soil properties can persist for decades, centuries, or even millennia after farmland abandonment. As a result of past land use, soil nutrient pools and successional trajectories may be altered, such that land-use legacies are also reflected in the composition and diversity of the resultant plant communities that form on these previously farmed areas, which in some cases may be evident some 2000 years after farmland abandonment.

10. The success of ecological restoration is increasingly being assessed in terms of ecosystem processes, such as nutrient cycling, soil structure, and carbon storage. Understanding how restoration efforts can recover important ecosystem processes is essential for achieving the long-term persistence and stability of the restored ecosystem, as well as for understanding the role of restoration in a broader landscape context.

11. Ecosystems are likely to receive resource inputs or spatial subsidies from other ecosystems, depending on their relative landscape positions and the degree to which they are connected by the flow of water, nutrients, organic matter, energy, or organisms. The productivity and food-web dynamics of a given ecosystem may thus be subsidized by the productivity of another. The concept of the metaecosystem, a system of interacting ecosystems, has emerged to study how spatial subsidies and the movements of organisms (mobile-link species) between ecosystems can affect their productivity and food-web dynamics.

12. Although the integration of landscape and food-web ecology has given rise to the study of spatial subsidies, resource subsidies have long been studied by ecosystem ecologists, particularly in the context of freshwater systems. Many aquatic systems rely on terrestrial subsidies for their productivity. Because water flows downhill,

terrestrial inputs into lakes and rivers tend to be greater than flows in the opposite direction.

13. According to the subsidy hypothesis, ecosystems that receive most of their resource inputs from outside sources (allochthonous inputs) are expected to experience stronger trophic cascades. In lakes, for example, terrestrial carbon sources represent a substantial subsidy to aquatic food webs, especially for higher trophic levels. Excessive nutrient loading, due to agricultural runoff from the surrounding landscape, increases the degree of internal (autochthonous) primary production, such that the aquatic food web is no longer dependent on terrestrial subsides. This could alter the strength of trophic interactions and radically alter nutrient dynamics and ecosystem functioning.

14. Spatial subsidies may also occur from freshwater or marine systems to terrestrial ones. Such cross-boundary subsidies reflect the interdependence of terrestrial and aquatic food webs, and illustrate that resource subsidies may be reciprocal, varying in magnitude and direction between ecosystems, depending upon location or time of year (e.g. seasonal pulses in nutrient fluxes).

15. Mobile-link species connect ecosystems through their provisioning or support of essential ecosystem functions. Mobile species may serve as resource linkers, by transporting organic material and nutrients among ecosystems; genetic linkers that carry genetic information among ecosystems (e.g. seed dispersers or pollinators); or as process linkers that provide or support an essential ecosystem process (e.g. grazers). These categories are not mutually exclusive, and a mobile species may fulfill more than one type of function.

16. By maintaining functional connectivity among different habitats or ecosystems, mobile-link species can enhance the resilience of these systems, by facilitating their recovery after a disturbance. Not all mobile-link species are equally good at providing or supporting a particular ecological function, however. Thus, ecosystem resilience is best achieved by maintaining functional heterogeneity within the community, that is, a high degree of variability in the functional capabilities of species. Functional heterogeneity enhances ecosystem resilience through a combination of functional redundancy (multiple species perform the same function) and response diversity, in which species having the same ecological function respond differently to a disturbance.

17. Source–sink metaecosystem dynamics result from asymmetries in the spatial flow of nutrients between ecosystems. The balance between inorganic nutrient flows (direct nutrient inputs) and organic matter flows (indirect nutrient inputs, requiring the breakdown of detritus) determine the net direction of nutrient flows between ecosystems with contrasting productivities, and thus affects whether an ecosystem is a net exporter (a source) or a net importer (a sink). Ecosystems with high productivity serve as organic matter sources to less-productive ecosystems (organic matter sinks), but because high productivity depletes the availability of limiting nutrients in those ecosystems, less-productive ecosystems might have higher nutrient concentrations and thus function as inorganic nutrient sources. The relative strength of these direct and indirect spatial flows of nutrients between ecosystems may be altered under different landscape configurations, as well as by loss or replacement of important trophic species in source ecosystems, and can even reverse the source–sink status of a given ecosystem (i.e. turning a sink into a source, or a source into a sink).

18. Ecosystems and landscapes both provide a variety of goods and services, which are generally assessed in terms of human benefits. Ecosystem services are categorized according to four types: (1) provisioning services, the products or goods obtained from ecosystems (e.g. food, fiber, fuel); (2) regulating services, the benefits obtained through the regulation of ecosystem processes (e.g. climate); (3) cultural services, the non-material benefits obtained from ecosystems (e.g. recreation, ecotourism), or sites that have aesthetic, cultural, spiritual, or religious significance; and (4) supporting services, which are necessary for the production of all other services (e.g. soil formation, nutrient cycling, productivity). Ecosystems that provide a variety of services are considered to be multifunctional.

19. Landscapes are also considered to be multifunctional because they provide provisioning, regulatory, and cultural services, in addition to various habitat functions. Habitat functions pertain to the maintenance of biological and evolutionary processes on the landscape, as well as to the effects of landscape structure and human land use on ecological functions. This emphasis on habitat functions and how landscape structure interacts with ecological or biophysical processes is what distinguishes landscape function from ecosystem function.

20. A disruption of landscape function can lead to landscape dysfunction. For example, chronic overgrazing and drought can truncate the patch-

size distribution of vegetation in arid grasslands, interfering with the ability of vegetation patches to intercept water, sediment, nutrients, litter, or seeds that runoff or blow across the landscape. Such 'leaky' landscapes are less productive and sustainable than functional landscapes that capture and retain water and nutrients, and which are thus more highly conserving. In its most severe form, a disruption of landscape function can manifest as a system state change (e.g. desertification), which in some cases may be irreversible.

21. Warning signs of impending landscape dysfunction may include: an alteration of landscape structure, such as in the patch-size distribution of vegetation or a reduction in landscape heterogeneity, especially if this occurs as a rapid or abrupt change (i.e. a threshold response); an alteration of landscape connectivity, which may also occur as a threshold response to landscape change; or system state changes, in which environmental conditions have been altered to such a degree that the system suddenly shifts to a different state (e.g. from grassland to desert).

22. Given that many types of state changes are undesirable, such as the desertification of grasslands or the eutrophication of lakes, it would be ideal from an ecosystem management standpoint if we could predict when such catastrophic state changes are imminent. Changes in the spatial structure or connectivity of the landscape (a disruption of landscape function) or in measures related to the rate of ecosystem recovery (critical slowing down), such as rising variability in some state variable as the system approaches the critical transition threshold, might provide early warning of impending state changes.

23. Recovery or restoration of an ecosystem will require far more than just a return to the environmental conditions that existed prior to the threshold. In other words, different critical thresholds define the initial state change versus the return of the system to its former state. The road to system recovery thus lies along a different trajectory than the one that led to its degradation, a condition referred to as hysteresis.

24. Sustainability represents the greatest challenge of the 21st century. Landscape ecology can contribute to the scientific basis for sustainable development, given that humans are the primary driver of landscape change and ecological processes are profoundly affected by landscape structure and dynamics. The landscape—as defined by the scale of human land-use activities—is thus the basic spatial unit or operational scale for managing and achieving sustainability.

25. Human land use is the key activity that determines to what degree landscapes can simultaneously support or provide a variety of environmental, social, and economic functions, that is, the extent to which landscapes are multifunctional. In human-dominated landscapes, such as agricultural production landscapes, landscape multifunctionality has traditionally been obtained through land sparing, in which separate areas are set aside for different purposes (e.g. crop production versus conservation areas). Conversely, the joint production of both agricultural commodities and biodiversity might be achieved through land sharing, by integrating production and conservation within the same heterogeneous landscape (i.e. by managing for multifunctional landscapes).

26. The goal of landscape sustainability is to provide landscape-specific ecosystem services that are essential for maintaining and improving human well-being in the face of environmental and socioeconomic changes. Landscape resilience, the ability of the landscape to tolerate disturbances without affecting its capacity to provide ecosystem services, is thus a key component of landscape sustainability. Because spatial heterogeneity enhances landscape resilience, it might be the key to achieving landscape sustainability.

Discussion Questions

1. Using your study system or an example provided by your instructor, try to identify the potential for spatial subsidies to occur between different habitats or ecosystems within the landscape. How might the magnitude and/or direction of these subsidies change over time (e.g. seasonally or from year to year)?

2. What mobile-link species occur within your study system, and what functions do they provide? If mobile-link species have a substantial effect on the structure and function of ecosystems, might they also be considered keystone species (**Chapter 3**)? What is the difference between a keystone species and a mobile-link species? (Hint: Think about relative differences in mobility, kinds of effects, top-down vs bottom-up controls on ecosystems, etc.).

3. Given the potential for topoedaphic effects on ecosystem structure and function, how might this

influence source–sink metaecosystem dynamics in hilly landscapes (e.g. **Figure 11.3**)? What role might mobile-link species play in metaecosystem dynamics? Which type of mobile-link species—a resource linker or process linker—would likely have the greatest effect on source–sink metaecosystem dynamics?

4. Make a table that identifies the different ecosystem services and landscape functions provided by your study system. In what ways might this landscape be considered multifunctional?

5. How would you advise monitoring landscape function in your study system? What indicators should an ecosystem manager use to give advance warning of a catastrophic state shift in this system, and why?

6. Explain how spatial heterogeneity is important to the resilience of the landscape or ecosystem in which you live or work. In what ways could sustainability be enhanced in this system?

7. This last one is more of an action item than a question for discussion: Locate the nearest regional chapter of the International Association for Landscape Ecology (www.landscape-ecology. org) and consider joining! Better yet, make plans to attend and to present your research at the next meeting of your regional chapter, or even, at the quadrennial IALE World Congress.

Glossary

4-cell neighborhood rule Neighborhood rule in which only the four cells that share an edge with the focal cell (the four nearest-neighbors) are considered adjacent; used to define patches in a **raster data** model. Synonymous with 4-cell rule.

8-cell neighborhood rule Neighborhood rule in which the four cells that share an edge (the four nearest-neighbors), along with the four cells that are diagonal to the focal cell (the next-nearest neighbors), are all considered adjacent; used to define patches in a **raster data** model. Synonymous with 8-cell rule.

active remote sensing Sensor that emits its own electromagnetic radiation to illuminate the Earth's surface. Examples include **lidar** and **synthetic aperture radar (SAR)**. Contrast **passive remote sensing**.

actual connectivity Realized connectivity of a landscape, assessed using detailed empirical information on movement responses to landscape structure (e.g. movement rates within different habitat types, response to habitat edges) or from evidence of gene flow. Compare **potential connectivity**.

adaptive loci See adaptive markers.

adaptive markers Loci or DNA segments (molecular markers) that directly influence individual fitness, and which are therefore assumed to be under selection.

added risk Proportional increase in the likelihood of an event or probability of occurrence relative to a reference condition.

afforestation Planting of trees in areas that have not supported forest in recent times.

agricultural extensification Clearing of natural land covers or habitats for the purposes of agricultural production, thereby extending the area under cultivation. Compare **agricultural intensification**.

agricultural intensification Increase in the productivity of existing agricultural lands, such as through the application of high-yielding crop varieties, irrigation, fertilizer, and pesticides. Compare **agricultural extensification**.

agri-environment schemes (AES) Incentives provided to farmers and landowners to restore or enhance natural and seminatural habitats within agricultural landscapes so as to conserve native species and **ecosystem services** (e.g. water and soil quality). In the EU, AES are the main tool and source of funding for conservation under the Common Agricultural Policy.

Allee effect In very small populations, the per capita growth rate (r) declines below some critical population density due to the greater difficulty in finding mates (mate limitation).

allele Alternate forms of a gene or homologous DNA sequence. Examples: A_1 and A_2 are different alleles for a gene; ACTGAA and ACTGAT are different alleles for a homologous DNA sequence.

allele frequencies Relative proportion of different **alleles** at a given **locus**.

allelic richness (A) Average number of **alleles** at a **locus**, standardized by sample size.

allogenic engineer Ecosystem engineers that change the environment by transforming living or non-living materials from one physical state to another. Example: beavers (*Custor canadensis* and *C. fiber*) gnaw and fell trees, which they use to construct dams to create ponds. Compare **autogenic engineer**.

allozyme Different structural forms of certain enzymes that are produced by changes in the DNA sequence that codes for that enzyme; used as a **molecular marker** to assess genetic variation within populations in early population **genetics** research.

alpha diversity (α diversity) **Species diversity** of a given community or site; local-scale diversity. Compare **beta diversity**; **gamma diversity**.

amplified fragment length polymorphisms (AFLPs) Type of **molecular marker** in which the entire DNA molecule is fragmented into different lengths using restriction enzymes; each fragment constitutes a **molecular marker**, whose presence varies among individuals. This approach produces tens to hundreds of markers per individual, resulting in a 'barcode' that depicts the presence or absence of the different fragments, giving each individual its own unique DNA fingerprint. Because the approach is based on the presence or absence of markers, it cannot be used to identify heterozygotes.

analytical hillshading **Topographic function** that uses a **digital elevation model** within a **geographical information system** to calculate the location of shadows based on the sun's position so as to illuminate the terrain and create a virtual relief of the landscape.

anisotropic Spatial pattern (landscape) that exhibits different properties depending on the direction of sampling or analysis. Contrast **isotropic**.

anthrome A human-dominated or anthropic landscape.

apparent competition Two species that compete for the same limiting resource may still be able to coexist if the superior competitor is more strongly limited by a shared natural enemy; because one competitor declines in the presence of the other, this might otherwise appear to be a competitive interaction, rather than one mediated by predation. Compare **competitive exclusion principle**.

area effects Ecological responses that are influenced by the size of the geographic or habitat patch area; proposed as an explanation for **elevational** and **latitudinal diversity gradients** (e.g. larger areas support more species than smaller areas). Compare **geographical area hypothesis**; **species–area relationship**.

area function Basic spatial measure featured in many landscape metrics and spatial analyses; a key function within the analysis subsystem of a **geographical information system**.

area-restricted search Behavioral response to a concentrated (patchy) resource distribution, in which individuals slow down and/or increase their turning frequency so as to remain within the resource patch.

aspect Direction a slope faces; one of the **topographic functions** that can be calculated from a **digital elevation model** in a **geographical information system**.

attribute data Non-spatial information that describes the characteristics of locations within a spatial data set. Attribute data comprises either **nominal**, **ordinal**, **interval**, or **ratio data**.

attribute error Type of error that affects the **data accuracy** of a spatial dataset, owing to the assignment of incorrect information to a particular cover type or landscape feature. May occur because of observer bias; sampling errors; problems associated with the instrumentation; or data-entry errors.

autogenic engineer **Ecosystem engineers** that change the environment through their own physical structure. Examples: forests, coral reefs, kelp forests. Compare **allogenic engineer**.

basic reproduction number (R_0) Number of secondary infections caused by a single infective individual during the period that it is infectious; epidemic spread is able to occur when $R_0 > 1$. See also **epidemic threshold**.

Bayesian genetic clustering An approach used in **landscape genetics** for delineating population genetic structure a posteriori, using the methods of Bayesian statistical inference. See also **spatial Bayesian cluster analysis**.

behavioral landscape ecology Field of study that investigates how individuals respond behaviorally to **landscape structure**, such as in their movement responses to habitat edges or propensity to move through different land covers of the landscape matrix.

beta diversity (β diversity) Degree to which species composition differs or changes among communities or sites; assayed either by the degree of **species turnover** between communities or via a statistical analysis of community similarity. Compare **alpha diversity**; **gamma diversity**.

biased correlated random walk **Correlated random walk**, in which there is a greater probability (bias) of moving in a particular direction, such as when orienting toward patches. See also **simple random walk**.

BIDE factors Demographic factors affecting population growth: birth rate (B), immigration rate (I), death rate (D), and emigration rate (E). Populations increase when $(B + I) > (D + E)$.

biocontrol Use of natural enemies (predators, parasitoids, pathogens) to control invasive or pest species. Portmanteau of biological control.

biodiversity Typically defined as the number of species in a given area, but also encompasses genetic and community or ecosystem (habitat) diversity; the sum total of all biological diversity across all levels (genetic, species, community, or ecosystem diversity). Portmanteau of biological diversity.

biogeography Discipline that studies the geographical distribution of plants and animals. Portmanteau of biology and geography.

biogeomorphology Discipline that studies how organisms contribute to landscape formation through their influence on erosional and depositional processes (e.g. by altering runoff and sedimentation rates). See **bioturbation**; compare **ecosystem engineer**.

biological legacies Organisms or their propagules that survive a disturbance, and which thus facilitate the recovery of an ecosystem following that disturbance. See **ecological succession**.

biotic disturbances Disturbances caused by organisms or their activities and which may contribute to landscape heterogeneity. Examples: grazing; burrowing; forest pest outbreaks. See **bioturbation**; **pyric herbivory**.

bioturbation Reworking of soils and sediments by animals or plants; transport of material from below ground to the surface, as by tunneling or burrowing animals. See **biogeomorphology**.

blocked-quadrat variance Type of moving-window analysis for detecting spatial structure across a range of scales. The variance among samples is calculated within non-overlapping blocks (boxes or windows) that are moved across the landscape; variance is calculated within blocks of different sizes (where block size increases as a power of 2), and a plot of variance versus block size (scale) can reveal the scale at which patchiness occurs (i.e. where variance peaks). Compare **local quadrat variance**; **two-term local quadrat variance**; **three-term local quadrat variance**.

Bray–Curtis dissimilarity coefficient (C_{BC}) A measure of **species turnover** that also incorporates the relative difference in species abundances between two communities; a measure of **beta diversity**. See also **Jaccard coefficient**; **Sørensen coefficient**.

broad scale A phenomenon or process that occurs across a large area or over a long time period. See **extent**.

buffering A type of **neighborhood function** that can be performed on **spatial data layers** in a **geographical information system**, in which a zone within a fixed distance (the buffer) of a landscape feature is created. For example, buffering may be used to delineate sensitive areas around wetlands or waterways for protection (e.g. riparian buffer zones).

candidate-gene approach **Allele frequencies** for a gene known a priori to be responsible for a given trait are analyzed to assess adaptive genetic variation among populations; the 'gene-first' approach.

carbon cycle Major global biogeochemical cycle in which carbon is exchanged among the atmosphere (CO_2), biosphere (all living organisms), soil, rocks (limestone), and the oceans through various processes and rates of uptake or release.

categorical data Data representing discrete categories or types (e.g. land-cover or land-use types). See **nominal data**.

categorical map A **spatial data layer** in which individual grid cells (**raster data** model) or polygons (**vector data** model) are classified into different categories (e.g. land-cover or land-use type). Synonymous with thematic map; chloropleth map.

catena Sequence of soils that occur along the length of a hillslope (i.e. from top to bottom).

catena concept Biogeochemical differences between upland and lowland sites are assumed to reflect the redistribution of soil constituents (mineral materials, organic matter) as a result of hillslope hydrology and erosional/depositional processes. For example, soils at the bottom of a hill tend to be deeper, heavier in clay content, richer in organic matter, and possess a higher soil moisture content than those at the ridgetop.

chloroplast DNA (cpDNA) Genetic information found in the chloroplasts of plant cells; usually maternally inherited, though some groups may exhibit paternal and even biparental modes of inheritance.

chronosequence In landscape ecology, a comparative analysis of landscape structure at different points in time in order to characterize the pattern of **landscape change**.

climate envelope model Type of **species distribution model** that depicts the geographic distribution of a species' climatic niche. The climatic niche includes such factors as temperature and precipitation that might limit a species' range. See also **ecological niche**.

climatic stability hypothesis One of the hypotheses for the **latitudinal diversity gradient**, which posits that greater diversity occurs in the tropics because this region has experienced greater climatic stability over evolutionary time relative to temperate regions, which have experienced repeated cycles of glaciation when large regions became uninhabitable for extended periods of time. See also **time-for-speciation hypothesis**.

climax state Mature vegetation assemblage that develops in a particular area over time; once thought to represent the natural end point of **ecological succession**.

common garden experiments To test if there is a genetic component to the phenotypic differences observed among populations or species found at different locations, individuals are moved from their native environments and raised within the same environment; any differences that emerge are then assumed to be under genetic rather than environmental control. These experiments have typically been performed with plants, which are transplanted into a 'common garden,' hence the name for this type of experiment.

competitive exclusion principle Species that share the same **ecological niche** should not be able to coexist, as the better competitor will displace the other(s) if species are competing for the same limiting resource. Often summarized as 'complete competitors cannot coexist.'

competitive reversal Superior competitor in one habitat type that is an inferior competitor in others; a form of **spatial niche differentiation** that permits coexistence among ecologically similar species at a landscape scale.

complete spatial randomness Realization of a spatial Poisson process (the simplest **spatial point process**), in which the locations of points are independent and follow a Poisson distribution (i.e. a random distribution, in which the variance is equal to the mean).

compounded disturbance Different types of disturbances may interact synergistically, possibly giving rise to unexpected events ('ecological surprises').

connectivity conservation Protection, management or restoration of ecological flows and functional linkages among habitat patches or metasystems (e.g. **metapopulation**; **metacommunity**; **metaecosystem or metalandscape**). See **landscape connectivity**.

coordinate thinning Data-processing task within a **geographical information system**, in which excessive detail (data points) is removed from lines and polygons in a **vector** dataset, giving shapes a smoother appearance.

correlated random walk Derivation of a **simple random walk**, in which individuals exhibit directional persistence over short distances. Although the direction of movement between successive steps is correlated, the overall direction of movement is still random. See also **biased correlated random walk**.

cradle of life hypothesis One of the hypotheses proposed for the **latitudinal diversity gradient**, which posits that there are so many species in the tropics because most groups of organisms originated there and so have been able to diversify over a longer period of time. See also the **time-for-speciation hypothesis**; **museum of biodiversity hypothesis**.

cross-K function Bivariate form of the **Ripley's K-function**, in which the joint distribution of two spatial point patterns is compared. See also **L-function**.

cross-scale interactions Processes at one spatial or temporal scale that interact with processes at another scale. Example: land-management practices operate at regional scales but may influence soil nutrient cycling at a local scale.

critical node In a landscape **graph**, a habitat or population **node** that is highly connected to other nodes and thus important for preserving overall network connectivity. See **landscape connectivity**.

critical slowing down As a system approaches a **critical threshold**, it may take longer to recover from disturbances or other perturbations (i.e. a **lagged response**); a possible early warning sign of an impending **system state change**.

critical threshold Point beyond which large, qualitative changes occur in the system of interest. May occur as a non-linear response, in which small changes in the vicinity of the threshold produce dramatic changes in system state or behavior. Example: **percolation threshold**. See also **regime shift**; **system state change**.

cross-boundary subsidies The productivity of a freshwater or marine ecosystem supports the productivity of a terrestrial ecosystem, or vice versa. See **spatial subsidies**; **mobile-link species**.

crown fire Most severe type of forest fire, in which fire burns through the upper canopy and thus has the potential to kill trees, potentially leading to stand replacement. See **fire severity**; compare **surface fire**.

cultural eutrophication Increased nutrient enrichment of aquatic ecosystems due to humans, such as from stormwater, wastewater, and agricultural runoff.

cultural services Non-material benefits obtained from **ecosystems**, such as opportunities for recreation, education or ecotourism, or sites that have aesthetic, cultural, spiritual, or religious significance; a type of **ecosystem service**.

data accuracy A measure of how close the recorded data are to their actual values. The accuracy of remote-sensing data is affected by both **attribute errors** and **positional errors**. Compare **data precision**.

data precision Repeatability or reproducibility of data obtained under similar conditions; the range of variability or uncertainty associated with the readings obtained for a given location. Compare **data accuracy**.

database management system (DBMS) Helps to streamline the editing, updating, and manipulation of both spatial and attribute data.

deme In genetics, a group of interbreeding individuals; the individual population or subpopulation.

demographic potential (k) Capacity of the population to increase, based on its reproductive output and dispersal abilities. Originally derived mathematically to give the fraction of habitat occupied by the **metapopulation** when the entire landscape is suitable (e.g. $k = 0.6$ means that 60% of suitable habitat is occupied when the landscape is entirely suitable).

deterministic interpolation methods Spatial interpolation method in which the weighting function is determined by some method, algorithm, or user-defined input. See **spatial moving average**; **inverse distance-weighting function**. Compare **geostatistical interpolation methods**.

diffusion coefficient (D) Mean squared displacement of individuals per unit time, assuming random diffusion. See **simple random walk**.

digital elevation model (DEM) A three-dimensional representation of a terrain's surface created from elevation data obtained from topographic maps produced via photogrammetric methods (e.g. the analysis of overlapping aerial or satellite images taken at slightly different locations) or **lidar**. A DEM is often used to correct terrain distortion in aerial and satellite imagery (i.e. **orthorectification**).

dilution effect In epidemiology, a reduction in the incidence or risk of disease as a result of high host diversity.

diploid Having two sets of homologous chromosomes, one set inherited from each parent. For example, humans are diploid organisms, with 23 pairs of homologous chromosomes (46 chromosomes in total).

directed dispersal That movement of individuals or their propagules toward a particular habitat or environment that is likely to be favorable for their survival or germination; may be facilitated by other dispersal agents (e.g. ants take seeds into their mounds, where they may subsequently germinate).

directed graph (digraph) Landscape **graph** in which linkages (**graph edges**) reflect the net direction of movement among **nodes** (habitat patches or populations).

disease nidality Distribution or areas of disease occurrence, resulting from the interaction of the pathogen, host, vector, or disease reservoir with the environment; the disease nidus is analogous to the **ecological niche** of a disease.

disease risk Likelihood of exposure to, or infection by, a particular pathogen or infectious disease agent.

disease risk mapping Application of statistical models and computerized mapping technologies (e.g. a **geographical information system**) to analyze and display spatial variation in disease risk, by deriving relationships between the distribution of vectors, reservoir species, hosts, or disease incidence and various landscape and environmental factors.

disease spillover Transmission of disease from a reservoir species, where the pathogen can attain high levels, to another host species.

dispersal Movement of individuals or their propagules from one habitat patch or population to another.

dispersal kernel **Probability density function** of dispersal distances for an individual or population. See **dispersal neighborhood**.

dispersal neighborhood Distance over which individuals can disperse or the local area over which propagules are distributed, which may be modeled as a distance-decay function of various types (e.g. negative exponential, leptokurtic, Weibull distribution). See **dispersal kernel**.

dispersal threshold Habitat level below which dispersal success abruptly declines.

distance function A basic spatial measure featured in many landscape metrics and spatial analyses; a function within the analysis subsystem of a **geographical information system**.

disturbance regime Spatial and temporal dimensions of a disturbance, which contribute to landscape heterogeneity and dynamics. See **spatial heterogeneity**; **temporal heterogeneity**.

diversity partitioning For hierarchical sampling schemes, in which samples are collected within a nested set of scales (e.g. sampling sites nested within habitat patches nested within landscapes), **species diversity** can be partitioned among these scales in either an additive or multiplicative fashion to determine whether or how diversity is spatially structured.

domain of scale Range of scales bounding a particular phenomenon or process in space or time (**grain**; **extent**); the range of scales over which a particular pattern-process relationship holds. See **variance staircase**.

drift-migration equilibrium Level of **migration** (**gene flow**) needed to offset genetic differentiation that would otherwise emerge in small populations via **genetic drift**.

drought Period of below-normal precipitation. There are for major types of drought: meteorological, agricultural, hydrological, and socioeconomic. Droughts are a type of disturbance and thus are defined by their intensity, spatial extent, and duration. See **drought intensity**; **drought spatial extent**; **drought duration**.

drought duration Number of years in which drought of a certain magnitude persists over a given area (**spatial extent**). See **drought intensity**; **drought spatial extent**.

drought intensity Magnitude of the precipitation, soil moisture, or water-storage deficit. Example: Palmer Drought Severity Index (PDSI).

drought spatial extent Area covered by drought; mapped using a measure of drought intensity (e.g. PDSI).

dynamic equilibrium As applied to landscapes, the relative constancy in a particular attribute (e.g. composition) over time despite ongoing disturbance dynamics. See **shifting-mosaic steady state**.

ecocline A physical gradient, such as in soil type, precipitation, or temperature, that gives rise to different ecosystems or habitat types. See **ecotone**.

ecological equivalent A species that fulfills the same or similar ecological role or ecosystem function as another. Example: domestic cattle are the ecological equivalents of bison (*Bison bison*) in many respects in the tallgrass prairie ecosystem.

ecological flows Movement of organisms, nutrients, matter, genes, or energy across a landscape.

ecological forecast/forecasting Use of ecological information or theory to predict the potential impact of certain factors (e.g. climate or land-use change) on future system structure or behavior, usually for the purposes of management or conservation. See **extrapolation**.

ecological niche All of the environmental conditions required by the species to persist; alternatively, the role of the species within its community or ecosystem. See **fundamental niche**; **realized niche**.

ecological niche model Type of **species distribution model** that attempts to describe the **ecological niche** (or at least the **realized niche**) of a species by relating its occurrence to various environmental features of the landscape. The ecological niche model is then projected onto the same or different landscape to identify regions that meet the species' niche requirements, thereby producing a hypothesized distribution for that species. See also **fundamental niche**.

ecological resilience Ability of an ecological system to recover from a disturbance. Sometimes defined as the amount of disturbance necessary to push the system from one state to another or as the ability to resist a **system state change** (**regime shift**). Compare **resistance**.

ecological succession Process by which communities develop and change over time, especially following a disturbance. Initially conceptualized as a series of developmental stages of increasing vegetation complexity toward a mature **climax state**. See **primary succession; secondary succession.**

ecological trap Habitat of marginal quality that is nevertheless preferred by a species. Compare **perceptual trap.**

ecosystem Ecological community and its abiotic environment, which are linked via the fluxes of matter and energy.

ecosystem engineer Organisms that directly or indirectly affect the availability of resources to other species, often modifying, creating, or maintaining habitat in the process. Compare **biogeomorphology.** See **allogenic engineer; autogenic engineer.**

ecosystem function Variously defined as the processes that regulate the fluxes of matter and energy within ecosystems; the extent to which those processes are influenced by species (the functional role of species; **functional diversity**); the functioning of the entire **ecosystem;** or **ecosystem services.**

ecosystem services Provisioning, regulating, supporting, or cultural benefits provided by **ecosystems.** See **provisioning services; regulating services; cultural services; supporting services.**

ecosystem structure Diversity of organisms that make up the ecological community as well as the physical features of the environment.

ecotone Naturally occurring boundary or transition between two ecosystems or habitat types. May reflect an underlying physical gradient. See **ecocline.**

edge effects A change in the environment or in an ecological response as a function of distance to a habitat edge or other ecotonal boundary. Although edge effects are often considered to be negative, owing to increased predation, parasitism, competition, or disease transmission along habitat edges, many species exhibit no response or can even have a **positive edge response.** See **edge species.**

edge matching A data-processing task within a **geographical information system,** which ensures that adjacent images are properly aligned (i.e. that spatial features along the edges match up).

edge species Species that show a positive response to habitat edges or other ecotonal boundaries. Contrast **interior species.** See also **edge effects; positive edge response.**

edge permeability Degree to which movement or other ecological flows can occur across a habitat patch boundary, assessed in terms of movement responses to the habitat structure or environmental conditions at the patch edge. See **hard edge; soft edge.**

edge tortuosity Complexity of the habitat patch-edge boundary, which may vary from simple (a straight line) to more irregular in shape, and thus may affect **edge permeability.**

effective evolutionary time hypothesis One of the hypotheses for the **latitudinal diversity gradient,** which posits that there are more species in the tropics because speciation (evolutionary) rates are effectively higher because the higher temperatures increase physiological processes that cause faster mutation rates as well as a more rapid response to selection pressures (e.g. shorter generation times). See also **metabolic theory of ecology.**

effective gene flow Movement and establishment of novel genes that are not currently present in the recipient population.

effective isolation Relative degree of isolation between two patches or populations, based on the likelihood of movement or dispersal through the intervening matrix. A patch or population is effectively isolated when the intervening matrix is hostile to dispersal, either because dispersers are unwilling or unable to traverse the matrix or because they suffer high mortality if they do. See **matrix resistance; patch connectivity; isolation by resistance.**

effective number of alleles (n_e) Number of **alleles** that would give the same level of **observed heterozygosity** at a given locus if all alleles were equally abundant.

effective population size (N_e) Size of an idealized population (one in **Hardy–Weinberg equilibrium**) that loses **heterozygosity** at the same rate as the observed population; N_e is usually less than the population size (N).

effective resistance A measure of **patch connectivity** based on the degree to which two patches or populations are isolated given the resistance of the intervening matrix to dispersal or gene flow and the number of alternative pathways between them. Synonymous with **resistance distance.** See **matrix resistance; effective isolation; isolation by resistance.**

effective species richness (S_e) Level of species richness expected if all species were equally abundant. See **evenness index.**

elevational climate model Mid-elevational peaks may be observed in the **elevational diversity gradient** for some taxa owing to the interaction between temperature (which decreases linearly with increasing elevation) and a unimodal water availability gradient (which peaks at middle elevations). According to the model, **species richness** should peak at mid-elevations in arid or temperate mountains because diversity is limited by water availability at low elevations but by extreme cold at high elevations.

elevational diversity gradient Species richness is expected to vary as a function of elevation, owing to changes in temperature and precipitation. See **elevational climate model.**

endemics–area relationship Species–area relationship based on species that are found only within that particular geographic region (i.e. endemic species).

enemy release Introduced species may become invasive in a new environment owing to decreased regulation by natural enemies (i.e. predators, parasitoids, or pathogens).

energy hypothesis Corollary of the **productivity hypothesis**, in which annual solar insolation (energy from the sun) is ultimately responsible for the greater diversity observed in the tropics relative to polar regions (i.e. the **latitudinal diversity gradient**).

Enhanced Thematic Mapper Plus (ETM+) Sensor carried aboard Landsat 7, which measures eight spectral bands that span the visible, near infrared, mid-infrared, and far infrared wavelengths, and provides spatial resolutions of 15 to 60 m. See also **Thematic Mapper; Multispectral Scanner System**.

epidemic threshold Threshold level above which disease spread can occur within a population. See **basic reproduction number**.

erase function In geographical information systems, a type of overlay function typically applied to **polygon-on-polygon overlays** that is used to identify areas that meet one criterion but not another; erase acts like the Boolean operator, NOT.

Euclidean nearest-neighbor distance (ENN) Straight-line distance between two habitat patches or populations, sometimes used as a measure of **patch connectivity**, although distance to the nearest occupied patch is a better measure. Synonymous with **nearest-neighbor distance**.

evenness index (E) Measure of **species diversity**, which quantifies how equal species are numerically within a community. A community in which all species are equally abundant has maximum evenness. An evenness index can be calculated by comparing a diversity index, such as the **Shannon diversity index (H′)**, to the expected value for that diversity index if all species were equally abundant (H'_{max}).

evolutionary landscape epidemiology Field that combines landscape epidemiology and evolutionary theory to understand how the ecology and evolution of host-pathogen interactions shape the distribution, dynamics, and severity of disease across heterogeneous landscapes.

evolutionary landscape genetics Study of how landscape heterogeneity shapes adaptive genetic variation.

exhaustive survey Comprehensive survey of all locations on a landscape.

expected heterozygosity (H_e) Proportion of heterozygotes expected to occur within a population that is in **Hardy–Weinberg equilibrium**. Compare **observed heterogeneity**.

exploitation/exploitative competition Indirect form of competition over a common limiting resource, in which the use of resources or space by one species depletes the amount available to others. Contrast **interference competition**.

extent Total area of coverage or duration of a study. See **spatial extent; temporal extent**.

extinction debt Eventual loss of species following habitat destruction or other disturbance, which may take many years, decades, or even centuries to play out; a type of **lagged response** to landscape change.

extinction threshold Minimum amount of habitat required for population or **metapopulation** persistence; the critical level of habitat at which the metapopulation goes extinct (i.e. $\hat{p} = 0$).

extrapolation Extending information beyond the spatial or temporal extent of the data set in order to make predictions about system structure or behavior. See **ecological forecasting; transferability**. Contrast **interpolation**.

F-statistic A measure of the relative difference between **observed** and **expected heterozygosity**; the F-statistic can be calculated at different levels (within versus between populations) to evaluate whether populations exhibit genetic structure. See F_{ST} **statistic; inbreeding coefficient**.

F_{ST} statistic Measures the average degree of **heterozygosity** observed among populations relative to that expected across all populations assuming no genetic structure. Compare **inbreeding coefficient** (F_{IS} statistic).

fault-block mountains Mountains created through the vertical movement of adjacent blocks of the Earth's crust along a fault line. Part of the block is thrust upward (the horst), while other portions of the fault block may 'drop' or subside (graben). See **rifting**.

fine scale Resolution of a study or data set (**grain**); a phenomenon or process that occurs frequently or locally.

finite population growth rate (λ) Relative change in population size from one year to the next (N_{t+1}/N_t). Also obtained as the dominant eigenvalue of the population matrix (see **matrix population model**). The population is increasing when $\lambda > 1$; stable when $\lambda = 1$; and decreasing when $\lambda < 1$. Compare the **intrinsic population growth rate, r**. Mathematically, $\lambda = e^r$, where e is the base of natural logarithms (e ~ 2.718).

fire-adapted species Species that have evolved life-history traits that enable them to withstand the effects of fire, and may even require fire to germinate, establish, or reproduce. Examples: big bluestem (*Andropogon gerardii*) of the tallgrass prairie ecosystem; serotinous conifer species (e.g. lodgepole pine, *Pinus contorta*). See **fire-dependent ecosystem**.

fire-dependent ecosystem Ecosystems that are especially prone to fire, whose structure and functioning are largely shaped by the **fire regime**, and which might even convert to some other ecosystem type in the absence of fire (i.e. a **regime shift**). Examples: tallgrass prairie; lodgepole pine forests. See **fire-adapted species.**

fire regime Disturbance regime of fire in a given ecosystem, characterized in terms of frequency, size, intensity, severity, and spatial pattern, among other parameters.

fire severity Impact of fire on an ecological system; degree to which different fuel types have been consumed and/or vegetation has been altered or killed. See **surface fire; crown fire**

first-passage time Measure of search effort within an area, determined by the time required to traverse a given distance.

flood pulse Period of peak water flow in rivers or streams, which may occur seasonally or in response to extreme precipitation events; plays a critical role in the structure and dynamics of riverine systems. See **pulse.**

focus Cumulative area or inference space represented by each data point in an analysis (e.g. whether each datum is a single sample or a mean value across samples).

fold-thrust belt Mountain ranges formed through a combination of geological folding and thrust faulting, in which younger rock layers may be pushed over older rock layers, along an orogenic belt where continental plates collide.

foraging hierarchy Different foraging behaviors may be expressed across a range of spatial and associated temporal scales in response to hierarchical (nested) patch structure.

forecast horizon Time period or spatial extent over which **ecological forecasts** are made.

forecast proficiency Accuracy and uncertainty associated with an **ecological forecast.**

fractal Patterns that are self-similar across a range of scales; a type of **power law** that gives a **fractal dimension** (note, however, that power laws by themselves do not indicate fractal structure).

fractal dimension (D) Patterns in nature exhibit complex structure across a range of scales and are thus not well-defined by simple Euclidean dimensions ($D = 1, 2,$ or 3). Instead, these patterns are often **fractal**, whose geometry lies somewhere in between the traditional Euclidean dimensions (i.e. the dimension is fractional), such as $D = 1.4$.

fractal landscape Natural landforms may exhibit **fractal** properties. May also refer to a class of **neutral landscape models**, in which landscapes are generated using fractal algorithms. Compare **random landscape; hierarchical random landscape.**

fuel mosaic Spatial pattern of flammable materials on the landscape; different fuel types (grass, leaf litter, downed woody debris) differ in their propensity to burn, and in conjunction with topography and weather, contribute to a heterogeneous fuel mosaic that influences the frequency, size, and severity of fire on the landscape. See **fire regime; fire severity.**

functional connectivity Assessment of **landscape connectivity**, based on the interaction between some type of ecological flow (e.g. movement of organisms) and the spatial distribution of habitat or land covers on a landscape. Contrast **structural connectivity.**

functional diversity Range of functional traits represented by species within a community (what species 'do' within **ecosystems**). Compare **species diversity.**

functional heterogeneity (ecosystems) Degree of variability among species in their ability to perform a particular **ecosystem function**. Example: frugivorous birds vary in their abilities to disperse seeds between habitats. Functional heterogeneity can enhance **ecological resilience** through a combination of **functional redundancy** and **response diversity**. See also **functional diversity.**

functional heterogeneity (landscapes) Environmental heterogeneity (spatial variation) that has an effect on some ecological process (e.g. movement behavior). See also **structural heterogeneity; functional connectivity.**

functional redundancy Multiple species performing the same **ecosystem function**; a component of **functional heterogeneity (ecosystems)** that enhances ecosystem resilience.

fundamental niche Full range of environmental conditions under which the species could persist. Contrast **realized niche.**

gamma diversity (γ diversity) Species diversity measured at a landscape or regional scale; the sum total of all species across all communities or sites surveyed. Compare **alpha diversity; beta diversity.**

gap-crossing ability Distance between patches (gaps) that a species is willing or able to traverse, as a function of its dispersal abilities, **perceptual range**, perceived predation risk, and nature of the intervening matrix. See also **matrix resistance.**

gap dynamics Disturbances create a **shifting mosaic** of sites (gaps) that undergo **ecological succession**, which contributes to landscape heterogeneity. See **secondary succession; patch dynamics.**

Geary's *c* Measure of **spatial autocorrelation**, derived as the sum of the squared differences in some variable between points at various lag distances (**spatial lag**), divided by the standard deviation of that variable across the landscape. Tends to be inversely related to **Moran's *I*.**

gene flow Movement of **alleles** or genes between populations via **dispersal**; **migration** in genetics. See also **effective gene flow.**

genetic distance A measure of the degree to which individuals or populations differ in **heterozygosity** or **allele frequencies** at a given **locus** (or averaged across loci). Genetic distance can be assayed using **pairwise** F_{ST} values or Nei's genetic distance.

genetic drift Random fluctuations in **allele** or **genotype** frequencies over time, which are especially pronounced in small or isolated populations, leading to a loss of **heterozygosity**.

genetic marker DNA sequence or gene with a known location on a chromosome that can be used to identify individuals or species; used to assess genetic variation. Synonymous with **molecular marker**. Examples: **allozyme; microsatellite; amplified fragment length polymorphisms; single nucleotide polymorphisms**.

genetics Study of genes, heredity, and genetic variation at an individual or population level. Example: population genetics. Compare **landscape genetics**.

genomics Study of the structure, function, and evolution of the **genome**, the full set of genes within an organism. See **population genomics; landscape genomics**.

genotype Specific **alleles** present at a given **locus**. Example: for a diploid organism that has two alleles at a particular locus (A_1 or A_2), there are three possible genotypes: A_1A_1, A_1A_2, and A_2A_2.

geographical area hypothesis One of the many hypotheses to explain the **latitudinal diversity gradient**, which posits that diversity is greater in the tropics because tropical regions previously had a much greater geographical **extent** than they do today, and large areas tend to support greater richness than smaller areas. See also the **species–area relationship**.

geographical information system (GIS) Integrated computer-based system of software and associated hardware, methods, and personnel involved in the input, management, manipulation, analysis, and display of spatial data.

georeference Assignment of spatial (map) coordinates to an object or landscape feature for the purposes of mapping and spatial analysis. Synonymous with **spatially referenced data**.

geostatistical interpolation methods Interpolation methods in which the spatial data are used to derive a model and set of parameters that are applied to produce a continuous landscape surface. See **kriging**. Compare **deterministic interpolation methods**.

genetic linkers Mobile-link **species** that carry genetic information among **ecosystems**. Examples: seed dispersers or pollinators.

giving-up density Resource or prey density at which optimal foragers should leave a patch in search of a more profitable one elsewhere. See **marginal value theorem**.

giving-up time Optimal amount of time that a foraging individual should remain in a patch, assessed as a trade-off between the costs of searching for an increasingly scarce resource versus the diminishing returns (e.g. energy intake) obtained from that resource. See **marginal value theorem**.

gliding-box method Type of moving-window analysis, in which a box (window) is moved across the landscape one sampling unit (e.g. grid cell) at a time, as opposed to one full box-size unit over, resulting in a more complete sampling of the spatial pattern. The use of different gliding-box sizes can be used to explore how spatial pattern changes as a function of scale (i.e. box size).

global spatial statistics Spatial statistics assessed across the entire landscape, which assumes that the assumption of **stationarity** has been met. Compare **local spatial statistics**.

grain Finest scale of measurement or resolution in a data set. See **spatial grain; temporal grain**.

graph From graph theory, a mathematical structure that describes the relationship among objects (**nodes**) as a function of the connections (**graph edges**) among them. See **landscape connectivity**.

graph diameter The 'longest shortest path' in a landscape **graph**; reflects the total inter-patch distance an organism would have to traverse to span the largest cluster. Compare **percolation cluster**.

graph edge Connections between **nodes** (habitat patches or populations) in a landscape **graph**. Connections may be structural (e.g. habitat corridors) or functional (i.e. the likelihood of movement between two patches). See **structural connectivity; functional connectivity**.

ground-truthing Verification of the classification or positional accuracy of spatial data from information obtained on-site in the field.

habitat-amount hypothesis Posits that **species richness** at a given sampling site (**alpha diversity**) should increase largely as a function of the total amount of habitat in the surrounding landscape rather than as a function of the local patch size in which that sampling site occurs; patch size and isolation are therefore important only to the extent that both factors influence the total amount of habitat in the surrounding landscape.

habitat fragmentation Pattern of **habitat loss** that results in an increased number of habitat patches on the landscape; patch sizes tend to decrease but inter-patch distances may or may not increase, depending on the pattern of fragmentation. Fragmentation also increases the amount of edge habitat on the landscape (see **edge effects**). Although habitat fragmentation typically occurs because of habitat loss, the two effects are not synonymous.

habitat heterogeneity hypothesis One of the hypotheses proposed to explain the **latitudinal diversity gradient**, which posits that diversity is greater in the tropics because **species richness** tends to be greater in more structurally complex habitats (e.g. rainforests); at a landscape scale, total species richness (**gamma diversity**) is expected to be greater in more heterogeneous landscapes than landscapes comprising a single habitat or land-cover type.

habitat loss Decrease in total habitat area on the landscape, which usually results in a reduction of the average patch size and an increase in inter-patch distances. Compare **habitat fragmentation**.

habitat selection Disproportionate use of habitat by a species or organism relative to the amount of that habitat on the landscape.

haploid Having one set of chromosomes, either maternally or paternally inherited; half the diploid number of chromosomes. Examples: gametes of sexually reproducing organisms; the gametophyte of plants, which is the dominant life stage in some groups (green algae, mosses); and male Hymenoptera (wasps, bees, ants).

hard edge Habitat patch boundary that is rarely or never traversed by a particular organism or ecological flow; a boundary that presents a physical (structural), physiological, or psychological barrier to movement. Contrast **soft edge**. See **edge permeability**.

Hardy–Weinberg equilibrium (HWE) Idealized population, in which **allele** and **genotype frequencies** remain constant over time; assumes a large, randomly mating population in which no mutation, natural selection, **genetic drift**, or **gene flow** occurs.

heterozygosity Proportion of individuals within a population that are **heterozygous** (having different **alleles**) at a given **locus**; a measure of genetic variation within populations. See **expected heterozygosity**; **observed heterozygosity**.

heterozygous Genotype in which the **alleles** at a given **locus** are different (e.g. A_1A_2).

hierarchical random landscape Type of **neutral landscape model**, generated using a **fractal** algorithm known as curdling, in which habitat is randomly distributed at each of several nested levels. Compare **fractal landscape**; **random landscape**.

hierarchy theory Theory of system organization in which hierarchical structure emerges as a consequence of interactions that occur among system components at lower levels, giving rise to higher levels of complexity; levels are characterized by both the magnitude and frequency of interactions. See **level**.

historical variability concept Management concept that embraces or tries to mimic the range of spatial and temporal variability inherent in the natural **disturbance regime**, especially that characteristic of the system prior to human settlement and our alteration or control of the natural disturbance regime. Sometimes used as a benchmark for evaluating the effects of human land-use and management practices.

home range Area traversed by an individual within a season, a year, or over a lifetime. When based on the **utilization distribution**, the home-range boundary is defined as the smallest area in which the individual spends 95% of its time. Compare **territory**.

homozygous Genotype in which **alleles** at a given **locus** are the same (e.g. A_1A_1 or A_2A_2).

human footprint In ecology, the degree of human influence on the landscape, quantified in terms of size, severity, and duration; a measure of anthropogenic disturbance.

hysteresis The **critical threshold** at which a system shifts from one state to another may differ depending on which direction the threshold is approached. For example, the amount of habitat needed for a species' recovery (the restoration threshold) may be far more than the amount of habitat at which the species first declined or went extinct (i.e. the **extinction threshold**).

ideal despotic distribution Form of preemptive habitat use, in which socially dominant individuals (despots) prevent subordinates from settling within high-quality habitat patches, resulting in a large discrepancy in fitness between dominant and subordinate individuals. Compare **ideal free distribution**.

ideal free distribution Form of density-dependent **habitat selection**, in which individuals distribute themselves among habitat patches that differ in quality so as to maximize their average fitness. Individuals should choose the better-quality patches up to the point where increasing population density in those patches would lower their expected gain in fitness, such that they would do just as well to settle in a lower quality patch. Compare **ideal despotic distribution**.

identity feature Within a **geographical information system**, a data attribute that is used to characterize or define a certain landscape feature of interest. Example: the boundaries of a nature reserve.

inbreeding coefficient (F_{IS}) The inbreeding coefficient, F_{IS}, measures the extent to which individuals within the population are homozygous ($F_{IS} = 1$) or heterozygous ($F_{IS} = -1$).

inbreeding depression Mating of closely related individuals may increase the frequency of deleterious recessive alleles expressed in the offspring, leading to a loss of fitness and thus genetic diversity (**heterozygosity**) over time. See **inbreeding coefficient** (F_{IS}).

incidence function model **Metapopulation** model that explicitly incorporates information on patch structure, such as the size and proximity of neighboring patches, to assess the probability that a given patch is occupied (i.e. incidence). Colonization and extinction rates are thus patch-specific. See **spatially explicit population model**.

integrodifference equation In ecology, a difference equation that includes population growth at each point on the landscape and an integral operator that accounts for the pattern of dispersal (i.e. a **dispersal kernel**).

interference competition A form of direct competition, in which one organism or species prevents others from using a resource via aggressive behavior or territoriality. Contrast **exploitative competition**.

interferogram In the case of **synthetic aperture radar** interferometry (InSAR), two SAR images of the same landscape are acquired simultaneously from different positions, either from separate systems aboard two flight platforms on different flight or orbital paths or by carrying two widely spaced radar systems on the same platform, to create signals that are slightly out of phase. The phase difference between the SAR images produces an interferogram that can then be analyzed and is used for topographic mapping and the production of **digital elevation models**.

interpolation Statistical procedure for predicting values or occurrences at new locations from known data within a sampled landscape. See **spatial interpolation**. Contrast **extrapolation**.

interpolation function Generates continuous landscape surfaces within a **geographical information system**, using various techniques to estimate values of unsampled locations based on the available spatial data (i.e. **spatial interpolation**). See **triangulated irregular networks; spatial moving average; inverse distance-weighting function; geostatistical interpolation methods; kriging**.

interior species Species that show a negative response to habitat edges and require large blocks of contiguous habitat. Contrast **edge species**.

intersect In **geographical information systems**, a type of overlay function typically applied to **polygon-on-polygon overlays** in which areas that meet two or more criteria are identified; intersect functions as the Boolean operator, AND.

interspersion Measure of how intermixed different patch types are on a landscape, based on the likelihood that cells of different patch types are adjacent (i.e. the unlike-adjacencies).

interval data Numeric data obtained from a measurement scale having an arbitrary (non-zero) reference point (e.g. elevation is measured as distance above mean sea level). See **attribute data**.

intrinsic rate of population increase (r) Difference between the birth (b) and death (d) rates within a population. If $r > 0$, the population is increasing ($b > d$); $r = 0$ when the population is stable ($b = d$); and $r < 0$ when the population is decreasing ($b < d$). Compare the **finite rate of population increase (λ)**. Mathematically, $r = \ln(\lambda)$.

inverse distance-weighting function Method of **interpolation**, in which greater influence is given to points that are near a given location than those that are farther away; a type of **interpolation function** within a **geographical information system**.

isolation by distance In population **genetics**, genetic differentiation among populations is expected to increase as a function of geographic (Euclidean) distance; used as the null hypothesis in **landscape genetics**. Contrast **isolation by resistance**.

isolation by resistance In **landscape genetics**, genetic differentiation among populations is expected to increase as a function of the resistance of the intervening matrix, rather than the geographic (Euclidean) distance between populations. Contrast **isolation by distance**. See **effective isolation; matrix resistance**.

isotropic Spatial pattern (landscape) that varies in a similar way in all directions. Contrast **anisotropic**.

Jaccard coefficient (C_J) Measure of **species turnover** that evaluates the degree of similarity between two communities; a measure of **beta diversity**. See also **Sørensen coefficient; Bray–Curtis dissimilarity coefficient**.

kernel Bivariate normal **probability density function**, used to give either the density of individuals as a function of distance from some source (a **dispersal kernel**) or the intensity of use over an animal's home range (kernel density estimation).

keystone process Ecological process or disturbance that has a disproportionate effect on an ecosystem, without which the system would be fundamentally different. Example: fire is a keystone process in tallgrass prairie ecosystems.

keystone species Species that has a disproportionate effect on community structure or ecosystem function that exceeds its own abundance or biomass. Examples: beaver (*Castor canadensis* or *C. fiber*); bison (*Bison bison*). See also **ecosystem engineer**.

kriging In geostatistics, a method of **spatial interpolation** that uses parameters obtained from the **theoretical variogram** (e.g. the **nugget, range**, and **sill**).

L-function Standardized form of the **Ripley's K-function**, whose value is used to assess whether **point-referenced data** are random, clumped, or evenly distributed. Synonymous with Ripley's L-function.

lacunarity index Index of the size distribution of gaps among habitat patches on the landscape (gap structure). A counterpart to the **fractal dimension**, it is a measure of **landscape texture**.

lacunarity threshold Critical habitat amount at which gaps sizes increase dramatically on the landscape, as assayed by the **lacunarity index**.

lagged response Delay in the population, community, or other ecological response to landscape change. Example: an **extinction debt**.

land sharing Integration of production or other land uses and conservation within the same area. Example: wildlife-friendly farming programs that provide incentives to plant wildflower strips or protect tree rows beside crop fields. See **landscape multifunctionality**. Contrast **land sparing**.

land sparing Separate areas in the landscape are set aside as wildlife habitat or for the protection of diversity (e.g. nature reserves) whereas other areas are devoted to intensive land use (e.g. farmland). Contrast **land sharing**.

land unit Component of a landscape defined in terms of its topography, soils, or vegetation. See **land-unit concept**.

land-unit concept Holistic view of landscapes as 'hierarchical wholes' made up of interacting **land units**; land-survey and mapping approach for characterizing different components of landscapes, especially in terms of their suitability for land use (e.g. crop production).

landform Unique feature of the landscape that contributes to its physical structure or geomorphology (i.e. topographic features). Examples: mountain, canyon, lake, or valley.

landscape change Variation in **landscape structure** or **landscape function** over time.

landscape complementation Landscapes in which two or more essential resources or habitats are abundant and in close proximity. For example, the proximity of foraging areas to breeding habitat contributes to high landscape complementation in many species.

landscape composition Number and relative abundance of different land covers or patch types on the landscape. Measures of landscape composition quantify the diversity or heterogeneity of the landscape. Compare **landscape configuration**.

landscape configuration Spatial arrangement of land covers or patch types on the landscape. Measures of landscape configuration assess the number, size, shape, aggregation, or interspersion of different patch types or land covers on the landscape. Compare **landscape composition**.

landscape connectivity Overall degree to which habitat, resources, or other spatial attributes are connected across the entire landscape; connectivity assessed at the scale of the entire landscape. Contrast **patch connectivity**. See also **structural connectivity**; **functional connectivity**.

landscape context Broader regional or land-use context in which the landscape occurs.

landscape design Application of landscape ecological principles to the planning and development of urban areas, with the goal of enhancing ecosystem services and sustainability to meet societal needs and values.

landscape-divergence hypothesis Posits that communities within a given landscape will exhibit similar dynamics and trajectories of change, which differ from those found in other landscapes of a similar type, causing communities from different landscapes to diverge over time.

landscape diversity Number and relative abundance of different patch or cover types on a landscape; a measure of **landscape composition** or heterogeneity. See **landscape evenness**; **landscape richness**.

landscape dysfunction Loss of one or more **landscape functions**, which may come about from the alteration of **landscape structure**, a disruption of **landscape connectivity**, or as a result of a **system state change** or **regime shift**.

landscape ecosystem ecology Study of how **landscape structure** influences ecosystem processes (i.e. **ecosystem function**).

landscape evenness (E) Proportion of maximum diversity attained by a landscape having a given number of cover types (i.e. richness); a measure of **landscape composition** or heterogeneity. Maximum evenness occurs when all cover types are equally abundant on the landscape ($E = 1$). See **landscape diversity**; **landscape richness**.

landscape evolution Formation and change in the physical or topographic features (terrain) of a landscape over geological time as a consequence of geomorphological processes (i.e. uplifting, rifting, depositional, and erosional processes).

landscape function Flow of energy, nutrients, organisms, or genetic information among spatial elements of a landscape. Landscape function may also refer to its **provisioning**, **regulating**, and **cultural services**, as well as its *habitat functions* related to the maintenance of biological and evolutionary processes, importance for human land use, and how the spatial structure of the landscape affects various ecological processes. See also **ecosystem function**; **multifunctional**.

landscape genetics Field that has emerged at the intersection of population **genetics**, landscape ecology, and spatial statistics to study how microevolutionary processes, such as **gene flow** and **genetic drift**, give rise to genetic variation within heterogeneous landscapes.

landscape genomics Field that builds upon **population genomics**, theoretical population **genetics**, and spatial/landscape analysis to study patterns of adaptive genetic variation across heterogeneous landscapes. The main goal of landscape genomics is to examine the relationship between genomic and environmental data to identify loci that have adaptive significance.

landscape limnology Spatially explicit study of lakes, streams, and wetlands across a range of spatial and temporal scales; the study of freshwater systems within a landscape context to determine how the surrounding terrestrial land covers or human land uses affect the physical, chemical, or biological characteristics of lakes, rivers, and streams.

landscape-mosaic concept View of landscapes as heterogeneous assemblages of different patch types, land covers, or land uses.

landscape multifunctionality Landscapes that are managed to provide a variety of **landscape functions**; typically applied to human-dominated landscapes, such as agricultural landscapes that are managed for the joint production of agricultural commodities and biodiversity.

landscape of fear Spatial variability in the perceived risk of predation, as a function of habitat, predator abundance, predator encounter rate, and the perceived lethality of those predators.

landscape resilience Capacity of a system to tolerate disturbance without exhibiting a **system state change**, which would then affect the capacity of that landscape to provide **ecosystem services** for current and future generations. See also **landscape sustainability; ecological resilience**.

landscape richness (R) Number of different patch or land-cover types on the landscape. Richness is a measure of **landscape composition** and provides a basic measure of **landscape diversity** or heterogeneity.

landscape structure Diversity or spatial distribution of landscape elements (e.g. habitat, soils, land covers, or land-use types).

landscape sustainability Capacity of a landscape to maintain its basic structure and to provide **ecosystem services** despite changing environmental, economic, and social conditions. See **sustainable development; sustainability science**.

lateral connectivity Flows that occur beyond the banks of the river, as a result of periodic flooding of the floodplain. Compare **longitudinal connectivity; vertical connectivity**.

latitudinal diversity gradient General biogeographic pattern in which the numbers of species increases from the poles to the equator; in other words, diversity tends to be greater in the tropics than in the polar regions.

Lefkovitch model Stage-structured **matrix population model**, in which demographic rates (birth and death rates) vary among different life stages.

Leslie model Age-structured **matrix population model**, in which demographic rates (birth and death rates) differ among different age classes within the population.

level In a hierarchically ordered system, this is a level of organization that emerges based on the magnitude and frequency of interactions occurring among lower-level system components; not to be confused with **scale**. See also **hierarchy theory**.

Lévy walk Mathematical model of movement, in which the frequency distribution of step lengths obtained from a movement pathway can be characterized by a power-law distribution (i.e. a **fractal**). Long-distance movements thus occur with greater frequency than expected if movement were completely random. Compare **simple random walk; correlated random walk**.

lidar Form of **active remote sensing** that uses light in the form of a pulsed laser to measure distance to the Earth's surface (the range). Possibly an acronym for 'LIght Detection And Ranging' or a portmanteau of 'light and 'radar.'

life-cycle graph **Graph** in which the **nodes** depict the different age classes or life stages of an organism, and **graph edges** depict the different transitions between them (e.g. stage-specific contributions to fecundity or the probability of surviving from one stage to the next). See also **matrix population model**.

life-zone concept Similar types of plant and animal communities form in response to the variable climatic conditions found along **elevational** or **latitudinal gradients** (life zones).

like-adjacencies In a **raster data** model, grid cells of the same type that share an edge, as defined by the **neighborhood rule**; used to define patches of a given land-cover type.

linearized F_{ST} Measure of **genetic distance** in which the F_{ST} statistic is divided by $(1-F_{ST})$ in order to make the relationship between F_{ST} and geographic (Euclidean) distance a linear one. See **isolation by distance**.

lines In a **vector data** model, features of the landscape that can be represented as one-dimensional or linked data points. Examples: streams, roads, elevation contour lines.

line-in-polygon overlay In a **geographical information system**, a type of **overlay function** applied to **vector data** that identifies which **line** segments intersect a particular type of **polygon** (e.g. roads that occur within the boundaries of a reserve).

local quadrat variance Method for calculating **blocked-quadrat variance**, which uses a range of block sizes (window or box sizes) and averages over all possible starting positions on the landscape. See also **two-term local quadrat variance; three-term local quadrat variance**.

local spatial statistics Statistical approaches in which the degree of spatial dependence is assessed only within a local neighborhood, usually because the **stationarity** assumption cannot be met. Compare **global spatial statistics**.

locus Position on a chromosome for a given gene or DNA sequence (plural loci).

longitudinal connectivity Flows that occur principally in one dimension, such as the flow of water and sediments in riverine networks; the degree to which upstream and downstream sections of a river are connected. Compare **lateral connectivity; vertical connectivity**.

macroecology Study of broad-scale patterns in the abundance and distribution of species over geographical and macroevolutionary scales. Macroecology seeks to uncover general principles that underlie the distribution of **biodiversity** at broad spatial and temporal scales, such as **latitudinal diversity gradients** and metabolic scaling relationships.

mainland-island metapopulation Type of **metapopulation** structure in which the source of colonists is assumed to be a large population that is resistant to extinction (the 'mainland' population), and thus colonization is not influenced by local patch occupancy (the 'island' populations).

majority rule In processing remotely sensed imagery, the spatial resolution can be coarsened by assigning the value of the major cover type to a group of pixels within a sampling window or box (e.g. a 2 × 2 group of cells) that is moved across the landscape image. See **resampling**.

Mantel test Test of the correlation between two distance matrices, such as the relationship between **genetic distance** and geographic distance (i.e. **isolation by distance**). Compare **partial Mantel test**.

mantle plume An upwelling of molten rock within the Earth's mantle, which may rise to the surface and create volcanic hotspots.

map scale Degree or ratio of reduction of a map. For example, a map scale of 1:10,000 (1/10,000) means one unit on the map is equivalent to 10,000 units on the ground (e.g. 1 cm = 10,000 cm). Maps with a small degree of reduction portray more detail and are considered *small scale*, whereas maps with a large degree of reduction depict less detail but cover a larger area and are thus *large scale*.

marginal value theorem Tenet of optimal foraging theory, which predicts that an individual should leave a resource patch at the point of diminishing returns (i.e. when the costs of searching exceed the benefits of finding additional resources or prey items). The marginal value theorem predicts that individuals should stay longer in patches of higher quality (greater profitability) and with increasing travel time (distance) to reach the patch. See also **giving-up time; giving-up density**.

matrix population model Population model in which demographic rates (birth and death rates) differ according to the age or life stage of individuals; demography is thus age- or stage-structured. See **Leslie model; Lefkovitch model**.

matrix resistance Relative degree to which the intervening habitat or land-use matrix resists the movement, dispersal, or gene flow of organisms between habitat patches or populations. See **effective isolation; isolation by resistance**.

mean squared displacement ($\bar{R}_t^2$) Average squared straight-line (Euclidean) distance traveled by a large population of random walkers, which is expected to increase linearly over time. See **simple random walk; diffusion coefficient**.

measured heterogeneity Environmental heterogeneity (spatial variation) that is quantified using landscape metrics or spatial statistics; **structural heterogeneity**. Compare **functional heterogeneity**.

metabolic theory of ecology The uptake and processing of energy by organisms (i.e. metabolism) is a fundamental biological constraint that controls ecological processes at every level of organization, from individuals to ecosystems, and therefore influences rates of biomass production, trophic dynamics, and patterns of species diversity (e.g. the **latitudinal diversity gradient**). See also **energy hypothesis; productivity hypothesis**.

metacommunity Collection of local communities linked by the dispersal of multiple, potentially interacting species.

metacommunity dynamics Spatial dynamics or regional properties (resulting from spatial dynamics) of communities occupying two or more interconnected patches.

metadata file File that contains information about a spatial data set, such as the data source, type of spatial data, method of collection, level of processing, date acquired, spatial resolution, **spatial extent**, data format (**raster** or **vector**), the projection or coordinate system used, and how to give credit for using the data.

metaecosystem Set of ecosystems connected by the spatial flows of energy, materials, and organisms.

metalandscape System of interacting landscape populations connected by dispersal; regional connectivity.

metapopulation System of interacting populations; a population of populations; a spatially distributed population in which populations are connected by dispersal and gene flow.

microgeographic adaptation Finer scale of adaptation than that typically associated with evolutionary divergence (e.g. speciation); local adaptations that emerge within the **dispersal neighborhood** of the organism.

microsatellite Short **nucleotide** sequence consisting of 2–6 DNA base pairs (motifs) that are repeated in tandem many times. Because the number of repeats is highly variable among individuals, this gives rise to numerous **alleles** within a population, and is useful as a **molecular marker** for quantifying genetic variation, as well as for detecting individual dispersal events and contemporary patterns of **gene flow** on the landscape. Heterozygotes can be distinguished from homozygotes, allowing a more robust estimate of genetic variation within populations. These markers are also neutral (**neutral makers**), and thus have been widely used in **landscape genetics**.

mid-domain effect Spatial constraints imposed by the base and summit of a mountain may limit the distribution of species, causing species with large and intermediate-sized ranges to overlap at middle elevations, resulting in a peak in species richness. The mid-domain effect was proposed to explain the hump-shaped distribution in the **elevational diversity gradients** of some taxa. See also **area effects**.

migration In **genetics**, **gene flow** between populations. Elsewhere in ecology, this is the seasonal movement of organisms, such as between their breeding and wintering ranges.

migratory connectivity Different landscapes, regions, or continents that are connected through the seasonal exchange of individuals of a given species through **migration**.

mitochondrial DNA (mtDNA) Genetic information contained within the mitochondria of cells in fungi, plants, and animals; usually maternally inherited, although paternal and biparental modes of inheritance have been found in some species (e.g. Mytilid mussels).

mobile-link species Organisms that move between habitats or **ecosystems** and which provide critical **ecosystem services** and/or support essential **ecosystem functions**, thereby contributing to greater **ecological resilience** as a consequence. See also **resource linkers**; **genetic linkers**; **process linkers**.

Modifiable Areal Unit Problem (MAUP) Analysis of landscape pattern is sensitive to the way in which its areal units (e.g. grid cells or patches) are defined; the areal units are considered modifiable because they can be reclassified or rescaled. See **reclassification**; **resampling**.

molecular marker See **genetic marker**.

monomorphic loci Loci having only one **allele**.

Moran's I Measure of global **spatial autocorrelation**, derived as the **spatial autocovariance** divided by the standard deviation of the measured variable.

Morisita's index (I_M) Measure of aggregation based on point count data (**point-referenced data with attributes**); similar to the **variance-to-mean ratio** but more sensitive to departures from randomness and not as strongly affected by sample size.

movement path Pattern of animal movement; a sequentially ordered series of **steps** between locations (**points**). Synonymous with movement pathway.

multifunctional Landscapes and ecosystems simultaneously provide many different types of functions or services. See **landscape function**; **ecosystem function**; **ecosystem services**; **landscape multifunctionality**.

Multispectral Scanner System (MSS) One of the sensors carried aboard the first five Landsat satellites. The MSS recorded 4–5 spectral bands: green, red, and two near-infrared bands, plus a fifth channel in the far infrared range starting with Landsat 3. See also **Thematic Mapper**; **Enhanced Thematic Mapper Plus**.

museum of biodiversity hypothesis One of the hypotheses proposed to explain the **latitudinal diversity gradient**, which posits that diversity is greater in the tropics because extinction rates are lower than at higher latitudes and so the tropics have simply accumulated more species over time. See also **cradle of life hypothesis**; **climatic stability hypothesis**; **time-for-speciation hypothesis**.

natural-color image In remote sensing, an image produced using a combination of three color spectra (red, green, blue), such that the land-cover types appear naturally colored to the human eye. Compare **false-color image**. Contrast **panchromatic image**.

natural flow regime Pattern of variability in water flow over time, characterized by the magnitude, frequency, duration, timing, and rate of change in water discharge levels; describes the flow dynamics of unregulated riverine systems. See **disturbance regime**.

natural variability concept See **historical variability concept**.

nearest-neighbor distance See **Euclidean nearest-neighbor distance**.

neighborhood function Function within the data manipulation and analysis subsystems of a **geographical information system** that incorporates the influence of neighboring points or cells on a given location, either for mapping purposes or as part of a proximity analysis. See **buffering**; **spatial filtering**.

neighborhood rule Number of cells around the focal cell that are considered adjacent (within the same neighborhood); used to define patches in a **raster data** model. See **4-cell neighborhood rule**; **8-cell neighborhood rule**; **like-adjacencies**.

network analysis Analysis of connectivity within landscape **graphs**. Connections (**graph edges**) may be physical (roads, waterways, habitat corridors) or functional linkages between **nodes**. See also **structural connectivity**; **functional connectivity**.

network robustness Resilience of a landscape **graph** to the removal of **nodes** or **graph edges** before overall connectivity is disrupted. See **critical node**.

network typology Characterization of a landscape **graph** structure to type. See **neutral landscape model**; **network analysis**.

neutral landscape model Theoretical representation of landscape structure, based on some mathematical, graphical, or statistical distribution, used to provide a statistical baseline for evaluating real landscape patterns (as a null model) or for exploring the effect of landscape pattern on ecological processes.

niche conservatism Tendency for species to retain their ancestral ecological traits (i.e. their **ecological niche**) over evolutionary time.

niche partitioning Ecologically similar species may be able to coexist if they use a shared resource in different ways, thereby limiting the degree of niche overlap and hence competition between them. See **competitive exclusion principle**.

nodes Points or polygons (e.g. patches or populations) in a landscape **graph**.

nucleotide Building blocks of nucleic acids (DNA, RNA), consisting of a nitrogenous base attached to a five-carbon sugar (e.g. deoxyribose), which in turn is attached to a phosphate group. In DNA, the nitrogenous bases are adenine (A), guanine (G), thymine (T), and cytosine (C). DNA sequencing thus determines the nucleotide sequence within a DNA segment (e.g. ATTCGT).

nucleotide diversity Average number of **nucleotide** differences within homologous DNA sequences; used as a measure of genetic variation for DNA sequence data.

neutral loci See neutral markers.

neutral markers Loci or molecular markers (regions of DNA) that are not under selection; **genetic drift** and **gene flow** thus have a greater influence on **allele frequencies** at these loci. Neutral markers are commonly used in population and **landscape genetics**. See **genetic marker**.

nominal data Type of data attribute represented by discrete categories (e.g. different land-cover or land-use classification types). See **attribute data**; **categorical data**. Compare **ordinal data**.

normalized difference vegetation index (NDVI) One of the most widely used indices for identifying vegetation from multispectral remote-sensing data, obtained as the ratio of the difference in reflectance between the near-infrared (NIR) and red bands relative to the total reflectance in the NIR and red bands. Healthy vegetation absorbs more strongly in the red part of the visible light spectrum, and thus NDVI > 0 for vegetated areas of a landscape, with NDVI $\rightarrow$ 1 for highly productive areas with greater plant biomass.

nugget In a **semivariogram**, the value of the **semivariance** at the y-intercept (lag distance = 0); although this should be zero in theory, the nugget is typically greater than 0 due to fine-scale heterogeneity that has not been measured, sampling errors, and/or measurement errors.

nutrient spiraling concept Cycling of nutrients (phosphorus, nitrogen) as they are carried downstream in rivers and streams.

observed heterozygosity (H_o) Proportion of heterozygotes that actually occurs within the population. Compare **expected heterozygosity**.

Operational Land Imager (OLI) One of the sensors aboard Landsat 8, which improves upon the sensor technology of previous Landsat missions and adds two new spectral bands in the deep blue and short-wave-infrared range, for a total of nine spectral bands. See **Multispectral Scanner System; Enhanced Thematic Mapper Plus; Thermal Infrared Sensor**.

ordinal data Type of data attribute that has a rank ordering (e.g. different stream orders or roads ordered by size and traffic volume). See **attribute data**.

orthoimage In remote sensing, a digital aerial photograph or satellite image in which the distortion created by topographic relief and camera tilt has been removed through the process of **orthorectification**. The resulting orthoimage has a uniform spatial scale, and thus combines the characteristics of the digital image with the spatial accuracy of a planimetric map. Synonymous with **orthophoto**.

orthophoto See orthoimage.

orthophoto map Map made by assembling a number of **orthophotos** into a single composite image. Orthophoto maps form the basis of many digital map products, such as those available within **geographical information systems**.

orthorectification Process of geometric correction, in which distortions due to topographic relief or camera tilt are removed from aerial photographs to produce an **orthoimage**. Orthorectification is necessary to maintain the geometric representation among mapped objects so that distances can be measured accurately. **Digital elevation models** are often used in orthorectification.

out of the tropics hypothesis One of the hypotheses proposed to explain the **latitudinal diversity gradient**, which posits that diversity is greater in the tropics because most lineages originated there and then spread outward from there to other regions of the world. See also **cradle of life hypothesis**.

outlier loci Loci with candidate genes under selection; **allele frequencies** (heterozygosity) at these loci thus differ greatly from those observed at neutral loci (i.e. they are statistical outliers). Outlier loci tend to have very high or low F_{ST} values compared to **neutral loci**.

overlay function Governs how **spatial data layers** are integrated within a **geographical information system**, such as through the use of mathematical operators or map algebra to combine **raster** or **vector** attribute data values among layers to produce a new output map layer or dataset. See also **point-in-polygon overlay; line-in-polygon overlay; polygon-on-polygon overlay**.

pairwise F_{ST} Genetic distance between each pair of populations, as assayed by the F_{ST} statistic.

panchromatic image In remote sensing, black-and-white images obtained via aerial photography or satellites, in which the imaging sensor spans a wide range of the visible light spectrum to create a high-resolution image of the landscape scene. Contrast **natural-color image; false-color image**.

partial Mantel test Test of the correlation between two distance matrices while controlling for the effects of a third. For example, we can evaluate the relationship between **genetic distance** and the **effective isolation** of populations while controlling for geographic distance in testing for **isolation by resistance**. See **Mantel test**.

passive remote sensing Sensor that measures the electromagnetic energy (wavelengths) that is reflected or emitted from the Earth's surface in either the visible light or infrared spectrum. Examples include airborne film or digital cameras and the multispectral scanners carried aboard Landsat, MODIS, and other Earth-orbiting satellites (e.g. **Multispectral Scanner System; Thematic Mapper; Enhanced Thematic Mapper Plus**). Contrast **active remote sensing**.

patch connectivity Relationship of a habitat patch to other patches on the landscape; connectivity assessed at the patch scale. Contrast **landscape connectivity**.

patch context Habitat or land-use matrix surrounding the patch, which may influence **patch permeability** and thus population and community dynamics within the patch.

patch dynamics Disturbance-mediated dynamics giving rise to a patch mosaic comprising different successional stages; the dynamics of populations or communities within patches.

patch isolation Measure of patch connectivity based on the relative distances among patches and (ideally) the degree to which patches are connected via dispersal, gene flow, or some other process. Also see **structural connectivity; functional connectivity**.

patch occupancy ($\hat{p}$) Proportion of suitable habitat patches occupied by a species on the landscape ($\hat{p}$). Alternatively, the probability that a given patch is occupied given its size, distance to neighboring patches, or other attributes. See **patch-occupancy model; incidence function model**.

patch-occupancy model Model that predicts the incidence or probability of occurrence of a given species within habitat patches as a function of patch and/or landscape attributes; a **metapopulation** model in which species persistence is assessed as the fraction of patches occupied ($\hat{p}$).

path (graph) Sequence of **graph edges** joining **nodes** in a landscape **graph**; may represent the possible route(s) an individual could take while moving through the landscape. Graphs are connected if there exists a path between each pair of nodes. See **landscape connectivity**.

pattern exploration In spatial analysis, the detection and description of spatial structure.

pedodiversity Variation in the types and relative abundance of soil types across a landscape or region.

perceptual extent Broadest scale, in both space and time, at which an organism perceives and responds to environmental heterogeneity.

perceptual grain Finest scale, in both space and time, at which an organism perceives and responds to environmental heterogeneity.

perceptual range Greatest distance from which an animal is able to detect habitat, as determined by its ability to orient towards a habitat edge.

perceptual trap High-quality habitat this is perceived to be of marginal quality, and thus avoided by the species. Compare **ecological trap**.

percolation cluster Single, large habitat patch that spans the entire landscape. In percolation-based applications, such as **neutral landscape models**, the probability that a percolation cluster occurs is used to assess **landscape connectivity**.

percolation threshold (p_c) In neutral landscape models, the critical proportion of habitat at which the landscape becomes disconnected; used as a measure of **landscape connectivity**.

phenotype Morphological, physiological, or behavioral expression of a particular gene or trait.

photogrammetry Science of extracting three-dimensional information and measuring distances between landscape features from two-dimensional aerial photographs; used in surveying and mapping.

phylogeography Discipline that combines phylogenetics and **biogeography** to make inferences about the effects of **migration** (**dispersal** and **gene flow**), population expansion, genetic bottlenecks, and vicariance on the evolutionary history of species.

points In a **vector data** model, objects that occur as discrete locations on the landscape. Examples: individual trees, small ponds, bird nests.

point-in-polygon overlay In a **geographical information system**, a type of **overlay function** applied to **vector data** to uncover associations between spatial **point** data and **polygons** (e.g. land covers).

point-referenced data Location-based **point** data, in which each point is referenced by its xy coordinates. See also **point-referenced data with attributes; spatially referenced data; spatial point pattern**.

point-referenced data with attributes Location-based **point** data that also contain information on the identity, magnitude, or intensity of the variable at that location (i.e. the attributes of the location). See also **spatially referenced data; spatial point pattern**.

pollination ecology Field of ecology that studies the reciprocal interactions between flowering plants and pollinators across a range of scales, from the effectiveness of a single pollinator for a given plant species, to the interactions among floral visitors and plants within a community, to regional and biogeographic patterns in the distribution of pollinators.

polygon In a **vector data** model, features of the landscape that can be represented as two-dimensional objects or as a sequence of points defining a bounded area. Examples: forest stands, agricultural fields, lakes, administrative areas, an animal's home range.

polygon-on-polygon overlay In a **geographical information system**, a type of **overlay function** in which different types of **polygons** can be combined in various ways. See **union**; **erase**; **intersect**.

polymorphic loci Loci having more than one **allele**.

polyploid Having more than two sets of homologous chromosomes; occurs in some cell types (e.g. mammalian liver cells) and in some organisms, especially plants. Example: tetraploids have four sets of chromosomes.

population genomics Area of population **genetics** that uses genome-wide sampling to investigate the roles played by different evolutionary processes (selection, **genetic drift**, **gene flow**) in giving rise to adaptive genetic variation within populations. In particular, population genomics seeks to distinguish the evolutionary processes that influence specific loci from those that influence the entire genome. See also **landscape genomics**.

population sink Population in which death rates exceed birth rates and immigration rates exceed emigration rates. Contrast **population source**.

population source Population in which birth rates exceed death rates, and emigration rates exceed immigration rates. Contrast **population sink**.

population viability Quantitative assessment of a population's current status and likelihood of extinction. Viability relates to the intrinsic capacity for increase ($r > 0$ or $\lambda > 1$) and ability to maintain a self-sustaining population without depending on immigration to sustain population numbers or growth.

positional error Incorrect or missing location data in a spatial dataset that affects **data accuracy**. May arise as a consequence of careless measurement or operator error in the field; data-entry errors; inadequate sensitivity of the equipment or sensor; registration errors; or other inaccuracies that creep in during the data conversion and processing stage.

positive edge response Ecological response in which **species richness** is greater along habitat edges, or for individual species, increased fitness or higher densities at or near habitat edges. See also **edge species**; **edge effects**.

potential connectivity An assessment of **functional connectivity**, based on the likely response of a species to landscape pattern as best determined given limited information on movement or dispersal, such as the maximum or average dispersal distance. Compare **actual connectivity**.

power law Relationship between two measured quantities in which one scales as a function of some power of the second. Example: area (A) scales as function of the square of length, l ($A = l^2$).

prediction bias Creation of a model that is too specific, such that it performs poorly when confronted with new data; a case of overfitting the model.

press Ongoing disturbance that persists for an extended period of time. Examples: grazing of grasslands; sediment loading of rivers following fire within forested watersheds. Contrast **pulse**; **ramp**.

primary succession Formation of ecological communities *de nouveau* in a barren area lacking vegetation and other organisms, such as occurs on lava flows or following a glacial retreat. See **ecological succession**. Compare **secondary succession**.

priority effects The first species to become established at a site may subsequently facilitate or inhibit the establishment of later-arriving species, thereby shaping successional trajectories. See **ecological succession**.

probability density function Shape of the frequency distribution for a continuous random variable or of a randomly generated process at different intervals (e.g. distances or locations in space). See **kernel**.

process linkers Mobile-link **species** that connect **ecosystems** by providing or supporting an essential ecosystem process (e.g. pest control), including some types of **biotic disturbances** (e.g. grazing).

productivity hypothesis Posits that more productive ecosystems should have more species; one of the explanations for the **latitudinal diversity gradient**, as to why species diversity is greater in the tropics (= greater productivity) than in polar regions. See also **energy hypothesis**.

propagule pressure Establishment of a species in a new area tends to increase with the number of individuals or propagules released.

provisioning services Products or goods obtained from **ecosystems**; a type of **ecosystem service**. Examples: food, fiber, fuel, genetic resources, or pharmaceuticals.

pseudoprediction Predictions that appear to have high predictability, but which are in fact illusory because the temporal scale is insufficient to capture the dynamics of the system.

pseudo-sink **Population source** that appears to be a sink because of high levels of immigration into the population. Unlike a true **population sink**, however, it is capable of persisting in the absence of immigration.

Public Land Survey System (PLSS) In the USA, a federal land survey based on a rectangular grid system that was implemented to plat newly acquired territory following the Treaty of Paris in 1783. Because 'witness trees' were recorded to locate the corners of each grid section, the PLSS has provided ecologists with a historical record of the forests that occurred in the pre-settlement landscape. See also **township**; **section**.

pulse Abrupt, short-term disturbance that abates before recurring. Example: a sudden increase in water flow within a river or stream following a heavy rainfall (a pulse flow). See **flood pulse**. Contrast **press**; **ramp**.

pyric herbivory Fire-mediated grazing; herbivores are attracted to, and concentrate their grazing activities within, recently burned areas of the landscape. Example: bison (*Bison bison*) in tallgrass prairie.

quantitative trait loci (QTL) Loci or DNA segments containing genes that underlie a quantitative trait (i.e. a trait that exhibits continuous variation, as opposed to only two or more discrete **phenotypes**). The analysis of QTLs links phenotypic data (trait measurements) with genotypic data (**molecular markers**) to explain the genetic basis of variation in complex traits.

query Search and retrieval of data within a **geographical information system**, which can sort and locate spatial features of interest based on user-specified attributes or by location.

rain shadow effect More rain falls on the windward side of mountain ranges than on the leeward side, which contributes to differences in the vegetation found at a particular elevation on opposite sides of the mountain range.

ramp Disturbance that gradually increases in severity or intensity over time. Example: drought is considered a 'creeping disaster' that steadily gets worse over time. Contrast **press**; **pulse**.

random landscape Type of **neutral landscape model**, generated as a simple random distribution of habitat within a lattice. Compare **hierarchical random landscape**; **fractal landscape**.

range (spatial statistics) In spatial statistics, the lag distance (**spatial lag**) over which the measured variable exhibits **spatial autocorrelation**; in a **semivariogram**, the scale at which **semivariance** levels off (i.e. the distance to the **sill**).

range contraction Decline in the abundance and distribution of a species, especially at the periphery of the range, leading to a reduction in the total geographic area occupied by the species. See also **range shift**.

range expansion Increase in the total geographic area occupied by a species, owing to an increase in the occurrence and distribution of the species beyond its current range limits. See also **range shift**.

range shift A change in the occurrence and distribution of a species, usually in response to environmental or climate changes, resulting in either an increase (**range expansion**) or decrease (**range contraction**) in the total area occupied by the species.

range-shift gaps Species unable to shift their ranges in response to climatic or other environmental changes may subsequently fail to occupy areas that fall within their new range, creating gaps in their distribution that may exacerbate their extinction risk. For example, many montane species in the tropics have narrow environmental tolerances and elevational ranges and are therefore unlikely to be able to shift their range in response to a warming climate, especially as climate change is likely to produce novel climatic conditions rather than just upward shifts in current climatic conditions.

rarefaction In ecology, a technique used to standardize sampling effort among studies or protocols that measure **species richness**. Rarefaction curves are produced by repeated resampling of the data and plotting the number of species expected if only 1, 2, 3,…N samples had been collected. See also **species-accumulation curve**.

raster data Data structure in which spatially referenced information is represented as a grid (lattice) of pixels (grid cells). Synonymous with raster data model. Compare **vector data**.

rasterization Data-processing task within a **geographical information system**, in which **vector data** are converted to **raster data**. Compare **vectorization**.

ratio data Numerical values obtained from a measurement scale with an absolute reference point of zero (e.g. tree density). See **attribute data**.

reaction-diffusion model In ecology, partial differential equations that include population growth (the reaction) and a diffusion coefficient to model invasive or disease spread in a homogeneous environment.

realized niche A reduced subset of environmental conditions in which a species can occur, owing to constraints on its distribution because of other species (competitors, predators) or geographic factors (barriers to dispersal). Contrast **fundamental niche**.

reclassification Data-processing task within a **geographical information system**, in which pixel values of a **raster** data set are changed to different values, usually to provide a broader classification scheme.

Red Queen hypothesis Species must continually evolve (or co-evolve) in response to other species with which they interact in order to maintain their evolutionary fitness; named for the fictional Red Queen from Lewis

Carroll's *Through the Looking-Glass*, in which she tells the main protagonist, a young girl named Alice, that 'it takes all the running you can do, to keep in the same place.'

regime shift Ecosystem transition from one state to another, usually as a consequence of an external forcing agent or perturbation. For example, overgrazing can cause the ecosystem to shift from a grassland to a shrubland state. Synonymous with **system state change**.

registration Data-processing task within a **geographical information system** to ensure that scanned images are digitally aligned and anchored to the appropriate geographic coordinates. Synonymous with **georeference**.

regulating services Non-material benefits obtained through the regulation of **ecosystem processes** involving climate, disease, or water; a type of **ecosystem service**.

Reid's paradox Rate at which most tree species expanded their range northward following the last glacial retreat is several orders of magnitude faster than should have been possible given their mean dispersal distances.

relational database One or more data sets (tables) that are linked based on their shared attributes. In a relational database, every table shares at least one attribute with another table in either a one-to-one, one-to-many, many-to-one, or many-to-many relationship. Example: **attribute data** are stored in a table that is linked to each location or spatial feature by a unique Feature Identification number (FID).

reprojection Data-processing task within a **geographical information system** to ensure that the map projection and coordinate systems are the same for all of the spatial data layers.

resampling Data-processing task within a **geographical information system**, which involves coarsening the resolution of a **raster** image to ensure that all images have the same spatial resolution.

rescue effect Immigration from a nearby source can help rescue populations from extinction. In the context of the **theory of island biogeography**, extinction rates for islands close to the mainland would thus be expected to have lower extinction rates than islands farther away. See also **mainland-island metapopulation; source-sink population dynamics**.

residence time Measure of search effort or habitat quality, in which individuals are assumed to spend more time (residence) in patches with a greater abundance or quality of resources. Synonymous with residence index.

resilience Ability of a system to return to its previous state following a perturbation. See also **ecological resilience**. Compare **resistance**.

resistance Ability of a system to maintain its current state in the face of a disturbance. Compare **resilience**.

resistance distance Derived from circuit theory, a measure of **patch connectivity** that includes information on the availability of alternative pathways (redundancy) as well as information on the shortest or least-cost path. Synonymous with **effective resistance**. See **effective isolation; graph theory; isolation by resistance**.

resource linkers Mobile-link species that transport organic material or nutrients among **ecosystems**. Example: the nutrient-rich guano of nesting seabirds, which is derived from a marine diet rich in fish, contributes to the terrestrial productivity of desert islands.

resource-selection function Type of **species distribution model** in which patterns of species occurrence are related to various measured attributes of the habitat; the function provides values that are proportional to the probability of that habitat being used by the species.

response diversity Measure of the degree to which species performing the same **ecosystem function** respond differently to a disturbance; a component of **functional heterogeneity** that enhances ecosystem resilience.

rifting Formation of fissures in the Earth's surface, which may contribute to elongated valleys (grabens). A form of extensional tectonics, continental rifting leads to divergent plate tectonics. See **fault-block mountains**.

Ripley's *K*-function Spatial statistic commonly used in the multiscale analysis of **point-referenced data**. See also ***L*-function; cross-*K* function**.

River Continuum Concept River sections are connected along their length via the flows of water, organisms, nutrients and sediments, producing a gradient in the relative importance of resource subsidies as a function of stream order or width. See **longitudinal connectivity**.

road ecology Field of ecology that studies the environmental and ecological consequences of roads; the application of ecological principles to the design and mitigation of negative road effects on species (e.g. constructing wildlife-crossing structures over or under highways to reduce vehicle collisions and road mortality).

scale In ecology, the dimensions bounding the data or observation set, which is defined by both the resolution (**grain**) at the finest scale and the **extent** at the broadest scale. Scale has units of measurement and encompasses both spatial and temporal dimensions. In geography and cartography, scale refers to the ratio of reduction of a map. See **map scale**.

scale invariance Pattern or system dynamic exhibiting a constant relationship regardless of scale. See **fractal**.

scale of effect Spatial scale at which an ecological response exhibits the greatest statistical effect size.

scope Relationship between the grain and extent of measurements within a study; the ratio of the **extent** to the resolution (**grain**) of the data set.

seascape ecology Study of coastal marine environments (seascapes), such as salt marshes, mangrove forests, seagrass meadows, and coral reefs, as spatially heterogeneous landscapes, in which patch structure is evident across a range of scales and can influence ecological processes within these benthic or pelagic marine environments.

secondary dispersal Dispersal by one or more agents following the initial dispersal of individuals or their propagules (i.e. after primary dispersal). Example: birds are the primary dispersal agents for the seeds of some fruiting trees, but ants may provide secondary dispersal services.

secondary succession Ecological succession that occurs within an already established ecosystem following a disturbance (e.g. after a fire or hurricane). See **ecological succession**. Compare **primary succession**.

section Under the United States **Public Land Survey System**, square **townships** (36 mi^2 = 93 km^2) are platted and subdivided into 36, 1 mi^2 sections. Each section covers 640 acres (2.6 km^2), and may be subdivided further for sale (e.g. into quarter-sections).

semivariance Distance-based function in which the calculation of spatial variance is derived as the sum of the squared pairwise differences, divided by 2 (i.e. the semivariance). Used for **spatial interpolation**, especially in the field of geostatistics.

semivariogram Plot of **semivariance** as a function of lag distance (**spatial lag**), used to identify the **range** in a spatial data set, along with other parameters, such as the **nugget** and **sill**, used in **kriging**. Synonymous with variogram. See also **theoretical variogram**.

Shannon diversity index (*H'*) Measure of **species diversity**, which incorporates information on the relative abundance of each species within a community. See also **evenness index**.

shifting mosaic The **disturbance regime** may create a heterogeneous and patchy landscape, whose composition changes over time owing to **ecological succession** within disturbance gaps. See **gap dynamics**; **secondary succession**; **patch dynamics**.

shifting-mosaic steady state Landscapes may exhibit a degree of constancy in their overall composition despite ongoing disturbance (i.e. relative constancy). See **shifting mosaic**.

simple random walk Mathematical model of movement in which movement is assumed to be both uncorrelated (direction of movement is independent of the previous direction moved) and unbiased (direction of movement is completely random); if an individual moves with equal probability in any direction, this is essentially diffusion. Compare **diffusion coefficient**; **correlated random walk**.

Single Nucleotide Polymorphisms (SNPs) Molecular marker involving a single **nucleotide** substitution of one base for another within a DNA sequence. Because these substitutions may occur within either coding or non-coding regions of the DNA molecule, they contribute to neutral as well as adaptive genetic variation within populations. Next-generation sequencing technologies are able to sequence hundreds or thousands of loci per individual, making SNPs a real boon to the study of **landscape genomics**.

slope Measure of the change in elevation over a certain distance (i.e. steepness of the terrain); one of the **topographic functions** typically calculated from a **digital elevation model** within a **geographical information system**.

slope curvature Measure of whether the terrain is concave (negative curvature, such as a valley) or convex (positive curvature, such as a ridge); one of the **topographic functions** that can be calculated from a **digital elevation model** within a **geographical information system**. Important for understanding hydrology, runoff, and erosion.

sill In a **semivariogram**, the lag distance (**spatial lag**) at which the **semivariance** reaches a plateau. Defines the **range** over which the variable exhibits **spatial autocorrelation**.

soft edge Habitat patch boundary that is readily crossed by individuals or some other type of ecological flow; a boundary that is highly permeable to movement. Contrast **hard edge**. See **edge permeability**.

Sørensen coefficient (*C_s*) Measure of **species turnover** that assesses the degree of similarity between two communities; a measure of **beta diversity**. Greater weight is given to species found in both communities than in the similar **Jaccard coefficient**. See also **Bray–Curtis dissimilarity coefficient**.

soundscape ecology Emerging field of ecology that studies how the interaction between natural and anthropogenic sounds shape the acoustic environment within a landscape (the soundscape), and how the acoustic environment in turn influences organisms living within that landscape.

source-sink inversion Former **population sinks** may convert to **population sources** if populations eventually adapt and do well in low-quality habitats. See **source-sink population dynamics**.

source-sink metapopulation System of interacting populations (a **metapopulation**) distributed among habitat patches that vary in quality, such that population growth rates are spatially dependent and dispersal from **population sources** helps to maintain populations in habitat sinks. See **source-sink population dynamics**.

source-sink population dynamics Population sinks are not intrinsically viable ($\lambda < 1$) but persist because of immigration from **population sources** ($\lambda > 1$). See **source-sink metapopulation**.

small island effect Species richness may vary independently of area below some critical patch or island size, where species establishment is determined by stochastic factors or environmental factors intrinsic to that particular island or habitat patch. The **species-area relationship** thus does not apply in this particular **domain of scale**.

space-for-time substitution Selection of a landscape series or set of study areas that collectively represent a range of conditions that characterize a particular dynamic, with the assumption that these replicates in space reflect how that landscape or system would change over time. Example: studying the effects of habitat loss by comparing contemporary landscapes that differ in their habitat amounts rather than observing the process over a decades-long study within individual landscapes.

space-time diagram Diagram depicting the spatial and temporal dimensions bounding particular biological, ecological, or physical phenomena or processes within a system. See **domains of scale**.

spatial autocorrelation Distance over which some attribute of the landscape is correlated, either positively or negatively. See **spatial heterogeneity**; **spatial correlogram**.

spatial autocovariance Statistic that measures the degree to which two locations within a certain distance covary with respect to some variable, calculated as the summed products of the deviations from the global mean for each member of the location pairs, obtained over the entire landscape for that variable.

spatial Bayesian cluster analysis Like **Bayesian genetic clustering**, but the geographic coordinates of individuals are used to infer the number of populations and to map population boundaries (the location of genetic discontinuities between populations).

spatial correlogram Plot of **spatial autocorrelation** as a function of lag distance (**spatial lag**); used to detect the scale(s) at which patchiness occurs within a spatially referenced dataset (i.e. **point-referenced data with attributes**).

spatial data layer In a geographic information system (GIS), spatial data sets are referred to as layers or coverages. Each layer provides different spatial information on the landscape of interest (e.g. location of roads, land uses, land-cover types, administrative boundaries, location of protected areas), which can then be combined and analyzed within a GIS.

spatial extent Broadest scale of coverage in a data set or study. Examples: size of a study area or the area encompassed by a remotely sensed image.

spatial filtering Type of **neighborhood function** that can be performed within a **geographical information system**, in which a data-processing window of a certain size (the spatial filter) is applied to the **spatial data layer**; a new value is then assigned to the focal location based on the mean, mode, or some other function within the spatial filter. See also **spatial moving average**.

spatial grain Finest resolution or scale of measurement in a data set. Examples: pixel size of a remotely sensed image or the dimensions of a sampling frame.

spatial heterogeneity Variously defined as either the variety of land-use/land-cover types present on a landscape (i.e. **landscape richness** or **landscape diversity**), or as spatial variation (patchiness) in a landscape feature of interest. Synonymous with landscape heterogeneity. Compare **structural heterogeneity**.

spatial interpolation Estimation and production of continuous landscape surfaces based on the available spatial data. See **interpolation**; **interpolation function**.

spatial lag In spatial analysis, the distance over which spatial dependence is being assessed between samples. Synonymous with lag distance.

spatial moving average Method of **spatial interpolation**, commonly performed in a **geographical information system**, in which a spatial filter of a certain size (a sampling window, box, or circle) is moved systematically over the landscape map and a mean value is assigned to each location based on the values of the sampled sites within that filter. Compare **spatial filtering**.

spatial niche differentiation Competitive coexistence can occur at a landscape scale if species exhibit **competitive reversals** in different habitat types, such that they are partitioning their use of habitats within a heterogeneous landscape. Compare **niche partitioning**.

spatial point pattern Collection of data **points** that are spatially referenced (**georeferenced**). See **point-referenced data**; **point-referenced data with attributes**.

spatial point process Any type of random process that generates a **spatial point pattern**.

spatial storage effect For species that exhibit **competitive reversals** in different habitat types, decreasing population growth rates in unfavorable habitats (where they are inferior competitors) are offset by positive population growth rates in favorable habitat where they are superior competitors. Species coexistence thus occurs as a result of **source-sink population dynamics**, in which species are buffered against spatial variation in population growth rates within heterogeneous landscapes.

spatial subsidies The productivity of one ecosystem supports the productivity of another owing to the asymmetrical flow of nutrients, materials, or organisms between them.

spatially explicit model Model in which the location or other properties (e.g. size, shape, quality) of individual patches or other landscape elements is specified. See **spatially explicit population model**.

spatially explicit population model Population model that incorporates information on the location or other properties of patches and/or populations on the landscape. See **incidence function model**; **spatially explicit population simulation model**. Contrast **spatially implicit population model**.

spatially explicit population simulation model Numerical simulation of dispersal and other population processes on a (usually grid-based) landscape; may be individual (agent) or population based. See **spatially explicit population model**.

spatially explicit simulation model Numerical simulation of ecological processes on a landscape, in which the location and other attributes of patches are explicitly considered and which may have a direct or indirect effect on the processes being modeled. See also **spatially explicit population simulation model**.

spatially implicit population model Population model in which patch structure is assumed to influence population processes (e.g. colonization–extinction dynamics) but the location and/or attributes of patches are not explicitly considered. Example: **mainland-island metapopulation**. Contrast **spatially explicit population model**.

spatially referenced data See **georeference**.

species-accumulation curve The cumulative number of species encountered tends to increase with increasing sampling effort (i.e. the number of samples collected or duration of sampling); sometimes referred to as a 'collector's curve.'

species–area relationship Scaling relationship (**power law**) in which the number of species (**species richness**) increases with the size of the area surveyed.

species distribution model Any of various methods that are used to develop statistical relationships between species occurrences (based on presence-absence or presence-only data) and environmental variables to describe or predict the geographic distribution of species in either the same or different landscape. See **climate envelope model**; **ecological niche model**; **resource-selection function**.

species diversity Richness and relative abundance of species within a community, assayed using metrics such as the **Shannon diversity index (H')**. Sometimes used synonymously with **species richness**.

species richness (S) Number of species that occur in a given area; a basic measure of community structure. Compare **species diversity**.

species turnover Change in species composition between two or more communities, sites, or habitats, or along an environmental gradient. See also **beta diversity**.

spectral signature In remote sensing, reflectance varies across a range of wavelengths in unique ways for different landscape features. The differences among these spectral response curves can be used to classify land-cover types in remotely sensed imagery, and can also aid in the identification of healthy versus stressed (diseased) vegetation (e.g. in crops).

stable isotopes Non-radioactive variants of chemical elements (e.g. ^{2}H, ^{13}C, ^{15}N, ^{34}S), whose analysis can be used to assess **migratory connectivity**, the trophic structure of communities, and energy flows among different ecosystems. See also **spatial subsidies**; **metaecosystem**.

stationarity In spatial statistics, the statistical parameters characterizing the spatial pattern (the mean, variance, or covariance) are assumed to be constant across the landscape.

statistical inference In spatial analysis, the estimation of parameters from a spatial distribution or the testing of hypotheses regarding spatial dependence.

stepping stones Small habitat patches that facilitate movement, dispersal, or gene flow between larger habitat fragments or nature reserves.

steps Segments of an animal movement pathway, corresponding to the linear distance traveled between two locations (**points**) in a fixed time interval, distance, or wherever the animal naturally stops. See **movement path**.

stratovolcano Composite volcano formed as a consequence of repeated eruptions over a long time period.

structural connectivity Assessment of **landscape connectivity**, based on the adjacency or physical relationships among habitat patches or other features of the landscape (**landscape structure**). Contrast **functional connectivity**.

structural heterogeneity Environmental heterogeneity (spatial variation) that can be measured, using landscape metrics or spatial analysis. Compare **functional heterogeneity (landscapes)**. See also **structural connectivity**.

subduction zone Convergent plate boundary, in which one or more tectonic plates is pushed beneath another. Mountain ranges and volcanoes typically form along subduction zones.

subgraph Connected region within a landscape **graph**.

subsidy hypothesis Posits that **ecosystems** receiving a larger fraction of their resource inputs from allochthonous sources (subsidies) should experience stronger tropic cascades (i.e. top-down regulation). See **spatial subsidies**.

supporting services Ecosystem processes and functions such as soil formation, productivity, and nutrient cycling that are necessary for the maintenance of all other **ecosystem services**.

surface fire Fire that consumes litter, grasses, and herbaceous plant cover, but does not burn into the soil organic layer or affect canopy trees. In forested ecosystems, surface fires are typically of low severity. See **fire severity**. Compare **crown fire**.

surface metrology Landscape metrics for the three-dimensional analysis of continuous surfaces. Synonymous with surface metrics.

sustainability science Discipline that addresses the dynamic relationship between nature and society across a range of scales in order to achieve the goal of **sustainable development**.

sustainable development Meeting human needs while sustaining the ability of natural systems to provide the natural resources and **ecosystem services** upon which society depends, and without compromising the ability of future generations to meet their own needs.

synthetic aperture radar (SAR) Form of **active remote sensing** in which the forward motion of the radar platform is used to synthesize a long antenna electronically. The antenna directs microwave pulses towards the Earth's surface and then measures the time delay associated with the return signal (the Doppler effect) to create high-resolution images of the terrain regardless of cloud cover or time of day.

system state change Often-abrupt change in the state of an **ecosystem** as a result of external forcing (e.g. eutrophication, overgrazing), which causes the system to shift from one stable state to another. Synonymous with **regime shift**.

temporal autocorrelation Degree and period over which events are correlated in time. See **temporal heterogeneity**.

temporal extent Duration of a study; duration or return interval of a disturbance; length of time over which a given process operates.

temporal grain Frequency of observation or measurement (data collection) within a study; frequency of a disturbance; rate at which a particular process operates.

temporal heterogeneity Pattern of change in the structure of a landscape over time, in terms of both its composition and configuration. See **landscape dynamics**.

territory An area that is defended by an individual or group of animals for breeding or foraging purposes. Compare **home range**.

Thematic Mapper (TM) One of the sensors aboard Landsat 4 and 5 satellites; basically, an enhanced **Multispectral Scanner System** that acquired data in seven spectral bands. See also **Enhanced Thematic Mapper Plus**.

thematic resolution Finest classification of land covers in a **spatial data layer** or **categorical map**.

theoretical variogram Theoretical model that is fit to a **semivariogram**, as a precursor to **kriging**. Examples: Gaussian model, spherical model, linear + sill model.

theory of island biogeography Equilibrium number of species on an island (or habitat patch) is predicted to represent a balance between immigration (colonization) and extinction rates. All else being equal, islands (or patches) that are close to the mainland (or some other source pool of species) are expected to have higher colonization rates than those that are farther away, and large islands (or patches) are expected to have lower extinction rates than small islands (or patches). The theory of island biogeography has been applied to predict the effects of habitat loss and fragmentation on species diversity.

Thermal Infrared Sensor (TIRS) One of the two sensors carried aboard Landsat 8, which has two channels that scan in the thermal infrared range (TIRS-1 and TIRS-2) with a 100 m resolution. See also **Operational Land Imager**.

three-term local quadrat variance (3TLQV) Method for calculating the **local quadrat variance**, which uses three adjacent blocks (boxes or windows of a given size) and averages over all possible starting locations to obtain the variance term. See also **two-term local quadrat variance**.

time-for-speciation hypothesis One of the hypotheses for the **latitudinal diversity gradient**, which posits that species richness is greater in the tropics because many groups of organisms originated there (**cradle of life hypothesis**) and so had longer to diversify, especially given the greater climatic stability of this region relative to higher latitudes. See also **climatic stability hypothesis**.

topographic function Used to calculate features that describe the terrain, such as from a **digital elevation model** within a **geographical information system**. See **slope**; **aspect**; **slope curvature**; **analytical hillshading**; **visibility analysis**.

township In the USA, a unit of land measure defined by the **Public Land Survey System**, encompassing 36 mi^2 (93 km^2). A survey township is referenced by a numbering system in relation to a principal meridian (north-south) and a base line (east-west). See also **section**.

transferability Application of a model to a different landscape or context from the one used to generate the model. A model that is transferable to many different situations has high generality. See **extrapolation**.

triangulated irregular network (TIN) Type of vector-based interpolation method used to produce a continuous landscape surface, which consists entirely of triangular facets (a network of vertices); TIN data structures may be used in **geographical information systems** and to represent terrain in a **digital elevation model**.

two-term local quadrat variance (TTLQV) Method for calculating the **local quadrat variance**, based on the squared difference between all adjacent pairs of blocks (boxes or windows of a given size) and averaged over all possible starting locations to obtain the variance term. See also **three-term local quadrat variance.**

union function In **geographical information systems,** a type of overlay function typically used for **polygon-on-polygon overlays** in which areas that meet at least one of some set criteria are identified; union behaves like the Boolean operator, OR.

urban adapter Species that do well in urban or suburban environments, such as early-successional or edge species, may benefit from moderate levels of disturbance or habitat fragmentation. Compare **urban avoider; urban exploiter.**

urban avoider Disturbance-sensitive species that are absent or rare from landscapes occupied at moderate-to-high density by people. Compare **urban adapter; urban exploiter.**

urban ecology Emerging field of ecology that studies organisms and their interactions within the urban environment.

urban exploiter Species that benefit from the various resources provided by humans in urban or suburban areas (e.g. vegetable gardens, bird feeders, pet food, trash). Compare **urban adapter; urban avoider.**

urban-rural gradient Gradient in human population and building density, both of which tend to decrease as one moves with increasing distance away from the urban core, and which is expected to be reflected in a corresponding ecological gradient. See also **urban adapter; urban avoider; urban exploiter.**

utilization distribution Representation of an animal's **home range** as a probability density surface reflecting the relative intensity of use within this area. The utilization distribution can be used to define the core area and boundary of the home range, as the smallest area encompassing 50% and 95% of total space use, respectively.

variance staircase Increasing the grain size of a data set results in a decrease in spatial variance with increasing scale, but rather than a smooth function this may manifest in abrupt transitions between different **domains of scale** (risers between the individual 'steps'), producing a relationship with the general appearance of a staircase.

variance-to-mean ratio (VMR) Simple measure of aggregation for point count data (**point-referenced data with attributes**), in which the sample variance (s^2) is compared to the sample mean ($\bar{x}$). The point pattern is random when the variance is equal to the mean (VMR = 1); clumped when the variance is greater than the mean (VMR > 1); and exhibits a regular distribution when the variance is less than the mean (VMR < 1).

vector data Data structure consisting of lines and polygons to represent spatial patterns. Synonymous with vector data model. Compare **raster data.**

vectorization Data-processing task within a **geographical information system,** in which **raster data** are converted to **vector data.** Compare **rasterization.**

vertical connectivity Exchanges that occur between the river, the atmosphere, and the groundwater or sediments below (the hyporheic zone). Compare **lateral connectivity; longitudinal connectivity.**

vicariance biogeography Speciation that occurs primarily because of the geographical isolation of subpopulations due to vicariance events, such as the formation of river drainages or mountain ranges, or the splitting of continents (**rifting**); allopatric speciation.

visibility analysis Topographic function that can be applied to a **digital elevational model** within a **geographical information system** that highlights what aspects of the terrain are visible from a particular location as a viewshed map.

Wahlund effect Reduction in **heterozygosity** (compared to **Hardy–Weinberg equilibrium** expectations) as a consequence of population subdivision.

Worldwide Reference System (WRS) Coordinate system used to index Landsat images, based on the orbital track of the satellite (its longitudinal path) that is subdivided into latitudinal rows corresponding to the center of each image; the combination of path and row uniquely identifies each scene. Because of a change in satellite orbit, two different reference systems are used to distinguish images obtained by Landsats 1–3 from subsequent missions (WRS-1 versus WRS-2, respectively).

zoonoses Infectious diseases that are transmitted from animals to humans.

References

Abbe, T. B. and D. R. Montgomery. 1996. Large woody debris jams, channel hydraulics and habitat formation in large rivers. *Regulated Rivers: Research and Management* 12: 201–221.

Abrams, M. D. and L. C. Hulbert. 1987. Effect of topographic position and fire on species composition in tallgrass prairie in northeast Kansas. *American Midland Naturalist* 117: 442–445.

Ackerman, M. W., W. D. Templin, J. E. Seeb, and L. W. Seeb. 2013. Landscape heterogeneity and local adaptation define the spatial genetic structure of Pacific salmon in a pristine environment. *Conservation Genetics* 14: 483–498.

Adams, M. S., C. N. Service, A. Bateman, M. Bourbonnais, K. A. Artelle, T. Nelson, P. C. Paquet, T. Levi, and C. T. Darimont. 2017. Intrapopulation diversity in isotopic niche over landscapes: Spatial patterns inform conservation of bear-salmon systems. *Ecosphere* 8(6): e01843.

Adler, F. R. and M. Kotar. 1999. Departure time versus departure rate: How to forage optimally when you are stupid. *Evolutionary Ecology Research* 1: 411–421.

Adler, P. B., and 57 others. 2011. Productivity is a poor predictor of plant species richness. *Science* 333: 1750–1753.

Aguilar, R., L. Ashworth, L. Galetto, and M. A. Aizen. 2006. Plant reproductive susceptibility to habitat fragmentation: Review and synthesis through a meta-analysis. *Ecology Letters* 9: 968–980.

Ahern, R. G., D. A. Landis, A. A. Reznicek, and D. W. Schemske. 2010. Spread of exotic plants in the landscape: The role of time, growth habit, and history of invasiveness. *Biological Invasions* 12: 3157–3169.

Ahrends, A., P. M. Hollingsworth, P. Beckschäfer, H. Chen, R. J. Zomer, L. Zhang, M. Wang, and J. Xu. 2017. China's fight to halt tree cover loss. *Proceedings of the Royal Society B* 284: 20162559.

Aikio, S., R. P. Duncan, and P. E. Hulme. 2010. Lag-phases in alien plant invasions: Separating the facts from the artefacts. *Oikos* 119: 370–378.

Alauddin, M. and J. Quiggin. 2008. Agricultural intensification, irrigation and the environment in South Asia: Issues and policy options. *Ecological Economics* 65: 111–124.

Alberti, M. 2008. *Advances in Urban Ecology: Integrating Human and Ecological Processes in Urban Ecosystems*. Springer, New York.

Alberti, M., J. M. Marzluff, E. Shulenberger, G. Bradley, C. Ryan, and C. Zumbrunnen. 2003. Integrating humans into ecology: Opportunities and challenges for studying urban ecosystems. *BioScience* 53: 1169–1179.

Allan, J. D. 2004. Landscapes and riverscapes: The influence of land use on stream ecosystems. *Annual Review of Ecology, Evolution, and Systematics* 35: 257–284.

Allan, J. D., D. L. Erickson, and J. Fay. 1997. The influence of catchment land use on stream integrity across multiple spatial scales. *Freshwater Biology* 37: 149–161.

Allen, C. D., and 19 others. 2010. A global overview of drought and heat-induced tree mortality reveals emerging climate change risks for forests. *Forest Ecology and Management* 259: 660–684.

Allen, C. D., D. D. Breshears, and N. G. McDowell. 2015. On underestimation of global vulnerability to tree mortality and forest die-off from hotter drought in the Anthropocene. *Ecosphere* 6(8):art129.

Allen, T. F. H. and T. B. Starr. 1982. *Hierarchy: Perspectives for Ecological Complexity*. Columbia University Press, New York.

Allen, T. F. H. and T. W. Hoekstra. 1992. *Toward a Unified Ecology*. Columbia University Press, New York.

Allred, B. W., S. D. Fuhlendorf, and R. G. Hamilton. 2011a. The role of herbivores in Great Plains conservation: Comparative ecology of bison and cattle. *Ecosphere* 2(3):art26.

Allred, B. W., S. D. Fuhlendorf, D. M. Engle, and R. D. Elmore. 2011b. Ungulate preference for burned patches reveals strength of fire-grazing interaction. *Ecology and Evolution* 1: 132–144.

Alsdorf, D. E., J. M. Melack, T. Dunne, L. A. K. Mertes, L. L. Hess, and L. C. Smith. 2000. Interferometric radar measurements of water level changes on the Amazon flood plain. *Nature* 404: 174–177.

Alves, D. S. 2002. Space-time dynamics of deforestation in Brazilian Amazônia. *International Journal of Remote Sensing* 23: 2903–2908.

Amarasekare, P. 2003. Competitive coexistence in spatially structured environments: A synthesis. *Ecology Letters* 6: 1109–1122.

AMS (American Meteorological Society). 1973. Policy statement of the American Meteorological Society on hurricanes. *Bulletin of the American Meteorological Society* 54: 46–47.

Amundson, R. 2001. The carbon budget in soils. *Annual Review of Earth and Planetary Sciences* 29: 535–562.

Amundson, R., Y. Guo, and P. Gong. 2003. Soil diversity and land use in the USA. *Ecosystems* 6: 470–482.

Andersen, C. F., and 13 others. 2007. *The New Orleans Hurricane Protection System: What Went Wrong and Why*. A Report by the American Society of Civil Engineers, Hurricane Katrina External Review Panel. American Society of Civil Engineers, Reston, Virginia.

Andersen, M. C., H. Adams, B. Hope and M. Powell. 2004. Risk assessment for invasive species. *Risk Analysis* 24: 787–793.

Anderson, C. D., B. K. Epperson, M.-J. Fortin, R. Holderegger, P. M. A. James, M. S. Rosenberg, K. T. Scribner, and S. Spear. 2010. Considering spatial and temporal scale in landscape-genetic studies of gene flow. *Molecular Ecology* 19: 3565–3575.

Anderson, D. L. and J. H. Natland. 2014. Mantle updrafts and mechanisms of oceanic volcanism. *Proceedings of the National Academy of Sciences of the United States of America* 111: E4298–E4304.

Anderson, K. and K. J. Gaston. 2013. Lightweight unmanned aerial vehicles will revolutionize spatial ecology. *Frontiers in Ecology and the Environment* 11: 138–146.

Anderson, M. J., K. E. Ellingsen, and B. H. McArdle. 2006. Multivariate dispersion as a measure of beta diversity. *Ecology Letters* 9: 683–693.

Anderson, M. J., and 13 others. 2011. Navigating the multiple meanings of β diversity: A roadmap for the practicing ecologist. *Ecology Letters* 14: 19–28.

Anderson, R. M. and R. M. May. 1978. Regulation and stability of host-parasite population interactions: I. Regulatory processes. *Journal of Animal Ecology* 47: 219–247.

Anderson, R. M. and R. M. May. 1979. Population biology of infectious diseases: Part I. *Nature* 280: 361–367.

Anderson, R. M., H. C. Jackson, R. M. May and A. M. Smith. 1981. Population dynamics of fox rabies in Europe. *Nature* 289: 765–771.

Anderson, W. B. and G. A. Polis. 1999. Nutrient fluxes from water to land: Seabirds affect plant nutrient status on Gulf of California islands. *Oecologia* 118: 324–332.

Andow, D. A., P. M. Kareiva, S. A. Levin, and A. Okubo. 1990. Spread of invading organisms. *Landscape Ecology* 4: 177–188.

Andreasen, A. M., K. M. Stewart, W. S. Longland, J. P. Beckmann, and M. L. Forister. 2012. Identification of source-sink dynamics in mountain lions of the Great Basin. *Molecular Ecology* 21: 5689–5701.

Andreassen, H. P. and R. A. Ims. 1998. The effects of experimental habitat destruction and patch isolation on space use and fitness parameters in female root vole *Microtus oeconomus*. *Journal of Animal Ecology* 67: 941–952.

Andreassen, H. P., S. Halle, and R.A. Ims. 1996. Optimal width of movement corridors for root voles: Not too narrow and not too wide. *Journal of Applied Ecology* 33: 63–70.

Andrén, H. 1994. Effects of habitat fragmentation on birds and mammals in landscapes with different proportions of suitable habitat: A review. *Oikos* 71: 355–366.

Andrén, H. 1995. Effects of landscape composition on predation rates at habitat edges. Pp. 225–255 in *Mosaic Landscapes and Ecological Processes* (L. Hansson, L. Fahrig, and G. Merriam, Editors). Chapman and Hall, London, UK.

Anselin, L. 1995. Local Indicators of Spatial Association—LISA. *Geographical Analysis* 27: 93–115.

Antonarakis, A. S., S. S Saatchi, R. L. Chazdon, and P. R. Moorcroft. 2011. Using Lidar and Radar measurements to constrain predictions of forest ecosystem structure and function. *Ecological Applications* 21: 1120–1137.

Antonovics, J. 1971. The effects of a heterogeneous environment on the genetics of natural populations. *American Scientist* 59: 593–599.

Antonovics, J. 2006. Evolution in closely adjacent plant populations X: Long-term persistence of prereproductive isolation at a mine boundary. *Heredity* 97: 33–37.

Antrop, M. 2007. Reflecting upon 25 years of landscape ecology. *Landscape Ecology* 22: 1441–1443.

Antrop, M., J. Brandt, T. Pinto Correia, I. Loupa Ramos, G. de Blust, M. Kozová, F. Papadimitriou, and V. Van Eetvelde. 2009. Why a European chapter of IALE? IALE Bulletin 27 (2), supplement: 1–5. Available online https://www.landscape-ecology.org

Araenas, M., N. Ray, M. Currat, and L. Excoffier. 2012. Consequences of range contractions and range shifts on molecular diversity. *Molecular Biology and Evolution* 29: 207–218.

Araújo, M. B. and M. Luoto. 2007. The importance of biotic interactions for modelling species distributions under climate change. *Global Ecology and Biogeography* 16: 743–753.

Araújo, M. B., M. Cabeza, W. Thuiller, L. Hannah, and P. H. Williams. 2004. Would climate change drive species out of reserves? An assessment of existing reserve-selection methods. *Global Change Biology* 10: 1618–1626.

Araújo, M. S. and R. Costa-Pereira. 2013. Latitudinal gradients in intraspecific ecological diversity. *Biology Letters* 9: 20130778.

Arkema, K. K., G. Guannel, G. Verutes, S. A. Wood, A. Guerry, M. Ruckelshaus, P. Kareiva, M. Lacayo, and J. M. Silver. 2013. Coastal habitats shield people and property from sea-level rise and storms. *Nature Climate Change* 3:913–918: 1–6.

Arnfield, A.J. 2003. Two decades of urban climate research: A review of turbulence, exchanges of energy and water, and the urban heat island. *International Journal of Climatology* 23: 1–26.

Aronson, R. B. and W. F. Precht. 1995. Landscape patterns of reef coral diversity: A test of the Intermediate Disturbance Hypothesis. *Journal of Experimental Marine Biology and Ecology* 192: 1–14.

Arrhenius, O. 1921. Species and area. *Journal of Ecology* 9: 95–99.

Ashley, M. V. 2010. Plant parentage, pollination, and dispersal: How DNA microsatellites have altered the landscape. *Critical Reviews in Plant Sciences* 29: 148–161.

Ashman, T.-L. and 11 others. 2004. Pollen limitation of plant reproduction: Ecological and evolutionary causes and consequences. *Ecology* 85: 2408–2421.

Askins, R. A. 1995. Hostile landscapes and the decline of migratory birds. *Science* 67: 1956–1957.

Asner, G. P., D. E. Knapp, E. N. Broadbent, P. J. C. Oliveira, M. Keller, and J. N. Silva. 2005. Selective logging in the Brazilian Amazon. *Science* 310: 480–482.

Asner, G. P., P. G. Brodrick, C. B. Anderson, N. Vaughn, D. E. Knapp, and R. E. Martin. 2016. Progressive forest canopy water loss during the 2012–2015 California drought. *Proceedings of the National Academy of Sciences of the United States of America* 113: E249–E255.

Atkinson, R. P., C. J. Rhodes, D. W. Macdonald, and R. M. Anderson. 2002. Scale-free dynamics in the movement patterns of jackals. *Oikos* 98: 134–140.

Austerlitz, F., C. W. Dick, C. Dutech, E. K. Klein, S. Oddou-Muratorio, P. E. Smouse, and V. L. Sork. 2004. Using genetic markers to estimate the pollen dispersal curve. *Molecular Ecology* 13: 937–954.

Avise, J. 2000. *Phylogeography*. Harvard University Press, Cambridge, Massachusetts.

Bachmaier, M. and M. Backes. 2008. Variogram or semivariogram? Understanding the variances in a variogram. *Precision Agriculture* 9: 173–175.

Baddeley, A., E. Rubak, and R. Turner. 2015. *Spatial Point Patterns: Methodology and Applications with R*. Chapman and Hall/CRC Press, London, UK.

Baggio, J. A., K. Salau, M. A. Janssen, M. L. Schoon, and Ö. Bodin. 2011. Landscape connectivity and predator-prey population dynamics. *Landscape Ecology* 26: 33–45.

Baguette, M. and H. Van Dyke 2007. Landscape connectivity and animal behavior: Functional grain as a key determinant for dispersal. *Landscape Ecology* 22: 1117–1129.

Bailey, D. J., W. Otten, and C. A. Gilligan. 2000. Saprotrophic invasion by the soil-borne fungal plant pathogen *Rhizoctonia solani* and percolation thresholds. *New Phytologist* 146: 535–544.

Bailey, H. and 14 others. 2012. Identification of distinct movement patterns in Pacific leatherback turtle populations influenced by ocean conditions. *Ecological Applications* 22: 735–747.

Bailey, R. G. 1996. *Ecosystem Geography*. Springer, New York.

Bak, P., C. Tang, and K. Wiesenfeld. 1988. Self-organized criticality. *Physical Review A* 38: 364–374.

Balaji, R., J. Bartram, D. Coates, R. Connor, J. Harding, M. Hellmuth, L. Leclerc, V. Pangare, and J. G. Shields. 2012. Beyond demand: Water's social and environmental benefits. Pp. 101–132 in *Managing Water Under Uncertainty and Risk*. United Nations World Water Development Report 4, Volume 1. UNESCO, Paris, France.

Balkenhol, N. and L. P. Waits. 2009. Molecular road ecology: Exploring the potential of genetics for investigating transportation impacts on wildlife. *Molecular Ecology* 18: 4151–4164.

Balkenhol, N. and E. L. Landguth. 2011. Simulation modelling in landscape genetics: On the need to go further. *Molecular Ecology* 20: 667–670.

Balkenhol, N., L. P. Waits, and R. J. Dezzani. 2009a. Statistical approaches in landscape genetics: An evaluation of methods for linking landscape and genetic data. *Ecography* 32: 818–830.

Balkenhol, N., F. Gugerli, S. A. Cushman, L. P. Waits, A. Coulon, J. W. Arntzen, R. Holderegger, H. H. Wagner, and the participants of the Landscape Genetics Research Agenda Workshop 2007. 2009b. Identifying future research needs in landscape genetics: Where to from here? *Landscape Ecology* 24: 455–463.

Balkenhol, N., S. A. Cushman, A. Storfer, and L. P. Waits. 2015. *Landscape Genetics: Concepts, Methods, Applications*. Wiley-Blackwell, West Sussex, UK.

Banavar, J. R., A. Maritan, and A. Rinaldo. 1999. Size and form in efficient transportation networks. *Nature* 399: 130–132.

Banks, J. E. 1997. Do imperfect trade-offs affect the extinction debt phenomenon? *Ecology* 78: 1597–1601.

Barbosa, P., Editor. 1998. *Conservation Biological Control*. Academic Press, San Diego, California.

Barlow, J. C. and S. N. Leckie. 2000. Eurasian Tree Sparrow (*Passer montanus*), *The Birds of North America Online* (A. Poole, Editor.). Cornell Lab of Ornithology, Ithaca, New York. Retrieved from the Birds of North America Online: https://birdsna.org/Species-Account/bna/species/eutspa/ (last accessed 2/15/19)

Barnes, J. H. and W. D. Sevon. 2002. *The Geological Story of Pennsylvania*, 3rd edition. Pennsylvania Geological Survey, Fourth Series, Educational Series 4, Harrisburg, Pennsylvania.

Barnett, T. P. and D. W. Pierce. 2008. When will Lake Mead go dry? *Water Resources Research* 44:W03201.

Barrett, R. D. H. and H. E. Hoekstra. 2011. Molecular spandrels: Tests of adaptation at the genetic level. *Nature Review Genetics* 12: 767–780.

Barrett, R. D. H., S. M. Rogers, and D. Schluter. 2008. Natural selection on a major armor gene in threespine stickleback. *Science* 322: 255–257.

Barros, G. S. C. 2014. Agricultura e indústria no desenvolvimento brasileiro. Pp. 79–116 in *O Mundo Rural no Brasil do Século 21*

(A. M. Buainain, E. Alves, J. J. Silveira, and Z. Navarro, Editors). EMBRAPA, Brasília, Brazil.

Bartels, P., J. Cucherousset, K. Steger, P. Eklöv, L. J. Tranvik, and H. Hellebrand. 2012. Reciprocal subsidies between freshwater and terrestrial ecosystems structure consumer resource dynamics. *Ecology* 93: 1173–1182.

Bartumeus, F., M. G. E. da Luz, G. M. Viswanathan, and J. Catalan. 2005. Animal search strategies: A quantitative random-walk analysis. *Ecology* 86: 2078–3087.

Bascompte, J. and R. V. Solé. 1996. Habitat fragmentation and extinction thresholds in spatially explicit models. *Journal of Animal Ecology* 65: 465–473.

Basnet, K., F. N. Scatena, G. E. Likens, and A. E. Lugo. 1993. Hurricane Hugo: Damage to a tropical rain forest in Puerto Rico. *Journal of Tropical Ecology* 8: 47–55.

Batáry, P., L. V. Dicks, D. Kleijn, and W. J. Sutherland. 2015. The role of agri-environment schemes in conservation and environmental management. *Conservation Biology* 29: 1006–1016.

Batista, W. B. and W. J. Platt. 2003. Tree population responses to hurricane disturbances: Syndromes in a south-eastern USA old-growth forest. *Journal of Ecology* 91:197–212.

Bayman, J. M. 2001. The Hohokam of southwest North America. *Journal of World Prehistory* 15: 257–311.

Beaumont, M. A. and B. Rannala. 2004. The Bayesian revolution in genetics. *Nature Reviews Genetics* 5: 251–261.

Bebber, D. P., M. A. T. Ramotowski, and S. J. Gurr. 2013. Crop pests and pathogens move polewards in a warming world. *Nature Climate Change* 3: 985–988.

Beerli, P. and J. Felsenstein. 2001. Maximum likelihood estimation of a migration matrix and effective populations sizes in *n* subpopulations by using a coalescent approach. *Proceedings of the National Academy of Sciences of the United States of America*: 98: 4563–4568.

Beier, P., and R. F. Noss. 1998. Do habitat corridors provide connectivity? *Conservation Biology* 12: 1241–1252.

Beier, P. D. R. Majka, and W. D. Spencer. 2008. Forks in the road: Choices in procedures for designing wildland linkages. *Conservation Biology* 22: 836–851.

Beissinger, S. R. and M. I. Westphal. 1998. On the use of demographic models of population viability in endangered species management. *Journal of Wildlife Management* 62: 821–841.

Bélisle, M. 2005. Measuring landscape connectivity: The challenge of behavioral landscape ecology. *Ecology* 86: 1988–1995.

Bellemare, J., G. Motzkin, and D. R. Foster. 2002. Legacies of the agricultural past in the forested present: An assessment of historical land-use effects on rich mesic forests. *Journal of Biogeography* 29: 1401–1420.

Bellingham, P. J. 1991. Landforms influence patterns of hurricane damage: Evidence from Jamaican montane forests. *Biotropica* 23: 427–433.

Bellingham, P. J., E. V. J. Tanner, and J. R. Healey. 1995. Damage and responsiveness of Jamaican montane tree species after disturbance by a hurricane. *Ecology* 76: 2562–2580.

Bellingham, P. J., E. V. J. Tanner, and J. R. Healey. 2005. Hurricane disturbance accelerates invasion by the alien tree *Pittosporum undulatum* in Jamaican montane rain forests. *Journal of Vegetation Science* 16: 675–684.

Ben-David, M., T. A. Hanley, and D. M. Schell. 1998. Fertilization of terrestrial vegetation by spawning Pacific salmon: The role of flooding and predator activity. *Oikos* 83: 47–55.

Bender, D. J. and L. Fahrig. 2005. Matrix structure obscures the relationship between interpatch movement and patch size and isolation. *Ecology* 86: 1023–1033.

Bender, D. J., T. A. Contreras, and L. Fahrig. 1998. Habitat loss and population decline: A meta-analysis of the patch size effect. *Ecology* 79: 517–533.

Bender, D. J., L. Tischendorf, and L. Fahrig. 2003a. Evaluation of patch isolation metrics for predicting animal movement in binary landscapes. *Landscape Ecology* 18: 17–39.

Bender, D. J., L. Tischendorf, and L. Fahrig. 2003b. Using patch isolation metrics to predict animal movement in binary landscapes. *Landscape Ecology* 18: 17–39.

Bender, M. A., T. R. Knutson, R. E. Tueya, J. J. Sirutis, G. A. Vecchi, S. T. Garner, and I. M. Held. 2010. Modeled impact of anthropogenic warming on the frequency of intense Atlantic hurricanes. *Science* 327: 454–458.

Benhamou, S. 2007. How many animals really do the Lévy walk? *Ecology* 88: 1962–1969.

Benhamou, S. and D. Cornélis. 2010. Incorporating movement behavior and barriers to improve kernel home range space use estimates. *Journal of Wildlife Management* 74: 1353–1360.

Benítez-López, A., R. Alkemade, and P. A. Verweij. 2010. The impacts of roads and other infrastructure on mammal and bird populations: A meta-analysis. *Biological Conservation* 143: 1307–1316.

Bennett, A. F., J. Q. Radford, and A. Haslem. 2006. Properties of land mosaics: Implications for nature conservation in agricultural environments. *Biological Conservation* 133: 250–264.

Benning, T. L. and T. R. Seastedt. 1995. Landscape-level interactions between topoedaphic features and nitrogen limitation in tallgrass prairie. *Landscape Ecology* 10: 337–348.

Bergen, K. M. and M. C. Dobson. 1999. Integration of remotely sensed radar imagery in modeling and mapping of forest biomass and net primary production. *Ecological Modelling* 122: 257–274).

Bernot, M. J., J. L. Tank, T. V. Royer, and M. B. David. 2006. Nutrient uptake in streams draining agricultural catchments of the midwestern United States. *Freshwater Biology* 51: 499–509.

Bhagwat, S. 2014. The history of deforestation and forest fragmentation: A global perspective. Pp. 5–19 in *Global Forest Fragmentation* (C. J. Kettle and L. P. Koh, Editors). CAB International, Oxfordshire, UK.

Bianchi, T. S. and M. A. Allison. 2009. Large-river delta-front estuaries as natural 'recorders' of global environmental change. *Proceedings of the National Academy of Sciences of the United States of America* 106: 8085–8092.

Biek, R. and L. A. Real. 2010. The landscape genetics of infectious disease emergence and spread. *Molecular Ecology* 19: 3515–3531.

Bissonette, J. A. and I. Storch, Editors. 2003. *Landscape Ecology and Resource Management: Linking Theory with Practice*. Island Press, Washington, D.C.

Bivand, R., B. Rowlingson, P. Diggle, G. Petris, and S. Eglen. 2017. Package 'splancs': Spatial and Space-Time Point Pattern Analysis, Version 2.01–40. Documentation available online: https://cran.r-project.org/web/packages/splancs/splancs.pdf

Black, W. C., C. F. Baer, M. F. Antolin, and N. M. DuTeau. 2001. Population genomics: Genome-wide sampling of insect populations. *Annual Review of Entomology* 46: 441–469.

Blackburn, T. M. and K. J. Gaston. 1997. The relationship between geographic area and the latitudinal gradient in species richness in New World birds. *Evolutionary Ecology* 11: 195–204.

Blair, C., D. E. Weigel, M. Balazik, A. T. H. Keeley, F. M. Walker, E. Landguth, S. Cushman, M. Murphy, L. Waits, and N. Balkenhol. 2012. A simulation-based evaluation of methods for inferring linear barriers to gene flow. *Molecular Ecology Resources* 12: 822–833.

Blair, J. M. 1997. Fire, N availability, and plant response in grasslands: A test of the transient maxima hypothesis. *Ecology* 78: 2359–2368.

Blanchong, J. A., M. D. Samuel, K. T. Scribner, B. V. Weckworth, J. A. Langenberg, and K. B. Filcek. 2008. Landscape genetics and the spatial distribution of chronic wasting disease. *Biology Letters* 4: 130–133.

Blevins, E., S. M. Wisely, and K. A. With. 2011. Historical processes and landscape context influence genetic structure in peripheral populations of the collared lizard (*Crotaphytus collaris*). *Landscape Ecology* 26: 1125–1136.

Blondel, J., D. W. Thomas, A. Charmantier, P. Perret, P. Bourgault, and M. M. Lambrechts. 2006. A thirty-year study of phenotypic and genetic variation of blue tits in Mediterranean habitat mosaics. *BioScience* 56: 661–673.

Blundell, G. M., J. A. K. Maier, and E. M. Debevec. 2001. Linear home ranges: Effects of smoothing, sample size, and autocorrelation on kernel estimates. *Ecological Monographs* 71: 469–489.

Boffey, P. M. 1976. International Biological Program: Was it worth the cost and effort? *Science* 193: 866–868.

Börger, L., N. Franconi, G. De Michele, A. Gantz, F. Meschi, A. Manica, S. Lovari, and T. Coulson. 2006. Effects of sampling regime on the mean and variance of home range size estimates. *Journal of Animal Ecology* 75: 1393–1405.

Börger, L., B. D. Dalziel, and J. M. Fryxell. 2008. Are there general mechanisms of animal home range behavior? A review and prospects for future research. *Ecology Letters* 11: 637–650.

Bolliger, J., J. C. Sprott, and D. J. Mladenoff. 2003. Self-organization and complexity in historical landscape patterns. *Oikos* 100: 541–553.

Bolliger, J., L. A. Shulte, S. N. Burrows, T. A. Sickley, and D. J. Mladenoff. 2004. Assessing ecological restoration potentials of Wisconsin (USA.) using historical landscape reconstructions. *Restoration Ecology* 12: 124–142.

Bolliger, J., T. Lander, and N. Balkenhol. 2014. Landscape genetics since 2003: Status, challenges and future directions. *Landscape Ecology* 29: 361–366.

Bolnick, D. I. and S. P. Otto. 2013. The magnitude of local adaptation under genotype-dependent dispersal. *Ecology and Evolution* 3: 4722–4735.

Bolnick, D. I., L. K. Snowberg, C. Patenia, W. E. Stutz, T. Ingram, and O. L Lau. 2009. Phenotype-dependent native habitat preference facilitates divergence between parapatric lake and stream stickleback. *Evolution* 63: 2004–2016.

Bolstad, P. V. and J. L. Smith. 1992. Errors in GIS: Assessing spatial data accuracy. *Journal of Forestry* 90: 21–29.

Bonaccorso, E., I. Kosh, and A. T. Peterson. 2006. Pleistocene fragmentation of Amazon species' ranges. *Diversity and Distributions* 12: 157–164.

Bond, W. J., F. I. Woodward, and G. F. Midgley. 2005. The global distribution of ecosystems in a world without fire. *New Phytologist* 165: 525–538.

Bonnell, T. R., R. R. Sengupta, C. A. Chapman, and T. L. Goldberg. 2010. An agent-based model of red colobus resources and disease dynamics implicates key resource sites as hot spots of disease transmission. *Ecological Modelling* 221: 2491–2500.

Bonney, R., C. B. Cooper, J. Dickinson, S. Kelling, T. Phillips, K. V. Rosenberg, and J. Shirk. 2009. Citizen science: A developing tool for expanding science knowledge and scientific literacy. *BioScience* 59: 977–984.

Boose, E. R., D. R. Foster, and M. Fluet. 1994. Hurricane impacts to tropical and temperate forest landscapes. *Ecological Monographs* 64: 369–400.

Borlaug, N. 2007. Feeding a hungry world. *Science* 318: 359.

Bormann, F. H. and G. E. Likens. 1979. *Patterns and Process in a Forested Ecosystem.* Springer-Verlag, New York.

Borowsky, R. L. 2001. Estimating nucleotide diversity from random amplified polymorphic DNA and amplified fragment length polymorphism data. *Molecular Phylogenetics and Evolution* 18: 143–148.

Boström, C., S. J. Pittman, C. Simenstad, and R. T. Kneib. 2011. Seascape ecology of coastal biogenic habitats: Advances, gaps, and challenges. *Marine Ecology Progress Series* 427: 191–217.

Botequilha Leitão, A. and J. Ahern. 2002. Applying landscape ecological concepts and metrics in sustainable landscape planning. *Landscape and Urban Planning* 59: 65–93.

Bouldin, J. 2008. Some problems and solutions in density estimation from bearing tree data: A review and synthesis. *Journal of Biogeography* 35: 2000–2011.

Boutin, S. and D. Hebert. 2002. Landscape ecology and forest management: Developing an effective partnership. *Ecological Applications* 12: 390–397.

Bowler, D. E. and T. G. Benton. 2005. Causes and consequences of animal dispersal strategies: Relating individual behavior to spatial dynamics. *Biological Reviews* 80: 205–225.

Bowman, J. 2003. Is dispersal distance of birds proportional to territory size? *Canadian Journal of Zoology* 81: 195–202.

Bowman, J., J. A. G. Jaeger, and L. Fahrig. 2002. Dispersal distance of mammals is proportional to home range size. *Ecology* 83: 2049–2055.

Boyce, M. S., P. R. Vernier, S. E. Nielsen, and F. K. A. Schmiegelow. 2002. Evaluating resource selection functions. *Ecological Modelling* 157: 281–300.

Boyles, J. G., P. M. Cryan, G. F. McCracken, and T. H. Kunz. 2011. Economic importance of bats in agriculture. *Science* 332: 41–42.

Bray, J. R. and J. T. Curtis. 1957. An ordination of upland forest communities of southern Wisconsin. *Ecological Monographs* 27: 325–349.

Breshears, D. D., and 12 others. 2005. Regional vegetation die-off in response to global-change-type drought. *Proceedings of the National Academy of Sciences of the United States of America* 102: 15144–15148.

Briggs, C. J. 2009. Host-parasitoid interactions. Pp. 213–219 in *The Princeton Guide to Ecology* (S. A. Levin, S. R. Carpenter, H. C. J. Godfray, A. P. Kinzig, M. Loreau, J. B. Losos, B. Walker, and D. S. Wilcove, Editors). Princeton University Press, Princeton, New Jersey.

Briggs, J. M. and A. K. Knapp. 1995. Interannual variability in primary production in tallgrass prairie: Climate, soil moisture, topographic position, and fire as determinants of aboveground biomass. *American Journal of Botany* 82: 1024–1030.

Briggs, J. M. and A. K. Knapp. 2001. Determinants of C_3 forb growth and production in a C_4 dominated grassland. *Plant Ecology* 152: 93–100.

Briggs J. M., A. K. Knapp, and B. L. Brock. 2002. Expansion of woody plants in tallgrass prairie: A fifteen-year study of fire and fire-grazing interactions. *American Midland Naturalist* 147: 287–294.

Briggs, J. M., G. A. Hoch, and L. C. Johnson. 2002. Assessing the rate, mechanisms, and consequences of the conversion of tallgrass prairie to *Juniperus virginiana* forest. *Ecosystems* 5: 578–586.

Briggs, J. M., A. K. Knapp, J. M. Blair, J. L. Heisler, G. A. Hoch, M. S. Lett, and J. K. McCarron. 2005. An ecosystem in transition: Causes and consequences of the conversion of mesic grassland to shrubland. *BioScience* 55: 243–254.

Bristow, C. S., K. A. Hudson-Edwards, and A. Chappell. 2010. Fertilizing the Amazon and equatorial Atlantic with West African dust. *Geophysical Research Letters* 37: L14807.

Britto, D. T. and H. J. Kronzucker. 2002. NH_4^+ toxicity in higher plants: A critical review. *Journal of Plant Physiology* 159: 567–584.

Broadbent, E. N., G. P. Asner, M. Keller, D. E. Knapp, P. J. C. Oliveira, and J. N. Silva. 2008. Forest fragmentation and edge effects from deforestation and selective logging in the Brazilian Amazon. *Biological Conservation* 141: 1745–1757.

Brock, W. A., S. R. Carpenter, and M. Scheffer. 2006. Regime shifts, environmental signals, uncertainty and policy choice. Pp. 180–206 in *A Theoretical Framework for Analyzing Social-Ecological Systems* (J. Norberg and G. Cumming, Editors). Columbia University Press, New York.

Brown, C. J., and 11 others. 2011. Quantitative approaches in climate change ecology. *Global Change Biology* 17: 3697–3713.

Brown, J. H. 1995. *Macroecology.* University of Chicago Press, Chicago, Illinois.

Brown, J. H. 2014. Why are there so many species in the tropics? *Journal of Biogeography* 41: 8–22.

Brown, J. H. and A. Kodric-Brown. 1977. Turnover rates in insular biogeography: Effect of immigration on extinction. *Ecology* 58: 445–449.

Brown, J. H., V. K. Gupta, B.-L. Li, B. T. Milne, C. Restrepo, and G. B. West. 2002. The fractal nature of nature: Power laws, ecological complexity and biodiversity. *Philosophical Transactions of the Royal Society of London, B:* 357: 619–626.

Brown, J. H., J. F. Gillooly, A. P. Allen, V. M. Savage, and G. B. West. 2004. Toward a metabolic theory of ecology. *Ecology* 85: 1771–1789.

Brown, J. S. 1988. Patch use as an indicator of habitat preference, predation risk, and competition. *Behavioral Ecology and Sociobiology* 22: 37–47.

Brown, J. S. 1999. Vigilance, patch use, and habitat selection: Foraging under predation risk. *Evolutionary Ecology Research* 1: 49–71.

Brown, K. A., S. Spector, and W. Wu. 2008. Multiscale analysis of species introductions: Combining landscape and demographic models to improve management decisions about non-native species. *Journal of Applied Ecology* 45: 1639–1648.

Brown, M. J. F. and R. J. Paxton. 2009. The conservation of bees: A global perspective. *Apidologie* 40: 410–416.

Bruggeman, D. J., T. Wiegand, and N. Fernández. 2010. The relative effects of habitat loss and fragmentation on population genetic variation in the red-cockaded woodpecker (*Picoides borealis*). *Molecular Ecology* 19: 3679–3691.

Brunsdon, C. and L. Comber. 2015. *An Introduction to R for Spatial Analysis and Mapping*. SAGE Publications, London, UK.

Buitenwerf, R., W. J. Bond, N. Stevens, and W. S. W. Trollope. 2012. Increased tree densities in South African savannas: >50 years of data suggest CO_2 as a driver. *Global Change Biology* 18: 675–684.

Bullock, J. M. and R. T. Clarke. 2000. Long distance seed dispersal by wind: Measuring and modelling the tail of the curve. *Oecologia* 124: 506–521.

Bullock, J. M., K. Shea, and O. Skarpaas. 2006. Measuring plant dispersal: An introduction to field methods and experimental design. *Plant Ecology* 186: 217–234.

Bunn, A. G., D. L. Urban, and T. H. Keitt. 2000. Landscape connectivity: A conservation application of graph theory. *Journal of Environmental Management* 59: 265–278.

Burdon, J. J. and P. H. Thrall. 1999. Spatial and temporal patterns in coevolving plant and pathogen associations. *American Naturalist* 153: S15–S33.

Burke, I. C. 1989. Control of nitrogen mineralization in a sagebrush steppe landscape. *Ecology* 70: 1115–1126.

Burney, J. A., S. J. Davis, and D. B. Lobell. 2010. Greenhouse gas mitigation by agricultural intensification. *Proceedings of the National Academy of Sciences of the United States of America* 107: 12052–12057.

Burt, W. H. 1943. Territoriality and home range as applied to mammals. *Journal of Mammalogy* 24: 346–352.

Busby, P. E., G. Motzkin, and E. R. Boose. 2008. Landscape-level variation in forest response to hurricane disturbance across a storm track. *Canadian Journal of Forest Research* 38: 2942–2950.

Butcher, G. S. and D. K. Niven. 2007. Combining data from the Christmas Bird Count and the Breeding Bird Survey to determine the continental status and trends of North America birds. National Audubon Society, New York. Available online: http://www.audubon.org/bird/stateofthebirds/CBID/content/Report.pdf (last accessed 3/3/19)

Butler, D. R. 1995. *Zoogeomorphology: Animals as Geomorphic Agents*. Cambridge University Press, Cambridge and New York.

Butler, D. R. and C. F. Sawyer. 2012. Introduction to the special issue—zoogeomorphology and ecosystem engineering. *Geomorphology* 157–158: 1–5.

Buyantuyev, A. and J. Wu. 2007. Effects of thematic resolution on landscape pattern analysis. *Landscape Ecology* 22: 7–13.

Cadbury, D. 2004. *Seven Wonders of the Industrial World*. Harper Perennial, London, UK.

Cadenasso, M. L., S. T. A. Pickett, and K. Schwarz. 2007. Spatial heterogeneity in urban ecosystems: Reconceptualizing land cover and a framework for classification. *Frontiers in Ecology* 5: 80–88.

Cadotte, M. W., D. V. Mai, S. Jantz, M. D. Collins, M. Keele, and J. A. Drake. 2006. On testing the competition-colonization trade-off in a multispecies assemblage. *American Naturalist* 168: 704–709.

Cagnolo, L., G. Valladares, A. Salvo, M. Cabido, and M. Zak. 2009. Habitat fragmentation and species loss across three interacting trophic levels: Effects of life-history and food-web traits. *Conservation Biology* 23: 1167–1175.

Cahill, J. F., Jr. and G. McNickle. 2011. The behavioral ecology of nutrient foraging by plants. *Annual Review of Ecology and Systematics* 42: 289–311.

Cain, D. H., K. Riitters, and K. Orvis. 1997. A multiscale analysis of landscape statistics. *Landscape Ecology* 12: 199–212.

Calabrese, J. M. and W. F. Fagan. 2004. A comparison-shopper's guide to connectivity metrics. *Frontiers in Ecology and the Environment* 2: 529–536.

Calcagno, V., N. Mouquet, P. Jarne, and P. David. 2006. Coexistence in a metacommunity: The competition-colonization trade-off is not dead. *Ecology Letters* 9: 897–907.

Calderone, N. W. 2012. Insect pollinated crops, insect pollinators and US agriculture: Trend analysis of aggregate data for the period 1992–2009. *PLoS One* 7(5): e37235.

Cale, W. G., G. M. Henebry, and J. A. Yeakley. 1989. Inferring process from pattern in natural communities. *BioScience* 39: 600–605.

Campbell, J. B. and R. H. Wynne. 2011. *Introduction to Remote Sensing*, 5th edition. Guilford Press, New York.

Canner, J. E. and M. Spence. 2011. A new technique using metal tags to track small seeds over short distances. *Ecological Research* 26: 233–236.

Cantwell, M. D. and R. T. T. Forman. 1993. Landscape graphs: Ecological modeling with graph theory to detect configurations common to diverse landscapes. *Landscape Ecology* 8: 239–255.

Carbonneau, P., M. A. Fonstad, W. A. Marcus, and S. J. Dugdale. 2012. Making riverscapes real. *Geomorphology* 137: 74–86.

Carpenter, S. R. and W. A. Brock. 2006. Rising variance: A leading indicator of ecological transition. *Ecology Letters* 9: 308–315.

Carpenter, S. R., S. W. Chisholm, C. J. Krebs, D. W. Schindler, and R. F. Wright. 1995. Ecosystem experiments. *Science* 269: 324–327.

Carpenter, S. R., N. F. Caraco, D. L. Correll, R. W. Howarth, A. N. Sharpley, and V. H. Smith. 1998. Nonpoint pollution of surface waters with phosphorus and nitrogen. *Ecological Applications* 8: 559–568.

Carpenter, S. R., J. J. Cole, M. L. Pace, M. Van de Bogert, D. L. Bade, D. Bastviken, C. M. Gille, J. R. Hodgson, J. F. Kitchell, and E. S. Kritzberg. 2005. Ecosystem subsidies: Terrestrial support of aquatic food webs from ^{13}C addition to contrasting lakes. *Ecology* 86: 2737–2750.

Castellarini, F., C. Provensal, and J. Polop. 2002. Effect of weather variables on the population fluctuation of muroid *Calomys venustus* in central Argentina. *Acta Oecologica* 23: 385–391.

Castro, C. N. 2015. Agriculture in Brazil's Midwest region: Limitations and future challenges to development. Discussion Paper 198, Institute for Applied Economic Research (IPEA), Brasília, Brazil.

Castro, M. C., R. L. Monte-Mór, D. O. Sawyer, and B. H. Singer. 2006. Malaria risk on the Amazon Frontier. *Proceedings of the National Academy of Sciences of the United States of America* 103: 2452–2457.

Caswell, H. 2000. *Matrix Population Models*, 2nd edition. Sinauer and Associates, Sunderland, Massachusetts.

Cavallero, L., E. Raffaele, and M. A. Aizen. 2013. Birds as mediators of passive restoration during early post-fire recovery. *Biological Conservation* 158: 242–350.

CCRIF (Caribbean Catastrophe Risk Insurance Facility). 2011. *A Snapshot of the Economics of Climate Adaptation Study in the Caribbean*. Grand Cayman, Cayman Islands. Available online: www.ccrif. org/publications/snapshot-economics-climate-adaptation-study-caribbean (last accessed 1/30/19).

Chace, J. F. and J. J. Walsh. 2006. Urban effects on native avifauna: A review. *Landscape and Urban Planning* 74: 46–69.

Chalfoun, A. D., F. R. Thompson III, and M. J. Ratnaswamy. 2002a. Nest predators and fragmentation: A review and meta-analysis. *Conservation Biology* 16: 306–318.

Chalfoun, A. D., M. J. Ratnaswamy, and F. R. Thompson III. 2002b. Songbird nest predators in forest-pasture edge and forest interior in a fragmented landscape. *Ecological Applications* 123: 858–867.

Chan, Y. K. and V. C. Koo. 2008. An introduction to synthetic aperture radar (SAR). *Progress in Electromagnetics Research B* 2: 27–60.

Chao, A., R. L. Chazdon, R. K. Colwell, and T.-J. Shen. 2005. A new statistical approach for assessing similarity of species composition with incidence and abundance data. *Ecology Letters* 8: 148–159.

Chao, A., R. L. Chazdon, R. K. Colwell, and T.-J. Shen. 2006. Abundance-based similarity indices and their estimation when there are unseen species in samples. *Biometrics* 62: 361–371.

Chapin, F. S., III, P. A. Matson, and P. Vitousek. 2011. *Principles of Terrestrial Ecosystem Ecology*. 2nd edition. Springer, New York.

Chaplin-Kramer, R., M. E. O'Rourke, E. J. Blitzer, and C. Kremen. 2011. A meta-analysis of crop pest and natural enemy response to landscape complexity. *Ecology Letters* 14: 922–932.

Chapman, C. A., M. D. Wasserman, T. R. Gillespie, M. L. Speirs, M. J. Lawes, T. L. Saj, and T. E. Ziegler. 2006. Do food availability, parasitism, and stress have synergistic effects on red colubus populations living in forest fragments? *American Journal of Physical Anthropology* 131: 525–534.

Charnov, E. L. 1976. Optimal foraging, the marginal value theorem. *Theoretical Population Biology* 9: 129–136.

Chase, J. M., and 11 others. 2005. Competing theories for competitive metacommunities. Pp. 335–354 in *Metacommunities: Spatial Dynamics and Ecological Communities* (M. Holyoak, M. A. Leibold, and R. D. Holt, Editors). Chicago University Press, Illinois.

Chen, C., E. Durand, F. Forbes, and O. François. 2007. Bayesian clustering algorithms ascertaining spatial population structure: A new computer program and a comparison study. *Molecular Ecology Notes* 7: 747–756.

Chen, I-C., J. K. Hill, R. Ohlemüller, D. B. Roy, and C. D. Thomas. 2011. Rapid range shifts of species associated with high levels of climate warming. *Science* 333: 1024–1026.

Chen, J., X. Zhu, J. E. Vogelmann, F. Gao, and S. Jin. 2011. A simple and effective method for filling gaps in Landsat ETM+ SLC-off images. *Remote Sensing of Environment* 115: 1053–1064.

Chen, N., C. Jayaprakash, K. Yu, and V. Guttal. 2018. Rising variability, not slowing down, as a leading indicator of a stochastically driven abrupt transition in a dryland ecosystem. *American Naturalist* 191: E1–E14.

Chen, X.-Y. and F. He. 2009. Speciation and endemism under the model of island biogeography. *Ecology* 90: 39–45.

Chesson, P. 1994. Multispecies competition in variable environments. *Theoretical Population Biology* 45: 227–276.

Chesson, P. 2000a. Mechanisms of maintenance of species diversity. *Annual Review of Ecology and Systematics* 31: 343–366.

Chesson, P. 2000b. General theory of competitive coexistence in spatially-varying environments. *Theoretical Population Biology* 58: 211–237.

Chesson, P. and M. Rosenzweig. 1991. Behavior, heterogeneity, and the dynamics of interacting species. *Ecology* 72: 1187–1195.

Cheville, N. F., D. R. McCullough, and L. R. Paulson. 1998. *Brucellosis in the Greater Yellowstone Area*. National Academy Press, Washington, D.C.

Chilès, J.-P. and P. Delfiner. 2012. *Geostatistics: Modeling Spatial Uncertainty*, 2nd edition. John Wiley & Sons, New Jersey.

Choat, B., and 23 others. 2012. Global convergence in the vulnerability of forests to drought. *Nature* 491: 752–755.

Chorowicz, J. 2005. The East African rift system. *Journal of African Earth Sciences* 43: 379–410.

Chown, S. L. and K. J. Gaston. 2000. Areas, cradles and museums: The latitudinal gradient in species richness. *Trends in Ecology and Evolution*: 14: 311–315.

Chown, S. L., and 15 others. 2012. Continent-wide risk assessment for the establishment of nonindigenous species in Antarctica. *Proceedings of the National Academy of Sciences of the United States of America* 109: 4938–4943.

Christensen, J. A., and 13 others. 2008. *Landscape Pattern Indicators for the Nation. A Report from the Heniz Center's Landscape Pattern Task Group*, H. John Heniz III Center for Science, Economics and the Environment, Washington, D.C.

Christensen, N. L., and 12 others. 1989. Interpreting the Yellowstone fires of 1988. *BioScience* 39: 678–685.

Christian, C. S. 1958. The concept of land units and land systems. *Proceedings of the Ninth Pacific Science Congress* 20: 74–81.

Christianini, A. V. and P. S. Oliveira. 2010. Birds and ants provide complementary seed dispersal in a neotropical savanna. *Journal of Ecology* 98: 573–582.

Chylek, P. and G. Lesins. 2008. Multidecadal variability of Atlantic hurricane activity: 1851–2007. *Journal of Geophysical Research Atmospheres* 113: D22106.

Ciannelli, L, P. Fauchald, K. S. Chan, V. N. Agostini, and G. E. Dingsør. 2008. Spatial fisheries ecology: Recent progress and future prospects. *Journal of Marine Systems* 71: 223–236.

Cilliers, S. S., N. S. G. Williams, and F. J. Barnard. 2008. Patterns of exotic plant invasions in fragmented urban and rural grasslands across continents. *Landscape Ecology* 23: 1243–1256.

Clark, J. S., and 12 others. 1998. Reid's paradox of rapid plant migration. *BioScience* 48: 13–24.

Clark, J. S., A. E. Gelfand, C. W. Woodall, and K. Zhu. 2014. More than the sum of their parts: Forest climate response from joint species distribution models. *Ecological Applications* 24: 990–999.

Clarke, K. R. 1993. Non-parametric multivariate analyses of changes in community structure. *Austral Ecology* 18: 117–143.

Clausen, J. and W. M. Hiesey. 1958. *Experimental Studies on the Nature of Species. IV. Genetic Structure of Ecological Races.* Publication #615, Carnegie Institution of Washington, D.C.

Clausen, J., D. D. Keck, and W. M. Hiesey. 1940. *Experimental Studies on the Nature of Species. I. Effects of Varied Environments on Western North American Plants.* Publication #520, Carnegie Institution of Washington, D.C.

Clements, F. E. 1916. *Plant Succession: An Analysis of the Development of Vegetation.* Carnegie Institution of Washington Publication Number 242, Washington, D.C.

Clevenger, A. P. and N. Waltho. 2005. Performance indices to identify attributes of highway crossing structures facilitating movement of large mammals. *Biological Conservation* 121: 453–464.

Clevenger, A. P., B. Chruszcz, and K. E. Gunson. 2003. Spatial patterns and factors influencing small vertebrate fauna road-kill aggregations. *Biological Conservation* 109: 15–26.

Codling, E. A., M. J. Plank, and S. Behamou. 2008. Random walk models in biology. *Journal of the Royal Society Interface* 5: 813–834.

Coffin, A. W. 2007. From roadkill to road ecology: A review of the ecological effects of roads. *Journal of Transport Geography* 15: 396–406.

Cogbill C. V., J. Burk, and G. Motzkin. 2002. The forests of presettlement New England, USA: Spatial and compositional patterns based on town proprietor surveys. *Journal of Biogeography* 29: 1279–1304.

Cohen, J. E., D. N. Schittler, D. G. Raffaelli, and D. C. Reuman. 2009. Food webs are more than the sum of their tritrophic parts. *Proceedings of the National Academy of Science of the United States of America* 106: 22335–22340.

Cohn, J. P. 2001. Resurrecting the dammed: A look at Colorado River restoration. *BioScience* 51: 998–1003.

Collinge, S. K. 2009. *Ecology of Fragmented Landscapes.* Johns Hopkins University Press, Baltimore, Maryland.

Collinge, S. K. and T. M. Palmer. 2002. The influences of patch shape and boundary contrast on insect response to fragmentation in California grasslands. *Landscape Ecology* 17: 647–656.

Collingham, Y. C. and B. Huntley. 2000. Impacts of habitat fragmentation and patch size upon migration rates. *Ecological Applications* 10: 131–144.

Collins, B. D. and D. R. Montgomery. 2002. Forest development, wood jams, and restoration of floodplain rivers in the Puget lowland, Washington. *Restoration Ecology* 10: 237–247.

Collins, S. L. 1992. Fire frequency and community heterogeneity in tallgrass prairie vegetation. *Ecology* 73: 2001–2006.

Collins, S. L. and L. B. Calabrese. 2012. Effects of fire, grazing and topographic variation on vegetation structure in tallgrass prairie. *Journal of Vegetation Science* 23: 563–575.

Colosimo, P. F., K. E. Hosemann, S. Balahadra, G. Villarreal Jr., M. Dickson, J. Grimwood, J. Schmutz, R. M. Myers, D. Schluter, and D. M. Kingsley. 2005. Widespread parallel evolution in sticklebacks by repeated fixation of Ectodysplasin alleles. *Science* 307: 1928–1933.

Colwell, R. K. and J. A. Coddington. 1994. Estimating terrestrial biodiversity through extrapolation. *Philosophical Transactions: Biological Sciences* 345: 101–118.

Colwell, R. K. and G. C. Hurtt. 1994. Non-biological gradients in species richness and a spurious Rapoport effect. *American Naturalist* 144: 570–595.

Colwell, R. K., C. Rahbek, and N. J. Gotelli. 2004. The mid-domain effect and species richness patterns: What have we learned so far? *American Naturalist* 163: E1–E23.

Colwell, R. K., G. Brehm, C. L Cardelús, A. C. Gilman, and J. T. Longino. 2008. Global warming, elevational range shifts, and lowland biotic attrition in the wet tropics. *Science* 322: 258–261.

Compton, B. W., K. McGarigal, S. A. Cushman, and L. R. Gamble. 2007. A resistant-kernel model of connectivity for amphibians that breed in vernal pools. *Conservation Biology* 21: 788–799.

Connell, J. H. 1978. Diversity in tropical rain forests and coral reefs. *Science* 199: 1302–1310.

Connell, J. H. 1980. Diversity and the coevolution of competitors, or the ghost of competition past. *Oikos* 35: 131–138.

Connell, J. H. 1997. Disturbance and recovery of coral assemblages. *Coral Reefs* 16: S101–S113.

Connell, J.H. and E. Orias. 1964. The ecological regulation of species diversity. *American Naturalist* 98: 399–414.

Conner, M. M., M. R. Ebinger, J. A. Blanchong, and P. C. Cross. 2008. Infectious disease in cervids of North America: Data, models, and management challenges. Pp. 146–172 in *Year in Ecology and Conservation Biology* (R. S. Ostfeld and W. H. Schlesinger, Editors.). Blackwell Publishing, Oxford, UK.

Connor, E. F. and E. D. McCoy. 1979. The statistics and biology of the species-area relationship. *American Naturalist* 113: 791–833.

Cook, B. I., R. L. Miller, and R. Seager. 2009. Amplification of the North American 'Dust Bowl' drought through human-induced land degradation. *Proceedings of the National Academy of Sciences of the United States of America* 106: 4997–5001.

Cook, B. I., K. J. Anchukaitis, J. O. Kaplan, M. J. Puma, M. Kelley, and D. Gueyffier. 2012. Pre-Columbian deforestation as an amplifier of drought in Mesoamerica. *Geophysical Research Letters* 39: L16706.

Cook, E. R., R. Seager, M. A. Cane, and D. W. Stahle. 2007. North American drought: Reconstructions, causes, and consequences. *Earth-Science Reviews* 81: 93–134.

Cook, E. R., R. Seager, R. R. Heim Jr., R. S. Vose, C. Herweijer, and C. Woodhouse. 2010. Megadroughts in North America: Placing IPCC projections of hydroclimatic change in a long-term palaeoclimate context. *Journal of Quaternary Science* 25: 48–61.

Cook, T. C. and D. T. Blumstein. 2013. The omnivore's dilemma: Diet explains variation in vulnerability to vehicle collision mortality. *Biological Conservation* 167: 310–315.

Coop, G., D. Witonsky, A. Di Rienzo, and J. K. Pritchard. 2010. Using environmental correlations to identify loci underlying local adaptation. *Genetics* 185: 1411–1423.

Corander, J., J. Sirén, and E. Arjas. 2008. Bayesian spatial modeling of genetic population structure. *Computational Statistics* 23: 111–129.

Coreau, A., G. Pinay, J. D. Thompson, P.-O. Cheptou, and L. Mermet. 2009. The rise of research on futures in ecology: Rebalancing scenarios and predictions. *Ecology Letters* 12: 1277–1286.

Costanza, R., and 12 others. 1997. The value of the world's ecosystem services and natural capital. *Nature* 387: 253–260.

Cote, D., D. G. Kehler, C. Bourne, and Y. F. Wiersma. 2009. A new measure of longitudinal connectivity for stream networks. *Landscape Ecology* 24: 101–113.

Cottenie, K. 2005. Integrating environmental and spatial processes in ecological community dynamics. *Ecology Letters* 8: 1175–1182.

Cousins, S. A. O. 2009. Extinction debt in fragmented grasslands: Paid or not? *Journal of Vegetation Science* 20: 3–7.

Cousins, S. A. O. and D. Vanhoenacker. 2011. Detection of extinction debt depends on scale and specialization. *Biological Conservation* 144: 782–787.

Coutts, S. R., R. D. van Klinken, H. Yokomizo, and Y. M. Buckley. 2011. What are the key drivers of spread in invasive plants: Dispersal, demography or landscape: And how can we use this knowledge to aid management? *Biological Invasions* 13: 1649–1661.

Couvreur, M., B. Christiaen, V. Veryheyen, and M. Hermy. 2004. Large herbivores as mobile links between isolated nature reserves through adhesive seed dispersal. *Applied Vegetation Science* 7: 229–236.

Cowie, R. J. 1977. Optimal foraging in great tits (*Parus major*). *Nature* 268: 137–139.

Craft, M. E., E. Volz, C. Packer, and L. A. Meyers. 2009. Distinguishing epidemic waves from disease spillover in a wildlife population. *Proceedings of the Royal Society* B 276: 1777–1785.

Cranmer, L., D. McCollin, and J. Ollerton. 2012. Landscape structure influences pollinator movements and directly affects plant reproductive success. *Oikos* 121: 562–568.

Creegan, H. P. and P. E. Osborne. 2005. Gap-crossing decisions of woodland songbirds in Scotland: An experimental approach. *Journal of Applied Ecology* 42: 678–687.

Cressie, N. and C. K. Wikle. 2011. *Statistics for Spatiotemporal Data.* John Wiley & Sons, New Jersey.

Crisafulli, C. M., F. J. Swanson, and V. H. Dale. 2005. Overview of ecological responses to the eruption of Mount St Helens—1980–2005. Pp. 287–299 in *Ecological Responses to the 1980 Eruption of Mount St Helens* (V. H. Dale, F. J. Swanson, and C. M. Crisafulli, Editors). Springer-Verlag, New York.

Crist, T. O. and J. A. Veech. 2006. Additive partitioning of rarefaction curves and species-area relationships: Unifying α-, β- and γ-diversity with sample size and habitat area. *Ecology Letters* 9: 923–932.

Crist, T. O., D. S. Guertin, J. A. Wiens, and B. T. Milne. 1992. Animal movements in heterogeneous landscapes: An experiment with *Eleodes* beetles in shortgrass prairie. *Functional Ecology* 6: 536–544.

Crist, T. O., J. A. Veech, J. C. Gering, and K. S. Summerville. 2003. Partitioning species diversity across landscapes and regions: A hierarchical analysis of α, β and γ diversity. *American Naturalist* 162: 734–743.

Crone, E. E. and C. B. Schultz. 2008. Old models explain new observations of butterfly movement at patch edges. *Ecology* 89: 2061–2067.

Cronin, J. T. 2003a. Matrix heterogeneity and host-parasitoid interactions in space. *Ecology* 84: 1506–1516.

Cronin, J. T. 2003b. Movement and spatial population structure of a prairie planthopper. *Ecology* 84: 1179–1188.

Cronin, J. T. 2004. Host-parasitoid extinction and colonization in a fragmented prairie landscape. *Oecologia* 139: 503–514.

Cronin, J. T. 2007. From population sources to sieves: The matrix alters host-parasitoid source-sink structure. *Ecology* 88: 2966–2976.

Cronin, J. T. and J. D. Reeve. 2014. An integrative approach to understanding host-parasitoid population dynamics in real landscapes. *Basic and Applied Ecology* 15: 101–113.

Crooks, K. R. and M. Sanjayan, Editors. 2006. *Connectivity Conservation*. Cambridge University Press, Cambridge, UK.

Crutzen, P. J. 2002. Geology of mankind. *Nature* 415: 23.

Cullingham, C. I., B. A. Pond, C. J. Kyle, E. E. Rees, R. C. Rosatte, and B. N. White. 2008. Combining direct and indirect genetic methods to estimate dispersal for informing wildlife disease management decisions. *Molecular Ecology* 17: 4874–4886.

Cullingham, C. I., C. J. Kyle, B. A. Pond, E. E. Rees, and B. N. White. 2009. Differential permeability of rivers to raccoon gene flow corresponds to rabies incidence in Ontario, Canada. *Molecular Ecology* 18: 43–53.

Cumming, G. S., P. Olsson, F. S. Chapin III, and C. S. Holling. 2013. Resilience, experimentation, and scale mismatches in social-ecological landscapes. *Landscape Ecology* 28: 1139–1150.

Cummins, K. W. 1974. Structure and function of stream ecosystems. *BioScience* 24: 631–641.

Cureton, A. N., H. J. Newbury, A. F. Raybould, and B. V. Ford-Lloyd. 2006. Genetic structure and gene flow in wild beet populations: The potential influence of habitat on transgene spread and risk assessment. *Journal of Applied Ecology* 43: 1203–1212.

Currie, D. J. 1991. Energy and large-scale patterns of animal- and plant-species richness. *American Naturalist* 137: 27–49.

Curtis, A. 1999. Using a spatial filter and a Geographic Information System to improve rabies surveillance data. *Emerging Infectious Diseases* 5: 603–606.

Curtis, J. T. 1956. The modification of mid-latitude grasslands and forest by man. Pp. 721–736 in *Man's Role in Changing the Face of the Earth* (W. L. Thomas, Jr., C. O. Sauer, M. Bates, and L. Mumford, Editors). University of Chicago Press, Chicago, Illinois.

Cushman, S. A. 2006. Effects of habitat loss and fragmentation on amphibians: A review and prospectus. *Biological Conservation* 128: 231–240.

Cushman, S. A. and E. L. Landguth. 2010. Spurious correlations and inference in landscape genetics. *Molecular Ecology* 19: 3592–2602.

Cushman, S. A., K. S. McKelvey, J. Hayden, and M. K. Schwartz. 2006. Gene flow in complex landscapes: Testing multiple hypotheses with causal modeling. *American Naturalist* 168: 486–499.

Cushman, S. A., K. McGarigal, and M. C. Neel. 2008. Parsimony in landscape metrics: Strength, universality, and consistency. *Ecological Indicators* 8: 691–703.

Cushman, S. A., K. S. McKelvey, and M. K. Schwartz. 2009. Use of empirically derived source-destination models to map regional conservation corridors. *Conservation Biology* 23: 368–376.

Cushman, S. A., A. Shirk, and E. L. Landguth. 2012. Separating the effects of habitat area, fragmentation and matrix resistance on genetic differentiation in complex landscapes. *Landscape Ecology* 27: 369–380.

Czajkowski, J., K. Simmons, and D. Sutter. 2011. An analysis of coastal and inland fatalities in landfalling U.S. hurricanes. *Natural Hazards* 59: 1513–1531.

D'Antonio, C. M. and P. M. Vitousek. 1992. Biological invasions by exotic grasses, the grass/fire cycle, and global change. *Annual Review of Ecology and Systematics* 23: 63–87.

Daehler, C. C. 2009. Short lag times for invasive tropical plants: Evidence from experimental plantings in Hawai'i. *PLoS ONE* 4(2): e4462.

Dai, A. 2011. Drought under global warming: A review. *Wiley Interdisciplinary Reviews (WIREs) Climate Change* 2: 45–65.

Daily, G. C. 1997. *Nature's Services: Societal Dependence on Natural Ecosystems*. Island Press, Washington, D.C.

Dakos, V., and 10 others. 2012. Methods for detecting early warnings of critical transitions in time series illustrated using simulated ecological data. *PLoS ONE* 7: e401010.

Dalan, R. A., G. R. Holley, W. I. Woods, H. W. Watters, Jr., and J. A. Koepke. 2003. *Envisioning Cahokia: A Landscape Perspective*. Northern Illinois University Press, Dekalb, Illinois.

Dale, M. R. T. 1999. *Spatial Pattern Analysis in Plant Ecology*. Cambridge University Press, Cambridge, UK.

Dale, M. R. T. 2000. Lacunarity analysis of spatial pattern: A comparison. *Landscape Ecology* 15: 467–478.

Dale, M. R. T. and M.-J. Fortin. 2014. *Spatial Analysis: A Guide for Ecologists*, 2nd edition. Cambridge University Press, Cambridge, UK.

Dale, M. R. T., P. Dixon, M.-J. Fortin, P. Legendre, D. E. Myers, and M. S. Rosenberg. 2002. Conceptual and mathematical relationships among methods for spatial analysis. *Ecography* 25: 558–577.

Dale, V. H., A. E. Lugo, J. A. MacMahon, and S. T. A. Pickett. 1998. Ecosystem management in the context of large, infrequent disturbances. *Ecosystems* 1: 546–557.

Dale, V. H., D. R. Campbell, W. M. Adams, C. M. Crisafulli, V. I. Dains, P. M. Frenzen, and R. F. Holland. 2005a. Plant succession on the Mount St Helens debris-avalanche deposit. Pp. 59–73 in *Ecological Responses to the 1980 Eruption of Mount St Helens* (V. H. Dale, F. J. Swanson, and C. M. Crisafulli, Editors). Springer-Verlag, New York.

Dale, V. H., C. M. Crisafulli, and F. J. Swanson. 2005b. 25 years of ecological change at Mount St Helens. *Science* 308: 961–962.

Damschen, E. I., N. M. Haddad, J. L. Orrock, J. J. Tewksbury, and D. J. Levey. 2006. Corridors increase plant species richness at large scales. *Science* 313: 1284–1286.

Davis, A.J., L. S. Jenkinson, J. L. Lawton, B. Shorrocks, and S. Wood. 1998. Making mistakes when predicting shifts in species range in response to global warming. *Nature* 391: 783–786.

Davis, M. A. 2009. *Invasion biology*. Oxford University Press, Oxford, UK.

Davis, S., P. Trapman, H. Leirs, M. Begon, and J. A. P. Heesterbeek. 2008. The abundance threshold for plague as a critical percolation phenomenon. *Nature* 454: 634–637.

de Groot, R. and L. Hein. 2007. Concept and valuation of landscape functions at different scales. Pp. 15–36 in *Multifunctional Land Use: Meeting Future Demands for Landscape Goods and Services* (Ü. Mander, H. Wiggering, and K. Helming, Editors). Springer, Berlin, Germany.

De Meester, L., A. Gómez, B. Okamura, and K. Schwenk. 2002. The Monopolization Hypothesis and the dispersal-gene flow paradox in aquatic organisms. *Acta Oecologica* 23: 121–135.

De Mita, S., A.-C. Thuillet A-C, L. Gay, N. Ahmadi, S. Manel, J. Ronfort, and Y. Vigouroux. 2013. Detecting selection along environmental gradients: Analysis of eight methods and their effectiveness for outbreeding and selfing populations. *Molecular Ecology* 22: 1383–1399.

de Smith, M. J., M. F. Goodchild, and P. A. Longley. 2018. *Geospatial Analysis: A Comprehensive Guide to Principles, Techniques and Software Tools*, 6th edition. Available online: www.spatialanalysisonline.com (last accessed 2/1/19)

DeAngelis, D. L. and J. C. Waterhouse. 1987. Equilibrium and non-equilibrium concepts in ecological models. *Ecological Monographs* 57: 1–21.

Debinski, D. M. and R. D. Holt. 2000. A survey and overview of habitat fragmentation experiments. *Conservation Biology* 14: 342–355.

del Moral, R., J. E. Sandler, and C. P. Muerdter. 2009. Spatial factors affecting primary succession on the Muddy River Lahar, Mount St Helens, Washington. *Plant Ecology* 202: 177–190.

Delang, C. O. and Z. Yuan. 2015. *China's Grain for Green Program*. Springer, Cham, Switzerland.

Delcourt, H. R. and P. A. Delcourt. 1988. Quaternary landscape ecology: Relevant scales in space and time. *Landscape Ecology* 2: 23–44.

Delcourt, H. R. and P. A. Delcourt. 1996. Pre-settlement landscape heterogeneity: Evaluating grain of resolution using General Land Office Survey data. *Landscape Ecology* 11: 363–381.

Delcourt, H. R., P. A. Delcourt, and T. Webb, III. 1983. Dynamic plant ecology: The spectrum of vegetational change in space and time. *Quaternary Science Reviews* 1: 153–175.

DeLuca, T. H. and C. A. Zabinski. 2011. Prairie ecosystems and the carbon problem. *Frontiers in Ecology and the Environment* 9: 407–413.

Deng, L., G. Liu, and Z. Shangguan. 2014. Land-use conversion and changing soil carbon stocks in China's 'Grain-for-Green' Program: A synthesis. *Global Change Biology* 20: 3544–3556.

Deng, L., G. Zhu, Z. Tang, and Z. Shangguan. 2016. Global patterns of the effects of land-use changes on soil carbon stocks. *Global Ecology and Conservation* 5: 127–138.

Dengler, J. 2009. Which function describes the species-area relationship best? A review and empirical evaluation. *Journal of Biogeography* 36: 728–744.

Denny, M. and L. Benedetti-Cecchi. 2012. Scaling up in ecology: Mechanistic approaches. *Annual Review of Ecology and Systematics* 43: 1–22.

Desrochers, A. and S. J. Hannon. 1997. Gap crossing decisions by forest songbirds during the post-fledging period. *Conservation Biology* 11: 1204–1210.

Deter, J., N. Charbonnel, and J.-F. Cosson. 2010. Evolutionary landscape epidemiology. Pp. 173–188 in *The Biogeography of Host-Parasite Interactions* (S. Morand and B. R. Krasnov, Editors). Oxford University Press, New York.

Dewhirst, S. and F. Lutscher. 2009. Dispersal in heterogeneous habitats: Thresholds, spatial scales, and approximate rates of spread. *Ecology* 90: 1138–1345.

Diamond, J. M. 1972. Biogeographic kinetics: Estimation of relaxation times for avifaunas of southwest Pacific islands. *Proceedings of the National Academy of Sciences of the United States of America* 69: 3199–3203.

Diamond, J. M. 1974. Colonization of exploded volcanic islands by birds: The supertramp strategy. *Science* 184: 803–806.

Dias, P. C. 1996. Sources and sinks in population biology. *Trends in Ecology and Evolution* 11: 326–330.

Díaz, S. and M. Cabido. 2001. Vive la différence: Plant functional diversity matters to ecosystem processes. *Trends in Ecology and Evolution* 16: 646–655.

Dicke, M. and P. A. Burrough. 1988. Using fractal dimensions for characterizing the tortuosity of animal trails. *Physiological Entomology* 13: 393–398.

Didham, R. K., J. M. Tylianakis, N. J. Gemmell, T. A. Rand, and R. M. Ewers. 2007. Interactive effects of habitat modification and species invasion on native species decline. *Trends in Ecology and Evolution* 22: 489–496.

Didham, R. K., V. Kapos, and R. M. Ewers. 2012. Rethinking the conceptual foundations of habitat fragmentation research. *Oikos* 121: 161–170.

Diniz-Filho, J. A. F., T. Siqueira, A. A. Padial, T. F. Rangel, V. L. Landeiro, and L. M. Bini. 2010. Spatial autocorrelation analysis allows disentangling the balance between neutral and niche processes in metacommunities. *Oikos* 121: 201–210.

Dixon, P. M. 2002. Ripley's K function. Pp. 1796–1803 in *Encyclopedia of Environmetrics* (A. H. El-Shaarawi and W. W. Piegorsch, Editors). John Wiley & Sons, Chichester, UK.

Dobzhansky, T. 1950. Evolution in the tropics. *American Scientist* 38: 209–221.

Doligez, B., C. Cadet, E. Danchin, and T. Boulinier. 2003. When to use public information for breeding habitat selection? The role of environmental predictability and density dependence. *Animal Behaviour* 66: 973–998.

Dolnik, C. and M. Breuer. 2008. Scale dependency in the species-area relationship of plant communities. *Folia Geobotanica* 43: 305–318.

Donnelly, C. A. and R. Woodroffe. 2012. Reduce uncertainty in UK badger culling. *Nature* 485: 582.

Donnelly, C. A., R. Woodroffe, D. R. Cox, J. Bourne, G. Gettinby, A. M. Le Fevre, J. P. McInerney, and W. I. Morrison. 2003. Impact of localized badger culling on tuberculosis incidence in British cattle. *Nature* 426: 834–837.

Donnelly, C. A., and 13 others. 2006. Positive and negative effects of widespread badger culling on tuberculosis in cattle. *Nature* 439: 843–846.

Donovan, T. M., F. R. Thompson III, J. Faaborg, and J. Probst. 1995. Reproductive success of neotropical migrant birds in habitat sources and sinks. *Conservation Biology* 9: 1380–1395.

Donovan, T. M., P. W. Jones, E. M. Annand, and F. R. Thompson III. 1997. Variation in local-scale edge effects: Mechanisms and landscape context. *Ecology* 78: 2064–2075.

Douglas, P. M. J., M. Pagani, M. A. Canuto, M. Brenner, D. A. Hodell, T. I. Eglinton, and J. H. Curtis. 2015. Drought, agricultural adaptation, and sociopolitical collapse in the Maya Lowlands. *Proceedings of the National Academy of Sciences of the United States of America*: 112: 5607–5612.

Downs, J. A. and M. W. Horner. 2008. Effects of point pattern shape on home-range estimates. *Journal of Wildlife Management* 72: 1813–1818.

Dray, S, and 16 others. 2012. Community ecology in the age of multivariate multiscale spatial analysis. *Ecological Monographs* 82: 257–275.

Druce, D. J., J. S. Brown, G. I. Kerley, B. P. Kotler, R. L. Mackey, and R. Slotow. 2009. Spatial and temporal scaling in habitat utilization by klipsringers (*Oreotragus oreotragus*) determined using giving-up densities. *Austral Ecology* 34: 577–587.

Duffield, W. A. 2005. *Volcanoes of Northern Arizona*. Grand Canyon Association, Arizona.

Dullinger, S., and 20 others. 2012. Extinction debt of high-mountain plants under twenty-first-century climate change. *Nature Climate Change* 2: 619–622.

Duncan, R. P., P. Cassey, and T. M. Blackburn. 2009. Do climate envelope models transfer? A manipulative test using dung beetle introductions. *Proceedings of the Royal Society* B 267: 1449–1457.

Dunning, J. B., B. J. Danielson, and H. R. Pulliam. 1992. Ecological processes that affect populations in complex landscapes. *Oikos* 65: 169–175.

Dunning, J. B., Jr., D. J. Stewart, B. J. Danielson, B. R. Noon, T. L. Root, R. H. Lamberson, and E. E. Stevens. 1995. Spatially explicit population models: Current forms and future uses. *Ecological Applications* 5: 3–11.

Dupouey, J. L., E. Bambrine, J. D. Laffite, and C. Moares. 2002. Irreversible impact of past land use on forest soils and biodiversity. *Ecology* 83: 2978–2984.

Durrett, R. and S. Levin. 1996. Spatial models for species-area curves. *Journal of Theoretical Biology* 179: 119–127.

Dusek, R. J., R. G. McLean, L. D. Kramer, S. R. Ubico, A. P. Dupuis II, G. D. Ebel, and S. C. Guptill. 2009. Prevalence of West Nile virus in migratory birds during spring and fall migration. *American Journal of Tropical Medicine and Hygiene* 81: 1151–1158.

Dyer, S. J., J. P. O'Neill, S. M. Wasel, and S. Boutin. 2002. Quantifying barrier effects of roads and seismic lines on movements of female woodland caribou in northeastern Alberta. *Canadian Journal of Zoology* 80: 839–845.

Eckert, A. J., J. van Heerwaarden, J. L. Wegrzyn, C. D. Nelson, J. Ross-Ibarra, S. C. González-Martínez, and D. B. Neale. 2010a. Patterns of population structure and environmental associations to aridity across the range of loblolly pine (*Pinus taeda* L., Pinaceae). *Genetics* 185: 969–982.

Eckert, A. J., A. D. Bower, S. C. González-Martínez, J. L.Wegrzyn, G. Coop, and D. B. Neale. 2010b. Back to nature: Ecological genomics of loblolly pine (*Pinus taeda*, Pinaceae). *Molecular Ecology* 19: 3789–3805.

EDDMapS. 2015. Early Detection & Distribution Mapping System. The University of Georgia—Center for Invasive Species and Ecosystem Health. Available online at https://www.eddmaps.org/ (last accessed 2/15/19).

Edwards, A. M., and 10 others. 2007. Revisiting Lévy flight search patterns of wandering albatrosses, bumblebees and deer. *Nature* 449: 1044–1048.

Egan, T. 2006. *The Worst Hard Time: The Untold Story of Those Who Survived the Great American Dust Bowl.* Mariner Books, Houghton Mifflin Company, New York.

Ehleringer, J. R., T. E. Cerling, and B. R. Helliker. 1997. C_4 photosynthesis, atmospheric CO_2, and climate. *Oecologia* 112: 285–299.

Eldridge, D. J., M. A. Bowker, F. T. Maestre, E. Roger, J. F. Reynolds, and W. G. Whitford. 2011. Impacts of shrub encroachment on ecosystem structure and functioning: Towards a global synthesis. *Ecology Letters* 14: 709–722.

Elith, J. and J. R. Leathwick. 2009. Species distribution models: Ecological explanation and prediction across space and time. *Annual Review of Ecology and Systematics* 40: 677–697.

Elith, J., C. H. Graham, and 25 others. 2006. Novel methods improve prediction of species' distributions from occurrence data. *Ecography* 29: 129–151.

Elith, J., M. Kearney, and S. Phillips. 2010. The art of modeling range-shifting species. *Methods in Ecology and Evolution* 1: 330–342.

Elith, J., S. J. Phillips, T. Hastie, M. Dudík, Y. E. Chee, and C. J. Yates. 2011. A statistical explanation of MaxEnt for ecologists. *Diversity and Distributions* 17: 43–57.

Ellis, A. M., T. Václavík, and R. K. Meentemeyer. 2010. When is connectivity important? A case study of the spatial pattern of sudden oak death. *Oikos* 119: 485–493.

Ellis, E. C. and N. Ramankutty. 2008. Putting people in the map: Anthropogenic biomes of the world. *Frontiers in Ecology and the Environment* 6: 439–447.

Ellison, A. M. 2010. Partitioning diversity. *Ecology* 91: 1962–1963.

Elmqvist, T, C. Folke, M. Nyström, G. Peterson, J. Bengtsson, B. Walker, and J. Norberg. 2003. Response diversity, ecosystem change, and resilience. *Frontiers in Ecology and the Environment* 1: 488–494.

Elsner, J. B., A. B. Kara, and M. A. Owens. 1999. Fluctuations in North Atlantic hurricane frequency. *Journal of Climate* 12: 427–437.

Elton, C. and M. Nicholson. 1942. The ten-year cycle in numbers of the lynx in Canada. *Journal of Animal Ecology* 11: 215–244.

Ensign, S. H. and M. W. Doyle. 2006. Nutrient spiraling in streams and river networks. *Journal of Geophysical Research* 111: G04009.

Epperson, B. K., and 10 others. 2010. Utility of computer simulations in landscape genetics. *Molecular Ecology* 19: 3549–3564.

Epstein, P. R. 2000. Is global warming harmful to health? *Scientific American* 283: 50–57.

Erens, H., M. Boudin, F. Mees, B. B. Mujinya, G. Baert, M. Van Strydonck, P. Boeckx, and E. Van Ranst. 2015. The age of large termite mounds—radiocarbon dating of *Macrotermes falciger* mounds of the Miombo woodland of Katanga, DR Congo. *Palaeogeography, Palaeoclimatology, Palaeoecology* 435: 265–271.

Erickson, R. O. 1945. The *Clematis fremontii* var. *Riehlii* population in the Ozarks. *Annals of the Missouri Botanical Garden* 32: 413-460.

Eswaran, H. and P. Reich. 2007. Human impact on land systems of the world. *Soil Horizons* 48: 11–15.

Eswaran, H., P. Reich and T. Vearasilp. 2005. A global assessment of land quality in the Anthropocene. Pp. 119–131 in *Proceedings of the International Conference on Soil, Water and Environmental Quality.* Indian Society of Soil Science, Calcutta, India.

Etienne, R. S., C. J. F. ter Braak, and C. C. Vos. 2004. Application of stochastic patch occupancy models to real metapopulations. Pp. 105–132 in *Ecology, Genetics and Evolution of Metapopulations* (I. Hanski and O. E. Gaggiotti, Editors). Elsevier Academic Press, Burlington, Massachusetts.

Evanno, G., S. Regnaut, and J. Goudet. 2005. Detecting the number of clusters of individuals using the software STRUCTURE: A simulation study. *Molecular Ecology* 14: 2611–2620.

Evans, M. R., K. J. Norris, and T. G. Benton. 2012. Predictive ecology: Systems approaches. *Philosophical Transactions of the Royal Society B* 367: 163–169.

Evju, M. and A. Sverdrup-Thygeson. 2016. Spatial configuration matters: A test of the habitat amount hypothesis for plants in calcareous grasslands. *Landscape Ecology* 31: 1891–1902.

Ewers, R. M. and R. K. Didham. 2006. Confounding factors in the detection of species responses to habitat fragmentation. *Biological Reviews* 81: 117–142.

Excoffier, L., P. E. Smouse, and J. M. Quattro. 1992. Analysis of molecular variance inferred from metric distances among DNA haplotypes: Application to human mitochondrial DNA restriction data. *Genetics* 131: 479–491.

Excoffier, L., M. Foll, and R. J. Petit. 2009. Genetic consequences of range expansions. *Annual Review of Ecology, Evolution and Systematics* 40: 481–501.

Ezard, T. H. G. and J. M. J. Travis. 2006. The impact of habitat loss and fragmentation on genetic drift and fixation time. *Oikos* 114: 367–375.

Fabiszewski, A. M., J. Umbanhowar, and C. E. Mitchell. 2010. Modeling landscape-scale pathogen spillover between domesticated and wild hosts: Asian soybean rust and kudzu. *Ecological Applications* 20: 582–592.

Fagan, W. F., R. S. Cantrell, and C. Cosner. 1999. How habitat edges change species interactions. *American Naturalist* 153: 165–182.

Fahrig, L. 1991. Simulation methods for developing general landscape-level hypotheses of single-species dynamics. Pp. 417–442 in *Quantitative Methods in Landscape Ecology* (M. G. Turner and R. H. Gardner, Editors). Springer-Verlag, New York.

Fahrig, L. 1997. Relative effects of habitat loss and fragmentation on population extinction. *Journal of Wildlife Management* 61: 603–610.

Fahrig, L. 2003. Effects of habitat fragmentation on biodiversity. *Annual Review of Ecology and Systematics* 34: 487–515.

Fahrig, L. 2013. Rethinking patch size and isolation effects: The habitat amount hypothesis. *Journal of Biogeography* 40: 1649–1663.

Fahrig, L. and T. Rytwinski. 2009. Effects of roads on animal abundance: An empirical review and synthesis. *Ecology and Society* 14: 21. Available online: www.ecologyandsociety.org/vol14/iss1/art21.

Fahrig, L., J. H. Pedlar, S. E. Pope, P. D. Taylor, and J. F. Wegner. 1995. Effect of road traffic on amphibian density. *Biological Conservation* 73: 177–182.

Falk, D. A., C. Miller, D. McKenzie, and A. E. Black. 2007. Cross-scale analysis of fire regimes. *Ecosystems* 10: 809–823.

Falk, D. A., E. K. Heyerdahl, P. M. Brown, C. Farris, P. Z. Fulé, D. McKenzie, T. W. Swetnam, A. H. Taylor, and M. L. Van Horne. 2011. Multiscale controls of historical forest-fire regimes: New insights from fire-scar networks. *Frontiers of Ecology and the Environment* 9: 446–454.

Falush, D., M. Stephens, and J. K. Pritchard. 2007. Inference of population structure using multilocus genotype data: Dominant markers and null alleles. *Molecular Ecology Notes* 7: 574–578.

FAO (Food and Agriculture Organization of the United Nations). 2010. *Global Forest Resources Assessment 2010: Main report.* FAO Forestry Paper No. 160. Rome, Italy. Available online: www.fao.org/forestry/fra/fra2010/en/ (last accessed on 1/30/19).

FAO. 2015. AQUASTAT website, Food and Agriculture Organization of the United Nations. Available online at: www.fao.org/nr/water/aquastat/water_use/index.stm (last accessed on 1/30/19).

Fauchald, P. and T. Tveraa. 2003. Using first-passage time in the analysis of area-restricted search and habitat selection. *Ecology* 84: 282–288.

Faulks, L. K., D. M. Gilligan, and L. B. Beheregaray. 2011. The role of anthropogenic vs. natural in-stream structures in determining connectivity and genetic diversity in an endangered freshwater fish, Macquarie perch (*Macquaria australasica*). *Evolutionary Applications* 4: 589–601.

Fausch, K. D., C. E. Torgersen, C. V. Baxter, and H. W. Li. 2002. Landscapes to riverscapes: Bridging the gap between research and conservation of stream fishes. *BioScience* 52: 483–498.

Fensham, R. J. and R. J. Fairfax. 1997. The use of land survey records to reconstruct pre-European vegetation patterns in the Darling Downs, Queensland, Australia. *Journal of Biogeography* 24: 827–836.

Fernández, R., M. Novo, M. Gutiérrez, A. Almodóvar, and D. J. Díaz Cosín. 2010. Life cycle and reproductive traits of the earthworm *Aporrectodea trapezoides* (Dugèsm 1828) in laboratory cultures. *Pedobiologia* 53: 295–299.

Ferrari, J. R., T. R. Lookingbill, and M. C. Neel. 2007. Two measures of landscape-graph connectivity: Assessment across gradients in area and configuration. *Landscape Ecology* 22: 1315–1323.

Ferraz, S. F. B., C. A. Vettorazzi, D. M. Theobald, and M. V. R. Ballester. 2005. Landscape dynamics of Amazonian deforestation between 1984 and 2002 in central Rondônia, Brazil: Assessment and future scenarios. *Forest Ecology and Management* 204: 67–83.

Fieberg, J. and S. P. Ellner. 2000. When is it meaningful to estimate an extinction probability? *Ecology* 81: 2040–2047.

Fischer, J., and 11 others. 2008. Should agricultural policies encourage land sparing or wildlife-friendly farming? *Frontiers in Ecology and the Environment* 6: 380–385.

Flaherty, E. A., W. P. Smith, S. Pyare, and M. Ben-David. 2008. Experimental trials of the northern flying squirrel (*Glaucomys sabrinus*) traversing managed rainforest landscapes: Perceptual range and fine-scale movements. *Canadian Journal of Zoology* 86: 1050–1058.

Flaspohler, D. J., S. A. Temple, and R. Rosenfield. 2001. The effects of forest edges on ovenbird demography in a managed forest landscape. *Conservation Biology* 15: 173–183.

Flather, C. H. and M. Bevers. 2002. Patchy reaction-diffusion and population abundance: The relative importance of habitat amount and aggregation. *American Naturalist* 159: 40–56.

Fletcher, R. J., Jr. 2006. Emergent properties of conspecific attraction in fragmented landscapes. *American Naturalist* 168: 207–219.

Fletcher, R. J., Jr., L. Ries, J. Battin, and A. D. Chalfoun. 2007. The role of habitat area and edge in fragmented landscapes: Definitively distinct or inevitably intertwined? *Canadian Journal of Zoology* 85: 1017–1030.

Flinn, K. M. and M. Vellend. 2005. Recovery of forest plant communities in post-agricultural landscapes. *Frontiers in Ecology and the Environment* 3: 243–250.

Focardi, S., P. Montanaro, and E. Pecchioli. 2010. Adaptive Lévy walks in foraging fallow deer. *PLoS ONE* 4: e6587.

Foley, J. A., and 18 others. 2005. Global consequences of land use. *Science* 309: 570–574.

Foley, J., D. Clifford, K. Castle, P. Cryan, and R. S. Ostfeld. 2010. Investigating and managing the rapid emergence of white-nose syndrome, a novel, fatal, infectious disease of hibernating bats. *Conservation Biology* 25: 223–231.

Folke, C., S. Carpenter, B. Walker, M. Scheffer, T. Elmqvist, L. Gunderson, and C. S. Holling. 2004. Regime shifts, resilience, and biodiversity in ecosystem management. *Annual Review of Ecology, Evolution, and Systematics* 35: 557–581.

Foppen, R. P. B., J. P. Chardon, and W. Liefveld. 2000. Understanding the role of sink patches in source-sink metapopulations: Reed warbler in an agricultural landscape. *Conservation Biology* 14: 1881–1892.

Forman, R. T. T. 1995a. *Land Mosaics. The Ecology of Landscapes and Regions.* Cambridge University Press, Cambridge, UK.

Forman, R. T. T. 1995b. Some general principles of landscape and regional ecology. *Landscape Ecology* 10: 133–142.

Forman, R. T. T. 2000. Estimate of the area affected ecologically by the road system in the USA. *Conservation Biology* 14: 31–35.

Forman, R. T. T. 2014. *Urban Ecology: Science of Cities.* Cambridge University Press, Cambridge, UK.

Forman, R. T. T. and M. Godron. 1981. Patches and structural components for a landscape ecology. *BioScience* 31: 733–740.

Forman, R. T. T. and M. Godron. 1986. *Landscape Ecology.* John Wiley and Sons, New York.

Forman, R. T. T. and L. E. Alexander. 1998. Roads and their major ecological effects. *Annual Review of Ecology and Systematics* 29: 207–231.

Forman, R. T. T., and 13 others. 2003. *Road Ecology: Science and Solutions.* Island Press, Washington, D.C.

Fornace, K. M., C. J. Drakeley, T. William, F. Espino, and J. Cox. 2014. Mapping infectious disease landscapes: Unmanned aerial vehicles and epidemiology. *Trends in Parasitology* 30: 514–519.

Forsman, E. D., and 26 others. 2011. Demography of the northern spotted owl. *Studies in Avian Biology*, Volume 40. University of California Press, Berkeley, California.

Fortin, D., R. Courtois, P. Etcheverry, C. Dussault, and A. Gingras. 2008. Winter selection of landscapes by woodland caribou: Behavioural response to geographical gradients in habitat attributes. *Journal of Applied Ecology* 45: 1392–1400.

Fortin, M.-J. and M. R. T. Dale. 2005. *Spatial Analysis: A Guide for Ecologists.* Cambridge University Press, Cambridge, UK.

Fortin, M.-J., T. H. Keitt, B. A. Mauer, M. L. Taper, D. M. Kaufman, and T. M. Blackburn. 2005. Species' geographic ranges and distributional limits: Pattern analysis and statistical issues. *Oikos* 108: 7–17.

Foster, D. R. 2002. Insights from historical geography to ecology and conservation: Lessons from the New England landscape. *Journal of Biogeography* 29: 1269–1275.

Foster, D. R. and E. R. Boose. 1992. Patterns of forest damage resulting from catastrophic wind in central New England, USA. *Journal of Ecology* 80: 79–98.

Foster, D. R., D. H. Knight, and J. F. Franklin. 1998a. Landscape patterns and legacies resulting from large, infrequent forest disturbances. *Ecosystems* 1: 497–510.

Foster, D. R., G. Motzkin, and B. Slater. 1998b. Land-use history as long-term broad-scale disturbance: Regional forest dynamics in central New England. *Ecosystems* 1: 96–119.

Foster, D., F. Swanson, J. Aber, I. Burke, N. Brokaw, D. Tilman, and A. Knapp. 2003. The importance of land-use legacies to ecology and conservation. *BioScience* 53: 77–88.

Foulger, G. R. 2010. *Plates vs. Plumes: A Geological Controversy*. Wiley-Blackwell, Chichester, UK.

Fowler, N. L. 1988. What is a safe site?: Neighbor, litter, germination date, and patch effects. *Ecology* 69: 947–961.

Fox, J. W. 2012. The intermediate disturbance hypothesis should be abandoned. *Trends in Ecology and Evolution* 28: 86–92.

Fox, M. D. and B. D. Fox. 1986. The susceptibility of communities to invasion. Pp. 97–105 in *Ecology of Biological Invasions: An Australian Perspective* (R. H. Groves and J. J. Burdon, Editors). Australian Academy of Science, Canberra, ACT, Australia.

Francis, C. D., C. P. Ortega, and A. Cruz. 2009. Noise pollution changes avian communities and species interactions. *Current Biology* 19: 1415–1419.

François, O. and E. Durand. 2010. Spatially explicit Bayesian clustering models in population genetics. *Molecular Ecology Resources* 10: 773–784.

Frankham, R., J. D. Ballou, and D. A. Briscoe. 2010. *Introduction to Conservation Genetics*, 2nd edition. Cambridge University Press, Cambridge, UK.

Franklin, J. 2010a. *Mapping Species Distributions: Spatial Inference and Prediction*. Cambridge University Press, Cambridge, UK.

Franklin, J. 2010b. Moving beyond static species distribution models in support of conservation and biogeography. *Diversity and Distributions* 16: 321–330.

Franklin, J. F. 1990. Biological legacies: A critical management concept from Mount St Helens. *Transactions of the North American Wildlife and Natural Resources Conference* 55: 216–219.

Franklin, J. F. and T. A. Spies. 1984. Characteristics of old-growth Douglas-fir forests. Pp. 328–334 in *Proceedings of the Society of American Foresters National Convention*. Society of American Foresters, Washington, D.C.

Franzén, M., O. Schweiger, and P.-E. Betzholtz. 2012. Species-area relationships are controlled by species traits. *PLoS ONE* 7: e37359.

Fraser, L. H., and 61 others. 2015. Worldwide evidence of a unimodal relationship between productivity and plant species richness. *Science* 349: 302–305.

Fraterrigo, J. M., M. G. Turner, S. M. Pearson, and P. Dixon. 2005. Effects of past land use on spatial heterogeneity of soil nutrients in southern Appalachian forests. *Ecological Monographs* 75: 215–230.

Freeman, C. C. and L. C. Hulbert. 1985. An annotated list of the vascular flora of Konza Prairie Research Natural Area, Kansas. *Transactions of the Kansas Academy of Science* 88: 84–115.

Fretwell, S. D. 1972. *Populations in a Seasonal Environment*. Princeton University Press, New Jersey.

Fretwell, S. D. and H. L. Lucas, Jr. 1970. On territorial behavior and other factors influencing habitat distribution in birds. I. Theoretical development. *Acta Biotheoretica* 19: 16–36.

Fridley, J. D., and 12 others. 2012. Comment on 'Productivity is a poor predictor of plant species richness.' *Science* 335: 1441.

Frissell, C. A., W. J. Liss, C. E. Warren, and M. D. Hurley. 1986. A hierarchical framework for stream habitat classification: Viewing streams in a watershed context. *Environmental Management* 10: 199–214.

Froese, R. 2006. Cube law, condition factor and weight-length relationships: History, meta-analysis and recommendations. *Journal of Applied Ichthyology* 22: 241–253.

Fu, B. J. and Y. H. Lu. 2006. The progress and perspectives of landscape ecology in China. *Progress in Physical Geography* 30: 232–244.

Fudickar, A. M., M. Wikelski, and J. Partecke. 2012. Tracking migratory songbirds: Accuracy of light-level loggers (geolocators) in forest habitats. *Methods in Ecology and Evolution* 3: 47-52.

Fuhlendorf, S. D. and D. M. Engle. 2001. Restoring heterogeneity on rangelands: Ecosystem management based on evolutionary grazing patterns. *BioScience* 51: 625–632.

Fuhlendorf, S. D. and D. M. Engle. 2004. Application of the fire-grazing interaction to restore a shifting mosaic on tallgrass prairie. *Journal of Applied Ecology* 41: 604–614.

Fuhlendorf, S. D., W. C. Harrell, D. M. Engle, R. G. Hamilton, C. A. Davis, and D. M. Leslie, Jr. 2006. Should heterogeneity be the basis for conservation? Grassland bird response to fire and grazing. *Ecological Applications* 16: 1706–1716.

Fuhlendorf, S. D., D. M. Engle, J. Kerby, and R. Hamilton. 2009. Pyric herbivory: Rewilding landscapes through the recoupling of fire and grazing. *Conservation Biology* 23: 588–598.

Fullerton, A. H., K. M. Burnett, E. A. Steel, R. L. Flitcroft, G. R. Pess, B. E. Feist, C. E. Torgersen, D. J. Miller and B. L. Sanderson. 2010. Hydrological connectivity for riverine fish: Measurement challenges and research opportunities. *Freshwater Biology* 55: 2215–2237.

Gabet, E. J., O. J. Reichman, and E. W. Seabloom. 2003. The effects of bioturbation on soil processes and sediment transport. *Annual Review of Earth and Planetary Sciences* 31: 249–273.

Gabet, E. J., J. T. Perron, and D. L. Johnson. 2014. Biotic origin for Mima mounds supported by numerical modelling. *Geomorphology* 206: 58–66.

Gallai, N., J.-M. Salles, J. Settele, and B. E. Vaissière. 2009. Economic valuation of the vulnerability of world agriculture confronted with pollinator decline. *Ecological Economics* 68: 810–821.

Gallardo, A., J. J. Rodríguez-Saucedo, F. Covelo, and R. Fernández-Alés. 2000. Soil nitrogen heterogeneity in a Dehesa ecosystem. *Plant and Soil* 222: 71–82.

Gálvez, R., M. A. Descalzo, I. Guerrero, G. Miró, and R. Molina. 2011. Mapping the current distribution and predicted spread of the leishmaniosis sand fly vector in the Madrid region (Spain) based on environmental variables and expected climate change. *Vector-Borne and Zoonotic Diseases* 11: 799–806.

Gandon, S. 2002. Local adaptation and the geometry of host-parasite coevolution. *Ecology Letters* 5: 246–256.

García, D. M., J. M. Herrera, and J. M. Morales. 2013. Functional heterogeneity in a plant-frugivore assemblage enhances seed dispersal resilience to habitat loss. *Ecography* 36: 197–2018.

Gardner, M. G., A. J. Fitch, T. Bertozzi, and A. J. Lowe. 2011. Rise of the machines—recommendations for ecologists when using next generation sequencing for microsatellite development. *Molecular Ecology Resources* 11: 1093–1101.

Gardner, R. H. 1992. Pattern, process, and the analysis of spatial scales. Pp. 17–35 in *Ecological Scale: Theory and Applications* (D. L. Peterson and V. T. Parker, Editors). Columbia University Press, New York.

Gardner, R. H. and R. V. O'Neill. 1991. Pattern, process, and predictability: The use of neutral models for landscape analysis. Pp. 289–307 in *Quantitative Methods in Landscape Ecology* (M. G. Turner and R. H. Gardner, Editors). Springer-Verlag, New York.

Gardner, R. H. and D. Urban. 2007. Neutral models for testing landscape hypotheses. *Landscape Ecology* 22: 15–29.

Gardner, R. H., B. T. Milne, M. G. Turner, and R. V. O'Neill. 1987. Neutral models for the analysis of broad-scale landscape pattern. *Landscape Ecology* 1: 19–28.

Gardner, T. A., I. M. Côté, J. A. Gill, A. Grant, and A. R. Watkinson. 2003. Long-term region-wide declines in Caribbean corals. *Science* 301: 958–960.

Gardner, T. A., I. M. Côté, J. A. Gill, A. Grant, and A. R. Watkinson. 2005. Hurricanes and Caribbean coral reefs: Impacts, recovery patterns, and role in long-term decline. *Ecology* 86: 174–184.

Garibaldi, L. A., and 22 others. 2011. Stability of pollination services decreases with isolation from natural areas despite honey bee visits. *Ecology Letters* 14: 1062–1072.

Garroway, C. J., J. Bowman, D. Carr, and P. J. Wilson. 2008. Applications of graph theory to landscape genetics. *Evolutionary Applications* 1: 620–630.

Gaston, K. J. 2000. Global patterns in biodiversity. *Nature* 405: 220–227.

Gaston, K. J. 2007. Latitudinal gradient in species richness. *Current Biology* 17(15): R574.

Gaston, K. J., Editor. 2010. *Urban Ecology*. Cambridge University Press, Cambridge, UK.

Gaston, K. J. and T. M. Blackburn. 1996. The tropics as a museum of biological diversity: An analysis of the New World avifauna. *Proceedings of the Royal Society of London B* 263: 63–68.

Gaston, K. J. and T. M. Blackburn. 2000. *Pattern and Process in Macroecology*. Blackwell Science, Malden, Massachusetts.

Gates, J. E. and L. W. Gysel. 1978. Avian nest dispersion and fledgling success in field-forest ecotones. *Ecology* 59: 871–883.

Gause, G. F. 1934. *The Struggle for Existence*. Williams & Wilkins, Baltimore, Maryland.

Gavier-Pizarro, G. I., V. C. Radeloff, S. I. Stewart, C. D. Huebner, and N. S. Keuler. 2010. Housing is positively associated with invasive exotic plant species richness in New England, USA. *Ecological Applications* 20: 1913–1925.

Geary, R. C. 1954. The contiguity ratio and statistical mapping. *Incorporated Statistician* 5: 115–145.

Gedalof, Z. 2011. Climate and spatial patterns of wildfire in North America. Pp. 89–115 in *The Landscape Ecology of Fire* (D. McKenzie, C. Miller, and D. A. Falk, Editors). Springer, Dordrecht, the Netherlands.

Gehlke, C. E. and K. Biehl. 1934. Certain effects of grouping upon the size of the correlation coefficient in census tract material. *Journal of the American Statistical Association* 29: 169–170.

Gende, S. M., R. T. Edwards, M. F. Willson, and M. S. Wipfli. 2002. Pacific salmon in aquatic and terrestrial ecosystems. *BioScience* 52: 917–928.

Gerlak, A. K., F. Zamora-Arroyo, and H. P. Kahler. 2013. A delta in repair: Restoration, binational cooperation, and the future of the Colorado River Delta. *Environment: Science and Policy for Sustainable Development* 55: 29–40.

Gerlanc, N. M. and G. A. Kaufman. 2003. Use of bison wallows by anurans on Konza Prairie. *American Midland Naturalist* 150: 158–168.

Getis, A. and J. Franklin. 1987. Second-order neighborhood analysis of mapped point patterns. *Ecology* 68: 473–477.

Getis, A. and J. K. Ord. 1992. The analysis of spatial association by use of distance statistics. *Geographical Analysis* 24: 189–206.

Getis, A. and J. K. Ord. 1996. Local spatial statistics: An overview. Pp. 261–277 in *Spatial Analysis: Modelling in a GIS Environment* (P. Longley and M. Batty, Editors). John Wiley & Sons, New York.

Getis. A. and D. A. Griffith. 2002. Comparative spatial filtering in regression analysis. *Geographical Analysis* 34: 130–140.

Getz, W. M. and C. C. Wilmers. 2004. A local nearest-neighbor convex-hull construction of home ranges and utilization distributions. *Ecography* 27: 489–505.

Getz, W. M., S. Fortmann-Roe, P. C. Cross, A. J. Lyons, J. Ryan, and C. C. Wilmers. 2007. LoCoH: Non-parametric kernel methods for constructing home ranges and utilization distributions. *PLoS ONE* 2: e207.

Getzin, S., R. S. Nuske, and K. Wiegand. 2014. Using unmanned aerial vehicles (UAV) to quantify spatial gap patterns in forests. *Remote Sensing* 6: 6988–7004.

Gibbs, H. K., A. S. Ruesch, F. Achard, M. K. Clayton, P. Holmgren, N. Ramankutty, and J. A. Foley. 2010. Tropical forests were the primary sources of new agricultural land in the 1980s and 1990s. *Proceedings of the National Academy of Sciences of the United States of America* 107: 16732–16737.

Gibbs, J. P. 1998. Amphibian movements in response to forest edges, roads, and streambeds in southern New England. *Journal of Wildlife Management* 62: 584–589.

Gibson, D. J. and L. C. Hulbert. 1987. Effects of fire, topography and year-to-year climatic variation on species composition in tallgrass prairie. *Vegetatio* 72: 175–185.

Gilbert, B. and J. M. Levine. 2013. Plant invasions and extinction debts. *Proceedings of the National Academy of Sciences of the United States of America* 110: 1744–1749.

Gill, L. and A. H. Taylor. 2009. Top-down and bottom-up controls on fire regimes along an elevational gradient on the east slope of the Sierra Nevada, California, USA. *Fire Ecology* 5: 57–75.

Gill, R. D. 2000. *The Great Maya Droughts: Water, Life, and Death*. University of New Mexico Press, Albuquerque, New Mexico.

Gillespie, T. R. and C. A. Chapman. 2007. Forest fragmentation, the decline of an endangered primate, and changes in host-parasite interactions relative to an unfragmented forest. *American Journal of Primatology* 70: 222–230.

Gillman, L. N. and S. D. Wright, S. D. 2014. Species richness and evolutionary speed: The influence of temperature, water and area. *Journal of Biogeography* 41: 39–51.

Gillson, L. 2004. Evidence of hierarchical patch dynamics in an East African savanna? *Landscape Ecology* 19: 883–894.

Gitzen, R. A., J. J. Millspaugh, and B. J. Kernohan. 2006. Bandwidth selection for fixed-kernel analysis of animal utilization distributions. *Journal of Wildlife Management* 70: 1334–1344.

Gleason, H. A. 1917. The structure and development of the plant association. *Bulletin of the Torrey Botanical Club* 43: 463–481.

Gleason, H. A. 1926. The individualistic concept of the plant association. *Bulletin of the Torrey Botanical Club* 53: 7–26.

Gleeson, T., Y. Wada, M. F. P. Bierkens, and L. P. H. van Beek. 2012. Water balance of global aquifers revealed by groundwater footprint. *Nature* 488: 197–200.

Glenn, E. P. and P. L. Nagler. 2005. Comparative ecophysiology of *Tamarix ramosissima* and native trees in the western U.S. riparian zones. *Journal of Arid Environments* 61: 419–446.

Godfray, H. C. J., J. R. Beddington, I. R. Crute, L. Haddad, D. Lawrence, J. F. Muir, J. Pretty, S. Robinson, S. M. Thomas, and C. Toulmin. 2010. Food security: The challenge of feeding 9 billion people. *Science* 327: 812–818.

Golley, F. B. 1987. Introducing landscape ecology. *Landscape Ecology* 1: 1–3.

Gomi, T., R. C. Sidle, and J. S. Richardson. 2002. Understanding processes and downstream linkages of headwater systems. *BioScience* 52: 905–916.

González, C., O. Wang, S. E. Strutz, C. González-Salazar, V. Sánchez-Cordero, and S. Sarkar. 2010. Climate change and risk of leishmaniasis in North America: Predictions from ecological niche models of vector and reservoir species. *PLoS Neglected Tropical Diseases* 4: e585.

González-Martínez, S. C., E. Ersoz, G. R. Brown, N. C. Wheeler, and D. B. Neale. 2006. DNA sequence variation and selection of tag SNPs at candidate genes for drought-stress response in *Pinus taeda* L. *Genetics* 172: 1915–1926.

González-Varo, J. P., C. S. Carvalho, J. M. Arroyo, and P. Jordano. 2017. Unravelling seed dispersal through fragmented landscapes: Frugivore species operate unevenly as mobile links. *Molecular Ecology* 26: 4309–4321.

Goodwin, B. J. 2003. Is landscape connectivity a dependent or independent variable? *Landscape Ecology* 18: 687–699.

Gordon, L. J., G. D. Peterson, and E. M. Bennett. 2008. Agricultural modifications of hydrological flows create ecological surprises. *Trends in Ecology and Evolution* 23: 211–219.

Gornall, J., R. Betts, E. Burke, R. Clark, J. Camp, K. Willett, and A. Wiltshire. 2010. Implications of climate change for agricultural productivity in the early twenty-first century. *Philosophical Transactions of the Royal Society B* 365: 2973–2989.

Gorte, R. W., C. H. Vincent, L. A. Hanson, and M. R. Rosenblum. 2012. *Federal Land Ownership: Overview and Data*. Congressional Research Service Report R42346, Washington, D.C.

Gotelli, N. J. and R. K. Colwell. 2001. Quantifying biodiversity: Procedures and pitfalls in the measurement and comparison of species richness. *Ecology Letters* 4: 379–391.

Graf, W. L. 2006. Downstream hydrologic and geomorphic effects of large dams on American rivers. *Geomorphology* 79: 336–360.

Graham, C. H., N. Silva, and J. Velásquez-Tibatá. 2010. Evaluating the potential causes of range limits of birds of the Colombian Andes. *Journal of Biogeography* 37: 1863–1875.

Graham, C. H., and 14 others. 2014. The origin and maintenance of montane diversity: Integrating evolutionary and ecological processes. *Ecography* 37: 711–719.

Gravel, D., F. Guichard, M. Loreau, and N. Mouquet. 2010. Source and sink dynamics in meta-ecosystems. *Ecology* 91: 2172–2184.

Graybill, D. A., D. A. Gregory, G. S. Funkhouser, and F. L. Nials. 2006. Long-term streamflow reconstructions, river channel morphology, and aboriginal irrigation systems along the Salt and Gila Rivers. Pp 69–123 in *Environmental Change and Human Adaptation in the Ancient American Southwest* (D. E. Doyel and J. S. Dean, Editors). University of Utah Press, Salt Lake City, Utah.

Green, D. G. 1994. Connectivity and complexity in landscapes and ecosystems. *Pacific Conservation Biology* 1: 194–200.

Green, P. T., D. J. O'Dowd, K. L. Abbott, M. Jeffery, K. Retallick, and R. MacNally. 2011. Invasional meltdown: Invader-invader mutualism facilitates a secondary invasion. *Ecology* 92: 1758–1768.

Green, R. E., S. J. Cornell, J. P. W. Scharlemann, and A. Balmford. 2005. Farming and the fate of wild nature. *Science* 307: 550–555.

Greenwood, P. J. 1980. Mating systems, philopatry and dispersal in birds and mammals. *Animal Behaviour* 7: 165–167.

Greig-Smith, P. 1952. The use of random and contiguous quadrats in the study of structure in plant communities. *Annals of Botany* 16: 293–316.

Grenouillet, G., S. Brosse, L. Tudesque, S. Lek, Y. Baraillé, and G. Loot. 2008. Concordance among stream assemblages and spatial autocorrelation along a fragmented gradient. *Diversity and Distributions* 14: 592–603.

Griffin, D. and K. J. Anchukaitis. 2014. How unusual is the 2012–2014 California drought? *Geophysical Research Letters* 41: 9017–9023.

Grime, J. P. 1973a. Competitive exclusion in herbaceous vegetation. *Nature* 242: 344–347.

Grime, J. P. 1973b. Control of species density in herbaceous vegetation. *Journal of Environmental Management* 1: 151–167.

Grimm, N. B., S. H. Faeth, N. E. Golubiewski, C. L. Redman, J. Wu, X. Bai, and J. M. Briggs. 2008. Global change and the ecology of cities. *Science* 319: 756–760.

Grinnell, J. 1917. The niche-relationships of the California Thrasher. *Auk* 34: 427–433.

Groffman, P. M., and 15 others. 2006. Ecological thresholds: The key to successful environmental management or an important concept with no practical application? *Ecosystems* 9: 1–13.

Groom, G., C. A. Mücher, M. Ihse, and T. Wrbka. 2006. Remote sensing in landscape ecology: Experiences and perspectives in a European context. *Landscape Ecology* 21: 391–408.

Groucutt, H. S., and 16 others. 2015. Rethinking the dispersal of *Homo sapiens* out of Africa. *Evolutionary Anthropology* 24: 149–164.

Grunwald, S., Editor. 2005. *Environmental Soil-Landscape Modeling: Geographic Information Technologies and Pedometrics*. CRC Press, Taylor & Francis Group, Boca Raton, Florida.

Grytnes, J. A. and O. R. Vetaas. 2002. Species richness and altitude: A comparison between null models and interpolated plant species richness along the Himalayan altitudinal gradient, Nepal. *American Naturalist* 159: 294–304.

Gu, W., J. Utzinger, and R. J. Novak. 2008. Habitat-based larval interventions: A new perspective for malaria control. *American Journal of Tropical Medicine and Hygiene* 78: 2–6.

Guichard, F. 2017. Recent advances in metacommunities and metaecosystem theories. *F1000 Research* 6: 610.

Guichoux, E., and 10 others. 2011. Current trends in microsatellite genotyping. *Molecular Ecology Resources* 11: 591–611.

Guillot, G. and F. Rousset. 2013. Dismantling the Mantel tests. *Methods in Ecology and Evolution* 4: 336–344.

Guillot, G., F. Mortier, and A. Estoup. 2005. GENELAND: A computer package for landscape genetics. *Molecular Ecology Notes* 5: 708–711.

Guillot, G., F. Santos, and A. Estoup. 2008. Analysing georeferenced population genetics data with GENELAND: A new algorithm to deal with null alleles and a friendly graphical user interface. *Bioinformatics* 24: 1406–1407.

Guisan, A. and N. E. Zimmermann. 2000. Predictive habitat distribution models in ecology. *Ecological Modelling* 135: 147–186.

Guisan, A., T. C. Edwards, Jr., and T. Hastie. 2002. Generalized linear and generalized additive models in studies of species distributions: Setting the scene. *Ecological Modelling* 157: 89–100.

Gundersen, G., E. Johannesen, H. P. Andreassen, and R. A. Ims. 2001. Source-sink dynamics: How sinks affect demography of sources. *Ecology Letters* 4: 14–21.

Gunderson, L. H. 2000. Ecological resilience—In theory and application. *Annual Review of Ecology and Systematics* 31: 425–439.

Guo, L. B. and R. M. Gifford. 2002. Soil carbon stocks and land use change: A meta analysis. *Global Change Biology* 8: 345–360.

Guo, Y., P. Gong, and R. Amundson. 2003. Pedodiversity in the United States of America. *Geoderma* 117: 99–115.

Gustafson, E. J. 1998. Quantifying landscape spatial pattern: What is the state of the art? *Ecosystems* 1: 143–156.

Gustafson, E. J. and G. R. Parker. 1992. Relationships between landcover proportion and indices of landscape spatial pattern. *Landscape Ecology* 7: 101–110.

Gustafson, E. J., and R. H. Gardner. 1996. The effect of landscape heterogeneity on the probability of patch colonization. *Ecology* 77: 94–107.

Guthrie, R. H. 2002. The effects of logging on frequency and distribution of landslides in three watersheds on Vancouver Island, British Columbia. *Geomorphology* 43: 273–292.

Gutiérrez, J. L., C. G. Jones, D. L. Strayer, and O. O. Iribarne. 2003. Mollusks as ecosystem engineers: The role of shell production in aquatic habitats. *Oikos* 101: 79–90.

Gutzwiller, K. J. and S. H. Anderson. 1992. Interception of moving organisms: Influences of patch shape, size, and orientation on community structure. *Landscape Ecology* 6: 293–303.

Haas, S. E., M. B. Hooten, D. M. Rizzo, and R. K. Meentemeyer. 2011. Forest species diversity reduces disease risk in a generalist plant pathogen invasion. *Ecology Letters* 14: 1108–1116.

Haase, P. 1995. Spatial pattern analysis in ecology based on Ripley's K-function: Introduction and methods of edge correction. *Journal of Vegetation Science* 6: 575–582.

Haberl, H., K. H. Erb, F. Krausmann, V. Gaube, A. Bondeau, C. Plutzar, S. Gingrich, W. Lucht, and M. Fischer-Kowalski. 2007. Quantifying and mapping the human appropriation of net primary production in earth's terrestrial ecosystems. *Proceedings of the National Academy of Sciences of the United States of America* 104: 12942–12947.

Haddad, N. M. 1999. Corridor use predicted from behaviors at habitat boundaries. *American Naturalist* 153: 215–227.

Haddad, N. M., and 11 others. 2014. Potential negative ecological effects of corridors. *Conservation Biology* 28: 1178–1187.

Haddad, N. M., and 23 others. 2015. Habitat fragmentation and its lasting impact on Earth's ecosystems. *Science Advances* 1: e1500052.

Hadley, A. S. and M. G. Betts. 2012. The effects of landscape fragmentation on pollination dynamics: Absence of evidence is not evidence of absence. *Biological Reviews* 87: 526–544.

Haffer, J. 1969. Speciation in Amazon forest birds. *Science* 165: 131–137.

Hagedorn, K. 2007. Towards an institutional theory of multifunctionality. Pp. 105–124 in *Multifunctional Land Use: Meeting Future Demands for Landscape Goods and Services* (Ü. Mander, H. Wiggering, and K. Helming, Editors), Springer, Berlin, Germany.

Hagenaars, T. J., C. A. Donnelly, and N. M. Ferguson. 2004. Spatial heterogeneity and the persistence of infectious diseases. *Journal of Theoretical Biology* 229: 349–359.

Hahn, M. B., R. E. Gangnon, C. Barcellos, G. P. Asner, and J. A. Patz. 2014. Influence of deforestation, logging, and fire on malaria in the Brazilian Amazon. *PLOS ONE* 9: e85725.

Haila, Y. 2002. A conceptual genealogy of fragmentation research: From island biogeography to landscape ecology. *Ecological Applications* 12: 321–334.

Haining, R. 2003. *Spatial Data Analysis: Theory and Practice*. Cambridge University Press, Cambridge, UK.

Hall, S. R., K. D. Lafferty, J. H. Brown, C. E. Cáceres, J. M. Chase, A. P. Dobson, R. D. Holt, C. G. Jones, S. E. Randolph, and P. Rohani. 2008. Is infectious disease just another type of predator-prey interaction? Pp. 223–241 in *Infectious Disease Ecology: Effects of Ecosystems on Disease and of Disease on Ecosystems*. (R. S. Osteld, F. Keesing, and V. T. Eviner, Editors). Princeton University Press, Princeton, New Jersey.

Hamilton, W. D. 1971. Geometry for the selfish herd. *Journal of Theoretical Biology* 31: 295–311.

Hammond, J. I., B. Luttbeg, T. Brodin, and A. Sih. 2012. Spatial scale influences the outcome of the predator-prey space race between tadpoles and predatory dragonflies. *Functional Ecology* 26: 522–531.

Hansen, M. J. and A. P. Clevenger. 2005. The influence of disturbance and habitat on the presence of non-native plant species along transport corridors. *Biological Conservation* 125: 249–259.

Hansen, M. M., I. Olivieri, D. M. Waller, E. E. Nielsen and The GeM Working Group. 2012. Monitoring adaptive genetic responses to environmental change. *Molecular Ecology* 21: 1311–1329.

Hanski, I. 1994. A practical model of metapopulation dynamics. *Journal of Animal Ecology* 63: 151–162.

Hanski, I. 1995. Effects of landscape pattern on competitive interactions. Pp. 203–224 in *Mosaic Landscapes and Ecological Processes*

(L. Hansson, L. Fahrig, and G. Merriam, Editors). Chapman & Hall, London, UK.

Hanski, I. 1999. *Metapopulation Ecology*. Oxford University Press, UK.

Hanski, I. 2011. Habitat loss, the dynamics of biodiversity, and a perspective on conservation. *Ambio* 40: 248–255.

Hanski, I. and O. Ovaskainen. 2000. The metapopulation capacity of a fragmented landscape. *Nature* 404: 755–758.

Hanski, I. and O. Ovaskainen. 2002. Extinction debt at extinction threshold. *Conservation Biology* 16: 666–673.

Hanski, I., G. A. Zurita, M. I. Bellocq, and J. Rybicki. 2013. Species-fragmented area relationship. *Proceedings of the National Academy of Sciences of the United States of America* 110: 12715–12720.

Hansson, L. 1991. Dispersal and connectivity in metapopulations. *Biological Journal of the Linnean Society* 42: 89–103.

Hardin, G. 1960. The competitive exclusion principle. *Science* 131: 1292–1297.

Harding, J. S., E. F. Benfield, P. V. Bolstad, G. S. Helfman, and E. B. D. Jones. 1998. Stream biodiversity: The ghost of land use past. *Proceedings of the National Academy of Sciences of the United States of America* 95: 14843–14847.

Hargis, C. D., J. A. Bissonette, and J. L. David. 1998. The behavior of landscape metrics commonly used in the study of habitat fragmentation. *Landscape Ecology* 13: 167–186.

Hargrove, W. W. and J. Pickering. 1992. Pseudo-replication: A *sine qua non* for regional ecology. *Landscape Ecology* 6: 251–258.

Harper, J. L., J. N. Clatworthy, I. H. McNaughton, and G. R. Sagar. 1961. The evolution and ecology of closely related species living in the same area. *Evolution* 15: 209–227.

Harper, J. L., P. H. Lovell, and K. G. Moore. 1970. The shapes and sizes of seeds. *Annual Review of Ecology and Systematics* 1: 327–356.

Harris, S., W. J. Cresswell, P. G. Forde, W. J. Trewhella, T. Woolard, and S. Wray. 1990. Home-range analysis using radio-tracking data—a review of problems and techniques particularly as applied to the study of mammals. *Mammal Review* 20: 97–123.

Harrison, R. G. 1980. Dispersal polymorphisms in insects. *Annual Review of Ecology and Systematics* 11: 95–118.

Harrison, S., D. D. Murphy, and P. R. Ehrlich. 1988. Distribution of the bay checkerspot butterfly, *Euphydryas editha bayensis*: Evidence for a metapopulation model. *American Naturalist* 132: 360–382.

Harte, J. and A. P. Kinzig. 1997. On the implications of species–area relationships for endemism, spatial turnover, and food web patterns. *Oikos* 80: 417–427.

Harte, J., A. B. Smith, and D. Storch. 2009. Biodiversity scales from plots to biomes with a universal species-area curve. *Ecology Letters* 12: 789–797.

Hastings, A. 1980. Disturbance, coexistence, history, and competition for space. *Theoretical Population Biology* 18: 363–373.

Hastings, A., and 13 others. 2005. The spatial spread of invasions: New developments in theory and evidence. *Ecology Letters* 8: 91–101.

Hastings, A., J. E. Byers, J. A. Crooks, K. Cuddington, C. G. Jones, J. G. Lambrinos, T. S. Talley, and W. G. Wilson. 2007. Ecosystem engineering in space and time. *Ecology Letters* 10: 153–164.

Haug, G. H., D. Günther, L. C. Peterson, D. M. Sigman, K. A. Hughen, and B. Aeschlimann. 2003. Climate and the collapse of the Maya Civilization. *Science* 299: 1731–1735.

Hawkes, C. 2009. Linking movement behavior, dispersal and population processes: Is individual variation a key? *Journal of Animal Ecology* 78: 894–906.

Hawkins, B. A. 2012. Eight (and a half) deadly sins of spatial analysis. *Journal of Biogeography* 39: 1–9.

Hawkins, B. A. and H. V. Cornell. 1994. Maximum parasitism rate and successful biological control. *Science* 262: 1886.

Hawkins, B. A., M. B. Thomas, and M. E. Hochberg. 1993. Refuge theory and biological control. *Science* 262: 1429–1432.

Hawkins, B. A., and 11 others. 2003. Energy, water, and broad-scale geographic patterns of species richness. *Ecology* 84: 3105–3117.

Haynes, G. 2012. Elephants (and extinct relatives) as earth-movers and ecosystem engineers. *Geomorphology* 157–158: 99–107.

Haynes, K. J. and J. T. Cronin. 2003. Matrix composition affects the spatial ecology of a prairie planthopper. *Ecology* 84: 2856–2866.

Haynes, K. J. and J. T. Cronin. 2006. Interpatch movement and edge effects: The role of behavioral responses to the landscape matrix. *Oikos* 113: 43–54.

He, F. and R. Condit. 2007. The distribution of species: Occupancy, scale, and rarity. Pp. 32–50 in *Scaling Biodiversity* (D. Storch, P. A. Marquet, and J. H. Brown, Editors), Cambridge University Press, Cambridge, UK.

He, F. and S. P. Hubbell. 2011. Species-area relationships always overestimate extinction rates from habitat loss. *Nature* 473: 368–371.

He, H. S., B. E. DeZonia, and D. J. Mladenoff. 2000. An aggregation index (AI) to quantify spatial patterns of landscapes. *Landscape Ecology* 15: 591–601.

He, Q., D. L. Edwards, and L. L. Knowles. 2013. Integrative testing of how environments from the past to the present shape genetic structure across landscapes. *Evolution* 67: 3386–3402.

Heaney, L. R. 2007. Is a new paradigm emerging for oceanic island biogeography? *Journal of Biogeography* 34: 753–757.

Hebblewhite, M. and D. T. Haydon. 2010. Distinguishing technology from biology: A critical review of the use of GPS telemetry data in ecology. *Philosophical Transactions of the Royal Society of London* B 365: 2303–2312.

Hedrick, P. W. 2005. A standardized genetic differentiation measure. *Evolution* 59: 1633–1638.

Heffernan, J. M., R. J. Smith, and L. M. Wahl. 2005. Perspectives on the basic reproductive ratio. *Journal of the Royal Society Interface* 2: 281–293.

Heilman, G. E., Jr., J. R. Strittholt, N. C. Slosser, and D. A. Dellasala. 2002. Forest fragmentation of the conterminous United States: Assessing forest intactness through road density and spatial characteristics. *BioScience* 52: 411–422.

Heim, C. E. 2001. Leapfrogging, urban sprawl, and growth management: Phoenix, 1950–2000. *American Journal of Economics and Sociology* 60: 245–283.

Heinz Center. 2008. *Landscape Pattern Indicators for the Nation. A Report from the Heinz Center's Landscape Pattern Task Group*, H. John Heinz III Center for Science, Economics and the Environment, Washington, D.C.

Heinz, S. K., C. Wissel, and K. Frank. 2006. The viability of metapopulations: Individual dispersal behavior matters. *Landscape Ecology* 21: 77–89.

Helfield, J. M. and R. J. Naiman. 2002. Salmon and alder as nitrogen sources to riparian forests in a boreal Alaskan watershed. *Oecologia* 133: 573–582.

Heller, R. and H. R. Siegismund. 2009. Relationship between three measures of genetic differentiation G_{ST}, D_{EST} and G'_{ST}: How wrong have we been? *Molecular Ecology* 18: 2080–2083.

Hemson, G., P. Johnson, A. South, R. Kenward, R. Ripley, and D. Macdonald. 2005. Are kernels the mustard? Data from global positioning system (GPS) collars suggests problems for kernel home-range analyses with least-squares cross-validation. *Journal of Animal Ecology* 74: 455–463.

Hendriks, A. J., B. J. C. Willers, H. J. R. Lenders, and R. S. E. W. Leuven. 2009. Towards a coherent allometric framework for individual home ranges, key population patches and geographic ranges. *Ecography* 32: 929–942.

Hereford, J. 2009. A quantitative survey of local adaptation and fitness trade-offs. *American Naturalist* 173: 579–588.

Hess, G. 1996. Disease in metapopulation models: Implications for conservation. *Ecology* 77: 1617–1632.

Hess, G. R. 1994. Conservation corridors and contagious disease: A cautionary note. *Conservation Biology* 8: 256–262.

Hesselbarth, M. H. K., M. Sciaini, J. Nowosad, and S. Hanß. 2019. Package 'landscapemetrics'. Available online: cran.r-project.org/web/packages/landscapemetrics/landscapemetrics.pdf

Hestbeck, J. B. 1982. Population regulation of cyclic mammals: The social fence hypothesis. *Oikos* 39: 157–163.

Hewitson, B. C. and R. G. Crane. 1996. Climate downscaling: Techniques and application. *Climate Research* 7: 85–95.

Hewitt, K., J. J. Clague, and J. F. Orwin. 2008. Legacies of catastrophic rock slope failures in mountain landscapes. *Earth-Science Reviews* 87: 1–38.

Heyerdahl, E. K., L. B. Brubaker, and J. K. Agee. 2001. Spatial controls of historical fire regimes: A multiscale example from the Interior West, USA. *Ecology* 82: 660–678.

Heywood, I., S. Cornelius, and S. Carver. 2011. *An Introduction to Geographical Information Systems*, 4th edition. Pearson Education Limited, Essex, UK.

Higgins, S. I. and D. M. Richardson. 1999. Predicting plant migration rates in a changing world: The role of long-distance dispersal. *American Naturalist* 153: 464–475.

Higgins, S. I. and M. L. Cain. 2002. Spatially realistic plant metapopulation models and the colonization-competition trade-off. *Journal of Ecology* 90: 616–626.

Hilderbrand, G. V., T. A. Hanley, C. T. Robbins, and C. C. Schwartz. 1999. Role of brown bears (*Ursus arctos*) in the flow of marine nitrogen into a terrestrial ecosystem. *Oecologia* 121: 546–550.

Hill, J. B., J. J. Clark, W. H. Doelle, and P. D. Lyons. 2004. Prehistoric demography in the Southwest: Migration, coalescence, and Hohokam population decline. *American Antiquity* 69: 689–716.

Hill, M. O. 1973a. Diversity and evenness: A unifying notation and its consequences. *Ecology* 54: 427–432.

Hill, M. O. 1973b. The intensity of spatial pattern in plant communities. *Journal of Ecology* 61: 225–235.

Hillebrand, H. 2004. On the generality of the latitudinal diversity gradient. *American Naturalist* 163: 192–211.

Hirzel, A. H. and G. Le Lay. 2008. Habitat suitability modelling and niche theory. *Journal of Applied Ecology* 45: 1372–1381.

Hobbie, S. E., J. C. Finlay, B. D. Janke, D. A. Nidzgorski, D. B. Millet, and L. A. Baker. 2017. Contrasting nitrogen and phosphorus budgets in urban watersheds and implications for managing urban water pollution. *Proceedings of the National Academy of Sciences of the United States of America* 114: 4177–4182.

Hobbs, N. T., D. S. Schimel, C. E. Owensby, and D. S. Ojima. 1991. Fire and grazing in the tallgrass prairie: Contingent effects on nitrogen budgets. *Ecology* 72: 1374–1382.

Hobbs, R. J. 2000. Land-use changes and invasion. Pp. 55–64 in *Invasive Species in a Changing World* (H. A. Mooney and R. J. Hobbs, Editors). Island Press, Washington, D.C.

Hobbs, R. J. and L. F. Huenneke. 1992. Disturbance, diversity, and invasion: Implications for conservation. *Conservation Biology* 6: 324–337.

Hobbs, R. J. and J. Wu. 2007. Perspectives and prospects of landscape ecology. Pp. 3–8 in *Key Topics in Landscape Ecology* (J. Wu and R. J. Hobbs, Editors). Cambridge University Press, Cambridge, UK.

Hobson, K. A. 1999. Tracing origins and migration of wildlife using stable isotopes: A review. *Oecologia* 120: 314–326.

Hobson, K. A. and L. I. Wassennar. 1997. Linking breeding and wintering grounds of neotropical migrant songbirds using stable hydrogen isotopic analysis of feathers. *Oecologia* 109: 142–148.

Hoekstra, T. W., T. F. H. Allen, and C. H. Flather. 1991. Implicit scaling in ecological research. *BioScience* 41: 148–154.

Holderegger, R. and H. W. Wagner. 2008. Landscape genetics. *BioScience* 58: 199–207.

Holderegger, R., R. Kamm, and F. Gugerli. 2006. Adaptive vs. neutral genetic diversity: Implications for landscape genetics. *Landscape Ecology* 21: 797–807.

Holderegger, R., D. Buehler, F. Gugerli, and S. Manel. 2010. Landscape genetics of plants. *Trends in Plant Science* 15: 675–683.

Holdridge, L. R. 1947. Determination of world plant formations from simple climatic data. *Science* 105: 367–368.

Holland, J. D., D. G. Bert, and L. Fahrig. 2004. Determining the spatial scale of species' responses to habitat. *BioScience* 54: 227–233.

Holling, C. S. 1973. Resilience and stability of ecological systems. *Annual Review of Ecology and Systematics* 4: 1–23.

Holling, C. S. 1996. Engineering resilience versus ecological resilience. Pp. 31–43 in *Engineering within Ecological Constraints* (P. C Schulze, Editor). National Academy Press, Washington, D.C.

Holsinger, K. E. and B. S. Weir. 2009. Genetics in geographically structured populations: Defining, estimating and interpreting F_{ST}. *Nature Reviews Genetics* 10: 639–650.

Holt, R. D. 1977. Predation, apparent competition, and the structure of prey communities. *Theoretical Population Biology* 12: 197–229.

Holt, R. D. 2001. Species coexistence. Pp. 413–426 in *Encyclopedia of Biodiversity*, Volume 5. Academic Press, Cambridge, Massachusetts.

Holt, R. D., J. Grover, and D. Tilman. 1994. Simple rules for interspecific dominance in systems with exploitative and apparent competition. *American Naturalist* 144: 741–771.

Holt, R. D., J. H. Lawton, G. A. Polis, and N. D. Martinez. 1999. Trophic rank and the species-area relationship. *Ecology* 80: 1495–1504.

Holyoak, M., M. A. Leibold, N. M. Mouquet, R. D. Holt, and M. F. Hoopes. 2005. Metacommunities: A framework for large-scale community ecology. Pp. 1–31 in *Metacommunities: Spatial Dynamics and Ecological Communities* (M. Holyoak, M. A. Leibold, and R. D. Holt, Editors). Chicago University Press, Chicago, Illinois.

Homer, C., C. Huang, L. Yang, B. Wylie, and M. Coan. 2004. Development of a 2001 National Land-Cover Database for the United States. *Photogrammetric Engineering and Remote Sensing* 70: 829–840.

Hood, G. A. and D. G. Larson. 2015. Ecological engineering and aquatic connectivity: A new perspective from beaver-modified wetlands. *Freshwater Biology* 60: 198–208.

Hook, P. B. and I. C. Burke. 2000. Biogeochemistry in a shortgrass landscape: Control by topography, soil texture, and microclimate. *Ecology* 81: 2686–2703.

Hooper, D. U. and 14 others. 2005. Effects of biodiversity on ecosystem functioning: A consensus of current knowledge. *Ecological Monographs* 75: 3–35.

Hoopes, M. F., R. D. Holt, and M. Holyoak. 2005. The effects of spatial processes on two-species interactions. Pp. 35–67 in *Metacommunities: Spatial Dynamics and Ecological Communities* (M. Holyoak, M. A. Leibold, and R. D. Holt, Editors). University of Chicago Press, Chicago, Illinois.

Hoorn, C., and 17 others. 2010. Amazonia through time: Andean uplift, climate change, landscape evolution, and biodiversity. *Science* 330: 927–931.

Hooten, M. B., E. M. Hanks, D. S. Johnson, and M. W. Alldredge. 2013. Reconciling resource utilization and resource selection functions. *Journal of Animal Ecology*: 82: 1146–1154.

Horn, H. S. and R. H. MacArthur. 1972. Competition among fugitive species in a harlequin environment. *Ecology* 54: 749–752.

Horne, J. K. and D. C. Schneider. 1995. Spatial variance in ecology. *Oikos* 74: 18–26.

Horreo, J. L., J. L. Martinez, F. Ayllon, I. G. Pola, J. A. Monteoliva, M. Héland, and E. Garcia-Vázquez. 2011. Impact of habitat fragmentation on the genetics of populations in dendritic landscapes. *Freshwater Biology* 56: 2567–2579.

Houghton, R. A. 1999. The annual net flux of carbon to the atmosphere from changes in land use 1850 to 1990. *Tellus* 50B: 298–313.

Houghton, R. A. 2007. Balancing the global carbon budget. *Annual Review of Earth and Planetary Sciences* 35: 13–47.

Houghton, R. A., J. I. House, J. Pongratz, G. R. van der Werf, R. S. DeFries, M. C. Hansen, C. LeQuéré, and N. Ramankutty. 2012. Carbon emissions from land use and land-cover change. *Biogeosciences* 9: 5125–5142.

Howard, J. B. 1992. Ancient engineers of the Arizona desert. *Geotimes* 37: 16–18.

Howitt, R. E., J. Medellin-Azuara, D. MacEwan, J. R. Lund, and D. A. Sumner. 2014. Economic analysis of the 2014 drought for California agriculture. Center for Watershed Sciences, University of California, Davis, California. Available online: https://watershed.ucdavis.edu/files/biblio/DroughtReport_23July2014_0.pdf (last accessed 1/30/19).

Hua, F., X. Wang, X. Zheng, B. Fisher, L. Wang, J. Zhu, Y. Tang, D. W. Yu, and D. S. Wilcove. 2016. Opportunities for biodiversity gains under the world's largest reforestation programme. *Nature Communications* 7: 12717.

Hua, F., L. Wang, B. Fischer. X. Zheng, X. Wang, D. W. Yu, Y. Tang, J. Zhu, and D. S. Wilcove. 2018. Tree plantations displacing native forests: The nature and drivers of apparent forest recovery on former croplands in Southwestern China from 2000 to 2015. *Biological Conservation* 222: 113–124.

Huang, C., E. L. Geiger, and J. A. Kupfer. 2006. Sensitivity of landscape metrics to classification scheme. *International Journal of Remote Sensing* 27: 2927–2948.

Hubbell, S. P. 2001. *The Unified Neutral Theory of Biodiversity and Biogeography.* Monographs in Population Biology 32. Princeton University Press, Princeton, New Jersey.

Huffaker, C. B. 1958. Experimental studies on predation: Dispersion factors and predator-prey oscillations. *Hilgardia* 27: 343–383.

Huffaker, C. B., K. P. Shea, and S. G. Herman. 1963. Experimental studies on predation: Complex dispersion and levels of food in an acarine predator-prey interaction. *Hilgardia* 34: 305–330.

Hughes, T. P. 1994. Catastrophes, phase shifts, and large-scale degradation of a Caribbean coral reef. *Science* 265: 1547–1551.

Hughes, T. P. and J. H. Connell. 1999. Multiple stressors on coral reefs: A long-term perspective. *Limnology and Oceanography* 44: 932–940.

Huijser, M. P., P. McGowen, J. Fuller, A. Hardy, A. Kociolek, A. P. Clevenger, D. Smith, and R. Ament. 2008. Wildlife-vehicle collision reduction study: Report to Congress. Report Number FHWA-HRT-08-034, Federal Highway Administration, Office of Safety Research and Development, McLean, Virginia.

Hulme, P. E. 2009. Trade, transport and trouble: Managing invasive species pathways in an era of globalization. *Journal of Applied Ecology* 46: 10–18.

Humphries, N. E., and 15 others. 2010. Environmental context explains Lévy and Brownian movement patterns of marine predators. *Nature* 465: 1066–1069.

Hunsaker, C. T., R. V. O'Neill, B. L. Jackson, S. P. Timmins, D. A. Levine, and D. J. Norton. 1994. Sampling to characterize landscape pattern. *Landscape Ecology* 9: 207–226.

Hurlbert, S. H. 1971. The nonconcept of species diversity: A critique and alternative parameters. *Ecology* 52: 577–586.

Hurlbert, S. H. 1990. Spatial distribution of the montane unicorn. *Oikos* 58: 257–271.

Hutchinson, G. E. 1959. Homage to Santa Rosalia or why are there so many kinds of animals? *American Naturalist* 93: 145–159.

Hutchinson, G. E. 1961. The paradox of the plankton. *American Naturalist* 95: 137–145.

Hutchison, D. W. 2003. Testing the central/peripheral model: Analyses of microsatellite variability in the eastern collared lizard (*Crotaphytus collaris collaris*). *American Midland Naturalist* 149: 148–162.

Ibáñez, J. J., S. De-Alba, F. F. Bermuidez, and A. Garcia-Alvarez. 1995. Pedodiversity: Concepts and measures. *Catena* 24: 215–232.

Ibáñez, J. J., S. De-Alba, A. Lobo, and V. Zucarello. 1998. Pedodiversity and global soil patterns at coarse scales. *Geoderma* 83: 171–192.

Iglesias, V., G. I. Yospin, and C. Whitlock. 2015. Reconstruction of fire regimes through integrated paleoecological proxy data and ecological modeling. *Frontiers in Plant Science* 5, 785.

Imaizumi, F. and R. C. Sidle. 2012. Effect of forest harvesting on hydrogeomorphic processes in steep terrain of central Japan. *Geomorphology* 169–170: 109–122.

Imhoff, M. L., T. D. Sisk, A. Milne, G. Morgan, and T. Orr. 1997. Remotely sensed indicators of habitat heterogeneity: Use of synthetic aperture radar in mapping vegetation structure and bird habitat. *Remote Sensing of Environment* 60: 217–227.

Ims, R. A. 1995. Movement patterns related to spatial structures. Pp. 85–109 in *Mosaic Landscapes and Ecological Processes* (L. Hansson, L. Fahrig, and G. Merriam, Editors). Chapman & Hall, London, UK.

Ims, R. A., J. Rolstad, and P. Wegge. 1993. Predicting space use responses to habitat fragmentation: Can voles *Microtus oeconomus* serve as an experimental model system (EMS) for capercaillie grouse *Tetrao urogallus* in boreal forest? *Biological Conservation* 63: 261–268.

Innis, H. A. 1999. *The fur trade in Canada: An introduction to Canadian economic history.* University of Toronto Press, Ontario, Canada.

IPCC (Intergovernmental Panel on Climate Change). 2012. *Managing the Risks of Extreme Events and Disasters to Advance Climate Change Adaptation: A Special Report of Working Groups I and II of the Intergovernmental Panel on Climate Change* (C.B. Field, V. Barros, T. F. Stocker, D. Qin, D. J. Dokken, K. L. Ebi, M. D. Mastrandrea, K. J. Mach, G.-K. Plattner, S. K. Allen, M. Tignor, and P. M. Midgley, Editors). Cambridge University Press, Cambridge, UK, and New York, New York, USA.

Isaaks, E. H. and R. M. Srivastava. 1989. *An Introduction to Applied Geostatistics.* Oxford University Press, New York.

IUSS (International Union of Soil Sciences) Working Group WRB (World Reference Base). 2006. *World Reference Base for Soil Resources 2006.* World Soil Resources Reports No. 103. Food and Agricultural Organization of the United Nations, Rome, Italy.

Iverson, L. R. 2007. Adequate data of known accuracy are critical to advancing the field of landscape ecology. Pp. 11-38 in *Key Topics in Landscape Ecology* (J. Wu and R. J. Hobbs, Editors). Cambridge University Press, Cambridge, UK.

Iverson, L. R. and A. M. Prasad. 1998. Predicting abundance of 80 tree species following climate change in the eastern United States. *Ecological Monographs* 68: 465–485.

Jablonski, D., K. Roy, and J. W. Valentine. 2006. Out of the tropics: Evolutionary dynamics of the latitudinal diversity gradient. *Science* 314: 102–106.

Jaccard, P. 1912. The distribution of the flora in the alpine zone. *New Phytologist* 11: 37–50.

Jackson, H. B. and L. Fahrig. 2012. What size is a biologically relevant landscape? *Landscape Ecology* 27: 929–941.

Jackson, S. M., F. Pinto, and J. R. Malcolm. 2000. A comparison of pre-European settlement (1857) and current (1981–1995) forest composition in central Ontario. *Canadian Journal of Forest Research* 30: 605–612.

Jakobsson, A. and O. Eriksson. 2000. A comparative study of seed number, seed size, seedling size and recruitment in grassland plants. *Oikos* 88: 494–501.

Jakobsson, A. and O. Eriksson. 2003. Trade-offs between dispersal and competitive ability: A comparative study of wind-dispersed Asteraceae forbs. *Evolutionary Ecology* 14: 233–246.

Jakobsson, M. and N. A. Rosenberg. 2007. CLUMPP: A cluster matching and permutation program for dealing with label switching and multimodality in analysis of population structure. *Bioinformatics* 23: 1801–1806.

Janzen, D. H. 1967. Why mountain passes are higher in the tropics. *American Naturalist* 101: 233–249.

Jauker, F., T. Diekötter, F. Schwarzbach, and V. Wolters. 2009. Pollinator dispersal in an agricultural matrix: Opposing responses of wild bees and hoverflies to landscape structure and distance from main habitat. *Landscape Ecology* 24: 547–555.

Jax, K. 2005. Function and 'functioning' in ecology: What does it mean? *Oikos* 111: 641–648.

Jelinski, D. E. and J. Wu. 1996. The modifiable areal unit problem and implications for landscape ecology. *Landscape Ecology* 11: 129–140.

Jenerette, G. D. and W. Shen. 2012. Experimental landscape ecology. *Landscape Ecology* 27: 1237–1248.

Jenkins, C. N, S. L. Pimm, and L. N. Joppa. 2013 Global patterns of terrestrial vertebrate diversity and conservation. *Proceedings of the National Academy of Sciences of the United States of America* 110: 11457–11462.

Jiggins, C. D. 2006. Sympatric speciation: Why the controversy? *Current Biology* 16: R333–R334.

Jobbágy, E. G. and R. B. Jackson. 2000. The vertical distribution of soil organic carbon and its relation to climate and vegetation. *Ecological Applications* 10: 423–436.

Johnson, A. R., B. T. Milne, and J. A. Wiens. 1992. Diffusion in fractal landscapes: Simulations and experimental studies of tenebrionid beetle movements. *Ecology* 73: 1968–1983.

Johnson, C. E., C. T. Driscoll, T. G. Siccama, and G. E. Likens. 2000. Element fluxes and landscape position in a northern hardwood forest watershed ecosystem. *Ecosystems* 3: 159–184.

Johnson, D. H. 1980. The comparison of usage and availability measurements for evaluating resource performance. *Ecology* 61: 65–71.

Johnson, L. C. and J. R. Matchett. 2001. Fire and grazing regulate belowground processes in tallgrass prairie. *Ecology* 82: 3377–3389.

Johnson, L. E., A. Ricciardi, and J. T. Carlton. 2001. Overland dispersal of aquatic invasive species: A risk assessment of transient recreational boating. *Ecological Applications* 11: 1789–1799.

Johnson, M. D. 2007. Measuring habitat quality: A review. *Condor* 109: 489–504.

Johnston, C. A. and R. J. Naiman. 1990a. Browse selection by beaver: Effects on riparian forest composition. *Canadian Journal of Forest Research* 20: 1036–1043.

Johnston, C. A. and R. J. Naiman. 1990b. The use of a geographic information system to analyze long-term landscape alteration by beaver. *Landscape Ecology* 4: 5–19.

Jones, A. G., C. M. Small, K. A. Paczolt, and N. L. Ratterman. 2010. A practical guide to methods of parentage analysis. *Molecular Ecology Resources* 10: 6–30.

Jones, C. G. 2012. Ecosystem engineers and geomorphological signatures in landscapes. *Geomorphology* 157–158: 75–87.

Jones, C. G., J. H. Lawton, and M. Shachak. 1994. Organisms as ecosystem engineers. *Oikos* 69: 373–386.

Jones, C. G., J. H. Lawton, and M. Shachak. 1997. Positive and negative effects of organisms as physical ecosystem engineers. *Ecology* 78: 1946–1957.

Jones, C. H., G. L. Farmer, B. Sageman, and S. Zhong. 2011. Hydrodynamic mechanism for the Laramide orogeny. *Geosphere* 7: 183–201.

Joost, S., A. Bonin, M. W. Bruford, L. Després, C. Conord, G. Erhardt, and P. Taberlet. 2007. A spatial analysis method (SAM) to detect candidate loci for selection: Towards a landscape genomics approach to adaptation. *Molecular Ecology* 16: 3955–3969.

Joost, S., S. Vuilleumier, J. D. Jensen, S. Schoville, K. Leempoel, S. Stucki, I. Widmer, C. Melodelima, J. Rolland, and S. Manel. 2013. Uncovering the genetic basis of adaptive change: On the intersection of landscape genomics and theoretical population genetics. *Molecular Ecology* 22: 3659–3665.

Joseph, F. 2010. *Advanced Civilizations of Prehistoric America: The Lost Kingdoms of the Adena, Hopewell, Mississippians, and Anasazi.* Bear and Company, Rochester, Vermont.

Jost, L. 2006. Entropy and diversity. *Oikos* 113: 363–375.

Jost, L. 2007. Partitioning diversity into independent alpha and beta components. *Ecology* 88: 2427–2439.

Jost, L. 2008. G_{ST} and its relatives do not measure differentiation. *Molecular Ecology* 17: 4015–4026.

Jost, L. 2010. Independence of alpha and beta diversities. *Ecology* 91: 1969–1974.

Jouquet, P., J. Dauber, J. Lagerlöf, P. Lavelle, and M. Lepage. 2006. Soil invertebrates as ecosystem engineers: Intended and accidental effects on soil and feedback loops. *Applied Soil Ecology* 32: 153–164.

Jules, E. S. 1998. Habitat fragmentation and demographic change for a common plant: *Trillium* in old-growth forest. *Ecology* 79: 1645–1656.

Jules, E. S., M. J. Kauffman, W. D. Ritts, and A. L. Carroll. 2002. Spread of an invasive pathogen over a variable landscape: A nonnative root rot on Port Orford cedar. *Ecology* 83: 3167–3181.

Junk, W. J., P. B. Bayley, and R. E. Sparks. 1989. The flood pulse concept in river-floodplain systems. *Canadian Special Publication of Fisheries and Aquatic Sciences* 106: 110–127.

Kalinowski, S. T. 2004. Counting alleles with rarefaction: Private alleles and hierarchical sampling design. *Conservation Genetics* 5: 539–543.

Kalinowski, S. T., M. L. Taper, and T. C. Marshall. 2007. Revising how the computer program CERVUS accommodates genotyping error increases success in paternity assignment. *Molecular Ecology* 16: 1099–1106.

Kaniewski, D., J. Guiot, and E. Van Campo. 2015. Drought and societal collapse 3200 years ago in the Eastern Mediterranean: A review. *Wiley Interdisciplinary Reviews (WIREs) Climate Change* 6: 369–382.

Kareiva, P. 1987. Habitat fragmentation and the stability of predator-prey interactions. *Nature* 326: 388–390.

Kareiva, P. 1994. Space: The final frontier for ecological theory. *Ecology* 75: 1.

Kareiva, P. and G. Odell. 1987. Swarms of predators exhibit 'preytaxis' if individual predators use area-restricted search. *American Naturalist* 130: 233–270.

Kareiva, P. M. and N. Shigesada. 1983. Analyzing insect movement as a correlated random walk. *Oecologia* 56: 234–238.

Kareiva, P. M. and M. Anderson. 1988. Spatial effects and species interactions: The wedding of models and experiments. pp. 33–52 in *Community Ecology* (A. Hastings, Editor). Springer-Verlag, New York.

Karl, T. R., J. M. Melillo, and T. C. Peterson, Editors. 2009. *Global Climate Change Impacts in the United States.* Cambridge University Press, New York, New York.

Kasischke, E. S., J. M. Melack, and M. C. Dobson. 1997. The use of imaging radars for ecological applications: A review. *Remote Sensing of Environment* 59: 141–156.

Kates, R. W. 2011. What kind of a science is sustainability science? *Proceedings of the National Academy of Sciences of the United States of America* 108: 19449–19450.

Kates, R. W., and 22 others. 2001. Sustainability science. *Science* 292: 641–642.

Kaufman, J. H., D. Brodbeck, and O. R. Melroy. 1998. Critical biodiversity. *Conservation Biology* 12: 521–532.

Kays, R., M. C. Crofoot, W. Jetz, and M. Wikelski. 2015. Terrestrial animal tracking as an eye on life and planet. *Science* 348: aaa2478.

Keating, K. A. and S. Cherry. 2009. Modeling utilization distributions in space and time. *Ecology* 90: 1971–1980.

Keeley, J. E. 2009. Fire intensity, fire severity and burn severity: A brief review and suggested usage. *International Journal of Wildland Fire* 18: 116–126.

Keeley, J. E., G. H. Aplet, N. L. Christensen, S. C. Conard, E. A. Johnson, P. N. Omi, D. L. Peterson, and T. W. Swetnam. 2009. Ecological foundations for fire management in North American forest and shrubland ecosystems. General Technical Report PNW-GTR-779, U.S. Department of Agriculture Forest Service, Pacific Northwest Research Station, Portland, Oregon.

Keeling, M. J. and P. Rohani. 2008. *Modeling Infectious Diseases in Humans and Animals.* Princeton University Press, New Jersey.

Keenan, R. J., G. A. Reams, F. Achard, J. V. de Freitas, A. Grainger, and E. Linquist. 2015. Dynamics of global forest area: Results from the FAO Global Forest Resources Assessment 2015. *Forest Ecology and Management* 352: 9–20.

Keesing, F., and 12 others. 2010. Impacts of biodiversity on the emergence and transmission of infectious diseases. *Nature* 468: 647–652.

Kéfi, S., M. Rietkerk, C. L. Alados, Y. Pueyo, V. P. Papanastasis, A. ElAich, and P. C. de Ruiter. 2007. Spatial vegetation patterns and imminent desertification in Mediterranean arid ecosystems. *Nature* 449: 213–217.

Kéfi, S., V. Guttal, W. A. Brock, S. R. Carpenter, A. M. Ellison, V. N. Livina, D. A. Seekell, M. Scheffer, E. H. van Nes, and V. Dakos. 2014. Early warning signals of ecological transitions: Methods for spatial patterns. *PLoS ONE* 9: e92097.

Keil, P., J. Belmaker, A. M. Wilson, P. Unitt, and W. Jetz. 2013. Downscaling of species distribution models: A hierarchical approach. *Methods in Ecology and Evolution* 4: 82–94.

Keinath, D. A., D. F. Doak, K. E. Hodges, L. R. Prugh, W. Fagan, C. H. Sekercioglu, S. H. M. Buchart, and M. Kauffman. 2017. A global analysis of traits predicting species sensitivity to habitat fragmentation. *Global Ecology and Biogeography* 26: 115–127.

Keith, S. A., and 10 others. 2012. What is macroecology? *Biology Letters* 8: 904–906.

Keitt, T. H. 2000. Spectral representation of neutral landscapes. *Landscape Ecology* 15: 479–494.

Keitt, T. H. 2009. Habitat conversion, extinction thresholds, and pollination services in agroecosystems. *Ecology* 19: 1561–1573.

Keitt, T. H., D. L. Urban, and B. T. Milne. 1997. Detecting critical scales in fragmented landscapes. *Conservation Ecology* [online]1(1): 4. Available online: http://www.consecol.org/vol1/iss1/art4/

Kelly, M. and R. K. Meentemeyer. 2002. Landscape dynamics of the spread of sudden oak death. *Photogrammetric Engineering and Remote Sensing* 68: 1001–1009.

Kenkel, N. C. 1988. Pattern of self-thinning in jack pine: Testing the random mortality hypothesis. *Ecology* 69: 1017–1024.

Kennedy, C. M., and 40 others. 2013. A global quantitative synthesis of local and landscape effects on wild bee pollinators in agroecosystems. *Ecology Letters* 16: 584–599.

Kennett D. J., and 17 others. 2012. Development and disintegration of Maya political systems in response to climate change. *Science* 338: 788–791.

Kernohan, B. J., R. A. Gitzen, and J. J. Millspaugh. 2001. High-tech behavioral ecology: Modeling the distribution of animal activities to better understand wildlife space use and resource selection. Pp. 309–326 in *Radio-tracking and Animal Populations* (J. J. Millspaugh and J. M. Marzluff, Editors). Academic Press, San Diego, California.

Kershaw, K. A. 1957. The use of cover and frequency in the detection of pattern in plant communities. *Ecology* 38: 291–299.

Keymer, J. E., P. A. Marquet, J. X. Velasco-Hernández, and S. A. Levin. 2000. Extinction thresholds and metapopulation persistence in dynamic landscapes. *American Naturalist* 156: 478–494.

Kier, G., J. Mutke, E. Dinerstein, T. H. Ricketts, W. Küper, H. Kreft, and W. Barthlott. 2005. Global patterns of plant diversity and floristic knowledge. *Journal of Biogeography* 32: 1–10.

Kindlmann, P. and F. Burel. 2008. Connectivity measures: A review. *Landscape Ecology* 23: 879–890.

Kindt, R. 2019. Package 'BiodiversityR': Package for Community Analysis and Suitability Analysis. R Package version 2.11-1. Available online: https://cran.r-project.org/web/packages/BiodiversityR/BiodiversityR.pdf

King, A. W. 1997. Hierarchy theory: A guide to system structure for wildlife biologists. Pp. 185–212 in *Wildlife and Landscape Ecology: Effects of Pattern and Scale* (J. A. Bissonette, Editor). Springer, New York.

King, A. W. and K. A. With. 2002. Dispersal success on spatially structured landscapes: When do spatial pattern and dispersal behavior really matter? *Ecological Modelling* 147: 23–39.

Kinzig, A. P. and J. Harte. 2000. Implications of endemics-area relationships for estimates of species extinctions. *Ecology* 81: 3305–3311.

Kitron, U. 1998. Landscape ecology and epidemiology of vector-borne diseases: Tools for spatial analysis. *Journal of Medical Entomology* 35: 435–445.

Kitron, U. 2000. Risk maps: Transmission and burden of vector-borne diseases. *Parasitology Today*: 15: 324–325.

Klein, A.-M., I. Steffan-Dewenter, and T. Tscharntke. 2003. Bee pollination and fruit set of *Coffea arabica* and *C. canephora* (Rubiaceae). *American Journal of Botany* 90: 153–157.

Klein, A.-M., B. E. Vaissière, J. H. Cane, I. Steffan-Dewenter, S. A. Cunningham, C. Kremen, and T. Tscharntke. 2007. Importance of pollinators in changing landscapes for world crops. *Proceedings of the Royal Society B* 274: 303–313.

Klink, C. A. and A. G. Moreira. 2002. Past and current human occupation, and land use. Pp. 69–88 in *The Cerrados of Brazil: Ecology and Natural History of a Neotropical Savanna* (P. S. Oliveira and R. J. Marquis, Editors). Columbia University Press, New York.

Klink, C. A. and R. B. Machado. 2005. Conservation of the Brazilian Cerrado. *Conservation Biology* 19: 707–713.

Klopatek, J. M. and R. H. Gardner, Editors. 1999. *Landscape Ecological Analysis: Issues and Applications*. Springer-Verlag, New York.

Kluser, S., P. Neumann, M.-P. Chauzat, and J. S. Pettis. 2010. Global honey bee colony disorders and other threats to insect pollinators. UNEP Emerging Issues, United Nations Environment Programme, Nairobi, Kenya. Available online: hdl.handle.net/20.500.11822/8544 (last accessed 2/21/19).

Knapp, A. K. and T. R. Seastedt. 1986. Detritus accumulation limits the productivity of tallgrass prairie. *BioScience* 36: 662–668.

Knapp, A. K., J. T. Fanhestock, S. P. Hamburg, L. B. Statland, T. R. Seastedt, and D. S. Schimel. 1993. Landscape patterns in soil-plant water relations and primary production in tallgrass prairie. *Ecology* 74: 549–560.

Knapp, A. K., J. M. Briggs, D. C. Hartnett, and S. L. Collins, Editors. 1998. *Grassland Dynamics: Long-term Ecological Research in Tallgrass Prairie*. Oxford University Press, New York.

Knapp, A. K., J. M. Blair, J. M. Briggs, S. L. Collins, D. C. Hartnett, L. C. Johnson, and E. G. Towne. 1999. The keystone role of bison in North American tallgrass prairie. *BioScience* 49:39–50.

Knapp, A. K., and 10 others. 2008. Shrub encroachment in North American grasslands: Shifts in growth form dominance rapidly alters control of ecosystem carbon inputs. *Global Change Biology* 14: 615–623.

Knapp, P. A. 1996. Cheatgrass (*Bromus tectorum L*) dominance in the Great Basin Desert: History, persistence and influences to human activities. *Global Environmental Change* 6: 37–52.

Kneitel, J. M. and J. M. Chase. 2004. Trade-offs in community ecology: Linking spatial scales and species coexistence. *Ecology Letters* 7: 69–80.

Knight, T. M., M. McCoy, J. M. Chase, K. A. McCoy, and R. D. Holt. 2005. Trophic cascades across ecosystems. *Nature* 437: 880–883.

Knutson, T. R., J. L. McBride, J. Chan, K. Emanuel, G. Holland, C. Landsea, I. Held, J. P. Kossin, A. K. Srivastava, and M. Sugi. 2010. Tropical cyclones and climate change. *Nature Geoscience* 3: 157–163.

Koen, E. L., J. Bowman, and A. A. Walpole. 2012. The effect of cost surface parameterization on landscape resistance estimates. *Molecular Ecology Resources* 12: 686–696.

Koh, L. P. and S. A. Wich. 2012. Dawn of drone ecology: Low-cost autonomous aerial vehicles for conservation. *Tropical Conservation Science* 5: 121–132.

Kolasa, J. and C. D. Rollo. 1991. Introduction: The heterogeneity of heterogeneity: A glossary. Pp. 1–23 in *Ecological Heterogeneity* (J. Kolasa and S. T. A. Pickett, Editors). Springer-Verlag, New York.

Koper, N., F. K. A. Schmiegelow, and E. H. Merrill. 2007. Residuals cannot distinguish between ecological effects of habitat amount and fragmentation: Implications for the debate. *Landscape Ecology* 22: 811–820.

Koren, I., Y. J. Kaufman, R. Washington, M. C. Todd, Y. Rudich, J. V. Martins, and D. Rosenfeld. 2006. The Bodélé depression: A single spot in the Sahara that provides most of the mineral dust to the Amazon forest. *Environmental Research Letters* 1: 014005.

Kormann, U., C. Scherber, T. Tscharntke, N. Klein, M. Larbig, J. J. Valente, A. S. Hadley, and M. G. Betts. 2016. Corridors restore animal-mediated pollination in fragmented tropical forest landscapes. *Proceedings of the Royal Society B* 283: 20152347.

Körner, C. 2007. The use of 'altitude' in ecological research. *Trends in Ecology and Evolution* 22: 569–574.

Kot, M., M. A. Lewis, and P. van den Driesshe. 1996. Dispersal data and the spread of invading organisms. *Ecology* 77: 2027–2042.

Kotliar, N. B. 2000. Application of the new keystone-species concept to prairie dogs: How well does it work? *Conservation Biology* 14: 1715–1721.

Kotliar, N. B. and J. A. Wiens. 1990. Multiple scales of patchiness and patch structure: A hierarchical framework for the study of heterogeneity. *Oikos* 59: 253–260.

Kraft, N. J. B., and 15 others. 2011. Disentangling the drivers of β diversity along latitudinal and elevational gradients. *Science* 333: 1755–1758.

Krauss, J., I. Steffan-Dewenter, and T. Tscharntke. 2003. How does landscape context contribute to effects of habitat fragmentation on diversity and population density of butterflies? *Journal of Biogeography* 30: 889–900.

Krebs, J. R. 1971. Territory and breeding density in the great tit, *Parus major* L. *Ecology* 52: 2–22.

Krebs, J. R., R. M. Anderson, T. Clutton-Brock, C. A. Donnelly, S. Frost, W. I. Morrison, R. Woodroffe, and D. Young. 1998. Badgers and bovine TB: Conflicts between conservation and health. *Science* 279: 817–818.

Kremen, C., N. M. Williams, R. L. Bugg, J. P. Fay, and R. W. Thorp. 2004. The area requirements of an ecosystem service: Crop pollination by native bee communities in California. *Ecology Letters* 7: 1109–1119.

Kremen, C., and 18 others. 2007. Pollination and other ecosystem services produced by mobile organisms: A conceptual framework for the effects of land-use change. *Ecology Letters* 10: 299–314.

Krige, D. G. 1966. Two-dimensional weighted moving average trend surfaces for ore-evaluation. *Journal of the South Africa Institute of Mining and Metallurgy* 66: 13–38.

Kronenfeld, B. J., and Y.-C. Wang. 2007. Accounting for surveyor inconsistency and bias in estimation of tree density from pre-settlement land survey records. *Canadian Journal of Forest Research* 37: 2365–2379.

Kruess, A. and T. Tscharntke. 1994. Habitat fragmentation, species loss, and biological control. *Science* 264: 1581–1584.

Krummel, J. R., R. H. Gardner, G. Sugihara, R. V. O'Neill, and P. R. Coleman. 1987. Landscape patterns in a disturbed environment. *Oikos* 48: 321–324.

Kunin, W. E. 1998. Extrapolating species abundance across spatial scales. *Science* 281: 151–1515.

Kupfer, J. A. 2006. National assessments of forest fragmentation in the US. *Global Environmental Change* 16: 73–82.

Kupfer, J. A. 2012. Landscape ecology and biogeography: Rethinking landscape metrics in a post-FRAGSTATS landscape. *Progress in Physical Geography* 36: 400–420.

Kupfer, J. A., G. P. Malanson, and S. B. Franklin. 2006. Not seeing the ocean for the islands: The mediating influence of matrix-based processes on forest fragmentation effects. *Global Ecology and Biogeography* 15: 8–20.

Kurz, W. A., C. C. Dymond, G. Stinson, G. J. Rampley, E. T. Neilson, A. L. Carroll, T. Ebata, and L. Safranyik. 2008. Mountain pine beetle and forest carbon feedback to climate change. *Nature* 452: 987–990.

Kuussaari, M., and 13 others. 2009. Extinction debt: A challenge for biodiversity conservation. *Trends in Ecology and Evolution* 24: 564–571.

LaDeau, S. L., G. E. Glass, N. T. Hobbs, A. Latimer, and R. S. Ostfeld. 2011. Data-model fusion to better understand emerging pathogens and improve infectious disease forecasting. *Ecological Applications* 21: 1443–1460.

Lafferty, K. D. 2009. The ecology of climate change and infectious diseases. *Ecology* 90: 888–900.

Lafferty, K. D. and A. M. Kuris. 2002. Trophic strategies, animal diversity and body size. *Trends in Ecology and Evolution* 11: 507–513.

Lahti, D. C. 2001. The 'edge effect on nest predation' hypothesis after twenty years. *Biological Conservation* 99: 365–374.

Lake, P. S. 2000. Disturbance, patchiness, and diversity in streams. *Journal of the North American Benthological Society* 19: 573–592.

Lal, R. 2004. Soil carbon sequestration to mitigate climate change. *Geoderma* 123: 1–22.

Lande, R. 1987. Extinction thresholds in demographic models of territorial populations. *American Naturalist* 130: 624–635.

Lande, R. 1988. Demographic models of the Northern Spotted Owl (*Strix occidentalis*). *Oecologia* 75: 601–607.

Lande, R. 1996. Statistics and partitioning of species diversity, and similarity among multiple communities. *Oikos* 76: 5–13.

Landguth, E. L., S. A. Cushman, M. K Schwartz, K. S. McKelvey, M. Murphy, and G. Luikart. 2010. Quantifying the lag time to detect barriers in landscape genetics. *Molecular Ecology* 19: 4179–4191.

Landres, P. B., P. Morgan, and F. J. Swanson. 1999. Overview of the use of natural variability concepts in managing ecological systems. *Ecological Applications* 9: 1179–1188.

Larsen, I. J. and D. R. Montgomery. 2012. Landslide erosion coupled to tectonics and river incision. *Nature Geoscience* 5: 468–473.

Laundré, J. W., L. Hernández, and K. B. Altendorf. 2001. Wolves, elk, and bison: Re-establishing the 'landscape of fear' in Yellowstone National Park, USA. *Canadian Journal of Zoology* 79: 1401–1409.

Laurance, W. F. 2008. Theory meets reality: How habitat fragmentation research has transcended island biogeographic theory. *Biological Conservation* 141: 1731–1744.

Laurance, W. F., T. E. Lovejoy, H. L. Vasconcelos, E. M. Bruna, R. K. Didham, and P. C. Stouffer, C. Gascon, R. O. Bierregaard, S. G. Laurance, and E. Sampaio. 2002. Ecosystem decay of Amazonian forest fragments: A 22-year investigation. *Conservation Biology* 16: 605–618.

Laurance, W. F., H. E. M. Nascimento, S. G. Laurance, A. Andrade, R. M. Ewers, K. E. Harms, R. C. C. Luizão, and J. E. Ribeiro. 2007. Habitat fragmentation, variable edge effects, and the landscape-divergence hypothesis. *PLoS ONE* 2: e1017.

Laurance, W. F., and 15 others. 2011. The fate of Amazonian forest fragments: A 32-year investigation. *Biological Conservation* 144: 56–67.

Laver, P. N. and M. J. Kelly. 2008. A critical review of home range studies. *Journal of Wildlife Management* 72: 290–298.

Lavorel, S., R. V. O'Neill, and R. H. Gardner. 1994. Spatiotemporal dispersal strategies and annual plant species coexistence in a structured landscape. *Oikos* 71: 75–88.

Lavorel, S., R. H. Gardner, and R. V. O'Neill. 1995. Dispersal of annual plants in hierarchically structured landscapes. *Landscape Ecology* 10: 277–289.

Law, R., J. Illian, D. F. R. P. Burslem, G. Gratzer, C. V. S. Gunatilleke, and I. A. U. N. Gunatilleke. 2009. Ecological information from spatial patterns of plants: Insights from point process theory. *Journal of Ecology* 97: 616–628.

Lawton, J. H. 1995. Ecological experiments with model systems. *Science* 269: 328–331.

Lawton, J. H. 1999. Are there general laws in ecology? *Oikos* 84: 177–192.

Laycock, W. A. 1991. Stable states and thresholds of range condition on North American rangelands: A viewpoint. *Journal of Range Management* 44: 427–433.

Le Rouzic, A., K. Østbye, T. O. Klepaker, T. F. Hansen, L. Bernatchez, D. Schluter, and L. A. Vøllestad. 2011. Strong and consistent natural selection associated with armour reduction in sticklebacks. *Molecular Ecology* 20: 2483–2493.

Leberg, P. L. 2002. Estimating allelic richness: Effects of sample size and bottlenecks. *Molecular Ecology* 11: 2445–2449.

Lee-Yaw, J. A., A. Davidson, B. H. McRae, and D. M. Green. 2009. Do landscape processes predict phylogeographic patterns in the wood frog? *Molecular Ecology* 18: 1863–1874.

Legendre, P. 1993. Spatial autocorrelation: Trouble or new paradigm? *Ecology* 74: 1659–1673.

Legendre, P., and M.-J. Fortin. 1989. Spatial pattern and ecological analysis. *Vegetatio* 80: 107–138.

Legendre, P., and M.-J. Fortin. 2010. Comparison of the Mantel test and alternative approaches for detecting complex multivariate relationships in the spatial analysis of genetic data. *Molecular Ecology Resources* 10: 831–844.

Legendre, P. and L. Legendre. 2012. *Numerical Ecology*, 3rd edition. Elsevier, Amsterdam, the Netherlands.

Legendre, P. and M. De Cáceres. 2013. Beta diversity as the variance of community data: Dissimilarity coefficients and partitioning. *Ecology Letters* 16: 951–963.

Leibold, M. A., and 11 others. 2004. The metacommunity concept: A framework for multi-scale community ecology. *Ecology Letters* 7: 601–613.

Lennartsson, T. 2002. Extinction thresholds and disrupted plant-pollinator interactions in fragmented plant populations. *Ecology* 83: 3060–3072.

Lenoir, J., J. C. Gégout, P. A. Marquet, P. de Ruffray, and H. Brisse. 2008. A significant upward shift in plant species optimum elevation during the 20th century. *Science* 320: 1768–1771.

Leopold, A. 1949. *A Sand County Almanac*. Oxford University Press, New York.

Lepš, J. 1990. Comparison of transect methods for the analysis of spatial pattern. Pp. 71–82 in *Spatial Processes in Plant Communities* (F. Krahulec, A.D.Q. Agnew, S. Agnew, and J. H. Willems, Editors). Academic Press, Prague, Czechoslovakia.

Leroux, S. J. and M. Loreau. 2008. Subsidy hypothesis and strength of trophic cascades across ecosystems. *Ecology Letters* 11: 1147–1156.

Leung, B., D. M. Lodge, D. Finnoff, J. F. Shogren, M. A. Lewis, and G. Lamberti. 2002. An ounce of prevention or a pound of cure: Bioeconomic risk analysis of invasive species. *Proceedings of the Royal Society of London* B 269: 2407–2413.

Levin, L. A., and 11 others. The function of marine critical transition zones and the importance of sediment biodiversity. *Ecosystems* 4: 420–451.

Levin, P. S., N. Tolimieri, M. Nicklin, and P. F. Sale. 2000. Integrating individual behavior and population ecology: the potential for habitat-dependent population regulation in a reef fish. *Behavioral Ecology* 11: 565–571.

Levin, S. A. 1992. The problem of pattern and scale in ecology. *Ecology* 73: 1943–1967.

Levin, S. A. and R. T. Paine. 1974. Disturbance, patch formation, and community structure. *Proceedings of the National Academy of Sciences of the United States of America* 71: 2744–2747.

Levine, J. M. and M. Rees. 2002. Coexistence and relative abundance in annual plant assemblages: The roles of competition and colonization. *American Naturalist* 160: 452–467.

Levins, R. 1969. Some demographic and genetic consequences of environmental heterogeneity for biological control. *Bulletin of the Entomological Society of America* 15: 237–240.

Levins, R. 1970. Extinction. Pp. 77–107 in *Some Mathematical Problems in Biology* (M. Gesternhaber, Editor), American Mathematical Society, Providence, Rhode Island.

Levins, R. and D. Culver. 1971. Regional coexistence of species and competition between rare species. *Proceedings of the National Academy of Sciences of the United States of America* 6: 1246–1248.

Lewis, S. L. and M. A. Maslin. 2015. Defining the Anthropocene. *Nature* 519: 171–180.

Li, B.-L. and S. Archer. 1997. Weighted mean patch size: A robust index for quantifying landscape structure. *Ecological Modelling* 102: 353–361.

Li, H. and J. F. Reynolds. 1993. A new contagion index to quantify spatial patterns of landscapes. *Landscape Ecology* 8: 155–162.

Li, H. and J. Wu. 2004. Use and misuse of landscape indices. *Landscape Ecology* 19: 389–399.

Li, P., L. Jiang, and Z. Feng. 2014. Cross-comparison of vegetation indices derived from Landsat-7 Enhanced Thematic Mapper Plus (ETM+) and Landsat-8 Operational Land Imager (OLI) sensors. *Remote Sensing* 6: 310–329.

Li, Y., W. Ye, M. Wang, and X. Yan. 2009. Climate change and drought: A risk assessment of crop-yield impacts. *Climate Research* 39: 31–46.

Lichti, N. and R. K. Swihart. 2011. Estimating utilization distributions with kernel versus local convex hull methods. *Journal of Wildlife Management* 75: 413–422.

Lidicker, W. Z., Jr. and W. D. Koenig. 1996. Responses of terrestrial vertebrates to habitat edges and corridors. Pp. 85–110 in *Metapopulations and Wildlife Conservation* (D. R. McCullough, Editor). Island Press, Washington D.C.

Likens, G. E. and E. H. Bormann. 1974. Linkages between terrestrial and aquatic ecosystems. *BioScience* 24: 447–456.

Lima, S. L. and L. M. Dill. 1990. Behavioral decisions made under the risk of predation: A review and prospectus. *Canadian Journal of Zoology* 68: 619–640.

Lima, S. L. and P. A. Zollner. 1996. Towards a behavioral ecology of ecological landscapes. *Trends in Ecology and Evolution* 11: 131–135.

Lin, N., K. Emanuel, M. Oppenheimer, and E. Vanmarcke. 2012. Physically based assessment of hurricane surge threat under climate change. *Nature Climate Change* 2: 462–467.

Lindenmayer, D. B. and H. P. Possingham. 1996. Ranking conservation and timber management options for Leadbeater's possum in southeastern Australia using population viability analysis. *Conservation Biology* 10: 235–251.

Lindenmayer, D. B. and J. Fischer. 2006. *Habitat Fragmentation and Landscape Change: An Ecological and Conservation Synthesis*. Island Press, Washington, D.C.

Lindenmayer, D. B., W. Blanchard, L. McBurney, D. Blair, S. C. Banks, D. Driscoll, A. L. Smith, and A. M. Gill. 2013. Fire severity and landscape context effects on arboreal marsupials. *Biological Conservation* 167: 137–148.

Lindsay, D. L, K. R. Barr, R. F. Lance, S. A. Tweddale, T. J. Hayden, and P. L. Leberg. 2008. Habitat fragmentation and genetic diversity of an endangered, migratory songbird, the golden-cheeked warbler (*Dendroica chrysoparia*). *Molecular Ecology* 17: 2122–2133.

Lipman, P.W. and D. R. Mullineaux, Editors. 1981. *The 1980 Eruptions of Mount St Helens*. U.S. Geological Survey Professional Paper 1250, Washington, D.C.

Lipman, P. W., J. M. Rhodes, and G. B. Dalrymple. 1990. The Ninole Basalt: Implications for the structural evolution of Mauna Loa volcano, Hawaii. *Bulletin of Volcanology* 53: 1–19.

Liu, F., X. B. Wu, E. Bai, T. W. Boutton, and S. R. Archer. 2010. Spatial scaling of ecosystem C and N in a subtropical savanna landscape. *Global Change Biology* 16: 2213–2223.

Liu, F., D. J. Mladenoff, N. S. Keuler, and L. S. Moore. 2011. Broadscale variability in tree data of the historical Public Land Survey and its consequences for ecological studies. *Ecological Monographs* 81: 259–275.

Liu, J. and W. W. Taylor, Editors. 2002. *Integrating Landscape Ecology into Natural Resource Management*. Cambridge University Press, Cambridge, UK.

Liu, J., S. Li, Z. Ouyang, C. Tam, and X. Chen. 2008. Ecological and socioeconomic effects of China's policies for ecosystem services. *Proceedings of the National Academy of Sciences of the United States of America* 105: 9477–9482.

Lockwood, J. L., P. Cassey, and T. Blackburn. 2005. The role of propagule pressure in explaining species invasions. *Trends in Ecology and Evolution* 20: 223–228.

Lockwood, J. L., M. F. Hoopes, and M. P. Marchetti. 2007. *Invasion Ecology*. Blackwell, Malden, Massachusetts.

Logan, J. A. and J. A. Powell. 2001. Ghost forests, global warming, and the mountain pine beetle (Coleoptera: Scolytidae). *American Entomologist* 47: 160–173.

Logue, J. B., N. Mouquet, H. Peter, H. Hillebrand, and the Metapopulation Working Group. 2011. Empirical approaches to metacommunities: A review and comparison with theory. *Trends in Ecology and Evolution* 26: 482–491.

Lomolino, M. V. 2000. Ecology's most general, yet protean pattern: The species–area relationship. *Journal of Biogeography* 27: 17–26.

Lomolino, M. V. 2001. The species-area relationship: New challenges for an old pattern. *Progress in Physical Geography* 25: 1–21.

Lomolino, M. V., B. R. Riddle and J. H. Brown. 2006. *Biogeography*, 3rd edition. Sinauer, Sunderland, Massachusetts.

Lomolino, M.V., B. R. Riddle, R. J. Whittaker, and J. H. Brown. 2010. *Biogeography*, 4th edition. Sinauer, Sunderland, Massachusetts.

Long, E. S., D. R. Diefenbach, C. S. Rosenberry, B. D. Wallingord, and M. R. D. Grund. 2005. Forest cover influences dispersal distance of white-tailed deer. *Journal of Mammalogy* 86: 623–629.

Long, J. C. 1991. The genetic structure of admixed populations. *Genetics* 127: 417–428.

Lorch, J. M., and 10 others. 2011. Experimental infection of bats with *Geomyces destructans* causes white-nose syndrome. *Nature* 480: 376–378.

Loreau, M., N. Mouquet, and R. D. Holt. 2003. Meta-ecosystems: A theoretical framework for a spatial ecosystem ecology. *Ecology Letters* 6: 673–679.

Lotka, A. J. 1925. *Elements of Physical Biology*, Williams & Wilkins, Baltimore, Maryland.

Lovell, S. T. and D. M. Johnston. 2009. Creating multifunctional landscapes: How can the field of ecology inform the design of the landscape? *Frontiers in Ecology and the Environment* 7: 212–220.

Lovett, G. M., C. G. Jones, M. G. Turner, and K. C. Weathers. 2005. Conceptual frameworks: Plan for a half-built house. Pp. 463–470 in *Ecosystem Function in Heterogeneous Landscapes* (G. M. Lovett, C. G. Jones, M. G. Turner, and K. C. Weathers, Editors). Springer, New York.

Lowe, S., M. Browne, S. Boudjelas, and M. De Poorter. 2000. *100 of the World's Worst Invasive Alien Species: A Selection from the Global Invasive Species Database*. Published by the Invasive Species Specialist Group (ISSG), Species Survival Commission (SSC), International Union for Conservation of Nature (IUCN), Auckland, New Zealand. Available online: http://www.issg.org/worst100_species.html (last accessed 1/30/19).

Lowry, D. B. 2010. Landscape evolutionary genomics. *Biology Letters* 6: 502–504.

Lowry, D. B., M. C. Hall, D. E. Salt, and J. H. Willis. 2009. Genetic and physiological basis of adaptive salt tolerance divergence between coastal and inland *Mimulus guttatus*. *New Phytologist* 183: 776–778.

Lozier, J. D., P. Aniello, and J. J. Hickerson. 2009. Predicting the distribution of Sasquatch in western North America: Anything goes with ecological niche modeling. *Journal of Biogeography* 36: 1623–1627.

Luck, G.W. 2005. An introduction to ecological thresholds. *Biological Conservation* 124: 299–300.

Ludwig, J. A. and D. J. Tongway. 1997. A landscape approach to rangeland ecology. Pp. 1–12 in *Landscape Ecology, Function, and Management: Principles from Australia's Rangelands* (J. A. Ludwig, D. J. Tongway, D. Freudenberger, J. C. Noble, and K. C. Hodgkinson, Editors), CSIRO, Collingwood, Australia.

Ludwig, J. A., B. P. Wilcox, D. D. Breshears, D. J. Tonway, and A. C. Imerson. 2005. Vegetation patches and runoff-erosion as interacting ecohydrological processes in semiarid landscapes. *Ecology* 86: 288–297.

Ludwig, J. A., G. N. Bastin, V. H. Chewings, R. W. Eager, and A. C. Liedloff. 2007. Leakiness: A new index for monitoring the health of arid and semiarid landscapes using remotely sensed vegetation cover and elevation data. *Ecological Indicators* 7: 442–454.

Luecke, D. F., J. Pitt, C. Congdon, E. Glenn, C. Valdés-Casillas, and M. Briggs. 1999. *A Delta Once More: Restoring Riparian and Wetland Habitat in the Colorado River Delta*. Environmental Defense Fund, Washington, D.C.

Lugo, A. E., S. L. Brown, R. Dodson, T. S. Smith, and H. H. Shugart. 1999. The Holdridge life zones of the conterminous United States in relation to ecosystem mapping. *Journal of Biogeography* 26: 1025–1038.

Lugo, A. E., C. S. Rogers, and S. W. Nixon. 2000. Hurricanes, coral reefs and rainforests: Resistance, ruin and recovery in the Caribbean. *Ambio* 29: 106–114.

Lugo-Fernandez, A., H. H. Roberts, and W. J. Wiseman. 1998. Tide effects on wave attenuation and wave set-up on a Caribbean coral reef. *Estuarine, Coastal and Shelf Science* 47: 385–393.

Luikart, G., P. R. England, D. Tallmon, S. Jordan, and P. Taberlet. 2003. The power and promise of population genomics: From genotyping to genome typing. *Nature Reviews Genetics* 4: 981–994.

Lundberg, J. and F. Moberg. 2003. Mobile link organisms and ecosystem functioning: Implications for ecosystem resilience and management. *Ecosystems* 6: 87–98.

MacArthur, R. 1965. Patterns of species diversity. *Biological Reviews* 40: 510–533.

MacArthur, R. and R. Levins. 1967. The limiting similarity, convergence, and divergence of coexisting species. *American Naturalist* 101: 377–385.

MacArthur, R. H. 1958. Population ecology of some warblers of northeastern coniferous forests. *Ecology* 39: 599–619.

MacArthur, R. H. and J. W. MacArthur. 1961. On bird species diversity. *Ecology* 42: 594–598.

MacArthur, R. H. and E. O. Wilson. 1963. An equilibrium theory of insular zoogeography. *Evolution* 17: 373–387.

MacArthur, R. H. and E. O. Wilson. 1967. *The Theory of Island Biogeography*. Princeton University Press, Princeton, New Jersey.

MacArthur, R. H., H. Recher, and M. Cody. 1966. On the relation between habitat selection and species diversity. *American Naturalist* 100: 319–322.

McCain, C. M. 2004. The mid-domain effect applied to elevational gradients: Species richness of small mammals in Costa Rica. *Journal of Biogeography* 31: 19–31.

McCain, C. M. 2005. Elevational gradients in diversity of small mammals. *Ecology* 86: 366–372.

McCain, C. M. 2007. Could temperature and water availability drive elevational species richness patterns? A global case study for bats. *Global Ecology and Biogeography* 16: 1–13.

McCain, C. M. 2009. Global analysis of bird elevational diversity. *Global Ecology and Biogeography* 18: 346–260.

McCain, C. M. and J.-A. Grytnes. 2010. Elevational gradients in species richness. *Encyclopedia of Life Sciences* (eLS). John Wiley & Sons, Chichester, UK.

McCarthy, M. A., S. J. Andelman, and H. P. Possingham. 2003. Reliability of relative predictions in population viability analysis. *Conservation Biology* 17: 982–989.

McClure, C. J. W., H. E. Ware, J. Carlisle, G. Kaltenecker, and J. R. Barber. 2013. An experimental investigation into the effects of traffic noise on distributions of birds: Avoiding the phantom road. *Proceedings of the Royal Society B* 280: 20132290.

MacDicken, K. G. 2015. Global Forest Resources Assessment 2015: What, why and how? *Forest Ecology and Management* 352: 3–8.

MacDicken, K. G., G. Reams, and J. de Freitas. 2015. Introduction to the changes in global forest resources from 1990 to 2015. *Forest Ecology and Management* 352: 1–2.

McDonnell, M. J. and S. T. A. Pickett. 1990. Ecosystem structure and function along urban-rural gradients: An unexploited opportunity for ecology. *Ecology* 71: 1232–1237.

McDonnell, M. J. and A. K. Hahs. 2008. The use of gradient analysis studies in advancing our understanding of the ecology of urbanizing landscapes: Current status and future directions. *Landscape Ecology* 10: 1143–1155.

McGarigal, K. and B. J. Marks. 1995. *FRAGSTATS: Spatial Pattern Analysis Program for Quantifying Landscape Structure*. General Technical Report PNW-GTR-351, USDA Forest Service, Pacific Northwest Research Station, Portland, OR.

McGarigal, K. and S. A. Cushman. 2002. Comparative evaluation of experimental approaches to the study of habitat fragmentation effects. *Ecological Applications* 12: 335–345.

McGarigal, K., S. Tagil, and S. A. Cushman. 2009. Surface metrics: An alternative to patch metrics for the quantification of landscape structure. *Landscape Ecology* 24: 433–450.

McGarigal, K., S. A. Cushman, and E. Ene. 2012. FRAGSTATS v4: Spatial Pattern Analysis Program for Categorical and Continuous Maps. Computer software program produced by the authors at the University of Massachusetts, Amherst. Available online: http://www.umass.edu/landeco/research/fragstats/fragstats.html (last accessed 2/3/19).

McGranahan, D. A., D. M. Engle, S. D. Fuhlendorf, S. J. Winter, J. R. Miller, and D. M. Debinski. 2012. Spatial heterogeneity across five rangelands managed with pyric-herbivory. *Journal of Applied Ecology* 49: 903–910.

McIntyre, N. E. and J. A. Wiens. 1999a. Interactions between landscape structure and animal behavior: The roles of heterogeneously distributed resources and food deprivation on movement patterns. *Landscape Ecology* 14: 437–447.

McIntyre, N. E. and J. A. Wiens. 1999b. Interactions between habitat abundance and configuration: Experimental validation of some predictions from percolation theory. *Oikos* 86: 129–137.

McKenzie, D. and M. C. Kennedy. 2012. Power laws reveal phase transitions in landscape controls of fire regimes. *Nature Communications* 3: 726.

McKenzie, D., C. Miller, and D. A. Falk. 2011. Toward a theory of landscape fire. Pp. 3–25 in *The Landscape Ecology of Fire* (D. McKenzie, C. Miller, and D. A. Falk, Editors). Springer, Dordrecht, the Netherlands.

MacKenzie, D. I., J. D. Nichols, G. B. Lachman, S. Droege, J. A. Royle, and C. A. Langtimm. 2002. Estimating site occupancy rates when detection probabilities are less than one. *Ecology* 83: 2248–2255.

MacKenzie, D. I., J. D. Nichols, J. A. Royle, K. H. Pollock, L. L. Bailey, and J. E. Hines. 2006. *Occupancy Estimation and Modeling: Inferring Patterns and Dynamics of Species Occurrence.* Elsevier Academic Press, Burlington, Massachusetts.

Mackey, R. L. and D. J. Currie. 2001. The diversity-disturbance relationship: Is it generally strong and peaked? *Ecology* 82: 3479–3492.

McKinley, D. C. and J. M. Blair. 2008. Woody plant encroachment by *Juniperus virginiana* in a mesic native grassland promotes rapid carbon and nitrogen accrual. *Ecosystems* 11: 454–468.

McKinney, M. L. 2002. Urbanization, biodiversity, and conservation. *BioScience* 52: 883–890.

McKinney, M. L. 2008. Effects of urbanization on species richness: A review of plants and animals. *Urban Ecosystems* 11: 161–176.

McLachlan, J. S., J. S. Clark, and P. S. Manos. 2005. Molecular indicators of tree migration capacity under rapid climate change. *Ecology* 86: 2088–2098.

McLoughlin, P. D., D. W. Morris, D. Fortin, E. Vander Wal, and A. L. Contasti. 2010. Considering ecological dynamics in resource selection functions. *Journal of Animal Ecology* 79: 4–12.

MacMahon, J. A., J. F. Mull, and T. O. Crist. 2000. Harvester ants (*Pogonomyrmex* spp.): Their community and ecosystem influences. *Annual Review of Ecology and Systematics* 31: 265–291.

McMillan, B. R., K. A. Pfeiffer, and D. W. Kaufman. 2011. Vegetation responses to an animal-generated disturbance (bison wallows) in tallgrass prairie. *American Midland Naturalist* 165: 60–73.

McNickle, G. G. and J. F. Cahill, Jr. 2009. Plant root growth and the marginal value theorem. *Proceedings of the National Academy of Sciences of the United States of America* 106: 4747–4751.

McRae, B. H. 2006. Isolation by resistance. *Evolution* 60: 1551–1561.

McRae, B. H. and P. Beier. 2007. Circuit theory predicts gene flow in plant and animal populations. *Proceedings of the National Academy of Sciences of the United States of America* 104: 19885–19890.

McRae, B. H., B. G. Dickson, T. H. Keitt, and V. B. Shah. 2008. Using circuit theory to model connectivity in ecology, evolution, and conservation. *Ecology* 89: 2712–2724.

McRae, B. H., V. B. Shah, and T.K. Mohapatra. 2013. Circuitscape 4 User Guide. The Nature Conservancy. Available online: https://www.circuitscape.org.

Magurran, A. E. 2004. *Measuring Biological Diversity.* Blackwell, Malden, Massachusetts.

Malamud, B. D., G., Morein, and, D. L. Turcotte. 1998. Forest fires: An example of self-organized criticality. *Science* 281: 1840–1842.

Malcolm, J. R., A. Markham, R. P. Neilson, and M. Garaci. 2002. Estimated migration rates under scenarios of global climate change. *Journal of Biogeography* 29: 835–849.

Mandelbrot, B. B. 1983. *The Fractal Geometry of Nature.* W. H. Freeman, New York.

Mander, Ü, H. Wiggering, and K. Helming, Editors. 2007. *Multifunctional Land Use: Meeting Future Demands for Landscape Goods and Services.* Springer, Berlin, Germany.

Manel, S. and R. Holderegger. 2013. Ten years of landscape genetics. *Trends in Ecology and Evolution* 28: 614–621.

Manel, S., M. K. Schwartz, G. Luikart, and P. Taberlet. 2003. Landscape genetics: Combining landscape ecology and population genetics. *Trends in Ecology and Evolution* 18: 189–197.

Manel, S., S. Joost, B. K. Epperson, R. Holderegger, A. Storfer, M. S. Rosenberg, K. T. Scribner, A. Bonin, and M.-J. Fortin. 2010. Perspectives on the use of landscape genetics to detect genetic adaptive variation in the field. *Molecular Ecology* 19: 3760–3772.

Manies, K. L. and D. L. Mladenoff. 2000. Testing methods to produce landscape-scale pre-settlement vegetation maps from the U.S. public land survey records. *Landscape Ecology* 15: 741–754.

Manies, K. L., D. J. Mladenoff, and E. V. Nordheim. 2001. Assessing large-scale surveyor variability in the historic forest data of the original U.S. Public Land Surveys. *Canadian Journal of Forest Research* 17:1719–1730.

Manly, B. F. J. 2006. *Randomization, Bootstrap and Monte Carlo Methods in Biology*, 3rd edition. Chapman & Hall/CRC Press, Florida.

Manly, B. F. J., L. L. McDonald, D. L. Thomas, T. L. McDonald, and W. P. Erickson. 2002. *Resource Selection in Animals: Statistical Design and Analysis for Field Studies*, 2nd edition. Kluwer, Norwell, Massachusetts.

Mannion, P. D., P. Upchurch, R. B. J. Benson, and A. Goswami. 2014. The latitudinal biodiversity gradient through deep time. *Trends in Ecology and Evolution* 29: 42–50.

Manrubia, S. and R. Solé. 1997. On forest spatial dynamics with gap formation. *Journal of Theoretical Biology* 187: 159–164.

Mantel, N. 1967. The detection of disease clustering and a generalized regression approach. *Cancer Research* 27: 209–220.

Marble, D. F. and D. J. Peuquet. 1983. Geographic information systems and remote sensing. Pp. 923–958 in *Manual of Remote Sensing*, Volume 1, 2nd edition (D. S. Simonett and F. T. Ulaby, Editors). American Society of Photogrammetry, Falls Church, Virginia.

Marcarelli, A. M., C. V. Baxter, M. M. Mineau, and R. O. Hall, Jr. 2011. Quantity and quality: Unifying food web and ecosystem perspectives on the role of resource subsidies in freshwaters. *Ecology* 92: 1215–1225.

Margosian, M. L., K. A. Garrett, J. M. S. Hutchinson, and K. A. With. 2009. Connectivity of the American agricultural landscape: Assessing the national risk of disease and crop pest spread. *BioScience* 59: 141–151.

Marini, L., E. Öckinger, K.-O. Bergman, B. Jauker, J. Krauss, M. Kuussaari, J. Pöyry, H. G. Smith, I. Steffan-Dewenter, and R. Bommarco. 2014. Contrasting effects of habitat area and connectivity on evenness of pollinator communities. *Ecography* 37: 544–551.

Marlon, J. R., and 11 others. 2012. Long-term perspective on wildfires in the western USA. *Proceedings of the National Academy of Sciences of the United States of America* 109: 3203–3204.

Martin, S. J., R. R. Funch, P. R. Hanson, and E.-H. Yoo. 2018. A vast 4,000-year-old spatial pattern of termite mounds. *Current Biology* 28: R1292–R1293.

Martin, T. E. 1993. Nest predation and nest sites. *BioScience* 43: 523–532.

Marzluff, J. M., E. Shulenberger, W. Endlicher, M. Alberti, G. Bradley, C. Ryan, U. Simon, and C. Zumbrunnen, Editors. 2008. *Urban Ecology: An International Perspective on the Interaction between Humans and Nature.* Springer, New York.

Massol, F., D. Gravel, N. Mouquet, M. W. Cadotte, T. Fukami, and M. A. Leibold. 2011. Linking community and ecosystem dynamics through spatial ecology. *Ecology Letters* 14: 313–323.

Mastrangelo, M. E., F. Weyland, S. H. Villarino, M. P. Barral, L. Nahuelhual, and P. Laterra. 2014. Concepts and methods for landscape multifunctionality and a unifying framework based on ecosystem services. *Landscape Ecology* 29: 345–358.

Mata, C., I. Hervás, J. Herranz, F. Suárez, and J. E. Malo. 2008. Are motorway wildlife passages worth building? Vertebrate use of road-crossing structures on a Spanish motorway. *Journal of Environmental Management* 88: 407–415.

Mathews, L. E. 2005. *Historical Imagery Holdings for the United States Department of Agriculture,* Aerial Photography Field Office (APFO). Available online: http://www.fsa.usda.gov/Internet/FSA_File/vault_holdings2.pdf (last accessed 2/3/19).

Mathiason, C. K., and 16 others. 2006. Infectious prions in the saliva and blood of deer with Chronic Wasting Disease. *Science* 314: 133–136.

Matson, P. A., W. J. Parton, A. G. Power, and M. J. Swift. 1997. Agricultural intensification and ecosystem properties. *Science* 277: 504–509.

Matthysen, E. 2005. Density-dependent dispersal in birds and mammals. *Ecography* 28: 403–416.

Maxwell, S. K., G. L. Schmidt, and J. C. Storey. 2007. A multiscale segmentation approach to filling gaps in Landsat ETM+ SLC-off images. *International Journal of Remote Sensing* 28: 5339–5356.

Mayer, A. L., B. Buma, A. Davis, S. A. Gagné, E. L. Loudermilk, R. M. Scheller, F. K. A. Schmiegelow, Y. F. Wiersma, and J. Franklin. 2016. How landscape ecology informs global land-change science and policy. *BioScience* 66: 458–469.

Mayor, S. J., D. C. Shneider, J. A. Schaefer, and S. P. Mahoney. 2009. Habitat selection at multiple scales. *Écoscience* 16: 238–247.

Mazerolle, M. J. and M.-A. Villard. 1999. Patch characteristics and landscape context as predictors of species presence and abundance: A review. *Écoscience* 6: 117–124.

Meagher, M. and M. E. Meyer. 1994. On the origin of brucellosis in bison of Yellowstone National Park—a review. *Conservation Biology* 8: 645–653.

Mech, S. G. and P. A. Zollner. 2002. Using body size to predict perceptual range. *Oikos* 98: 47–52.

Medeiros, A. C. and L. L. Loope. 1997. Status, ecology, and management of the invasive plant *Miconia calvescens* DC (Melastomataceae)

in the Hawaiian Islands. Records of the Hawaii Biological Survey for 1996, *Bishop Museum Occasional Papers* 48: 23–36.

Meentemeyer, R. K., N. E. Rank, B. L. Anacker, D. M. Rizzo, and J. H. Cushman. 2008. Influence of land-cover change on the spread of an invasive forest pathogen. *Ecological Applications* 18: 159–171.

Meentemeyer, R. K., M. A. Dorning, J. B. Vogler, D. Schmidt, and M. Garbelotto. 2015. Citizen science helps predict risk of emerging infectious disease. *Frontiers in Ecology and the Environment* 13: 189–194.

Meirmans, P. G. and P. W. Hedrick. 2011. Assessing population structure: F_{ST} and related measures. *Molecular Ecology Resources* 11: 5–18.

Melles, S. J., D. Badzinski, M.-J. Fortin, F. Csillag, and K. Lindsay. 2009. Disentangling habitat and social drivers of nesting patterns in songbirds. *Landscape Ecology* 24: 519–531.

Meng, X., N. Currit, and K. Zhao. 2010. Ground filtering algorithms for airborne LiDAR data: A review of critical issues. *Remote Sensing* 2: 833–860.

Mennechez, G., N. Schitckzelle, and M. Baguette. 2003. Metapopulation dynamics of the bog fritillary butterfly: Comparison of demographic parameters and dispersal between a continuous and a highly fragmented landscape. *Landscape Ecology* 18: 279–291.

Merriam, C. H. 1890. Results of a biological survey of the San Francisco Mountain region and desert of the Little Colorado in Arizona. North American Fauna 3, Bureau of Biological Survey, U.S. Department of Agriculture, Washington, D.C.

Merriam, G. 1984. Connectivity: A fundamental ecological characteristic of landscape pattern. Pp. 5–15 in *Proceedings of the First International Seminar on Methodology in Landscape Ecological Research and Planning*, Vol. I (J. Brandt and P. Agger, Editors). Roskilde Universitessforlag GeoRue, Roskilde, Denmark.

Merritt, D. M. and P. B. Shafroth. 2012. Edaphic, salinity, and stand structural trends in chronosequences of native and non-native dominated riparian forests along the Colorado River, USA. *Biological Invasions* 14: 2665–2685.

Metzger, J. P. and H. Décamps. 1997. The structural connectivity threshold: An hypothesis in conservation biology at the landscape scale. *Acta Œcologia* 18: 1–12.

Meyer, J. S., L. L. Irwin, and M. S. Boyce. 1998. Influence of habitat abundance and fragmentation on northern spotted owls in western Oregon. *Wildlife Monographs* 139: 3–51.

Meyer, J.-Y. 2010. The *Miconia* saga: 20 years of study and control in French Polynesia (1988–2008). Pp. 1–19 in *Proceedings of the International Miconia Conference*, Keanae, Maui, Hawaii, May 4–7, 2009 (L. L. Loope, J.-Y. Meyer, B.D. Hardesty and C.W. Smith, Editors), Maui Invasive Species Committee and Pacific Cooperative Studies Unit, University of Hawaii at Manoa. Available online: http://www.hear.org/conferences/miconia2009/pdfs/meyer.pdf (last accessed 1/30/19).

Meyer, K. M., K. Wiegand, D. Ward, and A. Moustakas. 2007. The rhythm of savanna patch dynamics. *Journal of Ecology* 95: 1306–1315.

Meysman, F. J. R., J. J. Middelburg, and C. H. R. Heip. 2006. Bioturbation: A fresh look at Darwin's last idea. *Trends in Ecology and Evolution* 21: 688–695.

Michener, C. D. 2000. *The Bees of the World*. Johns Hopkins Press, Baltimore, Maryland.

Michener, W. K., E. R. Blood, K. L. Bildstein, M. M. Brinson, and L. R. Gardner. 1997. Climate change, hurricanes and tropical storms, and rising sea level in coastal wetlands. *Ecological Applications* 7: 770–801.

Middleburg, J. J. 2014. Stable isotopes dissect aquatic food webs from the top to the bottom. *Biogeosciences* 11: 2357–2371.

Miles, C. M. and M. Wayne. 2008. Quantitative trait locus (QTL) analysis. *Nature Education* 1: 208. Available online: https://www.nature.com/scitable/topicpage/quantitative-trait-locus-qtl-analysis-53904

Millennium Ecosystem Assessment. 2003. *Ecosystems and Human Well-being: A Framework for Assessment*. Island Press, Washington, D.C.

Millennium Ecosystem Assessment. 2005. *Ecosystems and Human Well-being: Desertification Synthesis*. World Resources Institute, Washington, D.C. Available online: http://www.millenniumassessment.org/documents/document.355.aspx.pdf (last accessed 2/24/19)

Miller, C. and D. L. Urban. 2000. Connectivity of forest fuels and surface fire regimes. *Landscape Ecology* 15: 145–154.

Miller, J. D., H. D. Safford, M. Crimmins, and A. E. Thode. 2009. Quantitative evidence for increasing forest fire severity in the Sierra Nevada and southern Cascade mountains, California and Nevada, USA. *Ecosystems* 12: 16–32.

Miller, J. R., M. G. Turner, E. A. H. Smithwick, C. L. Dent, and E. H. Stanley. 2004. Spatial extrapolation: The science of predicting ecological patterns and processes. *BioScience* 54: 310–320.

Miller, M. A., and 10 others. 2002. Coastal freshwater runoff is a risk factor for *Toxoplasma gondii* infection of southern sea otters (*Enhydra lutris nereis*). *International Journal for Parasitology* 32: 997–1006.

Milliman, J. D. and R. H. Meade. 1983. Worldwide delivery of river sediment to the oceans. *Journal of Geology* 91: 1–21.

Mills, L. S., M. E. Soulé, and D. F. Doak. 1993. The keystone-species concept in ecology and conservation. *BioScience* 43: 219–224.

Millspaugh, J. J., R. M. Nielson, L. McDonald, J. M. Marzluff, R. A. Gitzen, C. D. Rittenhouse, M. W. Hubbard, and S. L. Sheriff. 2006. Analysis of resource selection using utilization distributions. *Journal of Wildlife Management* 70: 384–395.

Milne, B. T. 1997. Applications of fractal geometry in wildlife biology. Pp. 32–69 in *Wildlife and Landscape Ecology. Effects of Pattern and Scale*. (J.A. Bissonette, Editor). Springer-Verlag, New York.

Milne, B. T., M. G. Turner, J. A. Wiens, and A. R. Johnson. 1992. Interactions between the fractal geometry of landscapes and allometric herbivory. *Theoretical Population Biology* 41: 337–353.

Milne, G. 1935. Composite units for the mapping of complex soil associations. *Transactions of the Third International Congress of Soil Science* 1: 345–347.

Minor, E. S. and D. L. Urban. 2007. Graph theory as a proxy for spatially explicit population models in conservation planning. *Ecological Applications* 17: 1771–1782.

Minor, E. S. and D. L. Urban. 2008. A graph-theory framework for evaluating landscape connectivity and conservation planning. *Conservation Biology* 22: 297–307.

Minor, E. S., S. M. Tessel, K. A. M. Engelhardt, and T. R. Lookingbill. 2009. The role of landscape connectivity in assembling exotic plant communities: A network analysis. *Ecology* 90: 1802–1809.

Minshall, G. W. 1967. Role of allochthonous detritus in the trophic structure of a woodland springbrook community. *Ecology* 48: 139–149.

Mitchell, S. J. 2013. Wind as a natural disturbance agent in forests: A synthesis. *Forestry* 86: 147–157.

Mittelbach, G. G., and 21 others. 2007. Evolution and the latitudinal diversity gradient: Speciation, extinction and biogeography. *Ecology Letters* 10: 315–331.

Mladenoff, D. J., S. E. Dahir, E. V. Nordheim, L. A. Schulte, and G. G. Guntenspergen. 2002. Narrowing historical uncertainty: Probabilistic classification of ambiguously identified tree species in historical survey data. *Ecosystems* 5: 539–553.

Moilanen, A. and I. Hanski. 1998. Metapopulation dynamics: Effects of habitat quality and landscape structure. *Ecology* 79: 2503–2515.

Moilanen, A. and I. Hanski. 2001. On the use of connectivity measures in spatial ecology. *Oikos* 95: 147–151.

Moilanen, A. and M. Nieminen. 2002. Simple connectivity measures in spatial ecology. *Ecology* 83: 1131–1145.

Molden, D., Editor. 2007. *Water for Food, Water for Life: A Comprehensive Assessment of Water Management in Agriculture*. International Water Management Institute, Colombo, Sri Lanka. Earthscan, London, UK.

Moloney, K. A. and S. A. Levin. 1996. The effects of disturbance architecture on landscape-level population dynamics. *Ecology* 77: 375–394.

Montaggioni, L. F. 2005. History of Indo-Pacific coral reef systems since the last glaciation: Development patterns and controlling factors. *Earth-Science Reviews* 71: 1–75.

Monterrubio-Rico, T. C., K. Renton, J. M. Ortega-Rodríguez, A. Pérez-Arteaga, and R. Cacino-Murillo. 2010. The endangered yellow-headed parrot *Amazona oratrix* along the Pacific coast of Mexico. *Oryx* 44: 602–609.

Montoya, D., F. S. Albuquerque, M. Rueda, and M. A. Rodríguez. 2010. Species' response patterns to habitat fragmentation: Do trees support the extinction threshold hypothesis? *Oikos* 119: 1225–1343.

Moody, A. L., A. I. Houston, and J. M. McNamara. 1996. Ideal free distributions under predation risk. *Behavioral Ecology and Sociobiology* 38: 131–143.

Moody, M. E. and R. N. Mack. 1988. Controlling the spread of plant invasions: The importance of nascent foci. *Journal of Applied Ecology* 25: 1009–1021.

Moore, J. W., D. E. Schindler, and M. D. Scheuerell. 2004. Disturbance of freshwater habitats by anadromous salmon in Alaska. *Oecologia* 139: 298–308.

Moore, J. W., D. E. Schindler, J. L. Carter, J. Fox, J. Griffiths, and G. W. Holtgrieve. 2007. Biotic control of stream fluxes: Spawning salmon drive nutrient and matter export. *Ecology* 88: 1278–1291.

Moran, E. V. and J. S. Clark. 2012. Between-site differences in the scale of dispersal and gene flow in red oak. *PLoS ONE* 7(5): e36492.

Moran, P. A. P. 1950. Notes on continuous stochastic phenomena. *Biometrika* 37: 17–23.

Morelli, F., M. Beim, L. Jerzak, D. Jones, and P. Tryjanowski. 2014. Can roads, railways and related structures have positive effects on birds? A review. *Transportation Research Part D* 30: 21–31.

Morgan, J. K., J. Park, and C. A. Zelt. 2010. Rift zone abandonment and reconfiguration in Hawaii: Mauna Loa's Ninole rift zone. *Geology* 38: 471–474.

Morgan, J. L., S. E. Gergel, and N. C. Coops. 2010. Aerial photography: A rapidly evolving tool for ecological management. *BioScience* 60: 47–59.

Morgan, W. J. 1971. Convection plumes in the lower mantle. *Nature* 230: 42–43.

Morisita, M. 1962. I_δ-index, a measure of dispersion of individuals. *Researches on Population Ecology* 4: 1–7.

Morisita, M. 1971. Composition of the I_δ-index. *Researches on Population Ecology* 13: 1–27.

Morris, D. W. 1992. Scales and costs of habitat selection in heterogeneous landscapes. *Evolutionary Ecology* 6: 412–432.

Mouquet, N., M. F. Hoopes, and P. Amarasekare. 2005. The world is patchy and heterogeneous! Trade-off and source-sink dynamics in competitive metacommunities. Pp. 237–262 in *Metacommunities: Spatial Dynamics and Ecological Communities* (M. Holyoak, M. A. Leibold, and R. D. Holt, Editors). University of Chicago Press, Chicago, Illinois.

Mueller, U. G. and L. L. Wolfenbarger. 1999. AFLP genotyping and fingerprinting. *Trends in Ecology and Evolution* 14: 389–394.

Muller-Landau, H. C. 2010. The tolerance-fecundity trade-off and the maintenance of diversity in seed size. *Proceedings of the National Academy of Sciences of the United States of America* 107: 4242–4247.

Mumby, P. J., R. Vitolo, and D. S. Stephenson. 2011. Temporal clustering of tropical cyclones and its ecosystem impacts. *Proceedings of the National Academy of Sciences of the United States of America* 108: 17626–17630.

Mundt, C. C. 2002. Use of multiline cultivars and cultivar mixtures for disease management. *Annual Review of Phytopathology* 40: 381–410.

Mundt, C. C., K. E. Sackett, L. D. Wallace, C. Cowger, and J. P. Dudley. 2009a. Long-distance dispersal and accelerating waves of disease: Empirical relationships. *American Naturalist* 173: 456–466.

Mundt, C. C., K. E. Sackett, L. D. Wallace, C. Cowger, and J. P. Dudley. 2009b. Aerial dispersal and multiple-scale spread of epidemic disease. *EcoHealth* 6: 546–552.

Murdoch. W. W. and C. J. Briggs. 1996. Theory for biological control: Recent developments. *Ecology* 77: 2001–2013.

Murdoch, W. W., C. J. Briggs, R. M. Nisbet, W. S. C Gurney, and A. Stewart-Oaten. 1992. Aggregation and stability in metapopulation models. *American Naturalist* 140: 41–58.

Murphy, M. A., R. Dezzani, D. S. Pilliod, and A. Storfer. 2010. Landscape genetics of high mountain frog metapopulations. *Molecular Ecology* 19: 3634–3649.

Murray, J. D., E. A. Stanley, and D. L. Brown. 1986. On the spatial spread of rabies among foxes. *Proceedings of the Royal Society of London, Biology* 229:111–150.

Myers, N., R. A. Mittermeier, C. G. Mittermeier, G. A. B. da Fonesca, and J. Kent. 2000. Biodiversity hotspots for conservation priorities. *Nature* 402: 853–858.

Myster, R. W. and D. S. Fernández. 1995. Spatial gradients and patch structure on two Puerto Rican landslides. *Biotropica* 27: 149–159.

Myster, R. W. and L. R. Walker. 1997. Plant successional pathways on Puerto Rican landslides. *Journal of Tropical Ecology* 13: 165–173.

Myster, R. W., J. R. Thomlinson, and M. C. Larsen. 1997. Predicting landslide vegetation in patches on landscape gradients in Puerto Rico. *Landscape Ecology* 12: 299–307.

Naiman, R. J. and H. Décamps. 1997. The ecology of interfaces: Riparian zones. *Annual Review of Ecology and Systematics* 28: 621–658.

Naiman, R. J., C. A. Johnston, and J. C. Kelley. 1988. Alteration of North American streams by beaver. *BioScience* 38: 753–762.

Naiman, R. J., G. Pinay, C. A. Johnston, and J. Pastor. 1994. Beaver influences on the long-term biogeochemical characteristics of boreal forest drainage networks. *Ecology* 75: 905–921.

Naiman, R. J., R. E. Bilby, D. E. Schindler, and J. M. Helfield. 2002. Pacific salmon, nutrients, and the dynamics of freshwater and riparian ecosystems. *Ecosystems* 5: 399–417.

Nakano, S. and M. Murakami. 2001. Reciprocal subsidies: Dynamic interdependence between terrestrial and aquatic food webs. *Proceedings of the National Academy of Sciences of the United States of America* 98: 166–170.

Nams, V. O. 2005. Using animal movement paths to measure response to spatial scale. *Oecologia* 143: 179–188.

Nams, V. O. 2012. Shape of patch edges affects edge permeability for meadow voles. *Ecological Applications* 22: 1827–1837.

Nams, V. O. and M. Bourgeois. 2004. Fractal analysis measures habitat use at different spatial scales: An example with American marten. *Canadian Journal of Zoology* 82: 1738–1747.

Nanko, K., T. W. Giambelluca, R. A. Sutherland, R. G. Mudd. M. A. Nullet, and A. D. Ziegler. 2013. Erosion potential under *Miconia calvescens* stands on the island of Hawai'i. *Land Degradation & Development* 26: 218–226.

Nash, S. 2010. Making sense of Mount St Helens. *BioScience* 60: 571–575.

Nassauer, J. I. and P. Opdam. 2008. Design in science: Extending the landscape ecology paradigm. *Landscape Ecology* 23: 633–644.

Nathan, R., G. Perry, J. T. Cronin, A. E. Strand, and M. L. Cain. 2003. Methods for estimating long-distance dispersal. *Oikos* 103: 261–273.

National Research Council. 1999. *Our Common Journey: A Transition Toward Sustainability*. The National Academy Press, Washington, D.C.

Naveh, Z. 2007. Landscape ecology and sustainability. *Landscape Ecology* 22: 1437–1440.

Naveh, Z. and A. S. Lieberman. 1984. *Landscape Ecology: Theory and Application*. Springer-Verlag, New York.

Nee, S. and R. M. May. 1992. Dynamics of metapopulations: Habitat destruction and competitive coexistence. *Journal of Animal Ecology* 61: 37–40.

Neel, M. C., K. McGarigal, and S. A. Cushman. 2004. Behavior of class-level landscape metrics across gradients of class aggregation and area. *Landscape Ecology* 19: 435–455.

Nei, M. 1972. Genetic distance between populations. *American Naturalist* 106: 283–292.

Nei, M. 1973. Analysis of gene diversity in subdivided populations. *Proceedings of the National Academy of Sciences of the United States of America* 70: 3321–3323.

Nei, M. and W-H. Li. 1979. Mathematical model for studying genetic variation in terms of restriction endonucleases. *Proceedings of the National Academy of Sciences of the United States of America* 76: 5269–5273.

Nelson, M. C., K. Kintigh, D. R. Abbott, and J. M. Anderies. 2010. The cross-scale interplay between social and biophysical context and the vulnerability of irrigation-dependent societies: Archaeology's long-term perspective. *Ecology and Society* 15(3): 31. Available online: http://www.ecologyandsociety.org/vol15/iss3/art31/

Neteler, M., M. H. Bowman, M. Landa, and M. Metz. 2012. GRASS GIS: A multipurpose open source GIS. *Environmental Modelling & Software* 31: 124–130.

Newbold, J. D., J. W. Elwood, R. V. O'Neill, and W. Van Winkle. 1981. Measuring nutrient spiraling in streams. *Canadian Journal of Fisheries and Aquatic Sciences* 38: 860–863.

Newbold, T, and 40 others. 2015. Global effects of land use on local terrestrial biodiversity. *Nature* 520: 45–50.

Newton, A. C., R. A. Hill, C. Echeverría, D. Golicher, J. M. R. Benayas, L. Cayuela, and S. A. Hinsley. 2009. Remote sensing and the future of landscape ecology. *Progress in Physical Geography* 33: 528–546.

Nicholls, R. J., P. P. Wong, V. R. Burkett, J. O. Codignotto, J. E. Hay, R. F. McLean, S. Ragoonaden and C. D. Woodroffe. 2007. Coastal systems and low-lying areas. Pp. 315–356 in *Climate Change 2007: Impacts, Adaptation and Vulnerability. Contribution of Working Group II to the Fourth Assessment Report of the Intergovernmental Panel on Climate Change* (M. L. Parry, O. F. Canziani, J. P. Palutikof, P. J. van der Linden, and C. E. Hanson, Editors). Cambridge University Press, Cambridge, UK.

Nicholls, R. J., N. Marinova, J. A. Lowe, S. Brown, P. Vellinga, D. de Gusmão, J. Hinkel, and R. S. J. Tol. 2011. Sea-level rise and its possible impacts given a 'beyond 4°C world' in the twenty-first century. *Philosophical Transactions of the Royal Society A* 369: 161–181.

Nicholson, A. J. and V. A. Bailey. 1935. The balance of animal populations. Part I. *Journal of Zoology* 105: 551–598.

Nicolaus, M., S. P. M. Michler, K. M. Jalvingh, R. Ubels, M. van der Velde, J. Komdeur, C. Both, and J. M. Tinbergen. 2012. Social environment affects juvenile dispersal in great tits (*Parus major*). *Journal of Animal Ecology* 81: 827–837.

Niemelä, J., J. H. Breuste, T. Elmqvist, G. Guntenspergen, P. James, and N. E. McIntyre, Editors. 2011. *Urban Ecology: Patterns, Processes, and Applications*. Oxford University Press, New York.

NIFC (National Interagency Fire Center). 2017. Incident Management Situation Report, National Interagency Coordination Center, National Interagency Fire Center, United States. https://www.nifc.gov/nicc/

Nilsen, E. B., S. Pedersen, and J. D. C. Linnell. 2008. Can minimum convex polygon home ranges be used to draw biologically meaningful conclusions? *Ecological Research* 23: 635–639.

Nilsson, C. and K. Berggren. 2000. Alterations of riparian ecosystems caused by river regulation. *BioScience* 50: 783–792.

Nilsson, C., C. A. Reidy, M. Dynesius, and C. Revenga. 2005. Fragmentation and flow regulation of the world's large river systems. *Science* 308: 405–408.

Nogués-Bravo, D., M. B. Araújo, T. Romdal, and C. Rahbek. 2008. Scale effects and human impact on the elevational species richness gradients. *Nature* 453: 216–219.

Norris, M. D., J. M. Blair, and L. C. Johnson. 2007. Altered ecosystem nitrogen dynamics as a consequence of land cover change in tall-grass prairie. *American Midland Naturalist* 158: 432–445.

Nosil, P. 2008. Speciation with gene flow could be common. *Molecular Ecology* 17: 2103–2106.

Nosil, P., S. P. Egan, and D. J. Funk. 2008. Heterogeneous genomic differentiation between walking-stick ecotypes: 'Isolation by Adaptation' and multiple roles for divergent selection. *Evolution* 62: 316–336.

Núñez-Farfán, J. and C. D. Schlichting. 2001. Evolution in changing environments: The 'synthetic' work of Clausen, Keck, and Hiesey. *Quarterly Review of Biology* 76: 433–457.

Núñez-Farfán, J. and C. D. Schlichting. 2005. Natural selection in *Potentilla glandulosa* revisited. *Evolutionary Ecology Research* 7: 105–119.

Nystrand, M., M. Griesser, S. Eggers, and J. Ekman. 2010. Habitat-specific demography and source-sink dynamics in a population of Siberian jays. *Journal of Animal Ecology* 79: 266–274.

O'Dowd, D. J., P. T. Green, and P. S. Lake. 2003. Invasional 'meltdown' on an oceanic island. *Ecology Letters* 6: 812–817.

O'Neill, R. V., D. L. DeAngeles, J. B. Waide, and T. F. H. Allen. 1986. *A Hierarchical Approach to Ecosystems*. Princeton University Press, Princeton, New Jersey.

O'Neill R. V., and 11 others. 1988a. Indices of landscape pattern. *Landscape Ecology* 1: 153–162.

O'Neill, R. V., B. T. Milne, M. G. Turner, and R. H. Gardner. 1988b. Resource utilization scales and landscape pattern. *Landscape Ecology* 2: 63–69.

O'Neill, R. V., R. H. Gardner, and M. G. Turner. 1992. A hierarchical neutral model for landscape analysis. *Landscape Ecology* 7: 55–61.

O'Neill, R. V., C. T. Hunsaker, K. B. Jones, K. H. Riitters, J. D. Wickham, P. Schwarz, I. A. Goodman, B. Jackson, and W. S. Baillargeon. 1997. Monitoring environmental quality at the landscape scale. *BioScience* 47: 513–519.

Öckinger, E., and 10 others. 2010. Life-history traits predict species responses to habitat area and isolation: A cross-continental synthesis. *Ecology Letters* 13: 969–979.

Odum, E. P. 1953. *Fundamentals of Ecology*, 1st edition. W. B. Sanders, Philadelphia, Pennsylvania.

Odum, E. P. 1960. Organic production and turnover in old field succession. *Ecology* 41: 34–49.

Odum, E. P. 1964. The new ecology. *BioScience* 14: 14–16.

Odum, E. P. 1969. The strategy of ecosystem development. *Science* 164: 262–270.

OECD (Organization for Economic Cooperation and Development). 2015. Overview of the food and agricultural situation in Brazil. Pp. 27–49 in *Innovation, Agricultural Productivity and Sustainability in Brazil*, OECD Publishing, Paris, France. Available online: http://dx.doi.org/10.1787/9789264237056-5-en (last accessed 1/30/19).

Oglesby, R. J., T. L. Sever, W. Saturno, D. J. Erickson III, and J. Srikishen. 2010. Collapse of the Maya: Could deforestation have contributed? *Journal of Geophysical Research* 115: D12106.

Ohl, C. A. and S. Tapsell. 2000. Flooding and human health: The dangers posed are not always obvious. *British Medical Journal* 321: 1167–1168.

Ojima, D. S., D. S. Schimel, W. J. Parton, and C. E. Owensby. 1994a. Long- and short-term effects of fire on nitrogen cycling in tallgrass prairie. *Biogeochemistry* 24: 67–84.

Ojima, D. S., K. A. Galvin, and B. L. Turner, II. 1994b. The global impact of land-use change. *BioScience* 44: 300–304.

Okin, G. S., A. J. Parsons, J. Wainwright, J. E. Herick, B. T. Bestelmeyer, D. C. Peters, and E. L. Fredrickson. 2009. Do changes in connectivity explain desertification? *BioScience* 59: 237–244.

Oksanen, J., and 12 others. 2019. Package 'vegan': Community Ecology Package. R package version 2.5-4. Documentation available online: https://cran.r-project.org/web/packages/vegan/vegan.pdf

Okubo, A. 1980. *Diffusion and Ecological Problems: Mathematical Models.* Springer-Verlag, Berlin, Germany.

Okubo, A. and S. A. Levin. 2002. *Diffusion and Ecological Problems: Modern Perspectives,* 2nd edition. Springer-Verlag, New York.

Oldroyd, B. P. 2007. What's killing American honey bees? *PLoS Biology* 5(6): e168.

Oliveira, P. S. and R. J. Marquis, Editors. 2002. *The Cerrados of Brazil: Ecology and Natural History of a Neotropical Savanna.* Columbia University Press, New York.

Openshaw, S. 1983. *The Modifiable Areal Unit Problem.* Concepts and Techniques in Modern Geography No. 38. GeoBooks, Norwich, UK.

Openshaw, S. and P. Taylor. 1979. A million or so correlation coefficients: Three experiments on the modifiable areal unit problem. Pp. 127–144 in *Statistical Applications in the Spatial Sciences* (N. Wrigley, Editor). Pion, London, UK.

Oppenheimer, C. 2011. *Eruptions that Shook the World.* Cambridge University Press, Cambridge, UK.

Ord, J. K. and A. Getis. 1995. Local spatial autocorrelation statistics: Distributional issues and an application. *Geographical Analysis* 27: 286–306.

Orsini, L., J. Vanoverbeke, I. Swillen, J. Mergeay, and L. De Meester. 2013. Drivers of population genetic differentiation in the wild: Isolation by dispersal limitation, isolation by adaptation, and isolation by colonization. *Molecular Ecology* 22: 5983–5999.

Ortloff, C. R. 1995. Surveying and hydraulic engineering of the Pre-Columbian Chimú State: AD 900–1450. *Cambridge Archaeological Journal* 5: 55–74.

Ostertag, R., W. L. Silver, and A. E. Lugo. 2005. Factors affecting mortality and resistance to damage following hurricanes in a rehabilitated subtropical moist forest. *Biotropica* 37: 16–24.

Ostfeld, R. S. 2009. Climate change and the distribution and intensity of infectious diseases. *Ecology* 90: 903–905.

Ostfeld, R. S. and R. D. Holt. 2004. Are predators good for your health? Evaluating evidence for top-down regulation of zoonotic disease reservoirs. *Frontiers of Ecology and the Environment* 2: 13–20.

Ostfeld, R. S., G. E. Glass, and F. Keesing. 2005. Spatial epidemiology: An emerging (or re-emerging) discipline. *Trends in Ecology and Evolution* 20: 328–336.

Otero, X. L., S. De La Peña-Lastra, A. Pérez-Alberti, T. O. Ferreira, and M. A. Huerta-Diaz. 2018. Seabird colonies as important global drivers in the nitrogen and phosphorus cycles. *Nature Communications* 9: 246.

Ovaskainen, O. and I. Hanski. 2004. Metapopulation dynamics in highly fragmented landscapes. Pp. 73–103 in *Ecology, Genetics and Evolution of Metapopulations* (I. Hanski and O. E. Gaggiotti, Editors). Elsevier Academic Press, Burlington, Massachusetts.

Overpeck, J.T. 2013. Climate science: The challenge of hot drought. *Nature* 503: 350–351.

Owen-Smith, N., J. M. Fryxell, and E. H. Merrill. 2010. Foraging theory upscaled: The behavioural ecology of herbivore movement. *Philosophical Transactions of the Royal Society of London B* 365: 2267–2278.

Paine, R. T. 1969. A note on trophic complexity and community stability. *American Naturalist* 103: 91–93.

Paine, R. T., M. J. Tegner, and E. A. Johnson. 1998. Compounded perturbations yield ecological surprises. *Ecosystems* 1: 535–545.

Palmer, M. A. and C. M. Febria. 2012. The heartbeat of ecosystems. *Science* 336: 1393–1394.

Palmer, M. W. 1994. Variation in species richness: Towards a unification of hypotheses. *Folia Geobotanica et Phytotaxonomica* 29: 511–530.

Palmer, M. W. and P. S. White. 1994. Scale dependence and the species-area relationship. *American Naturalist* 144: 717–740.

Parisien, M.-A. and M. A. Moritz. 2009. Environmental controls on the distribution of wildfire at multiple spatial scales. *Ecological Monographs* 79: 127–154.

Parmesan, C. 1996. Climate and species' ranges. *Nature* 382: 765–766.

Parmesan, C. 2006. Ecological and evolutionary responses to recent climate change. *Annual Review of Ecology, Evolution and Systematics* 37: 637–669.

Parmesan, C., S. Gaines, L. Gonzalez, D. M. Kaufman, J. Kingsolver, A. T. Peterson, and R. Sagarin. 2005. Empirical perspectives on species borders: From traditional biogeography to global change. *Oikos* 108: 58–75.

Parsons, M., C. A. McLoughlin, K. A. Kotschy, K. H. Rogers, and M. W. Roundtree. 2005. The effects of extreme floods on the biophysical heterogeneity of river landscapes. *Frontiers in Ecology and the Environment* 3: 487–494.

Patten, M. A. and J. F. Kelly. 2010. Habitat selection and the perceptual trap. *Ecological Applications* 20: 2148–2156.

Pauchard, A. and P. B. Alaback. 2004. Influence of elevation, land use, and landscape context on patterns of alien plant invasions along roadsides in protected areas of south-central Chile. *Conservation Biology* 18: 239–248.

Paul, E. A. 1984. Dynamics of organic matter in soils. *Plant and Soil* 76: 275–285.

Pavlovski, E. N. 1966. *Natural Nidality of Transmissible Diseases, With Special Reference to the Landscape Epidemiology of Zooanthroponoses.* University of Illinois Press, Urbana, Illinois.

Pazzaglia, F. J. 2003. Landscape evolution models. Pp. 247–274 in *The Quaternary Period in the USA* (A. R. Gillespie, S.C. Porter, and B. F. Atwater, Editors). Developments in Quaternary Science, Volume 1, Elsevier, Amsterdam, the Netherlands.

Pearson, R. G. 2006. Climate change and the migration capacity of species. *Trends in Ecology and Evolution* 21: 111–113.

Pearson, R. G. and T. P. Dawson. 2003. Predicting the impacts of climate change on the distribution of species: Are bioclimate envelope models useful? *Global Ecology and Biogeography* 12: 361–371.

Pearson, S. M. and R. H. Gardner. 1997. Neutral models: Useful tools for understanding landscape patterns. Pp. 215–230 in *Wildlife and Landscape Ecology: Effects of Pattern and Scale* (J. A. Bissonette, Editor). Springer-Verlag, New York.

Pebesma, E. and B. Graeler. 2018. Package 'gstat': Spatial and spatiotemporal geostatistical modeling, prediction and simulation (version 1.1–6). Documentation available online: https://cran.r-project.org/web/packages/gstat/gstat.pdf

Pederson, G. T., S. T. Gray, C. A. Woodhouse, J. L. Betancourt, D. B. Fagre, J. S. Littell, E. Watson, B. H. Luckman, and L. J. Graumlich. 2011. The unusual nature of recent snowpack declines in the North American Cordillera. *Science* 333: 332–335.

Pedlowski, M. A., V. H. Dale, E. A. T. Matricardi, and E. P. S. Filho. 1997. Patterns and impacts of deforestation in Rondônia, Brazil. *Landscape and Urban Planning* 38: 149–157.

Pereira, C. G., D. P. Almenara, C. E. Winter, P. W. Fritsch, H. Lambers, and R. S. Oliveira. 2012. Underground leaves of *Philcoxia* trap and digest nematodes. *Proceedings of the National Academy of Sciences of the United States of America* 109: 1154–1158.

Pereira, H. M., and 22 others. 2010. Scenarios for global biodiversity in the 21st century. *Science* 330: 1496–1501.

Perera, A. H., L. J. Buse, and T. R. Crow, Editors. 2007. *Forest Landscape Ecology: Transferring Knowledge to Practice*. Springer, New York.

Perrins, C. M. 1965. Population fluctuations and clutch size in the great tit, *Parus major* L. *Journal of Animal Ecology* 34: 601–647.

Perron, J. T., P. W. Richardson, K. L. Ferrier, and M. Lapôtre. 2012. The root of branching river networks. *Nature* 492: 100–105.

Perry, G. L. W., B. P. Miller, and N. J. Enright. 2006. A comparison of methods for the statistical analysis of spatial point patterns in plant ecology. *Plant Ecology* 187: 59–82.

Perry, J. N., A. M. Liebhold, M. S. Rosenberg, J. Dungan, M. Miriti, A. Jakomulska, and S. Citron-Pousty. 2002. Illustrations and guidelines for selecting statistical methods for quantifying spatial pattern in ecological data. *Ecography* 25: 578–600.

Petchey, O. L. and K. J. Gaston. 2006. Functional diversity: Back to basics and looking forward. *Ecology Letters* 9: 741–758.

Petchey, O. L., and 17 others. 2015. The ecological forecast horizon, and examples of its uses and determinants. *Ecology Letters* 18: 597–611.

Peters, D. P. C., R. A. Pielke, Sr., B. T. Bestelmeyer, C. D. Allen, S. Munson-McGee, and K. M. Havstad. 2004. Cross-scale interactions, non-linearities, and forecasting catastrophic events. *Proceedings of the National Academy of Sciences of the United States of America* 101: 15130–15135.

Peters, D. P. C., B. T. Bestelmeyer, and M. G. Turner. 2007. Cross-scale interactions and changing pattern-process relationships: Consequences for system dynamics. *Ecosystems* 10: 790–796.

Peters, R. H. 1983. *The Ecological Implications of Body Size*. Cambridge University Press, New York, New York.

Peterson, A. T. 2003. Predicting the geography of species' invasions via ecological niche modeling. *Quarterly Review of Biology* 78: 419–560.

Peterson, A. T., M. A. Ortega-Huerta, J. Bartley, V. Sánchez-Cordero, J. Soberón, R. H. Buddemeier, and D. R. B. Stockwell. 2002. Future projections for Mexican faunas under global climate change scenarios. *Nature* 416: 626–629.

Peterson, A. T., M. Papeş, and M. Eaten. 2007. Transferability and model evaluation in ecological niche modeling: A comparison of GARP and Maxent. *Ecography* 30: 550–560.

Peterson, G., C. R. Allen, and C. S. Holling. 1998. Ecological resilience, biodiversity and scale. *Ecosystems* 1: 6–18.

Petit, R. J., A. El Mousadik, and O. Pons. 1998. Identifying populations for conservation on the basis of genetic markers. *Conservation Biology* 12: 844–855.

Petley, D. 2012. Global patterns of loss of life from landslides. *Geology* 40: 927–930.

Petren, K. 2013. The evolution of landscape genetics. *Evolution* 67: 3383–3385.

Pfannenstiel, R. S. and M. G. Ruder. 2015. Colonization of bison (*Bison bison*) wallows in tallgrass prairie by *Culicoides* spp (Diptera: Ceratopogonidae). *Journal of Vector Ecology* 40: 187–190.

Phillips, O. L., and 65 others. 2009. Drought sensitivity of the Amazon rainforest. *Science* 323: 1344–1347.

Phillips, S. J. 2008. Transferability, sample selection bias and background data in presence-only modelling: A response to Peterson et al. (2007). *Ecography* 31: 272–278.

Phillips, S. J., and Dudík, M. 2008. Modeling of species distributions with Maxent: New extensions and a comprehensive evaluation. *Ecography* 31: 161–175.

Phillips, S. J., R. P. Anderson, R. E. Schapire. 2006. Maximum entropy modeling of species geographic distributions. *Ecological Modelling* 190: 231–259.

Pianka, E. R. 1966. Latitudinal gradients in species diversity: A review of concepts. *American Naturalist* 100: 33–46.

Pickett, S. T. A. and J. N. Thompson. 1978. Patch dynamics and the design of nature reserves. *Biological Conservation* 13: 27–37.

Pickett, S. T. A. and P. S. White. 1985. *The Ecology of Natural Disturbance and Patch Dynamics*. Academic Press, San Diego, California.

Pickett, S. T. A. and M. L. Cadenasso. 1995. Landscape ecology: Spatial heterogeneity in ecological systems. *Science* 269: 331–334.

Pickett, S. T. A., and 13 others. 2011. Urban ecological systems: Scientific foundations and a decade of progress. *Journal of Environmental Management* 92: 331–362.

Pidgeon, A. M., V. C. Radeloff, and N. E. Mathews. 2003. Landscape-scale patterns of black-throated sparrow (*Amphispiza bilineata*) abundance and nest success. *Ecological Applications* 13: 530–542.

Pijanowski, B. C., L. J. Villanueva-Rivera, S. L. Dumyahn, A. Farina, B. L. Krause, B. M. Napoletano, S. H. Gage, and N. Pieretti. 2011. Soundscape ecology: The science of sound in the landscape. *BioScience* 61: 203–216.

Pimentel, D. 2006. Soil erosion: A food and environmental threat. *Environment, Development and Sustainability* 8: 119–137.

Pimm, S. L. and R. A. Askins. 1995. Forest losses predict bird extinctions in eastern North America. *Proceedings of the National Academy of Sciences of the United States of America* 92: 9343–9347.

Pittman, S. 2013. Seascape ecology. *Marine Scientist* 44: 20–23.

Pittman, S., R. Kneib, C. Simenstad, and I. Nagelkerken, Editors. 2011. Seascape ecology: Application of landscape ecology to the marine environment. *Marine Ecology Progress Series* 427: 187–302.

Plantegenest, M., C. Le May, and F. Fabre. 2007. Landscape epidemiology of plant diseases. *Journal of the Royal Society Interface* 4: 963–972.

Plotnick, R. E., R. H. Gardner, and R. V. O'Neill. 1993. Lacunarity indices as measures of landscape texture. *Landscape Ecology* 8: 201–211.

Poff, N. L. 1992. Why disturbances can be predictable: A perspective on the definition of disturbance in streams. *Journal of the North American Benthological Society* 11: 86–92.

Poff, N. L., J. D. Allan, M. B. Bain, J. R. Karr, K. L. Prestegaard, B. D. Richter, R. E. Sparks, and J. C. Stromberg. 1997. The natural flow regime. *BioScience* 47: 769–784.

Polis, G. A. and S. D. Hurd. 1995. Extraordinarily high spider densities on islands: Flow of energy from the marine to terrestrial food webs and the absence of predation. *Proceedings of the National Academy of Sciences of the United States of America* 92: 4382–4386.

Polis, G. A. and S. D. Hurd. 1996. Linking marine and terrestrial food webs: Allochthonous input from the ocean supports high secondary productivity on small islands and coastal land communities. *American Naturalist* 147: 396–423.

Polis, G. A., W. B. Anderson, and R. D. Holt. 1997. Toward an integration of landscape and food web ecology: The dynamics of spatially subsidized food webs. *Annual Review of Ecology and Systematics* 28: 289–316.

Polis, G. A., M. E. Power, and G. R. Huxel, Editors. 2004. *Food Webs at the Landscape Level.* University of Chicago Press, Chicago, Illinois.

Pope, L. C., R. K. Butlin, G. J. Wilson, R. Woodroffe, D. Erven, C. M. Conyers, T. Franklin, R. J. Delahay, C. L. Cheeseman, and T. Burke. 2007. Genetic evidence that culling increases badger movement: Implications for the spread of bovine tuberculosis. *Molecular Ecology* 16: 4919–4929.

Post, W. M. and K. C. Kwon. 2000. Soil carbon sequestration and land-use change: Processes and potential. *Global Change Biology* 6: 317–328.

Post, W. M., R. C. Izaurralde, T. O. West, M. A. Liebig, and A. W. King. 2012. Management opportunities for enhancing terrestrial carbon dioxide sinks. *Frontiers in Ecology and the Environment* 10: 554–561.

Potts, S. G., J. C. Biesmeijer, C. Kremen, P. Neumann, O. Schweiger, and W. E. Kunin. 2010. Global pollinator declines: Trends, impacts and drivers. *Trends in Ecology and Evolution* 25: 345–353.

Powell, A. G. and C. E. Mitchell. 2004. Pathogen spillover in disease epidemics. *American Naturalist* 164: S79–S89.

Powell, G. V. N. 1974. Experimental analysis of the social value of flocking by starlings (*Sturnus vulgaris*) in relation to predation and foraging. *Animal Behavior* 22: 501–505.

Powell, K. I., J. M. Chase, and T. M. Knight. 2013. Invasive plants have scale-dependent effects on diversity by altering species-area relationships. *Science* 339: 316–318.

Power, M. E., D. Tilman, J. A. Estes, B. A. Menge, W. J. Bond, L. S. Mills, G. Daily, J. C. Castilla, J. Lubchenco, and R. T. Paine. 1996. Challenges in the quest for keystones. *BioScience* 46: 609–620.

Preisler, H. K., J. A. Hicke, A. A. Ager, and J. L. Hayes. 2012. Climate and weather influences on spatial temporal patterns of mountain pine beetle populations in Washington and Oregon. *Ecology* 93: 2421–2434.

Pringle, C. M., R. J. Naiman, G. Bretschko, J. R. Karr, M. W. Oswood, J. R. Webster, R. L. Welcomme, and M. J. Winterbourn. 1988. Patch dynamics in lotic systems: The stream as a mosaic. *Journal of the North American Benthological Society* 7: 503–524.

Pritchard, J., M. Stephens, and P. Donnelly. 2000. Inference of population structure using multilocus genotype data. *Genetics* 155: 945–959.

Procheş, Ş., J. R. U. Wilson, R. Veldtman, J. M. Kalwij, D. M. Richardson, and S. L. Chown. 2005. Landscape corridors: Possible dangers? *Science* 310: 780–781.

Prugh, L. 2009. An evaluation of patch connectivity measures. *Ecological Applications* 19: 1300–1310.

Prugh, L. R., K. E. Hodges, A. R. E. Sinclair, and J. S. Brashares. 2008. Effect of habitat area and isolation on fragmented animal populations. *Proceedings of the National Academy of Sciences of the United States of America* 105: 20770–20775.

Pulliam, H. R. 1988. Sources, sinks, and population regulation. *American Naturalist* 132: 652–661.

Pulliam, H. R. and B. J. Danielson. 1991. Sources, sinks, and habitat selection: A landscape perspective on population dynamics. *American Naturalist* 137: S50–S66.

Puurtinen, M., M. Elo, M. Jalasvuori, A. Kahilainen, T. Ketola, J. S. Kotiaho, M. Mönkkönen, and O. T. Pentikäinen. 2016. Temperature-dependent mutational robustness can explain faster molecular evolution at warm temperatures, affecting speciation rate and global patterns of diversity. *Ecography* 39: 1025–1033.

Pyke, G. H. 1984. Optimal foraging theory: A critical review. *Annual Review of Ecology and Systematics* 15: 523–575.

Pywell, R. F., M. S. Heard, B. A. Woodcock, S. Hinsley, L. Ridding, M. Nowakowsi, and J. M. Bullock. 2015. Wildlife-friendly farming increases crop yield: Evidence for ecological intensification. *Proceedings of the Royal Society B* 282: 20141740.

Raddatz, R. L. 2007. Evidence for the influence of agriculture on weather and climate through the transformation and management of vegetation: Illustrated by examples from the Canadian Prairies. *Agricultural and Forest Meteorology* 142: 186–202.

Radeloff, V. C., D. J. Mladenoff, H. S. Hong, and M. S. Boyce. 1999. Forest landscape change in northwestern Wisconsin pine barrens from pre-European settlement to present. *Canadian Journal of Forest Research* 29: 1649–1659.

Raffa, K. F., B. H. Aukema, B. J. Bentz, A. L. Carroll, J. A. Hicke, M. G. Turner, and W. H. Romme. 2008. Cross-scale drivers of natural disturbances prone to anthropogenic amplification: The dynamics of bark beetle eruptions. *BioScience* 58: 501–517.

Raffel, T. R., L. B. Martin, and J. R. Rohr. 2008. Parasites as predators: Unifying natural enemy ecology. *Trends in Ecology and Evolution* 23: 610–618.

Rafferty, N. E. 2013. Pollination Ecology. In: *Oxford Bibliographies in Ecology* (D. J. Gibson, Editor). Oxford University Press, New York, New York.

Rahbek, C. 1995. The elevational gradient of species richness: A uniform pattern? *Ecography* 18: 200–205.

Rahbek, C. 2005. The role of spatial scale and the perception of large-scale species-richness patterns. *Ecology Letters* 8: 224–239.

Ramalho, C. E. and R. J. Hobbs. 2012. Time for a change: Dynamic urban ecology. *Trends in Ecology and Evolution* 27: 179–188.

Ramos-Fernández, G., J. L. Morales, O. Miramontes, G. Cosho, H. Larralde, and B. Ayala-Orozco. 2004. Lévy walk patterns in the foraging movements of spider monkeys (*Ateles geoffroyi*). *Behavioral Ecology and Sociobiology* 55: 223–230.

Rathcke, B. J. and E. S. Jules. 1993. Habitat fragmentation and plant-pollinator interactions. *Current Science* 65: 273–277.

Raufaste, N. and F. Rousset. 2001. Are partial Mantel tests adequate? *Evolution* 55: 1703–1705.

Rayfield, B., M.-J. Fortin, and A. Fall. 2010. The sensitivity of least-cost habitat graphs to relative cost surface values. *Landscape Ecology* 25: 519–532.

Raymond M. and F. Rousset. 1995. GENEPOP (version 1.2): Population genetics software for exact tests and ecumenicism. *Journal of Heredity* 86: 248–249.

Raynor, E. J., A. Joern, J. B. Nippert, and J. M. Briggs. 2016. Foraging decisions underlying restricted space use: Effects of fire and forage maturation on large herbivore nutrient uptake. *Ecology and Evolution* 6: 5843–5853.

Raynor, E. J., A. Joern, A. Skibbe, M. Sowers, J. M. Briggs, A. N. Laws, and D. Goodin. 2017. Temporal variability in large grazer space use in an experimental landscape. *Ecosphere* 8(1): e01674.

Real, L. A. and J. E. Childs. 2006. Spatial-temporal dynamics of rabies in ecological communities. Pp. 170–187 in *Disease Ecology* (S. K. Collinge and C. Ray, Editors.). Oxford University Press, New York.

Real, L. A. and R. Biek. 2007. Spatial dynamics and genetics of infectious diseases on heterogeneous landscapes. *Journal of the Royal Society Interface* 4: 935–948.

Reed, R. A., J. Johnson-Barnard, and W. A. Baker. 1996. Contribution of roads to forest fragmentation in the Rocky Mountains. *Conservation Biology* 10: 1098–1106.

Reed, W. J. and B. D. Hughes. 2002. From gene families to genera to incomes and internet file sizes: Why power laws are so common in nature. *Physical Review E* 66: 067103.

Reed, W. J. and K. S. McKelvey. 2002. Power-law behavior and parametric models for the size-distribution of forest fires. *Ecological Modelling* 150: 239–254.

Reeve, J. D, and J. T. Cronin. 2010. Edge behavior in a minute parasitic wasp. *Journal of Animal Ecology* 79: 483–490.

Reeve, J. D., J. T. Cronin, and K. J. Haynes. 2008. Diffusion models for animals in complex landscapes: Incorporating heterogeneity among substrates, individuals and edge behaviours. *Journal of Animal Ecology* 77: 898–904.

Reisen, W. K. 2010. Landscape epidemiology of vector-borne diseases. *Annual Review of Entomology* 55:461–483.

Remmel, T. K. and M.-J. Fortin. 2013. Categorical, class-focused map patterns: Characterization and comparison. *Landscape Ecology* 28: 1587–1599.

Ren, G., S. S. Young, L. Wang, W. Wang, Y. Long, R. Wu, J. Li, J. Zhu, and D. W. Yu. 2015. Effectiveness of China's National Forest Protection Program and nature reserves. *Conservation Biology* 29: 1368–1377.

Resasco, J., N. M. Haddad, J. L. Orrock, D. Shoemaker, L. A. Brudvig, E. I. Damschen, J. J. Tewksbury, and D. J. Levey. 2014. Landscape corridors can increase invasion by an exotic species and reduce diversity of native species. *Ecology* 95: 2033–2039.

Resetarits, W. J., Jr., C. A. Binckley, and D. R. Chalcraft. 2005. Habitat selection, species interactions, and processes of community assembly in complex landscapes: A metacommunity perspective. Pp. 374–398 in *Metacommunities: Spatial Dynamics and Ecological Communities* (M. Holyoak, M. A. Leibold, and R. D. Holt, Editors). Chicago University Press, Chicago, Illinois.

Resh, V. H., A. V. Brown, A. P. Covich, M. E. Gurtz, H. W. Li, G. W. Minshall, S. R. Reice, A. L. Sheldon, J. B. Wallace, and R. C. Wissmar. 1988. The role of disturbance in stream ecology. *Journal of the American Benthological Society* 7: 433–455.

Restrepo, C. and P. Vitousek. 2001. Landslides, alien species, and the diversity of a Hawaiian montane mesic ecosystem. *Biotropica* 33: 409–420.

Restrepo, C., P. Vitousek, and P. Neville. 2003. Landslides significantly alter land cover and the distribution of biomass: An example from the Ninole ridges of Hawai'i. *Plant Ecology* 166: 131–143.

Restrepo, C., and 12 others. 2009. Landsliding and its multiscale influence on mountainscapes. *BioScience* 59: 685–698.

Reutebuch, S. E., H-E. Anderson, and R. J. McGaughey. 2005. Light detection and ranging (LIDAR): An emerging tool for multiple resource inventory. *Journal of Forestry* 103: 286–292.

Reynolds, A. M. 2010. Bridging the gulf between correlated random walks and Lévy walks: Autocorrelation as a source of Lévy walk movement patterns. *Journal of the Royal Society Interface* 7: 1753–1758.

Reynolds, A. M., A. D. Smith, R. Menzel, U. Greggers, D. R. Reynolds, and J. R. Riley. 2007. Displaced honeybees perform optimal scale-free search flights. *Ecology* 88: 1955–1961.

Rhemtulla, J. M., D. J. Mladenoff, and M. K. Clayton. 2009. Legacies of historical land use on regional forest composition and structure in Wisconsin, USA (mid-1800s–1930s–2000s). *Ecological Applications* 19: 1061–1078.

Ribas, C. C., A. Aleixo, A. C. R. Nogueira, C. Y. Miyaki, and J. Cracraft. 2012. A palaeobiogeographic model for biotic diversification within Amazonia over the past three million years. *Proceedings of the Royal Society B* 279: 681–689.

Ribeiro, D. B., P. I. Prado, K. S. Brown Jr., and A. V. L. Freitas. 2008. Additive partitioning of butterfly diversity in a fragmented landscape: Importance of scale and implications for conservation. *Diversity and Distributions* 14: 961–968.

Ribeiro, Jr., P. J. and P. J. Diggle. 2018. Package 'geoR': Analysis of Geostatistical Data, version 1.7–5.2.1. Documentation available online: https://cran.r-project.org/web/packages/geoR/geoR.pdf

Richardson, D. M., N. Allsopp, C. M. D'Antonio, S. J. Milton, and M. Rejmánek. 2000. Plant invasions: The role of mutualisms. *Biological Reviews* 75: 65–93.

Richardson, J. L. and M. C. Urban. 2013. Strong selection barriers explain microgeographic adaptation in wild salamander populations. *Evolution* 67: 1729–1740.

Richardson, J. L., M. C. Urban, D. I. Bolnick, and D. K. Skelly. 2014. Microgeographic adaptation and the spatial scale of evolution. *Trends in Ecology and Evolution* 29: 165–176.

Richardson, J. S. and T. Sato. 2015. Resource subsidy flows across freshwater-terrestrial boundaries and influence on processes linking adjacent ecosystems. *Ecohydrology* 8: 406–415.

Ricketts, T. H. 2001. The matrix matters: Effective isolation in fragmented landscapes. *American Naturalist* 158: 87–99.

Ricketts, T. H., and 12 others. 2008. Landscape effects on crop pollination services: Are there general patterns? *Ecology Letters* 11: 499–515.

Ricotta, C., G. Avena, and M. Marchetti. 1999. The flaming sandpile: Self-organized criticality and wildfires. *Ecological Modelling* 119: 73–77.

Ries, L. and T. D. Sisk. 2004. A predictive model of edge effects. *Ecology* 85: 2917–2926.

Ries, L., R. J. Fletcher, Jr., J. Battin, and T. D. Sisk. 2004. Ecological responses to habitat edges: Mechanisms, models, and variability explained. *Annual Review of Ecology, Evolution, and Systematics* 35: 491–522.

Rignot, E., J. L. Bamber, M. R. van den Broeke, C. Davis, Y. Li, W. Jan van de Berg, and E. van Meijgaard. 2008. Recent Antarctic ice mass loss from radar interferometry and regional climate modeling. *Nature Geoscience* 1: 106–110.

Riitters, K. H. and J. D. Wickham. 2003. How far to the nearest road? *Frontiers in Ecology and the Environment* 1: 125–129.

Riitters, K. H., R. V. O'Neill, C. T. Hunsaker, J. D. Wickham, D. H. Yankee, S. P. Timmins, K. B. Jones, and B. L. Jackson. 1995. A factor analysis of landscape pattern and structure metrics. *Landscape Ecology* 10: 23–39.

Riitters, K. H., R. V. O'Neill, J. D. Wickham, and K. B. Jones. 1996. A note on contagion indices for landscape analysis. *Landscape Ecology* 11: 197–202.

Ripley, B. D. 1976. The second-order analysis of stationary processes. *Journal of Applied Probability* 13: 255–266.

Ripple, W. J. and R. L. Beschta. 2012. Trophic cascades in Yellowstone: The first 15 years after wolf introduction. *Biological Conservation* 145: 205–213.

Risser, P. G. 1995. The Allerton Park workshop revisited—a commentary. *Landscape Ecology* 10: 129–132.

Risser, P. G, J. R. Karr, and R. T. T. Forman. 1984. Landscape ecology: Directions and approaches. *Illinois Natural History Survey*, Number 2, Champaign, Illinois.

Rittenhouse, T. A. G. and R. D. Semlitsch. 2006. Grasslands as movement barriers for a forest-associated salamander: Migration behavior of adult and juvenile salamanders at a distinct habitat edge. *Biological Conservation* 131: 14–22.

Rizzo, D. M. and M. Garbelotto. 2003. Sudden oak death: Endangering California and Oregon forest ecosystems. *Frontiers in Ecology and the Environment* 1: 197–204.

Rizzo, D. M., M. Garbelotto, and E. M. Hansen. 2005. *Phytophthora ramorum*: Integrative research and management of an emerging pathogen in California and Oregon forests. *Annual Review of Phytopathology* 43: 309–335.

Robertson, B. A. and R. L. Hutto. 2006. A framework for understanding ecological traps and an evaluation of existing evidence. *Ecology* 87: 1075–1085.

Robinson, S. K., F. R. Thompson III, T. M. Donovan, D. R. Whitehead, and J. Faaborg. 1995. Regional forest fragmentation and the nesting success of migratory birds. *Science* 267: 1987–1990.

Rodríguez-Iturbe, I. and A. Rinaldo. 1997. *Fractal River Basins: Chance and Self-Organization*. Cambridge University Press, Cambridge, UK.

Roering, J. 2012. Landslides limit mountain relief. *Nature Geoscience* 5: 446–447.

Roger, E. and D. Ramp. 2009. Incorporating habitat use in models of fauna fatalities on roads. *Diversity and Distributions* 15: 221–231.

Roger, E., S. W. Laffan, and D. Ramp. 2007. Habitat selection by the common wombat (*Vombatus ursinus*) in disturbed environments: Implications for the conservation of a 'common' species. *Biological Conservation* 137: 437–449.

Roger, E., S. W. Laffan, and D. Ramp. 2011. Road impacts a tipping point for wildlife populations in threatened landscapes. *Population Ecology* 53: 215–227.

Roger, E., G. Bino, and D. Ramp. 2012. Linking habitat suitability and road mortalities across geographic ranges. *Landscape Ecology* 27: 1167–1181.

Rohde, K. 1992. Latitudinal gradients in species diversity: The search for the primary cause. *Oikos* 65: 514–527.

Roland, J. 1993. Large-scale forest fragmentation increases the duration of tent caterpillar outbreak. *Oecologia* 93: 25–30.

Roland, J. and P. D. Taylor. 1997. Insect parasitoid species respond to forest structure at different spatial scales. *Nature* 386: 710–713.

Romme, W. H. 1982. Fire and landscape diversity in subalpine forests of Yellowstone National Park. *Ecological Monographs* 52: 199–221.

Romme, W. H., E. H. Everham, L. E. Frelich, M. A. Moritz, and R. E. Sparks. 1998. Are large, infrequent disturbances qualitatively different from small, frequent disturbances? *Ecosystems* 1: 524–534.

Romme, W. H., M. S. Boyce, R. Gresswell, E. H. Merill, G. W. Minshall, C. Whitlock, and M. G. Turner. 2011. Twenty years after the 1988 Yellowstone fires: Lessons about disturbance and ecosystems. *Ecosystems* 14: 1196–1215.

Rosell, F., O. Bozsér, P. Collen, and H. Parker. 2005. Ecological impact of beavers *Castor fiber* and *Castor canadensis* and their ability to modify ecosystems. *Mammal Review* 35: 248–276.

Rosenberg, D. K., B. R. Noon, and E. C. Meslow. 1997. Biological corridors: Form, function, and efficacy. *BioScience* 47: 677–687.

Rosenzweig, M. L. 1995. *Species Diversity in Space and Time*. Cambridge University Press, Cambridge, UK.

Rossi, R. E., D. J. Mulla, A. G. Journel, and E. H. Franz. 1992. Geostatistical tools for modeling and interpreting ecological spatial dependence. *Ecological Monographs* 62: 277–314.

Roth, D., J. Roland, and T. Roslin. 2010. Parasitoids on the loose: Experimental lack of support of the parasitoid movement hypothesis. *Oikos* 115: 277–285.

Roubik, D. W. 2002. Tropical agriculture: The value of bees to the coffee harvest. *Nature* 417: 708.

Rousset, F. 1997. Genetic differentiation and estimation of gene flow from F-statistics under isolation by distance. *Genetics* 145: 1219–1228.

Row, J. R. and G. Blouin-Demers. 2006. Kernels are not accurate estimators of home-range size for herpetofauna. *Copeia* 4: 797–802.

Roy, D. P., J. Ju, K. Kline, P. L. Scaramuzza, V. Kovalskyy, M. Hansen, T. R. Loveland, E. F. Vermote, and C. Zhang. 2010. Web-enabled Landsat data (WELD): Landsat ETM+ composited mosaics of the conterminous United States. *Remote Sensing of Environment* 114: 35–49.

Royama, T. 1970. Factors governing hunting behavior and selection of food by great tits (*Parus major* L.). *Journal of Animal Ecology* 39: 619–668.

Royama, T. 1971. A comparative study of models for predation and parasitism. *Researches on Population Ecology* (Supplement) 1: 1–91.

Rubenstein, D. R. and K. A. Hobson. 2004. From birds to butterflies: Animal movement patterns and stable isotopes. *Trends in Ecology and Evolution* 19: 256–263.

Ruckelshaus, M., C. Hartway, and P. Kareiva. 1997. Assessing the data requirements of spatially explicit dispersal models. *Conservation Biology* 11: 1298–1306.

Ruddiman, W. F. 2003. The anthropogenic greenhouse era began thousands of years ago. *Climatic Change* 61: 261–293.

Ruddiman, W. F. 2013. The Anthropocene. *Annual Review of Earth and Planetary Sciences* 41: 45–68.

Ruddiman, W. F., E. C. Ellis, J. O. Kaplan, and D. Q. Fuller. 2015. Defining the epoch we live in. *Science* 348: 38–39.

Ruffell, J., C. Banks-Leite, and R. K. Didham. 2016. Accounting for the causal basis of collinearity when measuring the effects of habitat loss versus habitat fragmentation. *Oikos* 125: 117–125.

Ruggiero, A. and B. A. Hawkins. 2008. Why do mountains support so many species of birds? *Ecography* 31: 306–315.

Runge, J. P., M. C. Runge, and J. D. Nichols. 2006. The role of local populations within a landscape context: Defining and classifying sources and sinks. *American Naturalist* 167: 925–938.

Rutledge, L. Y., C. J. Garroway, K. M. Loveless, and B. R. Patterson. 2010. Genetic differentiation of eastern wolves in Algonquin Park despite bridging gene flow between coyotes and grey wolves. *Heredity* 105: 520–531.

Ryall, K. L. and L. Fahrig. 2006. Response of predators to loss and fragmentation of prey habitat: A review of theory. *Ecology* 87: 1086–1093.

Ryan, J. G., J. A. Ludwig, and C. A. McAlpine. 2007. Complex adaptive landscapes (CAL): A conceptual framework of multi-functional, non-linear ecohydrological feedback systems. *Ecological Complexity* 4: 113–127.

Rybicki, J. and I. Hanksi. 2013. Species-area relationships and extinctions caused by habitat loss and fragmentation. *Ecology Letters* 16: 27–38.

Rychert, C. A., G. Laske, N. Harmon, and P. M. Shearer. 2013. Seismic imaging of melt in a displaced Hawaiian plume. *Nature Geoscience* 6: 657–660.

Rytwinski, T. and L. Fahrig. 2013. Why are some animal populations unaffected or positively affected by roads? *Oecologia* 173: 1143–1156.

Sabo, J. L. and D. M. Post. 2008. Quantifying periodic, stochastic, and catastrophic environmental variation. *Ecological Monographs* 78: 19–40.

Sabo, J. L., K. Bestgen, W. Graf, T. Sinha, and E. E. Wohl. 2012. Dams in the Cadillac Desert: Downstream effects in a geomorphic context. *Annals of the New York Academy of Sciences* 1249: 227–246.

Safner, T., M. Miller, B. McRae, M.-J. Fortin, and S. Manel. 2011. Comparison of Bayesian clustering and edge detection methods for inferring boundaries in landscape genetics. *International Journal of Molecular Sciences* 12: 865–889.

Salau, K., M. L. Schoon, J. A. Baggio, and M. A. Janssen. 2012. Varying effects of connectivity and dispersal on interacting species dynamics. *Ecological Modelling* 242: 81–91.

Sale, P. F. 1977. Maintenance of high diversity in coral reef fish communities. *American Naturalist* 111: 337–359.

Sampson, F. and F. Knopf. 1994. Prairie conservation in North America. *BioScience* 44: 418–421.

Sandercock, B. K. 2006. Estimation of demographic parameters from live-encounter data: A summary review. *Journal of Wildlife Management* 70: 1504-1520.

Sanders, N. J. 2002. Elevational gradients in ant species richness: Area, geometry, and Rapoport's rule. *Ecography* 25: 25–32.

Sanderson, E. W., M. Jaiteh, M. A. Levy, K. H. Redford, A. V. Wannebo, and G. Woolmer. 2002. The human footprint and the last of the wild. *BioScience* 52: 891–904.

Sandin, L. and A. G. Solimini. 2009. Freshwater ecosystem structure-function relationships: From theory to application. *Freshwater Biology* 54: 2017–2024.

Saunders, D. A., R. J. Hobbs, and C. R. Margules. 1991. Biological consequences of ecosystem fragmentation: A review. *Conservation Biology* 5: 18–32.

Saura, S. and J. Martínez-Millán. 2001. Sensitivity of landscape pattern metrics to map spatial extent. *Photogrammetric Engineering & Remote Sensing* 67: 1027–2036.

Scanlon, B. R., C. C. Faunt, L. Longuevergne, R. C. Reedy, W. M. Alley, V. L. McGuire, and P. B. McMahon. 2012. Groundwater depletion and sustainability of irrigation in the U.S. High Plains and Central Valley. *Proceedings of the National Academy of Sciences of the United States of America* 109: 9320–9325.

Scanlon, T. M., K. K. Caylor, S. A. Levin, and I. Rodriguez-Iturbe. 2007. Positive feedbacks promote power-law clustering of Kalahari vegetation. *Nature* 449: 209–212.

Scarborough, V. L., and 10 others. 2012. Water and sustainable land use at the ancient tropical city of Tikal, Guatemala. *Proceedings of the National Academy of Sciences of the United States of America* 109: 12408–124123.

Schade, J. D., R. Sponseller, S. L. Collins, and A. Stiles. 2003. The influence of *Prosopis* canopies on understorey vegetation: Effects of landscape position. *Journal of Vegetation Science* 14: 743–750.

Scheffer, M., S. Carpenter, J. A. Foley, C. Folke, and B. Walker. 2001. Catastrophic shifts in ecosystems. *Nature* 413: 591–596.

Scheffer, M., J. Bascompte, W. A. Brock, V. Brovkin, S. R. Carpenter, V. Dakos, H. Held, E. H. van Nes, M. Rietkerk, and G. Sugihara. 2009. Early-warning signals for critical transitions. *Nature* 461: 53–59.

Scheffers, S. R., J. Haviser, T. Browne, and A. Scheffers. 2009. Tsunamis, hurricanes, the demise of coral reefs and shifts in prehistoric human populations in the Caribbean. *Quaternary International* 195: 69–87.

Scheiner, S. M. 2003. Six types of species–area curves. *Global Ecology and Biogeography* 12: 441–447.

Scheiner, S. M., S. B. Cox, M. Willig, G. G. Mittelbach, C. Osenberg, and M. Kaspari. 2000. Species richness, species–area curves and Simpson's paradox. *Evolutionary Ecology Research* 2: 791–802.

Scheiner, S. M., A. Chiarucci, G. A. Fox, M. R. Helmus, D. J. McGlinn, and M. R. Willig. 2011. The underpinnings of the relationship of species richness with space and time. *Ecological Monographs* 81: 195–213.

Schick, R. S. and S. T. Lindley. 2007. Directed connectivity among fish populations in a riverine network. *Journal of Applied Ecology* 44: 1116–1126.

Schick, R. S., S. R. Loarie, F. Colchero, B. D. Best, A. Boustany, D. A. Conde, P. N. Halpin, L. N. Joppa, C. M. McClellan, and J. S. Clark. 2008. Understanding movement data and movement processes: Current and emerging directions. *Ecology Letters* 11: 1338–1350.

Schimel, D. S. 1995. Terrestrial ecosystems and the carbon cycle. *Global Change Biology* 1: 77–91.

Schimel, D. S., T. G. F. Kittel, A. K. Knapp, T. R. Seastedt, W. J. Parton, and V. B. Brown. 1991. Physiological interactions along resource gradients in a tallgrass prairie. *Ecology* 72: 672–684.

Schippers, P., J. Verboom, J. P. Knappen, and R. C. van Apeldoorn. 1996. Dispersal and habitat connectivity in complex heterogeneous landscapes: An analysis with a GIS-based random walk model. *Ecography* 19: 97–106.

Schlenker, W. and M. J. Roberts. 2009. Non-linear temperature effects indicate severe damages to U.S. crop yields under climate change. *Proceedings of the National Academy of Sciences of the United States of America* 106: 15594–15598.

Schluter, D., K. B. Marchinko, R. D. H. Barrett, and S. M. Rogers. 2010. Natural selection and the genetics of adaptation in threespine stickleback. *Philosophical Transactions of the Royal Society B* 365: 2479–2486.

Schmitz, O. J. 2008. Herbivory from individuals to ecosystems. *Annual Review of Ecology, Evolution, and Systematics* 39: 133–152.

Schneider, D. C. 2001. The rise of the concept of scale in ecology. *BioScience* 51: 545–553.

Schoennagel, T., M. G. Turner, and W. H. Romme. 2003. The influence of fire interval and serotiny on postfire lodgepole pine density in Yellowstone National Park. *Ecology* 84: 1967–1978.

Scholl, A. E. and A. H. Taylor. 2010. Fire regimes, forest change, and self-organization in an old-growth mixed-conifer forest, Yosemite National Park, USA. *Ecological Applications* 20: 362–380.

Schooley, R. L. and J. A. Wiens. 2003. Finding habitat patches and directional connectivity. *Oikos* 102: 559–570.

Schreiber, K.-F. 1990. The history of landscape ecology in Europe. Pp. 21–33 in *Changing Landscapes: An Ecological Perspective* (I. S. Zonneveld and R. T. T. Forman, Editors). Springer-Verlag, New York.

Schrott, G. R., K. A. With, and A. W. King. 2005. On the importance of landscape history for assessing extinction risk. *Ecological Applications* 15: 493–506.

Schulte, L. A. and D. J. Mladenoff. 2001. The original US Public Land Survey records: Their use and limitations in reconstructing presettlement vegetation. *Journal of Forestry* 99: 5–10.

Schulte, L. A., D. J. Mladenoff, and E. V. Nordheim. 2002. Quantitative classification of a historic northern Wisconsin (U.S.A.) landscape: Mapping forests at regional scales. *Canadian Journal of Forest Research* 32: 1616–1638.

Schultz, C. B. 1998. Dispersal behavior and its implications for reserve design in a rare Oregon butterfly. *Conservation Biology* 12: 284–292.

Schumaker, N. H. 1996. Using landscape indices to predict habitat connectivity. *Ecology* 77: 1210–1225.

Schupp, E. W. and M. Fuentes. 1995. Spatial patterns of seed dispersal and the unification of plant population ecology. *Écoscience* 2: 267–275.

Sciaini, M., M. Fritsch, C. Simpkins, C. Scherer and S. Hanß. 2019. NLMR: Simulating Neutral Landscape Models. R package version 0.4.1. Documentation available online: https://cran.r-project.org/web/packages/NLMR/NLMR.pdf

Seaman, D. E. and R. A. Powell. 1996. An evaluation of the accuracy of kernel density estimators for home range analysis. *Ecology* 77: 2075–2085.

Seaman, D. E., J. J. Millspaugh, B. J. Kernohan, G. C. Brundige, K. J. Raedeke, and R. A. Gitzen. 1999. Effects of sample size on kernel home range estimates. *Journal of Wildlife Management* 63: 739–747.

Searle, K. R., T. Vandervelde, N. T. Hobbs, L. A. Shipley, and B. A. Wunder. 2006. Spatial context influences patch residence time in foraging hierarchies. *Oecologia* 148: 710–719.

Sears, A. L. W. and P. Chesson. 2007. New methods for quantifying the spatial storage effect: An illustration with desert annuals. *Ecology* 88: 2240–2247.

Seastedt, T. R., J. M. Briggs, and D. J. Gibson. 1991. Controls of nitrogen limitation in tallgrass prairie. *Oecologia* 87: 72–79.

Semmens, D. J., J. E. Diffendorfer, L. López-Hoffman, and C. D. Shapiro. 2011. Accounting for the ecosystem services of migratory species: Quantifying migration support and spatial subsidies. *Ecological Economics* 70: 2236–2242.

Senft, R. L., M. B. Coughenour, D. W. Bailey, L. R. Rittenhouse, O. E. Sala, and D. M. Swift. 1987. Large herbivore foraging and ecological hierarchies. *BioScience* 37: 789–799.

Seto, K. C., B. Güneralp, and L. R. Hutyra. 2012. Global forecasts of urban expansion to 2030 and direct impacts on biodiversity and carbon pools. *Proceedings of the National Academy of Sciences of the United States of America* 109: 16083–16088.

Seyfarth, R. M., D. L. Cheney, and P. Marler. 1980. Vervet monkey alarm calls: Semantic communication in a free-ranging primate. *Animal Behaviour* 28: 1070–1094.

Shackelford, N., and 16 others. 2017. Isolation predicts compositional change after discrete disturbances in a global meta-study. *Ecography* 40: 1256–1266.

Shen, W., Y. Lin, G. D. Jenerette, and J. Wu. 2011. Blowing litter across a landscape: Effects on ecosystem nutrient flux and implications for landscape management. *Landscape Ecology* 26: 629–644.

Shepard, D. B., A. R. Kuhns, M. J. Dreslik, and C. A. Phillips. 2008. Roads as barriers to animal movement in fragmented landscapes. *Animal Conservation* 11: 288–296.

Shigesada, N. and K. Kawasaki. 1997. *Biological Invasions*. Oxford University Press, Oxford.

Shmida, A. 1984. Whittaker's plant diversity sampling method. *Israel Journal of Botany* 33: 41–46.

Shmida, A. and M. V. Wilson. 1985. Biological determinants of species diversity. *Journal of Biogeography* 12: 1–20.

Shurin, J. B., E. T. Borer, E. W. Seabloom, K. Anderson, C. A. Blanchette, B. Broitman, S. D. Cooper, and B. S. Halpern. 2002. A cross-ecosystem comparison of the strength of trophic cascades. *Ecology Letters* 5: 785–791.

Sih, A. 2005. Predator–prey space use as an emergent outcome of a behavioral response race. Pp. 240–255 in *The Ecology of Predator–Prey Interactions* (P. Barbosa and I. Castellanos, Editors). Oxford University Press, Oxford, UK.

Silcock, J. L., T. P. Piddocke, and R. J. Fensham. 2013. Illuminating the dawn of pastoralism: Evaluating the record of European explorers to inform landscape change. *Biological Conservation* 159: 321–331.

Silverman, B. W. 1986. *Density Estimation for Statistics and Data Analysis*. Chapman and Hall, London, UK.

Simard, M., W. H. Romme, J. M. Griffin, and M. G. Turner. 2011. Do mountain pine beetle outbreaks change the probability of active crown fire in lodgepole pine forests? *Ecological Monographs* 81: 3–24.

Simberloff, D. 1976. Experimental zoogeography of islands: Effects of island size. *Ecology* 57: 629–648.

Simberloff, D. 2011. How common are invasion-induced ecosystem impacts? *Biological Invasions* 13: 1255–1268.

Simberloff, D. and B. Von Holle. 1999. Positive interactions of nonindigenous species: Invasional meltdown? *Biological Invasions* 1: 21–32.

Simberloff, D., J. A. Farr, J. Cox, and D. W. Mehlman. 1992. Corridors: Conservation bargains or poor investments? *Conservation Biology* 6: 493–504.

Simkin, T., L. Siebert, and R. Blong. 2001. Volcano fatalities—lessons from the historical record. *Science* 291: 255.

Simonti, C. N., and 20 others. 2016. The phenotypic legacy of admixture between modern humans and Neandertals. *Science* 351: 737–741.

Sims, D. W., and 17 others. 2008. Scaling laws of marine predator search behavior. *Nature* 451: 1098–1102.

Singer, A., and 14 others. 2016. Community dynamics under environmental change: How can next generation mechanistic models improve projections of species distributions? *Ecological Modelling* 326: 63–74.

Skellam, J. G. 1951. Random dispersal in theoretical populations. *Biometrika* 38: 196–218.

Sklar, L. S. and W. E. Dietrich. 2001. Sediment and rock strength controls on river incision into bedrock. *Geology* 29: 1087–1090.

Slatkin, M. 1987. Gene flow and the geographic structure of natural populations. *Science* 236: 787–792.

Slatkin, M. 1991. Inbreeding coefficients and coalescence times. *Genetical Research* 58: 167–175.

Slatkin, M. 1995. A measure of population subdivision based on microsatellite allele frequencies. *Genetics* 139: 457–462.

Small, C. and T. Naumann. 2001. The global distribution of human population and recent volcanism. *Environmental Hazards* 3: 93–109.

Smith, A. C., N. Koper, C. M. Francis, and L. Fahrig. 2009. Confronting collinearity: Comparing methods for disentangling the effects of habitat loss and fragmentation. *Landscape Ecology* 24: 1271–1285.

Smith, B. T., and 15 others. 2014. The drivers of tropical speciation. *Nature* 515: 406–409.

Smith, D. L., B. Lucey, L. A. Waller, J. E. Childs, and L. A Real. 2002. Predicting the spatial dynamics of rabies epidemics on heterogeneous landscapes. *Proceedings of the National Academy of Sciences of the United States of America* 99: 3668–3672.

Smith, G. J. D., and 12 others. 2009. Origins and evolutionary genomics of the 2009 swine-origin H1N1 influenza A epidemic. *Nature* 459: 1122–1125.

Smith, P. 2008. Land use change and soil organic carbon dynamics. *Nutrient Cycling in Agroecosystems* 81: 169–178.

Smithson, A. and M. R. MacNair. 1997. Density-dependent and frequency-dependent selection by bumblebees *Bombus terrestris* (L) (Hymenoptera: Apidae). *Biological Journal of the Linnean Society* 60: 401–417.

Smouse, P. E., J. C. Long, and R. R. Sokal. 1986. Multiple regression and correlation extensions of the Mantel test of matrix correspondence. *Systematic Zoology* 35: 627–632.

Sogge, M. K., S. J. Sferra, and E. H. Paxton. 2008. *Tamarix* as habitat for birds: Implications for riparian restoration in the Southwestern United States. *Restoration Ecology* 16: 146–154.

Solé, R. 2007. Scaling laws in the drier. *Nature* 449: 151–153.

Solé, R. and S. C. Manrubia. 1995. Are rainforests self-organized to a critical state? *Journal of Theoretical Biology* 173: 31–40.

Soranno, P. A., K. S. Cheruvelil, K. E. Webster, M. T. Bremigan, T. Wagner, and C. A. Stow. 2010. Using landscape limnology to classify freshwater ecosystems for multi-ecosystem management and conservation. *BioScience* 60: 440–454.

Sørensen, T. A. 1948. A method of establishing groups of equal amplitude in plant sociology based on similarity of species content, and its application to analyses of the vegetation on Danish commons. *Kongelige Danske Videnskabernes Selskabs Biologiske Skrifter* 5: 1–34.

Sork, V. L. 1984. Examination of seed dispersal and survival in red oak, *Quercus rubra* (Fagaceae), using metal-tagged acorns. *Ecology* 65: 1020–1022.

Sork, V. L. and P. E. Smouse. 2006. Genetic analysis of landscape connectivity in tree populations. *Landscape Ecology* 21: 821–836.

Sork, V. L. and L. P. Waits. 2010. Contributions of landscape genetics—approaches, insights, and future directions. *Molecular Ecology* 19: 3489–3495.

Sork, V. L., F. W. Davis, R. Westfall, A. Flint, M. Ikegami, H. Wang, and D. Grivet. 2010. Gene movement and genetic association with regional climate gradients in California valley oak (*Quercus lobata* Née) in the face of climate change. *Molecular Ecology* 19: 3806–3823.

Sork, V. L., S. N. Aitken, R. J. Dyer, A. J. Eckert, P. Legendre, and D. B. Neale. 2013. Putting the landscape into the genomics of trees: Approaches for understanding local adaptation and population responses to changing climate. *Tree Genetics and Genomes* 9: 901–911.

Soulé, M. E. 1985. What is conservation biology? *BioScience* 35: 727–734.

Spear, S. F., N. Balkenhol, M.-J. Fortin, B. H. McRae, and K. Scribner. 2010. Use of resistance surfaces for landscape genetic studies: Considerations for parameterization and analysis. *Molecular Ecology* 19: 3576–3591.

Spellerberg, I. F. and P. J. Fedor. 2003. A tribute to Claude Shannon (1916–2001) and a plea for more rigorous use of species richness, species diversity and the 'Shannon-Wiener' Index. *Global Ecology and Biogeography* 12: 177–179.

Stamps, J. A. 1988. Conspecific attraction and aggregation in territorial species. *American Naturalist* 131: 329–347.

Stamps, J. A. 2001. Habitat selection by dispersers: Integrating proximate and ultimate approaches. Pp. 230–242 in *Dispersal* (J. Clobert, E. Danchin, A. A. Dhondt, and J. D. Nichols, Editors). Oxford University Press, New York.

Stamps, J. A., M. Buechner, and V. V. Krishnan. 1987. The effects of edge permeability and habitat geometry on emigration from patches of habitat. *American Naturalist* 129: 533–552.

Stanners, D. A. and P. Bourdeau. 1995. Europe's Environment: The Dobříš Assessment. European Environment Agency, United Nations Economic Commission for Europe. Available online: www.eea.europa.eu/publications/92-826-5409-5 (last accessed 3/3/19)

Stebbins, G. L. 1974. *Flowering Plants: Evolution above the Species Level*. The Belknap Press of Harvard University Press, Cambridge, Massachusetts.

Steele, M. A., G. Turner, P. D. Smallwood, J. O. Wolff, and J. Radillo. 2001. Cache management by small mammals: Experimental evidence for the significance of acorn-embryo excision. *Journal of Mammalogy* 82: 35–42.

Steffan-Dewenter, I. 2003. Importance of habitat area and landscape context for species richness of bees and wasps in fragmented orchard meadows. *Conservation Biology* 17: 1036–1044.

Steffan-Dewenter, I. and T. Tscharntke. 1999. Effect of habitat isolation on pollinator communities and seed set. *Oecologia* 121: 432–440.

Steffan-Dewenter, I. and C. Westphal. 2008. The interplay of pollination diversity, pollination services and landscape change. *Journal of Applied Ecology* 45: 737–741.

Steffan-Dewenter, I., U. Münzenberg, C. Bürger, C. Thies, and T. Tscharntke. 2002. Scale dependent effects of landscape context on three pollinator guilds. *Ecology* 83: 1421–1432.

Steffan-Dewenter, I., S. G. Potts, and L. Packer. 2005. Pollinator diversity and crop pollination services are at risk. *Trends in Ecology and Evolution* 20: 651–652.

Steffan-Dewenter, I., A.-M. Klein, V. Gaebele, T. Alfert, and T. Tscharntke. 2006. Bee diversity and plant-pollinator interactions in fragmented landscapes. Pp. 287–407 in *Plant-Pollinator Interactions: From Specialization to Generalization* (N. M. Waser and J. Ollerton, Editors). University of Chicago Press, Chicago, Illinois.

Steiniger, S. and G. J. Hay. 2009. Free and open source geographic information tools for landscape ecology. *Ecological Informatics* 4: 183–195.

Steiniger, S. and A. J. S. Hunter. 2013. The 2012 free and open source GIS software map: A guide to facilitate research, development, and adoption. *Computers, Environment and Urban Systems* 39: 136–150.

Stephens, D. W. and J. R. Krebs. 1986. *Foraging Theory*. Princeton University Press, New Jersey.

Stephens, D. W., J. S. Brown, and R. C. Ydenberg. 2007. *Foraging: Behavior and Ecology*. University of Chicago Press, Illinois.

Stevens, G. C. 1989. The latitudinal gradient in geographical range: How so many species coexist in the tropics. *American Naturalist* 133: 240–256.

Stewart, K. J., P. Grogan, D. S. Coxson, and S. D. Siciliano. 2014. Topography as a key factor driving atmospheric nitrogen exchanges in arctic terrestrial ecosystems. *Soil Biology and Biochemistry* 70: 96–112.

Stockwell, D. R. B. and D. P. Peters. 1999. The GARP modeling system: Problems and solutions to automated spatial prediction. *International Journal of Geographic Information Systems* 13: 143–158.

Stockwell, D. R. B. and A. T. Peterson. 2002. Effects of sample size on accuracy of species distribution models. *Ecological Modelling* 148: 1–13.

Stoffel, M. and C. Hugel. 2012. Effects of climate change on mass movements in mountain environments. *Progress in Physical Geography* 36: 421–439.

Stohlgren, T. J. and M. Rejmánek. 2014. No universal scale-dependent impacts of invasive species on native plant species richness. *Biology Letters* 10: 20130939.

Stohlgren, T. J., M.B. Falkner, and L.D. Schell. 1995. A Modified-Whittaker nested vegetation sampling method. *Vegetatio* 117: 113–121.

Stohlgren, T. J., G. W. Chong, L. D. Schell, K. A. Rimar, Y. Otsuki, M. Lee, M. A. Kalkhan, and C. A. Villa. 2002. Assessing vulnerability to invasion by nonnative plant species at multiple spatial scales. *Environmental Management* 29: 566–577.

Stokstad, E. 2007. The case of the empty hives. *Science* 316: 971–972.

Stommel, H. 1963. Varieties of oceanographic experience. *Science* 139: 572–576.

Storfer, A., M. A. Murphy, J. S. Evans, C. S. Goldberg, S. Robinson, S. F. Spear, R. Dezzani, E. Delmelle, L. Vierling, and L. P. Waits. 2007. Putting the 'landscape' in landscape genetics. *Heredity* 98: 128–142.

Storfer, A., M. A. Murphy, S. F. Spear, R. Holderegger, and L. P. Waits. 2010. Landscape genetics: Where are we now? *Molecular Ecology* 19: 3496–3514.

Stürck, J. and P. H. Verburg. 2017. Multifunctionality at what scale? A landscape multifunctionality assessment for the European Union under conditions of land use change. *Landscape Ecology* 32: 481–500.

Stutchbury, B. J. M., S. A. Tarof, T. Done, E. Gow, P. M. Kramer, J. Tautin, J. W. Fox, and V. Afanasyev. 2009. Tracking long-distance songbird migration by using geolocators. *Science* 323:896.

Sumpter, D. J. T. 2010. *Collective Animal Behavior*. Princeton University Press, New Jersey.

Suttle, K. B., M. A. Thomsen, and M. E. Power. 2007. Species interactions reverse grassland responses to changing climate. *Science* 315: 640–642.

Svenning, J.-C. and F. Skov. 2004. Limited filling of the potential range in European tree species. *Ecology Letters* 7: 565–573.

Swetnam, T. L, P. D. Brooks, H. R. Barnard, A. A. Harpold, and E. L. Gallo. 2017. Topographically driven differences in energy and water constrain climatic control on forest carbon sequestration. *Ecosphere* 8(4): e01797.

Swetnam, T. W. and J. L. Betancourt. 1998. Mesoscale disturbance and ecological response to decadal climatic variability in the American Southwest. *Journal of Climate* 11: 3128–3147.

Swetnam, T. W., C. D. Allen, and J. L. Betancourt. 1999. Applied historical ecology: Using the past to manage for the future. *Ecological Applications* 9: 1189–1206.

Swift, T. L. and S. J. Hannon. 2010. Critical thresholds associated with habitat loss: A review of the concepts, evidence, and applications. *Biological Reviews* 85: 35–53.

Swihart, R. K. and N. A. Slade. 1985a. Influence of sampling interval on estimates of home range size. *Journal of Wildlife Management* 49: 1019–1025.

Swihart, R. K. and N. A. Slade. 1985b. Testing for independence of observations in animal movements. *Ecology* 66: 1176–1184.

Swindles, G. T., E. Watson, T. E. Turner, J. M. Galloway, T. Hadlari, J. Wheeler, and K. L. Bacon. 2015. Spheroidal carbonaceous particles are a defining stratigraphic marker for the Anthropocene. *Scientific Reports* 5: 10264.

Syvitski, J. P. M., C. V. Vörösmarty, A. J. Kettner, and P. Green. 2005. Impact of humans on the flux of terrestrial sediment to the global coastal ocean. *Science* 308: 376–380.

Tallaksen, L. M. and H. A. J. Van Lanen, Editors. 2004. Hydrological drought: Processes and estimation methods for streamflow and groundwater. *Developments in Water Science*, Volume 48. Elsevier, Amsterdam, the Netherlands.

Tanner, E. V. J. and P. J. Bellingham. 2006. Less diverse forest is more resistant to hurricane disturbance: Evidence from montane forests in Jamaica. *Journal of Ecology* 94: 1003–1010.

Tanner, J. E. 2003. Patch shape and orientation influences on seagrass epifauna are mediated by dispersal abilities. *Oikos* 100: 517–524.

Tansley, A. G. 1935. The use and abuse of vegetational concepts and terms. *Ecology* 16: 284–307.

Taylor, A. H. and C. N. Skinner. 2003. Spatial patterns and controls on historical fire regimes and forest structure in the Klamath Mountains. *Ecological Applications* 13: 704–719.

Taylor, P. D., L. Fahrig, K. Henein, and G. Merriam. 1993. Connectivity is a vital element of landscape structure. *Oikos* 68: 571–573.

Taylor, P. D., L. Fahrig, and K. A. With. 2006. Landscape connectivity: A return to the basics. Pp. 29–43 in *Connectivity Conservation* (K. R. Crooks and M. Sanjayan, Editors). Cambridge University Press, New York.

Templeton, A. R., R. J. Robertson, J. Brisson, and J. Strasburg. 2001. Disrupting evolutionary processes: The effect of habitat fragmentation on collared lizards in the Missouri Ozarks. *Proceedings of the National Academy of Sciences of the United States of America*: 98: 5426–5432.

Termorshuizen, J. W. and P. Opdam. 2009. Landscape services as a bridge between landscape ecology and sustainable development. *Landscape Ecology* 24: 1037–1052.

Tesch, F.-W. 2003. *The Eel*, 5th edition. Blackwell Science, Blackwell Publishing Company, Oxford, UK.

Tewksbury, J. J., D. J. Levey, N. M. Haddad, S. Sargent, J. L. Orrock, A. Weldon, B. J. Danielson, J. Brinkerhoff, E. I. Damschen, and P. Townsend. 2001. Corridors affect plants, animals, and their interactions in fragmented landscapes. *Proceedings of the National Academy of Sciences of the United States of America* 99: 12923–12926.

Tews, J., U. Brose, V. Grimm, K. Tielbörger, M. C. Wichmann, M. Schwager, and F. Jeltsch. 2004. Animal species diversity driven by habitat heterogeneity/diversity: The importance of keystone structures. *Journal of Biogeography* 31: 79–92.

The State of Canada's Forests. Annual Report 2011. 2011. Natural Resources Canada, Canadian Forest Service, Ottawa, Ontario. Available online: http://cfs.nrcan.gc.ca/pubwarehouse/pdfs/32683.pdf (last accessed 3/3/19)

Theobald, D. M. 2005. Landscape patterns of exurban growth in the USA from 1980 to 2020. *Ecology and Society* 10: 32. Available online: http://www.ecologyandsociety.org/vol10/iss1/art32/

Thies, C. and I. Steffan-Dewenter, and T. Tscharntke. 2003. Effects of landscape context on herbivory and parasitism at different spatial scales. *Oikos* 101: 18–25.

Thomas, C. D. and W. E. Kunin. 1999. The spatial structure of populations. *Journal of Animal Ecology* 68: 647–657.

Thomas, C. D., A. M. A. Franco, and J. K. Hill. 2006. Range retractions and extinction in the face of climate warming. *Trends in Ecology and Evolution* 21: 415–416.

Thompson, J. R., D. N. Carpenter, C. V. Cogbill, and D. R. Foster. 2013. Four centuries of change in northeastern United States. *PLoS ONE* 8: e72540.

Thomson, D. M. 2007. Do source-sink dynamics promote the spread of an invasive grass into a novel habitat? *Ecology* 88: 3126–3134.

Thomson, R. C., I. J. Wang, and J. R. Johnson. 2010. Genome-enabled development of DNA markers for ecology, evolution and conservation. *Molecular Ecology* 19: 2184–2195.

Tilman, D. 1994. Competition and biodiversity in spatially structured habitats. *Ecology* 75: 2–16.

Tilman, D. 1996. Biodiversity: Population versus ecosystem stability. *Ecology* 77: 350–363.

Tilman, D. and P. Kareiva. 1997. *Spatial Ecology: The Role of Space in Population Dynamics and Interspecific Interactions*. Princeton University Press, New Jersey.

Tilman, D., R. M. May, C. L. Lehman, and M. A. Nowak. 1994. Habitat destruction and the extinction debt. *Nature* 371: 65–66.

Tilman, D., C. Balzer, J. Hill, and B. L. Befort. 2011. Global food demand and the sustainable intensification of agriculture. *Proceedings of the National Academy of Sciences of the United States of America* 108: 20260–20264.

Tischendorf, L. and L. Fahrig. 2000a. On the usage and measurement of landscape connectivity. *Oikos* 90: 7–19.

Tischendorf, L. and L. Fahrig. 2000b. How should we measure landscape connectivity? *Landscape Ecology* 15: 633–641.

Tischendorf, L. and L. Fahrig. 2001. On the use of connectivity measures in spatial ecology: A reply. *Oikos* 95: 152–155.

Tischendorf, L., D. Bender, and L. Fahrig. 2003. Evaluation of patch-isolation metrics in mosaic landscapes for specialist vs. generalist dispersers. *Landscape Ecology* 18: 41–50.

Tjørve, E. 2009. Shapes and functions of species-area curves (II): A review of new models and parameterizations. *Journal of Biogeography* 36: 1435–1445.

Tobler, W. F. 1970. A computer movie simulating urban growth in the Detroit region. *Economic Geography* 46: 234–240.

Tockner, K. and J. A. Stanford. 2002. Riverine flood plains: Present state and future trends. *Environmental Conservation* 29: 308–330.

Tockner, K., F. Malard, and J. V. Ward. 2000. An extension of the flood pulse concept. *Hydrological Processes* 14: 2861–2883.

Tomlin, C. D. 1990. *Geographic Information Systems and Cartographic Modeling.* Prentice-Hall, Englewood-Cliffs, New Jersey.

Towne, E. G., D. C. Hartnett, and R. C. Cochran. 2005. Vegetation trends in tallgrass prairie from bison to cattle grazing. *Ecological Applications* 15: 1550–1559.

Townsend, A. K., A. B. Clark, K. J. McGowan, A. D. Miller, and E. L. Buckles. 2010. Condition, innate immunity and disease mortality of inbred crows. *Proceedings of the Royal Society* B 277: 2875–2883.

Townsend, C. R. 1989. The patch dynamics concept of stream community ecology. *Journal of the North American Benthological Society* 8: 36–50.

Townsend, P. A. and D. J. Levey. 2005. An experimental test of whether habitat corridors affect pollen transfer. *Ecology* 86: 466–475.

Traore, O., and 10 others. 2005. Processes of diversification and dispersion of rice yellow mottle virus inferred from large-scale and high-resolution phylogeographical studies. *Molecular Ecology* 14: 2097–2110.

Treml, E. A., P. N. Halpin, D. L. Urban, and L. F. Pratson. 2008. Modeling population connectivity by ocean currents, a graph-theoretic approach for marine conservation. *Landscape Ecology* 23: 19–36.

Triantis, K. A., K. Vardinoyannis, E. P. Tsolaki, I. Botsaris, K. Lika, and M. Mylonas. 2006. Re-approaching the small island effect. *Journal of Biogeography* 33: 914–923.

Triantis, K. A., F. Guilhaumon, and R. J. Whittaker. 2012. The island species-area relationship: Biology and statistics. *Journal of Biogeography* 39: 215–231.

Troll, C. 1939. Luftbildplan and okologische bodenforschung (Aerial photography and ecological studies of the earth). *Zeitschrift der Gesellschaft für Erdkunde,* Berlin: 241–298.

Troll, C. 1971. Landscape ecology (geoecology) and biogeocenology—a terminological study. *Geoforum* 2: 43–46.

Trombulak, S. C. and C. A. Frissell. 2000. Review of ecological effects of roads on terrestrial and aquatic communities. *Conservation Biology* 14: 18–30.

Trzcinski, M. K., L. Fahrig, and G. Merriam. 1999. Independent effects of forest cover and fragmentation on the distribution of forest breeding birds. *Ecological Applications* 9: 586–593.

Tscharntke, T., A. M. Klein, A. Kruess, I. Steffan-Dewenter, and C. Thies. 2005a. Landscape perspectives on agricultural intensification and biodiversity: Ecosystem service management. *Ecology Letters*: 8: 857–874.

Tscharntke, T., T. A. Rand, and F. J. J. A. Bianchi. 2005b. The landscape context of trophic interactions: Insect spillover across the crop-non-crop interface. *Annales Zoologici Fennici* 42: 421–432.

Tscharntke, T., R. Bommarco, Y. Clough, T. O. Crist, D. Kleijn, T. A. Rand, J. M. Tylianakis, S. van Nouhuys, and S. Vidal. 2007. Conservation biological control and enemy diversity on a landscape scale. *Biological Control* 43: 294–309.

Tsvetkov, N., O. Samson-Robert, K. Sood, H. S. Patel, D. A. Malena, P. H. Gajiwala, P. Maciukiewicz, V. Fournier, and A. Zayed. 2017.

Chronic exposure to neonicotinoids reduces honey bee health near corn crops. *Science* 356: 1395–1397.

Tucker, C. J. 1979. Red and photographic infrared linear combinations for monitoring vegetation. *Remote Sensing of Environment* 8: 127–150.

Tuomisto, H. 2010a. A diversity of beta diversities: Straightening up a concept gone awry. Part 1. Defining beta diversity as a function of alpha and gamma diversity. *Ecography* 33: 2–22.

Tuomisto, H. 2010b. A diversity of beta diversities: Straightening up a concept gone awry. Part 2. Quantifying beta diversity and related phenomena. *Ecography* 33: 23–45.

Turchin, P. 1991. Translating foraging movements in heterogeneous environments into the spatial distribution of foragers. *Ecology* 72: 1253–1266.

Turchin, P. 1998. *Quantitative Analysis of Movement: Measuring and Modeling Population Redistribution in Animals and Plants.* Sinauer Associates, Sunderland, Massachusetts.

Turnbull, L. A., M. Rees, and M. J. Crawley. 1999. Seed mass and the competition/colonization trade-off: A sowing experiment. *Journal of Ecology* 87: 899–912.

Turner, C. L., J. M. Blair, R. J. Schartz, and J. C. Neel. 1997. Soil N and plant responses to fire, topography, and supplemental N in tallgrass prairie. *Ecology* 78: 1832–1843.

Turner, M. G. 1989. Landscape ecology: The effect of pattern on process. *Annual Review of Ecology and Systematics* 20: 171–197.

Turner, M. G. and R. H. Gardner, Editors. 1991. *Quantitative Methods in Landscape Ecology: The Analysis and Interpretation of Landscape Heterogeneity.* Springer-Verlag, New York.

Turner, M. G. and W. H. Romme. 1994. Landscape dynamics in crown fire ecosystems. *Landscape Ecology* 9: 59–77.

Turner, M. G. and V. H. Dale. 1998. Comparing large, infrequent disturbances: What have we learned? *Ecosystems* 1: 493–496.

Turner, M. G. and F. S. Chapin, III. 2005. Causes and consequences of spatial heterogeneity in ecosystem function. Pp. 9–30 in *Ecosystem Function in Heterogeneous Landscapes* (G. M. Lovett, C. G. Jones, M. G. Turner and K. C. Weathers, Editors). Springer, New York.

Turner, M. G., R. H. Gardner, V. H. Dale, and R. V. O'Neill. 1989a. Predicting the spread of disturbance in heterogeneous landscapes. *Oikos* 55: 121–129.

Turner, M. G., R. V. O'Neill, R. H. Gardner, and B. T. Milne. 1989b. Effects of changing spatial scale on the analysis of landscape pattern. *Landscape Ecology* 3: 153–162.

Turner, M. G., W. H. Romme, R. H. Gardner, R. V. O'Neill, and T. K. Kratz. 1993. A revised concept of landscape equilibrium: Disturbance and stability on scaled landscapes. *Landscape Ecology* 9: 213–227.

Turner, M. G., W. W. Hargrove, R. H. Gardner, and W. H. Romme. 1994. Effects of fire on landscape heterogeneity in Yellowstone National Park, Wyoming. *Journal of Vegetation Science* 5: 731–742.

Turner, M. G., G. J. Arthaud, R. T. Engstrom, S. J. Heijl, J. Liu, S. Loeb, and K. McKelvey. 1995. Usefulness of spatially explicit population models in land management. *Ecological Applications* 5: 12–16.

Turner, M. G., V. H. Dale, and E. H. Everham, III. 1997. Fires, hurricanes, and volcanoes: Comparing large disturbances. *BioScience* 47: 758–768.

Turner, M. G., R. H. Gardner, and R. V. O'Neill. 2001. *Landscape Ecology in Theory and Practice: Pattern and Proces*s. Springer-Verlag, New York.

Turner, M. G., W. H. Romme, and D. B. Tinker. 2003. Surprises and lessons from the 1988 Yellowstone fires. *Frontiers in Ecology and Evolution* 1: 351–358.

Turner, M. G., D. C. Donato and W. H. Romme. 2013. Consequences of spatial heterogeneity for ecosystem services in changing forest landscapes: Priorities for future research. *Landscape Ecology* 28: 1081–1097.

Turner, W. R. and E. Tjørve. 2005. Scale-dependence in species–area relationships. *Ecography* 28: 721–730.

Underwood, N., P. Hambäck, and B. D. Inouye. 2005. Large-scale questions and small-scale data: Empirical and theoretical methods for scaling up in ecology. *Oecologia* 145: 177–178.

United Nations, Department of Economic and Social Affairs, Population Division. 2014. World Urbanization Prospects: The 2014 Revision, Highlights (ST/ESA/SER.A/352). Available online: http://esa.un.org/unpd/wup/Publications/Files/WUP2014-Highlights.pdf (last accessed 1/30/19).

Urban, D. L. 2005. Modeling ecological processes across scales. Ecology 86: 1996–2006.

Urban, D. and T. Keitt. 2001. Landscape connectivity: A graph-theoretic perspective. *Ecology* 82: 1205–1218.

Urban, D. L., R. V. O'Neill, and H. H. Shugart, Jr. 1987. Landscape ecology. *BioScience* 37: 119–127.

Urban, D. L., E. S. Minor, E. A. Treml, and R. S. Schick. 2009. Graph models of habitat mosaics. *Ecology Letters* 12: 260–273.

Urban, M. C. 2006. Road facilitation of trematode infections in snails of northern Alaska. *Conservation Biology* 20: 1143–1149.

USBR (United States Bureau of Reclamation). 2013a. Lower Colorado Region: Lake Mead High and Low Elevations (1935–2012). https://www.usbr.gov/lc/region/g4000/lakemead_line.pdf (last accessed 1/30/19).

USBR (United States Bureau of Reclamation). 2013b. Upper Colorado Region: Lake Powell Elevations (1963–2013). https://www.usbr.gov/rsvrWater/HistoricalApp.html (last accessed 1/30/19).

USDA (United States Department of Agriculture). 2014. 2012 *Census of Agriculture*. United States Department of Agriculture, National Agriculture Statistics Service, Washington, D.C. Available online: https://www.nass.usda.gov/Publications/AgCensus/2012/index.php (last accessed 3/3/19).

USDA-NRCS (United States Department of Agriculture Natural Resources Conservation Service). 2000. Global anthropic landscapes map. Retrieved from http://www.nrcs.usda.gov/wps/portal/nrcs/detail/national/nedc/training/soil/?cid=nrcs142p2_054001 (last accessed 3/3/19)

Valone, T. J. 1989. Group foraging, public information, and patch estimation. *Oikos* 56: 357-363.

Valone, T. J. 2007. From eavesdropping on performance to copying the behavior of others: A review of public information use. *Behavioral Ecology and Sociobiology* 62: 1-14.

van den Bosch, F., R. Hengeveld, and A. J. Mertz. 1992. Analyzing the velocity of animal range expansion. *Journal of Biogeography* 19: 135–150.

van der Ree, R., J. A. G. Jaeger, E. A. van der Grift, and A. P. Clevenger. 2011. Effects of roads and traffic on wildlife populations and landscape function: Road ecology is moving towards larger scales. *Ecology and Society* 16: 48. Available online: https://www.ecologyandsociety.org/vol16/iss1/art48/

Van Horne, B. 1983. Density as a misleading indicator of habitat quality. Journal of Wildlife Management 47: 893–901.

van Langevelde, F. 2000. Scale of habitat connectivity and colonization in fragmented nuthatch populations. *Ecography* 23: 614–622.

Van Loon, A. F. 2015. Hydrological drought explained. *Wiley Interdisciplinary Reviews (WIREs) Water* 2: 359–392.

van Mantgem, P. J., and 10 others. 2009. Widespread increase of tree mortality rates in western USA. *Science* 323: 521–524.

Van Noordwijk, C. G. E. (Toos), and 13 others. 2015. Species-area relationships are modulated by trophic rank, habitat affinity, and dispersal ability. *Ecology* 96: 518–531.

Van Valen, L. 1973. A new evolutionary law. *Evolutionary Theory* 1: 1–30.

Van Winkle, W. 1975. Comparison of several probabilistic home-range models. *Journal of Wildlife Management* 39: 118–123.

van Zanten, B. T., and 10 others. 2014. European agricultural landscapes, common agricultural policy and ecosystem services: A review. *Agronomy for Sustainable Development* 34: 309–325.

Vance, M. D., L. Fahrig, and C. H. Flather. 2003. Effect of reproductive rate on minimum habitat requirements of forest-breeding birds. *Ecology* 84: 2643–2653.

Vance, T. C. and R. E. Doel. 2010. Graphical methods and cold war scientific practice: The Stommel diagram's intriguing journey from the physical to the biological environmental sciences. *Historical Studies in the Natural Sciences* 40: 1–47.

Vannote, R. L., G. W. Minshall, K. W. Cummins, J. R. Sedell, and C. E. Cushing. 1980. The river continuum concept. *Canadian Journal of Fisheries and Aquatic Sciences* 37: 130–137.

Veech, J. A. and T. O. Crist. 2010a. Diversity partitioning without statistical independence of alpha and beta. *Ecology* 91: 1964–1969.

Veech, J. A. and T. O. Crist. 2010b. Toward a uniform view of diversity partitioning. *Ecology* 91: 1988–1992.

Veech, J. A., K. S. Summerville, T. O. Crist, and J. C. Gering. 2002. The additive partitioning of species diversity: Recent revival of an old idea. *Oikos* 99: 3–9.

Vejre, H., and 11 others. 2007. Multifunctional agriculture and multifunctional landscapes—land use as an interface. Pp. 93–104 in *Multifunctional Land Use: Meeting Future Demands for Landscape Goods and Services* (Ü. Mander, H. Wiggering, and K. Helming, Editors). Springer, Berlin, Germany.

Vellend, M. 2001. Do commonly used indices of β-diversity measure species turnover? *Journal of Vegetation Science* 12: 545–552.

Vellend, M. 2003. Habitat loss inhibits recovery of plant diversity as forests regrow. *Ecology* 84: 1158–1164.

Vellend, M., J. A. Myers, S. Gardescu, and P. L. Marks. 2003. Dispersal of *Trillium* seeds by deer: Implications for long-distance migration of forest herbs. *Ecology* 84: 1067–1072.

Vellend, M., K. Verheyen, H. Jackquemyn, A. Kolb, H. Van Calster, G. Peterken, and M. Hermy. 2006. Extinction debt of forest plants persists for more than a century following habitat fragmentation. *Ecology* 87: 542–548.

Vellend, M., C. D. Brown, H. M. Kharouba, J. L. McCune, and I. H. Myers-Smith. 2013. Historical ecology: Using unconventional data sources to test for effects of global environmental change. *American Journal of Botany* 100: 1294–1305.

Vermuelen, C., P. Lejeune, J. Lisein, P. Sawadogo, and P. Bouché. 2013. Unmanned aerial survey of elephants. *PLoS ONE* 8(2): e54700.

Vial, F. and C. A. Donnelly. 2012. Localized reactive badger culling increases risk of bovine tuberculosis in nearby cattle herds. *Biology Letters* 8: 50–53.

Vilà, M. and I. Ibáñez. 2011. Plant invasions in the landscape. *Landscape Ecology* 26: 461–472.

Viles, H. A., Editor. 1988. *Biogeomorphology*. Basil Blackwell, New York.

Villard, M.-A., M. K. Trzcinski, and G. Merriam. 1999. Fragmentation effects on forest birds: Relative influence of woodland cover and configuration on landscape occupancy. *Conservation Biology* 13: 774–783.

Villoria, N. B., D. Byerlee, and J. Stevenson. 2014. The effects of agricultural technological progress on deforestation: What do we really know? *Applied Economic Perspectives and Policy* 36: 211–237.

Viña, A., W. J. McConnell, H. Yang, Z. Xu, and J. Liu. 2016. Effects of conservation policy on China's forest recovery. *Science Advances* 2: e1500965.

Viswanathan, G. M., V. Afanasyev, S. V. Buldyrev, E. J. Murphy, P. A. Prince, and H. E. Stanley. 1996. Lévy flight search patterns of wandering albatrosses. *Nature* 381: 413–415.

Vitousek, P. M. and R. L. Stanford, Jr. 1986. Nutrient cycling in moist tropical forest. *Annual Review of Ecology and Systematics* 17: 137–167.

Vitousek, P. M., H. A. Mooney, J. Lubchenco, and J. M. Melillo. 1997. Human domination of Earth's ecosystems. *Science* 277: 494–499.

Vittor, A. Y., R. H. Gilman, J. Tielsch, G. Glass, T. Shields, W. Sánchez-Lozano, V. Pinedo-Cancino, and J. A. Patz. 2006. The effect of deforestation on the human-biting rate of *Anopheles darlingi*, the primary vector of falciparum malaria in the Peruvian Amazon. *American Journal of Tropical Medicine and Hygiene* 74: 3–11.

Vittor, A. Y., and 10 others. 2009. Linking deforestation to malaria in the Amazon: Characteristics of the breeding habitat of the principal malaria vector, *Anopheles darlingi*. *American Journal of Tropical Medicine and Hygiene* 81: 5–12.

Volterra, V. 1926. Variazioni e fluttuazioni del numero d'individui in specie animali conviventi (Variations and fluctuations of the number of individuals in animal species living together). *Memorie della Reale Accademia dei Lincei* 6 (2): 31–113.

von Schiller, D., and 26 others. 2017. River ecosystem processes: A synthesis of approaches, criteria of use and sensitivity to environmental stressors. *Science of the Total Environment* 596–597: 465–480.

Vos, C. C., J. Verboom, P. F. M. Opdam, and C. J. F. Ter Braak. 2001. Toward ecologically scaled landscape indices. *American Naturalist* 157: 24–41.

Vucetich, J. A. and T. A. Waite. 2000. Is one migrant per generation sufficient for the genetic management of fluctuating populations? *Animal Conservation* 3: 261–266.

Wagner, D., J. B. Jones, and D. M. Gordon. 2004. Development of harvester ant colonies alters soil chemistry. *Soil Biology and Biochemistry* 36: 797–804.

Wagner, H. H. and M.-J. Fortin. 2005. Spatial analysis of landscapes: Concepts and statistics. *Ecology* 86: 1975–1987.

Wagner, H. H. and M.-J. Fortin. 2013. A conceptual framework for the spatial analysis of landscape genetic data. *Conservation Genetics* 14: 253–261.

Walker, B. H. 1992. Biodiversity and ecological redundancy. *Conservation Biology* 6: 18–23.

Walker, B. H. 1995. Conserving biological diversity through ecosystem resilience. *Conservation Biology* 9: 747–752.

Wamsley, T. V., M. A. Cialone, J. M. Smith, J. H. Atkinson, and J. D. Rosati. 2010. The potential of wetlands in reducing storm surge. *Ocean Engineering* 37: 59–68.

Wang, I. J. and G. S. Bradburd. 2014. Isolation by environment. *Molecular Ecology* 23: 5649–5662.

Wang, I. J., W. K. Savage, and H. B. Shaffer. 2009. Landscape genetics and least-cost path analysis reveal unexpected dispersal routes in the California tiger salamander (*Ambystoma californiense*). *Molecular Ecology* 18: 1365–1374.

Ward, J. V. 1997. An expansive perspective of riverine landscapes: Pattern and process across scales. *River Ecosystems* 6: 52–60.

Ward, J. V., K. Tockner, D. B. Arscott, and C. Claret. 2002. Riverine landscape diversity. *Freshwater Biology* 47: 517–539.

Ware, H. E., C. J. W. McClure, J. D. Carlisle, and J. R. Barber. 2015. A phantom road experiment reveals traffic noise is an invisible source of habitat degradation. *Proceedings of the National Academy of Sciences of the United States of America* 112: 12105–12109.

Waring, R. H., J. B. Way, E. R. Hunt Jr., L. Morrissey, K. J. Ranson, J. F. Weishampel, R. Oren, and S. E. Franklin. 1995. Imaging radar for ecosystem studies. *BioScience* 45: 715–723.

Watkinson, A. R. and W. J. Sutherland. 1995. Sources, sinks and pseudo-sinks. *Journal of Animal Ecology* 64: 126–130.

Watt, A. S. 1947. Pattern and process in the plant community. *Journal of Ecology* 35: 1–22.

Webster, M. S., P. P. Marra, S. M. Haig, S. Bensch, and R. T. Holmes. 2002. Links between worlds: Unraveling migratory connectivity. *Trends in Ecology and Evolution* 17: 76–83.

Webster, R. and M. A. Oliver. 2007. *Geostatistics for Environmental Scientists*. Wiley, New York.

Wedding, L. M., C. A. Lepczyk, S. J. Pittman, A. M. Friedlander, and S. Jorgensen. 2011. Quantifying seascape structure: Extending terrestrial spatial pattern metrics to the marine realm. *Marine Ecology Progress Series* 427: 219–232.

Weintraub, S. R., P. G. Taylor, S. Porder, C. C. Cleveland, G. P. Asner, and A. R. Townsend. 2015. Topographic controls on soil nitrogen availability in a lowland tropical forest. *Ecology* 96: 1561–1574.

Weir, B. S. and C. C. Cockerham. 1984. Estimating *F*-statistics for the analysis of population structure. *Evolution* 38: 1358–1370.

Weir, J. T. and D. Schluter. 2007. The latitudinal gradient in recent speciation and extinction rates of birds and mammals. *Science* 314: 1574–1576.

Weiss, H. and R. S. Bradley. 2001. What drives societal collapse? *Science* 291: 609–610.

Weng, Y.-C. and C. P. S. Larsen. 2006. Do coarse resolution U.S. pre-settlement land survey records adequately represent the spatial pattern of individual tree species? *Landscape Ecology* 21: 1003–1017.

Wenny, D.G., T. L. DeVault, M. D. Johnson, D. Kelly, C. H. Sekercioglu, D. F. Tomback, and C. J. Whelan. 2011. The need to quantify ecosystem services provided by birds. *Auk* 128: 1–14.

West, G. B., J. H. Brown, and B. J. Enquist. 1997. A general model for the origin of allometric scaling laws in biology. *Science* 276: 122–126.

West, G. B., J. H. Brown, and B. J. Enquist. 1999. The fourth dimension of life: Fractal geometry and allometric scaling of organisms. *Science* 284: 1677–1679.

Westerling, A. L., H. G. Hidalgo, D. R. Cayan, and T. W. Swetnam. 2006. Warming and earlier spring increase western U.S. forest wildfire activity. *Science* 313: 940–943.

Westerling, A. L., M. G. Turner, E. A. H. Smithwick, W. H. Romme, and M. G. Ryan. 2011. Continued warming could transform Greater Yellowstone fire regimes by mid-21st century. *Proceedings of the National Academy of Sciences of the United States of America* 108: 13165–13170.

Westphal, C., I. Steffan-Dewenter, and T. Tscharntke. 2003. Mass flowering crops enhance pollinator densities at a landscape scale. *Ecology Letters* 6: 961–965.

Westphal, C., I. Steffan-Dewenter, and T. Tscharntke. 2006. Bumblebees experience landscapes at different spatial scales: Possible implications for coexistence. *Oecologia* 149: 289–300.

White, E. P. and J. H. Brown. 2005. The template: Patterns and processes of spatial variation. Pp. 31–47 in *Ecosystem Function in Heterogeneous Landscapes* (G. M. Lovett, C. G. Jones, M. G. Turner, and K. C. Weathers, Editors). Springer, New York.

White, G. C. and R. A. Garrott. 1990. *Analysis of Wildlife Radio-Tracking Data.* Academic Press, San Diego, California.

White, P. S. and S. T. A. Pickett. 1985. Natural disturbance and patch dynamics: An introduction. Pp. 3–13 in *The Ecology of Natural Disturbance and Patch Dynamics* (S. T. A. Pickett and P. S. White, Editors). Academic Press, San Diego, California.

White, R. E. 2006. *Principles and Practice of Soil Science: The Soil as a Natural Resource,* 4th edition. Blackwell, Malden, Massachusetts.

Whitlock, M. C. 2011. G'_{ST} and D do not replace F_{ST}. *Molecular Ecology* 20: 1083–1091.

Whitlock, M. C. and D. E. McCauley. 1999. Indirect measures of gene flow and migration: $F_{ST} \neq 1/(4Nm + 1)$. *Heredity* 82: 117–125.

Whitney, G. G. 1994. *From Coastal Wilderness to Fruited Plain: A History of Environmental Change in Temperate North America from 1500 to the Present.* Cambridge University Press, New York.

Whittaker, R. H. 1956. Vegetation of the Great Smoky Mountains. *Ecological Monographs* 26: 1–80.

Whittaker, R. H. 1960. Vegetation of the Siskiyou Mountains, Oregon and California. *Ecological Monographs* 30: 279–338.

Whittaker, R. H. 1967. Gradient analysis of vegetation. *Biological Reviews* 42: 207–264.

Whittaker, R. J. and K. A. Triantis. 2012. 'The species-area relationship: An exploration of that 'most general, yet protean pattern.' *Journal of Biogeography* 39: 623–626.

Whittaker, R. J., K. J. Willis, and R. Field. 2001. Scale and species richness: Towards a general, hierarchical theory of species diversity. *Journal of Biogeography* 28: 453–470.

Whittaker, R. J., K. A. Triantis, and R. J. Ladle. 2008. A general dynamic theory of oceanic island biogeography. *Journal of Biogeography* 35: 977–994.

Wiegand, T., S. Gunatilleke, N. Gunatilleke, and T. Okuda. 2007. Analyzing the spatial structure of a Sri Lankan tree species with multiple scales of clustering. *Ecology* 88: 3088–3102.

Wiens, J. A. 1976. Population responses to patchy environments. *Annual Review of Ecology and Systematics* 7: 81–120.

Wiens, J. A. 1977. On competition and variable environments. *American Scientist* 65: 590–597.

Wiens, J. A. 1989a. Spatial scaling in ecology. *Functional Ecology* 3: 385–397.

Wiens, J. A. 1989b. *The Ecology of Bird Communities.* Volume 2. Processes and variations. Cambridge University Press, Cambridge, UK.

Wiens, J. A. 1992. Ecological flows across landscape boundaries: A conceptual overview. Pp. 217–235 in *Landscape Boundaries: Consequences for Biotic Diversity and Ecological Flows* (A. J. Hansen & F. di Castri, Editors). Springer-Verlag, New York.

Wiens, J. A. 1995a. Habitat fragmentation: Island *v* landscape perspectives on bird conservation. *Ibis* 137: S97-S104.

Wiens, J. A. 1995b. Landscape mosaics and ecological theory. Pp. 1–26 in *Mosaic Landscapes and Ecological Processes* (L. Hansson, L. Fahrig, and G. Merriam, Editors). Chapman & Hall, London, UK.

Wiens, J. A. 1997. Metapopulation dynamics and landscape ecology. Pp. 43–62 in *Metapopulation Biology* (I. A. Hanski and M. E. Gilpin, Editors). Academic Press, San Diego, California.

Wiens, J. A. 1999. The science and practice of landscape ecology. Pp. 371–397 in *Landscape Ecological Analysis* (J. M. Klopatek and R. H. Gardner, Editors). Springer-Verlag, New York.

Wiens, J. A. 2002. Riverine landscapes: Taking landscape ecology into the water. *Freshwater Biology* 47: 501–515.

Wiens, J. A. 2005. Toward a unified landscape ecology. Pp. 365–373 in *Issues and Perspectives in Landscape Ecology* (J. A. Wiens and M. R Moss, Editors), Cambridge University Press, Cambridge, UK.

Wiens, J. A. 2008. Allerton Park 1983: The beginnings of a paradigm for landscape ecology? *Landscape Ecology* 23: 125–128.

Wiens, J. A. 2013. Is landscape sustainability a useful concept in a changing world? *Landscape Ecology* 28: 1047–1052.

Wiens, J. A. and B. T. Milne. 1989. Scaling of 'landscapes' in landscape ecology, or, landscape ecology from a beetle's perspective. *Landscape Ecology* 3: 87–96.

Wiens, J. A., J. T. Rotenberry, and B. Van Horne. 1986. A lesson in the limitations of field experiments: Shrubsteppe birds and habitat alteration. *Ecology* 67: 365–376.

Wiens, J. A., N. C. Stenseth, B. Van Horne, and R. A. Ims. 1993a. Ecological mechanisms and landscape ecology. *Oikos* 66: 369–380.

Wiens, J. A., T. O. Crist and B. T. Milne. 1993b. On quantifying insect movements. *Environmental Entomology* 22: 709–715.

Wiens, J. A., T. O. Crist, K. A. With, and B. T. Milne. 1995. Fractal patterns of insect movement in microlandscape mosaics. *Ecology* 76: 663–666.

Wiens, J. A., R. L. Schooley, and R. D. Weeks. 1997. Patchy landscapes and animal movements: Do beetles percolate? *Oikos* 78: 257–264.

Wiens, J. A., M. R. Moss, M. G. Turner and D. J. Mladenoff, Editors. 2007. *Foundation Papers in Landscape Ecology*. Columbia University Press, New York.

Wiens, J. A., D. Stralberg, D. Jongsomjit, C. A. Howell, and M. A. Snyder. 2009. Niches, models, and climate change: Assessing the assumptions and uncertainties. *Proceedings of the National Academy of Sciences of the United States of America* 106: 19729–19736.

Wiens, J. J. and M. J. Donoghue. 2004. Historical biogeography, ecology and species richness. *Trends in Ecology and Evolution* 19: 639–644.

Wiens, J. J. and C. H. Graham. 2005. Niche conservatism: Integrating evolution, ecology, and conservation biology. *Annual Review of Ecology, Evolution, and Systematics* 36: 519–539.

Wiens, J. J., and 13 others. 2010. Niche conservatism as an emerging principle in ecology and conservation biology. *Ecology Letters* 13: 1310–1324.

Wilcove, D. S., D. Rothstein, J. Dubrow, A. Phillips, and E. Losos. 1998. Quantifying threats to imperiled species in the United States. *BioScience* 48: 607–615.

Wilcox, K. R., J. M. Blair, and A. K. Knapp. 2016. Stability of grassland soil C and N pools despite 25 years of an extreme climatic and disturbance regime. *Journal of Geophysical Research: Biogeosciences* 121: 1934–1345.

Wildman, Jr., R. A. and N. A. Forde. 2012. Management of water shortage in the Colorado River Basin: Evaluating current policy and the viability of interstate water trading. *Journal of the American Water Resources Association* 48: 411–422.

Wiley, E. O. 1988. Vicariance biogeography. *Annual Review of Ecology and Systematics* 19: 513–542.

Wilkinson, D. M. 1999. The disturbing history of intermediate disturbance. *Oikos* 84: 145–147.

Willems, E. P. and R. A. Hill. 2009. Predator-specific landscapes of fear and resource distribution: Effects on spatial range use. *Ecology* 90: 546–555.

Williams, A. P., C. D. Allen, C. I. Millar, T. W. Swetnam, J. Michaelsen, C. J. Still, and S. W. Leavitt. 2010. Forest responses to increasing aridity and warmth in the southwestern United States. *Proceedings of the National Academy of Sciences of the United States of America* 107: 21289–21294.

Williams, A. P., and 14 others. 2013. Temperature as a potent driver of regional forest drought stress and tree mortality. *Nature Climate Change* 3: 292–297.

Williams, C. B. 1964. *Patterns in the Balance of Nature*. Academic Press, London, UK.

Williams, J. W., S. T. Jackson, and J. E. Kutzbach. 2007. Projected distributions of novel and disappearing climates by 2100 AD. *Proceedings of the National Academy of Sciences of the United States of America* 104: 5738–5742.

Williams, M. A. and W. L. Baker. 2010. Bias and error in using survey records for ponderosa pine landscape restoration. *Journal of Biogeography* 37: 707–721.

Williams, M. R., B. B. Lamont, and J. D. Henstridge. 2009. Species–area functions revisited. *Journal of Biogeography* 36: 1994–2004.

Williamson, M., K. J. Gaston, and W. M. Lonsdale. 2001. The species–area relationship does not have an asymptote! *Journal of Biogeography* 28: 827–830.

Williamson, M., K. Dehnen-Schmutz, I. Kuhn, M. Hill, S. Klotz, A. Milbau, J. Stout and P. Pyšek. 2009. The distribution of range sizes of native and alien plants in four European countries and the effects of residence time. *Diversity and Distributions* 15: 158–166.

Willig, M. R., D. M. Kaufman, and R. D. Stevens. 2003. Latitudinal gradients of biodiversity: Pattern, process, scale, and synthesis. *Annual Review of Ecology, Evolution, and Systematics* 34: 273–309.

Willis, A. J. 1997. The ecosystem: An evolving concept viewed historically. *Functional Ecology* 11: 268–271.

Willis, K. J. and R. J. Whittaker. 2002. Species diversity—scale matters. *Science* 295: 1245–1248.

Willson, M. E., S. M. Gende, and B. H. Marston. 1998. Fishes and forest. *BioScience* 48: 455–462.

Wilson, D.S. 1992. Complex interactions in metacommunities, with implications for biodiversity and higher levels of selection. *Ecology* 73: 1984–2000.

Wilson, J. P. and J. C. Gallant. 2000. *Terrain Analysis: Principles and Applications*. Wiley, New York.

Wilson, J. T. 1963. A possible origin of the Hawaiian Islands. *Canadian Journal of Physics* 41: 863–870.

Wilson, R. J., Z. G. Davies, and C. D. Thomas. 2009. Modelling the effect of habitat fragmentation on range expansion in a butterfly. *Proceedings of the Royal Society B* 276: 1421–1427.

Wilson, R. J., Z. G. Davies, and C. D. Thomas. 2010. Linking habitat use to range expansion rates in fragmented landscapes: A metapopulation approach. *Ecography* 33: 73–82.

Winemiller, K. O., A. S. Flecker, and D. J. Hoeinghaus. 2010. Patch dynamics and environmental heterogeneity in lotic ecosystems. *Journal of the North American Benthological Society* 29: 84–99.

Winfree, R., J. Dushoff, E. E. Crone, C. B. Schultz, R. V. Budny, N. M. Williams, and C. Kremen. 2005. Testing simple indices of habitat proximity. *American Naturalist* 165: 707–717.

Winfree, R., N. M. Williams, J. Dushoff, and C. Kremen. 2007. Native bees provide insurance against ongoing honey bee losses. *Ecology Letters* 10: 1105–1113.

Winfree, R., N. M. Williams, H. Gaines, J. S. Ascher, and C. Kremen. 2008. Wild bee pollinators provide the majority of crop visitation across land-use gradients in New Jersey and Pennsylvania, USA. *Journal of Applied Ecology* 45: 793–802.

Winfree, R., R. Aguilar, D. P. Vázquez, G. LeBuhn, and M. A. Aizen. 2009. A meta-analysis of bees' responses to anthropogenic disturbance. *Ecology* 90: 2068–2076.

Winsome, T., L. Epstein, P. F. Hendrix, and W. R. Horwath. 2006. Competitive interactions between native and exotic earthworm species as influenced by habitat quality in a California grassland. *Applied Soil Ecology* 32: 38–53.

Winter, M., D. H. Johnson, and J. Faaborg. 2000. Evidence for edge effects on multiple levels in tallgrass prairie. *Condor* 102: 256–266.

Wisz, M. S., R. J. Hijmans, J. Li, A. T. Peterson, C. H. Graham, NCEAS Predicting Species Distribution Working Group. 2008. Effects of sample size on the performance of species distribution models. *Diversity and Distributions* 14: 763–773.

With, K. A. 1994a. Using fractal analysis to assess how species perceive landscape structure. *Landscape Ecology* 9: 25–36.

With, K. A. 1994b. Ontogenetic shifts in how grasshoppers interact with landscape structure: An analysis of movement patterns. *Functional Ecology* 8: 477–485.

With, K. A. 1997. The application of neutral landscape models in conservation biology. *Conservation Biology* 11: 1069–1080.

With, K. A. 2002a. Using percolation theory to assess landscape connectivity and effects of habitat fragmentation. Pp. 105–130 in *Applying Landscape Ecology in Biological Conservation* (K. J. Gutzwiller, Editor). Springer-Verlag, New York.

With, K. A. 2002b. The landscape ecology of invasive spread. *Conservation Biology* 16: 1192–1203.

With, K. A. 2004. Assessing the risk of invasive spread in fragmented landscapes. *Risk Analysis* 24: 803–815.

With, K. A. 2007. Invoking the ghosts of landscapes past to understand the landscape ecology of the present…and the future. Pp. 43–58 in *Temporal Dimensions of Landscape Ecology: Wildlife Responses to Variable Resources* (J. A. Bissonette and I. Storch, Editors), Springer, New York.

With, K. A. 2015. How fast do migratory songbirds have to adapt to keep pace with rapidly changing landscapes? *Landscape Ecology* 30: 1351–1361.

With, K. A. 2016. Are landscapes more than the sum of their patches? *Landscape Ecology* 31: 969–980.

With, K. A. and T. O. Crist. 1995. Critical thresholds in species' responses to landscape structure. *Ecology* 76: 2446–2459.

With, K. A. and A. W. King. 1997. The use and misuse of neutral landscape models in ecology. *Oikos* 79: 219–229.

With, K. A. and A. W. King. 1999a. Extinction thresholds in fractal landscapes. *Conservation Biology* 13: 314–326.

With, K. A. and A. W. King. 1999b. Dispersal success in fractal landscapes: A consequence of lacunarity thresholds. *Landscape Ecology* 14: 73–82.

With, K. A. and A. W. King. 2001. Analysis of landscape sources and sinks: The effect of spatial pattern on avian demography. *Biological Conservation* 100: 75–88.

With, K. A. and D. M. Pavuk. 2011. Habitat area trumps fragmentation effects on arthropods in an experimental landscape system. *Landscape Ecology* 26: 1035–1048.

With, K. A. and D. M. Pavuk. 2012. Direct versus indirect effects of habitat fragmentation on community patterns in experimental landscapes. *Oecologia* 170: 517–528.

With, K. A., R. H. Gardner, and M. G. Turner. 1997. Landscape connectivity and population distributions in heterogeneous environments. *Oikos* 78: 151–169.

With, K. A., S. J. Cadaret, and C. Davis. 1999. Movement responses to patch structure in experimental fractal landscapes. *Ecology* 80: 1340–1353.

With, K. A., D. M. Pavuk, J. L. Worchuck, R. K. Oates, and J. L. Fisher. 2002. Threshold effects of landscape structure on biological control in agroecosystems. *Ecological Applications* 12: 52–65.

With, K. A., G. R. Schrott, and A. W. King. 2006. The implications of metalandscape connectivity for population viability in migratory songbirds. *Landscape Ecology* 21: 157–167.

With, K. A., A. W. King, and W. E. Jensen. 2008. Remaining large grasslands may not be sufficient to prevent grassland bird declines. *Biological Conservation* 141: 3152–3167.

Witt, A., B. D. Malamud, M. Rossi, F. Guzzetti, and S. Peruccacci. 2010. Temporal correlations and clustering of landslides. *Earth Surface Processes and Landforms* 35: 1138–1156.

Wolfe, C. J., S. C. Solomon, G. Laske, J. A. Collins, R. S. Detrick, J. A. Orcutt, D. Bercovici, and E. H. Hauri. 2009. Mantle shear-wave velocity structure beneath the Hawaiian hot spot. *Science* 326: 1388–1390.

Wood, C. C. 2007. Sockeye salmon ecotypes: Origin, vulnerability to human impacts, and conservation value. *American Fisheries Society Symposium* 54: 1–4.

Wood, D. M., D. Parry, R. D. Yanai, and N. E. Pitel. 2010. Forest fragmentation and duration of forest tent caterpillar (*Malacosoma disstria* Hübner) outbreaks in northern hardwood forests. *Forest Ecology and Management* 260: 1193–1197.

World Commission on Environment and Development. 1987. *Our Common Future.* Oxford University Press, New York.

World Factbook 2013–14. 2013. Central Intelligence Agency, Washington, D.C. Available online: https://www.cia.gov/library/publications/the-world-factbook (last accessed 1/30/19).

Wortley, L., J.-M. Hero, and M. Howes. 2013. Evaluating ecological restoration success: A review of the literature. *Restoration Ecology* 21: 537–543.

Worton, B. J. 1989. Kernel methods for estimating the utilization distribution in home-range studies. *Ecology* 70: 164–168.

Wright, J. P., C. G. Jones, and A. S. Flecker. 2002. An ecosystem engineer, the beaver, increases species richness at the landscape scale. *Oecologia* 132: 96–101.

Wright, S. 1921. Correlation and causation. *Journal of Agricultural Research* 20: 557–585.

Wright, S. 1931. Evolution in Mendelian populations. *Genetics* 16: 97–259.

Wright, S. 1943. Isolation by distance. *Genetics* 18: 114.

Wright, S. 1969. *Evolution and the Genetics of Populations*, Volume 2. *The Theory of Gene Frequencies.* University of Chicago Press, Chicago, Illinois.

Wright, S. 1978. *Evolution and the Genetics of Populations.* Volume 4. *Variability Within and Among Natural Populations.* University of Chicago Press, Chicago, Illinois.

Wu, J. 2004. Effects of changing scale on landscape pattern analysis: Scaling relations. *Landscape Ecology* 19: 125–138.

Wu, J. 2006. Landscape ecology, cross-disciplinarity, and sustainability science. *Landscape Ecology* 21: 1–4.

Wu, J. 2007a. Past, present and future of landscape ecology. *Landscape Ecology* 22: 1433–1435.

Wu, J. 2007b. Scale and scaling: A cross-disciplinary perspective. pp. 115–142 in *Key Topics in Landscape Ecology* (J. Wu and R. Hobbs, Editors). Cambridge University Press, New York.

Wu, J. 2012. A landscape approach for sustainability science. Pp. 59–77 in *Sustainability Science: The Emerging Paradigm and the Urban Environment* (M. P. Weinstein and R. E. Turner, Editors). Springer, New York.

Wu, J. 2013a. Key concepts and research topics in landscape ecology revisited: 30 years after the Allerton Park Workshop. *Landscape Ecology* 28: 1–11.

Wu, J. 2013b. Landscape sustainability science: Ecosystem services and human well-being in changing landscapes. *Landscape Ecology* 28: 999–1023.

Wu, J. and O. L. Loucks. 1995. From balance of nature to hierarchical patch dynamics: A paradigm shift in ecology. *Quarterly Review of Biology* 70: 439–466.

Wu, J. and R. J. Hobbs. 2007. Landscape ecology: The state-of-the-science. Pp. 271–287 in *Key Topics in Landscape Ecology*. Cambridge University Press, Cambridge, UK.

Wu, J., W. Shen, W. Sun, and P. T. Tueller. 2002. Empirical patterns of the effects of changing scale on landscape metrics. *Landscape Ecology* 17: 761–782.

Wulder, M. A., J. C. White, J. G. Masek, J. Dwyer, and D. P. Roy. 2011. Continuity of Landsat observations: Short term considerations. *Remote Sensing of Environment* 115: 747–751.

Yackulic, C. B. 2017. Competitive exclusion over broad spatial extents is a slow process: Evidence and implications for species distribution modeling. *Ecography* 40: 305–313.

Yahner, R. H. 1988. Changes in wildlife communities near edges. *Conservation Biology* 2: 333–339.

Yu, D. W. and H. B. Wilson. 2001. The competition-colonization trade-off is dead; long-live the competition-colonization trade-off. *American Naturalist* 138: 49–63.

Yue, T.-X., Z.-M. Fan, C.-F. Chen, X.-F. Sun, and B.-L. Li. 2011. Surface modeling of global terrestrial ecosystems under three climate change scenarios. *Ecological Modelling* 222: 2342–2361.

Zalasiewicz, J., and 25 others. 2015. When did the Anthropocene begin? A mid-twentieth century boundary level is stratigraphically optimal. *Quarternary International* 383: 196–203.

Zeller, K. A., K. McGarigal, and A. R. Whiteley. 2012. Estimating landscape resistance to movement: A review. *Landscape Ecology* 27: 777–797.

Zellmer, A. J. and L. L. Knowles. 2009. Disentangling the effects of historic vs. contemporary landscape structure on population genetic divergence. *Molecular Ecology* 18: 3593–3602.

Zhang, C., W. Li, and D. Civco. 2014. Application of geographically weighted regression to fill gaps in SLC-off Landsat ETM+ satellite imagery. *International Journal of Remote Sensing* 35: 7650–7672.

Zhang, P., G. Shao, G. Zhao, D. C. Le Master, G. R. Parker, J. B. Dunning Jr. and Q. Li. 2000. China's forest policy for the 21st century. *Science* 288: 2135–2136.

Zhang, X. and X. Cai. 2011. Climate change impacts on global agricultural land available. *Environmental Research Letters* 6: 014014.

Zollner, P. A. and S. L. Lima. 1997. Landscape-level perceptual abilities in white-footed mice: Perceptual range and the detection of forested habitat. *Oikos* 80: 51–60.

Zollner, P. A. and S. L. Lima. 1999. Illumination and the perception of remote habitat patches by white-footed mice. *Animal Behaviour* 58: 489–500.

Zollner, P. A. and S. L. Lima. 2005. Behavioral trade-offs when dispersing across a patchy landscape. *Oikos* 108: 219–230.

Zonneveld, I. S. 1972. *Land Evaluation and Land(scape) Science*. International Institute for Aerial Survey and Earth Sciences, Enschede, the Netherlands.

Zonneveld, I. S. 1989. The land unit—a fundamental concept in landscape ecology, and its applications. *Landscape Ecology* 3: 67–86.

Index

Note: boxes, figures, and tables are indicated by the suffixes *b*, *f*, and *t*. For example, 169*f* indicates a figure on page 169.